# MARTIN'S PHYSICAL PHARMACY AND PHARMACEUTICAL SCIENCES

## Physical Chemical and Biopharmaceutical Principles in the Pharmaceutical Sciences

**SIXTH EDITION**

### Editor

## PATRICK J. SINKO, PhD, RPh

Professor II (Distinguished)
Parke-Davis Chair Professor in Pharmaceutics and Drug Delivery
Ernest Mario School of Pharmacy
Rutgers, The State University of New Jersey
Piscataway, New Jersey

### Assistant Editor

## YASHVEER SINGH, PhD

Assistant Research Professor
Department of Pharmaceutics
Ernest Mario School of Pharmacy
Rutgers, The State University of New Jersey
Piscataway, New Jersey

Wolters Kluwer | Lippincott Williams & Wilkins
Health
Philadelphia · Baltimore · New York · London
Buenos Aires · Hong Kong · Sydney · Tokyo

*Editor:* David B. Troy
*Product Manager:* Meredith L. Brittain
*Vendor Manager:* Kevin Johnson
*Designer:* Holly McLaughlin
*Compositor:* Aptara®, Inc.

Sixth Edition

**Library of Congress Cataloging-in-Publication Data**

Martin's physical pharmacy and pharmaceutical sciences : physical chemical and biopharmaceutical principles in the pharmaceutical sciences.—6th ed. / editor, Patrick J. Sinko ; assistant editor, Yashveer Singh.
    p. ; cm.
 Includes bibliographical references and index.
 ISBN 978-1-6091-3402-0
 1. Pharmaceutical chemistry.    2. Chemistry, Physical and theoretical.
I. Martin, Alfred N. II. Sinko, Patrick J. III. Singh, Yashveer.
IV. Title: Physical pharmacy and pharmaceutical sciences.
 [DNLM: 1. Chemistry, Pharmaceutical.    2. Chemistry, Physical. QV 744
M386 2011]
 RS403.M34 2011
 615′.19—dc22                              2009046514

**DISCLAIMER**

Care has been taken to confirm the accuracy of the information present and to describe generally accepted practices. However, the authors, editors, and publisher are not responsible for errors or omissions or for any consequences from application of the information in this book and make no warranty, expressed or implied, with respect to the currency, completeness, or accuracy of the contents of the publication. Application of this information in a particular situation remains the professional responsibility of the practitioner; the clinical treatments described and recommended may not be considered absolute and universal recommendations.

The authors, editors, and publisher have exerted every effort to ensure that drug selection and dosage set forth in this text are in accordance with the current recommendations and practice at the time of publication. However, in view of ongoing research, changes in government regulations, and the constant flow of information relating to drug therapy and drug reactions, the reader is urged to check the package insert for each drug for any change in indications and dosage and for added warnings and precautions. This is particularly important when the recommended agent is a new or infrequently employed drug.

Some drugs and medical devices presented in this publication have Food and Drug Administration (FDA) clearance for limited use in restricted research settings. It is the responsibility of the health care providers to ascertain the FDA status of each drug or device planned for use in their clinical practice.

To purchase additional copies of this book, call our customer service department at (800) 638-3030 or fax orders to (301) 223-2320. International customers should call (301) 223-2300.

Visit Lippincott Williams & Wilkins on the Internet: at http://www.lww.com. Lippincott Williams & Wilkins customer service representatives are available from 8:30 am to 6:00 pm, EST.

*Dedicated to my parents Patricia and Patrick Sinko,*
*my wife Renee, and my children Pat, Katie (and Maggie)*

# DEDICATION

## ALFRED N. MARTIN (1919–2003)

This fiftieth anniversary edition of *Martin's Physical Pharmacy and Pharmaceutical Sciences* is dedicated to the memory of Professor Alfred N. Martin, whose vision, creativity, dedication, and untiring effort and attention to detail led to the publication of the first edition in 1960. Because of his national reputation as a leader and pioneer in the then emerging specialty of physical pharmacy, I made the decision to join Professor Martin's group of graduate students at Purdue University in 1960 and had the opportunity to witness the excitement and the many accolades of colleagues from far and near that accompanied the publication of the first edition of *Physical Pharmacy*. The completion of that work represented the culmination of countless hours of painstaking study, research, documentation, and revision on the part of Dr. Martin, many of his graduate students, and his wife, Mary, who typed the original manuscript. It also represented the fruition of Professor Martin's dream of a textbook that would revolutionize pharmaceutical education and research. *Physical Pharmacy* was for Professor Martin truly a labor of love, and it remained so throughout his lifetime, as he worked unceasingly and with steadfast dedication on the subsequent revisions of the book.

The publication of the first edition of *Physical Pharmacy* generated broad excitement throughout the national and international academic and industrial research communities in pharmacy and the pharmaceutical sciences. It was the world's first textbook in the emerging discipline of physical pharmacy and has remained the "gold standard" textbook on the application of physical chemical principles in pharmacy and the pharmaceutical sciences. *Physical Pharmacy*, upon its publication in 1960, provided great clarity and definition to a discipline that had been widely discussed throughout the 1950s but not fully understood or adopted. Alfred Martin's *Physical Pharmacy* had a profound effect in shaping the direction of research and education throughout the world of pharmaceutical education and research in the pharmaceutical industry and academia. The publication of this book transformed pharmacy and pharmaceutical research from an essentially empirical mix of art and descriptive science to a quantitative application of fundamental physical and chemical scientific principles to pharmaceutical systems and dosage forms. *Physical Pharmacy* literally changed the direction, scope,

focus, and philosophy of pharmaceutical education during the 1960s and the 1970s and paved the way for the specialty disciplines of biopharmaceutics and pharmacokinetics which, along with physical pharmacy, were necessary underpinnings of a scientifically based clinical emphasis in the teaching of pharmacy students, which is now pervasive throughout pharmaceutical education.

From the time of the initial publication of *Physical Pharmacy* to the present, this pivotal and classic book has been widely used both as a teaching textbook and as an indispensible reference for academic and industrial researchers in the pharmaceutical sciences throughout the world. This sixth edition of *Martin's Physical Pharmacy and Pharmaceutical Sciences* serves as a most fitting tribute to the extraordinary, heroic, and inspired vision and dedication of Professor Martin. That this book continues to be a valuable and widely used textbook in schools and colleges of pharmacy throughout the world, and a valuable reference to pharmaceutical scientists and researchers, is a most appropriate recognition of the life's work of Alfred Martin. All who have contributed to the thorough revision that has resulted in the publication of the current edition have retained the original format and fundamental organization of basic principles and topics that were the hallmarks of Professor Martin's classic first edition of this seminal book.

Professor Martin always demanded the best of himself, his students, and his colleagues. The fact that the subsequent and current editions of *Martin's Physical Pharmacy and Pharmaceutical Sciences* have remained faithful to his vision of scientific excellence as applied to understanding and applying the principles underlying the pharmaceutical sciences is indeed a most appropriate tribute to Professor Martin's memory. It is in that spirit that this fiftieth anniversary edition is formally dedicated to the memory of that visionary and creative pioneer in the discipline of physical pharmacy, Alfred N. Martin.

John L. Colaizzi, PhD
Rutgers, The State University of New Jersey
Piscataway, New Jersey
November 2009

Ever since the First Edition of *Martin's Physical Pharmacy* was published in 1960, Dr. Alfred Martin's vision was to provide a text that introduced pharmacy students to the application of physical chemical principles to the pharmaceutical sciences. This remains a primary objective of the Sixth Edition. *Martin's Physical Pharmacy* has been used by generations of pharmacy and pharmaceutical science graduate students for 50 years and, while some topics change from time to time, the basic principles remain constant, and it is my hope that each edition reflects the pharmaceutical sciences at that point in time.

## ORGANIZATION

As with prior editions, this edition represents an updating of most chapters, a significant expansion of others, and the addition of new chapters in order to reflect the applications of the physical chemical principles that are important to the Pharmaceutical Sciences today. As was true when Dr. Martin was at the helm, this edition is a work in progress that reflects the many suggestions made by students and colleagues in academia and industry. There are 23 chapters in the Sixth Edition, as compared with 22 in the Fifth Edition. All chapters have been reformatted and updated in order to make the material more accessible to students. Efforts were made to shorten chapters in order to focus on the most important subjects taught in Pharmacy education today. Care has been taken to present the information in "layers" from the basic to more in-depth discussions of topics. This approach allows the instructor to customize their course needs and focus their course and the students' attention on the appropriate topics and subtopics.

With the publication of the Sixth Edition, a Web-based resource is also available for students and faculty members (see the "Additional Resources" section later in this preface).

## FEATURES

Each chapter begins with a listing of **Chapter Objectives** that introduce information to be learned in the chapter. **Key Concept Boxes** highlight important concepts, and each **Chapter Summary** reinforces chapter content. In addition, illustrative **Examples** have been retained, updated, and expanded. **Recommended Readings** point out instructive additional sources for possible reference. **Practice Problems** have been

moved to the Web (see the "Additional Resources" section later in this preface).

## SIGNIFICANT CHANGES FROM THE FIFTH EDITION

Important changes include new chapters on Pharmaceutical Biotechnology and Oral Solid Dosage Forms. Three chapters were rewritten de novo on the basis of the valuable feedback received since the publication of the Fifth Edition. These include Chapter 1 ("Introduction"), which is now called Interpretive Tools; Chapter 20 ("Biomaterials"), which is now called Pharmaceutical Polymers; and Chapter 23 ("Drug Delivery Systems"), which is now called Drug Delivery and Targeting.

## ADDITIONAL RESOURCES

*Martin's Physical Pharmacy and Pharmaceutical Sciences*, Sixth Edition, includes additional resources for both instructors and students that are available on the book's companion Web site at thepoint.lww.com/Sinko6e.

### Instructors

Approved adopting instructors will be given access to the following additional resources:

■ Practice problems and answers to ascertain student understanding.

### Students

Students who have purchased *Martin's Physical Pharmacy and Pharmaceutical Sciences*, Sixth Edition, have access to the following additional resources:

■ A separate set of practice problems and answers to reinforce concepts learned in the text.

In addition, purchasers of the text can access the searchable Full Text Online by going to the *Martin's Physical Pharmacy and Pharmaceutical Sciences*, Sixth Edition, Web site at thePoint.lww.com/Sinko6e. See the inside front cover of this text for more details, including the passcode you will need to gain access to the Web site.

Patrick Sinko
*Piscataway, New Jersey*

**GREGORY E. AMIDON, PhD**
Research Professor
Department of Pharmaceutical Sciences
College of Pharmacy
University of Michigan
Ann Arbor, Michigan

**CHARLES RUSSELL MIDDAUGH, PhD**
Distinguished Professor
Department of Pharmaceutical Chemistry
University of Kansas
Lawrence, Kansas

**HOSSEIN OMIDIAN, PhD**
Assistant Professor
Department of Pharmaceutical Sciences
College of Pharmacy
Nova Southeastern University
Ft. Lauderdale, Florida

**KINAM PARK, PhD**
Showalter Distinguished Professor
Department of Biomedical Engineering
Professor of Pharmaceutics
Departments of Biomedical Engineering and Pharmaceutics
Purdue University
West Lafayette, Indiana

**TERUNA J. SIAHAAN, PhD**
Professor
Department of Pharmaceutical Chemistry
University of Kansas
Lawrence, Kansas

**YASHVEER SINGH, PhD**
Assistant Research Professor
Department of Pharmaceutics
Ernest Mario School of Pharmacy
Rutgers, The State University of New Jersey
Piscataway, New Jersey

**PATRICK J. SINKO, PhD, RPh**
Professor II (Distinguished)
Parke-Davis Chair Professor in Pharmaceutics and Drug Delivery
Ernest Mario School of Pharmacy
Rutgers, The State University of New Jersey
Piscataway, New Jersey

**HAIAN ZHENG, PhD**
Assistant Professor
Department of Pharmaceutical Sciences
Albany College of Pharmacy and Health Sciences
Albany, New York

# ■ACKNOWLEDGMENTS■

The Sixth Edition reflects the hard work and dedication of many people. In particular, I acknowledge Drs. Gregory Amidon (Ch 22), Russell Middaugh (Ch 21), Hamid Omidian (Chs 20 and 23), Kinam Park (Ch 20), Teruna Siahaan (Ch 21), and Yashveer Singh (Ch 23) for their hard work in spearheading the efforts to write new chapters or rewrite existing chapters de novo. In addition, Dr. Singh went beyond the call of duty and took on the responsibilities of Assistant Editor during the proofing stages of the production of the manuscripts. Through his efforts, I hope that we have caught many of the minor errors from the fourth and fifth editions. I also thank HaiAn Zheng, who edited the online practice problems for this edition, and Miss Xun Gong, who assisted him.

The figures and experimental data shown in Chapter 6 were produced by Chris Olsen, Yuhong Zeng, Weiqiang Cheng, Mangala Roshan Liyanage, Jaya Bhattacharyya, Jared Trefethen, Vidyashankara Iyer, Aaron Markham, Julian Kissmann and Sangeeta Joshi of the Department of Pharmaceutical Chemistry at the University of Kansas. The section on drying of biopharmaceuticals is based on a series of lectures and overheads presented by Dr. Pikal of the University of Connecticut in April of 2009 at the University of Kansas.

I would like to acknowledge Dr. Mayur Lodaya for his contributions to the continuous processing section of Chapter 22 on Oral Dosage forms.

Numerous graduate students contributed in many ways to this edition, and I am always appreciative of their insights, criticisms, and suggestions. Thanks also to Mrs. Amy Grabowski for her invaluable assistance with coordination efforts and support interactions with all contributors.

To all of the people at LWW who kept the project moving forward with the highest level of professionalism, skill, and patience. In particular, to David Troy for supporting our vision for this project and Meredith Brittain for her exceptional eye for detail and her persistent efforts to keep us on track.

And to my wonderful wife, Renee, who deserves enormous credit for juggling her hectic professional life as a pharmacist and her expert skill as the family organizer while maintaining a sense of calmness in what is an otherwise chaotic life.

Patrick Sinko
*Piscataway, New Jersey*

# CONTENTS

# 1 INTERPRETIVE TOOLS

**CHAPTER OBJECTIVES** At the conclusion of this chapter the student should be able to:

1. Understand the basic tools required to analyze and interpret data sets from the clinic, laboratory, or literature.
2. Describe the differences between classic dosage forms and modern drug delivery systems.
3. Use dimensional analysis.
4. Understand and apply the concept of significant figures.
5. Define determinant and indeterminant errors, precision, and accuracy.
6. Calculate the mean, median, and mode of a data set.
7. Understand the concept of variability.
8. Calculate standard deviation and coefficient of variation and understand when it is appropriate to use these parameters.
9. Use graphic methods to determine the slope of lines.
10. Interpret slopes of lines and how they relate to absorption and elimination from the body.

## INTRODUCTION

"One of the earmarks of evidence-based medicine is that the practitioner should not just accept the conventional wisdom of his/her mentor. Evidence-based medicine uses the scientific method of using observations and literature searches to form a hypothesis as a basis for appropriate medical therapy. This process necessitates education in basic sciences and an understanding of basic scientific principles."[1,2] Today more than ever before, the pharmacist and the pharmaceutical scientist are called upon to demonstrate a sound knowledge of biopharmaceutics, biochemistry, chemistry, pharmacology, physiology, and toxicology and an intimate understanding of the physical, chemical, and biopharmaceutical properties of medicinal products. Whether engaged in research and development, teaching, manufacturing, the practice of pharmacy, or any of the allied branches of the profession, the pharmacist must recognize the need to rely heavily on the basic sciences. This stems from the fact that pharmacy is an applied science, composed of principles and methods that have been culled from other disciplines. The pharmacist engaged in advanced studies must work at the boundaries between the various sciences and must keep abreast of advances in the physical, chemical, and biological fields in order to understand and contribute to the rapid developments in his or her profession. You are also expected to provide concise and practical interpretations of highly technical drug information to your patients and colleagues. With the abundance of information and misinformation that is freely and publicly available (e.g., on the Internet), having the tools and ability to provide meaningful interpretations of results is critical.

Historically, *physical pharmacy* has been associated with the area of pharmacy that dealt with the quantitative and theoretical principles of physicochemical science as they applied to the practice of pharmacy. Physical pharmacy attempted to integrate the factual knowledge of pharmacy through the development of broad principles of its own, and it aided the pharmacist and the pharmaceutical scientist in their attempt to predict the solubility, stability, compatibility, and biologic action of drug products. Although this remains true today, the field has become even more highly integrated into the biomedical aspects of the practice of pharmacy. As such, the field is more broadly known today as the *pharmaceutical sciences* and the chapters that follow reflect the high degree of integration of the biological and physical–chemical aspects of the field.

Developing new drugs and delivery systems and improving upon the various modes of administration are still the primary goals of the pharmaceutical scientist. A practicing pharmacist must also possess a thorough understanding of modern drug delivery systems as he or she advises patients on the best use of prescribed medicines. In the past, drug delivery focused nearly exclusively on *pharmaceutical technology* (in other words, the manufacture and testing of tablets, capsules, creams, ointments, solutions, etc.). This area of study is still very important today. However, the pharmacist needs to understand how these delivery systems perform in and respond to the normal and pathophysiologic states of the patient. The integration of physical–chemical and biological aspects is relatively new in the pharmaceutical sciences. As the field progresses toward the complete integration of these subdisciplines, the impact of the biopharmaceutical sciences and drug delivery will become enormous. The advent and commercialization of molecular, nanoscale, and microscopic drug delivery technologies is a direct result of the integration of the biological and physical–chemical sciences. In the past, a dosage (or dose) form and a drug delivery system were considered to be one and the same. *A dosage form* is the entity that is administered to patients so that they receive an effective dose of a drug. The traditional understanding of how an oral dosage form, such as a tablet, works is that a patient takes it by mouth with some fluid, the tablet disintegrates, and the drug dissolves in the stomach and is then absorbed through the intestines into the bloodstream. If the dose is

## KEY CONCEPT — PHARMACEUTICAL SCIENCES

Pharmacy, like many other applied sciences, has passed through a descriptive and empiric era. Over the past decade a firm scientific foundation has been developed, allowing the "art" of pharmacy to transform itself into a quantitative and mechanistic field of study. The integration of the biological, chemical, and physical sciences remains critical to the continuing evolution of the pharmaceutical sciences. The theoretical links between the diverse scientific disciplines that serve as the foundation for pharmacy are reflected in this book. The scientific principles of pharmacy are not as complex as some would believe, and certainly they are not beyond the understanding of the well-educated pharmacist of today.

too high, a lower-dose tablet may be prescribed. If a lower-dose tablet is not commercially available, the patient may be instructed to divide the tablet. However, a pharmacist who dispenses a nifedipine (Procardia XL) extended-release tablet or an oxybutynin (Ditropan XL) extended-release tablet to a patient would advise the patient not to bite, chew, or divide the "tablet." The reason for this is that the tablet dosage form is actually an elegant osmotic pump *drug delivery system* that looks like a conventional tablet (see Key Concept Box on Dosage Forms and Drug Delivery Systems). This creative and elegant approach solves numerous challenges to the delivery of pharmaceutical care to patients. On the one hand, it provides a sustained-release drug delivery system to patients so that they take their medication less frequently, thereby enhancing patient compliance and positively influencing the success rate of therapeutic regimens. On the other hand, patients see a familiar dosage form that they can take by a familiar route of administration. In essence, these osmotic pumps are delivery systems packaged into a dosage form that is familiar to the patient. The subtle differences between dose forms and delivery systems will become even more profound in the years to come as drug delivery systems successfully migrate to the molecular scale.

This course should mark the turning point in the study pattern of the student, for in the latter part of the pharmacy curriculum, emphasis is placed upon the application of scientific principles to practical professional problems. Although facts must be the foundation upon which any body of knowledge is built, the rote memorization of disjointed "particles" of knowledge does not lead to logical and systematic thought. This chapter provides a foundation for interpreting the observations and results that come from careful

## KEY CONCEPT — DOSAGE FORMS AND DRUG DELIVERY SYSTEMS

A Procardia XL extended-release tablet is similar in appearance to a conventional tablet. It consists, however, of a semipermeable membrane surrounding an osmotically active drug core. The core is divided into two layers: an "active" layer containing the drug and a "push" layer containing pharmacologically inert but osmotically active components. As fluid from the gastrointestinal (GI) tract enters the tablet, pressure increases in the osmotic layer and "pushes" against the drug layer, releasing drug through the precision laser-drilled tablet orifice in the active layer. Procardia XL is designed to provide nifedipine at an approximately constant rate over 24 hr. This controlled rate of drug delivery into the GI lumen is independent of pH or GI motility. The nifedepine release profile from Procardia XL depends on the existence of an osmotic gradient between the contents of the bilayer core and the fluid in the GI tract. Drug delivery is essentially constant as long as the osmotic gradient remains constant, and then gradually falls to zero. Upon being swallowed, the biologically inert components of the tablet remain intact during GI transit and are eliminated in the feces as an insoluble shell. The information that the pharmacist provides to the patient includes "Do not crush, chew, or break the extended-release form of Procardia XL. These tablets are specially formulated to release the medication slowly into the body. Swallow the tablets whole with a glass of water or another liquid. Occasionally , you may find a tablet form in the stool. Do not be alarmed, this is the outer shell of the tablet only,

the medication has been absorbed by the body." On examining the figure, you will notice how the osmotic pump tablet looks identical to a conventional tablet.

Remember, most of the time when a patient takes a tablet, it is also the delivery system. It has been optimized so that it can be mass-produced and can release the drug in a reproducible and reliable manner. Complete disintegration and deaggregation occurs and there is little, if any, evidence of the tablet dose form that can be found in the stool. However, with an osmotic pump delivery system, the "tablet" does not disintegrate even though all of the drug will be released. Eventually, the outer shell of the depleted "tablet" passes out of the body in the stool.

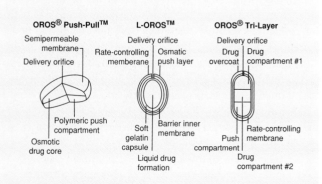

scientific study. At the conclusion of this chapter, you should have the ability to integrate facts and ideas into a meaningful whole and concisely convey a sense of that meaning to a third party. For example, if you are a pharmacy practitioner, you should be able to translate a complex scientific principle to a simple, practical, and useful recommendation for a patient.

The comprehension of course material is primarily the responsibility of the student. The teacher can guide and direct, explain, and clarify, but competence in solving problems in the classroom and the laboratory depends largely on the student's understanding of theory, recall of facts, ability to integrate knowledge, and willingness to devote sufficient time and effort to the task. Each assignment should be read and outlined, and assigned problems should be solved outside the classroom. The teacher's comments then will serve to clarify questionable points and aid the student to improve his or her judgment and reasoning abilities.

## MEASUREMENTS, DATA, PROPAGATION OF UNCERTAINTY

The goal of this chapter is to provide a foundation for the *quantitative reasoning* skills that are fundamental to the pharmacy practitioner and pharmaceutical scientist. "As mathematics is the language of science, statistics is the logic of science."[3] Mathematics and statistics are fundamental tools of the pharmaceutical sciences. You need to understand how and when to use these tools, and how to interpret what they tell us. You must also be careful not to overinterpret results. On the one hand, you may ask "do we really need to know how these equations and formulas were derived in order to use them effectively?" Logically, the answer would seem to be no. By analogy, you do not need to know how to build a computer in order to use one to send an e-mail message, do you? On the other hand, graphically represented data convey a sense dynamics that benefit from understanding a bit more about the fundamental equations behind the behavior. These equations are merely tools (that you should not memorize!) that allow for the transformation of a bunch of numbers into a behavior that you can interpret.

The mathematics and statistics covered in this chapter and this book are presented in a format to promote understanding and practical use. Therefore, many of the basic mathematical "tutorial" elements have been removed from the sixth edition, and in particular this chapter, because of the migration of numerous college-level topics to secondary school courses over the years. However, if you believe that you need a refresher in basic mathematical concepts, this information is still available in the online companion to this text (at thePoint.lww.com/Sinko6e). Statistical formulas and graphical method explanations have also been dramatically reduced in this edition. Depending on your personal goals and the philosophy of your program of study, you may well need an in-depth treatment of the subject matter. Additional detailed treatments can be found on the Web site and in the recommended readings.

## Data Analysis Tools

Readily available tools such as programmable calculators, computer spreadsheet programs (e.g., Microsoft Excel, Apple Numbers, or OpenOffice.org Calc), and statistical software packages (e.g., Minitab, SAS or SPSS) make the processing of data relatively easy. Spreadsheet programs have two distinct advantages: (1) data collection/entry is simple and can often be automated, reducing the possibility of errors in transcription, and (2) simple data manipulations and elementary statistical calculations are also easy to perform. In addition, many spreadsheet programs seamlessly interface with statistical packages when more robust statistical analysis is required. With very little effort, you can add data sets and generate pages of analysis. The student should appreciate that while it may be possible to automate data entry and have the computer perform calculations, the final interpretation of the results and statistical analysis is your responsibility! As you set out to analyze data keep in mind the simple acronym *GIGO—Garbage In, Garbage Out*. In other words, solid scientific results and sound methods of analysis will yield meaningful interpretations and conclusions. However, if the scientific foundation is weak, there are no known statistical tools that can make bad data significant.

## Dimensional Analysis

Dimensional analysis (also called the factor-label method or the unit factor method) is a problem-solving method that uses the fact that any number or expression can be multiplied by 1 without changing its value. For example, since 2.2 kg = 1 lb, you can divide both sides of the relationship by "1 lb," resulting in 2.2 kg/1 lb = 1. This is known as a *conversion factor*, which is a ratio of like-dimensioned quantities and is equal to the dimensionless unity (in other words, is equal to 1). On the face of it, the concept may seem a bit abstract and not very practical. However, dimensional analysis is very useful for any value that has a "unit of measure" associated with it, which is nearly everything in the pharmaceutical sciences. Simply put, this is a practical method for converting the units of one item to the units of another item.

## EXAMPLE 1–1

Solving problems using dimensional analysis is straightforward. You do not need to worry about the actual numbers until the very end. At first, simply focus on the units. Plug in all of the conversion factors that cancel out the units you do not want until you end up with the units that you do want. Only then do you need to worry about doing the calculation. If the units work out, you will get the right answer every time. In this example, the goal is to illustrate how to use the method for converting one value to another.

*Question*: How many seconds are there in 1 year?
*Conversion Factors*:
365 days = 1 year  24 hr = 1 day  60 min = 1 hr  60 sec = 1 min
*Rearrange Conversion Factors*:

$$\frac{365 \text{ days}}{1 \text{ year}} = 1 \quad \frac{24 \text{ hr}}{1 \text{ day}} = 1 \quad \frac{60 \text{ min}}{1 \text{ hr}} = 1 \quad \frac{60 \text{ sec}}{1 \text{ min}} = 1$$

Solve (arrange conversion factors so that the units that you do not want cancel out):

$$\frac{365 \text{ days}}{1 \text{ year}} \times \frac{24 \text{ hr}}{1 \text{ day}} \times \frac{60 \text{ min}}{1 \text{ hr}} \times \frac{60 \text{ sec}}{1 \text{ min}}$$

as you see the units become seconds.
*Calculate*: Now, plug the numbers carefully into your calculator and the resulting answer is 31,536,000 sec/year.

## EXAMPLE 1–2

This example will demonstrate the use of dimensional analysis in performing a calculation. How many calories are there in 3.00 joules? One should first recall a relationship or ratio that connects calories and joules. The relation 1 cal = 4.184 joules comes to mind. This is the key conversion factor required to solve this problem. The question can then be asked keeping in mind the conversion factor: If 1 cal equals 4.184 joules, how many calories are there in 3.00 joules? Write down the conversion factor, being careful to express each quantity in its proper units. For the unknown quantity, use an X.

$$X = \frac{3.00 \text{ joules} \times 1 \text{ cal}}{4.184 \text{ joules}}$$

$$X = 0.717 \text{ cal}$$

## EXAMPLE 1–3

How many gallons are equivalent to 2.0 liters? It would be necessary to set up successive proportions to solve this problem. In the method involving the identity of units on both sides of the equation, the quantity desired, X (gallons), is placed on the left and its equivalent, 2.0 liters, is set down on the right side of the equation. The right side must then be multiplied by known relations in ratio form, such as 1 pint per 473 mL, to give the units of gallons. Carrying out the indicated operations yields the result with its proper units:

$$X \text{ (in gallons)} = 2.0 \text{ liter} \times (1000 \text{ mL/liter})$$
$$\times (1 \text{ pint/473 mL}) \times (1 \text{ gallon/8 pints})$$
$$X = 0.53 \text{ gallon}$$

One may be concerned about the apparent disregard for the rules of significant figures in the equivalents such as 1 pint = 473 mL. The quantity of pints can be measured as accurately as that of milliliters, so that we assume 1.00 pint is meant here. The quantities 1 gallon and 1 liter are also exact by definition, and significant figures need not be considered in such cases.

## Significant Figures

A significant figure is any digit used to represent a magnitude or a quantity in the place in which it stands. The rules for interpreting significant figures and some examples are shown in **Table 1–1**. Significant figures give a sense of the accuracy of a number. They include all digits except leading and trailing zeros where they are used merely to locate the decimal point. Another way to state this is, the significant figures of a number include all certain digits plus the first uncertain digit. For example, one may use a ruler, the smallest subdivisions of which are centimeters, to measure the length of a piece of

## TABLE 1–1
## WRITING OR INTERPRETING SIGNIFICANT FIGURES IN NUMBERS

| Rule | Example |
| --- | --- |
| All nonzero digits are considered significant | 98.513 has five significant figures: 9, 8, 5, 1, and 3 |
| Leading zeros are not significant | 0.00361 has three significant figures: 3, 6, and 1 |
| Trailing zeros in a number containing a decimal point are significant | 998.100 has six significant figures: 9, 9, 8, 1, 0, and 0 |
| The significance of trailing zeros in a number not containing a decimal point can be ambiguous | The number of significant figures in numbers like 11,000 is uncertain because a decimal point is missing. If the number was written as 11,000, it would be clear that there are five significant figures |
| Zeros appearing anywhere between two nonzero digits are significant | 607.132 has six significant figures: 6, 0, 7, 1, 3, and 2 |

glass tubing. If one finds that the tubing measures slightly greater than 27 cm in length, it is proper to estimate the doubtful fraction, say 0.4, and express the number as 27.4 cm. A replicate measurement may yield the value 27.6 or 27.2 cm, so that the result is expressed as 27.4 ± 0.2 cm. When a value such as 27.4 cm is encountered in the literature without further qualification, the reader should assume that the final figure is correct to within about ±1 in the last decimal place, which is meant to signify the mean deviation of a single measurement. However, when a statement such as "not less than 99" is given in an official compendium, it means 99.0 and not 98.9.

## EXAMPLE 1–4

### How Many Significant Figures in the Number 0.00750?

The two zeros immediately following the decimal point in the number 0.00750 merely locate the decimal point and are not significant. However, the zero following the 5 is significant because it is not needed to write the number; if it were not significant, it could be omitted. Thus, the value contains three significant figures.

### How Many Significant Figures in the Number 7500?

The question of significant figures in the number 7500 is ambiguous. One does not know whether any or all of the zeros are meant to be significant or whether they are simply used to indicate the magnitude of the number. *Hint*: To express the significant figures of such a value in an unambiguous way, it is best to use exponential notation. Thus, the expression $7.5 \times 10^3$ signifies that the number contains two significant figures, and the zeros in 7500 are not to be taken as significant. In the value $7.500 \times 10^3$, both zeros are significant, and the number contains a total of four significant figures.

*Significant figures* are particularly useful for indicating the precision of a result. The proper interpretation of a value may be questioned specifically in cases when performing calculations (e.g., when spurious digits introduced by calculations carried out to greater accuracy than that of the original data) or when reporting measurements to a greater precision than the equipment supports. It is important to remember that the instrument used to make the measurement limits the precision of the resulting value that is reported. For example, a measuring rule marked off in centimeter divisions will not produce as great a precision as one marked off in 0.1 cm or mm. One may obtain a length of 27.4 ± 0.2 cm with the first ruler and a value of, say, 27.46 ± 0.02 cm with the second. The latter ruler, yielding a result with four significant figures, is obviously the more precise one. The number 27.46 implies

a precision of about 2 parts in 3000, whereas 27.4 implies a precision of only 2 parts in 300.

The absolute magnitude of a value should not be confused with its precision. We consider the number 0.00053 mole/liter as a relatively small quantity because three zeros immediately follow the decimal point. These zeros are not significant, however, and tell us nothing about the precision of the measurement. When such a result is expressed as $5.3 \times 10^{-4}$ mole/liter, or better as $5.3\,(\pm 0.1) \times 10^{-4}$ mole/liter, both its precision and its magnitude are readily apparent.

## EXAMPLE 1–5

The following example is used to illustrate *excessive precision*. If a faucet is turned on and 100 mL of water flows from the spigot in 31.47 sec, what is the average volumetric flow rate? By dividing the volume by time using a calculator, we get a rate of 3.17762948840165 mL/sec. Directly stating the uncertainty is the simplest way to indicate the precision of any result. Indicating the flow rate as 3.177 ± 0.061 mL/sec is one way to accomplish this. This is particularly appropriate when the uncertainty itself is important and precisely known. If the degree of precision in the answer is not important, it is acceptable to express trailing digits that are not known exactly, for example, 3.1776 mL/sec. If the precision of the result is not known you must be careful in how you report the value. Otherwise, you may overstate the accuracy or diminish the precision of the result.

In dealing with experimental data, certain rules pertain to the figures that enter into the computations:

1. In rejecting superfluous figures, increase by 1 the last figure retained if the following figure rejected is 5 or greater. Do not alter the last figure if the rejected figure has a value of less than 5.
2. Thus, if the value 13.2764 is to be rounded off to four significant figures, it is written as 13.28. The value 13.2744 is rounded off to 13.27.
3. In addition or subtraction include only as many figures to the right of the decimal point as there are present in the number with the least such figures. Thus, in adding 442.78, 58.4, and 2.684, obtain the sum and then round off the result so that it contains only one figure following the decimal point:

   This figure is rounded off to 503.9.

   Rule 2 of course cannot apply to the weights and volumes of ingredients in the monograph of a pharmaceutical preparation. The minimum weight or volume of each ingredient in a pharmaceutical formula or a prescription

## ⬛🠚KEY CONCEPT "WHEN SIGNIFICANT FIGURES DO NOT APPLY"

Since significant figure rules are based upon estimations derived from statistical rules for handling probability distributions, they apply only to *measured* values. The concept of significant figures does not pertain to values that are known to be exact. For example, integer counts (e.g., the number of tablets dispensed in a prescription bottle); legally defined conversions such as

1 pint = 473 mL; constants that are defined arbitrarily (e.g., a centimeter is 0.01 m); scalar operations such as "doubling" or "halving"; and mathematical constants, such as $\pi$ and $e$. However, physical constants such as Avogadro's number have a limited number of significant figures since the values for these constants are derived from measurements.

should be large enough that the error introduced is no greater than, say, 5 in 100 (5%), using the weighing and measuring apparatus at hand. Accuracy and precision in prescription compounding are discussed in some detail by Brecht.[5]

4. In multiplication or division, the rule commonly used is to retain the same number of significant figures in the result as appears in the value with the least number of significant figures. In multiplying 2.67 and 3.2, the result is recorded as 8.5 rather than as 8.544. A better rule here is to retain in the result the number of figures that produces a percentage error no greater than that in the value with the largest percentage uncertainty.

5. In the use of logarithms for multiplication and division, retain the same number of significant figures in the mantissa as there are in the original numbers. The characteristic signifies only the magnitude of the number and accordingly is not significant. Because calculations involved in theoretical pharmacy usually require no more than three significant figures, a four-place logarithm table yields sufficient precision for our work. Such a table is found on the inside back cover of this book. The calculator is more convenient, however, and tables of logarithms are rarely used today.

6. If the result is to be used in further calculations, retain at least one digit more than suggested in the rules just given. The final result is then rounded off to the last significant figure.

Remember, significant figures are not meant to be a perfect representation of uncertainty. Instead, they are used to prevent the loss of precision when rounding numbers. They also help you avoid stating more information than you actually know. Error and uncertainty are not the same. For example, if you perform an experiment in triplicate (in other words, you repeat the experiment three times), you will get a value that looks something like $4.351 \pm 0.076$. This does not mean that you made an error in the experiment or the collection of the data. It simply means that the outcome is naturally statistical.

## Data Types

The scientist is continually attempting to relate phenomena and establish generalizations with which to consolidate and interpret experimental data. The problem frequently resolves into a search for the relationship between two quantities that are changing at a certain rate or in a particular manner. The dependence of one property, the *dependent variable*, $y$, on the change or alteration of another measurable quantity, the *independent variable* $x$, is expressed mathematically as

$$y \propto x \qquad (1\text{--}1)$$

which is read "$y$ varies directly as $x$" or "$y$ is directly proportional to $x$." A proportionality is changed to an equation as follows. If $y$ is proportional to $x$ in general, then all pairs of

specific values of $y$ and $x$, say $y_1$ and $x_1$, $y_2$ and $x_2, \ldots,$ are proportional. Thus,

$$\frac{y_1}{x_1} = \frac{y_2}{x_2} = \ldots \qquad (1\text{--}2)$$

Because the ratio of any $y$ to its corresponding $x$ is equal to any other ratio of $y$ and $x$, the ratios are constant, or, in general

$$\frac{y}{x} = \text{Constant} \qquad (1\text{--}3)$$

Hence, it is a simple matter to change a proportionality to an equality by introducing a *proportionality constant*, $k$. To summarize, if

$$y \propto x$$

then

$$y = kx \qquad (1\text{--}4)$$

It is frequently desirable to show the relationship between $x$ and $y$ by the use of the more general notation

$$y = f(x) \qquad (1\text{--}5)$$

which is read "$y$ is some function of $x$." That is, $y$ may be equal, for example, to $2x$, to $27x^2$, or to $0.0051 + \log(a/x)$. The functional notation in equation (1–5) merely signifies that $y$ and $x$ are related in some way without specifying the actual equation by which they are connected.

As we begin to lay the foundation for the interpretation of data using descriptive statistics, some background information about the types of data that you will encounter in the pharmaceutical sciences is needed. In 1946, Stevens defined measurement as "the assignment of numbers to objects or events according to a rule."[4] He proposed a classification system that is widely used today to define data types. The first two, *intervals* and *ratios*, are categorized as continuous variables. These would include results of laboratory measurements for nearly all of the data that are normally collected in the laboratory (e.g., concentrations, weights). Only ratio or interval measurements can have units of measurement, and these variables are quantitative in nature. In other words, if you were given a set of "interval" data you would be able to calculate the exact differences between the different values. This makes this type of data "quantitative." Since the interval between measurements can be very small, we can also say that the data are "continuous." Another laboratory example of interval data measures is temperature. Think of the gradations on a common thermometer (in Celsius or Fahrenheit scale)—they are typically spaced apart by 1 degree with minor gradations at the 1/10th degree. The intervals could become even smaller; however, because of the physical limitations of common thermometers, smaller gradations are not possible since they cannot be read accurately. Of course, with digital thermometers the gradations (or intervals) could be much smaller but then the precision of the thermometer may become questionable. Another temperature scale that will be used in various sections of this text is the Kelvin scale, a thermodynamic temperature scale. By international agreement, the Kelvin and Celsius scales are related through the definition of absolute zero (in other words, 0 K = –273.15°C). Since the thermodynamic temperature is measured relative to absolute zero, the Kelvin scale is considered a ratio measurement. This also holds true for other physical quantities such as length or mass. The third common data type in the pharmaceutical sciences is *ordinal scale* measurements. Ordinal measurements represent the rank order of what is being measured. "Ordinals" are more subjective than interval or ratio measurements.

The final type of measurement is called *nominal* data. In this type of measurement, there is no order or sequence of the observations. They are merely assigned different groupings such as by name, make, or some similar characteristic. For example, you may have three groups of tablets: white tablets, red tablets, and yellow tablets. The only way to associate the various tablets is by their color. In clinical research, variables measured at a nominal level include sex, marital status, or race. There are a variety of ways to classify data types and the student is referred to texts devoted to statistics such as those listed in the recommended readings at the end of this chapter.[6,7]

# ERROR AND DESCRIBING VARIABILITY

If one is to maintain a high degree of accuracy in the compounding of prescriptions, the manufacture of products on a large scale, or the analysis of clinical or laboratory research results, one must know how to locate and eliminate constant and accidental errors as far as possible. Pharmacists must recognize, however, that just as they cannot hope to produce a perfect pharmaceutical product, neither can they make an absolute measurement. In addition to the inescapable imperfections in mechanical apparatus and the slight impurities that are always present in chemicals, perfect accuracy is impossible because of the inability of the operator to make a measurement or estimate a quantity to a degree finer than the smallest division of the instrument scale.

*Error* may be defined as a deviation from the absolute value or from the true average of a large number of results. Two types of errors are recognized: *determinate* (constant) and *indeterminate* (random or accidental).

## Determinate Errors

Determinate or constant errors are those that, although sometimes unsuspected, can be avoided or determined and corrected once they are uncovered. They are usually present in each measurement and affect all observations of a series in the same way. Examples of determinate errors are those inherent in the particular method used, errors in the calibration and the operation of the measuring instruments, impurities in the reagents and drugs, and biased personal errors that, for example, might recur consistently in the reading of a meniscus, in pouring and mixing, in weighing operations, in matching

colors, and in making calculations. The change of volume of solutions with temperature, although not constant, is a systematic error that can also be determined and accounted for once the coefficient of expansion is known.

Determinate errors can be reduced in analytic work by using a calibrated apparatus, using blanks and controls, using several different analytic procedures and apparatus, eliminating impurities, and carrying out the experiment under varying conditions. In pharmaceutical manufacturing, determinate errors can be eliminated by calibrating the weights and other apparatus and by checking calculations and results with other workers. Adequate corrections for determinate errors must be made before the estimation of indeterminate errors can have any significance.

## Indeterminate Errors

Indeterminate errors occur by accident or chance, and they vary from one measurement to the next. When one fires a number of bullets at a target, some may hit the bull's eye, whereas others will be scattered around this central point. The greater the skill of the marksman, the less scattered will be the pattern on the target. Likewise, in a chemical analysis, the results of a series of tests will yield a random pattern around an average or central value, known as the *mean*. Random errors will also occur in filling a number of capsules with a drug, and the finished products will show a definite variation in weight.

Indeterminate errors cannot be allowed for or corrected because of the natural fluctuations that occur in all measurements.

Those errors that arise from random fluctuations in temperature or other external factors and from the variations involved in reading instruments are not to be considered accidental or random. Instead, they belong to the class of determinate errors and are often called *pseudoaccidental* or *variable determinate* errors. These errors may be reduced by controlling conditions through the use of constant temperature baths and ovens, the use of buffers, and the maintenance of constant humidity and pressure where indicated. Care in reading fractions of units on graduates, balances, and other apparatus can also reduce pseudoaccidental errors. Variable determinate errors, although seemingly indeterminate, can thus be determined and corrected by careful analysis and refinement of technique on the part of the worker. Only errors that result from pure random fluctuations in nature are considered truly indeterminate.

## Precision and Accuracy

*Precision* is a measure of the agreement among the values in a group of data, whereas *accuracy* is the agreement between the data and the true value. Indeterminate or chance errors influence the precision of the results, and the measurement of the precision is accomplished best by statistical means. Determinate or constant errors affect the accuracy of data.

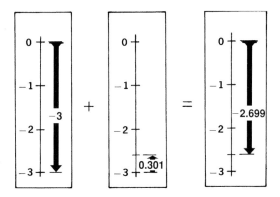

**Fig. 1–1.** The normal curve for the distribution of indeterminate errors.

The techniques used in analyzing the precision of results, which in turn supply a measure of the indeterminate errors, will be considered first, and the detection and elimination of determinate errors or inaccuracies will be discussed later.

Indeterminate or chance errors obey the laws of probability, both positive and negative errors being equally probable, and larger errors being less probable than smaller ones. If one plots a large number of results having various errors along the vertical axis against the magnitude of the errors on the horizontal axis, one obtains a bell-shaped curve, known as a *normal frequency distribution curve*, as shown in **Figure 1–1**. If the distribution of results follows the normal probability law, the deviations will be represented exactly by the curve for an infinite number of observations, which constitute the *universe* or *population*. Whereas the population is the whole of the category under consideration, the sample is that portion of the population used in the analysis.

# DESCRIPTIVE STATISTICS

Since the typical pharmacy student has sufficient exposure to descriptive statistics in other courses, this section will focus on introducing (or reintroducing) some of the key concepts that will be used numerous times in later chapters. The student who requires additional background in statistics is advised to seek out one of the many outstanding texts that have been published.[6,7] Descriptive statistics depict the basic features of a data set collected from an experimental study. They give summaries about the sample and the measures. However, viewing the individual data and tables of results alone is not always sufficient to understand the behavior of the data. Typically, a graphic analysis is paired with a tabular description to perform a quantitative analysis of the data set. The third component of descriptive statistics is "summary" statistics. These are single numbers that summarize the data. With interval data (e.g., the dose strength of individual tablets in a batch of 10,000 tablets), summary statistics focus on how big the value is and the variability among the values. The first of

these aspects relates to measures of "central tendency" (e.g., what is the average?), while the second refers to "dispersion" (in other words, the "variation" among a group of values).

## Central Tendency: Mean, Median, Mode

Central tendency can be described using a summary statistic (the mean, median, or mode) that gives an indication of the average value in the data set. The theoretical mean for a large number of measurements (the universe or population) is known as the *universe* or *population mean* and is given the symbol $\mu$ (mu).

The arithmetic mean $\overline{X}$ is obtained by adding together the results of the various measurements and dividing the total by the number $N$ of the measurements. In mathematical notation, the arithmetic mean for a small group of values is expressed as

$$\overline{X} = \frac{\sum(X_i)}{N} \qquad (1\text{–}6)$$

in which $\Sigma$ stands for "the sum of," $X_i$ is the $i$th individual measurement of the group, and $N$ is the number of values. $\overline{X}$ is an estimate of $\mu$ and approaches it as the number of measurements $N$ is increased. Remember, the "equations" used in all of the calculations are really a shorthand notation describing the various relationships that define some parameter.

**EXAMPLE 1–6**

A new student has just joined the lab and is being trained to pipette liquids correctly. She is using a 1-mL pipettor and is asked to withdraw 1 mL of water from a beaker and weigh it on a balance in a weighing boat. To determine her pipetting skill, she is asked to repeat this 10 times and take the average. What is the average volume of water that the student withdraws after 10 repeats? The density of water is 1 g/mL.

| Attempt | Weight (g) |
|---------|------------|
| 1 | 1.05 |
| 2 | 0.98 |
| 3 | 0.95 |
| 4 | 1.00 |
| 5 | 1.02 |
| 6 | 1.00 |
| 7 | 1.10 |
| 8 | 1.03 |
| 9 | 0.96 |
| 10 | 0.98 |

If $\Sigma X_i = 9.99$ and $N = 10$, so $9.99/10 = 0.999$. Given the number of significant figures, the average would be reported as 1.00 g, which equals 1 mL since the density of water is 1 g/mL.

The *median* is the middle value of a range of values when they are arranged in rank order (e.g., from lowest to highest). So, the median value of the list [1, 2, 3, 4, 5] is the number 3. In this case, the mean is also 3. So, which value is a better indicator of the central tendency of the data? The answer in this case is neither—both indicate central tendency equally well. However, the value of the median as a summary statistic becomes more obvious when the data set is skewed (in other words, when there are outliers or data points with values that are quite different from most of the others in the data set). For example, in the data set [1, 2, 2, 3, 10] the mean would be 3.6 but the median would be 2. In this case, the median is a better summary statistic than the mean because it gives a better representation of central tendency of the data set. Sometimes the median is referred to as a more "robust" statistic since it gives a reasonable outcome even with outlier results in the data set.

## EXAMPLE 1–7

As you have seen, calculating the median of a data set with an odd number of results is straightforward. But, what do you do when a data set has an even number of members? For example, in the data set [1, 2, 2, 3, 4, 10] you have 6 members to the data set. To calculate the median you need to find the two middle members (in this case, 2 and 3) then average them. So, the median would be 2.5.

Although it is human nature to want to "throw out" an outlying piece of data from a data set, it is not proper to do so under most circumstance or at least without rigorous statistical analysis. Using median as a summary statistic allows you to use all of the results in a data set and still get an idea of the central tendency of the results.

The *mode* is the value in the data set that occurs most often. It is not as commonly used in the pharmaceutical sciences but it has particular value in describing the most common occurrences of results that tend to center around more than one value (e.g., a bimodal distribution that has two commonly occurring values). For example, in the data set [1, 2, 4, 4, 5, 5, 5, 6, 9, 10] the mode value is equal to 5. However, we sometimes see a data set that has two "clusters" of results rather than one. For example, the data set [1, 2, 4, 4, 5, 5, 5, 6, 9, 10, 11, 11, 11, 11, 13, 14] is bimodal and thus has two modes (one mode is 5 and the other is 11). Taking the arithmetic mean of the data set would not give an indication of the bimodal behavior. Neither would the median.

## Variability: Measures of Dispersion

In order to fully understand the properties of the data set that you are analyzing, it is necessary to convey a sense of the dispersion or scatter around the central value. This is done so that an estimate of the variation in the data set can be calculated. This variability is usually expressed as the *range*, the *mean deviation*, or the *standard deviation*. Another useful measure of dispersion commonly used in the pharmaceutical sciences is the *coefficient of variation* (CV), which is a dimensionless parameter. Since much of this will be a review for many of the students using this text, only the most pertinent features will be discussed. The results obtained in the physical, chemical, and biological aspects of pharmacy have different characteristics. In the physical sciences, for example, instrument measurements are often not perfectly reproducible. In other words, variability may result from random measurement errors or may be due to errors in observations. In the biological sciences, however, the source of variation is viewed slightly differently since members of a population differ greatly. In other words, biological variations that we typically observe are intrinsic to the individual, organism, or biological process.

The *range* is the difference between the largest and the smallest value in a group of data and gives a rough idea of the dispersion. It sometimes leads to ambiguous results, however, when the maximum and minimum values are not in line with the rest of the data. The range will not be considered further.

The average distance of all the hits from the bull's eye would serve as a convenient measure of the scatter on the target. The average spread about the arithmetic mean of a large series of weighings or analyses is the mean deviation δ of the population. The sum of the positive and negative deviations about the mean equals zero; hence, the algebraic signs are disregarded to obtain a measure of the dispersion. The *mean deviation d* for a sample, that is, the deviation of an individual observation from the arithmetic mean of the sample, is obtained by taking the difference between each individual value $X_i$ and the arithmetic mean $\overline{X}$, adding the differences without regard to the algebraic signs, and dividing the sum by the number of values to obtain the average. The mean deviation of a sample is expressed as

$$d = \frac{\sum |X_i - \overline{X}|}{N} \qquad (1\text{–}7)$$

in which $\sum |X_i - \overline{X}|$ is the sum of the absolute deviations from the mean. The vertical lines on either side of the term in the numerator indicate that the algebraic sign of the deviation should be disregarded.

Youden[6] discourages the use of the mean deviation because it gives a biased estimate that suggests a greater precision than actually exists when a small number of values are used in the computation. Furthermore, the mean deviation of small subsets may be widely scattered around the average of the estimates, and accordingly, *d* is not particularly efficient as a measure of precision.

The *standard deviation* σ (the Greek lowercase letter sigma) is the square root of the mean of the squares of the deviations. This parameter is used to measure the dispersion or variability of a large number of measurements, for example, the weights of the contents of several million capsules. This set of items or measurements approximates the *population* and σ is, therefore, called the *population standard deviation*. Population standard deviations are shown in **Figure 1–1**.

As previously noted, any finite group of experimental data may be considered as a subset or sample of the population; the statistic or characteristic of a sample from the universe used to express the variability of a subset and supply an estimate of the standard deviation of the population is known as the *sample standard deviation* and is designated by the letter $s$. The formula is

$$s = \sqrt{\frac{\sum (X_i - \overline{X})^2}{N}} \qquad (1\text{--}8)$$

For a small sample, the equation is written

$$s = \sqrt{\frac{\sum (X_i - \overline{X})^2}{N - 1}} \qquad (1\text{--}9)$$

The term $(N - 1)$ is known as the *number of degrees of freedom*. It replaces $N$ to reduce the bias of the standard deviation $s$, which on the average is lower than the universe standard deviation.

The reason for introducing $(N - 1)$ is as follows. When a statistician selects a sample and makes a single measurement or observation, he or she obtains at least a rough estimate of the mean of the parent population. This single observation, however, can give no hint as to the degree of variability in the population. When a second measurement is taken, however, a first basis for estimating the population variability is obtained. The statistician states this fact by saying that two observations supply one *degree of freedom* for estimating variations in the universe. Three values provide two degrees of freedom, four values provide three degrees of freedom, and so on. Therefore, we do not have access to all $N$ values of a sample for obtaining an estimate of the standard deviation of the population. Instead, we must use 1 less than $N$, or $(N - 1)$, as shown in equation (1–9). When $N$ is large, say $N > 100$, we can use $N$ instead of $(N - 1)$ to estimate the population standard deviation because the difference between the two is negligible.

Modern statistical methods handle small samples quite well; however, the investigator should recognize that the estimate of the standard deviation becomes less reproducible and, on the average, becomes lower than the population standard deviation as fewer samples are used to compute the estimate. However, for many students studying pharmacy there is no compelling reason to view standard deviation in highly technical terms. So, we will simply refer to standard deviation as "SD" from this point forward.

A sample calculation involving the arithmetic mean, the mean deviation, and the estimate of the standard deviation follows.

### EXAMPLE 1–8

A pharmacist receives a prescription for a patient with rheumatoid arthritis calling for seven divided powders, each of which is to weigh 1.00 g. To check his skill in filling the powders, he removes the contents from each paper after filling the prescription by the block-and-divide method and then weighs the powders carefully. The results of the weighings are given in the first column of Table 1–2; the deviations of each value from the arithmetic mean, disregarding the sign, are given in column 2, and the squares of the deviations are shown in the last column. Based on the use of the mean deviation, the weight of the powders can be expressed as 0.98 ± 0.046 g. The variability of a single powder can also be expressed in

### TABLE 1–2
### STATISTICAL ANALYSIS OF DIVIDED POWDER COMPOUNDING TECHNIQUE

| Weight of Powder Contents (g) | Deviation (Sign Ignored), $\lvert X_i - \overline{X} \rvert$ | Square of the Deviation, $(X_i - \overline{X})^2$ |
|---|---|---|
| 1.00 | 0.02 | 0.0004 |
| 0.98 | 0.00 | 0.0000 |
| 1.00 | 0.02 | 0.0004 |
| 1.05 | 0.07 | 0.0049 |
| 0.81 | 0.17 | 0.0289 |
| 0.98 | 0.00 | 0.0000 |
| 1.02 | 0.04 | 0.0016 |
| Total $\Sigma = 6.84$ | $\Sigma = 0.32$ | $\Sigma = 0.0362$ |
| Average 0.98 | 0.046 | |

terms of percentage deviation by dividing the mean deviation by the arithmetic mean and multiplying by 100. The result is 0.98% ± 4.6%; of course, it includes errors due to removing the powders from the papers and weighing the powders in the analysis.

The standard deviation is used more frequently than the mean deviation in research. For large sets of data, it is approximately 25% larger than the mean deviation, that is, $\sigma = 1.25\delta$.

Statisticians have estimated that owing to chance errors, about 68% of all results in a large set will fall within one standard deviation on either side of the arithmetic mean, 95.5% within ±2 standard deviations, and 99.7% within ±3 standard deviations, as seen in **Figure 1–1**.

Goldstein[7] selected $1.73\delta$ as an equitable tolerance standard for prescription products, whereas Saunders and Fleming[8] advocated the use of ±3$\sigma$ as approximate limits of error for a single result. In pharmaceutical work, it should be considered permissible to accept ±2$s$ as a measure of the variability or "spread" of the data in small samples. Then, roughly 5% to 10% of the individual results will be expected to fall outside this range if only chance errors occur.

The estimate of the standard deviation in *Example 1–8* is calculated as follows:

$$s = \sqrt{\frac{0.0362}{(7 - 2)}} = 0.078 \text{ g}$$

and ±2$s$ is equal to ±0.156 g. That is, based upon the analysis of this experiment, the pharmacist should expect that roughly 90% to 95% of the sample values would fall within ±0.156 g of the sample mean.

The smaller the standard deviation estimate (or the mean deviation), the more *precise* is the operation. In the filling of capsules, precision is a measure of the ability of the pharmacist to put the same amount of drug in each capsule and to reproduce the result in subsequent operations. Statistical techniques for predicting the probability of occurrence of a specific deviation in future operations, although important in pharmacy, require methods that are outside the scope of this book. The interested reader is referred to treatises on statistical analysis.

Whereas the average deviation and the standard deviation can be used as measures of the *precision* of a method, the difference between the arithmetic mean and the *true* or *absolute* value expresses the error that can often be used as a measure of the *accuracy* of the method.

The true or absolute value is ordinarily regarded as the universe mean $\mu$—that is, the mean for an infinitely large set—because it is assumed that the true value is approached as the sample size becomes progressively larger. The universe mean does not, however, coincide with the true value of the quantity measured in those cases in which determinate errors are inherent in the measurements.

The difference between the sample arithmetic mean and the true value gives a measure of the accuracy of an operation; it is known as the *mean error*.

In *Example 1–8*, the true value is 1.00 g, the amount requested by the physician. The apparent error involved in compounding this prescription is

$$E = 1.0 - 0.98 = +0.02 \text{ g}$$

in which the positive sign signifies that the true value is greater than the mean value. An analysis of these results shows, however, that this difference is not statistically significant but rather is most likely due to accidental errors. Hence, the accuracy of the operation in *Example 1–8* is sufficiently great that no systemic error can be presumed. However, on further analysis it is found that one or several results are questionable. This possibility is considered later. If the arithmetic mean in *Example 1–8* were 0.90 instead of 0.98, the difference could be stated with assurance to have statistical significance because the probability that such a result could occur by chance alone would be small.

The mean error in this case is

$$1.00 - 0.90 = 0.10 \text{ g}$$

The *relative error* is obtained by dividing the mean error by the true value. It can be expressed as a percentage by multiplying by 100 or in parts per thousand by multiplying by 1000. It is easier to compare several sets of results by using the relative error rather than the absolute mean error. The relative error in the case just cited is

$$\frac{0.10 \text{ g}}{1.00 \text{ g}} \times 100 = 10\%$$

The reader should recognize that it is possible for a result to be precise without being accurate, that is, a constant error is present. If the capsule contents in *Example 1–8* had yielded an average weight of 0.60 g with a mean deviation of 0.5%, the results would have been accepted as precise. The degree of accuracy, however, would have been low because the aver-

age weight would have differed from the true value by 40%. Conversely, the fact that the result may be accurate does not necessarily mean that it is also precise. The situation can arise in which the mean value is close to the true value, but the scatter due to chance is large. Saunders and Fleming[8] observed, "it is better to be roughly accurate than precisely wrong."

A study of the individual values of a set often throws additional light on the exactitude of the compounding operations. Returning to the data of *Example 1–8* (**Table 1–2**), we note one rather discordant value, namely, 0.81 g. If the arithmetic mean is recalculated ignoring this measurement, we obtain a mean of 1.01 g. The mean deviation without the doubtful result is 0.02 g. It is now seen that the divergent result is 0.20 g smaller than the new average or, in other words, its deviation is 10 times greater than the mean deviation. A deviation greater than four times the mean deviation will occur purely by chance only about once or twice in 1000 measurements; hence, the discrepancy in this case is probably caused by some definite error in technique. Statisticians rightly question this rule, but it is a useful though not always reliable criterion for finding discrepant results.

Having uncovered the variable weight among the units, one can proceed to investigate the cause of the determinate error. The pharmacist may find that some of the powder was left on the sides of the mortar or on the weighing paper or possibly was lost during trituration. If several of the powder weights deviated widely from the mean, a serious deficiency in the compounder's technique would be suspected. Such appraisals as these in the college laboratory will aid the student in locating and correcting errors and will help the pharmacist become a safe and proficient compounder before entering the practice of pharmacy.

The CV is a dimensionless parameter that is quite useful. The CV relates the standard deviation to the mean and is defined as

$$CV = SD/mean \qquad (1\text{–}10)$$

It is valid only when the mean is nonzero. It is also commonly reported as a percentage (%CV is CV multiplied by 100). For example, if SD = 2 and mean = 3, then the %CV is 67%. The CV is useful because the standard deviation of data must always be understood in the context of the mean of the results. The CV should be used instead of the standard deviation to assess the difference between data sets with dissimilar units or very different means.

# VISUALIZING RESULTS: GRAPHIC METHODS, LINES

Scientists are not usually so fortunate as to begin each problem with an equation at hand relating the variables under study. Instead, the investigator must collect raw data and put them in the form of a table or graph to better observe the relationships. Constructing a graph with the data plotted in a manner so as to form a smooth curve often permits the investigator to observe the relationship more clearly and perhaps will allow expression of the connection in the form of a mathematical equation. The procedure of obtaining an empirical equation from a plot of the data is known as *curve fitting* and is treated in books on statistics and graphic analysis.

The magnitude of the independent variable is customarily measured along the horizontal coordinate scale, called the $x$ axis. The dependent variable is measured along the vertical scale, or the $y$ axis. The data are plotted on the graph, and a smooth line is drawn through the points. The $x$ value of each point is known as the $x$ coordinate or the *abscissa*; the $y$ value is known as the $y$ coordinate or the *ordinate*. The intersection of the $x$ axis and the $y$ axis is referred to as the *origin*. The $x$ and $y$ values may be either negative or positive.

We will first go through some of the technical aspects of lines and linear relationships. The simplest relationship between two variables, in which the variables contain no exponents other than 1, yields a straight line when plotted using rectangular coordinates. The straight-line or linear relationship is expressed as

$$y = a + bx \tag{1–11}$$

in which $y$ is the dependent variable, $x$ is the independent variable, and $a$ and $b$ are constants. The constant $b$ is the *slope* of the line; the greater the value of $b$, the steeper is the slope. It is expressed as the change in $y$ with the change in $x$, or $b = \frac{\Delta y}{\Delta x}$; $b$ is also the tangent of the angle that the line makes with the $x$ axis. The slope may be positive or negative depending on whether the line slants upward or downward to the right, respectively. When $b = 1$, the line makes an angle of $45°$ with the $x$ axis and the equation of the line can be written as follows:

$$y = a + x \tag{1–12}$$

When $b = 0$, the line is horizontal (in other words, parallel to the $x$ axis), and the equation reduces to

$$y = a \tag{1–13}$$

The constant $a$ is known as the $y$ *intercept* and denotes the point at which the line crosses the $y$ axis. If $a$ is positive, the line crosses the $y$ axis above the $x$ axis; if it is negative, the line intersects the $y$ axis below the $x$ axis. When $a$ is zero, equation (1–11) may be written as

$$y = bx \tag{1–14}$$

and the line passes through the origin.

The results of the determination of the refractive index of a benzene solution containing increasing concentrations of carbon tetrachloride are shown in **Table 1–3** The data are plotted in **Figure 1–2** and are seen to produce a straight line with a

### TABLE 1–3
### REFRACTIVE INDICES OF MIXTURES OF BENZENE AND CARBON TETRACHLORIDE

| Concentration of $CCl_4$ ($x$) (Volume %) | Refractive Index ($y$) |
|---|---|
| 10.0 | 1.497 |
| 25.0 | 1.491 |
| 33.0 | 1.488 |
| 50.0 | 1.481 |
| 60.0 | 1.477 |

negative slope. The equation of the line may be obtained by using the two-point form of the linear equation

$$y - y_1 = \frac{y_2 - y_1}{x_2 - x_1}(x - x_1) \tag{1–15}$$

The method involves selecting two widely separated points $(x_1, y_1)$ and $(x_2, y_2)$ on the line and substituting into the two-point equation.

### EXAMPLE 1–9

Referring to Figure 1–2, let 10.0% be $x_1$ and its corresponding $y$ value 1.497 be $y_1$; let 60.0% be $x_2$ and let 1.477 be $y_2$. The equation then becomes

$$y - 1.497 = \frac{1.477 - 1.497}{60.0 - 10.0}(x - 10.0)$$
$$y - 1.497 = -4.00 \times 10^{-4}(x - 10.0)$$
$$y = -4.00 \times 10^{-4}x + 1.501$$

The value $-4.00 \times 10^{-4}$ is the slope of the straight line and corresponds to $b$ in equation (1–11). A negative value for $b$ indicates that $y$ decreases with increasing values of $x$, as observed in **Figure 1–2**. The value 1.501 is the $y$ intercept and corresponds to $a$ in equation (1–11). It can be obtained

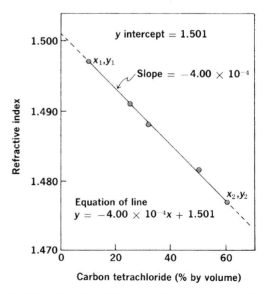

**Fig. 1–2.** Refractive index of the system benzene–carbon tetrachloride at 20°C.

### TABLE 1–4
### EMULSION STABILITY AS A FUNCTION OF EMULSIFIER CONCENTRATION

| Emulsifier (x) (% Concentration) | Oil Separation (y) (mL/month) | Logarithm of Oil Separation (log y) |
|---|---|---|
| 0.50 | 5.10 | 0.708 |
| 1.00 | 3.60 | 0.556 |
| 1.50 | 2.60 | 0.415 |
| 2.00 | 2.00 | 0.301 |
| 2.50 | 1.40 | 0.146 |
| 3.00 | 1.00 | 0.000 |

from the plot in **Figure 1–2** by *extrapolating* (extending) the line upward to the left until it intersects the y axis. It will also be observed that

$$\frac{y_2 - y_1}{x_2 - x_1} = \frac{\Delta y}{\Delta x} = b \qquad (1\text{–}16)$$

and this simple formula allows one to compute the slope of a straight line.

Not all experimental data form straight lines. Equations containing $x^2$ or $y^2$ are known as *second-degree* or *quadratic equations*, and graphs of these equations yield parabolas, hyperbolas, ellipses, and circles. The graphs and their corresponding equations can be found in standard textbooks on analytic geometry.

Logarithmic relationships occur frequently in scientific work. Data relating the amount of oil separating from an emulsion per month (dependent variable, y) as a function of the emulsifier concentration (independent variable, x) are collected in **Table 1–4**.

The data from this experiment may be plotted in several ways. In **Figure 1–3**, the oil separation y is plotted as ordinate against the emulsifier concentration x as abscissa on a rectangular coordinate grid. In **Figure 1–4**, the logarithm of the oil separation is plotted against the concentration. In

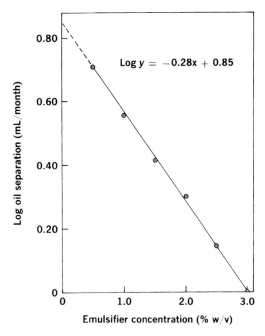

**Fig. 1–4.** A plot of the logarithm of oil separation of an emulsion versus concentration on a rectangular grid.

**Figure 1–5**, the data are plotted using semilogarithmic scale, consisting of a logarithmic scale on the vertical axis and a linear scale on the horizontal axis.

Although **Figure 1–3** provides a direct reading of oil separation, difficulties arise when one attempts to draw a smooth line through the points or to extrapolate the curve beyond the experimental data. Furthermore, the equation for the curve cannot be obtained readily from **Figure 1–3**. When the logarithm of oil separation is plotted as the ordinate, as in

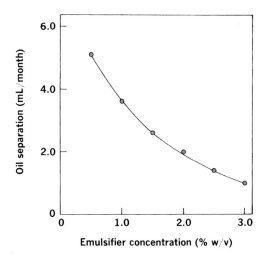

**Fig. 1–3.** Emulsion stability data plotted on a rectangular coordinate grid.

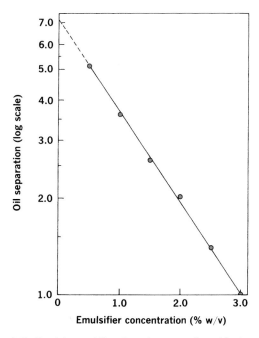

**Fig. 1–5.** Emulsion stability plotted on a semilogarithmic grid.

**Figure 1–4**, a straight line results, indicating that the phenomenon follows a logarithmic or exponential relationship. The slope and the *y* intercept are obtained from the graph, and the equation for the line is subsequently found by use of the two-point formula:

$$\log y = 0.85 - 0.28x$$

**Figure 1–4** requires that we obtain the logarithms of the oil-separation data before the graph is constructed and, conversely, that we obtain the antilogarithm of the ordinates to read oil separation from the graph. These inconveniences of converting to logarithms and antilogarithms can be overcome by plotting on a semilogarithmic scale. The *x* and *y* values of **Table 1–4** are plotted directly on the graph to yield a straight line, as seen in **Figure 1–5**. Although such a plot ordinarily is not used to obtain the equation of the line, it is convenient for reading the oil separation directly from the graph. It is well to remember that the ln of a number is simply 2.303 times the log of the number. Therefore, logarithmic graph scales may be used for ln as well as for log plots. In fact, today the natural logarithm is more commonly used than the base 10 log.

Not every pharmacy student will have the need to calculate slopes and intercepts of lines. In fact, once the basics are understood many of these operations can be performed quite easily with modern calculators. However, every pharmacy student should at least be able to look at a visual representation of data and get some sense of what it is telling you and why it is important. For example, the slope of a plasma drug concentration versus time curve is an approximation of the ratio of the input rate and the output rate of the drug in the body at that particular point in time (**Fig. 1–6**). At any given time point on that curve, the rate of change of drug in the body is equal to the rate of absorption (input) minus the rate of elimination (output or removal from the body). When the two rate processes are equal, the overall slope of the curve is zero. This is a very important $(x,y)$ point in pharmacokinetics because it is the time point where the peak blood levels occur $(C_{\max}, t_{\max})$. The rate of absorption is greater than the rate of elimination to left of the vertical line in **Figure 1–6**. This is called the *absorption phase*. When the rate of elimination is

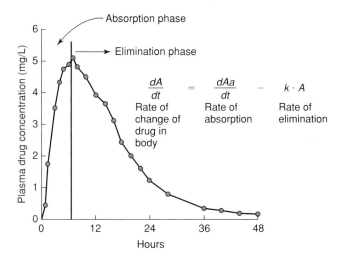

**Fig. 1–6.** A plot of the absorption and elimination phases of a typical plasma drug concentration versus time curve. During the absorption phase, the rate of input of drug into the body is greater than the rate of output (or elimination) of drug from the body. The opposite is true during the elimination phase. At the peak point ($T_{\max}$, $C_{\max}$), the rates of absorption and elimination are equal.

greater than the rate of absorption, it is called the *elimination phase*. This region falls to the right of the vertical line. The steepness of the slope is an indicator of the rate. For example, in **Figure 1–7**, three hypothetical products of the same drug are shown. As you can easily see, the rate of absorption of the drug into the bloodstream occurs most quickly from Product 1 and most slowly from Product 2 since the slope of the absorption phase is steepest for Product 1.

## Linear Regression Analysis

The data given in **Table 1–3** and plotted in **Figure 1–2** clearly indicate the existence of a linear relationship between the refractive index and the volume percent of carbon tetrachloride in benzene. The straight line that joins virtually all the points can be drawn readily on the figure by sighting the points along the edge of a ruler and drawing a line that can be extrapolated to the *y* axis with confidence.

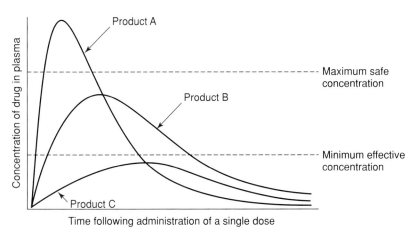

**Fig. 1–7.** A plot of plasma drug concentration versus time for three different products containing the same dose of a drug. Differences in the profiles are due to differences in the rate of absorption resulting from the three types of formulations. The slope of the absorption phase is equivalent to the rate of drug absorption. The steeper slope (Product A) equals a faster rate of absorption, whereas a less steep slope (Product C) has a slower absorption rate.

### TABLE 1–5

### REFRACTIVE INDICES OF MIXTURES OF BENZENE AND CARBON TETRACHLORIDE

| Concentration of $CCl_4$ ($x$) (Volume %) | Refractive Index ($y$) |
| --- | --- |
| 10.0 | 1.497 |
| 26.0 | 1.493 |
| 33.0 | 1.485 |
| 50.0 | 1.478 |
| 61.0 | 1.477 |

Let us suppose, however, that the person who prepared the solutions and carried out the refractive index measurements was not skilled and, as a result of poor technique, allowed indeterminate errors to appear. We might then be presented with the data given in **Table 1–5**. When these data are plotted on graph paper, an appreciable scatter is observed (**Fig. 1–8**) and we are unable, with any degree of confidence, to draw the line that expresses the relation between refractive index and concentration. It is here that we must employ better means of analyzing the available data.

The first step is to determine whether the data in **Table 1–5** should fit a straight line, and for this we calculate the *correlation coefficient, r*, using the following equation:

$$r = \frac{\sum(x - \bar{x})(y - \bar{y})}{\sqrt{\sum(x - \bar{x})^2 \sum(y - \bar{y})^2}} \quad (1\text{–}17)$$

When there is perfect correlation between the two variables (in other words, a perfect linear relationship), $r = 1$. When the two variables are completely independent, $r = 0$. Depending on the degrees of freedom and the chosen probability level, it is possible to calculate values of $r$ above which there is significant correlation and below which there is no significant correlation. Obviously, in the latter case, it is not profitable to proceed further with the analysis unless the data can be

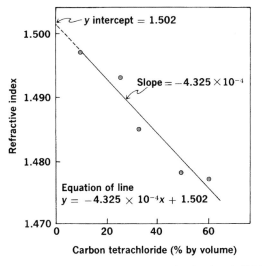

**Fig. 1–8.** Slope, intercept, and equation of line for data in Table 1–5 calculated by regression analysis.

plotted in some other way that will yield a linear relation. An example of this is shown in **Figure 1–4**, in which a linear plot is obtained by plotting the *logarithm* of oil separation from an emulsion against emulsifier concentration, as opposed to **Figure 1–3**, in which the raw data are plotted in the conventional manner.

Assuming that the calculated value of $r$ shows a significant correlation between $x$ and $y$, it is then necessary to calculate the slope and intercept of the line using the equation

$$b = \frac{\sum(x - \bar{x})(y - \bar{y})}{\sum(x - \bar{x})^2} \quad (1\text{–}18)$$

in which $\bar{b}$ is the *regression coefficient*, or slope. By substituting the value for $b$ in equation **(1–18)**, we can obtain the $y$ intercept:

$$\bar{y} = y + b(x - \bar{x}) \quad (1\text{–}19)$$

The following series of calculations, based on the data in **Table 1–5**, will illustrate the use of these equations.

### EXAMPLE 1–10

Using the data in Table 1–5, calculate the correlation coefficient, the regression coefficient, and the intercept on the $y$ axis.

Examination of equations (1–17) through (1–19) shows the various values we must calculate, and these are set up as follows:

| $x$ | $(x - \bar{x})$ | $(x - \bar{x})^2$ |
| --- | --- | --- |
| 10.0 | −26.0 | 676.0 |
| 26.0 | 10.0 | 100.0 |
| 33.0 | −3.0 | 9.0 |
| 50.0 | +14.0 | 196.0 |
| 61.0 | +25.0 | 625.0 |
| $\sum = 180.0$ | $\sum = 0$ | $\sum = 1606.0$ |
| $x = 36.0$ | | |

| $y$ | $(y - \bar{y})$ | $(y - \bar{y})^2$ |
| --- | --- | --- |
| 1.497 | +0.011 | 0.000121 |
| 1.493 | +0.007 | 0.000049 |
| 1.485 | −0.001 | 0.000001 |
| 1.478 | −0.008 | 0.000064 |
| 1.477 | −0.009 | 0.000081 |
| $\sum = 7.430$ | $\sum = 0$ | $\sum = 0.000316$ |
| $y \div 1.486$ | | |

| $(x - \bar{x})(y - \bar{y})$ |
| --- |
| −0.286 |
| 0.070 |
| +0.003 |
| −0.112 |
| −0.225 |
| $\sum = -0.690$ |

Substituting the relevant values into equation **(1–17)** gives

$$r = \frac{-0.690}{\sqrt{1606.0 \times 0.000316}} = -0.97$$

From equation **(1–18)**

$$b = \frac{-0.690}{1606.0} = -4.296 \times 10^4$$

and finally, from equation **(1–19)**

Intercept on the $y$ axis $= 1.486$

$$-4.315 \times 10^{-4}(0 - 36)$$
$$= +1.502$$

Note that for the intercept, we place $x$ equal to zero in equation (1–17). By inserting an actual value of $x$ into equation (1–19), we obtain the value of $y$ that should be found at that particular value of $x$. Thus, when $x = 10$,

$$y = 1.486 - 4.315 \times 10^{-4}(10 - 36)$$
$$= 1.486 - 4.315 \times 10^{-4}(-26)$$
$$= 1.497$$

The value agrees with the experimental value, and hence this point lies on the statistically calculated slope drawn in **Figure 1–8**.

## CHAPTER SUMMARY

Most of the statistical calculations reviewed in this chapter will be performed using a calculator or computer. The objective of this chapter was not to inundate you with statistical formulas or complex equations but rather to give the student a perspective on analyzing data as well as providing a foundation for the interpretation of results. Numbers alone are not dynamic and do not give a sense of the behavior of the results. In some situations, equations or graphic representations were used to give the more advanced student a sense of the dynamic behavior of the results.

**Practice problems for this chapter can be found at thePoint.lww.com/Sinko6e**

## References

1. D. L. Sackett, S. E. Straus, W. S. Richardson, W. Rosenberg, and R. B. Haynes, *Evidence-Based Medicine: How to Practice and Teach EBM*, 2nd Ed., Churchill Livingstone, Edinburgh, New York, 2000.
2. K. Skau, Am. J. Pharm. Educ. **71**, 11, 2007.
3. S. J. Ruberg, Teaching statistics for understanding and practical use. *Biopharmaceutical Report*, American Statistical Association, 1, 1992, pp. 14.
4. S. S. Stevens, Science, **103**, 677–680, 1946.
5. E. A. Brecht, in *Sprowls' American Pharmacy*, L. W. Dittert, Ed., 7th Ed., Lippincott Williams & Wilkins, Philadelphia, 1974, Chapter 2.
6. W. J. Youden, *Statistical Methods for Chemists*, R. Krieger, Huntington, New York, 1977, p. 9.
7. P. Rowe, in *Essential Statistics for the Pharmaceutical Sciences*, John Wiley & Sons, Ltd, West Sussex, England, 2007.
8. S. Bolton, *Pharmaceutical Statistics: Preclinical and Clinical Applications*, 3rd Ed., Marcel Dekker, Inc., New York, 1997.

## Recommended Readings

S. Bolton, *Pharmaceutical Statistics: Preclinical and Clinical Applications*, 3rd Ed., Marcel Dekker, Inc., New York, 1997.
P. Rowe, *Essential Statistics for the Pharmaceutical Sciences*, John Wiley & Sons, Inc., West Sussex, England, 2007.

### CHAPTER LEGACY

**Fifth Edition:** published as Chapter 1 (Introduction). Updated by Patrick Sinko.

**Sixth Edition:** published as Chapter 1 (Interpretive Tools). Updated by Patrick Sinko.

# 2 STATES OF MATTER

**CHAPTER OBJECTIVES** **At the conclusion of this chapter the student should be able to:**

**1** Understand the nature of the intra- and intermolecular forces that are involved in stabilizing molecular and physical structures.

**2** Understand the differences in these forces and their relevance to different types molecules.

**3** Discuss supercritical states to illustrate the utility of supercritical fluids for crystallization and microparticulate formulations.

**4** Appreciate the differences in the strengths of the intermolecular forces that are responsible for the stability of structures in the different states of matter.

**5** Perform calculations involving the ideal gas law, molecular weights, vapor pressure, boiling points, kinetic molecular theory, van der Waals real gases, the Clausius–Clapeyron equation, heats of fusion and melting points, and the phase rule equations.

**6** Understand the properties of the different states of matter.

**7** Describe the pharmaceutical relevance of the different states of matter to drug delivery systems by reference to specific examples given in the text boxes.

**8** Describe the solid state, crystallinity, solvates, and polymorphism.

**9** Describe and discuss key techniques utilized to characterize solids.

**10** Recognize and elucidate the relationship between differential scanning calorimetry, thermogravimetric, Karl Fisher, and sorption analyses in determining polymorphic versus solvate detection.

**11** Understand phase equilibria and phase transitions between the three main states of matter.

**12** Understand the phase rule and its application to different systems containing multiple components.

## BINDING FORCES BETWEEN MOLECULES

For molecules to exist as aggregates in gases, liquids, and solids, *inter*molecular forces must exist. An understanding of intermolecular forces is important in the study of pharmaceutical systems and follows logically from a detailed discussion of *intra*molecular bonding energies. Like *intra*molecular bonding energies found in covalent bonds, *inter*molecular bonding is largely governed by electron orbital interactions. The key difference is that covalency is not established in the *inter*molecular state. Cohesion, or the attraction of like molecules, and adhesion, or the attraction of unlike molecules, are manifestations of intermolecular forces. Repulsion is a reaction between two molecules that forces them apart. For molecules to interact, these forces must be balanced in an energetically favored arrangement. Briefly, the term energetically favored is used to describe the intermolecular distances and intramolecular conformations where the energy of the interaction is maximized on the basis of the balancing of attractive and repulsive forces. At this point, if the molecules are moved slightly in any direction, the stability of the interaction will change by either a decrease in attraction (when moving the molecules away from one another) or an increase in repulsion (when moving the molecules toward one another).

Knowledge of these forces and their balance (equilibrium) is important for understanding not only the properties of gases, liquids, and solids, but also interfacial phenomena, flocculation in suspensions, stabilization of emulsions,

compaction of powders in capsules, dispersion of powders or liquid droplets in aerosols, and the compression of granules to form tablets. With the rapid increase in biotechnology-derived products, it is important to keep in mind that these same properties are strongly involved in influencing biomolecular (e.g., proteins, DNA) secondary, tertiary, and quaternary structures, and that these properties have a profound influence on the stability of these products during production, formulation, and storage. Further discussion of biomolecular products will be limited in this text, but correlations hold between small-molecule and the larger biomolecular therapeutic agents due to the universality of the physical principles of chemistry.

### Repulsive and Attractive Forces

When molecules interact, both repulsive and attractive forces operate. As two atoms or molecules are brought closer together, the opposite charges and binding forces in the two molecules are closer together than the similar charges and forces, causing the molecules to attract one another. The negatively charged electron clouds of the molecules largely govern the balance (equilibrium) of forces between the two molecules. When the molecules are brought so close that the outer charge clouds touch, they repel each other like rigid elastic bodies.

Thus, attractive forces are necessary for molecules to cohere, whereas repulsive forces act to prevent the molecules from interpenetrating and annihilating each other.

Moelwyn-Hughes[1] pointed to the analogy between human behavior and molecular phenomena: Just as the actions of humans are often influenced by a conflict of loyalties, so too is molecular behavior governed by attractive and repulsive forces.

Repulsion is due to the interpenetration of the electronic clouds of molecules and increases exponentially with a decrease in distance between the molecules. At a certain equilibrium distance, about $(3-4) \times 10^{-8}$ cm (3–4 Å), the repulsive and attractive forces are equal. At this position, the potential energy of the two molecules is a minimum and the system is most stable (**Fig. 2–1**). This principle of minimum potential energy applies not only to molecules but also to atoms and to large objects as well. The effect of repulsion on the intermolecular three-dimensional structure of a molecule is well illustrated in considering the conformation of the two terminal methyl groups in butane, where they are energetically favored in the *trans*conformation because of a minimization of the repulsive forces. It is important to note that the arrangement of the atoms in a particular stereoisomer gives the *configuration* of a molecule. On the other hand, *conformation* refers to the different arrangements of atoms resulting from rotations about single bonds.

The various types of *attractive* intermolecular forces are discussed in the following subsections.

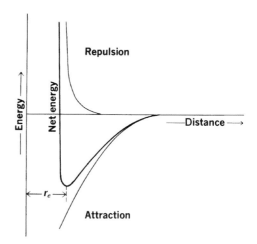

**Fig. 2–1.** Repulsive and attractive energies and net energy as a function of the distance between molecules. Note that a minimum occurs in the net energy because of the different character of the attraction and repulsion curves.

## Van der Waals Forces

Van der Waals forces relate to nonionic interactions between molecules, yet they involve charge–charge interactions (see Key Concept Box on van der Waals Forces). In organic

# KEY CONCEPT VAN DER WAALS FORCES

| Permanent dipole | Keesom forces | Debye forces | London forces |

van der Waal interactions are weak forces that involve the dispersion of charge across a molecule called a dipole. In a permanent dipole, as illustrated by the peptide bond, the electronegative oxygen draws the pair of electrons in the carbon–oxygen double bond closer to the oxygen nucleus. The bond then becomes polarized due to the fact that the oxygen atom is strongly pulling the nitrogen lone pair of electrons toward the carbon atom, thus creating a partial double bond. Finally, to compensate for valency, the nucleus of the nitrogen atom pulls the electron pair involved in the nitrogen–hydrogen bond closer to itself and creates a partial positive charge on the hydrogen. This greatly affects protein structure, which is beyond the scope of this discussion. In Keesom forces, the permanent dipoles interact with one another in an ionlike fashion. However, because the charges are partial, the strength of bonding is much weaker. Debye forces show the ability of a permanent dipole to polarize charge in a neighboring molecule. In London forces, two neighboring neutral molecules, for example, aliphatic hydrocarbons, induce partial charge distributions. If one conceptualizes the aliphatic chains in the lipid core of a membrane like a biologic membrane or a liposome, one can imagine the neighboring chains in the interior as inducing a network of these partial charges that helps hold the interior intact. Without this polarization, the membrane interior would be destabilized and lipid bilayers might break down. Therefore, London forces give rise to the fluidity and cohesiveness of the membrane under normal physiologic conditions.

chemistry, numerous reactions like nucleophilic substitutions are introduced where one molecule may carry a partial positive charge and be attractive for interaction with a partially negatively charged nucleophilic reactant. These partial charges can be permanent or be induced by neighboring groups, and they reflect the polarity of the molecule. The converse can be true for electrophilic reactants. The presence of these polarities in molecules can be similar to those observed with a magnet. For example, dipolar molecules frequently tend to align themselves with their neighbors so that the negative pole of one molecule points toward the positive pole of the next. Thus, large groups of molecules may be associated through weak attractions known as *dipole–dipole* or Keesom forces. Permanent dipoles are capable of inducing an electric dipole in nonpolar molecules (which are easily polarizable) to produce *dipole-induced dipole*, or Debye, interactions, and nonpolar molecules can induce polarity in one another by *induced dipole-induced dipole*, or London, attractions. This latter force deserves additional comment here.

The weak electrostatic force by which nonpolar molecules such as hydrogen gas, carbon tetrachloride, and benzene attract one another was first recognized by London in 1930. The *dispersion* or London force is sufficient to bring about the condensation of nonpolar gas molecules so as to form liquids and solids when molecules are brought quite close to one another. In all three types of van der Waals forces, the potential energy of attraction varies inversely with the distance of separation, $r$, raised to the sixth power, $r^6$. The potential energy of repulsion changes more rapidly with distance, as shown in **Figure 2–1**. This accounts for the potential energy minimum and the resultant equilibrium distance of separation, $r_e$.

A good conceptual analogy to illustrate this point is the interaction of opposite poles of magnets (**Fig. 2–2**). If two magnets of the same size are slid on a table so that the opposite poles completely overlap, the resultant interaction is attractive and the most energetically favored configuration

(**Fig. 2–2a**). If the magnets are slid further so that the poles of each slide into like-pole regions of the other (**Fig. 2–2b**), this leads to repulsion and a force that pushes the magnetic poles back to the energetically favored configuration (**Fig. 2–2a**). However, it must be noted that attractive (opposite-pole overlap) and repulsive (same-pole overlap) forces coexist. If the same-charged poles are slid into the proximity of one another, the resultant force is complete repulsion (**Fig. 2–2c**).

These several classes of interactions, known as van der Waals forces[*] and listed in **Table 2–1**, are associated with the condensation of gases, the solubility of some drugs, the formation of some metal complexes and molecular addition compounds, and certain biologic processes and drug actions. The energies associated with primary valence bonds are included for comparison.

## Orbital Overlap

An important *dipole–dipole* force is the interaction between pi-electron orbitals in systems. For example, aromatic–aromatic interactions can occur when the double-bonded pi-orbitals from the two rings overlap (**Fig. 2–3**).[2] Aromatic rings are dipolar in nature, having a partial negative charge in the pi-orbital electron cloud above and below the ring and partial positive charges residing at the equatorial hydrogens, as illustrated in **Figure 2–3a**. Therefore, a dipole–dipole interaction can occur between two aromatic molecules. In fact, at certain geometries aromatic–aromatic interaction can stabilize *inter-* and/or *intra*molecular interactions (**Fig. 2–3b** and *c*), with the highest energy interactions occurring when the rings are nearly perpendicular to one another.[2] This phenomena has been largely studied in proteins, where stacking of 50% to 60% of the aromatic side chains can often add stabilizing energy to secondary and tertiary structure (*inter*molecular) and may even participate in stabilizing quaternary interactions (*intra*molecular). Aromatic stacking can also occur in the solid state, and was first identified as a stabilizing force in the structure of small organic crystals.

It is important to point out that due to the nature of these interactions, repulsion is also very plausible and can be destabilizing if the balancing attractive force is changed. Finally, lone pairs of electrons on atoms like oxygen can also interact with aromatic pi orbitals and lead to attractive or repulsive interactions. These interactions are dipole–dipole in nature and they are introduced to highlight their importance. Students seeking additional information on this subject should read the excellent review by Meyer et al.[3]

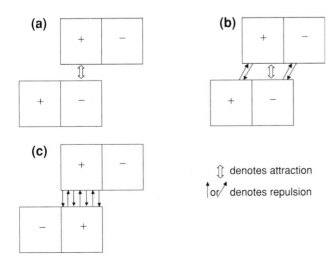

**(a)**

**(b)**

**(c)**

⇕ denotes attraction

↑ or ↗ denotes repulsion

**Fig. 2–2.** (*a*) The attractive, (*b*) partially repulsive, and (*c*) fully repulsive interactions of two magnets being brought together.

---

[*]The term van der Waals forces is often used loosely. Sometimes all combinations of intermolecular forces among ions, permanent dipoles, and induced dipoles are referred to as van der Waals forces. On the other hand, the London force alone is frequently referred to as the van der Waals force because it accounts for the attraction between nonpolar gas molecules, as expressed by the a/V2 term in the van der Waals gas equation. In this book, the three dipolar forces of Keesom, Debye, and London are called van der Waals forces. The other forces such as the ion-dipole interaction and the hydrogen bond (which have characteristics similar both to ionic and dipolar forces) are designated appropriately where necessary.

**TABLE 2–1**
## INTERMOLECULAR FORCES AND VALENCE BONDS

| Bond Type | Bond Energy (approximately) (kcal/mole) |
|---|---|
| Van der Waals forces and other intermolecular attractions | |
|     Dipole–dipole interaction, orientation effect, or Keesom force | |
|     Dipole-induced dipole interaction, induction effect, or Debye force | 1–10 |
|     Induced dipole–induced dipole interaction, dispersion effect, or London force | |
|     Ion–dipole interaction | |
|     Hydrogen bonds:     $O—H\cdots O$ | 6 |
|                      $C—H\cdots O$ | 2–3 |
|                      $O—H\cdots N$ | 4–7 |
|                      $N—H\cdots O$ | 2–3 |
|                      $F—H\cdots F$ | 7 |
| Primary valence bonds | |
|     Electrovalent, ionic, heteropolar | 100–200 |
|     Covalent, homopolar | 50–150 |

## Ion–Dipole and Ion-Induced Dipole Forces

In addition to the dipolar interactions known as van der Waals forces, other attractions occur between polar or nonpolar molecules and ions. These types of interactions account in part for the solubility of ionic crystalline substances in water; the cation, for example, attracts the relatively negative oxygen atom of water and the anion attracts the hydrogen atoms of the dipolar water molecules. Ion-induced dipole forces are presumably involved in the formation of the iodide complex,

$$I_2 + K^+I^- = K^+I_3^- \qquad (2\text{–}1)$$

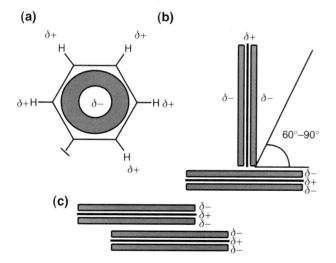

**Fig. 2–3.** Schematic depicting (*a*) the dipolar nature of an aromatic ring, (*b*) its preferred angle for aromatic–aromatic interactions between 60° and 90°, and (*c*) the less preferred planar interaction of aromatic rings. Although typically found in proteins, these interactions can stabilize states of matter as well. See the excellent review on this subject with respect to biologic recognition by Meyer et al.[3]

Reaction (2–1) accounts for the solubility of iodine in a solution of potassium iodide. This effect can clearly influence the solubility of a solute and may be important in the dissolution process.

## Ion–Ion Interactions

Another important interaction that involves charge is the ion–ion interaction. An ionic, electrovalent bond between two counter ions is the strongest bonding interaction and can persist over the longest distance. However, weaker ion–ion interactions, in particular salt formations, exist and influence pharmaceutical systems. This section focuses on those weaker ion–ion interactions. The ion–ion interactions of salts and salt forms have been widely discussed in prerequisite general chemistry and organic chemistry courses that utilize this text, but they will briefly be reviewed here.

It is well established that ions form because of valency changes in an atom. At neutrality, the number of protons and the number of electrons in the atom are equal. Imbalance in the ratio of protons to neutrons gives rise to a change in charge state, and the valency will dictate whether the species is cationic or anionic. Ion–ion interactions are normally viewed from the standpoint of attractive forces: A cation on one compound will interact with an anion on another compound, giving rise to an *inter*molecular association. Ion–ion interactions can also be repulsive when two ions of like charge are brought closely together. The repulsion between the like charges arises from electron cloud overlap, which causes the *inter*molecular distances to increase, resulting in an energetically favored dispersion of the molecules. The illustration of the magnetic poles in **Figure 2–2** offers an excellent corollary for the understanding of the attractive cationic (positive pole) and anionic (negative pole) interactions (panel A), as well as the need for proper distance to an energetically favored electrovalent interaction (panels A and B), and the repulsive forces that may occur between like charges (panels B and C). Ion–ion interactions may be *inter*molecular (e.g., a hydrochloride salt of a drug) or *intra*molecular (e.g., a salt-bridge interaction between counter ions in proteins).

Clearly, the strength of ion–ion interactions will vary according to the balancing of attractive and repulsive forces between the cation- and anion-containing species. It is important to keep in mind that ion–ion interactions are considerably stronger than many of the forces described in this section and can even be stronger than covalent bonding when an ionic bond is formed. The strength of ion–ion interactions has a profound effect on several physical properties of pharmaceutical agents including salt-form selection, solid-crystalline habit, solubility, dissolution, pH and p $K$ determination, and solution stability.

## Hydrogen Bonds

The interaction between a molecule containing a hydrogen atom and a strongly electronegative atom such as fluorine, oxygen, or nitrogen is of particular interest. Because of the small size of the hydrogen atom and its large electrostatic field, it can move in close to an electronegative atom and form an electrostatic type of union known as a *hydrogen bond* or *hydrogen bridge*. Such a bond, discovered by Latimer and Rodebush[4] in 1920, exists in ice and in liquid water; it accounts for many of the unusual properties of water including its high dielectric constant, abnormally low vapor pressure, and high boiling point. The structure of ice is an open but well ordered three-dimensional array of regular tetrahedra with oxygen in the center of each tetrahedron and hydrogen atoms at the four corners. The hydrogens are not exactly midway between the oxygens, as may be observed in **Figure 2–4**. Roughly one sixth of the hydrogen bonds of ice are broken when water passes into the liquid state, and essentially all the bridges are destroyed when it vaporizes. Hydrogen bonds can also exist between alcohol molecules, carboxylic acids, aldehydes, esters, and polypeptides.

The hydrogen bonds of formic acid and acetic acid are sufficiently strong to yield *dimers* (two molecules attached together), which can exist even in the vapor state. Hydrogen fluoride in the vapor state exists as a hydrogen-bonded polymer $(F-H\ldots)_n$, where $n$ can have a value as large as 6. This is largely due to the high electronegativity of the fluorine atom interacting with the positively charged, electropositive hydrogen atom (analogous to an ion–ion interaction). Several structures involving hydrogen bonds are shown in **Figure 2–4**. The dashed lines represent the hydrogen bridges. It will be noticed that *intra-* as well as *inter*molecular hydrogen bonds may occur (as in salicylic acid).

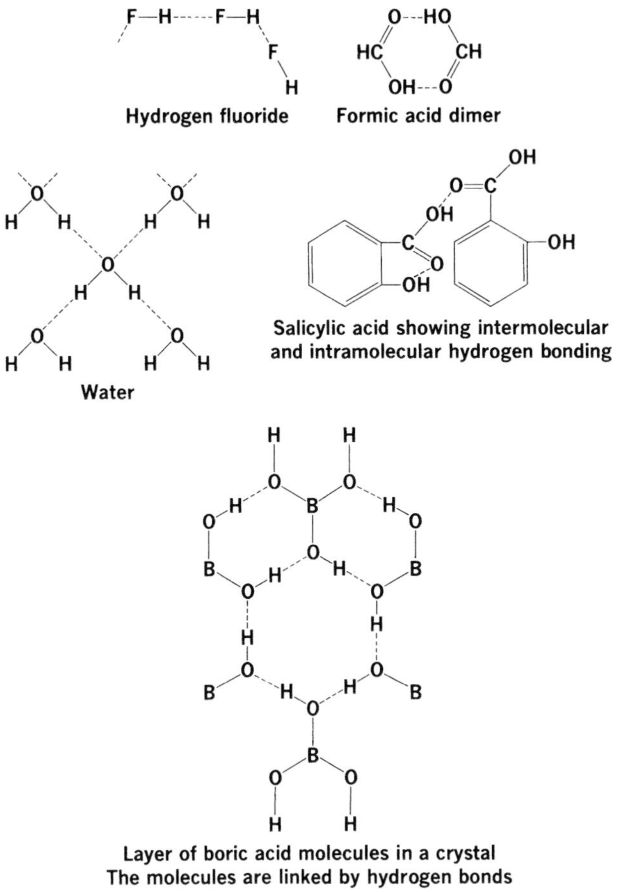

## Bond Energies

Bond energies serve as a measure of the strength of bonds. Hydrogen bonds are relatively weak, having a bond energy of about 2 to 8 kcal/mole as compared with a value of about 50

**Fig. 2–4.** Representative hydrogen-bonded structures.

to 100 kcal for the covalent bond and well over 100 kcal for the ionic bond. The metallic bond, representing a third type of primary valence, will be mentioned in connection with crystalline solids.

The energies associated with intermolecular bond forces of several compounds are shown in **Table 2–2**. It will be observed that the total interaction energies between molecules are contributed by a combination of orientation, induction, and dispersion effects. The nature of the molecules determines which of these factors is most influential in the attraction. In water, a highly polar substance, the orientation or dipole–dipole interaction predominates over the other two

## TABLE 2–2
### ENERGIES ASSOCIATED WITH MOLECULAR AND IONIC INTERACTIONS

| | Interaction (kcal/mole) | | | |
|---|---|---|---|---|
| Compound | Orientation | Induction | Dispersion | Total Energy |
| $H_2O$ | 8.69 | 0.46 | 2.15 | 11.30 |
| HCl | 0.79 | 0.24 | 4.02 | 5.05 |
| HI | 0.006 | 0.027 | 6.18 | 6.21 |
| NaCl | — | — | 3.0 | 183 |

forces, and solubility of drugs in water is influenced mainly by the orientation energy or dipole interaction. In hydrogen chloride, a molecule with about 20% ionic character, the orientation effect is still significant, but the dispersion force contributes a large share to the total interaction energy between molecules. Hydrogen iodide is predominantly covalent, with its intermolecular attraction supplied primarily by the London or dispersion force.

The ionic crystal sodium chloride is included in **Table 2–2** for comparison to show that its stability, as reflected in its large total energy, is much greater than that of molecular aggregates, and yet the dispersion force exists in such ionic compounds even as it does in molecules.

## STATES OF MATTER

Gases, liquids, and crystalline solids are the three primary states of matter or phases. The molecules, atoms, and ions in the solid state are held in close proximity by intermolecular, interatomic, or ionic forces. The atoms in the solid can oscillate only about fixed positions. As the temperature of a solid substance is raised, the atoms acquire sufficient energy to disrupt the ordered arrangement of the lattice and pass into the liquid form. Finally, when sufficient energy is supplied, the atoms or molecules pass into the gaseous state. Solids with high vapor pressures, such as iodine and camphor, can pass directly from the solid to the gaseous state without melting at room temperature. This process is known as *sublimation*, and the reverse process, that is, condensation to the solid state,

may be referred to as *deposition*. Sublimation will not be discussed in detail here but is very important in the freeze-drying process, as briefly detailed in the Key Concept Box on Sublimation in Freeze Drying (Lyophilization).

Certain molecules frequently exhibit a fourth phase, more properly termed a *mesophase* (Greek *mesos*, middle), which lies between the liquid and crystalline states. This so-called *liquid crystalline* state is discussed later. Supercritical fluids are also considered a mesophase, in this case a state of matter that exists under high pressure and temperature and has properties that are intermediate between those of liquids and gases. Supercritical fluids will also be discussed later because of their increased utilization in pharmaceutical agent processing.

## THE GASEOUS STATE

Owing to vigorous and rapid motion and resultant collisions, gas molecules travel in random paths and collide not only with one another but also with the walls of the container in which they are confined. Hence, they exert a *pressure*— a force per unit area—expressed in dynes/cm$^2$. Pressure is also recorded in atmospheres or in millimeters of mercury because of the use of the barometer in pressure measurement. Another important characteristic of a gas, its *volume*, is usually expressed in liters or cubic centimeters (1 cm$^3$ = 1 mL). The temperature involved in the gas equations is given according the absolute or Kelvin scale. Zero degrees on the centigrade scale is equal to 273.15 Kelvin (K).

## ➤ KEY CONCEPT ■ SUBLIMATION IN FREEZE DRYING (LYOPHILIZATION)

Freeze drying (lyophilization) is widely used in the pharmaceutical industry for the manufacturing of heat-sensitive drugs. Freeze drying is the most common commercial approach to making a sterilized powder. This is particularly true for injectable formulations, where a suspended drug might undergo rapid degradation in solution, and thus a dried powder is preferred. Many protein formulations are also prepared as freeze-dried powders to prevent chemical and physical instability processes that more rapidly occur in a solution state than in the solid state. As is implied by its name, freeze drying is a process where a drug suspended in water is frozen and then dried by a sublimation process. The following processes are usually followed in freeze drying: (*a*) The drug is formulated in a sterile buffer formulation and placed in a vial (it is important to note that there are different types of glass available and these types may have differing effects on solution stability; (*b*) a slotted stopper is partially inserted into the vial, with the stopper being raised above the vial so that air can get in and out of the vial; (*c*) the vials are loaded onto trays and placed in a lyophilizer, which begins the initial freezing; (*d*) upon completion of the primary freeze, which is conducted at a low temperature, vacuum is applied and the water sublimes

into vapor and is removed from the system, leaving a powder with a high water content (the residual water is more tightly bound to the solid powder); (*e*) the temperature is raised (but still maintaining a frozen state) to add more energy to the system, and a secondary freeze-drying cycle is performed under vacuum to pull off more of the tightly bound water; and (*f*) the stoppers are then compressed into the vials to seal them and the powders are left remaining in a vacuum-sealed container with no air exchange. These vials are subsequently sealed with a metal cap that is crimped into place. It is important to note that there is often residual water left in the powders upon completion of lyophilization. In addition, if the caps were not air tight, humidity could enter the vial and cause the powders to absorb atmospheric water (the measurement of the ability of a powder/solid material to absorb water is called its hygroscopicity), which could lead to greater instability. Some lyophilized powders are so hygroscopic that they will absorb enough water to form a solution; this is called deliquescence and is common in lyophilized powders. Finally, because the water is removed by sublimation and the compound is not crystalized out, the residual powder is commonly amorphous.

## The Ideal Gas Law

The student may recall from general chemistry that the gas laws formulated by Boyle, Charles, and Gay-Lussac refer to an ideal situation where no intermolecular interactions exist and collisions are perfectly elastic, and thus no energy is exchanged upon collision. Ideality allows for certain assumptions to be made to derive these laws. Boyle's law relates the volume and pressure of a given mass of gas at constant temperature,

$$P \propto \frac{1}{V}$$

or

$$PV = k \qquad (2\text{–}2)$$

The law of Gay-Lussac and Charles states that the volume and absolute temperature of a given mass of gas at constant pressure are directly proportional,

$$V \propto T$$
$$V = kT \qquad (2\text{–}3)$$

These equations can be combined to obtain the familiar relationship

$$\frac{P_1 V_1}{T_1} = \frac{P_2 V_2}{T_2} \qquad (2\text{–}4)$$

In equation (2–4), $P_1$, $V_1$, and $T_1$ are the values under one set of conditions and $P_2$, $V_2$, and $T_2$ the values under another set.

**EXAMPLE 2–1**

**The Effect of Pressure Changes on the Volume of an Ideal Gas**

In the assay of ethyl nitrite spirit, the nitric oxide gas that is liberated from a definite quantity of spirit and collected in a gas burette occupies a volume of 30.0 mL at a temperature of 20°C and a pressure of 740 mm Hg. Assuming the gas is ideal, what is the volume at 0°C and 760 mm Hg? Write

$$\frac{740 \times 30.0}{273 + 20} = \frac{760 \times V_2}{273}$$
$$V_2 = 27.2 \text{ mL}$$

From equation (2–4) it is seen that $PV/T$ under one set of conditions is equal to $PV/T$ under another set, and so on. Thus, one reasons that although $P$, $V$, and $T$ change, the ratio $PV/T$ is constant and can be expressed mathematically as

$$\frac{PV}{T} = R$$

or

$$PV = RT \qquad (2\text{–}5)$$

in which $R$ is the constant value for the $PV/T$ ratio of an ideal gas. This equation is correct only for 1 mole (i.e., 1 g molecular weight) of gas; for $n$ moles it becomes

$$PV = nRT \qquad (2\text{–}6)$$

Equation (2–6) is known as the *general ideal gas law*, and because it relates the specific conditions or state, that is, the pressure, volume, and temperature of a given mass of gas, it is called the *equation of state* of an ideal gas. Real gases do not interact without energy exchange, and therefore do not follow the laws of Boyle and of Gay-Lussac and Charles as ideal gases are assumed to do. This deviation will be considered in a later section.

The *molar gas constant R* is highly important in physical chemical science; it appears in a number of relationships in electrochemistry, solution theory, colloid chemistry, and other fields in addition to its appearance in the gas laws. To obtain a numerical value for $R$, let us proceed as follows. If 1 mole of an ideal gas is chosen, its volume under standard conditions of temperature and pressure (i.e., at 0°C and 760 mm Hg) has been found by experiment to be 22.414 liters. Substituting this value in equation (2–6), we obtain

$$1 \text{ atm} \times 22.414 \text{ liters} = 1 \text{ mole} \times R \times 273.16 \text{ K}$$
$$R = 0.08205 \text{ liter atm/mole K}$$

The molar gas constant can also be given in energy units by expressing the pressure in dynes/cm$^2$ (1 atm = 1.0133 × 10$^6$ dynes/cm$^2$) and the volume in the corresponding units of cm$^3$ (22.414 liters = 22,414 cm$^3$). Then

$$R = \frac{PV}{T} = \frac{(1.0133 \times 10^6) \times 22.414}{273.16}$$
$$= 8.314 \times 10^6 \text{ erg/mole K}$$

or, because 1 joule = 10$^7$ ergs,

$$R = 8.314 \text{ joules/mole K}$$

The constant can also be expressed in cal/mole deg, employing the equivalent 1 cal = 4.184 joules:

$$R = \frac{8.314 \text{ joules/mole deg}}{4.184 \text{ joules/cal}} = 1.987 \text{ cal/mole deg}$$

One must be particularly careful to use the value of $R$ commensurate with the appropriate units under consideration in each problem. In gas law problems, $R$ is usually expressed in liter atm/mole deg, whereas in thermodynamic calculations it usually appears in the units of cal/mole deg or joule/mole deg.

**EXAMPLE 2–2**

**Calculation of Volume Using the Ideal Gas Law**

What is the volume of 2 moles of an ideal gas at 25°C and 780 mm Hg?

$$(780 \text{ mm}/760 \text{ mm atm}^{-1}) \times V$$
$$= 2 \text{ moles} \times (0.08205 \text{ liter atm/mole deg}) \times 298 \text{ K}$$
$$V = 47.65 \text{ liters}$$

## Molecular Weight

The approximate molecular weight of a gas can be determined by use of the ideal gas law. The number of moles of gas $n$ is replaced by its equivalent $g/M$, in which $g$ is the number of grams of gas and $M$ is the molecular weight:

$$PV = \frac{g}{M}RT \qquad (2\text{--}7)$$

or

$$M = \frac{gRT}{PV} \qquad (2\text{--}8)$$

### EXAMPLE 2–3

**Molecular Weight Determination by the Ideal Gas Law**

If 0.30 g of ethyl alcohol in the vapor state occupies 200 mL at a pressure of 1 atm and a temperature of 100°C, what is the molecular weight of ethyl alcohol? Assume that the vapor behaves as an ideal gas. Write

$$M = \frac{0.30 \times 0.082 \times 373}{1 \times 0.2}$$

$$M = 46.0 \text{ g/mole}$$

The two methods most commonly used to determine the molecular weight of easily vaporized liquids such as alcohol and chloroform are the *Regnault* and *Victor Meyer* methods. In the latter method, the liquid is weighed in a glass bulb; it is then vaporized and the volume is determined at a definite temperature and barometric pressure. The values are finally substituted in equation (2–8) to obtain the molecular weight.

## Kinetic Molecular Theory

The equations presented in the previous section have been formulated from experimental considerations where the conditions are close to ideality. The theory that was developed to explain the behavior of gases and to lend additional support to the validity of the gas laws is called the *kinetic molecular theory*. Here are some of the more important statements of the theory:

1. Gases are composed of particles called atoms or *molecules*, the total volume of which is so small as to be negligible in relation to the volume of the space in which the molecules are confined. This condition is approximated in actual gases only at low pressures and high temperatures, in which case the molecules of the gas are far apart.
2. The particles of the gas do not attract one another, but instead move with complete independence; again, this statement applies only at low pressures.
3. The particles exhibit continuous random motion owing to their kinetic energy. The average kinetic energy, $E$, is directly proportional to the absolute temperature of the gas, or $E = (\frac{3}{2})RT$.
4. The molecules exhibit perfect elasticity; that is, there is no net loss of speed or transfer of energy after they collide with one another and with the molecules in the walls of the confining vessel, which latter effect accounts for the gas pressure. Although the net velocity, and therefore the average kinetic energy, does not change on collision, the speed

and energy of the individual molecules may differ widely at any instant. More simply stated, the net velocity can be an average velocity of many molecules; thus, a distribution of individual molecular velocities can be present in the system.

From these and other postulates, the following *fundamental kinetic equation* is derived:

$$PV = \frac{1}{3}nm\overline{c^2} \qquad (2\text{--}9)$$

where $P$ is the pressure and $V$ the volume occupied by any number $n$ of molecules of mass $m$ having an average velocity $\overline{c}$.

Using this fundamental equation, we can obtain the root mean square velocity $(\overline{c^2})^{1/2}$ (usually written $\mu$) of the molecules by an ideal gas.* Solving for $\overline{c^2}$ in equation (2–9) and taking the square root of both sides of the equation leads to the formula

$$\mu = \sqrt{\frac{3RV}{nm}} \qquad (2\text{--}10)$$

Restricting this case to 1 mole of gas, we find that $PV$ becomes equal to $RT$ from the equation of state (2–5), $n$ becomes Avogadro's number $N_A$, and $N_A$ multiplied by the mass of one molecule becomes the molecular weight $M$. The root mean square velocity is therefore given by

$$\mu = \sqrt{\frac{3RT}{M}} \qquad (2\text{--}11)$$

### EXAMPLE 2–4

**Calculation of Root Mean Square Velocity**

What is the root mean square velocity of oxygen (molecular weight, 32.0) at 25°C (298 K)?

$$\mu = \sqrt{\frac{3 \times 8.314 \times 10^7 \times 298}{32}} = 4.82 \times 10^4 \text{cm/sec}$$

Because the term $nm/V$ is equal to density, we can write equation (2–10) as

$$\mu = \sqrt{\frac{3P}{d}} \qquad (2\text{--}12)$$

Remembering that density is defined as a mass per unit volume, we see that the rate of diffusion of a gas is inversely proportional to the square root of its density. Such a relation confirms the early findings of Graham, who showed that a lighter gas diffuses more rapidly through a porous membrane than does a heavier one.

---

*Note that the root mean square velocity $(\overline{c^2})^{1/2}$ is not the same as the average velocity, $\overline{c}$. This can be shown by a simple example: Let $c$ have the three values 2, 3, and 4. Then $\overline{c} = (2 + 3 + 4)/3 = 3$, whereas $\mu = (\overline{c^2})^{1/2}$ is the square root of the mean of the sum of the squares, or $\sqrt{(2^2 + 3^2 + 4^2)/3} = \sqrt{9.67}$, and $\mu = 3.11$.

## The van der Waals Equation for Real Gases

The fundamental kinetic equation (2–9) is found to compare with the ideal gas equation because the kinetic theory is based on the assumptions of the ideal state. However, real gases are not composed of infinitely small and perfectly elastic nonattracting spheres. Instead, they are composed of molecules of a finite volume that tend to attract one another. These factors affect the volume and pressure terms in the ideal equation so that certain refinements must be incorporated if equation (2–5) is to provide results that check with experiment. A number of such expressions have been suggested, the *van der Waals equation* being the best known of these. For 1 mole of gas, the van der Waals equation is written as

$$\left(P + \frac{a}{V^2}\right)(V - b) = RT \qquad (2\text{–}13)$$

For the more general case of $n$ moles of gas in a container of volume $V$, equation (2–13) becomes

$$\left(P + \frac{an^2}{V^2}\right)(V - nb) = nRT \qquad (2\text{–}14)$$

The term $a/V^2$ accounts for the *internal pressure* per mole resulting from the intermolecular forces of attraction between the molecules; $b$ accounts for the incompressibility of the molecules, that is, the *excluded volume*, which is about four times the molecular volume. This relationship holds true for all gases; however, the influence of nonideality is greater when the gas is compressed. Polar liquids have high internal pressures and serve as solvents only for substances of similar internal pressures. Nonpolar molecules have low internal pressures and are not able to overcome the powerful cohesive forces of the polar solvent molecules. Mineral oil is immiscible with water for this reason.

When the volume of a gas is large, the molecules are well dispersed. Under these conditions, $a/V^2$ and $b$ become insignificant with respect to $P$ and $V$, respectively. Under these conditions, the van der Waals equation for 1 mole of gas reduces to the ideal gas equation, $PV = RT$, and at low pressures, real gases behave in an ideal manner. The values of $a$ and $b$ have been determined for a number of gases. Some of these are listed in **Table 2–3**. The weak van der Waals forces of attraction, expressed by the constant $a$, are those referred to in **Table 2–1**.

**TABLE 2–3**
**THE VAN DER WAALS CONSTANTS FOR SOME GASES**

| Gas | $a$ (liter$^2$ atm/mole$^2$) | $b$ (liter/mole) |
|---|---|---|
| $H_2$ | 0.244 | 0.0266 |
| $O_2$ | 1.360 | 0.0318 |
| $CH_4$ | 2.253 | 0.0428 |
| $H_2O$ | 5.464 | 0.0305 |
| $Cl_2$ | 6.493 | 0.0562 |
| $CHCl_3$ | 15.17 | 0.1022 |

**EXAMPLE 2–5**

**Application of the van der Waals Equation**

A 0.193-mole sample of ether was confined in a 7.35-liter vessel at 295 K. Calculate the pressure produced using (*a*) the ideal gas equation and (*b*) the van der Waals equation. The van der Waals *a* value for ether is 17.38 liter$^2$ atm/mole$^2$; the *b* value is 0.1344 liter/mole. To solve for pressure, the van der Waals equation can be rearranged as follows:

$$P = \frac{nRT}{V - nb} - \frac{an^2}{V^2}$$

(*a*)

$$P = \frac{0.193 \text{ mole} \times 0.0821 \text{ liter atm/deg mole} \times 295 \text{ deg}}{7.35 \text{ liter}}$$

$$= 0.636 \text{ atm}$$

(*b*)

$$P = \frac{0.193 \text{ mole} \times 0.0821 \text{ liter atm/deg mole} \times 295 \text{ deg}}{7.35 \text{ liter} - (0.193 \text{ mole}) \times (0.1344 \text{ liter/mole})}$$

$$- \frac{17.38 \text{ liter}^2 \text{ atm/mole}^2 (0.193 \text{ mole})^2}{(7.35 \text{ liter})^2}$$

$$= 0.626 \text{ atm}$$

**EXAMPLE 2–6**

**Calculation of the van der Waals Constants**

Calculate the pressure of 0.5 mole of $CO_2$ gas in a fire extinguisher of 1-liter capacity at 27°C using the ideal gas equation and the van der Waals equation. The van der Waals constants can be calculated from the critical temperature $T_c$ and the critical pressure $P_c$ (see the section Liquefaction of Gases for definitions):

$$a = \frac{27R^2 T_c^2}{64 P_c} \quad \text{and} \quad b = \frac{RT_c}{8 P_c}$$

The critical temperature and critical pressure of $CO_2$ are 31.0°C and 72.9 atm, respectively.

Using the ideal gas equation, we obtain

$$P = \frac{nRT}{V} = \frac{0.5 \text{ mole} \times 0.0821 \text{ liter atm/deg mole} \times 300.15 \text{ deg}}{1 \text{ liter}}$$

$$= 12.32 \text{ atm}$$

Using the van der Waals equation, we obtain

$$a = \frac{27 \times (0.0821 \text{ liter atm/deg mole})^2 \times (304.15 \text{ deg})^2}{64 \times 72.9 \text{ atm}}$$

$$= 3.608 \text{ liter}^2 \text{ atm/mole}^2$$

$$b = \frac{(0.0821 \text{ liter atm/deg mole}) \times 304.15 \text{ deg}}{8 \times 72.9 \text{ atm}}$$

$$= 0.0428 \text{ liter/mole}$$

$$P = \frac{nRT}{V - nb} - \frac{an^2}{V^2}$$

$$= \frac{(0.5 \text{ mole} \times 0.821 \text{ liter atm/deg mole}) \times 300.15 \text{ deg}}{1 \text{ liter} - (0.5 \text{ mole} \times 0.0428821 \text{ liter/mole})}$$

$$- \frac{(3.608 \text{ liter}^2 \text{ atm/mole}^2) \times 0.5 \text{ mole}^2}{(1 \text{ liter})^2}$$

$$= 11.69 \text{ atm}$$

Although it is beyond the scope of this text, it should be mentioned that to account for nonideality, the concept of fugacity was introduced by Lewis.[5] In general chemistry, the student learns about the concept of chemical potential. At equilibrium in an ideal homogeneous closed system, intermolecular interactions are considered to be nonexistent. However, in real gaseous states and in multiple-component systems, intermolecular interactions occur. Without going into great detail, one can say that these interactions can influence the chemical potential and cause deviations from the ideal state. These deviations reflect the activities of the component(s) within the system. Simply put, fugacity is a measurement of the activity associated with nonideal interactions. For further details pertaining to fugacity and its effects on gases, the student is directed to any introductory physical chemistry text.

# THE LIQUID STATE

## Liquefaction of Gases

When a gas is cooled, it loses some of its kinetic energy in the form of heat, and the velocity of the molecules decreases. If pressure is applied to the gas, the molecules are brought within the sphere of the van der Waals interaction forces and pass into the liquid state. Because of these forces, liquids are considerably denser than gases and occupy a definite volume. The transitions from a gas to a liquid and from a liquid to a solid depend not only on the temperature but also on the pressure to which the substance is subjected.

If the temperature is elevated sufficiently, a value is reached above which it is impossible to liquefy a gas irrespective of the pressure applied. This temperature, above which a liquid can no longer exist, is known as the *critical temperature*. The pressure required to liquefy a gas at its critical temperature is the *critical pressure*, which is also the highest vapor pressure that the liquid can have. The further a gas is cooled below its critical temperature, the less pressure is required to liquefy it. Based on this principle, all known gases have been liquefied. Supercritical fluids, where excessive temperature and pressure are applied, do exist as a separate/intermediate phase and will be discussed briefly later in this chapter.

The critical temperature of water is 374°C, or 647 K, and its critical pressure is 218 atm, whereas the corresponding values for helium are 5.2 K and 2.26 atm. The critical temperature serves as a rough measure of the attractive forces between molecules because at temperatures above the critical value, the molecules possess sufficient kinetic energy so that no amount of pressure can bring them within the range of attractive forces that cause the atoms or molecules to "stick" together. The high critical values for water result from the strong dipolar forces between the molecules and particularly the hydrogen bonding that exists. Conversely, only the weak London force attracts helium molecules, and, consequently, this element must be cooled to the extremely low temperature of 5.2 K before it can be liquefied. Above this critical temperature, helium remains a gas no matter what the pressure.

## Methods of Achieving Liquefaction

One of the most obvious ways to liquefy a gas is to subject it to intense cold by the use of freezing mixtures. Other methods depend on the cooling effect produced in a gas as it expands. Thus, suppose we allow an ideal gas to expand so rapidly that no heat enters the system. Such an expansion, termed an *adiabatic* expansion, can be achieved by carrying out the process in a Dewar, or vacuum, flask, which effectively insulates the contents of the flask from the external environment. The work done to bring about expansion therefore must come from the gas itself at the expense of its own heat energy content (collision frequency). As a result, the temperature of the gas falls. If this procedure is repeated a sufficient number of times, the total drop in temperature may be sufficient to cause liquefaction of the gas.

A cooling effect is also observed when a highly compressed *nonideal* gas expands into a region of low pressure. In this case, the drop in temperature results from the energy expended in overcoming the cohesive forces of attraction between the molecules. This cooling effect is known as the *Joule–Thomson effect* and differs from the cooling produced in adiabatic expansion, in which the gas does external work. To bring about liquefaction by the Joule–Thomson effect, it may be necessary to precool the gas before allowing it to expand. Liquid oxygen and liquid air are obtained by methods based on this effect.

## Aerosols

Gases can be liquefied under high pressures in a closed chamber as long as the chamber is maintained below the critical temperature. When the pressure is reduced, the molecules expand and the liquid reverts to a gas. This reversible change of state is the basic principle involved in the preparation of pharmaceutical aerosols. In such products, a drug is dissolved or suspended in a *propellant*, a material that is liquid under the pressure conditions existing inside the container but that forms a gas under normal atmospheric conditions. The container is so designed that, by depressing a valve, some of the drug–propellant mixture is expelled owing to the excess pressure inside the container. If the drug is nonvolatile, it forms a fine spray as it leaves the valve orifice; at the same time, the liquid propellant vaporizes off.

Chlorofluorocarbons and hydrofluorocarbons have traditionally been utilized as propellants in these products because of their physicochemical properties. However, in the face of increasing environmental concerns (ozone depletion) and legislation like the Clean Air Act, the use of chlorofluorocarbons and hydrofluorocarbons is tightly regulated. This has led researchers to identify additional propellants, which has led to the increased use of other gases such as nitrogen and carbon dioxide. However, considerable effort is being focused on finding better propellant systems. By varying the proportions of the various propellants, it is possible to produce pressures within the container ranging from 1 to 6 atm at room temperature. Alternate fluorocarbon propellants that do not deplete the ozone layer of the atmosphere are under investigation.[6]

The containers are filled either by cooling the propellant and drug to a low temperature within the container, which is then sealed with the valve, or by sealing the drug in the container at room temperature and then forcing the required amount of propellant into the container under pressure. In both cases, when the product is at room temperature, part of the propellant is in the gaseous state and exerts the pressure necessary to extrude the drug, whereas the remainder is in the liquid state and provides a solution or suspension vehicle for the drug.

The formulation of pharmaceuticals as aerosols is continually increasing because the method frequently offers distinct advantages over some of the more conventional methods of formulation. Thus, antiseptic materials can be sprayed onto abraded skin with a minimum of discomfort to the patient. One product, ethyl chloride, cools sufficiently on expansion so that when sprayed on the skin, it freezes the tissue and produces a local anesthesia. This procedure is sometimes used in minor surgical operations.

More significant is the increased efficiency often observed and the facility with which medication can be introduced into body cavities and passages. These and other aspects of aerosols have been considered by various researchers.[7,8] Byron and Clark[9a] studied drug absorption from inhalation aerosols and provided a rather complete analysis of the problem. The United States Pharmacopeia (USP)[9b] includes a discussion of metered-dose inhalation products and provides standards and test procedures (USP).

The identification of biotechnology-derived products has also dramatically increased the utilization of aerosolized formulations.[10] Proteins, DNA, oligopeptides, and nucleotides all demonstrate poor oral bioavailability due to the harsh environment of the gastrointestinal tract and their relatively large size and rapid metabolism. The pulmonary and nasal routes of administration enable higher rates of passage into systemic circulation than does oral administration.[11] It is important to point out that aerosol products are formulated under high pressure and stress limits. The physical stability of complex biomolecules may be adversely affected under these conditions (recall that pressure and temperature may influence the attractive and repulsive *inter-* and *intra*molecular forces present).

## Vapor Pressure of Liquids

Translational energy of motion (kinetic energy) is not distributed evenly among molecules; some of the molecules have more energy and hence higher velocities than others at any moment. When a liquid is placed in an evacuated container at a constant temperature, the molecules with the highest energies break away from the surface of the liquid and pass into the gaseous state, and some of the molecules subsequently return to the liquid state, or condense. When the rate of condensa-

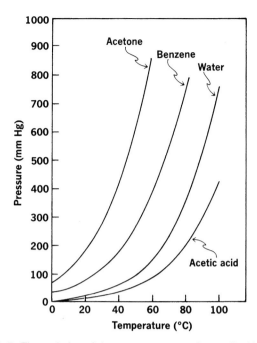

**Fig. 2–5.** The variation of the vapor pressure of some liquids with temperature.

tion equals the rate of vaporization at a definite temperature, the vapor becomes saturated and a dynamic equilibrium is established. The pressure of the saturated vapor* above the liquid is then known as the *equilibrium vapor pressure*. If a manometer is fitted to an evacuated vessel containing the liquid, it is possible to obtain a record of the vapor pressure in millimeters of mercury. The presence of a gas, such as air, above the liquid decreases the rate of evaporation, but it does not affect the equilibrium pressure of the vapor.

As the temperature of the liquid is elevated, more molecules approach the velocity necessary for escape and pass into the gaseous state. As a result, the vapor pressure increases with rising temperature, as shown in **Figure 2–5**. Any point on one of the curves represents a condition in which the liquid and the vapor exist together in equilibrium. As observed in the diagram, if the temperature of any of the liquids is increased while the pressure is held constant, or if the pressure is decreased while the temperature is held constant, all the liquid will pass into the vapor state.

## Clausius–Clapeyron Equation: Heat of Vaporization

The relationship between the vapor pressure and the absolute temperature of a liquid is expressed by the

---

*A gas is known as a *vapor* below its critical temperature. A less rigorous definition of a *vapor* is a substance that is a liquid or a solid at room temperature and passes into the gaseous state when heated to a sufficiently high temperature. A *gas* is a substance that exists in the gaseous state even at room temperature. Menthol and ethanol are vapors at sufficiently high temperatures; oxygen and carbon dioxide are gases.

Clausius–Clapeyron equation (the Clapeyron and the Clausius–Clapeyron equations are derived in Chapter 3):

$$\log \frac{p_2}{p_1} = \frac{\Delta H_v (T_2 - T_1)}{2.303 R T_1 T_2} \qquad (2\text{-}15)$$

where $p_1$ and $p_2$ are the vapor pressures at absolute temperatures $T_1$ and $T_2$, and $\Delta H_v$ is the *molar heat of vaporization*, that is, the heat absorbed by 1 mole of liquid when it passes into the vapor state. Heats of vaporization vary somewhat with temperature. For example, the heat of vaporization of water is 539 cal/g at 100°C; it is 478 cal/g at 180°C, and at the critical temperature, where no distinction can be made between liquid and gas, the heat of vaporization becomes zero. Hence, the $\Delta H_v$ of equation (2-15) should be recognized as an average value, and the equation should be considered strictly valid only over a narrow temperature range. The equation contains additional approximations, for it assumes that the vapor behaves as an ideal gas and that the molar volume of the liquid is negligible with respect to that of the vapor. These are important approximations in light of the nonideality of real solutions.

### EXAMPLE 2-7

**Application of the Clausius–Clapeyron Equation**

Compute the vapor pressure of water at 120°C. The vapor pressure $p_1$ of water at 100°C is 1 atm, and $\Delta H_v$ may be taken as 9720 cal/mole for this temperature range. Thus,

$$\log \frac{p_2}{1.0} = \frac{9720 \times (393 - 373)}{2.303 \times 1.987 \times 393 \times 373}$$

$$p_2 = 1.95 \text{ atm}$$

The Clausius–Clapeyron equation can be written in a more general form,

$$\log p = -\frac{\Delta H_v}{2.303 R} \frac{1}{T} + \text{constant} \qquad (2\text{-}16)$$

or in natural logarithms,

$$\ln p = -\frac{\Delta H_v}{R} \frac{1}{T} + \text{constant} \qquad (2\text{-}17)$$

from which it is observed that a plot of the logarithm of the vapor pressure against the reciprocal of the absolute temperature results in a straight line, enabling one to compute the heat of vaporization of the liquid from the slope of the line.

## Boiling Point

If a liquid is placed in an open container and heated until the vapor pressure equals the atmospheric pressure, the vapor will form bubbles that rise rapidly through the liquid and escape into the gaseous state. The temperature at which the vapor pressure of the liquid equals the external or atmospheric pressure is known as the *boiling point*. All the absorbed heat is used to change the liquid to vapor, and the temperature does not rise until the liquid is completely vaporized. The atmospheric pressure at sea level is approximately 760 mm Hg; at higher elevations, the atmospheric pressure decreases and the boiling point is lowered. At a pressure of 700 mm Hg, water boils at 97.7°C; at 17.5 mm Hg, it boils at 20°C.

The change in boiling point with pressure can be computed by using the Clausius–Clapeyron equation.

The heat that is absorbed when water vaporizes at the normal boiling point (i.e., the heat of vaporization at 100°C) is 539 cal/g or about 9720 cal/mole. For benzene, the heat of vaporization is 91.4 cal/g at the normal boiling point of 80.2°C. These quantities of heat, known as *latent heats of vaporization*, are taken up when the liquids vaporize and are liberated when the vapors condense to liquids.

The boiling point may be considered the temperature at which thermal agitation can overcome the attractive forces between the molecules of a liquid. Therefore, the boiling point of a compound, like the heat of vaporization and the vapor pressure at a definite temperature, provides a rough indication of the magnitude of the attractive forces.

The boiling points of normal hydrocarbons, simple alcohols, and carboxylic acids increase with molecular weight because the attractive van der Waals forces become greater with increasing numbers of atoms. Branching of the chain produces a less compact molecule with reduced intermolecular attraction, and a decrease in the boiling point results. In general, however, the alcohols boil at a much higher temperature than saturated hydrocarbons of the same molecular weight because of association of the alcohol molecules through hydrogen bonding. The boiling points of carboxylic acids are more abnormal still because the acids form dimers through hydrogen bonding that can persist even in the vapor state. The boiling points of straight-chain primary alcohols and carboxylic acids increase about 18°C for each additional methylene group. The rough parallel between the intermolecular forces and the boiling points or latent heats of vaporization is illustrated in **Table 2–4**. Nonpolar substances, the molecules of which are held together predominantly by the London force, have low boiling points and low heats of vaporization. Polar molecules, particularly those such as ethyl alcohol and water, which are associated through hydrogen bonds, exhibit high boiling points and high heats of vaporization.

### TABLE 2-4
**NORMAL BOILING POINTS AND HEATS OF VAPORIZATION**

| Compound | Boiling Point (°C) | Latent Heat of Vaporization (cal/g) |
|---|---|---|
| Helium | −268.9 | 6 |
| Nitrogen | −195.8 | 47.6 |
| Propane | −42.2 | 102 |
| Methyl chloride | −24.2 | 102 |
| Isobutane | −10.2 | 88 |
| Butane | −0.4 | 92 |
| Ethyl ether | 34.6 | 90 |
| Carbon disulfide | 46.3 | 85 |
| Ethyl alcohol | 78.3 | 204 |
| Water | 100.0 | 539 |

# SOLIDS AND THE CRYSTALLINE STATE

## Crystalline Solids

The structural units of crystalline solids, such as ice, sodium chloride, and menthol, are arranged in fixed geometric patterns or lattices. Crystalline solids, unlike liquids and gases, have definite shapes and an orderly arrangement of units. Gases are easily compressed, whereas solids, like liquids, are practically incompressible. Crystalline solids show definite melting points, passing rather sharply from the solid to the liquid state. Crystallization, as is sometimes taught in organic chemistry laboratory courses, occurs by precipitation of the compound out of solution and into an ordered array. Note that there are several important variables here, including the solvent(s) used, the temperature, the pressure, the crystalline array pattern, salts (if crystallization is occurring through the formation of insoluble salt complexes that precipitate), and so on, that influence the rate and stability of the crystal (see the section Polymorphism) formation. The various crystal forms are divided into six distinct crystal systems based on symmetry. They are, together with examples of each, cubic (sodium chloride), tetragonal (urea), hexagonal (iodoform), rhombic (iodine), monoclinic (sucrose), and triclinic (boric acid). The morphology of a crystalline form is often referred to as its *habit*, where the crystal habit is defined as having the same structure but different outward appearance (or alternately, the collection of faces and their area ratios comprising the crystal).

The units that constitute the crystal structure can be atoms, molecules, or ions. The sodium chloride crystal, shown in **Figure 2–6**, consists of a cubic lattice of sodium ions interpenetrated by a lattice of chloride ions, the binding force of the crystal being the electrostatic attraction of the oppositely charged ions. In diamond and graphite, the lattice units consist of atoms held together by covalent bonds. Solid carbon dioxide, hydrogen chloride, and naphthalene form crystals composed of molecules as the building units. In organic compounds, the molecules are held together by van der Waals forces, Coulombic forces, and hydrogen bonding, which account for the weak binding and for the low melting points of these crystals. Aliphatic hydrocarbons crystallize with their chains lying in a parallel arrangement, whereas fatty acids crystallize in layers of dimers with the chains lying parallel or tilted at an angle with respect to the base plane. Whereas ionic and atomic crystals in general are hard and brittle and have high melting points, molecular crystals are soft and have relatively low melting points.

Metallic crystals are composed of positively charged ions in a field of freely moving electrons, sometimes called the *electron gas*. Metals are good conductors of electricity because of the free movement of the electrons in the lattice. Metals may be soft or hard and have low or high melting points. The hardness and strength of metals depend in part on the kind of imperfections, or *lattice defects*, in the crystals.

## Polymorphism

Some elemental substances, such as carbon and sulfur, may exist in more than one crystalline form and are said to be *allotropic*, which is a special case of *polymorphism*. Polymorphs have different stabilities and may spontaneously convert from the metastable form at a temperature to the stable form. They also exhibit different melting points, x-ray crystal and diffraction patterns (see later discussion), and solubilities, even though they are chemically identical. The differences may not always be great or even large enough to "see" by analytical methods but may sometimes be substantial. Solubility and melting points are very important in pharmaceutical processes, including dissolution and formulation, explaining the primary reason we are interested in polymorphs. The formation of polymorphs of a compound may depend upon several variables pertaining to the crystallization process, including solvent differences (the packing of a crystal might be different from a polar versus a nonpolar solvent); impurities that may favor a metastable polymorph because of specific inhibition of growth patterns; the level of supersaturation from which the material is crystallized (generally the higher the concentration above the solubility, the more chance a metastable form is seen); the temperature at which the crystallization is carried out; geometry of the covalent bonds (are the molecules rigid and planar or free and flexible?); attraction and repulsion of cations and anions (later you will see how x-ray crystallography is used to define an electron density map of a compound); fit of cations into coordinates that are energetically favorable in the crystal lattice; temperature; and pressure.

Perhaps the most common example of polymorphism is the contrast between a diamond and graphite, both of which are composed of crystalline carbon. In this case, high pressure and temperature lead to the formation of a diamond from elemental carbon. When contrasting an engagement ring with a pencil, it is quite apparent that a diamond has a distinct crystal habit from that of graphite. It should be noted that a diamond is a less stable (*metastable*) crystalline form of carbon than is graphite. Actually, the imperfections in diamonds continue to occur with time and represent the diamond converting, very slowly at the low ambient temperature and pressure, into the more stable graphite polymorph.

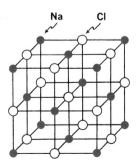

**Fig. 2–6.** The crystal lattice of sodium chloride.

Nearly all long-chain organic compounds exhibit polymorphism. In fatty acids, this results from different types of attachment between the carboxyl groups of adjacent molecules, which in turn modify the angle of tilt of the chains in the crystal. The triglyceride tristearin proceeds from the low-melting metastable alpha ($\alpha$) form through the beta prime ($\beta'$) form and finally to the stable beta ($\beta$) form, having a high melting point. The transition cannot occur in the opposite direction.

Theobroma oil, or cacao butter, is a polymorphous natural fat. Because it consists mainly of a single glyceride, it melts to a large degree over a narrow temperature range (34°C–36°C). Theobroma oil is capable of existing in four polymorphic forms: the unstable gamma form, melting at 18°C; the alpha form, melting at 22°C; the beta prime form, melting at 28°C; and the stable beta form, melting at 34.5°C. Riegelman[12] pointed out the relationship between polymorphism and the preparation of cacao butter suppositories. If theobroma oil is heated to the point at which it is completely liquefied (about 35°C), the nuclei of the stable beta crystals are destroyed and the mass does not crystallize until it is supercooled to about 15°C. The crystals that form are the metastable gamma, alpha, and beta prime forms, and the suppositories melt at 23°C to 24°C or at ordinary room temperature. The proper method of preparation involves melting cacao butter at the lowest possible temperature, about 33°C. The mass is sufficiently fluid to pour, yet the crystal nuclei of the stable beta form are not lost. When the mass is chilled in the mold, a stable suppository, consisting of beta crystals and melting at 34.5°C, is produced.

Polymorphism has achieved significance in last decade because different polymorphs exhibit different solubilities. In the case of slightly soluble drugs, this may affect the rate of dissolution. As a result, one polymorph may be more active therapeutically than another polymorph of the same drug. Aguiar et al.[13] showed that the polymorphic state of chloramphenicol palmitate has a significant influence on the biologic availability of the drug. Khalil et al.[14] reported that form II of sulfameter, an antibacterial agent, was more active orally in humans than form III, although marketed pharmaceutical preparations were found to contain mainly form III. Another case is that of the AIDS drug ritonavir, which was marketed in a dissolved formulation until a previously unknown, more stable and less soluble polymorph appeared. This resulted in a voluntary recall and reformulation of the product before it could be reintroduced to the market.

Polymorphism can also be a factor in suspension technology. Cortisone acetate exists in at least five different forms, four of which are unstable in the presence of water and change to a stable form.[15] Because this transformation is usually accompanied by appreciable caking of the crystals, these should all be in the form of the stable polymorph before the suspension is prepared. Heating, grinding under water, and suspension in water are all factors that affect the interconversion of the different cortisone acetate forms.[16]

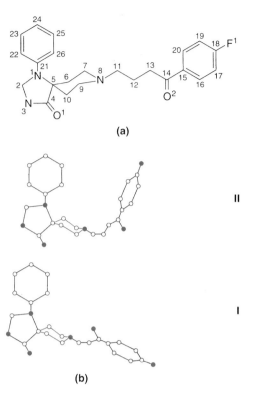

**Fig. 2–7.** (*a*) Structure and numbering of spiperone. (*b*) Molecular conformation of two polymorphs, I and II, of spiperone. (Modified from J. W. Moncrief and W. H. Jones, *Elements of Physical Chemistry*, Addison-Wesley, Reading, Mass., 1977, p. 93; R. Chang, *Physical Chemistry with Applications to Biological Systems*, 2nd Ed., Macmillan, New York, 1977, p. 162.) (From M. Azibi, M. Draguet-Brughmans, R. Bouche, B. Tinant, G. Germain, J. P. Declercq, and M. Van Meerssche, J. Pharm. Sci. **72**, 232, 1983. With permission.)

Although crystal structure determination has become quite routine with the advent of fast, high-resolution diffractometer systems as well as software allowing solution from powder x-ray diffraction data, it can be challenging to determine the crystal structure of highly unstable polymorphs of a drug. Azibi et al.[17] studied two polymorphs of spiperone, a potent antipsychotic agent used mainly in the treatment of schizophrenia. The chemical structure of spiperone is shown in **Figure 2–7a** and the molecular conformations of the two polymorphs, I and II, are shown in **Figure 2–7b**. The difference between the two polymorphs is in the positioning of the atoms in the side chains, as seen in **Figure 2–7b**, together with the manner in which each molecule binds to neighboring spiperone molecules in the crystal. The results of the investigation showed that the crystal of polymorph II is made up of dimers (molecules in pairs), whereas polymorph crystal I is constructed of nondimerized molecules of spiperone. In a later study, Azibi et al.[18] examined the polymorphism of a number of drugs to ascertain what properties cause a compound to exist in more than one crystalline form. Differences in intermolecular van der Waals forces and hydrogen bonds were found to produce different crystal structures

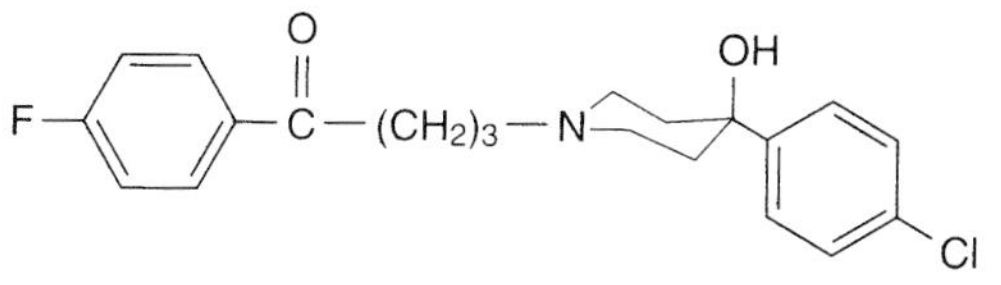

**Fig. 2–8.** Haloperidol.

in antipsychotic compounds such as haloperidol (**Fig. 2–8**) and bromperidol. Variability in hydrogen bonding also contributes to polymorphism in the sulfonamides.[19]

Goldberg and Becker[20] studied the crystalline forms of tamoxifen citrate, an antiestrogenic and antineoplastic drug used in the treatment of breast cancer and postmenopausal symptoms. The structural formula of tamoxifen is shown in **Figure 2–9**. Of the two forms found, the packing in the stable polymorph, referred to as form B, is dominated by hydrogen bonding. One carboxyl group of the citric acid moiety donates its proton to the nitrogen atom on an adjacent tamoxifen molecule to bring about the hydrogen-bonding network responsible for the stabilization of form B. The other polymorph, known as form A, is a metastable polymorph of tamoxifen citrate, its molecular structure being less organized than that of the stable B form. An ethanolic suspension of polymorph A spontaneously rearranges into polymorph B.

Lowes et al.[21] performed physical, chemical, and x-ray studies on carbamazepine. Carbamazepine is used in the treatment of epilepsy and trigeminal neuralgia (severe pain in the face, lips, and tongue). The $\beta$ polymorph of the drug can be crystallized from solvents of high dielectric constant, such as the aliphatic alcohols. The $\alpha$ polymorph is crystallized from solvents of low dielectric constant, such as carbon tetrachloride and cyclohexane. A rather thorough study of the two polymorphic forms of carbamazepine was made using infrared spectroscopy, thermogravimetric analysis (TGA), hot-stage microscopy, dissolution rate, and x-ray powder diffraction. The hydrogen-bonded structure of the $\alpha$ polymorph of carbamazepine is shown in **Figure 2–10a** and its molecular formula in **Figure 2–10b**.

Estrogens are essential hormones for the development of female sex characteristics. When the potent synthetic estrogen ethynylestradiol is crystallized from the solvents acetonitrile, methanol, and chloroform saturated with water, four different crystalline solvates are formed. Ethynylestradiol had been reported to exist in several polymorphic forms. However, Ishida et al.[22] showed from thermal analysis, infrared spectroscopy, and x-ray studies that these forms are crystals containing solvent molecules and thus should be classified

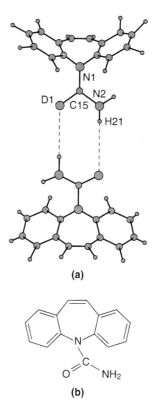

(a)

(b)

**Fig. 2–10.** (*a*) Two molecules of the polymorph $\alpha$-carbamazepine joined together by hydrogen bonds. (From M. M. J. Lowes, M. R. Caira, A. P. Lotter, and J. G. Van Der Watt, J. Pharm. Sci. **76,** 744, 1987. With permission.) (*b*) Carbamazepine.

as *solvates* rather than as polymorphs. Solvates are sometimes called *pseudopolymorphs.* Of course, solvates may also exhibit polymorphism as long as one compares "like" solvation-state crystal structures.[23] Other related estradiol compounds may exist in true polymeric forms.

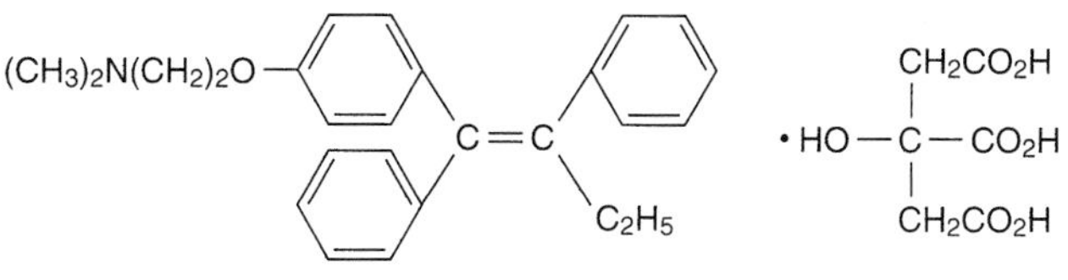

Behme et al.[24] reviewed the principles of polymorphism with emphasis on the changes that the polymorphic forms may undergo. When the change from one form to another is reversible, it is said to be *enantiotropic.* When the transition takes place in one direction only—for example, from a metastable to a stable form—the change is said to be *monotropic.* Enantiotropism and monotropism are important properties of polymorphs as described by Behme et al.[24]

The transition temperature in polymorphism is important because it helps characterize the system and determine the more stable form at temperatures of interest. At their transition temperatures, polymorphs have the same free energy (i.e., the forms are in equilibrium with each other), identical solubilities in a particular solvent, and identical vapor pressures. Accordingly, plots of logarithmic solubility of two polymorphic forms against $1/T$ provide the transition temperature at the intersection of the extrapolated curves. Often, the plots are nonlinear and cannot be extrapolated with accuracy. For dilute solutions, in which Henry's law applies, the logarithm of the solubility ratios of two polymorphs can be plotted

**Fig. 2–9.** Tamoxifen citrate.

against $1/T$, and the intersection at a ratio equal to unity gives the transition temperature.[25] This temperature can also be obtained from the phase diagram of pressure versus temperature and by using differential scanning calorimetry (DSC).[26]

## Solvates

Because many pharmaceutical solids are often synthesized by standard organic chemical methods, purified, and then crystallized out of different solvents, residual solvents can be trapped in the crystalline lattice. This creates a cocrystal, as described previously, termed a *solvate*. The presence of the residual solvent may dramatically affect the crystalline structure of the solid depending on the types of intermolecular interactions that the solvent may have with the crystalline solid. In the following sections we highlight the influence of solvates and how they can be detected using standard solid characterization analyses.

Biles[27] and Haleblian and McCrone[28] discussed in some detail the significance of polymorphism and solvation in pharmaceutical practice.

## Amorphous Solids

Amorphous solids as a first approximation may be considered supercooled liquids in which the molecules are arranged in a somewhat random manner as in the liquid state. Substances such as glass, pitch, and many synthetic plastics are amorphous solids. They differ from crystalline solids in that they tend to flow when subjected to sufficient pressure over a period of time, and they do not have definite melting points. In the Rheology chapter, a solid is characterized as any substance that must be subjected to a definite shearing force before it fractures or begins to flow. This force, below which the body shows elastic properties, is known as the *yield value*.

Amorphous substances, as well as cubic crystals, are usually *isotropic*, that is, they exhibit similar properties in all directions. Crystals other than cubic are *anisotropic*, showing different characteristics (electric conductance, refractive index, crystal growth, rate of solubility) in various directions along the crystal.

It is not always possible to determine by casual observation whether a substance is crystalline or amorphous. Beeswax and paraffin, although they appear to be amorphous, assume crystalline arrangements when heated and then allowed to cool slowly. Petrolatum contains both crystalline and amorphous constituents. Some amorphous materials, such as glass, may crystallize after long standing.

Whether a drug is amorphous or crystalline has been shown to affect its therapeutic activity. Thus, the crystalline form of the antibiotic novobiocin acid is poorly absorbed and has no activity, whereas the amorphous form is readily absorbed and therapeutically active.[29] This is due to the differences in the rate of dissolution. Once dissolved, the molecules exhibit no memory of their origin.

## X-Ray Diffraction

X-rays are a form of electromagnetic radiation (Chapter 4) having a wavelength on the order of interatomic distances (about 1.54 Å for most laboratory instruments using Cu $K\alpha$ radiation; the C—C bond is about 1.5 Å). X-rays are diffracted by the electrons surrounding the individual atoms in the molecules of the crystals. The regular array of atoms in the crystal (periodicity) causes certain directions to constructively interfere in some directions and destructively interfere in others, just as water waves interfere when you drop two stones at the same time into still water (due to the similarity of the wavelengths to the distance between the atoms or molecules of crystals mentioned). The x-ray diffraction pattern on modern instruments is detected on a sensitive plate arranged behind the crystal and is a "shadow" of the crystal lattice that produced it. Using computational methods, it is possible to determine the conformation of the molecules as well as their relationship to others in the structure. This results in a full description of the structure including the smallest building block, called the unit cell.

The electron density and, accordingly, the position of the atoms in complex structures, such as penicillin, may be determined from a comprehensive mathematical study of the x-ray diffraction pattern. The electron density map of crystalline potassium benzylpenicillin is shown in **Figure 2–11**. The elucidation of this structure by x-ray crystallography paved the way for the later synthesis of penicillin by organic chemists. Aspects of x-ray crystallography of pharmaceutical interest are reviewed by Biles[30] and Lien and Kennon.[31]

Where "single" crystals are unavailable or unsuitable for analysis, a powder of the substance may be investigated. The powder x-ray diffraction pattern may be thought of as a fingerprint of the single-crystal structure. Comparing the position and intensity of the lines (the same constructive interference discussed previously) on such a pattern with corresponding lines on the pattern of a known sample allows one to conduct a qualitative and a quantitative analysis. It is important to note that two polymorphs will provide two distinct powder x-ray diffraction patterns. The presence of a solvate will also influence the powder x-ray diffraction pattern because the solvate will have its own unique crystal structure. This may lead to a single polymorphic form appearing as changeable or two distinct polymorphs. One way to determine whether the presence of a change in a powder x-ray diffraction pattern is due to a solvate or is a separate polymorph is to measure the powder x-ray diffraction patterns at various temperatures. Because solvents tend to be driven out of the structure below the melting point, measuring the powder x-ray diffraction patterns at several temperatures may eliminate the solvent and reveal an unsolvated form. Lack of a change in the powder x-ray diffraction patterns at the different temperatures is a strong indication that the form is not really solvated, or minor changes may indicate a structure that maintains its packing motif without the solvent preset (see **Fig. 2–12** for spirapril). This can be confirmed by other methods as described later.

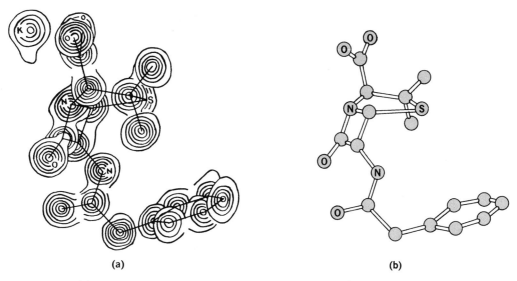

(a)                                          (b)

**Fig. 2–11.** (*a*) Electron density map of potassium benzylpenicillin. (Modified from G. L. Pitt, Acta Crystallogr. **5**, 770, 1952.) (*b*) A model of the structure that can be built from analysis of the electron density projection.

## Melting Point and Heat of Fusion

The temperature at which a liquid passes into the solid state is known as the *freezing point*. It is also the *melting point* of a pure crystalline compound. The freezing point or melting point of a pure crystalline solid is strictly defined as the temperature at which the pure liquid and solid exist in equilibrium. In practice, it is taken as the temperature of the equilibrium mixture at an external pressure of 1 atm; this is sometimes known as the *normal freezing* or *melting point*. The student is reminded that different intermolecular forces are involved in holding the crystalline solid together and that the addition of heat to melt the crystal is actually the addition

of energy. Recall that in a liquid, molecular motion occurs at a much greater rate than in a solid.

The heat (energy) absorbed when 1 g of a solid melts or the heat liberated when it freezes is known as the *latent heat of fusion*, and for water at $0°C$ it is about 80 cal/g (1436 cal/mole). The heat added during the melting process does not bring about a change in temperature until all of the solid has disappeared because this heat is converted into the potential energy of the molecules that have escaped from the solid into the liquid state. The normal melting points of some compounds are collected in **Table 2–5** together with the molar heats of fusion.

Changes of the freezing or melting point with pressure can be obtained by using a form of the Clapeyron equation, written as

$$\frac{\Delta T}{\Delta P} = T \frac{V_1 - V_s}{\Delta H_f} \qquad (2\text{–}18)$$

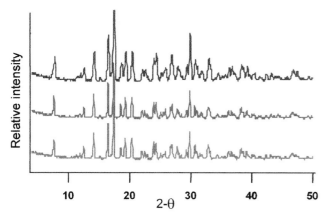

**Fig. 2–12.** Powder x-ray diffraction patterns for spirapril hydrochloride. The monohydrate (top) and the sample dehydrated at 75°C for 106 and 228 hr (middle and bottom, respectively) demonstrating that the structural motif is essentially unchanged in the "dehydrated hydrate." (From W. Xu, *Investigation of Solid State Stability of Selected Bioactive Compounds*, unpublished dissertation, Purdue University, Purdue, Ind., 1997. With permission.)

**TABLE 2–5**

**NORMAL MELTING POINTS AND MOLAR HEATS OF FUSION OF SOME COMPOUNDS**

| Substance | Melting Point (K) | Molar Heat of Fusion, $\Delta H_f$ (cal/mole) |
|---|---|---|
| $H_2O$ | 273.15 | 1440 |
| $H_2S$ | 187.61 | 568 |
| $NH_3$ | 195.3 | 1424 |
| $PH_3$ | 139.4 | 268 |
| $CH_4$ | 90.5 | 226 |
| $C_2H_6$ | 90 | 683 |
| $n\text{-}C_3H_8$ | 85.5 | 842 |
| $C_6H_6$ | 278.5 | 2348 |
| $C_{10}H_8$ | 353.2 | 4550 |

where $V_l$ and $V_s$ are the molar volumes of the liquid and solid, respectively. Molar volume (volume in units of $cm^3/mole$) is computed by dividing the gram molecular weight by the density of the compound. $\Delta H_f$ is the molar heat of fusion, that is, the amount of heat absorbed when 1 mole of the solid changes into 1 mole of liquid, and $\Delta T$ is the change of melting point brought about by a pressure change of $\Delta P$.

Water is unusual in that it has a larger molar volume in the solid state than in the liquid state ($V_s > V_l$) at the melting point. Therefore, $\Delta T / \Delta P$ is negative, signifying that the melting point is lowered by an increase in pressure. This phenomenon can be rationalized in terms of *Le Chatelier's principle*, which states that a system at equilibrium readjusts so as to reduce the effect of an external stress. Accordingly, if a pressure is applied to ice at $0°C$, it will be transformed into liquid water, that is, into the state of lower volume, and the freezing point will be lowered.

## EXAMPLE 2–8

### Demonstration of *Le Chatelier's Principle*

What is the effect of an increase of pressure of 1 atm on the freezing point of water (melting point of ice)?

At $0°C$, $T = 273.16$ K, $\Delta H_f \cong 1440$ cal/mole, the molar volume of water is 18.018, and the molar volume of ice is 19.651, or $V_l - V_s = -1.633$ $cm^3/mole$. To obtain the result in deg/atm using equation (2–18), we first convert $\Delta H_f$ in cal/mole into units of ergs/mole by multiplying by the factor $4.184 \times 10^7$ ergs/cal:

$$\Delta H_f = 6025 \times 10^7 \text{ ergs/mole} \quad \text{or} \quad 6025 \times 10^6 \text{ dyne cm/mole}$$

Then multiplying equation (2–18) by the conversion factor $(1.013 \times 10^6$ dynes/$cm^2$)/atm (which is permissible because the numerator and denominator of this factor are equivalent, so the factor equals 1) gives the result in the desired units:

$$\frac{\Delta T}{\Delta P} = \frac{273.16 \text{ deg} \times (-1.633 cm^3/mole)}{6025 \times 10^7 \text{ dynes cm/mole}}$$
$$\times (1.013 \times 10^6 \text{ dynes/}cm^2)/\text{atm}$$

$$\frac{\Delta T}{\Delta P} = -0.0075 \text{ deg/atm}$$

Hence, an increase of pressure of 1 atm lowers the freezing point of water by about 0.0075 deg, or an increase in pressure of about 133 atm would be required to lower the freezing point of water by 1 deg. (When the ice–water equilibrium mixture is saturated with air under a total pressure of 1 atm, the temperature is lowered by an additional 0.0023 deg.) Pressure has only a slight effect on the equilibrium temperature of *condensed systems* (i.e., of liquids and solids). The large molar volume or low density of ice (0.9168 $g/cm^3$ as compared with 0.9988 $g/cm^3$ for water at $0°C$) accounts for the fact that the ice floats on liquid water. The lowering of the melting point with increasing pressure is taken advantage of in ice-skating. The pressure of the skate lowers the melting point and thus causes the ice to melt below the skate. This thin layer of liquid provides lubricating action and allows the skate to glide over the hard surface. Of course, the friction of the skate also contributes greatly to the melting and lubricating action.

## EXAMPLE 2–9

### Pressure Effects on Freezing Points

According to *Example 2–8*, an increase of pressure of 1 atm reduces the freezing (melting) point of ice by 0.0075 deg. To what temperature is the melting point reduced when a 90-lb boy skates across the ice? The area of the skate blades in contact with the ice is 0.085 $cm^2$.

In addition to the atmospheric pressure, which may be disregarded, the pressure of the skates on the ice is the mass (90 lb = 40.8 kg) multiplied by the acceleration constant of gravity (981 cm/sec$^2$) and divided by the area of the skate blades (0.085 $cm^2$):

$$\text{Pressure} = \frac{40,800 \text{ g} \times (981 \text{ cm/sec}^2)}{0.085 cm^2}$$
$$= 4.71 \times 10^8 \text{dynes/}cm^2$$

Changing to atmospheres (1 atm = $1.01325 \times 10^6$ dynes/$cm^2$) yields a pressure of 464.7 atm. The change in volume $\Delta V$ from water to ice is $0.018 - 0.01963$ liter/mole, or $-0.00163$ liter/mole for the transition from ice to liquid water.

Use equation (2–18) in the form of a derivative:

$$\frac{dT}{dp} = T\frac{\Delta V}{\Delta H_f}$$

For a pressure change of 1 atm to 464.7 atm when the skates of the 90-lb boy touch the ice, the melting temperature will drop from 273.15 K ($0°C$) to $T$, the final melting temperature of the ice under the skate blades, which converts the ice to liquid water and facilitates the lubrication. For such a problem, we must put the equation in the form of an integral; that is, integrating between 273.15 K and $T$ caused by a pressure change under the skate blades from 1 atm to $(464.7 + 1)$ atm:

$$\int_{273.15 \text{ K}}^{T} \frac{1}{T}dT = \frac{\Delta V}{\Delta H_f} \int_{1 \text{ atm}}^{465.7 \text{ atm}} dP$$

$$\ln T - \ln(273.15) = \frac{-0.00163 \text{ liter/mole}}{1440 \text{ cal/mole}} \frac{24.2 \text{ cal}}{1 \text{ liter atm}}(P_2 - P_1)$$

In this integrated equation, 1440 cal/mole is the heat of fusion $\Delta H_f$ of water in the region of $0°C$, and 24.2 cal/liter atm is a conversion factor (see the front leaf of the book) for converting cal to liter atm. We now have

$$\ln T = (-274 \times 10^{-5}/\text{atm})(465.7 - 1 \text{ atm}) + \ln(273.15)$$
$$T = 269.69 \text{ K}$$

The melting temperature has been reduced from 273.15 to 269.69 K, or a reduction in melting point of 3.46 K by the pressure of the skates on the ice.

A simpler way to do the ice-skating problem is to realize that the small change in temperature, $-3.46$ K, occurs over a large pressure change of about 465 atm. Therefore, we need not integrate but rather may obtain the temperature change $\Delta T$ per unit atmosphere change, $\Delta P$, and multiply this value by the actual pressure, 464.7 atm. Of course, the heat of fusion of water, 1440 cal/mole, must be multiplied by the conversion factor, 1 liter atm/24.2 cal, to yield 59.504 liter atm. We have

$$\frac{\Delta T}{\Delta P} = \frac{T\Delta V}{\Delta H_f} = \frac{(273.15 \text{ K})(0.0180 - 0.0196) \text{ liter/mole}}{59.504 \text{ liter atm/mole}}$$

$$\frac{\Delta T}{\Delta P} = -0.00734 \text{ K/atm}$$

For a pressure change of 464.7 atm, the decrease in temperature is

$$\Delta T = -0.00734 \text{ K/atm} \times 464.7 \text{ atm}$$
$$= -3.41 \text{ K}$$

as compared with the more accurate value, $-3.46$ K.

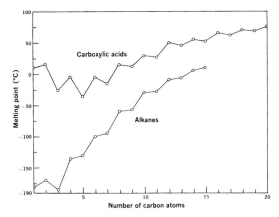

**Fig. 2–13.** The melting points of alkanes and carboxylic acids as a function of carbon chain length. (Modified from C. R. Noller, *Chemistry of Organic Compounds*, 2nd Ed., Saunders, Philadelphia, 1957, pp. 40, 149.)

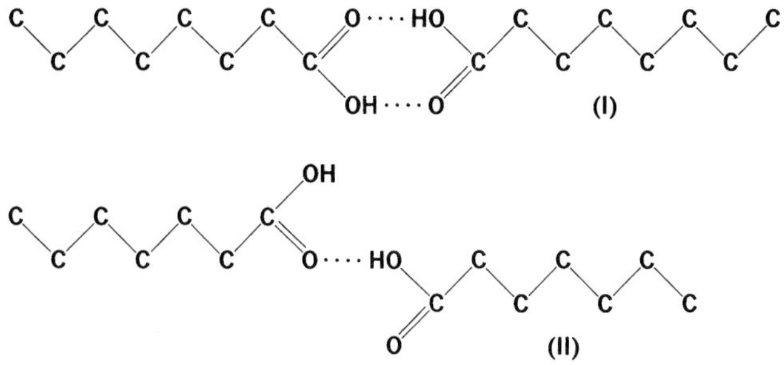

**Fig. 2–14.** Configuration of fatty acid molecules in the crystalline state. (Modified from A. E. Bailey, *Melting and Solidification of Fats*, Interscience, New York, 1950, p. 120.)

## Melting Point and Intermolecular Forces

The heat of fusion may be considered as the heat required to increase the interatomic or intermolecular distances in crystals, thus allowing melting (increased molecular motion) to occur. A crystal that is bound together by weak forces generally has a low heat of fusion and a low melting point, whereas one bound together by strong forces has a high heat of fusion and a high melting point.

Because polymorphic forms represent different molecular arrangements leading to different crystalline forms of the same compound, it is obvious that different intermolecular forces will account for these different forms. Then consider polymorph A, which is held together by higher attractive forces than is polymorph B. It is obvious that more heat will be required to break down the attractive forces in polymorph A, and thus its melting temperature will be higher than that of polymorph B.

Paraffins crystallize as thin leaflets composed of zigzag chains packed in a parallel arrangement. The melting points of normal saturated hydrocarbons increase with molecular weight because the van der Waals forces between the molecules of the crystal become greater with an increasing number of carbon atoms. The melting points of the alkanes with an even number of carbon atoms are higher than those of the hydrocarbons with an odd number of carbon atoms,

as shown in **Figure 2–13**. This phenomenon presumably is because alkanes with an odd number of carbon atoms are packed in the crystal less efficiently.

The melting points of normal carboxylic acids also show this alternation, as seen in **Figure 2–13**. This can be explained as follows. Fatty acids crystallize in molecular chains, one segment of which is shown in **Figure 2–14**. The even-numbered carbon acids are arranged in the crystal as seen in the more symmetric structure I, whereas the odd-numbered acids are arranged according to structure II. The carboxyl groups are joined at two points in the even-carbon compound; hence, the crystal lattice is more stable and the melting point is higher.

The melting points and solubilities of the xanthines of pharmaceutical interest, determined by Guttman and Higuchi,[32] further exemplify the relationship between melting point and molecular structure. Solubilities, like melting points, are strongly influenced by intermolecular forces. This is readily observed in **Table 2–6**, where the methylation of theophylline to form caffeine and the lengthening of the side chain from methyl (caffeine) to propyl in the 7 position result in a decrease of the melting point and an increase in solubility. These effects presumably are due to a progressive weakening of intermolecular forces.

## THE LIQUID CRYSTALLINE STATE

Three states of matter have been discussed thus far in this chapter: gas, liquid, and solid. A fourth state of matter is the

**TABLE 2–6**
**MELTING POINTS AND SOLUBILITIES OF SOME XANTHINES***

| Compound | Melting Point (°C Uncorrected) | Solubility in Water at 30°C (mole/liter $\times 10^2$) |
|---|---|---|
| Theophylline (R = H) | 270–274 | 4.5 |
| Caffeine (R = $CH_3$) | 238 | 13.3 |
| 7-Ethyltheophylline (R = $CH_2CH_3$) | 156–157 | 17.6 |
| 7-Propyltheophylline (R = $CH_2CH_2CH_3$ | 99–100 | 104.0 |

*From D. Guttman and T. Higuchi, J. Am. Pharm. Assoc. Sci. Ed. **46,** 4, 1957.

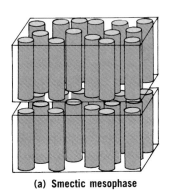

**(a) Smectic mesophase**

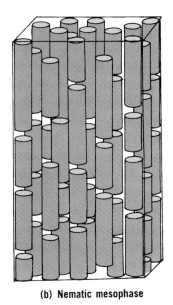

**(b) Nematic mesophase**

**Fig. 2–15.** Liquid crystalline state. (*a*) Smectic structure; (*b*) nematic structure.

*liquid crystalline* state or *mesophase*. The term *liquid crystal* is an apparent contradiction, but it is useful in a descriptive sense because materials in this state are in many ways intermediate between the liquid and solid states.

## Structure of Liquid Crystals

As seen earlier, molecules in the liquid state are mobile in three directions and can also rotate about three axes perpendicular to one another. In the solid state, on the other hand, the molecules are immobile, and rotations are not as readily possible.

It is not unreasonable to suppose, therefore, that intermediate states of mobility and rotation should exist, as in fact they do. It is these intermediate states that constitute the liquid crystalline phase, or mesophase, as the liquid crystalline phase is called.

The two main types of liquid crystals are termed *smectic* (soaplike or greaselike) and *nematic* (threadlike). In the smectic state, molecules are mobile in two directions and can rotate about one axis (**Fig. 2–15a**). In the nematic state, the molecules again rotate only about one axis but are mobile in three dimensions (**Fig. 2–15b**). A third type of crystal (*cholesteric*) exists but can be considered as a special case of the nematic type. In atherosclerosis, it is the incorporation of cholesterol and lipids in human subendothelial macrophages that leads to an insoluble liquid crystalline biologic membrane[33] that ultimately results in plaque formation.

The smectic mesophase is probably of most pharmaceutical significance because it is this phase that usually forms in ternary (or more complex) mixtures containing a surfactant, water, and a weakly amphiphilic or nonpolar additive.

In general, molecules that form mesophases (*a*) are organic, (*b*) are elongated and rectilinear in shape, (*c*) are rigid, and (*d*) possess strong dipoles and easily polarizable groups. The liquid crystalline state may result either from

the heating of solids (thermotropic liquid crystals) or from the action of certain solvents on solids (lyotropic liquid crystals). The first recorded observation of a thermotropic liquid crystal was made by Reinitzer in 1888 when he heated cholesteryl benzoate. At 145°C, the solid formed a turbid liquid (the thermotropic liquid crystal), which only became clear, to give the conventional liquid state, at 179°C.

## PROPERTIES AND SIGNIFICANCE OF LIQUID CRYSTALS

Because of their intermediate nature, liquid crystals have some of the properties of liquids and some of the properties of solids. For example, liquid crystals are mobile and thus can be considered to have the flow properties of liquids. At the same time they possess the property of being birefringent, a property associated with crystals. In birefringence, the light passing through a material is divided into two components with different velocities and hence different refractive indices.

Some liquid crystals show consistent color changes with temperature, and this characteristic has resulted in their being used to detect areas of elevated temperature under the skin that may be due to a disease process. Nematic liquid crystals may be sensitive to electric fields, a property used to advantage in developing display systems. The smectic mesophase has application in the solubilization of water-insoluble materials. It also appears that liquid crystalline phases of this type are frequently present in emulsions and may be responsible for enhanced physical stability owing to their highly viscous nature.

The liquid crystalline state is widespread in nature, with lipoidal forms found in nerves, brain tissue, and blood vessels. Atherosclerosis may be related to the laying down of lipid in the liquid crystalline state on the walls of blood vessels. The three components of bile (cholesterol, a bile acid salt, and water), in the correct proportions, can form a smectic mesophase, and this may be involved in the formation of gallstones. Bogardus[34] applied the principle of liquid crystal formation to the solubilization and dissolution of cholesterol, the major constituent of gallstones. Cholesterol is converted to a liquid crystalline phase in the presence of sodium oleate and water, and the cholesterol rapidly dissolves from the surface of the gallstones.

Nonaqueous liquid crystals may be formed from triethanolamine and oleic acid with a series of polyethylene glycols or various organic acids such as isopropyl myristate, squalane, squalene, and naphthenic oil as the solvents to replace the water of aqueous mesomorphs. Triangular plots or tertiary phase diagrams were used by Friberg et al.[35a,b] to show the regions of the liquid crystalline phase when either polar (polyethylene glycols) or nonpolar (squalene, etc.) compounds were present as the solvent.

Ibrahim[36] studied the release of salicylic acid as a model drug from lyotropic liquid crystalline systems across lipoidal barriers and into an aqueous buffered solution.

Finally, liquid crystals have structures that are believed to be similar to those in cell membranes. As such, liquid crystals may function as useful biophysical models for the structure and functionality of cell membranes.

Friberg wrote a monograph on liquid crystals.[35b] For a more detailed discussion of the liquid crystalline state, refer to the review by Brown,[37] which serves as a convenient entry into the literature.

## THE SUPERCRITICAL FLUID STATE

Supercritical fluids were first described more than 100 years ago and can be formed by many different normal gases such as carbon dioxide. Supercritical fluids have properties that are intermediate between those of liquids and gases, having better ability to permeate solid substances (gaslike) and having high densities that can be regulated by pressure (liquidlike). A *supercritical fluid* is a mesophase formed from the gaseous state where the gas is held under a combination of temperatures and pressures that exceed the *critical point* of a substance (**Fig. 2–16**). Briefly, a gas that is brought above its *critical temperature* $T_c$ will still behave as a gas irrespective of the applied pressure; the *critical pressure* ($P_c$) is the minimum pressure required to liquefy a gas at a given temperature. As the pressure is raised higher, the density of the gas can increase without a significant increase in the viscosity while the ability of the supercritical fluid to dissolve compounds also increases. A gas that may have little to no ability to dissolve a compound under ambient conditions can completely dissolve the compound under high pressure in the supercritical range. **Figure 2–17** illustrates this phenomenon with $CO_2$, where $CO_2$ held at the same temperature can dissolve different chemical classes from a natural product source when the

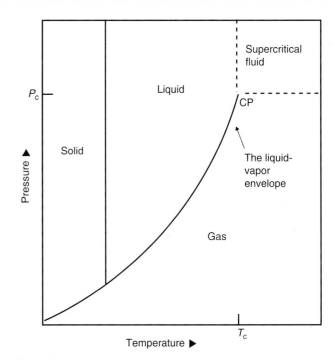

**Fig. 2–16.** Phase diagram showing the supercritical region of a compressed gas. Key: CP = critical point. (Modified from Milton Roy, Co., *Supercritical Fluid Extraction: Principles and Applications* [Bulletin 37.020], Milton Roy Co., Ivydale, Pa.)

pressure is increased.[38] It is also important to note that more than one gas or even the addition of a solvent (termed a cosolvent) such as water and/or ethanol can be made to increase the ability of the supercritical fluid to dissolve compounds.

Supercritical fluid applications in the pharmaceutical sciences were excellently reviewed by Kaiser et al.[39] There are several common uses for supercritical fluids including extraction,[40,41] crystallization,[42] and the preparation of formulations (they are increasingly being used to prepare polymer mixtures[43–45] and for the formation of micro- and nanoparticles[46]). Supercritical fluids offer several advantages over traditional methodologies: the potential for low-temperature extractions and purification of compounds (consider heat added for distillation procedures), solvent volatility under ambient conditions, selectivity of the extracted compounds (**Fig. 2–17**), and lower energy requirement and lower viscosity than solvents.[38] Most important is the reduced toxicity of the gases and the reduced need for hazardous solvents that require expensive disposal. For example, supercritical $CO_2$ can be simply disposed of by opening a valve and releasing the $CO_2$ into the atmosphere.

One of the best examples of the use of supercritical fluids is in the decaffeination of coffee.[39,47] Traditionally, solvents like methylene chloride have been used in the decaffeination process. This leads to great expense in the purchase and disposal of the residual solvents and increases the chance for toxicity. Supercritical $CO_2$ has now been utilized for the decaffeination of coffee and tea. Interestingly, initial supercritical $CO_2$ resulted in the removal of the caffeine and

**A**

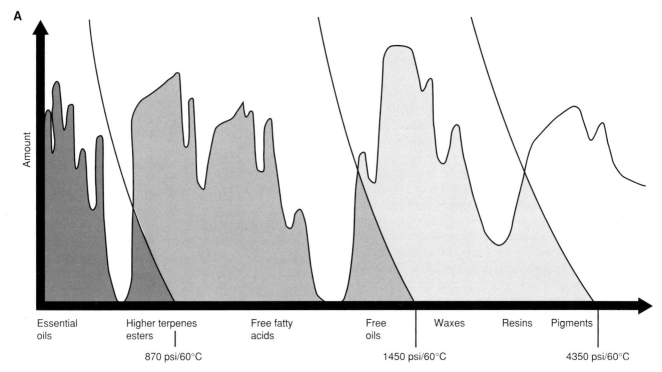

**Fig. 2–17.** The effect of pressure on the ability of supercritical fluids to selectively extract different compounds. (Modified from Milton Roy Co., *Supercritical Fluid Extraction: Principles and Applications* [Bulletin 37.020], Milton Roy Co., Ivyland, Pa.)

important flavor-adding compounds from coffee. The loss of the flavor-adding compounds resulted in a poor taste and an unacceptable product. Additional studies demonstrated that adding water to the supercritical $CO_2$ significantly reduced the loss of the flavor. However, in this process a sample run using coffee beans is performed through the system to saturate the water with the flavor-enhancing compounds. The supercritical $CO_2$ is passed over an extraction column upon release of the pressure and the residual caffeine it carries is collected on the column. The sample run of beans is removed and disposed of, and then the first batch of marketable coffee beans is passed through the system as the water is recirculated. Several batches of decaffeinated coffee are prepared in this manner before the water is discarded. The process leads to approximately 97% of caffeine being removed from the beans. This is an excellent example of how cosolvents may be added to improve the quality of the supercritical $CO_2$-processed material.

## THERMAL ANALYSIS

As noted earlier in this chapter, a number of physical and chemical effects can be produced by temperature changes, and methods for characterizing these alterations upon heating or cooling a sample of the material are referred to as *thermal analysis*. The most common types of thermal analysis are DSC, differential thermal analysis (DTA), TGA, and thermomechanical analysis (TMA). These methods have proved

to be valuable in pharmaceutical research and quality control for the characterization and identification of compounds, the determination of purity, polymorphism[20,21] solvent, and moisture content, amorphous content, stability, and compatibility with excipients.

In general, thermal methods involve heating a sample under controlled conditions and observing the physical and chemical changes that occur. These methods measure a number of different properties, such as melting point, heat capacity, heats of reaction, kinetics of decomposition, and changes in the flow (rheologic) properties of biochemical, pharmaceutical, and agricultural materials and food. The methods are briefly described with examples of applications. Differential scanning calorimetry is the most commonly used method and is generally a more useful technique because its measurements can be related more directly to thermodynamic properties. It appears that any analysis that can be carried out with DTA can be performed with DSC, the latter being the more versatile technique.

### Differential Scanning Calorimetry

In DSC, heat flows and temperatures are measured that relate to thermal transitions in materials. Typically, a sample and a reference material are placed in separate pans and the temperature of each pan is increased or decreased at a predetermined rate. When the sample, for example, benzoic acid, reaches its melting point, in this case 122.4°C, it remains at this temperature until all the material has passed into the liquid state because of the endothermic process of melting. A temperature difference therefore exists between benzoic acid and a reference, indium (melting point [mp] = 156.6°C), as the temperature of the two materials is raised gradually through the range 122°C to 123°C. A second temperature circuit is used in DSC to provide a heat input to overcome this temperature difference. In this way the temperature of the sample, benzoic acid, is maintained at the same value as that of the reference, indium. The difference is heat input to the sample, and the reference per unit time is fed to a computer and plotted as *dH/dt* versus the average temperature to which the sample and reference are being raised. The data collected in a DSC run for a compound such as benzoic acid are shown in the thermogram in **Figure 2–18**. There are a wide variety of features in DSCs such as autosamplers, mass flow controllers, and built-in computers. An example of a modern DSC, the Q200, is shown in **Figure 2–19**. The differential heat input is recorded with a sensitivity of $\pm0.2\ \mu$W, and the tempera-

ture range over which the instrument operates is $-180$°C to 725°C.

Differential scanning calorimetry is a measurement of heat flow into and out of the system. In general, an endothermic (the material is absorbing heat) reaction on a DSC arises from desolvations, melting, glass transitions, and, more rarely, decompositions. An exothermic reaction measured by DSC is usually indicative of a decomposition (energy is released from the bond breaking) process and molecular reorganizations such as crystallization. Differential scanning calorimetry has found increasing use in standardization of the lyophilization process.[48] Crystal changes and eutectic formation in the frozen state as well as amorphous character can be detected by DSC (and by DTA) when the instruments are operated below room temperature.

Although DSC is used most widely in pharmacy to establish identity and purity, it is almost as commonly used to obtain heat capacities and heats of fusion and capacities. It

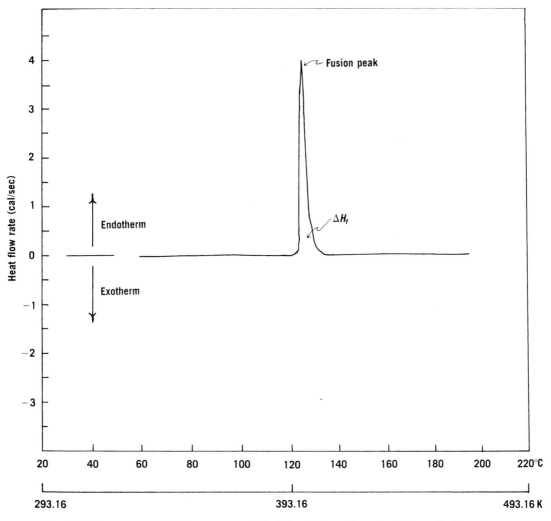

**Fig. 2–18.** Thermogram of a drug compound. Endothermic transitions (heat absorption) are shown in the upward direction and exothermic transitions (heat loss) are plotted downward. Melting is an endothermic process, whereas crystallization or freezing is an exothermic process. The area of the melting peak is proportional to the heat of fusion, $\Delta H_f$.

**Fig. 2–19.** Perkin Elmer differential scanning calorimeter 7 (DSC 7).

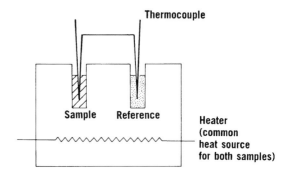

**Fig. 2–20.** Common heat source of differential thermal analyzer with thermocouples in contact with the sample and the reference material.

is also useful for constructing phase diagrams to study the polymorphs discussed in this chapter and for carrying out studies on the kinetics of decomposition of solids.

Differential scanning calorimetry and other thermal analytic methods have a number of applications in biomedical research and food technology. Guillory and associates[49,50] explored the applications of thermal analysis, DSC, and DTA in particular, in conjunction with infrared spectroscopy and x-ray diffraction. Using these techniques, they characterized various solid forms of drugs, such as sulfonamides, and correlated a number of physical properties of crystalline materials with interactions between solids, dissolution rates, and stabilities in the crystalline and amorphous states.

For additional references to the use of DSC in research and technology, contact the manufacturers of differential thermal equipment for complete bibliographies.

## Differential Thermal Analysis

In DTA, both the sample and the reference material are heated by a common heat source (**Fig. 2–20**) rather than the individual heaters used in DSC (**Fig. 2–21**). Thermocouples are placed in contact with the sample and the reference material in DTA to monitor the difference in temperature between the sample and the reference material as they are heated at a constant rate. The temperature difference between the sample and the reference material is plotted against time, and the endotherm as melting occurs (or exotherm as obtained during some decomposition reactions) is represented by a peak in the thermogram.

Although DTA is a useful tool, a number of factors may affect the results. The temperature difference, $\Delta T$, depends,

among other factors, on the resistance to heat flow, $R$. In turn, $R$ depends on temperature, nature of the sample, and packing of the material in the pans. Therefore, it is not possible to directly calculate energies of melting, sublimation, and decomposition, and DTA is used as a qualitative or semiquantitative method for calorimetric measurements. The DSC, although more expensive, is needed for accurate and precise results.

A related technique is dielectric analysis, also discussed in Chapter 4. The concept is that molecules move in response to an applied electric field according to their size, dipole moments, and environment (which changes with temperature). In this technique, the sample serves as the dielectric medium in a capacitor, and as the ac electric field oscillates, the motion of permanent dipoles is sensed as a phase shift relative to the timescale of the frequencies used. Each sample will have a characteristic response or permittivity at a given temperature and frequency. As the sample is heated or cooled, the response will vary because the mobility of the molecular dipoles will change. On going through a first-order transition such as melting, the mobility will exhibit a drastic frequency-independent change giving rise to a family of curves with different intensities but a common maximum occurring at the transition temperature. For glass transitions (pseudosecond order) where the transition occurs over a broad temperature range, there will be a distinct variation in both intensity and temperature, allowing extremely sensitive detection of such

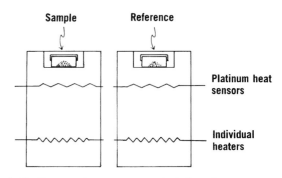

**Fig. 2–21.** Separate heat sources and platinum heat sensors used in differential scanning calorimetry.

## KEY CONCEPT PRACTICAL APPLICATION OF THERMAL ANALYSIS

Using TGA combined with DSC or DTA, we can classify endothermic and exothermic reactions. For example, if a DSC thermogram contains an endothermic reaction at 120°C, another endotherm at 190°C, and an exotherm at 260°C, the scientist must determine the cause of each of these heat exchange processes. With the utilization of a TGA instrument, the change in the weight of a sample is measured as a function of temperature. If the same material is subjected to TGA analysis, the transitions can be classified according to weight changes. Therefore, if a 4% weight loss is observed at 120°C, no weight change is observed at 190°C, and the weight is lost at 260°C, certain assumptions can be made.

The 4% weight loss was associated with an endothermic change, which can correspond to a desolvation, making it a likely choice for the endothermic response. The lack of a weight change at 190°C would most likely suggest a melt due to the endotherm observed on the DSC. The final loss of all of the remaining mass at 260°C is most likely due to decomposition; the exotherm on the DSC would support that conclusion. In addition, Karl Fisher analysis, described in the next subsection, could be used to determine whether the solid is a water solvate to determine whether the 4% loss in mass at 120°C is due to water.

events. Dielectric analysis detects the microscopic "viscosity" of the system and can yield information on activation energies of changes as well as the homogeneity of samples. The technique does, however, require significantly more data analysis than the other thermal methods discussed.

## Thermogravimetric and Thermomechanical Analyses

Changes in weight with temperature (thermogravimetric analysis, TGA) and changes in mechanical properties with temperature (thermomechanical analysis, TMA) are used in pharmaceutical engineering research and in industrial quality control. In TGA, a vacuum recording balance with a sensitivity of 0.1 $\mu$g is used to record the sample weight under pressures of $10^{-4}$ mm to 1 atm.

Thermogravimetric analysis instruments have now begun to be coupled with infrared or mass spectrometers to measure the chemical nature of the evolved gases being lost from the sample. The next subsection describes Karl Fisher analysis, which can also help in determining whether the desolvation may be attributed to water or residual solvents from chemical processing. The changes with temperature in hydrated salts such as calcium oxalate, $CaC_2O_4 \cdot H_2O$, are evaluated using TGA, as discussed by Simons and Newkirk.[51]

The characterization by TGA of bone tissue associated with dental structures was reported by Civjan et al.[52] Thermogravimetric analysis also can be used to study drug stability and the kinetics of decomposition.

Thermomechanical analysis measures the expansion and extension of materials or changes in viscoelastic properties and heat distortions, such as shrinking, as a function of temperature. By use of a probe assembly in contact with the test material, any motion due to expansion, melting, or other physical change delivers an electric signal to a recorder. The furnace, in which are placed a sample and a probe, controls the temperature, which can be programmed over a range from −150°C to 700°C. The apparatus serves essentially as a penetrometer, dilatometer, or tensile tester over a wide range of programmed temperatures. Humphries et al.[53] used TMA

in studies on the mechanical and viscoelastic properties of hair and the stratum corneum of the skin. Thermomechanical analysis is also widely used to look at polymer films and coatings used in pharmaceutical processes.

## Karl Fisher Method

The Karl Fisher method is typically performed as a potentiometric titration method commonly used to determine the amount of water associated with a solid material. The method follows the reaction of iodine (generated electrolytically at the anode in the reagent bath) and sulfur with water. One mole of iodine reacts with 1 mole of water, so the amount of water is directly proportional to the electricity produced. As mentioned, a DSC measurement may indicate an endothermic reaction at 120°C. This endothermic reaction may constitute an actual melt of the crystalline material or may be due to either desolvation or a polymorphic conversion. If one measured the same material using TGA and found a weight loss of about 4% at the same temperature as the endotherm, one could determine that the endotherm arose from a desolvation process.

Utilizing Karl Fisher analysis, one can add the solid material to the titration unit and determine the amount of water by mixing of reagents and the potentiometric electrodes. The Karl Fisher method is an aid in that it can determine whether the desolvation is all water (showing a 4% water content) or arises from the loss of a separate solvent trapped in the crystalline lattice. This method is routinely used for pharmaceutical applications, including the study of humidity effects in solids undergoing water sorption from the air and in quality control efforts to demonstrate the amount of water associated in different lots of manufactured solid products.

## Vapor Sorption/Desorption Analysis

This technique is similar to that of TGA in that it measures weight changes in solids as they are exposed to different solvent vapors and humidity and/or temperature conditions, although it is typically operated isothermally. Greenspan[54] published a definitive list of saturated salt solutions that can be used to control the relative humidity and are widely used to study many physicochemical properties of drugs. For example, if a selected saturated salt solution providing a high relative humidity is placed in a sealed container, the hygroscopicity of a drug can be assessed by determining the weight change in the solid under that humidity. A positive change in the weight would indicate that the solid material is absorbing (collectively called sorption) the solvent, in this case water, from the atmosphere inside the container. The ability of a solid to continuously absorb water until it goes into solution is called deliquescence. A weight loss could also be measured under low relative humidities controlled with different salts, which is termed desorption. Water vapor sorption/desorption can be used to study changes in the solvate state of a crystalline material.[55,56] Variations of commercial instruments also allow the determination of the sorption/desorption of other solvents. The degree of solvation of a crystalline form could have an adverse effect on its chemical stability[57] and/or its manufacturability.

Generally, the less sensitive a solid material or formulation is to changes in the relative humidity, the more stable will be the pharmaceutical shelf life and product performance. The pharmaceutical industry supplies products throughout the world, with considerable variation in climate. Therefore, the measurement of moisture sorption/desorption rates and extents is very important for the prediction of the stability of drugs.

# PHASE EQUILIBRIA AND THE PHASE RULE

The three primary phases (solid, liquid, and gaseous) of matter are often defined individually under different conditions, but in most systems we usually encounter phases in coexistence. For example, a glass of ice water on a hot summer day comprises three coexisting phases: ice (solid), water (liquid), and vapor (gaseous). The amount of ice in the drink depends heavily on several variables including the amount of ice placed in the glass, the temperature of the water in which it was placed, and the temperature of the surrounding air. The longer the drink is exposed to the warm air, the more the amount of ice in the drink will decrease, and the more water the melting ice will produce. However, evaporation of the water into vapor that is released into the large volume of air will also decrease the liquid volume. For this system, there is no establishment of equilibrium because the volume for vapor is infinite in contrast to the ice and liquid volumes.

If the ice water is sealed in a bottle, evaporation effects are limited to the available headspace, the ice melts to liquid, and evaporation becomes time and temperature dependent. For example, if the container is placed in a freezer, only one phase, ice, may be present after long-term storage. Heating of the container, provided the volume stays fixed, could potentially cause the formation of only a vapor phase. Opening and closing of the container would change the vapor phase composition and thus affect equilibrium. This one-component example can be extended to the two-component system of a drug suspension where solid drug is suspended and dissolved in solution and evaporation may take place in the headspace of the container. The suspended system will sit at equilibrium until the container is opened for administration of the drug, and then equilibrium would have to be reestablished for the new system. A new equilibrium or nonequilibrium state is established because dispensing of the suspension will decrease the volume of the liquid and solid in the container. Therefore, a new system is created after each opening, dispensing of the dose, and then resealing. This will be discussed in more detail later.

Before we get into detail about the individual phases, it is important to understand how the phases coexist, the rules that govern their coexistence, and the number of variables required to define the state(s) of matter present under defined conditions.

## The Phase Rule

In each of the examples just given, each phase can be defined by a series of independent variables (e.g., temperature) and their coexistence can only occur over a limited range. For example, ice does not last as long in boiling water as it does in cold water. Therefore, to understand and define the state of each phase, knowledge of several variables is required. J. Willard Gibbs formulated the *phase rule*, which is a relationship for determining the least number of intensive variables (independent variables that do not depend on the volume or size of the phase, e.g., temperature, pressure, density, and concentration) that can be changed without changing the equilibrium state of the system, or, alternately, the least number required to define the state of the system. This critical number is called **F**, the number of degrees of freedom of the system, and the rule is expressed as follows:

$$\mathbf{F} = C - P + 2 \qquad (2\text{--}19)$$

where $C$ is the number of components and $P$ is the number of phases present.

Looking at these terms in more detail, we can define a *phase* as a homogeneous, physically distinct portion of a system that is separated from other portions of the system by bounding surfaces. Thus, a system containing water and its vapor is a two-phase system. An equilibrium mixture of ice, liquid water, and water vapor is a three-phase system.

The *number of components* is the smallest number of constituents by which the composition of each phase in the system at equilibrium can be expressed in the form of a chemical formula or equation. The number of components in the equilibrium mixture of ice, liquid water, and water vapor is one because the composition of all three phases is described by the chemical formula $H_2O$. In the three-phase system $CaCO_3 = CaO + CO_2$, the composition of each phase can be expressed by a combination of any two of the chemical species present. For example, if we choose to use $CaCO_3$ and $CO_2$, we can

## TABLE 2–7

### APPLICATION OF THE PHASE RULE TO SINGLE-COMPONENT SYSTEMS*

| System | Number of Phases | Degrees of Freedom | Comments |
|---|---|---|---|
| Gas, liquid, or solid | 1 | $F = C - P + 2$ $= 1 - 1 + 2 = 2$ | System is *bivariant* ($F = 2$) and lies anywhere within the area marked vapor, liquid, or solid in Figure 2–22. We must fix two variables, e.g., $P_2$ and $t_2$, to define system D. |
| Gas–liquid, liquid–solid, or gas–solid | 2 | $F = C - P + 2$ $= 1 - 2 + 2 = 1$ | System is *univariant* ($F = 1$) and lies anywhere along a *line* between two-phase regions, i.e., AO, BO, or CO in Figure 2–22. We must fix one variable, e.g., either $P_1$ or $t_2$, to define system E. |
| Gas–liquid–solid | 3 | $F = C - P + 2$ $= 1 - 3 + 2 = 0$ | System is *invariant* ($F = 0$) and can lie only at the *point* of intersection of the lines bounding the three-phase regions, i.e., point O in Figure 2–22. |

*Key: $C$ = number of components; $P$ = number of phases.

write CaO as ($CaCO_3 - CO_2$). Accordingly, the number of components in this system is two.

The *number of degrees of freedom* is the *least* number of intensive variables that must be fixed/known to describe the system completely. Herein lies the utility of the phase rule. Although a large number of intensive properties are associated with any system, it is not necessary to report all of these to define the system. For example, let us consider a given mass of a gas, say, water vapor, confined to a particular volume. Using the phase rule only two independent variables are required to define the system, $F = 1 - 1 + 2 = 2$. Because we need to know two of the variables to define the gaseous system completely, we say that the system has two degrees of freedom. Therefore, even though this volume is known, it would be impossible for one to duplicate this system exactly (except by pure chance) unless the temperature, pressure, or another variable is known that may be varied *independent* of the volume of the gas. Similarly, if the temperature of the gas is defined, it is necessary to know the volume, pressure, or some other variable to define the system completely.

Next, consider a system comprising a liquid, say water, in equilibrium with its vapor. By stating the temperature, we define the system completely because the pressure under which liquid and vapor can coexist is also defined. If we decide to work instead at a particular pressure, then the temperature of the system is automatically defined. Again, this agrees with the phase rule because equation **(2–18)** now gives $F = 1 - 2 + 2 = 1$.

As a third example, suppose we cool liquid water and its vapor until a third phase (ice) separates out. Under these conditions, the state of the three-phase ice–water–vapor system is completely defined, and the rule gives $F = 1 - 3 + 2 = 0$; in other words, there are no degrees of freedom. If we attempt to vary the particular conditions of temperature or pressure necessary to maintain this system, we will lose a phase. Thus, if we wish to prepare the three-phase system of ice–water–vapor, we have no choice as to the temperature or pressure at which we will work; the combination is fixed and unique. This is known as the critical point, and later we discuss it in more detail.

The relation between the number of phases and the degrees of freedom in one-component systems is summarized in **Table 2–7**. The student should confirm these data by reference to **Figure 2–22**, which shows the phase equilibria of water at moderate pressures.

It is important to appreciate that as the number of components increases, so do the required degrees of freedom needed to define the system. Therefore, as the system becomes more complex, it becomes necessary to fix more variables to define the system. The greater the number of phases in equilibrium, however, the fewer are the degrees of freedom. Thus:

Liquid water + vapor
Liquid ethyl alcohol + vapor
Liquid water + liquid ethyl alcohol + vapor mixture
   (*Note*: Ethyl alcohol and water are completely miscible both as vapors and liquids.)
Liquid water + liquid benzyl alcohol + vapor mixture
   (*Note*: Benzyl alcohol and water form two separate liquid phases and one vapor phase. Gases are miscible in all proportions; water and benzyl alcohol are only partially miscible. It is therefore necessary to define the two variables in the completely miscible [one-phase] ethyl alcohol–water system but only one variable in the partially miscible [two-phase] benzyl–water system.)

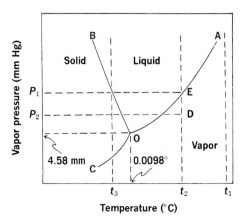

**Fig. 2–22.** Phase diagram for water at moderate pressures.

## Systems Containing One Component

We have already considered a system containing one component, namely, that for water, which is illustrated in **Figure 2–22** (not drawn to scale). The curve OA in the $P$–$T$ (pressure–temperature) diagram in **Figure 2–22** is known as the *vapor pressure curve*. Its upper limit is at the critical temperature, 374°C for water, and its lower end terminates at 0.0098°C, called the *triple point*. Along the vapor pressure curve, vapor and liquid coexist in equilibrium. This curve is analogous to the curve for water seen in **Figure 2–5**. Curve OC is the sublimation curve, and here vapor and solid exist together in equilibrium. Curve OB is the melting point curve, at which liquid and solid are in equilibrium. The negative slope of OB shows that the freezing point of water decreases with increasing external pressure, as we already found in *Example 2–8*.

The result of changes in pressure (at fixed temperature) or changes in temperature (at fixed pressure) becomes evident by referring to the phase diagram. If the temperature is held constant at $t_1$, where water is in the gaseous state above the critical temperature, no matter how much the pressure is raised (vertically along the dashed line), the system remains as a gas. At a temperature $t_2$ below the critical temperature, water vapor is converted into liquid water by an increase of pressure because the compression brings the molecules within the range of the attractive intermolecular forces. It is interesting to observe that at a temperature below the triple point, say $t_3$, an increase of pressure on water in the vapor state converts the vapor first to ice and then at higher pressure into liquid water. This sequence, vapor → ice → liquid, is due to the fact that ice occupies a larger volume than liquid water below the triple point. At the triple point, all three phases are in equilibrium, that is, the only equilibrium is at this pressure at this temperature of 0.0098°C (or with respect to the phase rule, **F** = 0).

As was seen in **Table 2–7**, in any one of the three regions in which pure solid, liquid, or vapor exists and $P = 1$, the phase rule gives

$$\mathbf{F} = 1 - 1 + 2 = 2$$

Therefore, we must fix two conditions, namely temperature and pressure, to specify or describe the system completely. This statement means that if we were to record the results of a scientific experiment involving a given quantity of water, it would not be sufficient to state that the water was kept at, say, 76°C. The pressure would also have to be specified to define the system completely. If the system were open to the atmosphere, the atmospheric pressure obtaining at the time of the experiment would be recorded. Conversely, it would not be sufficient to state that liquid water was present at a certain pressure without also stating the temperature. The phase rule tells us that the experimenter may alter two conditions without causing the appearance or disappearance of the liquid phase. Hence, we say that liquid water exhibits two degrees of freedom.

Along any three of the curves where two phases exist in equilibrium, **F** = 1 (see **Table 2–7**). Hence, only one condi-

tion need be given to define the system. If we state that the system contains both liquid water and water vapor in equilibrium at 100°C, we need not specify the pressure, for the vapor pressure can have no other value than 760 mm Hg at 100°C under these conditions. Similarly, only one variable is required to define the system along line OB or OC. Finally, at the triple point where the three phases—ice, liquid water, and water vapor—are in equilibrium, we saw that **F** = 0.

As already noted, the triple point for air-free water is 0.0098°C, whereas the freezing point (i.e., the point at which liquid water saturated with air is in equilibrium with ice at a total pressure of 1 atm) is 0°C. In increasing the pressure from 4.58 mm to 1 atm, we lower the freezing point by about 0.0075 deg (*Example 2–8*). The freezing point is then lowered an additional 0.0023 deg by the presence of dissolved air in water at 1 atm. Hence, the normal freezing point of water is 0.0075 deg + 0.0023 deg = 0.0098 deg below the triple point. In summary, the temperature at which a solid melts depends (weakly) on the pressure. If the pressure is that of the liquid and solid in equilibrium with the vapor, the temperature is known as the triple point; however, if the pressure is 1 atm, the temperature is the normal freezing point.

## CONDENSED SYSTEMS

We have seen from the phase rule that in a single-component system the maximum number of degrees of freedom is two. This situation arises when only one phase is present, that is, **F** = 1 − 1 + 2 = 2. As will become apparent in the next section, a maximum of three degrees of freedom is possible in a two-component system, for example, temperature, pressure, and concentration. To represent the effect of all these variables upon the phase equilibria of such a system, it would be necessary to use a three-dimensional model rather than the planar figure used in the case of water. Because in practice we are primarily concerned with liquid and/or solid phases in the particular system under examination, we frequently choose to disregard the vapor phase and work under normal conditions of 1 atm pressure. In this manner, we reduce the number of degrees of freedom by one. In a two-component system, therefore, only two variables (temperature and concentration) remain, and we are able to portray the interaction of these variables by the use of planar figures on rectangular-coordinate graph paper. Systems in which the vapor phase is ignored and only solid and/or liquid phases are considered are termed *condensed systems*. We shall see in the later discussion of three-component systems that it is again more convenient to work with condensed systems.

It is important to realize that in aerosol and gaseous systems, vapor cannot be ignored. Condensed systems are most appropriate for solid and liquid dosage forms. As will be discussed in this and later chapters, solids can also have liquid phase(s) associated with them, and the converse is true. Therefore, even in an apparently dry tablet form, small amounts of "solution" can be present. For example, it will be shown in the chapter on stability that solvolysis is a primary mechanism of solid drug degradation.

## Two-Component Systems Containing Liquid Phases

We know from experience that ethyl alcohol and water are miscible in all proportions, whereas water and mercury are, for all practical purposes, completely immiscible regardless of the relative amounts of each present. Between these two extremes lies a whole range of systems that exhibit partial miscibility (or immiscibility). One such system is phenol and water, and a portion of the condensed phase diagram is plotted in **Figure 2–23**. The curve *gbhci* shows the limits of temperature and concentration within which two liquid phases exist in equilibrium. The region outside this curve contains systems having but one liquid phase. Starting at the point *a*, equivalent to a system containing 100% water (i.e., pure water) at 50°C, adding known increments of phenol to a fixed weight of water, the whole being maintained at 50°C, will result in the formation of a single liquid phase until the point *b* is reached, at which point a minute amount of a second phase appears. The concentration of phenol and water at which this occurs is 11% by weight of phenol in water. Analysis of the second phase, which separates out on the bottom, shows it to contain 63% by weight of phenol in water. This phenol-rich phase is denoted by the point *c* on the phase diagram. As we prepare mixtures containing increasing quantities of phenol, that is, as we proceed across the diagram from point *b* to point *c*, we form systems in which the amount of the phenol-rich phase (B) continually increases, as denoted by the test tubes drawn in **Figure 2–23**. At the same time, the amount of the

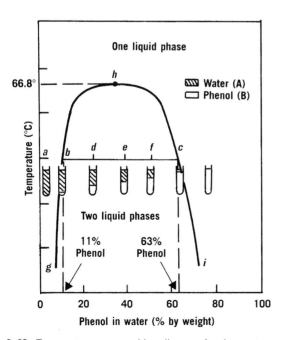

**Fig. 2–23.** Temperature–composition diagram for the system consisting of water and phenol. (From A. N. Campbell and A. J. R. Campbell, J. Am. Chem. Sec. **59**, 2481, 1937.)

water-rich phase (A) decreases. Once the total concentration of phenol exceeds 63% at 50°C, a single phenol-rich liquid phase is formed.

The maximum temperature at which the two-phase region exists is termed the *critical solution*, or *upper consolute*, *temperature*. In the case of the phenol–water system, this is 66.8°C (point *h* in **Fig. 2–23**). All combinations of phenol and water above this temperature are completely miscible and yield one-phase liquid systems.

The line *bc* drawn across the region containing two phases is termed a *tie line*; it is always parallel to the base line in two-component systems. An important feature of phase diagrams is that all systems prepared on a tie line, at equilibrium, will separate into phases of constant composition. These phases are termed *conjugate phases*. For example, any system represented by a point on the line *bc* at 50°C separates to give a pair of conjugate phases whose compositions are *b* and *c*. The *relative amounts* of the two layers or phases vary, however, as seen in **Figure 2–23**. Thus, if we prepare a system containing 24% by weight of phenol and 76% by weight of water (point *d*), at equilibrium we have two liquid phases present in the tube. The upper one, A, has a composition of 11% phenol in water (point *b* on the diagram), whereas the lower layer, B, contains 63% phenol (point *c* on the diagram). Phase B will lie below phase A because it is rich in phenol, and phenol has a higher density than water. In terms of the relative weights of the two phases, there will be more of the water-rich phase A than the phenol-rich phase B at point *d*. Thus:

$$\frac{\text{Weight of phase A}}{\text{Weight of phase B}} = \frac{\text{Length } dc}{\text{Length } bd}$$

The right-hand term might appear at first glance to be the reciprocal of the proportion one should write. The weight of phase A is greater than that of phase B, however, because point *d* is closer to point *b* than it is to point *c*. The lengths *dc* and *bd* can be measured with a ruler in centimeters or inches from the phase diagram, but it is frequently more convenient to use the units of percent weight of phenol as found on the abscissa of **Figure 2–23**. For example, because point *b* = 11%, point *c* = 63%, and point *d* = 24%, the ratio *dc/bd* = (63 − 24)/(24 − 11) = 39/13 = 3/1. In other words, for every 10 g of a liquid system in equilibrium represented by point *d*, one finds 7.5 g of phase A and 2.5 g of phase B. If, on the other hand, we prepare a system containing 50% by weight of phenol (point *f*, **Fig. 2–23**), the ratio of phase A to phase B is *fc/bf* = (63 − 50)/(50 − 11) = 13/39 = 1/3. Accordingly, for every 10 g of system *f* prepared, we obtain an equilibrium mixture of 2.5 g of phase A and 7.5 g of phase B. It should be apparent that a system containing 37% by weight of phenol will, under equilibrium conditions at 50°C, give equal weights of phase A and phase B.

Working on a tie line in a phase diagram enables us to calculate the *composition* of each phase in addition to the weight of the phases. Thus, it becomes a simple matter to calculate the distribution of phenol (or water) throughout the system as a whole. As an example, let us suppose that we mixed 24 g of phenol with 76 g of water, warmed the mixture to 50°C, and allowed it to reach equilibrium at this temperature. On separation of the two phases, we would find 75 g of phase A (containing 11% by weight of phenol) and 25 g of phase B (containing 63% by weight of phenol). Phase A therefore contains a total of $(11 \times 75)/100 = 8.25$ g of phenol, whereas phase B contains a total of $(63 \times 25)/100 = 15.75$ g of phenol. This gives a sum total of 24 g of phenol in the whole system. This equals the amount of phenol originally added and therefore confirms our assumptions and calculations. It is left to the reader to confirm that phase A contains 66.75 g of water and phase B 9.25 g of water. The phases are shown at *b* and *c* in **Figure 2–23**.

Applying the phase rule to **Figure 2–23** shows that with a two-component condensed system having one liquid phase, **F** = 3. Because the pressure is fixed, **F** is reduced to 2, and it is necessary to fix both temperature and concentration to define the system. When two liquid phases are present, **F** = 2; again, pressure is fixed. We need only define temperature to completely define the system because **F** is reduced to 1.* From **Figure 2–23**, it is seen that if the temperature is given, the compositions of the two phases are fixed by the points at the ends of the tie lines, for example, points *b* and *c* at 50°C. The compositions (relative amounts of phenol and water) of the two liquid layers are then calculated by the method already discussed.

The phase diagram is used in practice to formulate systems containing more than one component where it may be advantageous to achieve a single liquid-phase product. For example, the handling of solid phenol, a necrotic agent, is facilitated in the pharmacy if a solution of phenol and water is used. A number of solutions containing different concentrations of phenol are official in several pharmacopeias. Unless the freezing point of the phenol–water mixture is sufficiently low, however, some solidification may occur at a low ambient temperature. This will lead to inaccuracies in dispensing as well as a loss of convenience. Mulley[58] determined the relevant portion of the phenol–water phase diagram and suggested that the most convenient formulation of a single liquid phase solution was 80% w/v, equivalent to about 76% w/w. This mixture has a freezing point of about 3.5°C compared with liquefied phenol, USP, which contains approximately 90% w/w of phenol and freezes at about 17°C. It is not possible, therefore, to use the official preparation much below 20°C, or room temperature; the formulation proposed by Mulley from a consideration of the phenol–water phase diagram therefore is to be preferred. A number of other binary liquid systems of the same type as phenol and water have been studied, although few have practical application in pharmacy. Some of these are water–aniline, carbon disulfide–methyl alcohol, isopentane–phenol, methyl alcohol–cyclohexane, and isobutyl alcohol–water.

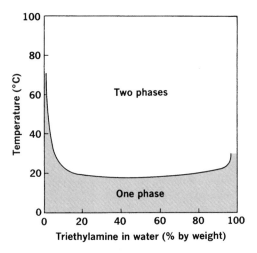

**Fig. 2–24.** Phase diagram for the system triethylamine–water showing lower consolute temperature.

**Figure 2–24** illustrates a liquid mixture that shows no upper consolute temperature, but instead has a *lower* consolute temperature below which the components are miscible in all proportions. The example shown is the triethylamine–water system. **Figure 2–25** shows the phase diagram for the nicotine–water system, which has both a lower and an upper consolute temperature. Lower consolute temperatures arise presumably because of an interaction between the components that brings about complete miscibility only at lower temperatures.

## Two-Component Systems Containing Solid and Liquid Phases: Eutectic Mixtures

We restrict our discussion, in the main, to those solid–liquid mixtures in which the two components are completely miscible in the liquid state and completely immiscible as solids, that is, the solid phases that form consist of pure crystalline

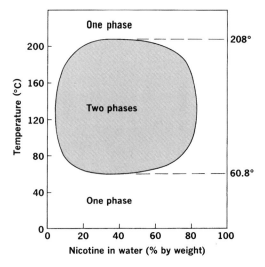

**Fig. 2–25.** Nicotine–water system showing upper and lower consolute temperatures.

---

*The number of degrees of freedom calculated from the phase rule if the system is not condensed is still the same. Thus, when one liquid phase and its vapor are present, **F** = 2 − 2 + 2 = 2; it is therefore necessary to define two conditions: temperature and concentration. When two liquids and the vapor phase exist, **F** = 2 − 3 + 2 = 1, and only temperature need be defined.

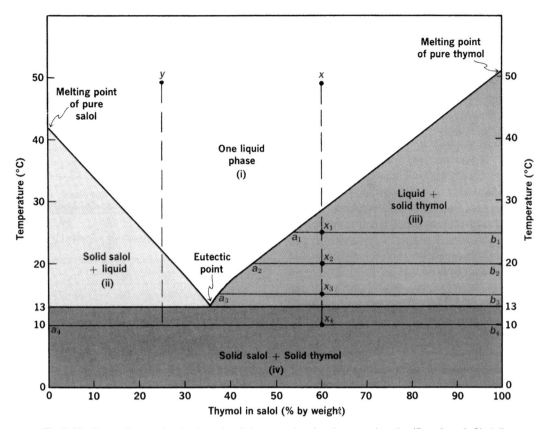

**Fig. 2–26.** Phase diagram for the thymol–salol system showing the eutectic point. (Data from A. Siedell, *Solubilities of Organic Compounds*, 3rd Ed., Vol. 2, Van Nostrand, New York, 1941, p. 723.)

components. Examples of such systems are salol–thymol, salol–camphor, and acetaminophen–propyphenazone.

The phase diagram for the salol–thymol system is shown in **Figure 2–26**. Notice that there are four regions: (i) a single liquid phase, (ii) a region containing solid salol and a conjugate liquid phase, (iii) a region in which solid thymol is in equilibrium with a conjugate liquid phase, and (iv) a region in which both components are present as pure solid phases. Those regions containing two phases (ii, iii, and iv) are comparable to the two-phase region of the phenol–water system shown in **Figure 2–23**. Thus it is possible to calculate both the composition and relative amount of each phase from knowledge of the tie lines and the phase boundaries.

Suppose we prepare a system containing 60% by weight of thymol in salol and raise the temperature of the mixture to 50°C. Such a system is represented by point $x$ in **Figure 2–26**. On cooling the system, we observe the following sequence of phase changes. The system remains as a single liquid until the temperature falls to 29°C, at which point a minute amount of solid thymol separates out to form a two-phase solid–liquid system. At 25°C (room temperature), system $x$ (denoted in **Fig. 2–26** as $x_1$) is composed of a liquid phase, $a_1$ (composition 53% thymol in salol), and pure solid thymol, $b_1$. The weight ratio of $a_1$ to $b_1$ is $(100 - 60)/(60 - 53) = 40/7$, that is, $a_1:b_1 = 5.71:1$. When the temperature is reduced to 20°C (point $x_2$), the composition of the liquid phase is $a_2$ (45% by weight of thymol in salol), whereas the solid phase is still pure thymol, $b_2$. The phase ratio is $a_2:b_2 = (100 - 60)/(60 - 45) =$

$40/15 = 2.67:1$. At 15°C (point $x_3$), the composition of the liquid phase is now 37% thymol in salol ($a_3$) and the weight ratio of liquid phase to pure solid thymol ($a_3:b_3$) is $(100 - 60)/(60 - 37) - 40/23 = 1.74:1$. Below 13°C, the liquid phase disappears altogether and the system contains two solid phases of pure salol and pure thymol. Thus, at 10°C (point $x_4$), the system contains an equilibrium mixture of pure solid salol ($a_4$) and pure solid thymol ($b_4$) in a weight ratio of $(100 - 60)/(60 - 0) = 40/60 = 0.67:1$. As system $x$ is progressively cooled, the results indicate that more and more of the thymol separates as solid. A similar sequence of phase changes is observed if system $y$ is cooled in a like manner. In this case, however, the solid phase that separates at 22°C is pure salol.

The lowest temperature at which a liquid phase can exist in the salol–thymol system is 13°C, and this occurs in a mixture containing 34% thymol in salol. This point on the phase diagram is known as the *eutectic point*. At the eutectic point, three phases (liquid, solid salol, and solid thymol) coexist. The eutectic point therefore denotes an invariant system because, in a condensed system, $F = 2 - 3 + 1 = 0$. The eutectic point is the point at which the liquid and solid phases have the same composition (*the eutectic composition*). The solid phase is an intimate mixture of fine crystals of the two compounds. The intimacy of the mixture gives rise to the phenomenon of "contact melting," which results in the lowest melting temperature over a composition range. Alternately explained, a eutectic composition is the composition of two or more compounds that exhibits a melting temperature lower than that of any other mixture of the compounds. Mixtures of salol and camphor show similar behavior. In this combination, the eutectic point occurs in a system containing 56% by weight of salol in camphor at a temperature of 6°C. Many other substances form eutectic mixtures (e.g., camphor, chloral hydrate, menthol, and betanaphthol). The primary criterion for eutectic formation is the mutual solubility of the components in the liquid or melt phase.

In the thermal analysis section in this chapter, we showed that calorimetry can be used to study phase transitions. Eutectic points are often determined by studying freezing point (melting point if one is adding heat) depression. Note that the freezing point in a one-component system is influenced simply by the temperature. In systems of two or more components, interactions between the components can occur, and depending on the concentrations of the components, the absolute freezing point may change. A eutectic point is the component ratio that exhibits the lowest observed melting point. This relationship is often used to provide information about how solutes interact in solution, with the eutectic point providing the favored composition for the solutes in solution, as illustrated in salol–thymol example. Lidocaine and prilocaine, two local anesthetic agents, form a 1:1 mixture having a eutectic temperature of 18°C. The mixture is therefore liquid at room temperature and forms a mixed local anesthetic that may be used for topical application. The liquid eutectic can be emulsified in water, opening the possibility for topical bioabsorption of the two local anesthetics.[59,60]

## Solid Dispersions

Eutectic systems are examples of solid dispersions. The solid phases constituting the eutectic each contain only one component and the system may be regarded as an intimate crystalline *mixture* of one component in the other. A second major group of solid dispersions are the *solid solutions*, in which each solid phase contains both components, that is, a solid solute is *dissolved* in a solid solvent to give a mixed crystal. Solid solutions are typically not stoichiometric, and the minor component or "guest" inserts itself into the structure of the "host" crystal taking advantage of molecular similarities and/or open spaces in the host lattice. Solid solutions may exhibit higher, lower, or unchanged melting behavior depending upon the degree of interaction of the guest in the crystal structure. A third common dispersion is the molecular dispersion of one component in another where the overall solid is amorphous. Such mixed amorphous or glass solutions exhibit an intermediate glass transition temperature between those of the pure amorphous solids. The dispersion of solid particles in semisolids is also a common dispersion strategy in which crystalline or amorphous solids are dispersed to aid delivery, as in some topical products (e.g., [tioconazole vaginal] Monistat-1).

There is widespread interest in solid dispersions because they may offer a means of facilitating the dissolution and frequently, therefore, the bioavailability of poorly soluble drugs

when combined with freely soluble "carriers" such as urea or polyethylene glycol. This increase in dissolution rate is achieved by a combination of effects, the most significant of which is reduction of particle size to an extent that cannot be readily achieved by conventional comminution approaches. Other contributing factors include increased wettability of the material, reduced aggregation and agglomeration, and a likely increase in solubility of the drug owing to the presence of the water-soluble carrier. Consult the reviews by Chiou and Riegelman[61] and Goldberg[62] for further details.

## Phase Equilibria in Three-Component Systems

In systems containing three components but only one phase, $F = 3 - 1 + 2 = 4$ for a noncondensed system. The four degrees of freedom are temperature, pressure, and the concentrations of two of the three components. Only two concentration terms are required because the sum of these subtracted from the total will give the concentration of the third component. If we regard the system as condensed and hold the temperature constant, then $F = 2$, and we can again use a planar diagram to illustrate the phase equilibria. Because we are dealing with a three-component system, it is more convenient to use triangular coordinate graphs, although it is possible to use rectangular coordinates.

The various phase equilibria that exist in three-component systems containing liquid and/or solid phases are frequently complex and beyond the scope of the present text. Certain typical three-component systems are discussed here, however, because they are of pharmaceutical interest. For example, several areas of pharmaceutical processing such as crystallization, salt form selection, and chromatographic analyses rely on the use of ternary systems for optimization.

## Rules Relating to Triangular Diagrams

Before discussing phase equilibria in ternary systems, it is essential that the reader becomes familiar with certain "rules" that relate to the use of triangular coordinates. It should have been apparent in discussing two-component systems that all concentrations were expressed on a weight–weight basis. This is because, although it is an easy and direct method of preparing dispersions, such an approach also allows the concentration to be expressed in terms of the mole fraction or the molality. The concentrations in ternary systems are accordingly expressed on a weight basis. The following statements should be studied in conjunction with **Figure 2–27**:

1. Each of the three corners or apexes of the triangle represent 100% by weight of one component (*A*, *B*, or *C*). As a result, that same apex will represent 0% of the other two components. For example, the top corner point in **Figure 2–27** represents 100% of *B*.
2. The three lines joining the corner points represent two-component mixtures of the three possible combinations of *A*, *B*, and *C*. Thus the lines *AB*, *BC*, and *CA* are used for

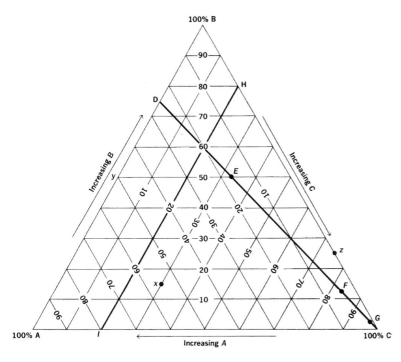

**Fig. 2–27.** The triangular diagram for three-component systems.

two-component mixtures of *A* and *B*, *B* and *C*, and *C* and *A*, respectively. By dividing each line into 100 equal units, we can directly relate the location of a point along the line to the percent concentration of one component in a two-component system. For example, point *y*, midway between *A* and *B* on the line *AB*, represents a system containing 50% of *B* (and hence 50% of *A* also). Point *z*, three fourths of the way along *BC*, signifies a system containing 75% of *C* in *B*.

In going along a line bounding the triangle so as to represent the concentration in a two-component system, it does not matter whether we proceed in a clockwise or a counterclockwise direction around the triangle, provided we are consistent. The more usual convention is clockwise and has been adopted here. Hence, as we move along *AB* in the direction of *B*, we are signifying systems of *A* and *B* containing increasing concentrations of *B*, and correspondingly smaller amounts of *A*. Moving along *BC* toward *C* will represent systems of *B* and *C* containing more and more of *C*; the closer we approach *A* on the line *CA*, the greater will be the concentration of *A* in systems of *A* and *C*.

**3.** The area within the triangle represents all the possible combinations of *A*, *B*, and *C* to give three-component systems. The location of a particular three-component system within the triangle, for example, point *x*, can be undertaken as follows.

The line *AC* opposite apex *B* represents systems containing *A* and *C*. Component *B* is absent, that is, *B* = 0. The horizontal lines running across the triangle parallel to *AC* denote increasing percentages of *B* from *B* = 0 (on line *AC*) to *B* = 100 (at point *B*). The line parallel to *AC* that cuts point *x* is equivalent to 15% *B*; consequently, the system contains 15% of *B* and 85% of *A* and *C* together.

Applying similar arguments to the other two components in the system, we can say that along the line *AB*, *C* = 0. As we proceed from the line *AB* toward *C* across the diagram, the concentration of *C* increases until at the apex, *C* = 100%. The point *x* lies on the line parallel to *AB* that is equivalent to 30% of *C*. It follows, therefore, that the concentration of *A* is 100 − (*B* + *C*) = 100 − (15 + 30) = 55%. This is readily confirmed by proceeding across the diagram from the line *BC* toward apex *A*; point *x* lies on the line equivalent to 55% of *A*.

**4.** If a line is drawn through any apex to a point on the opposite side (e.g., line *DC* in **Fig. 2–27**), then all systems represented by points on such a line have a constant ratio of two components, in this case *A* and *B*. Furthermore, the continual addition of *C* to a mixture of *A* and *B* will produce systems that lie progressively closer to apex *C* (100% of component *C*). This effect is illustrated in **Table 2–8**, in which increasing weights of *C* are added to a constant-weight mixture of *A* and *B*. Note that in all three systems, the ratio of *A* to *B* is constant and identical to that existing in the original mixture.

**5.** Any line drawn parallel to one side of the triangle, for example, line *HI* in **Figure 2–27**, represents ternary system in which the proportion (or percent by weight) of *one* component is constant. In this instance, all systems prepared along *HI* will contain 20% of *C* and varying concentrations of *A* and *B*.

## Ternary Systems with One Pair of Partially Miscible Liquids

Water and benzene are miscible only to a slight extent, and so a mixture of the two usually produces a two-phase system. The heavier of the two phases consists of water saturated with

**TABLE 2–8**

**EFFECT OF ADDING A THIRD COMPONENT (C) TO A BINARY SYSTEM OF A (5.0 G) AND B (15.0 G)**

| Weight of Third Component C Added (g) | Final System | | | Ratio of A to B | Location of System in Figure 2–27 |
| | Component | Weight (g) | Weight (%) | | |
| --- | --- | --- | --- | --- | --- |
| 10.0 | A | 5.0 | 16.67 | 3:1 | Point E |
| | B | 15.0 | 50.00 | | |
| | C | 10.0 | 33.33 | | |
| 100.0 | A | 5.0 | 4.17 | 3:1 | Point F |
| | B | 15.0 | 12.50 | | |
| | C | 100.0 | 83.33 | | |
| 1000.0 | A | 5.0 | 0.49 | 3:1 | Point G |
| | B | 15.0 | 1.47 | | |
| | C | 1000.0 | 98.04 | | |

benzene, while the lighter phase is benzene saturated with water. On the other hand, alcohol is completely miscible with both benzene and water. It is to be expected, therefore, that the addition of sufficient alcohol to a two-phase system of benzene and water would produce a single liquid phase in which all three components are miscible. This situation is illustrated in **Figure 2–28**, which depicts such a ternary system. It might be helpful to consider the alcohol as acting in a manner comparable to that of temperature in the binary phenol–water system considered earlier. Raising the temperature of the phenol–water system led to complete miscibility of the two conjugate phases and the formation of one liquid phase. The addition of alcohol to the benzene–water system achieves the same end but by different means, namely, a solvent effect in place of a temperature effect. There is a strong similarity between the use of heat to break cohesive forces

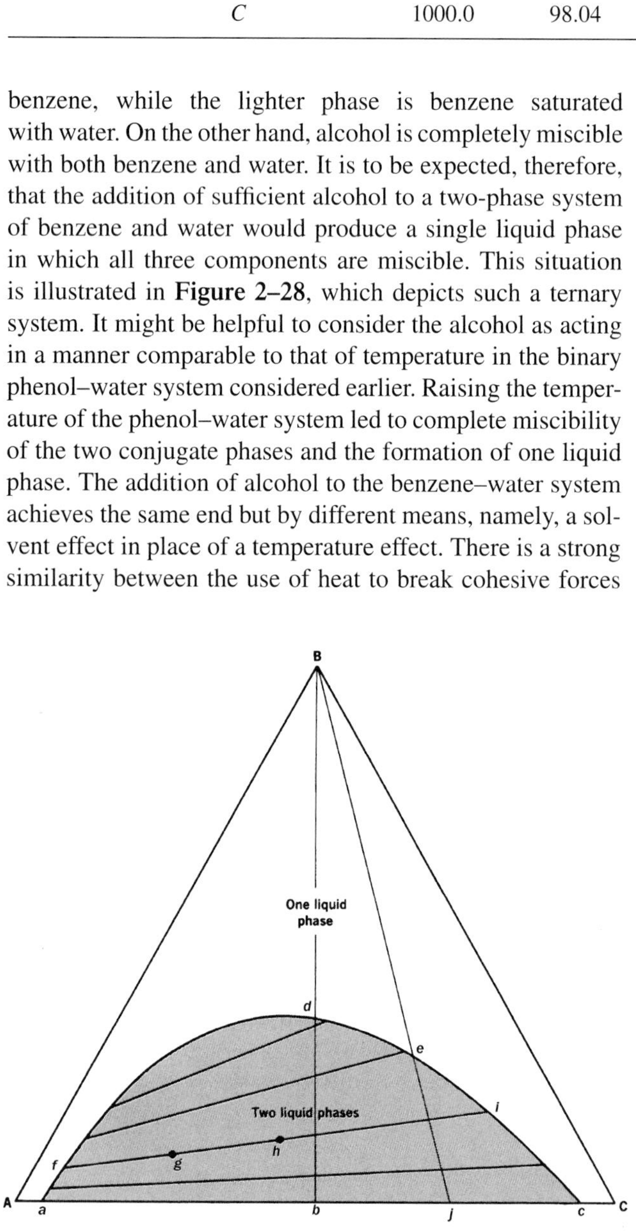

between molecules and the use of solvents to achieve the same result. The effect of alcohol will be better understood when we introduce dielectric constants of solutions and solvent polarity in later chapters. In this case, alcohol serves as an intermediate polar solvent that shifts the electronic equilibrium of the dramatically opposed highly polar water and nonpolar benzene solutions to provide solvation.

In **Figure 2–28**, let us suppose that A, B, and C represent water, alcohol, and benzene, respectively. The line AC therefore depicts binary mixtures of A and C, and the points a and c are the limits of solubility of C in A and of A in C, respectively, at the particular temperature being used. The curve afdeic, frequently termed a *binodal curve or binodal*, marks the extent of the two-phase region. The remainder of the triangle contains one liquid phase. The tie lines within the binodal are not necessarily parallel to one another or to the base line, AC, as was the case in the two-phase region of binary systems. In fact, the directions of the tie lines are related to the shape of the binodal, which in turn depends on the relative solubility of the third component (in this case, alcohol) in the other two components. Only when the added component acts equally on the other two components to bring them into solution will the binodal be perfectly symmetric and the tie lines run parallel to the baseline.

The properties of tie lines discussed earlier still apply, and systems g and h prepared along the tie line fi both give rise to two phases having the compositions denoted by the points f and i. The relative amounts, by weight, of the two conjugate phases will depend on the position of the original system along the tie line. For example, system g, after reaching equilibrium, will separate into two phases, f and i: The ratio of phase f to phase i, on a weight basis, is given by the ratio gi:fg. Mixture h, halfway along the tie line, will contain equal weights of the two phases at equilibrium.

The phase equilibria depicted in **Figure 2–28** show that the addition of component B to a 50 : 50 mixture of components A and C will produce a phase change from a two-liquid system to a one-liquid system at point d. With a 25:75 mixture of A

**Fig. 2–28.** A system of three liquids, one pair of which is partially miscible.

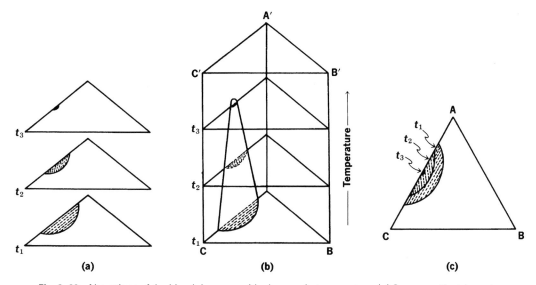

**Fig. 2–29.** Alterations of the binodal curves with changes in temperature. (*a*) Curves on the triangular diagrams at temperatures $t_1$, $t_2$, and $t_3$. (*b*) The three-dimensional arrangement of the diagrams in the order of increasing temperature. (*c*) The view one would obtain by looking down from the top of (*b*).

and *C*, shown as point *j*, the addition of *B* leads to a phase change at point *e*. Naturally, all mixtures lying along *dB* and *eB* will be one-phase systems.

As we saw earlier, **F** = 2 in a single-phase region, and so we must define two concentrations to fix the particular system. Along the binodal curve *afdeic*, **F** = 1, and we need only know one concentration term because this will allow the composition of one phase to be fixed on the binodal curve. From the tie line, we can then obtain the composition of the conjugate phase.

## Effect of Temperature

**Figure 2–29** shows the phase equilibria in a three-component system under isothermal conditions. Changes in temperature will cause the area of immiscibility, bounded by the binodal curve, to change. In general, the area of the binodal decreases as the temperature is raised and miscibility is promoted. Eventually, a point is reached at which complete miscibility is obtained and the binodal vanishes. To study the effect of

temperature on the phase equilibria of three-component systems, a three-dimensional figure, the triangular prism, is frequently used (**Fig. 2–29*b***). Alternatively, a family of curves representing the various temperatures may be used, as shown in **Figure 2–29*c***. The three planar sides of the prism are simply three-phase diagrams of binary-component systems. **Figure 2–29** illustrates the case of a ternary-component system containing one pair of partially immiscible liquids (A and C). As the temperature is raised, the region of immiscibility decreases. The volume outside the shaded region of the prism consists of a single homogeneous liquid phase.

## Ternary Systems with Two or Three Pairs of Partially Miscible Liquids

All the previous considerations for ternary systems containing one pair of partially immiscible liquids still apply. With two pairs of partially miscible liquids, there are two binodal curves. The situation is shown in **Figure 2–30*b***, in which A and C as well as B and C show partial miscibility; A and B

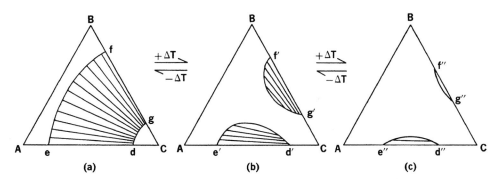

**Fig. 2–30.** Effect of temperature changes on the binodal curves representing a system of two pairs of partially miscible liquids.

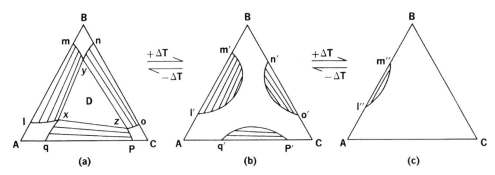

**Fig. 2–31.** Temperature effects on a system of three pairs of partially miscible liquids.

are completely miscible at the temperature used. Increasing the temperature generally leads to a reduction in the areas of the two binodal curves and their eventual disappearance (**Fig. 2–30c**). Reduction of the temperature expands the binodal curves, and, at a sufficiently low temperature, they meet and fuse to form a single band of immiscibility as shown in **Figure 2–30a**. Tie lines still exist in this region, and the usual rules apply. Nor do the number of degrees of freedom change—when $P = 1$, $\mathbf{F} = 2$; when $P = 2$, $\mathbf{F} = 1$.

Systems containing three pairs of partially miscible liquids are of interest. Should the three binodal curves meet (**Fig. 2–31a**), a central region appears in which *three* conjugate liquid phases exist in equilibrium. In this region, D, which is triangular, $\mathbf{F} = 0$ for a condensed system under isothermal conditions. As a result, *all* systems lying within this region consist of three phases whose compositions are always given by the points $x$, $y$, and $z$. The only quantities that vary are the relative amounts of these three conjugate phases. Increasing the temperature alters the shapes and sizes of the regions, as seen in **Figures 2–31b** and $c$.

The application and discussion of phase phenomena and their application in certain pharmaceutical systems will be discussed in later chapters.

## CHAPTER SUMMARY

As one of the foundational chapters of this text, many important subject areas have been covered from the examination of the binding forces between molecules to the various states of matter. Many of these subjects in this chapter are aimed at the more experienced pharmacy student or graduate student who is interested in understanding the fundamental physical aspects of the pharmaceutical sciences.

Practice problems for this chapter can be found at thePoint.lww.com/Sinko6e.

## References

1. E. A. Moelwyn-Hughes, *Physical Chemistry*, 2nd Ed., Pergamon, New York, 1961, p. 297.
2. L. Serrano, M. Bycroft, and A. Fersht, J. Mol. Biol. **218**, 465, 1991.
3. E. A. Meyer, R. K. Castellano, and F. Diederich, Angew. Chem. **42**, 1210, 2003.
4. W. M. Latimer and W. H. Rodebush, J. Am. Chem. Soc. **42**, 1419, 1920. According to P. Schuster, in P. Schuster, G. Zundel, and C. Sandorfy, (Eds.), *The Hydrogen Bond*, Vol. III, North-Holland, New York, 1976, the concept of the hydrogen bond was introduced by T. S. Moore and T. F. Winmill, J. Chem. Soc. **101**, 1635, 1912.
5. G. N. Lewis and M. Randall, *Thermodynamics*, Rev. K. S. Pitzer and L. Brewer, McGraw-Hill, New York, 1961.
6. P. R. Byron, J. Pharm. Sci. **75**, 433, 1986; R. N. Dalby, P. R. Byron, H. R. Shepherd, and E. Popadopoulous, Pharm. Technol. **14**, 26, 1990.
7. J. Pickthall, Pharm. J. **193**, 391, 1964.
8. J. J. Sciarra, J. Pharm. Sci. **63**, 1815, 1974.
9. (a) P. R. Byron and A. R. Clark, J. Pharm. Sci. **74**, 934, 939, 1985; (b) United States Pharmacopeia, *The United States Pharmacopeia*, 27th Rev., U.S. Pharmacopeial Convention, Rockville, Md., 2004.
10. C. Henry, Chem. Eng. News **78**, 49, 2000.
11. DeFelippis, M. R., Overcoming the challenges of noninvasive protein and peptide delivery. American Pharmaceutical Review **6**, 21, 2003.
12. S. Riegelman, in R. A. Lyman and J. B. Sprowls (Eds.), *American Pharmacy*, 4th Ed., Lippincott Williams & Wilkins, Philadelphia, 1955, Chapter 18.
13. A. J. Aguiar, J. Kro, A. W. Kinkle, and J. Samyn, J. Pharm. Sci. **56**, 847, 1967.
14. S. A. Khalil, M. A. Moustafa, A. R. Ebian, and M. M. Motawi, J. Pharm. Sci. **61**, 1615, 1972.
15. R. K. Callow and O. Kennard, J. Pharm. Pharmacol. **13**, 723, 1961.
16. J. E. Carless, M. A. Moustafa, and H. D. C. Rapson, J. Pharm. Pharmacol. **18**, 1905, 1966.
17. M. Azibi, M. Draguet-Brughmans, R. Bouche, B. Tinant, G. Germain, J. P. Declercq, and M. Van Meerssche, J. Pharm. Sci. **72**, 232, 1983.
18. M. Azibi, M. Draguet-Brughmans, and R. Bouche, J. Pharm. Sci. **73**, 512, 1984.
19. S. S. Yank and J. K. Guillory, J. Pharm. Sci. **61**, 26, 1972.
20. I. Goldberg and Y. Becker, J. Pharm. Sci. **76**, 259, 1987.
21. M. M. J. Lowes, M. R. Caira, A. P. Lotter, and J. G. Van Der Watt, J. Pharm. Sci. **76**, 744, 1987.
22. T. Ishida, M. Doi, M. Shimamoto, N. Minamino, K. Nonaka, and M. Inque, J. Pharm. Sci. **78**, 274, 1989.
23. M. Stoltz, A. P. Lötter, and J. G. Van Der Watt, J. Pharm. Sci. **77**, 1047, 1988.
24. R. J. Behme, D. Brooks, R. F. Farney, and T. T. Kensler, J. Pharm. Sci. **74**, 1041, 1985.
25. W. I. Higuchi, P. K. Lau, T. Higuchi, and J. W. Shell, J. Pharm. Sci. **52**, 150, 1963.
26. J. K. Guillory, J. Pharm. Sci. **56**, 72, 1967.
27. J. A. Biles, J. Pharm. Sci. **51**, 601, 1962.
28. J. Haleblian and W. McCrone, J. Pharm. Sci. **58**, 911, 1969; J. Haleblian, J. Pharm. Sci. **64**, 1269, 1975.
29. J. D. Mullins and T. J. Macek, J. Am. Pharm. Assoc. Sci. Ed. **49**, 245, 1960.
30. J. A. Biles, J. Pharm. Sci. **51**, 499, 1962.
31. E. J. Lien and L. Kennon, in *Remington's Pharmaceutical Sciences*, 17th Ed., Mack, Easton, Pa., 1975, pp. 176–178.

32. D. Guttman and T. Higuchi, J. Am. Pharm. Assoc. Sci. Ed. **46,** 4, 1957.
33. B. Lundberg, Atherosclerosis **56,** 93, 1985.
34. J. B. Bogardus, J. Pharm. Sci. **72,** 338, 1983.
35. (a) S. E. Friberg, C. S. Wohn, and F. E. Lockwood, J. Pharm. Sci. **74,** 771, 1985; (b) S. E. Friberg, Lyotropic Liquid Crystals, American Chemical Society, Washington, D.C., 1976.
36. H. G. Ibrahim, J. Pharm. Sci. **78,** 683, 1989.
37. G. H. Brown, Am. Sci. **60,** 64, 1972.
38. Milton Roy Co., *Supercritical Fluid Extraction: Principles and Applications* (Bulletin 37.020), Milton Roy Co., Ivyland, Pa.
39. C. S. Kaiser, H. Rompp, and P. C. Schmidt, Pharmazie **56,** 907, 2001.
40. D. E. LaCroix and W. R. Wolf, J. AOAC Int. **86,** 86, 2000.
41. C.-C. Chen and C.-T. Ho, J. Agric. Food Chem. **36,** 322, 1988.
42. H. H. Tong, B. Y. Shekunov, P. York, and A. H. Chow, Pharm. Res. **18,** 852, 2001.
43. V. Krukonis, Polymer News **11,** 7, 1985.
44. A. Breitenbach, D. Mohr, and T. Kissel, J. Control. Rel. **63,** 53, 2000.
45. M. Moneghini, I. Kikic, D. Voinovich, B. Perissutti, and J. Filipovic-Grcic, Int. J. Pharm. **222,** 129, 2001.
46. E. Reverchon, J. Supercritical Fluids **15,** 1, 1999.
47. H. Peker, M. P. Srinvasan, J. M. Smith, and B. J. McCoy, AIChE J. **38,** 761, 1992.
48. J. B. Borgadus, J. Pharm. Sci. **71,** 105, 1982; L. Gatlin and P. P. DeLuca, J. Parenter Drug Assoc. **34,** 398, 1980.
49. J. K. Guillory, S. C. Hwang, and J. L. Lach, J. Pharm. Sci. **58,** 301, 1969; H. H. Lin and J. K. Guillory, J. Pharm. Sci. **59,** 972, 1970.
50. S. S. Yang and J. K. Guillory, J. Pharm. Sci. **61,** 26, 1972; W. C. Stagner and J. K. Guillory, J. Pharm. Sci. **68,** 1005, 1979.
51. E. L. Simons and A. E. Newkirk, Talanta **11,** 549, 1964.
52. S. Civjan, W. J. Selting, L. B. De Simon, G. C. Battistone, and M. F. Grower, J. Dent. Res. **51,** 539, 1972.
53. W. T. Humphries, D. L. Millier, and R. H. Wildnauer, J. Soc. Cosmet. Chem. **23,** 359, 1972; W. T. Humphries and R. H. Wildnauer, J. Invest. Dermatol. **59,** 9, 1972.
54. L. Greenspan, J. Res. NBS **81A,** 89, 1977.
55. K. R. Morris, in H. G. Brittain (Ed.), *Polymorphism in Pharmaceutical Solids,* Marcel Dekker, New York, 1999, pp. 125–181.
56. R. V. Manek and W. M. Kolling, AAPS PharmSciTech. **5,** article 14, 2004. Available at www.aapspharmscitech.org/default/IssueView.asp? vol = 058issue = 01.
57. C. Ahlneck and G. Zografi, Int. J. Pharm. **62,** 87, 1990.
58. B. A. Mulley, Drug Stand. **27,** 108, 1959.
59. A. Brodin, A. Nyqvist-Mayer, T. Wadsten, B. Forslund, and F. Broberg, J. Pharm. Sci. **73,** 481, 1984.
60. A. Nyqvist-Mayer, A. F. Brodin, and S. G. Frank, J. Pharm. Sci. **74,** 1192, 1985.
61. W. L. Chiou and S. Riegelman, J. Pharm. Sci. **60,** 1281, 1971.
62. A. H. Goldberg, in L. J. Leeson and J. T. Carstensen (Eds.), *Dissolution Technology,* Academy of Pharmaceutical Science, Washington, D.C., 1974, Chapter 5.

## Recommended Readings

J. Liu, Pharm. Dev. Technol. **11,** 3–28, 2006.
D. Singhal and W. Curatolo, Adv. Drug Deliv. Rev. **56,** 335–347, 2004.

CHAPTER LEGACY

**Fifth Edition:** published as Chapter 2. Updated by Gregory Knipp, Susan Bogdanowich-Knipp and Kenneth Morris.

**Sixth Edition:** published as Chapter 2. Updated by Patrick Sinko.

# 3 ■ THERMODYNAMICS

Thermodynamics deals with the quantitative relationships of interconversion of the various forms of *energy*, including mechanical, chemical, electric, and radiant energy. Although thermodynamics was originally developed by physicists and engineers interested in the efficiencies of steam engines, the concepts formulated from it have proven to be extremely useful in the chemical sciences and related disciplines like pharmacy. As illustrated later in this chapter, the property called *energy* is broadly applicable, from determining the fate of simple chemical processes to describing the very complex behavior of biologic cells.

Thermodynamics is based on three "laws" or facts of experience that have never been proven in a direct way, in part due to the ideal conditions for which they were derived. Various conclusions, usually expressed in the form of mathematical equations, however, may be deduced from these three principles, and the results consistently agree with observations. Consequently, the laws of thermodynamics, from which these equations are obtained, are accepted as valid for systems involving large numbers of molecules.

It is useful at this point to distinguish the attributes of the three types of systems that are frequently used to describe thermodynamic properties. **Figure 3–1a** shows an *open* system in which energy and matter can be exchanged with the surroundings. In contrast, **Figure 3–1b** and *c* are examples of *closed* systems, in which there is no exchange of matter with the surroundings, that is, the system's mass is constant. However, energy can be transferred by *work* (**Fig. 3–1b**) or *heat* (**Fig. 3–1c**) through the *closed* system's boundaries. The

last example (**Fig. 3–1d**) is a system in which neither matter nor energy can be exchanged with the surroundings; this is called an *isolated* system.

For instance, if two immiscible solvents, water and carbon tetrachloride, are confined in a closed container and iodine is distributed between the two phases, each phase is an open system, yet the total system made up of the two phases is closed because it does not exchange matter with its surroundings.

## THE FIRST LAW OF THERMODYNAMICS

The first law is a statement of the conservation of energy. It states that, although energy can be transformed from one kind into another, it cannot be created or destroyed. Put in another way, the total energy of a system and its immediate surroundings remains constant during any operation. This statement follows from the fact that the various forms of energy are equivalent, and when one kind is formed, an equal amount of another kind must disappear. The relativistic picture of the universe expressed by Einstein's equation

$$\text{Energy} = (\text{Mass change}) \times (\text{Velocity of light})^2$$

suggests that matter can be considered as another form of energy, 1 g being equivalent to $9 \times 10^{13}$ joules. These enormous quantities of energy are involved in nuclear transformations but are not important in ordinary chemical reactions.

## ▶▶▶ KEY CONCEPT ■ BASIC DEFINITIONS OF THERMODYNAMICS

Before beginning with details about the origin and concepts involving these three laws, we define the language commonly used in thermodynamics, which has precise scientific meanings. A *system* in thermodynamics is a well-defined part of the universe that one is interested in studying. The system is separated from *surroundings*, the rest of the universe and from which the observations are made, by physical (or virtual) barriers defined

as *boundaries*. *Work* (*W*) and *heat* (*Q*) also have precise thermodynamic meanings. *Work* is a transfer of energy that can be used to change the height of a weight somewhere in the surroundings, and *heat* is a transfer of energy resulting from a temperature difference between the system and the surroundings. It is important to consider that both *work* and *heat* appear only at the system's *boundaries* where the *energy* is being transferred.

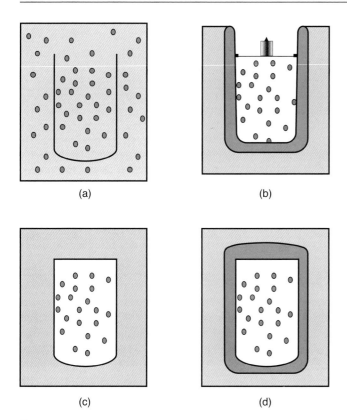

(a)                                  (b)

(c)                                  (d)

**Fig. 3–1.** Examples of thermodynamic systems. (*a*) An open system exchanging mass with its surroundings; (*b*) a closed system exchanging work with its surroundings; (*c*) a closed system exchanging heat with its surroundings; (*d*) an isolated system, in which neither work nor heat can be exchanged through boundaries.

According to the first law, the effects of $Q$ and $W$ in a given system during a transformation from an initial thermodynamic state to a final thermodynamic state are related to an intrinsic property of the system called the *internal energy*, defined as

$$\Delta E = E_2 - E_1 = Q + W \qquad (3\text{–}1)$$

where $E_2$ is the internal energy of the system in its final state and $E_1$ is the internal energy of the system in its initial state, $Q$ is the heat, and $W$ is the work. The change in internal energy $\Delta E$ is related to $Q$ and $W$ transferred between the system and its surroundings. Equation (3–1) also expresses the fact that *work* and *heat* are equivalent ways of changing the internal energy of the system.

The internal energy is related to the microscopic motion of the atoms, ions, or molecules of which the system is composed. Knowledge of its absolute value would tell us something about the microscopic motion of the vibrational, rotational, and translational components. In addition, the absolute value would also provide information about the kinetic and potential energies of their electrons and nuclear elements, which in practice is extremely difficult to attain. Therefore, change of internal energy rather than absolute energy value is the concern of thermodynamics.

**EXAMPLE 3–1**

**Thermodynamic State**

Consider the example of transporting a box of equipment from a camp in a valley to one at the top of a mountain. The main concern is with the potential energy rather than the internal energy of a system, but the principle is the same. One can haul the box to the top of the mountain by a block and tackle suspended from an overhanging cliff and produce little heat by this means. One can also drag the box up a path, but more work will be required and considerably more heat will be produced owing to the frictional resistance. The box can be carried to the nearest airport, flown over the appropriate spot, and dropped by parachute. It is readily seen that each of these methods involves a different amount of heat and work. The change in potential energy depends only on the difference in the height of the camp in the valley and the one at the top of the mountain, and it is independent of the path used to transport the box.

By using equation (3–1) (the first law), one can evaluate the change of internal energy by measuring $Q$ and $W$ during the change of state. However, it is useful to relate the change of internal energy to the *measurable properties* of the system: $P$, $V$, and $T$. Any two of these variables must be specified to define the internal energy. For an infinitesimal change in the energy $dE$, equation (3–1) is written as

$$dE = đq + đw \qquad (3\text{–}2)$$

where $đq$ is the heat absorbed and $đw$ is the work done during the infinitesimal change of the system. Capital letters $Q$ and $W$ are used for heat and work in equation (3–1) to signify finite changes in these quantities. The symbol $đ$ in equation (3–2) signifies infinitesimal changes of properties that depend on the "path," also called inexact differentials. Hence, $đq$ and $đw$ are not in these circumstances thermodynamic properties.

The infinitesimal change of any state property like $dE$, also called an exact differential, can be generally written, for instance, as a function of $T$ and $V$ as in the following equation for a closed system (i.e., constant mass):

$$dE = \left(\frac{\partial E}{\partial T}\right)_V dT + \left(\frac{\partial E}{\partial V}\right)_T dV \qquad (3\text{–}3)$$

The partial derivatives of the energy in equation (3–3) are important properties of the system and show the rate of change in energy with the change in $T$ at constant volume or with the change of $V$ at constant temperature. Therefore, it is useful to find their expression in terms of measurable properties. This can be done by combining equations (3–2) and (3–3) into

$$dq + dw = \left(\frac{\partial E}{\partial T}\right)_V dT + \left(\frac{\partial E}{\partial V}\right)_T dV \qquad (3\text{–}4)$$

This equation will be used later to describe some properties of $E$ as a function of $T$ and $V$.

## Isothermal and Adiabatic Processes

When the temperature is kept constant during a process, the reaction is said to be conducted *isothermally*. An isothermal reaction may be carried out by placing the system in a

## ▶ KEY CONCEPT ◀ THERMODYNAMIC STATE

The term *thermodynamic state* means the condition in which the measurable properties of the system have a definite value. The state of 1 g of water at $E_1$ may be specified by the conditions of, say, 1 atm pressure and 10°C, and the state $E_2$ by the conditions of 5 atm and 150°C. Hence, the states of most interest to the chemist ordinarily are defined by specifying any two of the three variables, temperature ($T$), pressure ($P$), and volume ($V$); however, additional independent variables sometimes are needed to specify the state of the system. Any equation relating the necessary variables—for example, $V = f(T,P)$—is an *equation of state*. The ideal gas law and the van der Waals equation described in Chapter 2 are equations of state. Thus, $V$, $P$, and $T$ are variables that define a state, so they are called *state variables*. The variables of a thermodynamic state are independent of how the state has been reached.

A feature of the change of internal energy, $\Delta E$, discovered by the first law is that it depends only on the initial and final thermodynamic states, that is, it is a variable of state; it is a thermodynamic property of the system. On the other hand, both $Q$ and $W$ depend on the manner in which the change is conducted. Hence, $Q$ and $W$ are not variables of state or thermodynamic properties; they are said to depend on the "path" of the transformation.

large constant-temperature bath so that heat is drawn from or returned to it without affecting the temperature significantly. When heat is neither lost nor gained during a process, the reaction is said to occur *adiabatically*. A reaction carried on inside a sealed Dewar flask or "vacuum bottle" is adiabatic because the system is thermally insulated from its surroundings. In thermodynamic terms, it can be said that an adiabatic process is one in which $dq = 0$, and the first law under adiabatic conditions reduces to

$$dw = dE \qquad (3-5)$$

According to equation (3–5), when work is done by the system, the internal energy decreases, and because heat cannot be absorbed in an adiabatic process, the temperature must fall. Here, the work done becomes a thermodynamic property dependent only on the initial and final states of the system.

### Work of Expansion Against a Constant Pressure

We first discuss the work term. Because of its importance in thermodynamics, initial focus is on the work produced by varying the volume of a system (i.e., expansion work or compression work) against a *constant* opposing external pressure, $P_{ex}$. Imagine a vapor confined in a hypothetical cylinder fitted with a weightless, frictionless piston of area $A$, as shown in **Figure 3–2**. If a constant external pressure $P_{ex}$ is exerted on the piston, the total force is $P_{ex} \times A$ because $P = $ Force/Area. The vapor in the cylinder is now made to expand by increasing the temperature, and the piston moves a distance $h$. The work done against the opposing pressure in one single stage is

$$W = -P_{ex} \times A \times h \qquad (3-6)$$

Now $A \times h$ is the increase in volume, $\Delta V = V_2 - V_1$, so that, at constant pressure,

$$W = -P_{ex}\Delta V = -P_{ex}(V_2 - V_1) \qquad (3-7)$$

### Reversible Processes

Now let us imagine the hypothetical case of water at its boiling point contained in a cylinder fitted with a weightless and frictionless piston (**Fig. 3–3a**). The apparatus is immersed in a constant-temperature bath maintained at the same temperature as the water in the cylinder. By definition, the vapor pressure of water at its boiling point is equal to the atmospheric pressure, represented in **Figure 3–3** by a set of weights equivalent to the atmospheric pressure of 1 atm; therefore, the temperature is 100°C. The process is an isothermal one, that is, it is carried out at constant temperature. Now, if the external pressure is decreased slightly by removing one of the infinitesimally small weights (**Fig. 3–3b**), the volume of the system increases and the vapor pressure falls infinitesimally. Water then evaporates to maintain the vapor pressure constant at its original value, and heat is extracted from the bath to keep the temperature constant and bring about the vaporization. During this process, a heat exchange between the system and the temperature bath will occur.

On the other hand, if the external pressure is increased slightly by adding an infinitesimally small weight (**Fig. 3–3c**), the system is compressed and the vapor pressure also rises infinitesimally. Some of the water condenses to reestablish the equilibrium vapor pressure, and the liberated heat

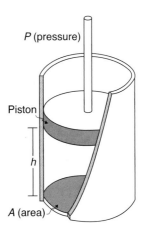

**Fig. 3–2.** A cylinder with a weightless and frictionless piston.

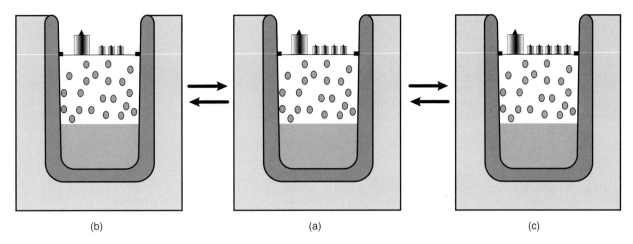

**Fig. 3–3.** A reversible process: evaporation and condensation of water at 1 atm in a closed system.
(*a*) System at equilibrium with $P_{ex} = 1$ atm; (*b*) expansion is infinitesimal; (*c*) compression is infinitesimal.

is absorbed by the constant-temperature bath. If the process could be conducted infinitely slowly so that no work is expended in supplying kinetic energy to the piston, and if the piston is considered to be frictionless so that no work is done against the force of friction, all the work is used to expand or compress the vapor. Then, because this process is always in a state of virtual thermodynamic equilibrium, being reversed by an infinitesimal change of pressure, it is said to be *reversible*. If the pressure on the system is increased or decreased rapidly or if the temperature of the bath cannot adjust instantaneously to the change in the system, the system is not in the same thermodynamic state at each moment, and the process is irreversible.

Although no real system can be made strictly reversible, some are nearly so. One of the best examples of reversibility is that involved in the measurement of the potential of an electrochemical cell using the potentiometric method.

## Maximum Work

The work done by a system in an isothermal expansion process is at a maximum when it is done reversibly. This statement can be shown to be true by the following argument. No work is accomplished if an ideal gas expands freely into a vacuum, where $P = 0$, because any work accomplished depends on the external pressure. As the external pressure becomes greater, more work is done by the system, and it rises to a maximum when the external pressure is infinitesimally less than the pressure of the gas, that is, when the process is reversible. Of course, if the external pressure is continually increased, the gas is compressed rather than expanded, and work is done *on* the system rather than *by* the system in an isothermal reversible process.

Then the maximum work done for a system that is expanded in reversible fashion is

$$W = \int_1^2 dw = -\int_{V_1}^{V_2} P\,dV \qquad (3\text{–}8)$$

where $P_{ex}$ was replaced by $P$ because the external pressure is only infinitesimally smaller than the pressure of the system. In similar fashion, it can be deduced that the minimum work in a reversible compression of the system will also lead to equation (**3–8**), because at each stage $P_{ex}$ is only infinitesimally larger than $P$. The right term in equation (**3–8**) is depicted in the shaded area in **Figure 3–4**, which represents the maximum expansion work or the minimum compression work in a reversible process.

**EXAMPLE 3–2**

A gas expands by 0.5 liter against a constant pressure of 0.5 atm at 25°C. What is the work in ergs and in joules done by the system?

$$W = P\Delta V$$
$$1\,\text{atm} = 1.013 \times 10^6\,\text{dynes/cm}^2$$
$$W = (0.507 \times 10^6\,\text{dynes/cm}^2) \times 500\,\text{cm}^3$$
$$= 2.53 \times 10^8\,\text{ergs} = 25.3\,\text{joules}$$

The following example demonstrates the kind of problem that can be solved by an application of the first law of thermodynamics.

The external pressure in equation (**3–8**) can be replaced by the pressure of an ideal gas, $P = nRT/V$, and by ensuring that the temperature of the gas remains constant during the change of state (isothermal process); then one can take $nRT$ outside the integral, giving the equation

$$W_{max} = \int dw_{max} = -nRT \int_{V_1}^{V_2} \frac{dV}{V} \qquad (3\text{–}9)$$

$$W_{max} = -nRT \ln \frac{V_2}{V_1} \qquad (3\text{–}10)$$

Note that in expansion, $V_2 > V_1$, and $\ln(V_2/V_1)$ is a positive quantity; therefore, the work is done by the system, so that its energy decreases (negative sign). When the opposite is true, $V_2 < V_1$, and $\ln(V_2/V_1)$ is negative due to gas compression, work is done by the system, so that its energy increases

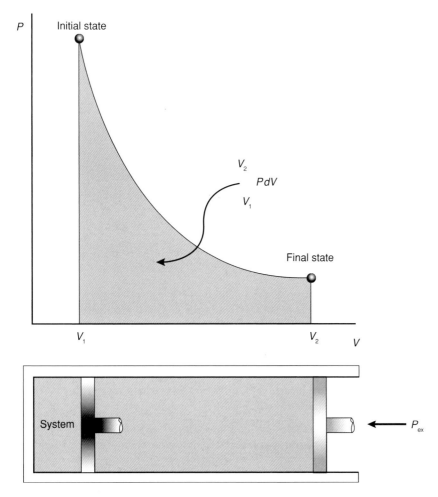

**Fig. 3–4.** Reversible expansion of a gas.

(positive sign). The process itself determines the sign of $W$ and $\Delta E$.

Equation **(3–10)** gives the maximum work done in the expansion as well as the heat absorbed, because $Q = \Delta E - W$, and, as will be shown later, $\Delta E$ is equal to zero for an ideal gas in an isothermal process. The maximum work in an isothermal reversible expansion may also be expressed in terms of pressure because, from Boyle's law, $V_2/V_1 = P_1/P_2$ at constant temperature. Therefore, equation **(3–10)** can be written as

$$W_{\max} = -nRT \ln \frac{P_1}{P_2} \qquad (3\text{–}11)$$

### EXAMPLE 3–3

One mole of water in equilibrium with its vapor is converted into steam at 100°C and 1 atm. The heat absorbed in the process (i.e., the heat of vaporization of water at 100°C) is about 9720 cal/mole. What are the values of the three first-law terms $Q$, $W$, and $\Delta E$?

The amount of heat absorbed is the heat of vaporization, given as 9720 cal/mole. Therefore,

$$Q = 9720 \,\text{cal/mole}$$

The work $W$ performed against the constant atmospheric pressure is obtained by using equation (3–10), $W = -nRT \ln(V_2/V_1)$. Now, $V_1$ is the volume of 1 mole of liquid water at 100°C, or about

0.018 liter. The volume $V_2$ of 1 mole of steam at 100°C and 1 atm is given by the gas law, assuming that the vapor behaves ideally:

$$V_2 = \frac{RT}{P} = \frac{0.082 \times 373}{1} = 30.6 \text{ liters}$$

It is now possible to obtain the work,

$$W = -(1 \text{ mole})(1.9872 \text{ cal/K mole})(398.15 \text{ K}) \ln(30.6/0.018)$$
$$W = -5883 \text{ cal}$$

The internal energy change $\Delta E$ is obtained from the first-law expression,

$$\Delta E = 9720 - 5883 = 3837 \,\text{cal}$$

Therefore, of the 9720 cal of heat absorbed by 1 mole of water, 5883 cal is employed in doing the work of expansion, or "$PV$ work," against an external pressure of 1 atm. The remaining 3837 cal increases the internal energy of the system. This quantity of heat supplies potential energy to the vapor molecules, that is, it represents the work done against the noncovalent forces of attraction.

### EXAMPLE 3–4

What is the maximum work done in the isothermal reversible expansion of 1 mole of an ideal gas from 1 to 1.5 liters at 25°C?

The conditions of this problem are similar to those of *Example 3–2*, except that equation (3–10) can now be used to obtain the (maximum) work involved in expanding reversibly this gas by 0.5 liters; thus,

$$W = -(1\,\text{mole})(8.3143\,\text{joules/K mole})(298.15\,\text{K})\ln(1.5/1.0)$$
$$W = -1005.3\,\text{joules}$$

When the solution to *Example 3–4* (work done in a reversible fashion) is contrasted with the results from *Example 3–3* (work done in one stage against a fixed pressure), it becomes apparent that the amount of work required for expansion between separate pathways can dramatically differ.

## Changes of State at Constant Volume

If the volume of the system is kept constant during a change of state, $dV = 0$, the first law can be expressed as

$$dE = đQ_v \qquad (3\text{–}12)$$

where the subscript $V$ indicates that volume is constant. Similarly, under these conditions the combined equation (3–4) is reduced to

$$dq_V = \left(\frac{\partial E}{\partial T}\right)_V dT \qquad (3\text{–}13)$$

This equation relates the heat transferred during the process at constant volume, $đQ_v$, with the change in temperature, $dT$. The ratio between these quantities defines the molar heat capacity at constant volume:

$$\overline{C}_v \equiv \frac{dq_V}{dT} = \left(\frac{\partial E}{\partial T}\right)_V \qquad (3\text{–}14)$$

## Ideal Gases and the First Law

An ideal gas has no internal pressure, and hence no work needs be performed to separate the molecules from their cohesive forces when the gas expands. Therefore, $đw = 0$, and the first law becomes

$$dE = đq \qquad (3\text{–}15)$$

Thus, the work done by the system in the isothermal expansion of an ideal gas is equal to the heat absorbed by the gas. Because the process is done isothermally, there is no temperature change in the surroundings, $dT = 0$, and $q = 0$. Equation (3–5) is reduced to

$$dE = \left(\frac{\partial E}{\partial V}\right)_T dV = 0 \qquad (3\text{–}16)$$

In this equation, $dV \neq 0$ because there has been an expansion, so that we can write

$$\left(\frac{\partial E}{\partial V}\right)_T = 0 \qquad (3\text{–}17)$$

Equation (3–17) suggests that the internal energy of an ideal gas is a function of the *temperature* only, which is one of the conditions needed to define an ideal gas in thermodynamic terms.

## Changes of State at Constant Pressure

When the work of expansion is done at *constant pressure*, $W = -P\,\Delta V = -P(V_2 - V_1)$ by equation (3–7), and under these conditions, the first law can be written as

$$\Delta E = Q_P - P(V_2 - V_1) \qquad (3\text{–}18)$$

where $Q_P$ is the heat absorbed at constant pressure. Rearranging the equation results in

$$Q_P = E_2 - E_1 + P(V_2 - V_1)$$
$$= (E_2 + PV_2) - (E_1 + PV_1) \qquad (3\text{–}19)$$

The term $E + PV$ is called the *enthalpy*, $H$. The increase in enthalpy, $\Delta H$, is equal to the heat absorbed at constant pressure by the system. It is the heat required to increase the internal energy and to perform the work of expansion, as seen by substituting $H$ in equation (3–19),

$$Q_P = H_2 - H_1 = \Delta H \qquad (3\text{–}20)$$

and writing equation (3–18) as

$$\Delta H = \Delta E + P\,\Delta V \qquad (3\text{–}21)$$

For an infinitesimal change, one can write as

$$đq_P = dH \qquad (3\text{–}22)$$

The heat absorbed in a reaction carried out at atmospheric pressure is independent of the number of steps and the mechanism of the reaction. It depends only on the initial and final conditions. This fact will be used in the section on thermochemistry.

It should also be stressed that $\Delta H = Q_P$ only when nonatmospheric work (i.e., work other than that against the atmosphere) is ruled out. When electric work, work against surfaces, or centrifugal forces are considered, one must write as

$$\Delta H = Q_P - W_{\text{nonatm}} \qquad (3\text{–}23)$$

The function $H$ is a composite of state properties, and therefore it is also a state property that can be defined as an exact differential. If $T$ and $P$ are chosen as variables, $dH$ can be written as

$$dH = \left(\frac{\partial H}{\partial T}\right)_P dT + \left(\frac{\partial E}{\partial P}\right)_T dP \qquad (3\text{–}24)$$

When the pressure is held constant, as, for example, when a reaction proceeds in an open container in the laboratory at essentially constant atmospheric pressure, equation (3–24) becomes

$$dH = \left(\frac{\partial H}{\partial T}\right)_P dT \qquad (3\text{–}25)$$

Because $đq_P = dH$ at constant pressure according to equation (3–22), the molar heat capacity $C_P$ at constant pressure is defined as

$$\overline{C}_P \equiv \frac{dq_P}{dT} = \left(\frac{\partial H}{\partial T}\right)_P \qquad (3\text{–}26)$$

**TABLE 3–1**
**MODIFIED FIRST-LAW EQUATIONS FOR PROCESSES OCCURRING UNDER VARIOUS CONDITIONS**

| Specified Conditions | Process | | Common Means for Establishing the Condition | Modification of the First Law, $dE = dq + dw$, Under the Stated Condition |
|---|---|---|---|---|
| (a) Constant heat | $dq = 0$ | Adiabatic | Insulated vessel, such as a Dewar flask | $dE = dw$ |
| (b) Reversible process at a constant temperature | $dT = 0$ | Isothermal | Constant-temperature bath | $dw = W_{max}$ |
| (c) Ideal gas at a constant temperature | $(\delta E/\delta V)_T = 0$ | Isothermal | Constant-temperature bath | $dE = 0, dq = -dw$ |
| (d) Constant volume | $dV = 0$ | Isometric (isochoric) | Closed vessel of constant volume, such as a bomb calorimeter | $dw = -PdV = 0, dE = Q_v$ |
| (e) Constant pressure | $dP = 0$ | Isobaric | Reaction occurring in an open container at constant (atmospheric) pressure | $dH = Q_P$ $dE = dH - PdV$ |

and for a change in enthalpy between products and reactants,

$$\Delta H = H_{products} - H_{reactants}$$

equation (3–26) may be written as

$$\left[\frac{\partial(\Delta H)}{\partial T}\right]_P = \Delta \overline{C}_P \qquad (3\text{–}27)$$

where $\Delta C_P = (C_P)_{products} - (C_P)_{reactants}$. Equation (3–27) is known as the *Kirchhoff equation*.

## Summary

Some of the special restrictions that have been placed on the first law up to this point in the chapter, together with the resultant modifications of the law, are brought together in **Table 3–1**. A comparison of the entries in **Table 3–1** with the material that has gone before will serve as a comprehensive review of the first law.

# THERMOCHEMISTRY

Many chemical and physical processes of interest are carried out at atmospheric (essentially constant) pressure. Under this condition, the heat exchanged during the process equals the change in enthalpy according to equation (3–20), $Q_P = \Delta H$. Thus, the change in enthalpy accompanying a chemical reaction remains a function only of temperature as stated in equation (3–27). A negative $\Delta H$ and $Q_P$ means that heat is released (exothermic); a positive value of $\Delta H$ and $Q_P$ means that heat is absorbed (endothermic). It is also possible that a reaction takes place in a closed container; in such a case the heat exchanged equals the change in internal energy (i.e., $Q_v = \Delta E$).

Thermochemistry deals with the heat changes accompanying isothermal chemical reactions at constant pressure or volume, from which values of $\Delta H$ or $\Delta E$ can be obtained. These thermodynamic properties are, of course, related by the definition of $H$, equation (3–21). In solution reactions, the $P \Delta V$ terms are not significant, so that $\Delta H \cong \Delta E$. This close approximation does not hold, however, for reactions involving gases.

## Heat of Formation

For any reaction represented by the chemical equation

$$aA + bB = cC + dD$$

the enthalpy change can be written as

$$\Delta H = \sum \overline{H}_{products} - \sum \overline{H}_{reactants} \qquad (3\text{–}28)$$

$$\Delta H = c\overline{H}_C + d\overline{H}_D - a\overline{H}_A - b\overline{H}_B \qquad (3\text{–}29)$$

where $\overline{H}$ = enthalpy per mole (called the molar enthalpy) and $a$, $b$, $c$, and $d$ are stoichiometric coefficients. It is known that only the molar enthalpies of compounds, either as reactants or products, contribute to the change of enthalpy of a chemical reaction. Enthalpies are all relative magnitudes, so it is useful to define the heat involved in the formation of chemical compounds.

One can, for instance, choose the reaction of formation of carbon dioxide from its elements,

$$C_{(s)} + O_{2(g)} = CO_{2(g)}; \quad \Delta H^{\circ}_{f(25^{\circ}C)} = -94,052 \text{ cal} \quad (3\text{–}30)$$

Here

$$\Delta H^{\circ}_{f(25^{\circ}C)} = \overline{H}(CO_2, g, 1 \text{ atm}) - \overline{H}(C, s, 1 \text{ atm})$$
$$-\overline{H}(O_2, g, 1 \text{ atm}) \qquad (3\text{–}31)$$

The subscripts represent the physical states, (s) standing for solid and (g) for gas. Additional symbols, (l) for liquid and (aq) for dilute aqueous solution, will be found in subsequent thermochemical equations.

**KEY CONCEPT** STANDARD
ENTHALPY

In thermodynamics, the convention of assigning zero enthalpy to all elements in their most stable physical state at 1 atm of pressure and 25°C is known as choosing a standard or reference state:

$$\overline{H}^\circ(\text{compound}) \equiv \Delta \overline{H}_f^\circ$$

If the $\overline{H}$ values for the elements $C_{(s)}$ and $O_{2(g)}$ are chosen arbitrarily to be zero, then according to equation (3–31), the molar enthalpy of the compound, $\overline{H}(CO_2, g, 1 \text{ atm})$, is equal to the enthalpy of the formation reaction, $\Delta H_{f(25°C)}^\circ$, for the process in equation (3–30) at 1 atm of pressure and 25°C.

The standard heat of formation of gaseous carbon dioxide is $\Delta H_{f(25°C)}^\circ = -94,052$ cal. The negative sign accompanying the value for $\Delta H$ signifies that heat is evolved, that is, the reaction is exothermic. The state of matter or allotropic form of the elements also must be specified in defining the standard state. Equation (3–30) states that when 1 mole of solid carbon (graphite) reacts with 1 mole of gaseous oxygen to produce 1 mole of gaseous carbon dioxide at 25°C, 94,052 cal is liberated. This means that the reactants contain 94,052 cal in excess of the product, so that this quantity of heat is evolved during the reaction. If the reaction were reversed and $CO_2$ were converted to carbon and oxygen, the reaction would be endothermic. It would involve the absorption of 94,052 cal, and $\Delta H$ would have a positive value.

The standard heats of formation of thousands of compounds have been determined, and some of these are given in **Table 3–2**.

**TABLE 3–2**
**STANDARD HEATS OF FORMATION AT 25°C\***

| Substance | $\Delta H^\circ$ (kcal/mole) | Substance | $\Delta H^\circ$ (kcal/mole) |
|---|---|---|---|
| $H_{2(g)}$ | 0 | | |
| $H_{(g)}$ | 52.09 | Methane$_{(g)}$ | −17.889 |
| $O_{2(g)}$ | 0 | Ethane$_{(g)}$ | −20.236 |
| $O_{(g)}$ | 59.16 | Ethylene$_{(g)}$ | 12.496 |
| $I_{2(g)}$ | 14.88 | Benzene$_{(g)}$ | 19.820 |
| $H_2O_{(g)}$ | −57.798 | Benzene$_{(l)}$ | 11.718 |
| $H_2O_{(l)}$ | −68.317 | Acetaldehyde$_{(g)}$ | −39.76 |
| $HCl_{(g)}$ | −22.063 | Ethyl alcohol$_{(l)}$ | −66.356 |
| $HI_{(g)}$ | 6.20 | Glycine$_{(g)}$ | −126.33 |
| $CO_{2(g)}$ | −94.052 | Acetic acid$_{(l)}$ | −116.4 |

\*From F. D. Rossini, K. S. Pitzer, W. J. Taylor, et al., *Selected Values of Properties of Hydrocarbons* (Circular of the National Bureau of Standards 461), U.S. Government Printing Office, Washington, D.C., 1947; F. D. Rossini, D. D. Wagman, W. H. Evans, et al., *Selected Values of Chemical Thermodynamic Properties* (Circular of the National Bureau of Standards 500), U.S. Government Printing Office, Washington, D.C., 1952.

## Hess's Law and Heat of Combustion

It is not possible to directly measure the heats of formation of every known compound as in equation (3–30). Incomplete or side reactions often complicate such determinations. However, as early as 1840, Hess showed that because $\Delta H$ depends only on the initial and final states of a system, thermochemical equations for several steps in a reaction could be added and subtracted to obtain the heat of the overall reaction. The principle is known as *Hess's law of constant heat summation* and is used to obtain heats of reaction that are not easily measured directly.

**EXAMPLE 3–5**

It is extremely important to use the heat of combustion, that is, the heat involved in the complete oxidation of 1 mole of a compound at 1 atm pressure, to convert the compound to its products. For instance, the formation of methane is written as

$$C_{(s)} + 2H_{2(g)} = CH_{4(g)}; \quad \Delta H_{f(25°C)}^\circ$$

and the combustion of methane is written as

$$CH_{4(g)} + 2O_{2(g)} = CO_{2(g)} + 2H_2O_{(l)};$$
$$\Delta H_{comb(25°C)}^\circ = -212.8 \text{ kcal}$$

The enthalpies of formation of both $CO_{2(g)}$ and $H_2O_{(l)}$ have been measured with extreme accuracy; therefore, $\Delta H_{f(25°C)}^\circ$ for methane gas can be obtained by subtracting $\Delta H_{f(25°C)}^\circ$ of $CO_{2(g)}$ and twice that of $H_2O_{(l)}$ from the heat of combustion:

$$CO_{2(g)} + 2H_2O_{(l)} = CH_{4(g)} + 2O_{2(g)};$$
$$\Delta H_{comb\,f(25°C)}^\circ = +212.8 \text{ kcal}$$

$$C_{(s)} + O_{2(g)} = CO_{2(g)}; \quad \Delta H_{f(25°C)}^\circ = -94.052 \text{ kcal}$$

$$2\,(H_{2(s)} + \tfrac{1}{2}O_{2(g)} = H_2O_{(l)}); \quad 2(\Delta H_{f(25°C)}^\circ = -68.317)$$

$$C_{(s)} + 2H_{2(g)} = CH_{4(g)}$$

$$\Delta H_{f(25°C)}^\circ(CH_4, g) = -\Delta H_{comb}^\circ + \Delta H_f(CO_2, g)$$
$$+ 2 \times \Delta H_f(H_2O, l)$$

$$= 212.8 \text{ kcal} + (-94.052 \text{ kcal}) + 2(-68.317 \text{ kcal})$$
$$= -17.886 \text{ kcal}$$

The heat of formation of $CH_{4(g)}$, methane, is reported in Table 3–2 as −17.889.

## Heats of Reaction from Bond Energies

In Chapter 2, energies of bonding were discussed in terms of binding forces that hold molecules (*intra*molecular bonding) or states of matter (*noncovalent forces*) together. For example, the energies associated with covalent bonding between atoms range from 50 to 200 kcal/mole. These figures encompass many of the types of bonding that you learned about in general and organic chemistry courses, including double and triple bonds that arise from $\pi$-electron orbital overlap, but they are weaker than covalent $\sigma$-electron orbital bonds. Electrovalent or ionic bonds occurring between atoms with opposing permanent charges can be much stronger than a covalent bond. In a molecule, the forces enabling a bond to form between atoms arise from a combination of attractive and repulsive energies associated with each atom in the molecule. These forces vary in strength because of interplay between attraction and repulsion. Therefore, the net bonding energy that holds a series of atoms together in a molecule is the additive result of all the individual bonding energies.

In a chemical reaction, bonds may be broken and new bonds may be formed to give rise to the product. The net energy associated with the reaction, the heat of reaction, can be estimated from the bond energies that are broken and the bond energies that are formed during the reaction process. Many of the common covalent and other bonding-type energies can be commonly found in books on thermodynamics, like the ones listed in the footnote in the opening of this chapter.

**EXAMPLE 3–6**

The covalent bonding energy in the reaction

$$H_2C{=}CH_2 + Cl{-}Cl \rightarrow \underset{\underset{H}{|}}{\overset{\overset{H}{|}}{Cl{-}C}} - \underset{\underset{H}{|}}{\overset{\overset{H}{|}}{C}}{-}Cl$$

can be calculated from knowing a C=C bond is broken (requiring 130 kcal), a Cl—Cl bond is broken (requiring 57 kcal), a C—C bond is formed (liberating 80 kcal), and two C—Cl bonds are formed (liberating 2 × 78 or 156 kcal). Thus, the energy $\Delta H$ of the reaction is

$$\Delta H = 130 + 57 - 80 - 156 = -49\,\text{kcal}$$

Because 1 cal = 4.184 joules, −49 kcal is expressed in SI units as $-2.05 \times 10^5$ joules.

The foregoing process is not an idealized situation; strictly speaking, it applies only to reactions in a gas phase. If there is a change from a gas into a condensed phase, then condensation, solidification, or crystallization heats must be involved in the overall calculations. This approach may find some use in estimating the energy associated with chemical instability.

## Additional Applications of Thermochemistry

Thermochemical data are important in many chemical calculations. Heat-of-mixing data can be used to determine whether a reaction such as precipitation is occurring during the mixing of two salt solutions. If no reaction takes place when dilute solutions of the salts are mixed, the heat of reaction is zero.

The constancy of the heats of neutralization obtained experimentally when dilute aqueous solutions of various strong acids and strong bases are mixed shows that the reaction involves only

$$H^+_{(aq)} + OH^-_{(aq)} = H_2O_{(l)}; \Delta H_{25°C} = -13.6\,\text{kcal} \quad (3-32)$$

No combination occurs between any of the other species in a reaction such as

$$HCl_{(aq)} + NaOH_{(aq)} = H_2O_{(l)} + Na^+_{(aq)} + Cl^-_{(aq)}$$

because HCl, NaOH, and NaCl are completely ionized in water. In the neutralization of a weak electrolyte by a strong acid or base, however, the reaction involves ionization in addition to neutralization, and the heat of reaction is no longer constant at about −13.6 kcal/mole. Because some heat is absorbed in the ionization of the weak electrolyte, the heat evolved falls below the value for the neutralization of completely ionized species. Thus, knowledge of $\Delta H$ of neutralization allows one to differentiate between strong and weak electrolytes.

Another important application of thermochemistry is the determination of the number of calories obtained from various foods. The subject is discussed in biochemistry texts as well as books on food and nutrition.

# THE SECOND LAW OF THERMODYNAMICS

In ordinary experience, most natural phenomena are observed as occurring only in one direction. For instance, heat flows spontaneously only from hotter to colder bodies, whereas two bodies having the same temperature do not evolve into two bodies having different temperatures, even though the first law of thermodynamics does not prohibit such a possibility. Similarly, gases expand naturally from higher to lower pressures, and solute molecules diffuse from a region of higher to one of lower concentration. These spontaneous processes will not proceed in reverse without the intervention of some external force to facilitate their occurrence. Although spontaneous processes are not thermodynamically reversible, they can be carried out in a nearly reversible manner by an outside force. Maximum work is obtained by conducting a spontaneous process reversibly; however, the frictional losses and the necessity of carrying out the process at an infinitely slow rate preclude the possibility of complete reversibility in real processes.

The first law of thermodynamics simply observes that energy must be conserved when it is converted from one form to another. It has nothing to say about the probability that a process will occur. The second law refers to the probability of the occurrence of a process based on the observed tendency of a system to approach a state of energy equilibrium.

The historical development of the thermodynamic property that explains the natural tendency of the processes to occur, now called *entropy*, has its origins in studies of the efficiency of steam engines, from which the following observation was made: "A steam engine can do work only with a fall in temperature and a flow of heat to the lower temperature. No useful work can be obtained from heat at constant temperature." This is one way to state the second law of thermodynamics.

## The Efficiency of a Heat Engine

An important consideration is that of the possibility of converting heat into work. Not only is heat isothermally unavailable for work, *it can never be converted completely into work*.

The spontaneous character of natural processes and the limitations on the conversion of heat into work constitute the second law of thermodynamics. Falling water can be made to do work owing to the difference in the potential energy at two different levels, and electric work can be done because of the difference in electric potential (emf). A heat engine (such as a steam engine) likewise can do useful work by using two heat reservoirs, a "source" and a "sink," at two different temperatures. Only part of the heat at the source is converted into work, with the remainder being returned to the sink (which, in practical operations, is often the surroundings) at the lower temperature. The fraction of the heat, $Q$, at the source converted into work, $W$, is known as the *efficiency* of the engine:

$$\text{Efficiency} \equiv \frac{W}{Q} \qquad (3\text{--}33)$$

The efficiency of even a hypothetical heat engine operating without friction cannot be unity because $W$ is always less than $Q$ in a continuous conversion of heat into work according to the second law of thermodynamics.

Imagine a hypothetical steam engine operating reversibly between an upper temperature $T_{hot}$ and a lower temperature $T_{cold}$. It absorbs heat $Q_{hot}$ from the hot boiler or source, and by means of the working substance, steam, it converts the quantity $W$ into work and returns heat $Q_{cold}$ to the cold reservoir or sink. Carnot, in 1824, proved that the efficiency of such an engine, operating reversibly at every stage and returning to its initial state (cyclic process), could be given by the expression

$$\frac{W}{Q_{hot}} = \frac{Q_{hot} - Q_{cold}}{Q_{hot}} \qquad (3\text{--}34)$$

It is known that heat flow in the operation of the engine follows the temperature gradient, so that the heat absorbed and rejected can be related directly to temperatures. Lord Kelvin used the ratios of the two heat quantities $Q_{hot}$ and $Q_{cold}$ of the Carnot cycle to establish the Kelvin temperature scale:

$$\frac{Q_{hot}}{Q_{cold}} = \frac{T_{hot}}{T_{cold}} \qquad (3\text{--}35)$$

By combining equations (3–33) through (3–35), we can describe the efficiency by

$$\text{Efficiency} = \frac{Q_{hot} - Q_{cold}}{Q_{hot}} = \frac{T_{hot} - T_{cold}}{T_{hot}} \qquad (3\text{--}36)$$

It is observed from equation (3–36) that the higher $T_{hot}$ becomes and the lower $T_{cold}$ becomes, the greater is the efficiency of the engine. When $T_{cold}$ reaches absolute zero on the Kelvin scale, the reversible heat engine converts heat completely into work, and its theoretical efficiency becomes unity. This can be seen by setting $T_{cold} = 0$ in equation (3–36). Because absolute zero is considered unattainable, however, an efficiency of 1 is impossible, and heat can never be completely converted to work. This statement can be written using the notation of limits as follows:

$$\lim_{T_{cold} \to 0} \frac{W}{Q} = 1 \qquad (3\text{--}37)$$

If $T_{hot} = T_{cold}$ in equation (3–36), the cycle is isothermal and the efficiency is zero, confirming the earlier statement that heat is isothermally unavailable for conversion into work.

### EXAMPLE 3–7

A steam engine operates between the temperatures of 373 and 298 K. (*a*) What is the theoretical efficiency of the engine? (*b*) If the engine is supplied with 1000 cal of heat $Q_{hot}$, what is the theoretical work in ergs?

(*a*)

$$\text{Efficiency} = \frac{W}{Q_{hot}} = \frac{373 - 298}{373} = 0.20, \quad \text{or} \quad 20\%$$

(*b*)

$$W = 1000 \times 0.20 = 200 \, \text{cal}$$
$$200 \, \text{cal} \times 4.184 \times 10^7 \, \text{ergs/cal} = 8.36 \times 10^9 \, \text{ergs}$$

## Entropy

In analyzing the properties of reversible cycles, one feature was unveiled; the sum of the quantities $Q_{rev,\,i}/T_i$ is zero for the cycle, that is,

$$\frac{Q_{hot}}{T_{hot}} + \frac{Q_{cold}}{T_{cold}} = 0 \qquad (3\text{--}38)$$

This behavior is similar to that of other state functions such as $\Delta E$ and $\Delta H$; Carnot recognized that when $Q_{rev}$, a path-dependent property, is divided by $T$, a *new* path-independent property is generated, called *entropy*. It is defined as

$$\Delta S = \frac{Q_{rev}}{T} \qquad (3\text{--}39)$$

and for an infinitesimal change,

$$dH = \left( \frac{\partial H}{\partial T} \right)_P dT \qquad (3\text{--}40)$$

Thus, the term $Q_{hot}/T_{hot}$ is known as the *entropy change* of the reversible process at $T_{hot}$, and $Q_{cold}/T_{cold}$ is the *entropy change* of the reversible process at $T_{cold}$. The entropy change $\Delta S_{hot}$ during the absorption of heat from $T_{hot}$ is positive, however, because $Q_{hot}$ is positive. At the lower temperature, $Q_{cold}$ is negative and the entropy change $\Delta S_{cold}$ is negative. The total entropy change $\Delta S_{cycle}$ in the reversible cyclic process is zero. Let us use the data from *Example 3–7* to demonstrate this. It can be seen that only part of the heat $Q_{hot}$ (1000 cal) is converted to work (200 cal) in the engine. The difference in energy (800 cal) is the heat $Q_{cold}$ that is returned to the sink at the lower temperature and is unavailable for work. When the entropy changes at the two temperatures are calculated, they are both found to equal 2.7 cal/deg:

$$\Delta S = \frac{Q_{hot,rev}}{T_{hot}} = \frac{1000 \, \text{cal}}{373} = 2.7 \, \text{cal/deg}$$

$$\Delta S = -\frac{Q_{cold,rev}}{T_{cold}} = -\frac{800 \, \text{cal}}{298} = -2.7 \, \text{cal/deg}$$

Therefore,

$$\Delta S_{cycle} = \Delta S_{hot} + \Delta S_{cold} = 0$$

It may also be noted that if $Q_{hot}$ is the heat absorbed by an engine at $T_{hot}$, then $-Q_{hot}$ must be the heat lost by the surroundings (the hot reservoir) at $T_{hot}$, and the entropy of the surroundings is

$$\Delta S_{surr} = -\frac{Q_{hot}}{T_{hot}} \qquad (3\text{--}41)$$

Hence, for any system and its surroundings or *universe*,

$$\Delta S_{univ} = \Delta S_{syst} + \Delta S_{surr} = 0 \qquad (3\text{--}42)$$

Therefore, there are two cases in which $\Delta S = 0$: (*a*) a system in a reversible cyclic process and (*b*) a system and its surroundings undergoing any reversible process.

In an *irreversible process*, the entropy change of the total system or universe (a system and its surroundings) is always positive because $\Delta S_{surr}$ is always less than $\Delta S_{syst}$ in an irreversible process. In mathematical symbols, we write,

$$\Delta S_{univ} = \Delta S_{syst} + \Delta S_{surr} > 0 \qquad (3\text{--}43)$$

and this can serve as a criterion of spontaneity of a real process.

Equations (3–42) and (3–43) summarize all of the possibilities for the entropy; for irreversible processes (real transformations) entropy always increases until it reaches its maximum at equilibrium, at which point it remains invariable (i.e., change equals zero).

Two examples of entropy calculations will now be given, first, a reversible process, and second, an irreversible process.

**EXAMPLE 3–8**

What is the entropy change accompanying the vaporization of 1 mole of water in equilibrium with its vapor at 25°C? In this reversible isothermal process, the heat of vaporization $\Delta H_v$ required to convert the liquid to the vapor state is 10,500 cal/mole.

The process is carried out at a constant pressure, so that $Q_P = \Delta H_v$, and because it is a reversible process, the entropy change can be written as

$$\Delta S = \frac{\Delta H_v}{T} = \frac{10,500}{298} = 35.2 \text{ cal/mole deg}$$

The entropy change involved as the temperature changes is often desired in thermodynamics; this relationship is needed for the next example. The heat absorbed at constant pressure is given by equation (3–26), $dq_P = C_P\, dT$, and for a reversible process

$$\frac{C_P dT}{T} = \frac{dq_{rev}}{T} = dS \qquad (3\text{--}44)$$

Integrating between $T_1$ and $T_2$ yields

$$\Delta S = C_P \ln \frac{T_2}{T_1} = 2.303\, C_P \log \frac{T_2}{T_1} \qquad (3\text{--}45)$$

**EXAMPLE 3–9**

Compute the entropy change in the (irreversible) transition of 1 mole of liquid water to crystalline water at $-10°C$, at constant pressure. The entropy is obtained by calculating the entropy changes for several *reversible* steps.

First consider the reversible conversion of supercooled liquid water at $-10°C$ to liquid water at $0°C$, then change this to ice at $0°C$, and finally cool the ice reversibly to $-10°C$. The sum of the entropy changes of these steps gives $\Delta S_{water}$. To this, add the entropy change undergone by the surroundings so as to obtain the total entropy change. If the process is spontaneous, the result will be a positive value. If the system is at equilibrium, that is, if liquid water is in equilibrium with ice at $-10°C$, there is no tendency for the transition to occur, and the total entropy change will be zero. Finally, if the process is not spontaneous, the total entropy change will be negative.

The heat capacity of water is 18 cal/deg and that of ice is 9 cal/deg within this temperature range.

The reversible change of water at $-10°C$ to ice at $-10°C$ is carried out as follows:

$$H_2O_{(l,-10°)} \rightarrow H_2O_{(l,0°)}; \quad \Delta S = C_P \ln \frac{T_{final}}{T_{initial}} = 0.67$$

$$H_2O_{(l,0°)} \rightarrow H_2O_{(s,0°)}; \quad \Delta S = \frac{q_{rev}}{T} = \frac{-1437}{273.2} = -5.26$$

$$H_2O_{(s,0°)} \rightarrow H_2O_{(s,-10°)}; \quad \Delta S = C_P \ln \frac{T_{final}}{T_{initial}} = -0.34$$

$$H_2O_{(l,-10°)} \rightarrow H_2O_{(s,-10°)}; \quad \Delta S_{H_2O} = -4.93 \text{ cal/mole deg}$$

The entropy of water decreases during the process because $\Delta S$ is negative, but the spontaneity of the process cannot be judged until one also calculates $\Delta S$ of the surroundings.

For the entropy change of the surroundings, the water needs to be considered to be in equilibrium with a large bath at $-10°C$, and the heat liberated when the water freezes is absorbed by the bath without a significant temperature increase. Thus, the reversible absorption of heat by the bath is given by

$$\Delta S_{surr} = -\frac{Q_{rev}}{T} = \frac{1343}{263.2} = 5.10 \text{ cal/mole, deg}$$

where 1343 cal/mole is the heat of fusion of water at $-10°C$. Thus,

$$\begin{aligned} \Delta S_{total\ system} &= \Delta S_{H_2O} + \Delta S_{bath} \\ &= -4.91 + 5.10 \\ &= 0.17 \text{ cal/mole deg} \end{aligned}$$

The process in *Example 3–8* is spontaneous because $\Delta S > 0$. This criterion of spontaneity is not a convenient one, however, because it requires a calculation of the entropy change both in the system and the surroundings. The free energy functions, to be treated in a later section, do not require information concerning the surroundings and are more suitable criteria of spontaneity.

## Entropy and Disorder

The second law provides a criterion for deciding whether a process follows the *natural* or *spontaneous* direction. Even though the causes for the preference for a particular direction of change of state or the impossibility for the reverse direction are unknown, the underlying reason is not that the reverse process is impossible, but rather that it is extraordinarily improbable.

Thermodynamic systems described by macroscopic properties such as $T$, $P$, or composition can also be described in terms of microscopic quantities such as molecular random motions. The microscopic or molecular interpretation of

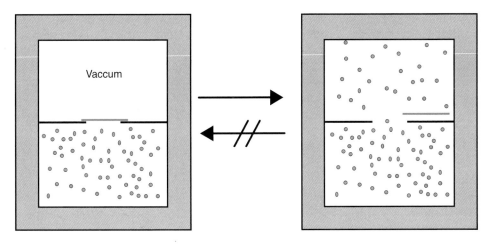

**Fig. 3–5.** Free expansion of a gas in a closed system.

entropy is commonly stated as a measure of disorder owing to molecular motion. As disorder increases, entropy will also increase. The impossibility of converting all thermal energy into work results from the "disorderliness" of the molecules existing in the system. Disorder can be seen as the number of ways the inside of a system can be arranged so that from the outside the system looks the same. A quantitative interpretation of entropy was given long before by Boltzmann; the equation for the entropy of a system in a given thermodynamic state is

$$S = k \ln O \qquad (3–46)$$

where $k$ is the Boltzmann constant and $O$ is the number of microscopic states or configuration that the system may adopt. Unlike the thermodynamic definition of entropy in equation (3–39) or (3–40), to understand what a configuration means (i.e., $O$), a structural model of the system is needed.

Let us imagine that the system consists of a gas in a closed container that is allowed to expand freely into a vacuum (**Fig. 3–5**). The initial state corresponds to $V_1$ and the final state corresponds to $V_2$, which is larger than $V_1$: $V_1 = \frac{1}{2}V_2$.

If only one gas particle is in the initial state, statistically there are two equal possibilities for a particle to occupy $V_2$, whereas only one exists for $V_1$. In other words, the probability $P$ for that particle to stay in $V_1$ is $P = \frac{1}{2}$ (**Fig. 3–6a**). Now consider two independent particles; the probability of finding both in $V_1$ after $V_2$ is available is $P = (\frac{1}{2})^2 = \frac{1}{4}$ (**Fig. 3–6b**). For three particles there are eight possible configurations, so the probability for all particles to stay in $V_1$ is only $\frac{1}{8}$. For a system composed of $N$ particles the probability becomes $P = (\frac{1}{2})^N$. Now, if $N$ is on the order of regular molecular quantities, that is, $\sim 10^{23}$ molecules, the probability of finding all of them in the initial state is extraordinarily small, $P = (\frac{1}{2})^{10^{23}}$.

Thus, in this model system, by changing from the initial state $V_1$ to a final state $V_2$, we significantly increase the number of distribution possibilities of the $N$ particles. Once the system has expanded, it is extremely improbable that they will be found by chance only in $V_1$. For this simple system, the probability expressed by the number of configurations

after an irreversible process from state 1 to state 2 is generally related to the volumes as

$$P = (O_1/O_2) = (V_1/V_2)^N \qquad (3–47)$$

and, therefore, the change of entropy for 1 mole of gas is given by

$$\Delta S = S_2 - S_1 = k \ln(O_1/O_2) = kN_A \ln(V_1/V_2) \qquad (3–48)$$

From the standpoint of this simple model, the isothermal expansion of an ideal gas increases the entropy because of

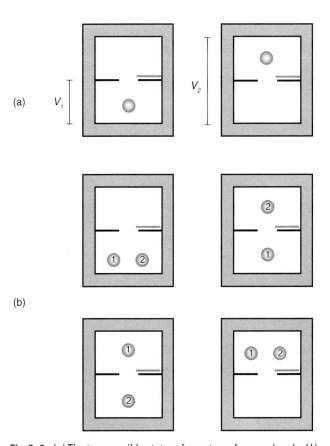

**Fig. 3–6.** (a) The two possible states of a system of one molecule. (b) The four possible states of a system of two independent molecules.

the enhanced number of configurations in a larger volume compared to a smaller one. The larger the number of configurations, the more disordered a system is considered. Thus, the increase in entropy with increasing number of configurations is nicely described by Boltzmann concept given in equation (3–46).

# THE THIRD LAW OF THERMODYNAMICS

The third law of thermodynamics states that the entropy of a pure crystalline substance is zero at absolute zero because the crystal arrangement must show the greatest orderliness at this temperature. In other words, a pure perfect crystal has only one possible configuration, and according to equation (3–46), its entropy is zero [i.e., $S = k \ln(1) = 0$]. As a consequence of the third law, the temperature of absolute zero (0 K) is not possible to reach even though sophisticated processes that use the orientation of electron spins and nuclear spins can reach very low temperatures of $2 \times 10^{-3}$ and $10^{-5}$ K, respectively.

The third law makes it possible to calculate the absolute entropies of pure substances from equation (3–44) rearranged as

$$S_T = \sum \frac{dq_{\text{rev},i}}{T_i} + S_0 \qquad (3\text{–}49)$$

where $S_0$ is the molar entropy at absolute zero and $S_T$ is the absolute molar entropy at any temperature. The magnitude $\text{đ}q_{\text{rev}}$ can be replaced by the corresponding $C_P \, dT$ values at constant pressure according to equation (3–26), so that $S_T$ may be determined from knowledge of the heat capacities and the entropy changes during a *phase change* as the temperature rises from 0 K to T. The following equation shows $S_T$ for a substance that undergoes two phase changes, melting (m) and vaporization (v):

$$S_T = \int_0^{T_m} \frac{C_P \, dT}{T} + \frac{\Delta H_m}{T_m} + \int_{T_m}^{T_v} \frac{C_P \, dT}{T}$$
$$+ \frac{\Delta H_v}{T_v} + \int_{T_v}^{T} \frac{C_P \, dT}{T}$$

where $S_0 = 0$ has been omitted. Each of these terms can be evaluated independently; in particular, the first integral is calculated numerically by plotting $C_P/T$ versus T. Below 20 K few data are available, so that it is customary to apply the Debye approximation $C_P \cong aT^3$, joules/mole K in this region.

# FREE ENERGY FUNCTIONS AND APPLICATIONS

The criterion for spontaneity given by the second law requires the knowledge of the entropy changes both in the system and the surroundings as stated in equation (3–43). It may be very

useful, however, to have a property that depends only on the system and nevertheless indicates whether a process occurs in the natural direction of change.

Let us consider an *isolated system* composed of a closed container (i.e., the system) in equilibrium with a temperature bath (i.e., the surroundings) as illustrated in **Figure 3–1d**, then according to equations (3–42) and (3–43), the resulting equation is

$$\Delta S_{\text{isolated system}} = \Delta S_{\text{syst}} + \Delta S_{\text{surr}} \geq 0 \qquad (3\text{–}50)$$

Equilibrium is a condition where the transfer of heat occurs reversibly; therefore, at constant temperature, $\Delta S_{\text{surr}} = -\Delta S_{\text{syst}}$ and

$$\Delta S_{\text{surr}} = -\frac{Q_{\text{rev}}}{T} \qquad (3\text{–}51)$$

Now, renaming $\Delta S_{\text{syst}} = \Delta S$ and $Q_{\text{rev}} = Q$, we write equation as

$$\Delta S - \frac{Q}{T} \geq 0 \qquad (3\text{–}52)$$

which can also be written in the following way:

$$Q - T\Delta S \leq 0 \qquad (\text{at } T = \text{const}) \qquad (3\text{–}53)$$

For a process at constant volume, $Q_v = \Delta E$, so equation (3–53) becomes

$$\Delta E - T\Delta S \leq 0 \qquad (\text{at } T = \text{const}, V = \text{const}) \qquad (3\text{–}54)$$

In the case of a process at constant pressure, $Q_P = \Delta H$, and

$$\Delta H - T\Delta S \leq 0 \qquad (\text{at } T = \text{const}, P = \text{const}) \qquad (3\text{–}55)$$

Equations (3–54) and (3–55) are combinations of the first and second laws of the thermodynamics and express both equilibrium (=) and spontaneity (<) conditions depending only on the properties of the system; therefore, they are of fundamental practical importance. The terms on the left of equations (3–54) and (3–55) are commonly defined as the *Helmholtz free energy or work function A*,

$$A = E - TS \qquad (3\text{–}56)$$

and the *Gibbs free energy G*,

$$G = H - TS \qquad (3\text{–}57)$$

respectively, which are two new state properties because they are composed of state variables E, H, T, and S. Thus, equilibrium and spontaneity conditions are reduced to only

$$\Delta A = 0 \qquad (\text{at } T = \text{const}, V = \text{const}) \qquad (3\text{–}58)$$

$$\Delta G = 0 \qquad (\text{at } T = \text{const}, P = \text{const}) \qquad (3\text{–}59)$$

## Maximum Net Work

The second law of the thermodynamics expressed in equation (3–53) for an isothermal process can be combined with the conservation of energy by substituting $Q = \Delta E - W$, yielding

$$\Delta E - W - T\Delta S = 0 \qquad (3\text{–}60)$$

**TABLE 3–3**
### CRITERIA FOR SPONTANEITY AND EQUILIBRIUM

| Function | Restrictions | Sign of Function | | |
| --- | --- | --- | --- | --- |
| | | **Spontaneous** | **Nonspontaneous** | **Equilibrium** |
| | Total system, | | | |
| $\Delta S_{universe}$ | $\Delta E = 0, \Delta V = 0$ | + or $>0$ | − or $<0$ | 0 |
| $\Delta G$ | $\Delta T = 0, \Delta P = 0$ | − or $<0$ | + or $>0$ | 0 |
| $\Delta A$ | $\Delta T = 0, \Delta V = 0$ | − or $<0$ | + or $>0$ | 0 |

In this case, $W$ represents all possible forms of work available and not only $PV$ work (i.e., expansion or compression of the system). The system does maximum work ($W_{max}$) when it is working under reversible conditions; therefore, only the equality in equation (**3–60**) applies. By including the definition for $A$ in equation (**3–60**), we obtain

$$\Delta A - W_{max} = 0 \quad \text{or} \quad \Delta A = W_{max} \quad \textbf{(3–61)}$$

The former equation explains the meaning of the *Helmholtz energy* (which also explains why it is called a *work function*), that is, the *maximum work* produced in an isothermal transformation is equal to the change in the Helmholtz energy.

Now if $W$ is separated into the $PV$ work and all other forms of work excluding expansion or compression, $W_a$ (also called *useful work*) for an isothermal process at constant pressure equation (**3–60**) can be written as

$$\Delta E - W_a = P\Delta V - T\Delta S = 0 \quad \textbf{(3–62)}$$

The definition of the Gibbs energy given in equation (**3–60**) can also be written as

$$G = E + PV - TS \quad \textbf{(3–63)}$$

For a reversible process equation, (**3–62**) transforms into

$$\Delta G - W_a = 0 \quad \text{or} \quad \Delta G = W_a \quad \textbf{(3–64)}$$

Equation (**3–64**) expresses the fact that the change in Gibbs free energy at constant temperature and pressure equals the *useful* or *maximum net work* ($W_a$) that can be obtained from the process. Under those circumstances in which the $PV$ term is insignificant (such as in electrochemical cells and surface tension measurements), the free energy change is approximately equal to the maximum work.

## Criteria of Equilibrium and Spontaneity

When net work can no longer be obtained from a process, $G$ is at a minimum, and $\Delta G = 0$. This statement signifies that the system is at equilibrium. On the other hand, from equation (**3–62**) a negative free energy change written as $\Delta G < 0$ signifies that the process is a spontaneous one. If $\Delta G$ is positive ($\Delta G > 0$), it indicates that net work must be absorbed for the reaction to proceed, and accordingly it is not spontaneous.

When the process occurs isothermally at constant volume rather than constant pressure, $\Delta A$ serves as the criterion for spontaneity and equilibrium. From equation (**3–58**) it is negative for a spontaneous process and becomes zero at equilibrium.

These criteria, together with the entropy criterion of equilibrium and spontaneity, are listed in **Table 3–3**.

The difference in sign between $\Delta G$ and $\Delta S_{universe}$ implies that the condition for a process being spontaneous has changed from an increase of the total entropy, $\Delta S_{universe} > 0$, to a decrease in Gibbs free energy, $\Delta G < 0$. However, Gibbs free energy is only a composite that expresses the total change in entropy in terms of the properties of the system alone. To prove that, let us combine equations (**3–42**) and (**3–51**) for a reversible isothermal process into

$$T\Delta S_{universe} = T\Delta S - Q_{rev} \quad \text{(at } T = \text{const)} \quad \textbf{(3–65)}$$

Then, if the pressure is constant,

$$T\Delta S_{universe} = T\Delta S - \Delta H \quad \text{(at } T = \text{const, } P = \text{const)} \quad \textbf{(3–66)}$$

For the same process,

$$\Delta G = \Delta H - T\Delta S \quad \textbf{(3–67)}$$

and therefore

$$\Delta G = -T\Delta S_{universe} \quad \textbf{(3–68)}$$

From the former equation it is clear that *the only criterion for spontaneous change is an increase in the entropy of the universe (i.e., the system and its surroundings).* Chemical reactions are usually carried out at constant temperature and constant pressure. Thus, equations involving $\Delta G$ are of particular interest to the chemist and the pharmacist.

It was once thought that at constant pressure a negative $\Delta H$ (evolution of heat) was itself proof of a spontaneous reaction. Many natural reactions do occur with an evolution of heat; the spontaneous melting of ice at $10°C$, however, is accompanied by absorption of heat, and a number of other examples can be cited to prove the error of this assumption. The reason $\Delta H$ is often thought of as a criterion of spontaneity can be seen from the familiar expression (**3–67**).

If $T\Delta S$ is small compared with $\Delta H$, a negative $\Delta H$ will occur when $\Delta G$ is negative (i.e., when the process is spontaneous). When $T\Delta S$ is large, however, $\Delta G$ may be negative, and the process may be spontaneous even though $\Delta H$ is positive.

The entropy of all systems, as previously stated, spontaneously tends toward randomness according to the second law, so that *the more disordered a system becomes, the higher is its probability and the greater its entropy.* Hence, equation (3–64) can be written as

$$\Delta G = \begin{bmatrix} \text{Difference in bond energies or} \\ \text{attractive energies between} \\ \text{products and reactants, } \Delta H \end{bmatrix}$$

$$- \begin{bmatrix} \text{Change in probability} \\ \text{during the process,} \\ T\,\Delta S \end{bmatrix} \quad (3\text{–}69)$$

One can state that $\Delta G$ will become negative and the reaction will be spontaneous either when the enthalpy decreases or the probability of the system increases at the temperature of the reaction.

Thus, although the conversion of ice into water at 25°C requires an absorption of heat of 1650 cal/mole, the reaction leads to a more probable arrangement of the molecules; that is, an increased freedom of molecular movement. Hence, the entropy increases, and $\Delta S = 6$ cal/mole deg is sufficiently positive to make $\Delta G$ negative, despite the positive value of $\Delta H$.

Many of the complexes of Chapter 10 form in solution with a concurrent absorption of heat, and the processes are spontaneous only because the entropy change is positive. The increase in randomness occurs for the following reason. The dissolution of solutes in water may be accompanied by a *decrease* in entropy because both the water molecules and the solute molecules lose freedom of movement as hydration occurs. In complexation, this highly ordered arrangement is disrupted as the separate ions or molecules react through coordination, and the constituents thus exhibit more freedom in the final complex than they had in the hydrated condition. The increase in entropy associated with this increased randomness results in a spontaneous reaction as reflected in the negative value of $\Delta G$.

Conversely, some association reactions are accompanied by a decrease in entropy, and they occur in spite of the negative $\Delta S$ only because the heat of reaction is sufficiently negative. For example, the Lewis acid–base reaction by which iodine is rendered soluble in aqueous solution,

$$I^-_{(aq)} + I_{2(aq)} = I^-_{3(aq)}; \quad \Delta H_{25°} = -5100\,\text{cal}$$

is accompanied by a $\Delta S$ of $-4$ cal/mole deg. It is spontaneous because

$$\Delta G = -5100 - [298 \times (-4)]$$
$$= -5100 + 1192 = -3908\,\text{cal/mole}$$

In the previous examples the reader should not be surprised to find a negative entropy associated with a spontaneous reaction. The $\Delta S$ values considered here are the changes in entropy of the *substance alone*, not of the total system, that is, the substance and its immediate surroundings. As stated before, when $\Delta S$ is used as a test of the spontaneity of a

reaction, the entropy change of the entire system must be considered. For reactions at constant temperature and pressure, which are the most common types, the change in free energy is ordinarily used as the criterion in place of $\Delta S$. It is more convenient because it eliminates the need to compute any changes in the surroundings.

By referring back to *Example 3–8*, it will be seen that $\Delta S$ was negative for the change from liquid to solid water at $-10°$C. This is to be expected because the molecules lose some of their freedom when they pass into the crystalline state. The entropy of water plus its surroundings increases during the transition, however, and it is a spontaneous process. The convenience of using $\Delta G$ instead of $\Delta S$ to obtain the same information is apparent from the following example, which may be compared with the more elaborate analysis required in *Example 3–8*.

**EXAMPLE 3–10**

$\Delta H$ and $\Delta S$ for the transition from liquid water to ice at $-10°$C and at 1 atm pressure are $-1343$ cal/mole and $-4.91$ cal/mole deg, respectively. Compute $\Delta G$ for the phase change at this temperature ($-10°$C $= 263.2$ K) and indicate whether the process is spontaneous. Write

$$\Delta G = -1343 - [263.2 \times (-4.91)] = -51\,\text{cal/mole}$$
$$= -213\,\text{joules}$$

The process is spontaneous, as reflected in the negative value of $\Delta G$.

## Pressure and Temperature Coefficients of Free Energy

By differentiating equation (3–63), one obtains several useful relationships between free energy and the pressure and temperature. Applying the differential of a product, $d(uv) = u\,dv + V\,du$, to equation (3–63), we obtain the following relationship:

$$dG = dE + P\,dV + V\,dP - T\,dS - S\,dT \quad (3\text{–}70)$$

Now, in a reversible process in which $dq_{rev} = T\,dS$, the first law, restricted to expansion work (i.e., $dE = dq_{rev} - P\,dV$), can be written as

$$dE = T\,dS - P\,dV \quad (3\text{–}71)$$

and substituting $dE$ of equation (3–74) into equation (3–70) gives

$$dG = T\,dS - P\,dV + P\,dV + V\,dP - T\,dS - S\,dT$$

or

$$dG = V\,dP - S\,dT \quad (3\text{–}72)$$

At constant temperature, the last term becomes zero, and equation (3–72) reduces to

$$dG = V\,dP \quad (3\text{–}73)$$

or

$$\left(\frac{\partial G}{\partial P}\right)_T = V \quad (3\text{–}74)$$

At constant pressure, the first term on the right side of equation (3–72) becomes zero, and

$$dG = -S\, dT \qquad (3\text{–}75)$$

or

$$\left(\frac{\partial G}{\partial T}\right)_P = -S \qquad (3\text{–}76)$$

To obtain the isothermal change of free energy, we integrate equation (3–73) between states 1 and 2 at constant temperature:

$$\int_{G_1}^{G_2} dG = \int_{P_1}^{P_2} V\, dP \qquad (3\text{–}77)$$

For an ideal gas, the volume $V$ is equal to $nRT/P$, thus allowing the equation to be integrated:

$$\Delta G = (G_2 - G_1) = nRT \int_{P_1}^{P_2} \frac{dP}{P}$$

$$\Delta G = nRT \ln \frac{P_2}{P_1} = 2.303 nRT \, \log \frac{P_2}{P_1} \qquad (3\text{–}78)$$

where $\Delta G$ is the free energy change of an *ideal gas* undergoing an *isothermal* reversible or irreversible alteration.

---

**EXAMPLE 3–11**

What is the free energy change when 1 mole of an ideal gas is compressed from 1 atm to 10 atm at 25°C? We write

$$\Delta G = 2.303 \times 1.987 \times 298 \times \log \frac{10}{1}$$

$$\Delta G = 1364 \text{ cal}$$

The change in free energy of a solute when the concentration is altered is given by the equation

$$\Delta G = 2.303 nRT \, \log \frac{a_2}{a_1} \qquad (3\text{–}79)$$

in which $n$ is the number of moles of solute and $a_1$ and $a_2$ are the initial and final activities of the solute, respectively.

---

**EXAMPLE 3–12**

Borsook and Winegarden[1] roughly computed the free energy change when the kidneys transfer various chemical constituents at body temperature (37°C or 310.2 K) from the blood plasma to the more concentrated urine. The ratio of concentrations was assumed to be equal to the ratio of activities in equation (3–79). They found

$$\Delta G = 2.303 \times 0.100 \times 1.987 \times 310.2 \times \log \frac{0.333}{0.00500}$$

$$\Delta G = 259 \text{ cal}$$

The concentration of urea in the plasma is 0.00500 mole/liter; the concentration in the urine is 0.333 mole/liter. Calculate the free energy change in transporting 0.100 mole of urea from the plasma to the urine.

This result means that 259 cal of work must be done on the system, or this amount of net work must be performed by the kidneys to bring about the transfer.

## Fugacity

For a reversible isothermal process restricted to $PV$ work, the Gibbs energy is described by equation (3–73) or (3–74). Because $V$ is always a positive magnitude, these relations indicate that at constant temperature the Gibbs energy varies proportional to the changes in $P$. Thus, from equation (3–73) it is possible to evaluate the Gibbs free energy change for a pure substance by integrating between $P^\circ$ and $P$:

$$\int_{P^\circ}^{P} dG = \int_{P^\circ}^{P} V\, dP$$

$$G - G^\circ = \int_{P^\circ}^{P} V\, dP \qquad (3\text{–}80)$$

For pure solids or liquids the volume has little dependence on pressure, so it can be approached as constant, and equation (3–80) is reduced to

$$G = G^\circ + V(P - P^\circ) \quad \text{(for solids and liquids)} \quad (3\text{–}81)$$

On the other hand, gases have a very strong dependence on pressure; by applying the ideal gas equation $V = nRT/P$, we find that equation (3–80) becomes

$$G - G^\circ = \int_{P^\circ}^{P} = \frac{nRT}{P} dP$$

$$= nRT \ln\left(\frac{P}{P^\circ}\right) \quad \text{(for ideal gas)} \qquad (3\text{–}82)$$

Relations (3–81) and (3–82) can be simplified by first assuming $P^\circ = 1$ atm as the reference state. Then, dividing by the amount of substance $n$ gives a new property ($G/n$) called *the molar Gibbs energy* or *chemical potential*, defined by the letter $\mu$,

$$\mu = \left(\frac{G}{n}\right) \qquad (3\text{–}83)$$

Thus, for an ideal gas we can write

$$\mu = \mu^\circ + RT \ln\left(\frac{P}{1 \text{ atm}}\right) \qquad (3\text{–}84)$$

where the integration constant $\mu^\circ$ depends only on the temperature and the nature of the gas and represents the chemical potential of 1 mole of the substance in the reference state where $P^\circ$ is equal to 1 atm. Note that $P$ in equation (3–84) is a finite number and not a function. When a real gas does not behave ideally, a function known as the *fugacity* ($f$) can be introduced to replace pressure, just as activities are introduced to replace concentration in nonideal solutions (see later discussion). Equation (3–84) becomes

$$\mu = \mu^\circ + RT \ln f \qquad (3\text{–}85)$$

## Open Systems

The systems considered so far have been closed. They exchange heat and work with their surroundings, but the processes involve no transfer of matter, so that the amounts of the components of the system remain constant.

The term *component* should be clarified before proceeding. A phase consisting of $w_2$ grams of NaCl dissolved in $w_1$ grams of water is said to contain two independently variable masses or two *components*. Although the phase contains the species $Na^+$, $Cl^-$, $(H_2O)_n$, $H_3O^+$, $OH^-$, and so on, they are not all independently variable. Because $H_2O$ and its various species, $H_3O^+$, $OH^-$, $(H_2O)_n$, and so on, are in equilibrium, the mass $m$ of water alone is sufficient to specify these species. All forms can be derived from the simple species $H_2O$. Similarly, all forms of sodium chloride can be represented by the single species NaCl, and the system therefore consists of just two components, $H_2O$ and NaCl. The *number of components* of a system is the smallest number of independently variable chemical substances that must be specified to describe the phases quantitatively.

In an open system in which the exchange of matter among phases also must be considered, any one of the extensive properties such as volume or free energy becomes a function of temperature, pressure, and the number of moles of the various components.

## Chemical Potential

Let us consider the change in Gibbs energy for an open system composed of a two-component phase (binary system). An infinitesimal reversible change of state is given by

$$dG = \left(\frac{\partial G}{\partial T}\right)_{P,n_1,n_2} dT + \left(\frac{\partial G}{\partial P}\right)_{T,n_1,n_2} dP$$
$$+ \left(\frac{\partial G}{\partial n_1}\right)_{T,P,n_2} dn_1 + \left(\frac{\partial G}{\partial n_2}\right)_{T,P,n_1} dn_2 \quad (3\text{–}86)$$

The partial derivatives $(\partial G/\partial n_1)_{T,P,n_2}$ and $(\partial G/\partial n_2)_{T,P,n_1}$ can be identified as the *chemical potentials* $(\mu)$ of the components $n_1$ and $n_2$, respectively, so that equation (3–86) is written more conveniently as

$$dG = \left(\frac{\partial G}{\partial T}\right)_{P,n_1,n_2} dT + \left(\frac{\partial G}{\partial P}\right)_{P,n_1,n_2} dP$$
$$+ \mu_1 dn_1 + \mu_2 dn_2 \quad (3\text{–}87)$$

Now, relationships (3–74) and (3–76), $(\partial G/\partial P)_T = V$ and $(\partial G/\partial T)_P = -S$, respectively, for a closed system also apply to an open system, so we can write equation (3–87) as

$$dG = -S\,dT + V\,dP + \mu_1\,dn_1 + \mu_2\,dn_2 + \cdots \quad (3\text{–}88)$$

The chemical potential, also known as the *partial molar free energy*, can be defined in terms of other extensive properties such as $E$, $H$, or $A$. However, what is most useful is the general definition given for Gibbs energy: At constant temperature and pressure, with the amounts of the other components ($n_j$) held constant, the chemical potential of a component $i$ is equal to the change in the free energy brought about by an infinitesimal change in the number of moles $n_i$ of the component:

$$\mu_i = \left(\frac{\partial G}{\partial n_i}\right)_{T,P,n_j} \quad (3\text{–}89)$$

It may be considered the change in free energy, for example, of an aqueous sodium chloride solution when 1 mole of NaCl is added to a large quantity of the solution so that the composition does not undergo a measurable change.

At constant temperature and pressure, the first two right-hand terms of equation (3–88) become zero, and

$$dG_{T,P} = \mu_1\,dn_1 + \mu_2\,dn_2 \quad (3\text{–}90)$$

or, in abbreviated notation,

$$dG_{T,P} = \sum \mu_i\,dn_i \quad (3\text{–}91)$$

which, upon integration, gives

$$G_{T,P,N} = \mu_1 n_1 + \mu_2 n_2 + \cdots \quad (3\text{–}92)$$

for a system of constant composition $N = n_1 + n_2 + \cdots$. In equation (3–92), the sum of the right-hand terms equals the total free energy of the system at constant pressure, temperature, and composition. Therefore, $\mu_1$, $\mu_2$, ..., $\mu_n$ can be considered as the contributions per mole of each component to the total free energy. The chemical potential, like any other partial molar quantity, is an *intensive* property; in other words, it is independent of the number of moles of the components of the system.

For a closed system at equilibrium and constant temperature and pressure, the free energy change is zero, $dG_{T,P} = 0$, and equation (3–91) becomes

$$\mu_1\,dn_1 + \mu_2\,dn_2 + \cdots = 0 \quad (3\text{–}93)$$

for all the phases of the overall system, which are closed.

## Equilibrium in a Heterogeneous System

We begin with an example suggested by Klotz.[2] For a two-phase system consisting of, say, iodine distributed between water and an organic phase, the overall system is a closed one, whereas the separate aqueous and organic solutions of iodine are open. The chemical potential of iodine in the aqueous phase is written as $\mu_{Iw}$ and that in the organic phase as $\mu_{Io}$. When the two phases are in equilibrium at constant temperature and pressure, the respective free energy changes $dG_w$ and $dG_o$ of the two phases must be equal because the free energy of the overall system is zero. Therefore, the chemical potentials of iodine in both phases are identical. This can be shown by allowing an infinitesimal amount of iodine to pass from the water to the organic phase, in which, at equilibrium, according to equation (3–93),

$$\mu_{Iw}\,dn_{Iw} + \mu_{Io}\,dn_{Io} = 0 \quad (3\text{–}94)$$

Now, a decrease of iodine in the water is exactly equal to an increase of iodine in the organic phase:

$$-dn_{Iw} = dn_{Io} \qquad (3-95)$$

Substituting equation (3–95) into (3–94) gives

$$\mu_{Iw}\, dn_{Iw} + \mu_{Io}(-dn_{Iw}) = 0 \qquad (3-96)$$

and finally

$$\mu_{Iw} = \mu_{Io} \qquad (3-97)$$

This conclusion may be generalized by stating that the chemical potential of a component is identical in all phases of a heterogeneous system when the phases are in equilibrium at a fixed temperature and pressure. Hence,

$$\mu_{i_\alpha} = \mu_{i_\beta} = \mu_{i_\gamma} = \cdots \qquad (3-98)$$

where $\alpha, \beta, \gamma, \cdots$ are various phases among which the substance $i$ is distributed. For example, in a saturated aqueous solution of sulfadiazine, the chemical potential of the drug in the solid phase is the same as its chemical potential in the solution phase.

When two phases are not in equilibrium at constant temperature and pressure, the total free energy of the system tends to decrease, and the substance passes spontaneously from a phase of higher chemical potential to one of lower chemical potential until the potentials are equal. Hence, the chemical potential of a substance can be used as a measure of the *escaping tendency* of the component from its phase. The concept of escaping tendency will be used in various chapters throughout the book. The analogy between chemical potential and electric or gravitational potential is evident, the flow in these cases always being from the higher to the lower potential and continuing until all parts of the system are at a uniform potential. For a phase consisting of a *single pure substance*, the chemical potential is the free energy of the substance per mole defined in equation (3–83).

For a two-phase system of a single component, for example, liquid water and water vapor in equilibrium at constant temperature and pressure, the *molar free energy G/n* is identical in all phases. This statement can be verified by combining equations (3–98) and (3–83).

## Clausius–Clapeyron Equation

If the temperature and pressure of a two-phase system of one component, for example, of liquid water (l) and water vapor (v) in equilibrium, are changed by a small amount, the molar free energy changes are equal and

$$dG_1 = dG_v \qquad (3-99)$$

In a phase change, the free energy changes for 1 mole of the liquid vapor are given by equation (3–72),

$$dG = V\, dP - T\, dS$$

Therefore, from equations (3–99) and (3–72),

$$V_1\, dP - S_1\, dT = V_v\, dP - S_v\, dT$$

or

$$\frac{dP}{dT} = \frac{S_v - S_1}{V_v - V_1} = \frac{\Delta S}{\Delta V} \qquad (3-100)$$

Now, at constant pressure, the heat absorbed in the reversible process (equilibrium condition) is equal to the molar heat of vaporization, and from the second law we have

$$\Delta S = \frac{\Delta H_v}{T} \qquad (3-101)$$

Substituting equation (3–101) into (3–100) gives

$$\frac{dP}{dT} = \frac{\Delta H_v}{T\, \Delta V} \qquad (3-102)$$

where $\Delta V = V_v - V_1$, the difference in the molar volumes in the two phases. This is the *Clapeyron equation.*

The vapor will obey the ideal gas law to a good approximation when the temperature is far enough away from the critical point, so that $V_v$ may be replaced by $RT/P$. Furthermore, $V_1$ is insignificant compared with $V_v$. In the case of water at $100°C$, for example, $V_v = 30.2$ liters and $V_1 = 0.0188$ liter.

Under these restrictive circumstances, equation (3–102) becomes

$$\frac{dP}{dT} = \frac{P\, \Delta H_v}{RT^2} \qquad (3-103)$$

which is known as the *Clausius–Clapeyron equation.* It can be integrated between the limits of the vapor pressures $P_1$ and $P_2$ and corresponding temperatures $T_1$ and $T_2$, assuming $\Delta H_v$ is constant over the temperature range considered:

$$\int_{P_1}^{P_2} \frac{dP}{P} = \frac{\Delta H}{R} \int_{T_1}^{T_2} T^{-2} dT \qquad (3-104)$$

$$[\ln P]_{P_1}^{P_2} = \frac{\Delta H_v}{R}\left[\left(-\frac{1}{T_2}\right) - \left(-\frac{1}{T_1}\right)\right] \qquad (3-105)$$

$$\ln P_2 - \ln P_1 = \frac{\Delta H_v}{R}\left(\frac{1}{T_1} - \frac{1}{T_2}\right) \qquad (3-106)$$

and finally

$$\ln \frac{P_2}{P_1} = \frac{\Delta H_v(T_2 - T_1)}{RT_1 T_2} = 0$$

or

$$\log \frac{P_2}{P_1} = \frac{\Delta H_v(T_2 - T_1)}{2.303\, RT_1 T_2} = 0 \qquad (3-107)$$

This equation is used to calculate the mean heat of vaporization of a liquid if its vapor pressure at two temperatures is available. Conversely, if the mean heat of vaporization and the vapor pressure at one temperature are known, the vapor pressure at another temperature can be obtained.

The Clapeyron and Clausius–Clapeyron equations are important in the study of various phase transitions and in the development of the equations of some colligative properties.

### EXAMPLE 3-13

The average heat of vaporization of water can be taken as about 9800 cal/mole within the range of 20°C to 100°C. What is the vapor pressure at 95°C? The vapor pressure $P_2$ at temperature $T_2 = 373$ K (100°C) is 78 cm Hg, and $R$ is expressed as 1.987 cal/deg mole. Write

$$\log \frac{78.0}{P_1} = \frac{9800}{2.303 \times 1.987} \left( \frac{373 - 368}{368 \times 373} \right)$$

$$P_1 = 65 \text{ cm Hg}$$

## Activities: Activity Coefficients

If the vapor above a solution can be considered to behave ideally, the chemical potential of the solvent in the vapor state in equilibrium with the solution can be written in the form of equation (3–84),

$$\mu = \mu^\circ + RT \ln P$$

If Raoult's law is now introduced for the solvent, $P_1 = P_1^\circ X_1$, equation (3–84) becomes

$$\mu_1 = \mu_1^\circ + RT \ln P_1^\circ + RT \ln X_1 \qquad (3\text{--}108)$$

Combining the first and second right-hand terms into a single constant gives

$$\mu_1 = \mu^\circ + RT \ln X_1 \qquad (3\text{--}109)$$

for an ideal solution. The reference state $\mu^\circ$ is equal to the chemical potential $\mu_1$ of the pure solvent (i.e., $X_1 = 1$). For nonideal solutions, equation (3–109) is modified by introducing the "effective concentration" or *activity* of the solvent to replace the mole fraction:

$$\mu_1 = \mu^\circ + RT \ln a_1 \qquad (3\text{--}110)$$

or, for

$$a = \gamma X \qquad (3\text{--}111)$$

and $\gamma$ is referred to as the *activity coefficient*, we have

$$\mu_1 = \mu^\circ + RT \ln \gamma_1 X_1 \qquad (3\text{--}112)$$

For the *solute* on the mole fraction scale,

$$\mu_2 = \mu^\circ + RT \ln a_2 \qquad (3\text{--}113)$$

$$\mu_2 = \mu^\circ + RT \ln \gamma_2 X_2 \qquad (3\text{--}114)$$

Based on the practical (molal and molar) scales

$$\mu_2 = \mu^\circ + RT \ln \gamma_m m \qquad (3\text{--}115)$$

$$\mu_2 = \mu^\circ + RT \ln \gamma_c c \qquad (3\text{--}116)$$

Equations (3–110) and (3–113) are frequently used as definitions of activity.

## Gibbs–Helmholtz Equation

For an isothermal process at constant pressure proceeding between the initial and final states 1 and 2, equation (3–57) yields

$$G_2 - G_1 = (H_2 - H_1) - T(S_2 - S_1)$$
$$\Delta G = \Delta H - T \Delta S \qquad (3\text{--}117)$$

Now, equation (3–76) may be written as

$$-\Delta S = -(S_2 - S_1) = \left( \frac{\partial G_2}{\partial T} \right)_P - \left( \frac{\partial G_1}{\partial T} \right)_P$$

or

$$-\Delta S = \left[ \frac{\partial (G_2 - G_1)}{\partial T} \right]_P = \left[ \frac{\partial (\Delta G)}{\partial T} \right]_P \qquad (3\text{--}118)$$

Substituting equation (3–118) into (3–119) gives

$$\Delta G = \Delta H + T \left[ \frac{\partial (\Delta G)}{\partial T} \right]_P \qquad (3\text{--}119)$$

which is one form of the Gibbs–Helmholtz equation

## Standard Free Energy and the Equilibrium Constant

Many of the processes of pharmaceutical interest such as complexation, protein binding, the dissociation of a weak electrolyte, or the distribution of a drug between two immiscible phases are systems at equilibrium and can be described in terms of changes of the Gibbs free energy ($\Delta G$).

Consider a closed system at constant pressure and temperature, such as the chemical reaction

$$a\text{A} + b\text{B} \rightleftharpoons c\text{C} + d\text{D} \qquad (3\text{--}120)$$

Because $G$ is a state function, the free energy change of the reaction going from reactants to products is

$$\Delta G = \sum \Delta G_{\text{products}} - \sum \Delta G_{\text{reactants}} \qquad (3\text{--}121)$$

Equation (3–120) represents a closed system made up of several components. Therefore, at constant $T$ and $P$ the total free energy change of the products and reactants in equation (3–121) is given as the sum of the chemical potential $\mu$ of each component times the number of moles [see equation (3–83)]:

$$\Delta G = (c\mu_\text{C} + d\mu_\text{D}) - (a\mu_\text{A} + b\mu_\text{B}) \qquad (3\text{--}122)$$

When the reactants and products are ideal gases, the chemical potential of each component is expressed in terms of partial pressure [equation (3–84)]. For nonideal gases, $\mu$ is written in terms of fugacities [equation (3–85)]. The corresponding expressions for solutions are given by equations (3–110) to (3–116). Let us use the more general expression that relates the chemical potential to the activity, equation

**(3–110)**. Substituting this equation for each component in equation **(3–121)** yields

$$\Delta G = c(\mu_C{}^\circ + RT \ln a_C) + d(\mu_D{}^\circ + RT \ln a_D)$$
$$- a(\mu_A{}^\circ + RT \ln a_A) - b(\mu_B{}^\circ + RT \ln a_B)$$
$$(3\text{–}123)$$

Rearranging equation **(3–123)** gives

$$\Delta G = c\mu_C{}^\circ + d\mu_D{}^\circ - a\mu_A{}^\circ - b\mu_B{}^\circ + RT\left(\ln a_C{}^c + \ln a_D{}^d\right)$$
$$- RT\left(\ln a_A{}^a + \ln a_B{}^b\right) \quad (3\text{–}124)$$

$\mu^\circ$ is the partial molar free energy change or chemical potential under standard conditions. Because it is multiplied by the number of moles in equation **(3–124)**, the algebraic sum of the terms involving $\mu^\circ$ represents the *total standard free energy change of the reaction* and is called $\Delta G^\circ$:

$$\Delta G^\circ = c\mu_C{}^\circ + d\mu_D{}^\circ - a\mu_A{}^\circ - b\mu_B{}^\circ \quad (3\text{–}125)$$

or, in general,

$$\Delta G^\circ = \sum n\mu^\circ(\text{products}) - \sum n\mu^\circ(\text{reactants}) \quad (3\text{–}126)$$

Using the rules of logarithms, we can express equation **(3–124)** as

$$\Delta G = \Delta G^\circ + RT \ln \left[\left(a_C{}^c a_D{}^d\right)/\left(a_A{}^a a_B{}^b\right)\right] \quad (3\text{–}127)$$

The products of activities in brackets are called the reaction quotients, defined as

$$Q = \left[\left(a_C{}^c a_D{}^d\right)/\left(a_A{}^a a_B{}^b\right)\right] \quad (3\text{–}128)$$

or, in general,

$$Q = \frac{\sum a_{\text{products}}^n}{\sum a_{\text{reactants}}^n} \quad (3\text{–}129)$$

Thus, equation **(3–127)** can be written as

$$\Delta G = \Delta G^\circ + RT \ln Q \quad (3\text{–}130)$$

Because $\Delta G^\circ$ is a constant at constant $P$ and constant $T$, $RT$ is also constant. The condition for equilibrium is $\Delta G = 0$, and therefore equation **(3–130)** becomes

$$0 = \Delta G^\circ + RT \ln K \quad (3\text{–}131)$$

or

$$\Delta G^\circ = -RT \ln K \quad (3\text{–}132)$$

where $Q$ has been replaced by $K$, the equilibrium constant.

Equation **(3–132)** is a very important expression, relating the standard free energy change of a reaction $\Delta G^\circ$ to the equilibrium constant $K$. This expression allows one to compute $K$ knowing $\Delta G^\circ$ and vice versa.

The equilibrium constant has been expressed in terms of activities. It also can be given as the ratio of partial pressures or fugacities (for gases) and as the ratio of the different concentration expressions used in solutions (mole fraction, molarity, molality). The equilibrium constant is dimensionless, the ratio of activities or concentration canceling the units.

However, the numerical value of $K$ differs depending on the units used (activity, mole fraction, fugacity, etc.).

**EXAMPLE 3–14**

Derive an expression for the free energies $\Delta G$ and $\Delta G^\circ$ of the reaction

$$Fe_{(s)} + H_2O_{(g)} = FeO_{(s)} + H_{2(g)}$$

Because the chemical potential of a solid is constant (it does not depend on concentration), the equilibrium constant depends only on the pressures (or fugacities) of the gases. Using pressures, we obtain

$$\Delta G = \mu^\circ{}_{FeO(s)} + \mu^\circ{}_{H_2(g)} - \mu^\circ{}_{Fe(s)} - \mu^\circ_{H_2O(g)} + RT \ln P_{H_2(g)}$$
$$- RT\, P_{H_2O_{(g)}} = 0$$
$$\Delta G = \Delta G^\circ + RT \ln P_{H_2(g)} - RT\, P_{H_2O_{(g)}} = 0$$

and

$$\Delta G^\circ = -RT \ln \frac{P_{H_2(g)}}{P_{H_2O_{(g)}}}$$

The magnitude and sign of $\Delta G^\circ$ indicate whether the reaction is spontaneous [see equation **(3–59)**], but only under *standard conditions*. When the reaction is not at equilibrium, $\Delta G \neq 0$ and the free energy change is described by equation **(3–130)**, where $Q$, like $K$, is the ratio of activities [equation **(3–129)**], fugacities, or concentration units of the products and reactants but under different conditions than those of equilibrium. $Q$ should not be confused with $K$, the ratio of activities, fugacities, and so on under *standard conditions at equilibrium*.

**EXAMPLE 3–15**

Sodium cholate is a bile salt that plays an important role in the dissolution or dispersion of cholesterol and other lipids in the body. Sodium cholate may exist either as monomer or as dimer (or higher *n-mers*) in aqueous solution. Let us consider the equilibrium monomer–dimer reaction[3]:

$$2(\text{Monomer}) \overset{K}{\rightleftharpoons} \text{Dimer}$$

which states that two moles (or molecules) of monomer form 1 mole (or molecule) of dimer.

(a)  If the molar concentration at 25°C of monomeric species is $4 \times 10^{-3}$ mole/liter and the concentration of dimers is $3.52 \times 10^{-5}$ mole/liter, what is the equilibrium constant and the standard free energy for the dimerization process? Write

$$K = \frac{[\text{Dimer}]}{[\text{Monomer}]^2} = \frac{3.52 \times 10^{-5}}{(4 \times 10^{-3})^2} = 2.20$$

$$\ln K = 0.788$$

$$\Delta G^\circ = -RT \ln K = -(1.9872 \times 298 \times 0.788)$$
$$= -466.6 \text{ cal/mole}$$

The process is spontaneous under standard conditions.

(b)  While keeping the concentration of monomer constant, suppose that one is able to remove part of the dimeric species by physical or chemical means so that its concentration is now four times less than the original dimer concentration. Compute the free energy change. What is the effect on the equilibrium?

The concentration of dimer is now

$$\frac{3.52 \times 10^{-5}}{4} = 8.8 \times 10^{-6} \text{ mole/liter}$$

Because the conditions are not at equilibrium, equation (3–130) should be used. First calculate $Q$:

$$Q = \frac{[\text{Dimer}]}{[\text{Monomer}]^2} = \frac{8.8 \times 10^{-6}}{(4 \times 10^{-3})^2} = 0.550; \ln Q = -0.598$$

and from equation (3–130),

$$\Delta G = 466.6 + [1.9872 \times 298 \times (-0.598)]$$
$$= -820.7 \text{ cal/mole}$$

$\Delta G$ is negative, $Q$ is less than $K$, and the reaction shifts to the right side of the equation with the formation of more dimer.

If the monomer is removed from the solution, the reaction is shifted to the left side, forming monomer, and $\Delta G$ becomes positive. Suppose that monomer concentration is now $1 \times 10^{-3}$ mole/liter and dimer concentration is $3.52 \times 10^{-5}$ mole/liter:

$$Q = \frac{3.52 \times 10^{-5}}{(1 \times 10^{-3})^2} = 35.2; \ln Q = 3.561$$

$$\Delta G = -466.6 + (1.9872 \times 298 \times 3.561) = +1642 \text{ cal/mole}$$

The positive sign of $\Delta G$ indicates that the reaction does not proceed forward spontaneously.

## The van't Hoff Equation

The effect of temperature on equilibrium constants is obtained by writing the equation

$$\ln K = \frac{\Delta G^\circ}{RT} \qquad (3\text{–}133)$$

and differentiating with respect to temperature to give

$$\frac{d \ln K}{dT} = -\frac{1}{R} \frac{d(\Delta G^\circ/T)}{dT} \qquad (3\text{–}134)$$

The Gibbs–Helmholtz equation may be written in the form (see one of the thermodynamics texts cited in the Recommended Readings section at the end of this chapter)

$$\frac{d(\Delta G/T)}{dT} = -\frac{\Delta H}{T^2} \qquad (3\text{–}135)$$

Expressing equation (3–135) in a form for the reactants and products in their standard states, in which $\Delta G$ becomes equal to $\Delta G^\circ$, and substituting into equation (3–134) yields

$$\frac{d \ln K}{dT} = \frac{\Delta H^\circ}{RT^2} \qquad (3\text{–}136)$$

where $\Delta H^\circ$ is the standard enthalpy of reaction. Equation (3–136) is known as the *van't Hoff equation*. It may be integrated, assuming $\Delta H^\circ$ to be constant over the temperature range considered; it becomes

$$\ln \frac{K_2}{K_1} = \frac{\Delta H^\circ}{R} \left( \frac{T_2 - T_1}{T_1 T_2} \right) \qquad (3\text{–}137)$$

Equation (3–137) allows one to compute the enthalpy of a reaction if the equilibrium constants at $T_1$ and $T_2$ are available. Conversely, it can be used to supply the equilibrium constant at a definite temperature if it is known at another temperature. Because $\Delta H^\circ$ varies with temperature and equation (3–137)

gives only an approximate answer, more elaborate equations are required to obtain accurate results. The solubility of a solid in an ideal solution is a special type of equilibrium, and it is not surprising that the solubility can be written as

$$\ln \frac{X_2}{X_1} = \frac{\Delta H_f}{R} \left( \frac{T_2 - T_1}{T_1 T_2} \right) \qquad (3\text{–}138)$$

which closely resembles equation (3–137). These expressions will be encountered in later chapters.

Combining equations (3–117) and (3–133) yields yet another form of the van't Hoff equation, namely

$$\ln K = -(\Delta H^\circ/R)1/T + \Delta S^\circ/R \qquad (3\text{–}139)$$

or

$$\log K = -[\Delta H^\circ/(2.303)R]1/T + \Delta S^\circ/(2.303R) \qquad (3\text{–}140)$$

where $\Delta S^\circ/R$ is the intercept on the $\ln K$ axis of a plot of $\ln K$ versus $1/T$.

Whereas equation (3–137) provides a value of $\Delta H^\circ$ based on the use of two $K$ values at their corresponding absolute temperatures $T_1$ and $T_2$, equations (3–139) and (3–140) give the values of $\Delta H^\circ$ and $\Delta S^\circ$, and therefore the value of $\Delta G^\circ = \Delta H^\circ - T\Delta S^\circ$. In the least squares linear regression equations (3–139) and (3–140), one uses as many $\ln K$ and corresponding $1/T$ values as are available from experiment.

### EXAMPLE 3–16

In a study of the transport of pilocarpine across the corneal membrane of the eye, Mitra and Mikkelson[4] presented a van't Hoff plot of the log of the ionization constant, $K_a$, of pilocarpine versus the reciprocal of the absolute temperature, $T^{-1} = 1/T$.

Using the data in Table 3–4, regress $K_a$ versus $T^{-1}$. With reference to the van't Hoff equation, equation (3–139), obtain the standard heat (enthalpy), $\Delta H^\circ$, of ionization for pilocarpine and the standard entropy for the ionization process. From $\Delta H^\circ$ and $\Delta S^\circ$

### TABLE 3–4
### IONIZATION CONSTANTS OF PILOCARPINE AT VARIOUS TEMPERATURES*,†

| $T(^\circ C)$ | $T(K)$ | $1/T \times 10^3$ | $K_a \times 10^7$ | $\log K_a$ |
|---|---|---|---|---|
| 15 | 288 | 3.47 | 0.74 | −7.13 |
| 20 | 293 | 3.41 | 1.07 | −6.97 |
| 25 | 298 | 3.35 | 1.26 | −6.90 |
| 30 | 303 | 3.30 | 1.58 | −6.80 |
| 35 | 308 | 3.24 | 2.14 | −6.67 |
| 40 | 313 | 3.19 | 2.95 | −6.53 |
| 45 | 318 | 3.14 | 3.98 | −6.40 |

*For the column headed $1/T \times 10^3$ the numbers are 1000 (i.e., $10^3$) times *larger* than the actual numbers. Thus, the first entry in column 3 has the value $3.47 \times 10^{-3}$ or 0.00347. Likewise, in the next column $K_a \times 10^7$ signifies that the number 0.74 and the other entries in this column are to be accompanied by the exponential value $10^{-7}$, not $10^{+7}$. Thus, the first value in the fourth column should be read as $0.74 \times 10^{-7}$ and the last value $3.98 \times 10^{-7}$.
†From A. K. Mitra and T. J. Mikkelson, J. Pharm. Sci. **77**, 772, 1988. With permission.

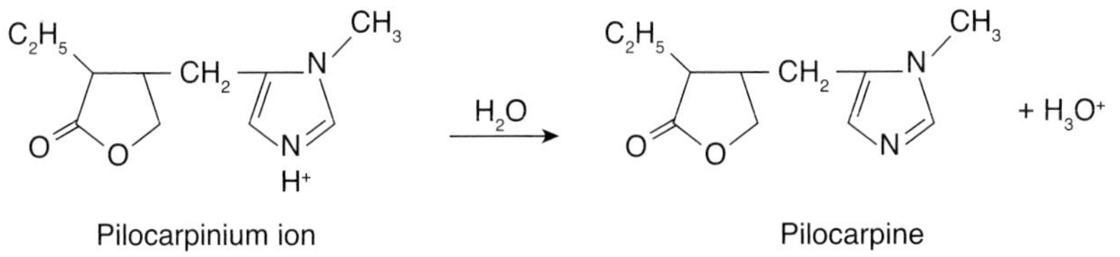

**Fig. 3–7.** Reaction of pilocarpinium ion to yield pilocarpine base.

calculate $\Delta G°$ at 25°C. What is the significance of the signs and the magnitudes of $\Delta H°$, $\Delta S°$, and $\Delta G°$?

Answers:

$$\Delta H° = 9784 \text{ cal/mole}$$
$$= 40.94 \text{ kJ/mole}$$
$$\Delta S° = 1.30 \text{ cal/mole deg}$$
$$\Delta G°_{25°} = \Delta H° - T\Delta S° = 9397 \text{ cal/mole}$$

These thermodynamic values have the following significance. $\Delta H°$ is a large positive value that indicates that the ionization of pilocarpine (as its conjugate acid) should increase as the temperature is elevated. The increasing values of $K_a$ in the table show this to be a fact. The standard entropy increase, $\Delta S° = 1.30$ entropy units, although small, provides a force for the reaction of the pilocarpinium ions to form pilocarpine (**Fig. 3–7**). The positively charged pilocarpine molecules, because of their ionic nature, are probably held in a more orderly arrangement than the predominantly nonionic pilocarpine in the aqueous environment. This increase in disorder in the dissociation process accounts for the increased entropy, which, however, is a small value: $\Delta S° = 1.30$ entropy units. Note that a positive $\Delta H°$ does not mean that the ionization will not occur; rather, it signifies that the equilibrium constant for the forward reaction (ionization) will have a small value, say $K_a \cong 1 \times 10^{-7}$, as observed in **Table 3–4**. A further explanation regarding the sign of $\Delta H°$ is helpful here. Mahan[5] pointed out that in the first stage of ionization of phosphoric acid, for example,

$$H_3PO_4 \rightarrow H^+ + H_2PO_4^-; \quad \Delta H° = -3.1 \text{ kcal/mole}$$

the hydration reaction of the ions being bound to the water molecules is sufficiently exothermic to produce the necessary energy for ionization, that is, enough energy to remove the proton from the acid, $H_3PO_4$. For this reason, $\Delta H°$ in the first stage of ionization is negative and $K_1 = 7.5 \times 10^{-3}$ at 25°C. In the second stage,

$$H_2PO_4^- \rightarrow H^+ + HPO_4^{2-}; \quad \Delta H° = 0.9 \text{ kcal/mole}$$

$\Delta H°$ is now positive, the reaction is *endothermic*, and $K_2 = 6.2 \times 10^{-8}$. Finally, in the third stage,

$$HPO_4^{2-} \rightarrow H^+ + PO_4^{3-}; \quad \Delta H° = 4.5 \text{ kcal/mole}$$

$\Delta H°$ is a relatively large positive value and $K_3 = 2.1 \times 10^{-13}$. These $\Delta H°$ and $K_a$ values show that increasing energy is needed to remove the positively charged proton as the negative charge increases in the acid from the first to the third stage of ionization. Positive $\Delta H°$ (endothermic reaction) values do not signal nonionization of the acid—that is, that the process is nonspontaneous—but rather simply show that the forward reaction, represented by its ionization constant, becomes smaller and smaller.

## CHAPTER SUMMARY

In this chapter, the quantitative relationships among different forms of energy were reviewed and expressed in the three laws of thermodynamics. Energy can be considered as the product of an intensity factor and a capacity factor; thus, the various types of energy may be represented as a product of an intensive property (i.e., independent of the quantity of material) and the differential of an extensive property that is proportional to the mass of the system. For example, mechanical work done by a gas on its surroundings is $P\,dV$. Some of the forms of energy, together with these factors and their accompanying units, are given in **Table 3–5**.

> **Practice problems for this chapter can be found at thePoint.lww.com/Sinko6e.**

**TABLE 3–5**

**INTENSITY AND CAPACITY FACTORS OF ENERGY**

| Energy Form | Intensity or Potential Factor (Intensive Property) | Capacity or Quantity Factor (Extensive Property) | Energy Unit Commonly Used |
|---|---|---|---|
| Heat (thermal) | Temperature (deg) | Entropy change (cal/deg) | Calories |
| Expansion | Pressure (dyne/cm$^2$) | Volume change (cm$^3$) | Ergs |
| Surface | Surface tension (dyne/cm) | Area change (cm$^2$) | Ergs |
| Electric | Electromotive force or potential difference (volts) | Quantity of electricity (coulombs) | Joules |
| Chemical | Chemical potential (cal/mole) | Number of moles | Calories |

## References

1. H. Borsook and H. M. Winegarden, Proc. Natl. Acad. Sci. USA **17,** 3, 1931.
2. J. M. Klotz, *Chemical Thermodynamics*, Prentice Hall, Englewood Cliffs, N.J., 1950, p. 226.
3. M. Vadnere and S. Lindenbaum, J. Pharm. Sci. **71,** 875, 1982.
4. A. K. Mitra and T. J. Mikkelson, J. Pharm. Sci. **77,** 771, 1988.
5. B. H. Mahan, *Elementary Chemical Thermodynamics*, Benjamin/Cummings, Menlo Park, Calif., 1963, pp. 139–142.

## Recommended Readings

W. Atkins and J. W. Locke, *Atkins's Physical Chemistry*, 7th Ed., Oxford University Press, Oxford, 2002.

G. W. Castellan, *Physical Chemistry*, 3rd Ed., Addison-Wesley, New York, 1983.

R. Chang, *Physical Chemistry for the Chemical and Biological Sciences*, 3rd Ed., University Science Books, Sausalito, Calif., 2000.

I. M. Klotz and R. M. Rosenberg, *Chemical Thermodynamics: Basic Theory and Methods*, 6th Ed., Wiley-Interscience, New York, 2000.

I. Tinoco, K. Sauer, J. C. Wang, and J. D. Puglisi, *Physical Chemistry: Principles and Applications in Biological Sciences*, 4th Ed., Prentice Hall, Englewood Cliffs, N.J., 2001.

### CHAPTER LEGACY

**Fifth Edition:** published as Chapter 3. Updated by Gregory Knipp, Hugo Morales-Rojas, and Dea Herrera-Ruiz.

**Sixth Edition:** published as Chapter 3. Updated by Patrick Sinko.

# 4  DETERMINATION OF THE PHYSICAL PROPERTIES OF MOLECULES

**CHAPTER OBJECTIVES**  **At the conclusion of this chapter the student should be able to:**

**1** Understand the nature of intra- and intermolecular forces that are involved in stabilizing molecular and physical structures.

**2** Understand the differences in the energetics of these forces and their relevance to different molecules.

**3** Understand the differences in energies between the vibrational, translational, and rotational levels and define their meaning.

**4** Understand the differences between atomic and molecular spectroscopic techniques and the information they provide.

**5** Appreciate the differences in the strengths of selected spectroscopic techniques used in the identification and detection of pharmaceutical agents.

**6** Define the electromagnetic radiation spectrum in terms of wavelength, wave number, frequency, and the energy associated with each range.

**7** Define and understand the relationships between atomic and molecular forces and their response to electromagnetic energy sources.

**8** Define and understand ultraviolet and visible light spectroscopy in terms of electronic structure.

**9** Define and understand fluorescence and phosphorescence in terms of electronic structure.

**10** Understand electron and nuclear precession in atoms subjected to electromagnetic radiation and its role in the determination of atomic structure in a molecule.

**11** Understand polarization of light beams and the ability to use polarized light to study chiral molecules.

**12** Understand fundamental principles of refraction of electron and neutron beams and how these beams are used to determine molecular properties.

## MOLECULAR STRUCTURE, ENERGY, AND RESULTING PHYSICAL PROPERTIES

An atom consists of a nucleus, made up of neutrons (neutral in charge) and protons (positively charged), with each particle carrying a weight of approximately 1 g/mole. In addition, electrons (negatively charged) exist in atomic orbits surrounding the nucleus and have a significantly lower weight. Charged atoms arise from an imbalance in the number of electrons and protons and can lead to ionic interactions (discussed in Chapter 2). The atomic mass is derived from counting the number of protons and neutrons in a nucleus. Isotopes may also exist for a given type of atom. For example, carbon has an atomic number of 6, which describes the number of protons, and there are several carbon isotopes with different numbers of neutrons in the nucleus: $^{11}C$ (with five neutrons), $^{12}C$ (with six neutrons), $^{13}C$ (with seven neutrons), $^{14}C$ (with eight neutrons), and $^{15}C$ (with nine neutrons).[1] Carbon-13, $^{13}C$, is a common isotope used in nuclear magnetic resonance (NMR) and kinetic isotope effect studies on rates of reaction, and $^{14}C$ is radioactive and used as a tracer for studies that require high sensitivity and for carbon dating. Both $^{11}C$ and $^{15}C$ are very short-lived, having half-lives of 20.3 min and 2.5 sec, respectively, and are not used in practical applications.

Molecules arise when interatomic bonding occurs. The molecular structure is reflected by the array of atoms within a molecule and is held together by bonding energy, which relies heavily on electron orbital orientation and overlap. This is illustrated in the Atomic Structure and Bonding Key Concept Box. Each bond in a complex molecule has an intrinsic energy and will have different properties, such as reactivity. The properties within a molecule depend on intramolecular interactions, and each molecule will possess a net energy of bonding that is defined by its unique composition of atoms. It is also important to note that for macromolecules like proteins or synthetic polymers, the presence of neighboring charges or dipole interactions, steric hindrance and repulsion, Debye forces, and so on, within the backbone or functional groups of the molecule can also give rise to distortions in the primary intramolecular bonding energies and can even give rise to additional covalent bonding, for example, disulfide bridges in a protein like insulin.

The more bonding that an atom participates in, the lower is the density of the electron cloud around the nucleus and the greater is the shared distribution of atomic electrons in the interatomic bonded space. For example, because of differences in atomic properties such as electronegativity, the nature of each interatomic bond can vary greatly. Here, the "nature" refers to the energy of the bond, the rotation around the bond, the vibration and motion of atoms in the bond, the rotation of a nucleus of an atom in a bond, and so on.

As an example of the nature of a bond, consider the carbonyl–amide bond between two amino acids in a protein

## KEY CONCEPT ATOMIC STRUCTURE AND BONDING

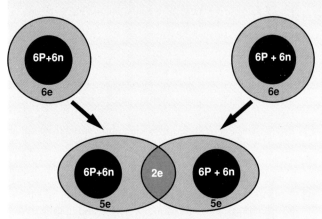

Consider two carbon atoms that are not bound. Each nucleus, carrying six protons (6P) and six neutrons (6n) and a net posi-

tive charge, is surrounded by a fairly uniform, six-electron (6e) cloud. Consider what happens when the two carbons form a covalent bond. Two electrons, one from each carbon, fill the hybrid orbital space between the atoms. The nucleus no longer is surrounded by a uniform cloud of equal charge. The nucleus is now surrounded closely by five electrons and has an electron that is shared in the bond. The charge distribution around each nucleus is similar to that of the atom alone, yet its shape is different. As we will see in this chapter, the difference in this distribution relates to the physical property of the molecule. When the atoms alone or in the molecule are exposed to external energy fields, they will behave differently. Therefore, changes in atomic order when molecular bonding occurs can be measured by different techniques using different levels of external energy.

(Fig. 4–1), as described by Mathews and van Holde.[3] The oxygen in the carbonyl is very electronegative and pulls electrons toward itself. The lone pair of electrons on the nitrogen remains unbound, but nitrogen has a considerably lower electronegativity than oxygen. Recall that the covalent bond occurs between the carbon (which also has a low electronega-

tivity) in the carbonyl group and the amine of the next amino acid. However, the electrons in the carbon–oxygen double bond in the carbonyl are pulled much closer to the oxygen due to its greater electronegativity, causing a partially negatively charged oxygen. The carbon then carries a partial positive charge and pulls the lone pair of electrons from the nitrogen. This forms a partial double bond, which causes the nitrogen to carry a partial positive charge, yielding a net dipole (see Permanent Dipole Moment of Polar Molecules section). Interestingly, in extended secondary structures, in particular $\alpha$-helices, the effective dipole of the peptide bond can have a strongly stabilizing effect on secondary structure.[4] This reduces rotation around the carbon–nitrogen bond, making only *cis/trans* isomers allowable. Although this is not discussed here, the student should recall discussions about energetically favored bond rotations in organic chemistry. The *trans* peptide bond orientation is energetically preferred, with the *cis* conformer appearing in only specialized cases, in particular, around the conformationally constrained proline. As discussed in general biochemistry courses, the secondary structure of a peptide is then determined by rotation around the $\phi$ and $\psi$ angles only. Finally, amino acids naturally occur in the L-conformer, with the exception of glycine. This chirality is followed in almost all peptides and proteins, resulting in a profound effect on function. An example of chirality with an L- and a D-amino acid is presented in the Key Concept Box Chirality of Amino Acids as a refresher.

## KEY CONCEPT CHIRALITY OF AMINO ACIDS

All $\alpha$-amino acids, with the exception of glycine, have a chiral $\alpha$-carbon. The L-conformer is preferred, yet the D-conformer is also found. To determine the conformer, Richardson[2] and others proposed the "corn crib" structure. In connection with the figure shown here, if one holds the $\alpha$-helix in front of one's eyes and looks down the bond with the $\alpha$-carbon, reading clockwise around the $\alpha$-carbon, one sees the word "CORN" pointing into the page for the L-amino acid conformer.

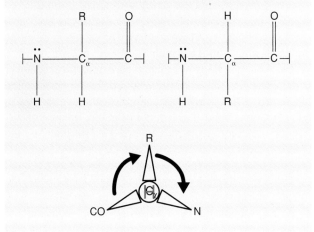

## ADDITIVE AND CONSTITUTIVE PROPERTIES

A study of the physical properties of drug molecules is a prerequisite for product formulation and leads to a better understanding of the relationship between a drug's molecular

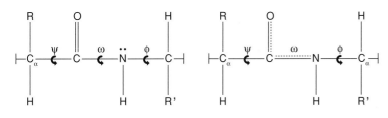

Amide bond is $\omega$        Amide bond is $\omega = 180\infty$

**Fig. 4–1.** The nature of an amide bond within a peptide/protein and its effect on structure. As mentioned in the text, the *trans*isomer of a peptide is energetically favored due to steric hinderance of the amino acids. Proline (tertiary amine group) and glycine (two hydrogens, little steric hinderence from the R group) can more readily adapt a *trans*isomer conformation in a peptide or protein than can the other amino acids.

and physicochemical properties and its structure and action. These properties come from the molecular bonding order of the atoms in the molecule and may be thought of as either *additive* (derived from the sum of the properties of the individual atoms or functional groups within the molecule) or *constitutive* (dependent on the structural arrangement of the atoms within the molecule). For example, mass is an additive property (the molecular weight is the sum of the masses of each atom in the molecule), whereas optical rotation may be thought of as a constitutive property because it depends on the chirality of the molecule, which is determined by atomic bonding order.

Many physical properties are constitutive and yet have some measure of additivity. The molar refraction of a compound, for example, is the sum of the refraction of the atoms and the array of functional groups making up the compound. Because the arrangements of atoms in each group are different, the refractive index of two molecules will also be different. That is, the individual groups in two different molecules contribute different amounts to the overall refraction of the molecules.

A sample calculation will clarify the principle of additivity and constitutivity. The molar refractions of the two compounds

$$\underset{C_2H_5-C-CH_3}{\overset{\overset{\displaystyle O}{\parallel}}{\phantom{C_2H_5-C-CH_3}}}$$

and

$$CH_3-CH=CH-CH_2-OH,$$

which have exactly the same number of carbon, hydrogen, and oxygen atoms, are calculated using **Table 4–1**. We obtain the following results:

$$\underset{C_2H_5-C-CH_3}{\overset{\overset{\displaystyle O}{\parallel}}{\phantom{C_2H_5-C-CH_3}}}$$

| | | | |
|---|---|---|---|
| 8H | $8 \times 1.100 =$ | 8.800 |
| 3C (single) | $3 \times 2.418 =$ | 7.254 |
| 1C (double) | $1 \times 1.733 =$ | 1.733 |
| 1O (C=O) | $1 \times 2.211 =$ | 2.211 |
| | | $19.998 = 20.0$ |

$$CH_3-CH=CH-CH_2-OH$$

| | | | |
|---|---|---|---|
| 8H | $8 \times 1.100 =$ | 8.800 |
| 2C (single) | $2 \times 2.418 =$ | 4.836 |
| 2C (double) | $2 \times 1.733 =$ | 3.466 |
| 1O (OH) | $1 \times 1.525 =$ | 1.525 |
| | | $18.627 = 18.7$ |

Thus, although these two compounds have the same number and type of atoms, their molar refractions are different. The molar refractions of the atoms are additive, but the carbon and oxygen atoms are constitutive in refraction. Refraction of a single-bonded carbon does not add equally to that of a double-bonded carbon, and a carbonyl oxygen (C–O) is not the same as a hydroxyl oxygen; therefore, the two compounds exhibit additive–constitutive properties and have different molar refractions. These will be more comprehensively discussed in this chapter and in other sections of the book.

Molecular perturbations result from external sources that exert energy on molecules. Actually, every molecule that exists does so under atmospheric conditions provided that ideality is excluded. As such, it exists in a perturbed natural state. Hence, physical properties encompass the specific relations between the atoms in molecules and well-defined forms of energy or other external "yardsticks" of measurement. For

**TABLE 4–1**

**ATOMIC AND GROUP CONTRIBUTIONS TO MOLAR REFRACTION***

| | |
|---|---|
| C– (single) | 2.418 |
| –C= (double) | 1.733 |
| –C≡ (triple) | 2.398 |
| Phenyl ($C_6H_5$) | 25.463 |
| H | 1.100 |
| O (C=O) | 2.211 |
| O (O–H) | 1.525 |
| O (ether, ester, C–O) | 1.643 |
| Cl | 5.967 |
| Br | 8.865 |
| L | 13.900 |

*These values are reported for the *D* line of sodium as the light source. From J. Dean (Ed.), *Lange's Handbook*, 12th Ed. McGraw-Hill, New York, 1979, pp. 10–94. See also Bower et al., in A. Weissberger (Ed.), *Physical Methods of Organic Chemistry*, Vol. 1, Part II, 3rd Ed., Wiley-Interscience, New York, 1960, Chapter 28.

example, the concept of weight uses the force of gravity as an external measure to compare the mass of objects, whereas that of optical rotation uses plane-polarized light to describe the optical rotation of molecules organized in a particular bonding pattern. Therefore, a physical property should be easily measured or calculated, and such measurements should be reproducible for an individual molecule under optimal circumstances.

When one carefully associates specific physical properties with the chemical nature of closely related molecules, one can (a) describe the spatial arrangement of atoms in drug molecules, (b) provide evidence for the relative chemical or physical behavior of a molecule, and (c) suggest methods for the qualitative and quantitative analysis of a particular pharmaceutical agent. The first and second of these often lead to implications about the chemical nature and potential action that are necessary for the creation of new molecules with selective pharmacologic activity. The third provides the researcher with tools for drug design and manufacturing and offers the analyst a wide range of methods for assessing the quality of drug products. The level of understanding of the physical properties of molecules has expanded greatly and, with increasing advances in technology, the development of computer-based computational tools to develop molecules with ideal physical properties has become commonplace.

Nowhere has the impact of computational modeling been of broader importance than in drug discovery and design, where computational tools offer great promise for enhancing the speed of drug design and selection. For example, computer modeling of therapeutic targets (e.g., HIV protease) is widely utilized. This approach is based on the screening of known or predicted physical properties (e.g., the protein conformation of the HIV protease) of a therapeutic target in a computer to elucidate area(s) where a molecule can be synthesized to optimally bind and either block (antagonist) or enhance (agonist) the actions of the target. This type of modeling usually reveals a limited number of atomic spatial arrangements and types, such as a functional group, that can be utilized in the design of a molecule to elicit the desired pharmacologic response. Other computational programs can then be utilized to compute the number of molecules that might make reasonable drugs based on these molecular restrictions.

It is important to note that these models are relatively new but offer great potential for the future of drug discovery and design. It is also important to note that all of these computational approaches are based solely on the physical properties of molecule(s) that have been determined through years of research. The ability to design and test these molecules using a computer and to rapidly synthesize chemical libraries has also dramatically increased the need for technologically superior equipment to measure the physical properties of these molecules to confirm that the actual properties match those that are predicted. Toward that end, chemical libraries can reach well over 10,000 individual molecules and can be tested for numerous molecular properties in modern *high-throughput systems*. In fact, "virtual" libraries routinely contain millions of compounds. This aids in the ability of scientists to select one lead compound whose potential to make it into clinical use is greater than all of the other molecules screened. Finally, the ability to measure the physical properties of a lead compound is critical to assuring that during the transition from computer to large-scale manufacturing, the initially identified molecule remains physically the same. This is a guideline that the Food and Drug Administration requires to ensure human safety and drug efficacy.

The remainder of this chapter describes some of the well-defined interactions that are important for determining the physical properties of molecules. All of the quantities presented are expressed in Système International (SI) units for all practical cases.

## DIELECTRIC CONSTANT AND INDUCED POLARIZATION

Electricity is related to the nature of charges in a dynamic state (e.g., electron flow). When a charged molecule is at rest, its properties are defined by electrostatics. For two charges separated by a distance $r$, their potential energy is defined by Coulomb's law,

$$U(r) = \frac{q_1 q_2}{4\pi \varepsilon_0 r} \qquad (4\text{--}1)$$

where the charges $q_1$ and $q_2$ are in coulombs (C), $r$ is in meters, $\varepsilon_0$ (the permittivity constant) $= 8.854 \times 10^{-12}$ $C^2$ $N^{-1}$ $m^{-2}$, and the potential energy is in joules. Coulomb's law can be used to describe both attractive and repulsive interactions. From equation (4–1), it is clear that electrostatic interactions also rely on the permittivity of the medium in which the charges exist. Different solvents have differing permittivities due to their chemical nature, polar or nonpolar. For a more comprehensive discussion, see Bergethon and Simons.[5]

When no permanent charges exist in molecules, a measure of their polarity (i.e., the electronic distribution within the molecule) is given by the property we called the dipole moment ($\mu$). In the simplest description, a dipole is a separation of two opposing charges over a distance $r$ (**Fig. 4–2**) and is generally described by a vector whose magnitude is given by $\mu = qr$. Dipoles do not have a net charge, but this charge separation can often create chargelike interactions and influence several physical and chemical properties. In complex molecules, the electron distribution can be approximated so that overall dipole moments can be estimated from partial atomic charges. A key feature of this calculation is the symmetry of the molecule. A symmetric molecule will display no dipole moment either because there is no charge separation (e.g., $H_2$, $O_2$, $N_2$) or as a consequence of cancellation of the dipole vectors (e.g., benzene, methane).

A molecule can maintain a separation of electric charge either through induction by an external electric field or by a permanent charge separation within a polar molecule. The

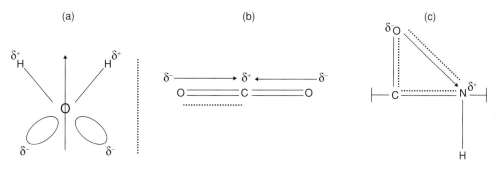

Fig. 4–2. Vectorial nature of permanent dipole moments for (*a*) water, (*b*) carbon dioxide, and (*c*) a peptide bond. The distance *r* is given by the dashed line for each molecule. The arrow represents the conventional direction that a dipole moment vector is drawn, from negative to positive.

dipolar nature of a peptide bond is an example of a fixed or permanent dipole, and its effects on the bonding structure were discussed earlier in this chapter. In proteins, the permanent dipolar nature of the peptide bond and side chains can stabilize secondary structures like $\alpha$-helices and can also influence the higher-order conformation of a protein. Induced dipoles in a protein can also influence hydrophobic core protein structures, so both types of dipole play a prominent role in the stabilization of protein-derived therapeutics. Permanent dipoles in a molecule are discussed in the next section.

To properly discuss dipoles and the effects of solvation, one must understand the concepts of polarity and dielectric constant. Placing a molecule in an electric field is one way to induce a dipole. Consider two parallel conducting plates, such as the plates of an electric condenser, which are separated by some medium across a distance *r*, and apply a potential across the plates (**Fig. 4–3**). Electricity will flow from the left plate to the right plate through the battery until the potential difference of the plates equals that of the battery supplying the initial potential difference. The *capacitance* (*C*, in farads [F]) is equal to the quantity of electric charge (*q*, in coulombs)

stored on the plates divided by the potential difference (*V*, in volts [V]) between the plates:

$$C = q/V \qquad (4\text{--}2)$$

The capacitance of the condenser in **Figure 4–3** depends on the type of medium separating the plates as well as on the thickness *r*. When a vacuum fills the space between the plates, the capacitance is $C_0$. This value is used as a reference to compare capacitances when other substances fill the space. If water fills the space, the capacitance is increased because the water molecule can orientate itself so that its negative end lies nearest the positive condenser plate and its positive end lies nearest the negative plate (see **Fig. 4–3**). This alignment provides additional movement of charge because of the increased ease with which electrons can flow between the plates. Thus, additional charge can be placed on the plates per unit of applied voltage.

The capacitance of the condenser filled with some material, $C_x$, divided by the reference standard, $C_0$, is referred to as the *dielectric constant*, $\varepsilon$:

$$\varepsilon = C_x/C_0 \qquad (4\text{--}3)$$

The dielectric constant ordinarily has no dimensions because it is the ratio of two capacitances. By definition, the dielectric constant of a vacuum is unity. Dielectric constants of some liquids are listed in **Table 4–2**. The dielectric constants of solvent mixtures can be related to drug solubility as described by Gorman and Hall,[6] and $\varepsilon$ for drug vehicles can be related to drug plasma concentration as reported by Pagay et al.[7]

The dielectric constant is a measure of the ability of molecule to resist charge separation. If the ratios of the capacitances are close, then there is greater resistance to a charge separation. As the ratios increase, so too does the molecule's ability to separate charges. Note that field strength and temperature play a role in these measurements. If a molecule has a permanent dipole, it will have minimal resistance to charge separation and will migrate in the field. However, the converse is also true. Consider a water molecule at the atomic level with its positively charged nuclei drawn toward the negative plate and its negatively charged electron cloud pulled and oriented toward the positive plate. Oxygen is strongly

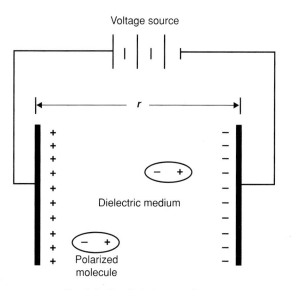

Voltage source

Dielectric medium

Polarized molecule

Fig. 4–3. Parallel-plate condenser.

## TABLE 4–2
## DIELECTRIC CONSTANTS OF SOME LIQUIDS AT 25°C

| Substance | Dielectric Constant, $\varepsilon$ | |
|---|---|---|
| N-Methylformamide | 182 | |
| Hydrogen cyanide | 114 | |
| Formamide | 110 | |
| Water | 78.5 | |
| Glycerol | 42.5 | |
| Methanol | 32.6 | |
| Tetramethylurea | 23.1 | |
| Acetone | 20.7 | |
| n-Propanol | 20.1 | |
| Isopropanol | 18.3 | |
| Isopentanol | 14.7 | |
| l-Pentanol | 13.9 | |
| Benzyl alcohol | 13.1 | |
| Phenol | 9.8 | (60°C) |
| Ethyl acetate | 6.02 | |
| Chloroform | 4.80 | |
| Hydrochloric acid | 4.60 | |
| Diethyl ether | 4.34 | (20°C) |
| Acetonitrile | 3.92 | |
| Carbon disulfide | 2.64 | |
| Triethylamine | 2.42 | |
| Toluene | 2.38 | |
| Beeswax (solid) | 2.8 | |
| Benzene | 2.27 | |
| Carbon tetrachloride | 2.23 | |
| 1,4-Dioxane | 2.21 | |
| Pentane | 1.84 | (20°C) |
| Furfural | 41 | (20°C) |
| Pyridine | 12.3 | |
| Methyl salicylate | 9.41 | (30°C) |

electronegative and will have a stronger attraction to the positive pole than will a less electronegative atom. Likewise, hydrogen atoms are more electropositive and will move further toward the positive plate than an atom that is more electronegative. Conceptually, this will reflect less resistance to the field.

A molecule placed in that field will align itself in the same orientation as the water molecules even though the extent of the alignment and induced charge separation will dramatically differ due to atomic structure. For this discussion, consider a molecule like pentane. Pentane is an aliphatic hydrocarbon that is not charged and relies on van der Waals interactions for its primary attractive energies. Pentane is not a polar molecule, and carbon is not strongly electronegative. Carbon–hydrogen bonds are much stronger (less acidic) than oxygen–hydrogen bonds; therefore, the electrons in the $\sigma$ bonds are more shared. The electronic structure would not favor a large charge separation in the molecule and should have a lower dielectric constant. If a polar functional group such as an alcohol moiety is added to generate 1-pentanol, an enhanced ability to undergo charge separation and an increased dielectric constant is the result. This is readily

## ⬛▶ KEY CONCEPT ◀ POLARIZABILITY

Polarizability is defined as the ease with which an ion or molecule can be polarized by any external force, whether it be an electric field or light energy or through interaction with another molecule. Large anions have large polarizabilities because of their loosely held outer electrons. Polarizabilities for molecules are given in the table. The units for $\alpha_p$ are $Å^3$ or $10^{-24}$ $cm^3$.

### POLARIZABILITIES

| Molecule | $\alpha_p \times 10^{24}$ ($cm^3$) |
|---|---|
| $H_2O$ | 1.68 |
| $N_2$ | 1.79 |
| HCl | 3.01 |
| HBr | 3.5 |
| HI | 5.6 |
| HCN | 5.9 |

observed in **Table 4–2**, where the dielectric constant for pentane (at 20°C) is 1.84, that for 1-pentanol (at 25°C) is 13.9, and that for isopentanol is 14.7 (at 25°C).

When nonpolar molecules like pentane are placed in a suitable solvent between the plates of a charged capacitor, an *induced polarization* of the molecules can occur. Again, this *induced dipole* occurs because of the separation of electric charge within the molecule due to the electric field generated between the plates. The displacement of electrons and nuclei from their original positions is the main result of the induction process. This temporary induced dipole moment is proportional to the field strength of the capacitor and the *induced polarizability*, $\alpha_p$, which is a characteristic property of the particular molecule.

From electromagnetic theory, it is possible to obtain the relationship

$$\frac{\varepsilon - 1}{\varepsilon + 2} = \frac{4}{3}\pi n \alpha_p \qquad (4\text{--}4)$$

where $n$ is the number of molecules per unit volume. Equation **(4–4)** is known as the *Clausius–Mossotti equation*. Multiplying both sides by the molecular weight of the substance, $M$, and dividing both sides by the solvent density, $\rho$, we obtain

$$\left(\frac{\varepsilon - 1}{\varepsilon + 2}\right)\frac{M}{\rho} = \frac{4}{3}\frac{\pi n M \alpha_p}{\rho} = \frac{4}{3}\pi N \alpha_p = P_i \qquad (4\text{--}5)$$

where $N$ is Avogadro's number, $6.023 \times 10^{23}$ mole$^{-1}$, and $P_i$ is known as the *induced molar polarization*. $P_i$ represents the induced dipole moment per mole of nonpolar substance when the electric field strength of the condenser, V/m in volts per meter, is unity.

**EXAMPLE 4–1**

### Polarizability of Chloroform

Chloroform has a molecular weight of 119 g/mole and a density of 1.43 g/cm$^3$ at 25°C. What is its induced molar polarizability? We have

$$P_i = \frac{(\varepsilon - 1)}{(\varepsilon + 2)} \times \frac{M}{\rho} = \frac{(4.8 - 1)}{(4.8 + 2)} \times \frac{119}{1.43} = 46.5 \, \text{cm}^3/\text{mole}$$

The concept of induced dipole moments can be extended from the condenser model just discussed to the model of a nonpolar molecule in solution surrounded by ions. In this case, an anion would repel molecular electrons, whereas a cation would attract them. This would cause an interaction of the molecule in relation to the ions in solution and produce an induced dipole. The distribution and ease of attraction or repulsion of electrons in the nonpolar molecule will affect the magnitude of this induced dipole, as would the applied external electric field strength.

## PERMANENT DIPOLE MOMENT OF POLAR MOLECULES

In a polar molecule, the separation of positively and negatively charged regions can be permanent, and the molecule will possess a *permanent dipole moment*, $\mu$. This is a nonionic phenomenon, and although regions of the molecule may possess partial charges, these charges balance each other so that the molecule does not have a net charge. Again, we can relate this to atomic structure by considering the electronegativity of the atoms in a bond. Water molecules possess a permanent dipole due to the differences in the oxygen and the hydrogen. The magnitude of the permanent dipole moment, $\mu$, is independent of any induced dipole from an electric field. It is defined as the vector sum of the individual charge moments within the molecule itself, including those from bonds and lone-pair electrons. The vectors depend on the distance of separation between the charges. **Figure 4–2** provides an illustration of dipole moment vectors for water, carbon dioxide, and a peptide bond. The unit of $\mu$ is the *debye*, with 1 debye equal to $10^{-18}$ electrostatic unit (esu) cm. The esu is the measure of electrostatic charge, defined as a charge in a vacuum that repels a like charge 1 cm away with a force of 1 dyne. In SI units, 1 debye = $3.34 \times 10^{-30}$ C-m. This is derived from the charge on the electron (about $10^{-10}$ esu) multiplied by the average distance between charged centers on a molecule (about $10^{-8}$ cm or 1 Å).

In an electric field, molecules with permanent dipole moments can also have induced dipoles. The polar molecule, however, tends to orient itself with its negatively charged centers closest to positively charged centers on other molecules *before* the electric field is applied. When the applied field is present, the orientation is in the direction of the field. Maximum dipole moments occur when the molecules are almost perfectly oriented with respect to the fields. Absolutely perfect orientation can never occur owing to thermal energy of the molecules, which contributes to molecular motion that opposes the molecular alignment. The *total* molar polarization, $P$, is the sum of induction and permanent dipole effects:

$$P = P_i + P_0 = \left(\frac{\varepsilon - 1}{\varepsilon + 2}\right)\frac{M}{\rho} \qquad (4-6)$$

where $P_0$ is the orientation polarization of the permanent dipoles. $P_0$ is equal to $4\pi N\mu^2/9\,kT$, in which $k$ is the Boltzmann constant, $1.38 \times 10^{-23}$ joule/K. Because $P_0$ depends on the temperature, $T$, equation (4–6) can be rewritten in a linear form as

$$P = P_i + A\frac{1}{T} \qquad (4-7)$$

where the slope $A$ is $4\pi \, N\mu^2/9k$ and $P_i$ is the $y$ intercept. If $P$ is obtained at several temperatures and plotted against $1/T$, the slope of the graph can be used to calculate $\mu$ and the intercept can be applied to compute $\alpha_p$. The values of $P$ can be obtained from equation (4–6) by measuring the dielectric constant and the density of the polar compound at various temperatures. The dipole moments of several compounds are listed in **Table 4–3**.

In solution, the permanent dipole of a solvent such as water can strongly interact with solute molecules. This interaction contributes to the solvent effect and is associated, in the case of water, with the hydration of ions and molecules. The symmetry of the molecule can also be associated with its dipole moment, which is observed with carbon dioxide (no net dipole) in **Figure 4–2**. Likewise, benzene and *p*-dichlorobenzene are symmetric planar molecules and have dipole moments of zero. Meta and ortho derivatives of benzene, however, are not symmetric and have significant dipole moments, as listed in **Table 4–3**.

The importance of dipole interactions should not be underestimated. For ionic solutes and nonpolar solvents, ion–induced dipole interactions have an essential role in solubility phenomena (Chapter 9). For drug–receptor binding, dipole–dipole interactions are essential noncovalent forces that contribute to enhance the pharmacologic effect, as described by Kollman.[8] For solids composed of molecules with permanent dipole moments, the dipole interactions contribute to the crystalline arrangement and overall structural nature of the solid. For instance, water molecules in ice crystals are organized through their dipole forces. Additional interpretations of the significance of dipole moments are given by Minkin et al.[9]

## ELECTROMAGNETIC RADIATION

Electromagnetic radiation is a form of energy that propagates through space as oscillating electric and magnetic fields at right angles to each other and to the direction of the propagation, shown in **Figure 4–4a**. Both electric and magnetic fields can be described by sinusoidal waves with characteristic amplitude, $A$, and frequency, $v$. The common representation of the electric field in two dimensions is shown in

**TABLE 4–3**

**DIPOLE MOMENTS OF SOME COMPOUNDS**

| Compound | Dipole Moment (Debye Units) |
|---|---|
| p-Dichlorobenzene | 0 |
| $H_2$ | 0 |
| Carbon dioxide | 0 |
| Benzene | 0 |
| 1,4-Dioxane | 0 |
| Carbon monoxide | 0.12 |
| Hydrogen iodide | 0.38 |
| Hydrogen bromide | 0.78 |
| Hydrogen chloride | 1.03 |
| Dimethylamine | 1.03 |
| Barbital | 1.10 |
| Phenobarbital | 1.16 |
| Ethylamine | 1.22 |
| Formic acid | 1.4 |
| Acetic acid | 1.4 |
| Phenol | 1.45 |
| Ammonia | 1.46 |
| m-Dichlorobenzene | 1.5 |
| Tetrahydrofuran | 1.63 |
| n-Propanol | 1.68 |
| Chlorobenzene | 1.69 |
| Ethanol | 1.69 |
| Methanol | 1.70 |
| Dehydrocholesterol | 1.81 |
| Water | 1.84 |
| Chloroform | 1.86 |
| Cholesterol | 1.99 |
| Ethylenediamine | 1.99 |
| Acetylsalicylic acid | 2.07 |
| o-Dichlorobenzene | 2.3 |
| Acetone | 2.88 |
| Hydrogen cyanide | 2.93 |
| Nitromethane | 3.46 |
| Acetanilide | 3.55 |
| Androsterone | 3.70 |
| Acetonitrile | 3.92 |
| Methyltestosterone | 4.17 |
| Testosterone | 4.32 |
| Urea | 4.56 |
| Sulfanilamide | 5.37 |

**Figure 4–4b.** This frequency, $v$, is the number of waves passing a fixed point in 1 sec. The wavelength, $\lambda$, is the extent of a single wave of radiation, that is, the distance between two successive maxima of the wave, and is related to frequency by the velocity of propagation, $v$:

$$v = v\lambda \qquad (4\text{–}8)$$

The frequency of the radiation depends on the source and remains constant; however, the velocity depends on the composition of the medium through which it passes. In a vacuum, the wave of radiation travels at its maximum, the speed of light, $c = 2.99792 \times 10^8$ m/sec. Thus, the frequency can be defined as

$$v = c/\lambda \qquad (4\text{–}9)$$

The wave number, $\bar{v}$, defined as the reciprocal of the wavelength, is another way of describing the electromagnetic radiation:

$$\bar{v} = \frac{1}{\lambda} = \frac{v}{\nu} \qquad (4\text{–}10)$$

The wave number (in cm$^{-1}$) represents the number of wavelengths found in 1 cm of radiation in a vacuum. It is widely used because it is directly proportional to the frequency, and thus to the energy of radiation, $E$, given by the famous and fundamental relationship of Planck and Einstein for the light quantum or energy of a photon:

$$E = hv = hc\bar{v} \qquad (4\text{–}11)$$

where $h$ is Planck's constant, which is equal to $6.6261 \times 10^{-34}$ joules.

The electromagnetic spectrum is classified according to its wavelength or its corresponding wave number, as illustrated in **Table 4–4**. The wavelength becomes shorter as the corresponding radiant energy increases. Most of our knowledge about atomic and molecular structure and properties comes from the interaction of electromagnetic radiation with matter. The electric field component is mostly responsible for phenomena such as transmission, reflection, refraction, absorption, and emission of electromagnetic radiation, which give rise to many of the spectroscopic techniques discussed in this chapter. The magnetic component is responsible for the absorption of energy in electron paramagnetic resonance (EPR) and nuclear magnetic resonance (NMR), techniques described at the end of the chapter.

## ATOMIC AND MOLECULAR SPECTRA

All atoms and molecules absorb electromagnetic radiation at certain characteristic frequencies. According to elementary quantum theory, ions and molecules can exist only in certain discrete states characterized by finite amounts of energy (i.e., energy levels or electronic states) because electrons are in motion around the positively charged nucleus in atoms. Thus, the radiant energy absorbed by a chemical species has only certain discrete values corresponding to the individual transitions from one energy state ($E_1$) to a second, upper energy state ($E_2$) that can occur in an atom or molecule (**Fig. 4–5**). The pattern of absorption frequencies is called an *absorption spectrum* and is produced if radiation of a particular wavelength is passed through a sample and measured. If the energy of the radiation is decreased upon exit, the change in the intensity of the radiation is due to electronic excitation.

In some instances, electromagnetic radiation can also be produced (*emission*) when the excited species (atoms, ions, or molecules excited by means of the heat of a flame, an electric

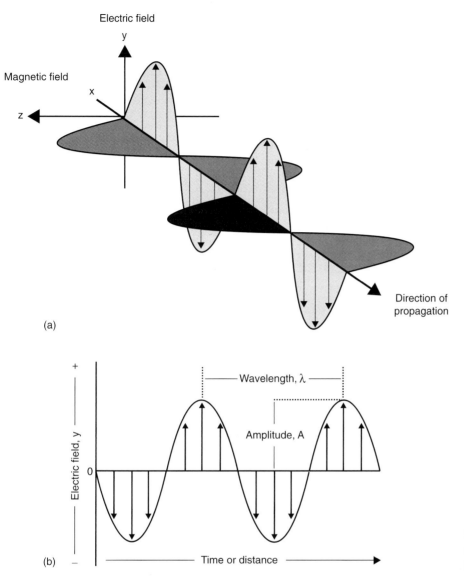

**Fig. 4–4.** The oscillating electric and magnetic fields associated with electromagnetic radiation.

current spark, absorption of light, or some other energy source) relax to lower energy levels by releasing energy as photons ($h\nu$). The emitted radiation as a function of the wavelength or frequency is called the emission spectrum.

Atomic spectra are the simplest to describe and usually present a series of lines corresponding to the frequencies of specific electronic transition states. An exact description of these transitions is possible only for the hydrogen atom, for which a complete quantum mechanical solution exists.[10] Even so, the absorption and emission spectra for the hydrogen atom also can be explained using a less sophisticated theory such as the Bohr's model. In this approach, the energy of an electron moving in a definite orbit around a positively charged nucleus in an atom is given by

$$E = -\frac{2\pi^2 Z^2 me^4}{n^2 h^2} \qquad \textbf{(4–12)}$$

where $Z$ is the atomic number, $m$ is the mass of the electron ($9.1 \times 10^{-31}$ kg), $n$ is an integer number corresponding to

the principal quantum number of the orbit, $e$ is the charge on the electron ($1.602 \times 10^{-19}$ coulomb or $1.519 \times 10^{-14}$ m$^{3/2}$ kg$^{1/2}$ sec$^{-1}$), and $h$ is Planck's constant.

Bohr's model also assumes that electrons moving in these orbits are stable (i.e., angular momentum is quantized), and therefore absorption or emission occurs only when electrons change orbit. Thus, the characteristic frequency of the photon absorbed or emitted corresponds to the absolute value of the difference of the energy states involved in an electronic transition in an atom. Thus, the difference between electron energy levels, $E_2 - E_1$, having respective quantum numbers $n_2$ and $n_1$, is given by the expression

$$E_2 - E_1 = \frac{2\pi^2 Z^2 me^4}{h^2} \left( \frac{1}{n_1^2} - \frac{1}{n_2^2} \right) \qquad \textbf{(4–13)}$$

The energy of a photon of electromagnetic radiation $E$ is related to the frequency of the radiation by equation **(4–12)**. Substituting the energy difference by the corresponding

## TABLE 4-4
## ELECTROMAGNETIC RADIATION RANGES ACCORDING TO WAVELENGTH, WAVE NUMBER, FREQUENCY, AND ENERGY, AND TECHNIQUES THAT UTILIZE THESE FORMS OF RADIATION*

| Electromagnetic Radiation | Wavelength $\lambda$ (m) | Wave Number $\bar{v}$ (cm$^{-1}$) | Frequency (Hz) | Energy (joule) | Representative Techniques (Not Inclusive) |
|---|---|---|---|---|---|
| Gamma rays | $<1 \times 10^{-11}$ | $>1 \times 10^{9}$ | $>3 \times 10^{19}$ | $>2 \times 10^{-14}$ | Microscopy, photography, radioactivity tracer |
| X-rays | $1 \times 10^{-11}$–$1 \times 10^{-8}$ | $1 \times 10^{9}$–$1 \times 10^{6}$ | $3 \times 10^{19}$–$3 \times 10^{16}$ | $2 \times 10^{-14}$–$2 \times 10^{-17}$ | Microscopy, photography, radioactivity tracer, diffraction, ionization |
| Vacuum ultraviolet | $1 \times 10^{-8}$–$2 \times 10^{-7}$ | $1 \times 10^{6}$–$5 \times 10^{4}$ | $3 \times 10^{16}$–$1.5 \times 10^{15}$ | $2 \times 10^{-17}$–$9.9 \times 10^{-19}$ | Spectrophotometry |
| Near-ultraviolet | $2 \times 10^{-7}$–$4 \times 10^{-7}$ | $5 \times 10^{4}$–$2.5 \times 10^{4}$ | $1.5 \times 10^{15}$–$7.5 \times 10^{14}$ | $9.9 \times 10^{-19}$–$5 \times 10^{-19}$ | Spectrophotometry, fluorescence, circular dichroism |
| Visible light | $4 \times 10^{-7}$–$7 \times 10^{-7}$ | $2.5 \times 10^{4}$–$1.4 \times 10^{4}$ | $7.5 \times 10^{14}$–$4 \times 10^{14}$ | $5 \times 10^{-19}$–$3 \times 10^{-19}$ | Spectrophotometry, fluorescence, light scattering, optical rotation |
| Near-infrared | $7 \times 10^{-7}$–$5 \times 10^{-6}$ | $1.4 \times 10^{4}$–$2 \times 10^{3}$ | $4 \times 10^{14}$–$6 \times 10^{13}$ | $3 \times 10^{-19}$–$4 \times 10^{-20}$ | Vibrational and rotational spectroscopy, solid-state characterization |
| Mid (intermediate) infrared | $5 \times 10^{-6}$–$3 \times 10^{-5}$ | $2 \times 10^{3}$–$3.33 \times 10^{2}$ | $6 \times 10^{13}$–$1 \times 10^{13}$ | $4 \times 10^{-20}$–$6.6 \times 10^{-21}$ | Vibrational and rotational spectroscopy |
| Far infrared | $3 \times 10^{-5}$–$1 \times 10^{-3}$ | $3.33 \times 10^{2}$–$10$ | $1 \times 10^{13}$–$3 \times 10^{11}$ | $6.6 \times 10^{-21}$–$2 \times 10^{-22}$ | Vibrational and rotational spectroscopy |
| Microwave | $1 \times 10^{-3}$–$0.1$ | $10$–$0.1$ | $3 \times 10^{11}$–$3 \times 10^{9}$ | $2 \times 10^{-22}$–$2 \times 10^{-24}$ | Electron paramagnetic resonance |
| Radio | $>0.1$ | $<0.1$ | $<3 \times 10^{9}$ | $<2 \times 10^{-24}$ | Nuclear magnetic resonance |

*Information on the values was derived in part from the NASA Web site http://imagine.gsfc.nasa.gov/docs/science/know_11/spectrum_chart.html and Campbell and Dweb.[12]

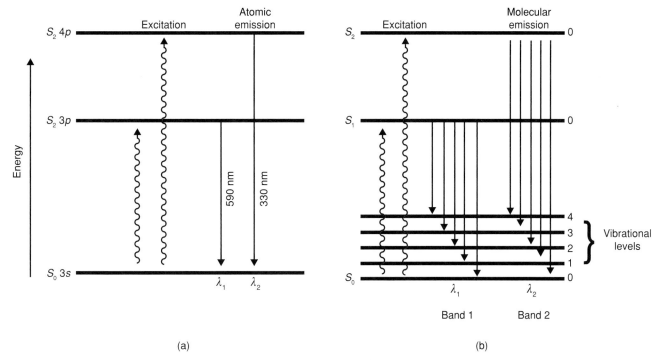

**Fig. 4–5.** Comparison of (a) atomic and (b) molecular energy levels.

characteristic wave number according to equation **(4–11)** in equation **(4–13)**, we obtain

$$\bar{v} = \frac{2\pi^2 Z^2 m e^4}{h^3 c} \left( \frac{1}{n_1^2} - \frac{1}{n_2^2} \right) \qquad (4\text{–}14)$$

where $n_1$ and $n_2$ are the principal quantum numbers for the orbital states involved in an electronic transition of the atom from level $E_1$ to $E_2$.

The first term on the right in equation **(4–14)** is composed of fundamental constants and the atomic number $Z$. For the hydrogen atom $Z = 1$, and substitution of the appropriate values affords a single constant $R_8$ of value 109,737.3 cm$^{-1}$. This value is calculated from the Bohr's theory and assumes that the nucleus of infinite mass was stationary with respect to the electron orbit. A more accurate analysis involves the replacement of the mass of the electron by the reduced mass of the electron in the hydrogen atom, $\mu_H = 0.9994558m$. The corrected theoretical value of $R_8$, now called $R_H$, is (109737.3) (0.9994558) = 109,677.6 cm$^{-1}$. The experimental value for this constant, also known as the *Rydberg constant*, is $R_H$ = 109,678.76 cm$^{-1}$. The remarkable agreement between the theoretical and experimental values of $R_H$ gave initial support to Bohr's theory, although it was later abandoned for the more complete quantum mechanical treatment.[10] Nevertheless, the absorption and emission spectra for the hydrogen atom are fully explained by Bohr's theory. Thus, introducing $R_H$ in equation **(4–14)**, we find the characteristic wave number only from the values of $n$, the principal quantum number of the electron's orbit for the transition, given by the simple relation

$$\bar{v} = R_H \left( \frac{1}{n_1^2} - \frac{1}{n_2^2} \right) \qquad (4\text{–}15)$$

**EXAMPLE 4–2**

**Energy of Excitation**

(a) What is the energy (in joules and cm$^{-1}$) of a quantum of radiation absorbed to promote the electron in the hydrogen atom from its ground state ($n_1 = 1$) to the second orbital, $n_2 = 2$?

We can solve this problem by using equation (4–13) and replacing the values of all physical constants involved and using $Z = 1$; we find

$$E_2 - E_1 = \frac{2 \times (3.14)^2 \times (1)^2 \times (9.1 \times 10^{-31}) \times (1.519 \times 10^{-14})^4}{(6.629 \times 10^{-34})^2}$$
$$\times \left( \frac{1}{(1)^2} - \frac{1}{(2)^2} \right)$$
$$= 1.63 \times 10^{-18} \text{ joule}$$

Note that joule = kg $\times$ m$^2$ $\times$ sec$^{-2}$.

To obtain the answer in cm$^{-1}$, we can use equation (4–11) to convert joules into wave number, or directly use equation (4–15):

$$\bar{v} = R_H \left( \frac{1}{n_1^2} - \frac{1}{n_2^2} \right)$$

Using this equation we obtain,

$$\bar{v} = (109{,}677.6 \text{ cm}^{-1}) \frac{1}{(1)^2} - \frac{1}{(2)^2} = 82{,}258.2 \text{ cm}^{-1}$$

and therefore, from equation (4–10), the wavelength of the spectral line when an electron passes from the $n = 1$ orbital to the $n = 2$ orbital is

$$\lambda = \frac{1}{\bar{v}} = \frac{1}{82{,}258.2} = 1.22 \times 10^{-5} \quad \text{or} \quad 122 \text{ nm}$$

This is the first line of the Lyman ultraviolet series of the atomic spectra for hydrogen. Note that, from equation **(4–12)**, the electron of the hydrogen atom in the ground state, $n = 1$, has a lower energy (–2.18 × 10$^{-18}$ joule) than that in the next highest electron state, $n = 2$ (–0.55 × 10$^{-18}$ joule).

When the electron acquires sufficient energy to leave the atom, it is regarded as infinitely distant from the nucleus, and the nucleus is considered no longer to affect the electron. The energy required for this process, which results in the ionization of the atom according to

$$H - 1e^- \rightarrow H^+$$

is also known as the *ionization potential*. If we consider this process as occurring when $n = \infty$, then the ionization potential from the ground state $(n = 1)$ to $n = \infty$ is

$$E_2 - E_1 = \frac{2\pi^2 Z^2 m e^4}{h^2} \left( \frac{1}{1} - \frac{1}{\infty} \right)$$

Because $1/\infty \cong 0$,

$$E_2 - E_1 = \frac{2\pi^2 Z^2 m e^4}{h^2}$$

This is equivalent to $-E$ for $n = 1$, according to equation (4–13). Thus, the ionization potential exactly equals the negative energy of the electron in the ground state with a value of $10,9677.6 \text{ cm}^{-1}$ (equals 13.6 eV).

Virtually all atoms but hydrogen have more than one electron. Therefore, approximate methods are used to evaluate electronic transition states for most atoms. In these atoms, spin pairing energy and electron–electron repulsion arise, so that description of the electronic states is not as straightforward as in the hydrogen atom. For practical purposes, most elements of the periodic table display a characteristic atomic absorption and emission spectrum associated with the electronic transition states that can be used to identify and quantify specific elements. Some of the more sensitive spectral wavelengths associated with particular atoms are given in **Table 4–5**. For instance, the simplified energy-level diagram for the sodium atom in **Figure 4–5a** illustrates the source of the lines in a typical emission spectrum of gas-phase sodium. The single-outer-electron ground state for sodium is located in the $3s$ orbital. After excitation (i.e., absorption of energy as shown by wavy lines), two emission lines (about 330 and 590 nm) result from the transitions $4p \rightarrow 3s$ and $3p \rightarrow 3s$,

**TABLE 4–5**

**SPECTRAL WAVELENGTHS ASSOCIATED WITH ELECTRONIC TRANSITIONS USED IN THE DETECTION OF PARTICULAR ELEMENTS**

| Element | Wavelength (nm) |
| --- | --- |
| As | 193.7 |
| CA | 422.7 |
| Na | 589.0 |
| Cu | 324.8 |
| Hg | 253.7 |
| Li | 670.8 |
| Pb | 405.8 |
| Zn | 213.9 |
| K | 766.5 |

respectively. Many types of atomic spectrometers are available depending on the methods used for atomization and introduction of the sample. Further detailed information can be found in the references cited in this chapter.

Atomic spectroscopy has pharmaceutical applications in analysis of metal ions from drug products and in the quality control of parenteral electrolyte solutions. For example, blood levels of lithium, used to treat bipolar disorder (manic depressive disorder), can be analyzed by atomic spectroscopy to check for overdosing of lithium salts.

In addition to having electronic states, molecules have quantized *vibrational states*, which are associated with energies due to interatomic vibrations (e.g., stretching and bending), and rotational states, which are related to the rotation of molecules around their center of gravity. These additional energy states available for electron transitions make the spectra of molecules more complex than those of atoms. In the case of vibration, the interatomic bonds may be thought of as springs between atoms (see **Fig. 4–14**) that can vibrate in various stretching or bending configurations depending on their energy levels. In rotation, the motion is similar to that of a top spinning according to its energy level. In addition, the molecule may have some kinetic energy associated with its translational (straight-line) motion in a particular direction.

The energy levels associated with these various transitions differ greatly from one another. The energy associated with movement of an electron from one orbital to another is typically about $10^{-18}$ joule (electronic transitions absorb in the ultraviolet and visible light region between 180 and 780 nm; see *Example 4–2*), where the energy involved in vibrational changes is about $10^{-19}$ to $10^{-20}$ joule (infrared region) depending on the atoms involved, and the energy for rotational change is about $10^{-21}$ joule. The energy associated with translational change is even smaller, about $10^{-35}$ joule. The precise energies associated with these individual transitions depend on the atoms and bonds that compose the molecule. Each electronic energy state of a molecule normally has several possible vibrational states, and each of these has several rotational states. The rotational energy levels are lower than the vibrational levels and are drawn in a similar manner to the quantized vibrational levels in the electronic states, as shown in **Figure 4–5b**. The translational states are so numerous and the energy levels between translational states are so small that they are normally considered a continuous form of energy and are not treated as quantized. The total energy of a molecule is the sum of its electronic, vibrational, rotational, and translational energies.

When a molecule absorbs electromagnetic radiation, it can undergo certain transitions that depend on the quantized amount of energy absorbed. In **Figure 4–5**, the absorption of radiation (wavy lines) leads to two different energy transitions, $\Delta E$, which result in the electronic transition from the lowest level of the ground state $(S_0)$ to an excited electronic state $(S_1$ or $S_2)$. Electronic transitions of molecules involve energies corresponding to ultraviolet or visible radiation.

Purely vibrational transitions may occur within the same electronic state (e.g., a change from level 1 to 2 in $S_0$) and involve near-infrared (IR) radiation. Rotational transitions (not shown in **Fig. 4–5**) are associated with low-energy radiation over the entire infrared wavelength region. The relatively large energy associated with an electronic transition usually leads to a variety of concurrent, different vibrational and rotational changes. Slight differences in the vibrational and rotational nature of the excited electronic state complicate the spectrum. These differences lead to broad bands, characteristic of the ultraviolet and visible regions, rather than the sharp, narrow lines characteristic of individual vibrational or rotational changes in the infrared region.

The energy absorbed by a molecule may be found only at a few discrete wavelengths in the ultraviolet, visible, and infrared regions, or the absorptions may be numerous and at longer wavelengths than originally expected. The latter case, involving longer-wavelength radiation, is normally found for molecules that have resonance structures, such as benzene, in which the bonds are elongated by the resonance and have lower energy transitions than would be expected otherwise. Electromagnetic energy may also be absorbed by a molecule from the microwave and radio wave regions (see **Table 4–4**). Low-energy transitions involve the spin of electrons in the microwave region and the spin of nuclei in the radio wave region. The study of these transitions constitutes the fields of EPR and NMR spectroscopy. These various forms of molecular spectroscopy are discussed in the following sections.

## ULTRAVIOLET AND VISIBLE SPECTROPHOTOMETRY

Electromagnetic radiation in the ultraviolet (UV) and visible (Vis) regions of the spectrum fits the energy of electronic transitions of a wide variety of organic and inorganic molecules and ions. Absorbing species are usually classified according to the type of molecular energy levels involved in the electronic transition, which depends on the electronic bonding within the molecule.[11,12] Commonly, covalent bonding occurs as a result of a pair of electrons moving around the nuclei in a way that minimizes both internuclear and interelectronic Coulombic repulsions. Combinations of atomic orbitals (i.e., overlap) give rise to *molecular orbitals*, which are locations in space with an associated energy in which bonding electrons within a molecule can be found.

For instance, the molecular orbitals commonly associated with single covalent bonds in organic species are called sigma orbitals ($\sigma$), whereas a double bond is usually described as containing two types of molecular orbitals: one sigma and one pi ($\pi$). Molecular orbital energy levels usually follow the order shown in **Figure 4–6**.

Thus, when organic molecules are exposed to light in the UV–Vis regions of the spectrum (see **Table 4–4**), they absorb light of particular wavelengths depending on the type of electronic transition that is associated with the absorption. For example, hydrocarbons that contain $\sigma$-type bonds can undergo only $\sigma \rightarrow \sigma^*$ electronic transitions from their ground state. The asterisk (*) indicates the antibonding molecular orbital occupied by the electron in the excited state after absorption of a quantized amount of energy. These electronic transitions occur at short-wavelength radiation in the vacuum ultraviolet region (wavelengths typically between 100 and 150 nm). If a carbonyl group is present in a molecule, however, the oxygen atom of this functional group possesses a pair of nonbonding ($n$) electrons that can undergo $n \rightarrow \pi^*$ or $n \rightarrow \sigma^*$ electronic orbital transitions. These transitions require a lower energy than do $\sigma \rightarrow \sigma^*$ transitions and therefore occur from the absorption of longer wavelengths of radiation (see **Fig. 4–6**). For acetone, $n \rightarrow \pi^*$ and $n \rightarrow \sigma^*$ transitions occur at 280 and 190 nm, respectively. For aldehydes and ketones, the region of the ultraviolet spectrum between 270 and 290 nm is associated with their carbonyl

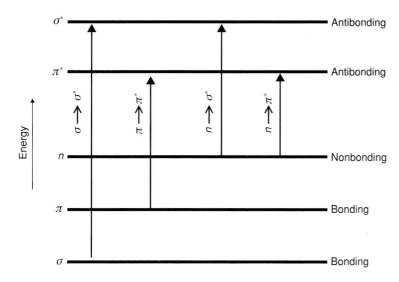

**Fig. 4–6.** Molecular orbital electronic transitions commonly found in ultraviolet–visible spectroscopy.

$n \rightarrow \pi^*$ electronic transitions, and this fact can be used for their identification. Thus, the types of electronic orbitals present in the ground state of the molecule dictate the region of the spectrum in which absorption can take place. Those parts of a molecule that can be directly associated with an absorption of ultraviolet or visible light, such as the carbonyl group, are called *chromophores*.

Most applications of absorption spectroscopy to organic molecules rely on transitions from $n$ or $\pi$ electrons to $\pi^*$ because the energies associated with these electronic transitions fall in an experimentally convenient region from 200 to 700 nm.

The amount of light absorbed by a sample is based on the measurement of the transmittance, $T$, or the absorbance, $A$, in transparent cuvettes or cells having a path length $b$ (in cm) according to equation (4–16):

$$A = -\log T = \log(I_0/I) = \varepsilon b C \qquad (4\text{–}16)$$

where $I_0$ is the intensity of the incident light beam and $I$ is in the intensity of light after it emerges from the sample.

Equation (4–16) is also known as *Beer's law*, and relates the amount of light absorbed ($A$) to the concentration of absorbing substance ($C$ in mole/liter), the length of the path of radiation passing through the sample ($b$ in cm), and the constant $\varepsilon$, known as the *molar absorptivity* for a particular absorbing species (in units of liter/mole cm). The molar absorptivity depends not only on the molecule whose absorbance is being determined but also on the type of solvent being used, the temperature, and the wavelength of light used for the analysis.

### EXAMPLE 4–3

**Absorbance Maxima Changes Due to Chemical Instability**

(a) A solution of $c = 2 \times 10^{-5}$ mole/liter of chlordiazepoxide dissolved in 0.1 N sodium hydroxide was placed in a fused silica cell having an optical path of 1 cm. The absorbance $A$ was found to be 0.648 at a wavelength of 260 nm. What is the molar absorptivity?
(b) If a solution of chlordiazepoxide had an absorbance of 0.298 in a 1-cm cell at 260 nm, what is its concentration?

(a)
$$\varepsilon = \frac{A}{bc} = \frac{6.48 \times 10^{-1}}{1 \times (2 \times 10^{-5})}$$
$$= 3.24 \times 10^4 \text{ liter/mole cm}$$

(b)
$$c = \frac{A}{b\varepsilon} = \frac{2.98 \times 10^{-1}}{1 \times (3.24 \times 10^4)} = 9.20 \times 10^{-6} \text{ mole/liter}$$

The large value of $\varepsilon$ indicates that chlordiazepoxide absorbs strongly at this wavelength. This molar absorptivity is characteristic of the drug dissolved in 0.1 N NaOH at this wavelength and is not the same as it would be in 0.1 N HCl. A lactam is formed from the drug under acid conditions that has an absorbance maximum at 245 rather than 260 nm and a correspondingly different $\varepsilon$ value. Large values for molar absorptivities are usually found in molecules having a high

degree of conjugation of chromophores, such as the conjugated double bonds in 1,3-butadiene, or the combination of carbonyl or carboxylic acids with double bonds as in $\alpha$, $\beta$ unsaturated ketones. In particular, highly conjugated molecules such as aromatic hydrocarbons and heterocycles typically display absorption bands characteristic of $\pi \rightarrow \pi^*$ transitions and have significantly higher $\varepsilon$ values.

### EXAMPLE 4–4

**Absorption Maxima Calculation**

Aminacrine is an anti-infective agent with the following molecular structure:

Its highly conjugated acridine ring produces a complex ultraviolet spectrum in dilute sulfuric acid that includes absorption maxima at 260, 313, 326, 381, 400, and 422 nm. The molar absorptivities of the absorbances at 260 and 313 nm are 63,900 and 1130 liter/mole cm, respectively. What is the minimum amount of aminacrine that can be detected at each of these two wavelengths?

If we assume an absorbance level of $A = 0.002$ corresponding to a minimum detectable concentration of the drug, then, at 260 nm,

$$c = \frac{A}{b \times \varepsilon} = \frac{0.002}{1 \times 63900} = 3.13 \times 10^{-8} \text{ mole/liter}$$

and at 381 nm

$$c = \frac{0.002}{1 \times 1130} = 1.77 \times 10^{-6} \text{ mole/liter}$$

Nearly 100 times greater sensitivity in detecting the drug is possible with the 260-nm absorption band, which is reflective of a higher number of electronic transitions that can occur at the higher energy. The absorbance level of 0.002 was chosen by judging this value to be a significant signal above instrumental noise (i.e., interference generated by the spectrophotometer in the absence of the drug). For a particular analysis, the minimum absorbance level for detection of a compound depends on both the instrumental conditions and the sample state, for example, the solvent chosen for dissolution of the sample. Instrumental conditions vary across spectrophotometers, and will be discussed later. With regard to the sample conditions, solvents may have dramatically different polarities, which are addressed in the Dielectric Constant and Induced Polarization section. The polarity of a solvent may influence the bonding state of a molecule by changing the attractive and repulsive forces in the molecule through solvation. For example, a polar compound dissolved in water will have different capacities to undergo electronic transitions than if it were dissolved in a nonpolar, aprotic solution.

As illustrated in *Example 4–4*, a molecule may have more than one characteristic absorption wavelength band, and the complete UV–Vis wavelength absorption spectrum can provide information for the positive identification of a compound. Each functional group has different associated molecular bonding and electronic structure, which will have different wavelengths that maximally absorb energy. Therefore, the resulting spectrum for a molecule can be used as a "molecular fingerprint" to determine the molecular structure. Changes in the order of bonding or in electronic structure will

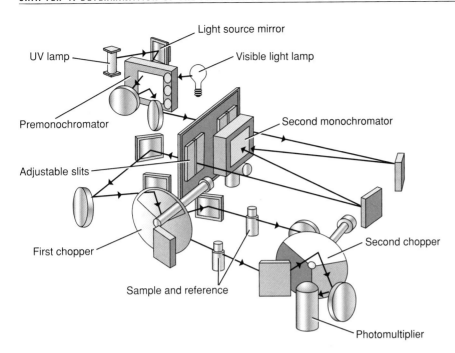

**Fig. 4–7.** A double-beam ultraviolet–visible spectrophotometer. (Courtesy of Varian, Inc., Palo Alto, CA.)

alter the UV–Vis spectrum, as illustrated with chlordiazepoxide and its lactam. As a result, many different molecular properties can be monitored by using a UV–Vis spectrophotometer including chemical reactions, complexation, and degradation, making it a powerful tool for pharmaceutical scientists.

Absorption spectroscopy is one of the most widely used methods for quantitative analysis due to its wide applicability to both organic and inorganic systems, moderate to high sensitivity and selectivity, and good accuracy and convenience. Modern spectroscopic instrumentation for measuring the complete molecular UV–Vis absorption spectra is commonplace in both industrial and academic settings. Two main types of spectrophotometers, usually coupled to personal computers for data analysis, are commercially available: double-beam and diode-array instruments.

A schematic diagram of a traditional double-beam UV–Vis spectrophotometer is shown in **Figure 4–7**. The beam of light from the source, usually a deuterium lamp, passes through a prism or grating monochromator to sort the light according to wavelength and spread the wavelengths over a wide range. This permits a particular wavelength region to be easily selected by passing it through the appropriate slits. The selected light is then split into two separate beams by a rotating mirror, or "chopper," with one beam passed through the reference, which is typically the blank solvent used to dissolve the sample, and the other through the sample cell containing the test molecule. After each beam passes through its respective cell, it is reflected onto a second mirror in another chopper assembly, which alternatively selects either the reference or the combined beams to focus onto the photomultiplier detector. The rapidly changing current signal from the detector is proportional to the intensity of the particular beam, and this is fed into an amplifier, which elec-

tronically separates the signals of the reference beam from those of the sample beam. The final difference in beam signals is automatically recorded. In addition, it is common practice to first put the solvent in the sample cell and set the spectrophotometer's absorbance to zero to serve as the baseline reference. The samples can then be placed in the sample cell and measured, with the difference from the baseline being reported. The sample data are reported as a plot of the intensity, usually as absorbance, against the wavelength, as shown for the chlordiazepam lactam in **Figure 4–8**.

**Figure 4–9** is an illustration of a typical diode-array spectrophotometer. Note that these instruments have simpler optical components, and, as a result, the radiation throughput is much higher than in traditional double-beam instruments. After the light beam passes the sample, the radiation is focused on an entrance slit and directed to a grating. The transducer is a diode array from which resolutions of 0.5 nm can be reached. A single scan from 200 to 1100 nm takes only 0.1 sec, which leads to a significant improvement in the signal-to-noise ratio by accumulating multiple scans in a short time.

Typically, a calibration curve from a series of standard solutions of known but varying concentration is used to generate standard curves for quantitative analysis. An absorbance spectrum can be used to determine one wavelength, typically an absorption maximum, where the absorbance of each sample can be efficiently measured. The absorbance for standards is measured at this point and plotted against the concentration, as shown in **Figure 4–10**, to obtain what is known as a Beer's-law plot. The concentration of an "unknown" sample can then be determined by interpolation from such a graph.

Spectrophotometry is a useful tool for studying chemical equilibria or determining the rate of chemical reactions. The chemical species participating in the equilibria must have

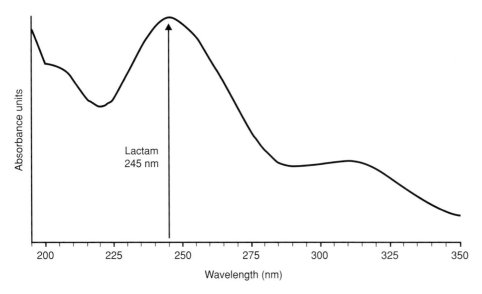

**Fig. 4–8.** The absorbance maxima of chlordiazepoxide lactam as a function of the wavelength. (Modified from http://chrom.tutms.tut.ac.jp/JINNO/DRUGDATA/17chlordia-zepoxide.html)

different absorption spectra due to changes in the electronic structure in the associated or dissociated states (Chapter 8), which influence the bonding structure. This change results in the variation in absorption at a representative wavelength for each species while the pH or other equilibrium variable is changed. If one determines the concentrations of the species from Beer's law and knows the pH of the solution, one can calculate an approximate $pK_a$ for a drug. For example, if the drug is a free acid (HA) in equilibrium with its base ($A^-$), then the $pK_a$ is defined by

$$pK_a = pH + \log\frac{[HA]}{[A^-]} \qquad (4\text{–}17)$$

When $[HA] = [A^-]$, as determined by their respective absorbances in the spectrophotometric determination, $pK_a \cong pH$.

### EXAMPLE 4–5

**Ionization State and Absorption**

Phenobarbital shows a maximum absorption of 240 nm as the monosodium salt ($A^-$), whereas the free acid (HA) shows no absorp-tion maxima in the wavelength region from 230 to 290 nm. If the free acid in water is slowly titrated with known volumes of dilute NaOH and the pH of the solution and the absorbance at 240 nm are measured after each titration, one reaches a maximum absorbance value at pH 10 after the addition of 10 mL of titrant. How can $pK_a$ be determined from this titration?

By plotting the absorbance against the pH over the titration range to pH = 10, one can obtain the midpoint in absorbance, where half the free acid has been titrated, and $[HA] = [A^-]$ (Fig. 4–8). The pH corresponding to this absorbance midpoint is approximately equal to the $pK_a$, namely, $pK_{a1}$ for the first ionization stage of phenobarbital. This midpoint occurs at a pH of 7.3; therefore, $pK_a \cong 7.3$. For more accurate $pK_a$ determinations, refer to the discussion in Chapter 8.

Reaction rates can also be measured when a particular reaction species has an absorption spectrum that is noticeably

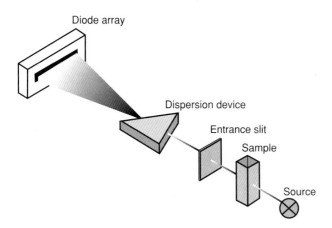

**Fig. 4–9.** Schematic of an HP-8453 diode-array ultraviolet–visible spectrophotometer. (Courtesy of Hewlett-Packard, Palo Alto, CA.)

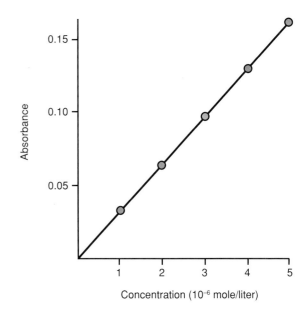

**Fig. 4–10.** A Beer's-law plot of absorbance against the concentration of chlordiazepoxide.

different from the spectra of other reactants or products. One can follow the rate of appearance or disappearance of the selected species by recording its absorbance at specific times during the reaction process. If no other reaction species absorbs at the particular wavelength chosen for this determination, the reaction rate will simply be proportional to the rate of change of absorbance with reaction time. Standard curves can be generated if the selected reactant is available in a pure form. It must be noted that the assumption that the other species or reaction intermediates do not interfere with the selected reactant can also be flawed, as discussed later. An example of the use of spectrophotometry for the determination of reaction rates in pharmaceutics is given by Jivani and Stella,[13] who used the disappearance of *para*-aminosalicylic acid from solution to determine its rate of decarboxylation.

Spectrophotometry can be used to study enzyme reactions and to evaluate the effects of drugs on enzymes. For example, the analysis of clavulanic acid can be accomplished by measuring the ultraviolet absorption of penicillin G at 240 nm, as described by Gutman et al.[14] Clavulanic acid inhibits the activity of $\beta$-lactamase enzymes, which convert penicillin G to penicilloic acid, where R is $C_6H_5CH_2-$.

The method first requires that the rate of absorbance change at 240 nm be measured with a solution containing penicillin G and a $\beta$-lactamase enzyme. Duplicate experiments are then performed with increasing standard concentrations of clavulanic acid. These show a decrease in absorbance change, equivalent to the enzyme inhibition from the drug, as the concentration of the drug increases. The concentration of an unknown amount of clavulanic acid is measured by comparing its rate of enzyme inhibition with that of the standards.

The primary weakness with a spectrophotometer used in such a manner is that it measures all of the species in a sample even though a single wavelength might be selected for its molecular detection specificity. UV–Vis spectra generated at a given wavelength(s) cannot properly detect changes in the species that arise during a reaction. For example, a reaction intermediate(s) may be generated that has stronger absorption at the wavelength selected than the reactant or the product being measured. Finally, there are many different reaction pathways that include multiple species with overlapping absorption, so the selection of a single wavelength for detection does not provide good selectivity. One way to correct this is to generate a UV–Vis absorption scan across a wide wavelength range and measure changes in several specific absorption maxima. However, the same problems can obfuscate the data in this approach as well. Therefore, in many cases, the complex reaction milieu is not readily adaptable to a single spectroscopic method. Thus, an ability to separate the individual species and quantitate their levels would provide enhanced sensitivity. UV–Vis spectrophotometers are also used in conjunction with many other methods for detecting molecules, for example, high-performance liquid chromatography (HPLC), which eliminates species interference by separating the compounds before detection occurs.

The major use of spectrophotometry is in the field of quantitative analysis, in which the absorbance of chromophores is determined. Various applications of spectrophotometry are discussed by Schulman and Vogt.[15]

# FLUORESCENCE AND PHOSPHORESCENCE

Luminescence is an emission of radiation in the ultraviolet, visible, or near-IR regions from electronically excited species. An electron in an atom or a molecule can be excited by means of absorbing energy, for instance a photon of light, to reach an electronic excited state (see prior discussion of orbital transitions). The excited state is relatively short-lived, and the electron can return to its ground state via radiative and nonradiative energy emission. If the preferred path to return to the ground state involves releasing energy through internal conversions by changes in vibrational states or through collisions with the environment (e.g., solvent molecules), then the molecule will not display luminescence. Many chemical species, however, emit radiation when returning to the ground state either as fluorescence or as phosphorescence, depending on the mechanism by which the electron finally returns to the ground state.

**Figure 4–11** shows a simplified energy diagram (also known as a Perrin–Jablonsky diagram) of the typical mechanisms that a chemical species undergoes after being electronically excited. The absorption of a photon usually excites an electron from the ground state toward its excited states without changing its spin (i.e., a singlet ground state will absorb into a singlet excited state, called a spin-allowed transition).

The triplet state usually cannot be achieved by excitation from the ground state, this being termed a "forbidden" transition according to quantum theory. It is usually reached through the process of *intersystem crossing* (ISC), which is a nonradiative transition between two isoenergetic vibrational levels belonging to electronic states of different multiplicities. For example, the excited singlet ($S_1$) in the 0 vibrational mode can move to the isoenergetic vibrational level of $T_n$ triplet state, then vibrational relaxation can bring it to the lowest vibrational level of $T_1$, with the concomitant energy loss (see **Fig. 4–11**). From $T_1$, radiative emission to $S_0$ can occur, which is called phosphorescence.

The excited triplet state ($T_1$) is usually considered more stable (i.e., having a longer lifetime) than the excited singlet state ($S_1$). The length of time during which light will be emitted after the molecule has become excited depends on the lifetime of the electronic transition. Therefore, we can expect phosphorescence to occur for a longer period after excitation than after fluorescence. Ordinarily, fluorescence occurs between $10^{-10}$ and $10^{-7}$ sec after excitation, whereas the lifetime for phosphorescence is between $10^{-7}$ and 1 sec. Because of its short lifetime, fluorescence is usually measured while the molecule is being excited. Phosphorescence uses a pulsed excitation source to allow enough time to detect the emission. It should be noted that measurements of

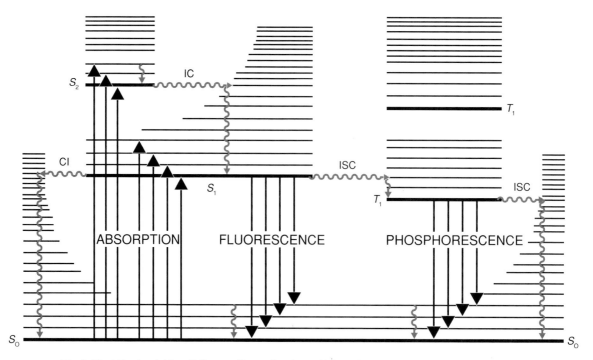

**Fig. 4–11.** A Perrin–Jablonski diagram illustrating the multiple pathways by which absorbed molecular energy can be released. Relaxation can occur through a nonradiative pathway or radiative emission of fluorescence or phosphorescence. The singlet (*S*) and triplet (*T*) states pertain to the electronic structure in the molecule after excitation.

fluorescence lifetimes on the order of femtoseconds have been demonstrated to be valuable in studying transition states, as described by Zewail.[16]

Fluorescence normally has a longer wavelength than the radiation used for the excitation phase, principally because of internal energy losses within the excited molecule before the fluorescent emission $S_1 \to S_0$ occurs (**Fig. 4–11**). The gap between the first absorption band and the maximum fluorescence is called the Stokes shift. Phosphorescence typically has even longer wavelengths than fluorescence owing to the energy difference that occurs in ISC as well as the loss of energy due to internal conversion over a longer lifetime.

A representative schematic of a typical fluorometer is shown in **Figure 4–12**. Generally, fluorescence intensity is measured in these instruments by placing the photomultiplier detector at right angles to the incident light beam that produces the excitation. The signal intensity is recorded as relative fluorescence against a standard solution. Because photoluminescence can occur in any direction from the sample, the detector will sense a part of the total emission at a characteristic wavelength and will not be capable of detecting radiation from the light beam used for excitation.

Fluorometry is a very sensitive technique, up to 1000 times more sensitive than spectrophotometry. This is because the fluorescence intensity is measured above a low background level, whereas in measuring low absorbances, two large signals that are slightly different are compared. Recently, advances in instrumentation have made it possible to detect fluorescence at a single-molecule level.[17]

Photoluminescence occurs only in those molecules that can undergo the specified photon emissions after excitation with consequent return to the ground state. Many molecules do not possess any photoluminescence, although they can largely absorb light. Most often, molecules that display fluorescence or phosphorescence contain a rigid conjugated structure such as aromatic hydrocarbons, rhodamines, coumarins, oxines, polyenes, and so on, or inorganic lanthanides ions like $Eu^{3+}$ or $Tb^{3+}$, which also show strong fluorescence. Many drugs (e.g., morphine), some natural amino acids and cofactors (e.g., tyrosine, tryptophan, nicotinamide adenine dinucleotide, reduced form flavin adenine dinucleotide, etc.) fluoresce. Some examples are given in **Table 4–6** along with their characteristic excitation and emission wavelengths, which can be used for qualitative and quantitative analyses.

Fluorescence detection and analysis has become a major tool in biopharmaceutical and chemical analysis, particularly in areas related to manipulation of biopolymers (nucleic acids and proteins) and imaging of biologic membranes and living organisms (cells, bacteria, etc.). For example, in peptides and proteins the UV absorbance for the $n \to \pi^*$ ($\varepsilon = 100$) and $\pi \to \pi^*$ ($\varepsilon = 7000$) transitions in the peptide bond occurs from about 210 to 220 and 190 nm, respectively.[12] The absorbance of several side-chain transitions can often be obfuscated absorbance from peptide bonds as well as other species present in the solution, and therefore does not offer good specificity. Tryptophan (280 nm), tyrosine (274), phenylalanine (257 nm), and cystine (250 nm) have electronic

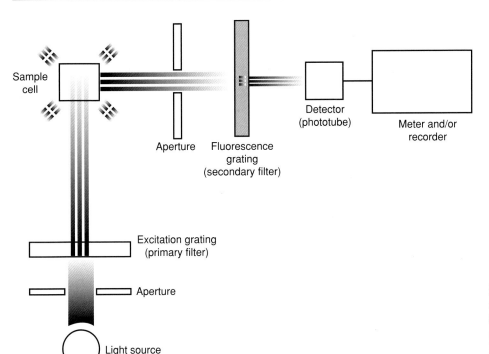

Fig. 4–12. Schematic diagram of a filter fluorometer. (From G. H. Schenk, *Absorption of Light and Ultraviolet Radiation*, Allyn and Bacon, Boston, 1973, p. 260. With permission.)

transitions that occur above the 220-nm range for the peptide bond, but these transitions can be obfuscated as well.[12] However, tryptophan, phenylalanine, and tyrosine can also undergo fluorescence that can be used to discriminate peptides or proteins from biologic matrices. The use of other dyes and fluorophores for biologic applications is also an exciting and rapidly developing area. The student is advised to seek out Web sites of companies such as Molecular Probes (www.probes.com) to look at the wide scope and use of fluorescent labels and probes as well as modern instrumentation in the field. Schulman and Sturgeon[18] give a thorough review of the applications of photoluminescence to the analysis of traditional pharmaceuticals.

# INFRARED SPECTROSCOPY

The study of the interaction of electromagnetic radiation with vibrational or rotational *resonances* (i.e., the harmonic oscillations associated with the stretching or bending of the bond) within a molecular structure is termed *infrared spectroscopy*. Normally, infrared radiation in the region from about 2.5 to 50 $\mu$m, equivalent to 4000 to 200 cm$^{-1}$ in wave number, is used in commercial spectrometers to determine most of the important vibration or vibration–rotation transitions. The individual masses of the vibrating or rotating atoms or functional groups, as well as the bond strength and molecular symmetry, determine the frequency (and, therefore, also the

## TABLE 4–6
### FLUORESCENCE OF SOME DRUGS

| Drug | Excitation Wavelength (nm) | Emission Wavelength (nm) | Solvent |
|---|---|---|---|
| Phenobarbital | 255 | 410–420 | 0.1 N NaOH |
| Hydroflumethiazide | 333 | 393 | 1 N HCl |
| Quinine | 350 | ~450 | 0.1 N $H_2SO_4$ |
| Thiamine | 365 | ~440 | Isobutanol, after oxidation with ferricyanide |
| Aspirin | 280 | 335 | 1% acetic acid in chloroform |
| Tetracycline hydrochloride | 330 | 450 | 0.05 N NaOH(aq) |
| Fluorescein | 493.5 | 514 | Water (pH 2) |
| Riboflavin | 455 | 520 | Ethanol |
| Hydralazine | 320 | 353 | Concentrated $H_2SO_4$ |

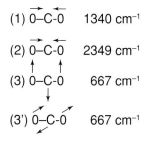

(1) O–C–O    1340 cm⁻¹

(2) O–C–O    2349 cm⁻¹

(3) O–C–O    667 cm⁻¹

(3') O–C–O    667 cm⁻¹

**Fig. 4–13.** The normal vibrational modes of $CO_2$ and their respective wave numbers, showing the directions of motion in reaching the extreme of the harmonic cycle. (Modified from W. S. Brey, *Physical Chemistry and Its Biological Applications*, Academic Press, New York, 1978, p. 316.)

wavelength) of the infrared absorption. The absorption of infrared radiation occurs only if the permanent dipole moment of the molecule changes with a vibrational or rotational resonance. The molecular symmetry relates directly to the permanent dipole moment, as already discussed. Bond stretching or bending may affect this symmetry, thereby shifting the dipole moment as found for the normal vibrational modes (2), (3), and (3') for $CO_2$ in **Figure 4–13**. Other resonances, such as (1) for $CO_2$ in **Figure 4–13**, do not affect the dipole moment and therefore do not produce infrared absorption. Resonances that *shift* the dipole moments can give rise to infrared absorption by molecules, even those considered to have no permanent dipole moment, such as benzene or $CO_2$. The frequencies of infrared absorption bands correspond closely to vibrations from particular parts of the molecule. The bending and stretching vibrations for acetaldehyde, together with the associated infrared frequencies of absorption, are shown in **Figure 4–14**. In addition to the fundamental absorption bands shown in this figure, each of which corresponds to a vibration or vibration–rotation resonance and a change in the dipole moment, weaker overtone bands may be observed for multiples of each of these frequencies (in wave numbers). For example, an overtone band may appear for acetaldehyde at 3460 cm⁻¹, which corresponds to twice the frequency (2 × 1730 cm⁻¹) for the carbonyl stretching band. Because the frequencies are simply associated with harmonic motion of the radiant energy, the overtones may be thought of as simple multiples that are exactly in phase with the fundamental frequency and can therefore "fit" into the same resonant vibration within the molecule.

Because the vibrational resonances of a complex molecule often can be attributed to particular bonds or groups, they behave as though they result from vibrations in a diatomic molecule. This means that vibrations produced by similar bonds and atoms are associated with infrared bands over a small frequency range even though these vibrations may occur in completely different molecules. A typical infrared spectrum of theophylline is shown in **Figure 4–15**. The spectrum "fingerprints" the drug and provides one method of verifying compounds. The individual bands can be associated

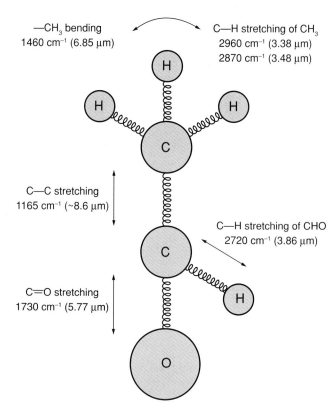

**Fig. 4–14.** Bending and stretching frequencies for acetaldehyde. (From H. H. Willard, L. L. Merritt, Jr., and J. A. Dean, *Instrumental Methods of Analysis*, 4th Ed., Van Nostrand, New York, 1968. With permission.)

with particular groups. For example, the band at 1660 cm⁻¹, *a* in **Figure 4–15**, is due to a carbonyl stretching vibration for theophylline.

Infrared spectra can be complex, and characteristic frequencies vary depending on the physical state of the molecule

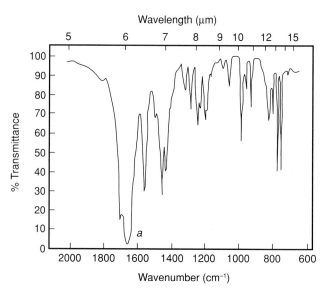

**Fig. 4–15.** Infrared spectrum of theophylline. (From E. G. C. Clarke, Ed., *Isolation and Identification of Drugs*, Pharmaceutical Press, London, 1969. With permission.)

being examined. For example, hydrogen bonding between sample molecules may change the spectra. For alcohols in dilute carbon tetrachloride solution, there is little intermolecular hydrogen bonding, and the hydroxyl stretching vibration occurs at about 3600 cm$^{-1}$. The precise position and shape of the infrared band associated with the hydroxyl group depend on the concentration of the alcohol and the degree of hydrogen bonding. Steric effects, the size and relative charge of neighboring groups, and phase changes can affect similar frequency shifts.

The use of infrared spectroscopy in pharmacy has centered on its applications for drug identification, as described by Chapman and Moss.[19] The development of Fourier transform infrared spectrometry[20] has enhanced infrared applications for both qualitative and quantitative analysis of drugs owing to the greater sensitivity and the enhanced ability to analyze aqueous samples with Fourier transform infrared instrumentation. Thorough surveys of the techniques and applications of infrared spectroscopy are provided by Smith[21] and Willard et al.[22]

## NEAR-INFRARED SPECTROSCOPY

Near-IR spectroscopy is rapidly becoming a valuable technique for analyzing pharmaceutical compounds. The near-IR region ranges from approximately 10,000 to 4000 cm$^{-1}$ and comprises mainly of mid-IR overtones and combinations of these overtones arising from heavy-atom electronic transitions, including O—H, C—H, and N—H stretching. Thus, near-IR spectroscopy is typically used for analyzing water, alcohols, and amines. In addition, near-IR bands arise from forbidden transitions; therefore, the bands are far less intense than the typical absorption bands for these atoms.[23]

Some advantages of using near-IR spectroscopy over other techniques include extremely fast analysis times, nondestructive analysis, lack of a need for sample preparation, qualitative and quantitative results, and the possibility of multicomponent analysis. Pharmaceutical applications include polymorph identification, water content analysis, and identification and/or monitoring of an active component within a tablet or other solid dosage form. Because of the nondestructive nature of the technique, online analysis for controlling large-scale processes is possible.

## ELECTRON PARAMAGNETIC AND NUCLEAR MAGNETIC RESONANCE SPECTROSCOPY

Now we enter into the description of two widely used spectroscopic techniques in the pharmaceutical sciences and related fields of chemistry, biology, and medicine. Electron paramagnetic resonance and NMR are based on the magnetic behavior of atomic particles such as electrons and nuclei in the presence of an external magnetic field. In the following,

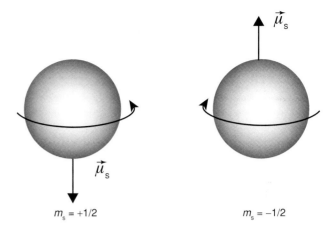

**Fig. 4–16.** Electron and nuclear rotation.

fundamentals of the techniques will be given, including differences between electrons and nuclei where applicable.

Charged particles such as electrons (negatively charged) and protons (positively charged) behave as if they were charged, spinning tops. Accordingly, they have magnetic moments associated with the movement around their axes, as shown in **Figure 4–16** (i.e., they behave as small magnets). In paired electrons the spins are opposite, so the magnetic moments cancel each other out. However, single unpaired electrons have a magnetic moment (Bohr magneton, $\mu_{\text{Bohr}}$) equal to

$$\mu_{\text{Bohr}} = \frac{eh}{4\pi m_{\text{e}}} = 9.274 \times 10^{-24} \text{ joule/T} \quad \textbf{(4–18)}$$

where $e$ is the elementary charge, $h$ is Planck's constant, and $m_{\text{e}}$ is the electron's mass.[*]

Similarly, nuclear motion in protons and certain nuclei also has an associated magnetic moment, whose magnitude can be calculated from an equation similar to **(4–18)** by replacing the mass of the particle; protons have a larger mass than electrons, so that the nuclear magneton is $\mu_{\text{N}} = 5.051 \times 10^{-27}$ joule/T.

When the electron (or the proton) is placed in an external magnetic field, its magnetic moment is oriented in two different directions relative to the magnetic field, $B$, parallel or antiparallel, giving rise to two new energy levels splitting from each other by a magnitude $\Delta E$, which depends on the applied field $B$ according to

$$\Delta E = g\,\mu_{\text{Bohr}}\,B \quad \textbf{(4–19)}$$

$g$ is termed the Landé *splitting factor*, or simply the $g$ factor. For organic free radicals, $g$ is nearly equal to its value for a free electron, 2.0023, which is the ratio of the electron's *spin* magnetic moment to its *orbital* magnetic moment. The $g$ *factor* is also characteristic for certain metal complexes

---

[*]The tesla, T, is the unit of magnetic flux density; 1 T induces a voltage of 1 V in a 1-m-long conductor that is moving at 1 m/sec. The tesla is equivalent to $10^4$ gauss, the unit that the tesla replaced.

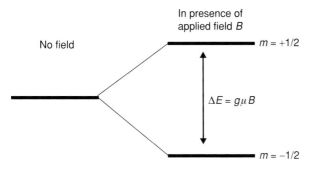

**Fig. 4–17.** Electron and nuclear spin splitting when subjected to an applied external magnetic field.

with unpaired electrons, from which information about their environment can be obtained.

For an electron placed in a magnetic field of $B = 1$ T, $\Delta E = 18.54 \times 10^{-24}$ joule; according to equation (**4–19**). Therefore, a simple calculation of the frequency that matches this energy difference, $v = \Delta E/h$, indicates that the electron can be excited with a radiation of 28 GHz, corresponding to the microwave region of the spectrum (see **Table 4–4**). A similar calculation for protons using

$$\Delta E = g_{N} \, \mu_{N} \, B \qquad (4\text{–}20)$$

with $g_{N} = 5.586$, affords an energy splitting of $2.82 \times 10^{-26}$ joule, corresponding to an electromagnetic frequency of 42.58 MHz, also in the microwave region.

Transitions between electron spin energy levels give rise to the phenomenon of EPR, as indicated in **Figure 4–17**; the corresponding property of the nuclear spin levels leads to NMR. The term "resonance" relates to the classical interpretation of the phenomenon, because transitions occur only when the frequency $v$ of the electromagnetic radiation matches the Larmor frequency of precession around the axis of the applied magnetic field, $v_{L}$, as displayed in **Figure 4–18**.

So far we have considered only electrons and protons, particles with spin $\frac{1}{2}$, which have only two possible energy levels. Several nuclei also have magnetic moments and, therefore, NMR signals. Nuclei with odd mass numbers have total spin of half-integral value, whereas nuclei with odd atomic numbers and even mass numbers have total spin of integral value. Examples include $^{13}$C, $^{19}$F, and $^{31}$P, which have spin $\frac{1}{2}$, and $^{2}$H and $^{14}$N, which have spin 1.

By considering equations (**4–19**) and (**4–20**), we can obtain the resonance signal by gradually varying the magnetic field strength, $B$, while keeping the radio frequency, $v$, constant. At some particular $B$ value, a spin transition will take place that flips the electron or nuclei from one spin state to another (e.g., from $I = -\frac{1}{2}$ to $+\frac{1}{2}$, as in **Figure 4–18**). In some spectrometers, the experiment can be done in the reverse fashion: keeping $B$ constant and varying $v$. This method is generally called *continuous-wave or field scan*, and the particular value of $v$ depends linearly on $B$, as illustrated in **Figure 4–19**.

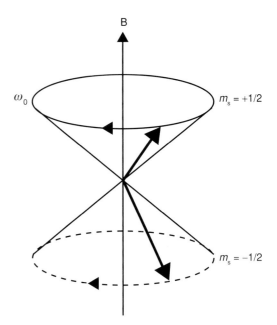

**Fig. 4–18.** Electronic or nuclear spin procession around an applied magnetic field.

In most modern spectrometers, particularly for NMR, all spins of one species in the sample are excited simultaneously by using a short, high-frequency pulse of electromagnetic radiation. Then their relaxation is measured as a function of time and transformed from the time domain to the frequency domain via fast Fourier transformation techniques. The signal-to-noise ratio can be significantly improved by the accumulation of a large number of measurements.

As we have seen, the theoretical basis of EPR spectroscopy is similar to that of NMR except that it is restricted to paramagnetic species (i.e., with unpaired electrons). Among thousands of types of substances constituting biologic systems with pharmacologic relevance, only very few, namely,

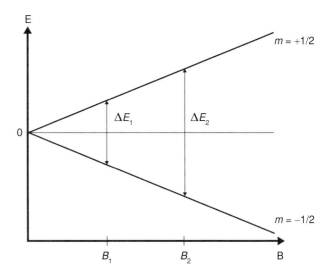

**Fig. 4–19.** Energy differences observed in electronic or nuclear spin processions as a function of magnetic field strengths.

free radicals, including molecular oxygen ($O_2$) and nitric oxide (NO), and a limited number of transition metal ions reveal paramagnetic properties and may be observed by EPR spectroscopy. Nevertheless, EPR has many potential uses in biochemistry and medicine by the intentional introduction into the system of exogenous EPR-detectable paramagnetic probes (spin labels). This method often eliminates the necessity to purify biologic samples. Thus, EPR spectroscopy applied to biologic tissues and fluids can identify the changes in redox processes that contribute to disease. Electron paramagnetic resonance may also be used to characterize certain plant-derived products as potentially important to biotechnology by increasing the level of free radicals and other reactive species produced during light-induced oxidative stress of the cell. Several other applications can be found in references at the end of the chapter.[24-26]

Nuclear magnetic resonance is a powerful experimental method in chemistry and biology because the resonance frequency of a nucleus within a molecule depends on the chemical environment. The local field experienced by an atom in a molecule is less than the external magnetic field by an amount given by

$$B_{effective} = B(1 - \sigma) \qquad (4\text{--}21)$$

where $\sigma$ is the shielding constant for a particular atom (i.e., a measure of its susceptibility to induction from a magnetic field) and $B$ is the external magnetic field strength. Instead of measuring the absolute value of $\sigma$, it is customary to measure the difference of $\sigma$ relative to a reference.

### EXAMPLE 4–6

**Nuclear Spin Energy**

(a) What is the energy change associated with the $^1H$ nuclear spin of chloroform, which has a shielding constant, $\sigma$, of $-7.25 \times 10^{-6}$ and a nuclear magnetic moment, $\mu_N$, of $5.050824 \times 10^{-27}$ joule/T in a magnetic field, $B$, of 1 T?

We can obtain $\Delta E = g_N \mu_N B_{local}$ for this particular nucleus, in which the corresponding local field experienced by the atom is given by equation (4–21). The local magnetic field $B_{local}$ is replaced, and we obtain

$$\Delta E = g_N \mu_N B(1-\sigma) = (5.585691)(5.050824 \times 10^{-27}\, joule/T)$$
$$\times (1\,T)(1 + 7.25 \times 10^{-6}) = 2.821254 \times 10^{-26}\, joule$$

(b) What is the radio frequency at which resonance will occur for this nuclear spin transition under the stated conditions?

$$\nu = \Delta E/h = (2.821254 \times 10^{-26}\, joule/6.626196$$
$$\times 10^{-34}\, joule\text{-}sec) = 4.2577289 \times 10^7\, Hz$$

Tetramethylsilane (TMS) is often used as a reference compound in proton NMR because the resonance frequency of its one proton signal from its four identical methyl groups is below that for most other compounds. In addition, TMS is relatively stable and inert.

### EXAMPLE 4–7

**Resonance of Tetramethylsilane**

What is the radio frequency at which resonance occurs for TMS in a magnetic field of 1 T?

The shielding constant $\sigma$ is 0.000, and $\Delta E = 2.821243 \times 10^{-26}$ joule for TMS, so

$$\nu = \Delta E/h = (2.82123 \times 10^{-26}\, joule/6.626196$$
$$\times 10^{-34}\, joule\text{-}sec) = 4.2576981 \times 10^7\, Hz$$

From the examples just given, it is important to note that the difference in the resonance frequency between chloroform and TMS at a constant magnetic field strength is only 308 Hz or waves/sec. This small radio frequency difference is equivalent to more than half the total range within which $^1H$ NMR signals are detected.

The value of the shielding constant, $\sigma$, for a particular nucleus will depend on local magnetic fields, including those produced by nearby electrons within the molecule. This effect is promoted by placing the molecule within a large external magnetic field, $B$. Greater shielding will occur with higher electron density near a particular nucleus, and this reduces the frequency at which resonance takes place. Thus, for TMS, the high electron density from the Si atom produces enhanced shielding and, therefore, a lower resonance frequency. The relative difference between a particular NMR signal and a reference signal (usually from TMS for proton NMR) is termed the *chemical shift*, $\delta$, given in parts per million (ppm). It is defined as

$$\delta = (\sigma_r - \sigma_s) \times 10^6 \qquad (4\text{--}22)$$

where $\sigma_r$ and $\sigma_s$ are the shielding constants for the reference and the sample nucleus, respectively.

If the separation between the sample and reference resonance is $\Delta H$ or $\Delta \nu$, then

$$\delta = \frac{\Delta H}{H_R} \times 10^6 = \frac{\Delta \nu}{\nu_R} \times 10^6 \qquad (4\text{--}23)$$

where $H_R$ and $\nu_R$ are the magnetic field strength and radio frequency for the nuclei, depending on whether field-swept or frequency-swept NMR is used.

### EXAMPLE 4–8

**Proton Shift Reference**

What is the chemical shift of the chloroform proton using TMS as a reference?

Substituting the frequencies obtained from *Examples 4–6b* and *4–7* into equation (4–23), we obtain the chemical shift:

$$\delta = \Delta \nu \times 10^6 / \nu_R = 308 \times 10^6 / 42.57 \times 10^6 = 7.23\, ppm$$

This is an approximate value owing to the relative accuracy used to determine each frequency in the example. The accepted experimental value for this chemical shift is 7.24 ppm. The chemical shift of a nucleus provides information about its local magnetic environment and therefore can "type" a nuclear species. **Table 4–7** lists some representative proton chemical shifts. **Figure 4–20** shows a proton NMR spectrum for benzyl acetate, $CH_3COOCH_2C_6H_5$, using TMS as a reference. Notice that each signal band represents a particular

### TABLE 4–7
### PROTON CHEMICAL SHIFTS FOR REPRESENTATIVE CHEMICAL GROUPS OR COMPOUNDS

| Compound or Group | ppm |
| --- | --- |
| Tetramethylsilane (TMS), $(CH_3)_4Si$ | 0.00 |
| Methane | 0.23 |
| Cyclohexane | 1.44 |
| Acetone | 2.08 |
| Methyl chloride | 3.06 |
| Chloroform | 7.25 |
| Benzene | 7.27 |
| Ethylene | 5.28 |
| Acetylene | 1.80 |
| R—OH (hydrogen bonded) | 0.5–5.0 |
| $R_2$—NH | 1.2–2.1 |
| Carboxylic acids (R—COOH) | 10–13 |
| $H_2O$ | ~4.7 |

type of proton, that is, the proton at $\delta = 2.0$ is from the $CH_3$ group, whereas that at 5.0 is due to $CH_2$, and the protons at about 7.3 are from the aromatic protons. The *integral* curve above the spectrum is the sum of the respective band areas, and its stepwise height is proportional to the number of protons represented by each band.

In **Figure 4–20**, the signal bands are of simple shape, with little apparent complexity. Such sharp single bands are known as *singlets* in NMR terminology. In most NMR spectra, the

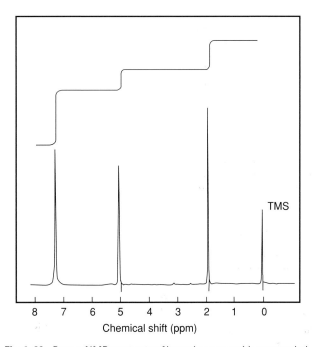

**Fig. 4–20.** Proton NMR spectrum of benzyl acetate with tetramethylsilane (TMS) as a reference. The TMS band appears at the far right. The upper curve is an integration of the spectral bands, the height of each step being proportional to the area under that band. (From W. S. Brey, *Physical Chemistry and Its Biological Applications*, Academic Press, New York, 1978, p. 498. With permission.)

bands are not as simple because each particular nucleus can be coupled by spin interactions to neighboring nuclei. If these neighboring nuclei are in different local magnetic environments, splitting of the bands can occur because of differences in electron densities. This leads to *multiplet* patterns with several lines for a single resonant nucleus. The pattern of splitting in the multiplet can provide valuable information concerning the nature of the neighboring nuclei.

**Figure 4–21** shows the proton NMR spectrum of acetaldehyde, $CH_3CHO$, in which the doublet of the $CH_3$ group (right side of the figure) is produced from coupling to the neighboring single proton, and the quartet of the lone proton in the CHO group (left side of the figure) is produced from coupling to the three methyl protons. In any molecule, if the representative coupled nuclei are in the proportion $A_X:A_Y$ on neighboring groups, then the resonance band for $A_X$ will be split into $(2_Y \times I + 1)$ lines, in which $I$ is the spin quantum number for the nuclei, whereas that for $A_Y$ will be split into $(2_X \times I + 1)$ lines, assuming the nuclei are in different local environments. For example, with acetaldehyde, $A_X:A_Y$ is $CH_3:CH$, and the resulting proton splitting pattern is $A_X$ ($CH_3$) as two lines and $A_Y$ (CH) as four lines. This splitting produces *first-order* spectra when the difference in ppm between all the lines of a multiplet (known as the *coupling constant, J*) is small compared with the difference in the chemical shift, $\delta$, between the coupled nuclei. First-order spectra produce simple multiplets with intensities determined by the coefficients of the binomial expansion, which can be obtained from Pascal's triangle:

$$
\begin{array}{cccccccccc}
n = 0 & & & & & 1 & & & & \\
n = 1 & & & & 1 & & 1 & & & \\
n = 2 & & & 1 & & 2 & & 1 & & \\
n = 3 & & 1 & & 3 & & 3 & & 1 & \\
n = 4 & 1 & & 4 & & 6 & & 4 & & 1
\end{array}
$$

where $n$ is the number of equivalent neighbor nuclei. Thus, a doublet presents intensities 1:1; a triplet, 1:2:1; and a quartet, 1:3:3:1. Further details of multiplicity and the interpretation of NMR spectra can be found in general texts on NMR.[27,28]

The typical range for NMR chemical shifts depends on the nucleus being observed: For protons, it is about 15 ppm with organic compounds, whereas it is about 400 ppm for $^{13}C$ and $^{19}F$ spectra. **Table 4–8** gives the basic NMR resonance for certain pure isotopes, together with their natural abundances. As the natural abundance of the isotope decreases, the relative sensitivity of NMR gets proportionally smaller.

A widely used nucleus in characterization of new chemical compounds is $^{13}C$. The development of $^{13}C$ NMR spectroscopy in recent years has been influenced by the application of *spin decoupling* techniques to intensify and simplify the otherwise complex $^{13}C$ NMR spectra. Decoupling of the proton spins is produced by continuously irradiating the entire proton spectral range with broadband radio frequency radiation. This decoupling produces the collapse of multiplet signals into simpler and more intense signals. It also produces an effect known as the *nuclear Overhauser effect* (NOE), in

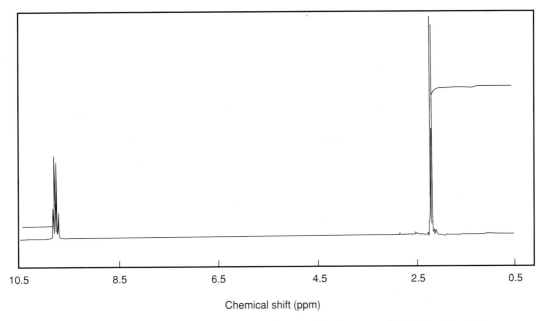

**Fig. 4–21.** Proton NMR spectrum of acetaldehyde. (From A. S. V. Burgen and J. C. Metcalfe, J. Pharm. Pharmacol. **22,** 156, 1970. With permission.)

which decoupling of the protons produces a dipole–dipole interaction and energy transfer to the carbon nuclei, resulting in greater relaxation (i.e., rate of loss of energy from nuclei in the higher spin state to the lower spin state) and, consequently, a greater population of carbon nuclei in the lower spin state (see **Fig. 4–13**). Because this greater population in the lower spin state permits a greater absorption signal in NMR, the NOE can increase the carbon nuclei signal by as much as a factor of 3. These factors, proton spin decoupling and the NOE, have enhanced the sensitivity of $^{13}$C NMR. They compensate for the low natural abundance of $^{13}$C as well as the smaller magnetic moment, $\mu$, of $^{13}$C compared with that for hydrogen.

A powerful extension of this enhancement of $^{13}$C spectra involves systematically decoupling only specific protons by irradiating at particular radio frequencies rather than by broadband irradiation. Such systematic decoupling permits

individual carbon atoms to show multiplet collapse and signal intensification, as mentioned previously, when protons coupled to the particular carbon atom are irradiated. These signal changes allow particular carbon nuclei in a molecular structure to be associated with particular protons and to produce what is termed *two-dimensional NMR spectrometry*. Farrar[29] describes the techniques involved in these experiments.

Nuclear magnetic resonance is a versatile tool in pharmaceutical research. Nuclear magnetic resonance spectra can provide powerful evidence for a particular molecular conformation of a drug, including the distinction between closely related isomeric structures. This identification is normally based on the relative position of chemical shifts as well peak multiplicity and other parameters associated with spin coupling. Drug–receptor interactions can be distinguished and characterized through specific changes in the NMR spectrum of the unbound drug after the addition of a suitable protein binder. These changes are due to restrictions in drug orientation. Burgen and Metcalfe[30] describe applications of NMR to problems involving drug–membrane and drug–protein interactions. Illustrations of these interactions are analyzed by Lawrence and Gill[31] using electron spin resonance (ESR) and by Tamir and Lichtenberg[32] using proton-NMR techniques. The ESR and proton-NMR results show that the psychotropic tetrahydrocannabinols reduce the molecular ordering in the bilayer of liposomes used as simple models of biologic membranes. These results suggest that the cannabinoids exert their psychotropic effects by way of a nonspecific interaction of the cannabinoid with lipid constituents, principally cholesterol, of nerve cell membranes. Rackham[33] reviewed the use of NMR in pharmaceutical research, with particular reference to analytical problems.

## TABLE 4–8

### BASIC NUCLEAR MAGNETIC RESONANCES (NMR) AND NATURAL ABUNDANCE OF SELECTED ISOTOPES*

| Isotope | NMR Frequency (MHz) at Given Field Strength | | Natural Abundance (%) |
|---|---|---|---|
| | At 1.0000 T | At 2.3487 T | |
| $^{1}_{1}$H | 42.57 | 100.00 | 99.985 |
| $^{13}_{6}$C | 10.71 | 25.14 | 1.108 |
| $^{15}_{7}$N | 4.31 | 10.13 | 0.365 |
| $^{19}_{9}$F | 40.05 | 94.08 | 100 |

*From A. J. Gordon and R. A. Ford, *The Chemist's Companion*, Wiley, New York, 1972, p. 314. With permission.

Nuclear magnetic resonance characterization has become routine, and specialized techniques have rapidly expanded its application breadth. Two particularly exciting applications rely on the determination of the three-dimensional structure of complex biomolecules[34] (proteins and nucleic acids) and the imaging of whole organisms (tomography). The paramount importance of these developments employing NMR has been recognized by the awarding of the Nobel Prize in Chemistry and Medicine in 2002 and 2003 for this work.

## REFRACTIVE INDEX AND MOLAR REFRACTION

Light passes more slowly through a substance than through a vacuum. As light enters a denser substance, the advancing waves interact with the atoms in the substance at the interface and throughout the thickness of the substance. These interactions modify the light waves by absorbing energy, resulting in the waves being closer together by reducing the speed and shortening the wavelength, as shown in **Figure 4–22**. If the light enters the denser substance at an angle, as shown, one part of the wave slows down more quickly as it passes the interface, and this produces a bending of the wave toward the interface. This phenomenon is called *refraction*. If light enters a less dense substance, it is refracted away from the interface rather than toward it. However, in both cases there is an important observation: The light travels in the same direction after interaction as before it. In reflection, the light travels in the opposite direction after incidence with the medium. The relative value of the effect of refraction between two substances is given by the *refractive index, n*:

$$n = \frac{\sin i}{\sin r} = \frac{\text{Velocity of light in the first substance}}{\text{Velocity of light in the second substance}} = \frac{c_1}{c_2}$$

(4–24)

where $\sin i$ is the sine of the angle of the incident ray of light, $\sin r$ is the sine of the angle of the refracted ray, and

$c_1$ and $c_2$ are the speeds of light of the respective media. Normally, the numerator is taken as the velocity of light in air and the denominator is that in the material being investigated. The refractive index, by this convention, is greater than 1 for substances denser than air. Theoretically, the reference state where $n = 1$ should be for light passing through a vacuum; however, the use of air as a reference produces a difference in $n$ of only 0.03% from that in a vacuum and is more commonly used.

The refractive index varies with the wavelength of light and the temperature because both alter the energy of interaction. Normally, these values are identified when a refractive index is listed; for example, $n_D{}^{20}$ signifies the refractive index using the $D$-line emission of sodium, at 589 nm, at a temperature of 20°C. Pressure must also be held constant in measuring the refractive index of gases. As discussed in Chapter 2, temperature and pressure are related. The refractive index can be used to identify a substance, to measure its purity, and to determine the concentration of one substance dissolved in another. Typically, a refractometer is used to determine refractive index. Refractometers can be an important tool for studying pharmaceutical compounds that do not undergo extensive UV–Vis absorption due to their electronic structure.

The *molar refraction*, $R_m$, is related to both the refractive index and the molecular properties of a compound being tested. It is expressed as

$$R_m = \frac{n^2 - 1}{n^2 + 2}\left(\frac{M}{\rho}\right)$$

(4–25)

where $M$ is the molecular weight and $\rho$ is the density of the compound. The $R_m$ value of a compound can often be predicted from the structural features of the molecule, and one should note the analogous equation to those utilized with dielectrics. Each constituent atom or group contributes a portion to the final $R_m$ value, as discussed earlier in connection with additive–constitutive properties (see **Table 4–1**). For example, acetone has an $R_m$ produced from three carbons ($R_m = 7.254$), six hydrogens (6.6), and a carbonyl oxygen (2.21) to give a total $R_m$ of 16.1 cm$^3$/mole. Because $R_m$ is independent of the physical state of the molecule, this value can often be used to distinguish between structurally different compounds, such as keto and enol tautomers.

Light incident on a molecule induces vibrating dipoles due to energy absorption at the interface, where the greater the refractive index at a particular wavelength, the greater the dipolar induction. Simply stated, the interaction of light photons with polarizable electrons of a dielectric molecule causes a reduction in the velocity of light. The dielectric constant, a measure of polarizability, is greatest when the resulting dipolar interactions with light are proportionally large. The refractive index for light of long wavelengths, $n_\infty$, is related to the dielectric constant for a nonpolar molecule, $\varepsilon$, by the expression

$$\varepsilon = n^2{}_\infty$$

(4–26)

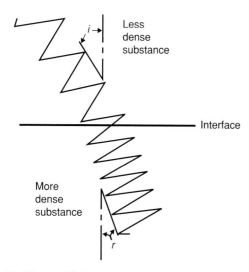

**Fig. 4–22.** Waves of light passing an interface between two substances of different density.

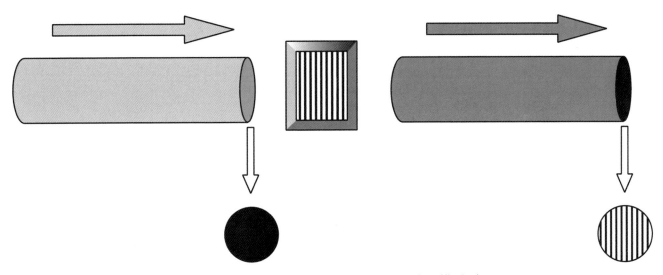

**Fig. 4–23.** The polarization of a dispersed light beam by a Nicol prism.

Molar polarization, $P_i$ [equation **(4–16)**], can be considered roughly equivalent to molar refraction, $R_m$, and can be written as

$$P_i = \left(\frac{n_\infty^2 - 1}{n_\infty^2 + 2}\right)\frac{M}{\rho} = \frac{4}{3}\pi N\alpha_p \qquad (4\text{–}27)$$

From this equation, the polarizability, $\alpha_p$, of a nonpolar molecule can be obtained from a measurement of refractive index. For practical purposes, the refractive index at a finite wavelength is used. This introduces only a relatively small error, approximately 5%, in the calculation.

## OPTICAL ROTATION

Electromagnetic radiation comprises two separate, sinusoidal-like wave motions that are perpendicular to one another, an electric wave and a magnetic wave (**Fig. 4–4**). Both waves have equal energy. A source will produce multiple waves of oscillating electromagnetic radiation at any given time, so that multiple electric and magnetic waves are emitted. These electromagnetic radiation waves travel in multiple orientations that are randomly dispersed in the resulting circular beam of radiation. Discussions of interactions with light sources are predominantly focused on the electric component of these waves, to which we restrict ourselves for the remainder of this section and in the Optical Rotatory Dispersion section and Circular Dichroism section.

Passing light through a polarizing prism such as a Nicol prism sorts the randomly distributed vibrations of electric radiation so that only those vibrations occurring in a single plane are passed (**Fig. 4–23**). The velocity of this *plane-polarized* light can become slower or faster as it passes through a sample, in a manner similar to that discussed for refraction. This change in velocity results in refraction of the polarized light in a particular direction for an *optically active* substance. A clockwise rotation in the planar light, as

observed looking into the beam of polarized light, defines a substance as *dextrorotatory*. When the plane of light is rotated by the sample in a counterclockwise manner, the sample is defined as a *levorotatory* substance. A dextrorotatory substance, which may be thought of as rotating the beam to the right, produces an *angle of rotation*, $\alpha$, that is defined as positive, whereas the levorotatory substance, which rotates the beam to the left, has an $\alpha$ that is defined as negative. Molecules that have an asymmetric center (chiral) and therefore lack symmetry about a single plane are optically active, whereas symmetric molecules (achiral) are optically inactive and consequently do not rotate the plane of polarized light. Optical activity can be considered as the interaction of a plane-polarized radiation with electrons in a molecule to produce electronic polarization. This interaction rotates the direction of vibration of the radiation by altering the electric field. A *polarimeter* is used to measure optical activity. **Figure 4–24** illustrates how a polarimeter can be used to measure these phenomena.

Optical rotation, $\alpha$, depends on the density of an optically active substance because each molecule provides an equal but small contribution to the rotation. The *specific rotation*, $\{\alpha_\lambda^t\}$ at a specified temperature $t$ and wavelength $\lambda$ (usually the D line of sodium), is characteristic for a pure, optically active substance. It is given by the equation

$$\{\alpha_\lambda^t\} = \frac{\alpha v}{lg} \qquad (4\text{–}28)$$

where $l$ is the length in decimeters (dm)* of the light path through the sample and $g$ is the number of grams of optically active substance in $v$ milliliter of volume. If the substance is dissolved in a solution, the solvent as well as the concentration should be reported with the specific rotation. The specific rotation of a molecule is indicative of its molecular structure

---

* The decimeter (dm) is chosen as the unit because of the long sample cells normally used in polarimeters; 1 dm = 10 cm = 0.1 m.

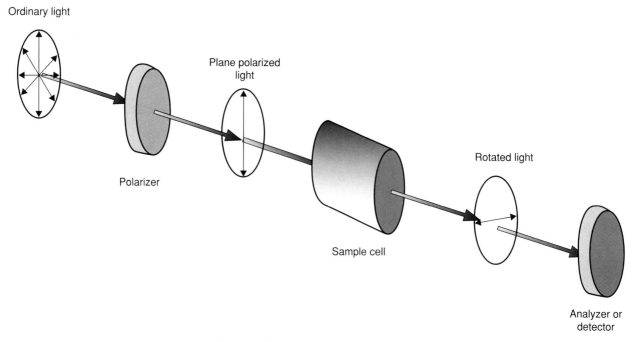

**Fig. 4–24.** Schematic view of a polarimeter.

and as such should not change when measured under the same conditions in any lab. Organic chemists use specific rotation as a tool to help confirm the identity of a synthesized compound whose molecular properties are already defined. Other measurements are often performed, but this is a quick test to check the identity and purity of a reaction product. The specific rotations of some drugs are given in **Table 4–9**. The subscript $D$ on $[\alpha]$ indicates that the measurement of specific rotation is made at a wavelength, $\lambda$, of 589 nm for sodium light. When the concentration is not specified, as in **Table 4–9**, the concentration is often assumed to be 1 g per milliliter of solvent. However, $[\alpha]$ values should be reported at specific temperatures and concentration including solvent. Data missing this information should be treated

with caution because the specific conditions are not known. The specific rotation of steroids, carbohydrates, amino acids, and other compounds of biologic importance are given in the *CRC Handbook.*[35]

# OPTICAL ROTATORY DISPERSION

Optical rotation changes as a function of the wavelength of light, and *optical rotatory dispersion* (ORD) is the measure of the angle of rotation as a function of the wavelength. Recall that this is light in the UV–Vis range and that different wavelengths of light have different energies, which may result in changes in absorption patterns of molecules due to their

**TABLE 4–9**
**SPECIFIC ROTATIONS**

| Drug | $[\alpha]_D$ (deg) | Temperature (°C) | Solvent |
|---|---|---|---|
| Ampicillin | +283 | 20 | Water |
| Aureomycin | +296 | 23 | Water |
| Benzylpenicillin | +305 | 25 | Water |
| Camphor | +41 to +43 | 25 | Ethanol |
| Colchicine | −121 | 17 | Chloroform |
| Cyanocobalamin | −60 | 23 | Water |
| Ergonovine | −16 | 20 | Pyridine |
| Nicotine | −162 | 20 | Pure liquid |
| Propoxyphene | +67 | 25 | Chloroform |
| Quinidine | +230 | 15 | Chloroform |
| Reserpine | −120 | 25 | Chloroform |
| Tetracycline hydrochloride | −253 | 24 | Methanol |
| *d*-Tubocurarine chloride | +190 | 22 | Water |
| Yohimbine | +51 to +62 | 20 | Ethanol |

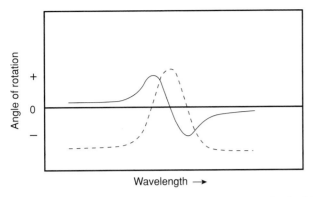

**Fig. 4–25.** The Cotton effect. Variation of the angle of rotation (solid line) in the vicinity of an absorption band of polarized light (dashed line). (Modified from W. S. Brey, *Physical Chemistry and Its Biological Applications*, Academic Press, New York, 1978, p. 330.)

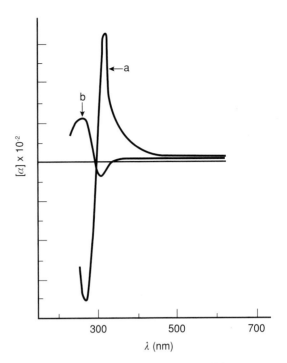

**Fig. 4–26.** Optical rotary dispersion curves for (a) *cis*-10-methyl-2-decalone and (b) *trans*-10-methyl-2-decalone. Note the positive effect in (a) and the negative effect in (b). (From C. Djerassi, *Optical Rotatory Dispersion*, McGraw-Hill, New York, 1960. With permission.)

electronic structure. Varying the wavelength of light changes the specific rotation for an optically active substance because of the electronic structure of the molecule. A graph of specific rotation versus wavelength shows an inflection and then passes through zero at the wavelength of maximum absorption of polarized light as shown in **Figure 4–25**. This change in specific rotation is known as the *Cotton effect*. By convention, compounds whose specific rotations show a maximum *before* passing through zero as the wavelength of polarized light becomes smaller are said to show a *positive Cotton effect*, whereas if {α} shows a maximum *after* passing through zero (under the same conditions of approaching shorter wavelengths), the compound shows a *negative Cotton effect*. Enantiomers can be characterized by the Cotton effect, as shown in **Figure 4–26**. In addition, ORD is often useful for the structural examination of organic compounds. Detailed discussions of ORD are given by Campbell and Dwek[12] and Crabbe.[36]

## CIRCULAR DICHROISM

Plane-polarized light is described as the vector sum of two circularly polarized components. Circularly polarized light has an electric vector that spirals around the direction of propagation. In plane-polarized light, there can be two such vectors spiraling in the opposite direction, a right-handed and a left-handed component. Handedness is used as a convention to conceptualize circularly polarized light. To visualize handedness, take both hands and hold them with the fingers pointed straight up and each thumb pointed toward your nose. Now curl the fingers around in a loose fist, holding the thumbs steady. The thumb on each hand indicates the light traveling toward your nose, and the curling of the fingers shows how the light vectors are spiraling into two separate vectors, the left and right circular directions. For an optically active substance, the values of the index of refraction, *n*, of the two vectors cannot be the same.

Because of chirality, one of the two vectors may be absorbed differently at one wavelength of light, whereas the converse may be true at another wavelength. This difference

changes the relative rate at which the polarized light spirals about its direction of propagation. Likewise, the speeds of the two components of polarized light become unequal as they pass through an optically active substance that is capable of absorbing light over a selected wavelength range. This is the same as saying that the two components of polarized light have different absorptivities at a particular wavelength of light. Conceptually, it is important to realize that at the source, the circularly polarized light is traveling at the same speed in a circle. When one component of light, say the left-handed component, has energy absorbed, it will slow down in contrast to the right-hand component, stretching the circle into an ellipse. The converse is true as well. This can also be illustrated by using the hand convention. When circularly polarized light is caused to become elliptically polarized, it is termed *circular dichroism* (CD).[12,37]

The Cotton effect is the name given to the unequal absorption (elliptic effect) of the two components of circularly polarized light by a molecule in the wavelength region near an absorption band. Circular dichroism spectra are plots of molar ellipticity, [θ], which is proportional to the difference in absorptivities between the two components of circularly polarized light, against the wavelength of light. By traditional convention, the molar ellipticity is given by

$$[\theta] = \frac{[\psi]M}{100} = 3300(\varepsilon_L - \varepsilon_R) \text{ deg liter mole}^{-1} \text{ dm}^{-1}$$

$$(4\text{--}29)$$

where [ψ] is the specific ellipticity, analogous to the specific rotation, *M* is the molecular weight, and $\varepsilon_L$ and $\varepsilon_R$ are the molar absorptivities for the left and right components of circularly polarized light at a selected wavelength. Perrin and Hart[38] review the applications of CD to interactions between small molecules and macromolecules, and the theory is described by Campbell and Dwek.[12] Circular dichroism has its greatest application in the characterization of chiral macromolecules, in particular proteins that are utilized in biotechnology-based therapeutics. Secondary structure in proteins can easily be defined by close investigation of CD spectra.[37,39–41] The magnitude of the absorption of the components of circularly polarized light is distinctly different for $\alpha$-helices, $\beta$-sheets, and $\beta$-turns. By measuring the CD spectrum of proteins whose secondary structure is known, one can compile database and use it to deconvolve the secondary structural composition of an unknown protein. The CD spectra for a protein can also serve as standard and be used to determine whether effects during production or formulation are altering the proteins' folding patterns, which could dramatically decrease its activity.

Circular dichroism has also been applied to the determination of the activity of penicillin, as described by Rasmussen and Higuchi.[42] The activity was measured as the change in the CD spectra of penicillin after addition of penicillinase, which enzymatically cleaves the $\beta$-lactam ring to form the penicillate ion, as shown for benzylpenicillin in **Figure 4–27**. Typical CD spectra for benzylpenicillin and its hydrolysis product are shown in **Figure 4–27**. The direct determination of penicillins by CD and the distinction of penicillins from cephalosporins by their CD spectra has been described by Purdie and Swallows.[43] The penicillins all have single positive CD bands with maxima at 230 nm, whereas the cephalosporins have two CD bands, a positive one with a maximum at 260 nm and a negative one with maximum at 230 nm (wavelengths for maxima are given to within $\pm 2$ nm). This permits easy differentiation between penicillins and cephalosporins by CD spectropolarimetry.

# ELECTRON AND NEUTRON SCATTERING AND EMISSION SPECTROSCOPY

Electrons can flow across media in response to charge separation, and do so at certain frequencies and wavelengths depending on the magnitude of the charge separation. Neutrons can also be generated and flow in a similar manner. The exact mathematical and theoretical description of this flow is well beyond the scope of this text. However, the concept that they can flow is important to the discussion of the following subjects. It must also be noted that whereas electron and neutron waves are not traditionally thought of as electromagnetic radiation, they can be used to determine important molecular properties and are included in this text. The main distinction between electrons and neutrons versus electromagnetic radiation is that electrons and neutrons have a finite resting mass, in contrast to electromagnetic radiation, which has a zero finite resting mass.[12]

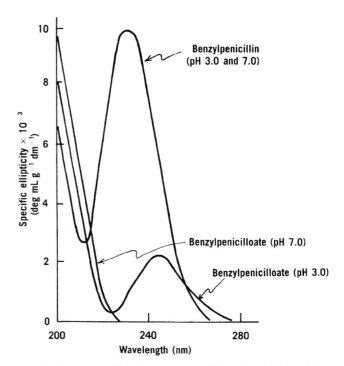

**Fig. 4–27.** Circular dichroism spectra of benzylpenicillin and its penicilloic acid derivative. (From C. E. Rassmussen and T. Higuchi, J. Pharm. Sci. **60,** 1616, 1971. With permission.)

Several techniques that are pharmaceutically relevant also employ electron or neutron beams, including small-angle neutron scattering and scanning and transmission electron microscopy (SEM and TEM, respectively). Scattering of energy waves results when a loss of energy occurs through interaction with a medium, as occurs in refraction. Scattering is usually measured at an angle $\theta$ from the incident beam of energy and does not measure transmittance of the energy wave. Both electron and neutron waves play an important role in identifying some of the molecular properties of compounds and excipients. They are discussed briefly here.

Scanning and transmission electron microscopy are two techniques that are routinely employed in pharmaceutical and biologic sciences to produce high-resolution images of surfaces. Let us consider SEM. In general, the image generated with SEM is produced by electron wave emission from a filament and bombardment of the sample, where the electrons undergo refraction after interaction with the sample surface. As the electrons pass through the sample surface, they continuously undergo refraction until they are refracted back out of the sample surface at a lower energy and are detected. In the case of refraction, the incident electron beam interacts with electrons in the sample's atomic electron cloud, which results in alteration of its path after collision. For heavier atoms, the electron cloud is dense, and the incident electrons from the beam undergo greater refraction. In SEM techniques, the sample is often sputter coated (sprayed) with gold to provide a greater electron density at the surface and prevent the beam from passing through the sample. In TEM, the electrons are transmitted through the sample. The refracted electrons generate a high-resolution snapshot of the sample surface and provide information about its contour and size; however, it does not offer molecular information.

Molecular (atomic) information can be obtained from monitoring the energy, which is in the form of x-rays emitted when the electrons interact with the molecules. In general, upon an inelastic collision the incidental electron beam can also knock out ground-state electrons in the molecules. This is known as Auger electron emission. The outer-shell electrons in each atom then relax to the ground state, which results in the emission of energy in the form of x-ray fluorescence. For each atom, the atomic structure is different, and the relaxation energies change on the basis of the individual atoms. The emission of these x-rays can be captured using techniques such as energy-dispersive x-ray analysis (EDAX) and wavelength-dispersive x-ray analysis. These techniques provide information about the molecular composition by detecting the different x-ray frequencies and wavelengths emitted during an SEM run. Energy-dispersive x-ray analysis captures the frequency patterns of the emitted x-ray waves, whereas wavelength-dispersive x-ray analysis measures the x-ray wavelengths from the atoms present. These emissions can be measured across a linear region of the sample or in the whole sample. They are qualitative but can be quantitative in nature, depending on the type of analysis performed.

A good example of the utility of this technique is the examination of a coated tablet that contains an iron-based drug in its core (**Fig. 4–28**). If an SEM and a linear EDAX analysis are performed across the surface of a split tablet, EDAX demonstrates that the iron is only present in the center of the tablet and not in the coated region. SEM provides a picture of the surface of the split tablet that can be used to look at formulation conditions. If a similar tablet is then placed in

water for 1 hr and subjected to the same analysis, one might observe a lower amount of iron in the core, and iron present in the coating, with an image that shows the difference in the tablet's core structure upon water exposure. This could provide very valuable information to a formulator, who may want to compare several different coatings to control the release of the iron-based drug or look at how different polymorphs (Chapter 2) of the compound might behave in the same formulation. In fact, this type of analysis is routinely performed for many types of formulations including micro- and nanospheres. The information about physical properties of state (Chapter 2) and molecular properties is very important. There are also many other techniques employing similar methods that are being widely incorporated into the industrial characterization and formulation development of drugs.

Small-angle neutron scattering is another technique that can be employed to illustrate the physical properties of a molecule. It is related to techniques like small-angle x-ray scattering and small-angle light scattering. Collectively, these techniques can be used to give information about the size (even molecular weights for large polymeric and protein molecules), shape, and even orientation of components in a sample.[44] Small-angle neutron scattering is primarily used for the characterization of polymers, in particular dendrimers, and colloids.

## CHAPTER SUMMARY

The goal of this chapter was to provide a foundation for understanding the physical properties of molecules including methods to make those determinations. Students should understand the nature of intra- and intermolecular forces, molecular energetics of these forces, and their relevance to different molecules. Another important aspect that was covered was the determination of these physical properties using a variety of atomic and molecular spectroscopic techniques. The student should also appreciate the differences in the techniques with respect to the identification and detection of pharmaceutical agents.

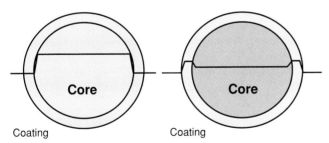

**Fig. 4–28.** A coated iron-containing tablet that was split, showing initially a smooth surface (*left*) and then, after exposure to water for 1 hr, erosion and a rougher surface (*right*). The line through the tablets shows how a hypothetical scanning electron microscope image combined with an energy-dispersive x-ray analysis (EDAX) linear profile focused on the iron x-ray fluorescence emission frequency is performed across the tablet. In the first tablet, the iron is uniformly contained in the center and the magnitude of the deviation of the line implies a higher iron concentration in the core. The image shows a smooth surface. After exposure to water for 1 hr, the iron and some of the excipients have migrated from the core, leaving a rough contour and a lower EDAX level. The coating has a higher level of iron by this scan, which could be due to dissolution and the effect of diffusion rates, as discussed in later chapters.

**Practice problems for this chapter can be found at thePoint.lww.com/Sinko6e.**

### References

1. Y. Bentor, *Chemical Element.com—Carbon.* Available at http://www.chemicalelements.com/elements/c.html. Accessed August 28, 2008.
2. J. S. Richardson, Adv. Protein Chem. **34,** 167, 1981.
3. C. K. Mathews and K. E. van Holde, *Biochemistry,* Benjamin Cummings, Redwood City, CA, 1990, Chapter 5.
4. W. G. Hol, Adv. Biophys. **19,** 133, 1985.
5. P. R. Bergethon and E. R. Simons, *Biophysical Chemistry: Molecules to Membranes,* Springer-Verlag, New York, 1990, Chapter 10.
6. W. G. Gorman and G. D. Hall, J. Pharm. Sci. **53,** 1017, 1964.
7. S. N. Pagay, R. I. Poust, and J. L. Colaizzi, J. Pharm. Sci. **63,** 44, 1974.

8. P. A. Kollman, in M. E. Wolff (Ed.), *The Basis of Medicinal Chemistry, Burger's Medicinal Chemistry, Part I*, Wiley, New York, 1980, Chapter 7.

9. V. I. Minkin, O. A. Osipov, and Y. A. Zhadanov, *Dipole Moments in Organic Chemistry*, Plenum Press, New York, 1970.

10. D. A. McQuarrie and J. D. Simon, *Physical Chemistry: A Molecular Approach*, University Science Books, Sausalito, CA, 1997.

11. W. S. Brey, *Physical Chemistry and Its Biological Applications*, Academic Press, New York, 1978, Chapter 9.

12. I. D. Campbell and R. A. Dwek, *Biological Spectroscopy*, Benjamin/Cummings, Menlo Park, CA, 1984.

13. S. G. Jivani and V. J. Stella, J. Pharm. Sci. **74**, 1274, 1985.

14. A. L. Gutman, V. Ribon, and J. P. Leblanc, Anal. Chem. **57**, 2344, 1985.

15. S. G. Schulman and B. S. Vogt, in J. W. Munson (Ed.), *Pharmaceutical Analysis: Modern Methods, Part B*, Marcel Dekker, New York, 1984, Chapter 8.

16. A. H. Zewail, J. Phys. Chem. A **104**, 5660, 2000.

17. B. Valeur, *Molecular Fluorescence Principles and Applications*, Wiley-VCH Verlag, Weinheim, Germany, 2002, Chapter 11.

18. S. G. Schulman and R. J. Sturgeon, in J. W. Munson (Ed.), *Pharmaceutical Analysis: Modern Methods, Part A*, Marcel Dekker, New York, 1984, Chapter 4.

19. D. I. Chapman and M. S. Moss, in E. G. C. Clarke (Ed.), *Isolation and Identification of Drugs*, Pharmaceutical Press, London, 1969, pp. 103–122.

20. J. R. Durig (Ed.), *Chemical, Biological, and Industrial Applications of Infrared Spectroscopy*, Wiley, New York, 1985; R. A. Hoult, *Development of the Model 1600 FT-IR Spectrophotometer*, Norwalk, Conn, 1983; R. J. Bell, *Introductory Fourier Transform Spectroscopy*, Academic Press, New York, 1972; P. R. Griffith, *Chemical Infrared Fourier Transform Spectroscopy*, Wiley, New York, 1975; R. J. Markovich and C. Pidgeon, Pharm. Res. **8**, 663, 1991.

21. A. L. Smith, *Applied Infrared Spectroscopy*, Wiley, New York, 1979.

22. H. H. Willard, L. L. Merritt, J. A. Dean, and F. A. Settle, *Instrumental Methods of Analysis*, 6th Ed., Van Nostrand, New York, 1981, Chapter 7.

23. C. D. Bugay and A. William, in H. G. Brittain (Ed.), *Physical Characterization of Pharmaceutical Solid*, Marcel Dekker, New York, 1995, pp. 65–79.

24. H. M. Swartz, J. R. Bolton, and D. C. Borg (Eds.), *Biological Applications of Electron Spin Resonance*, Wiley, New York, 1972.

25. P. M. Plonka and M. Elas, Curr. Top. Biophys. **26**, 175–189, 2002.

26. R. Galimzyanovich, L. Ivanovna, T. I. Vakul'skaya, and M. G. Voronkov, *Electron Paramagnetic Resonance in Biochemistry and Medicine*, Kluwer, Dordrecht, Netherlands, 2001.

27. H. Friebolin, *Basic One- and Two-Dimensional NMR Spectroscopy*, 2nd Ed., VCH, Weinheim, Germany, 1993.

28. H. Gunther, *NMR Spectroscopy*, 2nd Ed., Wiley, New York, 1998.

29. T. C. Farrar, Anal. Chem. **59**, 749A, 1987.

30. A. S. V. Burgen and J. C. Metcalfe, J. Pharm. Pharmacol. **22**, 153, 1970.

31. D. K. Lawrence and E. W. Gill, Mol. Pharmacol. **11**, 595, 1975.

32. J. Tamir and D. Lichtenberg, J. Pharm. Sci. **72**, 458, 1983.

33. D. M. Rackham, Talanta **23**, 269, 1976.

34. K. Wüthrich, *NMR of Proteins and Nucleic Acids*, Wiley, New York, 1986.

35. D. Lide (Ed.), *CRC Handbook of Chemistry and Physics*, 85th Ed., CRC Press, Boca Raton, Fl., 2004.

36. P. Crabbe, *ORD and CD in Chemistry and Biochemistry*, Academic Press, New York, 1972.

37. N. Berova, K. Nakanishi, and R. Woody, *Circular Dichroism: Principle and Applications*, VCH, New York, 2000.

38. J. H. Perrin and P. A. Hart, J. Pharm. Sci. **59**, 431, 1970.

39. M. C. Manning, J. Pharm. Biomed. Anal. **7**, 1103, 1989.

40. A. Perczel, K. Park, and G. D. Fasman, Anal. Biochem. **203**, 83, 1992.

41. G. D. Fasman (Ed.), Circular Dichroism and the Conformational Analysis of Biomolecules, Plenum Press, New York, 1996.

42. C. E. Rasmussen and T. Higuchi, J. Pharm. Sci. **60**, 1608, 1971.

43. N. Purdie and K. A. Swallows, Anal. Chem. **59**, 1349, 1987.

44. S. M. King, in R. A. Pethrick and J. V. Dawkins (Eds.), *Modern Techniques for Polymer Characterisation*, Wiley, New York, 1999, Chapter 7.

## Recommended Readings

C. R. Cantor and P. R. Schimmel, *Biophysical Chemistry Part II: Techniques for the Study of Biological Structure and Function*, W H Freeman, New York, 1980.

J. B. Lambert, H. F. Shurvell, D. Lightner, and R. G. Cooks, *Organic Structural Spectroscopy*, Prentice Hall, Upper Saddle River, NJ, 1998.

S. G. Schulman, *Fluorescence and Phosphorescence Spectroscopy, Physicochemical Principles and Practice*, Pergamon Press, Oxford, 1977.

R. M. Silverstein, F. X. Webster, and D. J. Kiemle, *Spectrometric Identification of Organic Compounds*, 7th Ed., John Wiley and Sons, Hoboken, NJ, 2005.

D. W. Turner, A. D. Baker, C. Baker, and C. R. Brundle, *Molecular Photoelectron Spectroscopy*, Wiley-Interscience, New York, 1970.

CHAPTER LEGACY

**Fifth Edition:** published as Chapter 4. Updated by Gregory Knipp, Dea Herrera-Ruiz, and Hugo Morales-Rojas.

**Sixth Edition:** published as Chapter 4. Updated by Patrick Sinko.

# 5      NONELECTROLYTES

CHAPTER OBJECTIVES **At the conclusion of this chapter the student should be able to:**

**1**   Identify and describe the four colligative properties of nonelectrolytes in solution.

**2**   Understand the various types of pharmaceutical solutions.

**3**   Calculate molarity, normality, molality, mole fraction, and percentage expressions.

**4**   Calculate equivalent weights.

**5**   Define ideal and real solutions using Raoult's and Henry's laws.

**6**   Use Raoult's law to calculate partial and total vapor pressure.

**7**   Calculate vapor pressure lowering, boiling point elevation, freezing point lowering, and pressure for solutions of nonelectrolytes.

**8**   Use colligative properties to determine molecular weight.

## INTRODUCTION

In this chapter, the student will begin to learn about pharmaceutical systems. In the pharmaceutical sciences, a *system* is generally considered to be a bounded space or an exact quantity of a material substance. Material substances can be mixed together to form a variety of pharmaceutical mixtures (or dispersions) such as true solutions, colloidal dispersions, and coarse dispersions. A *dispersion* consists of at least two phases with one or more dispersed (internal) phases contained in a single continuous (external) phase. The term *phase* is defined as a distinct homogeneous part of a system separated by definite boundaries from other parts of the system. Each phase may be consolidated into a contiguous mass or region, such as a single tea leaf floating in water. A *true solution* is defined as a mixture of two or more components that form a homogeneous molecular dispersion, in other words, a one-phase system. In a true solution, the suspended particles completely dissolve and are not large enough to scatter light but are small enough to be evenly dispersed resulting in a homogeneous appearance. The diameter of particles in *coarse dispersions* is greater than ~500 nm (0.5 $\mu$m). Two common pharmaceutical coarse dispersions are emulsions (liquid–liquid dispersions) and suspensions (solid–liquid dispersions). A colloidal dispersion represents a system having a particle size intermediate between that of a true solution and a coarse dispersion, roughly 1 to 500 nm. A colloidal dispersion may be considered as a two-phase (heterogeneous) system under some circumstances. However, it may also be considered as a one-phase system (homogeneous) under other circumstances. For example, liposomes or microspheres in an aqueous delivery vehicle are considered to be heterogeneous colloidal dispersions because they consist of distinct particles constituting a separate phase. On the other hand, a colloidal dispersion of acacia or sodium carboxymethylcellulose in water is homogeneous since it does not differ significantly from a solution of sucrose. Therefore, it may be considered as a single-phase system or true solution.[1] Another example

of a homogeneous colloidal dispersion that is considered to be a true solution is drug-polymer conjugates since they can completely dissolve in water.

This chapter focuses on molecular dispersions, which are also known as true solutions. A solution composed of only two substances is known as a *binary solution*, and the components or constituents are referred to as the *solvent* and the *solute*. Commonly, the terms *component* and *constituent* are used interchangeably to represent the pure chemical substances that make up a solution. The *number of components* has a definite significance in the phase rule. The constituent present in the greater amount in a binary solution is arbitrarily designated as the solvent and the constituent in the lesser amount as the solute. When a solid is dissolved in a liquid, however, the liquid is usually taken as the solvent and the solid as the solute, irrespective of the relative amounts of the constituents. When water is one of the constituents of a liquid mixture, it is usually considered the solvent. When dealing with mixtures of liquids that are miscible in all proportions,

## KEY CONCEPT   PHARMACEUTICAL DISPERSIONS

When two materials are mixed, one becomes dispersed in the other. To classify a pharmaceutical dispersion, only the size of the dispersed phase and not its composition is considered. The two components may become dispersed at the molecular level forming a true solution. In other words, the dispersed phase completely dissolves, cannot scatter light, and cannot be visualized using microscopy. If the dispersed phase is in the size range of 1 to 500 nm, it is considered to be a colloidal dispersion. Common examples of colloidal dispersions include blood, liposomes, and zinc oxide paste. If the particle size is greater than 500 nm (or 0.5 $\mu$m), it is considered to be a coarse dispersion. Two common examples of coarse dispersions are emulsions and suspensions.

such as alcohol and water, it is less meaningful to classify the constituents as solute and solvent.

# PHYSICAL PROPERTIES OF SUBSTANCES

The physical properties of substances can be classified as *colligative*, *additive*, and *constitutive*. Some of the constitutive and additive properties of molecules were considered in Chapter 4. In the field of thermodynamics, physical properties of systems are classified as *extensive* properties, which depend on the quantity of the matter in the system (e.g., mass and volume), and *intensive* properties, which are independent of the amount of the substances in the system (e.g., temperature, pressure, density, surface tension, and viscosity of a pure liquid).

*Colligative properties* depend mainly on the number of particles in a solution. The colligative properties of solutions are osmotic pressure, vapor pressure lowering, freezing point depression, and boiling point elevation. The values of the colligative properties are approximately the same for equal concentrations of different nonelectrolytes in solution regardless of the species or chemical nature of the constituents. In considering the colligative properties of solid-in-liquid solutions, it is assumed that the solute is nonvolatile and that the pressure of the vapor above the solution is provided entirely by the solvent.

*Additive properties* depend on the total contribution of the atoms in the molecule or on the sum of the properties of the constituents in a solution. An example of an additive property of a compound is the molecular weight, that is, the sum of the masses of the constituent atoms. The masses of the components of a solution are also additive, the total mass of the solution being the sum of the masses of the individual components.

*Constitutive properties* depend on the arrangement and to a lesser extent on the number and kind of atoms within a molecule. These properties give clues to the constitution of individual compounds and groups of molecules in a system. Many physical properties may be partly additive and partly constitutive. The refraction of light, electric properties, surface and interfacial characteristics, and the solubility of drugs are at least in part constitutive and in part additive properties; these are considered in other sections of the book.

## Types of Solutions

A solution can be classified according to the states in which the solute and solvent occur, and because three states of matter (gas, liquid, and crystalline solid) exist, nine types of homogeneous mixtures of solute and solvent are possible. These types, together with some examples, are given in **Table 5–1**.

When solids or liquids dissolve in a gas to form a gaseous solution, the molecules of the solute can be treated thermodynamically like a gas; similarly, when gases or solids dissolve in liquids, the gases and the solids can be considered to

**TABLE 5–1**
**TYPES OF SOLUTIONS**

| Solute | Solvent | Example |
|--------|---------|---------|
| Gas | Gas | Air |
| Liquid | Gas | Water in oxygen |
| Solid | Gas | Iodine vapor in air |
| Gas | Liquid | Carbonated water |
| Liquid | Liquid | Alcohol in water |
| Solid | Liquid | Aqueous sodium chloride solution |
| Gas | Solid | Hydrogen in palladium |
| Liquid | Solid | Mineral oil in paraffin |
| Solid | Solid | Gold—silver mixture, mixture of alums |

exist in the liquid state. In the formation of solid solutions, the atoms of the gas or liquid take up positions in the crystal lattice and behave like atoms or molecules of solids.

The solutes (whether gases, liquids, or solids) are divided into two main classes: *nonelectrolytes* and *electrolytes*. Nonelectrolytes are substances that do not ionize when dissolved in water and therefore do not conduct an electric current through the solution. Examples of nonelectrolytes are sucrose, glycerin, naphthalene, and urea. The colligative properties of solutions of nonelectrolytes are fairly regular. A 0.1-molar (M) solution of a nonelectrolyte produces approximately the same colligative effect as any other nonelectrolytic solution of equal concentration. Electrolytes are substances that form ions in solution, conduct electric current, and show apparent "anomalous" colligative properties; that is, they produce a considerably greater freezing point depression and boiling point elevation than do nonelectrolytes of the same concentration. Examples of electrolytes are hydrochloric acid, sodium sulfate, ephedrine, and phenobarbital.

Electrolytes may be subdivided further into *strong electrolytes* and *weak electrolytes* depending on whether the substance is completely or only partly ionized in water. Hydrochloric acid and sodium sulfate are strong electrolytes, whereas ephedrine and phenobarbital are weak electrolytes. The classification of electrolytes according to Arrhenius and the discussion of the modern theories of electrolytes are given later in the book.

# CONCENTRATION EXPRESSIONS

The concentration of a solution can be expressed either in terms of the quantity of solute in a definite *volume of solution* or as the quantity of solute in a definite *mass of solvent or solution*. The various expressions are summarized in **Table 5–2**.

## Molarity and Normality

Molarity and normality are the expressions commonly used in analytical work.[2] All solutions of the same molarity

**TABLE 5–2**
## CONCENTRATION EXPRESSIONS

| Expression | Symbol | Definition |
|---|---|---|
| Molarity | M,c | Moles (gram molecular weights) of solute in 1 liter of solution |
| Normality | N | Gram equivalent weights of solute in 1 liter of solution |
| Molality | $m$ | Moles of solute in 1000 g of solvent |
| Mole fraction | $X$,N | Ratio of the moles of one constituent (e.g., the solute) of a solution to the total moles of all constituents (solute and solvent) |
| Mole percent | | Moles of one constituent in 100 moles of the solution; mole percent is obtained by multiplying mole fraction by 100 |
| Percent by weight | % w/w | Grams of solute in 100 g of solution |
| Percent by volume | % v/v | Milliliters of solute in 100 mL of solution |
| Percent weight-in-volume | % w/v | Grams of solute in 100 mL of solution |
| Milligram percent | — | Milligrams of solute in 100 mL of solution |

contain the same number of solute molecules in a definite volume of solution. When a solution contains more than one solute, it may have different molar concentrations with respect to the various solutes. For example, a solution can be 0.001 M with respect to phenobarbital and 0.1 M with respect to sodium chloride. One liter of such a solution is prepared by adding 0.001 mole of phenobarbital (0.001 mole × 232.32 g/mole = 0.2323g) and 0.1 mole of sodium chloride (0.1 mole × 58.45 g/mole = 5.845 g) to enough water to make 1000 mL of solution.

Difficulties are sometimes encountered when one desires to express the molarity of an ion or radical in a solution. A molar solution of sodium chloride is 1 M with respect to both the sodium and the chloride ion, whereas a molar solution of $Na_2CO_3$ is 1 M with respect to the carbonate ion and 2 M with respect to the sodium ion because each mole of this salt contains 2 moles of sodium ions. A molar solution of sodium chloride is also 1 normal (1 N) with respect to both its ions; however, a molar solution of sodium carbonate is 2 N with respect to both the sodium and the carbonate ion.

Molar and normal solutions are popular in chemistry because they can be brought to a convenient volume; a volume aliquot of the solution, representing a known weight of solute, is easily obtained by the use of the burette or pipette.

Both molarity and normality have the disadvantage of changing value with temperature because of the expansion or contraction of liquids and should not be used when one wishes to study the properties of solutions at various temperatures. Another difficulty arises in the use of molar and normal solutions for the study of properties such as vapor pressure and osmotic pressure, which are related to the concentration of the solvent. The volume of the solvent in a molar or a normal solution is not usually known, and it varies for different solutions of the same concentration, depending upon the solute and solvent involved.

## Molality

A molal solution is prepared in terms of weight units and does not have the disadvantages just discussed; therefore,

molal concentration appears more frequently than molarity and normality in theoretical studies. It is possible to convert molality into molarity or normality if the final volume of the solution is observed or if the density is determined. In aqueous solutions more dilute than 0.1 M, it usually may be assumed for practical purposes that molality and molarity are equivalent. For example, a 1% solution by weight of sodium chloride with a specific gravity of 1.0053 is 0.170 M and 0.173 molal (0.173 $m$). The following difference between molar and molal solutions should also be noted. If another solute, containing neither sodium nor chloride ions, is added to a certain volume of a molal solution of sodium chloride, the solution remains 1 $m$ in sodium chloride, although the total volume and the weight of the solution increase. Molarity, of course, *decreases* when another solute is added because of the increase in volume of the solution.

Molal solutions are prepared by adding the proper weight of solvent to a carefully weighed quantity of the solute. The volume of the solvent can be calculated from the specific gravity, and the solvent can then be measured from a burette rather than weighed.

## Mole Fraction

Mole fraction is used frequently in experimentation involving theoretical considerations because it gives a measure of the relative proportion of moles of each constituent in a solution. It is expressed as

$$X_1 = \frac{n_1}{n_1 + n_2} \tag{5–1}$$

$$X_2 = \frac{n_2}{n_1 + n_2} \tag{5–2}$$

for a system of two constituents. Here $X_1$ is the mole fraction of constituent 1 (the subscript 1 is ordinarily used as the designation for the solvent), $X_2$ is the mole fraction of constituent 2 (usually the solute), and $n_1$ and $n_2$ are the numbers of moles of the respective constituents in the solution. The sum of the mole fractions of solute and solvent must equal unity. Mole fraction is also expressed in percentage terms by multiplying $X_1$ or $X_2$ by 100. In a solution containing 0.01 mole of solute and 0.04 mole of solvent, the mole fraction of the solute is $X_2 = 0.01/(0.04 + 0.01) = 0.20$. Because the mole fractions of the two constituents must equal 1, the mole fraction of the solvent is 0.8. The mole percent of the solute is 20%; the mole percent of the solvent is 80%.

The manner in which mole fraction is defined allows one to express the relationship between the number of solute and solvent molecules in a simple, direct way. In the example just given, it is readily seen that 2 of every 10 molecules in the solution are solute molecules, and it will be observed later that many of the properties of solutes and solvents are directly related to their mole fraction in the solution. For example, the partial vapor pressure above a solution brought about by the presence of a volatile solute is equal to the vapor pressure of the pure solute multiplied by the mole fraction of the solute in the solution.

## Percentage Expressions

The percentage method of expressing the concentration of pharmaceutical solutions is quite common. Percentage by weight signifies the number of grams of solute per 100 g of solution. A 10% by weight (% w/w) aqueous solution of glycerin contains 10 g of glycerin dissolved in enough water (90 g) to make 100 g of solution. Percentage by volume is expressed as the volume of solute in milliliters contained in 100 mL of the solution. Alcohol (United States Pharmacopeia) contains 92.3% by weight and 94.9% by volume of $C_2H_5OH$ at 15.56°C; that is, it contains 92.3 g of $C_2H_5OH$ in 100 g of solution or 94.9 mL of $C_2H_5OH$ in 100 mL of solution.

## Calculations Involving Concentration Expressions

The calculations involving the various concentration expressions are illustrated in the following example.

**EXAMPLE 5–1**

### Solutions of Ferrous Sulfate

An aqueous solution of exsiccated ferrous sulfate was prepared by adding 41.50 g of $FeSO_4$ to enough water to make 1000 mL of solution at 18°C. The density of the solution is 1.0375 and the molecular weight of $FeSO_4$ is 151.9. Calculate (*a*) the molarity; (*b*) the molality; (*c*) the mole fraction of $FeSO_4$, the mole fraction of water, and the mole percent of the two constituents; and (*d*) the percentage by weight of $FeSO_4$.

(*a*) Molarity:

$$\text{Moles of } FeSO_4 = \frac{\text{g of } FeSO_4}{\text{Molecular weight}}$$

$$= \frac{41.50}{151.9} = 0.2732$$

$$\text{Molarity} = \frac{\text{Moles of } FeSO_4}{\text{Liters of solution}} = \frac{0.2732}{1 \text{ liter}} = 0.2732 \text{ M}$$

(*b*) Molality:

$$\text{Grams of solution} = \text{Volume} \times \text{Density}$$

$$1000 \times 1.0375 = 1037.5 \text{ g}$$

$$\text{Grams of solvent} = \text{Grams of solution} - \text{Grams}$$

$$\text{of } FeSO_4 = 1037.5 - 41.5 = 996.0 \text{ g}$$

$$\text{Molality} = \frac{\text{Moles of } FeSO_4}{\text{kg of solvent}} = \frac{0.2732}{0.996} = 0.2743 \, m$$

(*c*) Mole fraction and mole percent:

$$\text{Moles of water} = \frac{996}{18.02} = 55.27 \text{ moles}$$

Mole fraction of $FeSO_4$:

$$X_2 = \frac{\text{Moles of } FeSO_4}{\text{Moles of water} + \text{Moles of } FeSO_4}$$

$$= \frac{0.2732}{55.27 + 0.2732} = 0.0049$$

Mole fraction of water:

$$X_1 = \frac{55.27}{55.27 + 0.2732} = 0.9951$$

Notice that
$X_1 + X_2 = 0.9951 + 0.0049 = 1.0000$
Mole percent of $FeSO_4 = 0.0049 \times 100 = 0.49\%$
Mole percent of water $= 0.9951 \times 100 = 99.51\%$

(*d*) Percentage by weight of $FeSO_4$:

$$= \frac{\text{g of } FeSO_4}{\text{g of solution}} \times 100$$

$$= \frac{41.50}{1037.5} \times 100 = 4.00\%$$

One can use the table of conversion equations, **Table 5–3**, to convert a concentration expression, say molality, into its value in molarity or mole fraction. Alternatively, knowing the weight, $w_1$, of a solvent, the weight, $w_2$, of the solute, and the molecular weight, $M_2$, of the solute, one can calculate the molarity, $c$, or the molality, $m$, of the solution. As an exercise, the reader should derive an expression relating $X_1$ to $X_2$ to the weights $w_1$ and $w_2$ and the solute's molecular weight, $M_2$. The data in *Example 5–1* are useful for determining whether your derived equation is correct.

## EQUIVALENT WEIGHTS

One gram atom of hydrogen weighs 1.008 g and consists of $6.02 \times 10^{23}$ atoms (Avogadro's number) of hydrogen. This gram atomic weight of hydrogen combines with $6.02 \times 10^{23}$ atoms of fluorine and with half of $6.02 \times 10^{23}$ atoms of oxygen. One gram atom of fluorine weighs 19 g, and 1 g atom of oxygen weighs 16 g. Therefore, 1.008 g of hydrogen combines with 19 g of fluorine and with half of 16 or 8 g

**TABLE 5–3**
## CONVERSION EQUATIONS FOR CONCENTRATION TERMS

A. Molality (moles of solute/kg of solvent, $m$) and mole fraction of solute ($X_2$):

$$X_2 = \frac{m}{m + \dfrac{1000}{M_1}}$$

$$m = \frac{1000X_2}{M_1(1 - X_2)}$$

$$= \frac{1000(1 - X_1)}{M_1 X_1}$$

B. Molarity (moles of solute/liter of solution, $c$) and mole fraction of solute ($X_2$):

$$X_2 = \frac{c}{c + \dfrac{1000\rho - cM_2}{M_1}}$$

$$c = \frac{1000\rho X_2}{M_1(1 - X_2) + M_2 X_2}$$

C. Molality ($m$) and molarity ($c$):

$$m = \frac{1000c}{1000\rho - M_2 c}$$

$$c = \frac{1000\rho}{\dfrac{1000}{m} + M_2}$$

D. Molality ($m$) and molarity ($c$) in terms of weight of solute, $w_2$, weight of solvent, $w_1$, and molecular weight, $M_2$, of solute:

$$m = \frac{w_2/M_2}{w_1/1000} = \frac{1000w_2}{w_1 M_2}$$

$$c = \frac{1000\rho w_2}{M_2(w_1 + w_2)}$$

Definition of terms:
$\rho$ = density of the solution (g/cm$^3$)
$M_1$ = molecular weight of the solvent
$M_2$ = molecular weight of the solute
$X_1$ = mole fraction of the solvent
$X_2$ = mole fraction of the solute
$w_1$ = weight of the solvent (g, mg, kg, etc.)
$w_2$ = weight of the solute (g, mg, kg, etc.)

of oxygen. The quantities of fluorine and oxygen combining with 1.008 g of hydrogen are referred to as the equivalent weight of the combining atoms. One equivalent (Eq) of fluorine (19 g) combines with 1.008 g of hydrogen. One equivalent of oxygen (8 g) also combines with 1.008 g of hydrogen.

It is observed that 1 equivalent weight (19 g) of fluorine is identical with its atomic weight. Not so with oxygen; its gram equivalent weight (8 g) is equal to half its atomic weight. Stated otherwise, the atomic weight of fluorine contains 1 Eq of fluorine, whereas the atomic weight of oxygen contains 2 Eq. The equation relating these atomic quantities is as

follows (the equation for molecules is quite similar to that for atoms, as seen in the next paragraph):

$$\text{Equivalent weight} = \frac{\text{Atomic weight}}{\substack{\text{Number of equivalents per} \\ \text{atomic weight (valence)}}} \quad (5\text{--}3)$$

The number of equivalents per atomic weight, namely, 1 for fluorine and 2 for oxygen, is the common *valence* of these elements. Many elements may have more than one valence and hence several equivalent weights, depending on the reaction under consideration. Magnesium will combine with two atoms of fluorine, and each fluorine can combine with one atom of hydrogen. Therefore, the valence of magnesium is 2, and its equivalent weight, according to equation (5–3), is one half of its atomic weight ($24/2 = 12$ g/Eq). Aluminum will combine with three atoms of fluorine; the valence of aluminum is therefore 3 and its equivalent weight is one third of its atomic weight, or $27/3 = 9$ g/Eq.

The concept of equivalent weights not only applies to atoms but also extends to molecules. The equivalent weight of sodium chloride is identical to its molecular weight, 58.5 g/Eq; that is, the equivalent weight of sodium chloride is the sum of the equivalent weights of sodium (23 g) and chlorine (35.5 g), or 58.5 g/Eq. The equivalent weight of sodium chloride is identical to its molecular weight, 58.5 g, because the valence of sodium and chlorine is each 1 in the compound. The equivalent weight of $Na_2CO_3$ is numerically half of its molecular weight. The valence of the carbonate ion, $CO_3^{2-}$, is 2, and its equivalent weight is $60/2 = 30$ g/Eq. Although the valence of sodium is 1, two atoms are present in $Na_2CO_3$, providing a weight of $2 \times 23$ g $= 46$ g; its equivalent weight is one half of this, or 23 g/Eq. The equivalent weight of $Na_2CO_3$ is therefore $30 + 23 = 53$ g, which is one half the molecular weight. The relationship of equivalent weight to molecular weight for molecules such as NaCl and $Na_2CO_3$ is [compare equation (5–3) for atoms]

$$\text{Equivalent weight (g/Eq)} = \frac{\text{molecular weight (g/mole)}}{\text{equivalent/mole}}$$

$$(5\text{--}4)$$

**EXAMPLE 5–2**

**Calculation of Equivalent Weight**

(*a*) What is the number of equivalents per mole of $K_3PO_4$, and what is the equivalent weight of this salt? (*b*) What is the equivalent weight of $KNO_3$? (*c*) What is the number of equivalents per mole of $Ca_3(PO_4)_2$, and what is the equivalent weight of this salt?

(*a*) $K_3PO_4$ represents 3 Eq/mole, and its equivalent weight is numerically equal to one third of its molecular weight, namely, (212 g/mole) $\div$ (3 Eq/mole) $= 70.7$ g/Eq.

(*b*) The equivalent weight of $KNO_3$ is also equal to its molecular weight, or 101 g/Eq.

(*c*) The number of equivalents per mole for $Ca_3(PO_4)_2$ is 6 (i.e., three calcium ions each with a valence of 2 or two phosphate ions each with a valence of 3). The equivalent weight of $Ca_3(PO_4)_2$ is therefore one sixth of its molecular weight, or $310/6 = 51.7$ g/Eq.

For a complex salt such as monobasic potassium phosphate (potassium acid phosphate), $KH_2PO_4$ (molecular weight, 136 g), the equivalent weight depends on how the compound is used. If it is used for its potassium content, the equivalent weight is identical to its molecular weight, or 136 g. When it is used as a buffer for its hydrogen content, the equivalent weight is one half of the molecular weight, $136/2 = 68$ g, because two hydrogen atoms are present. When used for its phosphate content, the equivalent weight of $KH_2PO_4$ is one third of the molecular weight, $136/3 = 45.3$ g, because the valence of phosphate is 3.

As defined in **Table 5–2**, the normality of a solution is the equivalent weight of the solute in 1 liter of solution. For NaF, $KNO_3$, and HCl, the number of equivalent weights equals the number of molecular weights, and normality is identical with molarity. For $H_3PO_4$, the equivalent weight is one third of the molecular weight, $98 g/3 = 32.67$ g/Eq, assuming complete reaction, and a 1 N solution of $H_3PO_4$ is prepared by weighing 32.67 g of $H_3PO_4$ and bringing it to a volume of 1 liter with water. For a 1 N solution of sodium bisulfate (sodium acid sulfate), $NaHSO_4$ (molecular weight 120 g), the weight of salt needed depends on the species for which the salt is used. If used for sodium or hydrogen, the equivalent weight would equal the molecular weight, or 120 g/Eq. If the solution were used for its sulfate content, $120/2 = 60$ g of $NaHSO_4$ would be weighed out and sufficient water added to make 1 liter of solution.

In electrolyte replacement therapy, solutions containing various electrolytes are injected into a patient to correct serious electrolyte imbalances. The concentrations are usually expressed as equivalents per liter or milliequivalents per liter. For example, the normal plasma concentration of sodium ions in humans is about 142 mEq/liter; the normal plasma concentration of bicarbonate ion, $HCO_3^-$, is 27 mEq/liter. Equation (5–4) is useful for calculating the quantity of salts needed to prepare electrolyte solutions in hospital practice. The moles in the numerator and denominator of equation (5–4) may be replaced with, say, liters to give

$$\text{Equivalent weight (in g/Eq)} = \frac{\text{Grams/liter}}{\text{Equivalents/liter}} \quad (5\text{–}5)$$

or

$$\text{Equivalent weight (in mg/mEq)} = \frac{\text{Milligrams/liter}}{\text{Milliequivalents/liter}}$$
$$(5\text{–}6)$$

Equivalent weight (analogous to molecular weight) is expressed in g/Eq, or what amounts to the same units, mg/mEq.

### EXAMPLE 5–3

**$Ca^{2+}$ in Human Plasma**

Human plasma contains about 5 mEq/liter of calcium ions. How many milligrams of calcium chloride dihydrate, $CaCl_2 \bullet 2H_2O$ (molecular weight 147 g/mole), are required to prepare 750 mL of a solution equal in $Ca^{2+}$ to human plasma? The equivalent weight of the dihydrate salt $CaCl_2 \bullet 2H_2O$ is half of its molecular weight, $147/2 = 73.5$ g/Eq, or 73.5 mg/mEq. Using equation (5–6), we obtain

$$73.5 \text{ mg/mEq} = \frac{\text{mg/liter}}{5 \text{ mEq/liter}}$$

$$73.5 \text{ mg/mEq} \times 5 \text{ mEq/liter} = 367.5 \text{ mg/liter}$$

For 750 cm$^3$, $367.5 \times \frac{750 \text{ mL}}{1000 \text{ mL}} = 275.6$ mg of $CaCl_2 \bullet 2H_2O$

### EXAMPLE 5–4

**Equivalent Weight and Molecular Weight**

Calculate the number of equivalents per liter of potassium chloride, molecular weight 74.55 g/mole, present in a 1.15% w/v solution of KCl.

Using equation (5–5) and noting that the equivalent weight of KCl is identical to its molecular weight, we obtain

$$74.55 \text{ g/Eq} = \frac{11.5 \text{ g/liter}}{\text{Eq/liter}}$$

$(11.5 \text{ g/liter})/(74.55 \text{ g/Eq}) = 0.154$ Eq/liter (or 154 mEq/liter)

### EXAMPLE 5–5

**Sodium Content**

What is the $Na^+$ content in mEq/liter of a solution containing 5.00 g of NaCl per liter of solution? The molecular weight and therefore the equivalent weight of NaCl is 58.5 g/Eq or 58.5 mg/mEq.

$$\text{mEq/liter} = \frac{\text{mg/liter}}{\text{Eq. wt.}} = \frac{5000 \text{ mg/liter}}{58.5 \text{ mg/mEq}}$$

$$= 85.47 \text{ mEq of } Na^+ \text{ per liter}$$

# IDEAL AND REAL SOLUTIONS

As stated earlier, the colligative properties of nonelectrolytes are ordinarily regular; on the other hand, solutions of electrolytes show apparent deviations. The remainder of this chapter relates to solutions of nonelectrolytes, except where comparison with an electrolyte system is desirable for clarity. Solutions of electrolytes are dealt with in Chapter 6.

An ideal gas is defined in Chapter 2 as one in which there is no attraction between the molecules, and it is found desirable to establish an ideal gas equation to which the properties of real gases tend as the pressure approaches zero. Consequently, the ideal gas law is referred to as a *limiting law*. It is convenient to define an *ideal solution* as one in which there is no change in the properties of the components, other than dilution, when they are mixed to form the solution. No heat is evolved or absorbed during the mixing process, and the final volume of the solution represents an additive property of the individual constituents. Stated another way, no shrinkage or expansion occurs when the substances are mixed. The constitutive properties, for example, the vapor pressure, refractive index, surface tension, and viscosity of the solution, are the weighted averages of the properties of the pure individual constituents.

KEY CONCEPT IDEALITY

Ideality in a gas implies the *complete absence* of attractive forces, and ideality in a solution means *complete uniformity* of attractive forces. Because a liquid is a highly condensed state, it cannot be expected to be devoid of attractive forces; nevertheless, if, in a mixture of A and B molecules, the forces between A and A, B and B, and A and B are all of the same order, the solution is considered to be ideal according to the definition just given.

Mixing substances with similar properties forms ideal solutions. For example, when 100 mL of methanol is mixed with 100 mL of ethanol, the final volume of the solution is 200 mL, and no heat is evolved or absorbed. The solution is nearly *ideal*.

When 100 mL of sulfuric acid is combined with 100 mL of water, however, the volume of the solution is about 180 mL at room temperature, and the mixing is attended by a considerable evolution of heat; the solution is said to be *nonideal*, or real. As with gases, some solutions are quite ideal in moderate concentrations, whereas others approach ideality only under extreme dilution.

## Escaping Tendency[3]

Two bodies are in thermal equilibrium when their temperatures are the same. If one body is heated to a higher temperature than the other, heat will flow "downhill" from the hotter to the colder body until both bodies are again in thermal equilibrium. This process is described in another way by using the concept of *escaping tendency*, and say that the heat in the hotter body has a greater escaping tendency than that in the colder one. Temperature is a quantitative measure of the escaping tendency of heat, and at thermal equilibrium, when both bodies finally have the same temperature, the escaping tendency of each constituent is the same in all parts of the system.

A quantitative measure of the escaping tendencies of material substances undergoing physical and chemical transformations is *free energy*. For a pure substance, the free energy per mole, or the *molar free energy*, provides a measure of escaping tendency; for the constituent of a solution, it is the *partial molar free energy* or *chemical potential* that is used as an expression of escaping tendency. Chemical potential is discussed in Chapter 3. The free energy of 1 mole of ice is greater than that of liquid water at 1 atm above 0°C and is spontaneously converted into water because

$$\Delta G = G_{liq} - G_{ice} < 0$$

At 0°C, at which temperature the system is in equilibrium, the molar free energies of ice and water are identical and $\Delta G = 0$. In terms of escaping tendencies, the escaping tendency of ice is greater than the escaping tendency of liquid water above 0°C, whereas at equilibrium, the escaping tendencies of water in both phases are identical.

## Ideal Solutions and Raoult's Law

The vapor pressure of a solution is a particularly important property because it serves as a quantitative expression of escaping tendency. In 1887, Raoult recognized that, in an ideal solution, the partial vapor pressure of each volatile constituent is equal to the vapor pressure of the pure constituent multiplied by its mole fraction in the solution. Thus, for two constituents A and B,

$$p_A = p_A{}^{\circ} X_A \qquad (5-7)$$

$$p_B = p_B{}^{\circ} X_B \qquad (5-8)$$

where $p_A$ and $p_B$ are the partial vapor pressures of the constituents over the solution when the mole fraction concentrations are $X_A$ and $X_B$, respectively. The vapor pressures of the pure components are $p_A{}^{\circ}$ and $p_B{}^{\circ}$, respectively. For example, if the vapor pressure of ethylene chloride in the pure state is 236 mm Hg at 50°C, then in a solution consisting of a mole fraction of 0.4 ethylene chloride and 0.6 benzene, the partial vapor pressure of ethylene chloride is 40% of 236 mm, or 94.4 mm. Thus, in an ideal solution, when liquid A is mixed with liquid B, the vapor pressure of A is reduced by dilution with B in a manner depending on the mole fractions of A and B present in the final solution. This will diminish the escaping tendency of each constituent, leading to a reduction in the rate of escape of the molecules of A and B from the surface of the liquid.

### EXAMPLE 5-6

**Partial Vapor Pressure**

**What is the partial vapor pressure of benzene and of ethylene chloride in a solution at a mole fraction of benzene of 0.6? The vapor pressure of pure benzene at 50°C is 268 mm, and the corresponding $p_A{}^{\circ}$ for ethylene chloride is 236 mm. We have**

$$p_B = 268 \times 0.6 = 160.8 \, \text{mm}$$

$$p_A = 236 \times 0.4 = 94.4 \, \text{mm}$$

If additional volatile components are present in the solution, each will produce a partial pressure above the solution, which can be calculated from Raoult's law. The total pressure is the sum of the partial pressures of all the constituents. In *Example 5–6*, the total vapor pressure $P$ is calculated as follows:

$$P = p_A + p_B = 160.8 + 94.4 = 255.2 \, \text{mm}$$

The vapor pressure–composition curve for the binary system benzene and ethylene chloride at 50°C is shown in **Figure 5–1**. The three lines represent the partial pressure of ethylene chloride, the partial pressure of benzene, and the total

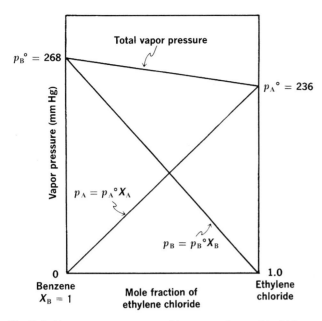

**Fig. 5–1.** Vapor pressure–composition curve for an ideal binary system.

pressure of the solution as a function of the mole fraction of the constituents.

## Aerosols and Raoult's Law

Aerosol dispensers have been used to package some drugs since the early 1950s. An aerosol contains the drug concentrated in a solvent or carrier liquid and a propellant mixture of the proper vapor characteristics. Chlorofluorocarbons (CFCs) were very popular propellants in aerosols until about 1989. Since that time they have been replaced in nearly every country because of the negative effects that CFCs have on the Earth's ozone layer. Today, two volatile hydrocarbons are commonly used as propellants in metered dose inhalers for treating asthma, hydrofluoroalkane 134a (1,1,1,2,-tetrafluoroethane) or hydrofluoroalkane 227 (1,1,1,2,3,3,3-heptafluoropropane) or combinations of the two. Early metered dose inhalers commonly used trichloromonofluoromethane (CFC propellant 11) and dichlorodifluoromethane (CFC propellant 12) as propellants. CFC 11 and CFC 12 were used in various proportions to yield the proper vapor pressure and density at room temperature. Although still used with drugs, these halogenated hydrocarbons are no longer used in cosmetic aerosols and have been replaced by nitrogen and unsubstituted hydrocarbons. Since propellants represent the vast majority (>99%) of the dose delivered by a metered dose inhaler, it must be nontoxic to the patient.

**EXAMPLE 5–7**

**Aerosol Vapor Pressure**

The vapor pressure of pure CFC 11 (molecular weight 137.4) at 21°C is $p_{11}° = 13.4$ lb/in² (psi) and that of CFC 12 (molecular weight 120.9) is $p_{12}° = 84.9$ psi. A 50:50 mixture by gram weight of the two propellants consists of $50 \div 137.4$ g/mole = 0.364 mole of CFC

11 and $50$ g $\div 120.9$ g/mole $= 0.414$ mole of CFC 12. What is the partial pressure of CFCs 11 and 12 in the 50:50 mixture, and what is the total vapor pressure of this mixture? We write

$$p_{11} = \frac{n_{11}}{n_{11} + n_{12}} p_{11}° = \frac{0.364}{0.364 + 0.414} (13.4) = 6.27 \text{ psi}$$

$$p_{12} = \frac{n_{12}}{n_{11} + n_{12}} p_{12}° = \frac{0.414}{0.364 + 0.414} (84.9) = 45.2 \text{ psi}$$

The total vapor pressure of the mixture is

$$6.27 + 45.2 = 51.5 \text{ psi}$$

To convert to gauge pressure (psig), one subtracts the atmospheric pressure of 14.7 psi:

$$51.5 - 14.7 = 36.8 \text{ psig}$$

The psi values just given are measured with respect to zero pressure rather than with respect to the atmosphere and are sometimes written psia to signify *absolute* pressure.

## Real Solutions

Ideality in solutions presupposes complete uniformity of attractive forces. Many examples of solution pairs are known, however, in which the "cohesive" attraction of A for A exceeds the "adhesive" attraction existing between A and B. Similarly, the attractive forces between A and B may be greater than those between A and A or B and B. This may occur even though the liquids are miscible in all proportions. Such mixtures are *real* or *nonideal*; that is, they do not adhere to Raoult's law throughout the entire range of composition. Two types of deviation from Raoult's law are recognized, *negative deviation* and *positive deviation*.

When the "adhesive" attractions between molecules of different species exceed the "cohesive" attractions between like molecules, the vapor pressure of the solution is less than that expected from Raoult's ideal solution law, and *negative deviation* occurs. If the deviation is sufficiently great, the total vapor pressure curve shows a minimum, as observed in **Figure 5–2**, where A is chloroform and B is acetone.

The dilution of constituent A by addition of B normally would be expected to reduce the partial vapor pressure of A; this is the simple dilution effect embodied in Raoult's law. In the case of liquid pairs that show negative deviation from the law, however, the addition of B to A tends to reduce the vapor pressure of A to a greater extent than can be accounted for by the simple dilution effect. Chloroform and acetone manifest such an attraction for one another through the formation of a hydrogen bond, thus further reducing the escaping tendency of each constituent. This pair forms a weak compound,

$$Cl_3C - H \cdots O = C(CH_3)_2$$

that can be isolated and identified. Reactions between dipolar molecules, or between a dipolar and a nonpolar molecule, may also lead to negative deviations. The interaction in these cases, however, is usually so weak that no definite compound can be isolated.

When the interaction between A and B molecules is less than that between molecules of the pure constituents,

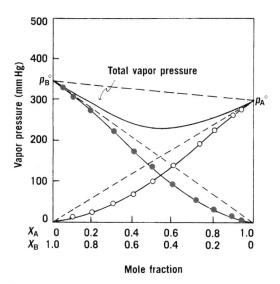

**Fig. 5–2.** Vapor pressure of a system showing negative deviation from Raoult's law.

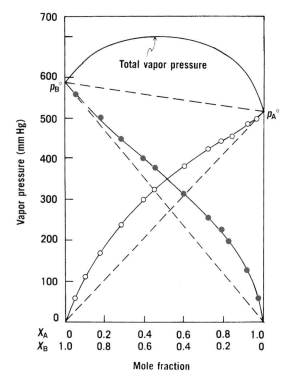

**Fig. 5–3.** Vapor pressure of a system showing positive deviation from Raoult's law.

the presence of B molecules reduces the interaction of the A molecules, and A molecules correspondingly reduce the B—B interaction. Accordingly, the dissimilarity of polarities or internal pressures of the constituents results in a greater escaping tendency of both the A and the B molecules. The partial vapor pressure of the constituents is greater than that expected from Raoult's law, and the system is said to exhibit *positive deviation.* The total vapor pressure often shows a maximum at one particular composition if the deviation is sufficiently large. An example of positive deviation is shown in **Figure 5–3.** Liquid pairs that demonstrate positive deviation are benzene and ethyl alcohol, carbon disulfide and acetone, and chloroform and ethyl alcohol.

Raoult's law does not apply over the entire concentration range in a nonideal solution. It describes the behavior of either component of a real liquid pair only when that substance is present in high concentration and thus is considered to be the solvent. Raoult's law can be expressed as

$$p_{\text{solvent}} = p_{\text{solvent}}^{\circ} X_{\text{solvent}} \qquad (5\text{–}9)$$

in such a situation, and it is valid only for the solvent of a nonideal solution that is sufficiently dilute with respect to the solute. It cannot hold for the component in low concentration, that is, the solute, in a dilute nonideal solution.

These statements will become clearer when one observes, in **Figure 5–2,** that the actual vapor pressure curve of chloroform (component A) approaches the ideal curve defined by Raoult's law as the solution composition approaches pure chloroform. Raoult's law can be used to describe the behavior of chloroform when it is present in high concentration (i.e., when it is the solvent). The ideal equation is not applicable to acetone (component B), however, which is present in low concentration in this region of the diagram, because the actual curve for acetone does not coincide with the ideal line. When one studies the left side of **Figure 5–2,** one observes that the conditions are reversed: Acetone is considered to be the

solvent here, and its vapor pressure curve tends to coincide with the ideal curve. Chloroform is the solute in this range, and its curve does not approach the ideal line. Similar considerations apply to **Figure 5–3.**

## Henry's Law

The vapor pressure curves for both acetone and chloroform as *solutes* are observed to lie considerably below the vapor pressure of an ideal mixture of this pair. The molecules of solute, being in relatively small number in the two regions of the diagram, are completely surrounded by molecules of solvent and so reside in a uniform environment. Therefore, the partial pressure or escaping tendency of chloroform at low concentration is in some way proportional to its mole fraction, but, as observed in **Figure 5–2,** the proportionality constant is not equal to the vapor pressure of the pure substance. The vapor pressure–composition relationship of the solute cannot be expressed by Raoult's law but instead by an equation known as *Henry's law:*

$$p_{\text{solute}} - k_{\text{solute}} X_{\text{solute}} \qquad (5\text{–}10)$$

where $k$ for chloroform is less than $p_{\text{CHCL}_3}^{\circ}$. Henry's law applies to the solute and Raoult's law applies to the solvent in dilute solutions of real liquid pairs. Of course, Raoult's law also applies over the entire concentration range (to both solvent and solute) when the constituents are sufficiently similar to form an ideal solution. Under any circumstance, when the partial vapor pressures of both of the constituents are directly proportional to the mole fractions over the entire range, the solution is said to be ideal; Henry's law becomes identical with Raoult's law, and $k$ becomes equal to $p°$. Henry's law is used for the study of gas solubilities discussed later in the book.

## Distillation of Binary Mixtures

The relationship between vapor pressure (and hence boiling point) and composition of binary liquid phases is the underlying principle in distillation. In the case of miscible liquids, instead of plotting vapor pressure versus composition, it is more useful to plot the boiling points of the various mixtures, determined at atmospheric pressure, against composition.

The higher the vapor pressure of a liquid—that is, the more volatile it is—the lower is the boiling point. Because the vapor of a binary mixture is always richer in the more volatile constituent, the process of distillation can be used to separate the more volatile from the less volatile constituent. **Figure 5–4** shows a mixture of a high-boiling liquid A and a low-boiling liquid B. A mixture of these substances having the composition $a$ is distilled at the boiling point $b$. The composition of the vapor $v_1$ in equilibrium with the liquid at this temperature is $c$; this is also the composition of the distillate when it is condensed. The vapor is therefore richer in B than the liquid from which it was distilled. If a fractionating column is used, A and B can be completely separated. The vapor rising in the column is met by the condensed vapor or downward-flowing liquid. As the rising vapor is cooled by contact with the liquid, some of the lower-boiling fraction condenses, and the vapor contains more of the volatile component than it did when it left the retort. Therefore, as the vapor proceeds up the fractionating column, it becomes progressively richer in the more volatile component B, and the liquid returning to the distilling retort becomes richer in the less volatile component A.

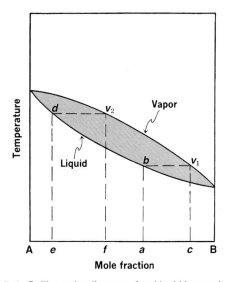

**Fig. 5–4.** Boiling point diagram of an ideal binary mixture.

**Figure 5–4** shows the situation for a pair of miscible liquids exhibiting ideal behavior. Because vapor pressure curves can show maxima and minima (see **Figs. 5–2** and **5–3**), it follows that boiling point curves will show corresponding minima and maxima, respectively. With these mixtures, distillation produces either pure A or pure B plus a mixture of constant composition and constant boiling point. This latter is known as an *azeotrope* (Greek: "boil unchanged") or *azeotropic mixture*. It is not possible to separate such a mixture completely into two pure components by simple fractionation. If the vapor pressure curves show a minimum (i.e., negative deviation from Raoult's law), the azeotrope has the highest boiling point of all the mixtures possible; it is therefore least volatile and remains in the flask, whereas either pure A or pure B is distilled off. If the vapor pressure curve exhibits a maximum (showing a positive deviation from Raoult's law), the azeotrope has the lowest boiling point and forms the distillate. Either pure A or pure B then remains in the flask.

When a mixture of HCl and water is distilled at atmospheric pressure, an azeotrope is obtained that contains 20.22% by weight of HCl and that boils at 108.58°C. The composition of this mixture is accurate and reproducible enough that the solution can be used as a standard in analytic chemistry. Mixtures of water and acetic acid and of chloroform and acetone yield azeotropic mixtures with maxima in their boiling point curves and minima in their vapor pressure curves. Mixtures of ethanol and water and of methanol and benzene both show the reverse behavior, namely, minima in the boiling point curves and maxima in the vapor pressure curves.

When a mixture of two practically *immiscible* liquids is heated while being agitated to expose the surfaces of both liquids to the vapor phase, each constituent independently exerts its own vapor pressure as a function of temperature as though the other constituent were not present. Boiling begins, and distillation may be effected when the sum of the partial pressures of the two immiscible liquids just exceeds the atmospheric pressure. This principle is applied in *steam distillation*, whereby many organic compounds insoluble in water can be purified at a temperature well below the point at which decomposition occurs. Thus, bromobenzene alone boils at 156.2°C, whereas water boils at 100°C at a pressure of 760 mm Hg. A mixture of the two, however, in any proportion, boils at 95°C. Bromobenzene can thus be distilled at a temperature 61°C below its normal boiling point. Steam distillation is particularly useful for obtaining volatile oils from plant tissues without decomposing the oils.

# COLLIGATIVE PROPERTIES

When a *nonvolatile solute* is combined with a *volatile solvent*, the vapor above the solution is provided solely by the solvent. The solute reduces the escaping tendency of the solvent, and, on the basis of Raoult's law, the vapor pressure of a solution

The freezing point, boiling point, and osmotic pressure of a solution also depend on the relative proportion of the molecules of the solute and the solvent. These are called *colligative properties* (Greek: "collected together") because they depend chiefly on the number rather than on the nature of the constituents.

In most solutions, changes in concentration are accompanied by linear and proportional changes in the cardinal colligative properties of the solvent—vapor pressure, freezing point, and boiling point. Measuring any of these properties provides an indirect indication of osmolality, but among them, only vapor pressure can be determined passively without a forced change in the sample's physical state.

containing a nonvolatile solute is lowered proportional to the relative number.

## Lowering of the Vapor Pressure

According to Raoult's law, the vapor pressure, $p_1$, of a solvent over a dilute solution is equal to the vapor pressure of the pure solvent, $p_1{}^\circ$, times the mole fraction of solvent in the solution, $X_1$. Because the solute under discussion here is considered to be nonvolatile, the vapor pressure of the solvent, $p_1$, is identical to the total pressure of the solution, $p$.

It is more convenient to express the vapor pressure of the solution in terms of the concentration of the solute rather than the mole fraction of the solvent, and this may be accomplished in the following way. The sum of the mole fractions of the constituents in a solution is unity:

$$X_1 + X_2 = 1 \qquad (5\text{--}11)$$

Therefore,

$$X_1 = 1 - X_2 \qquad (5\text{--}12)$$

where $X_1$ is the mole fraction of the solvent and $X_2$ is the mole fraction of the solute. Raoult's equation can be modified by substituting equation (5–12) for $X_1$ to give

$$p = p_1{}^\circ(1 - X_2) \qquad (5\text{--}13)$$

$$p_1{}^\circ - p = p_1{}^\circ X_2 \qquad (5\text{--}14)$$

$$\frac{p_1{}^\circ - p}{p_1{}^\circ} = \frac{\Delta p}{p_1{}^\circ} = X_2 = \frac{n_2}{n_1 + n_2} \qquad (5\text{--}15)$$

In equation (5–15), $\Delta p = p_1{}^\circ - p$ is the lowering of the vapor pressure and $\Delta p / p_1{}^\circ$ is the *relative vapor pressure lowering*. The relative vapor pressure lowering depends only on the mole fraction of the solute, $X_2$, that is, on the number of solute particles in a definite volume of solution. Therefore, the relative vapor pressure lowering is a *colligative property*.

**EXAMPLE 5–8**

**Relative Vapor Pressure Lowering of a Solution**

Calculate the relative vapor pressure lowering at 20°C for a solution containing 171.2 g of sucrose ($w_2$) in 100 g ($w_1$) of water. The

molecular weight of sucrose ($M_2$) is 342.3 and the molecular weight of water ($M_1$) is 18.02 g/mole. We have

$$\text{Moles of sucrose} = n_2 = \frac{w_2}{M_2} = \frac{171.2}{342.3} = 0.500$$

$$\text{Moles of water} = n_1 = \frac{w_1}{M_1} = 1000/18.02 = 55.5$$

$$\frac{\Delta p}{p_1{}^\circ} = X_2 = \frac{n_2}{n_1 + n_2}$$

$$\frac{\Delta p}{p_1{}^\circ} = \frac{0.50}{55.5 + 0.50} = 0.0089$$

Notice that in *Example 5–8*, the relative vapor pressure lowering is a dimensionless number, as would be expected from its definition. The result can also be stated as a percentage; the vapor pressure of the solution has been lowered 0.89% by the 0.5 mole of sucrose.

The mole fraction, $n_2/(n_1 + n_2)$, is nearly equal to, and may be replaced by, the mole ratio $n_2/n_1$ in a dilute solution such as this one. Then, the relative vapor pressure lowering can be expressed in terms of molal concentration of the solute by setting the weight of solvent $w_1$ equal to 1000 g. For an aqueous solution,

$$X_2 = \frac{\Delta P}{p_1{}^\circ} \cong \frac{n_2}{n_1} = \frac{w_2/M_2}{1000/M_1} = \frac{m}{55.5} = 0.018\,m \quad (5\text{--}16)$$

**EXAMPLE 5–9**

**Calculation of the Vapor Pressure**

Calculate the vapor pressure when 0.5 mole of sucrose is added to 1000 g of water at 20°C. The vapor pressure of water at 20°C is 17.54 mm Hg. The vapor pressure lowering of the solution is

$$\Delta p = p_1{}^\circ X_2 \cong p_1{}^\circ \times 0.018 \times m$$
$$= 17.54 \times 0.018 \times 0.5$$
$$= 0.158 \text{ mm} \cong 0.16 \text{ mm}$$

**The final vapor pressure is**

$$17.54 - 0.16 = 17.38 \text{ mm}$$

## Elevation of the Boiling Point

The normal boiling point is the temperature at which the vapor pressure of the liquid becomes equal to an external pressure of

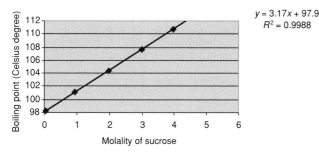

**Fig. 5–5.** Theoretical plot of the normal boiling point for water (solvent) as a function of molality in solutions containing sucrose (a nonvolatile solute) in increasing concentrations. Note that the normal boiling point of water increases as the concentration of sucrose increases. This is known as boiling point elevation.

### TABLE 5–4
**EBULLIOSCOPIC ($K_b$) AND CRYOSCOPIC ($K_f$) CONSTANTS FOR VARIOUS SOLVENTS**

| Substance | Boiling Point (°C) | $K_b$ | Freezing Point (°C) | $K_f$ |
|---|---|---|---|---|
| Acetic acid | 118.0 | 2.93 | 16.7 | 3.9 |
| Acetone | 56.0 | 1.71 | −94.82* | 2.40* |
| Benzene | 80.1 | 2.53 | 5.5 | 5.12 |
| Camphor | 208.3 | 5.95 | 178.4 | 37.7 |
| Chloroform | 61.2 | 3.54 | −63.5 | 4.96 |
| Ethyl alcohol | 78.4 | 1.22 | −114.49* | 3* |
| Ethyl ether | 34.6 | 2.02 | −116.3 | 1.79* |
| Phenol | 181.4 | 3.56 | 42.0 | 7.27 |
| Water | 100.0 | 0.51 | 0.00 | 1.86 |

*From G. Kortum and J. O'M. Bockris, *Textbook of Electrochemistry*, Vol. II, Elsevier, New York, 1951, pp. 618, 620.

760 mm Hg. A solution will boil at a higher temperature than will the pure solvent. This is the colligative property called boiling point elevation. As shown in **Figure 5–5**, the more of the solute that is dissolved, the greater is the effect. The boiling point of a solution of a nonvolatile solute is higher than that of the pure solvent owing to the fact that the solute lowers the vapor pressure of the solvent. This may be seen by referring to the curves in **Figure 5–6**. The vapor pressure curve for the solution lies below that of the pure solvent, and the temperature of the solution must be elevated to a value above that of the solvent in order to reach the normal boiling point. The elevation of the boiling point is shown in the figure as $T - T_o = \Delta T_b$. The ratio of the elevation of the boiling point, $\Delta T_b$, to the vapor pressure lowering, $\Delta p = p° - p$, at 100°C is approximately a constant at this temperature; it is written as

$$\frac{\Delta T_b}{\Delta p} = k' \qquad (5\text{–}17)$$

or

$$\Delta T_b = k' \Delta p \qquad (5\text{–}18)$$

Moreover, because $p°$ is a constant, the boiling point elevation may be considered proportional to $\Delta p/p°$, the relative

lowering of vapor pressure. By Raoult's law, however, the relative vapor pressure lowering is equal to the mole fraction of the solute; therefore,

$$\Delta T_b = k X_2 \qquad (5\text{–}19)$$

Because the boiling point elevation depends only on the mole fraction of the solute, it is a colligative property.

In dilute solutions, $X_2$ is equal approximately to $m/(1000/M_1)$ [equation (5–16)], and equation (5–19) can be written as

$$\Delta T_b = \frac{k M_1}{1000} m \qquad (5\text{–}20)$$

or

$$\Delta T_b = K_b m \qquad (5\text{–}21)$$

where $\Delta T_b$ is known as the *boiling point elevation* and $K_b$ is called the *molal elevation constant* or the *ebullioscopic constant*. $K_b$ has a characteristic value for each solvent, as seen in **Table 5–4**. It may be considered as the boiling point elevation for an ideal 1 $m$ solution. Stated another way, $K_b$ is the ratio of the boiling point elevation to the molal concentration in an extremely dilute solution in which the system is approximately ideal.

The preceding discussion constitutes a plausible argument leading to the equation for boiling point elevation. A more satisfactory derivation of equation (5–21), however, involves the application of the Clapeyron equation, which is written as

$$\frac{\Delta T_b}{\Delta p} = T_b \frac{V_v - V_1}{\Delta H_v} \qquad (5\text{–}22)$$

where $V_v$ and $V_1$ are the molar volume of the gas and the molar volume of the liquid, respectively, $T_b$ is the boiling point of the solvent, and $\Delta H_v$ is the molar heat of vaporization. Because $V_1$ is negligible compared to $V_v$, the equation becomes

$$\frac{\Delta T_b}{\Delta p} = T_b \frac{V_v}{\Delta H_v} \qquad (5\text{–}23)$$

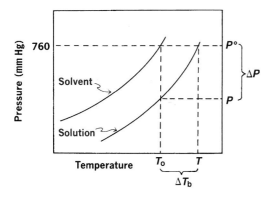

**Fig. 5–6.** Boiling point elevation of the solvent due to addition of a solute (not to scale).

and $V_v$, the volume of 1 mole of gas, is replaced by $RT_b/p°$ to give

$$\frac{\Delta T_b}{\Delta p} = \frac{RT_b^2}{p°\,\Delta H_v} \qquad (5\text{–}24)$$

or

$$\Delta T_b = \frac{RT_b^2}{\Delta H_v}\frac{\Delta p}{p°} \qquad (5\text{–}25)$$

From equation (5–16), $\Delta p/p_1° = X_2$, and equation (5–25) can be written as

$$\Delta T_b = \frac{RT_b^2}{\Delta H_v}X_2 = kX_2 \qquad (5\text{–}26)$$

which provides a more exact equation with which to calculate $\Delta T_b$.

Replacing the relative vapor pressure lowering $\Delta p/p_1°$ by $m/(1000/M_1)$ according to the approximate expression (5–16), in which $w_2/M_2 = m$ and $w_1 = 1000$ s, we obtain the formula

$$\Delta T_b = \frac{RT_b^2 M_1}{1000\,\Delta H_v}m = k_b m \qquad (5\text{–}27)$$

Equation (5–27) provides a less exact expression with which to calculate $\Delta T_b$.

For water at 100°C, we have $T_b = 373.2$ K, $\Delta H_v = 9720$ cal/mole, $M_1 = 18.02$ g/mole, and $R = 1.987$ cal/mole deg.

### EXAMPLE 5–10

**Calculation of the Elevation Constant**

A 0.200 $m$ aqueous solution of a drug gave a boiling point elevation of 0.103°C. Calculate the approximate molal elevation constant for the solvent, water. Substituting into equation (5–21) yields

$$K_b = \frac{\Delta T_b}{m} = \frac{0.103}{0.200} = 0.515\ \text{deg kg/mole}$$

The proportionality between $\Delta T_b$ and the molality is exact only at infinite dilution, at which the properties of real and ideal solutions coincide. The ebullioscopic constant, $K_b$, of a solvent can be obtained experimentally by measuring $\Delta T_b$ at various molal concentrations and extrapolating to infinite dilution ($m = 0$), as seen in **Figure 5–7**.

## Depression of the Freezing Point

The normal freezing point or melting point of a pure compound is the temperature at which the solid and the liquid phases are in equilibrium under a pressure of 1 atm. Equilibrium here means that the tendency for the solid to pass into the liquid state is the same as the tendency for the reverse process to occur, because both the liquid and the solid have the same escaping tendency. The value $T_0$, shown in **Figure 5–8**, for water saturated with air at this pressure is arbitrarily assigned a temperature of 0°C. The *triple point* of air-free water, at which solid, liquid, and vapor are in equilibrium, lies

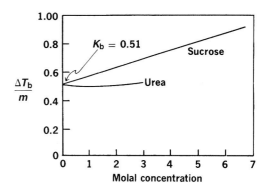

**Fig. 5–7.** The influence of concentration on the ebullioscopic constant.

at a pressure of 4.58 mm Hg and a temperature of 0.0098°C. It is not identical with the ordinary freezing point of water at atmospheric pressure but is rather the freezing point of water under the pressure of its own vapor. We shall use the triple point in the following argument because the depression $\Delta T_f$ here does not differ significantly from $\Delta T_f$ at a pressure of 1 atm. The two freezing point depressions referred to are illustrated in **Figure 5–7**. The $\Delta T_b$ of **Figure 5–6** is also shown in the diagram.

If a solute is dissolved in the liquid at the triple point, the escaping tendency or vapor pressure of the liquid solvent is lowered below that of the pure solid solvent. The temperature must drop to reestablish equilibrium between the liquid and the solid. Because of this fact, the freezing point of a solution is always lower than that of the pure solvent. It is assumed that the solvent freezes out in the pure state rather than as a *solid solution* containing some of the solute. When such a complication does arise, special calculations, not considered here, must be used.

The more concentrated the solution, the farther apart are the solvent and the solution curves in the diagram (see **Fig. 5–8**) and the greater is the freezing point depression. Accordingly, a situation exists analogous to that described

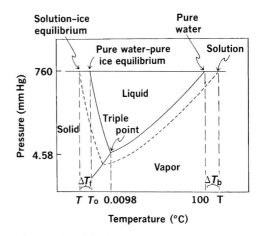

**Fig. 5–8.** Depression of the freezing point of the solvent, water, by a solute (not to scale).

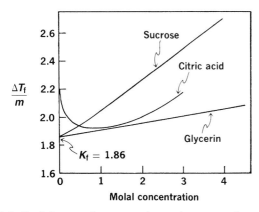

**Fig. 5–9.** The influence of concentration on the cryoscopic constant for water.

for the boiling point elevation, and the freezing point depression is proportional to the molal concentration of the solute. The equation is

$$\Delta T_f = K_f m \qquad (5\text{–}28)$$

or

$$\Delta T_f = K_f \frac{1000 w_2}{w_1 M_2} \qquad (5\text{–}29)$$

$\Delta T_f$ is the *freezing point depression*, and $K_f$ is the *molal depression constant* or the *cryoscopic constant*, which depends on the physical and chemical properties of the solvent.

The freezing point depression of a solvent is a function only of the number of particles in the solution, and for this reason it is referred to as a *colligative* property. The depression of the freezing point, like the boiling point elevation, is a direct result of the lowering of the vapor pressure of the solvent. The value of $K_f$ for water is 1.86. It can be determined experimentally by measuring $\Delta T_f/m$ at several molal concentrations and extrapolating to zero concentration. As seen in **Figure 5–9**, $K_f$ approaches the value of 1.86 for water solutions of sucrose and glycerin as the concentrations tend toward zero, and equation **(5–28)** is valid only in very dilute solutions. The apparent cryoscopic constant for higher concentrations can be obtained from **Figure 5–9**. For work in pharmacy and biology, the $K_f$ value of 1.86 can be rounded off to 1.9, which is good approximation for practical use with aqueous solutions, where concentrations are usually lower than 0.1 M. The value of $K_f$ for the solvent in a solution of citric acid is observed not to approach 1.86. This abnormal behavior is to be expected when dealing with solutions of electrolytes. Their irrationality will be explained in Chapter 6, and proper steps will be taken to correct the difficulty.

$K_f$ can also be derived from Raoult's law and the Clapeyron equation. For water at its freezing point, $T_f = 273.2$ K, $\Delta H_f$ is 1437 cal/mole, and

$$K_f = \frac{1.987 \times (273.2)^2 \times 18.02}{1000 \times 1437} = 1.86 \text{ deg kg/mole}$$

The cryoscopic constants, together with the ebullioscopic constants, for some solvents at infinite dilution are given in **Table 5–4**.

**EXAMPLE 5–11**

**Calculation of Freezing Point**

What is the freezing point of a solution containing 3.42 g of sucrose and 500 g of water? The molecular weight of sucrose is 342. In this relatively dilute solution, $K_f$ is approximately equal to 1.86. We have

$$\Delta T_f = K_f m = K_f \frac{1000 w_2}{w_1 M_2}$$

$$\Delta T_f = 1.86 \times \frac{1000 \times 3.42}{500 \times 342}$$

$$\Delta T_f = 0.037^\circ C$$

Therefore, the freezing point of the aqueous solution is $-0.037^\circ C$.

**EXAMPLE 5–12**

**Freezing Point Depression**

What is the freezing point depression of a 1.3 *m* solution of sucrose in water?

From the graph in Figure 5–8, one observes that the cryoscopic constant at this concentration is about 2.1 rather than 1.86. Thus, the calculation becomes

$$\Delta T_f = K_f \times m = 2.1 \times 1.3 = 2.73^\circ C$$

## Osmotic Pressure

If cobalt chloride is placed in a parchment sac and suspended in a beaker of water, the water gradually becomes red as the solute diffuses throughout the vessel. In this process of *diffusion*, both the solvent and the solute molecules migrate freely. On the other hand, if the solution is confined in a membrane permeable only to the solvent molecules, the phenomenon known as *osmosis* (Greek: "a push or impulse")[4] occurs, and the barrier that permits only the molecules of one of the components (usually water) to pass through is known as a *semipermeable membrane*. A thistle tube over the wide opening of which is stretched a piece of untreated cellophane can be used to demonstrate the principle, as shown in **Figure 5–10**. The tube is partly filled with a concentrated solution of sucrose, and the apparatus is lowered into a beaker of water. The passage of water through the semipermeable membrane into the solution eventually creates enough pressure to drive the sugar solution up the tube until the hydrostatic pressure of the column of liquid equals the pressure causing the water to pass through the membrane and enter the thistle tube. When this occurs, the solution ceases to rise in the tube. Osmosis is therefore defined as the passage of the solvent into a solution through a semipermeable membrane. This process tends to equalize the escaping tendency of the solvent on both sides of the membrane. Escaping tendency can be measured in terms of vapor pressure or the closely related colligative property *osmotic pressure*. It should be evident that osmosis can also

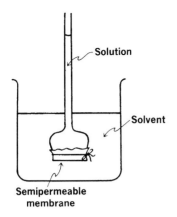

**Fig. 5–10.** Apparatus for demonstrating osmosis.

take place when a concentrated solution is separated from a less concentrated solution by a semipermeable membrane.

Osmosis in some cases is believed to involve the passage of solvent through the membrane by a distillation process or by dissolving in the material of the membrane in which the solute is insoluble. In other cases, the membrane may act as a sieve, having a pore size sufficiently large to allow passage of solvent but not of solute molecules.

In either case, the phenomenon of osmosis depends on the fact that the chemical potential (a thermodynamic expression of escaping tendency) of a solvent molecule in solution is less than exists in the pure solvent. Solvent therefore passes

spontaneously into the solution until the chemical potentials of solvent and solution are equal. The system is then at equilibrium. It may be advantageous for the student to consider osmosis in terms of the following sequence of events. (*a*) The addition of a nonvolatile solute to the solvent forms a solution in which the vapor pressure of the solvent is reduced (see Raoult's law). (*b*) If pure solvent is now placed adjacent to the solution but separated from it by a semipermeable membrane, solvent molecules will pass through the membrane into the solution in an attempt to dilute out the solute and raise the vapor pressure back to its original value (namely, that of the original solvent). (*c*) The osmotic pressure that is set up as a result of this passage of solvent molecules can be determined either by measuring the hydrostatic head appearing in the solution or by applying a known pressure that just balances the osmotic pressure and prevents any net movement of solvent molecules into the solution. The latter is the preferred technique. The osmotic pressure thus obtained is proportional to the reduction in vapor pressure brought about by the concentration of solute present. Because this is a function of the molecular weight of the solute, osmotic pressure is a colligative property and can be used to determine molecule weights.

An *osmotic pressure* osmometer (**Fig. 5–11**) is based on the same principle as the thistle tube apparatus shown in **Figure 5–10**. Once equilibrium has been attained, the height of the solution in the capillary tube on the solution side of the membrane is greater by the amount *h* than the height in

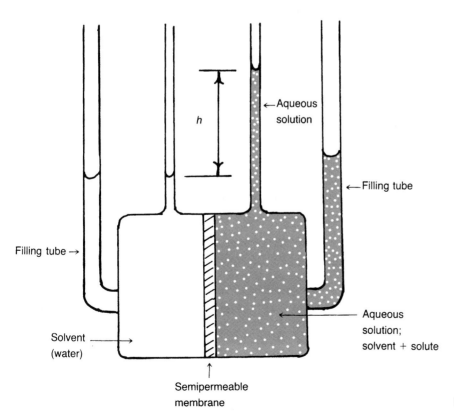

**Fig. 5–11.** Osmotic pressure osmometer.

the capillary tube on the solvent (water) side. The hydrostatic head, $h$, is related to the osmotic pressure through the expression osmotic pressure $\pi$ (atm) = Height $h$ × Solution density $\rho$ × Gravity acceleration. The two tubes of large bore are for filling and discharging the liquids from the compartments of the apparatus. The height of liquid in these two large tubes does not enter into the calculation of osmotic pressure. The determination of osmotic pressure is discussed in some detail in the next section.

## Measurement of Osmotic Pressure

The osmotic pressure of the sucrose solution referred to in the last section is not measured conveniently by observing the height that the solution attains in the tube at equilibrium. The concentration of the final solution is not known because the passage of water into the solution dilutes it and alters the concentration. A more exact measure of the osmotic pressure of the undiluted solution is obtained by determining the excess pressure on the solution side that just prevents the passage of solvent through the membrane. Osmotic pressure is defined as the excess pressure, or pressure greater than that above the pure solvent, that must be applied to the solution to prevent the passage of the solvent through a perfect semipermeable membrane. In this definition, it is assumed that a semipermeable sac containing the solution is immersed in the *pure* solvent.

In 1877, the botanist Wilhelm Pfeffer measured the osmotic pressure of sugar solutions, using a porous cup impregnated with a deposit of cupric ferrocyanide, $Cu_2Fe(CN)_6$, as the semipermeable membrane. The apparatus was provided with a manometer to measure the pressure. Although many improvements have been made through the years, including the attachment of sensitive pressure transducers to the membrane that can be electronically amplified to produce a signal,[5] the direct measurement of osmotic pressure remains difficult and inconvenient. Nevertheless, osmotic pressure is the colligative property best suited to the determination of the molecular weight of polymers such as proteins.

## van't Hoff and Morse Equations for Osmotic Pressure

In 1886, Jacobus van't Hoff recognized in Pfeffer's data proportionality between osmotic pressure, concentration, and temperature, suggested a relationship that corresponded to the equation for an ideal gas. van't Hoff concluded that there was an apparent analogy between solutions and gases and that the osmotic pressure in a dilute solution was equal to the pressure that the solute would exert if it were a gas occupying the same volume. The equation is

$$\pi V = nRT \qquad (5\text{--}30)$$

where $\pi$ is the osmotic pressure in atm, $V$ is the volume of the solution in liters, $n$ is the number of moles of solute, $R$ is the gas constant, equal to 0.082 liter atm/mole deg, and $T$ is the absolute temperature.

The student should be cautioned not to take van't Hoff's analogy too literally, for it leads to the belief that the solute molecules "produce" the osmotic pressure by exerting pressure on the membrane, just as gas molecules create a pressure by striking the walls of a vessel. It is more correct, however, to consider the osmotic pressure as resulting from the relative escaping tendencies of the *solvent* molecules on the two sides of the membrane. Actually, equation (5–30) is a limiting law applying to dilute solutions, and it simplifies into this form from a more exact expression [equation (5–36)] only after introducing a number of assumptions that are not valid for real solutions.

### EXAMPLE 5–13

**Calculating the Osmotic Pressure of a Sucrose Solution**

One gram of sucrose, molecular weight 342, is dissolved in 100 mL of solution at 25°C. What is the osmotic pressure of the solution? We have

$$\text{Moles of sucrose} = \frac{1.0}{342} = 0.0029$$

$$\pi \times 0.10 = 0.0029 \times 0.082 \times 298$$

$$\pi = 0.71 \, \text{atm}$$

Equation (5–30), the van't Hoff equation, can be expressed as

$$\pi = \frac{n}{V}RT = cRT \qquad (5\text{--}31)$$

where $c$ is the concentration of the solute in moles/liter (molarity). Morse and others have shown that when the concentration is expressed in molality rather than in molarity, the results compare more nearly with the experimental findings. The Morse equation is

$$\pi = RTm \qquad (5\text{--}32)$$

## Thermodynamics of Osmotic Pressure and Vapor Pressure Lowering

Osmotic pressure and the lowering of vapor pressure, both colligative properties, are inextricably related, and this relationship can be obtained from certain thermodynamic considerations.

We begin by considering a sucrose solution in the right-hand compartment of the apparatus shown in **Figure 5–12** and the pure solvent—water—in the left-hand compartment. A semipermeable membrane through which water molecules, but not sucrose molecules, can pass separates the two compartments. It is assumed that the gate in the air space connecting the solutions can be shut during osmosis. The external pressure, say 1 atm, above the pure solvent is $P_0$ and the pressure on the solution, provided by the piston in **Figure 5–12** and needed to maintain equilibrium, is $P$. The difference between the two pressures at equilibrium, $P - P_0$, or

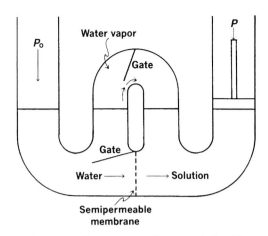

**Fig. 5–12.** Apparatus for demonstrating the relationship between osmotic pressure and vapor pressure lowering.

the excess pressure on the solution just required to prevent passage of water into the solution is the osmotic pressure $\pi$.

Let us now consider the alternative transport of water through the air space above the liquids. Should the membrane be closed off and the gate in the air space opened, water molecules would pass from the pure solvent to the solution by way of the vapor state by a distillation process. The space above the liquids actually serves as a "semipermeable membrane," just as does the real membrane at the lower part of the apparatus. The vapor pressure $p^\circ$ of water in the pure solvent under the influence of the atmospheric pressure $P_o$ is greater than the vapor pressure $p$ of water in the solution by an amount $p^\circ - p = \Delta p$. To bring about equilibrium, a pressure $P$ must be exerted by the piston on the solution to increase the vapor pressure of the solution until it is equal to that of the pure solvent, $p^\circ$. The excess pressure that must be applied, $P - P_o$, is again the osmotic pressure $\pi$. The operation of such an apparatus thus demonstrates the relationship between osmotic pressure and vapor pressure lowering.

By following this analysis further, it should be possible to obtain an equation relating osmotic pressure and vapor pressure. Observe that both the osmosis and the distillation process are based on the principle that the escaping tendency of water in the pure solvent is greater than that in the solution. By application of an excess pressure, $P - P_o = \pi$, on the solution side of the apparatus, it is possible to make the escaping tendencies of water in the solvent and solution identical. A state of equilibrium is produced; thus, the free energy of solvent on both sides of the membrane or on both sides of the air space is made equal, and $\Delta G = 0$.

To relate vapor pressure lowering and osmotic pressure, we must obtain the free energy changes involved in (a) transferring 1 mole of solvent from solvent to solution by a distillation process through the vapor phase and (b) transferring 1 mole of solvent from solvent to solution by osmosis. We have

**(a)**
$$\Delta G = RT \quad \ln \frac{p}{p^\circ} \qquad (5\text{–}33)$$

as the increase in free energy at constant temperature for the passage of 1 mole of water to the solution through the vapor phase, and

**(b)**
$$\Delta G = -V_1(P - P_o) = -V_1\pi \qquad (5\text{–}34)$$

as the increase in free energy at a definite temperature for the passage of 1 mole of water into the solution by osmosis. In equation (5–34), $V_1$ is the volume of 1 mole of solvent, or, more correctly, it is the *partial molar volume*, that is, the change in volume of the solution on the addition of 1 mole of solvent to a large quantity of solution.

Setting equations (5–33) and (5–34) equal gives

$$-\pi V_1 = RT \quad \ln \frac{p}{p^\circ} \qquad (5\text{–}35)$$

and eliminating the minus sign by inverting the logarithmic term yields

$$\pi = \frac{RT}{V_1} \ln \frac{p^\circ}{p} \qquad (5\text{–}36)$$

Equation (5–36) is a more exact expression for osmotic pressure than are equations (5–31) and (5–32), and it applies to concentrated as well as dilute solutions, provided that the vapor follows the ideal gas laws.

The simpler equation (5–32) for osmotic pressure can be obtained from equation (5–36), assuming that the solution obeys Raoult's law,

$$p = p^\circ X_1 \qquad (5\text{–}37)$$

$$\frac{p}{p^\circ} = X_1 = 1 - X_2 \qquad (5\text{–}38)$$

Equation (5–36) can thus be written

$$\pi V_1 = -RT \quad \ln(1 - X_2) \qquad (5\text{–}39)$$

and $\ln(1 - X_2)$ can be expanded into a series,

$$\ln(1 - X_2) = -X_2 - \frac{X_2^2}{2} - \frac{X_2^3}{3} \cdots - \frac{X_2^n}{2} \quad (5\text{–}40)$$

When $X_2$ is small, that is, when the solution is dilute, all terms in the expansion beyond the first may be neglected, and

$$\ln(1 - X_2) \cong -X_2 \qquad (5\text{–}41)$$

so that

$$\pi V_1 = RTX_2 \qquad (5\text{–}42)$$

For a dilute solution, $X_2$ equals approximately the mole ratio $n_2/n_1$, and equation (5–42) becomes

$$\pi \cong \frac{n_2}{n_1 V_1} RT \qquad (5\text{–}43)$$

where $n_1 V_1$, the number of moles of solvent multiplied by the volume of 1 mole, is equal to the total volume of solvent $V$ in liters. For a dilute aqueous solution, the equation becomes

$$\pi = \frac{n_2}{V} RT = RTm \qquad (5\text{–}44)$$

which is Morse's expression, equation (5–32).

**EXAMPLE 5–14**

**Compute $\pi$**

Compute $\pi$ for a 1 $m$ aqueous solution of sucrose using both equation (5–32) and the more exact thermodynamic equation (5–36). The vapor pressure of the solution is 31.207 mm Hg and the vapor pressure of water is 31.824 mm Hg at 30.0°C. The molar volume of water at this temperature is 18.1 cm³/mole, or 0.0181 liter/mole.

(a) By the Morse equation,

$$\pi = RTm = 0.082 \times 303 \times 1$$
$$\pi = 24.8 \text{atm}$$

(b) By the thermodynamic equation,

$$\pi = \frac{RT}{V_1} \ln \frac{p^\circ}{p}$$

$$\pi = \frac{0.082 \times 303}{0.0181} \times 2.303 \log \frac{31.824}{31.207}$$

$$= 27.0 \text{ atm}$$

The experimental value for the osmotic pressure of a 1 $m$ solution of sucrose at 30°C is 27.2 atm.

# MOLECULAR WEIGHT DETERMINATION

The four colligative properties that have been discussed in this chapter—vapor pressure lowering, freezing point lowering, boiling point elevation, and osmotic pressure—can be used to calculate the molecular weights of nonelectrolytes present as solutes. Thus, the lowering of the vapor pressure of a solution containing a nonvolatile solute depends only on the mole fraction of the solute. This allows the molecular weight of the solute to be calculated in the following manner.

Because the mole fraction of solvent, $n_1 = w_1/M_1$, and the mole fraction of solute, $n_2 = w_2/M_2$, in which $w_1$ and $w_2$ are the weights of solvent and solute of molecular weights $M_1$ and $M_2$ respectively, equation (5–15) can be expressed as

$$\frac{p_1{}^\circ - p_1}{p_1{}^\circ} = \frac{n_2}{n_1 + n_2} = \frac{w_2/M_2}{(w_1/M_1) + (w_2/M_2)} \quad (5\text{–}45)$$

In dilute solutions in which $w_2/M_2$ is negligible compared with $w_1/M_1$, the former term may be omitted from the denominator, and the equation simplifies to

$$\frac{\Delta p}{p_1{}^\circ} = \frac{w_2/M_2}{w_1/M_1} \quad (5\text{–}46)$$

The molecular weight of the solute $M_2$ is obtained by rearranging equation (5–46) to

$$M_2 = \frac{w_2 M_1 p_1{}^\circ}{w_1 \Delta p} \quad (5\text{–}47)$$

The molecular weight of a nonvolatile solute can similarly be determined from the boiling point elevation of the solution. Knowing $K_b$, the molal elevation constant, for the solvent and determining $T_b$, the boiling point elevation, one can

calculate the molecular weight of a nonelectrolyte. Because $1000w_2/w_1$ is the weight of solute per kilogram of solvent, molality (moles/kilogram of solvent) can be expressed as

$$m = \frac{w_2/M_2}{w_1} \times 1000 = \frac{1000w_2}{w_1 M_2} \quad (5\text{–}48)$$

and

$$\Delta T_b = K_b m \quad (5\text{–}49)$$

Then,

$$\Delta T_b = K_b \frac{1000w_2}{w_1 M_2} \quad (5\text{–}50)$$

or

$$M_2 = K_b \frac{1000w_2}{w_1 \Delta T_b} \quad (5\text{–}51)$$

**EXAMPLE 5–15**

**Determination of the Molecular Weight of Sucrose by Boiling Point Elevation**

A solution containing 10.0 g of sucrose dissolved in 100 g of water has a boiling point of 100.149°C. What is the molecular weight of sucrose? We write

$$M_2 = 0.51 \times \frac{1000 \times 10.0}{100 \times 0.149}$$
$$= 342 \text{ g/mole}$$

As shown in **Figure 5–8**, the lowering of vapor pressure arising from the addition of a nonvolatile solute to a solvent results in a depression of the freezing point. By rearranging equation (5–29), we obtain

$$M_2 = K_f \frac{1000w_2}{\Delta T_f w_1} \quad (5\text{–}52)$$

where $w_2$ is the number of grams of solute dissolved in $w_1$ grams of solvent. It is thus possible to calculate the molecular weight of the solute from cryoscopic data of this type.

**EXAMPLE 5–16**

**Calculating Molecular Weight Using Freezing Point Depression**

The freezing point depression of a solution of 2.000 g of 1,3-dinitrobenzene in 100.0 g of benzene was determined by the equilibrium method and was found to be 0.6095°C. Calculate the molecular weight of 1,3-dinitrobenzene. We write

$$M_2 = 5.12 \times \frac{1000 \times 2.000}{0.6095 \times 100.0} = 168.0 \text{ g/mole}$$

The van't Hoff and Morse equations can be used to calculate the molecular weight of solutes from osmotic pressure data, provided the solution is sufficiently dilute and ideal. The manner in which osmotic pressure is used to calculate the molecular weight of colloidal materials is discussed in Chapter 17.

**EXAMPLE 5–17**

**Determining Molecular Weight by Osmotic Pressure**

Fifteen grams of a new drug dissolved in water to yield 1000 mL of solution at 25°C was found to produce an osmotic pressure of 0.6 atm. What is the molecular weight of the solute? We write

$$\pi = cRT = \frac{c_g RT}{M_2} \tag{5–53}$$

where $c_g$ is in g/liter of solution. Thus,

$$\pi = \frac{15 \times 0.0821 \times 298}{M_2}$$

or

$$M_2 = \frac{15 \times 24.45}{0.6} = 612 \text{ g/mole}$$

## Choice of Colligative Properties

Each of the colligative properties seems to have certain advantages and disadvantages for the determination of molecular weights. The boiling point method can be used only when the solute is nonvolatile and when the substance is not decomposed at boiling temperatures. The freezing point method is satisfactory for solutions containing volatile solutes, such as alcohol, because the freezing point of a solution depends on the vapor pressure of the solvent alone. The freezing point method is easily executed and yields results of high accuracy for solutions of small molecules. It is sometimes inconvenient to use freezing point or boiling point methods, however, because they must be carried out at definite temperatures. Osmotic pressure measurements do not have this disadvantage, and yet the difficulties inherent in this method preclude its wide use. In summary, it may be said that the cryoscopic and newer vapor pressure techniques are the methods of choice, except for high polymers, in which instance the osmotic pressure method is used.

Because the colligative properties are interrelated, it should be possible to determine the value of one property from knowledge of any other. The relationship between vapor pressure lowering and osmotic pressure has already been shown. Freezing point depression and osmotic pressure can be related approximately as follows. The molality from the equation $m = \Delta T_f / K_f$ is substituted in the osmotic pressure equation, $\pi = RTm$, to give, at 0°C,

$$\pi = RT\frac{\Delta T_f}{K_f} = \frac{22.4}{1.86}\Delta T_f \tag{5–54}$$

or

$$\pi \cong 12\Delta T_f \tag{5–55}$$

Lewis[6] suggested the equation

$$\pi = 12.06\,\Delta T_f - 0.021\Delta T_f^2 \tag{5–56}$$

which gives accurate results.

**TABLE 5–5**

## APPROXIMATE EXPRESSIONS FOR THE COLLIGATIVE PROPERTIES

| Colligative Property | Expression | Proportionality Constant in Aqueous Solution |
|---|---|---|
| Vapor pressure lowering | $\Delta p = 0.018\,p_1°m$ | $0.018\,p_1° = 0.43$ at 25°C<br>$= 0.083$ at 0°C |
| Boiling point elevation | $\Delta T_b = K_b m$ | $K_b = 0.51$ |
| Freezing point depression | $\Delta T_f = K_f m$ | $K_f = 1.86$ |
| Osmotic pressure | $\pi = RTm$ | $RT = 24.4$ at 25°C<br>$= 22.4$ at 0°C |

**EXAMPLE 5–18**

**Osmotic Pressure of Human Blood Serum**

A sample of human blood serum has a freezing point of −0.53°C. What is the approximate osmotic pressure of this sample at 0°C? What is its more accurate value as given by the Lewis equation? We write

$$\pi = 12 \times 0.53 = 6.36 \text{ atm}$$

$$\pi = 12.06 \times 0.53 - 0.021(0.53)^2 = 6.39 \text{ atm}$$

Table 5–5 presents the equations and their constants in summary form. All equations are approximate and are useful only for dilute solutions in which the volume occupied by the solute is negligible with respect to that of the solvent.

## CHAPTER SUMMARY

This chapter focused on an important pharmaceutical mixture known as a molecular dispersion or true solution. Nine types of solutions, classified according to the states in which the solute and solvent occur, were defined. The concentration of a solution was expressed in two ways: either in terms of the quantity of solute in a definite volume of solution or as the quantity of solute in a definite mass of solvent or solution. You should be able to calculate molarity, normality, molality, mole fraction, and percentage expressions. Ideal and real solutions were described using Raoult's and Henry's laws. Finally, the colligative properties of solutions (osmotic pressure, vapor pressure lowering, freezing point depression, and boiling point elevation) were described. Colligative properties depend mainly on the number of particles in a solution and are approximately the same for equal concentrations of different nonelectrolytes in solution regardless of the species or chemical nature of the constituents.

**Practice problems for this chapter can be found at thePoint.lww.com/Sinko6e.**

## References

1. H. R. Kruyt, *Colloid Science*, Vol. II, Elsevier, New York, 1949, Chapter 1.
2. W. H. Chapin and L. E. Steiner, *Second Year College Chemistry*, Wiley, New York, 1943, Chapter 13; M. J. Sienko, *Stoichiometry and Structure*, Benjamin, New York, 1964.
3. G. N. Lewis and M. Randall, *Thermodynamics*, McGraw-Hill, New York, 1923, Chapter 16, J. M. Johlin, J. Biol. Chem. **91,** 551, 1931.
4. M. P. Tombs and A. R. Peacock, *The Osmotic Pressure of Biological Macromolecules*, Clarendon, Oxford, 1974, p. 1.
5. W. K. Barlow and P. G. Schneider, *Colloid Osmometry*, Wescor, Inc., Logan, Utah, 1978.
6. G. N. Lewis, J. Am. Chem. Soc. **30**, 668, 1908.

## Recommended Readings

P. W. Atkins and J. De Paula, *Atkins' Physical Chemistry*, 8th Ed., W. H. Freeman, New York, 2006.
J. H. Hildebrand and R. L. Scott, *The Solubility of Nonelectrolytes*, 3rd Ed., Dover Publications, New York, 1964.

### CHAPTER LEGACY

**Fifth Edition:** published as Chapter 5. Updated by Patrick Sinko.

**Sixth Edition:** published as Chapter 5. Updated by Patrick Sinko.

# 6 — ELECTROLYTE SOLUTIONS

The first satisfactory theory of ionic solutions was that proposed by Arrhenius in 1887. The theory was based largely on studies of electric conductance by Kohlrausch, colligative properties by van't Hoff, and chemical properties such as heats of neutralization by Thomsen. Arrhenius[1] was able to bring together the results of these diverse investigations into a broad generalization known as the theory of electrolytic dissociation.

Although the theory proved quite useful for describing weak electrolytes, it was soon found unsatisfactory for strong and moderately strong electrolytes. Accordingly, many attempts were made to modify or replace Arrhenius's ideas with better ones, and finally, in 1923, Debye and Hückel put forth a new theory. It is based on the principles that *strong electrolytes* are completely dissociated into ions in solutions of moderate concentration and that any deviation from complete dissociation is due to interionic attractions. Debye and Hückel expressed the deviations in terms of activities, activity coefficients, and ionic strengths of electrolytic solutions. These quantities, which had been introduced earlier by Lewis, are discussed in this chapter together with the theory of interionic attraction. Other aspects of modern ionic theory and the relationships between electricity and chemical phenomena are considered in following chapters.

This chapter begins with a discussion of some of the properties of ionic solutions that led to the Arrhenius theory of electrolytic dissociation.

## PROPERTIES OF SOLUTIONS OF ELECTROLYTES

### Electrolysis

When, under a potential of several volts, a direct electric current (dc) flows through an electrolytic cell (**Fig. 6–1**), a chemical reaction occurs. The process is known as *electrolysis*. Electrons enter the cell from the battery or generator at the *cathode* (road down); they combine with positive ions or *cations* in the solution, and the cations are accordingly reduced. The negative ions, or *anions*, carry electrons through the solution and discharge them at the *anode* (road up), and the anions are accordingly oxidized. *Reduction* is the addition of electrons to a chemical species, and *oxidation* is removal of electrons from a species. The current in a solution consists of a flow of positive and negative ions toward the electrodes, whereas the current in a metallic conductor consists of a flow of free electrons migrating through a crystal lattice of fixed positive ions. Reduction occurs at the cathode, where electrons enter from the external circuit and are added to a chemical species in solution. Oxidation occurs at the anode, where the electrons are removed from a chemical species in solution and go into the external circuit.

In the electrolysis of a solution of ferric sulfate in a cell containing platinum electrodes, a ferric ion migrates to the cathode, where it picks up an electron and is reduced:

$$Fe^{3+} + e = Fe^{2+} \qquad (6\text{–}1)$$

The sulfate ion carries the current through the solution to the anode, but it is not easily oxidized; therefore, hydroxyl ions of the water are converted into molecular oxygen, which escapes at the anode, and sulfuric acid is found in the solution around the electrode. The oxidation reaction at the anode is

$$OH^- = \frac{1}{4}O_2 + \frac{1}{2}H_2O + e \qquad (6\text{–}2)$$

Platinum electrodes are used here because they do not pass into solution to any extent. When *attackable* metals, such as copper or zinc, are used as the anode, their atoms tend to lose electrons, and the metal passes into solution as the positively charged ion.

In the electrolysis of cupric chloride between platinum electrodes, the reaction at the cathode is

$$\frac{1}{2}Cu^{2+} + e = \frac{1}{2}Cu \qquad (6\text{–}3)$$

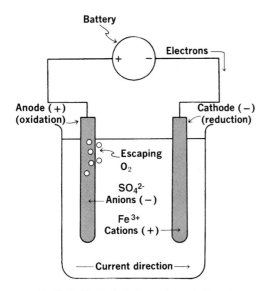

**Fig. 6–1.** Electrolysis in an electrolytic cell.

whereas at the anode, chloride and hydroxyl ions are converted, respectively, into gaseous molecules of chlorine and oxygen, which then escape. In each of these two examples, the net result is the transfer of one electron from the cathode to the anode.

## Transference Numbers

It should be noted that the flow of electrons through the solution from right to left in **Figure 6–1** is accomplished by the movement of cations to the right as well as anions to the left. The fraction of total current carried by the cations or by the anions is known as the *transport* or *transference number* $t_+$ or $t_-$:

$$t_+ = \frac{\text{Current carried by cations}}{\text{Total current}} \qquad (6\text{–}4)$$

$$t_- = \frac{\text{Current carried by anions}}{\text{Total current}} \qquad (6\text{–}5)$$

The sum of the two transference numbers is obviously equal to unity:

$$t_+ + t_- = 1 \qquad (6\text{–}6)$$

The transference numbers are related to the velocities of the ions, the faster-moving ion carrying the greater fraction of current. The velocities of the ions in turn depend on hydration as well as ion size and charge. Hence, the speed and the transference numbers are not necessarily the same for positive and for negative ions. For example, the transference number of the sodium ion in a 0.10 M solution of NaCl is 0.385. Because it is greatly hydrated, the lithium ion in a 0.10 M solution of LiCl moves more slowly than the sodium ion and hence has a lower transference number, 0.317.

## Electrical Units

According to Ohm's law, the strength of an electric current $I$ in amperes flowing through a metallic conductor is related to the difference in applied potential or voltage $E$ and the resistance $R$ in ohms, as follows:

$$I = \frac{E}{R} \qquad (6\text{–}7)$$

The current strength $I$ is the rate of flow of current or the quantity $Q$ of electricity (electronic charge) in coulombs flowing per unit time:

$$I = \frac{Q}{t} \qquad (6\text{–}8)$$

and

Quantity of electric charge, $Q$
$$= \text{Current, } I \times \text{Time, } t \qquad (6\text{–}9)$$

The quantity of electric charge is expressed in coulombs (1 coulomb $= 3 \times 10^9$ electrostatic units of charge, or esu), the current in amperes, and the electric potential in volts.

Electric energy consists of an intensity factor, electromotive force or voltage, and a quantity factor, coulombs.

$$\text{Electric energy} = E \times Q \qquad (6\text{–}10)$$

## Faraday's Laws

A univalent negative ion is an atom to which a valence electron has been added; a univalent positive ion is an atom from which an electron has been removed. Each gram equivalent of ions of any electrolyte carries Avogadro's number ($6.02 \times 10^{23}$) of positive or negative charges. Hence, from Faraday's laws, the passage of 96,500 coulombs of electricity results in the transport of $6.02 \times 10^{23}$ electrons in the cell. One faraday is an Avogadro's number of electrons, corresponding to the mole, which is an Avogadro's number of molecules. The passage of 1 faraday of electricity causes the electrolytic deposition of the following number of gram atoms or "moles" of various ions: 1 $Ag^+$, 1 $Cu^+$, ½ $Cu^{2+}$, ½ $Fe^{2+}$, $Fe^{3+}$. Thus, the number of positive charges carried by 1 g equivalent of $Fe^{3+}$ is $6.02 \times 10^{23}$, but the number of positive charges carried by 1 g atom or 1 mole of ferric ions is $3 \times 6.02 \times 10^{23}$.

Faraday's laws can be used to compute the charge on an electron in the following way. Because $6.02 \times 10^{23}$ electrons are associated with 96,500 coulombs of electricity, each electron has a charge $e$ of

$$e = \frac{96,500 \text{ coulombs}}{6.02 \times 10^{23}}$$
$$= 1.6 \times 10^{-19} \text{ coulomb} \qquad (6\text{--}11)$$

and because 1 coulomb $= 3 \times 10^9$ esu,

$$e = 4.8 \times 10^{-10} \text{ electrostatic unit of charge} \qquad (6\text{--}12)$$

## Electrolytic Conductance

The resistance, $R$, in ohms of any uniform metallic or electrolytic conductor is directly proportional to its length, $l$, in cm and inversely proportional to its cross-sectional area, $A$, in $cm^2$,

$$R = \rho \frac{l}{A} \qquad (6\text{--}13)$$

where $\rho$ is the resistance between opposite faces of a 1-cm cube of the conductor and is known as the *specific resistance*.

The *conductance*, $C$, is the reciprocal of resistance,

$$C = \frac{1}{R} \qquad (6\text{--}14)$$

and hence can be considered as a measure of the ease with which current can pass through the conductor. It is expressed in reciprocal ohms or *mhos*. From equation (6–13):

$$C = \frac{1}{R} = \frac{1}{\rho} \frac{A}{l} \qquad (6\text{--}15)$$

The *specific conductance* $\kappa$ is the reciprocal of specific resistance and is expressed in mhos/cm:

$$\kappa = \frac{1}{\rho} \qquad (6\text{--}16)$$

It is the conductance of a solution confined in a cube 1 cm on an edge as seen in **Figure 6–2**. The relationship between

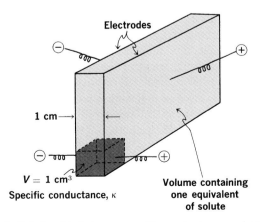

**Fig. 6–2.** Relationship between specific conductance and equivalent conductance.

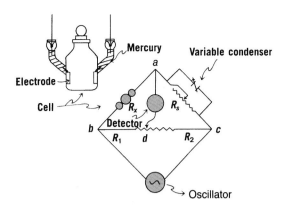

**Fig. 6–3.** Wheatstone bridge for conductance measurements.

specific conductance and conductance or resistance is obtained by combining equations (6–15) and (6–16):

$$\kappa = C \frac{1}{A} = \frac{1}{R} \frac{1}{A} \qquad (6\text{--}17)$$

## Measuring the Conductance of Solutions

The Wheatstone bridge assembly for measuring the conductance of a solution is shown in **Figure 6–3**. The solution of unknown resistance $R_x$ is placed in the cell and connected in the circuit. The contact point is moved along the slide wire $bc$ until at some point, say $d$, no current from the source of alternating current (oscillator) flows through the detector (earphones or oscilloscope). When the bridge is balanced, the potential at $a$ is equal to that at $d$, the sound in the earphones or the oscillating pattern on the oscilloscope is at a minimum, and the resistances $R_s$, $R_1$, and $R_2$ are read. In the balanced state, the resistance of the solution $R_x$ is obtained from the equation

$$R_x = R_s \frac{R_1}{R_2} \qquad (6\text{--}18)$$

The variable condenser across resistance $R_s$ is used to produce a sharper balance. Some conductivity bridges are calibrated in conductance as well as resistance values. The electrodes in the cell are platinized with platinum black by electrolytic deposition so that catalysis of the reaction will occur at the platinum surfaces and formation of a nonconducting gaseous film will not occur on the electrodes.

Water that is carefully purified by redistillation in the presence of a little permanganate is used to prepare the solutions. Conductivity water, as it is called, has a specific conductance of about $0.05 \times 10^{-6}$ mho/cm at 18°C, whereas ordinary distilled water has a value somewhat greater than $1 \times 10^{-6}$ mho/cm. For most conductivity studies, "equilibrium water" containing $CO_2$ from the atmosphere is satisfactory. It has a specific conductance of about $0.8 \times 10^{-6}$ mho/cm.

The specific conductance, $\kappa$, is computed from the resistance, $R_x$, or conductance, $C$, by use of equation (6–17). The quantity $l/A$, the ratio of distance between electrodes to the area of the electrode, has a definite value for each conductance cell; it is known as the *cell constant, K*. Equation (6–17) thus can be written as

$$\kappa = KC = K/R \qquad (6\text{–}19)$$

(The subscript *x* is no longer needed on *R* and is therefore dropped.) It would be difficult to measure *l* and *A*, but it is a simple matter to determine the cell constant experimentally. The specific conductance of several standard solutions has been determined in carefully calibrated cells. For example, a solution containing 7.45263 g of potassium chloride in 1000 g of water has a specific conductance of 0.012856 mho/cm at 25°C. A solution of this concentration contains 0.1 mole of salt per cubic decimeter (100 cm$^3$) of water and is known as a 0.1-*demal* solution. When such a solution is placed in a cell and the resistance is measured, the cell constant can be determined by use of equation (6–19).

**EXAMPLE 6–1**

**Calculating *K***

A 0.1-demal solution of KCl was placed in a cell whose constant *K* was desired. The resistance *R* was found to be 34.69 ohms at 25°C. Thus,

$$K = \kappa R = 0.012856 \text{ mho/cm} \times 34.69 \text{ ohms}$$
$$= 0.4460 \text{ cm}^{-1}$$

**EXAMPLE 6–2**

**Calculating Specific Conductance**

When the cell described in *Example 6–1* was filled with a 0.01 N Na$_2$SO$_4$ solution, it had a resistance of 397 ohms. What is the specific conductance? We write

$$\kappa = \frac{K}{R} = \frac{0.4460}{397} = 1.1234 \times 10^{-3} \text{ mho/cm}$$

## Equivalent Conductance

To study the dissociation of molecules into ions, independent of the concentration of the electrolyte, it is convenient to use equivalent conductance rather than specific conductance. All solutes of equal normality produce the same number of ions when completely dissociated, and equivalent conductance measures the current-carrying capacity of this given number of ions. Specific conductance, on the other hand, measures the current-carrying capacity of all ions in a unit volume of solution and accordingly varies with concentration.

*Equivalent conductance* $\Lambda$ is defined as the conductance of a solution of sufficient volume to contain 1 g equivalent of the solute when measured in a cell in which the electrodes are spaced 1 cm apart. The equivalent conductance $\Lambda_c$ at a concentration of *c* gram equivalents per liter is calculated from the product of the specific conductance $\kappa$ and the volume *V* in cm$^3$ that contains 1 g equivalent of solute. The cell may be imagined as having electrodes 1 cm apart and to be of

sufficient area so that it can contain the solution. The cell is shown in **Figure 6–2**. We have

$$V = \frac{1000 \text{ cm}^3/\text{liter}}{c \text{ Eq/liter}} = \frac{1000}{c} \text{ cm}^3/\text{Eq} \qquad (6\text{–}20)$$

The equivalent conductance is obtained when $\kappa$, the conductance per cm$^3$ of solution (i.e., the specific conductance), is multiplied by *V*, the volume in cm$^3$ that contains 1 g equivalent weight of solute. Hence, the equivalent conductance, $\Lambda_c$, expressed in units of mho cm$^2$/Eq, is given by the expression

$$\Lambda_c = \kappa \times V$$
$$= \frac{1000\,\kappa}{c} \text{ mho cm}^2/\text{Eq} \qquad (6\text{–}21)$$

If the solution is 0.1 N in concentration, then the volume containing 1 g equivalent of the solute will be 10,000 cm$^3$, and, according to equation (6–21), the equivalent conductance will be 10,000 times as great as the specific conductance. This is seen in the following example.

**EXAMPLE 6–3**

**Specific and Equivalent Conductance**

The measured conductance of a 0.1 N solution of a drug is 0.0563 ohm at 25°C. The cell constant at 25°C is 0.520 cm$^{-1}$. What is the specific conductance and what is the equivalent conductance of the solution at this concentration? We write

$$\kappa = 0.0563 \times 0.520 = 0.0293 \text{ mho/cm}$$
$$\Lambda_c = 0.0293 \times 1000/0.1$$
$$= 293 \text{ mho cm}^2/\text{Eq}$$

## Equivalent Conductance of Strong and Weak Electrolytes

As the solution of a strong electrolyte is diluted, the *specific conductance $\kappa$ decreases* because the number of ions per unit volume of solution is reduced. It sometimes goes through a maximum before decreasing. Conversely, the *equivalent conductance $\Lambda$* of a solution of a strong electrolyte steadily *increases* on dilution. The increase in $\Lambda$ with dilution is explained as follows. The quantity of electrolyte remains constant at 1 g equivalent according to the definition of equivalent conductance; however, the ions are hindered less by their neighbors in the more dilute solution and hence can move faster. The equivalent conductance of a weak electrolyte also increases on dilution, but not as rapidly at first.

Kohlrausch was one of the first investigators to study this phenomenon. He found that the equivalent conductance was a linear function of the square root of the concentration for strong electrolytes in dilute solutions, as illustrated in **Figure 6–4**. The expression for $\Lambda_c$, the equivalent conductance at a concentration *c* (Eq/L), is

$$\Lambda_c = \Lambda_0 - b\sqrt{c} \qquad (6\text{–}22)$$

where $\Lambda_0$ is the intercept on the vertical axis and is known as the *equivalent conductance at infinite dilution*. The constant

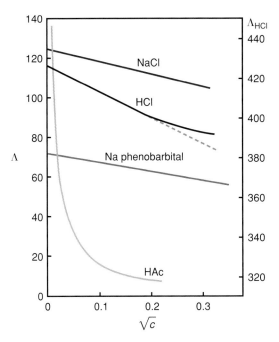

**Fig. 6–4.** Equivalent conductance of strong and weak electrolytes.

$b$ is the slope of the line for the strong electrolytes shown in **Figure 6–4**.

When the equivalent conductance of a weak electrolyte is plotted against the square root of the concentration, as shown for acetic acid in **Figure 6–4**, the curve cannot be extrapolated to a limiting value, and $\Lambda_0$ must be obtained by a method such as is described in the following paragraph. The steeply rising curve for acetic acid results from the fact that the dissociation of weak electrolytes increases on dilution, with a large increase in the number of ions capable of carrying the current.

Kohlrausch concluded that the ions of all electrolytes begin to migrate independently as the solution is diluted; the ions in dilute solutions are so far apart that they do not interact in any way. Under these conditions, $\Lambda_0$ is the sum of the equivalent conductances of the cations $l_c{}^o$ and the anions $l_a{}^o$ at infinite dilution

$$\Lambda_0 = l_c{}^o + l_a{}^o \qquad (6\text{–}23)$$

Based on this law, the known $\Lambda_0$ values for certain electrolytes can be added and subtracted to yield $\Lambda_0$ for the desired weak electrolyte. The method is illustrated in the following example.

**EXAMPLE 6–4**

**Equivalent Conductance of Phenobarbital**

What is the equivalent conductance at infinite dilution of the weak acid phenobarbital? The $\Lambda_0$ of the strong electrolytes HCl, sodium phenobarbital (NaP), and NaCl are obtained from the experimental results shown in Figure 6–4. The values are $\Lambda_{0,\text{HCl}} = 426.2$, $\Lambda_{0,\text{NaP}} = 73.5$, and $\Lambda_{0,\text{NaCl}} = 126.5$ mho cm$^2$/Eq.

Now, by Kohlrausch's law of the independent migration of ions,

$$\Lambda_{0,\text{HP}} = l_{H+}^o + l_P^o$$

and

$$\Lambda_{0,\text{HCl}} + \Lambda_{0,\text{NaP}} - \Lambda_{0,\text{NaCl}} = l_{H+}^o + l_{Cl-}^o + l_{Na+}^o + l_P^o - l_{Na+}^o - l_{Cl}^o$$

which, on simplifying the right-hand side of the equation, becomes

$$\Lambda_{0,\text{HCl}} + \Lambda_{0,\text{NaP}} - \Lambda_{0,\text{NaCl}} = l_{H+}^o + l_P^o$$

Therefore,

$$\Lambda_{0,\text{HP}} = \Lambda_{0,\text{HCl}} + \Lambda_{0,\text{NaP}} - \Lambda_{0,\text{NaCl}}$$

and

$$\begin{aligned}\Lambda_{0,\text{HP}} &= 426.2 + 73.5 - 126.5 \\ &= 373.2 \text{ mho cm}^2/\text{Eq}\end{aligned}$$

## Colligative Properties of Electrolytic Solutions and Concentrated Solutions of Nonelectrolytes

As stated in the previous chapter, van't Hoff observed that the osmotic pressure, $\pi$, of dilute solutions of nonelectrolytes, such as sucrose and urea, could be expressed satisfactorily by the equation $\pi = RT_c$ [equation **(5–34)**], where $R$ is the gas constant, $T$ is the absolute temperature, and $c$ is the concentration in moles/liter. van't Hoff found, however, that solutions of electrolytes gave osmotic pressures approximately two, three, and more times larger than expected from this equation, depending on the electrolyte investigated. Introducing a correction factor $i$ to account for the irrational behavior of ionic solutions, he wrote

$$\pi = iRTc \qquad (6\text{–}24)$$

By the use of this equation, van't Hoff was able to obtain calculated values that compared favorably with the experimental results of osmotic pressure. Van't Hoff recognized that $i$ approached the number of ions into which the molecule dissociated as the solution was made increasingly dilute.

The $i$ factor is plotted against the molal concentration of both electrolytes and nonelectrolytes in **Figure 6–5**. For nonelectrolytes, it is seen to approach unity, and for strong

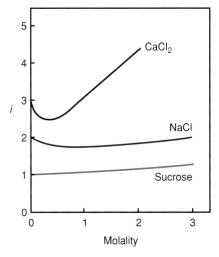

**Fig. 6–5.** van't Hoff $i$ factor of representative compounds.

electrolytes, it tends toward a value equal to the number of ions formed upon dissociation. For example, $i$ approaches the value of 2 for solutes such as NaCl and $CaSO_4$, 3 for $K_2SO_4$ and $CaCl_2$, and 4 for $K_3Fe(C)_6$ and $FeCl_3$.

The van't Hoff factor can also be expressed as the ratio of any colligative property of a real solution to that of an ideal solution of a nonelectrolyte because $i$ represents the number of times greater that the colligative effect is for a real solution (electrolyte or nonelectrolyte) than for an ideal nonelectrolyte.

The colligative properties in dilute solutions of electrolytes are expressed on the molal scale by the equations

$$\Delta p = 0.018 i p_1^\circ m \qquad (6\text{--}25)$$

$$\pi = iRTm \qquad (6\text{--}26)$$

$$\Delta T_f = iK_f m \qquad (6\text{--}27)$$

$$\Delta T_b = iK_b m \qquad (6\text{--}28)$$

Equation (6–25) applies only to aqueous solutions, whereas (6–26) through (6–28) are independent of the solvent used.

**EXAMPLE 6–5**

**Osmotic Pressure of Sodium Chloride**

What is the osmotic pressure of a 2.0 $m$ solution of sodium chloride at 20°C?

The $i$ factor for a 2.0 $m$ solution of sodium chloride as observed in Figure 6–5 is about 1.9. Thus,

$$\pi = 1.9 \times 0.082 \times 293 \times 2.0 = 91.3 \text{ atm}$$

# THEORY OF ELECTROLYTIC DISSOCIATION

During the period in which van't Hoff was developing the solution laws, the Swedish chemist Svante Arrhenius was preparing his doctoral thesis on the properties of electrolytes at the University of Uppsala in Sweden. In 1887, he published the results of his investigations and proposed the now classic theory of dissociation.[1] The new theory resolved many of the anomalies encountered in the earlier interpretations of electrolytic solutions. Although the theory was viewed with disfavor by some influential scientists of the nineteenth century, Arrhenius's basic principles of electrolytic dissociation were gradually accepted and are still considered valid today.

The theory of the existence of ions in solutions of electrolytes even at ordinary temperatures remains intact, aside from some modifications and elaborations that have been made through the years to bring it into line with certain stubborn experimental facts.

The original Arrhenius theory, together with the alterations that have come about as a result of the intensive research on electrolytes, is summarized as follows. When electrolytes are dissolved in water, the solute exists in the form of ions in the solution, as seen in the following equations:

$$H_2O + \underset{\text{[Ionic compound]}}{Na^+Cl^-} \rightarrow Na^+ + Cl^- + H_2O$$

$$\text{[Strong electrolyte]} \qquad (6\text{--}29)$$

$$H_2O + \underset{\substack{\text{[Covalent}\\\text{compound]}}}{HCl} \rightarrow H_3O^+ + Cl^-$$

$$\text{[Strong electrolyte]} \qquad (6\text{--}30)$$

$$H_2O + \underset{\substack{\text{[Covalent}\\\text{compound]}}}{CH_3COOH} \rightleftharpoons H_3O^+ + CH_3COO^-$$

$$\text{[Weak electrolyte]} \qquad (6\text{--}31)$$

The solid form of sodium chloride is marked with plus and minus signs in reaction (6–29) to indicate that sodium chloride exists as ions even in the crystalline state. If electrodes are connected to a source of current and are placed in a mass of fused sodium chloride, the molten compound will conduct the electric current because the crystal lattice of the pure salt consists of ions. The addition of water to the solid dissolves the crystal and separates the ions in solution.

Hydrogen chloride exists essentially as neutral molecules rather than as ions in the pure form and does not conduct electricity. When it reacts with water, however, it ionizes according to reaction [equation (6–30)]. $H_3O^+$ is the modern representation of the hydrogen ion in water and is known as the *hydronium or oxonium* ion. In addition to $H_3O^+$, other hydrated species of the proton probably exist in solution, but they need not be considered here.[2]

Sodium chloride and hydrochloric acid are *strong electrolytes* because they exist almost completely in the ionic

**KEY CONCEPT  van't HOFF FACTOR $i$**

The factor $i$ may also be considered to express the departure of concentrated solutions of nonelectrolytes from the laws of ideal solutions. The deviations of concentrated solutions of nonelectrolytes can be explained on the same basis as deviations of real solutions from Raoult's law, considered in the preceding chapter. They included differences of internal pressures of the solute and solvent, polarity, compound formation or complexation, and association of either the solute or the solvent. The departure of

electrolytic solutions from the colligative effects in ideal solutions of nonelectrolytes may be attributed—in addition to the factors just enumerated—to dissociation of weak electrolytes and to interaction of the ions of strong electrolytes. Hence, the van't Hoff factor $i$ accounts for the deviations of real solutions of nonelectrolytes and electrolytes, regardless of the reason for the discrepancies.

form in moderately concentrated aqueous solutions. Inorganic acids such as HCl, $HNO_3$, $H_2SO_4$, and HI; inorganic bases as NaOH and KOH of the alkali metal family and $Ba(OH)_2$ and $Ca(OH)_2$ of the alkaline earth group; and most inorganic and organic salts are highly ionized and belong to the class of strong electrolytes.

Acetic acid is a *weak electrolyte*, the oppositely directed arrows in equation **(6–31)** indicating that equilibrium between the molecules and the ions is established. Most organic acids and bases and some inorganic compounds, such as $H_3BO_3$, $H_2CO_3$, and $NH_4OH$, belong to the class of weak electrolytes. Even some salts (lead acetate, $HgCl_2$, HgI, and HgBr) and the complex ions $Hg(NH_3)_2{}^+$, $Cu(NH_3)_4{}^{2+}$, and $Fe(CN)_6{}^{3-}$ are weak electrolytes.

Faraday applied the term *ion* (Greek: "wanderer") to these species of electrolytes and recognized that the cations (positively charged ions) and anions (negatively charged ions) were responsible for conducting the electric current. Before the time of Arrhenius's publications, it was believed that a solute was not spontaneously decomposed in water but rather dissociated appreciably into ions only when an electric current was passed through the solution.

## Drugs and Ionization

Some drugs, such as anionic and cationic antibacterial and antiprotozoal agents, are more active when in the ionic state. Other compounds, such as the hydroxybenzoate esters (parabens) and many general anesthetics, bring about their biologic effects as nonelectrolytes. Still other compounds, such as the sulfonamides, are thought to exert their drug action both as ions and as neutral molecules.[3]

## Degree of Dissociation

Arrhenius did not originally consider strong electrolytes to be ionized completely except in extremely dilute solutions. He differentiated between strong and weak electrolytes by the fraction of the molecules ionized: the *degree of dissociation*, $\alpha$. A strong electrolyte was one that dissociated into ions to a high degree and a weak electrolyte was one that dissociated into ions to a low degree.

Arrhenius determined the degree of dissociation directly from conductance measurements. He recognized that the equivalent conductance at infinite dilution $\Lambda_0$ was a measure of the complete dissociation of the solute into its ions and that $\Lambda_c$ represented the number of solute particles present as ions at a concentration $c$. Hence, the fraction of solute molecules ionized, or the degree of dissociation, was expressed by the equation[4]

$$\alpha = \frac{\Lambda_c}{\Lambda_0} \qquad (6\text{–}32)$$

where $\Lambda_c / \Lambda_0$ is known as the *conductance ratio*.

---

**EXAMPLE 6–6**

**Degree of Dissociation of Acetic Acid**

The equivalent conductance of acetic acid at 25°C and at infinite dilution is 390.7 ohm $cm^2$/Eq. The equivalent conductance of a $5.9 \times 10^{-3}$ M solution of acetic acid is 14.4 ohm $cm^2$/Eq. What is the degree of dissociation of acetic acid at this concentration? We write

$$\alpha = \frac{14.4}{390.7} = 0.037 \text{ or } 3.7\%$$

---

The van't Hoff factor, $i$, can be connected with the degree of dissociation, $\alpha$, in the following way. The $i$ factor equals unity for an ideal solution of a nonelectrolyte; however, a term must be added to account for the particles produced when a molecule of an electrolyte dissociates. For 1 mole of calcium chloride, which yields three ions per molecule, the van't Hoff factor is given by

$$i = 1 + \alpha(3 - 1) \qquad (6\text{–}33)$$

or, in general, for an electrolyte yielding $v$ ions,

$$i = 1 + \alpha(v - 1) \qquad (6\text{–}34)$$

from which we obtain an expression for the degree of dissociation,

$$\alpha = \frac{i - 1}{v - 1} \qquad (6\text{–}35)$$

The cryoscopic method is used to determine $i$ from the expression

$$\Delta T_f = iK_f m \qquad (6\text{–}36)$$

or

$$i = \frac{\Delta T_f}{K_f m} \qquad (6\text{–}37)$$

---

**EXAMPLE 6–7**

**Degree of Ionization of Acetic Acid**

The freezing point of a 0.10 $m$ solution of acetic acid is –0.188°C. Calculate the degree of ionization of acetic acid at this concentration. Acetic acid dissociates into two ions, that is, $v = 2$. We write

$$i = \frac{0.188}{1.86 \times 0.10} = 1.011$$

$$\alpha = \frac{i - 1}{v - 1} = \frac{1.011 - 1}{2 - 1} = 0.011$$

---

In other words, according to the result of *Example 6–7*, the fraction of acetic acid present as free ions in a 0.10 $m$ solution is 0.011. Stated in percentage terms, acetic acid in 0.1 $m$ concentration is ionized to the extent of about 1%.

## THEORY OF STRONG ELECTROLYTES

Arrhenius used $\alpha$ to express the degree of dissociation of both strong and weak electrolytes, and van't Hoff introduced the factor $i$ to account for the deviation of strong and weak electrolytes and nonelectrolytes from the ideal laws of the colligative properties, regardless of the nature of these discrepancies. According to the early ionic theory, the degree of dissociation of ammonium chloride, a strong electrolyte, was calculated in the same manner as that of a weak electrolyte.

---

**EXAMPLE 6–8**

**Degree of Dissociation**

The freezing point depression for a 0.01 $m$ solution of ammonium chloride is 0.0367°C. Calculate the "degree of dissociation" of this electrolyte. We write

$$i = \frac{\Delta T_f}{K_f m} = \frac{0.0367°C}{1.86 \times 0.010} = 1.97$$

$$\alpha = \frac{1.97 - 1}{2 - 1} = 0.97$$

---

The Arrhenius theory is now accepted for describing the behavior only of weak electrolytes. The degree of dissociation of a weak electrolyte can be calculated satisfactorily from the conductance ratio $\Lambda_c / \Lambda_0$ or obtained from the van't Hoff $i$ factor.

Many inconsistencies arise, however, when an attempt is made to apply the theory to solutions of strong electrolytes. In dilute and moderately concentrated solutions, they dissociate almost completely into ions, and it is not satisfactory to write an equilibrium expression relating the concentration of the ions and the minute amount of undissociated molecules, as is done for weak electrolytes (Chapter 7). Moreover, a discrepancy exists between $\alpha$ calculated from the $i$ value and $\alpha$ calculated from the conductivity ratio for strong electrolytes in aqueous solutions having concentrations greater than about 0.5 M.

For these reasons, one does not account for the deviation of strong electrolyte from ideal nonelectrolyte behavior by calculating a degree of dissociation. It is more convenient to consider a strong electrolyte as completely ionized and to introduce a factor that expresses the deviation of the solute from 100% ionization. The *activity* and *osmotic coefficient*, discussed in subsequent paragraphs, are used for this purpose.

## Activity and Activity Coefficients

An approach that conforms well to the facts and that has evolved from a large number of studies on solutions of strong

electrolytes ascribes the behavior of strong electrolytes to an electrostatic attraction between the ions.

Not only are the ions interfered with in their movement by the "atmosphere" of oppositely charged ions surrounding them, they can also associate at high concentration into groups known as *ion pairs* (e.g., $Na^+Cl^-$) and ion triplets ($Na^+Cl^-Na^+$). Associations of still higher orders may exist in solvents of low dielectric constant, in which the force of attraction of oppositely charged ions is large.

Because of the electrostatic attraction and ion association in moderately concentrated solutions of strong electrolytes, the values of the freezing point depression and the other colligative properties are less than expected for solutions of unhindered ions. Consequently, a strong electrolyte may be *completely ionized, yet incompletely dissociated* into free ions.

One may think of the solution as having an "effective concentration" or, as it is called, an *activity*. The activity, in general, is less than the actual or stoichiometric concentration of the solute, not because the strong electrolyte is only partly ionized, but rather because some of the ions are effectively "taken out of play" by the electrostatic forces of interaction.

At infinite dilution, in which the ions are so widely separated that they do not interact with one another, the activity $a$ of an ion is equal to its concentration, expressed as molality or molarity. It is written on a molal basis at infinite dilution as

$$a = m \qquad (6\text{–}38)$$

or

$$\frac{a}{m} = 1 \qquad (6\text{–}39)$$

As the concentration of the solution is increased, the ratio becomes less than unity because the effective concentration or activity of the ions becomes less than the stoichiometric or molal concentration. This ratio is known as the *practical activity coefficient*, $\gamma_m$, on the molal scale, and the formula is written, for a particular ionic species, as

$$\frac{a}{m} = \gamma_m \qquad (6\text{–}40)$$

or

$$a = \gamma_m m \qquad (6\text{–}41)$$

On the molarity scale, another *practical activity coefficient*, $\gamma_c$, is defined as

$$a = \gamma_c c \qquad (6\text{–}42)$$

---

## KEY CONCEPT    WEAK AND STRONG ELECTROLYTES

The large number of oppositely charged ions in solutions of electrolytes influence one another through *interionic attractive forces*. Although this interference is negligible in dilute solutions, it becomes appreciable at moderate concentrations. In solutions of weak electrolytes, regardless of concentration, the

number of ions is small and the interionic attraction correspondingly insignificant. Hence, the Arrhenius theory and the concept of the degree of dissociation are valid for solutions of weak electrolytes but not for strong electrolytes.

and on the mole fraction scale, a *rational activity coefficient* is defined as

$$a = \gamma_x X \tag{6–43}$$

One sees from equations (6–41) through (6–43) that these coefficients are proportionality constants relating activity to molality, molarity, and mole fraction, respectively, for an ion. The activity coefficients take on a value of unity and are thus identical in infinitely dilute solutions. The three coefficients usually decrease and assume different values as the concentration is increased; however, the differences among the three activity coefficients may be disregarded in dilute solutions, in which $c \cong m < 0.01$. The concepts of activity and activity coefficient were first introduced by Lewis and Randall[5] and can be applied to solutions of nonelectrolytes and weak electrolytes as well as to the ions of strong electrolytes.

A cation and an anion in an aqueous solution may each have a different ionic activity. This is recognized by using the symbol $a_+$ when speaking of the activity of a cation and the symbol $a_-$ when speaking of the activity of an anion. An electrolyte in solution contains each of these ions, however, so it is convenient to define a relationship between the activity of the electrolyte $a_\pm$ and the activities of the individual ions. The activity of an electrolyte is defined by its *mean ionic activity*, which is given by the relation

$$a_\pm = (a_+{}^m a_-{}^n)^{1/(m+n)} \tag{6–44}$$

where the exponents $m$ and $n$ give the stoichiometric numbers of given ions that are in solution. Thus, an NaCl solution has a mean ionic activity of

$$a_\pm = (a_{Na^+} a_{Cl^-})^{1/2}$$

whereas an $FeCl_3$ solution has a mean ionic activity of

$$a_\pm = (a_{Fe^{+3}} a_{Cl^-}{}^3)^{1/4}$$

The ionic activities of equation (6–44) can be expressed in terms of concentrations using any of equations (6–41) to (6–43). Using equation (6–42), one obtains from equation (6–44) the expression

$$a_\pm = [(\gamma_+ c_+)^m (\gamma_- c_-)^n]^{1/(m+n)} \tag{6–45}$$

or

$$a_\pm = (\gamma_+{}^m \gamma_-{}^n)^{1/(m+n)} (c_+{}^m c_-{}^n)^{1/(m+n)} \tag{6–46}$$

The *mean ionic activity coefficient* for the electrolyte can be defined by

$$\gamma_\pm = (\gamma_+{}^m \gamma_-{}^n)^{1/(m+n)} \tag{6–47}$$

and

$$\gamma_\pm^{m+n} = \gamma_+{}^m \gamma_-{}^n \tag{6–48}$$

Substitution of equation (6–47) into equation (6–46) yields

$$a_\pm = \gamma_\pm (c_+{}^m c_-{}^n)^{1/(m+n)} \tag{6–49}$$

In using equation (6–49), it should be noted that the concentration of the electrolyte $c$ is related to the concentration of its ions by

$$c_+ = mc \tag{6–50}$$

and

$$c_- = nc \tag{6–51}$$

**EXAMPLE 6–9**

**Mean Ionic Activity of FeCl₃**

What is the mean ionic activity of a 0.01 M solution of $FeCl_3$? We write

$$a_\pm = \gamma_\pm (c_+ c_-{}^3)^{1/4} = \gamma_\pm [(0.01)(3 \times 0.01)^3]^{1/4}$$
$$= 2.3 \times 10^{-2} \gamma_\pm$$

It is possible to obtain the *mean ionic activity coefficient*, $\gamma_\pm$, of an electrolyte by several experimental methods as well as by a theoretical approach. The experimental methods include distribution coefficient studies, electromotive force measurements, colligative property methods, and solubility determinations. These results can then be used to obtain approximate activity coefficients for individual ions, where this is desired.[6]

Debye and Hückel developed a theoretical method by which it is possible to calculate the activity coefficient of a single ion as well as the mean ionic activity coefficient of a solute without recourse to experimental data. Although the theoretical equation agrees with experimental findings only in dilute solutions (so dilute, in fact, that some chemists have referred jokingly to such solutions as "slightly contaminated water"), it has certain practical value in solution calculations. Furthermore, the Debye–Hückel equation provides a remarkable confirmation of modern solution theory.

The mean ionic activity coefficients of a number of strong electrolytes are given in **Table 6–1**. The results of various investigators vary in the third decimal place; therefore, most of the entries in the table are recorded only to two places, providing sufficient precision for the calculations in this book. Although the values in the table are given at various molalities, we can accept these activity coefficients for problems involving molar concentrations (in which $m < 0.1$) because, in dilute solutions, the difference between molality and molarity is not great.

The mean values of **Table 6–1** for NaCl, $CaCl_2$, and $ZnSO_4$ are plotted in **Figure 6–6** against the square root of the molality. The reason for plotting the square root of the concentration is due to the form that the Debye–Hückel equation takes. The activity coefficient approaches unity with increasing dilution. As the concentrations of some of the electrolytes are increased, their curves pass through minima and rise again to values greater than unity. Although the curves for different electrolytes of the same ionic class coincide at lower concentrations, they differ widely at higher values. The initial decrease in the activity coefficient with increasing concentration is due to the interionic attraction, which causes the

**MEAN IONIC ACTIVITY COEFFICIENTS OF SOME STRONG ELECTROLYTES AT 25°C ON THE MOLAL SCALE**

| Molality ($m$) | HCl | NaCl | KCl | NaOH | CaCl$_2$ | H$_2$SO$_2$ | Na$_2$SO$_2$ | CuSO$_2$ | ZnSO$_2$ |
|---|---|---|---|---|---|---|---|---|---|
| 0.000 | 1.00 | 1.00 | 1.00 | 1.00 | 1.00 | 1.00 | 1.00 | 1.00 | 1.00 |
| 0.005 | 0.93 | 0.93 | 0.93 | – | 0.79 | 0.64 | 0.78 | 0.53 | 0.48 |
| 0.01 | 0.91 | 0.90 | 0.90 | 0.90 | 0.72 | 0.55 | 0.72 | 0.40 | 0.39 |
| 0.05 | 0.83 | 0.82 | 0.82 | 0.81 | 0.58 | 0.34 | 0.51 | 0.21 | 0.20 |
| 0.10 | 0.80 | 0.79 | 0.77 | 0.76 | 0.52 | 0.27 | 0.44 | 0.15 | 0.15 |
| 0.50 | 0.77 | 0.68 | 0.65 | 0.68 | 0.51 | 0.16 | 0.27 | 0.067 | 0.063 |
| 1.00 | 0.81 | 0.66 | 0.61 | 0.67 | 0.73 | 0.13 | 0.21 | 0.042 | 0.044 |
| 2.00 | 1.01 | 0.67 | 0.58 | 0.69 | 1.55 | 0.13 | 0.15 | – | 0.035 |
| 4.00 | 1.74 | 0.79 | 0.58 | 0.90 | 2.93 | 0.17 | 0.14 | – | – |

activity to be less than the stoichiometric concentration. The rise in the activity coefficient following the minimum in the curve of an electrolyte, such as HCl and CaCl$_2$, can be attributed to the attraction of the water molecules for the ions in concentrated aqueous solution. This *solvation* reduces the interionic attractions and increases the activity coefficient of the solute. It is the same effect that results in the salting out of nonelectrolytes from aqueous solutions to which electrolytes have been added.

## Activity of the Solvent

Thus far, the discussion of activity and activity coefficients has centered on the solute and particularly on electrolytes. It is customary to define the activity of the solvent on the mole fraction scale. When a solution is made infinitely dilute, it can be considered to consist essentially of pure solvent. Therefore, $X_1 \cong 1$, and the solvent behaves ideally in conformity

with Raoult's law. Under this condition, the mole fraction can be set equal to the activity of the solvent, or

$$a = X_1 = 1 \qquad (6–52)$$

As the solution becomes more concentrated in solute, the activity of the solvent ordinarily becomes less than the mole fraction concentration, and the ratio can be given, as for the solute, by the rational activity coefficient,

$$\frac{a}{X_1} = \gamma_x \qquad (6–53)$$

or

$$a = \gamma_x X_1 \qquad (6–54)$$

The activity of a volatile solvent can be determined rather simply. The ratio of the vapor pressure, $p_1$, of the solvent in a solution to the vapor pressure of pure solvent, $p_1°$, is approximately equal to the *activity* of the solvent at ordinary pressures: $a_1 = p_1/p°$.

**Calculating Escaping Tendency**

The vapor pressure of water in a solution containing 0.5 mole of sucrose in 1000 g of water is 17.38 mm and the vapor pressure of pure water at 20°C is 17.54 mm. What is the activity (or escaping tendency) of water in the solution? We write

$$a = \frac{17.38}{17.54} = 0.991$$

## Reference State

The assignment of activities to the components of solutions provides a measure of the extent of departure from ideal solution behavior. For this purpose, a *reference state* must be established in which each component behaves ideally. The reference state can be defined as the solution in which the concentration (mole fraction, molal or molar) of the component is equal to the activity,

$$\text{Activity} = \text{Concentration}$$

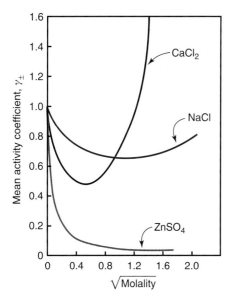

**Fig. 6–6.** Mean ionic activity coefficients of representative electrolytes plotted against the square root of concentration.

or, what amounts to the same thing, the activity coefficient is unity,

$$\gamma_i = \frac{\text{Activity}}{\text{Concentration}} = 1$$

The reference state for a solvent on the mole fraction scale was shown in equation (6–52) to be the pure solvent.

The reference state for the solute can be chosen from one of several possibilities. If a liquid solute is miscible with the solvent (e.g., in a solution of alcohol in water), the concentration can be expressed in mole fraction, and the pure liquid can be taken as the reference state, as was done for the solvent. For a liquid or solid solute having a limited solubility in the solvent, the reference state is ordinarily taken as the infinitely dilute solution in which the concentration of the solute and the ionic strength (see the following) of the solution are small. Under these conditions, the activity is equal to the concentration and the activity coefficient is unity.

## Standard State

The activities ordinarily used in chemistry are relative activities. It is not possible to know the absolute value of the activity of a component; therefore, a standard must be established just as was done in Chapter 1 for the fundamental measurable properties.

The *standard state* of a component in a solution is the state of the component at unit activity. The relative activity in any solution is then the ratio of the activity in that state to the value in the standard state. When defined in these terms, activity is a dimensionless number.

The pure liquid at 1 atm and at a definite temperature is chosen as the standard state of a solvent or of a liquid solute miscible with the solvent because, for the pure liquid, $a = 1$. Because the mole fraction of a pure solvent is also unity, mole fraction is equal to activity, and the reference state is identical with the standard state.

The standard state of the solvent in a solid solution is the pure solid at 1 atm and at a definite temperature. The assignment of $a = 1$ to pure liquids and pure solids will be found to be convenient in later discussions on equilibria and electromotive force.

The standard state for a solute of limited solubility is more difficult to define. The activity of the solute in an infinitely dilute solution, although equal to the concentration, is not unity, and the standard state is thus not the same as the reference state. The standard state of the solute is defined as a hypothetical solution of unit concentration (mole fraction, molal or molar) having, at the same time, the characteristics of an infinitely dilute or ideal solution. For complete understanding, this definition requires careful development, as carried out by Klotz and Rosenberg.[7]

## Ionic Strength

In dilute solutions of nonelectrolytes, activities and concentrations are considered to be practically identical because electrostatic forces do not bring about deviations from ideal behavior in these solutions. Likewise, for weak electrolytes that are present alone in solution, the differences between the ionic concentration terms and activities are usually disregarded in ordinary calculations because the number of ions present is small and the electrostatic forces are negligible.

However, for strong electrolytes and for solutions of weak electrolytes together with salts and other electrolytes, such as exist in buffer systems, it is important to use activities instead of concentrations. The activity coefficient, and hence the activity, can be obtained by using one of the forms of the Debye–Hückel equation (considered later) if one knows the ionic strength of the solution. Lewis and Randall[8] introduced the concept of *ionic strength*, $\mu$, to relate interionic attractions and activity coefficients. The ionic strength is defined on the molar scale as

$$\mu = \frac{1}{2}\left(c_1 z_1{}^2 + c_2 z_2{}^2 + c_3 z_3{}^2 + \cdots + c_j z_j{}^2\right) \quad \text{(6–55)}$$

or, in abbreviated notation,

$$\mu = \frac{1}{2}\sum_1^j c_i z_i{}^2 \quad \text{(6–56)}$$

where the summation symbol indicates that the product of $cz^2$ terms for all the ionic species in the solution, from the first one to the $j$th species, is to be added together. The term $c_i$ is the concentration in moles/liter of any of the ions and $z_i$ is its valence. Ionic strengths represent the contribution to the electrostatic forces of the ions of all types. It depends on the total number of ionic charges and not on the specific properties of the salts present in the solution. It was found that bivalent ions are equivalent not to two, but to four univalent ions; hence, by introducing the square of the valence, one gives proper weight to the ions of higher charge. The sum is divided by 2 because positive ion–negative ion pairs contribute to the total electrostatic interaction, whereas we are interested in the effect of each ion separately.

### EXAMPLE 6–11

**Calculating Ionic Strength**

**What is the ionic strength of (a) 0.010 M KCl, (b) 0.010 M BaSO₄, and (c) 0.010 M Na₂SO₄, and (d) what is the ionic strength of a solution containing all three electrolytes together with salicylic acid in 0.010 M concentration in aqueous solution?**

(a) KCl:

$$\mu = \frac{1}{2}[(0.01 \times 1^2) + (0.01 \times 1^2)]$$
$$= 0.010$$

(b) BaSO₄:

$$\mu = \frac{1}{2}[(0.01 \times 2^2) + (0.01 \times 2^2)]$$
$$= 0.040$$

(c) $Na_2 SO_4$:

$$\mu = \frac{1}{2}[(0.02 \times 1^2) + (0.01 \times 2^2)]$$

$$= 0.030$$

(d) The ionic strength of a 0.010 M solution of salicylic acid is 0.003 as calculated from a knowledge of the ionization of the acid at this concentration (using the equation $[H_3O^+] = \sqrt{K_a c}$). Unionized salicylic acid does not contribute to the ionic strength. The ionic strength of the mixture of electrolytes is the sum of the ionic strength of the individual salts. Thus,

$$\mu_{total} = \mu_{KCl} + \mu_{BaSO_4} + \mu_{Na_2SO_4} + \mu_{HSal}$$
$$= 0.010 + 0.040 + 0.030 + 0.003$$
$$= 0.083$$

---

**EXAMPLE 6–12**

**Ionic Strength of a Solution**

A buffer contains 0.3 mole of $K_2HPO_4$ and 0.1 mole of $KH_2PO_4$ per liter of solution. Calculate the ionic strength of the solution.

The concentrations of the ions of $K_2HPO_4$ are $[K^+] = 0.3 \times 2$ and $[HPO_4^{2-}] = 0.3$. The values for $KH_2PO_4$ are $[K^+] = 0.1$ and $[H_2PO_4^-] = 0.1$. Any contributions to $\mu$ by further dissociation of $[HPO_4^{2-}]$ and $[H_2PO_4^-]$ are neglected. Thus,

$$\mu = \frac{1}{2}[(0.3 \times 2 \times 1^2) + (0.3 \times 2^2) + (0.1 \times 1^2) + (0.1 \times 1^2)]$$

$$\mu = 1.0$$

---

It will be observed in *Example 6–11* that the ionic strength of a 1:1 electrolyte such as KCl is the same as the molar concentration; $\mu$ of a 1:2 electrolyte such as $Na_2SO_4$ is three times the concentration; and $\mu$ for a 2:2 electrolyte is four times the concentration.

The mean ionic activity coefficients of electrolytes should be expressed at various ionic strengths instead of concentrations. Lewis showed the uniformity in activity coefficients when they are related to ionic strength:

(a) The activity coefficient of a strong electrolyte is roughly constant in all dilute solutions of the same ionic strength, irrespective of the type of salts that are used to provide the additional ionic strength.

(b) The activity coefficients of all strong electrolytes of a single class, for example, all uni-univalent electrolytes, are approximately the same at a definite ionic strength, provided the solutions are dilute.

The results in **Table 6–1** illustrate the similarity of the mean ionic activity coefficients for 1:1 electrolytes at low concentrations (below 0.1 *m*) and the differences that become marked at higher concentrations.

Bull[9] pointed out the importance of the principle of ionic strength in biochemistry. In the study of the influence of pH on biologic action, the effect of the variable salt concentration in the buffer may obscure the results unless the buffer is adjusted to a constant ionic strength in each experiment. If the biochemical action is affected by the specific salts used, however, even this precaution may fail to yield satisfactory

results. Further use will be made of ionic strength in the chapters on ionic equilibria, solubility, and kinetics.

## The Debye–Hückel Theory

Debye and Hückel derived an equation based on the principles that strong electrolytes are completely ionized in dilute solution and that the deviations of electrolytic solutions from ideal behavior are due to the electrostatic effects of the oppositely charged ions. The equation relates the activity coefficient of a particular ion or the mean ionic activity coefficient of an electrolyte to the valence of the ions, the ionic strength of the solution, and the characteristics of the solvent. The mathematical derivation of the equation is not attempted here but can be found in Lewis and Randall's *Thermodynamics* as revised by Pitzer and Brewer.[10] The equation can be used to calculate the activity coefficients of drugs whose values have not been obtained experimentally and are not available in the literature.

According to the theory of Debye and Hückel, the activity coefficient, $\gamma_i$, of an ion of valence $z_i$ is given by the expression

$$\log \gamma_i = -A z_i^2 \sqrt{\mu} \qquad (6\text{–}57)$$

Equation (6–57) yields a satisfactory measure of the activity coefficient of an ion species up to an ionic strength, $\mu$, of about 0.02. For water at $25°C$, $A$, a factor that depends only on the temperature and the dielectric constant of the medium, is approximately equal to 0.51. The values of $A$ for various solvents of pharmaceutical importance are found in **Table 6–2**.

The form of the Debye–Hückel equation for a binary electrolyte consisting of ions with valences of $z_+$ and $z_-$ and present in a dilute solution ($\mu < 0.02$) is

$$\log \gamma_{\pm} = -A z_+ z_- \sqrt{\mu} \qquad (6\text{–}58)$$

The symbols $z_+$ and $z_-$ stand for the valences or charges, ignoring algebraic signs, on the ions of the electrolyte whose mean ionic activity coefficient is sought. The coefficient in equation (6–58) should actually be $\gamma_x$, the rational activity coefficient (i.e., $\gamma_{\pm}$ on the mole fraction scale), but in dilute solutions for which the Debye–Hückel equation is applicable, $\gamma_x$ can be assumed without serious error to be equal also to the practical coefficients, $\gamma_m$ and $\Lambda_c$, on the molal and molar scales.

**TABLE 6–2**

**VALUES OF *A* FOR SOLVENTS AT 25°C***

| Solvent | Dielectric Constant, $\varepsilon$ | $A_{calc}$ |
|---|---|---|
| Acetone | 20.70 | 3.76 |
| Ethanol | 24.30 | 2.96 |
| Water | 78.54 | 0.509 |

*$A_{calc} = (1.824 \times 10^6)/(\varepsilon \times T)^{3/2}$, where $\varepsilon$ is the dielectric constant and $T$ is the absolute temperature on the Kelvin scale.

**EXAMPLE 6–13**

**Mean Ionic Activity Coefficient**

Calculate the mean ionic activity coefficient for 0.005 M atropine sulfate (1:2 electrolyte) in an aqueous solution containing 0.01 M NaCl at 25°C. Because the drug is a uni-bivalent electrolyte, $z_1z_2 = 1 \times 2 = 2$. For water at 25°C, $A$ is 0.51. We have

$$\mu \text{ for atropine sulfate} = \frac{1}{2}[(0.005 \times 2 \times 1^2) + (0.005 \times 2^2)] = 0.015$$

$$\mu \text{ for NaCl} = \frac{1}{2}[(0.01 \times 1^2) + (0.01 \times 1^2)] = \underline{0.01}$$

$$\text{Total } \mu = 0.025$$

$$\log \gamma_\pm = -0.51 \times 2 \times \sqrt{0.025}$$

$$\log \gamma_\pm = -1.00 + 0.839 = -0.161$$

$$\gamma_\pm = 0.690$$

With the present-day accessibility of the handheld calculator, the intermediate step in this calculation (needed only when log tables are used) can be deleted.

---

Thus, one observes that the activity coefficient of a strong electrolyte in dilute solution depends on the total ionic strength of the solution, the valence of the ions of the drug involved, the nature of the solvent, and the temperature of the solution. Notice that although the ionic strength term results from the contribution of all ionic species in solution, the $z_1z_2$ terms apply only to the drug whose activity coefficient is being determined.

## Extension of the Debye–Hückel Equation to Higher Concentrations

The limiting expressions (6–57) and (6–58) are not satisfactory above an ionic strength of about 0.02, and equation (6–58) is not completely satisfactory for use in *Example 6–13*. A formula that applies up to an ionic strength of perhaps 0.1 is

$$\log \gamma_\pm = -\frac{Az_+z_-\sqrt{\mu}}{1 + a_iB\sqrt{\mu}} \quad (6\text{–}59)$$

The term $a_i$ is the mean distance of approach of the ions and is called the *mean effective ionic diameter* or the *ion size parameter*. Its exact significance is not known; however, it is somewhat analogous to the $b$ term in the van der Waals gas equation. The term $B$, like $A$, is a constant influenced only by the nature of the solvent and the temperature. The values of $a_i$ for several electrolytes at 25°C are given in **Table 6–3** and the values of $B$ and $A$ for water at various temperatures are shown in **Table 6–4**. The values of $A$ for various solvents, as previously mentioned, are listed in **Table 6–2**.

Because $a_i$ for most electrolytes equals 3 to $4 \times 10^{-8}$ and $B$ for water at 25°C equals $0.33 \times 10^8$, the product of $a_i$ and $B$ is approximately unity. Equation (6–59) then simplifies to

$$\log \gamma_\pm = -\frac{Az_+z_-\sqrt{\mu}}{1 + \sqrt{\mu}} \quad (6\text{–}60)$$

**TABLE 6–3**

**MEAN EFFECTIVE IONIC DIAMETER FOR SOME ELECTROLYTES AT 25°C**

| Electrolyte | $a_i$ (cm) |
|---|---|
| HCl | $5.3 \times 10^{-8}$ |
| NaCl | $4.4 \times 10^{-8}$ |
| KCl | $4.1 \times 10^{-8}$ |
| Methapyrilene HCl | $3.9 \times 10^{-8}$ |
| MgSO$_4$ | $3.4 \times 10^{-8}$ |
| K$_2$SO$_4$ | $3.0 \times 10^{-8}$ |
| AgNO$_3$ | $2.3 \times 10^{-8}$ |
| Sodium phenobarbital | $2.0 \times 10^{-8}$ |

**EXAMPLE 6–14**

**Comparing Activity Coefficients**

Calculate the activity coefficient of a 0.004 M aqueous solution of sodium phenobarbital at 25°C that has been brought to an ionic strength of 0.09 by the addition of sodium chloride. Use equations (6–58) through (6–60) and compare the results.

Equation (6–58): $\log \gamma_\pm = -0.51\sqrt{0.09}; \gamma_\pm = 0.70$

Equation (6–59): $\log \gamma_\pm =$

$$\frac{0.51\sqrt{0.09}}{1 + 2[(2 \times 10^8) \times (0.33 \times 10^3) \times \sqrt{0.09}]}; \gamma_\pm = 0.75$$

Equation (6–60): $\log \gamma_\pm = -\frac{0.51\sqrt{0.09}}{1 + \sqrt{0.09}}; \gamma_\pm = 0.76$

---

These results can be compared with the experimental values for some uni-univalent electrolytes in **Table 6–1** at a molal concentration of about 0.1.

For still higher concentrations, that is, at ionic strengths above 0.1, the observed activity coefficients for some electrolytes pass through minima and then increase with concentration; in some cases, they become greater than unity, as seen in **Figure 6–6**. To account for the increase in $\gamma_\pm$ at higher concentrations, an empirical term $C\mu$ can be added to the Debye–Hückel equation, resulting in the expression

$$\log \gamma_\pm = -\frac{Az_+z_-\sqrt{\mu}}{1 + a_iB\sqrt{\mu}} + C\mu \quad (6\text{–}61)$$

This equation gives satisfactory results in solutions of concentrations as high as 1 M. The mean ionic activity coefficient

**TABLE 6–4**

**VALUES OF *A* AND *B* FOR WATER AT VARIOUS TEMPERATURES**

| Temperature (°C) | *A* | *B* |
|---|---|---|
| 0 | 0.488 | $0.325 \times 10^8$ |
| 15 | 0.500 | $0.328 \times 10^8$ |
| 25 | 0.509 | $0.330 \times 10^8$ |
| 40 | 0.524 | $0.333 \times 10^8$ |
| 70 | 0.560 | $0.339 \times 10^8$ |
| 100 | 0.606 | $0.348 \times 10^8$ |

obtained from equation **(6–61)** is $\gamma_x$; however, it is not significantly different from $\gamma_m$ and $\gamma_c$ even at this concentration. Zografi et al.[11] used the extended Debye–Hückel equation [(equation **(6–61)**] in a study of the interaction between the dye orange II and quarternary ammonium salts.

Investigations have resulted in equations that extend the concentration to about 5 moles/liter.[12]

## COEFFICIENTS FOR EXPRESSING COLLIGATIVE PROPERTIES

### The $L$ Value

The van't Hoff expression $\Delta T_f = iK_f m$ probably provides the best single equation for computing the colligative properties of nonelectrolytes, weak electrolytes, and strong electrolytes. It can be modified slightly for convenience in dilute solutions by substituting molar concentration $c$ and by writing $iK_f$ as $L$, so that

$$\Delta T_f = Lc \qquad (6\text{–}62)$$

Goyan et al.[13] computed $L$ from experimental data for a number of drugs. It varies with the concentration of the solution. At a concentration of drug that is isotonic with body fluids, $L = iK_f$ is designated here as $L_{iso}$. It has a value equal to about 1.9 (actually 1.86) for nonelectrolytes, 2.0 for weak electrolytes, 3.4 for uni-univalent electrolytes, and larger values for electrolytes of high valences. A plot of $iK_f$ against the concentration of some drugs is presented in **Figure 6–7**, where each curve is represented as a band to show the variability of the $L$ values within each ionic class. The approximate $L_{iso}$ for each of the ionic classes can be obtained from the dashed line running through the figure.

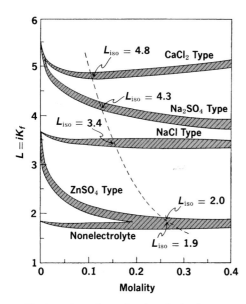

**Fig. 6–7.** $L_{iso}$ values of various ionic classes.

### Osmotic Coefficient

Other methods of correcting for the deviations of electrolytes from ideal colligative behavior have been suggested. One of these is based on the fact that as the solution becomes more dilute, $i$ approaches $\nu$, the number of ions into which an electrolyte dissociates, and at infinite dilution, $i = \nu$, or $i/\nu = 1$. Proceeding in the direction of more concentrated solutions, $i/\nu$ becomes less (and sometimes greater) than unity.

The ratio $i/\nu$ is designated as $g$ and is known as the *practical osmotic coefficient* when expressed on a molal basis. In the case of a weak electrolyte, it provides a measure of the degree of dissociation. For strong electrolytes, $g$ is equal to unity for complete dissociation, and the departure of $g$ from unity, that is, $1 - g$, in moderately concentrated solutions is an indication of the interionic attraction. Osmotic coefficients, $g$, for electrolytes and nonelectrolytes are plotted against ionic concentration, $\nu m$, in **Figure 6–8**. Because $g = 1/\nu$ or $i = g\nu$ in a dilute solution, the cryoscopic equation can be written as

$$\Delta T_f = g\nu K_f m \qquad (6\text{–}63)$$

The molal osmotic coefficients of some salts are listed in **Table 6–5**.

**EXAMPLE 6–15**

**Molality and Molarity**

The osmotic coefficient of LiBr at 0.2 $m$ is 0.944 and the $L_{iso}$ value is 3.4. Compute $\Delta T_f$ for this compound using $g$ and $L_{iso}$. Disregard the difference between molality and molarity. We have

$$\Delta T_f = g\nu K_f m = 0.944 \times 2 \times 1.86 \times 0.2$$
$$= 0.70°$$
$$\Delta T_f = L_{iso}c = 3.4 \times 0.2 = 0.68°$$

### Osmolality

Although osmotic pressure classically is given in atmospheres, in clinical practice it is expressed in terms of osmols (Osm) or milliosmols (mOsm). A solution containing 1 mole (1 g molecular weight) of a nonionizable substance in 1 kg of water (a 1 $m$ solution) is referred to as a 1-osmolal solution. It contains 1 osmol (Osm) or 1000 milliosmols (mOsm) of solute per kilogram of solvent. Osmolality measures the total number of particles dissolved in 1 kg of water, that is,

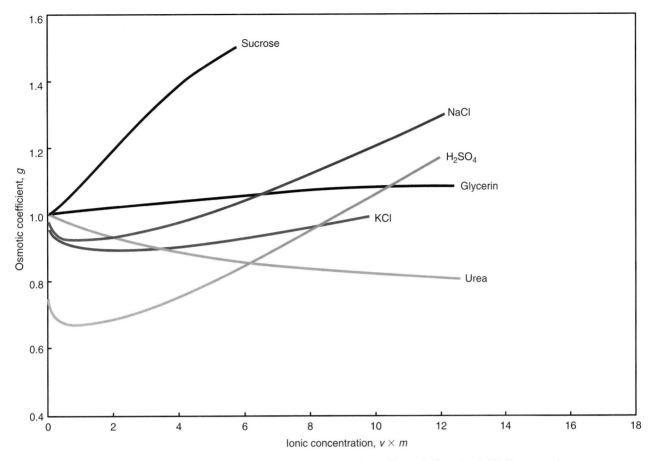

**Fig. 6–8.** Osmotic coefficient, *g*, for some common solutes. (From G. Scatchard, W. Hamer, and S. Wood, J. Am. Chem. Soc. **60,** 3061, 1938. With permission.)

the osmols per kilogram of water, and depends on the electrolytic nature of the solute. An ionic species dissolved in water will dissociate to form ions or "particles." These ions tend to associate somewhat, however, owing to their ionic interactions. The apparent number of "particles" in solution, as measured by osmometry or one of the other colligative methods, will depend on the extent of these interactions. An un-ionized material (i.e., a nonelectrolyte) is used as the reference solute for osmolality measurements, ionic interactions being insignificant for a nonelectrolyte. For an electrolyte that dissociates into ions in a dilute solution, osmolality or milliosmolality can be calculated from

$$\text{Milliosmolality (mOsm/kg)} = i \cdot mm \qquad \textbf{(6–64)}$$

where $i$ is approximately the number of ions formed per molecule and $mm$ is the millimolal concentration. If no ionic

**TABLE 6–5**
**OSMOTIC COEFFICIENT, *g*, AT 25°C***

| *m* | NaCl | KCl | H₂SO₄ | Sucrose | Urea | Glycerin |
|-----|------|-----|-------|---------|------|----------|
| 0.1 | 0.9342 | 0.9264 | 0.6784 | 1.0073 | 0.9959 | 1.0014 |
| 0.2 | 0.9255 | 0.9131 | 0.6675 | 1.0151 | 0.9918 | 1.0028 |
| 0.4 | 0.9217 | 0.9023 | 0.6723 | 1.0319 | 0.9841 | 1.0055 |
| 0.6 | 0.9242 | 0.8987 | 0.6824 | 1.0497 | 0.9768 | 1.0081 |
| 0.8 | 0.9295 | 0.8980 | 0.6980 | 1.0684 | 0.9698 | 1.0105 |
| 1.0 | 0.9363 | 0.8985 | 0.7176 | 1.0878 | 0.9631 | 1.0128 |
| 1.6 | 0.9589 | 0.9024 | 0.7888 | 1.1484 | 0.9496 | 1.0192 |
| 2.0 | 0.9786 | 0.9081 | 0.8431 | 1.1884 | 0.9346 | 1.0230 |
| 3.0 | 1.0421 | 0.9330 | 0.9922 | 1.2817 | 0.9087 | 1.0316 |
| 4.0 | 1.1168 | 0.9635 | 1.1606 | 1.3691 | 0.8877 | 1.0393 |
| 5.0 | 1.2000 | 0.9900 | – | 1.4477 | 0.8700 | 1.0462 |

*From G. Scatchard, W. G. Hamer, and S. E. Wood, J. Am. Chem. Soc. **60,** 3061, 1938. With permission.

interactions occurred in a solution of sodium chloride, $i$ would equal 2.0. In a typical case, for a 1:1 electrolyte in dilute solution, $i$ is approximately 1.86 rather than 2.0, owing to ionic interaction between the positively and negatively charged ions.

### EXAMPLE 6–16

#### Calculating Milliosmolality

What is the milliosmolality of a 0.120 $m$ solution of potassium bromide? What is its osmotic pressure in atmospheres?

For a 120 millimolal solution of KBr:

$$\text{Milliosmolality} = 1.86 \times 120 = 223 \text{ mOsm/kg}$$

A 1-osmolal solution raises the boiling point $0.52°C$, lowers the freezing point $1.86°C$, and produces an osmotic pressure of 24.4 atm at $25°C$. Therefore, a 0.223-Osm/kg solution yields an osmotic pressure of $24.4 \times 0.223 = 5.44$ atm.

Streng et al.[14] and Murty et al.[15] discuss the use of osmolality and osmolarity in clinical pharmacy. Molarity (moles of solute per liter of solution) is used in clinical practice more frequently than molality (moles of solute per kilogram of solvent). In addition, osmolarity is used more frequently than osmolality in labeling parenteral solutions in the hospital. Yet, osmolarity cannot be measured and must be calculated from the experimentally determined osmolality of a solution. As shown by Murty et al.,[15] the conversion is made using the relation

Osmolarity = (Measured osmolality)
$$\times \text{(Solution density in g/mL}$$
$$- \text{Anhydrous solute concentration in g/mL)} \quad (6–65)$$

According to Streng et al.,[14] osmolality is converted to osmolarity using the equation

mOsm/liter solution = mOsm/(kg $H_2O$)
$$\times [d_1°(1 - 0.001 \, \bar{v}_2°)] \quad (6–66)$$

where $d_1°$ is the density of the solvent and $\bar{v}_2°$ is the partial molal volume of the solute at infinite dilution.

### EXAMPLE 6–17

#### Converting Osmolality to Osmolarity

A 30-g/liter solution of sodium bicarbonate contains 0.030 g/mL of anhydrous sodium bicarbonate. The density of this solution was found to be 1.0192 g/mL at $20°C$ and its measured milliosmolality was 614.9 mOsm/kg. Convert milliosmolality to milliosmolarity. We have

Milliosmolarity = 614.9 mOsm/kg $H_2O$
$$\times (1.0192 \text{ g/mL} - 0.030 \text{ g/mL})$$
$$= 608.3 \text{ mOsm/L solution}$$

### EXAMPLE 6–18

#### Calculating Milliosmolarity

A 0.154 $m$ sodium chloride solution has a milliosmolality of 286.4 mOsm/kg (see *Example 6–19*). Calculate the milliosmolarity, mOsm/liter solution, using equation (6–66). The density of the

solvent—water—at $25°C$ is $d_1° = 0.9971$ g/cm³ and the partial molal volume of the solute—sodium chloride—is $\bar{v}_2° = 16.63$ mL/mole.

Milliosmolality = (286.4 mOsm/kg $H_2O$)
$$\times [0.9971(1 - 0.001(16.63))]$$
$$= 280.8 \text{ mOsm/L solution}$$

As noted, osmolarity differs from osmolality by only 1% or 2%. However, in more concentrated solutions of polyvalent electrolytes together with buffers, preservatives, and other ions, the difference may become significant. For accuracy in the preparation and labeling of parenteral solutions, osmolality should be measured carefully with a vapor pressure or freezing point osmometer (rather than calculated) and the results converted to osmolarity using equation (6–65) or (6–66).

Whole blood, plasma, and serum are complex liquids consisting of proteins, glucose, nonprotein nitrogenous materials, sodium, potassium, calcium, magnesium, chloride, and bicarbonate ions. The serum electrolytes, constituting less than 1% of the blood's weight, determine the osmolality of the blood. Sodium chloride contributes a milliosmolality of 275, whereas glucose and the other constituents together provide about 10 mOsm/kg to the blood.

Colligative properties such as freezing point depression are related to osmolality through equations (6–27) and (6–63):

$$\Delta T_f \cong K_f im \quad (6–67)$$

where $i = g\nu$ and $im = g\nu m$ is osmolality.

### EXAMPLE 6–19

#### Calculate Freezing Point Depression

Calculate the freezing point depression of (a) 0.154 $m$ solution of NaCl and (b) a 0.154 $m$ solution of glucose. What are the milliosmolalities of these two solutions?

(a) From Table 6–5, $g$ for NaCl at $25°C$ is about 0.93, and because NaCl ionizes into two ions, $i = \nu g = 2 \times 0.93 = 1.86$. From equation (6–64), the osmolality of a 0.154 $m$ solution is $im = 1.86 \times 0.154 = 0.2864$. The milliosmolality of this solution is therefore 286.4 mOsm/kg. Using equation (6–67), with $K_f$ also equal to 1.86, we obtain for the freezing point depression of a 0.154 $m$ solution—or its equivalent, a 0.2864-Osm/kg solution—of NaCl:

$$\Delta T_f = (1.86)(1.86)(0.154)$$
$$= (1.86)(0.2864) = 0.53°C$$

(b) Glucose is a nonelectrolyte, producing only one particle for each of its molecules in solution, and for a nonelectrolyte, $i = \nu = 1$ and $g = i/\nu = 1$. Therefore, the freezing point depression of a 0.154 $m$ solution of glucose is approximately

$$\Delta T_f = K_f im = (1.86)(1.00)(0.154)$$
$$= 0.286°C$$

which is nearly one half of the freezing point depression provided by sodium chloride, a 1:1 electrolyte that provides two particles rather than one particle in solution.

The osmolality of a nonelectrolyte such as glucose is identical to its molal concentration because osmolality = $i \times$ molality, and $i$

for a nonelectrolyte is 1.00. The milliosmolality of a solution is 1000 times its osmolality or, in this case, 154 mOsm/kg.

Ohwaki et al.[16] studied the effect of osmolality on the nasal absorption of secretin, a hormone used in the treatment of duodenal ulcers. They found that maximum absorption through the nasal mucosa occurred at a sodium chloride milliosmolarity of about 860 mOsm/liter (0.462 M), possibly owing to structural changes in the epithelial cells of the nasal mucosa at this high mOsm/liter value.

Although the osmolality of blood and other body fluids is contributed mainly by the content of sodium chloride, the osmolality and milliosmolality of these complex solutions by convention are calculated on the basis of $i$ for nonelectrolytes, that is, $i$ is taken as unity, and osmolality becomes equal to molality. This principle is best described by an example.

### EXAMPLE 6–20

#### Milliosmolarity of Blood

Freezing points were determined using the blood of 20 normal individuals and were averaged to –0.5712°C. This value of course is equivalent to a freezing point depression of +0.5712°C below the freezing point of water because the freezing point of water is taken as 0.000°C at atmospheric pressure. What is the average milliosmolality, $x$, of the blood of these subjects?

Using equation (6–67) with the arbitrary choice of $i = 1$ for body fluids, we obtain

$$0.5712 = (1.86)(1.00)x$$

$$x = 0.3071 \text{ Osm/kg}$$

$$= 307.1 \text{ mOsm/kg}$$

It is noted in *Example 6–20* that although the osmolality of blood and its freezing point depression are contributed mainly by NaCl, an $i$ value of 1 was used for blood rather than $g\nu = 1.86$ for an NaCl solution.

The milliosmolality for blood obtained by various workers using osmometry, vapor pressure, and freezing point depression apparatus ranges from about 250 to 350 mOsm/kg.[17] The normal osmolality of body fluids is given in medical handbooks[18] as 275 to 295 mOsm/kg, but normal values are likely to fall in an even narrower range of $286 \pm 4$ mOsm/kg.[19] Freezing point and vapor pressure osmometers are now used routinely in the hospital. A difference of 50 mOsm/kg or more from the accepted values of a body fluid suggests an abnormality such as liver failure, hemorrhagic shock, uremia, or other toxic manifestations. Body water and electrolyte balance are also monitored by measurement of milliosmolality.

## CHAPTER SUMMARY

In this chapter, solutions of electrolytes were introduced and discussed. Understanding the properties of electrolytes in solution is still very important today in the pharmaceutical sciences. You should be able to understand Faraday's law and

calculate the conductance in solutions. The rationale for using the concept of activity rather than concentrations is important in many chemical calculations since solutions that contain ionic solutes do not behave ideally even at very low concentrations. The information provided establishes a framework for calculating ionic strength, osmolality, and osmolarity. The successful student will not only be able to do these calculations but will understand the differences between osmolality and osmolarity.

 **Practice problems for this chapter can be found at thePoint.lww.com/Sinko6e.**

## References

1. S. Arrhenius, Z. Physik. Chem. **1,** 631, 1887.
2. H. L. Clever, J. Chem. Educ. **40,** 637, 1963; P. A. Giguère, J. Chem. Educ. **56,** 571, 1979; R. P. Bell, *The Proton in Chemistry*, 2nd Ed., Chapman and Hall, London, 1973, pp. 13–25.
3. A. Albert, *Selective Toxicity*, Chapman and Hall, London, 1979, pp. 344, 373–384.
4. S. Arrhenius, J. Am. Chem. Soc. **34,** 361, 1912.
5. G. N. Lewis and M. Randall, *Thermodynamics*, McGraw-Hill, New York, 1923, p. 255.
6. S. Glasstone, *An Introduction to Electrochemistry*, Van Nostrand, New York, 1942, pp. 137–139.
7. I. M. Klotz and R. M. Rosenberg, *Chemical Thermodynamics*, 3rd Ed., W. A. Benjamin, Menlo Park, Calif., 1973, Chapter 19.
8. G. N. Lewis and M. Randall, *Thermodynamics*, McGraw-Hill, New York, 1923, p. 373.
9. H. B. Bull, *Physical Biochemistry*, 2nd Ed., Wiley, New York, 1951, p. 79.
10. G. N. Lewis and M. Randall, *Thermodynamics, Rev.* K. S. Pitzer and L. Brewer, 2nd Ed., McGraw-Hill, New York, 1961, Chapter 23.
11. G. Zografi, P. Patel, and N. Weiner, J. Pharm. Sci. **53,** 545, 1964.
12. R. H. Stokes and R. A. Robinson, J. Am. Chem. Soc. **70,** 1870, 1948; M. Eigen and E. Wicke, J. Phys. Chem. **58,** 702, 1954.
13. F. M. Goyan, J. M. Enright, and J. M. Wells, J. Am. Pharm. Assoc. Sci. Ed. **33,** 74, 1944.
14. W. H. Streng, H. E. Huber, and J. T. Carstensen, J. Pharm. Sci. **67,** 384, 1978; H. E. Huber, W. H. Streng, and H. G. H. Tan, J. Pharm. Sci. **68,** 1028, 1979.
15. B. S. R. Murty, J. N. Kapoor, and P. O. DeLuca, Am. J. Hosp. Pharm. **33,** 546, 1976.
16. T. Ohwaki, H. Ando, F. Kakimoto, K. Uesugi, S. Watanabe, Y. Miyake, and M. Kayano, J. Pharm. Sci. **76,** 695, 1987.
17. C. Waymouth, In Vitro **6,** 109, 1970.
18. M. A. Kaupp, et al. (Eds.), *Current Medical Diagnosis and Treatment*, Lange, Los Altos, Calif., 1986, pp. 27, 1113.
19. L. Glasser, D. D. Sternalanz, J. Combie, and A. Robinson, Am. J. Clin. Pathol. **60,** 695–699, 1973.

## Recommended Reading

R. A. Robinson and R. H. Stokes, *Electrolyte Solutions: Second Revised Edition*, Dover Publications, Mineola NY, 2002.

CHAPTER LEGACY

**Fifth Edition:** published as Chapter 6 (Electrolyte Solutions). Updated by Patrick Sinko.

**Sixth Edition:** published as Chapter 6 (Electrolyte Solutions). Updated by Patrick Sinko.

# 7 ■ IONIC EQUILIBRIA

**CHAPTER OBJECTIVES**  **At the conclusion of this chapter the student should be able to:**

**1** Describe the Brönsted–Lowry and Lewis electronic theories.

**2** Identify and define the four classifications of solvents.

**3** Understand the concepts of acid–base equilibria and the ionization of weak acids and weak bases.

**4** Calculate dissociation constants $K_a$ and $K_b$ and understand the relationship between $K_a$ and $K_b$.

**5** Understand the concepts of pH, p$K$, and pOH and the relationship between hydrogen ion concentration and pH.

**6** Calculate pH.

**7** Define strong acid and strong base.

**8** Define and calculate acidity constants.

## INTRODUCTION

Arrhenius defined an acid as a substance that liberates hydrogen ions and a base as a substance that supplies hydroxyl ions on dissociation. Because of a need for a broader concept, Brönsted in Copenhagen and Lowry in London independently proposed parallel theories in 1923.[1] The *Brönsted–Lowry theory*, as it has come to be known, is more useful than the Arrhenius theory for the representation of ionization in both aqueous and nonaqueous systems.

### Brönsted–Lowry Theory

According to the Brönsted–Lowry theory, an acid is a substance, charged or uncharged, that is capable of donating a proton, and a base is a substance, charged or uncharged, that is capable of accepting a proton from an acid. The relative strengths of acids and bases are measured by the tendencies of these substances to give up and take on protons. Hydrochloric acid is a strong acid in water because it gives up its proton readily, whereas acetic acid is a weak acid because it gives up its proton only to a small extent. The strength of an acid or a base varies with the solvent. Hydrochloric acid is a weak acid

### ▶ KEY CONCEPT CLASSIFICATION OF SOLVENTS

Solvents can be classified as protophilic, protogenic, amphiprotic, and aprotic. A *protophilic* or basic solvent is one that is capable of accepting protons from the solute. Such solvents as acetone, ether, and liquid ammonia fall into this group. A *protogenic* solvent is a proton-donating compound and is represented by acids such as formic acid, acetic acid, sulfuric acid, liquid HCl, and liquid HF. *Amphiprotic* solvents act as both proton acceptors and proton donors, and this class includes water and the alcohols. *Aprotic* solvents, such as the hydrocarbons, neither accept nor donate protons, and, being neutral in this sense, they are useful for studying the reactions of acids and bases free of solvent effects.

in glacial acetic acid and acetic acid is a strong acid in liquid ammonia. Consequently, the strength of an acid depends not only on its ability to give up a proton but also on the ability of the solvent to accept the proton from the acid. This is called the *basic strength* of the solvent.

In the Brönsted–Lowry classification, acids and bases may be anions such as $HSO_4^-$ and $CH_3COO^-$, cations such as $NH_4^+$ and $H_3O^+$, or neutral molecules such as HCl and $NH_3$. Water can act as either an acid or a base and thus is amphiprotic. Acid–base reactions occur when an acid reacts with a base to form a new acid and a new base. Because the reactions involve a transfer of a proton, they are known as *protolytic reactions* or *protolysis*.

In the reaction between HCl and water, HCl is the acid and water the base:

$$\begin{array}{cccc} HCl & + H_2O & \rightarrow H_3O^+ & + Cl^- \\ Acid_1 & Base_2 & Acid_2 & Base_1 \end{array} \qquad (7\text{--}1)$$

$Acid_1$ and $Base_1$ stand for an *acid–base pair* or *conjugate pair*, as do $Acid_2$ and $Base_2$. Because the bare proton, $H^+$, is practically nonexistent in aqueous solution, what is normally referred to as the hydrogen ion consists of the hydrated proton, $H_3O^+$, known as the *hydronium ion*. Higher solvated forms can also exist in solution.[*] In an ethanolic solution, the "hydrogen ion" is the proton attached to a molecule of solvent, represented as $C_2H_5OH_2^+$. In equation (7–1), hydrogen chloride, the acid, has donated a proton to water, the base, to form the corresponding acid, $H_3O^+$, and the base, $Cl^-$.

The reaction of HCl with water is one of ionization. Neutralization and hydrolysis are also considered as acid–base reactions or proteolysis following the broad definitions of the Brönsted–Lowry concept. Several examples illustrate these types of reactions, as shown in **Table 7–1**. The displacement reaction, a special type of neutralization, involves the displacement of a weaker acid, such as acetic acid, from its salt as in the reaction shown later.

---

[*]Reports have appeared in the literature[2] describing the discovery of a polymer of the hydrogen ion consisting of 21 molecules of water surrounding one hydrogen ion, namely,

$$H^+ \cdot (H_2O)_{21}$$

**EXAMPLES OF ACID–BASE REACTIONS**

|  | Acid$_1$ | | Base$_2$ | | Acid$_2$ | | Base$_1$ |
|---|---|---|---|---|---|---|---|
| Neutralization | $NH_4^+$ | + | $OH^-$ | = | $H_2O$ | + | $NH_3$ |
| Neutralization | $H_3O^+$ | + | $OH^-$ | = | $H_2O$ | + | $H_2O$ |
| Neutralization | $HCl$ | + | $NH_3$ | = | $NH_4^+$ | + | $Cl^-$ |
| Hydrolysis | $H_2O$ | + | $CH_3COO^-$ | = | $CH_3COOH$ | + | $OH^-$ |
| Hydrolysis | $NH_4^+$ | + | $H_2O$ | = | $H_3O^+$ | + | $NH_3$ |
| Displacement | $HCl$ | + | $CH_3COO^-$ | = | $CH_3COOH$ | + | $Cl^-$ |

## Lewis Electronic Theory

Other theories have been suggested for describing acid–base reactions, the most familiar of which is the *electronic theory of Lewis*.[3]

According to the Lewis theory, an acid is a molecule or an ion that accepts an electron pair to form a covalent bond. A base is a substance that provides the pair of unshared electrons by which the base coordinates with an acid. Certain compounds, such as boron trifluoride and aluminum chloride, although not containing hydrogen and consequently not serving as proton donors, are nevertheless acids in this scheme. Many substances that do not contain hydroxyl ions, including amines, ethers, and carboxylic acid anhydrides, are classified as bases according to the Lewis definition. Two Lewis acid–base reactions are

$$H^+ \text{ (solvated)} + \; :\!\!\overset{\displaystyle H}{\underset{\displaystyle H}{N}}\!\!-\!H = \left[ H\!:\!\overset{\displaystyle \cdot\cdot}{\underset{\displaystyle \cdot\cdot}{N}}\!:\!H \right]^+ \quad (7\text{–}2)$$
$$\qquad\qquad\quad \text{Acid} \qquad \text{Base}$$

$$\overset{\displaystyle Cl}{\underset{\displaystyle Cl}{Cl\!-\!B}} + \; :\!O\!\!\overset{\displaystyle CH_3}{\underset{\displaystyle CH_3}{}} = \overset{\displaystyle Cl}{\underset{\displaystyle Cl}{Cl\!-\!B\!:\!O}}\!\!\overset{\displaystyle CH_3}{\underset{\displaystyle CH_3}{}}$$
$$\text{Acid} \qquad \text{Base} \qquad\qquad\qquad (7\text{–}3)$$

The Lewis system is probably too broad for convenient application to ordinary acid–base reactions, and those processes that are most conveniently expressed in terms of this electronic classification should be referred to simply as a form of electron sharing rather than as acid–base reactions.[4] The Lewis theory is finding increasing use for describing the mechanism of many organic and inorganic reactions. It will be mentioned again in the chapters on solubility and complexation. The Brönsted–Lowry nomenclature is particularly useful for describing ionic equilibria and is used extensively in this chapter.

## ACID–BASE EQUILIBRIA

The ionization or proteolysis of a weak electrolyte, acetic acid, in water can be written in the Brönsted–Lowry manner as

$$HAc + H_2O \; \rightleftharpoons \; H_3O^+ + Ac^-$$
$$\text{Acid}_1 \quad \text{Base}_2 \quad \text{Acid}_2 \quad \text{Base}_1 \qquad (7\text{–}4)$$

The arrows pointing in the forward and reverse directions indicate that the reaction is proceeding to the right and left simultaneously. According to the law of mass action, the velocity or rate of the forward reaction, $R_f$, is proportional to the concentration of the reactants:

$$R_f = k_1 \times [HAc]^1 \times [H_2O]^1 \qquad (7\text{–}5)$$

The speed of the reaction is usually expressed in terms of the decrease in the concentration of either the reactants per unit time. The terms rate, speed, and velocity have the same meaning here. The reverse reaction

$$R_r = k_2 \times [H_3O^+]^1 \times [Ac^-]^1 \qquad (7\text{–}6)$$

expresses the rate, $R_r$, of re-formation of un-ionized acetic acid. Because only 1 mole of each constituent appears in the

---

### KEY CONCEPT ■ EQUILIBRIUM

*Equilibrium* can be defined as a balance between two opposing forces or actions. This statement does not imply cessation of the opposing reactions, suggesting rather a dynamic equality between the velocities of the two. Chemical equilibrium maintains the concentrations of the reactants and products constant. Most chemical reactions proceed in both a forward and a reverse direction if the products of the reaction are not removed as they form. Some reactions, however, proceed nearly to completion and, for practical purposes, may be regarded as irreversible. The topic of chemical equilibria is concerned with truly reversible systems and includes reactions such as the ionization of weak electrolytes.

reaction, each term is raised to the first power, and the exponents need not appear in subsequent expressions for the dissociation of acetic acid and similar acids and bases. The symbols $k_1$ and $k_2$ are proportionality constants commonly known as *specific reaction rates* for the forward and the reverse reactions, respectively, and the brackets indicate concentrations. A better representation of the facts would be had by replacing concentrations with activities, but for the present discussion, the approximate equations are adequate.

## Ionization of Weak Acids

According to the concept of equilibrium, the rate of the forward reaction decreases with time as acetic acid is depleted, whereas the rate of the reverse reaction begins at zero and increases as larger quantities of hydrogen ions and acetate ions are formed. Finally, a balance is attained when the two rates are equal, that is, when

$$R_f = R_r \qquad (7-7)$$

The *concentrations* of products and reactants are not necessarily equal at equilibrium; the *speeds* of the forward and reverse reactions are what are the same. Because equation (7–7) applies at equilibrium, equations (7–5) and (7–6) may be set equal:

$$k_1 \times [HAc] \times [H_2O] = k_2 \times [H_3O^+] \times [Ac^-] \qquad (7-8)$$

and solving for the ratio $k_1/k_1$, one obtains

$$k = \frac{k_1}{k_2} = \frac{[H_3O^+][Ac^-]}{[HAc][H_2O]} \qquad (7-9)$$

In dilute solutions of acetic acid, water is in sufficient excess to be regarded as constant at about 55.3 moles/liter (1 liter $H_2O$ at 25°C weights 997.07 g, and 997.07/18.02 = 55.3). It is thus combined with $k_1/k_2$ to yield a new constant $K_a$, the *ionization constant* or the *dissociation constant* of acetic acid.

$$K_a = 55.3k = \frac{[H_3O^+][Ac^-]}{[HAc]} \qquad (7-10)$$

Equation (7–10) is the equilibrium expression for the dissociation of acetic acid, and the dissociation constant $K_a$ is an equilibrium constant in which the essentially constant concentration of the solvent is incorporated. In the discussion of equilibria involving charged as well as uncharged acids, according to the Brönsted–Lowry nomenclature, the term *ionization constant*, $K_a$, is not satisfactory and is replaced by the name *acidity constant*. Similarly, for charged and uncharged bases, the term *basicity constant* is now often used for $K_b$, to be discussed in the next section.

In general, the acidity constant for an uncharged weak acid HB can be expressed by

$$HB + H_2O \rightleftharpoons H_3O^+ + B^- \qquad (7-11)$$

$$K_a = \frac{[H_3O^+][B^-]}{[HB]} \qquad (7-12)$$

Equation (7–10) can be presented in a more general form, using the symbol $c$ to represent the initial molar concentration of acetic acid and $x$ to represent the concentration $[H_3O^+]$. The latter quantity is also equal to $[Ac^-]$ because both ions are formed in equimolar concentration. The concentration of acetic acid remaining at equilibrium $[HAc]$ can be expressed as $c - x$. The reaction [equation (7–4)] is

$$\begin{array}{ccc} HAc + H_2O \rightleftharpoons H_3O^+ + Ac^- \\ (c-x) \qquad\quad x \qquad x \end{array} \qquad (7-13)$$

and the equilibrium expression (7–10) becomes

$$K_a = \frac{x^2}{c - x} \qquad (7-14)$$

where $c$ is large in comparison with $x$. The term $c - x$ can be replaced by $c$ without appreciable error, giving the equation

$$K_a \cong \frac{x^2}{c} \qquad (7-15)$$

which can be rearranged as follows for the calculation of the hydrogen ion concentration of weak acids:

$$x^2 = K_a c$$
$$x = [H_3O^+] = \sqrt{K_a c} \qquad (7-16)$$

### EXAMPLE 7–1

In a liter of a 0.1 M solution, acetic acid was found by conductivity analysis to dissociate into $1.32 \times 10^{-3}$ g ions ("moles") each of hydrogen and acetate ion at 25°C. What is the acidity or dissociation constant $K_a$ for acetic acid?

According to equation (7–4), at equilibrium, 1 mole of acetic acid has dissociated into 1 mole each of hydrogen ion and acetate ion. The concentration of ions is expressed as moles/liter and less frequently as molality. A solution containing 1.0078 g of hydrogen ions in 1 liter represents 1 g ion or 1 mole of hydrogen ions. The molar concentration of each of these ions is expressed as $x$. If the original amount of acetic acid was 0.1 mole/liter, then at equilibrium the undissociated acid would equal $0.1 - x$ because $x$ is the amount of acid that has dissociated. The calculation according to equation (7–12) is

$$K_a = \frac{(1.32 \times 10^{-3})^2}{0.1 - (1.32 \times 10^{-3})}$$

It is of little significance to retain the small number, $1.32 \times 10^{-3}$, in the denominator, and the calculations give

$$K_a = \frac{(1.32 \times 10^{-3})^2}{0.1}$$

$$K_a = \frac{1.74 \times 10^{-6}}{1 \times 10^{-1}} = 1.74 \times 10^{-5}$$

The value of $K_a$ in *Example 7–1* means that, at equilibrium, the ratio of the product of the ionic concentrations to that of the undissociated acid is $1.74 \times 10^{-5}$; that is, the dissociation of acetic acid into its ions is small, and acetic acid may be considered as a weak electrolyte.

When a salt formed from a strong acid and a weak base, such as ammonium chloride, is dissolved in water, it dissociates completely as follows:

$$NH_4{}^+Cl^- \xrightarrow{H_2O} NH_4{}^+ + Cl^- \qquad (7\text{--}17)$$

The $Cl^-$ is the conjugate base of a strong acid, HCl, which is 100% ionized in water. Thus, the $Cl^-$ cannot react any further. In the Brönsted–Lowry system, $NH_4{}^+$ is considered to be a cationic acid, which can form its conjugate base, $NH_3$, by donating a proton to water as follows:

$$NH_4{}^+ + H_2O \rightleftharpoons H_3O^+ + NH_3 \qquad (7\text{--}18)$$

$$K_a = \frac{[H_3O^+][NH_3]}{[NH_4{}^+]} \qquad (7\text{--}19)$$

In general, for charged acids $BH^+$, the reaction is written as

$$BH^+ + H_2O \rightleftharpoons H_3O^+ + B \qquad (7\text{--}20)$$

and the acidity constant is

$$K_a = \frac{[H_3O^+][B]}{[BH^+]} \qquad (7\text{--}21)$$

## Ionization of Weak Bases

Nonionized weak bases B, exemplified by $NH_3$, react with water as follows:

$$B + H_2O \rightleftharpoons OH^- + BH^+ \qquad (7\text{--}22)$$

$$K_b = \frac{[OH^-][BH^+]}{[B]} \qquad (7\text{--}23)$$

which, by a procedure like that used to obtain equation (7–16), leads to

$$[OH^-] = \sqrt{K_b c} \qquad (7\text{--}24)$$

**EXAMPLE 7–2**

The basicity or ionization constant $K_b$ for morphine base is $7.4 \times 10^{-7}$ at $25°C$. What is the hydroxyl ion concentration of a 0.0005 M aqueous solution of morphine? We have

$$[OH^-] = \sqrt{7.4 \times 10^{-7} \times 5.0 \times 10^{-4}}$$

$$[OH^-] = \sqrt{37.0 \times 10^{-11}} = \sqrt{3.7 \times 10^{-10}}$$

$$x = [OH^-] = 1.92 \times 10^{-5} \text{ mole/liter}$$

Salts of strong bases and weak acids, such as sodium acetate, dissociate completely in aqueous solution to given ions:

$$Na^+CH_3COO^- \xrightarrow{H_2O} Na^+ + CH_3COO^- \qquad (7\text{--}25)$$

The sodium ion cannot react with water, because it would form NaOH, which is a strong electrolyte and would dissoci-

ate completely into its ions. The acetate anion is a Brönsted–Lowry weak base, and

$$CH_3COO^- + H_2O \rightleftharpoons OH^- + CH_3COOH$$

$$K_b = \frac{[OH^-][CH_3COOH]}{[CH_3COO^-]} \qquad (7\text{--}26)$$

In general, for an anionic base $B^-$,

$$B^- + H_2O \rightleftharpoons OH^- + HB$$

$$K_b = \frac{[OH^-][HB]}{[B^-]} \qquad (7\text{--}27)$$

## The Ionization of Water

The concentration of hydrogen or hydroxyl ions in solutions of acids or bases may be expressed as gram ions/liter or as moles/liter. A solution containing 17.008 g of hydroxyl ions or 1.008 g of hydrogen ions per liter is said to contain 1 g ion or 1 mole of hydroxyl or hydrogen ions per liter. Owing to the ionization of water, it is possible to establish a quantitative relationship between the hydrogen and hydroxyl ion concentrations of any aqueous solution.

The concentration of either the hydrogen or the hydroxyl ion in acidic, neutral, or basic solutions is usually expressed in terms of the hydrogen ion concentration or, more conveniently, in pH units.

In a manner corresponding to the dissociation of weak acids and bases, water ionizes slightly to yield hydrogen and hydroxyl ions. As previously observed, a weak electrolyte requires the presence of water or some other polar solvent for ionization. Accordingly, one molecule of water can be thought of as a weak electrolytic solute that reacts with another molecule of water as the solvent. This *autoprotolytic* reaction is represented as

$$H_2O + H_2O \rightleftharpoons H_3O^+ + OH^- \qquad (7\text{--}28)$$

The law of mass action is then applied to give the equilibrium expression

$$\frac{[H_3O^+][OH^-]}{[H_2O]^2} = k \qquad (7\text{--}29)$$

The term for molecular water in the denominator is squared because the reactant is raised to a power equal to the number of molecules appearing in the equation, as required by the law of mass action. Because molecular water exists in great excess relative to the concentrations of hydrogen and hydroxyl ions, $[H_2O]^2$ is considered as a constant and is combined with $k$ to give a new constant, $K_w$, known as the *dissociation constant*, the *autoprotolysis constant*, or the *ion product* of water:

$$K_w = k \times [H_2O]^2 \qquad (7\text{--}30)$$

The value of the ion product is approximately $1 \times 10^{-14}$ at $25°C$; it depends strongly on temperature, as shown in **Table 7–2**. In any calculations involving the ion product, one must be certain to use the proper value of $K_w$ for the temperature at which the data are obtained.

**TABLE 7–2**

**ION PRODUCT OF WATER AT VARIOUS TEMPERATURES***

| Temperature (°C) | $K_w \times 10^{14}$ | p$K_w$ |
|---|---|---|
| 0 | 0.1139 | 14.944 |
| 10 | 0.2920 | 14.535 |
| 20 | 0.6809 | 14.167 |
| 24 | 1.000 | 14.000 |
| 25 | 1.008 | 13.997 |
| 30 | 1.469 | 13.833 |
| 37 | 2.57 | 13.59 |
| 40 | 2.919 | 13.535 |
| 50 | 5.474 | 13.262 |
| 60 | 9.614 | 13.017 |
| 70 | 15.1 | 12.82 |
| 80 | 23.4 | 12.63 |
| 90 | 35.5 | 12.45 |
| 100 | 51.3 | 12.29 |
| 300 | 400 | 11.40 |

*From H. S. Harned and R. A. Robinson, Trans. Faraday Soc. **36**, 973, 1940.

Substituting equation (7–30) into (7–29) gives the common expression for the ionization of water:

$$[H_3O^+] \times [OH^-] = K_w \cong 1 \times 10^{-14} \text{ at } 25°C \quad (7\text{–}31)$$

In *pure* water, the hydrogen and hydroxyl ion concentrations are equal, and each has the value of approximately $1 \times 10^{-7}$ mole/liter at 25°C.*

$$[H_3O^+] = [OH^-] \cong \sqrt{1 \times 10^{-14}}$$
$$\cong 1 \times 10^{-7} \quad (7\text{–}32)$$

When an acid is added to pure water, some hydroxyl ions, provided by the ionization of water, must always remain. The increase in hydrogen ions is offset by a decrease in the hydroxyl ions so that $K_w$ remains constant at about $1 \times 10^{-14}$ at 25°C.

**EXAMPLE 7–3**

**Calculate [OH⁻]**

A quantity of HCl ($1.5 \times 10^{-3}$M) is added to water at 25°C to increase the hydrogen ion concentration from $1 \times 10^{-7}$ to $1.5 \times 10^{-3}$ mole/liter. What is the new hydroxyl ion concentration?

From equation (7–31),

$$[OH^-] = \frac{1 \times 10^{-14}}{1.5 \times 10^{-3}}$$
$$= 6.7 \times 10^{-12} \text{ mole/liter}$$

## Relationship Between $K_a$ and $K_b$

A simple relationship exists between the dissociation constant of a weak acid HB and that of its conjugate base B⁻, or

*Under laboratory conditions, distilled water in equilibrium with air contains about 0.03% by volume of $CO_2$, corresponding to a hydrogen ion concentration of about $2 \times 10^{-6}$ (pH $\cong$ 5.7).

between BH⁺ and B, when the solvent is amphiprotic. This can be obtained by multiplying equation (7–12) by equation (7–27):

$$K_a K_b = \frac{[H_3O^+][B^-]}{[HB]} \cdot \frac{[OH^-][HB]}{[B^-]}$$
$$= [H_3O^+][OH^-] = K_w \quad (7\text{–}33)$$

and

$$K_b = \frac{K_w}{K_a} \quad (7\text{–}34)$$

or

$$K_a = \frac{K_w}{K_b} \quad (7\text{–}35)$$

**EXAMPLE 7–4**

**Calculate $K_a$**

Ammonia has a $K_b$ of $1.74 \times 10^{-5}$ at 25°. Calculate $K_a$ for its conjugate acid, NH₄⁺. We have

$$K_a = \frac{K_w}{K_b} = \frac{1.00 \times 10^{-14}}{1.74 \times 10^{-5}}$$
$$= 5.75 \times 10^{-10}$$

## Ionization of Polyprotic Electrolytes

Acids that donate a single proton and bases that accept a single proton are called *monoprotic electrolytes*. A polyprotic (polybasic) acid is one that is capable of donating two or more protons, and a polyprotic base is capable of accepting two or more protons. A diprotic (dibasic) acid, such as carbonic acid, ionizes in two stages, and a triprotic (tribasic) acid, such as phosphoric acid, ionizes in three stages. The equilibria involved in the protolysis or ionization of phosphoric acid, together with the equilibrium expressions, are

$$H_3PO_4 + H_2O = H_3O^+ + H_2PO_4^- \quad (7\text{–}36)$$

$$\frac{[H_3O^+][H_2PO_4^-]}{[H_3PO_4]} = K_1 = 7.5 \times 10^{-3} \quad (7\text{–}37)$$

$$H_2PO_4^- + H_2O = H_3O^+ + HPO_4^{2-}$$

$$\frac{[H_3O^+][HPO_4^{2-}]}{[H_2PO_4^-]} = K_2 = 6.2 \times 10^{-8} \quad (7\text{–}38)$$

$$HPO_4^{2-} + H_2O = H_3O^+ + PO_4^{3-}$$

$$\frac{[H_3O^+][PO_4^{3-}]}{[HPO_4^{2-}]} = K_3 = 2.1 \times 10^{-13} \quad (7\text{–}39)$$

In any polyprotic electrolyte, the primary protolysis is greatest, and succeeding stages become less complete at any given acid concentration.

The negative charges on the ion $HPO_4^{2-}$ make it difficult for water to remove the proton from the phosphate ion, as reflected in the small value of $K_3$. Thus, phosphoric acid is weak in the third stage of ionization, and a solution of this acid contains practically no $PO_4^{3-}$ ions.

Each of the species formed by the ionization of a polyprotic acid can also act as a base. Thus, for the phosphoric acid system,

$$PO_4^{3-} + H_2O \rightleftharpoons HPO_4^{2-} + OH^- \quad (7\text{-}40)$$

$$K_{b1} = \frac{[HPO_4^{2-}][OH^-]}{[PO_4^{3-}]} = 4.8 \times 10^{-2} \quad (7\text{-}41)$$

$$HPO_4^{2-} + H_2O \rightleftharpoons H_2PO_4^- + OH^- \quad (7\text{-}42)$$

$$K_{b2} = \frac{[H_2PO_4^-][OH^-]}{[HPO_4^{2-}]} = 1.6 \times 10^{-7} \quad (7\text{-}43)$$

$$H_2PO_4^- + H_2O \rightleftharpoons H_3PO_4 + OH^- \quad (7\text{-}44)$$

$$K_{b3} = \frac{[H_3PO_4][OH^-]}{[H_2PO_4^-]} = 1.3 \times 10^{-12} \quad (7\text{-}45)$$

In general, for a polyprotic acid system for which the parent acid is $H_nA$, there are $n + 1$ possible species in solution:

$$H_nA + H_{n-j}A^{-j} + \cdots + HA^{-(n-1)} + A^{n-} \quad (7\text{-}46)$$

where $j$ represents the number of protons dissociated from the parent acid and goes from 0 to $n$. The total concentration of all species must be equal to $C_a$, or

$$[H_nA] + [H_{n-j}A^{-j}] + \cdots$$
$$+ [HA^{-(n-1)}] + [A^{n-}] = C_a \quad (7\text{-}47)$$

Each of the species pairs in which $j$ differs by 1 constitutes a conjugate acid–base pair, and in general

$$K_j K_{b(n+1-j)} = K_w \quad (7\text{-}48)$$

where $K_j$ represents the various acidity constants for the system. Thus, for the phosphoric acid system described by equations (7–37) to (7–45),

$$K_1 K_{b3} = K_2 K_{b2} = K_3 K_{b1} = K_w \quad (7\text{-}49)$$

### Ampholytes

In the preceding section, equations (7–37), (7–38), (7–41), and (7–43) demonstrated that in the phosphoric acid system, the species $H_2PO_4^-$ and $HPO_4^{2-}$ can function either as an acid or a base. A species that can function either as an acid or as a base is called an *ampholyte* and is said to be *amphoteric* in nature. In general, for a polyprotic acid system, all the species, with the exception of $H_nA$ and $A^{n-}$, are amphoteric.

Amino acids and proteins are ampholytes of particular interest in pharmacy. If glycine hydrochloride is dissolved in water, it ionizes as follows:

$$^+NH_3CH_2COOH + H_2O \rightleftharpoons$$
$$^+NH_3CH_2COO^- + H_3O^+ \quad (7\text{-}50)$$
$$^+NH_3CH_2COO^- + H_2O \rightleftharpoons$$
$$^+NH_2CH_2COO^- + H_3O^+ \quad (7\text{-}51)$$

The species $^+NH_3CH_2COO^-$ is amphoteric in that, in addition to reacting as an acid as shown in equation (7–51), it can react as a base with water as follows:

$$^+NH_3CH_2COO^- + H_2O \rightleftharpoons$$
$$^+NH_3CH_2COOH + OH^- \quad (7\text{-}52)$$

The amphoteric species $^+NH_3CH_2COO^-$ is called a *zwitterion* and differs from the amphoteric species formed from phosphoric acid in that it carries both a positive and a negative charge, and the whole molecule is electrically neutral. The pH at which the zwitterion concentration is a maximum is known as the *isoelectric point*. At the isoelectric point the net movement of the solute molecules in an electric field is negligible.

## SÖRENSEN'S pH

The hydrogen ion concentration of a solution varies from approximately 1 in a 1 M solution of a strong acid to about $1 \times 10^{-14}$ in a 1 M solution of a strong base, and the calculations often become unwieldy. To alleviate this difficulty, Sörensen[5] suggested a simplified method of expressing hydrogen ion concentration. He established the term *pH*, which was originally written as $p_H{}^+$, to represent the hydrogen ion potential, and he defined it as the common logarithm of the reciprocal of the hydrogen ion concentration:

$$pH = \log \frac{1}{[H_3O^+]} \quad (7\text{-}53)$$

According to the rules of logarithms, this equation can be written as

$$pH = \log 1 - \log[H_3O^+] \quad (7\text{-}54)$$

and because the logarithm of 1 is zero,

$$pH = -\log[H_3O^+] \quad (7\text{-}55)$$

Equations (7–53) and (7–55) are identical; they are acceptable for approximate calculations involving pH.

The pH of a solution can be considered in terms of a numeric scale having values from 0 to 14, which expresses in a quantitative way the degree of acidity (7 to 0) and alkalinity (7–14). The value 7 at which the hydrogen and hydroxyl ion concentrations are about equal at room temperature is referred to as the *neutral point*, or neutrality. The neutral pH at $0°C$ is 7.47, and at $100°C$ it is 6.15 (**Table 7–2**). The scale relating pH to the hydrogen and hydroxyl ion concentrations of a solution is given in **Table 7–3**.

### Conversion of Hydrogen Ion Concentration to pH

The student should practice converting from hydrogen ion concentration to pH and vice versa until he or she is proficient in these logarithmic operations. The following examples are given to afford a review of the mathematical operations

**TABLE 7-3**

**THE pH SCALE AND CORRESPONDING HYDROGEN AND HYDROXYL ION CONCENTRATIONS**

| pH | $[H^3O^+]$ (moles/liter) | $[OH^{-1}]$ (moles/liter) | |
|----|----|----|----|
| 0 | $10^0 = 1$ | $10^{-14}$ | |
| 1 | $10^{-1}$ | $10^{-13}$ | |
| 2 | $10^{-2}$ | $10^{-12}$ | |
| 3 | $10^{-3}$ | $10^{-11}$ | Acidic |
| 4 | $10^{-4}$ | $10^{-10}$ | |
| 5 | $10^{-5}$ | $10^{-9}$ | |
| 6 | $10^{-6}$ | $10^{-8}$ | |
| 7 | $10^{-7}$ | $10^{-7}$ | Neutral |
| 8 | $10^{-8}$ | $10^{-6}$ | |
| 9 | $10^{-9}$ | $10^{-5}$ | |
| 10 | $10^{-10}$ | $10^{-4}$ | |
| 11 | $10^{-11}$ | $10^{-3}$ | Basic |
| 12 | $10^{-12}$ | $10^{-2}$ | |
| 13 | $10^{-13}$ | $10^{-1}$ | |
| 14 | $10^{-14}$ | $10^0 = 1$ | |

involving logarithms. Equation (7–55) is more convenient for these calculations than equation (7–53).

**EXAMPLE 7-5**

**pH Calculation**

The hydronium ion concentration of a 0.05 M solution of HCl is 0.05 M. What is the pH of this solution? We write

$$pH = -\log(5.0 \times 10^{-2}) = -\log 10^{-2} - \log 5.0$$
$$= 2 - 0.70 = 1.30$$

A handheld calculator permits one to obtain pH simply by use of the log function followed by a change of sign.

A better definition of pH involves the activity rather than the concentration of the ions:

$$pH = -\log a_H{}^+ \qquad (7\text{-}56)$$

and because the activity of an ion is equal to the activity coefficient multiplied by the molal or molar concentration [equation (7–42)],

Hydronium ion concentration × Activity coefficient
$$= \text{Hydronium ion activity}$$

the pH may be computed more accurately from the formula

$$pH = -\log(\gamma_\pm \times c) \qquad (7\text{-}57)$$

**EXAMPLE 7-6**

**Solution pH**

The mean molar ionic activity coefficient of a 0.05 M solution of HCl is 0.83 at 25°C. What is the pH of the solution? We write

$$pH = -\log(0.83 \times 0.05) = 1.38$$

If sufficient NaCl is added to the HCl solution to produce a total ionic strength of 0.5 for this mixture of uni-univalent electrolytes,

the activity coefficient is 0.77. What is the pH of this solution? We write

$$pH = -\log(0.77 \times 0.05) = 1.41$$

Hence, the addition of a neutral salt affects the hydrogen ion activity of a solution, and activity coefficients should be used for the accurate calculation of pH.

**EXAMPLE 7-7**

**Solution pH**

The hydronium ion concentration of a 0.1 M solution of barbituric acid was found to be $3.24 \times 10^{-3}$ M. What is the pH of the solution? We write

$$pH = -\log(3.24 \times 10^{-3})$$
$$pH = 3 - \log 3.24 = 2.49$$

For practical purposes, activities and concentrations are equal in solutions of weak electrolytes to which no salts are added, because the ionic strength is small.

## Conversion of pH to Hydrogen Ion Concentration

The following example illustrates the method of converting pH to $[H_3O^+]$.

**EXAMPLE 7-8**

**Hydronium Ion Concentration**

If the pH of a solution is 4.72, what is the hydronium ion concentration? We have

$$pH = -\log[H_3O^+] = 4.72$$
$$\log[H_3O^+] = -4.72 = -5 + 0.28$$
$$[H_3O^+] = \text{antilog } 0.28 \times \text{antilog } (-5)$$
$$[H_3O^+] = 1.91 \times 10^{-5} \text{ mole/liter}$$

The use of a handheld calculator bypasses this two-step procedure. One simply enters −4.72 into the calculator and presses the key for antilog or $10^x$ to obtain $[H_3O^+]$.

## pK and pOH

The use of pH to designate the negative logarithm of hydronium ion concentration has proved to be so convenient that expressing numbers less than unity in "p" notation has become a standard procedure. The mathematician would say that "p" is a *mathematical operator* that acts on the quantity $[H^+]$, $K_a$, $K_b$, $K_w$, and so on to convert the value into the negative of its common logarithm. In other words, the term "p" is used to express the negative logarithm of the term following the "p." For example, pOH expresses $-\log[OH^-]$, $pK_a$ is used for $-\log K_a$, and $pK_w$ is $-\log K_w$. Thus, equations (7–31) and (7–33) can be expressed as

$$pH + pOH = pK_w \qquad (7\text{-}58)$$
$$pK_a + pK_b = pK_w \qquad (7\text{-}59)$$

where pK is often called the *dissociation exponent.*

The p$K$ values of weak acidic and basic drugs are ordinarily determined by ultraviolet spectrophotometry (95) and potentiometric titration (202). They can also be obtained by solubility analysis[6–8] (254) and by a partition coefficient method.[8]

## SPECIES CONCENTRATION AS A FUNCTION OF pH

As shown in the preceding sections, polyprotic acids, $H_nA$, can ionize in successive stages to yield $n + 1$ possible species in solution. In many studies of pharmaceutical interest, it is important to be able to calculate the concentration of all acidic and basic species in solution.

The concentrations of all species involved in successive acid–base equilibria change with pH and can be represented solely in terms of equilibrium constants and the hydronium ion concentration. These relationships can be obtained by defining all species in solution as fractions $\alpha$ of total acid, $C_a$, added to the system [see equation (7–47) for $C_a$]:

$$\alpha_0 = [H_nA]/C_a \tag{7–60}$$

$$\alpha_1 = [H_{n-1}A^{-1}]/C_a \tag{7–61}$$

and in general

$$\alpha_j = [H_{n-j}A^{-j}]/C_a \tag{7–62}$$

and

$$\alpha_n = [A^{-n}]/C_a \tag{7–63}$$

where $j$ represents the number of protons that have ionized from the parent acid. Thus, dividing equation (7–47) by $C_a$ and using equations (7–60) to (7–63) gives

$$\alpha_0 + \alpha_j + \cdots + \alpha_{n-1} + \alpha_n = 1 \tag{7–64}$$

All of the $\alpha$ values can be defined in terms of equilibrium constants $\alpha_0$ and $[H_3O^+]$ as follows:

$$K_1 = \frac{[H_{n-1}A^-][H_3O^+]}{[H_nA]} = \frac{\alpha_1 C_a [H_3O^+]}{\alpha_0 C_a} \tag{7–65}$$

Therefore,

$$\alpha_1 = K_1\alpha_0/[H_3O^+] \tag{7–66}$$

$$K_2 = \frac{[H_{n-2}A^{2-}][H_3O^+]}{[H_{n-1}A^-]} = \frac{[H_{n-2}A^{2-}][H_3O^+]^2}{K_1[H_nA]} \\ = \frac{\alpha_2 C_a [H_3O^+]^2}{\alpha_0 C_a K_1} \tag{7–67}$$

or

$$\alpha_2 = \frac{K_1 K_2 \alpha_0}{[H_3O^+]^2} \tag{7–68}$$

and, in general,

$$\alpha_j = (K_1 K_2 \dots K_j)\alpha_0/[H_3O^+]^j \tag{7–69}$$

Inserting the appropriate forms of equation (7–69) into equation (7–64) gives

$$\alpha_0 + \frac{K_1\alpha_0}{[H_3O^+]} + \frac{K_1 K_2 \alpha_0}{[H_3O^+]^2} + \cdots + \frac{K_1 K_2 \dots K_n \alpha_0}{[H_3O^+]^n} = 1 \tag{7–70}$$

Solving for $\alpha_0$ yields

$$\alpha_0 = [H_3O^+]^n/\{[H_3O^+]^n + K_1[H_3O]^{n-1} \\ + K_1 K_2[H_3O^+]^{n-2} + \cdots + K_1 K_2 \dots K_n\} \tag{7–71}$$

or

$$\alpha_0 = \frac{[H_3O^+]^n}{D} \tag{7–72}$$

where $D$ represents the denominator of equation (7–71). Thus, the concentration of $H_nA$ as a function of $[H_3O^+]$ can be obtained by substituting equation (7–60) into equation (7–72) to give

$$[H_nA] = \frac{[H_3O^+]^n C_a}{D} \tag{7–73}$$

Substituting equation (7–61) into equation (7–66) and the resulting equation into equation (7–72) gives

$$[H_{n-1}A^{-1}] = \frac{K_1[H_3O^+]^{n-1} C_a}{D} \tag{7–74}$$

In general,

$$[H_{n-j}A^{-j}] = \frac{K_1 \dots K_j [H_3O^+]^{n-j} C_a}{D} \tag{7–75}$$

and

$$[A^{-n}] = \frac{K_1 K_2 \dots K_n C_a}{D} \tag{7–76}$$

Although these equations appear complicated, they are in reality quite simple. The term $D$ in equations (7–72) to (7–76) is a power series in $[H_3O^+]$, each term multiplied by equilibrium constants. The series starts with $[H_3O^+]$ raised to the power representing $n$, the total number of dissociable hydrogens in the parent acid, $H_nA$. The last term is the product of all the acidity constants. The intermediate terms can be obtained from the last term by substituting $[H_3O^+]$ for $K_n$ to obtain the next to last term, then substituting $[H_3O^+]$ for $K_{n-1}$ to obtain the next term, and so on, until the first term is reached. The following equations show the denominators $D$ to be used in equations (7–72) to (7–76) for various types of polyprotic acids:

$$H_4A: \quad D = [H_3O^+]^4 + K_1[H_3O^+]^3 + K_1 K_2[H_3O^+]^2 \\ + K_1 K_2 K_3[H_3O^+] + K_1 K_2 K_3 K_4 \tag{7–77}$$

$$H_3A: \quad D = [H_3O^+]^3 + K_1[H_3O^+]^2 \\ + K_1 K_2[H_3O^+] + K_1 K_2 K_3 \tag{7–78}$$

$$H_2A: \quad D = [H_3O^+]^2 + K_1[H_3O^+] + K_1 K_2 \tag{7–79}$$

$$HA: \quad D = [H_3O^+] + K_a \tag{7–80}$$

In all instances, for a species in which $j$ protons have ionized, the numerator in equations (7–72) to (7–76) is $C_a$ multiplied by the term from the denominator $D$ that has $[H_3O^+]$ raised to the $n - j$ power. Thus, for the parent acid $H_2A$, the appropriate equation for $D$ is equation (7–79). The molar concentrations of the species $H_nA$ ($j = 0$), $HA^-$ ($j = 1$), and $A^{2-}$ ($j = 2$) can be given as

$$[H_2A] = \frac{[H_3O^+]^2 C_a}{[H_3O^+]^2 + K_1[H_3O^+] + K_1K_2} \quad (7\text{–}81)$$

$$[HA^-] = \frac{K_1[H_3O^+]C_a}{[H_3O^+]^2 + K_1[H_3O^+] + K_1K_2} \quad (7\text{–}82)$$

$$[A^{2-}] = \frac{K_1K_2C_a}{[H_3O^+]^2 + K_1[H_3O^+] + K_1K_2} \quad (7\text{–}83)$$

These equations can be used directly to solve for molar concentrations. It should be obvious, however, that lengthy calculations are needed for substances such as citric acid or ethylenediaminetetraacetic acid, requiring the use of a digital computer to obtain solutions in a reasonable time. Graphic methods have been used to simplify the procedure.[9]

# CALCULATION OF pH

## Proton Balance Equations

According to the Brönsted–Lowry theory, every proton donated by an acid must be accepted by a base. Thus, an equation accounting for the total proton transfers occurring in a system is of fundamental importance in describing any acid–base equilibria in that system. This can be accomplished by establishing a proton balance equation (PBE) for each system. In the PBE, the sum of the concentration terms for species that form by proton consumption is equated to the sum of the concentration terms for species that are formed by the release of a proton.

For example, when HCl is added to water, it dissociates completely into $H_3O^+$ and $Cl^-$ ions. The $H_3O^+$ is a species that is formed by the consumption of a proton (by water acting as a base), and the $Cl^-$ is formed by the release of a proton from HCl. In all aqueous solutions, $H_3O^+$ and $OH^-$ result from the dissociation of two water molecules according to equation (7–28). Thus, $OH^-$ is a species formed from the release of a proton. The PBE for the system of HCl in water is

$$[H_3O^+] = [OH^-] + [Cl^-]$$

Although $H_3O^+$ is formed from two reactions, it is included only once in the PBE. The same would be true for $OH^-$ if it came from more than one source.

The general method for obtaining the PBE is as follows:

(a) Always start with the species added to water.
(b) On the left side of the equation, place all species that can form when protons are consumed by the starting species.

(c) On the right side of the equation, place all species that can form when protons are released from the starting species.
(d) Each species in the PBE should be multiplied by the number of protons lost or gained when it is formed from the starting species.
(e) Add $[H_3O^+]$ to the left side of the equation and $[OH^-]$ to the right side of the equation. These result from the interaction of two molecules of water, as shown previously.

### EXAMPLE 7–9

**Proton Balance Equations**

What is the PBE when $H_3PO_4$ is added to water?
The species $H_2PO_4^-$ forms with the release of one proton. The species $HPO_4^{2-}$ forms with the release of two protons. The species $PO_4^{3-}$ forms with the release of three protons. We thus have

$$[H_3O^+] = [OH^-] + [H_2PO_4^-] + 2[HPO_4^{2-}] + 3[PO_4^{3-}]$$

### EXAMPLE 7–10

**Proton Balance Equations**

What is the PBE when $Na_2HPO_4$ is added to water?
The salt dissociates into two $Na^+$ and one $HPO_4^{2-}$; $Na^+$ is neglected in the PBE because it is not formed from the release or consumption of a proton; $HPO_4^{2-}$, however, does react with water and is considered to be the starting species.
The species $H_2PO_4^-$ results with the consumption of one proton.
The species of $H_3PO_4$ can form with the consumption of two protons.
The species $PO_4^{3-}$ can form with the release of one proton.
Thus, we have

$$[H_3O^+] + [H_2PO_4^-] + 2[H_3PO_4] = [OH^-] + [PO_4^{3-}]$$

### EXAMPLE 7–11

**Proton Balance Equations**

What is the PBE when sodium acetate is added to water?
The salt dissociates into one $Na^+$ and one $CH_3COO^-$ ion. The $CH^3COO^-$ is considered to be the starting species. The $CH_3COOH$ can form when $CH_3COO^-$ consumes one proton. Thus,

$$[H_3O^+] + [CH_3COOH] = [OH^-]$$

The PBE allows the pH of any solution to be calculated readily, as follows:

(a) Obtain the PBE for the solution in question.
(b) Express the concentration of all species as a function of equilibrium constants and $[H_3O^+]$ using equations (7–73) to (7–76).
(c) Solve the resulting expression for $[H_3O^+]$ using any assumptions that appear valid for the system.
(d) Check all assumptions.
(e) If all assumptions prove valid, convert $[H_3O^+]$ to pH.

If the solution contains a base, it is sometimes more convenient to solve the expression obtained in part (b) for $[OH^-]$, then convert this to pOH, and finally to pH by use of equation (7–58).

## Solutions of Strong Acids and Bases

Strong acids and bases are those that have acidity or basicity constants greater than about $10^{-2}$. Thus, they are considered to ionize 100% when placed in water. When HCl is placed in water, the PBE for the system is given by

$$[H_3O^+] = [OH^-] + [Cl^-] = \frac{K_w}{[H_3O^+]} + C_a \quad (7\text{--}84)$$

which can be rearranged to give

$$[H_3O^+]^2 - C_a[H_3O^+] - K_w = 0 \quad (7\text{--}85)$$

where $C_a$ is the total acid concentration. This is a quadratic equation of the general form

$$aX^2 + bX + c = 0 \quad (7\text{--}86)$$

which has the solution

$$X = \frac{-b \pm \sqrt{b^2 - 4ac}}{2a} \quad (7\text{--}87)$$

Thus, equation (7–85) becomes

$$[H_3O^+] = \frac{C_a + \sqrt{C_a^2 + 4K_w}}{2} \quad (7\text{--}88)$$

where only the positive root is used because $[H_3O^+]$ can never be negative.

When the concentration of acid is $1 \times 10^{-6}$ M or greater, $[Cl^-]$ becomes much greater than* $[OH^-]$ in equation (7–84) and $C_a^2$ becomes much greater than $4K_w$ in equation (7–88). Thus, both equations simplify to

$$[H_3O^+] \cong C_a \quad (7\text{--}89)$$

A similar treatment for a solution of a strong base such as NaOH gives

$$[OH^-] = \frac{C_b + \sqrt{C_b^2 + 4K_w}}{2} \quad (7\text{--}90)$$

and

$$[OH^-] \cong C_b \quad (7\text{--}91)$$

if the concentration of base is $1 \times 10^6$ M or greater.

## Conjugate Acid–Base Pairs

Use of the PBE enables us to develop one master equation that can be used to solve for the pH of solutions composed of weak acids, weak bases, or a mixture of a conjugate acid–base pair. To do this, consider a solution made by dissolving both a weak acid, HB, and a salt of its conjugate base, B$^-$, in water. The acid–base equilibria involved are

$$HB + H_2O \rightleftharpoons H_3O^+ + B^- \quad (7\text{--}92)$$
$$B^- + H_2O \rightleftharpoons OH^- + HB \quad (7\text{--}93)$$
$$H_2O + H_2O \rightleftharpoons H_3O^+ + OH^- \quad (7\text{--}94)$$

The PBE for this system is

$$[H_3O^+] + [HB] = [OH^-] + [B^-] \quad (7\text{--}95)$$

The concentrations of the acid and the conjugate base can be expressed as

$$[HB] = \frac{[H_3O^+]C_b}{[H_3O^+] + K_a} \quad (7\text{--}96)$$

$$[B^-] = \frac{K_aC_a}{[H_3O^+] + K_a} \quad (7\text{--}97)$$

Equation (7–96) contains $C_b$ (concentration of base added as the salt) rather than $C_a$ because in terms of the PBE, the species HB was generated from the species B$^-$ added in the form of the salt. Equation (7–97) contains $C_a$ (concentration of HB added) because the species B$^-$ in the PBE came from the HB added. Inserting equations (7–96) and (7–97) into equation (7–95) gives

$$[H_3O^+] + \frac{[H_3O^+]C_b}{[H_3O^+] + K_a}$$
$$= [OH^-] + \frac{K_aC_a}{[H_3O^+] + K_a} \quad (7\text{--}98)$$

which can be rearranged to yield

$$[H_3O^+] = K_a \frac{(C_a - [H_3O^+] + [OH^-])}{(C_b + [H_3O^+] - [OH^-])} \quad (7\text{--}99)$$

This equation is exact and was developed using no assumptions.* It is, however, quite difficult to solve. Fortunately, for real systems, the equation can be simplified.

## Solutions Containing Only a Weak Acid

If the solution contains only a weak acid, $C_b$ is zero, and $[H_3O^+]$ is generally much greater than $[OH^-]$. Thus, equation (7–99) simplifies to

$$[H_3O^+]^2 + K_a[H_3O^+] - K_aC_a = 0 \quad (7\text{--}100)$$

which is a quadratic equation with the solution

$$[H_3O^+] = \frac{-K_a + \sqrt{K_a^2 + 4K_aC_a}}{2} \quad (7\text{--}101)$$

In many instances, $C_a$ is much greater than $[H_3O^+]$, and equation (7–100) simplifies to

$$[H_3O^+] = \sqrt{K_aC_a} \quad (7\text{--}102)$$

**EXAMPLE 7–12**

**Calculate pH**

Calculate the pH of a 0.01 M solution of salicylic acid, which has a $K_a = 1.06 \times 10^{-3}$ at 25°C.

(a) Using equation (7–102), we find

$$[H_3O^+] = \sqrt{(1.06 \times 10^{-3}) \times (1.0 \times 10^{-2})}$$
$$= 3.26 \times 10^{-3} \text{ M}$$

The approximation that $C_a \gg H_3O^+$ is not valid.

---

*To adopt a definite and consistent method of making approximations throughout this chapter, the expression "much greater than" means that the larger term is at least 20 times greater than the smaller term.

*Except that, in this and all subsequent developments for pH equations, it is assumed that concentration may be used in place of activity.

(b) Using equation (7–101), we find

$$[H_3O^+] = -\frac{(1.06 \times 10^{-3})}{2}$$
$$+ \frac{\sqrt{(1.06 \times 10^{-3})^2 + 4(1.06 \times 10^{-3})(1.0 \times 10^{-2})}}{2}$$
$$= 2.77 \times 10^{-3} \text{ M}$$
$$pH = -\log(2.77 \times 10^{-3}) = 2.56$$

The example just given illustrates the importance of checking the validity of all assumptions made in deriving the equation used for calculating $[H_3O^+]$. The simplified equation (7–102) gives an answer for $[H_3O^+]$ with a relative error of 18% as compared with the correct answer given by equation (7–101).

### EXAMPLE 7–13

**Calculate pH**

Calculate the pH of a 1-g/100 mL solution of ephedrine sulfate. The molecular weight of the salt is 428.5, and $K_b$ for ephedrine base is $2.3 \times 10^{-5}$.

(a) The ephedrine sulfate, $(BH^+)_2SO_4$, dissociates completely into two $BH^+$ cations and one $SO_4^{2-}$ anion. Thus, the concentration of the weak acid (ephedrine cation) is twice the concentration, $C_s$, of the salt added.

$$C_a = 2C_s = \frac{2 \times 10 \text{ g/liter}}{428.5 \text{ g/mole}} = 4.67 \times 10^{-2} \text{ M}$$

(b)
$$K_a = \frac{1.00 \times 10^{-14}}{2.3 \times 10^{-5}} = 4.35 \times 10^{-10}$$

(c)
$$[H_3O^+] = \sqrt{(4.35 \times 10^{-10}) \times (4.67 \times 10^{-2})}$$
$$= 4.51 \times 10^{-6} \text{ M}$$

All assumptions are valid. We have

$$pH = -\log(4.51 \times 10^{-6}) = 5.35$$

## Solutions Containing Only a Weak Base

If the solution contains only a weak base, $C_a$ is zero, and $[OH^-]$ is generally much greater than $[H_3O^+]$. Thus, equation (7–99) simplifies to

$$[H_3O^+] = \frac{K_a[OH^-]}{C_b - [OH^-]} = \frac{K_aK_w}{[H_3O^+]C_b - K_w} \quad (7\text{–}103)$$

This equation can be solved for either $[H_3O^+]$ or $[OH^-]$. Solving for $[H_3O^+]$ using the left and rightmost parts of equation (7–103) gives

$$C_b[H_3O^+]^2 - K_w[H_3O^+] - K_aK_w = 0 \quad (7\text{–}104)$$

which has the solution

$$[H_3O^+] = \frac{K_w + \sqrt{K_w^2 + 4C_bK_aK_w}}{2C_b} \quad (7\text{–}105)$$

If $K_a$ is much greater than $[H_3O^+]$, which is generally true for solutions of weak bases, equation (7–100) gives

$$[H_3O^+] = \sqrt{\frac{K_aK_w}{C_b}} \quad (7\text{–}106)$$

Equation (7–103) can be solved for $[OH^-]$ by using the left and middle portions and converting $K_a$ to $K_b$ to give

$$[OH^-] = \frac{-K_b + \sqrt{K_b^2 + 4K_bC_b}}{2} \quad (7\text{–}107)$$

and if $C_b$ is much greater than $[OH^-]$, which generally obtains for solutions of weak bases,

$$[OH^-] = \sqrt{K_bC_b} \quad (7\text{–}108)$$

A good exercise for the student would be to prove that equation (7–106) is equal to equation (7–108). The applicability of both these equations will be shown in the following examples.

### EXAMPLE 7–14

**Calculate pH**

What is the pH of a 0.0033 M solution of cocaine base, which has a basicity constant of $2.6 \times 10^{-6}$? We have

$$[OH^-] = \sqrt{(2.6 \times 10^{-6}) \times (3.3 \times 10^{-3})}$$
$$= 9.26 \times 10^{-5} \text{ M}$$

All assumptions are valid. Thus,

$$pOH = -\log(9.26 \times 10^{-5}) = 4.03$$
$$pH = 14.00 - 4.03 = 9.97$$

### EXAMPLE 7–15

**Calculate pH**

Calculate the pH of a 0.165 M solution of sodium sulfathiazole. The acidity constant for sulfathiazole is $7.6 \times 10^{-8}$.

(a) The salt $Na^+B^-$ dissociates into one $Na^+$ and one $B^-$ as described by equations (7–24) to (7–27). Thus, $C_b = C_s = 0.165$ M. Because $K_a$ for a weak acid such as sulfathiazole is usually given rather than $K_b$ for its conjugate base, equation (7–106) is preferred over equation (7–108):

$$[H_3O^+] = \sqrt{\frac{(7.6 \times 10^{-8}) \times (1.00 \times 10^{-14})}{0.165}}$$
$$= 6.79 \times 10^{-11} \text{ M}$$

All assumptions are valid. Thus,

$$pH = -\log(6.79 \times 10^{-11}) = 10.17$$

## Solutions Containing a Single Conjugate Acid–Base Pair

In a solution composed of a weak acid and a salt of that acid (e.g., acetic acid and sodium acetate) or a weak base and a salt of that base (e.g., ephedrine and ephedrine hydrochloride), $C_a$ and $C_b$ are generally much greater than either $[H_3O^+]$ or $[OH^-]$. Thus, equation (7–99) simplifies to

$$[H_3O^+] = \frac{K_aC_a}{C_b} \quad (7\text{–}109)$$

**EXAMPLE 7–16**

**Calculate pH**

What is the pH of a solution containing acetic acid 0.3 M and sodium acetate 0.05 M? We write

$$[H_3O^+] = \frac{(1.75 \times 10^{-5}) \times (0.3)}{5.0 \times 10^{-2}}$$
$$= 1.05 \times 10^{-4} \text{ M}$$

All assumptions are valid. Thus,

$$pH = -\log(1.05 \times 10^{-4}) = 3.98$$

**EXAMPLE 7–17**

**Calculate pH**

What is the pH of a solution containing ephedrine 0.1 M and ephedrine hydrochloride 0.01 M? Ephedrine has a basicity constant of $2.3 \times 10^{-5}$; thus, the acidity constant for its conjugate acid is $4.35 \times 10^{-10}$.

$$[H_3O^+] = \frac{(4.35 \times 10^{-10}) \times (1.0 \times 10^{-2})}{1.0 \times 10^{-1}}$$
$$= 4.35 \times 10^{-11} \text{ M}$$

All assumptions are valid. Thus,

$$pH = -\log(4.35 \times 10^{-11}) = 10.36$$

Solutions made by dissolving in water both an acid and its conjugate base, or a base and its conjugate acid, are examples of buffer solutions. These solutions are of great importance in pharmacy and are covered in greater detail in the next two chapters.

## Two Conjugate Acid–Base Pairs

The Brönsted–Lowry theory and the PBE enable a single equation to be developed that is valid for solutions containing an ampholyte, which forms a part of two dependent acid–base pairs. An amphoteric species can be added directly to water or it can be formed by the reaction of a diprotic weak acid, $H_2A$, or a diprotic weak base, $A^{2-}$. Thus, it is convenient to consider a solution containing a diprotic weak acid, $H_2A$, a salt of its ampholyte, $HA^-$, and a salt of its diprotic base, $A^{2-}$, in concentrations $C_a$, $C_{ab}$, and $C_b$, respectively. The total PBE for this system is

$$[H_3O^+] + [H_2A]_{ab} + [HA^-]_b + 2[H_2A]_b$$
$$= [OH^-] + [HA^-]_a + 2[A^{2-}]_a$$
$$+ [A^{2-}]_{ab} \qquad (7-110)$$

where the subscripts refer to the source of the species in the PBE, that is, $[H_2A]_{ab}$ refers to $H_2A$ generated from the ampholyte and $[H_2A]_b$ refers to the $H_2A$ generated from the diprotic base. Replacing these species concentrations as a function of $[H_3O^+]$ gives

$$[H_3O^+] + \frac{[H_3O^+]^2 C_{ab}}{D} + \frac{K_1[H_3O^+]C_b}{D}$$
$$+ \frac{2[H_3O^+]^2 C_b}{D} = \frac{K_w}{[H_3O^+]}$$

$$+ \frac{K_1[H_3O^+]C_a}{D} + \frac{2K_1K_2C_a}{D}$$
$$+ \frac{K_1K_2C_{ab}}{D} \qquad (7-111)$$

Multiplying through by $[H_3O^+]$ and $D$, which is given by equation (7–79), gives

$$[H_3O^+]^4 + [H_3O^+]^3(K_1 + 2C_b + C_{ab})$$
$$+ [H_3O^+]^2[K_1(C_b - C_a) + K_1K_2 - K_w]$$
$$- [H_3O^+][K_1K_2(2C_a + C_{ab}) + K_1K_w]$$
$$- K_1K_2K_w = 0 \qquad (7-112)$$

This is a general equation that has been developed using no assumptions and that can be used for solutions made by adding a diprotic acid to water, adding an ampholyte to water, adding a diprotic base to water, and by adding combinations of these substances to water. It is also useful for tri- and quadriprotic acid systems because $K_3$ and $K_4$ are much smaller than $K_1$ and $K_2$ for all acids of pharmaceutical interest. Thus, these polyprotic acid systems can be handled in the same manner as a diprotic acid system.

## Solutions Containing Only a Diprotic Acid

If a solution is made by adding a diprotic acid, $H_2A$, to water to give a concentration $C_a$, the terms $C_{ab}$ and $C_b$ in equation (7–112) are zero. In almost all instances, the terms containing $K_w$ can be dropped, and after dividing through by $[H_3O^+]$, we obtain from equation (7–112)

$$[H_3O^+]^3 + [H_3O^+]^2 K_1 - [H_3O^+](K_1C_a - K_1K_2)$$
$$- 2K_1K_2C_a = 0 \qquad (7-113)$$

If $C_a \gg K_2$, as is usually true,

$$[H_3O^+]^3 + [H_3O^+]^2 K_1 - [H_3O^+]K_1C_a$$
$$- 2K_1K_2C_a = 0 \qquad (7-114)$$

If $[H_3O^+]$ is much greater than $2K_2$, the term $2K_1K_2C_a$ can be dropped, and dividing through by $[H_3O^+]$ yields the quadratic equation

$$[H_3O^+]^2 + [H_3O^+]K_1 - KC_a = 0 \qquad (7-115)$$

The assumptions $C_a$ is much greater than $K_2$ and $[H_3O^+]$ is much greater than $2K_2$ will be valid whenever $K_2$ is much less than $K_1$. Equation (7–115) is identical to equation (7–100), which was obtained for a solution containing a monoprotic weak acid. Thus, if $C_a$ is much greater than $[H_3O^+]$, equation (7–115) simplifies to equation (7–100).

**EXAMPLE 7–18**

**Calculate pH**

Calculate the pH of a $1.0 \times 10^{-3}$ M solution of succinic acid. $K_1 = 6.4 \times 10^{-5}$ and $K_2 = 2.3 \times 10^{-6}$.

(a) Use equation (7–102) because $K_1$ is approximately 30 times $K_2$:

$$[H_3O^+] = \sqrt{(6.4 \times 10^{-5}) \times (1.0 \times 10^{-3})}$$
$$= 2.53 \times 10^{-4} \text{ M}$$

The assumption that $C_a$ is much greater than $[H_3O^+]$ is not valid.

(b) Use the quadratic equation (7–115):

$$[H_3O^+] = -(6.4 \times 10^{-5})/2$$

$$+ \frac{\sqrt{(6.4 \times 10^{-5})^2 + 4(6.4 \times 10^{-5})(1.0 \times 10^{-3})}}{2}$$

$$= 2.23 \times 10^{-4} \text{ M}$$

Note that $C_a$ is much greater than $K_2$, and $[H_3O^+]$ is much greater than $2K_2$. Thus, we have

$$pH = -\log(2.23 \times 10^{-4}) = 3.65$$

## Solutions Containing Only an Ampholyte

If an ampholyte, $HA^-$, is dissolved in water to give a solution with concentration $C_{ab}$, the terms $C_a$ and $C_b$ in equation (7–112) are zero. For most systems of practical importance, the first, third, and fifth terms of equation (7–112) are negligible when compared with the second and fourth terms, and the equation becomes

$$[H_3O^+] = \sqrt{\frac{K_1 K_2 C_{ab} + K_1 K_w}{K_1 + C_{ab}}} \qquad (7\text{–}116)$$

The term $K_2 C_{ab}$ is generally much greater than $K_w$, and

$$[H_3O^+] = \sqrt{\frac{K_1 K_2 C_{ab}}{K_1 + C_{ab}}} \qquad (7\text{–}117)$$

If the solution is concentrated enough that $C_{ab}$ is much greater than $K_1$,

$$[H_3O^+] = \sqrt{K_1 K_2} \qquad (7\text{–}118)$$

### EXAMPLE 7–19

**Calculate pH**

Calculate the pH of a $5.0 \times 10^{-3}$ M solution of sodium bicarbonate at 25°C. The acidity constants for carbonic acid are $K_1 = 4.3 \times 10^{-7}$ and $K_2 = 4.7 \times 10^{-11}$.

Because $K_2 C_{ab}$ ($23.5 \times 10^{-14}$) is much greater than $K_w$ and $C_{ab}$ is much greater than $K_1$, equation (7–118) can be used. We have

$$[H_3O^+] = \sqrt{(4.3 \times 10^{-7}) \times (4.7 \times 10^{-11})}$$

$$= 4.5 \times 10^{-9} \text{ M}$$

$$pH = -\log(4.5 \times 10^{-9}) = 8.35$$

## Solutions Containing Only a Diacidic Base

In general, the calculations for solutions containing weak bases are easier to handle by solving for $[OH^-]$ rather than $[H_3O^+]$. Any equation in terms of $[H_3O^+]$ and acidity constants can be converted into terms of $[OH^-]$ and basicity constants by substituting $[OH^-]$ for $[H_3O^+]$, $K_{b1}$ for $K_1$, $K_{b2}$ for $K_2$, and $C_b$ for $C_a$. These substitutions are made into equation (7–112). Furthermore, for a solution containing only a diacidic base, $C_a$ and $C_{ab}$ are zero; all terms containing $K_w$ can be dropped; $C_b$ is much greater than $K_{b2}$; and

$[OH^-]$ is much greater than $2K_{b2}$. The following expression results:

$$[OH^-]^2 + [OH^-]K_{b1} - K_{b1}C_b = 0 \qquad (7\text{–}119)$$

If $C_b$ is much greater than $[OH^-]$, the equation simplifies to

$$[OH^-] = \sqrt{K_{b1}C_b} \qquad (7\text{–}120)$$

### EXAMPLE 7–20

**Calculate pH**

Calculate the pH of a $1.0 \times 10^{-3}$ M solution of $Na_2CO_3$. The acidity constants for carbonic acid are $K_1 = 4.31 \times 10^{-7}$ and $K_2 = 4.7 \times 10^{-11}$.

(a) Using equation (7–48), we obtain

$$K_{b1} = \frac{K_w}{K_2} = \frac{1.00 \times 10^{-14}}{4.7 \times 10^{-11}} = 2.1 \times 10^{-4}$$

$$K_{b2} = \frac{K_w}{K_1} = \frac{1.00 \times 10^{-14}}{4.31 \times 10^{-7}} = 2.32 \times 10^{-8}$$

(b) Because $K_{b2}$ is much greater than $K_{b2}$, one uses equation (7–120):

$$[OH^-] = \sqrt{(2.1 \times 10^{-4}) \times (1.0 \times 10^{-3})}$$

$$= 4.6 \times 10^{-4} \text{ M}$$

The assumption that $C_b$ is much greater than $[OH^-]$ is not valid, and equation (7–119) must be used. [See equations (7–86) and (7–87) for the solution of a quadratic equation.] We obtain

$$[OH^-] = -(2.1 \times 10^{-4})/2$$

$$+ \frac{\sqrt{(2.1 \times 10^{-4})^2 + 4(2.1 \times 10^{-4})(1.0 \times 10^{-3})}}{2}$$

$$= 3.7 \times 10^{-4} \text{ M}$$

$$pOH = -\log(3.7 \times 10^{-4}) = -3.4$$

$$pH = 14.00 - 3.4 = 10.6$$

Use of the simplified equation (7–120) gives an answer for $[OH^-]$ that has a relative error of 24% as compared with the correct answer given by equation (7–119). It is absolutely essential that all assumptions made in the calculation of $[H_3O^+]$ or $[OH^-]$ be verified!

## Two Independent Acid–Base Pairs

Consider a solution containing two independent acid–base pairs:

$$HB_1 + H_2O \rightleftharpoons H_3O^+ B_1^-$$

$$K_1 = \frac{[H_3O^+][B_1^-]}{[HB_1]} \qquad (7\text{–}121)$$

$$HB_2 + H_2O \rightleftharpoons H_3O^+ B_2^-$$

$$K_2 = \frac{[H_3O^+][B_2^-]}{[HB_2]} \qquad (7\text{–}122)$$

A general equation for calculating the pH of this type of solution can be developed by considering a solution made by adding to water the acids $HB_1$ and $HB_2$ in concentrations $C_{a1}$

and $C_{a2}$ and the bases $B_1{}^-$ and $B_2{}^-$ in concentrations $C_{b1}$ and $C_{b2}$. The PBE for this system is

$$[H_3O^+] + [HB_1]_{B1} + [HB_2]_{B2}$$
$$= [OH^-] + [B_1{}^-]_{A1} + [B_2{}^-]_{A2} \quad (7\text{–}123)$$

where the subscripts refer to the sources of the species in the PBE. Replacing these species concentrations as a function of $[H_3O^+]$ gives

$$[H_3O^+] + \frac{[H_3O^+]C_{b1}}{[H_3O^+] + K_1} + \frac{[H_3O^+]C_{b2}}{[H_3O^+] + K_2}$$
$$= \frac{K_w}{[H_3O^+]} + \frac{K_1 C_{a1}}{[H_3O^+] + K_1}$$
$$+ \frac{K_2 C_{a2}}{[H_3O^+] + K_2} \quad (7\text{–}124)$$

which can be rearranged to

$$[H_3O^+]^4 + [H_3O^+]^3(K_1 + K_2 + C_{b1} + C_{b2}) + [H_3O^+]^2$$
$$\times [K_1(C_{b2} - C_{a1}) + K_2(C_{b1} - C_{a2}) + K_1 K_2 - K_w]$$
$$- [H_3O^+][K_1 K_2(C_{a1} + C_{a2}) + K_w(K_1 + K_2)]$$
$$- K_1 K_2 K_w = 0 \quad (7\text{–}125)$$

Although this equation is extremely complex, it simplifies readily when applied to specific systems.

## Solutions Containing Two Weak Acids

In systems containing two weak acids, $C_{b1}$ and $C_{b2}$ are zero, and all terms in $K_w$ can be ignored in equation (7–125). For all systems of practical importance, $C_{a1}$ and $C_{a2}$ are much greater than $K_1$ and $K_2$, so the equation simplifies to

$$[H_3O^+]^2 + [H_3O^+](K_1 + K_2)$$
$$- (K_1 C_{a1} + K_2 C_{a2}) = 0 \quad (7\text{–}126)$$

If $C_{a1}$ and $C_{a2}$ are both greater than $[H_3O^+]$, the equation simplifies to

$$[H_3O^+] = \sqrt{K_1 C_{a1} + K_2 C_{a2}} \quad (7\text{–}127)$$

### EXAMPLE 7–21

**Calculate pH**

**What is the pH of a solution containing acetic acid, 0.01 mole/liter, and formic acid, 0.001 mole/liter? We have**

$$[H_3O^+] = \sqrt{(1.75 \times 10^{-5})(1.0 \times 10^{-2}) + (1.77 \times 10^{-4})(1.0 \times 10^{-3})}$$
$$= 5.93 \times 10^{-4} \text{ M}$$
$$pH = -\log(5.93 \times 10^{-4}) = 3.23$$

## Solutions Containing a Salt of a Weak Acid and a Weak Base

The salt of a weak acid and a weak base, such as ammonium acetate, dissociates almost completely in aqueous solution to yield $NH_4{}^+$ and $Ac^-$, the $NH_4{}^+$ is an acid and can be designated as $HB_1$, and the base $Ac^-$ can be designated as $B_2{}^-$ in

equations (7–121) and (7–122). Because only a single acid, $HB_1$, and a single base, $B_2{}^-$, were added to water in concentrations $C_{a1}$ and $C_{b2}$, respectively, all other stoichiometric concentration terms in equation (7–115) are zero. In addition, all terms containing $K_w$ are negligibly small and may be dropped, simplifying the equation to

$$[H_3O^+]^2(K_1 + K_2 + C_{b2})$$
$$+ [H_3O^+][K_1(C_{b2} - C_{a1}) + K_1 K_2]$$
$$- K_1 K_2 C_{a1} = 0 \quad (7\text{–}128)$$

In solutions containing a salt such as ammonium acetate, $C_{a1} = C_{b2} = C_s$, where $C_s$ is the concentration of salt added. In all systems of practical importance, $C_s$ is much greater than $K_1$ or $K_2$, and equation (7–128) simplifies to

$$[H_3O^+]^2 C_s + [H_3O^+]K_1 K_2 - K_1 K_2 C_s = 0 \quad (7\text{–}129)$$

which is a quadratic equation that can be solved in the usual manner. In most instances, however, $C_s$ is much greater than $[H_3O^+]$, and the quadratic equation reduces to

$$[H_3O^+] = \sqrt{K_1 K_2} \quad (7\text{–}130)$$

Equations (7–121) and (7–122) illustrate the fact that $K_1$ and $K_2$ are not the successive acidity constants for a single diprotic acid system, and equation (7–130) is not the same as equation (7–118); instead, $K_1$ is the acidity constant for $HB_1$ (Acid$_1$) and $K_2$ is the acidity constant for the conjugate acid, $HB_2$ (Acid$_2$), of the base $B_2{}^-$. The determination of Acid$_1$ and Acid$_2$ can be illustrated using ammonium acetate and considering the acid and base added to the system interacting as follows:

$$\begin{array}{cccc} NH_4{}^+ + AC^- & \rightleftharpoons & HAc + NH_3 \\ \text{Acid}_1 \quad \text{Base}_2 & & \text{Acid}_2 \quad \text{Base}_1 \end{array} \quad (7\text{–}131)$$

Thus, for this system, $K_1$ is the acidity constant for the ammonium ion and $K_2$ is the acidity constant for acetic acid.

### EXAMPLE 7–22

**Calculate pH**

**Calculate the pH of a 0.01 M solution of ammonium acetate. The acidity constant for acetic acid is $K_2 = K_a = 1.75 \times 10^{-5}$, and the basicity constant for ammonia is $K_b = 1.74 \times 10^{-5}$.**

*(a)* $K_1$ can be found by dividing $K_b$ for ammonia into $K_w$:

$$K_1 = \frac{1.00 \times 10^{-14}}{1.74 \times 10^{-5}} = 5.75 \times 10^{-10}$$
$$[H_3O^+] = \sqrt{(5.75 \times 10^{-10}) \times (1.75 \times 10^{-5})}$$
$$= 1.00 \times 10^{-7} \text{ M}$$

**Note that all of the assumptions are valid. We have**

$$pH = -\log(1.00 \times 10^{-7}) = 7.00$$

When ammonium succinate is dissolved in water, it dissociates to yield two $NH_4{}^+$ cations and one succinate ($S^{2-}$) anion. These ions can enter into the following acid–base equilibrium:

$$\begin{array}{cccc} NH_4{}^+ + S^{2-} & \rightleftharpoons & HS^- + NH_3 \\ \text{Acid}_1 \quad \text{Base}_2 & & \text{Acid}_2 \quad \text{Base}_1 \end{array} \quad (7\text{–}132)$$

In this system, $C_{b2} = C_s$ and $C_{a1} = 2C_s$, the concentration of salt added. If $C_s$ is much greater than either $K_1$ or $K_2$, equation (7–125) simplifies to

$$[H_3O^+]^2 - [H_3O^+]K_1 - 2K_1K_2 = 0 \quad (7\text{–}133)$$

and if $2K_2$ is much greater than $[H_3O^+]$,

$$[H_3O^+] = \sqrt{2K_1K_2} \quad (7\text{–}134)$$

In this example, equation (7–132) shows that $K_1$ is the acidity constant for the ammonium cation and $K_2$, referring to Acid$_2$, must be the acidity constant for the bisuccinate species $HS^-$ or the second acidity constant for succinic acid.

In general, when Acid$_2$ comes from a polyprotic acid $H_nA$, equation (7–128) simplifies to

$$[H_3O^+]^2 - [H_3O^+]K_1(n - 1) - nK_1K_2 = 0 \quad (7\text{–}135)$$

and

$$[H_3O^+] = \sqrt{nK_1K_2} \quad (7\text{–}136)$$

using the same assumptions that were used in developing equations (7–132) and (7–133).

It should be pointed out that in deriving equations (7–132) to (7–136), the base was assumed to be monoprotic. Thus, it would appear that these equations should not be valid for salts such as ammonium succinate or ammonium phosphate. For all systems of practical importance, however, the solution to these equations yields a pH value above the final $pK_a$ for the system. Therefore, the concentrations of all species formed by the addition of more than one proton to a polyacidic base will be negligibly small, and the assumption of only a one-proton addition becomes quite valid.

### EXAMPLE 7–23

**Calculate pH**

Calculate the pH of a 0.01 M solution of ammonium succinate. As shown in equation (7–132), $K_1$ is the acidity constant for the ammonium cation, which was found in the previous example to be $5.75 \times 10^{-10}$, and $K_2$ refers to the acid succinate ($HS^-$) or the second acidity constant for the succinic acid system. Thus, $K_2 = 2.3 \times 10^{-6}$. We have

$$[H_3O^+] = \sqrt{2(5.75 \times 10^{-10}) \times (2.3 \times 10^{-6})}$$
$$= 5.14 \times 10^{-8}$$
$$pH = -\log(5.14 \times 10^{-8}) = 7.29$$

## Solutions Containing a Weak Acid and a Weak Base

In the preceding section, the acid and base were added in the form of a single salt. They can be added as two separate salts or an acid and a salt, however, forming buffer solutions whose pH is given by equation (7–130). For example, consider a solution made by dissolving equimolar amounts of sodium acid phosphate, $NaH_2PO_4$, and disodium citrate, $Na_2HC_6H_5O_7$, in water. Both salts dissociate to give the

amphoteric species $H_2PO_4^-$ and $HC_6H_5O_7^{2-}$, causing a problem in deciding which species to designate as HB$_1$ and which to designate as B$_2^-$ in equations (7–121) and (7–122). This problem can be resolved by considering the acidity constants for the two species in question. The acidity constant for $H_2PO_4^-$ is 7.2 and that for the species $HC_6H_5O_7^{2-}$ is 6.4. The citrate species, being more acidic, acts as the acid in the following equilibrium:

$$\underset{\text{Acid}_1}{HC_6H_5O_7^{2-}} + \underset{\text{Base}_2}{H_2PO_4^-} \rightleftharpoons \underset{\text{Acid}_2}{H_3PO_4} + \underset{\text{Base}_1}{C_6H_5O_7^{3-}} \quad (7\text{–}137)$$

Thus, $K_1$ in equation (7–130) is $K_3$ for the citric acid system, and $K_2$ in equation (7–130) is $K_1$ for the phosphoric acid system.

### EXAMPLE 7–24

**Calculate pH**

What is the pH of a solution containing $NaH_2PO_4$ and disodium citrate (disodium hydrogen citrate) $Na_2HC_6H_5O_7$, both in a concentration of 0.01 M? The third acidity constant for $HC_6H_5O_7^{2-}$ is $4.0 \times 10^{-7}$, whereas the first acidity constant for phosphoric acid is $7.5 \times 10^{-3}$. We have

$$[H_3O^+] = \sqrt{(4.0 \times 10^{-7}) \times (7.5 \times 10^{-3})}$$
$$= 5.48 \times 10^{-5} \text{ M}$$

All assumptions are valid. We find

$$pH = -\log(5.48 \times 10^{-5}) = 4.26$$

The equilibrium shown in equation (7–137) illustrates the fact that the system made by dissolving $NaH_2PO_4$ and $Na_2HC_6H_5O_7$ in water is identical to that made by dissolving $H_3PO_4$ and $Na_3C_6H_5O_7$ in water. In the latter case, $H_3PO_4$ is HB$_1$ and the tricitrate is B$_2^-$, and if the two substances are dissolved in equimolar amounts, equation (7–130) is valid for the system.

A slightly different situation arises for equimolar combinations of substances such as succinic acid, $H_2C_4H_4O_4$, and tribasic sodium phosphate, $Na_3PO_4$. In this case, it is obvious that succinic acid is the acid, which can protonate the base to yield the species $HC_4H_4O_4^-$ and $HPO_4^{2-}$. The acid succinate ($pK_a$ 5.63) is a stronger acid than $HPO_4^{2-}$ ($pK_a$ 12.0), however, and an equilibrium cannot be established between these species and the species originally added to water. Instead, the $HPO_4^{2-}$ is protonated by the acid succinate to give $C_4H_4O_4^{2-}$ and $H_2PO_4^-$. This is illustrated in the following:

$$H_2C_4H_4O_4 + PO_4^{3-} \rightarrow$$
$$HC_4H_4O_4^- + HPO_4^{2-} \quad (7\text{–}138)$$

$$\underset{\text{Acid}_1}{HC_4H_4O_4^-} + \underset{\text{Base}_2}{HPO_4^{2-}} \rightleftharpoons \underset{\text{Base}_1}{C_4H_4O_4^{2-}} + \underset{\text{Acid}_2}{H_2PO_4^-} \quad (7\text{–}139)$$

Thus, $K_1$ in equation (7–139) is $K_2$ for the succinic acid system, and $K_2$ in equation (7–130) is actually $K_2$ from the phosphoric acid system.

**EXAMPLE 7–25**

**Calculate pH**

Calculate the pH of a solution containing succinic acid and tribasic sodium phosphate, each at a concentration of 0.01 M. The second acidity constant for the succinic acid system is $2.3 \times 10^{-6}$. The second acidity constant for the phosphoric acid system is $6.2 \times 10^{-8}$. Write

(a)
$$[H_3O^+] = \sqrt{(2.3 \times 10^{-6})(6.2 \times 10^{-8})}$$
$$= 3.78 \times 10^{-7} \text{ M}$$

All assumptions are valid. We have

$$pH = -\log(3.78 \times 10^{-7}) = 6.42$$

(b) Equation (7–130) can also be solved by taking logarithms of both sides to yield

$$pH = \frac{1}{2}(pK_1 + pK_2)$$
$$= \frac{1}{2}(5.63 + 7.21) = 6.42 \qquad (7\text{–}140)$$

Equations **(7–138)** and **(7–139)** illustrate the fact that solutions made by dissolving equimolar amounts of $H_2C_4H_4O_4$ and $Na_3PO_4$, $NaHC_4H_4O_4$ and $Na_2HPO_4$, or $Na_2C_4H_4O_4$ and $NaH_2PO_4$ in water all equilibrate to the same pH and are identical.

# ACIDITY CONSTANTS

One of the most important properties of a drug molecule is its acidity constant, which for many drugs can be related to physiologic and pharmacologic activity,[10–12] solubility, rate of solution,[13] extent of binding,[14] and rate of absorption.[15]

## Effect of Ionic Strength on Acidity Constants

In the preceding sections, the solutions were considered dilute enough that the effect of ionic strength on the acid–base equilibria could be ignored. A more exact treatment for the ionization of a weak acid, for example, would be

$$HB + H_2O \rightleftharpoons H_3O^+ + B$$
$$K = \frac{\alpha_{H_3O} + \alpha_B}{\alpha_{HB}} = \frac{[H_3O^+][B]}{[HB]} \cdot \frac{\gamma H_3O^+ \gamma_B}{\gamma_{HB}} \qquad (7\text{–}141)$$

where $K$ is the thermodynamic acidity constant, and the charges on the species have been omitted to make the equations more general. Equation **(7–141)** illustrates the fact that in solving equations involving acidity constants, both the concentration and the activity coefficient of each species must be considered. One way to simplify the problem would be to define the acidity constant as an apparent constant in terms of the hydronium ion activity and species concentrations and activity coefficients, as follows:

$$K = \alpha_{H_3O^+} \frac{[B]}{[HB]} \frac{\gamma_B}{\gamma_{HB}} = K' \frac{\gamma_B}{\gamma_{HB}} \qquad (7\text{–}142)$$

and

$$pK' = pK + \log \frac{\gamma_B}{\gamma_{HB}} \qquad (7\text{–}143)$$

The following form of the Debye–Hückel equation[16] can be used for ionic strengths up to about 0.3 M:

$$-\log \gamma_i = \frac{0.51 Z_i^2 \sqrt{\mu}}{1 + \alpha B \sqrt{\mu}} - K_s \mu \qquad (7\text{–}144)$$

where $Z_i$ is the charge on the species $i$. The value of the constant $\alpha B$ can be taken to be approximately 1 at 25°C, and $K_s$ is a "salting-out" constant. At moderate ionic strengths, $K_s$ can be assumed to be approximately the same for both the acid and its conjugate base.[16] Thus, for an acid with charge $Z$ going to a base with charge $Z - 1$,

$$pK' = pK + \frac{0.51(2Z - 1)\sqrt{\mu}}{1 + \sqrt{\mu}} \qquad (7\text{–}145)$$

**EXAMPLE 7–26**

**Calculate $pK'_2$**

Calculate $pK'_2$ for citric acid at an ionic strength of 0.01 M. Assume that $pK_2 = 4.78$. The charge on the acidic species is $-1$. We have

$$pK'_2 = 4.78 + \frac{0.51(-3)\sqrt{0.01}}{1 + \sqrt{0.01}}$$
$$= 4.78 - 1.53(0.091) = 4.64$$

If either the acid or its conjugate base is a zwitterion, it will have a large dipole moment, and the expression for its activity coefficient must contain a term $K_r$, the "salting-in" constant.[17] Thus, for the zwitterion $[+ -]$,

$$-\log \gamma_{+-} = (K_r - K_s)\mu \qquad (7\text{–}146)$$

The first ionization of an amino acid such as glycine hydrochloride involves an acid with a charge of $+1$ going to the zwitterion, $[+ -]$. Combining equations **(7–146)** and **(7–144)** with equation **(7–143)** gives

$$pK'_1 = pK_1 + \frac{0.51\sqrt{\mu}}{1 + \sqrt{\mu}} - K_r\mu \qquad (7\text{–}147)$$

The second ionization step involves the zwitterion going to a species with a charge of $-1$. Thus, using equations **(7–146)**, **(7–144)**, and **(7–143)** gives

$$pK'_2 = pK_2 - \frac{0.51\sqrt{\mu}}{1 + \sqrt{\mu}} + K_r\mu \qquad (7\text{–}148)$$

The "salting-in" constant, $K_r$, is approximately 0.32 for alpha-amino acids in water and approximately 0.6 for dipeptides.[17] Use of these values for $K_r$ enables equations **(7–147)** and **(7–148)** to be used for solutions with ionic strengths up to about 0.3 M.

The procedure to be used in solving pH problems in which the ionic strength of the solution must be considered is as follows:

(a) Convert all pK values needed for the problem into pK' values.

(b) Solve the appropriate equation in the usual manner.

**EXAMPLE 7–27**

**Calculate pH**

Calculate the pH of a 0.01 M solution of acetic acid to which enough KCl had been added to give an ionic strength of 0.01 M at 25°C. The $pK_a$ for acetic acid is 4.76.

(a)
$$pK_a' = 4.76 - \frac{0.51\sqrt{0.10}}{1 + \sqrt{0.10}}$$
$$= 4.76 - 0.12 = 4.64$$

(b) Taking logarithms of equation (7–99) gives

$$pH = \frac{1}{2}(pK_a' - \log C_a)$$

in which we now write $pK_a$ as $pK'_a$:

$$pH = \frac{1}{2}(4.64 + 2.00) = 3.32$$

**EXAMPLE 7–28**

**Calculate pH**

Calculate the pH of a $10^{-3}$ M solution of glycine at an ionic strength of 0.10 at 25°C. The $pK_a$ values for glycine are $pK_1 = 2.35$ and $pK_2 = 9.78$.

(a)
$$pK_1' = 2.35 + \frac{0.51\sqrt{0.10}}{1 + \sqrt{0.10}} - 0.32(0.10)$$
$$= 2.35 + 0.12 - 0.03 = 2.44$$

(b)
$$pK_2' = 9.78 - \frac{0.51\sqrt{0.10}}{1 + \sqrt{0.10}} + 0.32(0.10)$$
$$= 9.78 - 0.12 + 0.03 = 9.69$$

(c) Taking logarithms of equation (7–118) gives

$$pH = \frac{1}{2}(pK_1 + pK_2)$$
$$= \frac{1}{2}(2.44 + 9.69) = 6.07$$

# CHAPTER SUMMARY

In this chapter, the student is introduced to ionic equilibria in the pharmaceutical sciences. The Brönsted–Lowry and Lewis electronic theories are introduced. The four classes of solvents (protophilic, protogenic, amphiprotic, and aprotic) are described as well. The student should understand the concepts of acid–base equilibria and the ionization of weak acids and weak bases. Further, you should be able to use this theory in your practice. In other words, you should be able to calculate dissociation constants $K_a$ and $K_b$ and understand the relationship between $K_a$ and $K_b$. It is also very important to understand the concepts of pH, p$K$, and pOH and the relationship between hydrogen ion concentration and pH. Of course, you should be able to calculate pH. Finally, strong acids and strong bases were defined and described. You should strive for a working understanding of acidity constants.

Practice problems for this chapter can be found at thePoint.lww.com/Sinko6e.

# References

1. J. N. Brönsted, Rec. Trav. Chim. **42**, 718, 1923; Chem. Rev. **5**, 231, 1928; T. M. Lowry, J. Chem. Soc. **123**, 848, 1923.
2. A. W. Castleman Jr, J. Chem. Phys. **94**, 3268, 1991; Chem. Eng. News **69**, 47, 1991.
3. W. F. Luder and S. Zuffanti, *Electronic Theory of Acids and Bases*, Wiley, New York, 1947; G. N. Lewis, *Valency and the Structure of Atoms and Molecules*, Reinhold, New York, 1923.
4. R. P. Bell, *Acids and Bases*, Methuen, London, 1952, Chapter 7.
5. S. P. L. Sörensen, Biochem. Z. **21**, 201, 1909.
6. S. F. Kramer and G. L. Flynn, J. Pharm. Sci. **61**, 1896, 1972.
7. P. A. Schwartz, C. T. Rhodes and J. W. Cooper Jr, J. Pharm. Sci. **66**, 994, 1977.
8. J. Blanchard, J. O. Boyle and S. Van Wagenen, J. Pharm. Sci. **77**, 548, 1988.
9. T. S. Lee and L. Gunnar Sillen, *Chemical Equilibrium in Analytical Chemistry*, Interscience, New York, 1959; J. N. Butler, *Solubility and pH Calculations*, Addison–Wesley, Reading, Mass., 1964; A. J. Bard, *Chemical Equilibrium*, Harper & Row, New York, 1966.
10. P. B. Marshall, Br. J. Pharmacol. **10**, 270, 1955.
11. P. Bell and R. O. Roblin, J. Am. Chem. Soc. **64**, 2905, 1942.
12. I. M. Klotz, J. Am. Chem. Soc. **66**, 459, 1944.
13. W. E. Hamlin and W. I. Higuchi, J. Pharm. Sci. **55**, 205, 1966.
14. M. C. Meyer and D. E. Guttman, J. Pharm. Sci. **57**, 245, 1968.
15. B. B. Brodie, in T. B. Binns (Ed.), *Absorption and Distribution of Drugs*, Williams & Wilkins, Baltimore, 1964, pp. 16–48.
16. J. T. Edsall and J. Wyman, *Biophysical Chemistry*, Vol. 1, Academic Press, New York, 1958, p. 442.
17. J. T. Edsall and J. Wyman, *Biophysical Chemistry*, Vol. 1, Academic Press, New York, 1958, p. 443.

# Recommended Reading

J. N. Butler, *Ionic Equilibrium: Solubility and pH Calculations*, Wiley Interscience, Hoboken, NJ, 1998.

## CHAPTER LEGACY

**Fifth Edition:** published as Chapter 7 (Ionic Equilibria). Updated by Patrick Sinko.

**Sixth Edition:** published as Chapter 7 (Ionic Equilibria). Updated by Patrick Sinko.

# BUFFERED AND ISOTONIC SOLUTIONS

Buffers are compounds or mixtures of compounds that, by their presence in solution, resist changes in pH upon the addition of small quantities of acid or alkali. The resistance to a change in pH is known as *buffer action*. According to Roos and Borm,[1] Koppel and Spiro published the first paper on buffer action in 1914 and suggested a number of applications, which were later elaborated by Van Slyke.[2]

If a small amount of a strong acid or base is added to water or a solution of sodium chloride, the pH is altered considerably; such systems have no buffer action.

## THE BUFFER EQUATION

### Common Ion Effect and the Buffer Equation for a Weak Acid and Its Salt

The pH of a buffer solution and the change in pH upon the addition of an acid or base can be calculated by use of the

**KEY CONCEPT** ## WHAT IS A BUFFER?

A combination of a weak acid and its conjugate base (i.e., its salt) or a weak base and its conjugate acid acts as a buffer. If 1 mL of a 0.1 N HCl solution is added to 100 mL of pure water, the pH is reduced from 7 to 3. If the strong acid is added to a 0.01 M solution containing equal quantities of acetic acid and sodium acetate, the pH is changed only 0.09 pH units because the base $Ac^-$ ties up the hydrogen ions according to the reaction

$$Ac^- + H_3O^+ \rightleftharpoons HAc + H_2O \qquad (8\text{--}1)$$

If a strong base, sodium hydroxide, is added to the buffer mixture, acetic acid neutralizes the hydroxyl ions as follows:

$$HAc + OH^- \rightleftharpoons H_2O + Ac^- \qquad (8\text{--}2)$$

*buffer equation.* This expression is developed by considering the effect of a salt on the ionization of a weak acid when the salt and the acid have an ion in common.

For example, when sodium acetate is added to acetic acid, the dissociation constant for the weak acid,

$$K_a = \frac{[H_3O^+][Ac^-]}{[HAc]} = 1.75 \times 10^{-5} \qquad (8\text{--}3)$$

is momentarily disturbed because the acetate ion supplied by the salt increases the $[Ac^-]$ term in the numerator. To reestablish the constant $K_a$ at $1.75 \times 10^{-5}$, the hydrogen ion term in the numerator $[H_3O^+]$ is instantaneously decreased, with a corresponding increase in [HAc]. Therefore, the constant $K_a$ remains unaltered, and the equilibrium is shifted in the direction of the reactants. Consequently, the ionization of acetic acid,

$$HAc + H_2O \rightleftharpoons H_3O^+ + Ac^- \qquad (8\text{--}4)$$

is *repressed* upon the addition of the common ion, $Ac^-$. This is an example of the *common ion effect*. The pH of the final solution is obtained by rearranging the equilibrium expression for acetic acid:

$$[H_3O^+] = K_a \frac{[HAc]}{[Ac^-]} \qquad (8\text{--}5)$$

If the acid is weak and ionizes only slightly, the expression [HAc] may be considered to represent the total concentration of acid, and it is written simply as [Acid]. In the slightly ionized acidic solution, the acetate concentration $[Ac^-]$ can be considered as having come entirely from the salt, sodium acetate. Because 1 mole of sodium acetate yields 1 mole of acetate ion, $[Ac^-]$ is equal to the total salt concentration and is replaced by the term [Salt]. Hence, equation (8–5) is written as

$$[H_3O^+] = K_a \frac{[Acid]}{[Salt]} \qquad (8\text{--}6)$$

Equation (8–6) can be expressed in logarithmic form, with the signs reversed, as

$$-\log[H_3O^+] = -\log K_a - \log[Acid] + \log[Salt] \quad (8\text{–}7)$$

from which is obtained an expression, known as the *buffer equation* or the *Henderson–Hasselbalch equation*, for a weak acid and its salt:

$$pH = pK_a + \log\frac{[Salt]}{[Acid]} \quad (8\text{–}8)$$

The ratio [Acid]/[Salt] in equation (8–6) has been inverted by undertaking the logarithmic operations in equation (8–7), and it appears in equation (8–8) as [Salt]/[Acid]. The term $pK_a$, the negative logarithm of $K_a$, is called the *dissociation exponent*.

The buffer equation is important in the preparation of buffered pharmaceutical solutions; it is satisfactory for calculations within the pH range of 4 to 10.

### EXAMPLE 8–1

**pH Calculation**

What is the pH of 0.1 M acetic acid solution, $pK_a = 4.76$? What is the pH after enough sodium acetate has been added to make the solution 0.1 M with respect to this salt?

The pH of the acetic acid solution is calculated by use of the logarithmic form of equation (7–102):

$$pH = \frac{1}{2}pK_a - \frac{1}{2}\log c$$
$$pH = 2.38 + 0.50 = 2.88$$

The pH of the buffer solution containing acetic acid and sodium acetate is determined by use of the buffer equation (8–8):

$$pH = 4.76 + \log\frac{0.1}{0.1} = 4.76$$

It is seen from *Example 8–1* that the pH of the acetic acid solution has been *increased* almost 2 pH units; that is, the acidity has been *reduced* to about 1/100 of its original value by the addition of an equal concentration of a salt with a common ion. This example bears out the statement regarding the repression of ionization upon the addition of a common ion.

Sometimes it is desired to know the ratio of salt to acid in order to prepare a buffer of a definite pH. The following example demonstrates the calculation involved in such a problem.

### EXAMPLE 8–2

**pH and [Salt]/[Acid] Ratio**

What is the molar ratio, [Salt]/[Acid], required to prepare an acetate buffer of pH 5.0? Also express the result in mole percent.

$$5.0 = 4.76 + \log\frac{[Salt]}{[Acid]}$$

$$\log\frac{[Salt]}{[Acid]} = 5.0 - 4.76 = 0.24$$

$$\frac{[Salt]}{[Acid]} = \text{antilog } 0.24 = 1.74$$

Therefore, the mole ratio of salt to acid is 1.74/1. Mole percent is mole fraction multiplied by 100. The mole fraction of salt in the salt–acid mixture is $1.74/(1 + 1.74) = 0.635$, and in mole percent, the result is 63.5%.

## The Buffer Equation for a Weak Base and Its Salt

Buffer solutions are not ordinarily prepared from weak bases and their salts because of the volatility and instability of the bases and because of the dependence of their pH on $pK_w$, which is often affected by temperature changes. Pharmaceutical solutions—for example, a solution of ephedrine base and ephedrine hydrochloride—however, often contain combinations of weak bases and their salts.

The buffer equation for solutions of weak bases and the corresponding salts can be derived in a manner analogous to that for the weak acid buffers. Accordingly,

$$[OH^-] = K_b\frac{[Base]}{[Salt]} \quad (8\text{–}9)$$

and using the relationship $[OH^-] = K_w/[H_3O^+]$, the buffer equation is obtained

$$pH = pK_w - pK_b + \log\frac{[Base]}{[Salt]} \quad (8\text{–}10)$$

### EXAMPLE 8–3

**Using the Buffer Equation**

What is the pH of a solution containing 0.10 mole of ephedrine and 0.01 mole of ephedrine hydrochloride per liter of solution? Since the $pK_b$ of ephedrine is 4.64,

$$pH = 14.00 - 4.64 + \log\frac{0.10}{0.01}$$

$$pH = 9.36 + \log 10 = 10.36$$

## Activity Coefficients and the Buffer Equation

A more exact treatment of buffers begins with the replacement of concentrations by activities in the equilibrium of a weak acid:

$$K_a = \frac{a_{H_3O^+}a_{Ac^-}}{a_{HAc}} = \frac{(\gamma_{H_3O^+}c_{H_3O^+}) \times (\gamma_{Ac^-}c_{Ac^-})}{\gamma_{HAc}c_{HAc}} \quad (8\text{–}11)$$

The activity of each species is written as the activity coefficient multiplied by the molar concentration. The activity coefficient of the undissociated acid, $\gamma_{HAc}$, is essentially 1 and may be dropped. Solving for the hydrogen ion activity and pH, defined as $-\log a_{H_3O^+}$, yields the equations

$$a_{H_3O^+} = \gamma_{H_3O^+} \times c_{H_3O^+} = K_a\frac{c_{HAc}}{\gamma_{Ac^-}c_{Ac^-}} \quad (8\text{–}12)$$

$$pH = pK_a + \log\frac{[Salt]}{[Acid]} + \log\gamma_{Ac^-} \quad (8\text{–}13)$$

From the Debye–Hückel expression for an aqueous solution of a univalent ion at 25°C having an ionic strength not greater than about 0.1 or 0.2, we write

$$\log\gamma_{Ac^-} = \frac{-0.5\sqrt{\mu}}{1 + \sqrt{\mu}}$$

and equation **(8–13)** then becomes

$$pH = pK_a + \log\frac{[\text{Salt}]}{[\text{Acid}]} - \frac{0.5\sqrt{\mu}}{1 + \sqrt{\mu}} \qquad (8\text{–}14)$$

The general equation for buffers of polybasic acids is

$$pH = pK_n + \log\frac{[\text{Salt}]}{[\text{Acid}]} - \frac{A(2n-1)\sqrt{\mu}}{1 + \sqrt{\mu}} \qquad (8\text{–}15)$$

where $n$ is the stage of the ionization.

**EXAMPLE 8–4**

**Activity Coefficients and Buffers**

A buffer contains 0.05 mole/liter of formic acid and 0.10 mole/liter of sodium formate. The $pK_a$ of formic acid is 3.75. The ionic strength of the solution is 0.10. Compute the pH (*a*) with and (*b*) without consideration of the activity coefficient correction.

(*a*)

$$pH = 3.75 + \log\frac{0.10}{0.05} - \frac{0.5\sqrt{0.10}}{1 + \sqrt{0.10}}$$
$$= 3.93$$

(*b*)

$$pH = 3.75 + \log\frac{0.10}{0.05} = 4.05$$

## Some Factors Influencing the pH of Buffer Solutions

The addition of neutral salts to buffers changes the pH of the solution by altering the ionic strength, as shown in equation **(8–13)**. Changes in ionic strength and hence in the pH of a buffer solution can also be brought about by dilution. The addition of water in moderate amounts, although not changing the pH, may cause a small positive or negative deviation because it alters activity coefficients and because water itself can act as a weak acid or base. Bates[3] expressed this quantitatively in terms of a *dilution value*, which is the change in pH on diluting the buffer solution to one half of its original strength. Some dilution values for National Bureau of Standards buffers are given in **Table 8–1**. A positive dilution value signifies that the pH rises with dilution and a negative value signifies that the pH decreases with dilution of the buffer.

**TABLE 8–1**
**BUFFER CAPACITY OF SOLUTIONS CONTAINING EQUIMOLAR AMOUNTS (0.1 M) OF ACETIC ACID AND SODIUM ACETATE**

| Moles of NaOH Added | pH of Solution | Buffer Capacity, $\beta$ |
|---|---|---|
| 0 | 4.76 | |
| 0.01 | 4.85 | 0.11 |
| 0.02 | 4.94 | 0.11 |
| 0.03 | 5.03 | 0.11 |
| 0.04 | 5.13 | 0.10 |
| 0.05 | 5.24 | 0.09 |
| 0.06 | 5.36 | 0.08 |

Temperature also influences buffers. Kolthoff and Tekelenburg[4] determined the *temperature coefficient of* pH, that is, the change in pH with temperature, for a large number of buffers. The pH of acetate buffers was found to increase with temperature, whereas the pH of boric acid–sodium borate buffers decreased with temperature. Although the temperature coefficient of acid buffers was relatively small, the pH of most basic buffers was found to change more markedly with temperature, owing to $K_w$, which appears in the equation of basic buffers and changes significantly with temperature. Bates[3] referred to several basic buffers that show only a small change of pH with temperature and can be used in the pH range of 7 to 9. The temperature coefficients for the calomel electrode are given in the study by Bates.

## Drugs as Buffers

It is important to recognize that solutions of drugs that are weak electrolytes also manifest buffer action. Salicylic acid solution in a soft glass bottle is influenced by the alkalinity of the glass. It might be thought at first that the reaction would result in an appreciable increase in pH; however, the sodium ions of the soft glass combine with the salicylate ions to form sodium salicylate. Thus, there arises a solution of salicylic acid and sodium salicylate—a buffer solution that resists the change in pH. Similarly, a solution of ephedrine base manifests a natural buffer protection against reductions in pH. Should hydrochloric acid be added to the solution, ephedrine hydrochloride is formed, and the buffer system of ephedrine plus ephedrine hydrochloride will resist large changes in pH until the ephedrine is depleted by reaction with the acid. Therefore, a drug in solution may often act as its own buffer over a definite pH range. Such buffer action, however, is often too weak to counteract pH changes brought about by the carbon dioxide of the air and the alkalinity of the bottle. Additional buffers are therefore frequently added to drug solutions to maintain the system within a certain pH range. A quantitative measure of the efficiency or capacity of a buffer to resist pH changes will be discussed in a later section.

## pH Indicators

Indicators may be considered as weak acids or weak bases that act like buffers and also exhibit color changes as their degree of dissociation varies with pH. For example, methyl red shows its full alkaline color, yellow, at a pH of about 6 and its full acid color, red, at about pH 4.

The dissociation of an acid indicator is given in simplified form as

$$\underset{\substack{\text{Acid}_1 \\ \text{(Acid color)}}}{\text{HIn}} + \underset{\text{Base}_2}{\text{H}_2\text{O}} \rightleftharpoons \underset{\text{Acid}_2}{\text{H}_3\text{O}^+} + \underset{\substack{\text{Base}_1 \\ \text{(Alkaline color)}}}{\text{In}^-} \qquad (8\text{–}16)$$

The equilibrium expression is

$$\frac{[\text{H}_3\text{O}^+][\text{In}^-]}{[\text{HIn}]} = K_{\text{In}} \qquad (8\text{–}17)$$

**TABLE 8–2**
**COLOR, pH, AND INDICATOR CONSTANT, p$K_{In}$, OF SOME COMMON INDICATORS**

| | Color | | | |
| Indicator | Acid | Base | pH Range | p$K_{In}$ |
| --- | --- | --- | --- | --- |
| Thymol blue (acid range) | Red | Yellow | 1.2–2.8 | 1.5 |
| Methyl violet | Blue | Violet | 1.5–3.2 | – |
| Methyl orange | Red | Yellow | 3.1–4.4 | 3.7 |
| Bromcresol green | Yellow | Blue | 3.8–5.4 | 4.7 |
| Methyl red | Red | Yellow | 4.2–6.2 | 5.1 |
| Bromcresol purple | Yellow | Purple | 5.2–6.8 | 6.3 |
| Bromthymol blue | Yellow | Blue | 6.0–7.6 | 7.0 |
| Phenol red | Yellow | Red | 6.8–8.4 | 7.9 |
| Cresol red | Yellow | Red | 7.2–8.8 | 8.3 |
| Thymol blue (alkaline range) | Yellow | Blue | 8.0–9.6 | 8.9 |
| Phenolphthalein | Colorless | Red | 8.3–10.0 | 9.4 |
| Alizarin yellow | Yellow | Lilac | 10.0–12.0 | – |
| Indigo carmine | Blue | Yellow | 11.6–14 | – |

HIn is the un-ionized form of the indicator, which gives the acid color, and In$^-$ is the ionized form, which produces the basic color. $K_{In}$ is referred to as the *indicator constant*. If an acid is added to a solution of the indicator, the hydrogen ion concentration term on the right-hand side of equation (8–16) is increased, and the ionization is repressed by the common ion effect. The indicator is then predominantly in the form of HIn, the acid color. If base is added, [H$_3$O$^+$] is reduced by reaction of the acid with the base, reaction (8–16) proceeds to the right, yielding more ionized indicator In$^-$, and the base color predominates. Thus, the color of an indicator is a function of the pH of the solution. A number of indicators with their useful pH ranges are listed in **Table 8–2**.

The equilibrium expression (8–16) can be treated in a manner similar to that for a buffer consisting of a weak acid and its salt or conjugate base. Hence

$$[H_3O^+] = K_{In}\frac{[HIn]}{[In^-]} \tag{8–18}$$

and because [HIn] represents the acid color of the indicator and the conjugate base [In$^-$] represents the basic color, these terms can be replaced by the concentration expressions [Acid] and [Base]. The formula for pH as derived from equation (8–18) becomes

$$pH = pK_{In} + \log\frac{[Base]}{[Acid]} \tag{8–19}$$

**EXAMPLE 8–5**

**Calculate pH**

An indicator, methyl red, is present in its ionic form In$^-$, in a concentration of $3.20 \times 10^3$ M and in its molecular form, HIn, in an aqueous solution at 25°C in a concentration of $6.78 \times 10^3$ M. From Table 8–2 a p$K_{In}$ of 5.1 is observed for methyl red. What is the pH of this solution? We have

$$pH = 5.1 + \log\frac{3.20 \times 10^{-3}}{6.78 \times 10^{-3}} = 4.77$$

Just as a buffer shows its greatest efficiency when pH = p$K_a$, an indicator exhibits its *middle tint* when [Base]/[Acid] = 1 and pH = p$K_{In}$. The most efficient indicator range, corresponding to the effective buffer interval, is about 2 pH units, that is, p$K_{In} \pm 1$. The reason for the width of this color range can be explained as follows. It is known from experience that one cannot discern a change from the acid color to the salt or conjugate base color until the ratio of [Base] to [Acid] is about 1 to 10. That is, there must be at least 1 part of the basic color to 10 parts of the acid color before the eye can discern a change in color from acid to alkaline. The pH value at which this change is perceived is given by the equation

$$pH = pK_{In} + \log\frac{1}{10} = pK_{In} - 1 \tag{8–20}$$

Conversely, the eye cannot discern a change from the alkaline to the acid color until the ratio of [Base] to [Acid] is about 10 to 1, or

$$pH = pK_{In} + \log\frac{10}{1} = pK_{In} + 1 \tag{8–21}$$

Therefore, when base is added to a solution of a buffer in its acid form, the eye first visualizes a change in color at p$K_{In} - 1$, and the color ceases to change any further at p$K_{In} + 1$. The effective range of the indicator between its full acid and full basic color can thus be expressed as

$$pH = pK_{In} \pm 1 \tag{8–22}$$

Chemical indicators are typically compounds with chromophores that can be detected in the visible range and change color in response to a solution's pH. Most chemicals used as indicators respond only to a narrow pH range. Several indicators can be combined to yield so-called *universal indicators* just as buffers can be mixed to cover a wide pH range. A universal indicator is a pH indicator that displays different colors as the pH transitions from pH 1 to 12. A typical universal indicator will display a color range from red to purple.

For example, a strong acid (pH 0–3) may display as red in color, an acid (pH 3–6) as orange–yellow, neutral pH (pH 7) as green, alkaline pH (pH 8–11) as blue, and purple for strong alkaline pH (pH 11–14).

The colorimetric method for the determination of pH is probably less accurate and less convenient but is also less expensive than electrometric methods and it can be used in the determination of the pH of aqueous solutions that are not colored or turbid. This is particularly useful for the study of acid–base reactions in nonaqueous solutions. A note of caution should be added regarding the colorimetric method. Because indicators themselves are acids (or bases), their addition to unbuffered solutions whose pH is to be determined will change the pH of the solution. The colorimetric method is therefore not applicable to the determination of the pH of sodium chloride solution or similar unbuffered pharmaceutical preparations unless special precautions are taken in the measurement. Some medicinal solutions and pharmaceutical vehicles, however, to which no buffers have been added are buffered by the presence of the drug itself and can withstand the addition of an indicator without a significant change in pH. Errors in the result can also be introduced by the presence of salts and proteins, and these errors must be determined for each indicator over the range involved.

Recently, Kong et al.[5] reported on a rapid method for determining $pK_a$ based on spectrophotometric titration using a universal pH indicator. Historically, potentiometric titration, which typically uses pH electrodes, has been the most commonly used method for determining $pK_a$ values. This method takes time and requires the daily calibration of the pH electrode. Spectrophotometric titration has the advantage that less sample is required, it is not affected by $CO_2$ interference, and it can provide multiwavelength absorbance information. The method can be applied only to compounds with chromophores placed close to the titratable groups. The indicator spectra can then be used to calculate the pH value of a solution from the $pK_a$ values, concentration, and molar extinction coefficients of the indicator species. In contrast to pH electrodes, chemical indicators respond rapidly and do not require frequent calibration.

# BUFFER CAPACITY

Thus far it has been stated that a buffer counteracts the change in pH of a solution upon the addition of a strong acid, a strong base, or other agents that tend to alter the hydrogen ion concentration. Furthermore, it has been shown in a rather qualitative manner how combinations of weak acids and weak bases together with their salts manifest this buffer action. The resistance to changes of pH now remains to be discussed in a more quantitative way.

The magnitude of the resistance of a buffer to pH changes is referred to as the buffer capacity, $\beta$. It is also known as *buffer efficiency*, *buffer index*, and *buffer value*. Koppel and Spiro[1] and Van Slyke[2] introduced the concept of buffer

capacity and defined it as the ratio of the increment of strong base (or acid) to the small change in pH brought about by this addition. For the present discussion, the approximate formula

$$\beta = \frac{\Delta B}{\Delta pH} \qquad (8\text{–}23)$$

can be used, in which delta, $\Delta$, has its usual meaning, a *finite change*, and $\Delta B$ is the small increment in gram equivalents (g Eq)/liter of strong base added to the buffer solution to produce a pH change of $\Delta$ pH. According to equation (**8–23**), the buffer capacity of a solution has a value of 1 when the addition of 1 g Eq of strong base (or acid) to 1 liter of the buffer solution results in a change of 1 pH unit. The significance of this index will be appreciated better when it is applied to the calculation of the capacity of a buffer solution.

## Approximate Calculation of Buffer Capacity

Consider an acetate buffer containing 0.1 mole each of acetic acid and sodium acetate in 1 liter of solution. To this are added 0.01-mole portions of sodium hydroxide. When the first increment of sodium hydroxide is added, the concentration of sodium acetate, the [Salt] term in the buffer equation, increases by 0.01 mole/liter and the acetic acid concentration, [Acid], decreases proportionately because each increment of base converts 0.01 mole of acetic acid into 0.01 mole of sodium acetate according to the reaction

$$\underset{(0.1-0.01)}{HAc} + \underset{(0.01)}{NaOH} \rightleftharpoons \underset{(0.1+0.01)}{NaAc} + H_2O \qquad (8\text{–}24)$$

The changes in concentration of the salt and the acid by the addition of a base are represented in the buffer equation (**8–8**) by using the modified form

$$pH = pK_a + \log \frac{[Salt] + [Base]}{[Acid] - [Base]} \qquad (8\text{–}25)$$

Before the addition of the first portion of sodium hydroxide, the pH of the buffer solution is

$$pH = 4.76 + \log \frac{0.1 + 0}{0.1 - 0} = 4.76 \qquad (8\text{–}26)$$

The results of the continual addition of sodium hydroxide are shown in **Table 8–1**. The student should verify the pH values and buffer capacities by the use of equations (**8–25**) and (**8–23**), respectively.

As can be seen from **Table 8–1**, the buffer capacity is not a fixed value for a given buffer system but instead depends on the amount of base added. The buffer capacity changes as the ratio log([Salt]/[Acid]) increases with added base. With the addition of more sodium hydroxide, the buffer capacity decreases rapidly, and, when sufficient base has been added to convert the acid completely into sodium ions and acetate ions, the solution no longer possesses an acid reserve. The buffer has its greatest capacity before any base is added, where [Salt]/[Acid] = 1, and, therefore, according to equation (**8–8**), pH = $pK_a$. The buffer capacity is also influenced by an increase in the total concentration of the buffer constituents because, obviously, a great concentration of salt and acid provides a greater alkaline and acid reserve. The influence of concentration on buffer capacity is treated following the discussion of Van Slyke's equation.

## A More Exact Equation for Buffer Capacity

The buffer capacity calculated from equation (8–23) is only approximate. It gives the average buffer capacity over the increment of base added. Koppel and Spiro[1] and Van Slyke[2] developed a more exact equation,

$$\beta = 2.3C \frac{K_a[H_3O^+]}{(K_a + [H_3O^+])^2} \qquad (8\text{–}27)$$

where $C$ is the total buffer concentration, that is, the sum of the molar concentrations of the acid and the salt. Equation (8–27) permits one to compute the buffer capacity at any hydrogen ion concentration—for example, at the point where no acid or base has been added to the buffer.

### EXAMPLE 8–6

**Calculating Buffer Capacity**

At a hydrogen ion concentration of $1.75 \times 10^{-5}$ (pH = 4.76), what is the capacity of a buffer containing 0.10 mole each of acetic acid and sodium acetate per liter of solution? The total concentration, $C = [\text{Acid}] + [\text{Salt}]$, is 0.20 mole/liter, and the dissociation constant is $1.75 \times 10^{-5}$. We have

$$\beta = \frac{2.3 \times 0.20 \times (1.75 \times 10^{-5}) \times (1.75 \times 10^{-5})}{[(1.75 \times 10^{-5}) + (1.75 \times 10^{-5})]^2}$$

$$= 0.115$$

### EXAMPLE 8–7

**Buffer Capacity and pH**

Prepare a buffer solution of pH 5.00 having a capacity of 0.02. The steps in the solution of the problem are as follows:

(a) Choose a weak acid having a $pK_a$ close to the pH desired. Acetic acid, $pK_a = 4.76$, is suitable in this case.
(b) The ratio of salt and acid required to produce a pH of 5.00 was found in *Example 8–2* to be $[\text{Salt}]/[\text{Acid}] = 1.74/1$.
(c) Use the buffer capacity equation (8–27) to obtain the total buffer concentration, $C = [\text{Salt}] + [\text{Acid}]$:

$$0.02 = 2.3C \frac{(1.75 \times 10^{-5}) \times (1 \times 10^{-5})}{[(1.75 \times 10^{-5}) + (1 \times 10^{-5})]^2}$$

$$C = 3.75 \times 10^{-2} \text{ mole/liter}$$

(d) Finally from (b), $[\text{Salt}] = 1.74 \times [\text{Acid}]$, and from (c),

$$C = (1.74 \times [\text{Acid}]) + [\text{Acid}]$$

$$= 3.75 \times 10^{-2} \text{ mole/liter}$$

Therefore,

$$[\text{Acid}] = 1.37 \times 10^{-2} \text{ mole/liter}$$

and

$$[\text{Salt}] = 1.74 \times [\text{Acid}]$$

$$= 2.38 \times 10^{-2} \text{ mole/liter}$$

## The Influence of Concentration on Buffer Capacity

The buffer capacity is affected not only by the [Salt]/[Acid] ratio but also by the total concentrations of acid and salt. As shown in **Table 8–1**, when 0.01 mole of base is added to a 0.1 molar acetate buffer, the pH increases from 4.76 to 4.85, for a $\Delta$pH of 0.09.

If the concentration of acetic acid and sodium acetate is raised to 1 M, the pH of the original buffer solution remains at about 4.76, but now, upon the addition of 0.01 mole of base, it becomes 4.77, for a $\Delta$pH of only 0.01. The calculation, disregarding activity coefficients, is

$$\text{pH} = 4.76 + \log \frac{1.0 + 0.01}{1.0 - 0.01} = 4.77 \qquad (8\text{–}28)$$

Therefore, an increase in the concentration of the buffer components results in a greater buffer capacity or efficiency. This conclusion is also evident in equation (8–27), where an increase in the total buffer concentration, $C = [\text{Salt}] + [\text{Acid}]$, obviously results in a greater value of $\beta$.

## Maximum Buffer Capacity

An equation expressing the maximum buffer capacity can be derived from the buffer capacity formula of Koppel and Spiro[1] and Van Slyke,[2] equation (8–27). The maximum buffer capacity occurs where pH = $pK_a$, or, in equivalent terms, where $[H_3O^+] = K_a$. Substituting $[H_3O^+]$ for $K_a$ in both the numerator and the denominator of equation (8–27) gives

$$\beta_{max} = 2.303C \frac{[H_3O^+]^2}{(2[H_3O^+])^2} = \frac{2.303}{4}C$$

$$\beta_{max} = 0.576C \qquad (8\text{–}29)$$

where $C$ is the total buffer concentration.

### EXAMPLE 8–8

**Maximum Buffer Capacity**

What is the maximum buffer capacity of an acetate buffer with a total concentration of 0.020 mole/liter? We have

$$\beta_{max} = 0.576 \times 0.020$$

$$= 0.01152 \text{ or } 0.012$$

---

## ◤▶**KEY CONCEPT** ◼ BUFFER CAPACITY

The buffer capacity depends on (a) the value of the ratio [Salt]/[Acid], increasing as the ratio approaches unity, and (b) the magnitude of the individual concentrations of the buffer components, the buffer becoming more efficient as the salt and acid concentrations are increased.

## Neutralization Curves and Buffer Capacity

A further understanding of buffer capacity can be obtained by considering the titration curves of strong and weak acids when they are mixed with increasing quantities of alkali. The reaction of an equivalent of an acid with an equivalent of a base is called neutralization; it can be expressed according to the method of Brönsted and Lowry. The neutralization of a strong acid by a strong base and a weak acid by a strong base is written in the form

$$\begin{array}{cccc} \text{Acid}_1 & \text{Base}_2 & \text{Acid}_2 & \text{Base}_1 \end{array}$$

$$H_3O^+(Cl^-) + (Na^+)OH^- = H_2O + H_2O + Na^+ + Cl^-$$
$$HAc \quad\quad + (Na^+)OH^- = H_2O + (Na^+)Ac^-$$

where $(H_3O^+)(Cl^-)$ is the hydrated form of HCl in water. The neutralization of a strong acid by a strong base simply involves a reaction between hydronium and hydroxyl ions and is usually written as

$$H_3O^+ + OH^- = 2H_2O \qquad (8\text{--}30)$$

Because $(Cl^-)$ and $(Na^+)$ appear on both sides of the reaction equation just given, they may be disregarded without influencing the result. The reaction between the strong acid and the strong base proceeds almost to completion; however, the weak acid–strong base reaction is incomplete because $Ac^-$ reacts in part with water, that is, it hydrolyzes to regenerate the free acid.

The neutralization of 10 mL of 0.1 N HCl (curve I) and 10 mL of 0.1 N acetic acid (curve II) by 0.1 N NaOH is shown in **Figure 8–1**. The plot of pH versus milliliters of NaOH added produces the titration curve. It is computed as follows for HCl. Before the first increment of NaOH is added, the hydrogen ion concentration of the 0.1 N solution of HCl is $10^{-1}$ mole/liter, and the pH is 1, disregarding activities

and assuming HCl to be completely ionized. The addition of 5 mL of 0.1 N NaOH neutralizes 5 mL of 0.1 N HCl, leaving 5 mL of the original HCl in $10 + 5 = 15$ mL of solution, or $[H_3O^+] = 5/15 \times 0.1 = 3.3 \times 10^{-2}$ mole/liter, and the pH is 1.48. When 10 mL of base has been added, all the HCl is converted to NaCl, and the pH, disregarding the difference between activity and concentration resulting from the ionic strength of the NaCl solution, is 7. This is known as the equivalence point of the titration. Curve I in **Figure 8–1** results from plotting such data. It is seen that the pH does not change markedly until nearly all the HCl is neutralized. Hence, a solution of a strong acid has a high buffer capacity below a pH of 2. Likewise, a strong base has a high buffer capacity above a pH of 12.

The buffer capacity equations considered thus far have pertained exclusively to mixtures of weak electrolytes and their salts. The buffer capacity of a solution of a strong acid was shown by Van Slyke to be directly proportional to the hydrogen ion concentration, or

$$\beta = 2.303[H_3O^+] \qquad (8\text{--}31)$$

The buffer capacity of a solution of a strong base is similarly proportional to the hydroxyl ion concentration,

$$\beta = 2.303[OH^-] \qquad (8\text{--}32)$$

The total buffer capacity of a water solution of a strong acid or base at any pH is the sum of the separate capacities just given, equations **(8–31)** and **(8–32)**, or

$$\beta = 2.303([H_3O^+] + [OH^-]) \qquad (8\text{--}33)$$

### EXAMPLE 8–9

**Calculate Buffer Capacity**

What is the buffer capacity of a solution of hydrochloric acid having a hydrogen ion concentration of $10^{-2}$ mole/liter?

The hydroxyl ion concentration of such a solution is $10^{-12}$, and the total buffer capacity is

$$\beta = 2.303(10^{-2} + 10^{-12})$$
$$\beta = 0.023$$

The $OH^-$ concentration is obviously so low in this case that it may be neglected in the calculation.

Three equations are normally used to obtain the data for the titration curve of a weak acid (curve II of **Fig. 8–1**), although a single equation that is somewhat complicated can be used. Suppose that increments of 0.1 N NaOH are added to 10 mL of a 0.1 N HAc solution.

*(a)* The pH of the solution before any NaOH has been added is obtained from the equation for a weak acid,

$$\text{pH} = \frac{1}{2}pK_a - \frac{1}{2}\log c$$
$$= 2.38 - \frac{1}{2}\log 10^{-1} = 2.88$$

*(b)* At the equivalence point, where the acid has been converted completely into sodium ions and acetate ions, the

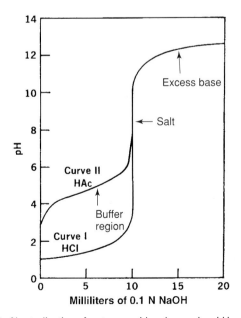

**Fig. 8–1.** Neutralization of a strong acid and a weak acid by a strong base.

pH is computed from the equation for a salt of a weak acid and strong base in log form:

$$pH = \frac{1}{2}pK_w + \frac{1}{2}pK_a + \frac{1}{2}\log c$$

$$= 7.00 + 2.38 + \frac{1}{2}\log(5 \times 10^{-2})$$

$$= 8.73$$

The concentration of the acid is given in the last term of this equation as 0.05 because the solution has been reduced to half its original value by mixing it with an equal volume of base at the equivalence point.

(c) Between these points on the neutralization curve, the increments of NaOH convert some of the acid to its conjugate base $Ac^-$ to form a buffer mixture, and the pH of the system is calculated from the buffer equation. When 5 mL of base is added, the equivalent of 5 mL of 0.1 N acid remains and 5 mL of 0.1 N $Ac^-$ is formed, and using the Henderson–Hasselbalch equation, we obtain

$$pH = pK_a + \log\frac{[Salt]}{[Acid]}$$

$$= 4.76 + \log\frac{5}{5} = 4.76$$

The slope of the curve is a minimum and the buffer capacity is greatest at this point, where the solution shows the smallest pH change per g Eq of base added. The buffer capacity of a solution is the reciprocal of the slope of the curve at a point corresponding to the composition of the buffer solution. As seen in **Figure 8–1**, the slope of the line is a minimum, and the buffer capacity is greatest at half-neutralization, where $pH = pK_a$.

The titration curve for a tribasic acid such as $H_3PO_4$ consists of three stages, as shown in **Figure 8–2**. These can be considered as being produced by three separate acids ($H_3PO_4$, $pK_1 = 2.21$; $H_2PO_4^-$, $pK_2 = 7.21$; and $HPO_4^{2-}$, $pK_3 = 12.67$)

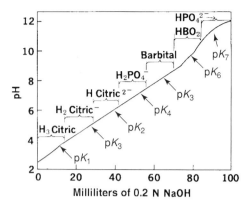

**Fig. 8–3.** Neutralization curve for a universal buffer. (From H. T. Britton, *Hydrogen Ions*, Vol. I, Van Nostrand, New York, 1956, p. 368.)

whose strengths are sufficiently different so that their curves do not overlap. The curves can be plotted by using the buffer equation and their ends joined by smooth lines to produce the continuous curve of **Figure 8–2**.

A mixture of weak acids whose $pK_a$ values are sufficiently alike (differing by no more than about 2 pH units) so that their buffer regions overlap can be used as a *universal buffer* over a wide range of pH values. A buffer of this type was introduced by Britton and Robinson.[6] The three stages of citric acid, $pK_1 = 3.15$, $pK_2 = 4.78$, and $pK_3 = 6.40$, are sufficiently close to provide overlapping of neutralization curves and efficient buffering over this range. Adding $Na_2HPO_4$, whose conjugate acid, $H_2PO_4^-$, has a $pK_2$ of 7.2, diethylbarbituric acid, $pK_1 = 7.91$, and boric acid, $pK_1 = 9.24$, provides a universal buffer that covers the pH range of about 2.4 to 12. The neutralization curve for the universal buffer mixture is linear between pH 4 and 8, as seen in **Figure 8–3**, because the successive dissociation constants differ by only a small value.

A titration curve depends on the ratio of the successive dissociation constants. Theoretically, when one $K$ is equal to or less than 16 times the previous $K$, that is, when successive $pK_s$ do not differ by greater than 1.2 units, the second ionization begins well before the first is completed, and the titration curve is a straight line with no inflection points. Actually, the inflection is not noticeable until one $K$ is about 50 to 100 times that of the previous $K$ value.

The buffer capacity of several acid–salt mixtures is plotted against pH in **Figure 8–4**. A buffer solution is useful within a range of about $\pm 1$ pH unit about the $pK_a$ of its acid, where the buffer capacity is roughly greater than 0.01 or 0.02, as observed in **Figure 8–4**. Accordingly, the acetate buffer should be effective over a pH range of about 3.8 to 5.8, and the borate buffer should be effective over a range of 8.2 to 10.2. In each case, the greatest capacity occurs where [Salt]/[Acid] = 1 and $pH = pK_a$. Because of interionic effects, buffer capacities do not in general exceed a value of 0.2. The buffer capacity of a solution of the strong acid HCl becomes marked below a pH of 2, and the buffer capacity of a strong base NaOH becomes significant above a pH of 12.

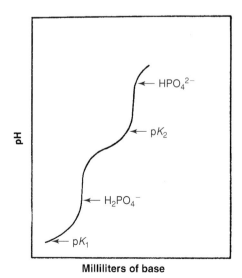

**Millimeters of base**

**Fig. 8–2.** Neutralization of a tribasic acid.

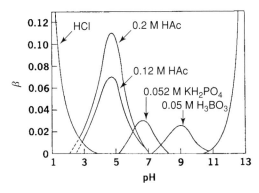

**Fig. 8–4.** The buffer capacity of several buffer systems as a function of pH. (Modified from R. G. Bates, *Electrometric pH Determinations*, Wiley, New York, 1954.)

The buffer capacity of a combination of buffers whose $pK_a$ values overlap to produce a universal buffer is plotted in **Figure 8–5**. It is seen that the total buffer capacity $\sum \beta$ is the sum of the $\beta$ values of the individual buffers. In this figure, it is assumed that the maximum $\beta$s of all buffers in the series are identical.

# BUFFERS IN PHARMACEUTICAL AND BIOLOGIC SYSTEMS

## In Vivo Biologic Buffer Systems

*Blood* is maintained at a pH of about 7.4 by the so-called primary buffers in the plasma and the secondary buffers in the erythrocytes. The plasma contains carbonic acid/bicarbonate and acid/alkali sodium salts of phosphoric acid as buffers. Plasma proteins, which behave as acids in blood, can combine with bases and so act as buffers. In the erythrocytes, the two buffer systems consist of hemoglobin/oxyhemoglobin and acid/alkali potassium salts of phosphoric acid.

The dissociation exponent $pK_1$ for the first ionization stage of carbonic acid in the plasma at body temperature and an ionic strength of 0.16 is about 6.1. The buffer equation for the carbonic acid/bicarbonate buffer of the blood is

$$pH = 6.1 + \log \frac{[HCO_3^-]}{[H_2CO_3]} \qquad (8\text{–}34)$$

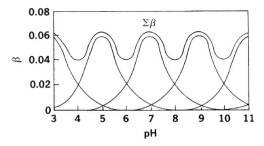

**Fig. 8–5.** The total buffer capacity of a universal buffer as a function of a pH. (From I. M. Kolthoff and C. Rosenblum, *Acid–Base Indicators*, Macmillan, New York, 1937, p. 29.)

where $[H_2CO_3]$ represents the concentration of $CO_2$ present as $H_2CO_3$ dissolved in the blood. At a pH of 7.4, the ratio of bicarbonate to carbonic acid in normal blood plasma is

$$\log \frac{[HCO_3^-]}{[H_2CO_3]} = 7.4 - 6.1 = 1.3$$

or

$$[HCO_3^-]/[H_2CO_3] = 20/1 \qquad (8\text{–}35)$$

This result checks with experimental findings because the actual concentrations of bicarbonate and carbonic acid in the plasma are about 0.025 M and 0.00125 M, respectively.

The buffer capacity of the blood in the physiologic range pH 7.0 to 7.8 is obtained as follows. According to Peters and Van Slyke,[7] the buffer capacity of the blood owing to hemoglobin and other constituents, exclusive of bicarbonate, is about 0.025 g equivalents per liter per pH unit. The pH of the bicarbonate buffer in the blood (i.e., pH 7.4) is rather far removed from the pH (6.1) where it exhibits maximum buffer capacity; therefore, the bicarbonate's buffer action is relatively small with respect to that of the other blood constituents. According to the calculation just given, the ratio $[NaHCO_3]/[H_2CO_3]$ is 20:1 at pH 7.4. Using equation (**8–27**), we find the buffer capacity for the bicarbonate system ($K_1 = 4 \times 10^{-7}$) at a pH of 7.4 ($[H_3O^+] = 4 \times 10^{-8}$) to be roughly 0.003. Therefore, the total buffer capacity of the blood in the physiologic range, the sum of the capacities of the various constituents, is $0.025 + 0.003 = 0.028$. Salenius[8] reported a value of $0.0318 \pm 0.0035$ for whole blood, whereas Ellison et al.[9] obtained a buffer capacity of about 0.039 g equivalents per liter per pH unit for whole blood, of which 0.031 was contributed by the cells and 0.008 by the plasma.

It is usually life-threatening for the pH of the blood to go below 6.9 or above 7.8. The pH of the blood in diabetic coma is as low as about 6.8.

*Lacrimal fluid*, or tears, have been found to have a great degree of buffer capacity, allowing a dilution of 1:15 with neutral distilled water before an alteration of pH is noticed.[10] In the terminology of Bates,[11] this would be referred to today as *dilution value* rather than buffer capacity. The pH of tears is about 7.4, with a range of 7 to 8 or slightly higher. It is generally thought that eye drops within a pH range of 4 to 10 will not harm the cornea.[12] However, discomfort and a flow of tears will occur below pH 6.6 and above pH 9.0.[12] Pure conjunctival fluid is probably more acidic than the tear fluid commonly used in pH measurements. This is because pH increases rapidly when the sample is removed for analysis because of the loss of $CO_2$ from the tear fluid.

## Urine

The 24-hr urine collection of a normal adult has a pH averaging about 6.0 units; it may be as low as 4.5 or as high as 7.8. When the pH of the urine is below normal values, hydrogen ions are excreted by the kidneys. Conversely, when the urine is above pH 7.4, hydrogen ions are retained by action of the kidneys in order to return the pH to its normal range of values.

## Pharmaceutical Buffers

Buffer solutions are used frequently in pharmaceutical practice, particularly in the formulation of ophthalmic solutions. They also find application in the colorimetric determination of pH and for research studies in which pH must be held constant.

Gifford[13] suggested two stock solutions, one containing boric acid and the other monohydrated sodium carbonate, which, when mixed in various proportions, yield buffer solutions with pH values from about 5 to 9.

Sörensen[14] proposed a mixture of the salts of sodium phosphate for buffer solutions of pH 6 to 8. Sodium chloride is added to each buffer mixture to make it isotonic with body fluids.

A buffer system suggested by Palitzsch[15] and modified by Hind and Goyan[16] consists of boric acid, sodium borate, and sufficient sodium chloride to make the mixtures isotonic. It is used for ophthalmic solutions in the pH range of 7 to 9.

The buffers of Clark and Lubs,[17] based on the original pH scale of Sörensen, have been redetermined at 25°C by Bower and Bates[18] so as to conform to the present definition of pH. Between pH 3 and 11, the older values were about 0.04 unit lower than the values now assigned, and at the ends of the scale, the differences were greater. The original values were determined at 20°C, whereas most experiments today are performed at 25°C.

The Clark–Lubs mixtures and their corresponding pH ranges are as follows:

(*a*) HCl and KCl, pH 1.2 to 2.2
(*b*) HCl and potassium hydrogen phthalate, pH 2.2 to 4.0
(*c*) NaOH and potassium hydrogen phthalate, pH 4.2 to 5.8
(*d*) NaOH and $KH_2PO_4$, pH 5.8 to 8.0
(*e*) $H_3BO_3$, NaOH, and KCl, pH 8.0 to 10.0

With regard to mixture (*a*), consisting of HCl and KCl and used for the pH range from 1.0 to 2.2, it will be recalled from the discussion of the neutralization curve I in **Figure 8–1** that HCl alone has considerable buffer efficiency below pH 2. KCl is a neutral salt and is added to adjust the ionic strength of the buffer solutions to a constant value of 0.10; the pH calculated from the equation $-\log a_{H}^{+} = -\log (y \pm c)$ corresponds closely to the experimentally determined pH. The role of the KCl in the Clark–Lubs buffer is sometimes erroneously interpreted as that of a salt of the buffer acid, HCl, corresponding to the part played by sodium acetate as the salt of the weak buffer acid, HAc. Potassium chloride is added to (*e*), the borate buffer, to produce an ionic strength comparable to that of (*d*), the phosphate buffer, where the pH of the two buffer series overlaps.

## KEY CONCEPT　PHOSPHATE BUFFERED SALINE

There are several variations in the formula for preparing PBS. Two common examples follow:

*Formula One*: Take 8 g NaCl, 0.2 g KCl, 1.44 g $Na_2HPO_4$, and 0.24 g $KH_2PO_4$ in 800 mL distilled water. Adjust pH to 7.4 using HCl. Add sufficient (*qs ad*) distilled water to achieve 1 liter.

*Formula Two*: Another variant of PBS. This one is designated as "10X PBS (0.1 M PBS, pH 7.2)" since it is much more concentrated than PBS and the pH is not yet adjusted to pH 7.4. Take 90 g NaCl, 10.9 g $Na_2HPO_4$, and 3.2 g $NaH_2PO_4$ in 1000 mL distilled water. Dilute 1:10 using distilled water and adjust pH as necessary.

Many buffers are available today. One of the most common biological buffers is phosphate buffered saline (PBS). Phosphate buffered saline contains sodium chloride (NaCl) and dibasic sodium phosphate ($Na_2PO_4$). It may also contain potassium chloride (KCl), monobasic potassium phosphate ($KH_2PO_4$), calcium chloride ($CaCl_2$), and magnesium sulfate ($MgSO_4$).

## General Procedures for Preparing Pharmaceutical Buffer Solutions

The pharmacist may be called upon at times to prepare buffer systems for which the formulas do not appear in the literature. The following steps should be helpful in the development of a new buffer.

(*a*) Select a weak acid having a $pK_a$ approximately equal to the pH at which the buffer is to be used. This will ensure maximum buffer capacity.
(*b*) From the buffer equation, calculate the ratio of salt and weak acid required to obtain the desired pH. The buffer equation is satisfactory for approximate calculations within the pH range of 4 to 10.
(*c*) Consider the individual concentrations of the buffer salt and acid needed to obtain a suitable buffer capacity. A *concentration* of 0.05 to 0.5 M is usually sufficient, and a *buffer capacity* of 0.01 to 0.1 is generally adequate.
(*d*) Other factors of some importance in the choice of a pharmaceutical buffer include availability of chemicals, sterility of the final solution, stability of the drug and buffer on aging, cost of materials, and freedom from toxicity. For example, a borate buffer, because of its toxic effects, certainly cannot be used to stabilize a solution to be administered orally or parenterally.
(*e*) Finally, determine the pH and buffer capacity of the completed buffered solution using a reliable pH meter. In some cases, sufficient accuracy is obtained by the use of

pH papers. Particularly when the electrolyte concentration is high, it may be found that the pH calculated by use of the buffer equation is somewhat different from the experimental value. This is to be expected when activity coefficients are not taken into account, and it emphasizes the necessity for carrying out the actual determination.

## Influence of Buffer Capacity and pH on Tissue Irritation

Friedenwald et al.[18] claimed that the pH of solutions for introduction into the eye may vary from 4.5 to 11.5 without marked pain or damage. This statement evidently would be true only if the buffer capacity were kept low. Martin and Mims[19] found that Sörensen's phosphate buffer produced irritation in the eyes of a number of individuals when used outside the narrow pH range of 6.5 to 8, whereas a boric acid solution of pH 5 produced no discomfort in the eyes of the same individuals. Martin and Mims concluded that a pH range of nonirritation cannot be established absolutely but instead depends upon the buffer employed. In light of the previous discussion, this apparent anomaly can be explained partly in terms of the low buffer capacity of boric acid as compared with that of the phosphate buffer and partly to the difference of the physiologic response to various ion species.

Riegelman and Vaughn[20] assumed that the acid-neutralizing power of tears when 0.1 mL of a 1% solution of a drug is instilled into the eye is roughly equivalent to 10 μL of a 0.01 N strong base. They pointed out that although in a few cases, irritation of the eye may result from the presence of the free base form of a drug at the physiologic pH, it is more often due to the acidity of the eye solution. For example, because only one carboxyl group of tartaric acid is neutralized by epinephrine base in epinephrine bitartrate, a 0.06 M solution of the drug has a pH of about 3.5. The prolonged pain resulting from instilling two drops of this solution into the eye is presumably due to the unneutralized acid of the bitartrate, which requires 10 times the amount of tears to restore the normal pH of the eye as compared with the result following two drops of epinephrine hydrochloride. Solutions of pilocarpine salts also possess sufficient buffer capacity to cause pain or irritation owing to their acid reaction when instilled into the eye.

Parenteral solutions for injection into the blood are usually not buffered, or they are buffered to a low capacity so that the buffers of the blood may readily bring them within the physiologic pH range. If the drugs are to be injected only in small quantities and at a slow rate, their solutions can be buffered weakly to maintain approximate neutrality.

According to Mason,[21] following oral administration, aspirin is absorbed more rapidly in systems buffered at low buffer capacity than in systems containing no buffer or in highly buffered preparations. Thus, the buffer capacity of the buffer should be optimized to produce rapid absorption and minimal gastric irritation of orally administered aspirin.

**KEY CONCEPT  PARENTERAL SOLUTIONS**

Solutions to be applied to tissues or administered parenterally are liable to cause irritation if their pH is greatly different from the normal pH of the relevant body fluid. Consequently, the pharmacist must consider this point when formulating ophthalmic solutions, parenteral products, and fluids to be applied to abraded surfaces. Of possible greater significance than the actual pH of the solution is its buffer capacity and the volume to be used in relation to the volume of body fluid with which the buffered solution will come in contact. The buffer capacity of the body fluid should also be considered. Tissue irritation, due to large pH differences between the solution being administered and the physiologic environment in which it is used, will be minimal (*a*) the lower is the buffer capacity of the solution, (*b*) the smaller is the volume used for a given concentration, and (*c*) the larger are the volume and buffer capacity of the physiologic fluid.

In addition to the adjustment of tonicity and pH for ophthalmic preparations, similar requirements are demanded for nasal delivery of drugs. Conventionally, the nasal route has been used for delivery of drugs for treatment of local diseases such as nasal allergy, nasal congestion, and nasal infections.[22] The nasal route can be exploited for the systemic delivery of drugs such as small molecular weight polar drugs, peptides and proteins that are not easily administered via other routes than by injection, or where a rapid onset of action is required. Examples include buserelin, desmopressin, and nafarelin.

## Stability versus Optimum Therapeutic Response

For the sake of completeness, some mention must be made at this point of the effect of buffer capacity and pH on the stability and therapeutic response of the drug being used in solution.

As will be discussed later, the undissociated form of a weakly acidic or basic drug often has a higher therapeutic activity than that of the dissociated salt form. This is because the former is lipid soluble and can penetrate body membranes readily, whereas the ionic form, not being lipid soluble, can penetrate membranes only with greater difficulty. Thus, Swan and White[23] and Cogan and Kinsey[24] observed an increase in therapeutic response of weakly basic alkaloids (used as ophthalmic drugs) as the pH of the solution, and hence concentration of the undissociated base, was increased. At a pH of about 4, these drugs are predominantly in the ionic form, and penetration is slow or insignificant. When the tears bring the pH to about 7.4, the drugs may exist to a significant degree in the form of the free base, depending on the dissociation constant of the drug.

**Mole Percent of Free Base**

The $pK_b$ of pilocarpine is 7.15 at 25°C. Compute the mole percent of free base present at 25°C and at a pH of 7.4. We have

$$C_{11}H_{16}N_2O_2 + H_2O \rightleftharpoons C_{11}H_{16}N_2O_2 + OH^-$$

| Pilocarpine | Pilocarpine |
|:---:|:---:|
| base | ion |

$$pH = pK_w - pK_b + \log \frac{[Base]}{[Salt]}$$

$$7.4 = 14.00 - 7.15 + \log \frac{[Base]}{[Salt]}$$

$$\log \frac{[Base]}{[Salt]} = 7.40 - 14.00 + 7.15 = 0.55$$

$$\frac{[Base]}{[Salt]} = \frac{3.56}{1}$$

$$\text{mole percent of base} = \frac{[Base]}{[Salt] + [Base]} \times 100$$

$$= [3.56/(1 + 3.56)] \times 100 = 78\%$$

Hind and Goyan[25] pointed out that the pH for maximum stability of a drug for ophthalmic use may be far below that of the optimum physiologic effect. Under such conditions, the solution of the drug can be buffered at a low buffer capacity and at a pH that is a compromise between that of optimum stability and the pH for maximum therapeutic action. The buffer is adequate to prevent changes in pH due to the alkalinity of the glass or acidity of $CO_2$ from dissolved air. Yet, when the solution is instilled in the eye, the tears participate in the gradual neutralization of the solution; conversion of the drug occurs from the physiologically inactive form to the undissociated base. The base can then readily penetrate the lipoidal membrane. As the base is absorbed at the pH of the eye, more of the salt is converted into base to preserve the constancy of $pK_b$; hence, the alkaloidal drug is gradually absorbed.

## pH and Solubility

Since the relationship between pH and the solubility of weak electrolytes is treated elsewhere in the book, it is only necessary to point out briefly the influence of buffering on the solubility of an alkaloidal base. At a low pH, a base is predominantly in the ionic form, which is usually very soluble in aqueous media. As the pH is raised, more undissociated base is formed, as calculated by the method illustrated in *Example 8–10*. When the amount of base exceeds the limited water solubility of this form, free base precipitates from solution. Therefore, the solution should be buffered at a sufficiently low pH so that the concentration of alkaloidal base in equilibrium with its salt is calculated to be less than the solu-

bility of the free base at the storage temperature. Stabilization against precipitation can thus be maintained.

# BUFFERED ISOTONIC SOLUTIONS

Reference has already been made to in vivo buffer systems, such as blood and lacrimal fluid, and the desirability for buffering pharmaceutical solutions under certain conditions. In addition to carrying out pH adjustment, pharmaceutical solutions that are meant for application to delicate membranes of the body should also be adjusted to approximately the same osmotic pressure as that of the body fluids. Isotonic solutions cause no swelling or contraction of the tissues with which they come in contact and produce no discomfort when instilled in the eye, nasal tract, blood, or other body tissues. Isotonic sodium chloride is a familiar pharmaceutical example of such a preparation.

The need to achieve isotonic conditions with solutions to be applied to delicate membranes is dramatically illustrated by mixing a small quantity of blood with aqueous sodium chloride solutions of varying tonicity. For example, if a small quantity of blood, defibrinated to prevent clotting, is mixed with a solution containing 0.9 g of NaCl per 100 mL, the cells retain their normal size. The solution has essentially the same salt concentration and hence the same osmotic pressure as the red blood cell contents and is said to be *isotonic* with blood. If the red blood cells are suspended in a 2.0% NaCl solution, the water within the cells passes through the cell membrane in an attempt to dilute the surrounding salt solution until the salt concentrations on both sides of the erythrocyte membrane are identical. This outward passage of water causes the cells to shrink and become wrinkled or *crenated*. The salt solution in this instance is said to be *hypertonic* with respect to the blood cell contents. Finally, if the blood is mixed with 0.2% NaCl solution or with distilled water, water enters the blood cells, causing them to swell and finally burst, with the liberation of hemoglobin. This phenomenon is known as *hemolysis*, and the weak salt solution or water is said to be *hypotonic* with respect to the blood.

The student should appreciate that the red blood cell membrane is not impermeable to all drugs; that is, it is not a perfect semipermeable membrane. Thus, it will permit the passage of not only water molecules but also solutes such as urea, ammonium chloride, alcohol, and boric acid.[26] A 2.0% solution of boric acid has the same osmotic pressure as the blood cell contents when determined by the freezing point method and is therefore said to be *isosmotic* with blood. The molecules of boric acid pass freely through the erythrocyte membrane, however, regardless of concentration. As a result, this solution acts essentially as water when in contact with blood cells. Because it is extremely hypotonic with respect to the blood, boric acid solution brings about rapid hemolysis. Therefore, a solution containing a quantity of drug calculated to be isosmotic with blood is isotonic *only* when

## KEY CONCEPT TONICITY

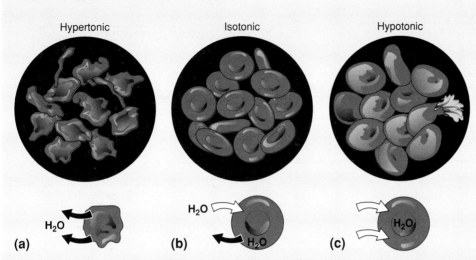

Osmolality and osmolarity are colligative properties that measure the concentration of the solutes independently of their ability to cross a cell membrane. Tonicity is the concentration of only the solutes that cannot cross the membrane since these solutes exert an osmotic pressure on that membrane. Tonicity is *not* the difference between the two osmolarities on opposing sides of the membrane. A solution might be hypertonic, isotonic, or hypotonic relative to another solution. For example, the relative tonicity of blood is defined in reference to that of the red blood cell (RBC) cytosol tonicity. As such, a hypertonic solution contains a higher concentration of impermeable solutes than the cytosol of the

RBC; there is a net flow of fluid out of the RBC and it shrinks (Panel A). The concentration of impermeable solutes in the solution and cytosol are equal and the RBCs remain unchanged, so there is no net fluid flow (Panel B). A hypotonic solution contains a lesser concentration of such solutes than the RBC cytosol and fluid flows into the cells where they swell and potentially burst (Panel C). In short, a solution containing a quantity of drug calculated to be isosmotic with blood is isotonic *only* when the blood cells are impermeable to the solute (drug) molecules and permeable to the solvent, water.

the blood cells are impermeable to the solute molecules and permeable to the solvent, water. It is interesting to note that the mucous lining of the eye acts as a true semipermeable membrane to boric acid in solution. Accordingly, a 2.0% boric acid solution serves as an isotonic ophthalmic preparation.

To overcome this difficulty, Husa[27] suggested that the term *isotonic* should be restricted to solutions having equal osmotic pressures with respect to a particular membrane. Goyan and Reck[28] felt that, rather than restricting the use of the term in this manner, a new term should be introduced that is defined on the basis of the sodium chloride concentration. These workers defined the term *isotonicity value* as the concentration of an aqueous NaCl solution having the same colligative properties as the solution in question. Although all solutions having an isotonicity value of 0.9 g of NaCl per 100 mL of solution need not *necessarily* be isotonic with respect to the living membranes concerned, many of them are roughly isotonic in this sense, and all may be considered isotonic across an ideal membrane. Accordingly, the term *isotonic* is used with this meaning throughout the present chapter. Only a few substances—those that penetrate animal membranes at a sufficient rate—will show exception to this classification.

The remainder of this chapter is concerned with a discussion of isotonic solutions and the means by which they can be buffered.

## Measurement of Tonicity

The tonicity of solutions can be determined by one of two methods. First, in the *hemolytic* method, the effect of various solutions of the drug is observed on the appearance of red blood cells suspended in the solutions. The various effects produced have been described in the previous section. Husa and his associates[27] used this method. In their later work, a quantitative method developed by Hunter[29] was used based on the fact that a hypotonic solution liberates oxyhemoglobin in direct proportion to the number of cells hemolyzed. By such means, the van't Hoff *i* factor can be determined and the value compared with that computed from cryoscopic data, osmotic coefficient, and activity coefficient.[30]

Husa found that a drug having the proper *i* value as measured by freezing point depression or computed from theoretical equations nevertheless may hemolyze human red blood cells; it was on this basis that he suggested restriction of the term *isotonic* to solutions having equal osmotic pressures with respect to a particular membrane.

**TABLE 8–3**

## AVERAGE $L_{iso}$ VALUES FOR VARIOUS IONIC TYPES*

| Type | $L_{iso}$ | Examples |
|---|---|---|
| Nonelectrolytes | 1.9 | Sucrose, glycerin, urea, camphor |
| Weak electrolytes | 2.0 | Boric acid, cocaine, phenobarbital |
| Di-divalent electrolytes | 2.0 | Magnesium sulfate, zinc sulfate |
| Uni-univalent electrolytes | 3.4 | Sodium chloride, cocaine hydrochloride, sodium phenobarbital |
| Uni-divalent electrolytes | 4.3 | Sodium sulfate, atropine sulfate |
| Di-univalent electrolytes | 4.8 | Zinc chloride, calcium bromide |
| Uni-trivalent electrolytes | 5.2 | Sodium citrate, sodium phosphate |
| Tri-univalent electrolytes | 6.0 | Aluminum chloride, ferric iodide |
| Tetraborate electrolytes | 7.6 | Sodium borate, potassium borate |

*From J. M. Wells, J. Am. Pharm. Assoc. Pract. Ed. **5**, 99, 1944.

The second approach used to measure tonicity is based on any of the methods that determine colligative properties earlier in the book. Goyan and Reck[28] investigated various modifications of the Hill–Baldes technique[31] for measuring tonicity. This method is based on a measurement of the slight temperature differences arising from differences in the vapor pressure of thermally insulated samples contained in constant-humidity chambers.

One of the first references to the determination of the freezing point of blood and tears (as was necessary to make solutions isotonic with these fluids) is that of Lumiere and Chevrotier,[32] in which the values of $-0.56°C$ and $-0.80°C$ were given, respectively, for the two fluids. Following work by Pedersen-Bjergaard and coworkers,[33,34] however, it is now well established that $-0.52°C$ is the freezing point of both human blood and lacrimal fluid. This temperature corresponds to the freezing point of a 0.90% NaCl solution, which is therefore considered to be isotonic with both blood and lacrimal fluid.

## Calculating Tonicity Using $L_{iso}$ Values

Because the freezing point depressions for solutions of electrolytes of both the weak and strong types are always greater than those calculated from the equation $\Delta T_f = K_f c$, a new factor, $L = i K_f$, is introduced to overcome this difficulty.[35] The equation, already discussed is

$$\Delta T_f = Lc \qquad (8-36)$$

The $L$ value can be obtained from the freezing point lowering of solutions of representative compounds of a given ionic type at a concentration $c$ that is isotonic with body fluids. This specific value of $L$ is written as $L_{iso}$.

The $L_{iso}$ value for a 0.90% (0.154 M) solution of sodium chloride, which has a freezing point depression of $0.52°C$ and is thus isotonic with body fluids, is 3.4: From

$$L_{iso} = \frac{\Delta T_f}{c} \qquad (8-37)$$

we have

$$L_{iso} = \frac{0.52°C}{0.154} = 3.4$$

The interionic attraction in solutions that are not too concentrated is roughly the same for all uni-univalent electrolytes regardless of the chemical nature of the various compounds of this class, and all have about the same value for $L_{iso}$, namely 3.4. As a result of this similarity between compounds of a given ionic type, a table can be arranged listing the $L$ value for each class of electrolytes at a concentration that is isotonic with body fluids. The $L_{iso}$ values obtained in this way are given in **Table 8–3**.

It will be observed that for dilute solutions of nonelectrolytes, $L_{iso}$ is approximately equal to $K_f$. **Table 8–3** is used to obtain the approximate $\Delta T_f$ for a solution of a drug if the ionic type can be correctly ascertained. A plot of $i K_f$ against molar concentration of various types of electrolytes, from which the values of $L_{iso}$ can be read, is shown in Figure 6–7 (in Chapter 6, "Electrolytes and Ionic Equilibria").

**EXAMPLE 8–11**

**Freezing Point Lowering**

What is the freezing point lowering of a 1% solution of sodium propionate (molecular weight 96)? Because sodium propionate is a uni-univalent electrolyte, its $L_{iso}$ value is 3.4. The molar concentration of a 1% solution of this compound is 0.104. We have

$$\Delta T_f = 3.4 \times 0.104 = 0.35°C \qquad (8-38)$$

Although 1 g/100 mL of sodium propionate is not the isotonic concentration, it is still proper to use $L_{iso}$ as a simple average that agrees with the concentration range expected for the finished solution. The selection of $L$ values in this concentration region is not sensitive to minor changes in concentration; no pretense to accuracy greater than about 10% is implied or needed in these calculations.

The calculation of *Example 8–11* can be simplified by expressing molarity c as grams of drug contained in a definite volume of solution. Thus,

$$\text{Molarity} = \frac{\text{Moles}}{\text{Liter}}$$

$$= \frac{\text{Weight in grams}}{\substack{\text{Molecular weight} \\ \text{in g/mole}}} \div \frac{\text{Volume in mL}}{1000\,\text{mL/liter}} \quad (8\text{--}39)$$

or

$$c = \frac{w}{\text{MW}} \times \frac{1000}{v} \quad (8\text{--}40)$$

where $w$ is the grams of solute, MW is the molecular weight of the solute, and $v$ is the volume of solution in milliliters. Substituting in equation **(8–36)** gives

$$\Delta T_f = L_{\text{iso}} \times \frac{w \times 1000}{\text{MW} \times v} \quad (8\text{--}41)$$

The problem in *Example 8–11* can be solved in one operation by the use of equation **(8–41)** without the added calculation needed to obtain the molar concentration:

$$\Delta T_f = 3.4 \times \frac{1 \times 1000}{96 \times 100} = 3.4 \times 0.104$$

$$= 0.35°C$$

The student is encouraged to derive expressions of this type; certainly equations **(8–40)** and **(8–41)** should not be memorized, for they are not remembered for long. The $L_{\text{iso}}$ values can also be used for calculating sodium chloride equivalents and Sprowls $V$ values, as discussed in subsequent sections of this chapter.

# METHODS OF ADJUSTING TONICITY AND pH

One of several methods can be used to calculate the quantity of sodium chloride, dextrose, and other substances that may be added to solutions of drugs to render them isotonic.

For discussion purposes, the methods are divided into two classes. In the class I methods, sodium chloride or some other substance is added to the solution of the drug to lower the freezing point of the solution to $-0.52°C$ and thus make it isotonic with body fluids. Under this class are included the *cryoscopic* method and the *sodium chloride equivalent* method. In the class II methods, water is added to the drug in a sufficient amount to form an isotonic solution. The preparation is then brought to its final volume with an isotonic or a buffered isotonic dilution solution. Included in this class are the *White–Vincent* method and the *Sprowls* method.

## Class I Methods

### Cryoscopic Method

The freezing point depressions of a number of drug solutions, determined experimentally or theoretically, are given in **Table 8–4**. According to the previous section, the freezing point depressions of drug solutions that have not been determined experimentally can be estimated from theoretical considerations, knowing only the molecular weight of the drug and the $L_{\text{iso}}$ value of the ionic class.

The calculations involved in the cryoscopic method are explained best by an example.

### EXAMPLE 8–12

**Isotonicity**

How much sodium chloride is required to render 100 mL of a 1% solution of apomorphine hydrochloride isotonic with blood serum?

From Table 8–4 it is found that a 1% solution of the drug has a freezing point lowering of 0.08°C. To make this solution isotonic with blood, sufficient sodium chloride must be added to reduce the freezing point by an additional 0.44°C (0.52°C − 0.08°C). In the freezing point table, it is also observed that a 1% solution of sodium chloride has a freezing point lowering of 0.58°C. By the method of proportion,

$$\frac{1\%}{X} = \frac{0.58°}{0.44°}; X = 0.76\%$$

Thus, 0.76% sodium chloride will lower the freezing point the required 0.44°C and will render the solution isotonic. The solution is prepared by dissolving 1.0 g of apomorphine hydrochloride and 0.76 g of sodium chloride in sufficient water to make 100 mL of solution.

### Sodium Chloride Equivalent Method

A second method for adjusting the tonicity of pharmaceutical solutions was developed by Mellen and Seltzer.[36] The *sodium chloride equivalent* or, as referred to by these workers, the "tonicic equivalent" of a drug is the amount of sodium chloride that is equivalent to (i.e., has the same osmotic effect as) 1 g, or other weight unit, of the drug. The sodium chloride equivalents $E$ for a number of drugs are listed in **Table 8–4**.

When the $E$ value for a new drug is desired for inclusion in **Table 8–4**, it can be calculated from the $L_{\text{iso}}$ value or freezing point depression of the drug according to formulas derived by Goyan et al.[37] For a solution containing 1 g of drug in 1000 mL of solution, the concentration $c$ expressed in moles/liter can be written as

$$c = \frac{1\,\text{g}}{\text{Molecular weight}} \quad (8\text{--}42)$$

and from equation **(8–36)**

$$\Delta T_f = L_{\text{iso}} \frac{1\,\text{g}}{\text{MW}}$$

Now, $E$ is the weight of NaCl with the same freezing point depression as 1 g of the drug, and for a NaCl solution containing $E$ grams of drug per 1000 mL,

$$\Delta T_f = 3.4 \frac{E}{58.45} \quad (8\text{--}43)$$

where 3.4 is the $L_{\text{iso}}$ value for sodium chloride and 58.45 is its molecular weight. Equating these two values of $\Delta T_f$ yields

$$\frac{L_{\text{iso}}}{\text{MW}} = 3.4 \frac{E}{58.45} \quad (8\text{--}44)$$

$$E \cong 17 \frac{L_{\text{iso}}}{\text{MW}} \quad (8\text{--}45)$$

**TABLE 8–4**
**ISOTONIC VALUES*,†**

| Substance | MW | E | V | $\Delta T_f^{1\%}$ | $L_{iso}$ |
|---|---|---|---|---|---|
| Alcohol, dehydrated | 46.07 | 0.70 | 23.3 | 0.41 | 1.9 |
| Aminophylline | 456.46 | 0.17 | 5.7 | 0.10 | 4.6 |
| Amphetamine sulfate | 368.49 | 0.22 | 7.3 | 0.13 | 4.8 |
| Antipyrine | 188.22 | 0.17 | 5.7 | 0.10 | 1.9 |
| Apomorphine hydrochloride | 312.79 | 0.14 | 4.7 | 0.08 | 2.6 |
| Ascorbic acid | 176.12 | 0.18 | 6.0 | 0.11 | 1.9 |
| Atropine sulfate | 694.82 | 0.13 | 4.3 | 0.07 | 5.3 |
| Diphenhydramine hydrochloride | 291.81 | 0.20 | 6.6 | 0.34 | 3.4 |
| Boric acid | 61.84 | 0.50 | 16.7 | 0.29 | 1.8 |
| Caffeine | 194.19 | 0.08 | 2.7 | 0.05 | 0.9 |
| Dextrose $\cdot$ $H_2O$ | 198.17 | 0.16 | 5.3 | 0.09 | 1.9 |
| Ephedrine hydrochloride | 201.69 | 0.30 | 10.0 | 0.18 | 3.6 |
| Ephedrine sulfate | 428.54 | 0.23 | 7.7 | 0.14 | 5.8 |
| Epinephrine hydrochloride | 219.66 | 0.29 | 9.7 | 0.17 | 3.7 |
| Glycerin | 92.09 | 0.34 | 11.3 | 0.20 | 1.8 |
| Lactose | 360.31 | 0.07 | 2.3 | 0.04 | 1.7 |
| Morphine hydrochloride | 375.84 | 0.15 | 5.0 | 0.09 | 3.3 |
| Morphine sulfate | 758.82 | 0.14 | 4.8 | 0.08 | 6.2 |
| Neomycin sulfate | – | 0.11 | 3.7 | 0.06 | – |
| Penicillin G potassium | 372.47 | 0.18 | 6.0 | 0.11 | 3.9 |
| Penicillin G Procaine | 588.71 | 0.10 | 3.3 | 0.06 | 3.5 |
| Phenobarbital sodium | 254.22 | 0.24 | 8.0 | 0.14 | 3.6 |
| Phenol | 94.11 | 0.35 | 11.7 | 0.20 | 1.9 |
| Potassium chloride | 74.55 | 0.76 | 25.3 | 0.45 | 3.3 |
| Procaine hydrochloride | 272.77 | 0.21 | 7.0 | 0.12 | 3.4 |
| Quinine hydrochloride | 396.91 | 0.14 | 4.7 | 0.08 | 3.3 |
| Sodium chloride | 58.45 | 1.00 | 33.3 | 0.58 | 3.4 |
| Streptomycin sulfate | 1457.44 | 0.07 | 2.3 | 0.04 | 6.0 |
| Sucrose | 342.30 | 0.08 | 2.7 | 0.05 | 1.6 |
| Tetracycline hydrochloride | 480.92 | 0.14 | 4.7 | 0.08 | 4.0 |
| Urea | 60.06 | 0.59 | 19.7 | 0.35 | 2.1 |
| Zinc chloride | 139.29 | 0.62 | 20.3 | 0.37 | 5.1 |

*The values were obtained from the data of E. R. Hammarlund and K. Pedersen-Bjergaard, J. Am. Pharm. Assoc. Pract. Ed. **19,** 39, 1958; J. Am. Pharm. Assoc. Sci. Ed. **47,** 107, 1958; and other sources. The values vary somewhat with concentration, and those in the table are for 1% to 3% solutions of the drugs in most instances. A complete table of E and $\Delta T_f$ values is found in the *Merck Index*, 11th Ed., Merck, Rahway, N. J., 1989, pp. MISC-79 to MISC-103. For the most recent results of Hammarlund, see J. Pharm. Sci. **70,** 1161, 1981; **78,** 519, 1989.
*Key:* MW = molecular weight of the drug; E = sodium chloride equivalent of the drug; V = volume in mL of isotonic solution that can be prepared by adding water to 0.3 g of the drug (the weight of drug in 1 fluid ounce of a 1% solution); $\Delta T_f^1\%$ = freezing point depression of a 1% solution of the drug; $L_{iso}$ = the molar freezing point depression of the drug at a concentration approximately isotonic with blood and lacrimal fluid.
†The full table is available at the book's companion website at thepoint.lww.com/Sinko6e.

**EXAMPLE 8–13**

**Sodium Chloride Equivalents**

Calculate the approximate E value for a new amphetamine hydrochloride derivative (molecular weight 187).

Because this drug is a uni-univalent salt, it has an $L_{iso}$ value of 3.4. Its E value is calculated from equation (8–45):

$$E = 17\frac{3.4}{187} = 0.31$$

Calculations for determining the amount of sodium chloride or other inert substance to render a solution isotonic (across an ideal membrane) simply involve multiplying the quantity of each drug in the prescription by its sodium chloride equivalent and subtracting this value from the concentration of sodium chloride that is isotonic with body fluids, namely, 0.9 g/100 mL.

**EXAMPLE 8–14**

**Tonicity Adjustment**

A solution contains 1.0 g of ephedrine sulfate in a volume of 100 mL. What quantity of sodium chloride must be added to make the solution isotonic? How much dextrose would be required for this purpose?

The quantity of the drug is multiplied by its sodium chloride equivalent, E, giving the weight of sodium chloride to which the quantity of drug is equivalent in osmotic pressure:

Ephedrine sulfate: **1.0 g × 0.23 = 0.23 g**

The ephedrine sulfate has contributed a weight of material osmotically equivalent to 0.23 g of sodium chloride. Because a total of 0.9 g of sodium chloride is required for isotonicity, 0.67 g (0.90 − 0.23 g) of NaCl must be added.

If one desired to use dextrose instead of sodium chloride to adjust the tonicity, the quantity would be estimated by setting up the following proportion. Because the sodium chloride equivalent of dextrose is 0.16,

$$\frac{1 \text{ g dextrose}}{0.16 \text{g NaCl}} = \frac{X}{0.67 \text{ g NaCl}}$$

$$X = 4.2 \text{ g of dextrose}$$

Other agents than dextrose can of course be used to replace NaCl. It is recognized that thimerosal becomes less stable in eye drops when a halogen salt is used as an "isotonic agent" (i.e., an agent like NaCl ordinarily used to adjust the tonicity of a drug solution). Reader[38] found that mannitol, propylene glycol, or glycerin—isotonic agents that did not have a detrimental effect on the stability of thimerosal—could serve as alternatives to sodium chloride. The concentration of these agents for isotonicity is readily calculated by use of the equation (see *Example 8–14*)

$$X = \frac{Y(\text{Additional amount of NaCl for isotonicity})}{E(\text{Grams of NaCl equivalent to 1 g of the isotonic agent})}$$

$$(8\text{–}46)$$

where $X$ is the grams of isotonic agent required to adjust the tonicity, $Y$ is the additional amount of NaCl for isotonicity over and above the osmotic equivalence of NaCl provided by the drugs in the solution, and $E$ is the sodium chloride equivalence of the isotonic agent.

**EXAMPLE 8–15**

**Isotonic Solutions**

Let us prepare 200 mL of an isotonic aqueous solution of thimerosal, molecular weight 404.84 g/mole. The concentration of this anti-infective drug is 1:5000, or 0.2 g/1000 mL. The $L_{iso}$ for such a compound, a salt of a weak acid and a strong base (a 1:1 electrolyte), is 3.4, and the sodium chloride equivalent $E$ is

$$E = 17\frac{L_{iso}}{\text{MW}} = 17\frac{3.4}{404.84} = 0.143$$

The quantity of thimerosal, 0.04 g for the 200-mL solution, multiplied by its $E$ value gives the weight of NaCl to which the drug is osmotically equivalent:

$$0.04 \text{ g thimerosal} \times 0.143 = 0.0057 \text{ g of NaCl}$$

Because the total amount of NaCl needed for isotonicity is 0.9 g/100 mL, or 1.8 g for the 200-mL solution, and because an equivalent of 0.0057 g of NaCl has been provided by the thimerosal, the additional amount of NaCl needed for isotonicity, $Y$, is

$$Y = 1.80 \text{ g NaCl needed} - 0.0057 \text{ g NaCl supplied by the drug}$$
$$= 1.794 \text{ g}$$

This is the additional amount of NaCl needed for isotonicity. The result, 1.8 g of NaCl, shows that the concentration of thimerosal is so small that it contributes almost nothing to the isotonicity of the solution. Thus, a concentration of 0.9% NaCl, or 1.8 g/200 mL, is required.

However, from the work of Reader[38] we know that sodium chloride interacts with mercury compounds such as thimerosal to reduce the stability and effectiveness of this preparation. Therefore, we replace NaCl with propylene glycol as the isotonic agent.

From equation (8–45) we calculate the $E$ value of propylene glycol, a nonelectrolyte with an $L_{iso}$ value of 1.9 and a molecular weight of 76.09 g/mole:

$$E = 17\frac{1.9}{76.09} = 0.42$$

Using equation (8–46), $X = Y/E$, we obtain

$$X = \frac{1.794}{0.42} = 4.3 \text{ g}$$

where $X = 4.3$ g is the amount of propylene glycol required to adjust the 200-mL solution of thimerosal to isotonicity.

## Class II Methods

### White–Vincent Method

The class II methods of computing tonicity involve the addition of water to the drugs to make an isotonic solution, followed by the addition of an isotonic or isotonic-buffered diluting vehicle to bring the solution to the final volume. Stimulated by the need to adjust the pH in addition to the tonicity of ophthalmic solutions, White and Vincent[39] developed a simplified method for such calculations. The derivation of the equation is best shown as follows.

Suppose that one wishes to make 30 mL of a 1% solution of procaine hydrochloride isotonic with body fluid. First, the weight of the drug, $w$, is multiplied by the sodium chloride equivalent, $E$:

$$0.3 \text{ g} \times 0.21 = 0.063 \text{ g} \qquad (8\text{–}47)$$

This is the quantity of sodium chloride osmotically equivalent to 0.3 g of procaine hydrochloride.

Second, it is known that 0.9 g of sodium chloride, when dissolved in enough water to make 100 mL, yields a solution that is isotonic. The volume, $V$, of isotonic solution that can be prepared from 0.063 g of sodium chloride (equivalent to 0.3 g of procaine hydrochloride) is obtained by solving the proportion

$$\frac{0.9 \text{ g}}{100 \text{ mL}} = \frac{0.063 \text{ g}}{V} \qquad (8\text{–}48)$$

$$V = 0.063 \times \frac{100}{0.9} \qquad (8\text{–}49)$$

$$V = 7.0 \text{ mL} \qquad (8\text{–}50)$$

In equation (8–49), the quantity 0.063 is equal to the weight of drug, $w$, multiplied by the sodium chloride equivalent, $E$, as seen in equation (8–47). The value of the ratio 100/0.9 is 111.1. Accordingly, equation (8–49) can be written as

$$V = w \times E \times 111.1 \qquad (8\text{–}51)$$

where $V$ is the volume in milliliters of isotonic solution that may be prepared by mixing the drug with water, $w$ is the weight in grams of the drug given in the problem, and $E$ is the sodium chloride equivalent obtained from **Table 8–4**. The constant, 111.1, represents the volume in milliliters of isotonic solution obtained by dissolving 1 g of sodium chloride in water.

The problem can be solved in one step using equation (8–51):

$$V = 0.3 \times 0.21 \times 111.1$$
$$V = 7.0 \, \text{mL}$$

To complete the isotonic solution, enough isotonic sodium chloride solution, another isotonic solution, or an isotonic-buffered diluting solution is added to make 30 mL of the finished product.

When more than one ingredient is contained in an isotonic preparation, the volumes of isotonic solution, obtained by mixing each drug with water, are additive.

### EXAMPLE 8–16

**Isotonic Solutions**

Make the following solution isotonic with respect to an ideal membrane:

Phenacaine hydrochloride ................................. 0.06 g
Boric acid .................................................. 0.30 g
Sterilized distilled water, enough to make .............. 100.0 mL

$$V = [(0.06 \times 0.20) + (0.3 \times 0.50)] \times 111.1$$
$$V = 18 \, \text{mL}$$

The drugs are mixed with water to make 18 mL of an isotonic solution, and the preparation is brought to a volume of 100 mL by adding an isotonic diluting solution.

### The Sprowls Method

A further simplification of the method of White and Vincent was introduced by Sprowls.[40] He recognized that equation (8–51) could be used to construct a table of values of $V$ when the weight of the drug, $w$, was arbitrarily fixed. Sprowls chose as the weight of drug 0.3 g, the quantity for 1 fluid ounce of a 1% solution. The volume, $V$, of isotonic solution that can be prepared by mixing 0.3 g of a drug with sufficient water can be computed for drugs commonly used in ophthalmic and parenteral solutions. The method as described by Sprowls[40] is further discussed in several reports by Martin and Sprowls.[41] The table can be found in the *United States Pharmacopeia*. A modification of the original table was made by Hammarlund and Pedersen-Bjergaard[42] and the values of $V$ are given in column 4 of **Table 8–4**, where the volume in milliliters of isotonic solution for 0.3 g of the drug, the quantity for 1 fluid ounce of a 1% solution, is listed. (The volume of isotonic solution in milliliters for 1 g of the drug can also be listed in tabular form if desired by multiplying the values in column 4 by 3.3.) The primary quantity of isotonic solution is finally brought to the specified volume with the desired isotonic or isotonic-buffered diluting solutions.

## CHAPTER SUMMARY

Buffers are compounds or mixtures of compounds that, by their presence in solution, resist changes in pH upon the addition of small quantities of acid or alkali. The resistance to a change in pH is known as *buffer action*. If a small amount of a strong acid or base is added to water or a solution of sodium chloride, the pH is altered considerably; such systems have no buffer action. In this chapter, the theory of buffers was introduced as were several formulas for making commonly used buffers. Finally, the important concept of tonicity was introduced. Pharmaceutical buffers must usually be made isotonic so that they cause no swelling or contraction of biological tissues, which would lead to discomfort in the patient being treated.

 **Practice problems for this chapter can be found at thePoint.lww.com/Sinko6e.**

## References

1. A. Roos and W. F. Borm, Respir. Physiol. **40**, 1980; M. Koppel and K. Spiro, Biochem. Z. **65**, 409, 1914.
2. D. D. Van Slyke, J. Biol. Chem. **52**, 525, 1922.
3. R. G. Bates, *Electrometric pH Determinations*, Wiley, New York, 1954, pp. 97–104, 116, 338.
4. I. M. Kolthoff and F. Tekelenburg, Rec. Trav. Chim. **46**, 33, 1925.
5. X. Kong, T. Zhou, Z. Liu, and R. C. Hider, J. Pharm. Sci. **96**, 2777, 2007.
6. H. T. S. Britton and R. A. Robinson, J. Chem. Soc. 458, 1931. DOI:10.1039/JR9310000458.
7. J. P. Peters and D. D. Van Slyke, *Quantitative Clinical Chemistry*, Vol. 1, Williams & Wilkins, Baltimore, 1931, Chapter 18.
8. P. Salenius, Scand. J. Clin. Lab. Invest. **9**, 160, 1957.
9. G. Ellison et al., Clin. Chem. **4**, 453, 1958.
10. G. N. Hosford and A. M. Hicks, Arch. Ophthalmol. **13**, 14, 1935; **17**, 797, 1937.
11. R. G. Bates, *Electrometric pH Determinations*, Wiley, New York, 1954, pp. 97–108.
12. D. M. Maurice, Br. J. Ophthalmol. **39**, 463–473, 1955.
13. S. R. Gifford, Arch. Ophthalmol. **13**, 78, 1935.
14. S. L. P. Sörensen, Biochem. Z. **21**, 131, 1909; **22**, 352, 1909.
15. S. Palitzsch, Biochem. Z. **70**, 333, 1915.
16. H. W. Hind and F. M. Goyan, J. Am. Pharm. Assoc. Sci. Ed. **36**, 413, 1947.
17. W. M. Clark and H. A. Lubs, J. Bacteriol. **2**, 1, 109, 191, 1917.
18. J. S. Friedenwald, W. F. Hughes, and H. Herrman, Arch. Ophthalmol. **31**, 279, 1944; V. E. Bower and R. G. Bates, J. Res. NBS **55**, 197, 1955.
19. F. N. Martin and J. L. Mims, Arch. Ophthalmol. **44**, 561, 1950; J. L. Mims, Arch. Ophthalmol. **46**, 644, 1951.
20. S. Riegelman and D. G. Vaughn, J. Am. Pharm. Assoc. Pract. Ed. **19**, 474, 1958.
21. W. D. Mason, J. Pharm. Sci. **73**, 1258, 1984.
22. L. Illum, J. Control. Release. **87**, 187, 2003.
23. K. C. Swan and N. G. White, Am. J. Ophthalmol. **25**, 1043, 1942.
24. D. G. Cogan and V. E. Kinsey, Arch. Ophthalmol. **27**, 466, 661, 696, 1942.
25. H. W. Hind and F. M. Goyan, J. Am. Pharm. Assoc. Sci. Ed. **36**, 33, 1947.
26. Y. Takeuchi, Y. Yamaoka, Y. Morimoto, I. Kaneko, Y. Fukumori, and T. Fukuda, J. Pharm. Sci. **78**, 3, 1989.
27. T. S. Grosicki and W. J. Husa, J. Am. Pharm. Assoc. Sci. Ed. **43**, 632, 1954; W. J. Husa and J. R. Adams, J. Am. Pharm. Assoc. Sci. Ed. **33**, 329, 1944; W. D. Easterly and W. J. Husa, J. Am. Pharm. Assoc. Sci. Ed. **43**, 750, 1954; W. J. Husa and O. A. Rossi, J. Am. Pharm. Assoc. Sci. Ed. **31**, 270, 1942.
28. F. M. Goyan and D. Reck, J. Am. Pharm. Assoc. Sci. Ed. **44**, 43, 1955.
29. F. R. Hunter, J. Clin. Invest. **19**, 691, 1940.
30. C. W. Hartman and W. J. Husa, J. Am. Pharm. Assoc. Sci. Ed. **46**, 430, 1957.

31. A. V. Hill, Proc. R. Soc. Lond. A **127**, 9, 1930; E. J. Baldes, J. Sci. Instrum. **11**, 223, 1934; A. Weissberger, *Physical Methods of Organic Chemistry*, 2nd Ed., Vol. 1, Part 1, Interscience, New York, 1949, p. 546.
32. A. Lumiere and J. Chevrotier, Bull. Sci. Pharmacol. **20**, 711, 1913.
33. C. G. Lund, P. Nielsen, and K. Pedersen-Bjergaard, *The Preparation of Solutions Isosmotic with Blood, Tears, and Tissue* (Danish Pharmacopeia Commission), Vol. 2, Einar Munksgaard, Copenhagen, 1947.
34. A. Krogh, C. G. Lund, and K. Pedersen-Bjergaard, Acta Physiol. Scand. **10**, 88, 1945.
35. F. M. Goyan, J. M. Enright, and J. M. Wells, J. Am. Pharm. Assoc. Sci. Ed. **33**, 74, 1944.
36. M. Mellen and L. A. Seltzer, J. Am. Pharm. Assoc. Sci. Ed. **25**, 759, 1936.
37. F. M. Goyan, J. M. Enright, and J. M. Wells, J. Am. Pharm. Assoc. Sci. Ed. **33**, 78, 1944.
38. M. J. Reader, J. Pharm. Sci. **73**, 840, 1984.
39. A. I. White and H. C. Vincent, J. Am. Pharm. Assoc. Pract. Ed. **8**, 406, 1947.
40. J. B. Sprowls, J. Am. Pharm. Assoc. Pract. Ed. **10**, 348, 1949.
41. A. Martin and J. B. Sprowls, Am. Profess. Pharm. **17**, 540, 1951; Pa Pharm. **32**, 8, 1951.
42. E. R. Hammarlund and K. Pedersen-Bjergaard, J. Am. Pharm. Assoc. Pract. Ed. **19**, 38, 1958.

## Recommended Readings

R. J. Benyon, *Buffer Solutions (The Basics)*, BIOS Scientific Publishers, Oxford, UK, 1996.

### CHAPTER LEGACY

**Fifth Edition:** published as Chapter 9 (Buffered and Isotonic Solutions). Updated by Patrick Sinko.

**Sixth Edition:** published as Chapter 8 (Buffered and Isotonic Solutions). Updated by Patrick Sinko.

# 9 SOLUBILITY AND DISTRIBUTION PHENOMENA

**CHAPTER OBJECTIVES** **At the conclusion of this chapter the student should be able to:**

**1** Define saturated solution, solubility, and unsaturated solution.

**2** Describe and give examples of polar, nonpolar, and semipolar solvents.

**3** Define complete and partial miscibility.

**4** Understand the factors controlling the solubility of weak electrolytes.

**5** Describe the influence of solvents and surfactants on solubility.

**6** Define thermodynamic, kinetic, and intrinsic solubility.

**7** Measure thermodynamic solubility.

**8** Describe what a distribution coefficient and partition coefficient are and their importance in pharmaceutical systems.

## GENERAL PRINCIPLES

### Introduction[1-4]

*Solubility* is defined in quantitative terms as the concentration of solute in a saturated solution at a certain temperature, and in a qualitative way, it can be defined as the spontaneous interaction of two or more substances to form a homogeneous molecular dispersion. Solubility is an *intrinsic* material property that can be altered only by chemical modification of the molecule.[1] In contrast to this, dissolution is an *extrinsic* material property that can be influenced by various chemical, physical, or crystallographic means such as complexation, particle size, surface properties, solid-state modification, or solubilization enhancing formulation strategies.[1] Dissolution is discussed in Chapter 13. Generally speaking, the solubility of a compound depends on the physical and chemical properties of the solute and the solvent as well as on such factors as temperature, pressure, the pH of the solution, and, to a lesser extent, the state of subdivision of the solute. Of the nine possible types of mixtures, based on the three states of matter, only liquids in liquids and solids in liquids are of everyday importance to most pharmaceutical scientists and will be considered in this chapter.

For the most part, this chapter will deal with the *thermodynamic solubility* of drugs (**Fig. 9–1**). The thermodynamic solubility of a drug in a solvent is the maximum amount of the most stable crystalline form that remains in solution in a given volume of the solvent at a given temperature and pressure under equilibrium conditions.[4] The equilibrium involves a balance of the energy of three interactions against each other: (1) solvent with solvent, (2) solute with solute, and (3) solvent and solute. Thermodynamic equilibrium is achieved when the overall lowest energy state of the system is achieved. This means that only the equilibrium solubility reflects the balance of forces between the solution and the most stable, lowest energy crystalline form of the solid. In practical terms, this means that one needs to be careful when evaluating a drug's solubility. For example, let us say that you want to determine the solubility of a drug and that you were not aware that it was not in its crystalline form. It is well known that a metastable solid form of a drug will have a higher apparent solubility. Given enough time, the limiting solubility of the most stable form will eventually dominate and since the most stable crystal form has the lowest solubility, this means that there will be excess drug in solution resulting in a precipitate. So, initially you would record a higher solubility but after a period of time the solubility that you measure would be significantly lower. As you can imagine, this could lead to serious problems. This was vividly illustrated by Abbott's antiviral drug ritonavir where the slow precipitation of a new stable polymorph from dosing solutions required the manufacturer to perform an emergency reformulation to ensure consistent drug release characteristics.[2,4]

## KEY CONCEPT SOLUTIONS AND SOLUBILITY

A *saturated solution* is one in which the solute in solution is in equilibrium with the solid phase. *Solubility* is defined in quantitative terms as the concentration of solute in a saturated solution at a certain temperature, and in a qualitative way, it can be defined as the spontaneous interaction of two or more substances to form a homogeneous molecular dispersion. An *unsaturated* or *subsaturated* solution is one containing the dissolved solute in a concentration below that necessary for complete saturation at a definite temperature. A *supersaturated solution* is one that contains more of the dissolved solute than it would normally contain at a definite temperature, were the undissolved solute present.

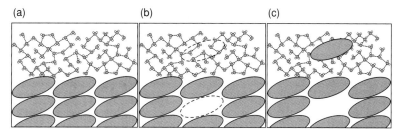

**Fig. 9–1.** The intermolecular forces that determine thermodynamic solubility. (*a*) Solvent and solute are segregated, each interacts primarily with other molecules of the same type. (*b*) To move a solute molecule into solution, the interactions among solute molecules in the crystal (lattice energy) and among solvent molecules in the space required to accommodate the solute (cavitation energy) must be broken. The system entropy increases slightly because the ordered network of hydrogen bonds among solvent molecules has been disrupted. (*c*) Once the solute molecule is surrounded by solvent, new stabilizing interactions between the solute and solvent are formed (solvation energy), as indicated by the dark purple molecules. The system entropy increases owing to the mingling of solute and solvent (entropy of mixing) but also decreases locally owing to the new short-range order introduced by the presence of the solute, as indicated by the light purple molecules."[4] (Adapted from Bhattachar et al. 2006.[4])

## Solubility Expressions

The solubility of a drug may be expressed in a number of ways. The *United States Pharmacopeia* (*USP*) describes the solubility of drugs as parts of solvent required for one part solute. Solubility is also quantitatively expressed in terms of molality, molarity, and percentage. The USP describes solubility using the seven groups listed in **Table 9–1**. The European Pharmacopeia lists six categories (it does not use the *practically insoluble* grouping). For exact solubilities of many substances, the reader is referred to standard reference works such as official compendia (e.g., USP) and the Merck Index.

## SOLVENT–SOLUTE INTERACTIONS

The pharmacist knows that water is a good solvent for salts, sugars, and similar compounds, whereas mineral oil is often a solvent for substances that are normally only slightly soluble in water. These empirical findings are summarized in the statement, "like dissolves like." Such a maxim is satisfying to most of us, but the inquisitive student may be troubled by this vague idea of "likeness."

## Polar Solvents

The solubility of a drug is due in large measure to the polarity of the solvent, that is, to its dipole moment. Polar solvents dissolve ionic solutes and other polar substances. Accordingly, water mixes in all proportions with alcohol and dissolves sugars and other polyhydroxy compounds. Hildebrand showed, however, that a consideration of dipole moments alone is not adequate to explain the solubility of polar substances in water. The ability of the solute to form hydrogen bonds is a far more significant factor than is the polarity as reflected in a high dipole moment. Water dissolves phenols, alcohols, aldehydes, ketones, amines, and other oxygen- and nitrogen-containing compounds that can form hydrogen bonds with water:

| **TABLE 9–1** | | | |
| --- | --- | --- | --- |
| **SOLUBILITY DEFINITION IN THE UNITED STATES PHARMACOPEIA** | | | |
| **Description Forms (Solubility Definition)** | **Parts of Solvent Required for One Part of Solute** | **Solubility Range (mg/mL)** | **Solubility Assigned (mg/mL)** |
| Very soluble (VS) | <1 | >1000 | 1000 |
| Freely soluble (FS) | From 1 to 10 | 100–1000 | 100 |
| Soluble | From 10 to 30 | 33–100 | 33 |
| Sparingly soluble (SPS) | From 30 to 100 | 10–33 | 10 |
| Slightly soluble (SS) | From 100 to 1000 | 1–10 | 1 |
| Very slightly soluble (VSS) | From 1000 to 10,000 | 0.1–1 | 0.1 |
| Practically insoluble (PI) | >10,000 | <0.1 | 0.01 |

$$R-O\cdots H-O\cdots$$

Alcohol

$$R-C=O\cdots H-O\cdots$$

Aldehyde

$$\begin{matrix} H_3C \\ \diagdown \\ \diagup \\ H_3C \end{matrix} C=O\cdots H-O\cdots$$

Ketone

$$R_3N\cdots H-O\cdots$$

Amine

A difference in acidic and basic character of the constituents in the Lewis electron donor–acceptor sense also contributes to specific interactions in solutions.

In addition to the factors already enumerated, the solubility of a substance also depends on structural features such as the ratio of the polar to the nonpolar groups of the molecule. As the length of a nonpolar chain of an aliphatic alcohol increases, the solubility of the compound in water decreases. Straight-chain monohydroxy alcohols, aldehydes, ketones, and acids with more than four or five carbons cannot enter into the hydrogen-bonded structure of water and hence are only slightly soluble. When additional polar groups are present in the molecule, as found in propylene glycol, glycerin, and tartaric acid, water solubility increases greatly. Branching of the carbon chain reduces the nonpolar effect and leads to increased water solubility. Tertiary butyl alcohol is miscible in all proportions with water, whereas *n*-butyl alcohol dissolves to the extent of about 8 g/100 mL of water at 20°C.

## Nonpolar Solvents

The solvent action of nonpolar liquids, such as the hydrocarbons, differs from that of polar substances. Nonpolar solvents are unable to reduce the attraction between the ions of strong and weak electrolytes because of the solvents' low dielectric constants. Nor can the solvents break covalent bonds and ionize weak electrolytes, because they belong to the group known as aprotic solvents, and they cannot form hydrogen bridges with nonelectrolytes. Hence, ionic and polar solutes are not soluble or are only slightly soluble in nonpolar solvents.

Nonpolar compounds, however, can dissolve nonpolar solutes with similar internal pressures through induced dipole interactions. The solute molecules are kept in solution by the weak van der Waals–London type of forces. Thus, oils and fats dissolve in carbon tetrachloride, benzene, and mineral oil. Alkaloidal bases and fatty acids also dissolve in nonpolar solvents.

## Semipolar Solvents

Semipolar solvents, such as ketones and alcohols, can *induce* a certain degree of polarity in nonpolar solvent molecules, so that, for example, benzene, which is readily polarizable, becomes soluble in alcohol. In fact, semipolar compounds can act as *intermediate solvents* to bring about miscibility of polar and nonpolar liquids. Accordingly, acetone increases the solubility of ether in water. Loran and Guth[5] studied the intermediate solvent action of alcohol on water–castor oil mixtures. Propylene glycol has been shown to increase the mutual solubility of water and peppermint oil and of water and benzyl benzoate.[6]

A number of common solvent types are listed in the order of decreasing "polarity" in **Table 9–2**, together with corresponding solute classes. The term *polarity* is loosely used here to represent not only the dielectric constants of the solvents and solutes but also the other factors enumerated previously.

## SOLUBILITY OF LIQUIDS IN LIQUIDS

Frequently two or more liquids are mixed together in the preparation of pharmaceutical solutions. For example, alcohol is added to water to form hydroalcoholic solutions of various concentrations; volatile oils are mixed with water to form dilute solutions known as aromatic waters; volatile oils are added to alcohol to yield spirits and elixirs; ether and alcohol are combined in collodions; and various fixed oils are blended into lotions, sprays, and medicated oils. Liquid–liquid systems can be divided into two categories according to the solubility of the substances in one another: (*a*) complete miscibility and (*b*) partial miscibility. The term *miscibility* refers to the mutual solubilities of the components in liquid–liquid systems.

## Complete Miscibility

Polar and semipolar solvents, such as water and alcohol, glycerin and alcohol, and alcohol and acetone, are said to be

TABLE 9-2
**TABLE 9-2**
**POLARITY OF SOME SOLVENTS AND THE SOLUTES THAT READILY DISSOLVE IN EACH CLASS OF SOLVENT**

| | Dielectric Constant of Solvent, $\varepsilon$ (Approximately) | Solvent | Solute | |
|---|---|---|---|---|
| Decreasing Polarity | 80 | Water | Inorganic salts, organic salts | Decreasing Water Solubility |
| ↓ | 50 | Glycols | Sugars, tannins | ↓ |
| | 30 | Methyl and ethyl alcohols | Caster oil, waxes | |
| | 20 | Aldehydes, ketones, and higher alcohols, ethers, esters, and oxides | Resins, volatile oils, weak electrolytes including barbiturates, alkaloids, and phenols | |
| | 5 | Hexane, benzene, carbon tetrachloride, ethyl ether, petroleum ether | Fixed oils, fats, petrolatum, paraffin, other hydrocarbons | |
| | 0 | Mineral oil and fixed vegetable oils | | |

completely miscible because they mix in all proportions. Nonpolar solvents such as benzene and carbon tetrachloride are also completely miscible. Completely miscible liquid mixtures in general create no solubility problems for the pharmacist and need not be considered further.

## Partial Miscibility

When certain amounts of water and ether or water and phenol are mixed, two liquid layers are formed, each containing some of the other liquid in the dissolved state. The phenol–water system has been discussed in detail in Chapter 2, and the student at this point should review the section dealing with the phase rule. It is sufficient here to reiterate the following points. (a) The mutual solubilities of partially miscible liquids are influenced by temperature. In a system such as phenol and water, the mutual solubilities of the two conjugate phases increase with temperature until, at the critical solution temperature (or upper consolute temperature), the compositions become identical. At this temperature, a homogeneous or single-phase system is formed. (b) From a knowledge of the phase diagram, more especially the tie lines that cut the binodal curve, it is possible to calculate both the composition of each component in the two conjugate phases and the amount of one phase relative to the other. *Example 9–1* gives an illustration of such a calculation.

**EXAMPLE 9-1**
**Component Weights**

A mixture of phenol and water at 20°C has a total composition of 50% phenol. The tie line at this temperature cuts the binodal at points equivalent to 8.4% and 72.2% w/w phenol. What is the weight of the aqueous layer and of the phenol layer in 500 g of the mixture and how many grams of phenol are present in each of the two layers?

Let $Z$ be the weight in grams of the aqueous layer. Therefore, $500 - Z$ is the weight in grams of the phenol layer, and the sum of the percentages of phenol in the two layers must equal the overall composition of 50%, or $500 \times 0.50 = 250$ g. Thus,

$$Z(8.4/100) + (500 - Z)(72.2/100) = 250$$
Weight of aqueous layer, $Z = 174$ g
Weight of phenol layer, $500 - Z = 326$ g
Weight of phenol in the aqueous layer, $174 \times 0.084 = 15$ g
Weight of phenol in the phenolic layer, $326 \times 0.722 = 235$ g

In the case of some liquid pairs, the solubility can increase as the temperature is lowered, and the system will exhibit a *lower consolute temperature*, below which the two members are soluble in all proportions and above which two separate layers form. Another type, involving a few mixtures such as nicotine and water, shows both an upper and a lower consolute temperature with an intermediate temperature region in which the two liquids are only partially miscible. A final type exhibits no critical solution temperature; the pair ethyl ether and water, for example, has neither an upper nor a lower consolute temperature and shows partial miscibility over the entire temperature range at which the mixture exists.

## Three-Component Systems

The principles underlying systems that can contain one, two, or three partially miscible pairs have been discussed in detail in Chapter 2. Further examples of three-component systems containing one pair of partially miscible liquids are water, $CCl_4$, and acetic acid; and water, phenol, and acetone. Loran and Guth[5] studied the three-component system consisting of water, castor oil, and alcohol and determined the proper proportions for use in certain lotions and hair preparations; a triangular diagram is shown in their report. A similar titration with water of a mixture containing peppermint oil and polyethylene glycol is shown in **Figure 9–2**.[6] Ternary diagrams have also found use in cosmetic formulations

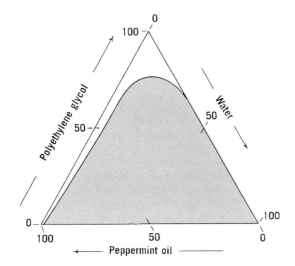

**Fig. 9–2.** A triangular diagram showing the solubility of peppermint oil in various proportions of water and polyethylene glycol.

involving three liquid phases.[7] Gorman and Hall[8] determined the ternary-phase diagram of the system of methyl salicylate, isopropanol, and water (**Fig. 9–3**).

# SOLUBILITY OF SOLIDS IN LIQUIDS

Systems of solids in liquids include the most frequently encountered and probably the most important type of pharmaceutical solutions. Many important drugs belong to the class of weak acids and bases. They react with strong acids and bases and, within definite ranges of pH, exist as ions that are ordinarily soluble in water.

Although carboxylic acids containing more than five carbons are relatively insoluble in water, they react with dilute sodium hydroxide, carbonates, and bicarbonates to form soluble salts. The fatty acids containing more than 10 carbon

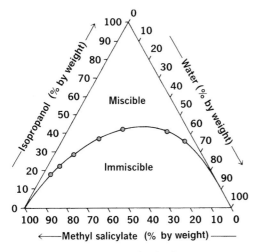

**Fig. 9–3.** Triangular phase diagram for the three-component system methyl salicylate–isopropanol–water. (From W. G. Gorman and G. D. Hall, J. Pharm. Sci. **53**, 1017, 1964. With permission.)

atoms form soluble soaps with the alkali metals and insoluble soaps with other metal ions. They are soluble in solvents having low dielectric constants; for example, oleic acid ($C_{17}H_{33}COOH$) is insoluble in water but is soluble in alcohol and in ether.

Hydroxy acids, such as tartaric and citric acids, are quite soluble in water because they are solvated through their hydroxyl groups. The potassium and ammonium bitartrates are not very soluble in water, although most alkali metal salts of tartaric acid are soluble. Sodium citrate is used sometimes to dissolve water-insoluble acetylsalicylic acid because the soluble acetylsalicylate ion is formed in the reaction. The citric acid that is produced is also soluble in water, but the practice of dissolving aspirin by this means is questionable because the acetylsalicylate is also hydrolyzed rapidly.

Aromatic acids react with dilute alkalies to form water-soluble salts, but they can be precipitated as the free acids if stronger acidic substances are added to the solution. They can also be precipitated as heavy metal salts should heavy metal ions be added to the solution. Benzoic acid is soluble in sodium hydroxide solution, alcohol, and fixed oils. Salicylic acid is soluble in alkalies and in alcohol. The OH group of salicyclic acid cannot contribute to the solubility because it is involved in an intramolecular hydrogen bond.

Phenol is weakly acidic and only slightly soluble in water but is quite soluble in dilute sodium hydroxide solution,

$$C_6H_5OH + NaOH \rightarrow C_6H_5O^- + Na^+ + H_2O$$

Phenol is a weaker acid than $H_2CO_3$ and is thus displaced and precipitated by $CO_2$ from its dilute alkali solution. For this reason, carbonates and bicarbonates cannot increase the solubility of phenols in water.

Many organic compounds containing a basic nitrogen atom in the molecule are important in pharmacy. These include the alkaloids, sympathomimetic amines, antihistamines, local anesthetics, and others. Most of these weak electrolytes are not very soluble in water but are soluble in dilute solutions of acids; such compounds as atropine sulfate and tetracaine hydrochloride are formed by reacting the basic compounds with acids. Addition of an alkali to a solution of the salt of these compounds precipitates the free base from solution if the solubility of the base in water is low.

The aliphatic nitrogen of the sulfonamides is sufficiently negative so that these drugs act as slightly soluble weak acids rather than as bases. They form water-soluble salts in alkaline solution by the following mechanism. The oxygens of the sulfonyl ($-SO_2-$) group withdraw electrons, and the resulting electron deficiency of the sulfur atom results in the electrons of the N:H bond being held more closely to the nitrogen atom. The hydrogen therefore is bound less firmly, and, in alkaline solution, the soluble sulfonamide anion is readily formed.

The sodium salts of the sulfonamides are precipitated from solution by the addition of a strong acid or by a salt of a strong acid and a weak base such as ephedrine hydrochloride.

The barbiturates, like the sulfonamides, are weak acids because the electronegative oxygen of each acidic carbonyl group tends to withdraw electrons and to create a positive carbon atom. The carbon in turn attracts electrons from the nitrogen group and causes the hydrogen to be held less firmly. Thus, in sodium hydroxide solution, the hydrogen is readily lost, and the molecule exists as a soluble anion of the weak acid. Butler et al.[9] demonstrated that, in highly alkaline solutions, the second hydrogen ionizes. The $pK_1$ for phenobarbital is 7.41 and the $pK_2$ is 11.77. Although the barbiturates are soluble in alkalies, they are precipitated as the free acids when a stronger acid is added and the pH of the solution is lowered.

## Calculating the Solubility of Weak Electrolytes as Influenced by pH

From what has been said about the effects of acids and bases on solutions of weak electrolytes, it becomes evident that the solubility of weak electrolytes is strongly influenced by the pH of the solution. For example, a 1% solution of phenobarbital sodium is soluble at pH values high in the alkaline range. The soluble ionic form is converted into molecular phenobarbital as the pH is lowered, and below 9.3, the drug begins to precipitate from solution at room temperature. On the other hand, alkaloidal salts such as atropine sulfate begin to precipitate as the pH is elevated.

To ensure a clear homogeneous solution and maximum therapeutic effectiveness, the preparations should be adjusted to an optimum pH. The pH below which the salt of a weak acid, sodium phenobarbital, for example, begins to precipitate from aqueous solution is readily calculated in the following manner.

Representing the free acid form of phenobarbital as HP and the soluble ionized form as $P^-$, we write the equilibria in a saturated solution of this slightly soluble weak electrolyte as

$$HP_{solid} \rightleftharpoons HP_{sol} \qquad (9\text{–}1)$$

$$HP_{sol} + H_2O \rightleftharpoons H_3O^+ + P^- \qquad (9\text{–}2)$$

Because the concentration of the un-ionized form in solution, $HP_{sol}$, is essentially constant, the equilibrium constant for the solution equilibrium, equation (9–1), is

$$S_o = [HP]_{sol} \qquad (9\text{–}3)$$

where $S_o$ is molar or intrinsic solubility. The constant for the acid–base equilibrium, equation (9–2), is

$$K_a = \frac{[H_3O^+][P^-]}{[HP]} \qquad (9\text{–}4)$$

or

$$[P^-] = K_a \frac{[HP]}{[H_3O^+]} \qquad (9\text{–}5)$$

where the subscript "sol" has been deleted from $[HP]_{sol}$ because no confusion should result from this omission.

The total solubility, $S$, of phenobarbital consists of the concentration of the undissociated acid, $[HP]$, and that of the conjugate base or ionized form, $[P^-]$:

$$S = [HP] + [P^-] \qquad (9\text{–}6)$$

Substituting $S_o$ for $[HP]$ from equation (9–3) and the expression from equation (9–5) for $[P^-]$ yields

$$S = S_o + K_a \frac{S_o}{[H_3O^+]} \qquad (9\text{–}7)$$

$$S = S_o \left(1 + \frac{K_a}{[H_3O^+]}\right) \qquad (9\text{–}8)$$

When the electrolyte is weak and does not dissociate appreciably, the solubility of the acid in water or acidic solutions is $S_o = [HP]$, which, for phenobarbital is approximately 0.005 mole/liter, in other words, 0.12%.

The solubility equation can be written in logarithmic form, beginning with equation (9–7). By rearrangement, we obtain

$$(S - S_o) = K_a \frac{S_o}{[H_3O^+]}$$

$$\log(S - S_o) = \log K_a + \log S_o - \log[H_3O^+]$$

and finally

$$pH_p = pK_a + \log \frac{S - S_o}{S_o} \qquad (9\text{–}9)$$

where $pH_p$ is the pH below which the drug separates from solution as the undissociated acid.

In pharmaceutical practice, a drug such as phenobarbital is usually added to an aqueous solution in the soluble salt form. Of the initial quantity of salt, sodium phenobarbital, that can be added to a solution of a certain pH, some of it is converted into the free acid, HP, and some remains in the ionized form, $P^-$ [equation (9–6)]. The amount of salt that can be added initially before the solubility $[HP]$ is exceeded is therefore equal to $S$. As seen from equation (9–9), $pH_p$ depends on the initial molar concentration, $S$, of salt added, the molar solubility of the undissociated acid, $S_o$, also known as the *intrinsic solubility*, and the $pK_a$. Equation (9–9) has been used to determine the $pK_a$ of sulfonamides and other drugs.[10,11] Solubility and pH data can also be used to obtain the $pK_1$ and $pK_2$ values of dibasic acids as suggested by Zimmerman[12] and Blanchard et al.[13]

### EXAMPLE 9–2

**Phenobarbital**

Below what pH will free phenobarbital begin to separate from a solution having an initial concentration of 1 g of sodium phenobarbital per 100 mL at 25°C? The molar solubility, $S_o$, of phenobarbital is 0.0050 and the $pK_a$ is 7.41 at 25°C. The secondary dissociation of phenobarbital, referred to previously, can ordinarily be disregarded. The molecular weight of sodium phenobarbital is 254.

The molar concentration of salt initially added is

$$\frac{g/liter}{mol.wt.} = \frac{10}{254} = 0.039 \text{ mole/liter}$$

$$pH_p = 7.41 + \log \frac{(0.039 - 0.005)}{0.005} = 8.24$$

An analogous derivation can be carried out to obtain the equation for the solubility of a weak base as a function of the pH of a solution. The expression is

$$pH_p = pK_w - pK_b + \log \frac{S_o}{S - S_o} \quad (9\text{–}10)$$

where $S$ is the concentration of the drug initially added as the salt and $S_o$ is the molar solubility of the free base in water. Here $pH_p$ is the pH *above* which the drug begins to precipitate from solution as the free base.

## The Influence of Solvents on the Solubility of Drugs

Weak electrolytes can behave like strong electrolytes or like nonelectrolytes in solution. When the solution is of such a pH that the drug is entirely in the ionic form, it behaves as a solution of a strong electrolyte, and solubility does not constitute a serious problem. However, when the pH is adjusted to a value at which un-ionized molecules are produced in sufficient concentration to exceed the solubility of this form, precipitation occurs. In this discussion, we are now interested in the solubility of nonelectrolytes and the undissociated molecules of weak electrolytes. The solubility of undissociated phenobarbital in various solvents is discussed here because it has been studied to some extent by pharmaceutical investigators.

Frequently, a solute is more soluble in a mixture of solvents than in one solvent alone. This phenomenon is known as *cosolvency*, and the solvents that, in combination, increase the solubility of the solute are called *cosolvents*. Approximately 1 g of phenobarbital is soluble in 1000 mL of water, in 10 mL of alcohol, in 40 mL of chloroform, and in 15 mL of ether at 25°C. The solubility of phenobarbital in water–alcohol–glycerin mixtures is plotted on a semilogarithm grid in **Figure 9–4** from the data of Krause and Cross.[14]

By drawing lines parallel to the abscissa in **Figure 9–4** at a height equivalent to the required phenobarbital concentration, it is a simple matter to obtain the relative amounts of the various combinations of alcohol, glycerin, and water needed to achieve solution. For example, at 22% alcohol, 40% glycerin, and the remainder water (38%), 1.5% w/v of phenobarbital is dissolved, as seen by following the vertical and horizontal lines drawn on **Figure 9–4**.

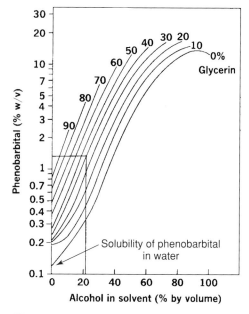

**Fig. 9–4.** The solubility of phenobarbital in a mixture of water, alcohol, and glycerin at 25°C. The vertical axis is a logarithmic scale representing the solubility of phenobarbital in g/100 mL. (From G. M. Krause and J. M. Cross, J. Am. Pharm. Assoc. Sci. Ed. **40**, 137, 1951. With permission.)

## Combined Effect of pH and Solvents

Stockton and Johnson[15] and Higuchi et al.[16] studied the effect of an increase of alcohol concentration on the dissociation constant of sulfathiazole, and Edmonson and Goyan[17] investigated the effect of alcohol on the solubility of phenobarbital.

Schwartz et al.[10] determined the solubility of phenytoin as a function of pH and alcohol concentration in various buffer systems and calculated the apparent dissociation constant. Kramer and Flynn[18] examined the solubility of hydrochloride salts of organic bases as a function of pH, temperature, and solvent composition. They described the determination of the $pK_a$ of the salt from the solubility profile at various temperatures and in several solvent systems. Chowhan[11] measured and calculated the solubility of the organic carboxylic acid naproxen and its sodium, potassium, calcium, and magnesium salts. The observed solubilities were in excellent agreement with the pH–solubility profiles based on equation **(9–9)**.

The results of Edmonson and Goyan[17] are shown in **Figure 9–5**, where one observes that the $pK_a$ of phenobarbital, 7.41, is raised to 7.92 in a hydroalcoholic solution containing

## ▶︎ KEY CONCEPT  SOLVENTS AND WEAK ELECTROLYTES

The solvent affects the solubility of a weak electrolyte in a buffered solution in two ways: (*a*) The addition of alcohol to a buffered aqueous solution of a weak electrolyte increases the solubility of the un-ionized species by adjusting the polarity of the solvent to a more favorable value. (*b*) Because it is less polar than water, alcohol decreases the dissociation of a weak electrolyte, and the solubility of the drug goes down as the dissociation constant is decreased ($pK_a$ is increased).

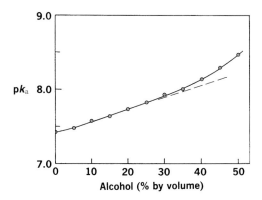

**Fig. 9–5.** The influence of alcohol concentration on the dissociation constant of phenobarbital. (From T. D. Edmonson and J. E. Goyan, J. Am. Pharm. Assoc. Sci. Ed. **47,** 810, 1958. With permission.)

30% by volume of alcohol. Furthermore, as can be seen in **Figure 9–4**, the solubility, $S_o$, of un-ionized phenobarbital is increased from 0.12 g/100 mL or 0.005 M in water to 0.64% or 0.0276 M in a 30% alcoholic solution. The calculation of solubility as a function of pH involving these results is illustrated in the following example.

**EXAMPLE 9–3**

**Minimum pH for Complete Solubility**

What is the minimum pH required for the complete solubility of the drug in a stock solution containing 6 g of phenobarbital sodium in 100 mL of a 30% by volume alcoholic solution? From equation (9–9),

$$pH_p = 7.92 + \log \frac{0.236 - 0.028}{0.028}$$
$$pH_p = 7.92 + 0.87 = 8.79$$

For comparison, the minimum pH for complete solubility of phenobarbital in an aqueous solution containing no alcohol is computed using equation (9–9):

$$pH_p = 7.41 + \log \frac{0.236 - 0.005}{0.005} = 9.07$$

From the calculations of *Example 9–3*, it is seen that although the addition of alcohol increases the $pK_a$, it also increases the solubility of the un-ionized form of the drug over that found in water sufficiently so that the pH can be reduced somewhat before precipitation occurs.

Equations **(9–9)** and **(9–10)** can be made more exact if activities are used instead of concentrations to account for interionic attraction effects. This refinement, however, is seldom required for practical work, where the values calculated from the approximate equations just given serve as satisfactory estimates.

## Influence of Complexation in Multicomponent Systems

Many liquid pharmaceutical preparations consist of more than a single drug in solution. Fritz et al.[19] showed that when several drugs together with pharmaceutical adjuncts interact in solution to form insoluble complexes, simple solubility profiles of individual drugs cannot be used to predict solubilities in mixtures of ingredients. Instead, the specific multicomponent systems must be studied to estimate the complicating effects of species interactions.

## Influence of Other Factors on the Solubility of Solids

The size and shape of small particles (those in the micrometer range) also affect solubility. Solubility increases with decreasing particle size according to the approximate equation

$$\log \frac{s}{s_o} = \frac{2\gamma V}{2.303 RTr} \qquad (9\text{–}11)$$

where $s$ is the solubility of the fine particles; $s_o$ is the solubility of the solid consisting of relatively large particles; $\gamma$ is the surface tension of the particles, which, for solids, unfortunately, is extremely difficult to obtain; $V$ is the molar volume (volume in $cm^3$ per mole of particles); $r$ is the final radius of the particles in cm; $R$ is the gas constant ($8.314 \times 10^7$ ergs/deg mole); and $T$ is the absolute temperature. The equation can be used for solid or liquid particles such as those in suspensions or emulsions. The following example is taken from the book by Hildebrand and Scott.[20]

**EXAMPLE 9–4**

**Particle Size and Solubility**

A solid is to be comminuted so as to increase its solubility by 10%, that is, $s/s_o$ is to become 1.10. What must be the final particle size, assuming that the surface tension of the solid is 100 dynes/cm and the volume per mole is 50 $cm^3$? The temperature is 27°C.

$$r = \frac{2 \times 100 \times 50}{2.303 \times 8.314 \times 10^7 \times 300 \times 0.0414}$$
$$= 4.2 \times 10^{-6} \text{cm} = 0.042 \ \mu m$$

The configuration of a molecule and the type of arrangement in the crystal also has some influence on solubility, and a symmetric particle can be less soluble than an unsymmetric one. This is because solubility depends in part on the work required to separate the particles of the crystalline solute. The molecules of the amino acid $\alpha$-alanine form a compact crystal with high lattice energy and consequently low solubility. The molecules of $\alpha$-amino-$n$-butyric acid pack less efficiently in the crystal, partly because of the projecting side chains, and the crystal energy is reduced. Consequently, $\alpha$-amino-$n$-butyric acid has a solubility of 1.80 moles/liter and $\alpha$-alanine has a solubility of only 1.66 moles/liter in water at 25°C, although the hydrocarbon chain is longer in $\alpha$-amino-$n$-butyric acid than in the other compound.

POOR AQUEOUS SOLUBILITY

"Poor aqueous solubility is caused by two main factors: high lipophilicity and strong intermolecular interactions, which make the solubilization of the solid energetically costly. What is meant by good and poorly soluble depends partly on the expected therapeutic dose and potency of the drug. As a rule of thumb from the delivery perspective, a drug with an average potency of 1 mg/kg should have a solubility of at least 0.1 g/L to be adequately soluble. If a drug with the same potency has a solubility of less than 0.01 g/L it can be considered poorly soluble."[3]

# DETERMINING THERMODYNAMIC AND "KINETIC" SOLUBILITY

## The Phase Rule and Solubility

Solubility can be described in a concise manner by the use of the Gibbs phase rule, which is described using

$$F = C - P + 2 \qquad (9\text{–}12)$$

where $F$ is the *number of degrees of freedom*, that is, the number of independent variables (usually temperature, pressure, and concentration) that must be fixed to completely determine the system, $C$ is the smallest number of components that are adequate to describe the chemical composition of each phase, and $P$ is the number of phases.

The Phase Rule can be used to determine the thermodynamic solubility of a drug substance. This method is based on the thermodynamic principles of heterogeneous equilibria that are among the soundest theoretical concepts in chemistry. It does not depend on any assumptions regarding kinetics or the structure of matter but is applicable to all drugs. The requirements for an analysis are simple, as the equipment needed is basic to most laboratories and the quantities of substances are small. Basically, drug is added in a specific amount of solvent. After equilibrium is achieved, excess drug is removed (usually by filtering) and then the concentration of the dissolved drug is measured using standard analysis techniques such as high-performance liquid chromatography.

A phase-solubility diagram for a pure drug substance is shown in **Figure 9–6**.[21] At concentrations below the saturation concentration there is only one degree of freedom since the studies are performed at constant temperature and pressure. In other words, only the concentration changes. This is represented in **Figure 9–6** by the segment A–B of the line. Once the saturation concentration is reached, the addition of more drug to the "system" does not result in higher solution concentrations (segment B–C). Rather, the drug remains in the solid state and the system becomes a two-phase system. Since the temperature, pressure, and solution concentration

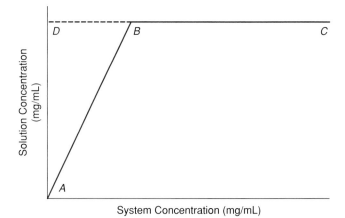

**Fig. 9–6.** Phase-solubility diagram for a pure drug substance. The line segment A–B represents one phase since the concentration of drug substance is below the saturation concentration. Line segment B–C represents a pure solid in a saturated solution at equilibrium. (From Remington, *The Science and Practice of Pharmacy*, 21st Ed., Lippincott Williams & Wilkins, 2006, p. 216. With permission.)

are constant at drug concentrations above the saturation concentration, the system has zero degree of freedom.

The situation in **Figure 9–6** is valid only for pure drug substances. What if the drug substance is not pure? This situation is described in **Figure 9–7**.[22] If the system has one impurity, the solution becomes saturated with the first component at point B. The situation becomes interesting at this point. In segment B–C of the line, the solution is saturated with component 1 (which is usually the major component such as the drug), so the drug would precipitate out of solution at concentrations greater than this. However, the impurity (the minor component or component 2) does not reach saturation

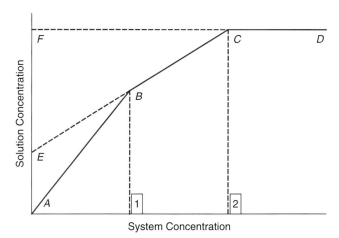

**Fig. 9–7.** Phase-solubility curve when the drug substance contains one impurity. At point B, the solution becomes saturated with component 1 (the drug). The segment B–C represents two phases—a solution phase saturated with the drug and some of the impurity and a solid phase of the drug. Segment C–D represents two phases—a liquid phase saturated with the drug and impurity and a solid phase containing the drug and the impurity. (From Remington, *The Science and Practice of Pharmacy*, 21st Ed., Lippincott Williams & Wilkins, 2006, p. 217. With permission.)

until it reaches point C on the line. The concentrations of the two components are saturated beyond point C (segment C–D) of the line. Once true equilibrium is achieved, one can extrapolate back to the $Y$ axis (solution concentration) to determine the solubility of the two components. Therefore, the thermodynamic solubility of the drug would be equal to the distance A–E and the solubility of the impurity would be equal to the distance represented by E–F. As one can see, this procedure can be used to measure the exact solubility of the pure drug without having a pure form of the drug to start with.

The practical aspect of measuring thermodynamic solubility is, on the surface, relatively simple but it can be quite time-consuming.[4] Some methods have been developed in attempt to reduce the time that it takes to get a result. Starting the experiment with a high purity crystalline form of the substance will give the best chance that the solubility measured after a reasonable incubation period will represent the true equilibrium solubility. However, this may still take several hours to several days. Also, there is still a risk that the incubation period will not be sufficient for metastable crystal forms to convert to the most stable form. This means that the measured concentration may represent the apparent solubility of a different crystal form. This risk must be taken into consideration when running a solubility experiment with material that is not known to be the most stable crystalline form.[4]

Bhattachar and colleagues[4] recently reviewed various aspects of solubility and they are summarized here. In practice, the stable crystalline form of the compound is not available in sufficient purity during early discovery and so the labor-intensive measurement of thermodynamic solubility is not commonly made. The amount of compound required to measure a thermodynamic solubility measurement depends on the volume of solvent used to make the saturated solution and the solubility of the compound in that solvent. Recent reports for miniaturized systems list compound requirements ranging from ~100 mg per measurement for poorly soluble compounds[23] to 3 to 10 mg for pharmaceutically relevant compounds.[24] Although early-stage solubility information is crucial to drug discovery teams, the number of compounds being assessed, the scarcity of compound, and questionable purity and crystallinity make it nearly impossible to assess thermodynamic solubility.

These challenges have been partially met using a high-throughput kinetic measurement of antisolvent precipitation

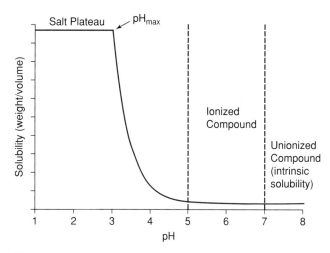

**Fig. 9–8.** pH–solubility profile for a compound with a single, basic $pK_a$ value of 5. The four regions of pH-dependent solubility are the salt plateau, $pH_{max}$, ionized compound, and un-ionized compound. (Adapted from Bhattachar et al. 2006,[4] with permission.)

commonly referred to as "*kinetic solubility*."[25–28] "Kinetic solubility is a misnomer, not because it is not kinetic, but because it measures a precipitation rate rather than solubility. Kinetic solubility methods are designed to facilitate high throughput measurements, using submilligram quantities of compound, in a manner that closely mimics the actual solubilization process used in biological laboratories. Typically, the compound is dissolved in dimethyl sulfoxide (because it is a strong organic solvent) to make a stock solution of known concentration. This stock is added gradually to the aqueous solvent of interest until the anti-solvent properties of the water drive the compound out of solution. The resulting precipitation is detected optically, and the kinetic solubility is defined as the point at which the aqueous component can no longer solvate the drug. Solubility results obtained from kinetic measurements might not match the thermodynamic solubility results perfectly; therefore, caution must be exercised such that the data from the kinetic solubility measurements are used only for their intended application. Since kinetic solubility is determined for compounds that have not been purified to a high degree or crystallized, the impurities and amorphous content in the material lead to a higher solubility than the thermodynamic solubility. Because kinetic solubility experiments begin with the drug in solution, there is a significant risk of achieving supersaturation of the aqueous solvent through precipitation of an amorphous or metastable crystalline form. This supersaturation can lead to a measured value that is significantly higher than the thermodynamic solubility, masking a solubility problem that will become apparent as soon as the compound is crystallized. Owing to the nature of kinetic solubility measurements, there is no time for equilibration of the compound in the aqueous solvent of measurement. Because the compounds tested are in dimethyl sulfoxide solutions, the energy required to break the crystal lattice is not factored into the solubility measurements."[4]

## KEY CONCEPT — EFFECT OF pH ON SOLUBILITY

Solubility must always be considered in the context of pH and $pK_a$. The relationship between pH and solubility is shown in Figure 9–8. If the measured solubility falls on the steep portion of the pH–solubility profile, small changes to the pH can have a marked effect on the solubility.[4]

## Some Limitations of Thermodynamic Solubility[3]

In a recent review, Faller and Ertl[3] have discussed some of the limitations of traditional methods for determining solubility. For example, if the traditional shake-flask method is used, adsorption to the vial or to the filter, incomplete phase separation, compound instability, and slow dissolution can affect the result. When the potentiometric method is used, inaccurate $pK_a$ determination, compound degradation during the titration, slow dissolution, or incorrect data analysis can affect the data quality. It is very important to define the experimental conditions well. The intrinsic solubility, $S_o$, needs to be distinguished from the solubility measured at a given pH value in a defined medium. Intrinsic solubility refers to the solubility of the unionized species. Artursson et al.[29] has shown that this parameter is relatively independent of the nature of the medium used. In contrast, solubility measured at a fixed pH value may be highly dependent on the nature and concentration of the counter ions present in the medium.[30] This is especially critical for poorly soluble compounds that are strongly ionized at the pH of the measurement. Finally, it is important to note that single pH measurements cannot distinguish between soluble monomers and soluble aggregates of drug molecules, which may range from dimers to micelles unless more sophisticated experiments are performed.[30]

## Computational Approaches

In addition to measuring solubility, computational approaches are widely used and were reviewed recently by Faller and Ertl.[3] Briefly, fragment-based models attempt to predict solubility as a sum of substructure contributions—such as contributions of atoms, bonds, or larger substructures. This approach is based on a general assumption that molecule properties are determined completely by molecular structure and may be approximated by the contributions of fragments in the molecule. The inverse relation between solubility and lipophilicity has also been recognized for a long time and empirical relationships between $\log S_o$ and $\log P$ have been reported. Finally, numerous other approaches for predicting water solubility have been reported. The array of possible molecular descriptors that can be used is nearly unlimited. The polar surface area, which characterizes molecule polarity and hydrogen bonding features, is one of the most useful descriptors. Polar surface area, defined as a sum of surfaces of polar atoms, is conceptually easy to understand and seems to encode in an optimal way a combination of hydrogen-bonding features and molecular polarity.

## DISTRIBUTION OF SOLUTES BETWEEN IMMISCIBLE SOLVENTS

If an excess of liquid or solid is added to a mixture of two immiscible liquids, it will distribute itself between the two phases so that each becomes saturated. If the substance is

### KEY CONCEPT HYDROPHOBIC PARAMETERS

Meyer in 1899[31] and Overton in 1901[32] showed that the pharmacologic effect of simple organic compounds was related to their oil/water partition coefficient, $P$. It later became clear that the partition coefficient was of little value for rationalizing specific drug activity (i.e., binding to a receptor) because specificity also relates to steric and electronic effects. However, in the early 1950s, Collander[33] showed that the rate of penetration of plant cell membranes by organic compounds was related to $P$. The partition coefficient, $P$, is a commonly used way of defining relative hydrophobicity (also known as lipophilicity) of compounds. For more about partition coefficients, see the text by Hansch and Leo.[34]

added to the immiscible solvents in an amount insufficient to saturate the solutions, it will still become distributed between the two layers in a definite concentration ratio.

If $C_1$ and $C_2$ are the equilibrium concentrations of the substance in Solvent$_1$ and Solvent$_2$, respectively, the equilibrium expression becomes

$$\frac{C_1}{C_2} = K \tag{9–13}$$

The equilibrium constant, $K$, is known as the *distribution ratio*, *distribution coefficient*, or *partition coefficient*. Equation (9–13), which is known as the *distribution law*, is strictly applicable only in dilute solutions where activity coefficients can be neglected.

### EXAMPLE 9–5

**Distribution Coefficient**

When boric acid is distributed between water and amyl alcohol at 25°C, the concentration in water is found to be 0.0510 mole/liter and in amyl alcohol it is found to be 0.0155 mole/liter. What is the distribution coefficient? We have

$$K = \frac{C_{H_2O}}{C_{alc}} = \frac{0.0510}{0.0155} = 3.29$$

No convention has been established with regard to whether the concentration in the water phase or that in the organic phase should be placed in the numerator. Therefore, the result can also be expressed as

$$K = \frac{C_{alc}}{C_{H_2O}} = \frac{0.0155}{0.0510} = 0.304$$

One should always specify, which of these two ways the distribution constant is being expressed.

Knowledge of partition is important to the pharmacist because the principle is involved in several areas of current pharmaceutical interest. These include preservation of oil–water systems, drug action at nonspecific sites, and the absorption and distribution of drugs throughout the body. Certain aspects of these topics are discussed in the following sections.

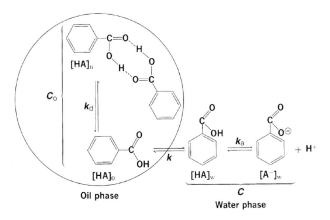

**Fig. 9–9.** Schematic representation of the distribution of benzoic acid between water and an oil phase. The oil phase is depicted as a magnified oil droplet in an oil-in-water emulsion.

## Effect of Ionic Dissociation and Molecular Association on Partition

The solute can exist partly or wholly as associated molecules in one of the phases or it may dissociate into ions in either of the liquid phases. The distribution law applies only to the concentration of the species common to both phases, namely, the *monomer* or simple molecules of the solute.

Consider the distribution of benzoic acid between an oil phase and a water phase. When it is neither associated in the oil nor dissociated into ions in the water, equation (9–13) can be used to compute the distribution constant. When association and dissociation occur, however, the situation becomes more complicated. The general case where benzoic acid associates in the oil phase and dissociates in the aqueous phase is shown schematically in **Figure 9–9**.

Two cases will be treated. *First*, according to Garrett and Woods,[35] benzoic acid is considered to be distributed between the two phases, peanut oil and water. Although benzoic acid undergoes dimerization (association to form two molecules) in many nonpolar solvents, it does not associate in peanut oil. It ionizes in water to a degree, however, depending on the pH of the solution. Therefore, in **Figure 9–9** for the case under consideration, $C_o$, the total concentration of benzoic acid in the oil phase, is equal to $[HA]_o$, the monomer concentration in the oil phase, because association does not occur in peanut oil.

The species common to both the oil and water phases are the unassociated and undissociated benzoic acid molecules. The distribution is expressed as

$$K = \frac{[HA]_o}{[HA]_w} = \frac{C_o}{[HA]_w} \qquad (9\text{--}14)$$

where $K$ is the *true distribution coefficient*, $[HA]_o = C_o$ is the molar concentration of the simple benzoic acid molecules in the oil phase, and $[HA]_w$ is the molar concentration of the undissociated acid in the water phase.

The total acid concentration obtained by analysis of the aqueous phase is

$$C_w = [HA]_w + [A^-]_w \qquad (9\text{--}15)$$

and the experimentally observed or *apparent distribution coefficient* is

$$K' = \frac{[HA]_o}{[HA]_w + [A^-]_w} = \frac{C_o}{C_w} \qquad (9\text{--}16)$$

As seen in **Figure 9–9**, the observed distribution coefficient depends on two equilibria: the distribution of the undissociated acid between the immiscible phases as expressed in equation (9–14) and the species distribution of the acid in the aqueous phase, which depends on the hydrogen ion concentration $[H_3O^+]$ and the dissociation constant $K_a$ of the acid, where

$$K_a \frac{[H_3O^+][A^-]_w}{[HA]_w} \qquad (9\text{--}17)$$

Association of benzoic acid in peanut oil does not occur, and $K_d$ (the equilibrium constant for dissociation of associated benzoic acid into monomer in the oil phase) can be neglected in this case.

Given these equations and the fact that the concentration, $C$, of the acid in the aqueous phase before distribution, assuming equal volumes of the two phases, is[*]

$$C = C_o + C_w \qquad (9\text{--}18)$$

one arrives at the combined result:[†]

$$\frac{K_a + [H_3O^+]}{C_w} = \frac{K_a}{C} + \frac{K+1}{C}[H_3O^+] \qquad (9\text{--}19)$$

Expression (9–19) is a linear equation of the form $y = a + bx$, and therefore a plot of $(K_a + [H_3O^+])/C_w$ against $[H_3O^+]$

---

[*]The meaning of $C$ in equation (9–18) is understood readily by considering a simple illustration. Suppose one begins with 1 liter of oil and 1 liter of water, and after benzoic acid has been distributed between the two phases, the concentration $C_o$ of benzoic acid in the oil is 0.01 mole/liter and the concentration $C_w$ of benzoic acid in the aqueous phase is 0.01 mole/liter. Accordingly, there is 0.02 mole/2 liter or 0.01 mole of benzoic acid per liter of total mixture after distribution equilibrium has been attained. Equation (9–18) gives

$$C = C_o + C_w = 0.01 \text{ mole/liter} + 0.01 \text{ mole/liter}$$
$$= 0.02 \text{ mole/liter}$$

The concentration, $C$, obviously is not the total concentration of the acid in the mixture at equilibrium but, rather, twice this value. $C$ is therefore seen to be the concentration of benzoic acid in the water phase (or the oil phase) before the distribution is carried out.

[†]Equation (9–19) is obtained as follows. Substituting for $[A^-]_w$ from equation (9–17) into equation (9–16) gives

$$K' = \frac{[HA]_o}{[HA]_w + \dfrac{K_a[HA]_w}{[H_3O^+]}} = \frac{[HA]_o[H_3O^+]}{[HA]_w (K_a + [H_3O^+])} \qquad (a)$$

Then $[HA]_w$ from equation (9–14) is substituted into $(a)$ to eliminate $[HA]_o$ from the equation:

$$K' = \frac{[HA]_o[H_3O^+]}{[HA]_o/K(K_a + [H_3O^+])} = \frac{K[H_3O^+]}{K_a + [H_3O^+]} \qquad (b)$$

yields a straight line with a slope $b = (K + 1)/C$ and an intercept $a = K_a/C$. The true distribution coefficient, $K$, can thus be obtained over the range of hydrogen ion concentration considered. Alternatively, the true distribution constant could be obtained according to equation (9–14) by analysis of the oil phase and of the water phase at a sufficiently low pH (2.0) at which the acid would exist completely in the un-ionized form. One of the advantages of equation (9–19), however, is that the oil phase need not be analyzed; only the hydrogen ion concentration and $C_w$, the total concentration remaining in the aqueous phase at equilibrium, need be determined.

### EXAMPLE 9–6

According to Garrett and Woods,[35] the plot of $(K_a + [H_3O^+])/C_w$ against $[H_3O^+]$ for benzoic acid distributed between equal volumes of peanut oil and a buffered aqueous solution yields a slope $b = 4.16$ and an intercept $a = 4.22 \times 10^{-5}$. The $K_a$ of benzoic acid is $6.4 \times 10^{-5}$. Compute the true partition coefficient, $K$, and compare it with the value $K = 5.33$ obtained by the authors. We have

$$b = (K + 1)/C$$

or

$$K = bC/1$$

Because

$$a = k_a/C \text{ or } c = \frac{K_a}{a}$$

the expression becomes

$$K = \frac{bk_a}{a}1 = \frac{bK_a - a}{a}$$

and

$$K = \frac{(4.16 \times 6.4 \times 10^{-5}) - 4.22 \times 10^{-5}}{4.22 \times 10^{-5}} = 5.31$$

*Second*, let us now consider the case in which the solute is associated in the organic phase and exists as simple molecules in the aqueous phase. If benzoic acid is distributed between benzene and acidified water, it exists mainly as associated molecules in the benzene layer and as undissociated molecules in the aqueous layer.

---

The apparent distribution constant is eliminated by substituting equation (b) into equation (9–16) to give

$$\frac{K[H_3O^+]}{K_a + [H_3O^+]} = \frac{C_o}{C_w} \qquad (c)$$

or

$$C_o = \frac{K[H_3O^+]C_w}{K_a + [H_3O^+]} \qquad (d)$$

$C_o$ is eliminated by substituting equation (c) into equation (9–18):

$$C = \frac{K[H_3O^+]C_w}{K_a + [H_3O^+]} + C_w$$

$$= \frac{K[H_3O^+]C_w + (K_a + [H_3O^+])C_w}{K_a + [H_3O^+]}$$

Rearranging equation (d) gives the final result:

$$= \frac{K_a + [H_3O^+]}{C_w} = \frac{[H_3O^+](K + 1) + K_a}{C}$$

The equilibrium between simple molecules HA and associated molecules $(HA)_n$ in

$$(HA)_n \rightleftharpoons n(HA)$$

Associated molecules      Simple molecules

and the equilibrium constant expressing the dissociation of associated molecules into simple molecules in this solvent is

$$K_d = \frac{[HA]_o{}^n}{[(HA)]_n} \qquad (9–20)$$

or

$$[HA]_o - \sqrt[n]{K_d} \sqrt[n]{[(HA)_n]} \qquad (9–21)$$

Because benzoic acid exists predominantly in the form of double molecules in benzene, $C_o$ can replace $[(HA)_2]$, where $C_o$ is the total molar concentration of the solute in the organic layer. Then equation (9–21) can be written approximately as

$$[HA]_o \cong constant \times \sqrt{C_o} \qquad (9–22)$$

In conformity with the distribution law as given in equation (9–14), the true distribution coefficient is always expressed in terms of simple species common to both phases, that is, in terms of $[HA]_w$ and $[HA]_o$. In the benzene–water system, $[HA]_o$ is given by equation (9–22), and the modified distribution constant becomes

$$K'' = \frac{[HA]_o}{[HA]_w} = \frac{\sqrt{C_o}}{[HA]_w} \qquad (9–23)$$

The results for the distribution of benzoic acid between benzene and water, as given by Glasstone,[36] are given in **Table 9–3**.

### Extraction

To determine the efficiency with which one solvent can extract a compound from a second solvent—an operation commonly employed in analytic chemistry and in organic chemistry—we follow Glasstone.[37] Suppose that $w$ grams of a solute is extracted repeatedly from $V_1$ mL of one solvent with successive portions of $V_2$ mL of a second solvent, which is immiscible with the first. Let $w_1$ be the weight of the solute remaining in the original solvent after extracting with the first portion of the other solvent. Then, the concentration

### TABLE 9–3

**DISTRIBUTION OF BENZOIC ACID BETWEEN BENZENE AND ACIDIFIED WATER AT 6°C[*],[†]**

| $[HA]_w$ | $C_o$ | $K'' - \sqrt{C_o}/[HA]_w$ |
|----------|-------|---------------------------|
| 0.00329 | 0.0156 | 38.0 |
| 0.00579 | 0.0495 | 38.2 |
| 0.00749 | 0.0835 | 38.6 |
| 0.0114 | 0.195 | 38.8 |

[*]The concentrations are expressed in mole/liter.
[†]From S. Glasstone, *Textbook of Physical Chemistry*, Van Nostrand, New York, 1946, p. 738.

of solute remaining in the first solvent is $(w_1/V_1)$ g/mL and the concentration of the solute in the extracting solvent is $(w - w_1)/V_2$ g/mL. The distribution coefficient is thus

$$K = \frac{\text{Concentration of solute in original solvent}}{\text{Concentration of solute in extracting solvent}}$$

$$K = \frac{w_1/V_1}{(w - w_1)V_2} \tag{9–24}$$

or

$$w_1 = w\frac{KV_1}{KV_1 + V_2} \tag{9–25}$$

The process can be repeated, and after $n$ extractions,[37]

$$w_n = w\left(\frac{KV_1}{KV_1 + V_2}\right)^n \tag{9–26}$$

By use of this equation, it can be shown that most efficient extraction results when $n$ is large and $V_2$ is small, in other words, when a large number of extractions are carried out with small portions of extracting liquid. The development just described assumes complete immiscibility of the two liquids. When ether is used to extract organic compounds from water, this is not true; however, the equations provide approximate values that are satisfactory for practical purposes. The presence of other solutes, such as salts, can also affect the results by complexing with the solute or by salting out one of the phases.

### EXAMPLE 9–7

#### Distribution Coefficient

The distribution coefficient for iodine between water and carbon tetrachloride at 25°C is $K = C_{H_2O}/C_{CCl_4} = 0.012$. How many grams of iodine are extracted from a solution in water containing 0.1 g in 50 mL by one extraction with 10 mL of $CCl_4$? How many grams are extracted by two 5-mL portions of $CCl_4$? We have

$$w_1 = 0.10 \times \frac{0.012 \times 50}{(0.012 \times 50) + 10}$$

$$= 0.0057 \text{ g remains or } 0.0943 \text{ g is extracted}$$

$$w_2 = 0.10 \times \left(\frac{0.012 \times 50}{(0.012 \times 50) + 5}\right)^2$$

$$= 0.0011 \text{ g of iodine}$$

Thus, 0.0011 g of iodine remains in the water phase, and the two portions of $CCl_4$ have extracted 0.0989 g.

## Solubility and Partition Coefficients

Hansch et al.[38] observed a relationship between aqueous solubilities of nonelectrolytes and partitioning. Yalkowsky and Valvani[39] obtained an equation for determining the aqueous solubility of liquid or crystalline organic compounds:

$$\log S = -\log K$$
$$-1.11\frac{\Delta S_f(\text{mp} - 25)}{1364} + 0.54 \tag{9–27}$$

where $S$ is aqueous solubility in moles/liter, $K$ is the octanol–water partition coefficient, $\Delta S_f$ is the molar entropy of fusion,

and mp is the melting point of a solid compound on the centigrade scale. For a liquid compound, mp is assigned a value of 25 so that the second right-hand term of equation (9–27) becomes zero.

The entropy of fusion and the partition coefficient can be estimated from the chemical structure of the compound. For rigid molecules, $\Delta S_f = 13.5$ entropy units (eu). For molecules with $n$ greater than five nonhydrogen atoms in a flexible chain,

$$\Delta S_f = 13.5 + 2.5(n - 5)\text{eu} \tag{9–28}$$

Leo et al.[38] provided partition coefficients for a large number of compounds. When experimental values are not available, group contribution methods[38,40] are available for estimating partition coefficients.

### EXAMPLE 9–8

#### Molar Aqueous Solubility

Estimate the molar aqueous solubility of heptyl $p$-aminobenzoate, mp 75°C, at 25°C:

It is first necessary to calculate $\Delta S_f$ and $\log K$.

There are nine nonhydrogens in the flexible chain (C, O, and the seven carbons of $CH_3$). Using equation (9–28), we obtain

$$\Delta S_f = 13.5 + 2.5(9 - 5) = 23.5 \text{ eu}$$

For the partition coefficient, Leo et al.[38] give for $\log K$ of benzoic acid a value of 1.87, the contribution of $NH_2$ is $-1.16$, and that of $CH_2$ is 0.50, or $7 \times 0.50 = 3.50$ for the seven carbon atoms of $CH_3$ in the chain:

$$\log K(\text{heptyl } p\text{-aminobenzoate}) = 1.87 - 1.16 + 3.50 = 4.21$$

We substitute these values into equation (9–27) to obtain

$$\log S = -4.21 - 1.11\left(\frac{23.5(75 - 25)}{1364}\right) + 0.54$$
$$\log S = -4.63$$
$$S_{(\text{calc})} = 2.36 \times 10^{-5} \text{ M}$$
$$S_{(\text{obs})} = 2.51 \times 10^{-5} \text{ M}$$

The oil–water partition coefficient is an indication of the lipophilic or hydrophobic character of a drug molecule. Passage of drugs through lipid membranes and interaction with macromolecules at receptor sites sometimes correlate well with the octanol–water partition coefficient of the drug. In the last few sections, the student has been introduced to the distribution of drug molecules between immiscible solvents together with some important applications of partitioning; a number of useful references are available for further study on the subject.[41–44] Three excellent books[45–47] on solubility in the pharmaceutical sciences will be of interest to the serious student of the subject.

## CHAPTER SUMMARY

The concept of *solubility* was presented in this chapter. As described, solubility is defined in quantitative terms as the concentration of solute in a saturated solution at a certain temperature, and in a qualitative way, it can be defined as the spontaneous interaction of two or more substances to form a homogeneous molecular dispersion. Solubility is an *intrinsic* material property that can be altered only by chemical modification of the molecule. Solubilization was not covered in this chapter. In order to determine the true solubility of a compound, one must measure the thermodynamic solubility. However, given the constraints that were discussed an alternate method, kinetic solubility determination, was presented that offers a more practical alternative given the realities of the situation. Distribution phenomena were also discussed in some detail. The distribution behavior of drug molecules is important to many pharmaceutical processes including physicochemical (e.g., when formulating drug substances) and biological (e.g., absorption across a biological membrane) processes.

 **Practice problems for this chapter can be found at thePoint.lww.com/Sinko6e.**

# References

1. S. Stegemann, F. Leveiller, D. Franchi, H. de Jong, and H. Linden, Eur. J. Pharm. Sci. **31,** 249, 2007.
2. J. Bauer, S. Spanton, R. Henry, J. Quick, W. Dziki, and W. Porter, J. Morris, Pharm. Res. **18,** 859, 2001.
3. B. Faller and P. Ertl, Adv. Drug Deliv. Rev. **59,** 535, 2007.
4. S. N. Bhattachar, L. A. Deschenes, and J. A. Wesley, Drug Discov. Today **11,** 1012, 2006.
5. M. R. Loran and E. P. Guth, J. Am. Pharm. Assoc. Sci. Ed. **40,** 465, 1951.
6. W. J. O'Malley, L. Pennati, and A. Martin, J. Am. Pharm. Assoc. Sci. Ed. **47,** 334, 1958.
7. R. J. James and R. L. Goldemberg, J. Soc. Cosm. Chem. **11,** 461, 1960.
8. W. G. Gorman and G. D. Hall, J. Pharm. Sci. **53,** 1017, 1964.
9. T. C. Butler, J. M. Ruth, and G. F. Tucker, J. Am. Chem. Soc. **77,** 1486, 1955.
10. P. A. Schwartz, C. T. Rhodes, and J. W. Cooper, Jr., J. Pharm. Sci. **66,** 994, 1977.
11. Z. T. Chowhan, J. Pharm. Sci. **67,** 1257, 1978.
12. I. Zimmerman, Int. J. Pharm. **31,** 69, 1986.
13. J. Blanchard, J. O. Boyle, and S. Van Wagenen, J. Pharm. Sci. **77,** 548, 1988.
14. G. M. Krause and J. M. Cross, J. Am. Pharm. Assoc. Sci. Ed. **40,** 137, 1951.
15. J. R. Stockton and C. R. Johnson, J. Am. Chem. Soc. **33,** 383, 1944.
16. T. Higuchi, M. Gupta, and L. W. Busse, J. Am. Pharm. Assoc. Sci. Ed. **42,** 157, 1953.
17. T. D. Edmonson and J. E. Goyan, J. Am. Pharm. Assoc. Sci. Ed. **47,** 810, 1958.
18. S. F. Kramer and G. L. Flynn, J. Pharm. Sci. **61,** 1896, 1972.
19. B. Fritz, J. L. Lack, and L. D. Bighley, J. Pharm. Sci. **60,** 1617, 1971.
20. J. H. Hildebrand and R. L. Scott, *Solubility of Nonelectrolytes*, Dover, New York, 1964, p. 417.
21. Remington, *The Science and Practice of Pharmacy*, 21st Ed., Lippincott Williams & Wilkins, Baltimore, MD, 2006, p. 216.
22. Remington, *The Science and Practice of Pharmacy*, 21st Ed., Lippincott Williams & Wilkins, Baltimore, MD, 2006, p. 217.
23. A. Glomme, J. März, and J. B. Dressman, J. Pharm. Sci. **94,** 1, 2005.
24. S. N. Bhattachar, J. A. Wesley, and C. Seadeek, J. Pharm. Biomed. Anal. **41,** 152, 2006.
25. A. Avdeef, *Absorption and Drug Development Solubility, Permeability and Charge State*, John Wiley & Sons, Hoboken, NJ, 2003.
26. K. A. Dehring, J. Pharm. Biomed. Anal. **36,** 447, 2004.
27. T. M. Chen, H. Shen, and C. Zhu, Comb. Chem. High Throughput Screen, **5,** 575, 2002.
28. C. A. Lipinski, F. Lombardo, B. W. Dominy, and P. J. Feeney, Adv. Drug Del. Rev. **46,** 3, 2001.
29. C. A. S. Bergstrom, K. Luthman, and P. Artursson, Eur. J. Pharm. Sci. **22,** 387, 2004.
30. A. Avdeef, D. Voloboy, and A. Foreman, Dissolution and Solubility, in J. B. Taylor and D. J. Triggle (Eds.), *Comprehensive Medicinal Chemistry II*, Vol. 5, Elsevier, Oxford, UK, ISBN: 978-0-08-044513-7, 2007, pp. 399–423.
31. H. Meyer, Naunyn Schmiedebergs Arch. Pharmacol. **42,** 109, 1899.
32. E. Overton, *Studien über die Narkose*, Fischer, Jena, Germany, 1901.
33. R. Collander, Annu. Rev. Plant Physiol. **8,** 335, 1957.
34. C. Hansch and A. Leo, *Substituent Constants for Correlation Analysis in Chemistry and Biology*, Wiley, New York, 1979.
35. E. R. Garrett and O. R. Woods, J. Am. Pharm. Assoc. Sci. Ed. **42,** 736, 1953.
36. S. Glasstone, *Textbook of Physical Chemistry*, Van Nostrand, New York, 1946, p. 738.
37. S. Glasstone, *Textbook of Physical Chemistry*, Van Nostrand, New York, 1946, pp. 741–742.
38. C. Hansch, J. E. Quinlan, and G. L. Lawrence, J. Org. Chem. **33,** 347, 1968; A. J. Leo, C. Hansch, and D. Elkin, Chem. Rev. **71,** 525, 1971; C. Hansch and A. J. Leo, *Substituent Constants for Correlation Analysis in Chemistry and Biology*, Wiley, New York, 1979.
39. S. H. Yalkowsky and S. C. Valvani, J. Pharm. Sci. **69,** 912, 1980; G. Amidon, Int. J. Pharmaceutics, **11,** 249, 1982.
40. R. F. Rekker, *The Hydrophobic Fragmental Constant*, Elsevier, New York, 1977; G. G. Nys and R. F. Rekker, Eur. J. Med. Chem. **9,** 361, 1974.
41. C. Hansch and W. J. Dunn III, J. Pharm. Sci. **61,** 1, 1972; C. Hansch and J. M. Clayton, J. Pharm. Sci. **62,** 1, 1973; R. N. Smith, C. Hansch, and M. M. Ames, J. Pharm. Sci. **64,** 599, 1975.
42. K. C. Yeh and W. I. Higuchi, J. Pharm. Soc. **65,** 80, 1976.
43. R. D. Schoenwald and R. L. Ward, J. Pharm. Sci. **67,** 787, 1978.
44. W. J. Dunn III, J. H. Block, and R. S. Pearlman (Eds.), *Partition Coefficient, Determination and Estimation*, Pergamon Press, New York, 1986.
45. K. C. James, *Solubility and Related Properties*, Marcel Dekker, New York, 1986.
46. D. J. W. Grant and T. Higuchi, *Solubility Behavior of Organic Compounds*, Wiley, New York, 1990.
47. S. H. Yalkowsky and S. Banerjee, *Aqueous Solubility*, Marcel Dekker, New York, 1992.

# Recommended Readings

S. N. Bhattachar, L. A. Deschenes, and J. A. Wesley, Drug Discov. Today **11,** 1012–1018, 2006.
B. Faller and P. Ertl, Adv. Drug Del. Rev. **59,** 533–545, 2007.
C. A. Lipinski, F. Lombardo, B. W. Dominy, and P. J. Feeney, Adv. Drug Del. Rev. **46,** 3–26, 2001.
S. H. Yalkowski, *Solubility and Solubilization in Aqueous Media*, American Chemical Society, Washington DC, 1999.

CHAPTER LEGACY

**Fifth Edition:** published as Chapter 10 (Solubility and Distribution Phenomena). Updated by Patrick Sinko.

**Sixth Edition:** published as Chapter 8 (Solubility and Distribution Phenomena). Updated by Patrick Sinko.

# 10 COMPLEXATION AND PROTEIN BINDING

**At the conclusion of this chapter the student should be able to:**

**1** Define the three classes of complexes (coordination compounds) and identify pharmaceutically relevant examples.

**2** Describe chelates, their physically properties, and what differentiates them from organic molecular complexes.

**3** Describe the types of forces that hold together organic molecular complexes and give examples.

**4** Describe the forces involved in polymer–drug complexes used for drug delivery and situations where reversible or irreversible complexes may be advantageous.

**5** Discuss the uses and give examples of cyclodextrins in pharmaceutical applications.

**6** Determine the stoichiometric ratio and stability constant for complex formation.

**7** Describe the methods of analysis of complexes and their strengths and weaknesses.

**8** Discuss the ways that protein binding can influence drug action.

**9** Describe the equilibrium dialysis and ultrafiltration methods for determining protein binding.

**10** Understand the factors affecting complexation and protein binding.

**11** Understand the thermodynamic basis for the stability of complexes.

Complexes or coordination compounds, according to the classic definition, result from a donor–acceptor mechanism or Lewis acid–base reaction between two or more different chemical constituents. Any nonmetallic atom or ion, whether free or contained in a neutral molecule or in an ionic compound, that can donate an electron pair can serve as the donor. The acceptor, or constituent that accepts a share in the pair of electrons, is frequently a metallic ion, although it can be a neutral atom. Complexes can be divided broadly into two classes depending on whether the acceptor component is a metal ion or an organic molecule; these are classified according to one possible arrangement in **Table 10–1**. A third class, the inclusion/occlusion compounds, involving the entrapment of one compound in the molecular framework of another, is also included in the table.

Intermolecular forces involved in the formation of complexes are the van der Waals forces of dispersion, dipolar, and induced dipolar types. Hydrogen bonding provides a significant force in some molecular complexes, and coordinate covalence is important in metal complexes. Charge transfer and hydrophobic interaction are introduced later in the chapter.

## METAL COMPLEXES

A satisfactory understanding of metal ion complexation is based upon a familiarity with atomic structure and molecular forces, and the reader would do well to consult texts on inorganic and organic chemistry to study those sections dealing with electronic structure and hybridization before proceeding.

## Inorganic Complexes

The ammonia molecules in hexamminecobalt (III) chloride, as the compound $[Co(NH_3)_6]^{3+} Cl_3^-$ is called, are known as the *ligands* and are said to be *coordinated* to the cobalt ion. The *coordination number* of the cobalt ion, or number of ammonia groups coordinated to the metal ions, is six. Other complex ions belonging to the inorganic group include $[Ag(NH_3)_2]^+$, $[Fe(CN)_6]^{4-}$, and $[Cr(H_2O)_6]^{3+}$.

Each ligand donates a pair of electrons to form a coordinate covalent link between itself and the central ion having an incomplete electron shell. For example,

$$Co^{3+} + 6\ddot{N}H_3 = [Co(NH_3)_6]^{3+}$$

Hybridization plays an important part in coordination compounds in which sufficient bonding orbitals are not ordinarily available in the metal ion. The reader's understanding of hybridization will be refreshed by a brief review of the argument advanced for the quadrivalence of carbon. It will be recalled that the ground-state configuration of carbon is

1s    2s     2p

This cannot be the bonding configuration of carbon, however, because it normally has four rather than two valence electrons. Pauling[1] suggested the possibility of *hybridization* to account for the quadrivalence. According to this mixing process, one of the 2s electrons is promoted to

**TABLE 10–1**
**CLASSIFICATION OF COMPLEXES***

**I.** Metal ion complexes
  A. Inorganic type
  B. Chelates
  C. Olefin type
  D. Aromatic type
    1. Pi ($\pi$) complexes
    2. Sigma ($\sigma$) complexes
    3. "Sandwich" compounds
**II.** Organic molecular complexes
  A. Quinhydrone type
  B. Picric acid type
  C. Caffeine and other drug complexes
  D. Polymer type
**III.** Inclusion/occlusion compounds
  A. Channel lattice type
  B. Layer type
  C. Clathrates
  D. Monomolecular type
  E. Macromolecular type

*This classification does not pretend to describe the mechanism or the type of chemical bonds involved in complexation. It is meant simply to separate out the various types of complexes that are discussed in the literature. A highly systematized classification of electron donor–acceptor interactions is given by R. S. Mulliken, J. Phys. Chem. **56**, 801, 1952.

the available 2p orbital to yield four equivalent bonding orbitals:

$$1s \quad 2s \quad 2p$$

These are directed toward the corners of a tetrahedron, and the structure is known as an $sp^3$ hybrid because it involves one s and three p orbitals. In a double bond, the carbon atom is considered to be $sp^2$ hybridized, and the bonds are directed toward the corners of a triangle.

Orbitals other than the 2s and 2p orbitals can become involved in hybridization. The transition elements, such as iron, copper, nickel, cobalt, and zinc, seem to make use of their 3d, 4s, and 4p orbitals in forming hybrids. These hybrids account for the differing geometries often found for the complexes of the transition metal ions. **Table 10–2** shows some compounds in which the central atom or metal ion is hybridized differently and the geometry that results.

Ligands such as $H_2\ddot{O}$ $H_3\ddot{N}$, $C\ddot{N}^-$, or $\ddot{Cl}^-$ donate a pair of electrons in forming a complex with a metal ion, and the electron pair enters one of the unfilled orbitals on the metal ion. A useful but not inviolate rule to follow in estimating the type of hybridization in a metal ion complex is to select that complex in which the metal ion has its 3d levels filled or that can use the lower-energy 3d and 4s orbitals primarily in the hybridization. For example, the ground-state electronic configuration of $Ni^{2+}$ can be given as

$$3d \qquad 4s \qquad 4p$$

In combining with $4CN^-$ ligands to form $[Ni(CN)_4]^{2-}$, the electronic configuration of the nickel ion may become either

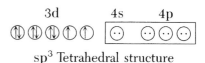

$sp^3$ Tetrahedral structure

or

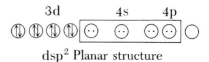

$dsp^2$ Planar structure

in which the electrons donated by the ligand are shown as dots. The $dsp^2$ or square planar structure is predicted to be the complex formed because it uses the lower-energy 3d orbital. By the preparation and study of a number of complexes, Werner deduced many years ago that this is indeed the structure of the complex.

Similarly, the trivalent cobalt ion, Co(III), has the ground-state electronic configuration

$$3d \qquad 4s \qquad 4p$$

and one may inquire into the possible geometry of the complex $[Co(NH_3)_6]^{3+}$. The electronic configuration of the metal ion leading to filled 3d levels is

$$3d \qquad 4s \qquad 4p$$

$d^2sp^3$ Octahedral

and thus the $d^2sp^3$ or octahedral structure is predicted as the structure of this complex. Chelates (see following section) of octahedral structure can be resolved into optical isomers, and in an elegant study, Werner[2] used this technique to prove that cobalt complexes are octahedral.

In the case of divalent copper, Cu(II), which has the electronic configuration

$$3d \qquad 4s \qquad 4p$$

the formation of the complex $[Cu(NH_3)_4]^{2+}$ requires the promotion of one d electron of $Cu^{2+}$ to a 4p level to obtain a filled 3d configuration in the complexed metal ion, and a $dsp^2$ or planar structure is obtained:

$$3d \qquad 4s \qquad 4p$$

Although the energy required to elevate the d electron to the 4p level is considerable, the formation of a planar complex

## TABLE 10–2
## BOND TYPES OF REPRESENTATIVE COMPOUNDS

| Coordination Number | Orbital Configuration | Bond Geometry | Example | |
|---|---|---|---|---|
| | | | Formula | Structure |
| 2 | sp | Linear | $O_2$ | O—O |
| 3 | $sp^2$ | Trigonal | $BCl_3$ | |
| 4 | $sp^3$ | Tetrahedral | $CH_4$ | |
| 4 | $dsp^2$ | Square planar | $Cu(NH_3)_4^2$ | |
| 5 | $dsp^3$ | Bipyramidal | $PF_5$ | |
| 6 | $d^2sp^3$ | Octahedral | $Co(NH_3)_6^3$ | |

having the 3d levels filled entirely more than "pays" for the expended energy.

The metal ion Fe(III) has the ground-state configuration

and in forming the complex $[Fe(CN)_6]^{3-}$, no electron promotion takes place,

because no stabilization is gained over that which the $d^2sp^3$ configuration already possesses. Compounds of this type, in which the ligands lie "above" a partially filled orbital, are termed *outer-sphere complexes*; when the ligands lie "below" a partially filled orbital, as in the previous example, the com-

pound is termed an *inner-sphere complex*. The presence of unpaired electrons in a metal ion complex can be detected by *electron spin resonance spectroscopy*.

## Chelates

Chelation places stringent steric requirements on both metal and ligands. Ions such as Cu(II) and Ni(II), which form square planar complexes, and Fe(III) and Co(III), which form octahedral complexes, can exist in either of two geometric forms. As a consequence of this isomerism, only *cis-coordinated ligands*—ligands adjacent on a molecule—will be readily replaced by reaction with a chelating agent. Vitamin $B_{12}$ and the hemoproteins are incapable of reacting with chelating agents because their metal is already coordinated in such a way that only the *trans*coordination positions of the metal are available for complexation. In contrast, the metal ion in certain enzymes, such as alcohol dehydrogenase, which contains

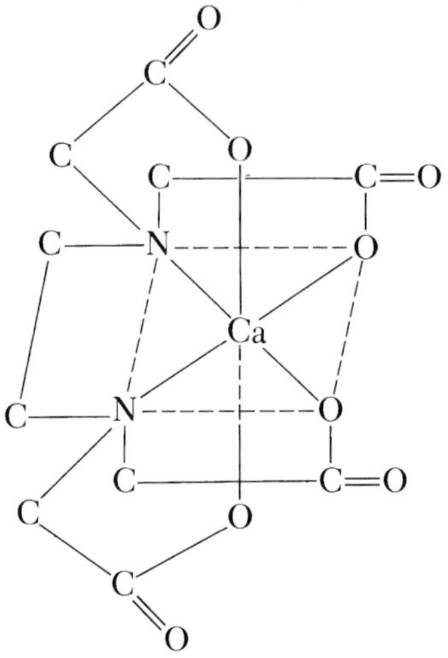

**Fig. 10–1.** Calcium ions sequestered by ethylendediaminetetraacetic acid.

zinc, can undergo chelation, suggesting that the metal is bound in such a way as to leave two *cis* positions available for chelation.

Chlorophyll and hemoglobin, two extremely important compounds, are naturally occurring chelates involved in the life processes of plants and animals. Albumin is the main carrier of various metal ions and small molecules in the blood serum. The amino-terminal portion of human serum albumin binds Cu(II) and Ni(II) with higher affinity than that of dog serum albumin. This fact partly explains why humans are less susceptible to copper poisoning than are dogs. The binding of copper to serum albumin is important because this metal is possibly involved in several pathologic conditions.[3] The synthetic chelating agent ethylenediaminetetraacetic acid (**Fig. 10–1**) has been used to tie up or *sequester* iron and copper ions so that they cannot catalyze the oxidative degradation of ascorbic acid in fruit juices and in drug preparations. In the process of sequestration, the chelating agent and metal ion form a water-soluble compound. Ethylenediaminetetraacetic acid is widely used to sequester and remove calcium ions from hard water.

Chelation can be applied to the assay of drugs. A calorimetric method for assaying procainamide in injectable solutions is based on the formation of a 1:1 complex of procainamide with cupric ion at pH 4 to 4.5. The complex absorbs visible radiation at a maximum wavelength of 380 nm.[4] The many uses to which metal complexes and chelating agents can be put are discussed by Martell and Calvin.[5]

## ORGANIC MOLECULAR COMPLEXES

An organic coordination compound or molecular complex consists of constituents held together by weak forces of the donor–acceptor type or by hydrogen bonds.

The difference between complexation and the formation of organic compounds has been shown by Clapp.[6] The compounds dimethylaniline and 2,4,6-trinitroanisole react in the cold to give a molecular complex:

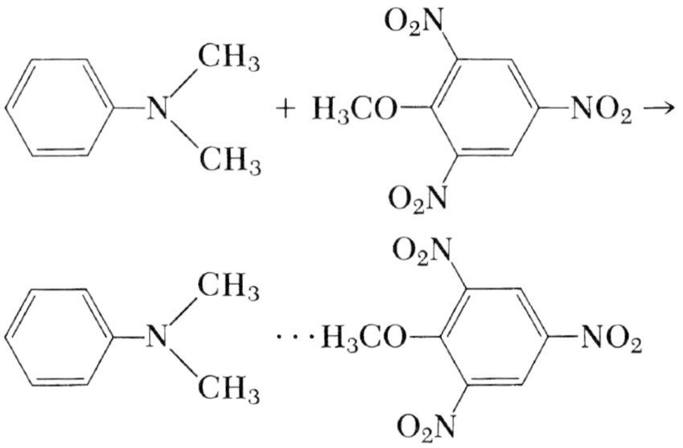

**(10–1)**

On the other hand, these two compounds react at an elevated temperature to yield a salt, the constituent molecules of which are held together by primary valence bonds:

The dotted line in the complex of equation **(10–1)** indicates that the two molecules are held together by a weak secondary valence force. It is not to be considered as a clearly defined bond but rather as an overall attraction between the two aromatic molecules.

The type of bonding existing in molecular complexes in which hydrogen bonding plays no part is not fully understood,

### KEY CONCEPT CHELATES

A substance containing two or more donor groups may combine with a metal to form a special type of complex known as a *chelate* (Greek: "kelos, claw"). Some of the bonds in a chelate may be ionic or of the primary covalent type, whereas others are coordinate covalent links. When the ligand provides one group for attachment to the central ion, the chelate is called *monodentate.* Pilocarpine behaves as a monodentate ligand toward Co(II), Ni(II), and Zn(II) to form chelates of pseudotetrahedral geometry.[7]

The donor atom of the ligand is the pyridine-type nitrogen of the imidazole ring of pilocarpine. Molecules with two and three donor groups are called *bidentate* and *tridentate*, respectively. Ethylenediaminetetraacetic acid has six points for attachment to the metal ion and is accordingly *hexadentate*; however, in some complexes, only four or five of the groups are coordinated.

but it may be considered for the present as involving an electron donor–acceptor mechanism corresponding to that in metal complexes but ordinarily much weaker.

$$(10\text{–}2)$$

Many organic complexes are so weak that they cannot be separated from their solutions as definite compounds, and they are often difficult to detect by chemical and physical means. The energy of attraction between the constituents is probably less than 5 kcal/mole for most organic complexes. Because the bond distance between the components of the complex is usually greater than 3 Å, a covalent link is not involved. Instead, one molecule polarizes the other, resulting in a type of ionic interaction or charge transfer, and these molecular complexes are often referred to as *charge transfer complexes*. For example, the polar nitro groups of trinitrobenzene induce a dipole in the readily polarizable benzene molecule, and the electrostatic interaction that results leads to complex formation:

(Donor)  (Acceptor)

Electron drift or partial electron
transfer by polarization ($\pi$ bonding)

X-ray diffraction studies of complexes formed between trinitrobenzene and aniline derivatives have shown that one of the nitro groups of trinitrobenzene lies over the benzene ring of the aniline molecule, the intermolecular distance between the two molecules being about 3.3 Å. This result strongly suggests that the interaction involves $\pi$ bonding between the $\pi$ electrons of the benzene ring and the electron-accepting nitro group.

A factor of some importance in the formation of molecular complexes is the steric requirement. If the approach and close association of the donor and acceptor molecules are hindered by steric factors, the complex is not likely to form. Hydrogen

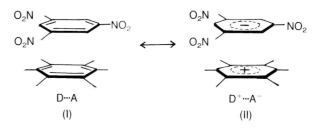

**Fig. 10–2.** Resonance in a donor–acceptor complex of trinitrobenzene (acceptor, top) and hexamethylbenzene (donor, bottom). (From F. Y. Bullock, in M. Florkin and E. H. Stotz (Eds.), *Comprehensive Biochemistry*, Elsevier, New York, 1967, pp. 82–85. With permission.)

bonding and other effects must also be considered, and these are discussed in connection with the specific complexes considered on the following pages.

The difference between a *donor–acceptor* and a *charge transfer* complex is that in the latter type, resonance makes the main contribution to complexation, whereas in the former, London dispersion forces and dipole–dipole interactions contribute more to the stability of the complex. A resonance interaction is shown in **Figure 10–2** as depicted by Bullock.[8] Trinitrobenzene is the acceptor, A, molecule and hexamethylbenzene is the donor, D. On the left side of the figure, weak dispersion and dipolar forces contribute to the interaction of A and D; on the right side of the figure, the interaction of A and D results from a significant transfer of charge, making the electron acceptor trinitrobenzene negatively charged ($A^-$) and leaving the donor, hexamethylbenzene, positively charged ($D^+$). The overall donor–acceptor complex is shown by the double-headed arrow to resonate between the uncharged D $\cdots$ A and the charged $D^+ \cdots A^-$ moieties. If, as in the case of hexamethylbenzene–trinitrobenzene, the resonance is fairly weak, having an intermolecular binding energy $\Delta G$ of about $-4700$ calories, the complex is referred to as a *donor–acceptor complex*. If, on the other hand, resonance between the charge transfer structure ($D^+ \cdots A^-$) and the uncharged species (D $\cdots$ A) contributes greatly to the binding of the donor and acceptor molecule, the complex is called a *charge transfer complex*. Finally, those complexes bound together by van der Waals forces, dipole–dipole interactions, and hydrogen bonding but lacking charge transfer are known simply as *molecular complexes*. In both charge transfer and donor–acceptor complexes, new absorption bands occur in the spectra, as shown later in **Figure 10–13**. In this book we do not attempt to separate the first two classes, but rather refer to all interactions that produce absorption bands as charge transfer or as electron donor–acceptor complexes without distinction. Those complexes that do not show new bands are called molecular complexes.

Charge transfer complexes are of importance in pharmacy. Iodine forms 1:1 charge transfer complexes with the drugs disulfiram, chlomethiazole, and tolnaftate. These drugs have recognized pharmacologic actions of their own:

Disulfiram is used against alcohol addiction, clomethiazole is a sedative–hypnotic and anticonvulsant, and tolnaftate is an antifungal agent. Each of these drugs possesses a nitrogen–carbon–sulfur moiety (see the accompanying structure of tolnaftate), and a complex may result from the transfer of charge from the pair of free electrons on the nitrogen and/or sulfur atoms of these drugs to the antibonding orbital of the iodine atom. Thus, by tying up iodine, molecules containing the N—C=S moiety inhibit thyroid action in the body.[9]

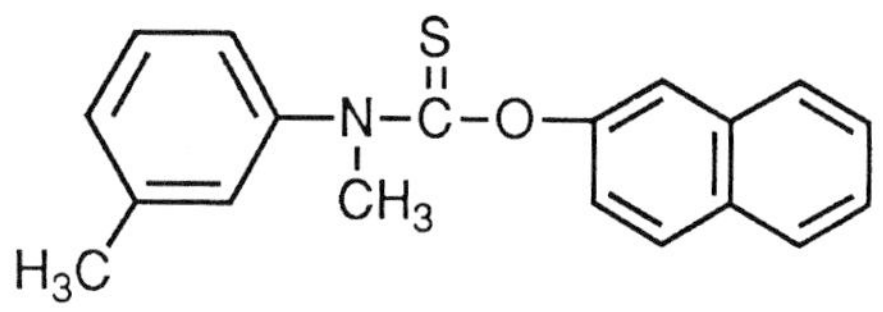

Tolnaftate (Tinactin)

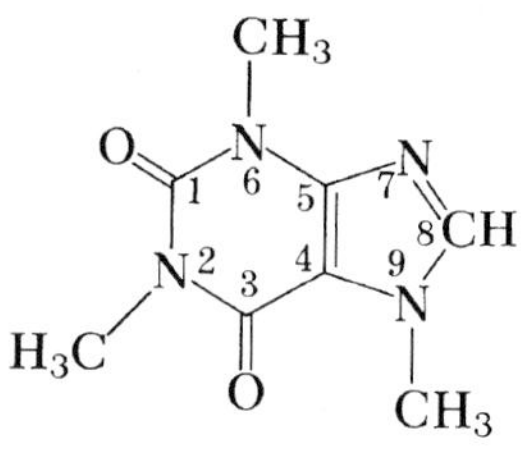

## Drug Complexes

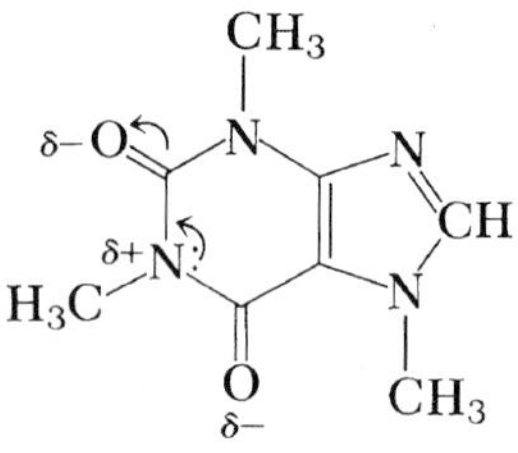

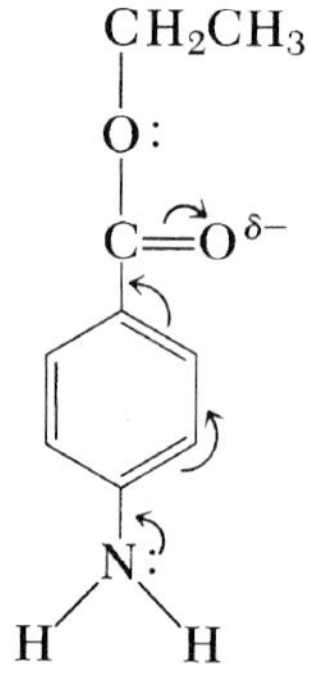

Higuchi and his associates[10] investigated the complexing of caffeine with a number of acidic drugs. They attributed the interaction between caffeine and a drug such as a sulfonamide or a barbiturate to a dipole–dipole force or hydrogen bonding between the polarized carbonyl groups of caffeine and the hydrogen atom of the acid. A secondary interaction probably occurs between the nonpolar parts of the molecules, and the resultant complex is "squeezed out" of the aqueous phase owing to the great internal pressure of water. These two effects lead to a high degree of interaction.

The complexation of esters is of particular concern to the pharmacist because many important drugs belong to this class. The complexes formed between esters and amines, phenols, ethers, and ketones have been attributed to the hydrogen bonding between a nucleophilic carbonyl oxygen and an active hydrogen. This, however, does not explain the complexation of esters such as benzocaine, procaine, and tetracaine with caffeine, as reported by Higuchi et al.[11] There are no activated hydrogens on caffeine; the hydrogen in the number 8 position (formula I) is very weak ($K_a = 1 \times 10^{-14}$) and is not likely to enter into complexation. It might be suggested that, in the caffeine molecule, a relatively positive center exists that serves as a likely site of complexation. The caffeine molecule is numbered in formula I for convenience in the discussion. As observed in formula II, the nitrogen at the 2 position presumably can become strongly electrophilic or acidic just as it is in an imide, owing to the withdrawal of electrons by the oxygens at positions 1 and 3. An ester such as benzocaine also becomes polarized (formula III) in such a way that the carboxyl oxygen is nucleophilic or basic. The complexation can thus occur as a result of a dipole–dipole interaction between the nucleophilic carboxyl oxygen of benzocaine and the electrophilic nitrogen of caffeine.

I

II

III

Caffeine forms complexes with organic acid *anions* that are more soluble than the pure xanthine, but the complexes formed with organic acids, such as gentisic acid, are less soluble than caffeine alone. Such insoluble complexes provide caffeine in a form that masks its normally bitter taste and should serve as a suitable state for chewable tablets. Higuchi and Pitman[12] synthesized 1:1 and 1:2 caffeine–gentisic acid complexes and measured their equilibrium solubility and rates of dissolution. Both the 1:1 and 1:2 complexes were less soluble in water than caffeine, and their dissolution rates were also less than that of caffeine. Chewable tablets formulated from these complexes should provide an extended-release form of the drug with improved taste.

York and Saleh[13] studied the effect of sodium salicylate on the release of benzocaine from topical vehicles, it being recognized that salicylates form molecular complexes with benzocaine. Complexation between drug and complexing agents can improve or impair drug absorption and bioavailability; the authors found that the presence of sodium salicylate significantly influenced the release of benzocaine, depending on the type of vehicle involved. The largest increase in absorption was observed for a water-miscible polyethylene glycol base.

TABLE 10–3
## TABLE 10–3
### SOME MOLECULAR ORGANIC COMPLEXES OF PHARMACEUTICAL INTEREST*

| Agent | Compounds That Form Complexes with the Agent Listed in the First Column |
|---|---|
| Polyethylene glycols | m-Hydroxybenzoic acid, p-hydroxybenzoic acid, salicylic acid, o-phthalic acid, acetylsalicylic acid, resorcinol, catechol, phenol, phenobarbital, iodine (in $I_2 \cdot KI$ solutions), bromine (in presence of HBr) |
| Povidone (polyvinyl-pyrrolidone, PVP) | Benzoic acid, m-hydroxybenzoic acid, p-hydroxybenzoic acid, salicylic acid, sodium salicylate, p-aminobenzoic acid, mandelic acid, sulfathiazole, chloramphenicol, phenobarbital |
| Sodium carboxymethylcellulose | Quinine, benadryl, procaine, pyribenzamine |
| Oxytetracycline and tetracycline | N-Methylpyrrolidone, N,N-dimethylacetamide, γ-valerolactone, γ-butyrolactone, sodium p-aminobenzoate, sodium salicylate, sodium p-hydroxybenzoate, sodium saccharin, caffeine |

*Compiled from the results of T. Higuchi et al., J. Am. Pharm. Assoc. Sci. Ed. **43**, 393, 398, 456, 1954; **44**, 668, 1955; **45**, 157, 1956; **46**, 458, 587, 1957; and J. L. Lach et al., Drug Stand. **24**, 11, 1956. An extensive table of acceptor and donor molecules that form aromatic molecular complexes was compiled by L. J. Andrews, Chem. Rev. **54**, 713, 1954. Also refer to T. Higuchi and K. A. Connors, Phase Solubility Techniques. *Advances in Analytical Chemistry and Instrumentation*, in C. N. Reilley (Ed.), Wiley, New York, 1965, pp. 117–212.

## Polymer Complexes

Polyethylene glycols, polystyrene, carboxymethylcellulose, and similar polymers containing nucleophilic oxygens can form complexes with various drugs. The incompatibilities of certain polyethers, such as the Carbowaxes, Pluronics, and Tweens with tannic acid, salicylic acid, and phenol, can be attributed to these interactions. Marcus[14] reviewed some of the interactions that may occur in suspensions, emulsions, ointments, and suppositories. The incompatibility may be manifested as a precipitate, flocculate, delayed biologic absorption, loss of preservative action, or other undesirable physical, chemical, and pharmacologic effects.

Plaizier-Vercammen and De Nève[15] studied the interaction of povidone (PVP) with ionic and neutral aromatic compounds. Several factors affect the binding to PVP of substituted benzoic acid and nicotine derivatives. Although ionic strength has no influence, the binding increases in phosphate buffer solutions and decreases as the temperature is raised.

Crosspovidone, a cross-linked insoluble PVP, is able to bind drugs owing to its dipolar character and porous structure. Frömming et al.[16] studied the interaction of crosspovidone with acetaminophen, benzocaine, benzoic acid, caffeine, tannic acid, and papaverine hydrochloride, among other drugs. The interaction is mainly due to any phenolic groups on the drug. Hexylresorcinol shows exceptionally strong binding, but the interaction is less than 5% for most drugs studied (32 drugs). Crosspovidone is a disintegrant in pharmaceutical granules and tablets. It does not interfere with gastrointestinal absorption because the binding to drugs is reversible.

Solutes in parenteral formulations may migrate from the solution and interact with the wall of a polymeric container. Hayward et al.[17] showed that the ability of a polyolefin container to interact with drugs depends linearly on the octanol–water partition coefficient of the drug. For parabens and drugs that exhibit fairly significant hydrogen bond donor properties, a correction term related to hydrogen-bond formation is needed. Polymer–drug container interactions may result in loss of the active component in liquid dosage forms.

Polymer–drug complexes are used to modify biopharmaceutical parameters of drugs; the dissolution rate of ajmaline is enhanced by complexation with PVP. The interaction is due to the aromatic ring of ajmaline and the amide groups of PVP to yield a dipole–dipole-induced complex.[18]

Some molecular organic complexes of interest to the pharmacist are given in **Table 10–3**. (Complexes involving caffeine are listed in **Table 10–6**.)

## INCLUSION COMPOUNDS

The class of addition compounds known as *inclusion* or *occlusion* compounds results more from the architecture of molecules than from their chemical affinity. One of the constituents of the complex is trapped in the open lattice or cagelike crystal structure of the other to yield a stable arrangement.

### Channel Lattice Type

The *cholic acids* (bile acids) can form a group of complexes principally involving deoxycholic acid in combination with paraffins, organic acids, esters, ketones, and aromatic compounds and with solvents such as ether, alcohol, and dioxane. The crystals of deoxycholic acid are arranged to form a channel into which the complexing molecule can fit (**Fig. 10–3**). Such stereospecificity should permit the resolution of optical isomers. In fact, camphor has been partially resolved by complexation with deoxycholic acid, and *dl*-terpineol has been resolved by the use of digitonin, which

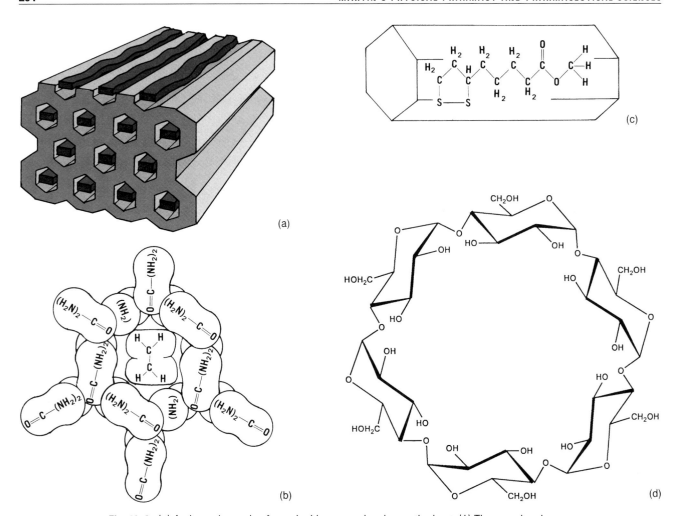

**Fig. 10–3.** (*a*) A channel complex formed with urea molecules as the host. (*b*) These molecules are packed in an orderly manner and held together by hydrogen bonds between nitrogen and oxygen atoms. The hexagonal channels, approximately 5 Å in diameter, provide room for guest molecules such as long-chain hydrocarbons, as shown here. (From J. F. Brown, Jr., Sci. Am. **207**, 82, 1962. Copyright © 1962 by Scientific American, Inc. All rights reserved.) (*c*) A hexagonal channel complex (adduct) of methyl α-lipoate and 15 g of urea in methanol prepared with gentle heating. Needle crystals of the adduct separated overnight at room temperature. This inclusion compound or adduct begins to decompose at 63°C and melts at 163°C. Thiourea may also be used to form the channel complex. (From H. Mina and M. Nishikawa, J. Pharm. Sci. **53**, 931, 1964. With permission.). (*d*) Cyclodextrin (cycloamylose, Schardinger dextrin). (See *Merck Index*, 11th Ed., Merck, Rahway, N.J., 1989, p. 425.)

occludes certain molecules in a manner similar to that of deoxycholic acid:

Deoxycholic acid

Urea and thiourea also crystallize in a channel-like structure permitting enclosure of unbranched paraffins, alcohols, ketones, organic acids, and other compounds, as shown in **Figure 10–3***a* and *b*. The well-known starch–iodine solution is a channel-type complex consisting of iodine molecules entrapped within spirals of the glucose residues.

Forman and Grady[19] found that monostearin, an interfering substance in the assay of dienestrol, could be extracted easily from dermatologic creams by channel-type inclusion in urea. They felt that urea inclusion might become a general approach for separation of long-chain compounds in assay methods. The authors reviewed the earlier literature on urea inclusion of straight-chain hydrocarbons and fatty acids.

## Layer Type

Some compounds, such as the clay montmorillonite, the principal constituent of bentonite, can trap hydrocarbons, alcohols, and glycols between the layers of their lattices.[20] Graphite can also intercalate compounds between its layers.

## Clathrates[21]

The clathrates crystallize in the form of a cagelike lattice in which the coordinating compound is entrapped. Chemical bonds are not involved in these complexes, and only the molecular size of the encaged component is of importance. Ketelaar[22] observed the analogy between the stability of a clathrate and the confinement of a prisoner. The stability of a clathrate is due to the strength of the structure, that is, to the high energy that must be expended to decompose the compound, just as a prisoner is confined by the bars that prevent escape.

Powell and Palin[23] made a detailed study of clathrate compounds and showed that the highly toxic agent hydroquinone (quinol) crystallizes in a cagelike hydrogen-bonded structure, as seen in **Figure 10–4**. The holes have a diameter of 4.2 Å and permit the entrapment of one small molecule to about every two quinol molecules. Small molecules such as methyl alcohol, $CO_2$, and HCl may be trapped in these cages, but smaller molecules such as $H_2$ and larger molecules such as ethanol cannot be accommodated. It is possible that clathrates may be used to resolve optical isomers and to bring about other processes of molecular separation.

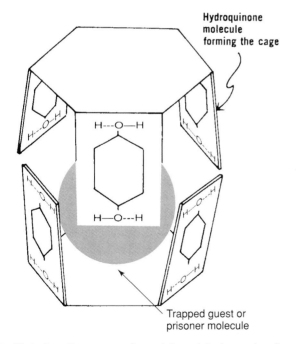

Hydroquinone
molecule
forming the cage

Trapped guest or
prisoner molecule

**Fig. 10–4.** Cagelike structure formed through hydrogen bonding of hydroquinone molecules. Small molecules such as methanol are trapped in the cages to form the clathrate. (Modified from J. F. Brown, Jr., Sci. Am. **207**, 82, 1962. Copyright © 1962 by Scientific American, Inc. All rights reserved.)

One official drug, warfarin sodium, United States Pharmacopeia, is a clathrate of water, isopropyl alcohol, and sodium warfarin in the form of a white crystalline powder.

## Monomolecular Inclusion Compounds: Cyclodextrins

Inclusion compounds were reviewed by Frank.[24a] In addition to channel- and cage-type (clathrate) compounds, Frank added classes of *mono-* and *macromolecular* inclusion compounds. Monomolecular inclusion compounds involve the entrapment of a single guest molecule in the cavity of one host molecule. Monomolecular host structures are represented by the cyclodextrins (CD). These compounds are cyclic oligosaccharides containing a minimum of six D-(+)-glucopyranose units attached by $\alpha$-1,4 linkages produced by the action on starch of *Bacillus macerans* amylase. The natural $\alpha$-, $\beta$-, and $\gamma$-cyclodextrins ($\alpha$-CD, $\beta$-CD, and $\gamma$-CD, respectively) consist of six, seven, and eight units of glucose, respectively.

Their ability to form inclusion compounds in aqueous solution is due to the typical arrangement of the glucose units (see **Fig. 10–3d**). As observed in cross section in the figure, the cyclodextrin structure forms a torus or doughnut ring. The molecule actually exists as a truncated cone, which is seen in **Figure 10–5a**; it can accommodate molecules such as mitomycin C to form inclusion compounds (**Fig. 10–5b**). The interior of the cavity is relatively hydrophobic because of the $CH_2$ groups, whereas the cavity entrances are hydrophilic owing to the presence of the primary and secondary hydroxyl groups.[25,26] $\alpha$-CD has the smallest cavity (internal diameter almost 5 Å). $\beta$-CD and $\gamma$-CD are the most useful for pharmaceutical technology owing to their larger cavity size (internal diameter almost 6 Å and 8 Å, respectively). Water inside the cavity tends to be squeezed out and to be replaced by more hydrophobic species. Thus, molecules of appropriate size and stereochemistry can be included in the cyclodextrin cavity by

### KEY CONCEPT CYCLODEXTRINS

According to Davis and Brewster,[24b] "Cyclodextrins are cyclic oligomers of glucose that can form water-soluble inclusion complexes with small molecules and portions of large compounds. These biocompatible, cyclic oligosaccharides do not elicit immune responses and have low toxicities in animals and humans. Cyclodextrins are used in pharmaceutical applications for numerous purposes, including improving the bioavailability of drugs. Of specific interest is the use of cyclodextrin-containing polymers to provide unique capabilities for the delivery of nucleic acids." Davis and Brewster[24b] discuss cyclodextrin-based therapeutics and possible future applications, and review the use of cyclodextrin-containing polymers in drug delivery.

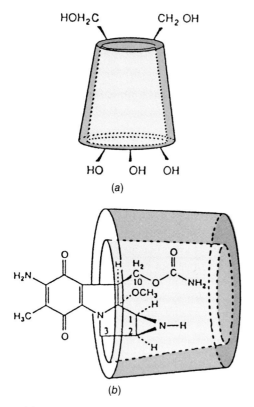

**Fig. 10–5.** (*a*) Representation of cyclodextrin as a truncated cone. (*b*) Mitomycin C partly enclosed in cyclodextrin to form an inclusion complex. (From O. Beckers, Int. J. Pharm. **52,** 240, 247, 1989. With permission.)

hydrophobic interactions. Complexation does not ordinarily involve the formation of covalent bonds. Some drugs may be too large to be accommodated totally in the cavity. As shown in **Figure 10–5***b*, mitomycin C interacts with $\gamma$-CD at one side of the torus. Thus, the aziridine ring

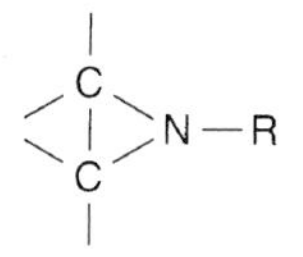

of mitomycin C is protected from degradation in acidic solution.[27] Bakensfield et al.[28] studied the inclusion of indomethacin with $\beta$-CD using a [1]H-NMR technique. The *p*-chlorobenzoyl part of indomethacin (shaded part of **Fig. 10–6**) enters the $\beta$-CD ring, whereas the substituted indole moiety (the remainder of the molecule) is too large for inclusion and rests against the entrance of the CD cavity.

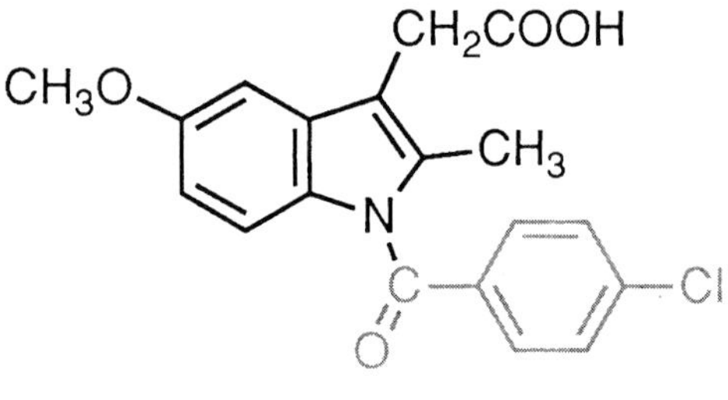

**Fig. 10–6.** Indomethacin (Indocin).

Cyclodextrins are studied as solubilizing and stabilizing agents in pharmaceutical dosage forms. Lach and associates[29] used cyclodextrins to trap, stabilize, and solubilize sulfonamides, tetracyclines, morphine, aspirin, benzocaine, ephedrine, reserpine, and testosterone. The aqueous solubility of retinoic acid (0.5 mg/liter), a drug used topically in the treatment of acne,[30] is increased to 160 mg/liter by complexation with $\beta$-CD. Dissolution rate plays an important role in bioavailability of drugs, fast dissolution usually favoring absorption. Thus, the dissolution rates of famotidine,[31] a potent drug in the treatment of gastric and duodenal ulcers, and that of tolbutamide, an oral antidiabetic drug, are both increased by complexation with $\beta$-cyclodextrin.[32]

Cyclodextrins may increase or decrease the reactivity of the guest molecule depending on the nature of the reaction and the orientation of the molecule within the CD cavity. Thus, $\alpha$-cyclodextrin tends to favor pH-dependent hydrolysis of indomethacin in aqueous solution, whereas $\beta$-cyclodextrin inhibits it.[28] Unfortunately, the water solubility of $\beta$-CD (1.8 g/100 mL at 25°C) is often insufficient to stabilize drugs at therapeutic doses and is also associated with nephrotoxicity when CD is administered by parenteral routes.[33] The relatively low aqueous solubility of the cyclodextrins may be due to the formation of intramolecular hydrogen bonds between the hydroxyl groups (see **Fig. 10–3***d*), which prevent their interaction with water molecules.[34]

Derivatives of the natural crystalline CD have been developed to improve aqueous solubility and to avoid toxicity. Partial methylation (alkylation) of some of the OH groups in CD reduces the intermolecular hydrogen bonding, leaving some OH groups free to interact with water, thus increasing the aqueous solubility of CD.[34] According to Müller and Brauns,[35] a low degree of alkyl substitution is preferable. Derivatives with a high degree of substitution lower the surface tension of water, and this has been correlated with the hemolytic activity observed in some CD derivatives. Amorphous derivatives of $\beta$-CD and $\gamma$-CD are more effective as solubilizing agents for sex hormones than the parent cyclodextrins. Complexes of testosterone with amorphous hydroxypropyl $\beta$-CD allow an efficient transport of hormone into the circulation when given sublingually.[36] This route avoids both metabolism of the drug in the intestines and rapid *first-pass* decomposition in the liver (see Chapter 15), thus improving bioavailability.

In addition to hydrophilic derivatives, hydrophobic forms of $\beta$-CD have been found useful as sustained-release drug carriers. Thus, the release rate of the water-soluble calcium antagonist diltiazem was significantly decreased by complexation with ethylated $\beta$-CD. The release rate was controlled by mixing hydrophobic and hydrophilic derivatives of cyclodextrins at several ratios.[37] Ethylated $\beta$-CD has also been used to retard the delivery of isosorbide dinitrate, a vasodilator.[38]

Cyclodextrins may improve the organoleptic characteristics of oral liquid formulations. The bitter taste of suspensions of femoxetine, an antidepressant, is greatly suppressed by complexation of the drug with $\beta$-cyclodextrin.[39]

## Molecular Sieves

Macromolecular inclusion compounds, or *molecular sieves* as they are commonly called, include zeolites, dextrins, silica gels, and related substances. The atoms are arranged in three dimensions to produce cages and channels. Synthetic zeolites may be made to a definite pore size so as to separate molecules of different dimensions, and they are also capable of ion exchange. See the review article by Frank[24a] for a detailed discussion of inclusion compounds.

# METHODS OF ANALYSIS[40]

A determination of the *stoichiometric ratio* of ligand to metal or donor to acceptor and a quantitative expression of the *stability constant* for complex formation are important in the study and application of coordination compounds. A limited number of the more important methods for obtaining these quantities is presented here.

## Method of Continuous Variation

Job[41] suggested the use of an additive property such as the spectrophotometric extinction coefficient (dielectric constant or the square of the refractive index may also be used) for the measurement of complexation. If the property for two species is sufficiently different and if no interaction occurs when the components are mixed, then the value of the property is the weighted mean of the values of the separate species in the mixture. This means that if the additive property, say dielectric constant, is plotted against the mole fraction from 0 to 1 for one of the components of a mixture where no complexation occurs, a linear relationship is observed, as shown by the dashed line in **Figure 10–7**. If solutions of two species $A$ and $B$ of equal molar concentration (and hence of a fixed total con-

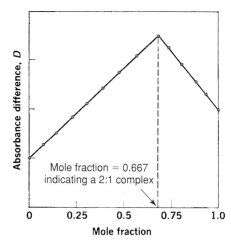

**Fig. 10–8.** A plot of absorbance difference against mole fraction showing the result of complexation.

centration of the species) are mixed and if a complex forms between the two species, the value of the additive property will pass through a maximum (or minimum), as shown by the upper curve in **Figure 10–7**. For a constant total concentration of $A$ and $B$, the complex is at its greatest concentration at a point where the species $A$ and $B$ are combined in the ratio in which they occur in the complex. The line therefore shows a break or a change in slope at the mole fraction corresponding to the complex. The change in slope occurs at a mole fraction of 0.5 in **Figure 10–7**, indicating a complex of the 1:1 type.

When spectrophotometric absorbance is used as the physical property, the observed values obtained at various mole fractions when complexation occurs are usually subtracted from the corresponding values that would have been expected had no complex resulted. This difference, $D$, is then plotted against mole fraction, as shown in **Figure 10–8**. The molar ratio of the complex is readily obtained from such a curve. By means of a calculation involving the concentration and the property being measured, the stability constant of the formation can be determined by a method described by Martell and Calvin.[42] Another method, suggested by Bent and French,[43] is given here.

If the magnitude of the measured property, such as absorbance, is proportional only to the concentration of the complex $MA_n$, the molar ratio of ligand $A$ to metal $M$ and the stability constant can be readily determined. The equation for complexation can be written as

$$M + nA = MA_n \qquad (10\text{–}3)$$

and the stability constant as

$$K = \frac{[MA_n]}{[M][A]^n} \qquad (10\text{–}4)$$

or, in logarithmic form,

$$\log [MA_n] = \log K + \log [M] + n \log [A] \quad (10\text{–}5)$$

where $[MA_n]$ is the concentration of the complex, $[M]$ is the concentration of the uncomplexed metal, $[A]$ is the

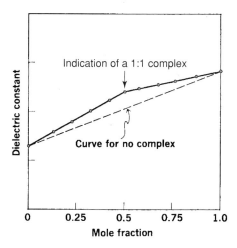

**Fig. 10–7.** A plot of an additive property against mole fraction of one of the species in which complexation between the species has occurred. The dashed line is that expected if no complex had formed. (Refer to C. H. Giles et al., J. Chem. Soc., 1952, 3799, for similar figures.)

concentration of the uncomplexed ligand, $n$ is the number of moles of ligand combined with 1 mole of metal ion, and $K$ is the equilibrium or *stability* constant for the complex. The concentration of a metal ion is held constant while the concentration of ligand is varied, and the corresponding concentration, $[MA_n]$, of complex formed is obtained from the spectrophotometric analysis.[41] Now, according to equation **(10–5)**, if log $[MA_n]$ is plotted against log $[A]$, the slope of the line yields the stoichiometric ratio or the number $n$ of ligand molecules coordinated to the metal ion, and the intercept on the vertical axis allows one to obtain the stability constant, $K$, because $[M]$ is a known quantity.

Job restricted his method to the formation of a single complex; however, Vosburgh et al.[44] modified it so as to treat the formation of higher complexes in solution. Osman and Abu-Eittah[45] used spectrophotometric techniques to investigate 1:2 metal–ligand complexes of copper and barbiturates. A greenish-yellow complex is formed by mixing a blue solution of copper (II) with thiobarbiturates (colorless). By using the Job method, an apparent stability constant as well as the composition of the 1:2 complex was obtained.

## pH Titration Method

This is one of the most reliable methods and can be used whenever the complexation is attended by a change in pH. The chelation of the cupric ion by glycine, for example, can be represented as

$$Cu^{2+} 2NH_3^+ CH_2 COO^- = Cu(NH_2CH_2COO)_2 + 2H^+$$
$$\text{(10–6)}$$

Because two protons are formed in the reaction of equation **(10–6)**, the addition of glycine to a solution containing cupric ions should result in a decrease in pH.

Titration curves can be obtained by adding a strong base to a solution of glycine and to another solution containing glycine and a copper salt and plotting the pH against the equivalents of base added. The results of such a potentiometric titration are shown in **Figure 10–9**. The curve for

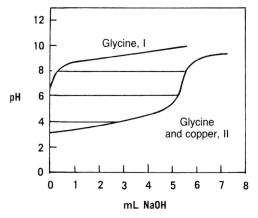

**Fig. 10–9.** Titration of glycine and of glycine in the presence of cupric ions. The difference in pH for a given quantity of base added indicates the occurrence of a complex.

the metal–glycine mixture is well below that for the glycine alone, and the decrease in pH shows that complexation is occurring throughout most of the neutralization range. Similar results are obtained with other zwitterions and weak acids (or bases), such as $N,N'$-diacetylethylenediamine diacetic acid, which has been studied for its complexing action with copper and calcium ions.

The results can be treated quantitatively in the following manner to obtain stability constants for the complex. The two successive or stepwise equilibria between the copper ion or metal, $M$, and glycine or the ligand, $A$, can be written in general as

$$M + A = MA; \quad K_1 = \frac{[MA]}{[M][A]} \quad \text{(10–7)}$$

$$MA + A = MA_2; \quad K_2 = \frac{[MA_2]}{[MA][A]} \quad \text{(10–8)}$$

and the overall reaction, **(10–7)** and **(10–8)**, is

$$M + 2A = MA_2; \quad \beta = K_1 K_2 = \frac{[MA_2]}{[M][A]^2} \quad \text{(10–9)}$$

Bjerrum[46] called $K_1$ and $K_2$ the *formation constants*, and the equilibrium constant, $\beta$, for the overall reaction is known as the *stability constant*. A quantity $n$ may now be defined. It is the number of ligand molecules bound to a metal ion. The *average* number of ligand groups bound per metal ion present is therefore designated $\tilde{n}$ ($n$ bar) and is written as

$$\overline{n} = \frac{\text{Total concentration of ligand bound}}{\text{Total concentration of metal ion}} \quad \text{(10–10)}$$

or

$$\overline{n} = \frac{[MA] + 2[MA_2]}{[M] + [MA] + [MA_2]} \quad \text{(10–11)}$$

Although $n$ has a definite value for each species of complex (1 or 2 in this case), it may have any value between 0 and the largest number of ligand molecules bound, 2 in this case. The numerator of equation **(10–11)** gives the total concentration of ligand species bound. The second term in the numerator is multiplied by 2 because two molecules of ligand are contained in each molecule of the species, $MA_2$. The denominator gives the total concentration of metal present in all forms, both bound and free. For the special case in which $\tilde{n} = 1$, equation **(10–11)** becomes

$$[MA] + 2[MA_2] = [M] + [MA] + [MA_2]$$
$$[MA_2] = [M] \quad \text{(10–12)}$$

Employing the results in equations **(10–9)** and **(10–12)**, we obtain the following relation:

$$\beta = K_1 K_2 = \frac{1}{[A]^2} \quad \text{or} \quad \log \beta = -2 \log[A]$$

and finally

$$p[A] = \frac{1}{2} \log \beta \text{ at } \overline{n} = 1 \quad \text{(10–13)}$$

where p[A] is written for $-\log[A]$. Bjerrum also showed that, to a first approximation,

$$p[A] = \log K_1 \text{ at } \bar{n} = \frac{1}{2} \qquad (10\text{–}14)$$

$$p[A] = \log K_2 \text{ at } \bar{n} = \frac{3}{2} \qquad (10\text{–}15)$$

It should now be possible to obtain the individual complex formation constants, $K_1$ and $K_2$, and the overall stability constant, $\beta$, if one knows two values: $\bar{n}$ and p[A].

Equation (10–10) shows that the concentration of bound ligand must be determined before $\tilde{n}$ can be evaluated. The horizontal distances represented by the lines in **Figure 10–9** between the titration curve for glycine alone (curve I) and for glycine in the presence of $Cu^{2+}$ (curve II) give the amount of alkali used up in the following reactions:

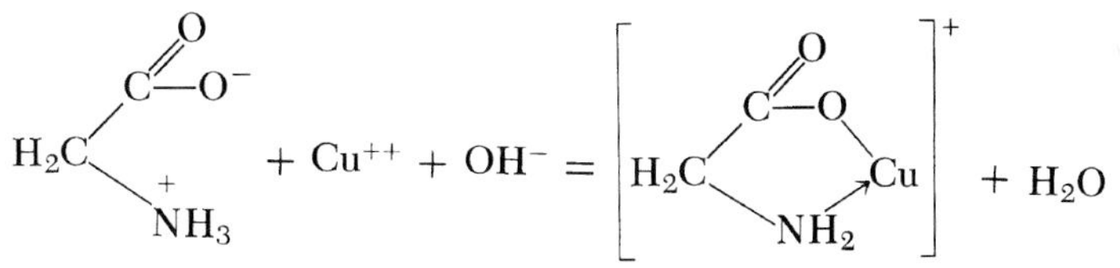

$$(10\text{–}16)$$

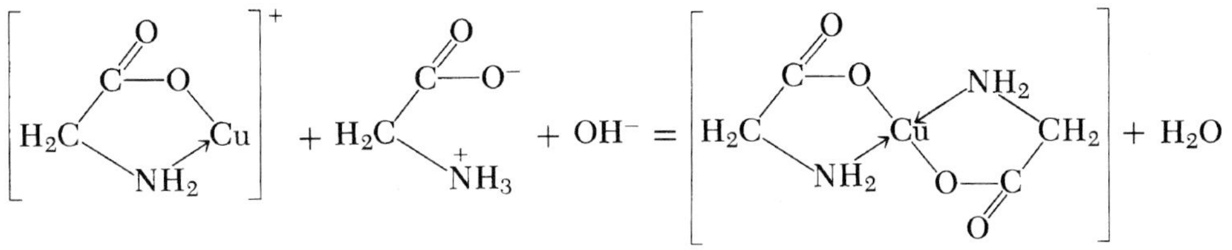

$$(10\text{–}17)$$

This quantity of alkali is exactly equal to the concentration of ligand bound at any pH, and, according to equation (10–10), when divided by the total concentration of metal ion, gives the value of $\bar{n}$.

The concentration of free glycine [A] as the "base" $NH_2CH_2COO^-$ at any pH is obtained from the acid dissociation expression for glycine:

$$NH_3^+CH_2COO^- + H_2O = H_3O^+ + NH_2CH_2COO^-$$

$$K_a = \frac{[H_3O^+][NH_2CH_2COO^-]}{[NH_3^+CH_2COO^-]} \qquad (10\text{–}18)$$

or

$$[NH_2CH_2COO^-], = [A] = \frac{K_a[HA]}{[H_3O^+]} \qquad (10\text{–}19)$$

The concentration $[NH_3^+CH_2COO^-]$, or $[HA]$, of the acid species at any pH is taken as the difference between the initial concentration, $[HA]_{init}$, of glycine and the concentration, $[NaOH]$, of alkali added. Then

$$[A] = K_a \frac{[HA]_{init} - [NaOH]}{[H_3O^+]} \qquad (10\text{–}20)$$

or

$$-\log, [A] = p[A] = pK_a - pH - \log, ([HA]_{init} - [NaOH]) \qquad (10\text{–}21)$$

where [A] is the concentration of the ligand glycine.

### EXAMPLE 10–1

**Calculate Average Number of Ligands**

If 75-mL samples containing $3.34 \times 10^{-2}$ mole/liter of glycine hydrochloride alone and in combination with $9.45 \times 10^{-3}$ mole/liter of cupric ion are titrated with 0.259 N NaOH, the two curves I and II, respectively, in Figure 10–9 are obtained. Compute $\tilde{n}$ and p[A] at pH 3.50 and pH 8.00. The $pK_a$ of glycine is 9.69 at 30°C.

(a) From Figure 10–9, the horizontal distance at pH 3.50 for the 75-mL sample is 1.60 mL NaOH or $2.59 \times 10^{-4}$ mole/mL × 1.60 $= 4.15 \times 10^{-4}$ mole. For a 1-liter sample, the value is 5.54 × $10^{-3}$ mole. The total concentration of copper ion per liter is $9.45 \times 10^{-3}$ mole, and from equation (10–10), $\bar{n}$ is given by

$$\bar{n} = \frac{5.54 \times 10^{-3}}{9.45 \times 10^{-3}} = 0.59$$

From equation (10–21),

$$p[A] = 9.69 - 3.50 - \log[(3.34 \times 10^{-2})$$
$$- (5.54 \times 10^{-3})] = 7.75$$

(b) At pH 8.00, the horizontal distance between the two curves I and II in Figure 10–9 is equivalent to 5.50 mL of NaOH in the 75-mL sample, or $2.59 \times 10^{-4} \times 5.50 \times 1000/75 = 19.0 \times 10^{-3}$ mole/liter. We have

$$\bar{n} = \frac{19.0 \times 10^{-3}}{9.45 \times 10^{-3}} = 2.01$$

$$p[A] = 9.69 - 8.00 - \log[(3.34 \times 10^{-2})$$
$$- (1.90 \times 10^{-2})] = 3.53$$

The values of $\bar{n}$ and p[A] at various pH values are then plotted as shown in **Figure 10–10**. The curve that is obtained is known as a *formation curve*. It is seen to reach a limit at $\bar{n} = 2$, indicating that the maximum number of glycine molecules that can combine with one atom of copper is 2. From this curve at $\bar{n} = 0.5$, at $\bar{n} = 3/2$, and at $\bar{n} = 1.0$, the approximate values for $\log K_1$, $\log K_2$, and $\log \beta$, respectively, are obtained.

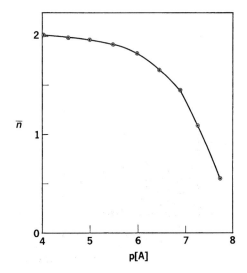

**Fig. 10–10.** Formation curve for the copper–glycine complex.

**TABLE 10–4**

**POTENTIOMETRIC TITRATION OF GLYCINE HYDROCHLORIDE ($3.34 \times 10^{-2}$ MOLE/LITER, $pK_a$, 9.69) AND CUPRIC CHLORIDE ($9.45 \times 10^{-3}$ MOLE/LITER) IN 75-mL SAMPLES USING 0.259 N NaOH AT 30°C***

| pH | $\Delta$(mL NaOH) (per 75-mL Sample) | Moles OH⁻, MA Complexed (mole/liter) | $\bar{n}$ | p[A] |
|---|---|---|---|---|
| 3.50 | 1.60 | $5.54 \times 10^{-3}$ | 0.59 | 7.66 |
| 4.00 | 2.90 | $10.1 \times 10^{-3}$ | 1.07 | 7.32 |
| 4.50 | 3.80 | $13.1 \times 10^{-3}$ | 1.39 | 6.85 |
| 5.00 | 4.50 | $15.5 \times 10^{-3}$ | 1.64 | 6.44 |
| 5.50 | 5.00 | $17.3 \times 10^{-3}$ | 1.83 | 5.98 |
| 6.00 | 5.20 | $18.0 \times 10^{-3}$ | 1.91 | 5.50 |
| 6.50 | 5.35 | $18.5 \times 10^{-3}$ | 1.96 | 5.02 |
| 7.00 | 5.45 | $18.8 \times 10^{-3}$ | 1.99 | 4.53 |
| 7.50 | 5.50 | $19.0 \times 10^{-3}$ | 2.03 | 4.03 |
| 8.00 | 5.50 | $19.0 \times 10^{-3}$ | 2.01 | 3.15 |

*From the data in the last two columns, the formation curve, **Figure 10–10**, is plotted, and the following results are obtained from the curve: $\log K_1 = 7.9$, $\log K_2 = 6.9$, and $\log \beta = 14.8$ (average $\log \beta$ from the literature at 25°C is about 15.3).

A typical set of data for the complexation of glycine by copper is shown in **Table 10–4**. Values of $\log K_1$, $\log K_2$, and $\log \beta$ for some metal complexes of pharmaceutical interest are given in **Table 10–5**.

Pecar et al.[47] described the tendency of pyrrolidone 5-hydroxamic acid to bind the ferric ion to form mono, bis, and tris chelates. These workers later studied the thermodynamics of these chelates using a potentiometric method to determine stability constants. The method employed by Pecar et al. is known as the *Schwarzenbach method* and can be used instead of the potentiometric method described here when complexes are unusually stable. Sandmann and Luk[48] measured the stability constants for lithium catecholamine complexes by potentiometric titration of the free lithium ion. The results demonstrated that lithium forms complexes with

the zwitterionic species of catecholamines at pH 9 to 10 and with deprotonated forms at pH values above 10. The interaction with lithium depends on the dissociation of the phenolic oxygen of catecholamines. At physiologic pH, the protonated species show no significant complexation. Some lithium salts, such as lithium carbonate, lithium chloride, and lithium citrate, are used in psychiatry.

Agrawal et al.[49] applied a pH titration method to estimate the average number of ligand groups per metal ion, $\bar{n}$, for several metal–sulfonamide chelates in dioxane–water. The maximum $\bar{n}$ values obtained indicate 1:1 and 1:2 complexes.

**TABLE 10–5**

**SELECTED CONSTANTS FOR COMPLEXES BETWEEN METAL IONS AND ORGANIC LIGAND***

| Organic Ligand | Metal Ion | $\log K_1$ | $\log K_2$ | $\log, \beta = \log K_1 K_2$ |
|---|---|---|---|---|
| Ascorbic acid | $Ca^{2+}$ | 0.19 | — | — |
| Nicotinamide | $Ag^+$ | — | — | 3.2 |
| Glycine (aminoacetic acid) | $Cu^{2+}$ | 8.3 | 7.0 | 15.3 |
| Salicylaldehyde | $Fe^{2+}$ | 4.2 | 3.4 | 7.6 |
| Salicylic acid | $Cu^{2+}$ | 10.6 | 6.3 | 16.9 |
| p-hydroxybenzoic acid | $Fe^{3+}$ | 15.2 | — | — |
| Methyl salicylate | $Fe^{3+}$ | 9.7 | — | — |
| Diethylbarbituric acid (barbital) | $Ca^{2+}$ | 0.66 | — | — |
| 8-Hydroxyquinoline | $Cu^{2+}$ | 15 | 14 | 29 |
| Pteroylglutamic acid (folic acid) | $Cu^{2+}$ | — | — | 7.8 |
| Oxytetracycline | $Ni^{2+}$ | 5.8 | 4.8 | 10.6 |
| Chlortetracycline | $Fe^{3+}$ | 8.8 | 7.2 | 16.0 |

*From J. Bjerrum, G. Schwarzenback, and L. G. Sillen, *Stability Constants, Part I, Organic Ligands*, The Chemical Society, London, 1957.

The linear relationship between the $pK_a$ of the drugs and the log of the stability constants of their corresponding metal ion complexes shows that the more basic ligands (drugs) give the more stable chelates with cerium (IV), palladium (II), and copper (II). A potentiometric method was described in detail by Connors et al.[50] for the inclusion-type complexes formed between $\alpha$-cyclodextrin and substituted benzoic acids.

## Distribution Method

The method of distributing a solute between two immiscible solvents can be used to determine the stability constant for certain complexes. The complexation of iodine by potassium iodide may be used as an example to illustrate the method. The equilibrium reaction in its simplest form is

$$I_2 + I^- \rightleftharpoons I_3^- \qquad (10\text{--}22)$$

Additional steps also occur in polyiodide formation; for example, $2I^- + 2I_2 \rightleftharpoons I_6^{2-}$ may occur at higher concentrations, but it need not be considered here.

### EXAMPLE 10–2

**Free and Total Iodine**

When iodine is distributed between water (w) at $25°C$ and carbon disulfide as the organic phase (o), as depicted in Figure 10–11, the distribution constant $K(o/w) = C_o/C_w$ is found to be 625. When it is distributed between a 0.1250 M solution of potassium iodide and carbon disulfide, the concentration of iodine in the organic solvent is found to be 0.1896 mole/liter. When the aqueous KI solution is analyzed, the concentration of iodine is found to be 0.02832 mole/liter.

In summary, the results are as follows:

Total concentration of $I_2$ in the aqueous

layer (free + complexed iodine): 0.02832, mole/liter

Total concentration of KI in the aqueous

layer (free + complexed KI): 0.1250 mole/liter

Concentration of $I_2$ in the $CS_2$ layer (free): 0.1896 mole/liter

Distribution coefficient, $K(o/w) = [I_2]_o/[I_2]_w = 625$

The species common to both phases is the free or uncomplexed iodine; the distribution law expresses only the concentration of *free* iodine, whereas a chemical analysis yields the *total* concentration of

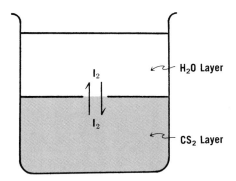

**Fig. 10–11.** The distribution of iodine between water and carbon disulfide.

iodine in the aqueous phase. The concentration of free iodine in the aqueous phase is obtained as follows:

$$[I_2]_w = \frac{[I_2]_o}{K(o/w)} = \frac{0.1896}{625} = 3.034 \times 10^{-4} \text{ mole/liter}$$

To obtain the concentration of iodine in the complex and hence the concentration of the complex, $[I_3^-]$, one subtracts the free iodine from the total iodine of the aqueous phase:

$$
\begin{aligned}
[I_2]_{complexed} &= [I_2]_{w,\,total} - [I_2]_{w,\,free} \\
&= 0.02832 - 0.000303 \\
&= 0.02802 \text{ mole/liter}
\end{aligned}
$$

According to equation (10–22), $I_2$ and KI combine in equimolar concentrations to form the complex. Therefore,

$$[KI]_{complexed} = [I_2]_{complexed} = 0.02802 \text{ mole/liter}$$

KI is insoluble in carbon disulfide and remains entirely in the aqueous phase. The concentration of *free* KI is thus

$$
\begin{aligned}
[KI]_{free} &= [KI]_{total} - [KI]_{complexed} \\
&= 0.1250 - 0.02802 \\
&= 0.09698 \text{ mole/liter}
\end{aligned}
$$

and finally

$$
\begin{aligned}
K &= \frac{[\text{Complex}]}{[I_2]_{free}\,[KI]_{free}} \\
&= \frac{0.02802}{0.000303 \times 0.09698} = 954
\end{aligned}
$$

Higuchi and his associates investigated the complexing action of caffeine, polyvinylpyrrolidone, and polyethylene glycols on a number of acidic drugs, using the partition or distribution method. According to Higuchi and Zuck,[51] the reaction between caffeine and benzoic acid to form the benzoic acid–caffeine complex is

Benzoic acid + Caffeine = (Benzoic acid-Caffeine)

$$(10\text{--}23)$$

and the stability constant for the reactions at $0°C$ is

$$K = \frac{[\text{Benzoic acid-Caffeine}]}{[\text{Benzoic acid}][\text{Caffeine}]} = 37.5 \quad (10\text{--}24)$$

The results varied somewhat, the value 37.5 being an average stability constant. Guttman and Higuchi[52] later showed that caffeine exists in aqueous solution primarily as a monomer, dimer, and tetramer, which would account in part for the variation in $K$ as observed by Higuchi and Zuck.

## Solubility Method

According to the solubility method, excess quantities of the drug are placed in well-stoppered containers, together with a solution of the complexing agent in various concentrations, and the bottles are agitated in a constant-temperature bath until equilibrium is attained. Aliquot portions of the supernatant liquid are removed and analyzed.

Higuchi and Lach[53] used the solubility method to investigate the complexation of $p$-aminobenzoic acid (PABA) by caffeine. The results are plotted in **Figure 10–12**. The point A at which the line crosses the vertical axis is the solubility of

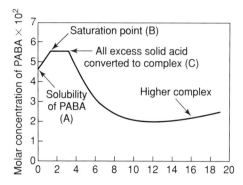

**Fig. 10–12.** The solubility of *para*-aminobenzoic acid (PABA) in the presence of caffeine. (From T. Higuchi and J. L. Lack, J. Am. Pharm. Assoc. Sci. Ed. **43**, 525, 1954.)

the drug in water. With the addition of caffeine, the solubility of PABA rises linearly owing to complexation. At point B, the solution is saturated with respect to the complex and to the drug itself. The complex continues to form and to precipitate from the saturated system as more caffeine is added. At point C, all the excess solid PABA has passed into solution and has been converted to the complex. Although the solid drug is exhausted and the solution is no longer saturated, some of the PABA remains uncomplexed in solution, and it combines further with caffeine to form higher complexes such as (PABA-2 caffeine) as shown by the curve at the right of the diagram.

**EXAMPLE 10–3**

**Stoichiometric Complex Ratio**

The following calculations are made to obtain the stoichiometric ratio of the complex. The concentration of caffeine, corresponding to the plateau BC, equals the concentration of caffeine entering the complex over this range, and the quantity of *p*-aminobenzoic acid entering the complex is obtained from the undissolved solid remaining at point B. It is computed by subtracting the acid in solution at the saturation point B from the total acid initially added to the mixture, because this is the amount yet undissolved that can form the complex.

The concentration of caffeine in the plateau region is found from Figure 10–12 to be $1.8 \times 10^{-2}$ mole/liter. The free, undissolved solid PABA is equal to the total acid minus the acid in solution at point B, namely, $7.3 \times 10^{-2} - 5.5 \times 10^{-2}$, or $1.8 \times 10^{-2}$ mole/liter, and the stoichiometric ratio is

$$\frac{\text{Caffeine in complex}}{\text{PABA in complex}} = \frac{1.8 \times 10^{-2}}{1.8 \times 10^{-2}}$$

The complex formation is therefore written as

$$\text{PABA} + \text{Caffeine} \equiv \text{PABA-Caffeine} \qquad (10\text{–}25)$$

and the stability constant for this 1:1 complex is

$$K = \frac{[\text{PABA-Caffeine}]}{[\text{PABA}][\text{Caffeine}]} \qquad (10\text{–}26)$$

*K* may be computed as follows. The concentration of the complex [PABA-Caffeine] is equal to the total acid concentration at saturation less the solubility [PABA] of the acid in water. The concentration [Caffeine] in the solution at equilibrium is equal to the caffeine added to the system less the concentration that has been converted to the complex. The total acid concentration of saturation is $4.58 \times 10^{-2}$ mole/liter when no caffeine is added (solubility of PABA) and

is $5.312 \times 10^{-2}$ mole/liter when $1.00 \times 10^{-2}$ mole/liter of caffeine is added. We have

$$[\text{PABA-Caffeine}] = (5.31 \times 10^{-2}) - (4.58 \times 10^{-2})$$
$$= 0.73 \times 10^{-2}[\text{PABA}] = 4.58 \times 10^{-2}$$
$$[\text{Caffeine}] = (1.00 \times 10^{-2}) - (0.73 \times 10^{-2}) = 0.27 \times 10^{-2}$$

Therefore,

$$K = \frac{\text{PABA-Caffeine}}{[\text{PABA}][\text{Caffeine}]} = \frac{0.73 \times 10^{-2}}{(4.58 \times 10^{-2})(0.27 \times 10^{-2})} = 59$$

The stability constants for a number of caffeine complexes obtained principally by the distribution and the solubility methods are given in **Table 10–6**. Stability constants for a number of other drug complexes were compiled by Higuchi and Connors.[54] Kenley et al.[55] studied water-soluble complexes of various ligands with the antiviral drug acyclovir using the solubility method.

## Spectroscopy and Change Transfer Complexation

Absorption spectroscopy in the visible and ultraviolet regions of the spectrum is commonly used to investigate electron donor–acceptor or charge transfer complexation.[56,57] When iodine is analyzed in a noncomplexing solvent such as $CCl_4$, a curve is obtained with a single peak at about 520 nm. The solution is violet. A solution of iodine in benzene exhibits a maximum shift to 475 nm, and a new peak of considerably higher intensity for the charge-shifted band appears at 300 nm. A solution of iodine in diethyl ether shows a still greater shift to lower wavelength and the appearance of a new maximum. These solutions are red to brown. Their curves are shown in **Figure 10–13**. In benzene and ether, iodine is the electron acceptor and the organic solvent is the donor; in

**TABLE 10–6**

**APPROXIMATE STABILITY CONSTANTS OF SOME CAFFEINE COMPLEXES IN WATER AT 30°C**

| Compound Complexed with Caffeine | Approximate Stability Constant |
| --- | --- |
| Suberic acid | 3 |
| Sulfadiazine | 7 |
| Picric acid | 8 |
| Sulfathiazole | 11 |
| *o*-Phthalic acid | 14 |
| Acetylsalicylic acid | 15 |
| Benzoic acid (monomer) | 18 |
| Salicylic acid | 40 |
| *p*-aminobenzoic acid | 48 |
| Butylparaben | 50 |
| Benzocaine | 59 |
| *p*-hydroxybenzoic acid | >100 |

*Compiled from T. Higuchi et al., J. Am. Pharm. Assoc. Sci. Ed. **42**, 138, 1953; **43**, 349, 524, 527, 1954; **45**, 290, 1956; **46**, 32, 1957. Over 500 such complexes with other drugs are recorded by T. Higuchi and K. A. Connors. Phase solubility Techniques, in C. N. Reilley (Ed.), *Advances in Analytical Chemistry and Instrumentation*, Wiley, Vol. 4, New York, 1965, pp. 117–212.

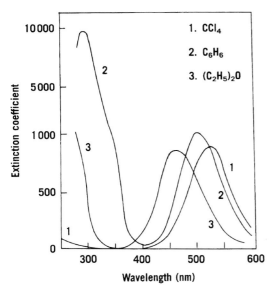

**Fig. 10–13.** Absorption curve of iodine in the noncomplexing solvent (1) carbontetrachloride and the complexing solvents (2) benzene and (3) diethyl ether. (From H. A. Benesi and J. A. Hildebrand, J. Am. Chem. Soc. **70**, 2832, 1948.)

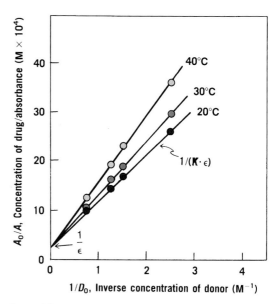

**Fig. 10–14.** A Benesi–Hildebrand plot for obtaining the stability constant, *K*, from equation (10–28) for charge transfer complexation. (From M. A. Slifkin, Biochim. Biophys. Acta **109**, 617, 1965.)

$CCl_4$, no complex is formed. The shift toward the ultraviolet region becomes greater as the electron donor solvent becomes a stronger electron-releasing agent. These spectra arise from the transfer of an electron from the donor to the acceptor in close contact in the excited state of the complex. The more easily a donor such as benzene or diethyl ether releases its electron, as measured by its ionization potential, the stronger it is as a donor. Ionization potentials of a series of donors produce a straight line when plotted against the frequency maximum or charge transfer energies (1 nm = 18.63 cal/mole) for solutions of iodine in the donor solvents.[56,57]

The complexation constant, *K*, can be obtained by use of visible and ultraviolet spectroscopy. The association between the donor *D* and acceptor *A* is represented as

$$D + A \underset{k_{-1}}{\overset{k_1}{\rightleftharpoons}} DA \qquad (10\text{–}27)$$

where $K = k_1/k_{-1}$ is the equilibrium constant for complexation (stability constant) and $k_1$ and $k_{-1}$ are the interaction rate constants. When two molecules associate according to this scheme and the absorbance *A* of the charge transfer band is measured at a definite wavelength, *K* is readily obtained from the *Benesi–Hildebrand equation*[58]:

$$\frac{A_0}{A} = \frac{1}{\epsilon} + \frac{1}{K\epsilon} \frac{1}{D_0} \qquad (10\text{–}28)$$

$A_0$ and $D_0$ are initial concentrations of the acceptor and donor species, respectively, in mole/liter, $\epsilon$ is the molar absorptivity of the charge transfer complex at its particular wavelength, and *K*, the stability constant, is given in liter/mole or $M^{-1}$. A plot of $A_0/A$ versus $1/D_0$ results in a straight line with a slope of $1/(K\epsilon)$ and an intercept of $1/\epsilon$, as observed in **Figure 10–14**.

Borazan et al.[59] investigated the interaction of nucleic acid bases (electron acceptors) with catechol, epinephrine, and isoproterenol (electron donors). Catechols have low ionization potentials and hence a tendency to donate electrons. Charge transfer complexation was evident as demonstrated by ultraviolet absorption measurements. With the assumption of 1:1 complexes, the equilibrium constant, *K*, for charge transfer interaction was obtained from Benesi–Hildebrand plots at three or four temperatures, and $\Delta H°$ was obtained at these same temperatures from the slope of the line as plotted in **Figure 10–15**. The values of *K* and the thermodynamic parameters $\Delta G°$, $\Delta H°$, and $\Delta S°$ are given in **Table 10–7**.

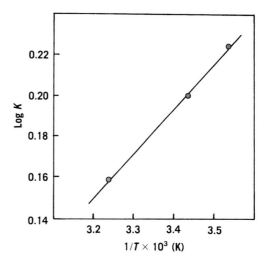

**Fig. 10–15.** Adenine–catechol stability constant for charge transfer complexation measured at various temperatures at a wavelength of 340 nm. (From F. A. Al-Obeidi and H. N. Borazan, J. Pharm. Sci. **65**, 892, 1976. With permission.)

### TABLE 10–7
### STABILITY CONSTANT, *K*, AND THERMODYNAMIC PARAMETERS FOR CHARGE TRANSFER INTERACTION OF NUCLEIC ACID BASES WITH CATECHOL IN AQUEOUS SOLUTION*

| Temperature (°C) | $K$ ($M^{-1}$) | $\Delta G°$ (cal/mole) | $\Delta H°$ (cal/mole) | $\Delta S°$ (cal/deg mole) |
|---|---|---|---|---|
| | | Adenine–catechol | | |
| 9 | 1.69 | −294 | | |
| 18 | 1.59 | −264 | −1015 | −2.6 |
| 37 | 1.44 | −226 | | |
| | | Uracil–catechol | | |
| 6 | 0.49 | 396 | | |
| 18 | 0.38 | 560 | −3564 | −14 |
| 25 | 0.32 | 675 | | |
| 37 | 0.26 | 830 | | |

*From F. A. Al-Obeidi and H. N. Borazan, J. Pharm. Sci. **65,** 892, 1976. With permission.

### EXAMPLE 10–4

**Calculate Molar Absorptivity**

When $A_0/A$ is plotted against $1/D_0$ for catechol (electron-donor) solutions containing uracil (electron acceptor) in 0.1 N HCl at 6°C, 18°C, 25°C, and 37°C, the four lines were observed to intersect the vertical axis at 0.01041. Total concentration, $A_0$, for uracil was $2 \times 10^{-2}$ M, and $D_0$ for catechol ranged from 0.3 to 0.8 M. The slopes of the lines determined by the least-squares method were as follows:

| 6°C | 18°C | 25°C | 37°C |
|---|---|---|---|
| 0.02125 | 0.02738 | 0.03252 | 0.04002 |

Calculate the molar absorptivity and the stability constant, $K$. Knowing $K$ at these four temperatures, how does one proceed to obtain $\Delta H°$, $\Delta G°$, and $\Delta S°$?

The intercept, from the Benesi–Hildebrand equation, is the reciprocal of the molar absorptivity, or $1/0.01041 = 96.1$. The molar absorptivity, $\epsilon$, is a constant for a compound or a complex, independent of temperature or concentration. $K$ is obtained from the slope of the four curves:

$$(1)\ 0.02125 = 1/(K \times 96.1); K = 0.49\ M^{-1}$$
$$(2)\ 0.02738 = 1/(K \times 96.1); K = 0.38\ M^{-1}$$
$$(3)\ 0.03252 = 1/(K \times 96.1); K = 0.32\ M^{-1}$$
$$(4)\ 0.04002 = 1/(K \times 96.1); K = 0.26\ M^{-1}$$

These $K$ values are then plotted as their logarithms on the vertical axis of a graph against the reciprocal of the four temperatures, converted to kelvin. This is a plot of equation (10–49) and yields $\Delta H°$ from the slope of the line. $\Delta G°$ is calculated from log $K$ at each of the four temperatures using equation (10–48), in which the temperature, $T$, is expressed in kelvin. $\Delta S°$ is finally obtained using relation: $\Delta G° = \Delta H° - T\Delta S°$. The answers to this sample problem are given in Table 10–7.

Webb and Thompson[60] studied the possible role of electron donor–acceptor complexes in drug–receptor binding using quinoline and naphthalene derivatives as model electron donors and a trinitrofluorene derivative as the electron acceptor. The most favorable arrangement for the donor 8-aminoquinoline (heavy lines) and the acceptor

9-dicyanomethylene trinitrofluorene (light lines), as calculated by a quantum chemical method, is shown below:

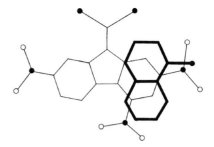

Filled circles are nitrogen atoms and open circles oxygen atoms. The donor lies above the acceptor molecule at an intermolecular distance of about 3.35 Å and is attached by a binding energy of –5.7 kcal/mole. The negative sign signifies a positive binding force.

## Other Methods

A number of other methods are available for studying the complexation of metal and organic molecular complexes. They include NMR and infrared spectroscopy, polarography, circular dichroism, kinetics, x-ray diffraction, and electron diffraction. Several of these will be discussed briefly in this section.

Complexation of caffeine with L-tryptophan in aqueous solution was investigated by Nishijo et al.[61] using $^1$H-NMR spectroscopy. Caffeine interacts with L-tryptophan at a molar ratio of 1:1 by parallel stacking. Complexation is a result of polarization and $\pi - \pi$ interactions of the aromatic rings. A possible mode of parallel stacking is shown in **Figure 10–16**. This study demonstrates that tryptophan, which is presumed to be the binding site in serum albumin for certain drugs, can interact with caffeine even as free amino acid. However,

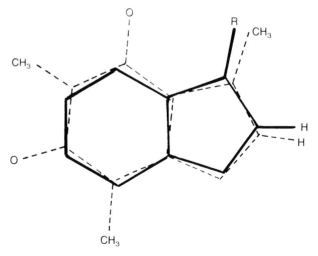

**Fig. 10–16.** Stacking of L-tryptophan (solid line) overlying caffeine (dashed line). The benzene ring of tryptophan is located above the pyrimidine ring of caffeine, and the pyrrole ring of L-tryptophan is above the imidazole ring of caffeine. (From J. Nishijo, I. Yonetami, E. Iwamoto, et al., J. Pharm. Sci. **79,** 18, 1990. With permission.)

caffeine does not interact with other aromatic amino acids such as L-valine or L-leucine.

Borazan and Koumriqian[62] studied the coil–helix transition of polyadenylic acid induced by the binding of the catecholamines norepinephrine and isoproterenol, using circular dichroism. Most mRNA molecules contain regions of polyadenylic acid, which are thought to increase the stability of mRNA and to favor genetic code translation. The change of the circular dichroism spectrum of polyadenylic acid was interpreted as being due to intercalative binding of catecholamines between the stacked adenine bases. These researchers suggested that catecholamines may exert a control mechanism through induction of the coil-to-helix transition of polyadenylic acid, which influences genetic code translation.

De Taeye and Zeegers-Huyskens[63] used infrared spectroscopy to investigate the hydrogen-bonded complexes involving polyfunctional bases such as proton donors. This is a very precise technique for determining the thermodynamic parameters involved in hydrogen-bond formation and for characterizing the interaction sites when the molecule has several groups available to form hydrogen-bonded. Caffeine forms hydrogen-bonded complexes with various proton donors: phenol, phenol derivatives, aliphatic alcohols, and water. From the infrared technique, the preferred hydrogen-bonding sites are the carbonyl functions of caffeine. Seventy percent of the complexes are formed at the C=O group at position 6 and 30% of the complexes at the C=O group at position 2 of caffeine. El-Said et al.[64] used conductometric and infrared methods to characterize 1:1 complexes between uranyl acetate and tetracycline. The structure suggested for the uranyl–tetracycline complex is shown below.

## KEY CONCEPT DRUG–PROTEIN BINDING

The binding of drugs to proteins contained in the body can influence their action in a number of ways. Proteins may (a) facilitate the distribution of drugs throughout the body, (b) inactivate the drug by not enabling a sufficient concentration of free drug to develop at the receptor site, or (c) retard the excretion of a drug. The interaction of a drug with proteins may cause (a) the displacement of body hormones or a coadministered agent, (b) a configurational change in the protein, the structurally altered form of which is capable of binding a coadministered agent, or (c) the formation of a drug–protein complex that itself is biologically active. These topics are discussed in a number of reviews.[65,66] Among the plasma proteins, albumin is the most important owing to its high concentration relative to the other proteins and also to its ability to bind both acidic and basic drugs. Another plasma protein, $\alpha_1$-acid glycoprotein, has been shown to bind numerous drugs; this protein appears to have greater affinity for basic than for acidic drug molecules.

## PROTEIN BINDING

A complete analysis of protein binding, including the multiple equilibria that are involved, would go beyond our immediate needs. Therefore, only an abbreviated treatment is given here.

### Binding Equilibria

We write the interaction between a group or free receptor $P$ in a protein and a drug molecule $D$ as

$$P + D \rightleftharpoons PD \qquad (10\text{--}29)$$

The equilibrium constant, disregarding the difference between activities and concentrations, is

$$K = \frac{[PD]}{[P][D_f]} \qquad (10\text{--}30)$$

or

$$K[P][D_f] = [PD] \qquad (10\text{--}31)$$

where $K$ is the association constant, $[P]$ is the concentration of the protein in terms of free binding sites, $[D_f]$ is the concentration, usually given in moles, of free drug, sometimes called the ligand, and $[PD]$ is the concentration of the protein–drug complex. $K$ varies with temperature and would be better represented as $K(T)$; $[PD]$, the symbol for bound drug, is sometimes written as $[D_b]$, and $[D]$, the free drug, as $[D_f]$.

If the total protein concentration is designated as $[P_t]$, we can write

$$[P_t] = [P] + [PD]$$

or

$$[P] = [P_t] - [PD] \qquad (10\text{--}32)$$

Substituting the expression for $[P]$ from equation (10–32) into (10–31) gives

$$[PD] = K[D_f]([P_t] - [PD]) \qquad (10\text{--}33)$$

$$[PD] = K[D_f][PD] = K[D_f][P_t] \qquad (10\text{--}34)$$

$$\frac{[PD]}{P_t} = \frac{K[D_f]}{1 + K[D_f]} \qquad (10\text{--}35)$$

Let $r$ be the number of moles of drug bound, $[PD]$, per mole of total protein, $[P_t]$; then $r = [PD]/[P_t]$, or

$$r = \frac{K[D_f]}{1 + K[D_f]} \qquad (10\text{–}36)$$

The ratio $r$ can also be expressed in other units, such as milligrams of drug bound, $x$, per gram of protein, $m$. Equation (10–36) is one form of the Langmuir adsorption isotherm. Although it is quite useful for expressing protein-binding data, it must not be concluded that obedience to this formula necessarily requires that protein binding be an adsorption phenomenon. Expression (10–36) can be converted to a linear form, convenient for plotting, by inverting it:

$$\frac{1}{r} = \frac{1}{K[D_f]} + 1 \qquad (10\text{–}37)$$

If $v$ independent binding sites are available, the expression for $r$, equation (10–36), is simply $v$ times that for a single site, or

$$r = v\frac{K[D_f]}{1 + K[D_f]} \qquad (10\text{–}38)$$

and equation (10–37) becomes

$$\frac{1}{r} = \frac{1}{vK}\frac{1}{[D_f]} + \frac{1}{v} \qquad (10\text{–}39)$$

Equation (10–39) produces what is called a *Klotz reciprocal plot*.[67]

An alternative manner of writing equation (10–38) is to rearrange it first to

$$r + rK[D_f] = vK[D_f] \qquad (10\text{–}40)$$

$$\frac{r}{[D_f]} = vK - rK \qquad (10\text{–}41)$$

Data presented according to equation (10–41) are known as a *Scatchard plot*.[67,68] The binding of bishydroxycoumarin to human serum albumin is shown as a Scatchard plot in **Figure 10–17**.

Graphical treatment of data using equation (10–39) heavily weights those experimental points obtained at low concentrations of free drug, $D$, and may therefore lead to misinterpretations regarding the protein-binding behavior at high concentrations of free drug. Equation (10–41) does not have this disadvantage and is the method of choice for plotting data. Curvature in these plots usually indicates the existence of more than one type of binding site.

Equations (10–39) and (10–41) cannot be used for the analysis of data if the nature and the amount of protein in the experimental system are unknown. For these situations, Sandberg et al.[69] recommended the use of a slightly modified form of equation (10–41):

$$\frac{[D_b]}{[D_f]} = -K[D_b] + vK[P_t] \qquad (10\text{–}42)$$

where $[D_b]$ is the concentration of bound drug. Equation (10–42) is plotted as the ratio $[D_b]/[D_f]$ versus $[D_b]$, and

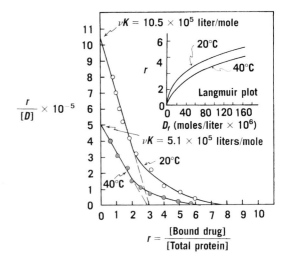

**Fig. 10–17.** A Scatchard plot showing the binding of bishydroxycoumarin to human serum albumin at 20°C and 40°C plotted according to equation (10–41). Extrapolation of the two lines to the horizontal axis, assuming a single class of sites with no electrostatic interaction, gives an approximate value of 3 for $v$. (From M. J. Cho, A. G. Mitchell, and M. Pernarowski, J. Pharm. Sci. **60**, 196, 1971; **60**, 720, 1971. With permission.) The inset is a Langmuir adsorption isotherm of the binding data plotted according to equation (10–36).

in this way $K$ is determined from the slope and $vK[P_t]$ is determined from the intercept.

The Scatchard plot yields a straight line when only one class of binding sites is present. Frequently in drug-binding studies, $n$ classes of sites exist, each class $i$ having $v_i$ sites with a unique association constant $K_i$. In such a case, the plot of $r/[D_f]$ versus $r$ is not linear but exhibits a curvature that suggests the presence of more than one class of binding sites. The data in **Figure 10–17** were analyzed in terms of one class of sites for simplification. The plots at 20°C and 40°C clearly show that multiple sites are involved. Blanchard et al.[70] reviewed the case of multiple classes of sites. Equation (10–38) is then written as

$$r = \frac{v_1 K_1[D_f]}{1 + K_1[D_f]} + \frac{v_2 K_2[D_f]}{1 + K_2[D_f]} + \cdots \frac{v_n K_n[D_f]}{1 + K_n[D_f]} \qquad (10\text{–}43)$$

or

$$r = \sum_{i=1}^{n} \frac{v_i K_i[D_f]}{1 + K_i[D_f]} \qquad (10\text{–}44)$$

As previously noted, only $v$ and $K$ need to be evaluated when the sites are all of one class. When $n$ classes of sites exist, equations (10–43 and 10–44) can be written as

$$r = \sum_{i=1}^{n-1} \frac{v_i K_i[D_f]}{1 + K_i[D_f]} + v_n K_n[D_f] \qquad (10\text{–}45)$$

The binding constant, $K_n$, in the term on the right is small, indicating extremely weak affinity of the drug for the sites, but this class may have a large number of sites and so be considered unsaturable.

## Equilibrium Dialysis (ED) and Ultrafiltration (UF)

A number of methods are used to determine the amount of drug bound to a protein. Equilibrium dialysis, ultrafiltration, and electrophoresis are the classic techniques used, and in recent years other methods, such as gel filtration and nuclear magnetic resonance, have been used with satisfactory results. We shall discuss the equilibrium dialysis, ultrafiltration, and kinetic methods.

The equilibrium dialysis procedure was refined by Klotz et al.[71] for studying the complexation between metal ions or small molecules and macromolecules that cannot pass through a semipermeable membrane.

According to the equilibrium dialysis method, the serum albumin (or other protein under investigation) is placed in a Visking cellulose tubing (Visking Corporation, Chicago) or similar dialyzing membrane. The tubes are tied securely and suspended in vessels containing the drug in various concentrations. Ionic strength and sometimes hydrogen ion concentration are adjusted to definite values, and controls and blanks are run to account for the adsorption of the drug and the protein on the membrane.

If binding occurs, the drug concentration in the sac containing the protein is greater at equilibrium than the concentration of drug in the vessel outside the sac. Samples are removed and analyzed to obtain the concentrations of free and complexed drug.

Equilibrium dialysis is the classic technique for protein binding and remains the most popular method. Some potential errors associated with this technique are the possible binding of drug to the membrane, transfer of substantial amounts of drug from the plasma to the buffer side of the membrane, and osmotic volume shifts of fluid to the plasma side. Tozer et al.[72] developed mathematical equations to calculate and correct for the magnitude of fluid shifts. Briggs et al.[73] proposed a modified equilibrium dialysis technique to minimize experimental errors for the determination of low levels of ligand or small molecules.

Ultrafiltration methods are perhaps more convenient for the routine determination because they are less time-consuming. The ultrafiltration method is similar to equilibrium dialysis in that macromolecules such as serum albumin are separated from small drug molecules. Hydraulic pressure or centrifugation is used in ultrafiltration to force the solvent and the small molecules, unbound drug, through the membrane while preventing the passage of the drug bound to the protein. This ultrafiltrate is then analyzed by spectrophotometry or other suitable technique.

The concentration of the drug that is free and unbound, $D_f$, is obtained by use of the Beer's law equation:

$$A = \epsilon bc \qquad (10\text{–}46)$$

where $A$ is the spectrophotometric absorbance (dimensionless), $\epsilon$ is the molar absorptivity, determined independently for each drug, $c$ ($D_f$ in binding studies) is the concentration of the free drug in the ultrafiltrate in moles/liter, and $b$ is the

optical path length of the spectrophotometer cell, ordinarily 1 cm. The following example outlines the steps involved in calculating the Scatchard $r$ value and the percentage of drug bound.

### EXAMPLE 10–5

#### Binding to Human Serum Albumin

The binding of sulfamethoxypyridazine to human serum albumin was studied at 25°C, pH 7.4, using the ultrafiltration technique. The concentration of the drug under study, $[D_t]$, is $3.24 \times 10^{-5}$ mole/liter and the human serum albumin concentration, $[P_t]$, is $1.0 \times 10^{-4}$ mole/liter. After equilibration the ultrafiltrate has an absorbance, $A$, of 0.559 at 540 nm in a cell whose optical path length, $b$, is 1 cm. The molar absorptivity, $\epsilon$, of the drug is $5.6 \times 10^4$ liter/mole cm. Calculate the Scatchard $r$ value and the percentage of drug bound.

The concentration of free (unbound) drug, $[D_f]$, is given by

$$[D_f] = \frac{A}{b\epsilon} = \frac{0.559}{(5.6 \times 10^4)1} = 0.99 \times 10^{-5} \text{ mole/liter}$$

The concentration of bound drug, $[D_b]$, is given by

$$\begin{aligned}[D_b] &= [D_t] - [D_f] \\ &= (3.24 \times 10^{-5}) - (0.99 \times 10^{-5}) \\ &= 2.25 \times 10^{-5} \text{ mole/liter}\end{aligned}$$

The $r$ value is

$$r = \frac{[D_b]}{[P_t]} = \frac{2.25 \times 10^{-5}}{1.0 \times 10^{-4}} = 0.225$$

The percentage of bound drug is $[D_b]/[D_t] \times 100 = 69\%$.

A potential error in ultrafiltration techniques may result from the drug binding to the membrane. The choice between ultrafiltration and equilibrium dialysis methods depends on

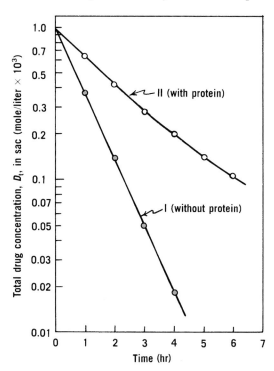

**Fig. 10–18.** The dynamic dialysis plot for determining the concentration of bound drug in a protein solution (From M. C. Meyer and D. E. Guttman, J. Pharm. Sci. **57**, 1627, 1968. With permission).

## ⬛▬KEY CONCEPT▬ PROTEIN BINDING

Protein binding (PB) plays an important role in the pharmacokinetics and pharmacodynamics of a drug. The extent of PB in the plasma or tissue controls the volume of distribution and affects both hepatic and renal clearance. In many cases, the free drug concentration, rather than the total concentration in plasma, is correlated to the effect. Drug displacement from drug–protein complex can occur by direct competition of two drugs for the same binding site and is important with drugs that are highly bound (>95%), for which a small displacement of bound drug can greatly increase the free drug concentration in the plasma. In order to measure free fraction or PB of a drug, ultrafiltration (UF), ultracentrifugation, equilibrium dialysis (ED), chromatography, spectrophotometry, electrophoresis, etc. have been used. Essential methodologic aspects of PB study include the selection of assay procedures, devices, and materials. The most commonly used method for PB measurement is ED, which is believed to be

less susceptible to experimental artifacts. However, it is time consuming and is not suitable for unstable compounds because it requires substantial equilibration time (3–24 hr) depending on drugs, membrane materials, and devices. Many researchers have used UF centrifugal devices for PB measurement. UF is a simple and rapid method in which centrifugation forces the buffer containing free drugs through the size exclusion membrane and achieves a fast separation of free from protein-bound drug. However, the major disadvantage of this method is nonspecific binding of drugs on filter membranes and plastic devices. When the drug binds extensively to the filtration membrane, the ultrafiltrate concentration may deviate from the true free concentration. (From K.-J. Lee, R. Mower, T. Hollenbeck, J. Castelo, N. Johnson, P. Gordon, P. J. Sinko, K. Holme, and Y.-H. Lee, Pharm. Res. **20**, 1015, 2003. With permission.)

---

the characteristics of the drug. The two techniques have been compared in several protein-binding studies.[74–76]

### Dynamic Dialysis

Meyer and Guttman[77] developed a kinetic method for determining the concentrations of bound drug in a protein solution. The method has found favor in recent years because it is relatively rapid, economical in terms of the amount of protein required, and readily applied to the study of competitive inhibition of protein binding. It is discussed here in some detail. The method, known as *dynamic dialysis*, is based on the rate of disappearance of drug from a dialysis cell that is proportional to the concentration of unbound drug. The apparatus consists of a 400-mL jacketed (temperature-controlled) beaker into which 200 mL of buffer solution is placed. A cellophane dialysis bag containing 7 mL of drug or drug–protein solution is suspended in the buffer solution. Both solutions are stirred continuously. Samples of solution external to the dialysis sac are removed periodically and analyzed spectrophotometrically, and an equivalent amount of buffer solution is returned to the external solution. The dialysis process follows the rate law

$$\frac{-d[D_t]}{dt} = k[D_f] \qquad (10\text{–}47)$$

where $[D_t]$ is the total drug concentration, $[D_f]$, is the concentration of free or unbound drug in the dialysis sac, $-d[D_t]/dt$ is the rate of loss of drug from the sac, and $k$ is the first-order rate constant (see Chapter 13) representative of the diffusion process. The factor $k$ can also be referred to as the apparent permeability rate constant for the escape of drug from the sac. The concentration of unbound drug, $[D_f]$, in the sac (protein compartment) at a total drug concentration $[D_t]$ is calculated using equation (10–45), knowing $k$ and the rate $-d[D_t]/dt$ at a particular drug concentration, $[D_t]$. The rate constant, $k$,

is obtained from the slope of a semilogarithmic plot of $[D_t]$ versus time when the experiment is conducted in the absence of the protein.

**Figure 10–18** illustrates the type of kinetic plot that can be obtained with this system. Note that in the presence of protein, curve II, the rate of loss of drug from the dialysis sac is slowed compared with the rate in the absence of protein, curve I. To solve equation (10–47) for free drug concentration, $[D_f]$, it is necessary to determine the slope of curve II at various points in time. This is not done graphically, but instead it is accurately accomplished by first fitting the time-course data to a suitable empirical equation, such as the following, using a computer.

$$[D_t] = C_1 e^{-C_2 t} + C_3 e^{-C_4 t} + C_5 e^{-C_6 t} \qquad (10\text{–}48)$$

The computer fitting provides estimates of $C_1$ through $C_6$. The values for $d[D_t]/dt$ can then be computed from equation (10–49), which represents the first derivative of equation (10–48):

$$-\frac{d[D_t]}{dt} = C_1 C_2 e^{-C_2 t} + C_3 C_4 e^{-C_4 t} + C_5 C_6 e^{-C_6 t} \qquad (10\text{–}49)$$

Finally, once we have a series of $[D_f]$ values computed from equations (10–49) and (10–47) corresponding to experimentally determined values of $[D_t]$ at each time $t$, we can proceed to calculate the various terms for the Scatchard plot.

---

**EXAMPLE 10–6***

**Calculate Scatchard Values**

Assume that the kinetic data illustrated in Figure 10–18 were obtained under the following conditions: initial drug concentration, $[D_{t0}]$, is $1 \times 10^{-3}$ mole/liter and protein concentration is $1 \times 10^{-3}$ mole/liter. Also assume that the first-order rate constant, $k$, for the control

---

*Example 10–6 was prepared by Prof. M. Meyer of the University of Tennessee.

(curve I) was determined to be 1.0 hr$^{-1}$ and that fitting of curve II to equation (10–48) resulted in the following empirical constants: $C_1 = 5 \times 10^{-4}$ mole/liter, $C_2 = 0.6$ hr$^{-1}$, $C_3 = 3 \times 10^{-4}$ mole/liter, $C_4 = 0.4$ hr$^{-1}$, $C_5 = 2 \times 10^{-4}$ mole/liter, and $C_6 = 0.2$ hr$^{-1}$.

Calculate the Scatchard values (the Scatchard plot was discussed in the previous section) for $r$ and $r[D_f]$ if, during the dialysis in the presence of protein, the experimentally determined value for $[D_t]$ was $4.2 \times 10^{-4}$ mole/liter at 2 hr. Here, $r = [D_b]/P_t$, where $[D_b]$ is drug bound and $P_t$ is total protein concentration. We have

Using equation (10–49),

$$-\frac{d[D_t]}{dt} = k[d_f]$$
$$= (5 \times 10^{-4})(0.6)e^{-0.6(2)} + (3 \times 10^{-4})(0.4)e^{-0.4(2)}$$
$$+ (2 \times 10^{-4})(0.2)e^{-0.2(2)}$$

where the (2) in the exponent stands for 2 hr. Thus,

$$[D_f]_{2\ hr} = \frac{1.7 \times 10^{-4}\ \text{mole/liter hr}^{-1}}{1.0\ \text{hr}^{-1}} = 1.7 \times 10^{-4}\ \text{mole/liter}$$

It follows that at 2 hr,

$$[D_b] = [D_t] - [D_f]$$
$$= 4.2 \times 10^{-4}\ \text{mole/liter} - 1.7 \times 10^{-4}\ \text{mole/liter}$$
$$= 2.5 \times 10^{-4}\ \text{mole/liter}$$
$$r = [D_b]/[P_t] = (2.5 \times 10^{-4})/(1 \times 10^{-3}) = 0.25$$
$$(r)/[D_f] = (0.25)/(1.7 \times 10^{-4}) = 1.47 \times 10^3\ \text{liter/mole}$$

Additional points for the Scatchard plot would be obtained in a similar fashion, using the data obtained at various points throughout the dialysis. Accordingly, this series of calculations permits one to prepare a Scatchard plot (see Fig. 10–17).

Judis[78] investigated the binding of phenol and phenol derivatives by whole human serum using the dynamic dialysis technique and presented the results in the form of Scatchard plots.

## Hydrophobic Interaction

Hydrophobic "bonding," first proposed by Kauzmann,[79] is actually not bond formation at all but rather the tendency of hydrophobic molecules or hydrophobic parts of molecules to avoid water because they are not readily accommodated in the hydrogen-bonding structure of water. Large hydrophobic species such as proteins avoid the water molecules in an aqueous solution insofar as possible by associating into micelle-like structures (Chapter 15) with the nonpolar portions in contact in the inner regions of the "micelles," the polar ends facing the water molecules. This attraction of hydrophobic species, resulting from their unwelcome reception in water, is known as hydrophobic bonding, or, better, *hydrophobic interaction*. It involves van der Waals forces, hydrogen bonding of water molecules in a three-dimensional structure, and other interactions. Hydrophobic interaction is favored thermodynamically because of an increased disorder or entropy of the water molecules that accompanies the association of the nonpolar molecules, which squeeze out the water. Globular proteins are thought to maintain their ball-like structure in water because of the hydrophobic effect. Hydrophobic interaction is depicted in **Figure 10–19**.

Nagwekar and Kostenbauder[80] studied hydrophobic effects in drug binding, using as a model of the protein a copolymer of vinylpyridine and vinylpyrrolidone. Kristiansen et al.[81] studied the effects of organic solvents in decreasing complex formation between small organic molecules in aqueous solution. They attributed the interactions of the organic species to a significant contribution by both hydrophobic bonding and the unique effects of the water structure. They suggested that some nonclassic "donor–acceptor" mechanism may be operating to lend stability to the complexes formed.

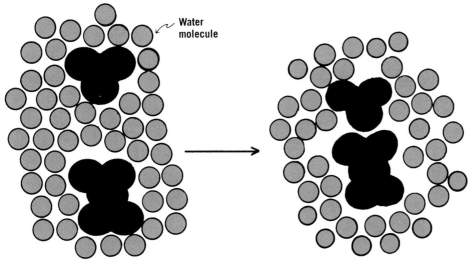

**(a)** Ordered water structure surrounding two large nonpolar molecules.

**(b)** Hydrophobic interaction of the nonpolar molecules with the "squeezing out" of water molecules into a more randomized structure.

**Fig. 10–19.** Schematic view of hydrophobic interaction. (*a*) Two hydrophobic molecules are separately enclosed in cages, surrounded in an orderly fashion by hydrogen-bonded molecules of water (*open circles*). The state at (*b*) is somewhat favored by breaking of the water cages of (*a*) to yield a less ordered arrangement and an overall entropy increase of the system. Van der Waals attraction of the two hydrophobic species also contributes to the hydrophobic interaction.

Feldman and Gibaldi[82] studied the effects of urea, methylurea, and 1,3-dimethylurea on the solubility of benzoic and salicylic acids in aqueous solution. They concluded that the enhancement of solubility by urea and its derivatives was a result of hydrophobic bonding rather than complexation. Urea broke up the hydrogen-bonded water clusters surrounding the nonpolar solute molecules, increasing the entropy of the system and producing a driving force for solubilization of benzoic and salicylic acids. It may be possible that the ureas formed channel complexes with these aromatic acids as shown in **Figure 10–3** *a*, *b*, and *c*.

The interaction of drugs with proteins in the body may involve hydrophobic bonding at least in part, and this force in turn may affect the metabolism, excretion, and biologic activity of a drug.

## Self-Association

Some drug molecules may self-associate to form dimers, trimers, or aggregates of larger sizes. A high degree of association may lead to formation of micelles, depending on the nature of the molecule (Chapter 16). Doxorubicin forms dimers, the process being influenced by buffer composition and ionic strength. The formation of tetramers is favored by hydrophobic stacking aggregation.[83] Self-association may affect solubility, diffusion, transport through membranes, and therapeutic action. Insulin shows concentration-dependent self-association, which leads to complications in the treatment of diabetes. Aggregation is of particular importance in long-term insulin devices, where insulin crystals have been observed. The initial step of insulin self-association is a hydrophobic interaction of the monomers to form dimers, which further associate into larger aggregates. The process is favored at higher concentrations of insulin.[84] Addition of urea at nontoxic concentrations (1.0–3 mg/mL) has been shown to inhibit the self-association of insulin. Urea breaks up the "icebergs" in liquid water and associates with structured water by hydrogen bonding, taking an active part in the formation of a more open "lattice" structure.[85]

Sodium salicylate improves the rectal absorption of a number of drugs, all of them exhibiting self-association. Touitou and Fisher[86] chose methylene blue as a model for studying the effect of sodium salicylate on molecules that self-associate by a process of stacking. Methylene blue is a planar aromatic dye that forms dimers, trimers, and higher aggregates in aqueous solution. The workers found that sodium salicylate prevents the self-association of methylene blue. The inhibition of aggregation of porcine insulin by sodium salicylate results in a 7875-fold increase in solubility.[87] Commercial heparin samples tend to aggregate in storage depending on factors such as temperature and time in storage.[88]

## Factors Affecting Complexation and Protein Binding

Kenley et al.[55] investigated the role of hydrophobicity in the formation of water-soluble complexes. The logarithm of the ligand partition coefficient between octanol and water was chosen as a measure of hydrophobicity of the ligand. The authors found a significant correlation between the stability constant of the complexes and the hydrophobicity of the ligands. Electrostatic forces were not considered as important because all compounds studied were uncharged under the conditions investigated. Donor–acceptor properties expressed in terms of orbital energies (from quantum chemical calculations) and relative donor–acceptor strengths correlated poorly with the formation constants of the complex. It was suggested that ligand hydrophobicity is the main contribution to the formation of water-soluble complexes. Coulson and Smith[89] found that the more hydrophobic chlorobiocin analogues showed the highest percentage of drug bound to human serum albumin. These workers suggested that chlorobiocin analogues bind to human albumin at the same site as warfarin. This site consists of two noncoplanar hydrophobic areas and a cationic group. Warfarin, an anticoagulant, serves as a model drug in protein-binding studies because it is extensively but weakly bound. Thus, many drugs are able to compete with and displace warfarin from its binding sites. The displacement may result in a sudden increase of the free (unbound) fraction in plasma, leading to toxicity, because only the free fraction of a drug is pharmacologically active. Diana et al.[90] investigated the displacement of warfarin by nonsteroidal anti-inflammatory drugs. **Table 10–8** shows the variation of the stability constant, $K$, and the number of binding sites, $n$, of the complex albumin–warfarin after addition of competing drugs. Azapropazone markedly decreases the $K$ value, suggesting that both drugs, warfarin and azapropazone, compete for the same binding site on albumin. Phenylbutazone also competes strongly for the binding site on albumin. Conversely, tolmetin may increase $K$, as suggested by the authors, by a conformational change in the albumin molecule that favors warfarin binding. The other drugs (see **Table 10–8**) decrease the $K$ value of warfarin to a lesser extent, indicating that they do not share exclusively the same binding site as that of warfarin.

Plaizier-Vercammen[91] studied the effect of polar organic solvents on the binding of salicylic acid to povidone. He

### TABLE 10–8

**BINDING PARAMETERS ($\pm$ STANDARD DEVIATION) FOR WARFARIN IN THE PRESENCE OF DISPLACING DRUGS***

|                      | Racemic Warfarin | |
| Competing Drug       | $n$           | $K \times 10^{-5}$ M$^{-1}$ |
| -------------------- | ------------- | --------------------------- |
| None                 | $1.1 \pm 0.0$ | $6.1 \pm 0.2$               |
| Azapropazone         | $1.4 \pm 0.1$ | $0.19 \pm 0.02$             |
| Phenylbutazone       | $1.3 \pm 0.2$ | $0.33 \pm 0.06$             |
| Naproxen             | $0.7 \pm 0.0$ | $2.4 \pm 0.2$               |
| Ibuprofen            | $1.2 \pm 0.2$ | $3.1 \pm 0.4$               |
| Mefenamic acid       | $0.9 \pm 0.0$ | $3.4 \pm 0.2$               |
| Tolmetin             | $0.8 \pm 0.0$ | $12.6 \pm 0.6$              |

*From F. J. Diana, K. Veronich, and A. L. Kapoor, J. Pharm. Sci. **78**, 195, 1989. With permission.

found that in water–ethanol and water–propylene glycol mixtures, the stability constant of the complex decreased as the dielectric constant of the medium was lowered. Such a dependence was attributed to hydrophobic interaction and can be explained as follows. Lowering the dielectric constant decreases polarity of the aqueous medium. Because most drugs are less polar than water, their *affinity to the medium increases* when the dielectric constant decreases. As a result, the binding to the macromolecule is reduced.

Protein binding has been related to the solubility parameter $\delta$ of drugs. Bustamante and Selles[92] found that the percentage of drug bound to albumin in a series of sulfonamides showed a maximum at $\Delta = 12.33$ cal$^{1/2}$ cm$^{-3/2}$. This value closely corresponds to the $\delta$ value of the postulated binding site on albumin for sulfonamides and suggests that the closer the solubility parameter of a drug to the $\delta$ value of its binding site, the greater is the binding.

# CHAPTER SUMMARY

Complexation is widely used in the pharmaceutical sciences to improve properties such as solubility. The three classes of complexes or coordination compounds were discussed in the context to identify pharmaceutically relevant examples. The physical properties of chelates and what differentiates them from organic molecular complexes were also described. The types of forces that hold together organic molecular complexes also play an important role in determining the function and use of complexes in the pharmaceutical sciences. One widely used complex system, the cyclodextrins, was described in detail with respect to pharmaceutical applications. The stoichiometry and stability of complexes was described as well as methods of analysis to determine their strengths and weaknesses. Protein binding is important for many drug substances. The ways in which protein binding could influence drug action were discussed. Also, methods such as the equilibrium dialysis and ultrafiltration were described for determining protein binding.

 **Practice problems for this chapter can be found at thePoint.lww.com/Sinko6e.**

# References

1. L. Pauling, *The Nature of the Chemical Bond*, Cornell University Press, Ithaca, New York, 1940, p. 81.
2. A. Werner, Vjschr. nuturf. Ges. Zurich, **36**, 129, 1891.
3. P. Mohanakrishnan, C. F. Chignell, and R. H. Cox, J. Pharm. Sci. **74**, 61, 1985.
4. J. E. Whitaker and A. M. Hoyt, Jr., J. Pharm. Sci. **73**, 1184, 1984.
5. A. E. Martell and M. Calvin, *Chemistry of the Metal Chelate Compounds*, Prentice Hall, New York, 1952.
6. L. B. Clapp, in J. R. Bailar, Jr. (Ed.), *The Chemistry of the Coordination Compounds*, Reinhold, New York, 1956, Chapter 17.
7. G. Canti, A. Scozzafava, G. Ciciani, and G. Renzi, J. Pharm. Sci. **69**, 1220, 1980.
8. F. J. Bullock, in M. Florkin and E. H. Stotz (Eds.), *Comprehensive Biochemistry*, Elsevier, New York, 1967, pp. 82–85.
9. J. Buxeraud, A. C. Absil, and C. Raby, J. Pharm. Sci. **73**, 1687, 1984.
10. T. Higuchi and J. L. Lach, J. Am. Pharm. Assoc. Sci. Ed. **43**, 349, 525, 527, 1954; T. Higuchi and D. A. Zuck, J. Am. Pharm. Assoc. Sci. Ed. **42**, 132, 1953.
11. T. Higuchi and L. Lachman, J. Am. Pharm. Assoc. Sci. Ed. **44**, 521, 1955; L. Lachman, L. J. Ravin, and T. Higuchi, J. Am. Pharm. Assoc. Sci. Ed. **45**, 290, 1956; L. Lachman and T. Higuchi, J. Am. Pharm. Assoc. Sci. Ed. **46**, 32, 1957.
12. T. Higuchi and I. H. Pitman, J. Pharm. Sci. **62**, 55, 1973.
13. P. York and A. Saleh, J. Pharm. Sci. **65**, 493, 1976.
14. A. Marcus, Drug Cosmet. Ind. **79**, 456, 1956.
15. J. A. Plaizier-Vercammen and R. E. De Nève, J. Pharm. Sci. **70**, 1252, 1981; J. A. Plaizier-Vercammen, J. Pharm. Sci. **76**, 817, 1987.
16. K. H. Frömming, W. Ditter, and D. Horn, J. Pharm. Sci. **70**, 738, 1981.
17. D. S. Hayward, R. A. Kenley, and D. R. Jenke, Int. J. Pharm. **59**, 245, 1990.
18. T. Hosono, S. Tsuchiya, and H. Matsumaru, J. Pharm. Sci. **69**, 824, 1980.
19. B. J. Forman and L. T. Grady, J. Pharm. Sci. **58**, 1262, 1969.
20. H. van Olphen, *Introduction to Clay Calloid Chemistry*, 2nd Ed., Wiley, New York, 1977, pp. 66–68.
21. *Merck Index*, 11th Ed., Merck, Rahway, N.J., 1989, p. 365.
22. J. A. A. Ketelaar, *Chemical Constitution*, Elsevier, New York, 1958, p. 365.
23. H. M. Powell and D. E. Palin, J. Chem. Soc. **208**, 1947; J. Chem. Soc. **61**, 571, 815, 1948; J. Chem. Soc. **298**, 300, 468, 1950; J. Chem. Soc. 2658, 1954.
24. (a) S. G. Frank, J. Pharm. Sci. **64**, 1585, 1975; (b) M. E. Davis and M. E. Brewster, Nat. Rev. Drug Discov. **3**, 1023, 2004.
25. K. Cabrera and G. Schwinn, Am. Lab. **22**, 22, 24, 26, 28, 1990.
26. D. Duchêne and D. Wouessidjewe, Pharm. Tech. **14**(6), 26, 28, 32, 34, 1990; **14**(8), 22, 24, 26, 28, 30, 1990.
27. O. Beckers, J. H. Beijnen, E. H. G. Bramel, M. Otagiri, A. Bult, and W. J. M. Underberg, Int. J. Pharm. **52**, 239, 1989.
28. T. Bakensfield, B. W. Müller, M. Wiese, and J. Seydel, Pharm. Res. **7**, 484, 1990.
29. J. L. Lach and J. Cohen, J. Pharm. Sci. **52**, 137, 1963; W. A. Pauli and J. Lach, J. Pharm. Sci. **54**, 1945, 1965; J. Lach and W. A. Pauli, J. Pharm. Sci. **55**, 32, 1966.
30. D. Amdidouche, H. Darrouzet, D. Duchêne, and M.-C. Poelman, Int. J. Pharm. **54**, 175, 1989.
31. M. A. Hassan, M. S. Suleiman, and N. M. Najib, Int. J. Pharm. **58**, 19, 1990.
32. F. Kedzierewicz, M. Hoffman, and P. Maincent, Int. J. Pharm. **58**, 22, 1990.
33. M. E. Brewster, K. S. Estes, and N. Bodor, Int. J. Pharm. **59**, 231, 1990; M. E. Brewster, K. S. Estes, T. Loftsson, R. Perchalski, H. Derendorf, G. Mullersman, and N. Bodor, J. Pharm. Sci. **77**, 981, 1988.
34. A. R. Green and J. K. Guillory, J. Pharm. Sci. **78**, 427, 1989.
35. B. W. Müller and U. Brauns, J. Pharm. Sci. **75**, 571, 1986.
36. J. Pitha, E. J. Anaissie, and K. Uekama, J. Pharm. Sci. **76**, 788, 1987.
37. Y. Horiuchi, F. Hiriyama, and K. Uekama, J. Pharm. Sci. **79**, 128, 1990.
38. F. Hirayama, N. Hirashima, K. Abe, K. Uekama, T. Ijitsu, and M. Ueno, J. Pharm. Sci. **77**, 233, 1988.
39. F. M. Andersen, H. Bungaard, and H. B. Mengel, Int. J. Pharm. **21**, 51, 1984.
40. K. A. Connors, in *A Textbook of Pharmaceutical Analysis*, Wiley, New York, 1982, Chapter 4; K. A. Connors, *Binding Constants*, Wiley, New York, 1987.
41. P. Job, Ann. Chim. **9**, 113, 1928.
42. A. E. Martell and M. Calvin, *Chemistry of the Metal Chelate Compounds*, Prentice Hall, New York, 1952, p. 98.
43. M. E. Bent and C. L. French, J. Am. Chem. Soc. **63**, 568, 1941; R. E. Moore and R. C. Anderson, J. Am. Chem. Soc. **67**, 168, 1945.
44. W. C. Vosburgh, et al., J. Am. Chem. Soc. **63**, 437, 1941; J. Am. Chem. Soc. **64**, 1630, 1942.
45. A. Osman and R. Abu-Eittah, J. Pharm. Sci. **69**, 1164, 1980.
46. J. Bjerrum, *Metal Amine Formation in Aqueous Solution*, Haase, Copenhagen, 1941.
47. M. Pecar, N. Kujundzic, and J. Pazman, J. Pharm. Sci. **66**, 330, 1977; M. Pecar, N. Kujundzic, B. Mlinarevic, D. Cerina, B. Horvat, and M. Veric, J. Pharm. Sci. **64**, 970, 1975.

48. B. J. Sandmann and H. T. Luk, J. Pharm. Sci. **75**, 73, 1986.

49. Y. K. Agrawal, R. Giridhar, and S. K. Menon, J. Pharm. Sci. **76**, 903, 1987.

50. K. A. Connors, S.-F. Lin, and A. B. Wong, J. Pharm. Sci. **71**, 217, 1982.

51. T. Higuchi and A. A. Zuck, J. Am. Pharm. Assoc. Sci. Ed. **42**, 132, 1953.

52. D. Guttman and T. Higuchi, J. Am. Pharm. Assoc. Sci. Ed. **46**, 4, 1957.

53. T. Higuchi and J. L. Lach, J. Am. Pharm. Assoc. Sci. Ed. **43**, 525, 1954.

54. T. Higuchi and K. A. Connors, in C. N. Reilley (Ed.), *Advances in Analytical Chemistry and Instrumentation*, Vol. 4, Wiley, New York, 1965, pp. 117–212.

55. R. A. Kenley, S. E. Jackson, J. S. Winterle, Y. Shunko, and G. C. Visor, J. Pharm. Sci. **75**, 648, 1986.

56. R. Foster (Ed.), *Molecular Complexes*, Vol. 1, Elek, London, 1973.

57. M. A. Slifkin, *Charge Transfer Interactions of Biomolecules*, Academic Press, New York, 1971, Chapters 1 and 2.

58. H. A. Benesi and J. H. Hildebrand, J. Am. Chem. Soc. **70**, 2832, 1948.

59. F. A. Al-Obeidi and H. N. Borazan, J. Pharm. Sci. **65**, 892, 1976; J. Pharm. Sci. **65**, 982, 1976; H. M. Taka, F. A. Al-Obeidi, and H. N. Borazan, J. Pharm. Sci. **68**, 631, 1979; N. I. Al-Ani and H. N. Borazan, J. Pharm. Sci. **67**, 1381, 1978; H. N. Borazan and Y. H. Ajeena, J. Pharm. Sci. **69**, 990, 1980; J. Pharm. Sci. **77**, 544, 1988.

60. N. E. Webb and C. C. Thompson, J. Pharm. Sci. **67**, 165, 1978.

61. J. Nishijo, I. Yonetami, E. Iwamoto, S. Tokura, and K. Tagahara, J. Pharm. Sci. **79**, 14, 1990.

62. H. N. Borazan and S. N. Koumriqian, J. Pharm. Sci. **72**, 1450, 1983.

63. J. De Taeye and Th. Zeegers-Huyskens, J. Pharm. Sci. **74**, 660, 1985.

64. A. El-Said, E. M. Khairy, and A. Kasem, J. Pharm. Sci. **63**, 1453, 1974.

65. A Goldstein, Pharmacol. Rev. **1**, 102, 1949.

66. J. J. Vallner, J. Pharm. Sci. **66**, 447, 1977.

67. C. K. Svensson, M. N. Woodruff, and D. Lalka, in W. E. Evans, J. J. Schentag, and W. J. Jusko (Eds.), *Applied Pharmacokinetics*, 2nd Ed., Applied Therapeutics, Spokane, Wash., 1986, Chapter 7.

68. G. Scatchard, Ann. N. Y. Acad. Sci. **51**, 660, 1949.

69. A. A. Sandberg, H. Rosenthal, S. L. Schneider, and W. R. Slaunwhite, in T. Nakao, G. Pincus, and J. F. Tait (Eds.), *Steroid Dynamics*, Academic Press, New York, 1966, p. 33.

70. J. Blanchard, W. T. Fink, and J. P. Duffy, J. Pharm. Sci. **66**, 1470, 1977.

71. I. M. Klotz, F. M. Walker, and R. B. Puvan, J. Am. Chem. Soc. **68**, 1486, 1946.

72. T. N. Tozer, J. G. Gambertoglio, D. E. Furst, D. S. Avery, and N. H. G. Holford, J. Pharm. Sci. **72**, 1442, 1983.

73. C. J. Briggs, J. W. Hubbard, C. Savage, and D. Smith, J. Pharm. Sci. **72**, 918, 1983.

74. A. Zini, J. Barre, G. Defer, J. P. Jeanniot, G. Houin, and J. P. Tillement, J. Pharm. Sci. **74**, 530, 1984.

75. E. Okezaki, T. Teresaki, M. Nakamura, O. Nagata, H. Kato, and A. Tsuji, J. Pharm. Sci. **78**, 504, 1989.

76. J. W. Melten, A. J. Wittebrood, H. J. J. Williams, G. H. Faber, J. Wemer, and D. B. Faber, J. Pharm. Sci. **74**, 692, 1985.

77. M. C. Meyer and D. E. Guttman, J. Pharm. Sci. **57**, 1627, 1968.

78. J. Judis, J. Pharm. Sci. **71**, 1145, 1982.

79. W. Kauzmann, Adv. Protein Chem. **14**, 1, 1959.

80. J. B. Nagwekar and H. B. Kostenbauder, J. Pharm. Sci. **59**, 751, 1970.

81. H. Kristiansen, M. Nakano, N. I. Nakano, and T. Higuchi, J. Pharm. Sci. **59**, 1103, 1970.

82. S. Feldman and M. Gibaldi, J. Pharm. Sci. **56**, 370, 1967.

83. M. Menozzi, L. Valentini, E. Vannini, and F. Arcamone, J. Pharm. Sci. **73**, 6, 1984.

84. S. Sato, C. D. Ebert, and S. W. Kim, J. Pharm. Sci. **72**, 228, 1983.

85. A. M. Saleh, A. R. Ebian, and M. A. Etman, J. Pharm. Sci. **75**, 644, 1986.

86. E. Touitou and P. Fisher, J. Pharm. Sci. **75**, 384, 1986.

87. E. Touitou, F. Alhaique, P. Fisher, A. Memoli, F. M. Riccieri, and E. Santucci, J. Pharm. Sci. **76**, 791, 1987.

88. T. J. Racey, P. Rochon, D. V. C. Awang, and G. A. Neville, J. Pharm. Sci. **76**, 314, 1987.

89. J. Coulson and V. J. Smith, J. Pharm. Sci. **69**, 799, 1980.

90. F. J. Diana, K. Veronich, and A. Kapoor, J. Pharm. Sci. **78**, 195, 1989.

91. J. A. Plaizier-Vercammen, J. Pharm. Sci. **72**, 1042, 1983.

92. P. Bustamante and E. Selles, J. Pharm. Sci. **75**, 639, 1986.

## Recommended Readings

H. Dodziuk, *Cyclodextrins and Their Complexes: Chemistry, Analytical Methods, Applications*, Wiley-VCH, Weinheim, Germany, 2006.

J. A. Goodrich and J. F. Kugel, *Binding and Kinetics for Molecular Biologists*, Cold Spring Harbor Laboratory Press, New York, 2007.

## CHAPTER LEGACY

**Fifth Edition:** published as Chapter 11 (Complexation and Protein Binding). Updated by Patrick Sinko.

**Sixth Edition:** published as Chapter 9 (Complexation and Protein Binding). Updated by Patrick Sinko.

# 11 DIFFUSION

## INTRODUCTION

The fundamentals of diffusion are discussed in this chapter. Free diffusion of substances through liquids, solids, and membranes is a process of considerable importance in the pharmaceutical sciences. Topics of mass transport phenomena applying to the pharmaceutical sciences include the release and dissolution of drugs from tablets, powders, and granules; lyophilization, ultrafiltration, and other mechanical processes; release from ointments and suppository bases; passage of water vapor, gases, drugs, and dosage form additives through coatings, packaging, films, plastic container walls, seals, and caps; and permeation and distribution of drug molecules in living tissues. This chapter treats the fundamental basis for diffusion in pharmaceutical systems.

There are several ways that a solute or a solvent can traverse a physical or biologic membrane. The first example (**Fig. 11–1**) depicts the flow of molecules through a physical barrier such as a polymeric membrane. The passage of matter through a solid barrier can occur by simple molecular permeation or by movement through pores and channels. Molecular diffusion or permeation through nonporous media depends on the solubility of the permeating molecules in the bulk membrane (**Fig. 11–1a**), whereas a second process can involve passage of a substance through solvent-filled pores of a membrane (**Fig. 11–1b**) and is influenced by the relative size of the penetrating molecules and the diameter and shape of the pores. The transport of a drug through a polymeric membrane involves dissolution of the drug in the matrix of the membrane and is an example of simple molecular diffusion. A second example relates to drug and solvent transport across the skin. Passage through human skin of steroidal molecules substituted with hydrophilic groups may predominantly involve transport through hair follicles, sebum ducts, and sweat pores in the epidermis (**Fig. 11–19**). Perhaps a better representation of a membrane on the molecular scale is a matted arrangement of polymer strands with branching and intersecting channels as shown in **Figure 11–1c**. Depending on the size and shape of the diffusing molecules, they may pass through the tortuous pores formed by the overlapping strands of polymer. If it is too large for such channel transport, the diffusant may dissolve in the polymer matrix and pass through the film by simple diffusion. Diffusion also plays an important role in drug and nutrient transport in biologic membranes in the brain, intestines, kidneys, and liver. In addition to diffusion through the lipoidal membrane, several other transport

## KEY CONCEPT DIFFUSION

*Diffusion* is defined as a process of mass transfer of individual molecules of a substance brought about by random molecular motion and associated with a driving force such as a concentration gradient. The mass transfer of a solvent (e.g., water) or a solute (e.g., a drug) forms the basis for many important phenomena in the pharmaceutical sciences. For example, diffusion of a drug across a biologic membrane is required for a drug to be absorbed into and eliminated from the body, and even for it to get to the site of action within a particular cell. On the negative side, the shelf life of a drug product could be significantly reduced if a container or closure does not prevent solvent or drug loss or if it does not prevent the absorption of water vapor into the container. These and many more important phenomena have a basis in diffusion. Drug release from a variety of drug delivery systems, drug absorption and elimination, dialysis, osmosis, and ultrafiltration are some of the examples covered in this and other chapters.

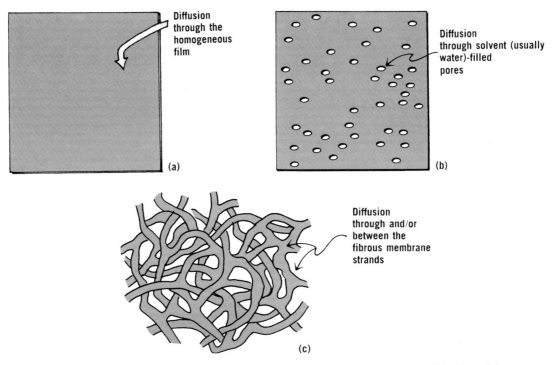

**Fig. 11–1.** (*a*) Homogeneous membrane without pores. (*b*) Membrane of dense material with straight-through pores, as found in certain filler barriers such as Nucleopore. (*c*) Cellulose membrane used in the filtration process, showing the intertwining nature of the fibers and the tortuous channels.

mechanisms have been characterized in biologic membranes including energy-dependent and energy-independent carrier-mediated transport as well as diffusion through the paracellular spaces between cells. The multitude of mechanisms of transport across various mucosal membranes will be introduced later in this chapter. Several pharmaceutically important diffusion-based processes are covered in this and subsequent chapters.

## Drug Absorption and Elimination

Diffusion through biologic membranes is an essential step for drugs entering (i.e., absorption) or leaving (i.e., elimination) the body. It is also an important component along with convection for efficient drug distribution throughout the body and into tissues and organs. Diffusion can occur through the lipoidal bilayer of cells. This is termed *transcellular diffusion*. On the other hand, paracellular diffusion occurs through the spaces between adjacent cells. In addition to diffusion, drugs and nutrients also traverse biologic membranes using membrane transporters and, to a lesser extent, cell surface receptors. Membrane transporters are specialized proteins that facilitate drug transport across biologic membranes. The interactions between drugs and transporters can be classified as energy dependent (i.e., active transport) or energy independent (i.e., facilitated diffusion). Membrane transporters are located in every organ responsible for the absorption, distribution, metabolism, and excretion (ADME) of drug substances. Specialized membrane trans-

port mechanisms were covered in more detail in Chapter 12 (Biopharmaceutics) and Chapter 13 (Drug Release and Dissolution).

## Elementary Drug Release

Elementary drug release is an important process that literally affects nearly every person in everyday life. Drug release is a multistep process that includes diffusion, disintegration, deaggregation, and dissolution. These processes are described in this and other chapters. Common examples are the release of steroids such as hydrocortisone from topical over-the-counter creams and ointments for the treatment of skin rashes and the release of acetaminophen from a tablet that is taken by mouth. Drug release must occur before the drug can be pharmacologically active. This includes pharmaceutical products such as capsules, creams, liquid suspensions, ointments, tablets, and transdermal patches.

## Osmosis

*Osmosis* was originally defined as the passage of both solute and solvent across a membrane but now refers to an action in which only the solvent is transferred. The solvent passes through a semipermeable membrane to dilute the solution containing solute and solvent. The passage of solute together with solvent is now called *diffusion* or *dialysis*. Osmotic drug release systems use osmotic pressure as a driving force for the controlled delivery of drugs. A simple osmotic pump consists of an osmotic core (containing drug with or without an osmotic agent) and is coated with a semipermeable membrane. The semipermeable membrane has an orifice for drug release from the "pump." The dosage form, after coming in contact with the aqueous fluids, imbibes water at a rate determined by the fluid permeability of the membrane and osmotic pressure of core formulation. The osmotic imbibition of water results in high hydrostatic pressure inside the pump, which causes the flow of the drug solution through the delivery orifice.

## Ultrafiltration and Dialysis

Ultrafiltration is used to separate colloidal particles and macromolecules by the use of a membrane. Hydraulic pressure is used to force the solvent through the membrane, whereas the microporous membrane prevents the passage of large solute molecules. Ultrafiltration is similar to a process called *reverse osmosis*, but a much higher osmotic pressure is developed in reverse osmosis, which is used in desalination of brackish water. Ultrafiltration is used in the pulp and paper industry and in research to purify albumin and enzymes. *Microfiltration*, a process that employs membranes of slightly larger pore size, 100 nm to several micrometers, removes bacteria from intravenous injections, foods, and drinking water.[1] Hwang and Kammermeyer[2] defined *dialysis* as a separation process based on unequal rates of passage of solutes and solvent through microporous membranes, carried out in batch or continuous mode. *Hemodialysis* is used in treating kidney malfunction to rid the blood of metabolic waste products (small molecules) while preserving the high-molecular-weight components of the blood. In ordinary osmosis as well as in dialysis, separation is spontaneous and does not involve the high-applied pressures of ultrafiltration and reverse osmosis.

Diffusion is caused by random molecular motion and, in relative terms, is a slow process. In a classic text on diffusion, E. L. Cussler stated, "In gases, diffusion progresses at a rate of about 10 cm in a minute; in liquids, its rate is about 0.05 cm/min; in solids, its rate may be only about 0.0001 cm/min."[3] A relevant question to ask at this point is, Can such a slow process be meaningful to the pharmaceutical sciences? The answer is a resounding "yes." Although the rate of diffusion appears to be quite slow, other factors such as the distance that a diffusing molecule must traverse are also very important. For example, a typical cell membrane is approximately 5-nm thick. If it is assumed that a drug will diffuse into a cell at a rate of 0.0005 cm/min, then it takes only a fraction of a second for that drug molecule to enter the cell. On the other hand, the thickest biomembrane is skin, with an average thickness of 3 $\mu$m (**Fig. 11–19**). For the same rate of diffusion, it would take 600 times longer for the same drug molecule to diffuse through the skin. The time difference in the appearance of the drug on the other side of the skin is known as the lag time. An even more extreme example is the levonorgestrel-releasing implant.[4] This long-acting contraceptive has been approved for 5 years of continuous use in human patients. To achieve low constant diffusion rates, six matchstick-sized, flexible, closed capsules made of silicon rubber tubing are inserted into the upper arm of patients. Annual pregnancy rates of Norplant users are below 1 per 100 throughout 7 years of continuous use. Levonorgestrel implants provide low progestogen doses: 40 to 50 $\mu$g/day at 1 year of use, decreasing to 25 to 30 $\mu$g/day in the fifth year. Serum levels of levonorgestrel at 5 years are 60% to 65% of those levels measured at 1 month of use.[4] Although diffusion plays an important role in the successful delivery of levonorgestrel from the Norplant system, drug release from long-acting delivery systems is a function of many other factors as well.

Another pharmaceutically relevant example of diffusion relates to the mixing of a drug in solution with intestinal contents immediately prior to drug absorption across the intestinal mucosa. At first glance, mixing appears to be a simple process; however, several molecular- and macroscopic-level processes must occur in parallel for efficient mixing to occur. It is important to remember that diffusion depends on random molecular motions that take place over small molecular distances. Therefore, other processes are responsible for the movement of molecules over much larger distances and are required for mixing to occur. These processes are called macroscopic processes and include convection, dispersion, and stirring. After the macroscopic movement of molecules occurs, diffusion mixes newly adjacent portions of the intestinal fluid. Diffusion and the macroscopic processes all contribute to mixing, and, qualitatively, the effects are similar. In 1860, Maxwell was one of the first to recognize this when he stated, "Mass transfer is due partly to the motion of translation and partly to that of agitation."[5] Unlike many other phenomena, diffusion in a solution always occurs in parallel with convection. Convection is the bulk movement of fluid accompanied by the transfer of heat (energy) in the presence of agitation. An example of convection relevant to intestinal absorption of drugs is fluid flow down the intestine. Dispersion is also relevant to intestinal flow and is related to diffusion. "The relation exists on two very different levels. First, dispersion is a form of mixing, and so on a microscopic level it involves diffusion of molecules. Second, dispersion and diffusion are described with very similar mathematics."[3] Although it is somewhat difficult to assess intestinal dispersion patterns in humans, they are most likely characterized as "turbulent." In certain experimental models, such as the single-pass intestinal perfusion procedure[6] that is used to estimate the intestinal permeability of drugs in rats, flow conditions are optimized to obtain laminar flow hydrodynamics. Laminar flow conditions are a special example of the coupling of flow and diffusion. In contrast to turbulent flow, when operating under laminar flow conditions a dispersion coefficient can be accurately predicted. In this system, mass transport occurs by radial diffusion (i.e., movement toward the intestinal mucosa) and axial convection (i.e., flow down the length of the intestine).

# STEADY-STATE DIFFUSION

## Thermodynamic Basis

Mass transfer is the movement of molecules in response to an applied driving force. Convective and diffusive mass transfer is important to many pharmaceutical science applications. Diffusive mass transfer is the subject of this chapter, but convective mass transfer will not be covered in detail, and the student is referred to other texts.[7–9] Mass transfer is a kinetic process, occurring in systems that are not in equilibrium.[7] To better understand the thermodynamic basis of mass transfer, consider an isolated system consisting of two sections separated by an imaginary membrane (**Fig. 11–2**).[7] At equilibrium, the temperatures, $T$, pressures, $P$, and chemical potentials, $\mu$, of each of two species A and B are equal in the two sections. If this isolated system is unperturbed, it will remain at this thermodynamic equilibrium indefinitely. Suppose that the chemical potential of one of the species, A, is now increased in section I so that $\mu_{A,I} > \mu_{A,II}$. Because the chemical potential of A is related to its concentration, the ideality of the solution, and the temperature, this perturbation of the system can be achieved by increasing the concentration of A in section I. The system will respond to this perturbation by establishing a new thermodynamic equilibrium. Although it could reestablish the equilibrium by altering any of the three variables in the system ($T$, $P$, or $\mu$), let us assume that it will reequilibrate the chemical potentials, leaving $T$ and $P$ unaffected. If the membrane separating the two sections will allow for the passage of species A, then equilibrium will be reestablished by the movement of species A from section I to section II until the chemical potentials of section I and II are once again equal. The movement of mass in response to a spatial gradient in chemical potential as a result of random molecular motion (i.e., Brownian motion) is called diffusion. Although the thermodynamic basis for diffusion is best described using chemical potentials, it is mathematically simpler to describe it using concentration, a more experimentally practical variable.

## Fick's Laws of Diffusion

In 1855, Fick recognized that the mathematical equation of heat conduction developed by Fourier in 1822 could be applied to mass transfer. These fundamental relationships govern diffusion processes in pharmaceutical systems. The amount, $M$, of material flowing through a unit cross section, $S$, of a barrier in unit time, $t$, is known as the flux, $J$:

$$J = \frac{dM}{S \cdot dt} \qquad (11\text{–}1)$$

The flux, in turn, is proportional to the concentration gradient, $dC/dx$:

$$J = -D\frac{dC}{dx} \qquad (11\text{–}2)$$

where $D$ is the *diffusion coefficient* of a penetrant (also called the *diffusant*) in cm$^2$/sec, $C$ is its concentration in g/cm$^3$, and $x$ is the distance in centimeter of movement perpendicular to the surface of the barrier. In equation (**11–1**), the mass, $M$, is usually given in grams or moles, the barrier surface area, $S$, in cm$^2$, and the time, $t$, in seconds. The units of $J$ are g/cm$^2$ sec. The SI units of kilogram and meter are sometimes used, and the time may be given in minutes, hours, or days. The negative sign of equation (**11–2**) signifies that diffusion occurs in a direction (the positive $x$ direction) opposite to that of increasing concentration. That is, diffusion occurs in the direction of decreasing concentration of diffusant; thus, the flux is always a positive quantity. Diffusion will stop when the concentration gradient no longer exists (i.e., when $dC/dx = 0$).

Although the *diffusion coefficient, D*, or *diffusivity*, as it is often called, appears to be a proportionality constant, it does not ordinarily remain constant. $D$ is affected by concentration, temperature, pressure, solvent properties, and the chemical nature of the diffusant. Therefore, $D$ is referred to more correctly as a *diffusion coefficient* rather than as a constant. Equation (**11–2**) is known as *Fick's first law*.

## Fick's Second Law

Fick's second law of diffusion forms the basis for most mathematical models of diffusion processes. One often wants to examine the rate of change of diffusant concentration at a point in the system. An equation for mass transport that emphasizes the change in *concentration* with time at a definite location rather than the *mass* diffusing across a unit area of barrier in unit time is known as *Fick's second law*. This diffusion equation is derived as follows. The concentration, $C$, in a particular volume element (**Figs. 11–3** and **11–4**) changes only as a result of net flow of diffusing molecules into or out of the region. A difference in concentration results from a difference in input and output. The concentration of diffusant in the volume element changes with time, that is, $\Delta C/\Delta t$, as

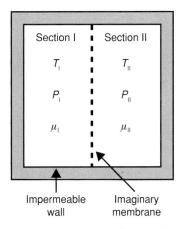

**Fig. 11–2.** Isolated system consisting of two sections separated by an imaginary permeable membrane. At equilibrium, the temperatures (*T*), pressures (*P*), and chemical properties (*μ*) of each of the species in the system are equal in the two sections. (Modified from G. L. Amidon, P. I. Lee, and E. M. Topp (Eds.), *Transport Processes in Pharmaceutical Systems*, Marcel Dekker, New York, 2000, p. 13.)[84]

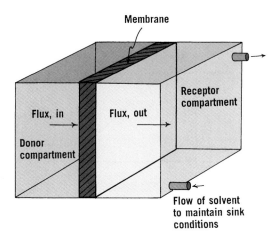

**Fig. 11–3.** Diffusion cell. The donor compartment contains diffusant at concentration $C$.

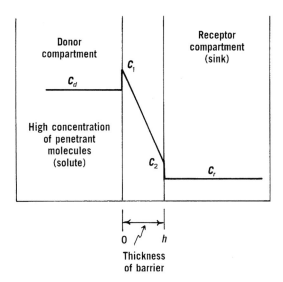

**Fig. 11–4.** Concentration gradient of diffusant across the diaphragm of a diffusion cell. It is normal for the concentration curve to increase or decrease sharply at the boundaries of the barrier because, in general, $C_1$ is different from $C_d$, and $C_2$ is different from $C_r$. The concentration $C_1$ would be equal to $C_d$, for example, only if $K - C_1/C_d$ had a value of unity.

the flux or amount diffusing changes with distance, $\Delta J/\Delta x$, in the $x$ direction, or*

$$\frac{\partial C}{\partial t} = -\frac{\partial J}{\partial x} \qquad (11\text{–}3)$$

Differentiating the first-law expression, equation **(11–2)**, with respect to $x$, one obtains

$$-\frac{\partial J}{\partial x} = D\frac{\partial^2 c}{\partial x^2} \qquad (11\text{–}4)$$

Substituting $\partial C/\partial t$ from equation **(11–3)** into equation **(11–4)** results in Fick's second law, namely,

$$\frac{\partial C}{\partial t} = D\frac{\partial^2 c}{\partial x^2} \qquad (11\text{–}5)$$

Equation **(11–5)** represents diffusion only in the $x$ direction. If one wishes to express concentration changes of diffusant in three dimensions, Fick's second law is written in the general form

$$\frac{\partial C}{\partial t} = D\left(\frac{\partial^2 C}{\partial x^2} + \frac{\partial^2 C}{\partial y^2} + \frac{\partial^2 C}{\partial z^2}\right) \qquad (11\text{–}6)$$

This expression is not usually needed in pharmaceutical problems of diffusion, however, because movement in one direction is sufficient to describe most cases. Fick's second law states that the change in concentration with time in a particular region is proportional to the change in the concentration gradient at that point in the system.

## Steady State

An important condition in diffusion is that of the *steady state*. Fick's first law, equation **(11–2)**, gives the flux (or rate of diffusion through unit area) in the steady state of flow. The second law refers in general to a change in concentration of diffusant with time at any distance, $x$ (i.e., a nonsteady state

*Concentration and flux are often written as $C(x, t)$ and $J(x, t)$, respectively, to emphasize that these parameters are functions of both distance, $x$, and time, $t$.

of flow). Steady state can be described, however, in terms of the second law, equation **(11–5)**. Consider the diffusant originally dissolved in a solvent in the left-hand compartment of the chamber shown in **Figure 11–3**. Solvent alone is placed on the right-hand side of the barrier, and the solute or penetrant diffuses through the central barrier from solution to solvent side (donor to receptor compartment). In diffusion experiments, the solution in the receptor compartment is constantly removed and replaced with fresh solvent to keep the concentration at a low level. This is referred to as "sink conditions," the left compartment being the source and the right compartment the sink.

Originally, the diffusant concentration will fall in the left compartment and rise in the right compartment until the system comes to equilibrium, based on the rate of removal of diffusant from the sink and the nature of the barrier. When the system has been in existence a sufficient time, the concentration of diffusant in the solutions at the left and right of the barrier becomes constant with respect to time but obviously not the same in the two compartments. Then, within each diffusional slice perpendicular to the direction of flow, the rate of change of concentration, $dC/dt$, will be zero, and by Fick's second law,

$$\frac{dC}{dt} = D\frac{d^2 C}{dx^2} = 0 \qquad (11\text{–}7)$$

$C$ is the concentration of the permeant in the barrier expressed in mass/cm³. Equation **(11–7)** demonstrates that because $D$ is not equal to zero, $d^2C/dx^2 = 0$. When a second derivative such as this equals zero, one concludes that there is no change in $dC/dx$. In other words, the concentration gradient across the membrane, $dC/dx$, is constant, signifying a linear relationship between concentration, $C$, and distance, $x$. This is shown in **Figure 11–4** (in which the distance $x$ is equal to $h$) for drug diffusing from left to right in the cell of **Figure 11–3**. Concentration will not be rigidly constant, but rather is likely to vary slightly with time, and then $dC/dt$ is not exactly zero. The conditions are referred to as a "quasistationary" state, and little error is introduced by assuming steady state under these conditions.

## Diffusion Driving Forces

There are numerous diffusional driving forces in pharmaceutical systems. Up to this point the discussion has focused on "ordinary diffusion," which is driven by a concentration gradient.[8] However, other driving forces include pressure, temperature, and electric potential. Examples of driving forces in pharmaceutical systems are shown in **Table 11–1**.

# DIFFUSION THROUGH MEMBRANES

## Steady Diffusion Across a Thin Film and Diffusional Resistance

Yu and Amidon[18] concisely developed an analysis for steady diffusion across a thin film as it relates to diffusional resistance. **Figure 11–4** depicts steady diffusion across a thin film of thickness $h$. In this case, the diffusion coefficient is considered constant because the solutions on both sides of the film are dilute. The concentrations on both sides of the film, $C_d$ and $C_r$, are kept constant and both sides are well mixed. Diffusion occurs in the direction from the higher concentration ($C_d$) to the lower concentration ($C_r$). After sufficient time, steady state is achieved and the concentrations are constant at all points in the film as shown in **Figure 11–5**. At steady state ($dC/dt = 0$), Fick's second law becomes

$$D\frac{\partial^2 e}{\partial z^2} = 0 \qquad (11\text{–}8)$$

![KEY CONCEPT] **MEMBRANES AND BARRIERS**

Flynn et al.[19] differentiated between a membrane and a barrier. A *membrane* is a biologic or physical "film" separating the phases, and material passes by passive, active, or facilitated transport across this film. The term *barrier* applies in a more general sense to the region or regions that offer resistance to passage of a diffusing material, the total barrier being the sum of individual resistances of membranes.

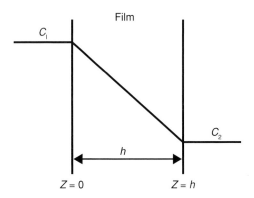

**Fig. 11–5.** Diffusion across a thin film. The solute molecules diffuse from the well-mixed higher concentration, $C_1$, to the well-mixed lower concentration, $C_2$. The concentrations on both sides of the film are kept constant. At steady state, the concentrations remain constant at all points in the film. The concentration profile inside the film is linear, and the flux is constant.

Integrating equation (**11–8**) twice using the conditions that at $z = 0$, $C = C_d$ and at $z = h$, $C = C_r$, yields the following equation:

$$J = \frac{D}{h}(C_1 - C_2) \qquad (11\text{–}9)$$

The term $h/D$ is often called the diffusional resistance, denoted by $R$. The flux equation can then be written as

$$J = \frac{C_1 - C_2}{R} \qquad (11\text{–}10)$$

Although resistance to diffusion is a fundamental scientific principle, permeability is a term that is used more often in the pharmaceutical sciences. Resistance and permeability are inversely related. In other words, the higher the resistance to diffusion, the lower is the permeability of the diffusing substance. In the next few sections the concepts of permeability and series resistance will be introduced.

## Permeability

Fick adapted the two diffusion equations (**11–2**) and (**11–5**) to the transport of matter from the laws of heat conduction. Equations of heat conduction are found in the book by Carslaw.[20] General solutions to these differential equations yield complex expressions; simple equations are used here for the most part, and worked examples are provided so that the reader should have no difficulty in following the discussion of dissolution and diffusion.

If a membrane separates the two compartments of a diffusion cell of cross-sectional area $S$ and thickness $h$, and if the concentrations in the membrane on the left (donor) and on the right (receptor) sides are $C_1$ and $C_2$, respectively (**Figure 11–4**), the first law of Fick can be written as

$$J = \frac{dM}{S\,dt} = D\left(\frac{C_1 - C_2}{h}\right) \qquad (11\text{–}11)$$

## TABLE 11–1
## DRIVING FORCES IN PHARMACEUTICAL SYSTEMS

| Driving Force | Example | Description | References |
|---|---|---|---|
| Concentration | Passive diffusion | Passive diffusion is a process of mass transfer of individual molecules of a substrate brought about by random molecular motion and associated with a concentration gradient | 3 |
| | Drug dissolution | Drug "dissolution" occurs when a tablet is introduced into a solution and is usually accompanied by disintegration and deaggregation of the solid matrix followed by drug diffusion from the remaining small particles | 10 |
| Pressure | Osmotic drug release | Osmotic drug release systems utilize osmotic pressure as the driving force for controlled delivery of drugs; a simple osmotic pump consists of an osmotic core (containing drug with or without an osmotic agent) coated with a semipermeable membrane; the semipermeable membrane has an orifice for drug release from the pump; the dosage form, after contacting with the aqueous fluids, imbibes water at a rate determined by the fluid permeability of the membrane and osmotic pressure of core formulation; this osmotic imbibition of water results in high hydrostatic pressure inside the pump, which causes the flow of the drug solution through the delivery orifice | 11 |
| | Pressure-driven jets for drug delivery | Pressure-driven jets are used for drug delivery; a jet injector produces a high-velocity jet ($>100$ m/sec) that penetrates the skin and delivers drugs subcutaneously, intradermally, or intramuscularly without the use of a needle; the mechanism for the generation of high-velocity jets includes either a compression spring or compressed air | 12 |
| Temperature | Lyophilization | Lyophilization (freeze-drying) of a frozen aqueous solution containing a drug and a inner-matrix building substance involves the simultaneous change in receding boundary with time, phase transition at the ice–vapor interface governed by the Clausius–Clapeyron pressure–temperature relationship, and water vapor diffusion across the pore path length of the dry matrix under low temperature and vacuum conditions | 13 |
| | Microwave-assisted extraction | Microwave-assisted extraction (MAE) is a process of using microwave energy to heat solvents in contact with a sample in order to partition analytes from the sample matrix into the solvent; the ability to rapidly heat the sample solvent mixture is inherent to MAE and is the main advantage of this technique; by using closed vessels, the extraction can be performed at elevated temperatures, accelerating the mass transfer of target compounds from the sample matrix | 14 |
| Electrical potential | Iontophoretic dermal drug delivery | Iontophoresis is used to enhance transdermal delivery of drugs by applying a small current through a reservoir that contains ionized drugs; one electrode (positive electrode to deliver positively charged ions and negative electrode to deliver negatively charged ions) is placed between the drug reservoir and the skin; the other electrode with opposite charge is placed a short distance away to complete the circuit, and the electrodes are connected to a power supply; when the current flows, charged ions are transported across the skin through a pore | 15, 16 |
| | Electrophoresis | Electrophoresis involves the movement of charged particles through a liquid under the influence of an applied potential difference; an electrophoresis cell fitted with two electrodes contains dispersion; when a potential is applied across the electrodes, the particles migrate to the oppositely charged electrode; capillary electrophoresis is widely used as an analytical tool in the pharmaceutical sciences | 17 |

where $(C_1 - C_2)/h$ approximates $dC/dx$. The gradient $(C_1 - C_2)/h$ within the diaphragm must be assumed to be constant for a quasistationary state to exist. Equation (11–11) presumes that the aqueous boundary layers (so-called static or unstirred aqueous layers) on both sides of the membrane do not significantly affect the total transport process. The potential influence of multiple resistances on diffusion such as those introduced by aqueous boundary layers (i.e., multi-layer diffusion) is covered later in this chapter.

The concentrations $C_1$ and $C_2$ within the membrane ordinarily are not known but can be replaced by the partition coefficient multiplied by the concentration $C_d$ on the donor side or $C_r$ on the receiver side, as follows. The distribution or partition coefficient, $K$, is given by

$$K = \frac{C_1}{C_d} = \frac{C_2}{C_r} \qquad (11\text{–}12)$$

Hence,

$$\frac{dM}{dt} = \frac{DSK(C_d - C_r)}{h} \qquad (11\text{–}13)$$

and, if sink conditions hold in the receptor compartment, $C_r \cong 0$,

$$\frac{dM}{dt} = \frac{DSKc_d}{h} = PSC_d \qquad (11\text{–}14)$$

where

$$P = \frac{DK}{h} (\text{cm/sec}) \qquad (11\text{–}15)$$

It is noteworthy that the permeability coefficient, also called the permeability, $P$, has units of linear velocity.*

In some cases, it is not possible to determine $D$, $K$, or $h$ independently and thereby to calculate $P$. It is a relatively simple matter, however, to measure the rate of barrier permeation and to obtain the surface area, $S$, and concentration, $C_d$, in the donor phase and the amount of permeant, $M$, in the receiving sink. One can then obtain $P$ from the slope of a linear plot of $M$ versus $t$:

$$M = PSC_d t \qquad (11\text{–}16)$$

provided that $C_d$ remains relatively constant throughout time. If $C_d$ changes appreciably with time, one recognizes that $C_d = M_d/V_d$, the amount of drug in the donor phase divided by the donor phase volume, and then one obtains $P$ from the slope of log $C_d$ versus $t$:

$$\log C_d = \log C_d(0) - \frac{PSt}{2.303 V_d} \qquad (11\text{–}17)$$

**EXAMPLE 11–1**

**Simple Drug Diffusion Through a Membrane**

A newly synthesized steroid is allowed to pass through a siloxane membrane having a cross-sectional area, $S$, of 10.36 $cm^2$ and a thickness, $h$, of 0.085 cm in a diffusion cell at 25°C. From the horizontal

intercept of a plot of $Q = M/S$ versus $t$, the lag time, $t_L$, is found to be 47.5 min. The original concentration $C_0$ is 0.003 mmole/$cm^3$. The amount of steroid passing through the membrane in 4.0 hr is $3.65 \times 10^{-3}$ mmole.

(a) Calculate the parameter $DK$ and the permeability, $P$. We have

$$Q = \frac{3.65 \times 10^{-3}\ \text{mmole}}{10.36\ cm^2} = 0.35 \times 10^{-3}\ \text{mmole/}cm^2$$

$$= DK \left( \frac{0.003\ \text{mmole/}cm^3}{0.085\ \text{cm}} \right) \left[ 4.0\ \text{hr} - \left( \frac{47.5}{60} \right) \text{hr} \right]$$

$$DK = 0.0031\ cm^2/\text{hr} = 8.6 \times 10^{-7}\ cm^2/\text{sec}$$

$$P = DK/h = (8.6 \times 10^{-7} cm^2/\text{sec})/0.085\ \text{cm}$$

$$= 1.01 \times 10^{-5}\ \text{cm/sec}$$

(b) Using the lag time $t_L = h^2/6D$, calculate the diffusion coefficient. We have

$$D = \frac{h^2}{6t_L} = \frac{(0.085)^2\ cm^2}{6 \times 47.5\ \text{min}}$$

$$= 25.4 \times 10^{-6}\ cm^2/\text{min}$$

or

$$= 4.23 \times 10^{-7}\ cm^2/\text{min}$$

(c) Combining the permeability, Equation (11–15), with the value of $D$ from (b), calculate the partition coefficient, $K$. We have

$$K = \frac{Ph}{D} = \frac{(1.01 \times 10^{-5}\ \text{cm/sec})(0.085\ \text{cm})}{4.23 \times 10^{-7}\ cm^2/\text{sec}} = 2.03$$

Partition coefficients have already been discussed in the chapter on solubility.

## Examples of Diffusion and Permeability Coefficients

Diffusivity is a fundamental material property of the system and is dependent on the solute, the temperature, and the medium through which diffusion occurs.[20] Gas molecules diffuse rapidly through air and other gases. Diffusivities in liquids are smaller, and in solids still smaller. Gas molecules pass slowly and with great difficulty through metal sheets and crystalline barriers. Diffusivities are a function of the molecular structure of the diffusant as well as the barrier material. Diffusion coefficients for gases and liquids passing through water, chloroform, and polymeric materials are given in **Table 11–2**. Approximate diffusion coefficients and permeabilities for drugs passing from a solvent in which they are dissolved (water, unless otherwise specified) through natural and synthetic membranes are given in **Table 11–3**. In the chapter on colloids, we will see that the molecular weight and the radius of a spherical protein can be obtained from knowledge of its diffusivity.

## Multilayer Diffusion

There are many examples of multilayer diffusion in the pharmaceutical sciences. Diffusion across biologic barriers may involve a number of layers consisting of separate membranes, cell contents, and fluids of distribution. The passage

---

*Confusion arises when the permeability coefficient is defined by $P = DK$ ($cm^2/\text{sec}$) as used when $D$ and $K$ are not independently known. Equation (11–11), including $h$ in the denominator, is the conventional definition of permeability.

## TABLE 11–2
## DIFFUSION COEFFICIENTS OF COMPOUNDS IN VARIOUS MEDIA*

| Diffusant | Partial Molar Volume (cm³/mole) | $D \times 10^6$ (cm²/sec) | Medium or Barrier (Temperature, °C) |
|---|---|---|---|
| Ethanol | 40.9 | 12.4 | Water (25) |
| n-Pentanol | 89.5 | 8.8 | Water (25) |
| Formamide | 26 | 17.2 | Water (25) |
| Glycine | 42.9 | 10.6 | Water (25) |
| Sodium lauryl sulfate | 235 | 6.2 | Water (25) |
| Glucose | 116 | 6.8 | Water (25) |
| Hexane | 103 | 15.0 | Chloroform (25) |
| Hexadecane | 265 | 7.8 | Chloroform (25) |
| Methanol | 25 | 26.1 | Chloroform (25) |
| Acetic acid dimer | 64 | 14.2 | Chloroform (25) |
| Methane | 22.4 | 1.45 | Natural rubber (40) |
| n-Pentane | — | 6.9 | Silicone rubber (50) |
| Neopentane | — | 0.002 | Ethycellulose (50) |

*From G. L. Flynn, S. H. Yalkowsky, and T. J. Roseman, J. Pharm. Sci. **63**, 507, 1974. With permission.

of gaseous or liquid solutes through the walls of containers and plastic packaging materials is also frequently treated as a case of multilayer diffusion. Finally, membrane permeation studies using Caco-2 or MDCK cell monolayers on permeable supports such as polycarbonate filters are other common examples of multilayer diffusion.

Higuchi[32] considered the passage of a topically applied drug from its vehicle through the lipoidal and lower hydrous layers of the skin. Two barriers in series, the lipoidal and the hydrous skin layers of thickness $h_1$ and $h_2$, respectively, are shown in **Figure 11–6**. The resistance, $R$, to diffusion in each layer is equal to the reciprocal of the permeability coefficient, $P_i$, of that particular layer. Permeability, $P$, was defined earlier [equation (11–15)] as the diffusion coefficient, $D$, multiplied by the partition coefficient, $K$, and divided by the membrane thickness, $h$. For a particular lamina i,

$$P_i = D_i K_i / h_i \qquad (11–18)$$

and

$$R_i = 1 / P_i = h_i / D_i K_i \qquad (11–19)$$

where $R_i$ is the resistance to diffusion. The total resistance, $R$, is the reciprocal of the total permeability, $P$, and is additive for a series of layers. It is written in general as

$$R = R_1 + R_2 + \cdots R_n \qquad (11–20a)$$

$$1/P = 1/P_i + 1/P_2 \cdots + 1/P_n \qquad (11–20b)$$

$$R = 1/P = h_1/D_1 K_1$$
$$+ h_2/D_2 K_2 + \cdots + h_n/D_n K_n \qquad (11–20c)$$

where $K_i$ is the distribution coefficient for layer i relative to the next corresponding layer, i + 1, of the system.[19] The total permeability for the two-ply model of the skin is obtained by taking the reciprocal of equation (11–20c), expressed in terms of two layers, to yield

$$P = \frac{D_1 K_1 D_2 K_2}{h_1 D_2 K_2 + h_2 D_1 K_1} \qquad (11–21)$$

The lag time to steady state for a two-layer system is

$$t_L = \frac{\dfrac{h_1^2}{D_1}\left(\dfrac{h_1}{6D_1 K_1} + \dfrac{h_2}{2D_2 K_2}\right) + \dfrac{h_2^2}{D_2}\left(\dfrac{h_1}{2D_1 K_1} + \dfrac{h_2}{6D_2 K_2}\right)}{(h_1/D_1 K_1 + h_2/D_2 K_2)} \qquad (11–22)$$

When the partition coefficients, $K_i$, of the two layers are essentially the same and one of the $h/D$ terms, say 1, is much larger than the other, however, the time lag equation for the bilayer skin system reduces to the simple time lag expression

$$t_L = h_1^2/6D_1 \qquad (11–23)$$

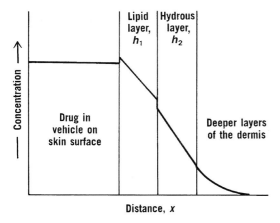

**Fig. 11–6.** Passage of a drug on the skin's surface through a lipid layer, $h_1$, and a hydrous layer, $h_2$, and into the deeper layers of the dermis. The curve of concentration against the distance changes sharply at the two boundaries because the two partition coefficients have values other than unity.

**TABLE 11–3**

**DRUG DIFFUSION AND PERMEABILITY COEFFICIENTS***

| Drug | Membrane Diffusion Coefficient (cm²/sec) | Membrane Permeability Coefficient (cm/sec) | Pathway | References |
|---|---|---|---|---|
| Amiloride | — | $1.63 \times 10^{-4}$ | Absorption from human jejunum | 21 |
| Antipyrine | — | $4.5 \times 10^{-4}$ | Absorption from human jejunum | 22 |
| Atenolol | — | $0.2 \times 10^{-4}$ | Absorption from human jejunum | 22 |
| Benzoic acid | — | $36.6 \times 10^{-4}$ | Absorption from rat jejunum | 23 |
| Carbamazepine | — | $4.3 \times 10^{-4}$ | Absorption from human jejunum | 22 |
| Chloramphenicol | — | $1.87 \times 10^{-6}$ | Through mouse skin | 24 |
| Cyclosporin A | $4.3 \times 10^{-6}$ | — | Diffusion across cellulose membrane | 25 |
| Desipramine · HCl | — | $4.4 \times 10^{-4}$ | Absorption from human jejunum | 22 |
| Enalaprilat | — | $0.2 \times 10^{-4}$ | Absorption from human jejunum | 22 |
| Estrone | — | $20.7 \times 10^{-4}$ | Absorption from rat jejunum | 23 |
| Furosemide | — | $0.05 \times 10^{-4}$ | Absorption from human jejunum | 22 |
| Glucosamine | $9.0 \times 10^{-6}$ | — | Diffusion across cellulose membrane | 25 |
| Glucuronic acid | $9.0 \times 10^{-6}$ | — | Diffusion across cellulose membrane | 25 |
| Hydrochlorothiazide | — | $0.04 \times 10^{-4}$ | Absorption from human jejunum | 22 |
| Hydrocortisone | — | $0.56 \times 10^{-4}$ | Absorption from rat jejunum | 23 |
|  | — | $5.8 \times 10^{-5}$ | Absorption from rabbit vaginal tract | 26 |
| Ketoprofen | $2.1 \times 10^{-3}$ | — | Diffusion across abdominal skin from a hairless male rat | 27 |
| Ketoprofen | — | $8.4 \times 10^{-4}$ | Absorption from human jejunum | 22 |
| Mannitol | $8.8 \times 10^{-6}$ | — | Diffusion across cellulose membrane | 25 |
| Mannitol | — | $0.9 \times 10^{-4}$ | Diffusion across excised bovine nasal mucosa | 28 |
| Metoprolol · $^1/_2$ tartrate | — | $1.3 \times 10^{-4}$ | Absorption from human jejunum | 22 |
| Naproxen | — | $8.3 \times 10^{-4}$ | Absorption from human jejunum | 22 |
| Octanol | — | $12 \times 10^{-4}$ | Absorption from rat jejunum | 23 |
| PEG 400 | — | $0.58 \times 10^{-4}$ | Absorption from human jejunum | 29 |
| Piroxicam | — | $7.8 \times 10^{-4}$ | Absorption from human jejunum | 22 |
| Progesterone | — | $7 \times 10^{-4}$ | Absorption from rat jejunum | 23 |
| Propranolol | — | $3.8 \times 10^{-4}$ | Absorption from human jejunum | 29 |
| Salycylates | $1.69 \times 10^{-6}$ | — | Absorption from rabbit vaginal tract | 30 |
| Salicylic acid | — | $10.4 \times 10^{-4}$ | Absorption from rat jejunum | 23 |
| Terbutaline · $^1/_2$ sulfate | — | $0.3 \times 10^{-4}$ | Absorption from human jejunum | 22 |
| Testosterone | $7.6 \times 10^{-6}$ | — | Diffusion across cellulose membrane | 25 |
| Testosterone | — | $20 \times 10^{-4}$ | Absorption from rat jejunum | 23 |
| Verapamil · HCl | — | $6.7 \times 10^{-4}$ | Absorption from human jejunum | 22 |
| Water | $2.8 \times 10^{-10}$ | $2.78 \times 10^{-7}$ | Diffusion into human skin layers | 31 |

*All at 37°C.

**EXAMPLE 11–2**

**Series Resistances in Cell Culture Studies**

Cell culture models are increasingly used to study drug transport; however, in many instances only the effective permeability, $P_{eff}$, is calculated. For very hydrophobic drugs, interactions with the filter substratum or the aqueous boundary layer (ABL) may provide more resistance to drug transport than the cell monolayer itself. Because the goal of the study is to assess the cell transport properties of drugs, $P_{eff}$ may be inherently biased due to drug interactions with the substratum or ABL. Reporting $P_{eff}$ is of value only if the monolayer is the rate-limiting transport barrier. Therefore, prior to reporting the $P_{eff}$ of a compound, the effect of each of these barriers should be evaluated to ensure that the permeability relates to that across the cell monolayer. In cell culture systems the resistance to drug transport, $P_{eff}$, is composed of a series of resistances including those of the ABL ($R_{aq}$), the cell monolayer ($R_{mono}$), and the filter resistance

($R_f$) (Fig. 11–7). Total resistance is additive for a series of layers:

$$R_{eff} = R_{aq} + R_{mono} + R_f \tag{11–24}$$

This can be written in terms of the reciprocal of the total permeability:

$$\frac{1}{P_{eff}} = \frac{1}{P_{aq}} + \frac{1}{P_{mono}} + \frac{1}{P_f}$$

$$P_{eff} = \frac{1}{\dfrac{1}{P_{aq}} + \dfrac{1}{P_{mono} + \dfrac{1}{P_f}}} \tag{11–25}$$

where $P_{eff}$ is the measured effective permeability, $P_{aq}$ is the total permeability of the ABL (adjacent to both the apical surface of the cell monolayer and the free surface of the filter), and $P_f$ is the permeability of the supporting microporous filter.

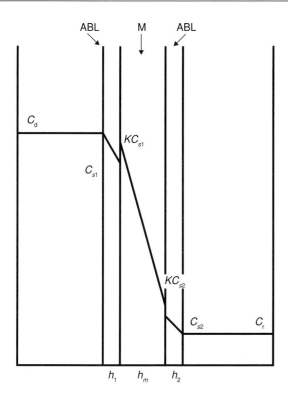

**Fig. 11–7.** Diffusion of drug across the aqueous boundary layer (ABL) and cell monolayer (M) in a cell culture system.

Permeability across the filter, $P_f$, can be obtained experimentally by measuring the $P_{eff}$ across blank filters:

$$\frac{1}{P_{eff}^{blank}} = \frac{1}{P_{aq}} + \frac{1}{P_f} \qquad (11\text{–}26)$$

Because $P_{aq}$ is dependent on the flow rate,

$$P_{aq} = KV^n \qquad (11\text{–}27)$$

(where $k$ is a hybrid constant that takes into account the diffusivity of the compound, kinematic viscosity, and geometric factors of the chamber; $V$ is the stirring rate in mL/min; and $n$ is an exponent that varies between 0 and 1 depending on the hydrodynamic conditions in the diffusion chamber), $P_f$ can be calculated using nonlinear regression by obtaining $P_{eff}$ across blank filters at various flow rates.

Similarly, $1/P_f + 1/P_{mono}$ can be determined by measuring the $P_{eff}$ through the cell monolayer at various flow rates and by using nonlinear regression and the equation.

$$P_{eff} = \frac{1}{\dfrac{1}{P_f} + \dfrac{1}{P_{mono}} + \dfrac{1}{KV^n}} \qquad (11\text{–}28)$$

The implicit assumption of this method is that each resistance in series is independent of the other barriers. Therefore, $P_{mono}$ is calculated by difference, using the independently determined $P_f$. Because $P_{aq}$ is independent of the presence of the monolayer, $P_{mono}$ can be calculated as follows:

$$\frac{1}{P_{mono}} = \frac{1}{P_{eff}} - \frac{1}{P_{eff}^{blank}} \qquad (11\text{–}29)$$

Because the contributions of $R_f$ and $R_{aq}$ vary depending on the nature of the drug, it is important to correct for these biases by reporting $P_{mono}$. The deviation between $P_{mono}$ and $P_{eff}$ becomes more significant if the flow rate is low (i.e., $R_{aq}$ is high) or if the filter has low effective porosity (i.e., $R_f$ is high). In addition, the permeability of the drug also plays a major role such that the deviation

between $P_{mono}$ and $P_{eff}$ becomes more significant for highly permeable compounds.[33]

## Membrane Control and Diffusion Layer Control

A multilayer case of special importance is that of a membrane between two aqueous phases with stationary or stagnant solvent layers in contact with the donor and receptor sides of the membrane (**Fig. 11–8**).

The permeability of the total barrier, consisting of the membrane and two static aqueous diffusion layers, is

$$P = \frac{1}{R} = \frac{D_m K D_a}{h_m D_a + 2h_a D_m K} = \frac{1}{h_m/D_m K + 2h_a/D_a} \qquad (11\text{–}30)$$

This expression is analogous to equation (**11–21**). In equation (**11–30**), however, only one partition coefficient, $K$, appears that giving the ratio of concentrations of the drug in the membrane and in the aqueous solvent, $K = C_3/C_4 = C_3/C_2$. The flux $J$ through this three-ply barrier is simply equal to the permeability, $P$, multiplied by the concentration gradient, $(C_1 - C_5)$, that is, $J = P(C_1 - C_5)$. The receptor serves as a sink (i.e., $C_5 = 0$), and the donor concentration $C_1$ is assumed to be constant, providing a steady-state flux.[34] We thus have

$$J = \frac{1}{S}\frac{dM}{dt} = \frac{D_m K D_a C_1}{h_m D_a + 2h_a D_m K} \qquad (11\text{–}31)$$

In equations (**11–30**) and (**11–31**), $D_m$ and $D_a$ are membrane and aqueous solvent diffusivities, $h_m$ is the membrane thickness, and $h_a$ is the thickness of the aqueous diffusion layer, as shown in **Figure 11–8**. $M$ is the amount of permeant reaching the receptor, and $S$ is the cross-sectional area of the barrier. It is important to realize that $h_a$ is physically influenced by the hydrodynamics in the bulk aqueous phases. The higher the degree of stirring, the thinner is the stagnant aqueous

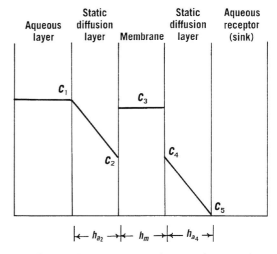

**Fig. 11–8.** Schematic of a multilayer (three-ply) barrier. The membrane is found between two static aqueous diffusion layers. (From G. L. Flynn, O. S. Carpenter, and S. H. Yalkowsky, J. Pharm. Sci. **61**, 313, 1972. With permission.)

diffusion layer; the slower the stirring, the thicker is this aqueous layer.

Equation (**11–31**) is the starting point for considering two important cases of multilayer diffusion, namely, diffusion under *membrane control* and diffusion under *aqueous diffusion layer control*.

### Membrane Control

When the membrane resistance to diffusion is much greater than the resistances of the aqueous diffusion layers, that is, $R_m$ is greater than $R_a$ by a factor of at least 10, or correspondingly, $P_m$ is much less than $P_a$, the rate-determining step (slowest step) is diffusion across the membrane. This is reflected in equation (**11–31**) when $h_m D_a$ is much greater than $2h_a D_m$. Thus, equation (**11–31**) reduces to

$$J = \left( \frac{K D_m}{h_m} \right) C_1 \qquad (11\text{–}32)$$

Equation (**11–32**) represents the simplest case of membrane control of flux.

### Aqueous Diffusion Layer Control

When $2h_a K D_m$ is much greater than $h_m D_a$, equation (**11–31**) becomes

$$J = \left( \frac{D_a}{2h_a} \right) C_1 \qquad (11\text{–}33)$$

and it is now said that the rate-determining barriers to diffusional transport are the stagnant aqueous diffusion layers. This statement means that the concentration gradient that controls the flux now resides in the aqueous diffusion layers rather than in the membrane. From the relationship $2h_a K D_m \gg h_m D_a$, it is observed that membrane control shifts to diffusion layer control when the partition coefficient $K$ becomes sufficiently large.

### EXAMPLE 11–3

**Transfer from Membrane to Diffusion-Layer Control**

Flynn and Yalkowsky[34] demonstrated a transfer from membrane to diffusion-layer control in a homologous series of *n*-alkyl *p*-aminobenzoates (PABA esters). The concentration gradient is almost entirely within the silicone rubber membrane for the short-chain PABA esters. As the alkyl chain of the ester is lengthened proceeding from butyl to pentyl to hexyl, the concentration no longer drops across the membrane. Instead, the gradient is now found in the aqueous diffusion layers, and diffusion-layer control takes over as the dominant factor in the permeation process. The steady-state flux, $J$, for hexyl *p*-aminobenzoate was found to be $1.60 \times 10^{-7}$ mmole/cm$^2$ sec. $D_a$ is $6.0 \times 10^{-6}$ cm$^2$/sec and the concentration of the PABA ester, $C$, is 1.0 mmole/liter. The system is in diffusion-layer control, so equation (**11–33**) applies. Calculate the thickness of the static diffusion layer, $h_a$. We have

$$J = \left( \frac{D_a}{2h_a} \right) C \quad \text{or} \quad h_a = \left( \frac{D_a}{2J} \right) C$$

$$h_a = \frac{6.0 \times 10^{-6} \text{ cm}^2/\text{sec}}{2(1.60 \times 10^{-7} \text{ mmole/cm}^2 \text{ sec})}$$

$$\times (1.0 \times 10^{-3} \text{ mmole/cm}^3) = 0.019 \text{ cm}$$

One observes from equations (**11–32**) and (**11–33**) that, under sink conditions, steady-state flux is proportional to concentration, $C$, in the donor phase whether the flux-determining mechanism is under membrane or diffusion-layer control. Equation (**11–33**)

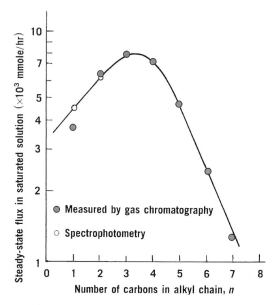

**Fig. 11–9.** Steady-state flux of a series of *p*-aminobenzoic acid esters. Maximum flux occurs between the esters having three and four carbons and is due to a change from membrane to diffusion-layer control, as explained in the text. (From G. L. Flynn and S. H. Yalkowsky, J. Pharm. Sci. **61**, 838, 1972. With permission.)

shows that the flux is independent of membrane thickness, $h_m$, and other properties of the membrane when under static diffusion layer control.

The maximum flux obtained in a membrane preparation depends on the solubility, or limiting concentration, of the PABA homologue. The maximum flux can therefore be obtained using equation (**11–31**) in which $C$ is replaced by $C_s$, the solubility of the permeating compound:

$$J_{max} = \frac{D_m K D_a}{h_m D_a + 2h_a K D_m} C_s \qquad (11\text{–}34)$$

The maximum steady-state flux, $J_{max}$, for saturated solutions of the PABA esters is plotted against the ester chain length in Figure 11–9.[34] The plot exhibits peak flux between $n = 3$ and $n = 4$ carbons, that is, between propyl and butyl *p*-aminobenzoates. The peak in Figure 11–9 suggests in part the solubility characteristics of the PABA esters but primarily reflects the change from membrane to static diffusion-layer control of flux. For the methyl, ethyl, and propyl esters, the concentration gradient in the membrane gradually decreases and shifts, in the case of the longer-chain esters, to a concentration gradient in the diffusion layers.

By using a well-characterized membrane such as siloxane of known thickness and a homologous series of PABA esters, Flynn and Yalkowsky[34] were able to study the various factors: solubility, partition coefficient, diffusivity, diffusion lag time, and the effects of membrane and diffusion-layer control. From such carefully designed and conducted studies, it is possible to predict the roles played by various physicochemical factors as they relate to diffusion of drugs through plastic containers, influence release rates from sustained-delivery forms, and influence absorption and excretion processes for drugs distributed in the body.

## Lag Time Under Diffusion-Layer Control

Flynn et al.[19] showed that the lag time for ultrathin membranes under conditions of diffusion-layer control can be represented as

$$t_L = \frac{\left( \sum h_a \right)^2}{6 D_a} \qquad (11\text{–}35)$$

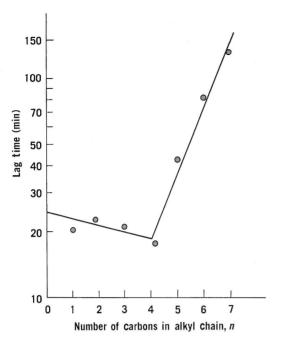

Fig. 11–10. Change in lag time of *p*-aminobenzoic acid esters with alkyl chain length. (From G. L. Flynn and S. H. Yalkowsky, J. Pharm. Sci. **61**, 838, 1972. With permission.)

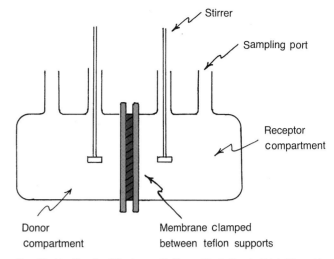

Fig. 11–11. Simple diffusion cell. (From M. G. Karth, W. I. Higuchi, and J. L. Fox, J. Pharm. Sci. **74**, 612, 1985. With permission.)

where $\sum h_a$ is the sum of the thicknesses of the aqueous diffusion layers on the donor and receptor sides of the membrane. The correspondence between $t_L$ in equation (11–35) with that for systems under membrane control, equation (11–32), is evident. The lag time for *thick* membranes operating under diffusion layer control is

$$t_L = \frac{h_m h_{a1} h_{a2} K}{(h_{a1} + h_{a2}) D_a} \tag{11–36}$$

When the diffusion layers, $h_{a1}$ and $h_{a2}$, are of the same thickness, the lag time reduces to

$$t_L = \frac{h_m h_a K}{2 D_a} \tag{11–37}$$

The partition coefficient, which was shown earlier to be instrumental in converting the flux from membrane to diffusion-layer control, now appears in the numerator of the lag-time equation. A large $K$ signifies lipophilicity of the penetrating drug species. As one ascends a homologous series of PABA esters, for example, the larger lipophilicity increases the onset time for steady-state behavior; in other words, lengthening of the ester molecule increases the lag time once the system is in diffusion-layer control. The sharp increase in lag time for PABA esters with alkyl chain length beyond $C_4$ is shown in Figure 11–10.[34]

## PROCEDURES AND APPARATUS FOR ASSESSING DRUG DIFFUSION

A number of experimental methods and diffusion chambers have been reported in the literature. Examples of those used mainly in pharmaceutical and biologic transport studies are introduced here.

Diffusion chambers of simple construction, such as the one reported by Karth et al.[35] (Fig. 11–11), are probably best for diffusion work. They are made of glass, clear plastic, or polymeric materials, are easy to assemble and clean, and allow visibility of the liquids and, if included, a rotating stirrer. They may be thermostated and lend themselves to automatic sample collection and assay. Typically, the donor chamber is filled with drug solution. Samples are collected from the receiver compartment and subsequently assayed using a variety of analytical methods such as liquid scintillation counting or high-performance liquid chromatography with a variety of detectors (e.g., ultraviolet, fluorescence, or mass spectrometry). Experiments may be run for hours under these controlled conditions.

Biber and Rhodes[36] constructed a Plexiglas three-compartment diffusion cell for use with either synthetic or isolated biologic membranes. The drug was allowed to diffuse from the two outer donor compartments in a central receptor chamber. Results were reproducible and compared favorably with those from other workers. The three-compartment design created greater membrane surface exposure and improved analytic sensitivity.

The permeation through plastic film of water vapor and of aromatic organic compounds from aqueous solution can be investigated in two-chamber glass cells similar in design to those used for studying drug solutions in general. Nasim et al.[37] reported on the permeation of 19 aromatic compounds from aqueous solution through polyethylene films. Higuchi and Aguiar[38] studied the permeability of water vapor through enteric coating materials, using a glass diffusion cell and a McLeod gauge to measure changes in pressure across the film.

The sorption of gases and vapors can be determined by use of a microbalance enclosed in a temperature-controlled and evacuated vessel that is capable of weighing within a sensitivity of $\pm 2 \times 10^{-6}$ g. The gas or vapor is introduced at controlled pressures into the glass chamber containing the

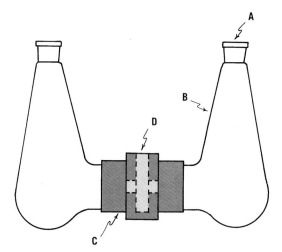

**Fig. 11–12.** Diffusion cell for permeation through stripped skin layers. The permeant may be in the form of a gas, liquid, or gel. Key: *A*, glass stopper; *B*, glass chamber; *C*, aluminum collar; *D*, membrane and sample holder. (From D. E. Wurster, J. A. Ostrenga, and L. E. Matheson, Jr., J. Pharm. Sci. **68**, 1406, 1410, 1979. With permission.)

polymer or biologic film of known dimensions suspended on one arm of the balance. The mass of diffusant sorbed at various pressures by the film is recorded directly.[39] The rate of approach to equilibrium sorption permits easy calculation of the diffusion coefficients for gases and vapors.

In studying percutaneous absorption, animal or human skin, ordinarily obtained by autopsy, is employed. Scheuplein[31] described a cell for skin penetration experiments made of Pyrex and consisting of two halves, a donor and a receptor chamber, separated by a sample of skin supported on a perforated plate and securely clamped in place. The liquid in the receptor was stirred by a Teflon-coated bar magnet. The apparatus was submerged in a constant-temperature bath, and samples were removed periodically and assayed by appropriate means. For compounds such as steroids, penetration was slow, and radioactive methods were found to be necessary to determine the low concentrations.

Wurster et al.[40] developed a permeability cell to study the diffusion through stratum corneum (stripped from the human forearm) of various permeants, including gases, liquids, and gels. The permeability cell is shown in **Figure 11–12**. During diffusion experiments it was kept at constant temperature and gently shaken in the plane of the membrane. Samples were withdrawn from the receptor chamber at definite times and analyzed for the permeant.

The kinetics and equilibria of liquid and solute absorption into plastics, skin, and chemical and other biologic materials can be determined simply by placing sections of the film in a constant-temperature bath of the pure liquid or solution. The sections are retrieved at various times, excess liquid is removed with absorbant tissue, and the film samples are accurately weighed in tared weighing bottles. A radioactive-counting technique can also be used with this method to

analyze for drug remaining in solution and, by difference, the amount sorbed into the film.

Partition coefficients are determined simply by equilibrating the drug between two immiscible solvents in a suitable vessel at a constant temperature and removing samples from both phases, if possible, for analysis.[41] Equilibrium solubilities of drug solutes are also required in diffusion studies, and these are obtained as described earlier (Chapter 8).

Addicks et al.[42] described a flowthrough cell and Addicks et al.[43] designed a cell that yields results more comparable to the diffusion of drugs under clinical conditions. Grass and Sweetana[44] proposed a side-by-side acrylic diffusion cell for studying tissue permeation. In a later paper, Hidalgo et al.[45] developed and validated a similar diffusion chamber for studying permeation through cultured cell monolayers. These chambers (**Fig. 11–13** *a* and *b*), derived from the Ussing chamber, have the advantage of employing laminar flow conditions across the tissue or cell surface allowing for an assessment of the aqueous boundary layer and calculation of intrinsic membrane drug permeability.

## BIOLOGIC DIFFUSION

**EXAMPLE 11–4**

**Intestinal Drug Absorption and Secretion**

The apparent permeability, $P_{app}$, of Taxol across a monolayer of Caco-2 cells is $4.4 \times 10^{-6}$ cm/sec in the apical to basolateral direction (i.e., absorptive direction) and is $31.8 \times 10^{-6}$ cm/sec from basolateral to apical direction (i.e., secretory direction). Assuming that both absorptive and secretory drug transport occurs under sink conditions ($C_r \ll C_d$), what is the amount of Taxol absorbed through the intestinal wall by 2 hr after administering an oral dose? Assume that the Taxol concentration in the intestinal fluid is 0.1 mg/mL, and following intravenous administration, the initial Taxol concentration in the plasma is 10 $\mu$g/mL. How much Taxol will be secreted into the feces 2 hr after dosing? Assume that the effective area for intestinal absorption and secretion is 1 m.[2,46,47] We have

$$\frac{dJ}{dt} = \frac{dM}{dt\,A} = P_{app}C_d$$

For intestinal absorption,

$$M_a = P_{app}C_d At = (4.4 \times 10^{-6}\ \text{cm/sec})(0.1\ \text{mg/mL})(1\ \text{m}^2)(2\ \text{hr})$$
$$= 3.17\ \mu\text{g}$$

For intestinal secretion,

$$M_e = P_{app}C_d At = (31.8 \times 10^{-6}\ \text{cm/sec})(10\ \mu\text{g/mL})(1\ \text{m}^2)(2\ \text{hr})$$
$$= 2.29\ \mu\text{g}$$

## Gastrointestinal Absorption of Drugs

Drugs pass through living membranes according to two main classes of transport, passive and carrier mediated. Passive transfer involves a simple diffusion driven by differences in drug concentration on the two sides of the membrane. In intestinal absorption, for example, the drug travels in most cases by passive transport from a region of high concentration

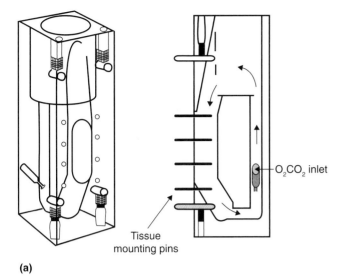

(a)

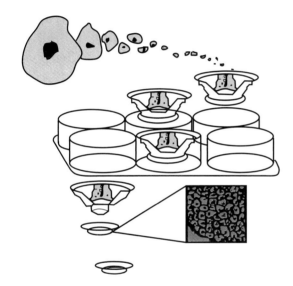

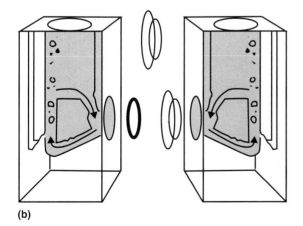

(b)

**Fig. 11–13.** (*a*) Sweetana/Grass diffusion cell. Tissue is mounted between acrylic half-cells. Buffer is circulated by gas lift ($O_2/CO_2$) at the inlet and flows in the direction of arrows, parallel to the tissue surface. Temperature control is maintained by a heating block.

in the gastrointestinal tract to a region of low concentration in the systemic circulation. Given the instantaneous dilution of absorbed drug once it reaches the bloodstream, sink conditions are essentially maintained at all times.

Carrier-mediated transport can be classified as active transport (i.e., requires an energy source) or as facilitated diffusion (i.e., does not depend on an energy source such as adenosine triphosphate). In active transport the drug can proceed from regions of *low* concentration to regions of *high* concentration through the "pumping action" of these biologic transport systems. Facilitative-diffusive carrier proteins cannot transport drugs or nutrients "uphill" or against a concentration gradient. We will make limited use of specialized carrier systems in this chapter and will concentrate attention mainly on passive diffusion.

Many drugs are weakly acidic or basic, and the ionic character of the drug and the biologic compartments and membranes have an important influence on the transfer process. From the Henderson–Hasselbalch relationship for a weak acid,

$$pH = pK_a + \log \frac{[A^-]}{[HA]}$$

where [HA] is the concentration of the nonionized weak acid and $[A^-]$ is the concentration of its conjugate base. For a weak base, the equation is

$$pH = pK_a + \log \frac{[B]}{[BH^+]}$$

where [B] is the concentration of the base and $[BH^+]$ that of its conjugate acid. $pK_a$ is the dissociation exponent for the weak acid in each case. For the weak base, $pK_a = pK_w - pK_b$.

The percentage ionization of a weak acid is the ratio of concentration of drug in the ionic form, $I$, to total concentration of drug in ionic, $I$, and undissociated, $U$, form, multiplied by 100:

$$\% \text{ Ionized} = \frac{I}{I + U} \times 100 \qquad (11\text{--}38)$$

Therefore, the Henderson–Hasselbalch equation for weak acids can be written as

$$\frac{U}{I} = 10^{(pK_a - pH)} = \text{ antilog}(pK_a - pH) \quad (11\text{--}39)$$

or

$$U = I \text{ antilog}(pK_a - pH) \qquad (11\text{--}40)$$

Substituting $U$ into the equation for percentage ionization yields

$$\% \text{ Ionized } = \frac{100}{1 + \text{antilog}(pK_a - pH)} \qquad (11\text{--}41)$$

Similarly, for a weak molecular base,

$$\% \text{ Ionized } = \frac{100}{1 + \text{antilog}(pH - pK_a)} \qquad (11\text{--}42)$$

### PERCENTAGE SULFISOXAZOLE, $pK_a \cong 5.0$, DISSOCIATED AND UNDISSOCIATED AT pH VALUES

| pH | Percentage Dissociated | Percentage Undissociated |
|-----|-----|-----|
| 2.0 | 0.100 | 99.900 |
| 4.0 | 9.091 | 90.909 |
| 5.0 | 50.000 | 50.000 |
| 6.0 | 90.909 | 9.091 |
| 8.0 | 99.900 | 0.100 |
| 10.0 | 99.999 | 0.001 |

In equation (11–41), $pK_a$ refers to the weak acid, whereas in (11–42), $pK_a$ signifies the acid that is conjugate to the weak base.

The percentage ionization at various pH values of the weak acid sulfisoxazole, $pK_a \cong 5.0$, is given in **Table 11–4**. At a point at which the pH is equal to the drug's $pK_a$, equal amounts are present in the ionic and molecular forms.

The molecular diffusion of drugs across the intestinal mucosa was long thought to be the major pathway for drug absorption into the body. Drug absorption by means of diffusion through intestinal cells (i.e., enterocytes) or in between those cells (i.e., paracellular diffusion) is governed by the state of ionization of the drug, its solubility and concentration in the intestine, and its membrane permeability.

## pH-Partition Hypothesis

Biologic membranes are predominantly lipophilic, and drugs penetrate these barriers mainly in their molecular, undissociated form. Brodie and his associates[48] were the first workers to apply the principle, known as the *pH-partition hypothesis*, that drugs are absorbed from the gastrointestinal tract by passive diffusion depending on the fraction of undissociated drug at the pH of the intestines. It is reasoned that the partition coefficient between membranes and gastrointestinal fluids is large for the undissociated drug species and favors transport of the molecular form from the intestine through the mucosal wall and into the systemic circulation.

The pH-partition principle has been tested in a large number of in vitro and in vivo studies, and it has been found to be only partly applicable in real biologic systems.[48,49] In many cases, the ionized as well as the un-ionized form partitions into, and is appreciably transported across, lipophilic membranes. It is found for some drugs, such as sulfathiazole, that the in vitro permeability coefficient for the ionized form may actually exceed that for the molecular form of the drug.

Transport of a drug by diffusion across a membrane such as the gastrointestinal mucosa is governed by Fick's law:

$$-\frac{dM}{dt} = \frac{D_{\mathrm{m}}SK}{h}(C_{\mathrm{g}} - C_{\mathrm{p}}) \qquad (11\text{–}43)$$

where $M$ is the amount of drug in the gut compartment at time $t$, $D_{\mathrm{m}}$ is diffusivity in the intestinal membrane, $S$ is the area of the membrane, $K$ is the partition coefficient between membrane and aqueous medium in the intestine, $h$ is the membrane thickness, $C_{\mathrm{g}}$ is the concentration of drug in the intestinal compartment, and $C_{\mathrm{p}}$ is the drug concentration in the plasma compartment at time $t$. The gut compartment is kept at a high concentration and has a large enough volume relative to the plasma compartment so as to make $C_{\mathrm{g}}$ a constant. Because $C_{\mathrm{p}}$ is relatively small, it can be omitted. Equation (11–43) then becomes

$$-\frac{dM}{dt} = \frac{D_{\mathrm{m}}SKC_{\mathrm{g}}}{h} \qquad (11\text{–}44)$$

The left-hand side of (11–44) is converted into concentration units, $C$ (mass/unit volume) $\times$ $V$ (volume). On the right-hand side of (11–44), the diffusion constant, membrane area, partition coefficient, and membrane thickness are combined to yield a *permeability coefficient*. These changes lead to the pair of equations

$$-V\frac{dC_{\mathrm{g}}}{dt} = P_{\mathrm{g}}C_{\mathrm{g}} \qquad (11\text{–}45)$$

$$-V\frac{dC_{\mathrm{p}}}{dt} = P_{\mathrm{p}}C_{\mathrm{g}} \qquad (11\text{–}46)$$

where $C_{\mathrm{g}}$ and $P_{\mathrm{g}}$ of equation (11–45) are the concentration and permeability coefficient, respectively, for drug passage from intestine to plasma. In equation (11–46), $C_{\mathrm{p}}$ and $P_{\mathrm{p}}$ are corresponding terms for the reverse passage of drug from plasma to intestine. Because the gut volume, $V$, and the gut concentration, $C_{\mathrm{g}}$, are constant, dividing (11–45) by (11–46) yields

$$\frac{dC_{\mathrm{g}}/dt}{dC_{\mathrm{p}}/dt} = \frac{P_{\mathrm{g}}}{P_{\mathrm{p}}} \qquad (11\text{–}47)$$

Equation (11–47) demonstrates that the ratio of absorption rates in the intestine-to-plasma and the plasma-to-intestine directions equals the ratio of permeability coefficients.

The study by Turner et al.[49] showed that undissociated drugs pass freely through the intestinal membrane in either direction by simple diffusion, in agreement with the pH-partition principle. Drugs that are partly ionized show an increased permeability ratio, indicating favored penetration from intestine to plasma. Completely ionized drugs, either negatively or positively charged, show permeability ratios $P_{\mathrm{g}}/P_{\mathrm{p}}$ of about 1.3, that is, a greater passage from gut to plasma than from plasma to gut. This suggests that penetration of ions is associated with sodium ion flux. Their forward passage, $P_{\mathrm{g}}$, is apparently due to a coupling of the ions with sodium transport, which mechanism then ferries the drug ions across the membrane, in conflict with the simple pH-partition hypothesis.

Colaizzi and Klink[50] investigated the pH-partition behavior of the tetracyclines, a class of drugs having three separate $pK_a$ values, which complicates the principles of pH partition. The lipid solubility and relative amounts of the ionic forms of a tetracycline at physiologic pH may have a bearing on the

biologic activity of the various tetracycline analogues used in clinical practice.

## Modification of the pH-Partition Principle

Ho and coworkers[51] also showed that the pH-partition principle is only approximate, assuming as it does that drugs are absorbed through the intestinal mucosa in the nondissociated form alone. Absorption of relatively small ionic and nonionic species through the aqueous pores and the aqueous diffusion layer in front of the membrane must be considered.[23] Other complicating factors, such as metabolism of the drug in the gastrointestinal membrane, absorption and secretion by carrier-mediated processes, absorption in micellar form, and enterohepatic circulatory effects, must also be accounted for in any model that is proposed to reflect in vivo processes.

Ho, Higuchi, and their associates[23] investigated the gastrointestinal absorption of drugs using diffusional principles and a knowledge of the physiologic factors involved. They employed an in situ preparation, as shown in **Figure 11–14**, known as the modified Doluisio method for in situ rat intestinal absorption. (The original rat intestinal preparation[52] employed two syringes without the mechanical pumping modification.)

The model used for the absorption of a drug through the mucosal membrane of the small intestine is shown in **Figure 11–15**. The aqueous boundary layer is in series with the biomembrane, which is composed of lipid regions and aqueous pores in parallel. The final reservoir is a sink consisting of the blood. The flux of a drug permeating the mucosal membrane is

$$J = P_{app}(C_b - C_{blood}) \qquad (11\text{–}48)$$

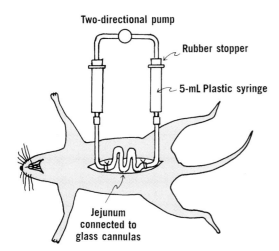

Two-directional pump

Rubber stopper

5-mL Plastic syringe

Jejunum connected to glass cannulas

**Fig. 11–14.** Modified Doluisio technique for in situ rat intestinal absorption. (From N. F. H. Ho, J. Y. Park, G. E. Amidon, et al., in A. J. Aguiar (Ed.), *Gastrointestinal Absorption of Drugs*, American Pharmaceutical Association, Academy of Pharmaceutical Sciences, Washington, D. C., 1981. With permission.)

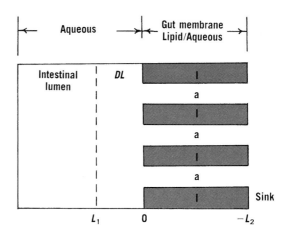

**Fig. 11–15.** Model for the absorption of a drug through the mucosa of the small intestine. The intestinal lumen is on the left, followed by a static aqueous diffusion layer (DL). The gut membrane consists of aqueous pores (a) and lipoidal regions (l). The distance from the membrane wall to the systemic circulation (sink) is marked off from 0 to $-L_2$; the distance through the diffusion layer is 0 to $L_1$. (From N. F. Ho, W. I. Higuchi, and J. Turi, J. Pharm. Sci. **61**, 192, 1972. With permission.)

or, because the blood reservoir is a sink, $C_{blood} \cong 0$, and

$$J = P_{app}C_b \qquad (11\text{–}49)$$

where $P_{app}$ is the apparent permeability coefficient (cm/sec) and $C_b$ is the total drug concentration in bulk solution in the lumen of the intestine. The apparent permeability coefficient is given by

$$P_{app} = \frac{1}{\dfrac{1}{P_{aq}} + \dfrac{1}{P_m}} \qquad (11\text{–}50)$$

where $P_{aq}$ is the permeability coefficient of the drug in the aqueous boundary layer (cm/sec) and $P_m$ is the effective permeability coefficient for the drug in the lipoidal and polar aqueous regions of the membrane (cm/sec).

The flux can be written in terms of drug concentration, $C_b$, in the intestinal lumen by combining with it a term for the volume, or

$$J = -\frac{V}{S} \cdot \frac{dC_b}{dt} \qquad (11\text{–}51)$$

where $S$ is the surface area and $V$ is the volume of the intestinal segment. The first-order disappearance rate, $K_u$ (sec$^{-1}$), of the drug in the intestine appears in the expression

$$\frac{dC_b}{dt} = -K_u C_b \qquad (11\text{–}52)$$

Substituting equation (11–52) into (11–51) gives

$$J = \frac{V}{S} \cdot K_u C_b \qquad (11\text{–}53)$$

## ⬛▶KEY CONCEPT◀ TRANSPORT PATHWAYS

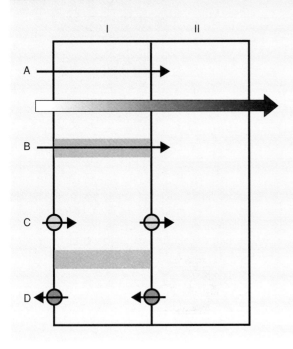

*Parallel* transport pathways are all potential pathways encountered during a particular absorption step. Although many pathways are potentially available for drug transport across biologic membranes, drugs will traverse the particular absorption step by the path of least resistance.

For transport steps in *series* (i.e., one absorption step must be traversed before the next one), the slower absorption step is always the rate-determining process.

and from equations **(11–49)** and **(11–50)**, together with **(11–53)**, we find

$$P_{app} = \frac{1}{\dfrac{1}{P_{aq}} + \dfrac{1}{P_m}} = \frac{V}{S}K_u \qquad \textbf{(11–54)}$$

or

$$K_u = \frac{S}{V} \cdot \frac{P_{aq}}{1 + \dfrac{P_{aq}}{P_m}} \qquad \textbf{(11–55)}$$

Consideration of two cases, (*a*) aqueous boundary layer control and (*b*) membrane control, results in simplification of equation **(11–55)**.

(*a*)  When the permeability coefficient of the intestinal membrane (i.e., the velocity of drug passage through the membrane in centimeter per second) is much greater than that of the aqueous layer, the aqueous layer will cause a slower passage of the drug and become a rate-limiting barrier. Therefore, $P_{aq}/P_m$ will be much less than unity, and equation **(11–55)** reduces to

$$K_{u,max} = (S/V)P_{aq} \qquad \textbf{(11–56\textit{a})}$$

$K_u$ is now written as $K_{u,max}$ because the maximum possible diffusional rate constant is determined by passage across the aqueous boundary layer.

(*b*)  If, on the other hand, the permeability of the aqueous boundary layer is much greater than that of the membrane, $P_{aq}/P_m$ will become much larger than unity, and equation **(11–55)** reduces to

$$K_u = (S/V)P_m \qquad \textbf{(11–56\textit{b})}$$

The rate-determining step for transport of drug across the membrane is now under membrane control. When neither $P_{aq}$ nor $P_m$ is much larger than the other, the process is controlled by the rate of drug passage through both the stationary aqueous layer and the membrane. **Figures 11–16** and **11–17** show the absorption studies of *n*-alkanol and *n*-alkanoic acid homologues, respectively, that concisely illustrate the biophysical interplay of pH, p$K_a$, solute lipophilicity via carbon chain length, membrane permeability of the lipid and aqueous pore

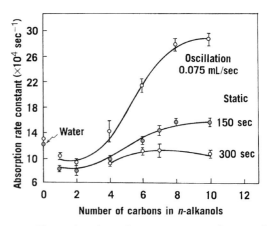

**Fig. 11–16.** First-order absorption rate constant for a series of *n*-alkanols under various hydrodynamic conditions (static or low stirring rates and oscillation or high stirring fluid at 0.075 mL/sec) in the jejunum, using the modified Doluisio technique. (From N. F. H. Ho, J. Y. Park, W. Morozowich, and W. I. Higuchi, in E. B. Roche (Ed.), *Design of Biopharmaceutical Properties Through Prodrugs and Analogs*, American Pharmaceutical Association, Academy of Pharmaceutical Sciences, Washington, D. C., 1977, p. 148. With permission.)

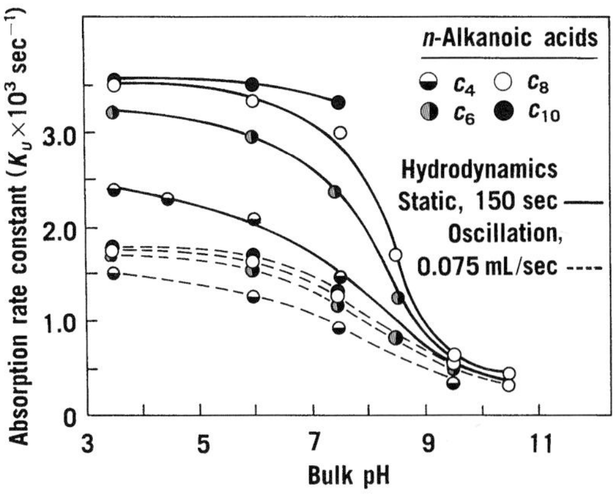

**Fig. 11–17.** First-order absorption rate constants of alkanoic acids versus buffered pH of the bulk solution of the rat gut lumen, using the modified Doluisio technique. Hydrodynamic conditions are shown in the figure. (From N. F. H. Ho, J. Y. Park, W. Morozowich, and W. I. Higuchi, in E. B. Roche, (Ed.), *Design of Biopharmaceutical Properties Through Prodrugs and Analogs*, American Pharmaceutical Association, Academy of Pharmaceutical Sciences, Washington, D. C., 1977, p. 150. With permission.)

pathways, and permeability of the aqueous diffusion layer as influenced by the hydrodynamics of the stirred solution.

**Small Intestinal Transport of a Small Molecule**

Calculate the first-order rate constant, $K_u$, for transport of an aliphatic alcohol across the mucosal membrane of the rat small intestine if $S/V = 11.2$ cm$^{-1}$, $P_{aq} = 1.5 \times 10^{-4}$ cm/sec, and $P_m = 1.1 \times 10^{-4}$ cm/sec. We have

$$K_u = (11.2)\frac{1.5 \times 10^{-4} \text{ cm/sec}}{1 + \dfrac{1.5 \times 10^{-4} \text{ cm/sec}}{1.1 \times 10^{-4} \text{ cm/sec}}} = 11.2 \left(\frac{1.5 \times 10^{-4}}{2.3636}\right)$$

$$K_u = 7.1 \times 10^{-4} \text{ sec}^{-1}$$

For a weak electrolytic drug, the absorption rate constant, $Ku$, is[23]

$$K_u = \frac{S}{V} \cdot \frac{P_{aq}}{1 + \dfrac{P_{aq}}{P_0 X_s + P_p}} \quad (11\text{–}57)$$

where $P_m$ of the membrane is now separated into a term $P_0$, the permeability coefficient of the lipoidal pathway for nondissociated drug, and a term $P_p$, the permeability coefficient of the polar or aqueous pathway for both ionic and nonionic species:

$$P_m = P_0 X_s + P_p \quad (11\text{–}58)$$

The fraction of nondissociated drug species, $X_s$, at the pH of the membrane surface in the aqueous boundary is

$$X_s = \frac{[H^+]_s}{[H^+]_s + K_a} = \frac{1}{1 + \text{antilog}(pH_s - pK_a)} \quad (11\text{–}59)$$

for weak acids, and

$$X_s = \frac{K_a}{[H^+]_s + K_a} = \frac{1}{1 + \text{antilog}(pK_a - pH_s)} \quad (11\text{–}60)$$

for weak bases. Note the relationship between equations (11–59) and (11–41) and between (11–60) and (11–42). $K_a$ is the dissociation constant of a weak acid or of the acid conjugate to a weak base, and $[H^+]_s$ is the hydrogen ion concentration at the membrane surface, where $s$ stands for surface. The surface $pH_s$ is not necessarily equal to the pH of the buffered drug solution[23] because the membrane of the small intestine actively secretes buffer species (principally $CO_2{}^{2-}$ and $HC_3{}^-$). It is only at a pH of about 6.5 to 7.0 that the surface pH is equal to the buffered solution pH. One readily recognizes that for nonelectrolytes, $X_s$ becomes unity, and also that for large molecules such as steroids, $P_p$ is insignificant.

**Duodenal Absorption Rate Constant**

A weakly acidic drug having a $K_a$ value of $1.48 \times 10^{-5}$ is placed in the duodenum in a buffered solution of pH 5.0. Assume $[H^+]_s = 1 \times 10^{-5}$ in the duodenum, $P_{aq} = 5.0 \times 10^{-4}$ cm/sec, $P_0 = 1.14 \times 10^{-3}$ cm/sec, $P_p = 2.4 \times 10^{-5}$ cm/sec, and $S/V = 11.20$ cm$^{-1}$. Calculate the absorption rate constant, $K_u$, using equation (11–57).

First, from equation (11–58), we have

$$X_s = \frac{1 \times 10^{-5}}{1 \times 10^{-5} + 1.48 \times 10^{-5}} = 0.403$$

Then,

$$K_u = (11.2)\frac{5.0 \times 10^{-4}}{1 + \dfrac{5.0 \times 10^{-4}}{(1.14 \times 10^{-3})\,0.403 + 2.4 \times 10^{-5}}}$$

$$K_u = 2.75 \times 10^{-3} \text{ sec}^{-1}$$

**Transcorneal Permeation of Pilocarpine**

In gastrointestinal absorption (*Example 11–5*) the permeability coefficient is divided into $P_0$ for the lipoidal pathway for undissociated drug and $P_p$ for the polar pathway for both ionic and nonionic species. In an analogous way, $P$ can be divided for corneal penetration of a weak base into two permeation coefficients: $P_B$ for the un-ionized species and $P_{BH+}$ for its ionized conjugated acid. The following example demonstrates the use of these two permeability coefficients.

Mitra and Mikkelson[53] studied the transcorneal permeation of pilocarpine using an in vitro rabbit corneal preparation clamped into a special diffusion cell. The permeability (permeability coefficient) $P$ as determined experimentally is given at various pH values in Table 11–5.

**TABLE 11–5**

**PERMEABILITY COEFFICIENTS AT VARIOUS pH VALUES**

| pH, donor solution | 4.67 | 5.67 | 6.24 | 6.40 | 6.67 | 6.91 | 7.04 | 7.40 |
|---|---|---|---|---|---|---|---|---|
| $P \times 10^6$ cm/sec | 4.72 | 5.44 | 6.11 | 6.81 | 7.06 | 7.66 | 6.79 | 8.85 |

(a) Compute the un-ionized fraction, $f_B$, of pilocarpine at the pH values found in the table, using equation (11–60). The p$K_a$ of pilocarpine (actually the p$K_a$ of the conjugate acid of the weak base, pilocarpine, and known as the pilocarpinium ion) is 6.67 at 34°C.

(b) The relationship between the permeability $P$ and the un-ionized fraction $f_B$ of pilocarpine base over this range of pH values is given by the equation

$$P = P_B f_B + P_{BH^+} f_{BH^+} \qquad (11\text{–}61)$$

where B stands for base and $BH^+$ for its ionized or conjugate acid form. Noting that $f_{BH^+} = 1 - f_B$, we can write equation (11–61) as

$$P = P_{BH^+} + (P_B - P_{BH^+})f_B \qquad (11\text{–}62)$$

Obtain the permeability for the protonated species, $P_{BH^+}$, and the uncharged base, $P_B$, using least-squares linear regression on equation (11–62) in which $P$, the total permeability, is the dependent variable and $f_B$ is the independent variable.

(c) Obtain the ratio of the two permeability coefficients, $P_{BH^+}/P_B$.

Answers:

(a) The calculated $f_B$ values are given at the various pH values in the following table:

| pH, donor solution | 4.67 | 5.67 | 6.24 | 6.40 | 6.67 | 6.91 | 7.04 | 7.40 |
|---|---|---|---|---|---|---|---|---|
| $f_B$ | 0.01 | 0.09 | 0.27 | 0.35 | 0.50 | 0.64 | 0.70 | 0.84 |

(b) Upon linear regression, equation (11–62) gives

$$P = 4.836 \times 10^{-6} + 4.897 \times 10^{-6} f_B$$

$$\text{Intercept} = 4.836 \times 10^{-6} = P_{BH^+}$$

$$\text{Slope} = 4.897 \times 10^{-6} = (P_B - 4.836 \times 10^{-6})$$

$$P_B = 9.733 \times 10^{-6} \text{ cm/sec}$$

(c) The ratio $P_B/P_{BH^+} \cong 2$. The permeability of the un-ionized form is seen to be about twice that of the ionized form.

The reader should now be in a position to explain the result under (c) based on the pH-partition hypothesis.

## Percutaneous Absorption

Percutaneous penetration, that is, passage through the skin, involves (a) dissolution of a drug in its vehicle, (b) diffusion of solubilized drug (solute) from the vehicle to the surface of the skin, and (c) penetration of the drug through the layers of the skin, principally the stratum corneum. **Figure 11–18** shows the various structures of the skin involved in percutaneous absorption. The slowest step in the process usually involves passage through the stratum corneum; therefore, this is the rate that limits or controls the permeation.*

Scheuplein[54] found that the average permeability constant, $P_s$, for water into skin is $1.0 \times 10^{-3}$ cm/hr and the average diffusion constant, $D_s$, is $2.8 \times 10^{-10}$ cm²/sec (the subscript s on $D$ stands for skin). Water penetration into the stratum corneum appears to alter the barrier only slightly, primarily by its effect on the pores of the skin. The stratum corneum is considered to be a dense homogeneous film.

Small polar nonelectrolytes penetrate into the bulk of the stratum corneum and bind strongly to its components; diffusion of most substances through this barrier is quite slow. Diffusion, for the most part, is transcellular rather than occurring through channels between cells or through sebaceous pores and sweat ducts (**Fig. 11–18**, mechanism A rather than B, C, or E). Stratum corneum, normal and even hydrated, is the most impermeable biologic membrane; this is one of its important features in living systems.

It is an oversimplification to assume that one route prevails under all conditions.[54] Yet after steady-state conditions have been established, transdermal diffusion through the stratum corneum most likely predominates. In the early stages of penetration, diffusion through the appendages (hair follicles, sebaceous and sweat ducts) may be significant. These *shunt* pathways are even important in steady-state diffusion in the case of large polar molecules, as noted in the following.

Scheuplein et al.[55] investigated the percutaneous absorption of a number of steroids. They found that the skin's main barrier to penetration by steroid molecules is the stratum corneum. The diffusion coefficient, $D_s$, for these compounds is approximately $10^{-11}$ cm²/sec, several orders of magnitude smaller than for most nonelectrolytes. This small value of $D_s$ results in low permeability of the steroids. The addition of polar groups to the steroid molecule reduces the diffusion constant still more. For the polar steroids, sweat and sebaceous ducts appear to play a more important part in percutaneous absorption than diffusion through the bulk stratum corneum.

The studies of Higuchi and coworkers[56] demonstrated the methods used to characterize the permeability of different sections of the skin. Distinct protein and lipid domains appear to have a role in the penetration of drugs into the stratum corneum. The uptake of a solute may depend on the characteristics of the protein region, the lipid pathway, or a combination of these two domains in the stratum corneum and depends on the lipophilicity of the solute. The lipid content of the stratum corneum is important in the uptake of lipophilic solutes but is not involved in the attraction of hydrophilic drugs.[57]

The proper choice of vehicle is important in ensuring bioavailability of topically applied drugs. Turi et al.[58] studied the effect of solvents—propylene glycol in water and polyoxypropylene 15 stearyl ether in mineral oil—on the penetration of diflorasone diacetate (a steroid ester) into the skin. The percutaneous flux of the drug was reduced by the presence of excess solvent in the base. Optimum solvent concentrations were determined for products containing both 0.05% and 0.1% diflorasone diacetate.

The important factors influencing the penetration of a drug into the skin are (a) concentration of dissolved drug, $C_s$, because penetration rate is proportional to concentration; (b) the partition coefficient, $K$, between the skin and the vehicle, which is a measure of the relative affinity of the drug for skin and vehicle; and (c) diffusion coefficients, which represent the resistance of drug molecule movement through vehicle,

---

*Appreciable amounts of some drugs, such as steroids, may also penetrate the skin through sebaceous ducts ordinarily associated with hairs on the skin surface (transfollicular absorption).

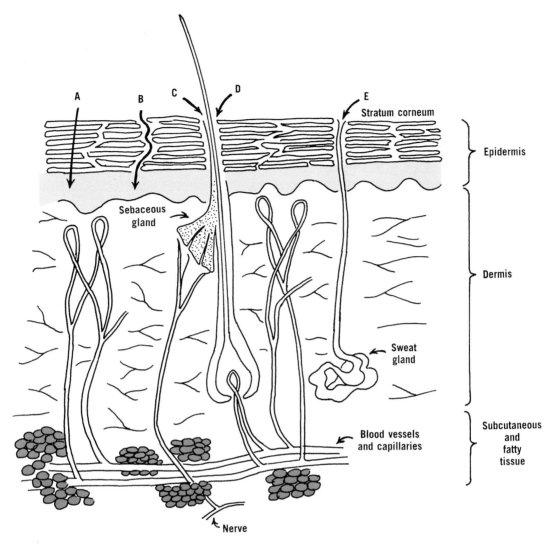

**Fig. 11–18.** Skin structures involved in percutaneous absorption. Thickness of layers is not drawn to scale. Key to sites of percutaneous penetration: A, transcellular; B, diffusion through channels between cells; C, through sebaceous ducts; D, transfollicular; E, through sweat ducts.

$D_v$, and skin, $D_s$, barriers. The relative magnitude of the two diffusion coefficients, $D_v$ and $D_s$, determines whether release from vehicle or passage through the skin is the rate-limiting step.[58,59]

For diflorasone diacetate in propylene glycol–water (a highly polar base) and in polyoxypropylene 15 stearyl ether in mineral oil (a nonpolar base), the skin was found to be the rate-limiting barrier. The diffusional equation for this system is

$$-\frac{dC_v}{dt} = \frac{SK_{vs}D_sC_v}{Vh} \qquad (11\text{–}63)$$

where $C_v$ is the concentration of dissolved drug in the vehicle ($g/cm^3$), $S$ is the surface area of application ($cm^2$), $K_{sv}$ is the skin–vehicle partition coefficient of diflorasone diacetate, $D_s$ is the diffusion coefficient of the drug in the skin ($cm^2/sec$), $V$ is the volume of the drug product applied ($cm^3$), and $h$ is the thickness of the skin barrier (cm).

The diffusion coefficient and the skin barrier thickness can be replaced by a resistance, $R_s$, to diffusion in the skin:

$$R_s = h/D_s \qquad (11\text{–}64)$$

and equation **(11–63)** becomes

$$-\frac{dC_v}{dt} = \frac{SK_{vs}C_v}{VR_s} \qquad (11\text{–}65)$$

In a percutaneous experimental procedure, Turi et al.[58] measured the drug in the receptor rather than in the donor compartment of an in vitro diffusion apparatus, the barrier of which consisted of hairless mouse skin. At steady-state penetration,

$$-V\frac{dC_v}{dt} = V_R \cdot \frac{dC_R}{dt} \qquad (11\text{–}66)$$

The rate of loss of drug from the vehicle in the donor compartment is equal to the rate of gain of drug in the receptor

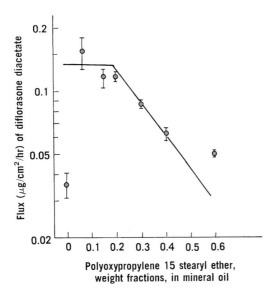

**Fig. 11–19.** Steady-state flux of diflorasone diacetate in a mixture of polyoxypropylene 15 stearate ether in mineral oil. (From J. S. Turi, D. Danielson, and W. Wolterson, J. Pharm. Sci. **68**, 275, 1979. With permission.)

compartment. With this change, equation **(11–65)** is integrated to yield

$$M_R = \left(\frac{SK_{vs}C_v}{R_s}\right)t \qquad (11\text{--}67)$$

where $M_R$ is the amount of diflorasone diacetate in the receptor solution at time $t$. The flux, $J$, is

$$J = \frac{M_R}{St} = \frac{K_{vs}C_v}{R_s} \qquad (11\text{--}68)$$

The steady-state flux for a 0.05% diflorasone diacetate formulation containing various proportions (weight fractions) of polyoxypropylene 15 stearyl ether in mineral oil is shown in **Figure 11–19**. The skin–vehicle partition coefficient was measured for each vehicle formulation. The points represent the experimental values obtained with the diffusion apparatus; the line was calculated using equation **(11–68)**. The point at 0 weight fraction of the ether cosolvent is due to low solubility and slow dissolution rate of the drug in mineral oil and can be disregarded. Beyond a critical concentration, about 0.2 weight fraction of polyoxypropylene 15 stearyl ether, penetration rate decreases. The results[69] indicated that one application of the topical steroidal preparation per day was adequate and that the 0.05% concentration was as effective as the 0.1% preparation.

## EXAMPLE 11–8

**Diflorasone Diacetate Permeation of Hairless Mouse Skin**

A penetration study of $5.0 \times 10^{-3}$ g/cm³ diflorasone diacetate solution was conducted at 27°C in the diffusion cell of Turi et al.[58] using a solvent of 0.4 weight fraction of polyoxypropylene 15 stearyl ether in mineral oil. The partition coefficient, $K_{vs}$, for the drug distributed between hairless mouse skin and vehicle was found to be 0.625. The resistance, $R_s$, of the drug in the mouse skin was determined to be

6666 hr/cm. The diameter of a circular section of mouse skin used as the barrier in the diffusion cell was 1.35 cm.* Calculate (*a*) the flux, $J = M_R/St$, in g/cm, and (*b*) the amount, $M_R$ in $\mu$g, of diflorasone diacetate that diffused through the hairless mouse skin in 8 hr.

Using equation (11–68), we obtain

(*a*)

$$J = \frac{K_{vs}C_v}{R_s} = \frac{(0.625)(5.0 \times 10^{-3}\ \text{g/cm}^3)}{6666\ \text{hr/cm}}$$

$$J = 4.69 \times 10^{-7}\ \text{g/cm}^2/\text{hr}$$

(*b*)

$$M_R = J \times S \times t$$

$$M_R = (4.69 \times 10^{-7}\ \text{g/cm}^2/\text{hr}) \times \frac{\pi}{4}(1.35\ \text{cm})^2\ (8\ \text{hr})$$

$$M_R = 5.37 \times 10^{-6}\ \text{g} = 5.37\ \mu\text{g}$$

Ostrenga and his associates[60] studied the nature and composition of topical vehicles as they relate to the transport of a drug through the skin. The varied $D_s$, $K_{vs}$, and $C_v$ to improve skin penetration of two topical steroids, fluocinonide and fluocinolone acetonide, incorporated into various propylene glycol–water gels. In vivo penetration and in vitro diffusion using abdominal skin removed at autopsy were studied. It was concluded that clinical efficacy of topical steroids can be estimated satisfactorily from in vitro data regarding release, diffusion, and the physical chemical properties of drug and vehicle.

The diffusion, $D_s$, of the drug in the skin barrier can be influenced by components of the vehicle (mainly solvents and surfactants), and an optimum partition coefficient can be obtained by altering the affinity of the vehicle for the drug.

The in vitro rate of skin penetration of the drug, $dQ/dt$, at 25°C is obtained experimentally at definite times, and the cumulative amount penetrating (measured in radioactive disintegrations per minute) is plotted against time in minutes or hours. After steady state has been attained, the slope of the straight line yields the rate, $dM/dt$. The lag time is obtained by extrapolating the steady-state line to the time axis.

In vitro penetration of human cadaver skin and in vivo penetration of fluocinolone acetonide from propylene glycol gels into living skin are compared in **Figure 11–20**. It is observed that the shapes and peaks of the two curves are approximately similar. Thus, in vitro studies using human skin sections should serve as a rough guide to the formulation of acceptable bases for these steroidal compounds.

Ostrenga et al.[60] were able to show a relationship between release of the steroid from its vehicle, in vitro penetration through human skin obtained at autopsy, and in vivo vasoconstrictor activity of the drug depending on compositions of the vehicle. The correlations obtained suggest that information obtained from diffusion studies can assist in the design of effective topical dosage forms. Some useful guidelines are (*a*) all the drug should be in solution in the vehicle, (*b*) the

---

*The area, $S$, of a circle, expressed in terms of its diameter, $d$, is $S = \frac{1}{4}\pi\ d^2$. (See inside front cover.)

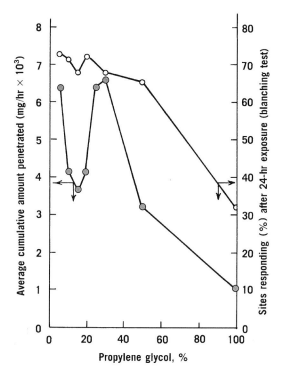

**Fig. 11–20.** Comparison of in vitro penetration of steroid through a skin section and in vivo skin blanching test. Key: •, in vitro method; ○, in vivo method. (From J. Ostrenga, C. Steinmetz, and B. Poulsen, J. Pharm. Sci. **60,** 1177, 1971. With permission.)

solvent mixtures must maintain a favorable partition coefficient so that the drug is soluble in the vehicle and yet have a great affinity for the skin barrier into which it penetrates, and (c) the components of the vehicle should favorably influence the permeability of the stratum corneum.

Sloan and coworkers[61] studied the effect of vehicles having a range of solubility parameters, d, on the diffusion of salicylic acid and theophylline through hairless mouse skin. They were able to correlate the partition coefficient, K, for the drugs between the vehicle and skin calculated from solubility parameters and the permeability coefficient, P, obtained experimentally from the diffusion data. The results obtained with salicylic acid, a soluble molecule, and with theophylline, a poorly soluble molecule with quite different physical chemical properties, were practically the same.

In the studies of skin permeation described thus far, efforts were made to increase percutaneous absorption processes. It is important, however, that some compounds not be absorbed. Pharmaceutical adjuvants such as antimicrobial agents, antioxidants, coloring agents, and drug solubilizers, although they should remain in the vehicle on the skin's surface, can penetrate the stratum corneum.

Parabens, typical preservatives incorporated into cosmetics and topical dosage forms, may cause allergic reactions if absorbed into the dermis. Komatsu and Suzuki[62] studied the in vitro percutaneous absorption of butylparaben (butyl p-hydroxybenzoate) through guinea pig skin. Disks of dorsal skin were placed in a diffusion cell between a donor and receptor chamber, and the penetration of $^{32}$C-butylparaben was determined by the fractional collection of samples from the cell's receptor side and measurement of radioactivity in a liquid scintillation counter.

When butylparaben was incorporated into various vehicles containing polysorbate 80, propylene glycol, and polyethylene glycol 400, a constant diffusivity was obtained averaging 3.63 ($\pm$ 0.47 SD) $\times 10^{-4}$ cm$^2$/hr.

The partition coefficient, $K_{vs}$, for the paraben between vehicle and skin changed markedly depending upon the vehicle. For a 0.015% (w/v) aqueous solution of butyl paraben, $K_{vs}$ was found to be 2.77. For a 0.1% w/v solution of the preservative containing 2% (w/v) of polysorbate 80 and 10% (w/v) propylene glycol in water, the partition coefficient dropped to 0.18. There was no apparent complexation between these solubilizers and butylparaben, according to the authors.

The addition of either propylene glycol or polyethylene glycol 400 to water was found to increase the solubility of paraben in the vehicle and to reduce its partition coefficient between vehicle and skin. By this means, skin penetration of butylparaben could be retarded, maintaining the preservative in the topical vehicle where it was desired.

In the case of polysorbate 80, Komatsu and Suzuki[62] found that this surfactant also reduced preservative absorption, maintaining the antibacterial action of the paraben in the vehicle. These workers concluded that the action of polysorbate 80 was a balance of complex factors that is difficult for the product formulator to predict and manage.

**Buccal Absorption**

Using a wide range of organic acids and bases as drug models, Beckett and Moffat[63] studied the penetration of drugs into the lipid membrane of the mouths of humans. In harmony with the pH-partition hypothesis, absorption was related to the p$K_a$ of the compound and its lipid–water partition coefficient.

Ho and Higuchi[64] applied one of the earlier mass transfer models[65] to the analysis of the buccal absorption of n-alkanoic acids.[66] They utilized the aqueous–lipid phase model in which the weak acid species are transported across the aqueous diffusion layer and, subsequently, only the nonionized species pass across the lipid membrane. Unlike the intestinal membrane, the buccal membrane does not appear to possess significant aqueous pore pathways, and the surface pH is essentially the same as the buffered drug solution pH. Buccal absorption is assumed to be a first-order process owing to the nonaccumulation of drug on the blood side:

$$\ln \frac{C}{C_0} = -K_u t \qquad (11\text{–}69)$$

where $C$ is the aqueous concentration of the n-alkanoic acid in the donor or mucosal compartment. The absorption rate constant, $K_u$, is

$$K_u = \frac{S}{V} \cdot \frac{P_{aq}}{1 + \dfrac{P_{aq}}{P_0 X_s}} \qquad (11\text{–}70)$$

the terms of which have been previously defined. Recall that $X_s = 1/(1 + 10^{pH_s - pK_a})$, or, by equation **(11–59)**, $X_s = 1/[1 + \text{antilog}(pH_s - pK_a)]$ and is the fraction of un-ionized weak acid at $pH_s$.

With $S = 100$ cm$^2$, $V = 25$ cm$^3$, $P_{aq} = 1.73 \times 10^{-3}$ cm/sec, $P_0 = 2.27 \times 10^{-3}$ cm/sec, $pK_a = 4.84$, and $pH_s = 4.0$, equations **(11–59)** and **(11–70)** yield for caproic acid an absorption rate constant

$$K_u = \frac{100}{25} \left( \frac{1.73 \times 10^{-3}}{1 + \dfrac{1.73 \times 10^{-3}}{2.27 \times 10^{-3} \times 0.874}} \right) = 3.7 \times 10^{-3} \text{ sec}^{-1}$$

Buccal absorption rate constants constructed according to the model of Ho and Higuchi agree well with experimental values. The study shows an excellent correspondence between diffusional theory and in vivo absorption and suggests a fruitful approach for structure–activity studies not only for buccal membrane permeation but also for bioabsorption in general.

## Uterine Diffusion

Drugs such as progesterone and other therapeutic and contraceptive compounds may be delivered in microgram amounts into the uterus by means of diffusion-controlled forms (intrauterine device). In this way the patient is automatically and continuously provided medication or protected from pregnancy for days, weeks, or months.[67]

Yotsuyanagi et al.[68] performed in situ vaginal drug absorption studies using the rabbit doe as an animal model to develop more effective uterine drug delivery systems. A solution of a model drug was perfused through a specially constructed cell and implanted in the vagina of the doe (**Fig. 11–21**), and the drug disappearance was monitored. The drug release followed first-order kinetics, and the results permitted the

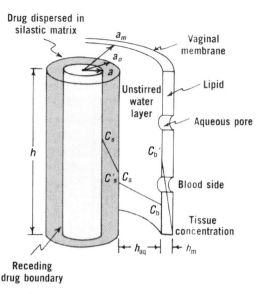

**Fig. 11–22.** Contraceptive drug in water-insoluble silicone polymer matrix. Dimensions and sections of the matrix are shown together with concentration gradients across the drug release pathway. (From S. Hwang, E. Owada, T. Yotsuyanagi, et al., J. Pharm. Sci. **65**, 1578, 1976. With permission.)

calculation of apparent permeability coefficient and diffusion layer thickness.

The drug may also be implanted in the vagina in a silicone matrix (**Fig. 11–22**), and drug release at any time can be calculated using a quadratic expression,[68]

$$\left( \frac{1}{2\pi h a_0^2 A} \right) M^2 + \frac{D_e K_s}{a_0} \left( \frac{1}{P_{aq}} + \frac{1}{P_m} \right) M$$
$$- (2\pi h D_e C_s t) = 0 \quad \textbf{(11–71)}$$

The method of calculation can be shown, using the data of Hwang et al.,[69] which are given in **Table 11–6**. When the

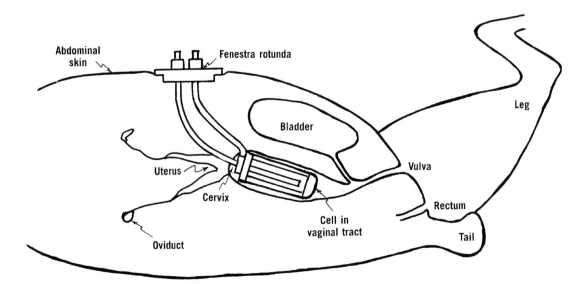

**Fig. 11–21.** Implanted rib-cage cell in the vaginal tract of a rabbit. (From T. Yotsuyanagi, A. Molakhia, et al., J. Pharm. Sci. **64**, 71, 1975. With permission.)

**PHYSICAL PARAMETERS FOR THE RELEASE OF PROGESTERONE AND HYDROCORTISONE FROM A SILICONE MATRIX FOR VAGINAL ABSORPTION IN THE RABBIT[69]**

|  | Progesterone | Hydrocortisone |
|---|---|---|
| Solubility in matrix, $C_s$ (mg/cm³) | 0.572 | 0.014 |
| Diffusion coefficient in matrix, $D_e$ (cm²/sec) | $4.5 \times 10^{-7}$ | $4.5 \times 10^{-7}$ |
| Silicone–water partition coefficient, $K_s$ | 50.2 | 0.05 |
| Permeability coefficient of rabbit vaginal membrane, $P_m$ (cm/sec) | $7 \times 10^{-4}$ | $5.8 \times 10^{-5}$ |
| $P_{aq}$ (when $h_{aq} = 100\ \mu m$) | $7 \times 10^{-4}$ | $7 \times 10^{-4}$ |
| $P_{aq}$ (when $h_{aq} = 1000\ \mu m$) | $0.7 \times 10^{-4}$ | $0.7 \times 10^{-4}$ |

aqueous diffusion layer, $h_{aq}$, is 100 $\mu$m, the aqueous permeability coefficient, $P_{aq}$, is $7 \times 10^{-4}$ cm/sec; this value is used in the following example. The length, $h$, of the silastic cylinder (**Fig. 11–21**) is 6 cm, its radius, $a_0$, is 1.1 cm, and the initial amount of drug per unit volume of plastic cylinder, or loading concentration, $A$, is 50 mg/cm³.

Equation **(11–71)** is of the quadratic form $aM^2 + bM + c = 0$, where, for progesterone,

$$a = \frac{1}{2\pi h a_0^2 A} = \frac{1}{(2)(3.1416)(6\ \text{cm})(1.1\ \text{cm})^2(50\ \text{mg/cm}^3)}$$

$$= 0.000438\ \text{mg}^{-1}$$

$$b = \frac{D_e K_s}{a_0}\left(\frac{1}{P_{aq}} + \frac{1}{P_m}\right) = \frac{(4.5 \times 10^{-7}\ \text{cm}^2/\text{sec})(50.2)}{1.1\ \text{cm}}$$

$$\left(\frac{1}{7 \times 10^{-4}\ \text{cm/sec}} + \frac{1}{7 \times 10^{-4}\ \text{cm/sec}}\right)$$

$$= 0.0587\ (\text{dimensionless})$$

$$c = -2\pi h D_e C_s t = -2(3.1416)(6\ \text{cm})$$

$$\times (4.5 \times 10^{-7}\ \text{cm}^2/\text{sec})(0.572\ \text{mg/cm}^3)$$

$$\times \left(\frac{86,400\ \text{sec}}{\text{day}}\right)(t\ \text{days}) = -0.8384 \times t\ (\text{days})$$

How much progesterone is released in 5 days? In 20 days? The quadratic formula to be used here is

$$M = \frac{-b + \sqrt{b^2 - 4ac}}{2a}$$

After 5 days,

$$c = -0.8384\ \text{mg/day} \times 5\ \text{days} = -4.1920\ \text{mg},\ \text{and}$$

$$M = \frac{-0.0587 + \sqrt{(0.0587)^2 - (4)(0.000438)(-4.1920)}}{2(0.000438)}$$

$$= 51.6\ \text{mg}$$

After 20 days, $C = -0.8384$ mg/day $\times$ 20 days $= -16.77$ mg, and

$$M = \frac{-0.0587 + \sqrt{(0.0587)^2 - (4)(0.000438)(-16.77)}}{2(0.000438)}$$

$$= 139.8\ \text{mg}$$

Okada et al.[70] carried out detailed studies on the vaginal absorption of hormones.

# ELEMENTARY DRUG RELEASE

Release from dosage forms and subsequent bioabsorption are controlled by the physical chemical properties of drug and delivery form and the physiologic and physical chemical properties of the biologic system. Drug concentration, aqueous solubility, molecular size, crystal form, protein binding, and p$K_a$ are among the physical chemical factors that must be understood to design a delivery system that exhibits controlled or sustained-release characteristics.[71]

The release of a drug from a delivery system involves factors of both dissolution and diffusion. As the reader has already observed in this chapter, the foundations of diffusion and dissolution theories bear many resemblances. Dissolution rate is covered in great detail in the next chapter.

## Zero-Order Drug Release

The flux, $J$, of equation **(11–11)** is actually proportional to a gradient of thermodynamic activity rather than concentration. The activity will change in different solvents, and the diffusion rate of a solvent at a definite concentration may vary widely depending on the solvent employed. The thermodynamic activity of a drug can be held constant ($a = 1$) in a delivery form by using a saturated solution in the presence of excess solid drug. Unit activity ensures constant release of the drug at a rate that depends on the membrane permeability and the geometry of the dosage form. **Figure 11–23** shows the

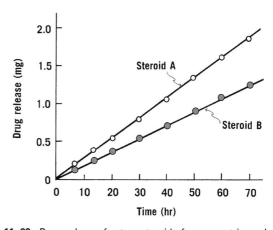

**Fig. 11–23.** Drug release for two steroids from a matrix or device providing zero-order release. (After R. W. Baker and H. K. Lonsdale, in, A. C. Tanquary and R. E. Lacey (Eds.), *Controlled Release of Biologically Active Agents*, Plenum Press, New York, 1974, p. 30.)

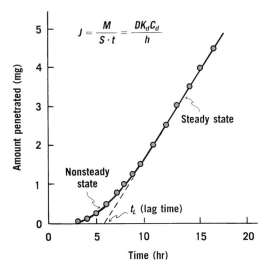

$$J = \frac{M}{S \cdot t} = \frac{DK_d C_d}{h}$$

**Fig. 11–24.** Butyl paraben diffusing through guinea pig skin from aqueous solution. Steady-state and nonsteady-state regions are shown. (From H. Komatsu and M. Suzuki, J. Pharm. Sci. **68**, 596, 1979. With permission.)

rate of delivery of two steroids from a device providing constant drug activity and what is known as "zero-order release." For more information about zero-order processes the reader is referred to the chapter on kinetics (Chapter 14). If excess solid is not present in the delivery form, the activity decreases as the drug diffuses out of the device, the release rate falls exponentially, and the process is referred to as first-order release, analogous to the well-known reaction in chemical kinetics. First-order release from dosage forms is discussed by Baker and Lonsdale.[72]

## Lag Time

A constant-activity dosage form may not exhibit a steady-state process from the initial time of release. **Figure 11–24** is a plot of the amount of butylparaben penetrating through guinea pig skin from a dilute aqueous solution of the penetrant. It is observed that the curve of **Figure 11–23** is convex with respect to the time axis in the early stage and then becomes linear. The early stage is the nonsteady-state condition. At later times, the rate of diffusion is constant, the curve is essentially linear, and the system is at steady state. When the steady-state portion of the line is extrapolated to the time axis, as shown in **Figure 11–24**, the point of intersection is known as the *lag time*, $t_L$. This is the time required for a penetrant to establish a uniform concentration gradient within the membrane separating the donor from the receptor compartment.

In the case of a time lag, the straight line of **Figure 11–24** can be represented by a modification of equation (**11–13**):

$$M = \frac{SDKC_d}{h}(t - t_L) \tag{11–72}$$

The lag time, $t_L$, is given by

$$t_L = \frac{h^2}{6D} \tag{11–73}$$

and its measurement provides a means of calculating the diffusivity, $D$, presuming a knowledge of the membrane thickness, $h$. Also, knowing $P$, one can calculate the thickness, $h$, from

$$t_L = \frac{h}{6P} \tag{11–74}$$

## Drugs in Polymer Matrices

A powdered drug is homogeneously dispersed throughout the matrix of an erodible tablet. The drug is assumed to dissolve in the polymer matrix and to diffuse out from the surface of the device. As the drug is released, the distance for diffusion becomes increasingly greater. The boundary that forms between drug and empty matrix recedes into the tablet as drug is eluted. A schematic illustration of such a device is shown in **Figure 11–25a**. **Figure 11–25b** shows a granular matrix with interconnecting pores or capillaries. The drug is leached out of this device by entrance of the surrounding medium. **Figure 11–25c** depicts the concentration profile and shows the receding depletion zone that moves to the center of the tablet as the drug is released.

Higuchi[32] developed an equation for the release of a drug from an ointment base and later[73] applied it to diffusion of solid drugs dispersed in homogeneous and granular matrix dosage systems (**Fig. 11–25**).

Fick's first law,

$$\frac{dM}{S\,dt} = \frac{dQ}{dt} = \frac{DC_s}{h} \tag{11–75}$$

can be applied to the case of a drug embedded in a polymer matrix, in which $dQ/dt^*$ is the rate of drug released per unit area of exposed surface of the matrix. Because the boundary between the drug matrix and the drug-depleted matrix recedes with time, the thickness of the empty matrix, $dh$, through which the drug diffuses also increases with time.

Whereas $C_s$ is the solubility or saturation concentration of drug in the matrix, $A$ is the total concentration (amount per unit volume), dissolved and undissolved, of drug in the matrix.

As drug passes out of a homogeneous matrix (**Fig. 11–25a**), the boundary of drug (represented by the dashed vertical line in **Fig. 11–25c**) moves to the left by an infinitesimal distance, $dh$. The infinitesimal amount, $dQ$, of drug released because of this shift of the front is given by the approximate linear expression

$$dQ = A\,dh - \tfrac{1}{2}C_s\,dh \tag{11–76}$$

Now $dQ$ of equation (**11–76**) is substituted into equation (**11–75**), integration is carried out, and the resulting equation

---

*$dM$ is amount of drug diffusing; $dQ$ is introduced here to represent $dM/S$, where $S$ is surface area of the boundary.

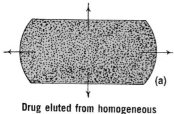

Drug eluted from homogeneous
polymer matrix

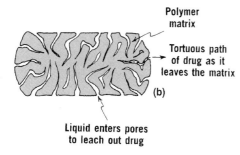

Polymer
matrix

Tortuous path
of drug as it
leaves the matrix

(b)

Liquid enters pores
to leach out drug

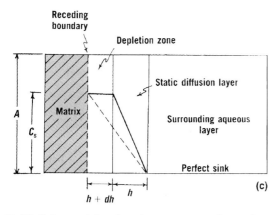

Receding
boundary

Depletion zone

Static diffusion layer

Matrix

Surrounding aqueous
layer

Perfect sink

(c)

$h + dh$   $h$

**Fig. 11–25.** Release of drug from homogenous and granular matrix dosage forms. (a) Drug eluted from a homogenous polymer matrix. (b) Drug leached from a heterogeneous or granular matrix. (c) Schematic of the solid matrix and its receding boundary as drug diffuses from the dosage form. (From T. Higuchi, J. Pharm. Sci. **50,** 874, 1961. With permission.)

is solved for $h$. The steps of the derivation as given by Higuchi[32] are

$$\left(A - \tfrac{1}{2}C_s\right) dh = \frac{DC_s}{h} dt \qquad (11\text{–}77)$$

$$\frac{2A - C_s}{2DC_s} \int h\, dh = \int dt \qquad (11\text{–}78)$$

$$t = \frac{(2A - C_s)}{4DC_s} h^2 + C \qquad (11\text{–}79)$$

The integration constant, $C$, can be evaluated at $t = 0$, at which $h = 0$, giving

$$t = \frac{(2A - C_s)h^2}{4DC_s} \qquad (11\text{–}80)$$

$$h = \left(\frac{4DC_s t}{2A - C_s}\right)^{1/2} \qquad (11\text{–}81)$$

The amount of drug depleted per unit area of matrix, $Q$, at time $t$ is obtained by integrating equation (11–76) to yield

$$Q = hA - 1/2\, hC_s \qquad (11\text{–}82)$$

Substituting equation (11–81) into (11–82) produces the result

$$Q = \left(\frac{DC_s t}{2A - C_s}\right)^{1/2} (2A - C_s) \qquad (11\text{–}83)$$

which is known as the Higuchi equation:

$$Q = [D(2A - C_s)C_s t]^{1/2} \qquad (11\text{–}84)$$

The instantaneous rate of release of a drug at time $t$ is obtained by differentiating equation (11–84) to yield

$$\frac{dQ}{dt} = \frac{1}{2}\left[\frac{D(2A - C_s)C_s}{t}\right]^{1/2} \qquad (11\text{–}85)$$

Ordinarily, $A$ is much greater than $C_s$, and equation (11–84) reduces to

$$Q = (2ADC_s t)^{1/2} \qquad (11\text{–}86)$$

and equation (11–85) becomes

$$\frac{dQ}{dt} = \left(\frac{ADC_s}{2t}\right)^{1/2} \qquad (11\text{–}87)$$

for the release of a drug from a homogeneous polymer matrix–type delivery system. Equation (11–86) indicates that the amount of drug released is proportional to the square root of $A$, the total amount of drug in unit volume of matrix; $D$, the diffusion coefficient of the drug in the matrix; $C_s$, the solubility of drug in polymeric matrix; and $t$, the time.

The rate of release, $dQ/dt$, can be altered by increasing or decreasing the drug's solubility, $C_s$, in the polymer by complexation. The total concentration, $A$, of drug that the physician prescribes is also seen to affect the rate of drug release.

---

**EXAMPLE 11–9**

**Classic Drug Release: Higuchi Equation**

(a) What is the amount of drug per unit area, $Q$, released from a tablet matrix at time $t = 120$ min? The total concentration of drug in the homogeneous matrix, $A$, is 0.02 g/cm³. The drug's solubility $C_s$ is $1.0 \times 10^{-3}$ g/cm³ in the polymer. The diffusion coefficient, $D$, of the drug in the polymer matrix at 25°C is $6.0 \times 10^{-6}$ cm²/sec, or $360 \times 10^{-6}$ cm²/min.

We use equation (11–86):

$$Q = [2(0.02\ \text{g/cm}^3)(360 \times 10^{-6}\ \text{cm}^2/\text{min})$$
$$\times (1.0 \times 10^{-3}\ \text{g/cm}^3)(120\ \text{min})]^{1/2}$$
$$= 1.3 \times 10^{-3}\ \text{g/cm}^2$$

(b) What is the instantaneous rate of drug release occurring at 120 min? We have

$$dQ/dt = \left[\frac{(0.02)(360 \times 10^{-6})(1.0 \times 10^{-3})}{2 \times 120}\right]^{1/2}$$
$$= 5.5 \times 10^{-6}\ \text{g/cm}^2\ \text{min}$$

## Release from Granular Matrices: Porosity and Tortuosity

The release of a solid drug from a granular matrix (**Fig. 11–25b**) involves the simultaneous penetration of the surrounding liquid, dissolution of the drug, and leaching out of the drug through interstitial channels or pores. A granule is, in fact, defined as a porous rather than a homogeneous matrix. The volume and length of the opening in the matrix must be accounted for in the diffusional equation, leading to a second form of the Higuchi equation,

$$Q = \left[ \frac{D\epsilon}{\tau}(2A - \epsilon C_s)C_s t \right]^{1/2} \qquad (11\text{–}88)$$

where $\epsilon$ is the porosity of the matrix and $\tau$ is the tortuosity of the capillary system, both parameters being dimensionless quantities.

Porosity, $\epsilon$, is the fraction of matrix that exists as pores or channels into which the surrounding liquid can penetrate. The porosity term, $\epsilon$, in equation (**11–88**) is the total porosity of the matrix after the drug has been extracted. This is equal to the initial porosity, $\epsilon_0$, due to pores and channels in the matrix before the leaching process begins and the porosity created by extracting the drug. If $A$ g/cm$^3$ of drug is extracted from the matrix and the drug's specific volume or reciprocal density is $1/\rho$ cm$^3$/g, then the drug's concentration, $A$, is converted to volume fraction of drug that will create an additional void space or porosity in the matrix once it is extracted. The total porosity of the matrix, $\epsilon$, becomes

$$\epsilon = \epsilon_0 + A(1/\rho) \qquad (11\text{–}89)$$

The initial porosity, $\epsilon_0$, of a compressed tablet may be considered to be small (a few percent) relative to the porosity $A/\rho$ created by the dissolution and removal of the drug from the device. Therefore, the porosity frequently is calculated conveniently by disregarding $\epsilon_0$ and writing

$$\epsilon \cong (A/\rho) \qquad (11\text{–}90)$$

Equation (**11–88**) differs from equation (**11–84**) only in the addition of $\epsilon$ and $\tau$. Equation (**11–84**) is applicable to release from a homogeneous tablet that gradually erodes and releases the drug into the bathing medium. Equation (**11–88**) applies instead to a drug-release mechanism based upon entrance of the surrounding medium into a polymer matrix, where it dissolves and leaches out the soluble drug, leaving a shell of polymer and empty pores. In equation (**11–88**), diffusivity is multiplied by porosity, a fractional quantity, to account for the decrease in $D$ brought about by empty pores in the matrix. The apparent solubility of the drug, $C_s$, is also reduced by the volume fraction term, which represents porosity.

Tortuosity, $\tau$, is introduced into equation (**11–88**) to account for an increase in the path length of diffusion due to branching and bending of the pores as compared to the shortest "straight-through" pores. Tortuosity tends to reduce the amount of drug release in a given interval of time, and

so it appears in the denominator under the square root sign. A straight channel has a tortuosity of unity, and a channel through spherical beads of uniform size has a tortuosity of 2 or 3. At times, an unreasonable value of, say, 1000 is obtained for $\tau$, as Desai et al.[74] noted. When this occurs, the pathway for diffusion evidently is not adequately described by the concept of tortuosity, and the system must be studied in more detail to determine the factors controlling matrix permeability. Methods for obtaining diffusivity, porosity, tortuosity, and other quantities required in an analysis of drug diffusion are given by Desai et al.[75]

Equation (**11–88**) has been adapted to describe the kinetics of lyophilization,[13] commonly called *freeze-drying*, of a frozen aqueous solution containing drug and an inert matrix-building substance (e.g., mannitol or lactose). The process involves the simultaneous change in the receding boundary with time, phase transition at the ice–vapor interface governed by the Clausius–Clapeyron pressure–temperature relationship, and water vapor diffusion across the pore path length of the dry matrix under low temperature and vacuum conditions.

## Soluble Drugs in Topical Vehicles and Matrices

The original Higuchi model[32,73] does not provide a fit to experimental data when the drug has a significant solubility in the tablet or ointment base. The model can be extended to drug release from homogeneous solid or semisolid vehicles, however, using a quadratic expression introduced by Bottari et al.,[76]

$$Q^2 + 2DRA^*Q - 2DA^*C_s t = 0 \qquad (11\text{–}91)$$

where

$$A^* = A - \frac{1}{2}(C_s + C_v) \qquad (11\text{–}92)$$

$Q$ is the amount of drug released per unit area of the dosage form, $D$ is an effective diffusivity of the drug in the vehicle, $A$ is the total concentration of drug, $C_s$ is the solubility of drug in the vehicle, $C_v$ is the concentration of drug at the vehicle–barrier interface, and $R$ is the diffusional resistance afforded by the barrier between the donor vehicle and the receptor phase. $A^*$ is an effective $A$ as defined in equation (**11–92**) and is used when $A$ is only about three or four times greater than $C_s$.

When

$$Q^2 \gg 2DRA^*Q \qquad (11\text{–}93)$$

equation (**11–91**) reduces to one form of the Higuchi equation [equation (**11–86**)]:

$$Q^2 = (2A^*DC_s t)^{1/2} \qquad (11\text{–}94)$$

Under these conditions, resistance to diffusion, $R$, is no longer significant at the interface between vehicle and receptor phase. When $C_s$ is not negligible in relation to $A$, the vehicle-controlled model of Higuchi becomes

$$Q = [D(2A - C_s)C_s t]^{1/2} \qquad (11\text{--}95)$$

The quadratic expression of Bottari, equation (11–91), should allow one to determine diffusion of drugs in ointment vehicles or homogeneous polymer matrices when $C_s$ becomes significant in relation to $A$. The approach of Bottari et al.[76] follows.

Because it is a second-degree power series in $Q$, equation (11–91) can be solved using the well-known quadratic approach. One writes

$$aQ^2 + bQ + c = 0 \qquad (11\text{--}96)$$

where, with reference to equation (11–91), $a = 1$, $b = 2$ $DRA^*$, and $C = -2DA^*C_s t$. Equation (11–96) has the well-known solution

$$Q = \frac{-b \pm \sqrt{b^2 - 4ac}}{2a} \qquad (11\text{--}97)$$

or

$$Q = \frac{-2DRA^* + \sqrt{(2DRA^*)^2 + (2DA^*C_s t)}}{2} \qquad (11\text{--}98)$$

in which the positive root is taken for physical significance. If a lag time occurs, $t$ in equation (11–98) is replaced by $(t - t_L)$ for the steady-state period. Bottari et al.[76] obtained satisfactory value for $b$ and $c$ by use of a least-square fit of equation (11–91) involving the release of benzocaine from suspension-type aqueous gels. The diffusional resistance, $R$, is determined from steady-state permeation, and $C_v$ is then obtained from the expression

$$C_v = R(dQ/dt) \qquad (11\text{--}99)$$

The application of equation (11–91) is demonstrated in the following example.

### EXAMPLE 11–10

**Benzocaine Release from an Aqueous Gel**

(a) Calculate $Q$, the amount in milligrams of micronized benzocaine released per square centimeter of surface area, from an aqueous gel after 9000 sec (2.5 hr) in a diffusion cell. Assume that the total concentration, $A$, is 10.9 mg/mL, the solubility, $C_s$, is 1.31 mg/mL, $C_v = 1.05$ mg/mL, the diffusional resistance, $R$, of a silicone rubber barrier separating the gel from the donor compartment is $8.10 \times 10^3$ sec/cm, and the diffusivity, $D$, of the drug in the gel is $9.14 \times 10^{-6}$ cm²/sec. From equation (11–92) we have

$$A^* = 10.9 \text{ mg/mL} - \frac{1}{2}(1.31 + 1.05) \text{ mg/mL} = 9.72 \text{ mg/mL}$$

Then,

$$DRA^* = (9.14 \times 10^{-6} \text{ cm}^2/\text{sec}) \times (8.10 \times 10^3 \text{ sec/cm})$$
$$\times (9.72 \text{ mg/mL})$$
$$= 0.7196 \text{ mg cm}^{-2}$$
$$DA^*C_s t = (9.14 \times 10^{-6})(9.72)(1.31)(9000) = 1.047 \text{ mg}^2/\text{cm}^4$$
$$Q = -0.7196 + [(0.7196)^2 + 2(1.047)]^{1/2} \text{ mg/cm}^2$$
$$= -0.7196 + [1.616] = 0.90 \text{ mg/cm}^2$$

The $Q_{(calc)}$ of 0.90 mg/cm² compares well with $Q_{(obs)} = 0.88$ mg/cm².

A slight increase in accuracy can be obtained by replacing $t = 9000$ sec with $t = (9000 - 405)$ sec, in which the lag time $t = 405$ sec is obtained from a plot of experimental $Q$ values versus $t^{1/2}$. This correction yields a $Q_{(calc)} = 0.87$ mg/cm².

(b) Calculate $Q$ using equation (11–95) and compare the result with that obtained in equation (11–94). We have

$$Q = \{(9.14 \times 10^{-6})[(2 \times 10.9) - 1.31](1.31)(9000)\}^{1/2}$$
$$= 1.49 \text{ mg/cm}^2$$

Paul and coworkers[77] studied cases in which $A$, the matrix loading of drug per unit volume in a polymeric dosage form, may be greater than, equal to, or less than the equilibrium solubility, $C_s$, of the drug in a matrix. The model is a refinement of the original Higuchi approach,[32,73] providing an accurate set of equations that describe release rates of drugs, fertilizers, pesticides, antioxidants, and preservatives in commercial and industrial applications over the entire range of ratios of $A$ to $C_s$.

---

A silastic capsule, as depicted in **Figure 11–26a**, has been used to sustain and control the delivery of drugs in pharmacy and medicine.[78–80] The release of a drug from a silastic capsule is shown schematically in **Figure 11–26b**. The molecules of the crystalline drug lying against the inside wall of the capsule leave their crystals, pass into the polymer wall by a dissolution process, diffuse through the wall, and pass into the liquid diffusion layer and the medium surrounding the capsule. The concentration differences across the polymer wall of thickness $h_m$ and the stagnant diffusion layer of thickness $h_a$ are represented by the lines $C_p - C_m$ and $C_s - C_b$, respectively. $C_p$ is the solubility of the drug in the polymer and $C_m$ is the concentration at the polymer–solution interface, that is, the concentration of drug in the polymer in contact with the solution. $C_s$, on the other hand, is the concentration of the drug in the solution at the polymer–solution interface, and it

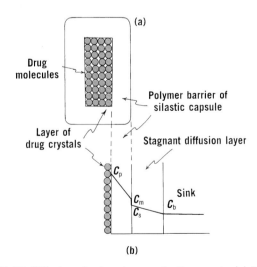

**Fig. 11–26.** Diffusion of a drug from a silastic capsule. (a) Drug in the capsule surrounded by a polymer barrier; (b) diffusion of the drug through the polymer wall and stagnant aqueous diffusion layer and into the receptor compartment at sink conditions. (After Y. W. Chien, in J. R. Robinson (Ed.), *Sustained and Controlled Release Drug Delivery Systems*, Marcel Dekker, New York, 1978, p. 229; and Y. W. Chien, Chem. Pharm. Bull. **24**, 147, 1976.)

is seen in **Figure 11–25b** to be somewhat below the solubility of drug in polymer at the interface. There is a real difference between the solubility of the drug in the polymer and in the solution, although both exist at the interface. Finally, $C_b$ is the concentration of the drug in the bulk solution surrounding the capsule.

To express the rate of drug release under sink conditions, Chien[78] used the following expression:

$$Q = \frac{K_r D_a D_m}{K_r D_a h_m + D_m h_a} C_p t \qquad (11\text{--}100)$$

which is an integrated form analogous to equation **(11–31)**. In equation **(11–100)**, $Q$ is the amount of drug released per unit surface area of the capsule and $K_r$ is the partition coefficient, defined as*

$$K_r = C_s/C_p \qquad (11\text{--}101)$$

When diffusion through the capsule membrane or film is the limiting factor in drug release, that is, when $K_r D_a h_m$ is much greater than $D_m h_a$, equation **(11–100)** reduces to

$$Q = \left(\frac{D_m}{h_m}\right) C_p t \qquad (11\text{--}102)$$

and when the limiting factor is passage through the diffusion layer ($D_m h_a \gg K_r D_a h_m$),

$$Q = \left(\frac{D_a}{h_a}\right) C_s t = \left(\frac{K_r D_a}{h_a}\right) C_p t \qquad (11\text{--}103)$$

The right-hand expression can be written because $C_s = K_r C_p$, as defined earlier in equation **(11–101)**.

The rate of drug release, $Q/t$, for a polymer-controlled process can be calculated from the slope of a linear $Q$ versus $t$ plot and from equation **(11–102)** is seen to equal $C_p D_m/h_m$. Likewise, $Q/t$, for the diffusion-layer–controlled process, resulting from plotting $Q$ versus $t$, is found to be $C_s D_a/h_a$. Furthermore, a plot of the release rate, $Q/t$, versus $C_s$, the solubility of the drug in the surrounding medium, should be linear with a slope of $D_a/h_a$.

### EXAMPLE 11–11

**Progesterone Diffusion out of a Silastic Capsule**

The partition coefficient, $K_r = C_s/C_p$, of progesterone is 0.022; the solution diffusivity, $D_a$, is $4.994 \times 10^{-2}$ cm²/day; the silastic membrane diffusivity $D_m$, is $14.26 \times 10^{-2}$ cm²/day; the solubility of progesterone in the silastic membrane, $C_p$, is 513 $\mu$g/cm³; the thickness of the capsule membrane, $h_m$, is 0.080 cm, and that of the diffusion layer, $h_a$, as estimated by Chien, is 0.008 cm.

Calculate the rate of release of progesterone from the capsule and express it in $\mu$g/cm² per day. Compare the calculated result with

the observed value, $Q/t = 64.50$ $\mu$g/cm² per day. Using equation **(11–100)**, we obtain

$$\frac{Q}{t} = \frac{C_p K_r D_a D_m}{K_r D_a h_m + D_m h_a}$$

$$\frac{Q}{t} = \frac{\begin{array}{c}(513\ \mu\text{g/cm}^3)(0.022)(4.994 \times 10^{-2}\ \text{cm}^2/\text{day}) \\ \times (14.26 \times 10^{-2}\ \text{cm}^2/\text{day})\end{array}}{\begin{array}{c}(0.022)(4.994 \times 10^{-2}\ \text{cm}^2/\text{day})(0.008\ \text{cm}) \\ + (14.26 \times 10^{-2}\ \text{cm}^2/\text{day})(0.008\ \text{cm})\end{array}}$$

$$\frac{Q}{t} = \frac{0.08037}{0.00123} = 65.34\ \mu\text{g/cm}^2/\text{day}$$

In the example just given, (a) is $K_r D_a h_m$ much greater than $D_m h_a$ or (b) is $D_m h_a$ much greater than $KD_a h_m$? (c). What conclusion can be drawn regarding matrix or diffusion-layer control? First, we have

$$K_r D_a h_m = 8.79 \times 10^{-5};\ D_m h_a = 1.14 \times 10^{-3}$$

$$D_m h_a/(K_r D_a h_m + D_m h_a) = (1.14 \times 10^{-3})/[(8.79 \times 10^{-5})$$

$$+ (1.14 \times 10^{-3})] = 0.93$$

Therefore, $D_m h_a$ is much greater than $K_r D_a h_m$, and the system is 93% under aqueous diffusion-layer control. It should thus be possible to use the simplified equation **(11–103)**:

$$\frac{Q}{t} = \frac{K_r D_a C_p}{h_a} = \frac{(0.022)(4.994 \times 10^{-2})(513)}{0.008} = 70.45\ \mu\text{g/cm}^2/\text{day}$$

Although $D_m h_a$ is larger than $K_r D_a h_m$ by about one order of magnitude (i.e., $D_m h_a/KD_a h_m = 13$), it is evident that a considerably better result is obtained by using the full expression, equation **(11–100)**.

### EXAMPLE 11–12

**Contraceptive Release from Polymeric Capsules**

Two new contraceptive steroid esters, A and B, were synthesized, and the parameters determined for release from polymeric capsules are as follows[78]:

| | $K_r$ | $D_a$ (cm²/day) | $D_m$ (cm²/day) | $C_p$ ($\mu$g/cm³) | $h_a$(cm) | $Q/t$(obs) ($\mu$g/cm²/ day) |
|---|---|---|---|---|---|---|
| A | 0.15 | $25 \times 10^{-2}$ | $2.6 \times 10^{-2}$ | 100 | 0.008 | 24.50 |
| B | 0.04 | $4.0 \times 10^{-2}$ | $3.0 \times 10^{-2}$ | 85 | 0.008 | 10.32 |

Using equation **(11–100)** and the quantities given in the table, calculate values of $h_m$ in centimeter for these capsule membranes. First, we write

$$\frac{Q}{t} = \frac{C_p K_r D_a D_m}{K_r D_a h_m + D_m h_a}$$

$$\frac{Q}{t}(K_r D_a h_m + D_m h_a) = C_p K_r D_a D_m$$

$$\frac{Q}{t}(K_r D_a h_m) = C_p K_r D_a D_m - D_m h_a(Q/t)$$

$$h_m = \frac{C_p K_r D_a D_m - D_m h_a(Q/t)}{(Q/t) K_r D_a}$$

For capsule A,

$$h_m = \frac{(100)(0.15)(25 \times 10^{-2})(2.6 \times 10^{-2}) - (2.6 \times 10^{-2})(0.008)(24.50)}{(24.50)(0.15)(25 \times 10^{-2})}$$

$$h_m = \frac{0.0924}{0.9188}\ \text{cm} = 0.101\ \text{cm}$$

---

*Contrary to convention, the partition coefficient, $K_r$, in equation **(11–101)** is defined as the water/lipid distribution coefficient. Thus, $K_r$ decreases here as the compound becomes more hydrophobic. The $K$ as used previously, for example, in equations **(11–19)**, **(11–30)**, and **(11–36)**, is the lipid/water distribution coefficient such that $K$ increases with increasing hydrophobicity of the compound. $C_m$ varies with time and eventually equals $C_p$. Therefore, the equilibrium partition coefficient does not involve $C_m$, but, rather, the equilibrium values $C_p$ and $C_s$.

Note that all units cancel except centimeter in the equation for $h_m$. The reader should carry out the calculations for compound B. (*Answer*: 0.097 cm.)

# FICK'S SECOND LAW AS A STARTING POINT

Fick's first law, equation **(11–2)**, has been used throughout this chapter as a starting point in the development of equations to describe the diffusion of drugs through natural and polymeric membranes. However, there are many diffusion problems in which the first law of Fick is not applicable, and the second law, equation **(11–6)**,

$$\frac{\partial C}{\partial t} = D\frac{\partial^2 c}{\partial x^2} \qquad (11\text{–}104)$$

must be used. Here we use $u$ instead of $C$ to express concentration. The symbol $\partial$ indicates that *partial derivatives* are being used because $u$ is a function of both $t$ and $x$. The second law is used to express diffusion in cylinders and spheres as well as through flat plates. The simplest form of the second-law diffusion equation is

$$\frac{\partial u}{\partial t} = D\left(\frac{\partial^2 u}{\partial r^2} + \frac{1}{r}\frac{\partial u}{\partial r}\right) \qquad (11\text{–}105)$$

for symmetric diffusion outward from the axis of a cylinder of radius $r$.

For diffusion proceeding symmetrically about the center of a sphere of radius $r$, the partial differential equation representing Fick's second law in its simplest form is

$$r\frac{\partial u}{\partial t} = D\left(r\frac{\partial^2 u}{\partial r^2} + 2\frac{\partial u}{\partial r}\right) \qquad (11\text{–}106)$$

The equations for diffusion in cylinders and spheres are discussed by Crank[81] and Jacobs.[82]

Although the derivation of equations based on Fick's second law is in most cases beyond the mathematical scope of this book, it is of value to present some equations and obtain their solutions. Such exercises give the student practice in calculations for diffusion problems that are more complicated than those derived from Fick's first law.

## Diffusion in a Closed System

### Determination of *D*

A simple apparatus (**Fig. 11–27**) was used by Graham (1861), one of the pioneers in diffusion studies, to obtain the diffusion coefficient, $D$, for solutes in various solvents. The coefficients for some solutes diffusing through various media are listed in **Table 11–2**. In the apparatus depicted in **Figure 11–27**, the height of the solution is $h$, the combined height of solution and solvent is $H$, and the distance traversed by the solute is $x$. The concentration of solute at a position $x$ and time $t$ in the solution is $u$ and its initial concentration is $u_0$. From the experimental values of $u$, $x$, and $t$, it is possible to determine the diffusion coefficient, $D$, for the solute in the solvent.

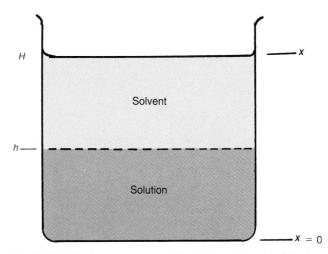

**Fig. 11–27.** Simple apparatus used by Graham for early diffusion studies. (From M. H. Jacobs, *Diffusion Processes*, Springer-Verlag, New York, 1976, p. 24. With permission.)

Initially—that is, at time $t = 0$ sec—the concentration $u$ is equal to $u_0$ (moles or grams per $cm^3$) in the cell from position $x = 0$ to $x = h$ (cm) and $u = 0$ from $x = h$ to $x = H$. These statements are known as *initial conditions*. In a case in which $h$ is taken equal to be equal to $H/2$, that is, both solution and solvent are of equal volume, the equation for $u$ is[82]

$$u = \frac{u_0}{2} + \frac{2u_0}{\pi}\left[\left(\cos\frac{\pi x}{H}\right)e^{-\pi^2 Dt/H^2}\right.$$
$$\left. - \frac{1}{3}\left(\cos\frac{3\pi x}{H}\right)e^{-9\pi^2 Dt/H^2} + \cdots\right] \qquad (11\text{–}107)$$

Equation **(11–107)** is simplified if we choose $x$, the position of sampling in the cell, to be $H/6$; the second cosine term in the parenthesis of equation **(11–107)** becomes $\cos(\pi/2) = \cos 90° = 0$. This leaves only the first cosine term, $\cos(\pi/6) = \cos 30° = 0.866$. Thus, taking $x = H/6$, we have

$$u = \frac{u_0}{2} + \frac{2u_0}{\pi}\left(0.866 e^{-\pi^2 Dt/H^2}\right) \qquad (11\text{–}108)$$

Recall that with trigonometric functions such as $\cos(\pi/6)$, $\pi$ is given in degrees, that is, $\pi = 180°$ and $\pi/6 = 30°$, whereas in terms such as $2u_0/\pi$ and $e^{-\pi^2 Dt/H^2}$, the value of $\pi$ is $3.14159\ldots$ .

## EXAMPLE 11–13

**Determination of an Aqueous Diffusion Coefficient**

A new water-soluble drug, corazole, is placed in a Graham diffusion cell (see Fig. 11–27) at an initial concentration of $u_0 = 0.030$ mmole/$cm^3$ to determine its diffusion coefficient in water at 25°C. The height of the solution, $h$, in the cell is 2.82 cm and the total height of aqueous solution and overlying water, $H$, is 5.64 cm. A sample is taken at a depth of $x = H/6$ cm at time, $t$, of 4.3 hr (15,480 sec) and is found by spectrophotometric analysis to have a

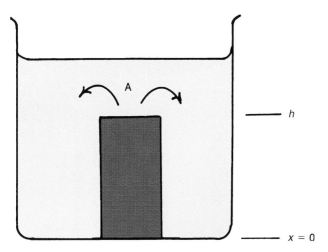

**Fig. 11–28.** Diffusion apparatus with one open and one closed boundary. (From M. H. Jacobs, *Diffusion Processes*, Springer-Verlag, New York, 1976, p. 47. With permission.)

concentration, $u$, of 0.0225 mmole/cm$^3$. $D$ is obtained by rearranging equation (11–108):

$$D = -\left[\ln\left(\frac{u - u_0/2}{0.866(2u_0/\pi)}\right)\right] \cdot \frac{H^2}{\pi^2 t}$$

$$= (-0.79113)\frac{(-31.8096)}{(9.8696)(15480)}$$

$$= 16.47 \times 10^{-5} \text{ cm}^2/\text{sec}$$

## Diffusion in Systems with One Open Boundary

The Graham cell for the determination of diffusion coefficients is an example of a closed system. In pharmaceutics, physiology, and biochemistry, systems with one or two open boundaries are of more interest than the closed-boundary system. In 1850, Graham introduced a system with one open and one closed boundary, as shown in **Figure 11–28**. Insignificant mixing occurs between the solution and the water because of differences in density. The condition at the interface between the solution and the water layer, known as a *boundary condition*, is expressed as "$u = 0$ when $x = h$." A second boundary condition states that the change in concentration, $u$, with the change in position, $x$, is zero, or, in mathematical notation, $\partial u/\partial x = 0$. This occurs at the bottom of the cell, because the solute cannot pass out through the bottom. In addition to the two boundary conditions, it is useful to specify an *initial condition*, as was done for the closed cell treated earlier. The initial condition is often taken as uniformity of concentration within the solution in the inner vessel of the cell, that is, $u = u_0$ at $t = 0$.

For a system with one open and one closed surface, the amount, $M_{0,t}$, of solute escaping between time 0 and time $t$ is expressed by the equation[82]*

$$M_{0,t} = u_0 Ah \left[1 - \frac{8}{\pi^2}\left(e^{-\pi^2 Dt/4h^2} + \frac{1}{9}e^{-9\pi^2 Dt/4h^2} + \cdots\right)\right]$$

(11–109)

*We use the symbol $M$ to replace Jacobs's term $Q$.

where $A$ is the cross-sectional area of the inner cell of height $h$ (see **Fig. 11–28**), and the other terms have been defined in connection with equations (11–107) and (11–108).

**Drug Diffusion from an Open Boundary**

Calculate the total amount, $M_{0,t}$, of the new drug corazole that escapes between times $t = 0$ and $t = 2.70$ hr (9720 sec) from the cell with one open boundary (Fig. 11–28). The area, $A$, of the cell is 8.27 cm$^2$ and its height, $h$, is 2.65 cm. The original concentration, $u_0$, of the drug in the cell is 0.0437 g/cm$^3$. The total amount of drug, $M$, in the cell is the concentration in g/cm$^3$ multiplied by $A \times h$, the volume of the cell: 0.0437 g/cm$^3$ × 8.27 cm$^2$ × 2.65 cm = 0.9577 g. The diffusion coefficient, $D$, of the drug corazole in water at 25°C is $16.5 \times 10^{-5}$ cm$^2$/sec, as found in *Example 11–13*.

Inserting these values into equation (11–109) yields

$$M_{0,t} = (0.0437 \text{ g/cm}^3 \times 827 \text{ cm}^2 \times 2.65 \text{ cm})$$

$$\times \left[1 - \frac{8}{\pi^2}\left(e^{-\frac{\pi^2(16.5\times10^{-5} \text{ cm}^2/\text{sec})(9720\,\text{sec})}{4\times7.0225\text{ cm}^2}}\right.\right.$$

$$\left.\left. + \frac{1}{9}e^{-\frac{9\pi^2(16.5\times10^{-5} \text{ cm}^2/\text{sec})(9720\ \text{sec})}{4\times7.0225\text{ cm}^2}} + \cdots\right)\right].$$

$$M_{0,t} = 0.9577 \times 0.53805 = 0.51529 \text{ g}$$

Thus, we arrive at the result that in a cell containing 0.0437 g/cm$^3$ or 0.9577 g of total drug, 0.5153 g diffuses out in 2.7 hr.

The diffusion of macromolecules, such as proteins, is discussed in the chapter on colloids.

## Osmotic Drug Release[11]

Osmotic drug release systems use osmotic pressure as driving force for the controlled delivery of drugs. A simple osmotic pump consists of an osmotic core containing drug with or without an osmotic agent coated with a semipermeable membrane. The semipermeable membrane has an orifice for drug release from the pump. The dosage form, after coming in contact with aqueous fluids, imbibes water at a rate determined by the fluid permeability of the membrane and osmotic pressure of core formulation. This osmotic imbibition of water results in high hydrostatic pressure inside the pump, which causes the flow of the drug solution through the delivery orifice. A lag time of 30 to 60 min is observed in most of the cases as the system hydrates. Approximately 60% to 80% of drug is released at a constant rate (zero order) from the pump.

The drug release rate from a simple osmotic pump can be described by the following mathematic equation:

$$dM/dt = AK/h(\Delta\pi - \Delta p)C \qquad (11\text{–}110)$$

where $dM/dt$ is drug release rate, $A$ is the membrane area, $K$ is the membrane permeability, $h$ is the membrane thickness, $\Delta\pi$ and $\Delta p$ are the osmotic and hydrostatic pressure differences between the inside and outside of the system, respectively, and $C$ is the drug concentration inside the pump (i.e., dispensed fluid). If the size of the delivery orifice is sufficiently large, the hydrostatic pressure inside the system is minimized and $\Delta\pi$ is much greater than $\Delta p$. When the osmotic pressure in an environment is negligible, such as the gastrointestinal fluids, as compared to that of core, $\pi$ can be safely substituted for $\Delta\pi$. Therefore, equation (11–110) can be simplified to

$$dM/dt = AK/h\,\pi C \qquad (11\text{–}111)$$

When all the parameters on the right-hand side of equation (11–111) remain constant, the drug release rate from an osmotic device is constant. This can be achieved by carefully designing the formulation and selecting the semipermeable membrane to achieve a saturated drug solution inside the pump so that $\pi$ and $C$ remain constant.

Drug release from osmotic systems is governed by various formulation factors such as the solubility and osmotic pressure of the core component(s), size of the delivery orifice, and nature of the rate-controlling membrane.

## Solubility

The kinetics of osmotic drug release is directly related to the solubility of the drug within the core. Assuming a tablet core of pure drug, we find the fraction of the core released with zero-order kinetics from.

$$F(z) = 1 - S/\rho \qquad (11\text{–}112)$$

where $F(z)$ is the fraction released by zero-order kinetics, $S$ is the drug's solubility (g/mL), and $\rho$ is the density (g/mL) of the core tablet. Drugs with low solubility ($\leq 0.05$ g/mL) can easily reach saturation and would be released from the core through zero-order kinetics. However, according to equation (11–112), the zero-order release rate would be slow due to the small osmotic pressure gradient and low drug concentration. Conversely, highly water-soluble drugs would demonstrate a high release rate that would be zero order for a small percentage of the initial drug load. Thus, the intrinsic water solubility of many drugs might preclude them from incorporation into an osmotic pump. However, it is possible to modulate the solubility of drugs within the core and thus extend this technology to the delivery of drugs that might otherwise have been poor candidates for osmotic delivery.

## Osmotic Pressure

Osmotic pressure, like vapor pressure and boiling point, is a colligative property of a solution in which a nonvolatile solute is dissolved in a volatile solvent. The osmotic pressure of a solution is dependent on the number of discrete entities of solute present in the solution. From equation (11–111), it is evident that the release rate of a drug from an osmotic system is directly proportional to the osmotic pressure of the core formulation. For controlling drug release from these systems, it is important to optimize the osmotic pressure gradient between the inside compartment and the external environment. It is possible to achieve and maintain a constant osmotic pressure by maintaining a saturated solution of osmotic agent in the compartment. If a drug does not possess sufficient osmotic pressure, an osmotic agent can be added to the formulation.

## Delivery Orifice

Osmotic delivery systems contain at least one delivery orifice in the membrane for drug release. The size of the delivery orifice must be optimized to control the drug release from osmotic systems. If the size of delivery orifice is too small, zero-order delivery will be affected because of the development of hydrostatic pressure within the core. This hydrostatic pressure may not be relieved because of the small orifice size and may lead to deformation of the delivery system, thereby resulting in unpredictable drug delivery. On the other hand, the size of the delivery orifice should not be too large, for otherwise solute diffusion from the orifice may take place. To optimize the size of the orifice, we can use the equation

$$A_s = 8\pi(LV/t)(\eta/\Delta P)^{1/2} \qquad (11\text{–}113)$$

where $A_s$ is the cross-sectional area, $\pi = 3.14\ldots$, $L$ is the diameter of the orifice, $V/t$ is the volume release per unit time, $\eta$ is the viscosity of the drug solution, and $\Delta P$ is the difference in hydrostatic pressure.

## Semipermeable Membrane

The choice of a rate-controlling membrane is an important aspect in the formulation development of oral osmotic systems. The semipermeable membrane should be biocompatible with the gastrointestinal tract. The membrane should also be water permeable and provide effective isolation from the dissolution process in the gut environment. Therefore, drug release from osmotic systems is independent of the pH and agitational intensity of the gastrointestinal tract. To ensure that the coating is able to resist the pressure within the device, the thickness of the membrane is usually kept between 200 and 300 $\mu$m. Selecting membranes that have high water permeability can ensure high hydrostatic pressure inside the osmotic device and hence permit rapid drug release flow through the orifice.

In summary, designing a drug with suitable solubility and selecting a semipermeable membrane with favorable water permeability and orifice size are the key factors for ensuring a sustained and constant drug release rate through an osmotic drug delivery system.

### EXAMPLE 11–15

**Osmotic Release of Potassium Chloride**

Five hundred mg of potassium chloride was pressed into 0.25 mL of water; the semipermeable membrane thickness is 0.025 cm with an area of 2.2 cm². The drug solubility is 330 mg/mL. The density of the solution is 2 g/mL. Here $K\pi = 0.686 \times 10^{-3}$ cm²/hr, and the diffusion coefficient, $D$, is $0.122 \times 10^{-3}$ cm²/hr. What is the release rate of potassium chloride in this osmotic delivery system?[83]

Assuming the osmotic pressure is the main driving force of the system, we obtain, using equation (11–111),

$$dM/dt = A/h\,K\,\pi\,C = (2.2\ \text{cm}^2/0.025\ \text{cm})(0.686 \times 10^{-3}\ \text{cm}^2/\text{hr})$$
$$(330\ \text{mg/mL}) = 19.92\ \text{mg/hr}$$

Correcting for the contribution of diffusion, we obtain

$$dM/dt = A/hC(K\pi + D) = (2.2\ \text{cm}^2/0.025\ \text{cm})$$
$$(330\ \text{mg/mL})(0.686 \times 10^{-3}\ \text{cm}^2/\text{hr} + 0.122$$
$$\times 10^{-3}\ \text{cm}^2/\text{hr}) = 23.5\ \text{mg/hr}$$

# CHAPTER SUMMARY

The fundamentals of diffusion were discussed in this chapter. Free diffusion of substances through liquids, solids, and membranes is a process of considerable importance to the pharmaceutical sciences. A fundamental understanding of the processes of dialysis, osmosis, and ultrafiltration is essential for pharmaceutical sciences. The mechanisms of transport in pharmaceutical systems were described in some detail. Fick's laws of diffusion were also defined and their application described. Important parameters such as diffusion coefficient, permeability, and lag time were discussed and sample calculations were performed to illustrate their use. The various driving forces behind diffusion, drug absorption, and elimination were described as well as elementary drug diffusion. Although many of the treatments in this chapter appear to be highly mathematical because of the extensive use of equations, the equations and their derivations are useful as the student learns about these important pharmaceutical processes at the mechanistic level.

 **Practice problems for this chapter can be found at thePoint.lww.com/Sinko6e.**

# References

1. V. T. Stannett, et al., in *Advances in Polymer Science,* **Vol. 32,** Springer-Verlag, Berlin, 1979, pp. 71–122; D. R. Paul and G. Morel, in R. E. Kirk and D. F. Othmer, *Encyclopedia of Chemical Technology,* 3rd Ed., **Vol. 15,** Wiley, New York, 1981, pp. 92–131.

2. S. T. Hwang and K. Kammermeyer, *Membranes in Separations*, Wiley-Interscience, New York, 1975, pp. 23–28.

3. E. L. Cussler, *Diffusion: Mass Transfer in Fluid Systems*, 2nd Ed., Cambridge University Press, New York, 1997.

4. I. Sivin, Drug Saf. **26,** 303, 2003.

5. J. C. Maxwell, Phil. Mag. **19,** 20, 1860.

6. D. A. Johnson and G. L. Amidon, J. Theor. Biol. **131,** 93, 1988.

7. E. M. Topp, in G. L. Amidon, P. I. Lee, and E. M. Topp (Eds.), *Transport Processes in Pharmaceutical Systems*, Marcel Dekker, New York, 2000.

8. R. B. Bird, W. E. Stewart, and E. N. Lightfoot, *Transport Phenomena*, Wiley, New York, 1960.

9. J. Crank, *The Mathematics of Diffusion*, 2nd Ed., Clarendon Press, Oxford, 1983.

10. D. E. Wurster and P. W. Tylor, J. Pharm. Sci. **54,** 169, 1965.

11. R. K. Verma et al., J. Control. Release **79,** 7, 2002.

12. J. Schramm and S. Mitragotri, Pharm. Res. **19,** 1673, 2002.

13. N. F. H. Ho and T. J. Roseman, J. Pharm. Sci. **68,** 1170, 1979.

14. C. S. Eskilsson and E. Bjorklund, J. Chromatogr. A, **902,** 227, 2000.

15. A. K. Banga, S. Bose, and T. K. Ghosh, Int. J. Pharm. **179,** 1, 1999.

16. P. Tyle, Pharm. Res. **3,** 318, 1986.

17. P. Gebauer, J. L. Beckers, and P. Bocek, Electrophoresis, **23,** 1779, 2002.

18. Yu and G. L. Amidon, in G. L. Amidon, P. I. Lee, and E. M. Topp (Eds.), *Transport Processes in Pharmaceutical Systems*, Marcel Dekker, New York, 2000, p. 29.

19. G. L. Flynn, S. H. Yalkowsky, and T. J. Roseman, J. Pharm. Sci. **63,** 479, 1974.

20. H. Carslaw, *Mathematical Theory of the Conduction of Heat*, Macmillan, New York, 1921.

21. H. Lennernas, L. Knutson, T. Knutson, A. Hussain, L. Lesko, T. Salmonson, and G. L. Amidon, Eur. J. Pharm. Sci. **15,** 271, 2002.

22. S. Winiwarter, N. M. Bonham, F. Ax, A. Hallberg, H. Lennernas, and A. Karlen. J. Med. Chem. **41,** 4939, 1998.

23. N. F. H. Ho, J. Y. Park, G. E. Amidon, et al., in A. J. Aguiar (Ed.), *Gastrointestinal Absorption of Drugs*, American Pharmaceutical Association, Academy of Pharmaceutical Sciences, Washington, D. C., 1981; W. I. Higuchi, N. F. H. Ho, J. Y. Park, and I. Komiya, in L. F. Prescott and W. S. Nimmo (Eds.), *Drug Absorption*, Adis Press, Balgowlah, Australia, 1981, p. 35; N. F. H. Ho, J. Y. Park, W. Morozowich, and W. I. Higuchi, in E. B. Roche (Ed.), *Design of Biopharmaceutical Properties Through Prodrugs and Analogs*, American Pharmaceutical Association, Academy of Pharmaceutical Sciences, Washington, D. C., 1977, Chapter 8.

24. A. J. Aguiar and M. A. Weiner. J. Pharm. Sci. **58,** 210, 1969.

25. A. W. Larhed, P. Artursson, J. Grasjo, and E. Bjork. J. Pharm. Sci. **86,** 660, 1997.

26. N. F. Ho, L. Suhardja, S. Hwang, E. Owada, A. Molokhia, G. L. Flynn, W. I. Higuchi, and J. Y. Park. J. Pharm. Sci. **65,** 1578, 1976.

27. M. Fujii, N. Hori, K. Shiozawa, K. Wakabayashi, E. Kawahara, and M. Matsumoto, Int. J. Pharm. **205,** 117, 2000.

28. S. Lang, B. Rothen-Rutishauser, J. C. Perriard, M. C. Schmidt, and H. P. Merkle, Peptides **19,** 599, 1998.

29. N. Takamatsu, O. N. Kim, L. S. Welage, N. M. Idkaidek, Y. Hayashi, J. Barnett, R. Yamamoto, E. Lipka, H. Lennernas, A. Hussain, A. Lesko, and G. L. Amidon, Pharm. Res. **18,** 742, 2001.

30. K. F. Farng and K. G. Nelson, J. Pharm. Sci. **66,** 1611, 1977.

31. R. Scheuplein, J. Invest. Dermatol. **45,** 334, 1965.

32. T. Higuchi, J. Soc. Cosm. Chem. **11,** 85, 1960.

33. H. Yu and P. J. Sinko, J. Pharm. Sci. **86,** 1448, 1997.

34. G. L. Flynn and S. H. Yalkowsky, J. Pharm. Sci. **61,** 838, 1972.

35. M. G. Karth, W. I. Higuchi, and J. L. Fox, J. Pharm. Sci. **74,** 612, 1985.

36. M. Z. Biber and C. T. Rhodes, J. Pharm. Sci. **65,** 564, 1976.

37. K. Nasim, N. C. Meyer, and J. Autian, J. Pharm. Sci. **61,** 1775, 1972.

38. T. Higuchi and A. Aguiar, J. Am. Pharm. Assoc. Sci. Ed. **48,** 574, 1959.

39. J. Crank and G. S. Park, *Diffusion in Polymers*, Academic Press, New York, 1968, p. 20.

40. D. E. Wurster, J. A. Ostrenga, and L. E. Matheson, Jr., J. Pharm. Sci. **68,** 1406, 1410, 1979.

41. Y. C. Martin, *Quantitative Drug Design*, Marcel Dekker, New York, 1978, p. 76.

42. W. J. Addicks, G. L. Flynn, and N. Weiner, Pharm. Res. **4,** 337, 1987.

43. W. J. Addicks, G. L. Flynn, N. Weiner, and C.-M. Chiang, Pharm. Res. **5,** 377, 1988.

44. G. M. Grass and S. A. Sweetana, Pharm. Res. **5,** 372, 1988.

45. I. J. Hidalgo, G. M. Grass, K. M. Hillgren, and R. T. Borchardt, In Vitro Cell. Dev. Biol. **28 A,** 578, 1992.

46. U. K. Walle and T. Walle, Drug Metab. Dispos, **26,** 343, 1998.

47. A. Sparreboom et al., Proc. Natl. Acad. Sci. USA **94,** 2031, 1997.

48. B. B. Brodie and C. A. M. Hogben, J. Pharm. Pharmacol. **9,** 345, 1957; P. A. Shore, B. B. Brodie, and C. A. M. Hogben, J. Pharmacol. Exp. Ther. **119,** 361, 1957; L. S. Shanker, D. J. Tocco, B. B. Brodie, and C. A. M. Hogben, J. Pharmacol. Exp. Ther. **123,** 81, 1958; C. A. M. Hogben, D. J. Tocco, B. B. Brodie, and L. S. Shanker, J. Pharmacol. Exp. Ther. **125,** 275, 1959; L. S. Shanker, J. Med. Chem. **2,** 343, 1960; L. S. Shanker, Annu. Rev. Pharmacol. **1,** 29, 1961.

49. R. H. Turner, C. S. Mehta, and L. Z. Benet, J. Pharm. Sci. **59,** 590, 1970.

50. J. L. Colaizzi and P. R. Klink, J. Pharm. Sci. **58,** 1184, 1969.

51. A. Suzuki, W. I. Higuchi, and N. F. H. Ho, J. Pharm. Sci. **59,** 644, 651, 1970; N. F. H. Ho, W. I. Higuchi, and J. Turi, J. Pharm. Sci. **61,** 192, 1972; N. F. H. Ho and W. I. Higuchi, J. Pharm. Sci. **63,** 686, 1974.

52. J. T. Doluisio, F. N. Billups, L. W. Dittert, E. T. Sugita, and V. J. Swintosky, J. Pharm. Sci. **58,** 1196, 1969.

53. A. K. Mitra and T. J. Mikkelson, J. Pharm. Sci. **77,** 771, 1988.

54. R. J. Scheuplein, J. Invest. Dermatol. **45,** 334, 1965; R. J. Scheuplein, J. Invest. Dermatol. **48,** 79, 1967; R. J. Scheuplein and I. H. Blank, Physiol. Rev. **51,** 702, 1971.

55. R. J. Scheuplein, I. H. Blank, G. J. Brauner, and D. J. MacFarlane, J. Invest. Dermatol. **52,** 63, 1969.

56. C. D. Yu, J. L. Fox, N. F. H. Ho, and W. I. Higuchi, J. Pharm. Sci. **68,** 1341, 1347, 1979; J. Pharm. Sci. **69,** 772, 1980; H. Durrheim, G. L. Flynn, W. I. Higuchi, and C. R. Behl, J. Pharm. Sci. **69,** 781, 1980; G. L. Flynn, H. Durrheim, and W. I. Higuchi, J. Pharm. Sci. **70,** 52, 1981; C. R. Behl, G. L. Flynn, T. Kurihara, N. Harper, W. Smith, W. I. Higuchi,

N. F. H. Ho, and C. L. Pierson, J. Invest. Dermatol. **75,** 346, 1980.

57. P. V. Raykar, M.-C. Fung, and B. D. Anderson, Pharm. Res. **5,** 140, 1988.

58. J. S. Turi, D. Danielson, and W. Wolterson, J. Pharm. Sci. **68,** 275, 1979.

59. S. H. Yalkowsky and G. L. Flynn, J. Pharm. Sci. **63,** 1276, 1974.

60. J. Ostrenga, C. Steinmetz, and B. Poulsen, J. Pharm. Sci. **60,** 1175, 1180, 1971.

61. K. B. Sloan, K. G. Siver, and S. A. M. Koch, J. Pharm. Sci. **75,** 744, 1986.

62. H. Komatsu and M. Suzuki, J. Pharm. Sci. **68,** 596, 1979.

63. A. H. Beckett and A. C. Moffat, J. Pharm. Pharmacol. **20,** 239 S, 1968.

64. N. F. H. Ho and W. I. Higuchi, J. Pharm. Sci. **60,** 537, 1971.

65. A. Suzuki, W. I. Higuchi, and N. F. H. Ho, J. Pharm. Sci. **59,** 644, 651, 1970.

66. G. L. Flynn, N. F. H. Ho, S. Hwang, E. Owada, A. Motokhia, C. R. Behl, W. I. Higuchi, T. Yotsuyanagi, Y. Shah, and J. Park, in D. R. Paul and F. W. Harris (Eds.), *Controlled Release Polymeric Formulations,* American Chemical Society, Washington, D. C., 1976, p. 87.

67. S. K. Chandrasekaran, H. Benson, and J. Urquhart, in J. R. Robinson (Ed.), *Sustained and Controlled Release Drug Delivery Systems,* Marcel Dekker, New York, 1978, p. 572.

68. T. Yotsuyanagi, A. Molakhia, S. Hwang, N. F. H. Ho, C. L. Flynn, and W. I. Higuchi, J. Pharm. Sci. **64,** 71, 1975.

69. S. Hwang, E. Owada, T. Yotsuyanagi, L. Suhardja, Jr., N. F. H. Ho, G. L. Flynn, W. I. Higuchi, and J. V. Park, J. Pharm. Sci. **65,** 1574, 1578, 1976.

70. H. Okada, I. Yamazaki, Y. Ogawa, S. Hirai, T. Yashiki, and H. Mima, J. Pharm. Sci. **71,** 1367, 1982; H. Okada, I. Yamazaki, T. Yashiki, and H. Mima, J. Pharm. Sci. **72,** 75, 1983; H. Okada, T. Yashiki, and H. Mima, J. Pharm. Sci. **72,** 173, 1983; H. Okada, I. Yamazaki, T. Yashiki, T. Shimamoto, and H. Mima, J. Pharm. Sci. **73,** 298, 1984.

71. J. R. Robinson, *Sustained and Controlled Release Drug Delivery Systems,* Marcel Dekker, New York, 1978.

72. R. W. Baker and H. K. Lonsdale, in A. C. Tanquary and R. E. Lacey (Eds.), *Controlled Release of Biologically Active Agents,* Plenum Press, New York, 1974, p. 30.

73. T. Higuchi, J. Pharm. Sci. **50,** 874, 1961.

74. S. J. Desai, A. P. Simonelli, and W. I. Higuchi, J. Pharm. Sci. **54,** 1459, 1965.

75. S. J. Desai, P. Singh, A. P. Simonelli, and W. I. Higuchi, J. Pharm. Sci. **55,** 1224, 1230, 1235, 1966.

76. F. Bottari, G. DiColo, E. Nannipieri, et al., J. Pharm. Sci. **63,** 1779, 1974; J. Pharm. Sci. **66,** 927, 1977.

77. D. R. Paul and S. K. McSpadden, J. Memb. Sci. **1,** 33, 1976; S. K. Chandrasekaran and D. R. Paul, J. Pharm. Sci. **71,** 1399, 1982.

78. Y. W. Chien, in J. R. Robinson (Ed.), *Sustained and Controlled Release Drug Delivery Systems,* Marcel Dekker, New York, 1978, Chapter 4; Y. W. Chien, Chem. Pharm. Bull. **24,** 147, 1976.

79. C. K. Erickson, K. I. Koch, C. S. Metha, and J. W. McGinity, Science **199,** 1457, 1978; J. W. McGinity, L. A. Hunke, and A. Combs, J. Pharm. Sci. **68,** 662, 1979.

80. Y. W. Chien, *Novel Drug Delivery Systems: Biomedical Applications and Theoretical Basis,* Marcel Dekker, New York, 1982, Chapter 9.

81. J. Crank, *The Mathematics of Diffusion,* 2nd Ed., Oxford University Press, Oxford, 1975, Chapters 5 and 6.

82. M. H. Jacobs, *Diffusion Processes,* Springer-Verlag, New York, 1967, pp. 43, 49, equation 39.

83. F. Theeuwes, J. Pharm. Sci. **64,** 1987, 1975.

84. G. L. Amidon, P. I. Lee, and E. M. Topp (Eds.), *Transport Processes in Pharmaceutical Systems,* Marcel Dekker, New York, 2000, p. 13.

## Recommended Readings

J. Crank, *The Mathematics of Diffusion,* 2nd Ed., Clarendon Press, Oxford, 1983.

E. L. Cussler, *Diffusion: Mass Transfer in Fluid Systems,* 2nd Ed., Cambridge University Press, New York, 1997.

G. L. Flynn, S. H. Yalkowsky, and T. J. Roseman, J. Pharm. Sci. **63,** 479, 1974.

CHAPTER LEGACY

**Fifth Edition:** published as Chapter 12 (Diffusion). Updated by Patrick Sinko.

**Sixth Edition:** published as Chapter 11 (Diffusion). Updated by Patrick Sinko.

# 12 BIOPHARMACEUTICS

The purpose of this chapter is to provide the student with a biopharmaceutical foundation for studying the contemporary pharmaceutical sciences. The intended audience is predoctoral pharmacy (PharmD) and pharmaceutical science (PhD) graduate students. This information will be valuable to the practicing pharmacist because he or she is in a unique position to integrate and interpret the vast amounts of biologic, chemical, and physical information regarding drugs and drug products. The pharmacist can then convey practical advice to patients regarding the potential for drug interactions and for managing complex multidrug-treatment regimens. The pharmaceutical scientist also requires an understanding of this subject matter because of its emerging importance in drug discovery and development, including basic and applied research as well as in fields such as regulatory affairs.

The pharmaceutical sciences have been undergoing a revolution of sorts over the last two decades. A transition has occurred from focusing solely on the physical aspects of pharmacy such as dissolution, solubility, and compaction physics to the integration of these important disciplines with the biopharmaceutical sciences. First, in the 1970s and 1980s, there was an explosion of activity in the discovery and characterization of the cytochromes P-450 (CYP450s), an important group of enzymes responsible for metabolizing many endogenous and exogenous substances,[1] including 40% to 50% of all medications.[2] CYP450s transform drugs and other xenobiotics into more hydrophilic substances in order to facilitate their elimination from the body. More recently, it has become increasingly recognized that membrane transporters play an important role in the absorption and elimination of drugs. Drug transporters are membrane-spanning proteins that facilitate the movement of endogenous or exogenous molecules across biologic membranes. Included in this

broad category are proteins involved in active (i.e., energy- or adenosine triphosphate [ATP]-dependent) transport, facilitated transport, and ion channels. Membrane transporters and metabolizing enzymes are found throughout the human body and in all organs involved in the absorption and disposition of drugs. Whereas membrane transporters facilitate the movement of drugs and their metabolites into and out of specific organs, groups of organs make up subsystems in the human body with discrete functions. For example, "enterohepatic cycling" involves the movement of drugs from the intestine into the liver, back out into the bile, and then back into the intestine.

Although the complex interplay between these systems evolved for various physiologic purposes, the impact on drug–blood levels and the resulting therapeutic effect can be quite significant. We will explore the basic biopharmaceutical foundation for these complex systems and present the practical implications for pharmacotherapy throughout the chapter.

In an editorial in *Molecular Pharmaceutics*,[3] G. L. Amidon concisely reflected the sentiment that was the basis for preparing this chapter: "Traditional scientific endeavors in drug delivery and drug product development have been rather phenomenological, more descriptive, and somewhat based on trial and error. That has been primarily due to a lack of tools and our limited understanding of the mechanisms involved at a cellular and molecular level. The rapid advances in the field of biological sciences, cell and molecular biology, and genomics and proteomics, in particular, have penetrated more than just the drug discovery phase of the pharmaceutical sciences. They are now rapidly changing the views and strategies in pharmaceutics and the pharmaceutical development sciences. While the impact is particularly evident in the membrane transporter and metabolism fields, advances in the physical and material sciences, in parallel, are altering the pharmaceutical properties of drug candidates and delivery systems in new and innovative ways. The computational tool of bioinformatics, molecular property, biopharmaceutical property, and metabolism estimation have advanced rapidly in the past decade and are now having a significant impact on drug discovery and drug development. Traditional pharmacokinetics did not have molecular tools for understanding membrane transport and metabolism, or the effect of genetic polymorphism. Prodrug design and synthesis could not readily consider where and what enzymes convert the prodrug to the active drug. Receptors were primarily investigated for designing new chemical entities until molecular pharmaceutical scientists recognized that ligand-receptor interactions could be used to target drug to the receptor-expressing cells. Moreover, traditional dosage forms have not had to deal with high-molecular-weight 'biopharmaceuticals,' which generally have more complex pharmaceutical properties and sites of action hidden deep within the target cells." In this chapter, the student will be introduced to the biopharmaceutical considerations of the pharmaceutical sciences.

# FUNDAMENTALS

## Absorption, Distribution, Metabolism, and Excretion

The molecular processes, tissues, and organs that control the absorption and disposition of drugs form the basis for the study of biopharmaceutics. The two primary processes relate to input into the body, that is, absorption and output from the body (i.e., disposition).

*Absorption* relates to the mechanisms of drug input into the body and into a tissue or an organ within the body. *Disposition* can be broken down into distribution and elimination. After a drug enters the systemic circulation, it is distributed to the body's tissues. *Distribution* depends on many factors, including blood perfusion, cell membrane permeability, and tissue binding. The penetration of a drug into a tissue depends on the rate of blood flow to the tissue, partition characteristics between blood and tissue, and tissue mass. When entry and exit rates are the same, distribution equilibrium between blood and tissue is reached. It is reached more rapidly in richly vascularized areas than in poorly perfused areas unless diffusion across membrane barriers is the rate-limiting step. After equilibrium is attained, bound and unbound drug concentrations in tissues and in extracellular fluids are reflected by plasma concentrations. *Elimination* relates to the chemical transformation and/or physical removal of drug from the body. Hence, elimination is the sum of the processes related to drug loss from the body, namely, *metabolism* and *excretion*. Metabolism and excretion occur simultaneously with distribution, making the process dynamic and complex. Excretion is the process by which a drug or metabolite is removed from the body without further chemical modification. Three primary routes of excretion occur through the bile (i.e., biliary excretion), intestine, and kidney (i.e., renal excretion).

## Terminology

The following definitions are taken primarily from a variety of sources, which are indicated immediately after the defined term. In some cases, the original source has an expanded discussion on the topic, so the student is encouraged to utilize these sources for additional insights.

### Bioavailability[4]

According to the U.S. Food and Drug Administration, bioavailability describes the rate and extent to which the active drug ingredient is absorbed from a drug product and becomes available at the site of drug action. Because pharmacologic response is generally related to the concentration of drug at the site of drug action, the availability of a drug from a dosage form is a critical element of a drug product's clinical efficacy. However, drug concentrations usually cannot be readily measured directly at the site of action. Therefore, most bioavailability studies involve the determination of drug concentration in the blood or urine. This is based on the premise that the drug at the site of action is in equilibrium with drug in the blood. This does not mean that the drug concentrations in blood and tissues are equal. Instead, it assumes that equilibrium is maintained and that blood concentrations are proportional to tissue and active-site concentrations. It is therefore possible to obtain an indirect measure of drug response by monitoring drug levels in the blood or urine. Thus, bioavailability is concerned with how quickly and how much of a drug appears in the blood after a specific dose is administered. The bioavailability of a drug product often determines the therapeutic efficacy of that product because it affects the onset, intensity, and duration of therapeutic response of the drug.

In most cases, one is concerned with the extent of absorption of drug (i.e., the fraction of the dose that actually reaches the bloodstream) because this represents the "effective dose" of a drug. This is generally less than the amount of drug that is actually administered in the dosage form.

In some cases, notably those where acute conditions are being treated, one is also concerned with the rate of absorption of a drug because rapid onset of pharmacologic action

---

**KEY CONCEPT  BIOPHARMACEUTICAL PROCESS: ADME**

- Absorption
- Disposition
  - Distribution
  - Elimination
    - Metabolism
    - Excretion

# KEY CONCEPT BIOAVAILABILITY VARIABILITY AND THERAPEUTIC INDEX

Therapeutic problems (e.g., toxicity, lack of efficacy) are encountered most frequently during long-term therapy when a patient who is stabilized on one formulation is given a nonequivalent substitute. This is well known for drugs like digoxin or phenytoin. Sometimes therapeutic equivalence may be achieved despite differences in bioavailability. For example, the therapeutic index (ratio of the maximum tolerated dose to the minimum effective dose) of amoxicillin is so wide that moderate blood concentration differences due to bioavailability differences in amoxicillin products may not affect therapeutic efficacy or safety. In contrast, bioavailability differences are important for a drug with a relatively narrow therapeutic index (e.g., digoxin).

is desired. Food can slow drug absorption and result in lower blood levels. This is particularly important for drugs that depend on certain levels for maximum effectiveness. Good examples of this are antibiotics that need to achieve minimum inhibitory concentrations to be effective. Conversely, there are instances where a slower rate of absorption is desired, either to avoid adverse effects or to produce a prolonged duration of action.

## Absolute Bioavailability[4]

"Absolute" bioavailability, $F$, is the fraction of an administered dose that actually reaches the systemic circulation and ranges from $F = 0$ (i.e., no drug absorption) to $F = 1$ (i.e., complete drug absorption). Because the total amount of drug reaching the systemic circulation is directly proportional to the area under the curve (AUC) of plasma drug concentration versus time, $F$ is determined by comparing the respective AUCs of the test product and the drug administered intravenously.

Bioequivalence[5] refers to chemical equivalents (i.e., drug products that contain the same compound in the same amount and that meet current official standards) that, when administered to the same person in the same dosage regimen, result in equivalent concentrations of drug in blood and tissues. Therapeutic equivalence refers to drug products that, when administered to the same person in the same dosage regimen, provide essentially the same therapeutic effect or toxicity. Bioequivalent products are expected to be therapeutically equivalent.

## Genotype[6]

Genotype is the "internally coded, inheritable information" carried by all living organisms. This stored information is used as a "blueprint" or set of instructions for building and maintaining a living creature. These instructions are found within almost all cells (hence the word "internal" in the definition), are written in a coded language (the genetic code), are copied at the time of cell division or reproduction, and are passed from one generation to the next ("inheritable"). These instructions are intimately involved with all aspects of the life of a cell or an organism. They control everything from the formation of protein macromolecules to the regulation of metabolism and synthesis.

## Membrane Permeability

Membrane permeability relates to the velocity with which a drug molecule moves across a membrane. The units of mea-

surement for permeability are distance per time (e.g., cm/sec). Permeability is inversely related to the resistance of transport across membranes or tissues. Therefore, the higher the permeability, the lower is the resistance to movement across the membrane. Drugs can permeate membranes by passive diffusion through the cell membrane or between cells and by using transporters that "carry" drugs across the membrane. Passive permeability across membranes is determined by the solubility of the permeating molecule in the membrane, diffusion across the membrane into the cell, and the thickness of the barrier. This is covered in more detail in Chapter 13.

Permeability as it relates to drug transporters is covered later in this chapter.

## Phenotype[6]

Phenotype is the "outward, physical manifestation" of the organism. These are the physical parts, the sum of the atoms, molecules, macromolecules, cells, structures, metabolism, energy utilization, tissues, organs, reflexes, and behaviors: anything that is part of the observable structure, function, or behavior of a living organism. Rogers et al.[7] defined phenotype as it relates to drug metabolism: "Phenotype is the observed characteristic (as influenced by dietary intake and environmental exposure) of a patient's enzyme activity, and includes such designations as 'poor metabolizer,' 'intermediate metabolizer,' 'extensive metabolizer,' and 'ultrarapid extensive metabolizer.'" Patients who express dysfunctional or inactive enzymes are considered poor metabolizers.[8] Prodrugs, which require biotransformation to an active metabolite to elicit a therapeutic effect, are often not effective in these patients.

## EXAMPLE 12–1

### Codeine and Analgesia

Drug toxicity can be observed in patients who are poor metabolizers because of impaired clearance of medications requiring biotransformation for elimination and excretion. Intermediate metabolizers are patients who demonstrate decreased enzyme activity and have diminished drug metabolism.[2] Extensive metabolizers are patients who express enzymes that have normal activity,[9] in whom the anticipated medication response would be seen with standard doses of drugs. Ultrarapid extensive metabolizers are patients who have higher quantities of expressed enzymes because of gene duplication.[8] Normal doses of drugs in these patients may result in reduced or no efficacy (or toxicity with prodrugs) because of rapid metabolism.[8] In this example, codeine, a commonly used analgesic for postoperative pain, will be examined. It is well known that the quality of pain management varies among patients. Codeine is thought to be an effective analgesic because it is metabolized by the cytochrome P-450 2D6 (CYP2D6) pathway to morphine. This product is then quickly glucuronidated to morphine-3-glucuronide (M3G) and morphine-6-glucuronide (M6G), active analgesics. However, it has been shown that CYP2D6 is polymorphic in a number of different alleles, potentially causing a slow-metabolism phenotype, and may even be overexpressed (i.e., ultrarapid metabolizing phenotype) in certain people. This polymorphism results in patients having differing abilities to utilize that pathway for creating an active and effective analgesic. How common are these variations in drug metabolism? The frequency of poor metabolizers is 6% of the U.S. population, whereas only 1% of the Asian population is considered to be poor metabolizers. Similarly, the ultrarapid-metabolizing phenotype is also found worldwide, the greatest percentage being in Ethiopia (29%). In the case of codeine, a poor metabolizer will not receive as much pain management from a typical dose as a normal patient, and an ultrarapid-metabolizing patient may overdose at a similar dose.

### Relationship Between Genotype and Phenotype

The "internally coded, inheritable information," or genotype, carried by all living organisms holds the critical instructions that are used and interpreted by the cellular machinery of the cells to produce the "outward, physical manifestation," or phenotype, of the organism.[6]

### Relative Bioavailability[4]

"Relative" bioavailability refers to the availability of a drug product as compared to another dosage form or product of the same drug given in the same dose. These measurements determine the effects of formulation differences on drug absorption. The relative bioavailability of one product compared to that of another, both products containing the same dose of the same drug, is obtained by comparing their respective AUCs.

### Pharmacokinetics[4]

Pharmacokinetics is the mathematics of the time course of absorption, distribution, metabolism, and excretion (ADME) of drugs in the body. The biologic, physiologic, and physicochemical factors that influence the transfer processes of drugs in the body also influence the rate and extent of ADME of those drugs in the body. In many cases, the pharmacologic action and the toxicologic action are related to the plasma concentration of drugs. Through the study and application of pharmacokinetics, the pharmacist can individualize therapy for the patient.

### Pharmacodynamics

Pharmacodynamics is the study of the biochemical and physiologic effects of drugs and their mechanisms of action.

### Pharmacogenetics

Pharmacogenetics is the study of how genetic variations affect drug response.

Omics[10]. The burgeoning fields of genomics and proteomics are spawning multiple "omic" subdisciplines and related areas. The suffix generally refers to the study of a complete grouping or system of biomolecules, such as a genome, containing all of an organism's genes, or its proteome, containing all of its proteins. For example, genomics is the scientific study of a genome and the roles that genes play, alone and together, in directing growth and development and in controlling and determining biologic structure and function. As the field has grown, it has been broken down into several major branches. Structural genomics focuses on the physical aspects of the genome through the construction and comparison of gene maps and sequences as well as gene discovery, localization, and characterization. At the same time, functional genomics attempts to move data from structural genomics toward biologic function by understanding what genes do, how they are regulated, and their activity. Pharmacogenomics looks at genetic makeup or genetic variations and their connection to drug response. Variations in drug targets, usually proteins, and target pathways are studied to understand how the variations are manifested and how they influence response. The term pharmacogenetics is sometimes used instead, but it can also refer specifically to genetic profiles or tests that predict drug response.

## MOLECULAR AND CELLULAR BIOPHARMACEUTICS

### Introduction*

A biologic membrane is a lipid bilayer, typically embedded with proteins, that acts as a barrier within or surrounds the components of a cell. The membrane that separates a cell from the surrounding medium is called a plasma membrane. Such membranes also define most organelles (i.e., structures with specialized functions suspended in the cytoplasm) within cells. The typical structure of a cell membrane is shown in **Figure 12–1**. The membrane is characterized by a lipid bilayer that is typically about 5-nm thick. The lipid bilayer is composed of two opposing layers of lipid molecules arranged so that their hydrocarbon tails face one another to form the oily bilayer core, whereas their electrically charged or polar heads face the watery or "aqueous" solutions on either side of the membrane. Most of the proteins found in biologic membranes are integral membrane proteins (i.e., they are anchored to the cytoskeleton). Examples of the functions that integral membrane proteins serve include the identification of the cell for recognition by other cells, the anchoring of one cell to another or to surrounding media, the initiation of intracellular responses to external molecules, and the transport of molecules across the membrane. In 1899, Overton[12,13] concluded that the entry of any molecule into a cell is the result of its "selective solubility" in the cell's boundary, and that the more soluble in lipids the molecule is, the greater is its permeability, a discovery that has since been called the Overton rule.[11] Overton's studies led to the hypothesis that cell membranes are composed of lipid domains, which mediate transport of lipophilic molecules, and protein pores, which transport hydrophilic molecules. Eventually, these data were unified in the hypothesis that cell

---

*Much of this section is adapted from an article by Al-Awqati,[11] which the student is encouraged to read for a more in-depth analysis.

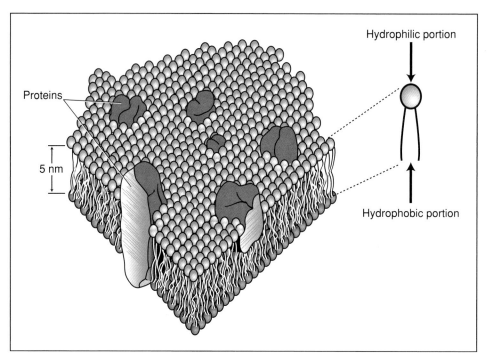

**Fig. 12–1.** Typical structure of a cell membrane. It is composed of a lipid bilayer that is 5-nm thick. Various types of proteins are part of the bilayer and serve a variety of physiological functions.

membranes are mosaics composed of lipid domains, through which lipophilic molecules permeate, and (water-filled) pore regions, presumably made up of proteins that allow the transport of hydrophilic molecules.[11–16]

Overton also suggested that ions must use a different pathway across the membrane because the low dielectric constant of lipids prevented solvation of charged particles. However, large ions have significant lipid solubility because their charge is spread over a much larger area, providing an explanation for the rapid permeability of large charged particles, including many drugs. Because many drugs are organic acids or bases, the role of ionization in drug absorption has been the subject of much study over the years. Historically, in the pharmaceutical sciences, the role of ionization in membrane permeation has been described by the pH-partition hypothesis.[17,18] As a general rule, the pH-partition hypothesis states that nonionized (i.e., lipid-soluble) drugs pass quickly through membranes, whereas ionized species are too polar to pass easily. Thus, it was expected that the rate of permeation of most drugs, which are organic acids and bases, is determined by the gradient for the nonionized form. The pH-partition hypothesis is covered in detail in Chapter 11.

Over the last several years, the mechanisms by which drugs and other xenobiotics are transported across biologic membranes have been reevaluated in light of the vast amount of information recently discovered during the mapping of the human genome and the identification of a vast number of proteins that may be involved in moving drugs across membranes. With the identification of aquaporins (in other words, water-conducting, protein-based channels), lipid transporters capable of transporting lipids such as the nonionized form of

short-chain fatty acids, and drug transporters that can transport water and lipid-soluble drugs, the view of membrane transport is rapidly changing. In the words of Al-Awqati, in evaluating Overton's landmark work, "Needless to say, the [current analysis] . . . neither reduces the importance of Overton's insight into the lipid structure of the cell membrane nor nullify the likelihood that a few molecules may indeed travel through the lipid bilayer. However, what is certain today is that most molecules of physiological or pharmacological significance are transported into or out of cells by proteins rather than by a 'passive' solubility into the lipid layer and diffusion through it."[11] This adequately sums up the changes in our thinking about the membrane transport of drugs. The next section introduces drug-transporting proteins.

## Drug Transporters, Cells, and Transport Pathways

Transporters are membrane proteins whose function is to facilitate the movement of molecules across cellular membranes. Although their primary function is to transport nutrients or other endogenous substances, many transporters also translocate drugs. For example, PepT1 is a transporter located at the brush border membrane of the human intestine responsible for the uptake of di- and tripeptides. However, it is able to transport many different drugs, such as valacyclovir, the L-valine ester prodrug of the acyclic nucleoside acyclovir,[19] angiotensin-converting enzyme inhibitors,[20] and cephalosporin antibiotics.[21] The human genome sequence suggests that there are more than 700 known transport/carrier genes,[22] and it has been estimated that at least 4% to 5% of the human proteome could be transporters[23] (**Fig. 12–2**).

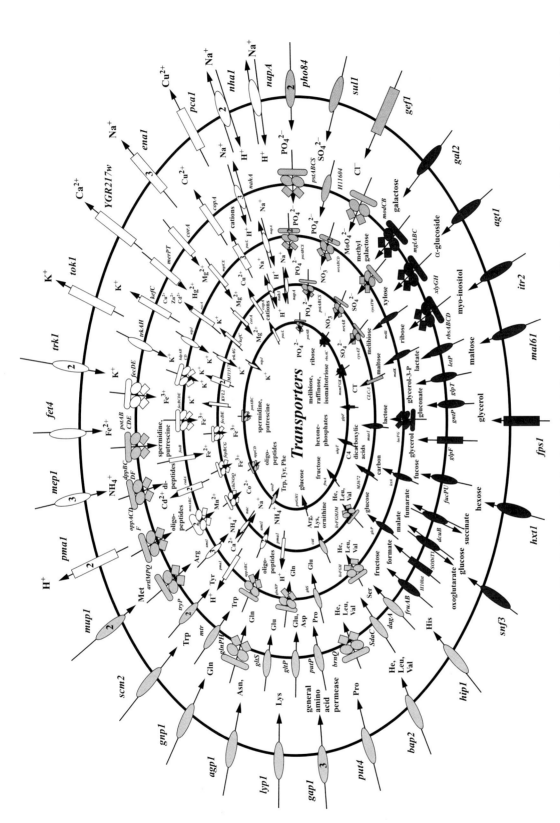

**Fig. 12–2.** Membrane transport functions identified by analysis of coding regions in five complete genomes. Circular representations from the center to the outer ring: *Mycoplasma genitalium, Methanococcus jannaschii, Synechocystis PCC6803, Haemophilus influenzae,* and *Saccharomyces cerevisiae.* Colors represent the four role categories: (1) amino acids, peptides and amines (light purple); (2) carbohydrates, organic alcohols and acids (dark purple); (3) cations (white); and (4) anions (gray). Ion-coupled permeases are designated by ovals, ABC transporters are shown as composites (circles, diamonds and ovals), and all other transporters are represented by rectangles. Arrows that point outward indicate efflux from the cell; those that point inward designate solute uptake from the environment. (From R. A. Clayton, O. White, K. A. Ketchum, and J. C. Venter, The first genome from the third domain of life, Nature **387**(6632), 459–462, 1997. With permission.)

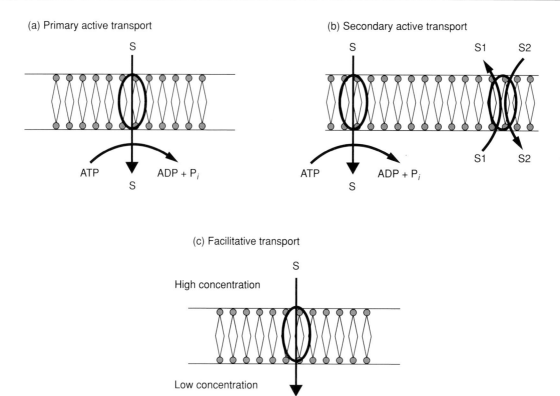

**Fig. 12–3.** Drug transport mechanisms across cell membranes fall into three general categories. In panel (A) primary active transport is depicted. In primary active transport, a drug or nutrient (S) is translocated across the membrane by means of a transport protein that spans the membrane. Energy in the form of ATP is required to drive the process. In panel (B) secondary active transport is depicted. In this case the drug or nutrient (S) crosses the membrane in the same manner as in (A). However, a second substrate (S1 or S2) is also moved into or out of the cell. In panel (C) facilitative transport is shown. Although transport is facilitated by proteins, the process is not energy dependent.

Drug transport mechanisms fall into three categories based on energetics and cotransport of other substances. These are primary and secondary active transport and facilitative transport. These mechanisms are depicted in **Figure 12–3**. Active transport involves the use of energy, usually ATP, to transport substrates across a biologic membrane. By using ATP, active transporters can move substrates to areas of high or low concentration. P-Glycoprotein (P-gp) is an example of a primary active transporter. Secondary active transport involves the cotransport of another substance such as an ion (e. g., $H^+$ or $Na^+$) along with the substrate. If the cotransported substance is transported in the same direction as the substrate, the process is called symport. If the cotransported substance is moved in the opposite direction, it is called antiport. An example of a symporter is the oligopeptide transporter PepT1. PepT1 transports a $H^+$ and a small peptide, typically a di- or tripeptide, into cells. Numerous drugs are also substrates for PepT1 including valacyclovir,[19] angiotensin-converting enzyme inhibitors,[20] and cephalosporin antibiotics.[21] Glucose and $Na^+$ transport by means of the glucose transporter is another classic example of symport. An example of an antiporter is $Na^+/K^+$-ATPase, which transports $Na^+$ and $K^+$

in opposite directions. Facilitative transport (also known as facilitated diffusion) is a non–energy-dependent transporter-mediated mechanism. Because the transport mechanism is not energy dependent, these transporters cannot move substrates against a concentration gradient. In other words, substrates can only move from areas of high concentration to areas of low concentration. An example of a facilitative transport mechanism involves the equilibrative nucleoside transporters *es* and *ei*.

In this chapter, two major transporter superfamilies, the ATP-binding cassette (ABC) and the solute carrier family (SLC), will be introduced. The ABC transporter superfamily is the largest transporter gene family. ABC transporters directly use ATP hydrolysis as the driving force to pump substrates out of cells or prevent them from entering cells. The genes encoding ABC transporters are widely dispersed in the genome and show a high degree of amino acid sequence identity among eukaryotes.[24,25] Using phylogenetic analysis, we can divide the human ABC superfamily into seven subfamilies with more than 40 members. Several well-characterized drug transporting–related members are listed in **Table 12–1**.[26] Delineation of the topology of a transporter is

## TABLE 12–1
## ATP-BINDING CASSETTE FAMILY TRANSPORTERS

| | Symbol | Alias | Rodent Orthologue | Tissue Distribution | Subcellular Localization | Functions |
|---|---|---|---|---|---|---|
| Subfamily A | ABCA1 | ABC-1 | Abca1 | Many tissues | – | A major regulator of cellular cholesterol and phospholipid homeostasis. It effluxes phospholipids (PS) and cholesterol from macrophages to apoA-I, reversing foam cell formation. Likely not involved in hepatic cholesterol secretion and intestinal apical cholesterol transport. |
| Subfamily B | ABCB1 | P-gp, MDR1 | Abcb1b | Many tissues (especially those with barrier functions such as L, BBB, P, K, I) | Apical | Efflux pump for xenobiotic compounds with broad substrate specificity, which is responsible for decreased drug accumulation in multidrug-resistant cells and often mediates the development of resistance to anticancer drugs. |
| | ABCB4 | MDR3 | Abcb1a | L | Apical | Most likely involved in biliary phosphatidylcholine secretion from hepatocytes in a bile salt–dependent manner. |
| Subfamily C | ABCC1 | MRP1 | Abcc1a | Lu, T, I | Lateral | MRP1 transports glucuronides and sulfate-conjugated steroid hormones and bile salts. It also transports drugs and other hydrophobic compounds in presence of glutathione. |
| | ABCC2 | MRP2, CMOAT | Abcc2 | L, I, K | Apical | MRP2 excretes glucuronides and sulfate-conjugated steroid hormones and bile salts into bile. Other substrates include anticancer drugs such as vinblastine and anti-HIV drugs such as saquinavir. Contributes to drug resistance. |
| | ABCC3 | MRP3 | Abcc3 | I, K | Lateral | MRP3 is inducible transporter in the biliary and intestinal excretion of organic anions. |
| | ABCC4 | MRP4 | | Many tissues (especially L) | Basolateral | MRP4 transports prostaglandins out of hepatocytes back to blood circulation. It also transports cyclic nucleotides and some nucleoside monophosphate analogues including nucleoside-based antiviral drugs. |
| | ABCC5 | | | L | | Similar substrate specificity with MRP4 |
| | ABCC6 | | | K and L | | MRP6 transports glutathione conjugates. |
| Subfamily G | ABCG2 | BCRP, MXR, ABCP | Abcg2 | P, B, L, I | Apical | BCRP functions as a xenobiotic transporter, which contributes to multidrug resistance. It serves as a cellular defense mechanism in response to mitoxantrone and anthracycline exposure. It also transports organic anions, steroids (cholesterol, estradiol, progesterone, testosterone), and certain chlorophyll metabolites. |

Key: L = liver; Lu = lung; T = testis; I = intestine; P = placenta; B = brain; K = kidney; BBB = blood–brain barrier.

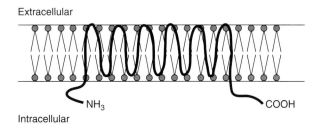

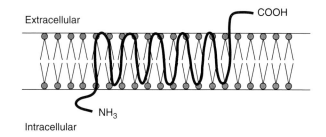

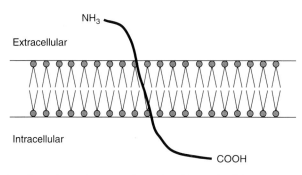

**Fig. 12–4.** Topologies of membrane-integrated proteins. (Drawn by G. You. With permission.)

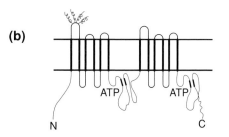

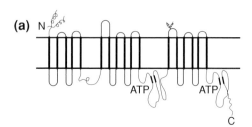

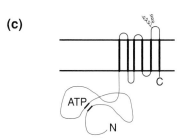

very important to gaining an understanding of its physiologic functions and substrate specificity. Knowing the transporter's structure may also enable the design of useful molecules to manipulate its transport activity and optimize the drug's pharmacokinetic behavior through enhanced absorption and targeted delivery. Membrane transporters may have various configurations across membranes. These configurations are shown in **Figure 12–4** and are referred to as topologies. The topology of a membrane transporter relates to its physiologic function. Because of difficulties in crystallizing membrane proteins, the topology of ABC transporters has been proposed on the basis of computational simulations and confirmed by experimental data. In general, ABC transporters contain two ATP-binding domains, also known as nucleotide-binding domains, which are located intracellularly, and 12 membrane–spanning $\alpha$-helices, which associate with each other to become specific membrane-spanning domains. Some transporters in the ABC superfamily, such as breast cancer resistance protein (BCRP, ABCG2), contain only one membrane-spanning domain and nucleotide-binding domain and are believed to associate with other proteins themselves to become functional. As a result of this, BCRP is also known as a half transporter.

The commonly agreed topologies of three well-studied ABC transporters, P-gp, BCRP, and multidrug-resistance protein-1 (MRP1), are shown in **Figure 12–5**. It is important to realize that these structures are highly educated guesses and will remain somewhat controversial until the exact crystal structures can be determined.

**Fig. 12–5.** Predicted structure of (*a*) multidrug resistance protein-1, (*b*) P-glycoprotein, and (*c*) breast cancer resistance protein. Shown is the linear secondary structure with putative transmembrane helices and the two ATP-binding domains. The potential glycosylation sites are denoted as C. (From T. Litman, T. E. Druley, W. D. Stein, and S. E. Bates, Cell. Mol. Life Sci. **58** (7), 931, 2001.

## TABLE 12–2
## SOLUTE CARRIER FAMILY TRANSPORTERS

|  | Symbol | Alias | Rodent Orthologue | Tissue Distribution | Subcellular Localization | Functions |
|---|---|---|---|---|---|---|
| SLC15 | SLC15A1 | hPepT1 | Slc15a1 | I, K | – | Proton-coupled uptake of oligopeptides of 2–4 amino acids, beta-lactam antibiotics |
| SLC21 | SLC21A3 | OATP, OATP-A | Slc21a7 | Many tissues (B, I, L, P, K) | Lateral | Mediates cellular uptake of organic ions in the liver. Its substrates include bile acids, bromosulfophthalein, some steroidal compounds, and fexofenadin. |
|  | SLC21A6 | OATP2, OATP-C, LST-1 |  | L | Basolateral | Mediates the Na⁺-independent transport of organic anions such as pravastatin, taurocholate, methotrexate, dehydroepiandrosterone sulfate, 17-beta-glucuronosyl estradiol, estrone sulfate, prostaglandin e2, thromboxane b2, leukotriene c3, leukotriene e4, thyroxine, and triiodothyronine. It may play an important role in the hepatic clearance of bile acids and organic anions. |
| SLC22 | SLC22A1 | OCT1 | Slc22a1 | L, K, I | Basolateral | Play a critical role in the elimination of many endogenous small organic cations as well as a wide array of drugs and environmental toxins. |
|  | SLC22A6 | OAT1, PAHT | Slc22a6 | K, B, P | Basolateral | Involved in the sodium-dependent transport and excretion of endogenous organic anions, such as $p$-aminohippurate, cyclic nucleotides, dicarboxylates, and neurotransmitter metabolites, and xenobiotics such as $\beta$-lactam antibiotics, nonsteroidal anti-inflammatory drugs, and anti-HIV and antitumor drugs. |

Key: L = liver; T = testis; I = intestine; P = placenta; B = brain; K = kidney.

ABC transporters interact with a wide variety of substrates, including sugars, amino acids, metal ions, peptides, proteins, and a large number of hydrophobic compounds and metabolites. ABC transporters play very important roles in many cellular processes, and several human genetic disorders such as cystic fibrosis, neurologic disease, retinal degeneration, cholesterol and bile transport defects, and anemia are the result of transporter dysfunction. ABC transporters are among the key players in multidrug resistance, a frequently observed phenomenon in cancer therapy, and they also significantly influence the pharmacokinetic behavior of many drugs. Because of this, drug transporters are becoming increasingly implicated in determining therapeutic outcomes.

Another transporter family that is involved in drug absorption and disposition is known as the SLC. More than 20 SLC subfamilies have been identified; they are responsible for transporting a variety of endogenous and exogenous substances, such as amino acids, glucose, oligopeptides, antibiotics, and nonsteroidal anti-inflammatory, antitumor, and anti-HIV drugs. In this chapter, only those well-characterized and drug transport–relevant SLC members are described. Their tissue distribution, subcellular localization, functions, and substrates are listed in **Table 12–2**. Computer modeling based on hydropathy analysis has enabled us to predict the linear secondary structures of organic anion transporters (OATs), organic cation transporters (OCTs), and organic anion-transporting polypeptides (OATPs) (**Fig. 12–6**). Unlike ABC transporters, these membrane-integrated transporters do not carry potential ATP-binding sites or an ATPase domain that can hydrolyze ATP. Most SLC members transport substrates with another ion in the same or opposite direction as the substrate. For example, the transport of oligopeptides and peptidomimetics by means of PepT1, the small peptide transporter (a symporter), has been shown to be proton and

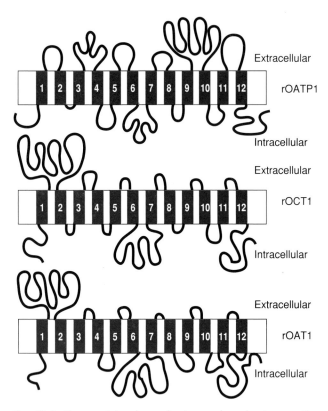

**Fig. 12–6.** Proposed topology of rat organic anion-transporting peptide-1 (rOATP1), organic cation transporter-1 (rOCT1), and organic ion transporter-1 (rOAT1). All transporters have 12 transmembrane helices. rOATP1 has a large extracellular loop between transmembrane helices 9 and 10, whereas rOCT1 and rOAT1 carry the large extracellular loop between helices 1 and 2.

membrane potential dependent on the apical surface of an epithelial cell (**Fig. 12–7**). Briefly, the $Na^+/H^+$ exchanger, an antiporter, located at the apical cell surface generates a lower pH in the microclimate of intestinal villi. Substrates are then taken up into epithelial cells by PepT1 coupled with the influx of proton. The influx of protons acts as the driving force for PepT1-mediated uptake, and this transport system is known as proton dependent.[27] It was demonstrated that the lower pH in the lumen (pH 5–6.5) and the inside negative membrane potential established by $Na^+/K^+$-ATPase were critical factors in the unidirectional uptake of substrates into cells by PepT1. Therefore, PepT1 functions as a net absorptive influx transporter rather than as secretory transporter in intestine. The effects of the proton on the orientation of PepT1 and symport events are depicted in **Figure 12–8**. In a similar manner, $Na^+$ and glucose cotransport effect the orientation and function of the glucose transporter (SLC2A1). These events are depicted in **Figure 12–9**.

The elimination of organic anions from blood into urine by OAT also involves the translocation of multiple ions across the basolateral membrane in the proximal tubule epithelium in the kidney. As shown in **Figure 12–10**, $Na^+/K^+$-ATPase establishes an inwardly directed $Na^+$ gradient. This $Na^+$ gradient then moves dicarboxylates into the cells via a Na-dicarboxylate cotransporter (SDCT2), which produces an outwardly directed dicarboxylate gradient. Finally, OAT transfers organic anions such as *para*-aminohippurate (PAH) into cells using the coupled efflux of dicarboxylate as the driving force.

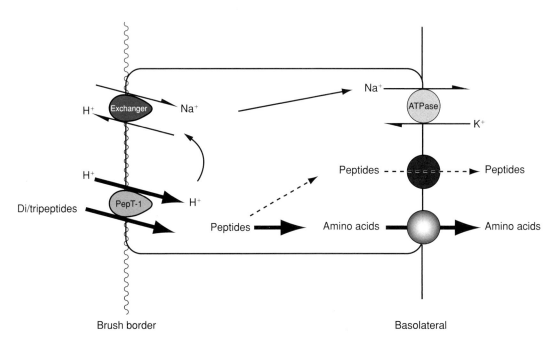

**Fig. 12–7.** A proposed model of PepT1-mediated transport. Di- or tripeptides enter the cells together with protons via PepT1 at the brush border membrane, and the proton and $Na^+$ gradients are maintained by the $H^+$–$Na^+$ exchanger and $Na^+/K^+$-ATPase, respectively.

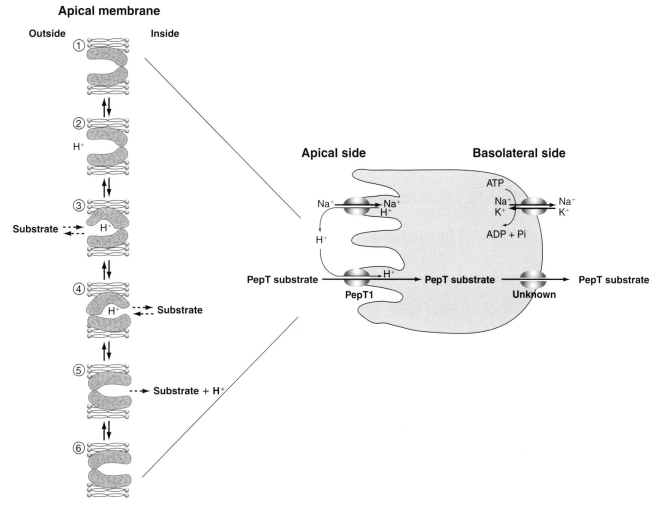

**Fig. 12–8.** Proposed mechanism of substrate transport by PepT1, the intestinal peptide transporter. (From D. R. Herrera-Ruiz and G. T. Knipp, J. Pharm. Sci. **92,** 691, 2003. With permission.)

In addition to transporting a wide variety of endogenous substance such as PAH, urate, cAMP, cGMP, tetraethyl ammonium, aliphatic quaternary ammonium compounds, and bile acids, SLC transporters also transport many clinically useful drugs including antibiotics and antiallergy, anti-HIV, and antitumor medications. A large body of data has shown that SLC transporters play a very important role in drug absorption and disposition and may be at the heart of numerous and significant drug–drug interactions.

Most drug transporters are located in tissues with barrier functions such as intestine, kidney, liver, and the brain barriers. The cells at the border of these barriers are usually polarized. In other words, the plasma membrane of these cells is organized into at least two discrete regions with different compositions and functions. **Figure 12–11** shows an example of a polarized cell and the transport pathways through and between cells. Enterocytes (i.e., intestinal absorptive cells) at the brush border membrane of intestine and epithelial cells

at the renal proximal tubule have an apical domain (AP) facing the lumen and basolateral domain (BL) facing the blood circulation; hepatocytes are polarized into a canalicular (AP) membrane facing the bile duct and a sinusoidal (BL) membrane facing the blood circulation; syncytiotrophoblasts at the maternal–fetal interface of placenta have apical domain facing the maternal blood and a basolateral domain facing the fetus. The brain capillary endothelial cells that function as the blood–brain barrier (BBB) are also polarized into luminal and antiluminal membranes. In most cases, the expression of a drug transporter is usually restricted to one side, the apical or basolateral domain, of polarized cells (e.g., PepT1 is located only on the apical membrane).

Drug transporters can be categorized into efflux or influx transporters according to the direction that they transport substrates across cell membranes. Under this definition, transporters that pump substrates out of the cells are called efflux transporters and those that transfer substrates into the cells are

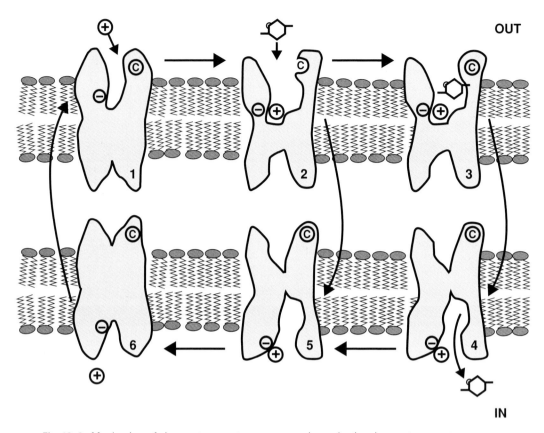

**Fig. 12–9.** Mechanism of glucose transport across a membrane by the glucose transporter.

called influx transporters. This definition is widely used when drug transport studies are performed at the cellular level. For example, P-gp and multidrug-resistant proteins (MRPs) belong to the efflux transporter group because they pump substrates out of the cytosol and into the extracellular environment. On the other hand, PepT1, OCTs, OATs, and OATPs are categorized as influx transporters due to their ability to bring

substrates into cells. Another way of classifying drug transporters is from a pharmacokinetic point of view. Based on this terminology system, transporters that transfer their substrates in the direction of the systemic circulation from outside the

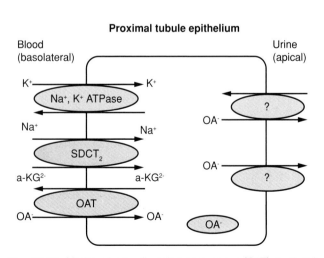

**Fig. 12–10.** Model of organic anion transporter (OAT)-mediated organic anion transport. Key: SDCT2 = Na$^+$-coupled dicarboxylate cotransporter-2; OA$^-$ = organic anion.

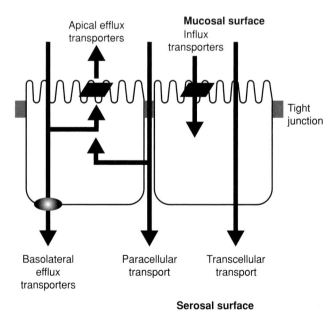

**Fig. 12–11.** Pathways across a cell monolayer. Drugs can cross between cells (i.e., paracellularly) or through cells (i.e., transcellularly). Drug transport out of cells is termed efflux and into cells is called influx.

body or into organs like the brain or the liver are called absorptive transporters, whereas transporters that transport drugs out of an organ or from the blood circulation into bile, urine, and gut lumen are called secretory transporters. For example, MRP1 (ABCC1) is an efflux transporter that can pump drugs such as saquinavir, an HIV protease inhibitor, out of cells. However, considering that MRP1 expression in enterocytes in the intestine is restricted to the basolateral membrane, efflux of saquinavir by MRP1 in the intestine leads to the movement of drug into the blood circulation. Therefore, MRP1 is considered an absorptive efflux transporter. Similarly, influx transporters could function as either absorptive or secretory transporters depending on the tissue and membrane domain where they are expressed. For example, intestinally expressed organic anion-transporting polypeptide-A (OATP-A) is localized on the apical domain of enterocytes. Orally administered fexofenadine, a histamine $H_1$-receptor antagonist, is transported into intestinal cells by OATP-A and then into the blood stream; therefore, OATP-A is considered an absorptive influx transporter.[28] An influx transporter could also act as a secretory pump. For example, studies at the cellular level have demonstrated that organic anion transporter-1 (OAT1) is an influx transporter with substrates such as PAH, prostaglandin E2 (PGE2), decarboxylates, and various anionic drugs. In the kidney, OAT1 is found on the basolateral membrane of tubular epithelial cells responsible for eliminating certain endogenous and exogenous substances and their metabolites from the blood into the urine. Therefore, kidney OAT1 is thought to be a secretory influx transporter. In other words, OAT1 takes up substrates into kidney cells, but the process is oriented toward moving them out of the body.[29]

---

**EXAMPLE 12–2**

**Concerted Transport Across Cells[12]**

Concerted drug or xenobiotic transport occurs when the transport of a compound is facilitated across both membrane domains of a polarized cell by membrane transporters. Concerted transport implies that the transporters on the two domains move drugs in the same direction (Fig. 12–12). In this example, Madin–Darby canine kidney (MDCK) cells were cultured in a specialized device called a Transwell. A Transwell is a porous support made of a polymer and is used to grow cells as a continuous monolayer of polarized cell membranes (apical and basolateral) and functional cell–cell tight junctions. In this example, the human MRP2 or OATP8 genes were heterologously expressed in MDCK cells, and their expression was restricted to the apical and basolateral membranes, respectively (Fig. 12–13). When the basolateral-to-apical (B → A) transport of estradiol-17-$\beta$-glucuronide, a metabolite of a sex hormone, was measured, it was found that the rank order of B → A permeability of estradiol-17-$\beta$-glucuronide is MDCK–MRP2/OATP8 (MDCK expressing both MRP2 and OATP8) ≫ MDCK–MRP2 (MDCK expressing MRP2) > MDCK (Fig. 12–14). This observation indicated that estradiol-17-$\beta$-glucuronide was taken up by OATP8 at the basolateral membrane and extruded by MRP2 at the apical membrane in a concerted manner. In other words, OATP8 and MRP2 transport their common substrates in the same direction to achieve more efficient substrate movement across the cell monolayer. When MRP2 was not present, translocation (i.e., net transport) was reduced, suggesting that

without the MRP2 transporter, estradiol-17-$\beta$-glucuronide would accumulate inside the cells. Because many drugs are substrates for more than one transporter, it is likely that there are numerous concerted transport pathways for drugs across each membrane domain.

---

## Drug Metabolism

### Introduction[5]

The first human metabolism study was performed in 1841 by Alexander Ure, who observed the conversion of benzoic acid to hippuric acid and proposed the use of benzoic acid for the treatment of gout.[30] Much has been learned about drug metabolism since that time, and the purpose of this section is to provide the biopharmaceutical background needed for the student to better understand drug metabolism. Because lipophilic drugs are efficiently deposited in tissues and cells, are readily reabsorbed across renal tubular cells, and tend to be highly bound to plasma proteins such as albumin, their clearance is very low. This can be partially explained by the fact that lipophilic drugs rapidly diffuse into hepatocytes or other cells containing various metabolic enzymes and have easy access to cytochrome P-450 anchored to endoplasmic reticulum (**Fig. 12–15**). To facilitate drug elimination and maintain homeostasis after the exposure to xenobiotics including drugs and environmental toxins, numerous biochemical transformations occur. These transformations are facilitated by two major groups of enzymes, and the process is called "drug metabolism." Metabolic reactions generally have the effect of converting drugs into more polar metabolites than the parent drug. The conversion to a more polar form has important biologic consequences because it enhances the ability of the body to eliminate drugs.

Drug metabolism involves a wide range of chemical reactions, including oxidation, reduction, hydrolysis, hydration, conjugation, condensation, and isomerization. The enzymes involved are present in many tissues but generally are more concentrated in the liver. For many drugs, metabolism occurs in two apparent phases. Phase I reactions involve the formation of a new or modified functional group or a cleavage (oxidation, reduction, hydrolysis); these are nonsynthetic reactions. Phase II reactions involve conjugation with an endogenous compound (e.g., glucuronic acid, sulfate, glycine) and are therefore synthetic reactions. Metabolites formed in synthetic reactions are more polar and more readily excreted by the kidneys (in urine) and the liver (in bile) than those formed in nonsynthetic reactions. Some drugs undergo either phase I or phase II reactions; thus, phase numbers reflect functional rather than sequential classification. Drugs are metabolized to various degrees by oxidation, reduction, hydrolysis, and conjugation in the body. Some drugs are eliminated without any structural changes occurring at all. The process of elimination of a compound from the body without further chemical modification is known as excretion. Williams[32] classified all known metabolic

# ⟹ KEY CONCEPT   CONCERTED TRANSPORT[12,31]

Given the presence of various transporters on the two domains of polarized cells and the possible differences in transport direction (i.e., influx vs. efflux), it is natural to consider how these transporters may work with each other (i.e., in "concert") or against each other. Recently, the phenomenon of "concerted transport," when membrane transporters on the AP and BL domains of a cell transport a drug substrate in the same direction (i.e., absorptive or secretory direction), has been studied. The liver is an important organ involved in the metabolism and clearance of endo- and xenobiotics. As seen in the accompanying figure,

drugs are taken up by hepatocytes in the liver directly by passive diffusion or by means of influx transporters at the sinusoidal membrane (BL). They are then converted intracellularly to pharmacologically inactive, active, or sometimes toxic metabolites by the cytochromes P450 (CYPs). The metabolites are then conjugated with various endogenous compounds such as glucuronide and sulfate; consequently, they are excreted into the bile passively by diffusion or by means of transporters such as the MRP family at the canalicular membrane (AP). It has been found that influx transporters at the sinusoidal membrane (e.g., OATP-C and OATP8) and efflux transporters at the canalicular membrane (e.g., MRP2 and P-gp) work in concert to transport drugs and other substances into the bile. Therefore, the alliance between influx transporters on the basolateral/sinusoidal membrane and efflux transporters at apical/canalicular membranes of hepatocytes can efficiently eliminate endogenous wastes or toxic xenobiotics into bile.

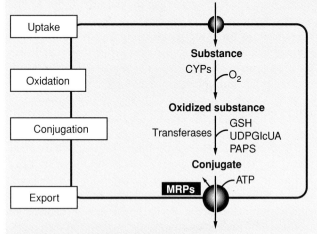

Uptake, biotransformation, and multidrug-resistance protein (MRP)–mediated export of endogenous substances, drugs, and carcinogens. Key: CYPs, cytochrome P-450s; GSH = ...; UDPGlc UA = ...; PAPS = ...; ATP = adenosine triphosphate. (From J. König et al., Biochim. Biophys. Acta **1461**, 377, 1999. With permission.)

reactions as either phase I or phase II reactions. In recent years, a third phase of drug metabolism has been classified and is commonly referred to as phase III metabolism. The three "phases" of drug metabolism are shown in **Figure 12–16**. Phase I reactions include oxidation, reduction, or hydrolysis of the drug. In a phase II reaction, the drug or its polar metabolite is coupled to an endogenous substrate such as uridine diphosphate (UDP) glucuronic acid, sulfate, acetate, or amino acid. The third phase of drug metabolism involves transporting the drug, metabolite, or conjugated metabolite across a biologic membrane and out of the body. For example, one such mechanism, originally called phase III detoxification,[33] utilizes the GS-X pump to transport xenobiotic metabolites out of the body. Because phase III reactions (i.e., membrane transporters) were covered in the previous section, only phase I and II reactions will be discussed here.

## Phase I Reactions

A major class of oxidative transformations was initially characterized by O. Hayaishi in Japan[34] and H. S. Mason in

the United States.[35] This class of oxygenases had requirements for both an oxidant (molecular oxygen) and a reductant (reduced nicotinamide-adenine dinucleotide phosphate [NADP]) and hence was given the name "mixed-function oxidases." An understanding of the biochemical nature of these reactions grew out of early studies on liver pigments by Garfinkel[36] and Klingenberg,[37] who observed in liver microsomes an unusual carbon monoxide–binding pigment with an absorbance maximum at 450 nm. Omura and Sato[38] ultimately characterized this pigment as a cytochrome. The function of this unique cytochrome, called P-450 (CYP450), was initially revealed in 1963 in studies by Estabrook et al.,[39] using microsomes from the adrenal cortex for the catalysis of the hydroxylation of 17-hydroxyprogesterone to deoxycorticosterone.

The most actively studied drug metabolism reaction is the CYP450-mediated reaction because the CYP450 family represents key enzymes in phase I reactions with several unique properties. This vast family is composed of more than 57 isoforms in humans alone (**Table 12–3**), mediates multiple oxidative reactions, and has very broad substrate

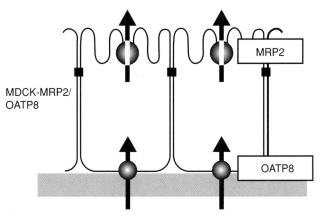

**Fig. 12–12.** Vectorial transport by human uptake transporter SLC21A8 (OATP8) and the apical export pump ABCC2 (MRP2). Key: MDCK = Madin–Darby canine kidney cell; MRP2 = multidrug-resistance protein-2; OATP8, organic anion-transporting peptide-8. (From Y. Cui, J. Konig, and D. Keppler, Mol. Pharmacol. **60** (5), 934, 2001. With permission.)

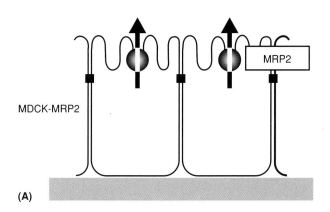

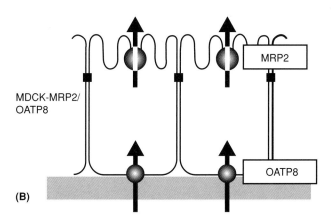

**Fig. 12–13.** Vectorial transport by (A) human uptake transporter SLC21A8 (OATP8) or (B) the apical export pump ABCC2 (MRP2). Key: MDCK = Madin–Darby canine kidney cell, MRP2 = multidrug-resistance protein-2; OATP8, organic anion-transporting peptide-8. (From Y. Cui, J. Konig, and D. Keppler, Mol. Pharmacol. **60** (5), 934, 2001. With permission.)

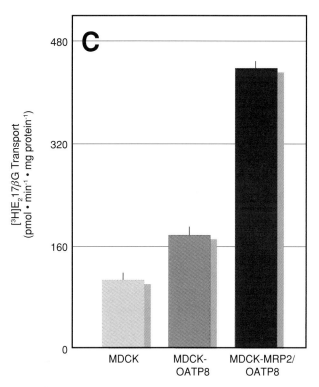

**Fig. 12–14.** Transcellular transport (from basolateral to apical) of estradiol 17-$\beta$-gluronide (E$_2$ 17$\beta$G). (From Y. Cui, J. Konig, and D. Keppler, Mol. Pharmacol. **60** (5), 934, 2001. With permission.)

specificity (**Table 12–4**). Phase I reactions introduce a functional group (—OH, —NH$_2$, —SH, or —COOH) to drugs and usually results in a small increase in hydrophilicity. Various molecules with very diverse chemical structures and different molecular weights, ranging from ethylene (28 g/mole) to cyclosporine (1201 g/mole) are known to be substrates and/or inhibitors of CYP450. Catalysis by CYP450 is very slow compared to that of other enzymes such as catalase, superoxide dismutase, and peroxidase.

CYP450 actually consists of two enzymes, catalyzing two separate but coupled reactions:

| Reductive half: | $O_2 + NADPH + H^+$ | $\rightarrow$ | $[O] + H_2O + NADP^+$ |
|---|---|---|---|
| Oxidative half: | $RH + [O]$ | $\rightarrow$ | $ROH$ |
| Coupled: | $RH + O_2 + NADPH + H^+ \rightarrow ROH + H_2O + NADP^+$ | | |

More than 60 reactions are catalyzed by CYP450 (**Table 12–5**), and even a single CYP450 isoform can generate the several metabolites from a single substrate. The function of CYP depends largely on the presence of molecular oxygen and/or drugs as substrates. For example, CYP450 can act as both an oxidative and a reducing enzyme (**Fig. 12–17**). These complex enzyme systems or mixed-function oxidases require NADPH, molecular oxygen, CYP450, NADPH–CYP450

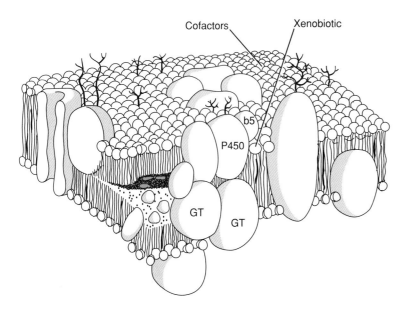

**Fig. 12–15.** Phase I and phase II enzymes integrated within or spanning across the lipid bilayer of the smooth endoplasmic reticulum (SER). Cytochrome P-450 enzyme complex, including cytochrome b5, and reduced nicotinamide-adenine dinucleotide phosphate (NADPH) as a cofactor are present in SER. Also present are glucuronyltransferase enzymes (GT). (From A. S. Kane, University of Maryland, College Park, Md., http://aquaticpath.umd.edu/appliedtox/metabolism.pdf, 2003.)

reductase, and phospholipids. An overall scheme showing the catalytic cycle of CYP450 is shown in **Figure 12–18**.

## Phase II Reactions

The first reference to phase II drug metabolism was made more than 150 years ago, when Stadeler[40] referred to the presence of conjugated phenol in urine, which was later isolated and characterized as phenyl sulfate by Baumann. A glucuronide conjugate was first discovered by Jeffe,[41] who found that o-nitrotoluene gave rise to o-nitrobenzyl alcohol excreted as a conjugate in the dog: a few other sugar conjugates were also reported in the 1870s by von Mering's group[42] and Schmiedeberg's group.[43] Although the mercapturic acids were found in 1879 by Baumann and Preusse,[44] the full mechanism of glutathione conjugation was not known until Barnes et al.[45] characterized the relationship between glutathione and mercapturic acid in 1959. The enzymes responsible for phase II reactions are UDP-glucuronosyltransferases (UGTs) for glucuronidation, sulfotransferases (SULTs) for sulfation, and glutathione S-transferases (GSTs) for glutathione conjugation. Glucuronidation, the most common phase II reaction, is the only one that occurs in the liver microsomal enzyme system.[5]

There are numerous enzyme families, and a variety of isoforms within families and different types of isozymes are found in the various animal species and humans. Three well-documented and important phase II enzyme families are shown in **Table 12–6**. Phase II reactions involve the conjugation of certain functional groups using conjugating cofactors as shown in **Table 12–7**. Polar groups introduced by phase I reactions are used as attachment sites for conjugation (phase II) reactions. For instance, a hydroxyl group added during a phase I reaction is a good target for glucuronide or sulfate conjugation. Conjugation reactions greatly increase the hydrophilicity and promote the excretion of drugs. Hydrophilic conjugates of drugs are typically less active than the parent compounds, with some notable exceptions, such as morphine-6-glucuronide, N-(4-hydroxylphenyl) retinamide glucuronide, and minoxidil sulfate, where the metabolites are more potent than their respective parent drugs. Glucuronides are secreted in bile and eliminated in urine. Chloramphenicol, meprobamate, and morphine are metabolized this way. Amino acid conjugation with glutamine or glycine produces conjugates (e.g., salicyluric acid formed from salicylic acid and glycine) that are readily excreted in urine but are not extensively secreted in bile. Acetylation is the primary metabolic pathway for sulfonamides. Hydralazine, isoniazid, and procainamide are also acetylated. Sulfoconjugation is the reaction between phenolic or alcoholic groups and inorganic sulfate, which is partially derived from sulfur-containing amino acids (e.g., cysteine). The sulfate esters formed are polar and readily

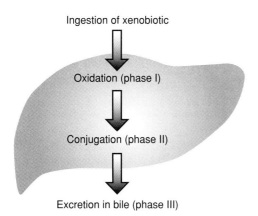

**Fig. 12–16.** Phase III elimination reactions in bile begin with oxidation in the liver (phase I), followed by conjugation (phase II). (From M. Vore, Environ. Health Perspect. **102**, 5, 1994. With permission.)

**TABLE 12–3**

## ISOFORMS OF CYTOCHROME P-450 ENZYME IN ANIMALS AND HUMANS

| Isoform | Mouse | Rat | Rabbit | Dog | Human |
|---------|-------|-----|--------|-----|-------|
| CYP1A | 1, 2 | 1, 2 | 1, 2 | 1, 2 | 1, 2 |
| CYP2A | 4, 5, 12 | 1–3 | 10, 11 | ? | 6, 7, 14 |
| CYP2B | 9, 10, 14, 19, 20 | 1–3, 8, 12, 15, 22–24 | 4, 5 | 11 | 6, 7 |
| CYP2C | 9, 10, 29, 37–40 | 6, 7, 11–14, 23, 24 | 1–5, 11–16, 30 | 21, 41, 42 | 8, 9, 18, 19 |
| CYP2D | 9–12 | 1–5, 18 | 23, 24 | 15 | 6 |
| CYP2E | 1 | 1 | 1, 2 | 1 | 1 |
| CYP3A | 11, 14, 16, 25 | 1, 2, 9, 18, 23 | 6 | 12, 26 | 3, 4, 5, 7, 43 |

excreted in urine. Drugs that form sulfate conjugates include acetaminophen, estradiol, methyldopa, minoxidil, and thyroxine. Methylation is a major metabolic pathway for inactivation of some catecholamines. Niacinamide and thiouracil are also methylated.[5]

### Enzyme Induction and Inhibition

Most marketed drugs are metabolized by more than one CYP450 isoform, which means it is highly likely that drug–drug interactions are possible. Pharmacokinetic interactions related to drug metabolism can be categorized as either enzyme induction or inhibition.

Induction. Most drug-metabolizing enzymes are expressed constitutively, that is, they are synthesized in the absence of any discernible external stimulus. It has been shown that an increase in the activity of hepatic microsomal enzymes such as CYP450 can occur after exposure to structurally diverse drugs and xenobiotics. The stimulation of enzyme activity in response to an environmental signal is referred to as enzyme induction. Enzyme induction involves multiple mechanisms and usually occurs at the gene transcriptional level. Inducing agents may increase the rate of their own metabolism as well as those of other unrelated drugs by inducing various phase I and phase II enzymes (**Table 12–8**).

**TABLE 12–4**

## MARKER SUBSTRATES, REACTIONS, AND TYPICAL INHIBITORS FOR CYTOCHROME P-450 ISOZYMES

| CYP | Substrate (Reaction) | Inhibitor (Mechanism) |
|-----|----------------------|------------------------|
| 1A2 | 7-Methoxyresorufin ($O$-demethylation) | Furafylline (mechanism based) |
| | Caffeine ($N$3-demethylation) | 7,8-Benzoflavone (competitive) |
| | Phenacetin ($O$-deethylation) | Fluvoxamine (competitive) |
| 2A6 | Coumarin (7-hydroxylation) | Methoxalen (mechanism based) |
| | | Tryptamine (competitive) |
| | | Trancylcypromine (competitive) |
| 2B6 | 7-Benzoxyresorufin ($O$-debenzylation) | Orphenadrine (competitive) |
| | ($S$)-Mephenytoin ($N$-demethylation) | |
| 2C8 | Paclitaxel ($6\alpha$-hydroxylation) | Quercetin |
| 2C9 | Tolbutamide (methyl hydroxylation) | Sulfaphenazole (competitive) |
| | Phenytoin ($4\alpha$-hydroxylation) | Tienilic acid (mechanism based) |
| | Diclofenac ($4\alpha$-hydroxylation) | |
| | ($S$)-Warfarin (7-hydroxylation) | |
| 2C19 | ($S$)-Mephenytoin ($4\alpha$-hydroxylation) | Ticlopidine (competitive) |
| | Omeprazole (oxidation) | Omeprazole |
| 2D6 | Debrisoquine (4-hydroxylation) | Quinidine (competitive) |
| | Bufuralol (1-hydroxylation) | Fluoxetine (competitive) |
| | Dextromethorphan ($O$-demethylation) | Paroxetine (competitive) |
| 2E1 | Chlorzoxazone (6-hydroxylation) | Diethyldithiocarbamate (mechanism based) |
| | 4-Nitrophenol (3-hydroxylation) | 4-Methylpyrazole (competitive) |
| | $N$-Nitrosodimethylamine ($N$-demethylation) | Disulfiram (mechanism based) |
| | Aniline (4-hydroxylation) | Pyridine |
| 3A4 | Nifedipine (oxidation) | Troleandomycin (metabolic intermediate complex) |
| | Erythromycin ($N$-demethylation) | Erythromycin (metabolic intermediate complex) |
| | Testosterone ($6\alpha$-hydroxylation) | Ketoconazole, itraconazole (competitive) |
| | Midazolam (1-hydroxylation) | Gestodene (mechanism based) |

## TABLE 12–5
## PHASE I DRUG-METABOLIZING REACTIONS

| Oxidation | Reduction | Hydrolysis |
|---|---|---|
| Aromatic hydroxylation | Nitro reduction | Amidine hydrolysis |
| Alipatic hydroxylation | Azo reduction | Ester hydrolysis |
| N-Oxidation (formation of N-oxide and N-OH) | Ketone reduction | Amide hydrolysis |
| S-oxidation (sulfoxidation) | | |
| N-, O-, S-dealkylation | Reduction of $\alpha$, $\beta$-unsaturated ketones | |
| Oxidation of cyclic amines to lactams | Aldehyde reduction | |
| Oxidative deamination | N-, S-oxide reduction | |
| Oxidation of methyl to carboxyl group | | |
| Epoxidation | | |
| Alcohol oxidation (conversion to aldehyde or carboxylic acid) | | |
| Dehydrogenation, $\beta$-oxidation | | |

### EXAMPLE 12–3

**Clinically Significant Enzyme Induction[46]**

Administration of two or more drugs together may lead to serious drug interactions as a result of enzyme induction. Triazolam is a short-acting hypnotic drug that is extensively metabolized by CYP3A4. Rifampin is used with other medicines to treat tuberculosis and is known to be a potent inducer of CYP3A4. Coadministration of rifampin markedly reduces plasma concentrations and the pharmacologic effects of many drugs including triazolam. To potentially induce CYP3A4 enzymes, 600 mg of rifampin or placebo was administered once daily to 10 healthy volunteers for 5 days. On the sixth day, 0.5 mg of triazolam was orally administered, and the plasma concentration profile of triazolam was monitored for 10 hr (Fig. 12–19). As expected, a significant drug–drug interaction between rifampin and triazolam was observed. The area under the plasma concentration versus time curve of triazolam in the rifampin phase was only 5% of that in the placebo phase and the maximum plasma concentration of triazolam was 12.4% of the control value. The conclusion of this study was that triazolam becomes pharmacologically ineffective after long-term rifampin treatment because the induction of microsomal enzymes by rifampin causes an increase in the metabolism of triazolam and a marked decrease in the plasma concentration and the efficacy of triazolam.

Inhibition.  Enzyme inhibition generally occurs without delay and can result in the immediate loss of activity for one or more enzymes. Many drugs and xenobiotics are capable of inhibiting drug metabolism. With metabolism decreases, drug accumulation often occurs, leading to prolonged drug action and possibly serious adverse effects. Enzyme inhibition by drugs or xenobiotics can occur in several ways, including competitive inhibition, the destruction of preexisting enzymes, interference with enzyme synthesis, and inactivation of the drug-metabolizing enzymes by complexation. Drugs containing imidazole, pyridine, or quinoline groups, such as ketoconazole, metyrapone, and quinidine, are well-known reversible inhibitors. Inactivation of metabolizing enzymes by complexation is called quasi-irreversible inhibition and occurs when a noncovalent tight bond is formed between the metabolite and CYP450. Macrolide antibiotics such as troleandomycin and erythromycin, hydrazines such as isoniazid, and methylenedioxybenzenes such as isosafrole are all known as quasi-irreversible inhibitors. Lastly, xenobiotics containing specific functional groups can be metabolized by CYP450 to reactive intermediates that bind to the enzyme covalently. For example, compounds that contain olefins and acetylenes can alkylate the heme. It is also known that some S- or N-containing compounds such as tienilic and cyclopropylamine covalently bind to the apoprotein.

### EXAMPLE 12–4

**Clinically Significant Enzyme Inhibition: Grapefruit Juice[47]**

Many commonly consumed foods, drinks, and natural products or dietary supplements are known to alter the disposition of drugs. One particularly well-studied case, grapefruit juice, is the subject of this example. Grapefruit juice is known to cause a considerable

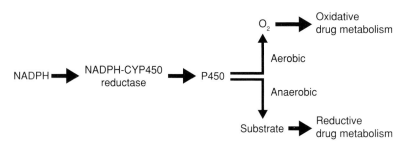

**Fig. 12–17.** Dual function of cytochrome P-450. Key: NADP = reduced nicotinamide-adenine dinucleotide phosphate. (From A. Y. H. Lu, Rutgers University, New Brunswick, N. J., Lecture Note, 2003.)

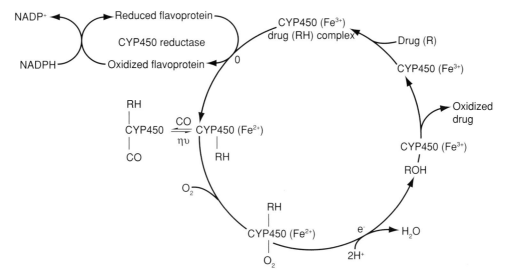

**Fig. 12–18.** Catalytic cycle of cytochrome P-450. Key: NADP = nicotinamide-adenine dinucleotide phosphate. $\eta\upsilon$ represents energy. RH represents the drug substrate. (From A. P. Alvares and W. B. Pratt, in W. B. Pratt and P. Taylor (Eds.), *Principles of Drug Action*, 3rd Ed., Churchill Livingstone, New York, 1990, p. 469. With permission.)

increase in the oral bioavailability of many drugs. This is because grapefruit juice is an inhibitor of CYP3A, a major player in the intestinal and hepatic metabolism of drugs, and when coadministered with substrates of CYP3A it causes a significant increase in drug–blood levels. The main mechanism of this drug–drug interaction is attributed to a decrease in intestinal CYP3A in humans (i.e., CYP3A4) by approximately 45% to 65%. In contrast, a single "dose" of grapefruit juice does not affect hepatic CYP3A4. Long-term multiple administrations of grapefruit juice inhibit CYP3A4 not only in the intestine but also in the liver. Verapamil is a calcium-channel blocker that undergoes extensive metabolism mainly by CYP3A4 in human. Verapamil was administered to 24 volunteers as a CYP3A4 substrate with grapefruit juice for 7 days. Grapefruit juice caused about a 50% increase in steady-state plasma concentrations of verapamil, showing a significant food–drug interaction (Fig. 12–20).

### Enzyme Inhibition Kinetics

Although a thorough discussion of enzyme kinetics is presented in Chapter 16, some pertinent points are highlighted here. If needed as a refresher, the student is referred to basic biochemistry textbooks for relevant background information. Enzyme inhibition (**Fig. 12–21**) can be classified by enzyme kinetics expressed by a change in $K_m$ (the Michaelis–Menten constant) and/or $V_m$ (maximal velocity). In competitive inhi-

bition, the inhibitor binds to a free binding site on the enzyme. An increase in the inhibitor concentration results in a lower chance of binding between the substrate and enzyme, and thus the $K_m$ value increases without a corresponding change in $V_m$,

$$V = \frac{S \cdot V_m}{S + K_m(1 + I/K_i)} \tag{12–1}$$

where $I$ is concentration of inhibitor and $K_i$ is inhibitory constant. In the case of noncompetitive inhibition, the inhibitor binds to its own binding site regardless of whether the substrate binding site is occupied. Because the degree of enzyme inhibition is dependent on the inhibitor concentration and is independent of substrate binding, the $V_m$ values decrease with increasing inhibitor concentrations without changing $K_m$,

$$V = \frac{S \cdot V_m/(1 + I/K_i)}{S + K_m} \tag{12–2}$$

Uncompetitive inhibition is observed when the inhibitor binds only to the enzyme–substrate complex. This results in a change in both $K_m$ and $V_m$. Mixed-type enzyme inhibition

**TABLE 12–6**
**ISOFORMS OF PHASE II ENZYMES IN ANIMALS AND HUMANS**

| Enzyme | Mouse | Rat | Rabbit | Dog | Human |
|---|---|---|---|---|---|
| UGT1A | 1, 5, 6, 9 | 1–3, 4P, 5–8, 9P, 10 | ? | 6 | 1, 2P, 3–10, 11P, 12P |
| UGT2A | 1 | 1 | 2 | ? | 1 |
| UGT2B | 5 | 1–3, 6, 8, 12 | 13, 14, 16 | ? | 4, 7, 10, 11, 15, 17 |
| SULT1 | A, B, C, E | A, B, C, E | ? | A | A, B, C, E |
| SULT2 | A | A | ? | ? | A, B |
| GST | A, P | A, M, S | ? | ? | A, M, P, T, Z |

### TABLE 12–7
## PHASE II DRUG-METABOLIZING REACTIONS AND COFACTORS

| Reactions | Cofactors |
|---|---|
| Glucuronidation<br>  N-, O-, S-, C-Glucuronidation<br>  Carbamic acid glucuronide | UDP-Glucuronic acid<br>  (UDPGA) |
| Sulfate conjugation | 3-Phosphoadenosine<br>  5′-phosphosulfate (PAPS) |
| Glycine conjugation | Glycine |
| Acetylation | Acetyl CoA |
| Methylation<br>  N-, O-, S-Methylation | S-adenosyl-L-methionine<br>  (SAM) |
| Glutathione conjugation | Glutathione |

can also lead to changes in both parameters, and that could confuse the interpretation of the results.

### Metabolism and Drug Disposition

Orally dosed drug molecules are absorbed from the gastrointestinal (GI) tract through the GI wall and pass through the liver prior to reaching the systemic circulation. During the absorption process, drug molecules are exposed to various dispositional processes such as intestinal metabolism, intestinal secretion, hepatic metabolism, and biliary secretion. Among these processes, intestinal and hepatic metabolism are lumped together and are commonly referred to as "first-pass metabolism." In the past, first-pass metabolism and hepatic first-pass metabolism were considered synonymous because of the dominating role of the liver in drug metabolism. However, recent studies have demonstrated that intestinal metabolism can be significant, especially if the role and potential impact of intestinal drug secretion is considered, and so it is best to refer to the process as "first-pass metabolism" unless specific mechanistic information is available that

### TABLE 12–8
## DRUGS THAT INDUCE METABOLISM CLINICALLY*

| Inducing Agent | Induced Enzyme |
|---|---|
| Tobacco | CYP1A2 |
| Phenobarbital, rifampin | CYP2B6 |
| Rifampin, secobarbital | CYP2C9 |
| Ethanol, isoniazid | CYP2E1 |
| Carbamazepine, troglitazone,<br>  phenobarbital, phenytoin,<br>  rifabutin, rifampin, St. John's<br>  wort (hyperforin) | CYP3A4,5,7 |
| Phenobarbital | Glucuronide transferase |
| Red grape (ellagic acid), garlic<br>  oil, rosemary, soy, cabbage,<br>  brussels sprouts† | Glutathione-S-transferase and<br>  glucuronide transferase |

*Modified from Department of Medicine, Indiana University, Bloomington, Ind., http://medicine.iupui.edu/flockhart/clinlist.htm, 2003.
†In vitro or animal data.

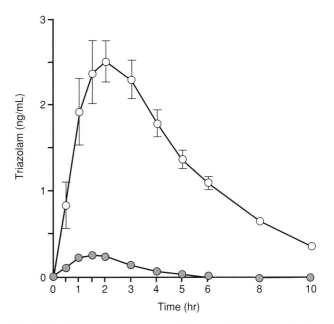

**Fig. 12–19.** Plasma concentration of triazolam in 10 individuals after 0.5 mg of oral triazolam, after pretreatment with 600 mg of rifampin once daily (●) or placebo (○) for 5 days.

defines the role of the intestine and/or liver. The degree of the metabolism may vary greatly with each drug, and the resulting oral bioavailability can be very low. If a drug is completely absorbed and is not secreted, the oral fraction of dose absorbed or bioavailability, $F$, indicates the portion of drug that is absorbed intact:

$$F = \frac{\text{AUC}_{\text{oral}}}{\text{AUC}_{\text{IV}}} \qquad (12\text{–}3)$$

To get the precise value of first-pass metabolism from the $F$ value, all of the foregoing assumptions must be satisfied. The fraction lost to metabolism would be equal to $1 - F$. Of course, if absorption were incomplete, then $1 - F$ would represent the fraction of drug not absorbed due to incomplete absorption and/or lost to metabolism. Poor absorption could be due to many factors including low intestinal permeability; binding to intestinal tissue, mucus, or debris; or instability in the GI tract. The extraction ratio (ER) is commonly used to directly measure drug removal from the intestine or liver:

$$\text{ER} = \frac{C_a - C_v}{C_a} \qquad (12\text{–}4)$$

where $C_a$ is the drug concentration in the blood entering the organ and $C_v$ is the drug concentration leaving the organ. The relationship between bioavailability and intestinal and hepatic extraction ratios is expressed by the equation

$$F = F_{\text{ABS}}(1 - \text{ER}_{\text{GI}})(1 - \text{ER}_{\text{H}}) \qquad (12\text{–}5)$$

where $F_{\text{ABS}}$ is the fraction of the dose absorbed through the intestinal mucosal membrane into the portal vein and $\text{ER}_{\text{GI}}$ and $\text{ER}_{\text{H}}$ are the gut and hepatic extraction ratios, respectively. When absorption is complete (i.e., $F_{\text{ABS}} = 1$) and

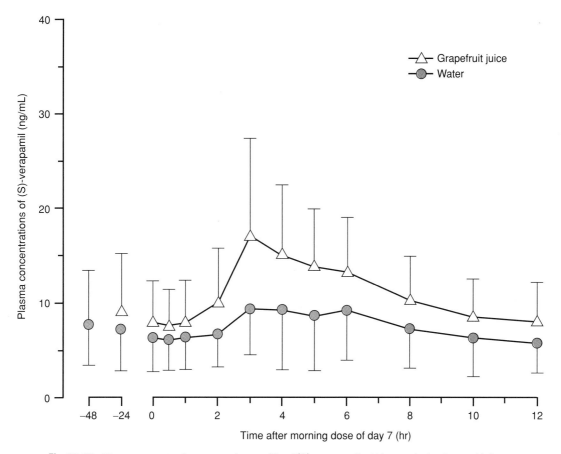

**Fig. 12–20.** Mean concentration versus time profile of (S)-verapamil within one dosing interval following oral administration of 120 mg of verapamil twice daily for 7 days during coadministration of water or grapefruit juice starting 48 hr prior to the dosing interval.

intestinal extraction is negligible ($ER_{GI} \cong 0$), the equation can be simplified to

$$F = 1 - ER_H \qquad (12\text{–}6)$$

Substituting equation (12–6) into equation (12–3) and rearranging results in

$$ER_H = 1 - \frac{AUC_{oral}}{AUC_{IV}} \qquad (12\text{–}7)$$

which is used as an estimation of the liver extraction ratio.

**Metabolism and Protein Binding**

Hepatic clearance, $CL_H$, can be related to liver blood flow, $Q$, and the intrinsic clearance, $CL_{int}$, of liver using the following equation:

$$CL_H = Q\frac{CL_{int}}{Q + CL_{int}} \qquad (12\text{–}8)$$

where the intrinsic clearance is the ability of the liver to remove drug without flow limitations. The Michaelis–Menten equation can be rearranged to give

$$\frac{V}{S} = \frac{V_m}{K_m + S} = CL_{int} \qquad (12\text{–}9)$$

where the value of the rate of reaction, $V$, divided by the drug concentration, $S$, is conceptually the same as the intrinsic clearance, $CL_{int}$. Because the hepatic metabolizing enzymes are rarely saturated in a clinical situation, one can assume that $K_m$ is much greater than $S$, and equation (12–9) reduces to

$$CL_{int} = \frac{V}{S} = \frac{V_m}{K_m} \qquad (12\text{–}10)$$

In other words, the intrinsic clearance is constant, assuming the metabolizing enzymes are not saturated and that the protein binding of the drug is constant. Protein binding affects the intrinsic clearance of drugs because intrinsic clearance contains the free drug fraction, $f_u$, and the intrinsic clearance of free drug, $CL'_{int}$:

$$CL_{int} = f_u \cdot CL'_{int} \qquad (12\text{–}11)$$

Equation (12–8) can be rewritten as

$$CL_H = Q\frac{f_u \cdot CL'_{int}}{Q + f_u \cdot CL'_{int}} \qquad (12\text{–}12)$$

When the intrinsic clearance is much greater than hepatic blood flow, the hepatic clearance is dependent only on blood

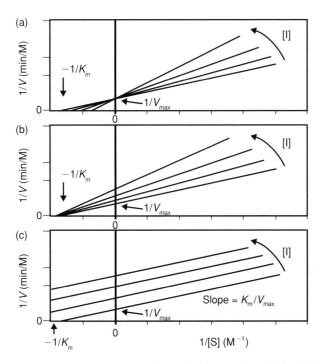

**Fig. 12–21.** The Lineweaver–Burk plot is used to distinguish the types of enzyme inhibition. Key: [I] = inhibitor concentration, [S] = substrate concentration. (*a*) Competitive inhibition, (*b*) noncompetitive inhibition, (*c*) uncompetitive inhibition. (Modified from D. P. Goldenberg, University of Utah, Salt Lake City, Utah. Lecture Slides inhibitors II. pdf, http://courses.biology.utah.edu/goldenberg/biol.3515/index.html, 2003.)

flow. On the other hand, if the intrinsic clearance is much lower than hepatic blood flow, then hepatic clearance is dependent only on the intrinsic clearance. These two extremes are called flow-limited and metabolism-limited extraction, respectively. Drug protein binding does not affect hepatic clearance for drugs that demonstrate high intrinsic clearance. However, low-extraction drugs may be affected by protein binding, depending on the free fraction of drug. The studies of Blaschke[48] demonstrate the relationship between protein binding and hepatic clearance (**Fig. 12–22**). Hepatic clearance of low extracted and medium- or low-binding drugs (less than 75% to about 80%) are not greatly affected by the changes in protein binding. These drugs are categorized as capacity-limited, binding-insensitive drugs. High-protein-binding and low-extraction drugs are considered capacity-limited, binding-sensitive drugs because a small change in the bound portion usually means large changes in the free drug fraction.

### Drug Metabolism at the Subcellular Level

The endoplasmic reticulum (ER) is one of the most important cellular organelles in drug metabolism. Other fractions such as the mitochondria or cytosolic fraction play an important role in some cases (**Table 12–9**). The subcellular fractions comprising the S9 fraction, microsomes, and cytosolic fraction are the most widely used in vitro systems for

studying drug metabolism. These fractions can be isolated by differential centrifugation techniques, which have permitted important advances in studies of drug metabolism. Because microsomal enzymes can oxidize a large portion of xenobiotics, incubation of a drug with liver microsomes is a widely used in vitro technique. Enzyme preparations have another advantage in that they are easy to prepare and can be obtained from small amounts of tissue. The level of drug-metabolizing enzymes is also readily determined. Enzyme kinetic parameters obtained from liver microsomes can be used to predict in vivo clearance. For instance, intrinsic clearance, $CL_{int}$, is calculated from $K_m$ and $V_m$ in equation **(12–8)**, and this intrinsic clearance is used again in calculating in vivo $CL_H$ using equation **(12–6)** or **(12–10)** when the major elimination route of the drug is hepatic metabolism. Cellular fractions, like all experimental systems, have some limitations.

## ORGAN-LEVEL BIOPHARMACEUTICS

Organ-level biopharmaceutics is an important aspect of pharmacokinetics because the many transport and metabolism components that have already been introduced work together in a dynamic environment. In essence, it serves as a link between the molecular/cellular-level aspects and the "intact" system studied in pharmacokinetics. In this section, various organ systems and groups of organs that define key biopharmaceutical and pharmacokinetic processes will be described in detail. The key organ systems that will be discussed are the brain and choroid plexus, intestine, kidney, and liver. The kidney and liver are the primary organs of drug excretion (**Fig. 12–23**). Because the lung and the skin are minor organs of drug excretion, they will not be covered in depth in this chapter. Examples of groups of organs include the "first-pass" organ system, which includes the intestine, liver, and lungs, and the "enterohepatic" recirculation organs, which include the intestine, liver, and gallbladder. This chapter will not deal extensively with species differences in organ-level biopharmaceutics. However, for reference, **Tables 12–10** through **12–12** are provided to show differences in organ weights, volumes, and blood flow for mice, rats, rabbits, rhesus monkeys, dogs, and humans. These values are used in correlative studies using physiologically based pharmacokinetic (PBPK) models (see the next section) in order to predict species differences and human dosing based on preclinical animal results.

### Brain-Barrier Systems

In 1885, Paul Ehrlich,[49] a German scientist, observed that many dyes can be distributed widely into body tissues but fail to stain brain parenchyma. This was attributed to a brain-barrier system. In fact, because the central nervous system is so well perfused, permeability is generally the major determinant of the drug distribution rate into the brain. Drugs reach the central nervous system by means of brain capillaries and the cerebral spinal fluid (CSF). Although the brain receives about one sixth of the cardiac output, distribution of drugs to brain tissue is highly restricted. The restricted brain exposure to drugs and other xenobiotics is the result of two brain-barrier systems: (*a*) the BBB, which is formed by brain capillary endothelial cells, and (*b*) the blood–CSF barrier or the choroid plexus.

The brain-barrier systems are shown in **Figure 12–24**. **Figure 12–24***a* depicts a brain capillary. It is composed of four kinds of cells: endothelial cells, pericytes, astrocytes, and neurons. The endothelial cells of the brain capillaries are more tightly joined to one another than are those of other capillaries. Another barrier to water-soluble drugs is the glial connective tissue cells (astrocytes), which form an astrocytic sheath close to the basement membrane of the capillary endothelium. The capillary endothelium and the astrocytic sheath form the BBB. Because the capillary wall rather than the parenchymal cell forms the barrier, the brain's permeability characteristics differ from those of other tissues. Drugs can also enter ventricular CSF directly by means of the choroid plexus. **Figure 12–24***b* shows the choroid plexus. It acts as a BBB in the brain parenchyma because this capillary has a tight junction. There are two kinds of blood–CSF barriers. One is the arachnoid membrane and the other is formed by the epithelial cells of choroid plexus. Because these capillaries are permeable, only the arachnoid membrane and epithelial cells of choroid plexus function as a brain barrier.

Polar compounds cannot enter the brain by passive diffusion but can enter the interstitial fluids of most other tissues. Because membrane transporters are known to play a major role in the uptake of many compounds, it is likely that they also play a major role in the blood–brain and blood–CSF barriers. For example, drug uptake into brain endothelial cells is likely to be assisted by membrane transporters as described earlier in this chapter. However, secretory efflux transporters like P-gp may ultimately play a major role in limiting drug uptake into the brain parenchyma. As shown in **Figure 12–25**, a drug may permeate the apical membrane and be taken up into the brain endothelial cell. However, efflux transporters like P-gp are able to move the drugs back across the apical membrane and into the blood, protecting the brain from toxic substances or preventing drug absorption into the brain tissue. So, in addition to having a physically tight endothelium, membrane transporters play a major role in the brain's barrier properties. The following example with the HIV protease inhibitor amprenavir provides a good demonstration of the important role that membrane transporters play in the function of the BBB.

### EXAMPLE 12–5

**Amprenavir Brain Penetration**

To examine the role of P-gp in the effectiveness of the blood–brain barrier, Polli et al.[50] examined the brain uptake of the HIV protease inhibitor amprenavir in mice. They examined the effect of the coadministration of ritonavir, another HIV protease inhibitor (GF120918), a specific P-gp inhibitor, or "genetic" P-gp–knockout (mdr1a/1b double knockout) mice on the brain uptake of amprenavir. Using whole-body autoradiography, they were able to visualize the brain uptake of amprenavir under these three conditions. In mice treated with GF120918 and in the genetic knockout mice, they observed a 14- and 27-fold increase in brain amprenavir concentrations, respectively, due to the lack of P-gp. This can be visualized nicely in Figure 12–26. In Figure 12–26*a* the brain and CSF are shaded gray, indicating that amprenavir was able to penetrate the brain when P-gp was inhibited by GF120918. The control animal in Figure 12–26*b* shows is no amprenavir in the brain when P-gp is active and functional, suggesting that membrane transporters such as P-gp are an effective part of the blood–brain barrier. Ritonavir did not have an effect on amprenavir brain concentrations.

## Gastrointestinal Tract[51]

The GI tract is depicted in **Figure 12–27**. The role of GI tract in drug absorption is clearly evident. However, more recently the realization that the GI tract and, more specifically, the intestine play a role in drug metabolism and excretion has occurred. The stomach provides several major functions that affect the bioavailability of orally administered drugs. It processes food into chyme by vigorous contractions that mix ingested contents with gastric secretions and assist intestinal absorption. It regulates input of the liquefied nutrients into the small intestine. The stomach is a major site of chemical and enzymatic degradation. Because the stomach controls the rate of input into the intestine, where the majority of

## KEY CONCEPT  BLOOD LEVELS AND RATES OF ABSORPTION/ELIMINATION

The rates of absorption and elimination of drugs ultimately determine the resulting blood levels of drug that are achieved in the blood circulation, organs, tissues, and cells. Each point on a blood/plasma/serum drug concentration versus time curve reflects the rates of absorption and elimination at that time point (Fig. 12–28). From now on these curves will be referred to as plasma level versus time curves (PLTCs). If the rate of absorption is greater than the rate of elimination at that time point, the slope of the PLTC will be positive and the plasma concentrations are increasing. If the rate of absorption is slower than the rate of elimination at a given time point, the slope of the PLTC will be negative and the plasma concentrations are decreasing. When the rates of absorption and elimination are equal, the slope is zero and the corresponding (*x, y*) time point is known as ($T_{max}$, $C_{max}$). As shown in Figure 12–28a, when the net rate of input into the body decreases, the slope of the absorptive phase also decreases and there is a shift of $T_{max}$ (to a larger value) and $C_{max}$ (to a lower value) as well. A slower input rate would result from a lower permeability, lower solubility, or slower gastric emptying rate. When the input rate is held constant but the rate of elimination is varied, a similar situation occurs (Fig. 12–28b). As the elimination rate constant increases, there is a shift in $T_{max}$ to the left (i.e., shorter) and a decrease in $C_{max}$.

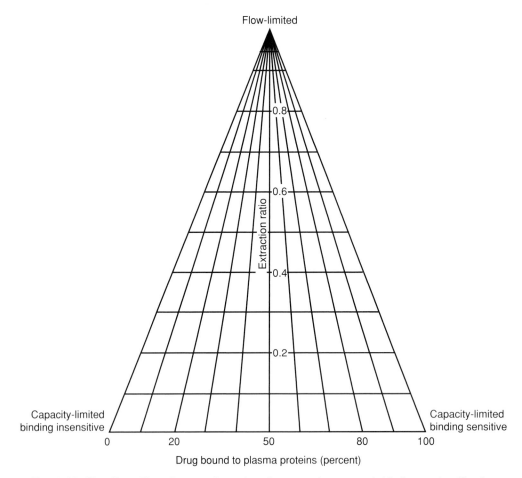

**Fig. 12–22.** The effect of hepatic extraction ratio and percent plasma protein binding on classification of hepatic clearance (flow limited; capacity-limited, binding sensitive and capacity-limited, binding insensitive). Any drug metabolized primarily by the liver can be plotted on the triangular graph. The closer a drug falls to a corner of the triangle, the more likely it is to have the characteristic changes in disposition in liver disease. (From T. F. Blaschke, Clin. Pharmacokinet. **2**, 40, 1977. With permission.)

drug and nutrient absorption takes place, it has a considerable impact on the pharmacokinetics of drugs. This is because it controls the concentration of drugs at the most important site of GI absorption—the small intestine. Therefore, if a drug's permeability and solubility are not low and do not limit their absorption, the gastric emptying rate will essentially control the blood concentration–time profile of the drug. If gastric emptying is slower, then the net absorption rate will be slower, the peak blood levels, $C_{max}$, will be lower, and the time to peak blood levels, $T_{max}$, will be longer.

### EXAMPLE 12–6[52]

**Gastric Emptying**

Gastric emptying rate significantly affects drug absorption and the appearance of drug in the blood. Acetaminophen is a high-permeability and high-solubility drug. Therefore, the appearance of acetaminophen in the blood is strictly related to its emptying from the stomach and presentation to the absorbing site, the small intestine. In this study, acetaminophen was administered to five healthy male individuals. To stimulate gastric emptying, they also received metoclopramide. To reduce the gastric emptying rate, they received propantheline. As can be seen in Figure 12–29, altering the gas-

tric emptying rate significantly altered the rate of acetaminophen absorption. In fact, Figure 12–29 is very similar to the theoretical expectation as seen in the simulation in Figure 12–28a. Although this example demonstrates the role of gastric emptying, one should keep in mind that reduced permeability, reduced solubility, or even slower release from a drug product would result in qualitatively similar behavior. Examination of PLTCs has limited value because it is easy to conclude from Figure 12–29 that absorption rate was slowed, but the cause of the reduced absorption rate cannot be understood without further information about the drug or its biopharmaceutics.

The stomach can be thought of as a two-part system, the upper part consisting of the fundus and upper body and the lower part consisting of the antrum and lower body. These two sections affect the motility of gastric contents and are very different. The upper section acts as a reservoir that can expand to accommodate ingested materials. This expansion does not cause a significant increase in internal pressure and helps generate a pressure gradient between the stomach and the small intestine. Gastric emptying is controlled by a gastric pacemaker, a group of smooth muscle cells in the midcorpus on the greater curvature of the stomach. Neural control of

TABLE 12–9
## SUBCELLULAR FRACTIONS AND METABOLIC REACTIONS

| Fraction | Centrifugation | Metabolic Reaction |
|---|---|---|
| Nuclei and cell debris | $500 \times g$ | Little metabolic activity |
| Mitochondira | $8000 \times g$ | Glycine conjugation, fatty acid $\beta$-oxidation, monoamine oxidase |
| Lysosomes | $15,000 \times g$ | Ester hydrolysis, not so much involved in drug metabolism |
| Microsomes | $100,000 \times g$ | Most of phase I reaction, glucuronidation, N-, O-methylation |
| Cytosol | $100,000 \times g$ supernatant | Hydrolysis, alcohol and aldehyde dehydrogenase, sulfate and glutathione conjugation, acetylation |

gastric emptying occurs by means of extrinsic and intrinsic innervation. Contractions occur at a basal rate of three to four cycles per minute or as peristaltic waves initiated by the entry of solids into the stomach.[53] Emptying occurs at a constant rate because the antrum maintains a relatively constant volume.[54] The proximal stomach controls the emptying of liquids. It is directly related to the gastroduodenal pressure gradient.[55] Noncaloric liquids such as sodium chloride empty from the stomach in a monoexponential pattern, the rate decreasing as intragastric volume and pressure decrease. If the intragastric fluid is caloric, acidic, or nonisotonic, initial emptying is retarded and then follows a more linear pattern.[56] The lower section of the stomach acts as a forceful grinder by developing powerful peristaltic contractions. These waves of contraction increase in force as they near the pylorus. When these forceful waves reach the pylorus, the membrane that separates the stomach from the duodenum is opened, and the contents of the stomach are administered as spurts of chyme.

Gastric motility is controlled by a very complex set of neural and hormonal signals. For instance, the system has a feedback loop in case the chyme is too acidic. Whereas gastrin is a hormone that stimulates gastric acid secretion, motilin is associated with housekeeping waves of motility that occur in the fasted condition. The fasted gastric motility cycle serves two functions and occurs as four "phases." This cycle repeats about every 2 hr during the fasted state. Phase I typically lasts 40 to 60 min and consists of a gentle mixing period due to smooth muscle quiescence, during which there are only rare contractions. Phase II follows with peristaltic contractions occurring with an increase in frequency for approximately 25 to 40 min. These waves of activity originate in the stomach and propagate through the small intestine. Phase III is sometimes referred to as the "housekeeper" wave because the pylorus remains open to allow indigestible particles that are less than 12 mm in size to pass into the small intestine. Particles that are larger than 12 mm are rejected by the pylorus and remain in the stomach until they become small enough to pass. Phase III, which lasts 15 to 25 min, is characterized by powerful peristaltic contractions that occur three times a minute and empty the stomach into the small intestine. Phase IV lasts up to 7 min and is a transition between the forceful contractions of phase III and the gentle mixing contractions of phase I. The pH of fasting healthy adults is approximately 2 to 3, whereas fed-state pH is considerably higher, in the range of pH 5 to 6.

The volume and composition of ingested food determines the rate of gastric emptying.[57] Gastric emptying of liquids is

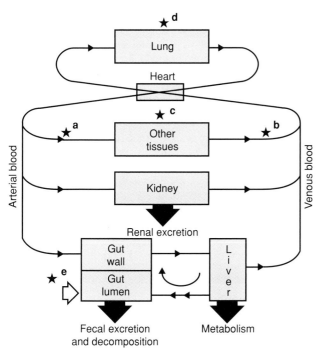

**Fig. 12–23.** Once absorbed from any of the many sites of administration, drug is conveyed by blood to all sites within the body including the eliminating organs. Sites of administration include: a, artery; b, peripheral vein; c, muscle and subcutaneous tissue; d, lung; and e, gastrointestinal tract. The dark- and light-colored lines with arrows refer to the mass movement of drug in blood and in bile, respectively. The movement of virtually any drug can be followed from site of administration to site of elimination. (From M. Rowland and T. N. Tozer, *Clinical Pharmacokinetics: Concepts and Application*, 3rd Ed., Lippincott Williams & Wilkins, Baltimore, MD, 1995, p. 12.)

**TABLE 12–10**

**WEIGHTS OF VARIOUS ORGANS IN THE MOUSE, RAT, RABBIT, MONKEY, DOG, AND HUMAN***

| Organ | Mouse (0.02 kg) | Rat (0.25 kg) | Rabbit (2.5 kg) | Rhesus Monkey (5 kg) | Dog (10 kg) | Human (70 kg) |
|---|---|---|---|---|---|---|
| Brain | 0.36 | 1.8 | 14 | 90 | 80 | 1400 |
| Liver | 1.75 | 10.0 | 77 | 150 | 320 | 1800 |
| Kidneys | 0.32 | 2.0 | 14 | 25 | 50 | 310 |
| Heart | 0.08 | 1.0 | 5 | 18.5 | 80 | 330 |
| Spleen | 0.1 | 0.75 | 1 | 8 | 25 | 180 |
| Adrenals | 0.004 | 0.05 | 0.5 | 1.2 | 1 | 14 |
| Lung | 0.12 | 1.5 | 18 | 33 | 100 | 1000 |

*Organ weights are given in grams.
From B. Davies and T. Morris, Pharm. Res. **10,** 1093, 1993.

rapid (half-time is about 12 min so that 95% is emptied within 1 hr).[58] An increase in caloric content generally slows gastric emptying so that the rate of delivery of calories into the duodenum is relatively constant.[59] It is estimated that nearly 50% of a solid meal remains in the stomach after ~2 hr. The temperature of the ingested meal is not important for liquids, which conduct heat rapidly, but may delay the emptying of hot or cold semisolid or solid meals, which have a higher thermal inertia. Gastric emptying occurs more rapidly in the morning than in the evening.[60] Gastric emptying is slightly slower in healthy individuals older than 70 years of age of both sexes,[61] even though the absorption of orally administered drugs does not seem to vary with age.[61,62] The results of studies of the effect of body weight on gastric emptying of solids and liquids are inconsistent. Accelerated,[63] delayed,[64] and unchanged gastric emptying[65] have all been reported. The differences in emptying rates are difficult to explain, but it appears that moderate obesity is not a major modifying factor, although the emptying of solids may be delayed in obese individuals who are at least 63% in excess of ideal weight.[66] The influence of gender on gastric emptying is controversial. Whereas some authors have found similar gastric emptying rates for men and women,[61,66] others have found slower gastric emptying in women than in men.[67,68] The difference could be attributed to the phase of the menstrual cycle at the study time because the rate of solid gastric emptying decreases linearly during the menstrual cycle toward the luteal phase (days 19–28). The emptying of liquids does not differ between the two phases of the cycle.[66,69] Pregnancy is believed to delay gastric emptying. However, the majority of studies have not shown delayed gastric emptying of liquids in women presenting during the first or second trimester for terminations of pregnancy, at elective caesarean section,[70] and at first and third postpartum days.[71] Absorption may occur from the stomach, but, typically, absorption is minimal. Although nonionized lipophilic molecules of moderate size may be absorbed, and even though the duration of exposure to epithelium is short, there is very little absorption

**TABLE 12–11**

**VOLUMES OF VARIOUS BODY FLUIDS AND ORGANS IN THE MOUSE, RAT, RABBIT, MONKEY, DOG, AND HUMAN***

| | Mouse (0.02 kg) | Rat (0.25 kg) | Rabbit (2.5 kg) | Rhesus Monkey (5 kg) | Dog (10 kg) | Human (70 kg) |
|---|---|---|---|---|---|---|
| Brain | – | 1.2 | – | – | 72 | 1450 |
| Liver | 1.3 | 19.6 | 100 | 145 | 480 | 1690 |
| Kidneys | 0.34 | 3.7 | 15 | 30 | 60 | 280 |
| Heart | 0.095 | 1.2 | 6 | 17 | 120 | 310 |
| Spleen | 0.1 | 1.3 | 1 | – | 36 | 192 |
| Lungs | 0.1 | 2.1 | 17 | – | 120 | 1170 |
| Gut | 1.5 | 11.3 | 120 | 230 | 480 | 1650 |
| Muscle | 10.0 | 245 | 1450 | 2500 | 5530 | 35,000 |
| Adipose | – | 10.0 | 120 | – | – | 10,000 |
| Skin | 2.9 | 40.0 | 110 | 500 | – | 7800 |
| Blood | 1.7 | 14.5 | 165 | 367 | 900 | 5200 |
| Total body water | 14.5 | 167 | 1790 | 3465 | 6036 | 42,000 |
| Intracellular fluid | – | 92.8 | 1165 | 2425 | 3276 | 23,800 |
| Extracellular fluid | – | 74.2 | 625 | 1040 | 2760 | 18,200 |
| Plasma volume | 1.0 | 7.8 | 110 | 224 | 515 | 3000 |

*Organ and other volumes are given in milliliters.
From B. Davies and T. Morris, Pharm. Res. **10,** 1093, 1993.

### TABLE 12–12
### FLOW OF BLOOD THROUGH THE MAJOR ORGANS AND FLOW OF OTHER FLUIDS IN THE MOUSE, RAT, RABBIT, MONKEY, DOG, AND HUMAN*

|  | Mouse (0.02 kg) | Rat (0.25 kg) | Rabbit (2.5 kg) | Rhesus Monkey (5 kg) | Dog (10 kg) | Human (70 kg) |
|---|---|---|---|---|---|---|
| Brain | – | 1.3 | – | 72 | 45 | 700 |
| Liver | 1.8 | 14.8 | 177 | 218 | 309 | 1450 |
| Kidneys | 1.3 | 9.2 | 80 | 148 | 216 | 1240 |
| Heart | 0.28 | 3.9 | 16 | 60 | 54 | 240 |
| Spleen | 0.09 | 0.63 | 9 | 21 | 25 | 77 |
| Gut | 1.5 | 7.5 | 111 | 125 | 216 | 1100 |
| Muscle | 0.91 | 7.5 | 155 | 90 | 250 | 750 |
| Adipose | – | 0.4 | 32 | 20 | 35 | 260 |
| Skin | 0.41 | 5.8 | – | 54 | 100 | 300 |
| Hepatic artery | 0.35 | 2.0 | 37 | 51 | 79 | 300 |
| Portal vein | 1.45 | 9.8 | 140 | 167 | 230 | 1150 |
| Cardiac output | 8.0 | 74.0 | 530 | 1086 | 1200 | 5600 |
| Urine flow | 1.0 | 50.0 | 150 | 375 | 300 | 1400 |
| Bile flow | 2.0 | 22.5 | 300 | 125 | 120 | 350 |
| GFR | 0.28 | 1.31 | 7.8 | 10.4 | 61.3 | 125 |

*All blood flows are in mL/min; urine and bile flows and glomerular filtration rate (GFR) are in mL/day.

because of the small epithelial surface area and the physically thick, viscous mucus layer.

Absorption of drugs, fluid, and nutrients can occur from each section of the small intestine and colon. The absorption of fluids, nutrients, electrolytes, and xenobiotics occurs as chyme moves through the GI tract. The small intestine is partitioned into three sections of different sizes and function, the duodenum, the jejunum, and the ileum. Water is able to flow into or out of the lumen to maintain the isotonicity of the luminal contents with plasma. Approximately 8 to 9 liters

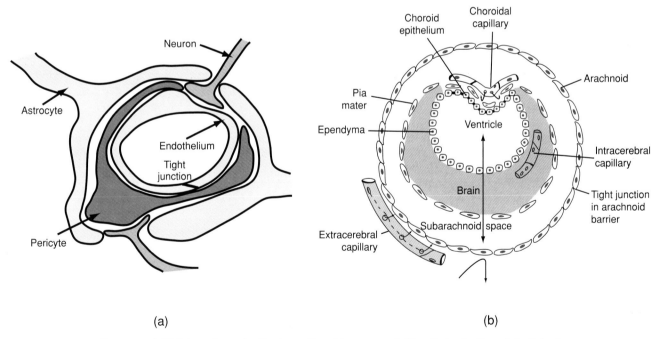

(a)                                                                (b)

**Fig. 12–24.** (*a*) This is a schematic of brain capillary. It is composed of four kinds of cells; endothelial cell, pericyte, astrocyte, and neurons. Because of the close anatomical proximity of the cells, they stimulate endothelial cells to proliferate and differentiate. (*b*) This is a schematic of the brain-barrier system. The capillary has a tight junction. Hence, it acts as a blood–brain barrier in the brain parenchyma. There are two kinds of blood–CSF barriers. One is the arachnoid membrane and the other is the epithelial cell of choroid plexus. These capillaries are permeable. So, in these areas only arachnoid membrane and epithelial cells of choroid plexus can function as a barrier. (From B. Schlosshauer, Bioessays **15**, 341, 1993. With permission.)

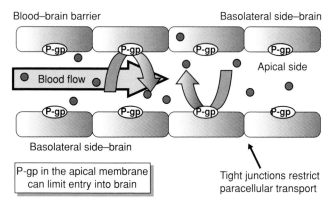

**Fig. 12–25.** P-glycoprotein, an efflux secretory transporter, is widely thought to limit entry of drugs into the brain, testis, intestines, and other organs and tissues. Drugs enter cells but are effluxed out of the cell by P-gp before they can enter the brain. This mechanism is responsible for minimizing brain exposure to toxic chemicals.

of fluid enter the upper GI tract every day—approximately 7 liters of secreted juices and 1.5 liters of ingested fluid. About 1 liter of fluid enters the colon, and only 100 mL of water leaves the body in the feces. For most drugs, the duodenum and the proximal jejunum are the best sites of absorption

because they have the highest absorptive surface area. In general, absorptive surface area decreases as one travels down the intestine.[72] The ratio of the absolute surface area of the human stomach to that of the small intestine is 1 to 3800; this shows why absorption of substances by the stomach is generally neglected. Similarly, the 570-fold difference[73] between the small intestine and the colon suggests that the majority of absorption occurs in the small intestine. However, although this takes into account the surface area, the transit time of the colon is 4 to 24 times longer (i.e., 12–72 hr as compared to 3–4 hr) than in the small intestine. Therefore, a longer residence time could offset a lower absorptive surface area, making the colon as good site for the absorption of drugs as the small intestine.

Small intestinal absorption is also dramatically affected by regional differences in the distribution of transporters, metabolic enzymes, and so on. The practical implication of this is that even though the absorptive surface area in the duodenum is higher than in the ileum, absorption from the ileum is not necessarily lower for drugs and nutrients. For example, intestinal reabsorption of bile salts plays a crucial role in human health and disease. The small intestine absorbs 90% to 95% of the bile salts. Of the remaining bile salts, the colon converts the salts of deoxycholic acid and

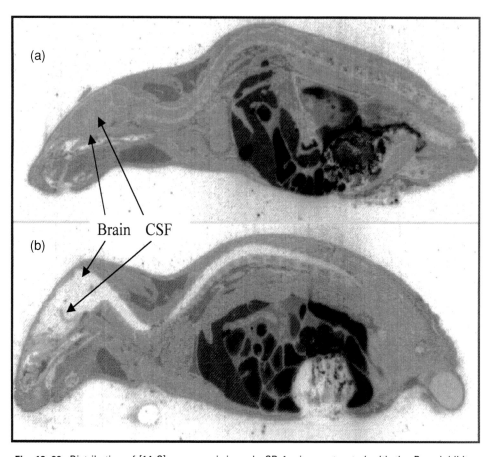

**Fig. 12–26.** Distribution of [14 C]-amprenavir in male CD-1 mice pretreated with the P-gp inhibitor GF120918. Animals treated with GF120918 (*a*) had a 13-fold increase in brain and 3.3-fold increase in CSF levels of amprenavir-related material over vehicle-treated mice (*b*).

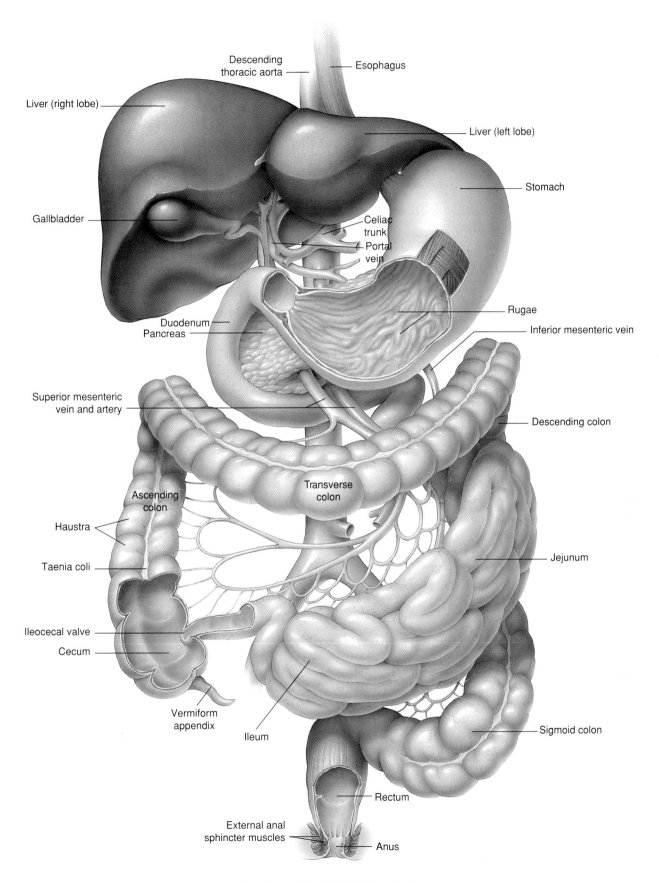

**Fig. 12–27.** Gastrointestinal system.

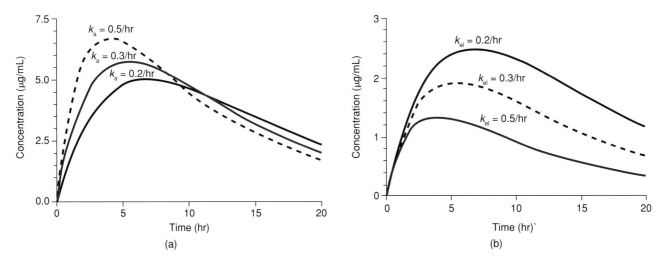

**Fig. 12–28.** Effect of absorption and elimination rate constant on the plasma concentration versus time profile. (*a*) Absorption (input) rate is increased from 0.2/hr to 0.5/hr while holding the elimination rate constant resulting in an increase in the slope of the absorptive phase. (*b*) The input rate is held constant while elimination rate constant is changed. Note that absorptive phase is unchanged (i.e., slopes are equal).

lithocholic acid. Only 1% of the lithocholate is absorbed, and the colon excretes the rest. The bile salts lost to excretion in the colon are replaced by synthesis of new ones in the liver at a rate of 0.2 to 0.4 g/day, with a total bile salt amount of 3.5 g, which is constantly recycled by enterohepatic circulation.

Enterohepatic circulation is discussed later in this chapter. Bile acid reabsorption is primarily localized in the terminal ileum and is mediated by a 48-kd sodium-dependent bile acid cotransporter known as ASBT, which is given the molecular designation SLC10A2.[74]

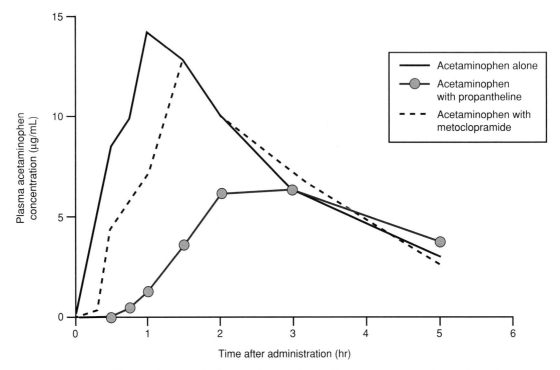

**Fig. 12–29.** Effect of propantheline and metoclopramide on acetaminophen absorption. Acetaminophen absorption is very rapid and is only limited by its introduction into the intestine by the stomach. Metoclopramide increases the rate of acetaminophen gastric emptying resulting in a faster rate of absorption, higher $C_{max}$ and shorter $t_{max}$. Propantheline has the opposite effect, slowing gastric emptying rate and delaying absorption. (From J. Nimmo, R. C. Heading, P. Tothill, and L. F. Prescott, Br. Med. J. **1,** 587, 1973. With permission.)

Region-specific absorption has been reported in animals and humans for a variety of drugs including allopurinol,[75] amoxicillin,[76] benzoate,[77] lefradafiban,[78] oxyprenolol,[79] talinolol,[80] and thymidine analogues.[81] Regional variation in the distribution of drug transporters also brings variatiability in absorption. Regional distribution of apical absorptive transporters including the apical bile acid transporter,[82,83] monocarboxylic acid transporter-1,[84] a nucleoside transporter,[85] OATP3,[86] and the peptide transporter PEPT1[87] has been reported. Segmental variability is also known to occur for metabolic enzymes and efflux/secretory transporters as well. The cytochrome P-450 3 A,[88–94] SULTs, GSTs, and the UDP-glucuronosyltransferases[95–99] are higher at the proximal than at the distal intestine. MRP2 intestinal secretion follows the distribution of the cytochrome P-450s and conjugation enzymes,[100,101] whereas P-glycoprotein[93,102–105] is higher in the jejunum/ileum than other parts of the intestine. Basolateral MRP3, in contrast to MRP2, is more prevalent in the ileum and colon.[106]

The impact of the varied regional distributions of drug transporters and metabolizing enzymes is difficult to predict because drugs can be substrates for numerous transporters and enzymes. For example, saquinavir, an HIV protease inhibitor, is known to be a substrate for the transporters P-gp, MRP1, MRP2, OATP-A, OATP-C, and the metabolic enzyme CYP3A.[106–110] In addition to regional intestinal distribution, substrate affinity, enzyme/transporter capacity, turnover rate, and other factors ultimately determine the segmental absorption behavior and pharmacokinetics of drugs. Changes also occur in the characteristics of the paracellular spaces throughout the intestine. Intestinal pH is relatively constant and ranges from about pH 5 in the duodenal bathing region of the upper small intestine to pH 6.5 to 7.2 in other areas of the intestine and colon.

## Kidney

Excretion is the process by which a drug or a metabolite is eliminated from the body without further chemical change. The kidneys, which transport water-soluble substances out of the body, are the major organs of excretion. The kidney performs two critical functions in the distribution and excretion of drug molecules. They excrete the metabolites formed by the liver or other organs/tissues and control the concentrations of many of the molecules found in the blood stream. The kidney does this by filtration of the blood. A depiction of a kidney is shown in **Figure 12–30**. Blood enters the glomerulus through the afferent arteriole and leaves through the efferent arteriole. About one fifth of the plasma reaching the glomerulus is filtered through pores in the glomerular endothelium; the remainder passes through the efferent arterioles surrounding the renal tubules. Drugs bound to plasma proteins are not filtered; only unbound drug is contained in the filtrate.[5] After filtration in the glomerulus, the blood and waste/filtrate streams continue to be processed by the nephron, the individual working unit of the kidney. There are approximately

1 million nephrons in each kidney. The glomerular filtrate has essentially the same composition as the plasma that entered the glomerulus without a significant amount of protein and no red blood cells. Filtration occurs in the glomerulus by size and charge exclusion. However, secretion and reabsorption occurring in the tubules occur because of the permeability of the molecule being transported. The pore size of the glomerulus is large enough to allow molecules that are up to 8 nm in diameter to pass through.

As seen in **Table 12–13**, there is a steep molecular weight dependence on permeability in the kidney. The permeability of the solute is affected by size and charge if it is transported by passive diffusion; however, in the kidney, solutes are transported out of the tubules by active transport. The primary reason is that passive diffusion occurs from regions of higher solute concentrations to lower concentrations. Because typical solute concentrations in the blood will be more dilute than those in the collecting duct and urine, diffusion out will not be favored. Another important factor is that the pores are lined with proteoglycans that have a very strong negative charge. It is this electrostatic repulsion that keeps albumin, which is only 6 nm in diameter, and most other proteins greater than molecular weight 69,000 from being filtered in the glomerulus. The kidney has a blood flow of 1200 mL/min, which creates a flow from the glomerulus into the proximal tubule of 125 mL/min. The bulk of this fluid flow is water, and if water was not actively reabsorbed, 180 liters of water would be lost each day. Fortunately, more than 99% of the water and varying amounts of its solutes are normally reabsorbed into the blood by way of the proximal tubules. This concentrates the filtrate greatly. The filtrate that comes from the glomerulus passes through the proximal tubule, where conservation of ions, glucose, and amino acids occurs by active and passive transport. In the proximal tubule, these molecules are reabsorbed from the glomerulus filtrate by the blood in the efferent arteriole. About 65% of the glomerular filtrate is reabsorbed before reaching the loop of Henle. The filtrate continues moving through the loop of Henle and distal tubule, where it is continuously reabsorbed. The maximal rate of reabsorption for various substances is shown in **Tables 12–14** and **12–15**. These values indicate the maximum rate at which a species can be reabsorbed. The transport rate, however, may not be linear in concentration. This occurs when a system undergoes saturation kinetics. Although the efferent arteriole is in the process of reabsorbing water and other vital ions and solutes from the tubules, it also secretes molecules into the tubules. The remaining substance in the tubules enters a collecting duct and is considered urine. The ability of the kidneys to clean or clear the plasma of various substances is defined as plasma clearance:

Plasma clearance (mL/min) =

$$\frac{\text{Urine flow (mL/min)} \times \text{Concentration in urine}}{\text{Concentration in plasma}} \quad (12\text{–}13)$$

There are several substances that are routinely measured to determine kidney function: creatinine, inulin, and

Cross section

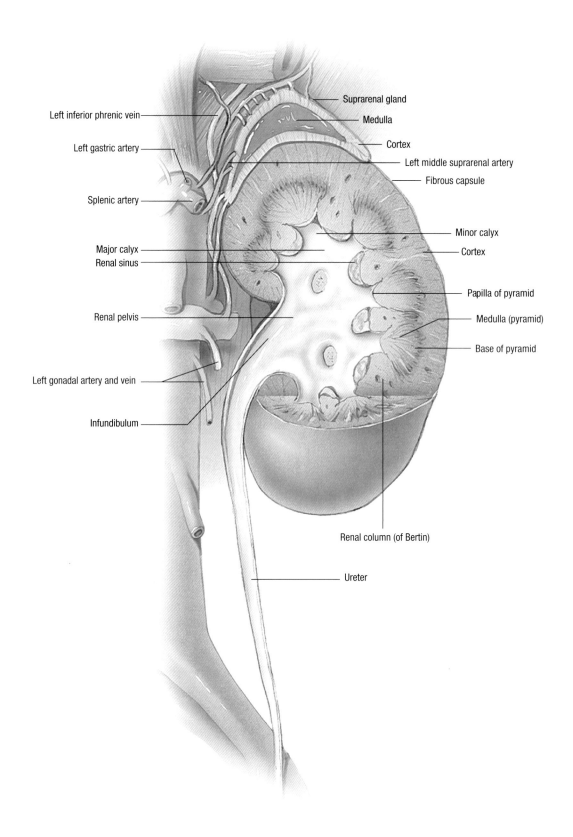

**Fig. 12–30.** Left kidney and adrenal gland.

## TABLE 12–13
### PERMEABILITY AND MOLECULAR SIZE

| Molecular Weight | Permeability Compared to Water | Example Substance |
|---|---|---|
| 5200 | 1.00 | Inulin |
| 30,000 | 0.5 | Very small protein |
| 69,000 | 0.005 | Albumin |

## TABLE 12–15
### TUBULAR TRANSPORT MAXIMUMS OF IMPORTANT SUBSTANCES SECRETED INTO RENAL TUBULES

| | |
|---|---|
| Creatinine | 16 mg/min |
| p-Aminohippurate | 80 mg/min |
| Glomerular filtrate | 125 mL/min |
| Flowing into the loops of Henle | 45 mL/min |
| Flowing into the distal tubules | 25 mL/min |
| Flowing into the collecting tubules | 12 mL/min |
| Flowing into the urine | 1 mL/min |

p-aminohippurate (PAH). These measure glomerular filtration rate and plasma flow through the kidneys. Inulin is not reabsorbed from the tubules and is not actively secreted into the tubules; therefore, any inulin found in the urine comes from glomerular filtration. As shown in **Table 12–16**, inulin is filtered in the glomerulus as easily as water is. Therefore, the plasma clearance of inulin is equal to the glomerular filtration rate. In terms of ADME and pharmacokinetics, the kidney is a primary organ of drug excretion. Drugs may be filtered by the glomerulus, reabsorbed into the blood stream by the proximal tubule, or secreted from the blood stream into the distal tubule. For proteins, general rules for glomerular filtration are as follows: (a) If the protein is bigger than immunoglobulin G (150 kd, 55-Å radius), it is rarely excreted; (b) if the protein is smaller than 40 kd and has a radius less than 30 Å, it is almost completely excreted; (c) negatively charged molecules are retained, even if small, because of charge repulsion with Bowman's space; and (d) elongated molecules have higher clearance than spherical molecules.

The limited ability of the kidneys to clear large materials from the body has been used as a way to increase the circulation time and decrease the clearance of drugs. Numerous studies have been published showing that polymers of similar size to proteins are cleared in a similar manner. For example, the molecular-weight threshold limiting glomerular filtration of an HPMA (N-(2-hydroxypropyl)methacrylamide) copolymer was found to be about 45 kd in rats.[111] In mice, the molecular-weight cutoff was found to be about 30 kd for poly(ethylene glycol) (PEG).[112] This size limitation has been exploited in drug delivery, with PEG being the most

common polymer employed to date. PEG is advantageous as a protein-modifying agent because it is inert, water-soluble, nontoxic, and modular in size. Pegylation (i.e., chemically adding a PEG to a therapeutic agent) is now a well-established method of modifying the pharmacologic properties of a protein by, for example, delaying clearance and reducing protein immunogenicity.[110–115] Among the various disease states that have been targeted for the study of drugs incorporating pegylation technology, the treatment of chronic hepatitis C with interferon-based compounds offers significant potential for clinical impact. Two compounds, peginterferon alfa-2a (PEGASYS) and peginterferon alfa-2b (PEG-Intron), are both approved for use alone and in combination with ribavirin for the treatment of chronic hepatitis C. However, the different PEG moieties attached to the native protein, the site of attachment, and the type of bond involved lead to vast differences with respect to the pharmacokinetics and pharmacodynamics of these two compounds.

## EXAMPLE 12–7

### The Design and Development of Pegfilgrastim (PEG-rmetHuG-CSF, Neulasta)

The following is the abstract of a paper by Molineux[116]:

> The addition of a polyethylene glycol (PEG) moiety to filgrastim (rmetHu-G-CSF, Neupogen) resulted in the development of pegfilgrastim. Pegfilgrastim is a long-acting form of filgrastim that requires only once-per-cycle administration for the management of chemotherapy-induced neutropenia. Pegylation increases the size of filgrastim so that it becomes too large for renal clearance. Consequently, neutrophil-mediated clearance predominates in elimination of the drug. This extends the median serum half-life of pegfilgrastim to 42 hr, compared with between 3.5 and 3.8 hr for Filgrastim, though in fact the half-life is variable, depending on the absolute neutrophil count, which in turn reflects the ability of pegfilgrastim to sustain production of those same cells. The clearance of the molecule is thus dominated by a self-regulating mechanism. Pegfilgrastim retains the same biological activity as filgrastim and binds to the same G-CSF receptor, stimulating the proliferation, differentiation, and activation of neutrophils. Once-per-chemotherapy cycle administration of pegfilgrastim reduces the duration of severe neutropenia as effectively as daily treatment with filgrastim. In clinical trials, patients receiving pegfilgrastim also had a lower observed incidence of febrile neutropenia than patients receiving filgrastim.

## TABLE 12–14
### TUBULAR TRANSPORT MAXIMUMS OF IMPORTANT SUBSTANCES REABSORBED FROM RENAL TUBULES

| Substance | Value | Units |
|---|---|---|
| Glucose | 320 | mg/min |
| Phosphate | 0.1 | mm/min |
| Sulfate | 0.06 | mm/min |
| Amino acids | 1.5 | mm/min |
| Urate | 15 | mg/min |
| Plasma protein | 30 | mg/min |
| Hemoglobin | 1 | mg/min |
| Lactate | 75 | mg/min |
| Acetoacetate | Variable, ~30 | mg/min |

**TABLE 12–16**

**RELATIVE CONCENTRATIONS OF SUBSTANCES IN THE GLOMERULAR FILTRATE AND IN THE URINE**

| Substance[*] | Glomerular Filtrate (125 mL/min) | | Urine (1 mL/min) | | Urine/Concentration in Plasma (plasma clearance per minute) |
|---|---|---|---|---|---|
| | Quantity/min (mEq) | Concentration (mEq/liter) | Quantity/min (mEq) | Concentration (mEq/liter) | |
| $Na^+$ | 17.7 | 142 | 0.128 | 128 | 0.9 |
| $K^+$ | 0.63 | 5 | 0.06 | 60 | 12 |
| $Ca^{2+}$ | 0.5 | 4 | 0.0048 | 4.8 | 1.2 |
| $Mg^{2+}$ | 0.38 | 3 | 0.015 | 15 | 5.0 |
| $Cl^-$ | 12.9 | 103 | 0.144 | 144 | 1.3 |
| $HCO_3^-$ | 3.5 | 28 | 0.014 | 14 | 0.5 |
| $H_2PO_4^-$ | 0.25 | 2 | 0.05 | 50 | 25 |
| $HPO_4^{2-}$ | 0.25 | 2 | 0.05 | 50 | 25 |
| $SO_4^{2-}$ | 0.09 | 0.7 | 0.033 | 33 | 47 |
| Glucose | 125[†] | 100[†] | 0[†] | 0[†] | 0 |
| Urea | 33 | 26 | 18.2 | 1820 | 70 |
| Uric acid | 3.8 | 3 | 0.42 | 42 | 14 |
| Creatinine | 1.4 | 1.1 | 1.96 | 196 | 140 |
| Inulin | – | – | – | – | 125 |
| PAH | – | – | – | – | 585 |

[*]PAH = *p*-Aminohippurate.

[†]Units for glucose are mg for quantity and mg/dL for concentration.

From *Textbook of Medical Physiology*, 8th Ed., W. B. Saunders, Philadelphia, PA, 1991, p. 304.

## Liver

The liver is an extremely important organ in biopharmaceutics and pharmacokinetics. After drug is absorbed from the gut, it potentially undergoes metabolism in the liver, secretion from the liver into bile, or reaches the systemic circulation intact. Of course, metabolites may also be secreted into bile, further metabolized, or make it into the systemic circulation. The liver is unique in its blood supply because it receives oxygenated blood from the hepatic artery and nutrient-rich but deoxygenated blood from the stomach, intestine, and spleen. The split between the two streams is approximately one fifth oxygenated and the remainder is nutrient rich. In most cases, the liver is thought of as containing lobules serviced/drained by a central vein in the center of each. However, the liver can functionally be thought of as being organized into acini, with two input streams, the hepatic artery and the portal vein, passing through the sinusoids and leaving through a terminal hepatic vein (**Fig. 12–31**). The sinusoids are lined with unique epithelial cells called hepatocytes. The hepatocytes have distinct polarity. Their basolateral side is lined with microvilli to take up nutrients, proteins, and xenobiotics. The apical side forms the canalicular membranes through which bile components are secreted. It is the hepatocytes that perform functions essential for life. These functions include the production of bile and its carriers (bile acids, cholesterol, lecithin, phospholipids), the synthesis of essential serum proteins (albumin, carrier proteins, coagulation factors, many hormonal and growth factors), the regulation of nutrients and metabolism, and the conjugation of lipophilic compounds (bilirubin, cations, drugs) for excretion in the bile or the urine.[117]

Earlier in the chapter, there was a discussion about metabolizing enzymes and transporters. These two systems are found in abundance within liver and play a major role in drug distribution and elimination. The liver is a major site of metabolism in the body, and it works with the kidney in removing waste from the blood stream. As mentioned previously, there are

## KEY CONCEPT  IMPORTANCE OF EQUILIBRIA IN ADME

It is very important to realize that ADME is filled with a number of dynamic equilibria that occur in a variety of organs and tissues. The net result of all of these processes is the observed plasma concentration versus time profiles. From the moment that a drug enters the body, the drug molecule strives to be in equilibrium between the tissues and blood. Other equilibria that occur are blood: active site concentration, parent: metabolite, blood: bile, blood: urine, and bound drug: unbound drug. These are just some of the equilibria that occur and the ones that play an important role in the blood or plasma concentration level measured in the study of pharmacokinetics. The rates of absorption, distribution, and elimination control drug–blood concentrations and are discussed further in the next section.

**Cross section of liver lobule**

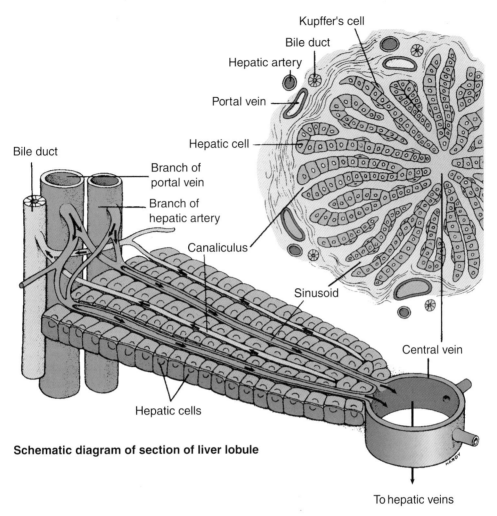

**Schematic diagram of section of liver lobule**

**Fig. 12–31.** A section of liver lobule showing the location of hepatic veins, hepatic cells, liver sinusoids, and branches of the portal vein and hepatic artery. (From S. C. Smeltzer and B. G Bare, *Textbook of Medical-Surgical Nursing*, 9th Ed., Lippincott Williams & Wilkins, Philadelphia, 2000.)

three phases of drug metabolism. Phases I and II are involved in the biotransformation of drugs and, therefore, are strictly related to the ADME process of metabolism. In addition to metabolism, the liver also plays an important role in drug and metabolite excretion out of the body. This process is also known as phase III metabolism and involves the transport of drugs and metabolites out of cells by means of membrane transporters. In the liver this process is known as enterohepatic cycling and occurs by biliary excretion from the gall bladder and intestinal reabsorption of a solute (i.e., drug or metabolite), sometimes with hepatic conjugation (see phase II discussion earlier in the chapter) and intestinal deconjugation. Therefore, the liver's role in drug distribution or excretion occurs in conjunction with the intestine and the gall bladder. Drug is absorbed from the intestine and enters the liver, where the drug or metabolites can be secreted into the bile of the gall bladder. The gall bladder secretes bile, usually in conjunction with meals, and the drugs and metabolites

reenter the intestinal tract. Therefore, the biliary "system" contributes to excretion to the degree that drug is not reabsorbed from the GI tract. In other words, the drug or metabolite is considered eliminated from the body as long as it is not reabsorbed from the intestine. On the other hand, the biliary system also contributes to drug distribution to the extent that intact secreted drug is reabsorbed from the intestine. In a fairly unique set of circumstances, in the enterohepatic cycling system even metabolized drug, usually a terminal step, can be reversed, adding to the distribution phase of drug disposition. For example, it has been shown that phase II metabolism (i.e., conjugation), particularly with glucuronic acid, typically leads to biliary excretion. Drug conjugates secreted into the intestine also undergo enterohepatic cycling when they are hydrolyzed and the drug becomes available for reabsorption. Metabolism is usually considered to be part of the elimination process (i.e., permanent removal from the body). However, the process of enterohepatic recycling could also be considered as distribution because the metabolism step is reversible and drug can be absorbed over and over again into the body. Once again, secreted conjugates that are not converted back to the original drug and are excreted in the feces are considered to be "eliminated."

Drugs and their metabolites that are extensively excreted in bile are transported across the biliary epithelium against a concentration gradient requiring active secretory transport by membrane transporters. Secretory transport may approach an upper limit at high plasma concentrations of a drug, and substances with similar physicochemical properties may compete for excretion via the same mechanism. The drug transporters responsible for this behavior are those found in the liver and will not be reviewed here. Factors affecting biliary excretion include drug characteristics (chemical structure, polarity, and molecular size), transport across sinusoidal plasma membrane and cannicular membranes, biotransformation, and possible reabsorption from intrahepatic bile ductules. Intestinal reabsorption to complete the enterohepatic cycle may depend on hydrolysis of a drug conjugate by gut bacteria. Larger drugs (i.e., with a molecular weight greater than 300–500 g/mole) with both polar and lipophilic groups are more likely to be excreted in bile. Smaller molecules are generally excreted only in negligible amounts. The renal and hepatic excretion pathways are complementary to each other. In other words, a compound with high renal excretion, which is typical for a low-molecular-weight compound, will have low biliary excretion and vice versa.[118,119] These values can be species dependent. For example, the excretion of organic anions greater than 500 g/mole is found to occur in the bile in humans, whereas the values are slightly lower for rats, guinea pigs, and rabbits, ranging from 325 to 475 g/mole.[120] Additionally, compounds which are usually excreted into the bile are more lipophilic; contain charged groups, such as carboxylic acid, sulfonic acid, or quaternary ammonium groups, are highly protein-bound anions bound to albumin, whereas cations are mainly bound to orosomucoid or a1-acid glycoprotein and have a high molecular weight. The opposite is true for substrates of renal excretion. These broad classifications should serve only as a guide. Levofloxacin, ofloxacin (Floxin), and ciprofloxacin are broad-spectrum antimicrobial agents for oral administration and are part of a class of fluorinated carboxyquinolones. These drugs are primarily excreted in the urine, yet they are carboxylic acids. For example, only 4% to 8% of Floxin (molecular weight 361.4) is excreted in the feces,[121] which would disprove the rule of carboxylic acids always being excreted in the feces.

Finally, dose dependencies are expected for enterohepatic circulation because membrane transporters play a major role and saturation at high doses or inhibition by competing substances may occur. This could lead to excretion by an alternative pathway or reduced drug excretion and significantly higher blood levels and, possibly, toxicity. In general, enterohepatic cycling may prolong the pharmacologic effect of certain drugs and drug metabolites. The pharmacokinetics (i.e., apparent volume of distribution and clearance) of a drug that undergoes enterohepatic cycling may be substantially altered. Enterohepatic cycling is also associated with the occurrence of multiple drug–blood level peaks and a longer apparent half-life in the plasma concentration–time profile. Of particular importance is the potential amplifying effect of enterohepatic variability in defining differences in the bioavailability. Bioavailability is also affected by the extent of intestinal absorption, gut-wall P-glycoprotein efflux, and gut-wall metabolism. Recently, there has been a considerable increase in our understanding of the role of transporters, gene expression of intestinal and hepatic enzymes, and hepatic zonation. Drugs, disease, and genetics may result in induced or inhibited activity of transporters and metabolizing enzymes. Reduced expression of one transporter, for example, hepatic canalicular multidrug resistance-associated protein-2 (MRP2), is often associated with enhanced expression of others, for example, the usually quiescent basolateral efflux MRP3, to limit hepatic toxicity.

## EXAMPLE 12–8

### Biliary Excretion

Although the first impression about biliary excretion may be that it plays a role in orally absorbed medications, this example shows that drugs introduced into the body by other routes (e.g., intravenously) may also be excreted into the bile. These drugs may have poor oral absorption properties, so enterohepatic cycling is probably minimal for them. Both P-glycoprotein and MRP2, ATP-dependent membrane transporters, exist in a variety of normal tissues and play important roles in the disposition of various drugs. Sugie et al.[122] studied the contribution of P-glycoprotein and/or MRP2 to the disposition of azithromycin in rats. The disappearance of azithromycin from plasma after intravenous administration was significantly delayed in rats treated with intravenous injection of cyclosporine, a P-glycoprotein inhibitor, but was normal in rats pretreated with an intraperitoneal injection of erythromycin, a CYP3A4 inhibitor. When rats received an infusion of azithromycin with cyclosporine and probenecid, an MRP2 inhibitor, a significant decrease in the steady-state biliary clearance of azithromycin of 5% and 40% of the corresponding control values was observed, respectively. However, neither inhibitor altered the renal clearance of azithromycin, suggesting the lack of renal tubular secretion of azithromycin. Tissue distribution experiments showed that azithromycin is distributed largely into liver, kidney, and lungs, whereas neither inhibitor altered the tissue-to-plasma concentration ratio of azithromycin. Significant reduction in the biliary excretion of azithromycin was observed in Eisai hyperbilirubinemic rats, which have a hereditary deficiency in MRP2. These results suggest that azithromycin is a substrate for both P-glycoprotein and MRP2 and that the biliary and intestinal excretion of azithromycin is mediated via these two drug transporters.

# INTRODUCTORY PHARMACOKINETICS

## Introduction

This section is not meant to replace pharmacokinetics textbooks, but rather to link the basic biopharmaceutical concepts introduced in this chapter to the simplest pharmacokinetic models, parameters, and behavior that relate to drug input and output into/from the body. We will cover the correlation

## ▶ KEY CONCEPT APPARENT VOLUME OF DISTRIBUTION[5]

The volume of fluid into which a drug appears to be distributed or diluted is called the apparent volume of distribution (i.e., the fluid volume required to contain the drug in the body at the same concentration as in plasma). This parameter provides a reference for the plasma concentration expected for a given dose and for the dose required to produce a given concentration. However, it provides little information about the specific pattern of distribution. Each drug is uniquely distributed in the body. Some drugs go into fat, others remain in the extracellular fluid, and still others are bound avidly to specific tissues, commonly liver or kidney. Many acidic drugs (e.g., warfarin, salicylic acid) are highly protein bound and thus have a small apparent volume of distribution. Many basic drugs (e.g., amphetamine, meperidine) are avidly taken up by tissues and thus have an apparent volume of distribution larger than the volume of the entire body.

between in vitro and in vivo data using compartment models, permeability, and intrinsic clearance. Pharmacokinetics is the kinetic study of the ADME of drugs in the body. The compartment model assumes that the body is a simplified system of compartments and that drug transfer and elimination rates between/from compartments occur by a first-order process. Other transfer and elimination functions (e.g., nonlinear functions) have also been used in compartment models but will not be the focus of this chapter. A one-compartment model is the simplest and best-studied pharmacokinetic model even though few drugs truly follow these simplified kinetics. A number of in vitro and in situ models have been employed to predict in vivo drug absorption, including the parallel artificial membrane permeability assay, human colon carcinoma cells (Caco-2), Madin–Darby canine kidney (MDCK) cells, Ussing chamber using animal intestinal tissues, and in situ intestinal perfusion. The permeability data from these models, such as apparent permeability, $P_a$, and effective permeability, $P_{eff}$, can be used in the calculation of an absorption rate constant, $K_a$, in the one-compartment model. $P_a$ and $P_{eff}$ are typically synonymous terms and are considered "lumped" permeability coefficients because they represent a measure of all of the transport and metabolism processes occurring at a particular time. In other words, the apparent or effective permeability is the net permeability due to permeability by all pathways in the intended direction but also accounting for loss due to degradation, metabolism, binding, or transport in the opposite direction. In this section, we will also link the basic biopharmaceutical processes to the elimination rate constant, $K_{el}$, which can be calculated using the intrinsic clearance, $CL_{int}$, from in vitro metabolism experiments. The basic assumptions for each type of correlation will be listed and explained in this section with brief introduction of the one compartment model.

## Compartmental Models and $K_a/K_{el}$

In the first model, we will not consider drug absorption but rather drug elimination. In the one-compartment model with rapid intravenous injection, a drug distributes into the body according to one-compartment-model "behavior." In other words, drug distribution in a one-compartment model is

complete and instantaneous. The drug is eliminated by a first-order process,

$$\frac{dX}{dt} = -k_{el}X \tag{12–14}$$

where $X$ represents the amount of drug in the body at time $t$ after administration and $k_{el}$ is the elimination rate constant. Integration of equation (12–14) gives the following expression:

$$X = X_0 E^{-k_{el}t} \quad \text{or} \quad \log X = \log X_0 - \frac{k_{el}t}{2.303} \tag{12–15}$$

where $X_0$ is the initial drug dose. The elimination rate constant, $k_{el}$, can be calculated from two fundamental pharmacokinetic parameters, total body clearance, $CL_t$, and apparent volume of distribution, $V_d$:

$$k_{el} = \frac{CL_t}{V_d} \tag{12–16}$$

$CL_t$ is defined as the volume of plasma or blood that is completely cleared of drug per unit time:

$$CL_t = \frac{-\left(\frac{dX}{dt}\right)}{C} \tag{12–17}$$

$V_d$ is a theoretical volume factor relating the amount of drug in the body and the concentration of drug in the plasma or blood:

$$V_d = \frac{X}{C} \tag{12–18}$$

where $C$ is the drug concentration in plasma or blood.

The elimination rate constant represents the sum of two processes:

$$k_{el} = k_m + k_e \tag{12–19}$$

where $k_m$ is the elimination rate constant by metabolism and $k_e$ is the elimination rate constant by excretion. If the metabolism is dominant over excretion during elimination, the elimination constant can be replaced by $k_m$. If (a) the liver is the major metabolic organ, (b) hepatic drug metabolism shows no enzymatic saturation, and (c) the intrinsic clearance, $CL_{int}$, is much smaller than liver blood flow, $Q$, then total body clearance, $CL_t$, values, calculated with the one-compartment

## ▰▰▰KEY CONCEPT▰ DRUG CLEARANCE AND ORGAN BLOOD FLOW

Two situations arise that show the relationship between drug clearance and liver blood flow. Organ clearance, CL, is given by $CL = QCL_{int}/(Q + CL_{int})$. In the case when $Q$ much greater than $CL_{int}$, organ $CL = CL_{int}$. This occurs for drugs such as antipyrine, barbiturates, antiepileptics, and cumarin derivatives. In the second case, when $Q$ is much less than $CL_{int}$, organ $CL = Q$. This occurs for various analgesics, tricyclic antidepressants, and beta-blockers. Protein binding may also have an effect, so, considering the free fraction of drug, $f_B$, one should use $f_B \cdot CL_{int}$ instead of $CL_{int}$ in this situation. Liver blood flow in humans is 20.7 mL/min/kg or 1450 mL/min for a 70-kg person.

**Example 1:** $Q$ much greater than $CL_{int}$. Antipyrine is negligibly bound to plasma proteins, eliminated exclusively through hepatic metabolism, and more than 99% of a given dose is excreted into urine as metabolites.[123] The intrinsic clearance of antipyrine is 12.8 mL/min/person, which was calculated from in vitro intrinsic clearance, $1.62 \times 10^{-4}$ mL/min/mg protein,[118] and total liver

microsomal protein, $7.88 \times 10^{-4}$ mg/person.[90] Human hepatic blood flow is reported as 1450 mL/min/70-kg person.[124] In vivo systemic clearance of antipyrine is reported as 13.5 (9.3–22.8) mL/min/person in patients with liver cirrhosis and 49.3 (31.1–103) mL/min/person in healthy individuals. Calculated in vitro intrinsic clearance of antipyrine is close to the values for patients with liver cirrhosis, probably because in vitro experiments were done with liver samples obtained from patients who underwent partial hepatectomy.

**Example 2:** $CL_{int}$ much greater than $Q$. In vivo systemic clearance propranolol is $1.21 \pm 0.15$ liter/min for (+)-propranolol and $1.03 \pm 0.12$ liter/min for (−)-propranolol.[125] The intrinsic clearance of racemic propranolol was 4180 mL/min/person, which was calculated from in vitro intrinsic clearance, 0.053 mL/min/mg protein,[126] and total liver microsomal protein, $7.88 \times 10^4$ mg/person.[90] Human hepatic blood flow is 1450 mL/min/70-kg person as shown in *Example-1*.

intravenous model, correlates well with the intrinsic clearance, $CL_{int}$.

In a one-compartment model with a drug absorption step such as oral administration, the drug enters the body by a first-order process. In this case, absorption is slower than the instantaneous injection that occurs during intravenous administration. Distribution of the absorbed drug molecules is instantaneous and elimination occurs according to one-compartment-model behavior as described previously:

$$\frac{dX_a}{dt} = -k_a X_a \qquad (12\text{–}20)$$

$$\frac{dX}{dt} = k_a X_a - k_{el} X \qquad (12\text{–}21)$$

where $X_a$ is the amount of drug in the absorption site at time $t$ after administration and $k_a$ is the absorption rate constant. Integration of equations (12–20) and (12–21) gives the following expression:

$$X = \frac{k_a F X_0}{V_d(k_a - k)}(e^{-k_{el}t} - e^{-k_{el}t}) \qquad (12\text{–}22)$$

where $F$ is the fraction of the dose, $X_0$, absorbed following oral administration.

The absorption rate constant as well as the elimination rate constant can be calculated from in vitro or in situ data in the oral absorption model. The absorption rate constant, $k_a$, can be related to the effective permeability,

$$k_a = \frac{SA \cdot P_{eff}}{V} = \frac{2P_{eff}}{r} \qquad (12\text{–}23)$$

where SA is the surface area, $V$ is the volume of the intestinal segment, and $r$ is the intestinal radius. If one assumes that a cylinder can be used to estimate the intestinal shape, then the SA/V ratio simplifies to $2/r$. Others have examined the effect of other, more realistic intestinal geometries.[127] However, for

the purposes of this example, assuming cylindrical geometry keeps the mathematics straightforward.

One can "build" a model of the human body absorption and disposition of drugs by using compartmental models. Each compartment can represent an organ, tissue, or set of organs or tissues (**Fig. 12–23**). For example, sometimes a two-compartment model is appropriate. Here, fast-perfused and slow-perfused tissues are grouped together into separate compartments. Typically, when organs or tissues are lumped together it is difficult to examine the behavior of specific individual organ systems. When the goal is to examine specific organ systems, PBPK models are constructed (**Fig. 12–23**). Using flow rates (e.g., blood flow, intestinal transit), volumes, and input and output rate constants, one can construct a PK model of an organ system. The PBPK models have a long and rich history that is covered in much more detail in a course in pharmacokinetics. We leave those details to those courses.

## BIOAVAILABILITY[4]

### Introduction

The words absorption and bioavailability are used in many ways. The purpose of this section is to introduce the student to the biopharmaceutical basis and practical meanings of the word bioavailability. "Bioavailability," as defined by the U.S. Food and Drug Administration in the *Code of Federal Regulations* (21 CFR 320.1[a]), means the rate and extent to which the active ingredient or active moiety is absorbed from a drug product and becomes available at the site of action. Because pharmacologic response is generally related to the concentration of drug at the receptor site, the availability of a drug from a dosage form is a critical element of a drug product's clinical efficacy. However, drug concentrations usually cannot be readily measured directly at the site of action. Therefore, most bioavailability studies involve the determination of drug concentrations in the blood or the urine. This is based on the premise that the drug at the site of action is in equilibrium with the drug in the blood. It is therefore possible to obtain an indirect measure of drug response by monitoring drug levels in the blood or the urine. Thus, bioavailability is concerned with how quickly and how much of a drug appears in the blood after a specific dose is administered. The bioavailability of a drug product often determines the therapeutic efficacy of that product because it affects the onset, intensity, and duration of therapeutic response of the drug. In most cases one is concerned with the extent of absorption of drug (i.e., the fraction of the dose that actually reaches the blood stream) because this represents the "effective dose" of a drug. This is generally less than the amount of drug actually administered in the dosage form. In some cases, notably those where acute conditions are being treated, one is also concerned with the rate of absorption of a drug because rapid onset of pharmacologic action is desired. Conversely, these are instances where a slower rate of absorption is desired, either to avoid adverse effects or to produce a prolonged duration of action.

## Causes of Low Bioavailability[5]

When a drug rapidly dissolves and readily crosses membranes, absorption tends to be complete, but absorption of orally administered drugs is not always complete. Before reaching the vena cava, a drug must move down the GI tract and pass through the gut wall and liver, common sites of drug metabolism; thus, a drug may be metabolized (first-pass metabolism) before it can be measured in the systemic circulation. Many drugs have low oral bioavailability because of extensive first-pass metabolism. For such drugs (e.g., isoproterenol, norepinephrine, testosterone), extraction in these tissues is so extensive that bioavailability is virtually zero. For drugs with an active metabolite, the therapeutic consequence of first-pass metabolism depends on the contributions of the drug and the metabolite to the desired and undesired effects. Intestinal secretion of drugs by transporters such as MRP2 and P-gp and enterohepatic recirculation may also cause low oral bioavailability. Low bioavailability is most common with oral dosage forms of poorly water-soluble, slowly absorbed drugs. More factors can affect bioavailability when absorption is slow or incomplete than when it is rapid and complete, so slow or incomplete absorption often leads to variable therapeutic responses. Insufficient time in the GI tract is a common cause of low bioavailability. Ingested drug is exposed to the entire GI tract for no more than 1 to 2 days and to the small intestine for only 2 to 4 hr. If the drug does not dissolve readily or cannot penetrate the epithelial membrane (e.g., if it is highly ionized and polar), time at the absorption site may be insufficient. In such cases, bioavailability tends to be highly variable as well as low. Age, gender, activity, genetic phenotype, stress, disease (e.g., achlorhydria, malabsorption syndromes), or previous GI surgery can affect drug

bioavailability. Reactions that compete with absorption can reduce bioavailability. They include complex formation (e.g., between tetracycline and polyvalent metal ions), hydrolysis by gastric acid or digestive enzymes (e.g., penicillin and chloramphenicol palmitate hydrolysis), conjugation in the gut wall (e.g., sulfoconjugation of isoproterenol), adsorption to other drugs (e.g., digoxin and cholestyramine), and metabolism by luminal microflora.

## CHAPTER SUMMARY

A shift has occurred in the pharmaceutical sciences from focusing solely on the physical and chemical aspects of pharmacy such as dissolution, solubility, and compaction physics to the integration of these important disciplines with the biopharmaceutical sciences. The purpose of this chapter was to provide the student with a biopharmaceutical foundation for studying the contemporary pharmaceutical sciences. At this point you should be able to define ADME and understand the differences between the two possibilities for Ds (distribution and disposition) and Es (excretion and elimination) in ADME. Two major membrane transporter superfamilies play an important role in ADME and if they work together (concerted drug transport), drugs and metabolites can be moved into or out of the body with great efficiency. Phase 1, 2, and "3" drug metabolism was also introduced in this chapter. The student should have a good understanding of the concepts of inhibition and induction as they relate to drug transporters, metabolizing enzymes, ADME, and pharmacokinetics. Graphical representations of the rates of absorption, disposition, metabolism, and elimination and blood (plasma) level versus time curves were also introduced. The very important concept of bioavailability was introduced. Finally, various organ systems were covered to give the student a better understanding of the complexity of ADME and the interplay of molecular, cellular, and organ level functions on pharmacokinetics.

 **Practice problems for this chapter can be found at thePoint.lww.com/Sinko6e.**

## References

1. P. R. Ortiz de Montellano, in T. F. Woolf (Ed.), *Handbook of Drug Metabolism*, Marcel Dekker, New York, 1999, p. 109.
2. M. Ingelman-Sundberg, M. Oscarson, and R. A. McLellan, Trends Pharmacol. Sci. **20**, 342, 1999.
3. G. L. Amidon, Mol. Pharm. **1**, 1, 2004.
4. M. Makoid, P. Vuchetich, and U. Banakar, Basic Pharmacokinetics on the Web, available at http://pharmacy.creighton.edu/pha443/pdf(Default.asp.
5. M. H. Beers and R. Berkow (Eds.), Drug Input and Disposition, in *The Merck Manual of Diagnosis and Therapy*, 17th Ed., Section 22: Clinical Pharmacology, Chapter 298, Wiley, New York, 1999.
6. http://www.brooklyn.cuny.edu/bc/ahp/BioInfo/GP/Definition.html.
7. J. F. Rogers, A. N. Nafziger, and J. S. Bertino, Am. J. Med. **114**, 746, 2002.
8. R. M. Norton, Drug Discov. Today, **6**, 180, 2001.

9. S. C. Bursen, Pharm. Lett, **15,** 31, 1999.
10. C. M. Henry, Chem. Eng. News **81,** 20, 2003.
11. Q. Al-Awqati, Nat. Cell Biol. **1,** E201, 1999.
12. Y. Cui, J. Konig, and D. Keppler, Mol. Pharmacol. **60,** 934, 2002.
13. E. Overton, Vierteljahrsschr. Naturforsch. Ges. Zurich **44,** 88, 1899.
14. R. Höber, Physiol. Rev. **16,** 52, 1936.
15. H. Davson and J. F. Danielli, *The Permeability of Natural Membranes,* Cambridge University Press, Cambridge, 1943.
16. S. J. Singer and G. L. Nicolson, Science **175,** 720, 1972.
17. P. A. Shore, B. B. Brodie, and C. A. M. Hogben, J. Pharmacol. Exp. Ther. **119,** 361, 1957.
18. C. A. M. Hogben, D. J. Tocco, B. B. Brodie, and L. S. Schanker, J. Pharmal. Exp. Ther. **125,** 275, 1959.
19. P. V. Balimane, I. Tamai, A. L. Guo, T. Nakanishi, H. Kitada, F. H. Leibach, A. Tsuji, and P. J. Sinko, Biochem. Biophys. Res. Commun. **250,** 246, 1998.
20. T. Zhu, X.-Z. Chen, A. Steel, M. A. Hediger, and D. E. Smith, Pharm. Res. **17,** 526, 2000.
21. M. E. Ganapathy, P. D. Prasad, B. Mackenzie, V. Ganapathy, and F. H. Leibach, Biochim. Biophys. Acta **1424,** 296, 1997.
22. J. C. Ventner, et al., Science **291** (5507), 1404, 2001.
23. H. P. Klenk, et al., Nature **390,** 364–370, 1997.
24. M. Dean, Y. Hamon, and G. Chimini, J. Lipid Res. **42,** 1007, 2001.
25. P. M. Jones and A. M. George, Eur. J. Biochem. **267,** 5298, 2000.
26. M. Muller and P. L. M. Jansen, Am. J. Physiol. Gastrointest. Liver Physiol. **35,** G1285, 1997.
27. S. A. Adibi, Gastroenterology **114,** 332, 1997.
28. G. K. Dresser, D. G. Bailey, B. F. Leake, U. I. Schwarz, P. A. Dawson, D. J. Freeman, and R. B. Kim, Clin. Pharmacol. Ther. **71,** 11, 2002.
29. G. You, Med. Res. Rev. **22,** 602, 2002.
30. A. Ure, Pharm. J. Trans. **1,** 24, .
31. M. Sasaki, H. Suzuki, K. Ito, T. Abe, and Y. Sugiyama, J Biol Chem. **277,** 6497, 2002.
32. R. T. Williams, *Detoxication Mechanisms,* Wiley, New York, 1959.
33. T. Ishikawa, Trends Biochem. Sci. **17,** 463, 1992.
34. O. Hayaishi, in *Proceedings Plenary Session, 6th International Congress of Biochemistry,* New York, 1964, p. 31.
35. H. S. Mason, Adv. Enzymol. **19,** 74, 1957.
36. D. Garfinkel, Arch. Biochem. Biophys. **77,** 493, 1958.
37. M. Klingenberg, Arch. Biochem. Biophys. **75,** 376, 1958.
38. T. Omura and R. Sato, J. Biol. Chem. **239,** 2370, 1964.
39. R. W. Estabrook, D. Y. Cooper, and O. Rosenthal, Biochem. Z. **338,** 741, 1963.
40. G. Stadeler, Ann. Chem. Liebigs **77,** 17, 1851.
41. M. Jeffe, Z. Physiol. Chem. **2,** 47, 1878.
42. J. von Mering and F. Musculus, Ber. Dtsch. Chem. Ges. **8,** 662, 1875.
43. O. Schmiedeberg and H. Meyer, Z. Physiol. Chem. **3,** 422, 1879.
44. E. Baumann and C. Preusse, Ber. Dtsch. Chem. Ges. **12,** 806, 1879.
45. M. M. Barnes, S. P. James, and P. B. Wood, Biochem. J. **71,** 680, 1959.
46. V. Villikka, K. T. Kivisto, J. T. Backman, K. T. Olkkola, and P. J. Neuvonen, Clin. Pharmacol. Ther. **61,** 8, 1997.
47. U. Fuhr, H. Muller-Peltzer, R. Kern, P. Lopez-Rojas, M. Junemann, S. Harder, and A. H. Staib, Eur. J. Clin. Pharmacol. **58,** 45, 2002.
48. T. F. Blaschke, Clin. Pharmacokinet. **2,** 32, 1977.
49. P. Ehrlich, *Das Sauerstoff-bedurfnis des Organismus. Eine farbenan-alytische Studie.* Hirschwald, Berlin, 1885.
50. J. W. Polli et al., Pharm. Res. **16,** 1206, 1999.
51. O. U. Petring and D. W. Blake, Anaesth. Intensive Care **21,** 774, 1993.
52. J. Nimmo, R. C. Heading, P. Tothill, and L. F. Prescott, Br. Med. J. **1,** 587, 1973.
53. C. H. Kim, A. R. Zinsmeister, and J.-R. Malagelada, Gastroenterology **92,** 993, 1987.
54. J. G. Moore, P. E. Christian, and R. E. Coleman, Dig. Dis. Sci. **26,** 16, 1981.
55. J. E. Valenzuela and D. P. Liu, Scand. J. Gastroenterol. **17,** 293, 1982.
56. C. P. Dooley, J. B. Reznick, and J. E. Valenzuela. Gastroenterology **87,** 1114, 1984.
57. M. G. Velchik, J. C. Reynolds, and A. Alavi, J. Nucl. Med. **30,** 1106, 1989.
58. C. J. Cote, in R. K. Stoelting (Ed.), *Advances in Anesthesia,* **Vol. 9,** Mosby Year Book, St. Louis, Mo., 1992, p. 1.
59. J. N. Hunt and D. F. Stubbs, J. Physiol. **245,** 209, 1975.
60. R. H. Goo, J. G. Moore, E. Greenberg, and N. Alazraki, Gastroenterology **93,** 515, 1987.
61. M. Horowitz, G. J. Maddern, B. E. Chatterton, P. J. Collins, P. E. Harding, and D. J. C. Shearman, Clin. Sci. **67,** 214, 1984.
62. M. Divoll, B. Ameer, D. R. Abernathy, and D. J. Greenblatt, J. Am. Geriatr. Soc. **30,** 240, 1982.
63. R. A. Wright, S. Krinsky, C. Fleeman, J. Trujillo, and E. Teague, Gastroenterology **84,** 747, 1983.
64. M. Horowitz, P. J. Collins, D. J. Cook, P. E. Harding, and D. J. C. Shearman, Int. J. Obesity **7,** 415, 1983.
65. H. Sasaki, M. Nagulesparan, A. Dubois, I. M. Samloff, E. Straus, M. L. Sievers, and R. H. Unger, Int. J. Obesity **8,** 183, 1984.
66. O. U. Petring and H. Flachs, Br. J. Clin. Pharmacol. **29,** 703, 1990.
67. F. L. Datz, P. E. Christian, and J. Moore, Gastroenterology **92,** 1443, 1989.
68. W. Hutson, R. Roehrkasse, D. Kaslewicz, B. Stoney, and A. Wald, Gastroenterology **92,** 1443, 1989.
69. R. C. Gill, D. Murphy, H. R. Hooper, K. L. Bowes, and Y. J. Kingma, Digestion **36,** 168, 1987.
70. A. G. Macfie, A. D. Magides, M. N. Richmond, and C. S. Reilly, Br. J. Anaesth. **67,** 54, 1991.
71. T. Gin, A. M. W. Cho, J. K. L. Lew, G. S. N. Lau, P. M. Yuen, A. J. H. Critchley, and T. E. Oh, Anaesth. Intensive Care **19,** 521, 1991.
72. D. F. Magee and A. F. Dalley II, *Digestion and the Structure and Function of the Gut,* Karger, Basel, Switzerland, 1986.
73. J. M. DeSesso and C. F. Jacobson, Food Chem. Toxicol. **39,** 209, 2001.
74. B. L. Shneider, J. Pediatr. Gastroenterol. Nutr. **32,** 407, 2001.
75. V. S. Patel and W. G. Kramer, J. Pharm. Sci. **75,** 275, 1986.
76. W. H. Barr, E. M. Zola, E. L. Candler, S. M. Hwang, A. V. Tendolkar, R. Shamburek, B. Parker, and M. D. Hilty, Clin. Pharmacol. Ther. **56,** 279, 1994.
77. D. Cong, A. K. Y. Fong, R. Lee, and K. S. Pang, Drug Metab. Dispos. **29,** 1539, 2001.
78. J. Drewe, H. Narjes, G. Heinzel, R. S. Brickl, A. Rohr, and C. Beglinger, J. Clin. Pharmacol. **50,** 69, 2000.
79. J. Godbillon, N. Vidon, R. Palma, A. Pfeiffer, C. Franchisseur, M. Bovet, G. Gosset, J. J. Bernier, and J. Hirtz, Br. J. Clin. Pharmacol. **24,** 335, 1987.
80. T. Grámatte, R. Oertel, B. Terhaag, and W. Kirch, Clin. Pharmacol. Ther. **59,** 541, 1996.
81. G. B. Park and A. K. Mitra, Pharm. Res. (N.Y.) **9,** 326, 1992.
82. B. L. Shneider, P. A. Dawson, D. M. Christie, W. Hardikar, M. H. Wong, and F. J. Suchy, J. Clin. Invest. **95,** 745, 1995.
83. R. Aldini, M. Montagnani, A. Roda, S. Hrelia, P. L. Biagi, and E. Roda, Gastroenterology **110,** 459, 1996.
84. I. Tamai, Y. Sai, A. Ono, Y. Kido, H. Yabuuchi, H. Takanaga, E. Satoh, T. Ogihara, O. Amano, S. Izeki, and A. Tsuji, J. Pharm. Pharmacol. **51,** 1114, 1999.
85. L. Y. Ngo, S. D. Patil, and J. D. Unadkat, Am. J. Physiol. **280,** G475, 2001.
86. H. C. Walters, A. L. Craddock, H. Fusegawa, D. C. Willingham, and P. A. Dawson, Am. J. Physiol. **279,** G1188, 2000.
87. Y. J. Fei, Y. Kanai, S. Nussberger, V. Ganapathy, F. H. Leibach, M. F. Romero, S. K. Singh, W. F. Boron, and M. A. Hediger, Nature (Lond.) **368,** 563, 1994.
88. H. Hoensch, C. H. Woo, S. B. Raffin, and R. Schmid, Gastroenterology **70,** 1063, 1976.
89. H. L. Bonkovsky, H. P. Hauri, U. Martï, R. Gasser, and U. A. Meyer, Gastroenterology **88,** 458, 1985.
90. M. F. Paine, M. Khalighi, J. M. Fisher, D. D. Shen, K. L. Kunze, C. L. Marsh, J. D. Perkins, and K. E. Thummel, J. Pharmacol. Exp. Ther. **283,** 1552, 1997.
91. M. F. Paine, D. D. Shen, K. L. Kunze, J. D. Perkins, C. L. Marsh, J. P. McVicar, D. M. Barr, B. S. Gilles, and K. E. Thummel, Clin. Pharmacol. Ther. **60,** 14, 1996.
92. K. E. Thummel, D. O'Shea, M. F. Paine, D. D. Shen, K. L. Kunze, J. D. Perkins, and G. R. Wilkinson, Clin. Pharmacol. Ther. **59,** 491, 1996.

93. K. S. Lown, R. R. Mayo, A. B. Leichtman, H.-L. Hsiao, D. K. Turgeon, P. Schmiedlin-Ren, M. B. Brown, W. Guo, S. J. Rossi, L. Z. Benet, and P. B. Watkins, Clin. Pharmacol. Ther. **62,** 248, 1997.

94. L. Y. Li, G. L. Amidon, J. S. Kim, T. Heimbach, F. Kesisoglou, J. T. Topliss, and D. Fleisher, J. Pharmacol. Exp. Ther. **301,** 586, 2002.

95. G. Clifton and N. Kaplowitz, Cancer Res. **37,** 788, 1977.

96. L. M. Pinkus, J. N. Ketley, and W. B. Jakoby, Biochem. Pharmacol. **26,** 2359, 1977.

97. L. R. Schwarz and M. Schwenk, Biochem. Pharmacol. **33,** 3353, 1984.

98. A. S. Koster, A. C. Frankhuijzen-Sierevogel, and J. Noordhoek, Biochem. Pharmacol. **34,** 3527, 1985.

99. B. F. Coles, G. Chen, F. F. Kadlubar, and A. Radominska-Pandya, Arch. Biochem. Biophys. **403,** 270, 2002.

100. Y. Gotoh, H. Suzuki, S. Kinoshita, T. Hirohashi, Y. Kato, and Y. Sugiyama, J. Pharmacol. Exp. Ther. **292,** 433, 2000.

101. A. D. Mottino, T. Hoffman, L. Jennes, and M. Vore, J. Pharmacol. Exp. Ther. **293,** 717, 2000.

102. V. D. Makhey, A. Guo, D. A. Norris, P. Hu, Y. Yan, and P. J. Sinko, Pharm. Res. **15,** 1160, 1998.

103. A. Collett, N. B. Higgs, E. Sims, M. Rowland, and G. Warhurst, J. Pharmacol. Exp. Ther. **288,** 171, 1999.

104. A. Nakayama, H. Saitoh, M. Oda, M. Takada, and B. J. Aungst, Eur. J. Pharm. Sci. **11,** 317, 2000.

105. R. H. Stephens, C. A. O'Neill, A. Warhurst, G. L. Carlson, M. Rowland, and G. Warhurst, J. Pharmacol. Exp. Ther. **296,** 584, 2001.

106. D. Rost, S. Mahner, Y. Sugiyama, and W. Stremmel, Am. J. Physiol. **282,** G720, 2002.

107. G. C. Williams, A. Liu, G. Knipp, and P. J. Sinko, Antimicrob. Agents Chemother. **46,** 3456, 2002.

108. Y. Su, X. Zhang, and P. J. Sinko, Mol. Pharm. **1,** 49, 2004.

109. M. P. McRae, et al., JPET **318,** 1068, 2006.

110. S. J. Mouly, M. F. Paine, and P. B. Watkins, J. Pharmacol. Exp. Ther. **308,** 941, 2004.

111. L. W. Seymour, R. Duncan, J. Strohalm, and J. Kopecek, J. Biomed. Mater. Res. **21,** 1441, 1987.

112. T. Yamaoka, Y. Tabata, and Y. Ikada, J. Pharm. Sci. **83,** 601, 1994.

113. C. Delgado, G. E. Francis, and D. Fisher, Crit. Rev. Ther. Drug Carrier Syst. **9,** 249, 1992.

114. X.-H. He, P.-C. Shaw, and S.-C. Tam, Life Sci. **65,** 355, 1999.

115. Y. Inada, M. Furukawa, H. Sasaki, Y. Kodera, M. Hiroto, H. Nishimura, and A. Matsushima, Trends Biotechnol. **14,** 86, 1995.

116. G. Molineux, Curr. Pharm. Des. **10,** 1235, 2004.

117. M. Ghany and J. H. Hoofnagle, in E. Braunwald, A. S. Fauci, D. L. Kasper, S. L. Hauser, D. L. Longo, and J. L. Jameson, (Eds.), *Harrison's Principles of Internal Medicine,* 15th Ed., McGraw-Hill, New York, 2001, p. 1707.

118. J. E. van Montfoort, B. Hagenbuch, G. M. M. Groothuis, H. Koepsell, P. J. Meier, and D. K. F. Meijer, Curr. Drug Metab. **4,** 185, 2003.

119. P. C. Hirom, P. Millburn, and R. L. Smith, Xenobiotica **6,** 55, 1976.

120. P. C. Hirom, P. Millburn, R. L. Smith, and R. T. Williams, Biochem. J. **129,** 1071, 1972.

121. Medical Economics Staff and PDR Staff (Eds.), *Physicians' Desk Reference,* 57th Ed., Medical Economics, Oradell, 2002, NJ.

122. M. Sugie, E. Asakura, Y. L. Zhao, S. Torita, M. Nadai, K. Baba, K. Kitaichi, K. Takagi, and T. Hasegawa, Antimicrob. Agents Chemother. **48,** 809, 2004.

123. G. Engel, U. Hofmann, H. Heidemann, J. Cosme, and M. Eichelbaum, Clin. Pharmacol. Ther. **59,** 614, 1996.

124. B. Davies and T. Morris, Pharm. Res. **10,** 1093, 1993.

125. L. S. Olanoff, T. Walle, U. K. Walle, T. D. Cowart, and T. E. Gaffney, Clin. Pharmacol. Ther. **35,** 755, 1984.

126. R. S. Obach, Drug Metab. Dispos. **25,** 1459, 1997.

127. W. L. Chiou, Int. J. Clin. Pharmacol. Ther. **32,** 474, 1994.

## Recommended Readings

Q. Al-Awqati, Nat. Cell Biol. **1,** E201, 1999.

P. A. Shore, B. B. Brodie, and C. A. M. Hogben, J. Pharmacol. Exp. Ther. **119,** 361, 1957.

H. van de Waterbeend, B, Testa, R. Mannhold, H. Kubinyi, and G. Folkers (eds), *Drug Bioavailability: Estimation of Solubility, Permeability, Absorption and Bioavailability,* 2nd Ed., Wiley-VCH, Weinheim, Germany, 2008.

CHAPTER LEGACY

**Fifth Edition:** published as Chapter 14 (Biopharmaceutics). Updated by Patrick Sinko.

**Sixth Edition:** published as Chapter 12 (Biopharmaceutics). Updated by Patrick Sinko.

# 13 | DRUG RELEASE AND DISSOLUTION

**CHAPTER OBJECTIVES** At the conclusion of this chapter the student should be able to:

**1** Define dissolution and describe relevant examples in the pharmaceutical sciences and practice of pharmacy.

**2** Understand the differences among immediate-, modified-, delayed-, extended-, and controlled-release delivery systems.

**3** Differentiate between zero-order and first-order release kinetics.

**4** Define and understand intrinsic dissolution rate and define the driving force for dissolution.

**5** Understand the effect of surface area on dissolution rate.

**6** Differentiate the Hixson–Crowell, Noyes–Whitney, and Higuchi models of dissolution and release.

**7** Understand the concept of sink conditions.

**8** Define the Biopharmaceutics Classification System and discuss the role of permeability and solubility.

**9** Understand how media properties can affect dissolution, for example, viscosity, pH, lipids, surfactants.

**10** Describe and understand the mechanics of the most commonly used dissolution apparatuses.

## INTRODUCTION*

Disintegration tests, official in the *United States Pharmacopeia* (USP) since 1950, are only indirectly related to drug bioavailability and product performance.[1] In 1962, dissolved drug was known to be necessary for physiologic action and it was becoming increasingly recognized that capsule and tablet monographs in which the drug substance had a solubility of less than 1% in aqueous media should include a dissolution requirement. In 1968, the *USP/National Formulary* (NF) recommended the adoption of a basket-stirred-flask test apparatus (USP apparatus 1) to determine the dissolution of solid oral dosage forms.[1] With the introduction of USP XIX/NF XIV in 1975, it was shown that a compendial in vivo bioavailability standard was not required, provided that a satisfactory in vitro–in vivo correlation (IVIVC) could be established. In 1978, the USP paddle apparatus was officially adopted and was found to be advantageous for disintegrating dosage forms. Today, the quality control of many drug products is based on the kinetics of drug release in vitro.[2–5] Drug release testing is also routinely used to predict how formulations or drug products are expected to perform in patients. These two distinct areas of dissolution and drug release testing have evolved over the past decade and will be described in this chapter.

The five types of dosage forms that can be characterized by release in vitro are: (*a*) solid oral dosage forms, (*b*) rectal dosage forms such as suppositories, (*c*) pulmonary (lung delivery) dosage forms, (*d*) modified-release dosage forms, and (*e*) semisolid products such as ointments, creams, and transdermal products. Over the last several years, the pharmaceutical industry, pharmaceutical scientists, and the Food and Drug Administration have worked together to improve the guidance available for classifying, studying, and documenting postapproval changes to manufacturing processes.[5] The first round of this effort resulted in the publication of several SUPAC (Scale Up Post Approval Change) guidances, including the initial guidance, SUPAC-IR,[3] for immediate-release drug products, followed by SUPAC-MR,[5] for modified-release drug products, SUPAC-SS,[4] for semisolids, and PAC-ATLS, for analytical lab changes. A number of additional SUPAC documents are in various stages of development. Parallel with these efforts, the explicit link between physicochemical properties such as drug dissolution and bioavailability was becoming formally recognized as reflected in the Biopharmaceutics Classification System (BCS), introduced in 1995.[6] The BCS proposed a straightforward classification of drug products on the basis of their

## KEY CONCEPT DRUG RELEASE

Drug release is the process by which a drug leaves a drug product and is subjected to absorption, distribution, metabolism, and excretion, eventually becoming available for pharmacologic action. Drug release is described in several ways. Immediate release refers to the instantaneous availability of drug for absorption or pharmacologic action. *Immediate-release* drug products allow drugs to dissolve with no intention of delaying or prolonging dissolution or absorption of the drug. *Modified-release* dosage forms include both delayed- and extended-release drug products. *Delayed release* is defined as the release of a drug at a time other than immediately following administration. *Extended-release* products are formulated to make the drug available over an extended period after administration. Finally, *controlled release* includes extended-release and pulsatile-release products. *Pulsatile release* involves the release of finite amounts (or pulses) of drug at distinct time intervals that are programmed into the drug product.

---

*Much of this opening section was adapted from Cohen et al.[1]

## KEY CONCEPT DISSOLUTION

Dissolution refers to the process by which a solid phase (e.g., a tablet or powder) goes into a solution phase such as water. In essence, when a drug "dissolves," solid particles separate and mix molecule by molecule with the liquid and appear to become part of that liquid. Therefore, drug dissolution is the process by which drug molecules are liberated from a solid phase and enter into a solution phase. If particles remain in the solid phase once they are introduced into a solution, a pharmaceutical suspension results. Suspensions are covered in Chapters 17 and 18. In the vast majority of circumstances, only drugs in solution can be absorbed, distributed, metabolized, excreted, or even exert pharmacologic action. Thus, dissolution is an important process in the pharmaceutical sciences.

solubility and permeability characteristics. Beginning in 1998, interest in studying the link between dissolution testing and bioavailability was revived and continues through today.[7–11] Standard pharmacopeial dissolution monographs were typically designed as quality control procedures to ensure that batch-to-batch drug product variability is kept within acceptable scientific and regulatory standards. However, with the widespread adoption of the BCS, the possibility of substituting dissolution tests for clinical studies has called the conditions of established compendial dissolution tests into question because there is now a need to better predict the in vivo performance of drug products. In this chapter, we cover the basic theoretical and analytical background for performing drug release and dissolution calculations, the release testing of oral drug products, the BCS and biorelevant dissolution conditions, and dissolution testing methods and apparatus, and provide numerous examples to help the student gain an understanding of dissolution and drug release.

## TERMINOLOGY*

**Drug Product:** A drug product is a finished dosage form (e.g., tablet and capsule) that contains a drug substance, generally, but not necessarily in association with one or more other ingredients (21 *Code of Federal Regulations* 314.3(b)). A solid oral dosage form includes but is not limited to tablets, chewable tablets, enteric-coated tablets, capsules, caplets, encapsulated beads, and gelcaps.

**Drug Substance:** An active ingredient that is intended to furnish pharmacologic activity or other direct effect in the diagnosis, cure, mitigation, treatment, or prevention of a disease, or to affect the structure of any function of the human body, but does not include intermediates used in the synthesis of such ingredient (21 *Code of Federal Regulations* 314.3(b)).

**Enteric Coated:** Intended to delay the release of the drug (or drugs) until the dosage form has passed through the stomach. Enteric-coated products are delayed-release dosage forms.

**Extended Release:** Extended-release products are formulated to make the drug available over an extended period after ingestion. This allows a reduction in dosing frequency compared to a drug presented as a conventional dosage form (e.g., as a solution or an immediate-release dosage form).

**Modified-Release Dosage Forms:** Dosage forms whose drug-release characteristics of time course and/or location are chosen to accomplish therapeutic or convenience objectives not offered by conventional dosage forms such as a solution or an immediate-release dosage form. Modified-release solid oral dosage forms include both delayed- and extended-release drug products.

**Immediate Release:** Allows the drug to dissolve in the gastrointestinal contents with no intention of delaying or prolonging the dissolution or absorption of the drug.

**In Vitro–in Vivo Correlation:** A predictive mathematical model describing the relationship between an in vitro property of an oral dosage form (usually the rate or extent of drug dissolution or release) and a relevant in vivo response (e.g., plasma drug concentration or amount of drug absorbed).

## THE BASICS

Biopharmaceutics (Chapter 12) and the design of modern drug delivery systems (Chapter 23), as dealt with later, are based partly on principles of diffusion and dissolution theory. This chapter lays a foundation for the study of these topics by way of presenting concepts, illustrations, and worked examples. Drug release is introduced first because it is largely based on diffusion, which was introduced in Chapter 11. We then cover drug dissolution with examples from the literature and with applications of both subjects to pharmaceutical problems.

Drug dissolution and release patterns commonly fall into two groups: zero- and first-order release. Typically in the pharmaceutical sciences, zero-order release is achieved from nondisintegrating dosage forms such as topical or transdermal delivery systems, implantable depot systems, or oral controlled-release delivery systems. Because many of these delivery systems are covered in Chapter 23 on drug delivery systems and Chapter 20 on pharmaceutical polymers, the mathematical basis will be introduced in this chapter. In

---

*This section was adapted with some modification from reference 12.

## KEY CONCEPT ZERO-ORDER RELEASE KINETICS

Zero-order release kinetics refers to the process of constant drug release from a drug delivery device such as oral osmotic tablets, transdermal systems, matrix tablets with low-soluble drugs, and other delivery systems. "Constant" release is defined in this context as the same amount of drug release per unit of time. In its simplest form, zero-order drug release can be represented as

$$Q = Q_0 + K_0 t \qquad (13\text{--}1)$$

where $Q$ is the amount of drug released or dissolved (assuming that release occurs rapidly after the drug dissolves), $Q_0$ is the initial amount of drug in solution (it is usually zero), and $K_0$ is the zero-order release constant.

these cases, drug "dissolution" is commonly referred to as drug "release" because it is dependent on diffusion (Chapter 11). The advanced student can find a more in-depth treatment of the mathematical models of dissolution from the review by Costa and Sousa Lobo.[13] The following sections review basic models with an emphasis on understanding the conceptual basis for drug release and dissolution. The student should view each equation as a short-hand way of describing the relationships among the parameters/factors that affect the process that is being described.

## Dissolution

When a tablet or other solid drug form is introduced into a beaker of water or into the gastrointestinal tract, the drug begins to pass into solution from the intact solid. Unless the tablet is a contiguous polymeric device, the solid matrix also disintegrates into granules, and these granules deaggregate in turn into fine particles. Disintegration, deaggregation, and dissolution may occur simultaneously with the release of a

drug from its delivery form. These steps are separated for clarification as depicted in **Figure 13–1**.

The effectiveness of a tablet in releasing its drug for systemic absorption depends somewhat on the rate of disintegration of the dosage forms and deaggregation of the granules. Ordinarily of more importance, however, is the dissolution rate of the solid drug. Frequently, dissolution is the limiting or rate-controlling step in the absorption of drugs with low solubility because it is often the slowest of the various stages involved in release of the drug from its dosage form and passage into systemic circulation. Classical dissolution has been reviewed by Wurster and Taylor,[14] Wagner,[15] and Leeson and Carstensen.[16] Release rate processes in general are discussed by Higuchi.[17] This has been an active area of research for many years, and reviews have appeared recently on numerous aspects of drug dissolution, including the influence of physicochemical properties of drugs on dissolution[18] and on the modeling and comparison of dissolution profiles.[13] Articles such as these will provide the student with a thorough yet broad overview of the current status of the field.

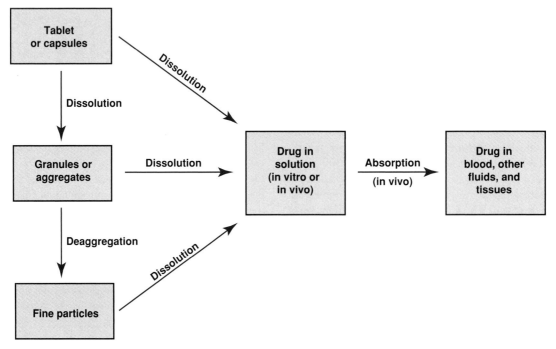

**Fig. 13–1.** Disintegration, deaggregation, and dissolution stages as a drug leaves a tablet or granular matrix. (From J. G. Wagner, *Biopharmaceutics and Relevant Pharmacokinetics*, Drug Intelligence Publications, Hamilton, IL, 1971, p. 99. With permission).

Several theories have been used to build mathematical models that describe drug dissolution from immediate- and modified-release dosage forms. In this chapter, the focus will be on drug dissolution from solid dosage forms. Because dissolution is a kinetic process, the rate of dissolution reflects the amount of drug dissolved over a given time period. In certain cases, an equation can be exactly derived that describes the dissolution time dependence. This is called an analytical mathematical solution. However, in many cases, an analytical solution cannot be derived and an empirical relationship is used. Several common mathematical models will be covered in the following sections. The pharmacy student should keep in mind that the most important lesson to be learned at this stage is not how to derive these equations but rather how to use them as short-hand formulas to understand the different factors that affect dissolution rate and how dissolution patterns can vary and ultimately influence the efficacy of therapeutic regimens in patients.

The rate at which a solid dissolves in a solvent was proposed in quantitative terms by Noyes and Whitney[19] in 1897 and elaborated subsequently by other workers. The equation can be written as

$$\frac{dM}{dt} = \frac{DS}{h}(C_s - C) \qquad (13-2)$$

or

$$\frac{dC}{dt} = \frac{DS}{Vh}(C_s - C) \qquad (13-3)$$

where $M$ is the mass of solute dissolved in time $t$, $dM/dt$ is the mass rate of dissolution (mass/time), $D$ is the diffusion coefficient of the solute in solution, $S$ is the surface area of the exposed solid, $h$ is the thickness of the diffusion layer, $C_s$ is the solubility of the solid (i.e., concentration of a saturated solution of the compound at the surface of the solid and at the temperature of the experiment), and $C$ is the concentration of solute in the bulk solution and at time $t$. The quantity $dC/dt$ is the dissolution rate, and $V$ is the volume of solution.

In dissolution or mass transfer theory, it is assumed that an *aqueous diffusion layer* or *stagnant liquid film* of thickness $h$ exists at the surface of a solid undergoing dissolution, as observed in **Figure 13–2**. This thickness, $h$, represents a stationary layer of solvent in which the solute molecules exist in concentrations from $C_s$ to $C$. Beyond the static diffusion layer, at $x$ greater than $h$, mixing occurs in the solution, and

The Noyes–Whitney[19] equation is

$$dC/dt = K(C_s - C) \qquad (13-4)$$

where $K$ is the "first-order" proportionality constant. The Hixson and Crowell[20] equation further considers the surface area of the dissolving solid:

$$dM/dt = KS(C_s - C) \qquad (13-5)$$

where $K = D/h$

the drug is found at a uniform concentration, $C$, throughout the bulk phase.

At the solid surface–diffusion layer interface, $x = 0$, the drug in the solid is in equilibrium with drug in the diffusion layer. The gradient, or change in concentration with distance across the diffusion layer, is constant, as shown by the straight downward-sloping line. This is the gradient represented in equations (13–2) and (13–3) by the term $(C_s - C)/h$. The similarity of the Noyes–Whitney[19] equation to Fick's first law (Chapter 11) is evident in equation (13–2).

Therefore, when $C$ is considerably less than the drug's solubility, $C_s$, the system is represented by *sink conditions*, and concentration $C$ can be eliminated from equations (13–2) and (13–3). Equation (13–2) then becomes

$$dM/dt = DSC_s/h \qquad (13-7)$$

In the derivation of equations (13–2) and (13–3), it was assumed that $h$ and $S$ were constant, but this is not the case. The static diffusion layer thickness is altered by the force of agitation at the surface of the dissolving tablet and will be referred to later. The surface area, $S$, obviously does not remain constant as a powder, granule, or tablet dissolves, and it is difficult to obtain an accurate measure of $S$ as the process continues. In experimental studies of dissolution, the surface may be controlled by placing a compressed pellet in a holder that exposes a surface of constant area. Although this ensures better adherence to the requirements of equations (13–2) through (13–7) and provides valuable information on the drug, it does not simulate the actual dissolution of the material in practice.

The saturation solubility of a drug is a key factor in the Noyes–Whitney[19] equation. The driving force for dissolution is the concentration gradient across the boundary layer. Therefore, the driving force depends on the thickness of the boundary layer and the concentration of drug that is already dissolved. When the concentration of dissolved drug, $C$, is less than 20% of the saturation concentration, $C_s$, the system is said to operate under "sink conditions." The driving force for dissolution is greatest when the system is under sink conditions. Under sink conditions, equation (13–5) can be written in a simplified form:

$$Q_t = Q_0 e^{-Kt} \qquad \text{or} \qquad \ln(Q_t / Q_0) = Kt \qquad (13-6)$$

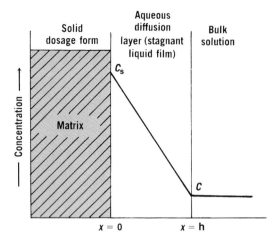

**Fig. 13–2.** Dissolution of a drug from a solid matrix, showing the stagnant diffusion layer between the dosage form surface and the bulk solution.

In calculating the diffusion coefficient and dissolution rate constant, the application of equations **(13–2)** through **(13–7)** is demonstrated by way of the following two examples.

### EXAMPLE 13–1

**Calculate Dissolution Rate Constant**

A preparation of drug granules weighing 0.55 g and having a total surface area of 0.28 m² (0.28 × 10⁴ cm²) is allowed to dissolve in 500 mL of water at 25°C. After the first minute, 0.76 g has passed into solution. The quantity $D/h$ can be referred to as a dissolution rate constant, $k$.

If the solubility, $C_s$, of the drug is 15 mg/mL at 25 °C, what is $k$? From equation (12–7), $M$ changes linearly with $t$ initially, and

$$\frac{dM}{dt} = \frac{760 \text{ mg}}{60 \text{ sec}} = 12.67 \text{ mg/sec}$$

$$12.67 \text{ mg/sec} = k \times 0.28 \times 10^4 \text{ cm}^2 \times 15 \text{ mg/cm}^3$$

$$k = 3.02 \times 10^{-4} \text{ cm/sec}$$

In this example, 0.760 g dissolved in 500 mL after a time of 1 min, or 760 mg/500 mL = 1.5 mg/cm³. This value is one tenth of the drug's solubility and can be omitted from equation **(13–2)** without introducing significant error, shown by employing the full equation **(13–2)**:

$$k = \frac{12.67 \text{ mg/sec}}{(0.28 \times 10^4 \text{ cm}^2)(15 \text{ mg/cm}^3 - 1.5 \text{ mg/cm}^3)}$$

$$k = 3.35 \times 10^{-4} \text{ cm/sec}$$

When this result is compared with 3.02 × 10⁻⁴ cm/sec, obtained using the less exact expression, it shows that "sink conditions" are in effect, and that the concentration term, $C$, can be omitted from the rate equation.

### EXAMPLE 13–2

**Hixson–Crowell Cube-Root Law[20]**

The diffusion layer thickness in *Example 13–1* is estimated to be 5 × 10⁻³ cm. Calculate $D$, the diffusion coefficient, using the relation $k = D/h$.

We have

$$D = (3.35 \times 10^{-4} \text{ cm/sec}) \times (5 \times 10^{-3} \text{ cm})$$

$$= 1.68 \times 10^{-6} \text{ cm}^2/\text{sec}$$

If a dosage form's dimensions diminish proportionally in such a manner that the geometric shape of the dosage form stays constant as dissolution is occurring, then dissolution occurs in planes that are parallel to the dosage form surface and we use the Hixson–Crowell[20] cube-root model to understand its behavior. It is thought that tablet dissolution occurs in this manner. For a drug powder consisting of uniformly sized particles, it is possible to derive an equation that expresses the rate of dissolution based on the cube root of the weight of the particles.

The radius of the particle is not assumed to be constant. The particle (sphere) shown in **Figure 13–3** has a radius $r$ and a surface area $4\pi r^2$.

Through dissolution, the radius is reduced by $dr$, and the infinitesimal volume of this section lost is

$$dV = 4\pi r^2 \, dr \qquad (13\text{–}8)$$

For $N$ such particles, the volume loss is

$$dV = 4N\pi r^2 \, dr \qquad (13\text{–}9)$$

The surface area of $N$ particles is

$$S = 4N\pi r^2 \qquad (13\text{–}10)$$

Now, the infinitesimal mass change as represented by the Noyes–Whitney law,[19] equation **(13–2)**, is

$$-dM = kSC_s \, dt \qquad (13\text{–}11)$$

where $k$ is used for $D/h$ as in *Example 13–1*. The drug's density multiplied by the infinitesimal volume change, $\rho \, dV$, can be set equal to $dM$, or

$$-\rho \, dV = kSC_s \, dt \qquad (13\text{–}12)$$

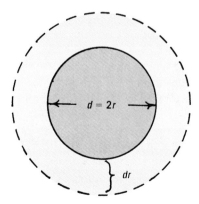

**Fig. 13–3.** Schematic of a particle, showing the change in surface area and volume as the particle dissolves. The volume, $dV$, dissolved in $dt$ seconds is given by Thickness × Surface area = $dr \times 4\pi r^2$. (From J. T. Carstensen, *Pharmaceutics of Solids and Solid Dosage Forms*, Wiley, New York, 1977, p. 75. With permission.)

Equations **(13–9)** and **(13–10)** are substituted into equation **(13–12)** to yield

$$-4\rho N\pi r^2 dr = 4N\pi r^2 kC_s dt \qquad (13–13)$$

Equation **(13–13)** is divided through by $4N\pi r^2$ to give

$$-\rho \, dr = kC_s \, dt \qquad (13–14)$$

Integration with $r = r_0$ at $t = 0$ produces the expression

$$r = r_0 - \frac{kC_s t}{\rho} \qquad (13–15)$$

The radius of spherical particles can be replaced by the mass of $N$ particles by using the relationship (see inside front cover for the volume of a sphere)

$$M = N\rho(\pi/6)d^3 \qquad (13–16)$$

where $d = 2r$, the diameter of the particle. Taking the cube root of equation **(13–16)** yields

$$M^{1/3} = [N\rho(\pi/6)]^{1/3}d \qquad (13–17)$$

The diameter, $d$, from equation **(13–17)** is substituted for $2r$ into equation **(13–15)**, giving

$$M_0^{1/3} - M^{1/3} = \kappa t \qquad (13–18)$$

where

$$\kappa = [N\rho(\pi/6)]^{1/3}\frac{2kC_s}{\rho} = \frac{M_0^{1/3}}{d}\frac{2kC_s}{\rho} \qquad (13–19)$$

$M_0$ is the original mass of the drug particles. Equation **(13–18)** is known as the Hixson–Crowell cube-root law,[20] and $\kappa$ is the cube-root dissolution rate constant.

### EXAMPLE 13–3

#### Calculate Dissolution Rate Constant

A specially prepared tolbutamide powder of fairly uniformly sized particles with a diameter of 150 $\mu$m weighed 75 mg. Dissolution of the drug was determined in 1000 mL of water at 25°C as a function of time. Determine the value of $\kappa$, the cube-root dissolution rate constant, at each time interval and calculate the average value of $\kappa$. The data and results are set forward in the accompanying table.

#### DISSOLUTION OF TOLBUTAMIDE POWDER*

| Time (min) | Concentration Dissolved (mg/mL) | Weight Undissolved, $M$(g) | $M_0^{1/3} - M^{1/3}$ | $\kappa$ (g$^{1/3}$/min) |
|---|---|---|---|---|
| 0 | 0 | 0.0750($M_0$) | 0 | — |
| 10 | 0.01970 | 0.0553 | 0.0406 | 0.0041 |
| 20 | 0.0374 | 0.0376 | 0.0866 | 0.0043 |
| 30 | 0.0510 | 0.0240 | 0.1332 | 0.0044 |
| 40 | 0.0595 | 0.0155 | 0.1724 | 0.0043 |
| 50 | 0.0650 | 0.0100 | 0.2063 | 0.0041 |

$$\kappa_{av} = \frac{\sum \kappa}{5} = \frac{0.0212}{5} = 0.00424 \, \text{g}^{1/3}/\text{min}$$

*Based on M. J. Miralles, M. S. Thesis, University of Texas, Austin, 1980.

In the situation in which the aqueous diffusion layer thickness of a spherical particle is comparable to or larger than the size of the sphere (e.g., micronized particles less than 50 $\mu$m in diameter), the change in particle radius with time becomes

$$r^2 = r_0^2 - \frac{2DC_s t}{\rho} \qquad (13–20)$$

and the estimated time for complete dissolution, $\tau$ (i.e., when $r^2 = 0$), is

$$\tau = \frac{\rho r_0^2}{2DC_s} \qquad (13–21)$$

### EXAMPLE 13–4

#### Dissolution Time

In clinical practice, diazepam injection (a sterile solution of diazepam in a propylene glycol–ethanol–water cosolvent system) is often diluted manyfold with normal saline injection. An incipient precipitation of diazepam occurs invariably upon addition of saline followed by complete dissolution within 1 min upon shaking. With $C_s$ in water equal to 3 mg/mL, $\rho$ about equal to 1.0 g/mL, and $D$ equal to $5 \times 10^{-6}$ cm$^2$/sec, calculate the time for complete dissolution when $r_0 = 10 \, \mu$m ($10 \times 10^{-4}$ cm).

We have

$$\tau = \frac{(1 \, \text{g/mL})(10 \times 10^{-4} \, \text{cm})^2}{2(5 \times 10^{-6} \, \text{cm}^2/\text{sec})(3 \times 10^{-3} \, \text{g/mL})}$$

$$= 33 \, \text{sec}$$

If $r_0 = 25 \, \mu$m, $\tau = 208$ sec

### More Complex Models of Dissolution: Convective Diffusion

Convection, the transfer of heat (energy) and the presence of agitation accompanying the movement of a fluid, can be combined with diffusion to provide a *convective diffusion model* for the study of dissolution.[21] The convective diffusion model, unlike the simpler Noyes–Whitney[19] and Nernst–Brünner[22] approaches, takes into consideration such factors as flow rate, mixing (agitation), and the dimensions of the dosage form. Nelson and Shah[23] investigated the convective diffusion model for the dissolution of alkyl $p$-aminobenzoates as test compounds. De Smidt et al.[24] also used a convective diffusion model in the study of the dissolution kinetics of griseofulvin in solutions of the solubilizing agent sodium dodecylsulfate.

### EXAMPLE 13–5

#### Drug Dissolution

De Smidt et al.[24] introduced a drug dissolution rate approach to model the rates of dissolution of alkyl $p$-aminobenzoates in a specially designed diffusion cell. The model is based on convective diffusion, the equations of which can be used to calculate $R$, the rate of diffusion or permeation rate:

$$R = 0.808D^{2/3}C_s\alpha^{1/3}bL^{2/3} \qquad (13–22)$$

for a rectangular tablet surface of width $b$ and length $L$ in the direction of flow, and

$$R = 2.157D^{2/3}C_s\alpha^{1/3}r^{5/3} \qquad (13–23)$$

for a circular tablet surface of radius $r$. In these equations, $D$ is the diffusivity or diffusion coefficient, $C_s$ is the solubility, and $\alpha$ is the rate of shear as the solvent is pumped over the dissolving surface. The rate of shear is calculated from $\alpha = 6Q/H^2W$, where $Q$ is the flow rate and $H$ and $W$ are the height and width, respectively, of a channel in the diffusion cell to allow the flow of solvent (water) over the dissolving tablet.

Experiments on dissolution rate, $R$, were carried out at 37°C with rectangular tablet surfaces containing the drug model ethyl $p$-aminobenzoate. The long axis of the rectangular surface was 25.4 mm and the short axis 3.175 mm.

(*a, b*) Compute the rate of dissolution, $R$, with the long axis, $L$, placed perpendicular to the direction of flow, and then with the long axis placed parallel to the direction of flow. The flow rate, $Q$, is 14.9 mL/min; the diffusivity and the solubility of the drug are $D = 9.86 \times 10^{-6}$ cm²/sec and $C_s = 7.27 \times 10^{-6}$ mole/cm³, respectively; and $H^2W = 0.3506$ cm³.

(*c*) The experiment is repeated but using a disk with a circular surface of area equal to the surface area of the rectangle referred to in (*a*). Compute $R$, expressing the results in mole/min.

(*d*) What differences do you find between this model and the classic stagnant or unstirred diffusion layer model? You can refer to Nelson and Shah[23] to check the answers given here.

(*a*) The rate of shear is

$$\alpha = 6Q/H^2W = 6 \times 14.9 \text{ cm}^3 \text{ min}^{-1}/0.3506 \text{ cm}^3 = 255.0 \text{ min}^{-1}$$

For the long axis perpendicular to flow, $b = 2.54$ cm and $L = 0.3175$ cm. Then,

$$R = 0.808(9.86 \times 10^{-6} \text{ cm}^2/\text{sec} \times 60 \text{ sec/min})^{2/3}$$
$$\times (7.27 \times 10^{-6} \text{ mole/cm}^3) \times (255.0 \text{ min}^{-1})^{1/3}$$
$$\times (2.54 \text{ cm}) \times (0.3175 \text{ cm})^{2/3}$$
$$= 3.10 \times 10^{-7} \text{ mole/min}$$

(*b*) For the long axis parallel to the flow, $b = 0.3175$ cm and $L = 2.54$ cm. Then,

$$R = 0.808(9.86 \times 10^{-6} \text{ cm}^2/\text{sec} \times 60 \text{ sec/min})^{2/3}$$
$$\times (7.27 \times 10^{-6} \text{ mole/cm}^3) \times (255.0 \text{ min}^{-1})^{1/3}$$
$$\times (0.3175 \text{ cm}) \times (2.54 \text{ cm})^{2/3}$$
$$= 1.55 \times 10^{-7} \text{ mole/min}$$

For the long axis perpendicular to the flow, $R$ is twice the value when the long axis is parallel to the flow, as observed in (*a*) and (*b*):

$$R = 3.10 \times 10^{-7}/(1.55 \times 10^{-7}) = 2.0$$

(*c*) The surface area of the rectangular tablet is 2.54 cm $\times$ 0.3175 cm = 0.806 cm², which is also the surface area of the circular tablet or disk. Therefore, the radius, $r$, of the circular surface is $\pi r^2 = 0.806$, or $r = 0.507$ cm, and the rate, $R$, of diffusion or permeation for a tablet of circular surface [equation (13–23)] is

$$R = 2.157(9.86 \times 10^{-6} \times 60 \text{ cm}^2/\text{min})^{2/3}$$
$$\times (7.27 \times 10^{-6} \text{ mole/cm})^3 \times (255 \text{ min}^{-1})^{1/3}$$
$$\times (0.507 \text{ cm})^{5/3}$$
$$= 2.26 \times 10^{-7} \text{ mole/min}$$

(*d*) The convective diffusion (CD) model, which takes into account fluid flow as well as diffusion, has several parameters in common with the classic diffusion model. These include the solubility, $C_s$, diffusion coefficient or diffusivity, $D$, and the dimensions of a rectangular or circular surface, $b$, $L$, and $r$. In the classic model, $R$ is proportional to $D$, where in the CD model, $R$ is proportional to $D^{2/3}$. In the classic model, $R$ is proportional to the surface area, $S$, of a rectangle or disk; in the CD model, $R$ is proportional to a reduced function of surface area, that is, $bL^{2/3}$ or $r^{5/3}$. A new parameter, $\alpha$, the rate of shear over the dissolving surface, is introduced in the CD model; it is calculated from the flow rate and the dimensions of the diffusion cell.

## Drug Release

Release from dosage forms and subsequent absorption are controlled by the physical chemical properties of drug and delivery system and the physiologic and physical chemical properties of the biologic system. Drug concentration, aqueous solubility, molecular size, crystal form, protein binding, and p$K_a$ are among the physical chemical factors that must be understood to design a delivery system that exhibits controlled- or sustained-release characteristics.[25]

### The Higuchi (Equation) Model[26,27]

Higuchi developed a theoretical model for studying the release of water-soluble and poorly soluble drugs from a variety of matrices, including semisolid and solids. We will cover the factors that control drug release from solid dosage forms later in this chapter. A powdered drug is homogeneously dispersed throughout the matrix of an erodible tablet. The drug is assumed to dissolve in the polymer matrix and to diffuse out from the surface of the device. As the drug is released, the distance for diffusion becomes increasingly greater. The boundary that forms between drug and empty matrix therefore recedes into the tablet as drug is eluted. A schematic illustration of such a device is shown in **Figure 13–4a**. **Figure 13–4b** shows a granular matrix with interconnecting pores or capillaries. The drug is leached out of this device by entrance of the surrounding medium. **Figure 13–4c** depicts the concentration profile and shows the receding depletion zone that moves to the center of the tablet as the drug is released.

Higuchi[26] developed an equation for the release of a drug from an ointment base and later[27] applied it to diffusion of solid drugs dispersed in homogeneous and granular matrix dosage systems (**Fig. 13–4**).

Recall that Fick's first law (Chapter 11),

$$\frac{dM}{S\,dt} = \frac{dQ}{dt} = \frac{DC_s}{h} \tag{13–24}$$

can be applied to the case of a drug embedded in a polymer matrix, where $dQ/dt$ is the rate of drug released per unit area of exposed surface of the matrix. Because the boundary between the drug matrix and the drug-depleted matrix recedes with time, the thickness of the empty matrix, $dh$, through which the drug diffuses also increases with time.

Whereas $C_s$ is the solubility or saturation concentration of drug in the matrix, $A$ is the total concentration (amount per unit volume), dissolved and undissolved, of drug in the matrix.

As drug passes out of a homogeneous matrix (**Fig. 13–4a**), the boundary of drug (represented by the dashed vertical line in **Fig. 13–4c**) moves to the left by an infinitesimal distance, $dh$. The infinitesimal amount, $dQ$, of drug released because of this shift of the front is given by the approximate linear expression

$$dQ = A\,dh - \frac{1}{2}C_s\,dh \tag{13–25}$$

Now, $dQ$ of equation (**13–35**) is substituted into equation (**13–34**), integration is carried out, and the resulting equation

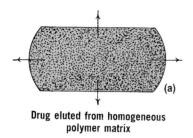

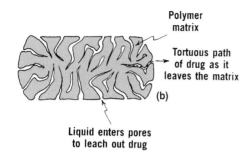

Drug eluted from homogeneous
polymer matrix

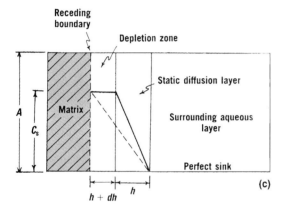

Fig. 13–4. Release of drug from homogeneous and granular matrix dosage forms. (*a*) Drug eluted from a homogeneous polymer matrix. (*b*) Drug leached from a heterogeneous or granular matrix. (*c*) Schematic of the solid matrix and its receding boundary as drug diffuses from the dosage form. (From T. Higuchi, J. Pharm. Sci. **50**, 874, 1961. With permission.)

is solved for *h*. The steps of the derivation as given by Higuchi[26] are

$$\left(A - \frac{1}{2}C_s\right) dh = \frac{DC_s}{h}dt \qquad (13\text{–}26)$$

$$\frac{2A - C_s}{2DC_s}\int h\,dh = \int dt \qquad (13\text{–}27)$$

$$t = \frac{(2A - C_s)}{4DC_s}h^2 + C \qquad (13\text{–}28)$$

The integration constant, *C*, can be evaluated at $t = 0$, at which $h = 0$, giving

$$t = \frac{(2A - C_s)h^2}{4DC_s} \qquad (13\text{–}29)$$

$$h = \left(\frac{4DC_s t}{2A - C_s}\right)^{1/2} \qquad (13\text{–}30)$$

The amount of drug depleted per unit area of matrix, *Q*, at time *t* is obtained by integrating equation (13–25) to yield

$$Q = hA - \frac{1}{2}hC_s \qquad (13\text{–}31)$$

Substituting equation (13–30) into (13–31) produces the result

$$Q = \left(\frac{DC_s t}{2A - C_s}\right)^{1/2}(2A - C_s) \qquad (13\text{–}32)$$

which is known as the Higuchi equation:

$$Q = [D(2A - C_s)C_s t]^{1/2} \qquad (13\text{–}33)$$

The instantaneous rate of release of a drug at time *t* is obtained by differentiating equation (13–33) to yield

$$\frac{dQ}{dt} = \frac{1}{2}\left[\frac{D(2A - C_s)C_s}{t}\right]^{1/2} \qquad (13\text{–}34)$$

Ordinarily, *A* is much greater than $C_s$, and equation (13–33) reduces to

$$Q = (2ADC_s t)^{1/2} \qquad (13\text{–}35)$$

and equation (13–34) becomes

$$\frac{dQ}{dt} = \left(\frac{ADC_s}{2t}\right)^{1/2} \qquad (13\text{–}36)$$

for the release of a drug from a homogeneous polymer matrix–type delivery system. Equation (13–35) indicates that the amount of drug released is proportional to the square root of *A*, the total amount of drug in unit volume of matrix; *D*, the diffusion coefficient of the drug in the matrix; $C_s$, the solubility of drug in polymeric matrix; and *t*, the time.

The rate of release, $dQ/dt$, can be altered by increasing or decreasing the drug's solubility, $C_s$, in the polymer by complexation. The total concentration, *A*, of drug that the physician prescribes is also seen to affect the rate of drug release.

### EXAMPLE 13–6

**Drug Release**

(*a*) What is the amount of drug per unit area, *Q*, released from a tablet matrix at time $t = 120$ min? The total concentration of drug in the homogeneous matrix, *A*, is 0.02 g/cm$^3$. The drug's solubility, $C_s$, is $1.0 \times 10^{-3}$ g/cm$^3$ in the polymer. The diffusion coefficient, *D*, of the drug in the polymer matrix at 25°C is $6.0 \times 10^{-6}$ cm$^2$/sec or $360 \times 10^{-6}$ cm$^2$/min.

We use equation (13–35):

$$Q = [2(0.02 \text{ g/cm}^3)(360 \times 10^{-6} \text{ cm}^2/\text{min})$$
$$\times (1.0 \times 10^{-3} \text{ g/cm}^3)(120 \text{ min})]^{1/2}$$
$$= 1.3 \times 10^{-3} \text{ g/cm}^2$$

(*b*) What is the instantaneous rate of drug release occurring at 120 min?

We have

$$dQ/dt = \left[\frac{(0.02)(360 \times 10^{-6})(1.0 \times 10^{-3})}{2 \times 120}\right]^{1/2}$$
$$= 5.5 \times 10^{-6} \text{ g cm}^{-2}\text{min}^{-1}$$

## Release from Granular Matrices: Porosity and Tortuosity

The release of a solid drug from a granular matrix (Fig. 13–4b) involves the simultaneous penetration of the surrounding liquid, dissolution of the drug, and leaching out of the drug through interstitial channels or pores. A granule is, in fact, defined as a porous rather than a homogeneous matrix. The volume and length of the opening in the matrix must be accounted for in the diffusional equation, leading to a second form of the Higuchi equation:

$$Q = \left[\frac{D\varepsilon}{\tau}(2A - \varepsilon C_s)C_s t\right]^{1/2} \quad (13\text{--}37)$$

where $\varepsilon$ is the porosity of the matrix and $\tau$ is the tortuosity of the capillary system, both parameters being dimensionless quantities.

Porosity, $\varepsilon$, is the fraction of matrix that exists as pores or channels into which the surrounding liquid can penetrate. The porosity term, $\varepsilon$, in equation (13–37) is the total porosity of the matrix after the drug has been extracted. This is equal to the initial porosity, $\varepsilon_0$, due to pores and channels in the matrix before the leaching process begins and the porosity created by extracting the drug. If $A$ g/cm³ of drug is extracted from the matrix and the drug's specific volume or reciprocal density is $1/\rho$ cm³/g, then the drug's concentration, $A$, is converted to volume fraction of drug that will create an additional void space or porosity in the matrix once it is extracted. The total porosity of the matrix, $\varepsilon$, becomes

$$\varepsilon = \varepsilon_0 + A(1/\rho) \quad (13\text{--}38)$$

The initial porosity, $\varepsilon_0$, of a compressed tablet can be considered to be small (a few percent) relative to the porosity, $A/\rho$, created by the dissolution and removal of the drug from the device. Therefore, the porosity frequently is calculated conveniently by disregarding $\varepsilon_0$ and writing

$$\varepsilon \cong A/\rho \quad (13\text{--}39)$$

Tablet porosity and its measurement and applications in pharmacy are discussed in more detail in sections Capsule-Type Devices and Dissolution and Release from Oral Drug Products.

Equation (13–37) differs from equation (13–33) only in the addition of $\varepsilon$ and $\tau$. Equation (13–33) is applicable to release from a homogeneous tablet that gradually erodes and releases the drug into the bathing medium. Equation (13–37) applies instead to a drug-release mechanism based on entrance of the surrounding medium into a polymer matrix, where it dissolves and leaches out the soluble drug, leaving a shell of polymer and empty pores. In equation (13–37), diffusivity is multiplied by porosity, a fractional quantity, to account for the decrease in $D$ brought about by empty pores in the matrix. The apparent solubility of the drug, $C_s$, is also reduced by the volume fraction term, which represents porosity.

Tortuosity, $\tau$, is introduced into equation (13–37) to account for an increase in the path length of diffusion due to branching and bending of the pores, as compared to the shortest "straight-through" pores. Tortuosity tends to reduce

the amount of drug release in a given interval of time, and so it appears in the denominator under the square root sign. A straight channel has a tortuosity of unity, and a channel through spherical beads of uniform size has a tortuosity of 2 or 3. At times, an unreasonable value of, say, 1000 is obtained for $\tau$, as Desai et al[28a] noted. When this occurs, the pathway for diffusion evidently is not adequately described by the concept of tortuosity, and the system must be studied in more detail to determine the factors controlling matrix permeability. Methods for obtaining diffusivity, porosity, tortuosity, and other quantities required in an analysis of drug diffusion are given by Desai et al.[28b]

Equation (13–37) has been adapted to describe the kinetics of lyophilization,[29] commonly called *freeze-drying*, of a frozen aqueous solution containing drug and an inert matrix-building substance (e.g., mannitol or lactose). The process involves the simultaneous change in the receding boundary with time, phase transition at the ice–vapor interface governed by the Clausius–Clapeyron pressure–temperature relationship, and water vapor diffusion across the pore path length of the dry matrix under low-temperature and vacuum conditions.

## Soluble Drugs in Topical Vehicles and Matrices

The original Higuchi model[26,27] does not provide a fit to experimental data when the drug has a significant solubility in the tablet or ointment base. The model can be extended to drug release from homogeneous solid or semisolid vehicles, however, using a quadratic expression introduced by Bottari et al.[30]:

$$Q^2 + 2DRA^*Q - 2DA^*C_s t = 0 \quad (13\text{--}40)$$

Here,

$$A^* = A - \frac{1}{2}(C_s + C_v) \quad (13\text{--}41)$$

$Q$ is the amount of drug released per unit area of the dosage form, $D$ is an effective diffusivity of the drug in the vehicle, $A$ is the total concentration of drug, $C_s$ is the solubility of drug in the vehicle, $C_v$ is the concentration of drug at the vehicle–barrier interface, and $R$ is the diffusional resistance afforded by the barrier between the donor vehicle and the receptor phase. $A$ is an effective area as defined in equation (13–41) and is used when $A$ is only about three or four times greater than $C_s$.

When

$$Q^2 \gg 2DRA^*Q \quad (13\text{--}42)$$

equation (13–40) reduces to one form of the Higuchi equation [equation (13–35)]:

$$Q = (2A^*DC_s t)^{1/2} \quad (13\text{--}42a)$$

Under these conditions, resistance to diffusion, $R$, is no longer significant at the interface between vehicle and receptor phase. When $C_s$ is not negligible in relation to $A$, the vehicle-controlled model of Higuchi becomes

$$Q = [D(2A - C_s)C_s t]^{1/2} \quad (13\text{--}42b)$$

as derived earlier equation (13–33).

The quadratic expression of Bottari, equation (13–40), should allow one to determine diffusion of drugs in ointment vehicles or homogeneous polymer matrices when $C_s$ becomes significant in relation to $A$. The approach of Bottari et al.[30] follows.

Because it is a second-degree power series in $Q$, equation (13–40) can be solved using the well-known quadratic approach. One writes

$$aQ^2 + bQ + c = 0 \qquad (13–43)$$

where, with reference to equation (13–40), $a$ is unity, $b = 2DRA$, and $c = -2DAC_st$. Equation (13–43) has the well-known solution

$$Q = \frac{-b \pm \sqrt{b^2 - 4ac}}{2a} \qquad (13–44)$$

or

$$Q = \frac{-2DRA^* + \sqrt{(2DRA^*)^2 + (2DA^*C_st)}}{2} \qquad (13–45)$$

where the positive root is taken for physical significance. If a lag time occurs, $t$ in equation (13–45) is replaced by $(t - t_L)$ for the steady-state period. Bottari et al.[30] obtained satisfactory values for $b$ and $c$ by use of a least-square fit of equation (13–40) involving the release of benzocaine from suspension-type aqueous gels. $R$, the diffusional resistance, is determined from steady-state permeation, and $C_v$ is then obtained from the expression

$$C_v = R(dQ/dt) \qquad (13–46)$$

The application of equation (13–40) is demonstrated in the following example.

## EXAMPLE 13–7

### Calculate $Q$

Calculate $Q$, the amount in milligrams, of micronized benzocaine released per $cm^2$ of surface area from an aqueous gel after 9000 sec (2.5 hr) in a diffusion cell. Assume that the total concentration, $A$, is 10.9 mg/mL; the solubility, $C_s$, is 1.31 mg/mL; $C_v = 1.05$ mg/mL; the diffusional resistance, $R$, of a silicone rubber barrier separating the gel from the donor compartment is $8.10 \times 10^3$ sec/cm; and the diffusivity, $D$, of the drug in the gel is $9.14 \times 10^{-6}$ $cm^2$/sec. From equation (13–41),

$$A^* = 10.9 \text{ mg/mL} - \frac{1}{2}(1.31 + 1.05) \text{ mg/mL} = 9.72 \text{ mg/mL}$$

Then,

$$DRA^* = (9.14 \times 10^{-6} \text{ cm}^2/\text{sec}) \times (8.10 \times 10^3 \text{ sec/cm})$$
$$\times (9.72 \text{ mg/mL}) = 0.7196 \text{ mg/cm}^2$$

$$DA^*C_st = (9.14 \times 10^{-6})(9.72)(1.31)(9000) = 1.047 \text{ mg}^2/\text{cm}^4$$
$$Q = -0.7196 + [(0.7196)^2 + 2(1.047)]^{1/2} \text{ mg/cm}^2$$
$$= -0.7196 + [1.616] = 0.90 \text{ mg/cm}^2$$

The $Q_{(calc)}$ of 0.90 mg/cm$^2$ compares well with $Q_{(obs)} = 0.88$ mg/cm$^2$.

A slight increase in accuracy can be obtained by replacing $t = 9000$ sec with $t = (9000 - 405)$ sec, in which the lag time $t = 405$ sec is obtained from a plot of experimental $Q$ values versus $t^{1/2}$. This correction yields a $Q_{(calc)} = 0.87$ mg/cm$^2$.

(b) Calculate $Q$ using equation (13–42b) and compare the result with that obtained in (a).

We have

$$Q = \{(9.14 \times 10^{-6})[(2 \times 10.9) - 1.31](1.31)(9000)\}^{1/2}$$
$$= 1.49 \text{ mg/cm}^2$$

Paul and coworkers[31] studied cases in which $A$, the matrix loading of drug per unit volume in a polymeric dosage form, may be greater than, equal to, or less than the equilibrium solubility, $C_s$, of the drug in a matrix. The model is a refinement of the original Higuchi approach,[26,27] providing an accurate set of equations that describe release rates of drugs, fertilizers, pesticides, antioxidants, and preservatives in commercial and industrial applications over the entire range of ratios of $A$ to $C_s$.

### Capsule-Type Device

A silastic capsule, as depicted in **Figure 13–5a**, has become a popular sustained and controlled delivery form in pharmacy and medicine.[32–34] The release of a drug from a silastic capsule is shown schematically in **Figure 13–5b**. The molecules of the crystalline drug lying against the inside wall of the capsule leave their crystals, pass into the polymer wall by a dissolution process, diffuse through the wall, and pass into the liquid diffusion layer and the medium surrounding the capsule. The concentration differences across the polymer wall of thickness, $h_m$, and the stagnant diffusion layer of thickness, $h_a$, are represented by the lines $C_p - C_m$ and $C_s - C_b$, respectively, where $C_p$ is the solubility of the drug in the polymer and $C_m$ is the concentration at the polymer–solution interface, that is, the concentration of drug in the polymer in contact with the solution. On the other hand, $C_s$ is the concentration of the drug in the solution at the polymer–solution interface,

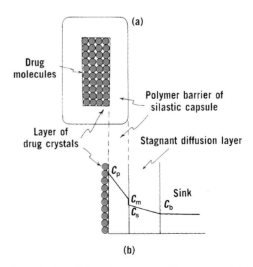

**Fig. 13–5.** Diffusion of drug from an elastic capsule. (a) Drug in the capsule surrounded by a polymer barrier; (b) diffusion of drug through the polymer wall and stagnant aqueous diffusion layer and into the receptor compartment at sink conditions. (From Y. W. Chien, in J. R. Robinson (Ed.), *Sustained and Controlled Release Drug Delivery Systems*, Marcel Dekker, New York, 1978, p. 229; Y. W. Chien, Chem. Pharm. Bull. **24**, 147, 1976.)

and it is seen in **Figure 13–5b** to be somewhat below the solubility of drug in polymer at the interface. There is a real difference between the solubility of the drug in the polymer and that in the solution, although both exist at the interface. Finally, $C_b$ is the concentration of the drug in the bulk solution surrounding the capsule.

To express the rate of drug release under sink conditions, Chien[32] used the following expression:

$$Q = \frac{K_r D_a D_m}{K_r D_a h_m + D_m h_a} C_p t \qquad (13\text{–}47)$$

In equation (13–47), $Q$ is the amount of drug released per unit surface area of the capsule and $K_r$ is the partition coefficient, defined as:

$$K_r = C_s/C_p \qquad (13\text{–}48)$$

When diffusion through the capsule membrane or film is the limiting factor in drug release, that is, when $K_r D_a h_m$ is much greater than $D_m h_a$, equation (13–47) reduces to:

$$Q = \frac{D_m}{h_m} C_p t \qquad (13\text{–}49)$$

and when the limiting factor is passage through the diffusion layer $D_m h_a \gg K_r D_a h_m$),

$$Q = \frac{D_a}{h_a} C_s t = \frac{K_r D_a}{h_a} C_p t \qquad (13\text{–}50)$$

The right-hand expression can be written because $C_s = K_r C_p$ as defined earlier, in equation (13–48).

The rate of drug release, $Q/t$, for a polymer-controlled process can be calculated from the slope of a linear $Q$-versus-$t$ plot, and from equation (13–49) is seen to equal $C_p D_m/h_m$. Likewise, $Q/t$, for the diffusion-layer–controlled process, resulting from plotting $Q$ versus $t$, is found to be $C_s D_a/h_a$. Furthermore, a plot of the release rate, $Q/t$, versus $C_s$, the solubility of the drug in the surrounding medium, should be linear with a slope of $D_a/h_a$.

### EXAMPLE 13–8

**Progesterone Release Rate**

The partition coefficient, $K_r = C_s/C_p$, of progesterone is 0.022; the solution diffusivity, $D_a$, is $4.994 \times 10^{-2}$ cm$^2$/day; the silastic membrane diffusivity, $D_m$, is $14.26 \times 10^{-2}$ cm$^2$/day; the solubility of progesterone in the silastic membrane, $C_p$, is 513 $\mu$g/cm$^3$; the thickness of the capsule membrane, $h_m$, is 0.080 cm; and that of the diffusion layer, $h_a$, as estimated by Chien,[30] is 0.008 cm.

Calculate the rate of release of progesterone from the capsule and express it in $\mu$g/cm$^2$ per day. Compare the calculated result with the observed value, $Q/t = 64.50$ $\mu$g/cm$^2$ per day.

Using equation (13–47), we find

$$Q/t = \frac{C_p K_r D_a D_m}{K_r D_a h_m + D_m h_a}$$

$$Q/t = \frac{\begin{array}{c}(513\ \mu\text{g/cm}^3)(0.022)(4.994 \times 10^{-2}\ \text{cm}^2/\text{day}) \\ \times (14.26 \times 10^{-2}\ \text{cm}^2/\text{day})\end{array}}{\begin{array}{c}(0.022)(4.994 \times 10^{-2}\ \text{cm}^2/\text{day})(0.080\ \text{cm}) \\ + (14.26 \times 10^{-2}\ \text{cm}^2/\text{day})(0.008\ \text{cm})\end{array}}$$

$$Q/t = \frac{0.08037}{0.00123} = 65.34\ \mu\text{g/cm}^2\ \text{per day}$$

In the example just given, (a) is $K_r D_a h_m \gg D_m h_a$ or (b) is $D_m h_a \gg K D_a h_m$? (c). What conclusion can be drawn regarding matrix or diffusion-layer control?

We have

$$K_r D_a h_m = 8.79 \times 10^{-5}; \quad D_m h_a = 1.14 \times 10^{-3}$$

$$D_m h_a/(K_r D_a h_m + D_m h_a)$$
$$= (1.14 \times 10^{-3})/[(8.79 \times 10^{-5})$$
$$+ (1.14 \times 10^{-3})] = 0.93$$

Therefore, $D_m h_a \gg K_r D_a h_m$, and the system is 93% under aqueous diffusion-layer control. It should thus be possible to use the simplified equation (13–50):

$$Q/t = \frac{K_r D_a C_p}{h_a} = \frac{(0.022)(4.994 \times 10^{-2})(513)}{0.008}$$

$$= 70.45\ \mu g/\text{cm}^2\ \text{per day}$$

Although $D_m h_a$ is larger than $K_r D_a h_m$ by about one order of magnitude (i.e., $D_m h_a/K D_a h_m = 13$), it is evident that a considerably better result is obtained by using the full expression, equation (13–47).

### EXAMPLE 13–9

**Calculate Membrane Thickness**

Two new contraceptive steroid esters, A and B, were synthesized, and the parameters determined for release from polymeric capsules are as follows[32]:

| | $K_r$ | $D_a$ (cm$^2$/day) | $D_m$ (cm$^2$/day) | $C_p$ ($\mu$g/cm$^3$) | $h_a$ (cm) | $Q/t_{(obs)}$ ($\mu$g/cm$^2$ per day) |
|---|---|---|---|---|---|---|
| A | 0.15 | $25 \times 10^{-2}$ | $2.6 \times 10^{-2}$ | 100 | 0.008 | 24.50 |
| B | 0.04 | $4.0 \times 10^{-2}$ | $3.0 \times 10^{-2}$ | 85 | 0.008 | 10.32 |

Using equation (13–47) and the quantities in the table, calculate values of $h_m$ in centimeter for these capsule membranes.

We have

$$Q/t = \frac{C_p K_r D_a D_m}{K_r D_a h_m + D_m h_a}$$

$$(Q/t)(K_r D_a h_m + D_m h_a) = C_p K_r D_a D_m$$

$$(Q/t)(K_r D_a h_m) = C_p K_r D_a D_m - D_m h_a(Q/t)$$

$$h_m = \frac{C_p K_r D_a D_m - D_m h_a(Q/t)}{(Q/t)K_r D_a}$$

For capsule A,

$$h_m = \frac{\begin{array}{c}(100)(0.15)(25 \times 10^{-2})(2.6 \times 10^{-2}) \\ -(2.6 \times 10^{-2})(0.008)(24.50)\end{array}}{(24.50)(0.15)(25 \times 10^{-2})}$$

$$h_m = \frac{0.0924}{0.9188}\ \text{cm} = 0.101\ \text{cm}$$

Note that all units cancel except centimeter in the equation for $h_m$. The reader should carry out the calculations for compound B. (*Answer:* 0.097 cm)

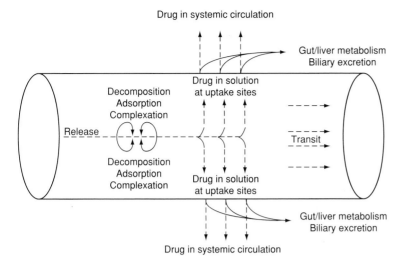

**Fig. 13–6.** A schematic representation of the factors that determine the fraction of drug that is absorbed from a drug product across the intestinal mucosa. Decomposition, adsorption to intestinal components, or complexation can reduce the amount of drug available for absorption. Drug uptake is controlled by the drug's permeability through the intestinal mucosa and the length of time that it stays at the absorption site (i.e., residence time). The longer it stays within the "absorption window" and the higher the permeability, the more is the drug absorbed across the intestinal mucosa. (From J. B. Dressman, G. L. Amidon, C. Reppas, and V. P. Shah, Pharm. Res. **15**, 11, 1998. With permission.)

# DISSOLUTION AND RELEASE FROM ORAL DRUG PRODUCTS

After a solid dosage form such as a tablet is administered by mouth to a patient, it must first disintegrate into larger clusters of particles known as aggregates. Deaggregation then occurs and individual particles are liberated. Finally, particles dissolve, releasing the active drug into solution. Dissolution is a time-dependent (or kinetic) process that represents the final step of drug release, which is ultimately required before a drug can be absorbed or exert a pharmacologic effect. For immediate-release dosage forms, the rate of drug release and dissolution relative to the rate of transit through the intestine and the permeability profile of the small intestine to the drug determines the rate and the extent of drug absorption (**Fig. 13–6**). If drug dissolution is slow compared with drug absorption, less drug may be absorbed, especially if the drug is absorbed preferentially in certain locations ("absorption windows") of the gastrointestinal tract. Slower absorption due to slower dissolution can also result in lower peak drug blood levels. On the other hand, semisolid dosage forms such as topical drug products are applied to the skin and remain in the area of application. As described in the SUPAC-SS Guidance,[4b] semisolid dosage forms are complex formulations having complex structural elements. Often they are composed of two phases (oil and water), one of which is a continuous (external) phase, the other of which is a dispersed (internal) phase. The active ingredient is often dissolved in one phase even though occasionally the drug is not fully soluble in the system and is dispersed in one or both phases, thus creating a three-phase system. The physical properties of the dosage form depend on various factors: the size of the dispersed particles, the interfacial tension between the phases, the partition coefficient of the active ingredient between the phases, and the product rheology. These factors combine to determine the release characteristics of the drug as well as other characteristics, such as viscosity.

The BCS[6] categorizes drugs into four types (**Table 13–1**), depending on their solubility and permeability characteristics. Solubility is covered in Chapter 9 and permeability in Chapter 11. For the purposes of this chapter, it would be helpful to give some perspective on the role of solubility, permeability, and drug release on the availability of drug in the human body after oral administration. In most situations, only drug that dissolves and is released from the drug product will be available for absorption through the intestinal tissues and into the blood stream of patients. Therefore, the rate at which the drug dissolves (in other words, dissolution rate) and its solubility become important factors, and these have already been discussed in some detail. Permeability is a measure of how rapidly a drug can penetrate a biologic tissue such as the intestinal mucosa and appears on the other side (e.g., the blood side). Therefore, a drug must be soluble and permeable

| **TABLE 13–1** | | | |
| --- | --- | --- | --- |
| **THE BIOPHARMACEUTICS CLASSIFICATION SYSTEM (BCS)\*,†** | | | |
| Class I | Class II | Class III | Class IV |
| High solubility, high permeability | Low solubility, high permeability | High solubility, low permeability | Low solubility, low permeability |

\*From G. L. Amidon, H. Lennernas, V. P. Shah, and J. R. Crison, Pharm. Res. **12**, 413, 1995.
†The goal of the system is to provide guidance as to when in vitro studies may be used in lieu of clinical studies to establish the bioequivalence of two products.

## KEY CONCEPT　THE ROLE OF DISSOLUTION TESTING

In a 1998 review article, Dressman and colleagues[7] summarized the situation well:

> Dissolution tests are used for many purposes in the pharmaceutical industry: in the development of new products, for quality control, and to assist with the determination of bioequivalence. Recent regulatory developments such as the Biopharmaceutics Classification Scheme have highlighted the importance of dissolution in the regulation of postapproval changes and introduced the possibility of substituting dissolution tests for clinical studies in some cases. Therefore, there is a need to develop dissolution tests that better predict the in vivo performance of drug

products. This could be achieved if the conditions of the gastrointestinal tract were successfully reconstructed in vivo.

Numerous factors need to be considered if dissolution tests are to be considered biorelevant. They are the composition, hydrodynamics (fluid flow patterns), and volume of the contents in the gastrointestinal tract. Biorelevant media considerations are covered in this section and the apparatus used to measure dissolution are covered in the next. Other aspects are also covered throughout the book. The student who wants to study this in more detail is referred to the original review article.

for absorption to occur. To classify drugs according to these two important factors, the BCS was proposed. According to the BCS, class I drugs are well absorbed (more than 90% absorbed) because they are highly permeable and go rapidly into solution. Poor absorption of class I drugs is only expected if they are unstable or if they undergo reactions (such as binding or complexation) in the intestine that inactivate them. Bioavailability could also be low if they are metabolized in the intestine or liver or are subject to secretory processes such as intestinal or enterohepatic cycling. Class II drugs are those with solubilities too low to be consistent with complete absorption even though they are highly membrane permeable. Class III drugs have good solubility but low permeability. In other words, they are unable to permeate the gut wall quickly enough for absorption to be complete. Class IV drugs have neither sufficient solubility nor permeability for absorption to be complete. Class IV drugs tend to be the most problematic, although there are numerous examples of class IV drugs that are successfully used in the clinic. The student should keep in mind that even though class IV drugs do not possess optimal properties, some drugs in this category may still be absorbed well enough so that oral administration is a viable option.

Typically, the BCS is used to build an IVIVC. According to the Food and Drug Administration Guidance document, an IVIVC is "a predictive mathematical model describing the relationship between an in vitro property of an oral dosage form (usually the rate or extent of drug dissolution or release) and a relevant in vivo response (e.g., plasma drug concentration or amount of drug absorbed)."[5] Because the focus of this chapter is drug dissolution and release, we will focus on aspects of IVIVCs related only to these phenomena. Correlation of in vivo results with dissolution tests is likely to be best for class II drugs because dissolution rate is the principal limiting characteristic to absorption. Another case where good IVIVCs are often obtained is when a class I drug is formulated as an extended-release product. This is because the release profile controls the rate of absorption and absorption profile. In the first case, drug dissolution (and solubility) are the rate-limiting step for absorption, whereas in the second

case, the drug has adequate solubility and permeability, so its ability to be absorbed is controlled by its availability in the lumen of the gastrointestinal tract. Therefore, release from the dosage form is the key process. These examples highlight the practical differences between the processes of drug dissolution and release. Controlled-release products work using a combination of mechanisms and are covered in Chapter 21.

There are several physicochemical and physiologic factors that control the dissolution of drug products in humans and need to be considered when designing dissolution tests. They are the composition, mixing patterns (i.e., hydrodynamics), and volume of the contents in the gastrointestinal tract. These are reviewed in detail by Dressman et al.[7] The student must keep in mind that the gastrointestinal tract is an organ with a multitude of functions and drug products; foods and nutrients can remain in the gastrointestinal tract for up to 24 to 30 hr if they are not completely absorbed. The conditions of the gastrointestinal tract vary with the location of the segment of interest. To link the three key factors that control the dissolution of drugs in the gastrointestinal tract with the mathematical understanding developed earlier in the chapter, let us reexamine the Noyes–Whitney equation[19] with some commonly used modifications[7] as introduced by Levich[21] and Nernst–Brunner[22]

$$\frac{dX_d}{dt} = \frac{AD}{\delta}\left(C_s \frac{Xd}{V}\right) \qquad (13\text{–}51)$$

where $A$ is the effective surface area of the drug, $D$ is the diffusion coefficient of the drug, $\delta$ is the diffusion boundary layer thickness adjacent to the dissolving surface, $C_s$ is the saturation solubility of the drug under intestinal conditions, $X_d$ is the amount of drug already in solution, and $V$ is the volume of the dissolution media. To manipulate the effective surface area of a drug, a formulator may attempt to reduce the particle size to increase wettability. In the intestine, however, there are natural surface tension–reducing agents that promote drug dissolution. Most of these natural surfactants are found in the secretions that come from the stomach and bile. If one were to design a biorelevant dissolution test for drugs with poor dissolution characteristics, it would be important to account for the natural surfactants located in the stomach and the intestine. The maximum solubility of the drug in the intestine is influenced by many factors (solubility is covered in Chapter 9) such as the buffer capacity, pH, the presence of food, and natural surfactants such as bile salts and lecithin. Dressman et al.[35] showed that the presence of a meal in humans immediately raised the pH of the stomach from its normally acidic state (pH 1.5–3.0) to pH 5.5 to 7.0. This could dramatically affect the solubility of drugs, which, in turn, can affect the oral bioavailability. The diffusivity of the drug, or its natural ability to diffuse through the intestinal contents, will be a function of the viscosity of the intestinal contents. Viscosity will depend on the level of natural secretions, which may vary in the fed and fasted states, and the presence of food. The boundary layer thickness around the dissolving particle will depend on how vigorous local mixing is. In other words, if motility or mixing is higher, then the stagnant layer surrounding the particle will be smaller. If mixing is reduced, then the stagnant layer becomes larger and may alter dissolution. For dissolution to proceed, there must be a driving force. As shown in equation (13–51), the "driving force" is represented by the term $C_s - X_d/V$. If the difference between $C_s$ and $X_d/V$ is great, then the rate of drug dissolution, $X_d/dt$, will be greater. As the concentration of dissolved drug ($X_d/V$) becomes larger, the driving force is reduced. So, the relevant question is, "How can drug dissolution in the gastrointestinal tract ever be complete?" To maximize drug dissolution, the concentration of dissolved drug must be minimized. Of course, this happens when the drug is absorbed through the intestinal wall and into the blood. The rate of absorption is related to the permeability of the drug (Chapter 15) and intestinal drug concentration, $X_d/V$. For passively absorbed drugs, the greater the intestinal drug concentration, the faster is the rate of absorption. Therefore, dissolved drug concentrations in the intestine can be kept low, which enhances dissolution. When $X_d/V$ is no greater than 20% of $C_s$, this condition is met and is known as "sink conditions." Maintaining sink conditions in a dissolution test is another matter altogether, but is a very important concern.

## Biorelevant Media

Based on all of the previous considerations, biorelevant media have been proposed. The rationale for proposing the various components was provided in the previous comments. Because of the significant difference between the stomach and the intestine, media representative of the gastric and intestinal environments is commonly used. The major differences between gastric and intestinal media are the pH and presence of bile. Another important consideration is the absence or presence of food in the stomach. When food is absent, conditions between patients do not vary too much. Because the stomach is acidic (<pH 3) in most patients in the fasted state, the main variables are the type and volume

### TABLE 13–2
### COMPOSITION OF DISSOLUTION MEDIA FOR IN VITRO DISSOLUTION TESTING

| Medium | Composition | Amount |
|---|---|---|
| Simulated gastric fluid pH 1.2 (SGFsp), USP 26 | NaCl | 2.0 g |
| | Concentrated HCl | 7.0 mL |
| | Deionized water to | 1.0 L* |
| Simulated intestinal fluid pH 6.8 (SIFsp), USP 26 | $KH_2PO_4$ | 68.05 g |
| | NaOH | 8.96 g |
| | Deionized water to | 10.0 L† |

*Add 3.2 g of pepsin for SGF.
†Add X g of pancreatin for SIF.

of liquid administered with the dosage form. If water is the administered fluid, the buffer capacity is low, and this would not be a factor in dissolution testing. Although it is known that the surface tension of gastric contents is reduced, the exact physiologic agents that are responsible are not known. Therefore, sodium lauryl sulfate is commonly used in dissolution testing to achieve this effect. The composition of simulated fasted-state gastric fluid (pH 1.2) is rather simple and is listed in Table 13–2. In the fed state, the conditions of the stomach are highly dependent on the type and quantity of meal ingested. Simulated intestinal fluid (SIF) is described in the 26th edition of the *United States Pharmacopeia* as a 0.05 M buffer solution containing potassium dihydrogen phosphate (Table 13–2). The pH of this buffer is 6.8 and falls within the range of normal intestinal pH. Pancreatin may also be added if a more biorelevant form of the medium is required. Pancreatin is a mixture of the fat-dissolving enzyme lipase, the protein-degrading enzymes called proteases, and those that break down carbohydrates, like amylase. If SIF does not contain pancreatin, it is indicated using the notation SIFsp, where the "sp" means "sans pancreatin" or "without pancreatin." Some of the parameters that can profoundly influence the dissolution rate of drug products such as the buffer capacity, pH, and surfactant concentrations and how they can be introduced into a biorelevant dissolution test have been discussed. Other considerations are the volume of the contents in the stomach or intestinal segment and the duration of the test as it related to residence time in the stomach or intestinal segment.

## METHODS AND APPARATUS

The objective of most pharmacopeial dissolution monographs is to establish procedures for evaluating batch-to-batch consistency in the dissolution of drug products. Similar dissolution characteristics for different batches of the same drug product imply similar performance of the product in humans. Although there are many customized and original dissolution testing devices reported in the literature, the purpose of this section is to introduce the basic apparatus used in compendial testing of immediate- and modified-release oral dosage forms.

## USP Methods I and II for Dissolution[36–38]

The most commonly used methods for evaluating dissolution first appeared in the 13th edition of the *United States Pharmacopeia* in early 1970. These methods are known as the USP basket (method I) and paddle (method II) methods and are referred to as "closed-system" methods because a fixed volume of dissolution medium is used. In practice, a rotating basket or paddle provides a steady stirring motion in a large vessel with 500 to 1000 mL of fluid that is immersed in a temperature-controlled water bath (**Fig. 13–7a**)[36,39]

Variants of these two standard apparatuses have been reported and are depicted in **Figure 13–7b** (see Shiu[36] for a complete discussion). The devices are very simple, robust, and easily standardized. Descriptions for apparatus specifications are detailed in the current version of the USP. The USP basket and paddle methods are the methods of choice for dissolution testing of immediate-release oral solid dosage forms. The use of alternative dissolution methods should be considered only after USP methods I and II are found to be unsatisfactory. Biorelevant dissolution media were discussed in the previous section. Other commonly used media include (*a*) water,

Basket apparatus
No stirring

Paddle apparatus
No stirring

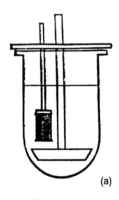

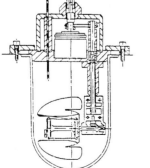

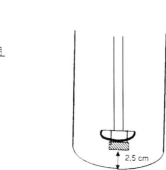

(a)                          (b)

(c)                          (d)

**Fig. 13–7.** (Top) Pictures of USP basket and paddle apparatus. Note the dye coming from the tablet in the basket in the left panel. Types of dissolution apparatus include (*a*) a stationary basket-rotating paddle for immediate-release oral solid dosage forms, (*b*) a modified stationary basket rotating paddle for suppositories, (*c*) a rotating dialysis cell, and (*d*) a rotating paddle–rotating basket. (From G. K. Shiu, Drug Inf. J. **30**, 1045, 1996. With permission.)

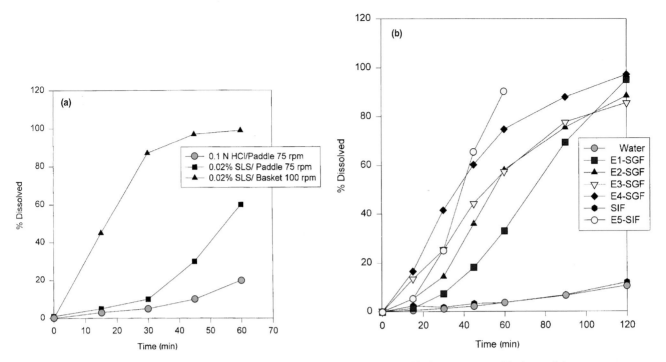

**Fig. 13–8.** (*a*) Dissolution profiles of norethindrone acetate of a norethindrone acetate: ethinyl estradiol combination tablet by various dissolution media and methods. (*b*) Dissolution profiles of theophylline from an aged theophylline soft gelatin product under various dissolution media by USP basket method at 100 rpm rotating speed. E1, E2, E3, and E4 are pepsin with increasing enzyme activity. E5 is a commercially available intestinal enzyme, pancreatin. (From G. K. Shiu, Drug Inf. J. **30**, 1045, 1996. With permission.)

(*b*) 0.1 N HCl, (*c*) buffer solutions, (*d*) water or buffers with surfactants, and (*e*) low-content alcoholic aqueous solutions. The temperature of the medium is usually maintained at body temperature (37°C) for dissolution testing. Although water is one of the most commonly listed dissolution media found in USP monographs, it may not be physiologically relevant due to the lack of buffering capacity. In the following examples, a variety of conditions are used that result in significantly different dissolution profiles, suggesting that the appropriate selection of dissolution conditions must be made.

**EXAMPLE 13–10**

**Dissolution Profiles of Norethindrone Acetate (Fig. 13–8a)**

Before 1990, water was used as the dissolution medium for testing combination of oral contraceptive drug products. Water was used for norethindrone (NE): ethyl estradiol (EE) tablets and 3% isopropanol was used for NE: mestranol (ME) tablets. The dissolution data for norethindrone are shown in Figure 13–8a. An example of an acceptable dissolution profile is seen when the dissolution medium is 0.1 N HCl with 0.02% sodium lauryl sulfate in the USP basket method at 100 rpm. (Example taken from Shiu[36]; original data in Nguyen et al.[40])

**EXAMPLE 13–11**

**Dissolution Profiles of Theophylline from Soft Gelatin Capsules[36] (Fig. 13–8b)**

This example shows the role of biorelevant dissolution media. Inclusion of the appropriate enzymes in the dissolution medium has been

considered appropriate because they are found naturally in the gastrointestinal tract. We often assume that the dosage form assists in improving bioavailability. This example shows that this is not always the case; this example is consistent with the report of a workshop published in 1996,[41] where it was recognized that discrepancies between dissolution and bioavailability occur because the gel that comprises soft and hard gelatin capsules becomes cross-linked. In this example, the dissolution of theophylline from aged soft gelatin capsules was studied in a variety of media: water, increasing amounts of pepsin in simulated gastric fluid (SGF) (E1 through E4), SIF, and SIF with pancreatin (Fig. 13–8b). The highest dissolution rates were observed in the media with enzymes present, and for pepsin, an increase in dissolution was observed with increasing pepsin activity, showing the role of the soft gelatin capsule dosage form in hindering the release of theophylline.

## Special Considerations for Modified-Release Dosage Forms: USP Apparatuses 3 and 4*

Modified-release delivery systems are similar in size and shape to conventional immediate-release dosage forms. For example, shown in **Figure 13–9** are nifedipine (Procardia® XL) "tablets," which are actually nondisintegrating osmotic pumps (Chapter 21). The mechanisms for controlling the release of the drugs are becoming very sophisticated, and special consideration must be given to how drug release is

---

*Modified from Crison.[42]

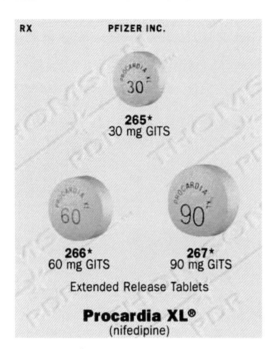

**Fig. 13-9.** Procardia® XL tablets. These tablets are orally delivered osmotic pumps that release drug through a laser-drilled orifice. Although the tablets look like conventional tablets, they behave differently. For example, they do not disintegrate and are excreted in the stool intact. (From *Physicians' Desk Reference*, 58th Ed., Thomson PDR, Montvale, N. J., 2004. With permission.)

evaluated. Regulatory guidances recommend four dissolution apparatuses for modified-release dosage forms. Although the existing apparatuses are adequate for the intended purpose, equipment may require either modifications or completely new designs to accommodate these new release mechanisms. For example, nondisintegrating dosage forms (e.g., Procardia XL) requiring a delivery orifice for drug release may dictate a special design or modification of the dissolution apparatus so that the orifice is not blocked. In contrast, disintegrating or eroding delivery systems pose the challenge of transferring the dosage form to different media without losing any of the pieces. In general, methods of agitation, changing the medium, and holding the dosage form in the medium without obstructing the release mechanism are relevant to drug testing. A challenging component of a dissolution test for a modified-release delivery system is changing the media to obtain a pH gradient or to simulate fed and fasted conditions. The ability to easily change the medium is the focus of commercially available dissolution equipment targeted for modified-release delivery systems, and several equipment designs are available. The USP Apparatus 3, a reciprocating cylinder, dips a transparent cylinder containing the dosage form at a rate determined by the operator.[43,44] The tubes have a mesh base to allow the medium to drain into a sampling reservoir as the tube moves up and down, thus creating convective forces for dissolution. The cylinders can also be transferred to different media at specified times, automatically. A second design is the rotating bottle apparatus, which also allows for changing of medium to simulate a pH gradient or fed and fasted conditions. The USP Apparatus 4 is a flowthrough cell containing the dosage form that is fed with dissolution medium from a reservoir. Directing the fluid through a porous glass plate or a bed of beads produces a dispersed flow of medium. Turbulent or laminar flow can be achieved by changing the bottom barrier. As with Apparatus 3, the medium can be changed to provide a pH gradient, surfactants, and other medium components.

## CHAPTER SUMMARY

Dissolution and drug release are fundamental concepts that affect the practice of pharmacy on a daily basis. Examples include patients who now have to take only one tablet daily instead of one tablet three times daily because they are taking the osmotic pump form of the medication. Not only does drug delivery improve convenience for patients but it also improves compliance as they adhere to treatment regimens that may have been too complex. At this point the student should understand these concepts and understand the differences among immediate-, modified-, delayed-, extended-, and controlled-release delivery systems. You should also be able to differentiate between zero-order and first-order release kinetics as well as understand intrinsic dissolution rate and the driving force for dissolution. Understanding the effect of surface area and sink conditions on dissolution rate is critical and helps explain why drugs are so well absorbed after oral administration. The BCS was discussed and the important role of permeability and solubility was demonstrated. Finally, the student should have an appreciation for the different roles that dissolution testing plays in the pharmaceutical sciences (a quality control versus predictive role) and understand how media properties such as viscosity, pH, lipids, and surfactants can affect dissolution.

**Practice problems for this chapter can be found at thePoint.lww.com/Sinko6e.**

### References

1. J. L. Cohen, B. B. Hubert, L. J. Leeson, C. T. Rhodes, J. R. Robinson, T. J. Roseman, and E. Shefter, Pharm. Res. **7,** 983, 1990.
2. J. M. Aiache, J. Pharm. Biomed. Anal. **8,** 499, 1990.
3. U. S. Food and Drug Administration, SUPAC IR Guidance.
4. (a) G. L. Flynn, V. P. Shah, S. N. Tenjarla, M. Corbo, D. DeMagistris, T. G. Feldman, T. J. Franz, D. R. Miran, D. M. Pearce, J. A. Sequeira, J. Swarbrick, J. C. Wang, A. Yacobi, and J. L. Zatz, Pharm. Res. **16,** 1325, 1999; (b) U. S. Food and Drug Administration, SUPAC-SS.
5. U. S. Department of Health and Human Services, Food and Drug Administration Center for Drug Evaluation and Research (CDER), *Guidance for Industry SUPAC-MR: Modified Release Solid Oral Dosage Forms*, September 1997, CMC 8.
6. G. L. Amidon, H. Lennernas, V. P. Shah, and J. R. Crison, Pharm. Res. **12,** 413, 1995.
7. J. B. Dressman, G. L. Amidon, C. Reppas, and V. P. Shah, Pharm. Res. **15,** 11, 1998.
8. E. Nicolaides, E. Galia, C. Efthymiopoulos, J. B. Dressman, and C. Reppas, Pharm. Res. **16,** 1876, 1999.

9. J. B. Dressman and C. Reppas, Eur. J. Pharm. Sci. **11** (Suppl 2), S73, 2000.

10. E. Nicolaides, M. Symillides, J. B. Dressman, and C. Reppas, Pharm. Res. **18**, 380–388, 2001.

11. E. S. Kostewicz, U. Brauns, R. Becker, and J. B. Dressman, Pharm. Res. **19**, 345, 2002.

12. C-P. Milne, Drug Inf. J. **34**, 681, 2000.

13. P. Costa and J. M. Sousa Lobo, Eur. J. Pharm. Sci. **13**, 123, 2001.

14. D. E. Wurster and P. W. Taylor, J. Pharm. Sci. **54**, 169, 1965.

15. J. G. Wagner, *Biopharmaceutics and Relevant Pharmacokinetics,* Drug Intelligence Publications, Hamilton, IL, 1971, pp. 98–147.

16. L. J. Leeson and J. T. Carstensen, *Dissolution Technology,* Academy of Pharmaceutical Sciences, Washington, D. C., 1974.

17. W. I. Higuchi, J. Pharm. Sci. **56**, 315, 1967.

18. D. Horter and J. B. Dressman, Adv. Drug Del. Rev. **46**, 75, 2001.

19. A. S. Noyes and W. R. Whitney, J. Am. Chem. Soc. **19**, 930, 1897.

20. A. Hixson and J. Crowell, Ind. Eng. Chem. **23**, 923, 1931.

21. V. Levich, *Physicochemical Hydrodynamics,* Prentice Hall, Englewood Cliffs, N. J., 1982.

22. W. Nernst and E. Brunner, Z. Physik. Chem. **47**, 52, 1904.

23. K. G. Nelson and A. C. Shah, J. Pharm. Sci. **64**, 610–1518, 1975.

24. J. H. de Smidt, J. C. A. Offringa, and D. J. A. Crommelin, J. Pharm. Sci. **76**, 711, 1987.

25. J. R. Robinson, *Sustained and Controlled Release Drug Delivery Systems,* Marcel Dekker, New York, 1978.

26. T. Higuchi, J. Soc. Cosm. Chem. **11**, 85, 1960.

27. T. Higuchi, J. Pharm. Sci. **50**, 874, 1961.

28. (a) S. J. Desai, A. P. Simonelli, and W. I. Higuchi, J. Pharm. Sci. **54**, 1459, 1965; (b) S. J. Desai, P. Singh, A. P. Simonelli, and W. I. Higuchi, J. Pharm. Sci. **55**, 1224,1230, 1235, 1966.

29. N. F. H. Ho and T. J. Roseman, J. Pharm. Sci. **68**, 1170, 1979.

30. F. Bottari, G. DiColo, E. Nannipieri, M. F. Sactonne, and M. F. Serafini, J. Pharm. Sci. **63**, 1779, 1974; ibid. **66**, 927, 1977.

31. D. R. Paul and S. K. McSpadden, J. Membr. Sci. **1**, 33, 1976; S. K. Chandrasekaran and D. R. Paul, J. Pharm. Sci. **71**, 1399, 1982.

32. Y. W. Chien, in J. R. Robinson (Ed.), *Sustained and Controlled Release Drug Delivery Systems,* Marcel Dekker, New York, 1978, Chapter 4; Y. W. Chien, Chem. Pharm. Bull. **24**, 147, 1976.

33. C. K. Erickson, K. I. Koch, C. S. Metha, and J. W. McGinity, Science **199**, 1457, 1978; J. W. McGinity, L. A. Hunke, and A. Combs, J. Pharm. Sci. **68**, 662, 1979.

34. Y. W. Chien, *Novel Drug Delivery Systems: Biomedical Applications and Theoretical Basis,* Marcel Dekker, New York, 1982, Chapter 9.

35. J. B. Dressman, R. B. Berardi, L. C. Dermentzoglou, T. L. Russell, S. P. Schmaltz, J. L. Barnett, and K. M. Jarvenpaa, Pharm. Res. **7**, 756, 1990.

36. G. K. Shiu, Drug Inf. J. **30**, 1045, 1996.

37. J. T. Carstensen, T. Lai, and V. K. Prasad, J. Pharm. Sci. **66**, 607, 1977.

38. J. T. Carstensen, T. Y.-F. Lai, and V. K. Prasad, J. Pharm. Sci. **67**, 1303, 1978.

39. J. Mauger, J. Ballard, R. Brockson, S. De, V. Gray, and D. Robinson, Dissolution Technol. **10**, 6, 2003.

40. H. T. Nguyen, G. K. Shiu, W. H. Worsley, and J. P. Skelly, J. Pharm. Sci. **79**, 163, 1990.

41. I. J. McGilveray, Drug Inf. J. **30**, 1029, 1996.

42. John R. Crison, Developing Dissolution Tests for Modified Release Dosage Forms: General Considerations. Available at: http://www.dissolutiontech.com/DTresour/299articles/299Crison.htm.

43. *United States Pharmacopeia 23/National Formulary 18,* USPC, Inc., Rockville, Md., 1995.

44. William A. Hanson, *Handbook of Dissolution Testing,* 2nd Ed. Rev., Aster, Eugene, Ore., 1991, Chapter 3.

## Recommended Readings

G. L. Amidon, H. Lennernas, V. P. Shah, and J. R. Crison, Pharm. Res. **12**, 413, 1995.

J. B. Dressman, G. L. Amidon, C. Reppas, and V. P. Shah, Pharm. Res. **15**, 11, 1998.

CHAPTER LEGACY

**Fifth Edition:** published as Chapter 13 (Drug Release and Dissolution). Updated by Patrick Sinko.

**Sixth Edition:** published as Chapter 13 (Drug Release and Dissolution). Updated by Patrick Sinko.

# CHEMICAL KINETICS AND STABILITY

**At the conclusion of this chapter the student should be able to:**

**1** Define reaction rate, reaction order, and molecularity.

**2** Understand and apply apparent zero-order kinetics to the practice of pharmacy.

**3** Calculate half-life and shelf life of pharmaceutical products and drugs.

**4** Understand Michaelis–Menten (nonlinear) kinetic behavior and linearization techniques.

**5** Interpret pH–rate profiles and kinetic data.

**6** Understand the basis for transition-state theory and its application to chemical kinetics.

**7** Describe the influence of temperature, ionic strength, solvent, pH, and dielectric constant on reaction rates.

**8** Calculate the increase in rate constant as a function of temperature ($Q_{10}$).

**9** Describe the factors that influence solid-state chemical kinetics.

**10** Identify and describe methods for the stabilization of pharmaceutical agents.

**11** Understand stability-testing protocols and regulatory requirements.

---

The purpose of stability testing is to provide evidence on how the quality of a drug substance or drug product varies with time under the influence of a variety of environmental factors, such as temperature, humidity, and light, and to establish a retest period for the drug substance or a shelf life for the drug product and recommended storage conditions. Although the pharmaceutical scientist plays a critical role in determining the stability of pharmaceuticals, practicing pharmacists should be able to interpret this information for their patients. This chapter introduces the rates and mechanisms of reactions with particular emphasis on decomposition and stabilization of drug products. It is essential for pharmacists and pharmaceutical scientists to study, understand, and interpret conditions of instability of pharmaceutical products as well as to be able to offer solutions for the stabilization of these products. Pharmaceutical manufacturers routinely utilize the principles covered in this chapter; however, with the resurgence of pharmaceutical compounding, it is essential for practicing pharmacists to understand drug product stability as well. If a community pharmacist is asked to compound a prescription product, there are many factors that he or she must consider. The pharmacist must recognize that alterations in stability may occur when a drug is combined with other ingredients. For example, if thiamine hydrochloride, which is most stable at a pH of 2 to 3 and is unstable above pH 6, is combined with a buffered vehicle of, say, pH 8 or 9, the vitamin is rapidly inactivated.[1] Knowing the rate at which a drug deteriorates at various hydrogen ion concentrations allows one to choose a vehicle that will retard or prevent the degradation. Patients expect that products will have a reasonable shelf life. Even though pharmaceutical manufacturers label prescription and over-the-counter drug products with expiration dating to guide the patient/consumer in these matters, patients may store these products in a bathroom medicine cabinet where the humidity and temperature are higher than the typical storage place for medications. How does this affect the shelf life of the product? A community pharmacy practitioner should be able to understand this and advise patients on these matters.

The experimental investigation of the possible breakdown of new drugs is not a simple matter. Applications of chemical kinetics in pharmacy result in the production of more stable drug preparations, the dosage and rationale of which may be established on sound scientific principles. Thus, as a result of current research involving the kinetics of drug systems, the pharmacist is able to assist the physician and patient regarding the proper storage and use of medicinal agents. This chapter brings out a number of factors that bear on the formulation, stabilization, and administration of drugs. Concentration, temperature, light, pH, and catalysts are important in relation to the speed and the mechanism of reactions and will be discussed in turn.

## FUNDAMENTALS AND CONCENTRATION EFFECTS

### Rates, Order, and Molecularity of Reactions

The rate, velocity, or speed of a reaction is given by the expression $dc/dt$, where $dc$ is the increase or decrease of concentration over an infinitesimal time interval $dt$. According to the law of mass action, the rate of a chemical reaction is proportional to the product of the molar concentration of the reactants each raised to a power usually equal to the number of molecules, $a$ and $b$, of the substances $A$ and $B$, respectively, undergoing reaction. In the reaction

$$aA + bB + \cdots = \text{Products} \qquad (14\text{–}1)$$

the rate of the reaction is

$$\text{Rate} = \frac{1}{a}\frac{d[A]}{dt}$$

$$= \frac{1}{b}\frac{d[B]}{dt} = \cdots k[A]^a[B]^b \cdots \qquad (14\text{–}2)$$

where $k$ is the *rate constant*.

The overall *order* of a reaction is the sum of the exponents [$a + b$, e.g., in equation (**14–2**)] of the concentration terms, $A$ and $B$. The order with respect to one of the reactants, $A$ or $B$, is the exponent $a$ or $b$ of that particular concentration term. In the reaction of ethyl acetate with sodium hydroxide in aqueous solution, for example,

$$CH_3COOC_2H_5 + NaOH_{soln} \rightarrow CH_3COONa + C_2H_5OH$$

the rate expression is

$$Rate = \frac{d[CH_3COOC_2H_5]}{dt}$$
$$= -\frac{d[NaOH]}{dt} = k[CH_3COOC_2H_5]^1 \, [NaOH]^1$$
$$(14-3)$$

The reaction is first order ($a = 1$) with respect to ethyl acetate and first order ($b = 1$) with respect to sodium hydroxide solution; overall the reaction is second order ($a + b = 2$).

Suppose that in this reaction, sodium hydroxide as well as water was in great excess and ethyl acetate was in a relatively low concentration. As the reaction proceeded, ethyl acetate would change appreciably from its original concentration, whereas the concentrations of NaOH and water would remain essentially unchanged because they are present in great excess. In this case, the contribution of sodium hydroxide to the rate expression is considered constant and the reaction rate can be written as

$$\frac{d[CH_3COOC_2H_5]}{dt} = k'[CH_3COOC_2H_5] \quad (14-4)$$

where $k' = k[NaOH]$. The reaction is then said to be a *pseudo–first-order* reaction because it depends only on the first power ($a = 1$) of the concentration of ethyl acetate. In general, when one of the reactants is present in such great excess that its concentration may be considered constant or nearly so, the reaction is said to be of *pseudo-order*.

### EXAMPLE 14–1

**Reaction Order**

**In the reaction of acetic anhydride with ethyl alcohol to form ethyl acetate and water,**

$$(CH_3CO)_2O + 2C_2H_5OH \rightarrow 2CH_3CO_2C_2H_5 + H_2O$$

**the rate of reaction is**

$$Rate = -\frac{d[(CH_3CO)_2O]}{dt}$$
$$= k[(CH_3CO)_2O][C_2H_5OH]^2 \quad (14-5)$$

What is the order of the reaction with respect to acetic anhydride? With respect to ethyl alcohol? What is the overall order of the reaction?

If the alcohol, which serves here as the solvent for acetic anhydride, is in large excess such that a small amount of ethyl alcohol is used up in the reaction, write the rate equation for the process and state the order.

*Answer:* The reaction appears to be first order with respect to acetic anhydride, second order with respect to ethyl alcohol, and

overall third order. However, because alcohol is the solvent, its concentration remains essentially constant, and the rate expression can be written as

$$-\frac{d[(CH_3CO)_2O]}{dt} = k'[(CH_3CO)_2O] \quad (14-6)$$

Kinetically the reaction is therefore a pseudo–first-order reaction, as noted by Glasstone.[2]

## Molecularity

A reaction whose overall order is measured can be considered to occur through several steps or elementary reactions. Each of the elementary reactions has a stoichiometry giving the number of molecules taking part in that step. Because the order of an elementary reaction gives the number of molecules coming together to react in the step, it is common to refer to this order as the *molecularity* of the elementary reaction. If, on the other hand, a reaction proceeds through several stages, the term molecularity is not used in reference to the observed rate law: One step may involve two molecules, a second step only one molecule, and a subsequent step one or two molecules. Hence, order and molecularity are ordinarily identical only for elementary reactions. Bimolecular reactions may or may not be second order.

In simple terms, molecularity is the number of molecules, atoms, or ions reacting in an elementary process. In the reaction

$$Br_2 \rightarrow 2Br$$

the process is *unimolecular* because the single molecule, $Br_2$, decomposes to form two bromine atoms. In the single-step reaction

$$H_2 + I_2 \rightarrow 2HI$$

the process is *bimolecular* because two molecules, one of $H_2$ and one of $I_2$, must come together to form the product HI. *Termolecular* reactions, that is, processes in which three molecules must come together simultaneously, are rare.

Chemical reactions that proceed through more than one step are known as *complex reactions*. The overall order determined kinetically may not be identical with the molecularity because the reaction consists of several steps, each with its own molecularity. For the overall reaction

$$2NO + O_2 \rightarrow 2NO_2$$

the order has been found experimentally to be 2. The reaction is not termolecular, in which two molecules of NO would collide simultaneously with one molecule of $O_2$. Instead, the mechanism is postulated to consist of two elementary steps, each being bimolecular:

$$2NO \rightarrow N_2O_2$$

$$N_2O_2 + O_2 \rightarrow 2NO_2$$

## Rate Constants, Half-Life, Shelf Life, and Apparent or Pseudo-order

### Specific Rate Constant

The constant, $k$, appearing in the rate law associated with a single-step (elementary) reaction is called the *specific rate constant* for that reaction. Any change in the conditions of the reaction, for example, in temperature or solvent, or a slight change in one of the reacting species, will lead to a rate law having a different value for the specific rate constant. Experimentally, a change of specific rate constant corresponds simply to a change in the slope of the line given by the rate equation. Variations in the specific rate constant are of great physical significance because a change in this constant necessarily represents a change at the molecular level as a result of a variation in the reaction conditions.

Rate constants derived from reactions consisting of a number of steps of different molecularity are functions of the specific rate constants for the various steps. Any change in the nature of a step due to a modification in the reaction conditions or in the properties of the molecules taking part in this step could lead to a change in the value of the overall rate constant. At times, variations in an overall rate constant can be used to provide useful information about a reaction, but quite commonly, anything that affects one specific rate constant will affect another; hence, it is quite difficult to attach significance to variations in the overall rate constant for these reactions.

### Units of the Basic Rate Constants

To arrive at units for the rate constants appearing in zero-, first-, and second-order rate laws, the equation expressing the law is rearranged to have the constant expressed in terms of the variables of the equation. Thus, for a zero-order reaction,

$$k = -\frac{dA}{dt} = \frac{\text{moles/liter}}{\text{second}}$$
$$= \frac{\text{moles}}{\text{liter second}} = \text{moles liter}^{-1} \text{second}^{-1}$$

for a first-order reaction,

$$k = -\frac{dA}{dt}\frac{1}{A} = \frac{\text{moles/liter}}{\text{second-moles/liter}}$$
$$= \frac{1}{\text{second}} = \text{second}^{-1}$$

**HALF-LIFE AND SHELF LIFE**

The *half-life* is the time required for one-half of the material to disappear; it is the time at which $A$ has decreased to $\frac{1}{2} A$. The shelf life is the time required for 10% of the material to disappear; it is the time at which $A$ has decreased to 90% of its original concentration (i.e., $0.9 A$).

and for a second-order reaction,

$$k = \frac{dA}{dt}\frac{1}{A^2} = \frac{\text{moles/liter}}{\text{second (moles/liter)}^2}$$
$$= \frac{\text{liter}}{\text{moles-second}} = \text{liter second}^{-1} \text{mole}^{-1}$$

where $A$ is the molar concentration of the reactant. It is an easy matter to replace the units moles/liter by any other units (e.g., pressure in atmospheres) to obtain the proper units for the rate constants if quantities other than concentration are being measured.

### Zero-Order Reactions

Garrett and Carper[3] found that the loss in color of a multisulfa product (as measured by the decrease of spectrophotometric absorbance at a wavelength of 500 nm) followed a zero-order rate. The rate expression for the change of absorbance, $A$, with time is therefore

$$-\frac{dA}{dt} = k_0 \qquad (14-7)$$

where the minus sign signifies that the absorbance is decreasing (i.e., the color is fading). The velocity of fading is seen to be constant and independent of the concentration of the colorant used. The rate equation can be integrated between the initial absorbance, $A_0$, corresponding to the original color of the preparation at $t = 0$, and $A_t$, the absorbance after $t$ hours:

$$\int_{A_0}^{A_t} dA = -k_0 \int_0^t dt$$
$$A_t - A_0 = -k_0 t$$

or

$$A_t = A_0 - k_0 t \qquad (14-8)$$

The initial concentration corresponding to $A_0$ is ordinarily written as $a$ and the concentration remaining at time $t$ as $c$.

**APPARENT OR PSEUDO-ORDER**

"Apparent" or "pseudo"-order describes a situation where one of the reactants is present in large excess or does not effect the overall reaction and can be held constant. For example, many hydrolysis decomposition reactions of drug molecules are second order. Usually the amount of water present is in excess of what is needed for the reaction to proceed. In other words, the concentration of water is essentially constant throughout the reaction. In this case, the second-order reaction behaves like a first-order reaction and is called an apparent or pseudo–first-order reaction.

When this linear equation is plotted with $c$ on the vertical axis against $t$ on the horizontal axis, the slope of the line is equal to $-k_0$. Garrett and Carper[3] obtained a value for $k$ of 0.00082 absorbance decrease per hour at 60°C, indicating that the color was fading at this constant rate independent of concentration.

Because the *half-life* is the time required for one-half of the material to disappear, in the present case $A_0 = 0.470$ and $\frac{1}{2}A_0 = 0.235$:

$$t_{1/2} = \frac{\frac{1}{2}A_0}{k_0} = \frac{0.235}{8.2 \times 10^{-4}} = 2.9 \times 10^2 \text{ hr}$$

### Suspensions. Apparent Zero-Order Kinetics[4]

Suspensions are another case of zero-order kinetics, in which the concentration in solution depends on the drug's solubility. As the drug decomposes in solution, more drug is released from the suspended particles so that the concentration remains constant. This concentration is, of course, the drug's equilibrium solubility in a particular solvent at a particular temperature. The important point is that the amount of drug in solution remains constant despite its decomposition with time. The reservoir of solid drug in suspension is responsible for this constancy.

The equation for an ordinary solution, with no reservoir of drug to replace that depleted, is the first-order expression, equation (14–11):

$$\frac{-d[A]}{dt} = k[A]$$

where $[A]$ is the concentration of drug remaining undecomposed at time $t$, and $k$ is known as a first-order rate constant. When the concentration $[A]$ is rendered constant, as in the case of a suspension, we can write

$$k[A] = k_0 \qquad (14\text{–}9)$$

so that the first-order rate law (14–11) becomes

$$-\frac{d[A]}{dt} = k_0 \qquad (14\text{–}10)$$

Equation (14–10) obviously is a zero-order equation. It is referred to as an *apparent zero-order equation*, being zero order only because of the suspended drug reservoir, which ensures constant concentration. Once all the suspended particles have been converted into drug in solution, the system changes to a first-order reaction.

### EXAMPLE 14–2

#### Shelf Life of an Aspirin Suspension

A prescription for a liquid aspirin preparation is called for. It is to contain 325 mg/5 mL or 6.5 g/100 mL. The solubility of aspirin at 25°C is 0.33 g/100 mL; therefore, the preparation will definitely be a suspension. The other ingredients in the prescription cause the product to have a pH of 6.0. The first-order rate constant for aspirin degradation in this solution is $4.5 \times 10^{-6}$ sec$^{-1}$. Calculate the zero-order rate constant. Determine the shelf life, $t_{90}$, for the liquid prescription, assuming that the product is satisfactory until the time at which it has decomposed to 90% of its original concentration (i.e., 10% decomposition) at 25°C.

*Answer*: $k_0 = k \times$ [Aspirin in solution], from equation (14–9).

Thus,

$$k_0 = (4.5 \times 10^{-6} \text{ sec}^{-1}) \times (0.33 \text{ g/100 mL})$$

$$k_0 = 1.5 \times 10^{-6} \text{ g/100 mL sec}^{-1}$$

$$t_{90} = \frac{0.10[A]_0}{k_0} = \frac{(0.10)(6.5 \text{ g/100 mL})}{(1.5 \times 10^{-6} \text{ g/100 mL sec}^{-1})}$$

$$= 4.3 \times 10^5 \text{ sec} = 5.0 \text{ days}$$

### First-Order Reactions

In 1918, Harned[5] showed that the decomposition rate of hydrogen peroxide catalyzed by 0.02 M KI was proportional to the concentration of hydrogen peroxide remaining in the reaction mixture at any time. The data for the reaction

$$2H_2O_2 \rightarrow 2H_2O + O_2$$

are given in **Table 14–1**. Although two molecules of hydrogen peroxide appear in the stoichiometric equation as just written, the reaction was found to be first order. The rate equation is written as

$$-\frac{dc}{dt} = kc \qquad (14\text{–}11)$$

where $c$ is the concentration of hydrogen peroxide remaining undecomposed at time $t$ and $k$ is the first-order velocity constant. Integrating equation (14–11) between concentration $c_0$ at time $t = 0$ and concentration $c$ at some later time, $t$, we have

$$\int_{c_0}^{c} \frac{dc}{c} = -k \int_{0}^{t} dt$$

$$\ln c - \ln c_0 = -k(t - 0)$$

$$\ln c = \ln c_0 - kt \qquad (14\text{–}12)$$

## KEY CONCEPT  SHELF LIFE AND EXPIRATION DATING

*Shelf life* (also referred to as the expiration dating period) is the time period during which a drug product is expected to remain within the approved specification for use, provided that it is stored under the conditions defined on the container label.

*Expiration date* is the date placed on the container label of a drug product designating the time prior to which a batch of the product is expected to remain within the approved shelf-life specification if stored under defined conditions and after which it must not be used.

TABLE 14–1
**DECOMPOSITION OF HYDROGEN PEROXIDE AT 25°C IN AQUEOUS SOLUTION CONTAINING 0.02 M KI\***

| $t$ (min) | $a - x$ | $K$ (min$^{-1}$) |
|-----------|---------|------------------|
| 0         | 57.90   | –                |
| 5         | 50.40   | 0.0278           |
| 10        | 43.90   | 0.0277           |
| 25        | 29.10   | 0.0275           |
| 45        | 16.70   | 0.0276           |
| 65        | 9.60    | 0.0276           |
| $\infty$  | 0       | –                |

*Based on H. S. Harned, J. Am. Chem. Soc. **40**, 1462, 1918.

Converting to common logarithms yields

$$\log c = \log c_0 - kt/2.303 \qquad (14\text{–}13)$$

or

$$k = \frac{2.303}{t} \log \frac{c_0}{c} \qquad (14\text{–}14)$$

In exponential form, equation (14–12) becomes

$$c = c_0 e^{-kt} \qquad (14\text{–}15)$$

and equation (14–13) becomes

$$c = c_0 10^{-kt/2.303} \qquad (14\text{–}16)$$

Equations (14–15) and (14–16) express the fact that, in a first-order reaction, the concentration decreases exponentially with time. As shown in **Figure 14–1**, the concentration begins at $c_0$ and decreases as the reaction becomes progressively slower. The concentration asymptotically approaches a final value $c_\infty$ as time proceeds toward infinity.

Equation (14–14) is often written as

$$k = \frac{2.303}{t} \log \frac{a}{a - x} \qquad (14\text{–}17)$$

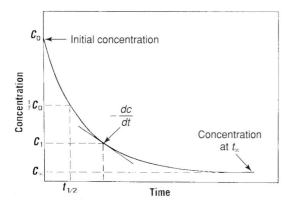

**Fig. 14–1.** Fall in concentration of a decomposing drug with time. In addition to $C_0$ and $C_\infty$, $\frac{1}{2}C_0$ and the corresponding time, $t_{1/2}$, are shown. The rate of decrease of concentration with time, $-dC/dt$, at an arbitrary concentration, $C_1$, is also shown.

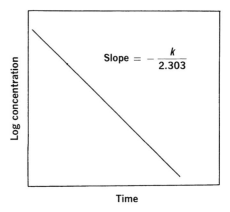

**Fig. 14–2.** A linear plot of log $C$ versus time for a first-order reaction.

where the symbol $a$ is customarily used to replace $c_0$, $x$ is the decrease of concentration in time $t$, and $a - x = c$.

The specific reaction rates listed in **Table 14–1** were calculated by using equation (14–17). Probably the best way to obtain an average $k$ for the reaction is to plot the logarithm of the concentration against the time, as shown in **Figure 14–2**. The linear expression in equation (14–13) shows that the slope of the line is $-k/2.303$, from which the rate constant is obtained. If a straight line is obtained, it indicates that the reaction is first order. Once the rate constant is known, the concentration of reactant remaining at a definite time can be computed as demonstrated in the following examples.

**EXAMPLE 14–3**

**Decomposition of Hydrogen Peroxide**

The catalytic decomposition of hydrogen peroxide can be followed by measuring the volume of oxygen liberated in a gas burette. From such an experiment, it was found that the concentration of hydrogen peroxide remaining after 65 min, expressed as the volume in milliliters of gas evolved, was 9.60 from an initial concentration of 57.90.

    (a)  Calculate $k$ using equation (14–14).
    (b)  How much hydrogen peroxide remained undecomposed after 25 min?

($a$)

$$k = \frac{2.303}{65} \log \frac{57.90}{9.60} = 0.0277 \, \text{min}^{-1}$$

($b$)

$$0.0277 = \frac{2.303}{25} \log \frac{57.90}{c}; c = 29.01$$

**EXAMPLE 14–4**

**First-Order Half-Life**

A solution of a drug contained 500 units/mL when prepared. It was analyzed after 40 days and was found to contain 300 units/mL. Assuming the decomposition is first order, at what time will the drug have decomposed to one-half of its original concentration?
    We have

$$k = \frac{2.303}{40} \log \frac{500}{300} = 0.0128 \, \text{day}^{-1}$$

$$t = \frac{2.303}{0.0128} \log \frac{500}{250} = 54.3 \, \text{days}$$

## Half-Life

The period of time required for a drug to decompose to one-half of the original concentration as calculated in *Example 14–3* is the half-life, $t_{1/2}$, for a first-order reaction:

$$t_{1/2} = \frac{2.303}{k} \log \frac{500}{250} = \frac{2.303}{k} \log 2$$

$$t_{1/2} = \frac{0.693}{k} \qquad (14\text{–}18)$$

In *Example 14–4*, the drug has decomposed by 250 units/mL in the first 54.3 days. Because the half-life is a constant, independent of the concentration, it remains at 54.3 days regardless of the amount of drug yet to be decomposed. In the second half-life of 54.3 days, half of the remaining 250 units/mL, or an additional 125 units/mL, are lost; in the third half-life, 62.5 units/mL are decomposed, and so on.

The student should now appreciate the reason for stating the half-life rather than the time required for a substance to decompose completely. Except in a zero-order reaction, theoretically it takes an infinite period of time for a process to subside completely, as illustrated graphically in **Figure 14–1**. Hence, a statement of the time required for complete disintegration would have no meaning. Actually, the rate ordinarily subsides in a finite period of time to a point at which the reaction may be considered to be complete, but this time is not accurately known, and the half-life, or some other fractional-life period, is quite satisfactory for expressing reaction rates.

The same drug may exhibit different orders of decomposition under various conditions. Although the deterioration of hydrogen peroxide catalyzed with iodine ions is first order, it has been found that decomposition of concentrated solutions stabilized with various agents may become zero order. In this case, in which the reaction is independent of drug concentration, decomposition is probably brought about by contact with the walls of the container or some other environmental factor.

## Second-Order Reactions

The rates of bimolecular reactions, which occur when two molecules come together,

$$A + B \rightarrow \text{Products}$$

are frequently described by the second-order equation. When the speed of the reaction depends on the concentrations of $A$ and $B$ with each term raised to the first power, the rate of decomposition of $A$ is equal to the rate of decomposition of $B$, and both are proportional to the product of the concentrations of the reactants:

$$-\frac{d[A]}{dt} = \frac{d[B]}{dt} = k[A][B] \qquad (14\text{–}19)$$

If $a$ and $b$ are the initial concentrations of $A$ and $B$, respectively, and $x$ is the concentration of each species reacting in time $t$, the rate law can be written as

$$\frac{dx}{dt} = k(a-x)(b-x) \qquad (14\text{–}20)$$

where $dx/dt$ is the rate of reaction and $a - x$ and $b - x$ are the concentrations of $A$ and $B$, respectively, remaining at time $t$. When, in the simplest case, both $A$ and $B$ are present in the same concentration so that $a = b$,

$$\frac{dx}{dt} = k(a-x)^2 \qquad (14\text{–}21)$$

Equation **(14–21)** is integrated, using the conditions that $x = 0$ at $t = 0$ and $x = x$ at $t = t$.

$$\int_0^x \frac{dx}{(a-x)^2} = k \int_0^t dt$$

$$\left(\frac{1}{a-x}\right) - \left(\frac{1}{a-0}\right) = kt$$

$$\frac{x}{a(a-x)} = kt \qquad (14\text{–}22)$$

or

$$k = \frac{1}{at}\left(\frac{x}{a-x}\right) \qquad (14\text{–}23)$$

When, in the general case, $A$ and $B$ are not present in equal concentrations, integration of equation **(14–20)** yields

$$\frac{2.303}{a-b} \log \frac{b(a-x)}{a(b-x)} = kt \qquad (14\text{–}24)$$

or

$$k = \frac{2.303}{t(a-b)} \log \frac{b(a-x)}{a(b-x)} \qquad (14\text{–}25)$$

It can be seen by reference to equation **(14–22)** that when $x/a(a-x)$ is plotted against $t$, a straight line results if the reaction is second order. The slope of the line is $k$. When the initial concentrations $a$ and $b$ are not equal, a plot of $\log[b(a-x)/a(b-x)]$ against $t$ should yield a straight line with a slope of $(a-b)k/2.303$. The value of $k$ can thus be obtained. It is readily seen from equation **(14–23)** or **(14–25)** that the units in which $k$ must be expressed for a second-order reaction are $1/(\text{mole/liter}) \times 1/\text{sec}$ where the concentrations are given in mole/liter and the time in seconds. The rate constant, $k$, in a second-order reaction therefore has the dimensions liter/(mole sec) or liter mole$^{-1}$ sec$^{-1}$.

### EXAMPLE 14–5

**Saponification of Ethyl Acetate**

Walker[6] investigated the saponification of ethyl acetate at 25°C:

$$CH_3COOC_2H_5 + NaOH \rightarrow CH_3COONa + C_2H_5OH$$

The initial concentrations of both ethyl acetate and sodium hydroxide in the mixture were 0.01000 M. The change in concentration, $x$, of alkali during 20 min was 0.000566 mole/liter; therefore, $(a - x) = 0.01000 - 0.00566 = 0.00434$.

Compute (*a*) the rate constant and (*b*) the half-life of the reaction.

(*a*) Using equation (14–23), we obtain

$$k = \frac{1}{0.01 \times 20} \frac{0.00566}{0.00434} = 6.52 \text{ liter mole}^{-1} \text{ min}^{-1}$$

(*b*) The half-life of a second-order reaction is

$$t_{1/2} = \frac{1}{ak} \qquad (14\text{–}26)$$

It can be computed for the reaction only when the initial concentrations of the reactants are identical. In the present example,

$$t_{1/2} = \frac{1}{0.01 \times 6.52} = 15.3\,\text{min}$$

## Determination of Order

The order of a reaction can be determined by several methods.

Substitution Method. The data accumulated in a kinetic study can be substituted in the integrated form of the equations that describe the various orders. When the equation is found in which the calculated $k$ values remain constant within the limits of experimental variation, the reaction is considered to be of that order.

Graphic Method. A plot of the data in the form of a graph as shown in **Figure 14–2** can also be used to ascertain the order. If a straight line results when concentration is plotted against $t$, the reaction is zero order. The reaction is first order if $\log(a - x)$ versus $t$ yields a straight line, and it is second order if $1/(a - x)$ versus $t$ gives a straight line (in the case in which the initial concentrations are equal). When a plot of $1/(a - x)^2$ against $t$ produces a straight line with all reactants at the same initial concentration, the reaction is third order.

Half-Life Method. In a zero-order reaction, the half-life is proportional to the initial concentration, $a$, as observed in **Table 14–2**. The half-life of a first-order reaction is independent of $a$; $t_{1/2}$ for a second-order reaction, in which $a = b$, is proportional to $1/a$; and in a third-order reaction, in which $a = b = c$, it is proportional to $1/a^2$. The relationship between these results shows that, in general, the half-life of a reaction in which the concentrations of all reactants are identical is

$$t_{1/2} \propto \frac{1}{a^{n-1}} \qquad (14\text{–}27)$$

where $n$ is the order of the reaction. Thus, if two reactions are run at different initial concentrations, $a_1$ and $a_2$, the respective half-lives $t_{1/2(2)}$ and $t_{1/2(2)}$ are related as follows:

$$\frac{t_{1/2(1)}}{t_{1/2(2)}} = \frac{(a_2)^{n-1}}{(a_1)^{n-1}} = \left(\frac{a_2}{a_1}\right)^{n-1} \qquad (14\text{–}28)$$

### TABLE 14–2
### RATE AND HALF-LIFE EQUATIONS

| Order | Integrated Rate Equation | Half-Life Equation |
|---|---|---|
| 0 | $x = kt$ | $t_{1/2} = \dfrac{a}{2k}$ |
| 1 | $\log\dfrac{a}{a - x} = \dfrac{k}{2.303}t$ | $t_{1/2} = \dfrac{0.693}{k}$ |
| 2 | $\dfrac{x}{a(a - x)} = kt$ | $t_{1/2} = \dfrac{1}{ak}$ |
| 3 | $\dfrac{2ax - x^2}{a^2(a - x)^2} = 2kt$ | $t_{1/2} = \dfrac{3}{2}\dfrac{1}{a^2 k}$ |

or, in logarithmic form,

$$\log\frac{t_{1/2(1)}}{t_{1/2(2)}} = (n - 1)\log\frac{a_2}{a_1} \qquad (14\text{–}29)$$

and finally

$$n = \log\frac{\left(t_{1/2(1)}/t_{1/2(2)}\right)}{\log(a_2/a_1)} + 1 \qquad (14\text{–}30)$$

The half-lives are obtained graphically by plotting $a$ versus $t$ at two different initial concentrations and reading the time at $1/2a_1$ and $1/2a_2$. The values for the half-lives and the initial concentrations are then substituted into equation **(14–30)**, from which the order $n$ is obtained directly. Rather than using different initial concentrations, one can take two concentrations during a single run as $a_1$ and $a_2$ and determine the half-lives $t_{1/2(1)}$ and $t_{1/2(2)}$ in terms of these. If the reaction is first order, $t_{1/2(1)} = t_{1/2(2)}$ because the half-life is independent of concentration in a first-order reaction. Then $\log(t_{1/2(1)}/t_{1/2(2)}) = \log 1 = 0$, and one can see from equation **(14–30)** that

$$n = 0 + 1 = 1$$

## Complex Reactions

Many reactions cannot be expressed by simple zero-, first-, and second-, or third-order equations. They involve more than one-step or elementary reactions and accordingly are known as *complex reactions*. These processes include reversible, parallel, and consecutive reactions.

(*a*) Reversible reaction:

$$A + B \underset{k_{-1}}{\overset{k_1}{\rightleftharpoons}} C + D$$

(*b*) Parallel or side reactions:

$$A \begin{array}{c} \overset{k_1}{\longrightarrow} B \\ \underset{k_2}{\longrightarrow} C \end{array}$$

(*c*) Series or consecutive reactions:

$$A \xrightarrow{k_1} B \xrightarrow{k_2} C$$

Reversible Reactions. The simplest reversible reaction is one in which both the forward and the reverse steps are first-order processes:

$$A \underset{k_r}{\overset{k_f}{\rightleftharpoons}} B$$

Although at first this equation appears to be that for an equilibrium between $A$ and $B$, it must be pointed out that an equilibrium situation requires that the concentrations of $A$ and $B$ do not change with time. Because this expression is intended to explain a kinetic process, it must follow that the equation describes the approach to equilibrium. That is, the situation represented is one in which $A$ decreases to form $B$ and some of the product $B$ reverts back to $A$. According to this description, the *net* rate at which $A$ decreases will be given by the rate at which $A$ decreases in the forward step less the rate at which $A$ increases in the reverse step:

$$-\frac{dA}{dt} = k_f A - k_r B \qquad (14\text{–}31)$$

This rate law can be integrated by noting that

$$A_0 - A = B \qquad (14\text{–}32)$$

Substitution of equation (14–32) into equation (14–31) affords, upon integration,

$$\ln \frac{k_f A_0}{(k_f + k_r)A - k_r A_0} = (k_f + k_r)t \qquad (14\text{–}33)$$

Equation (14–33) can be simplified by introducing the equilibrium condition

$$k_f A_{eq} = k_r B_{eq} \qquad (14\text{–}34)$$

where

$$A_0 - A_{eq} = B_{eq} \qquad (14\text{–}35)$$

Equations (14–34) and (14–35) can be used to solve for the equilibrium concentration in terms of the starting concentration:

$$A_{eq} = \frac{k_r}{k_f + k_r} A_0 \qquad (14\text{–}36)$$

Use of equation (14–36) in equation (14–33) enables us to give a simple form of the rate law:

$$\ln \frac{A_0 - A_{eq}}{A - A_{eq}} = (k_f + k_r)t \qquad (14\text{–}37)$$

or

$$\log \frac{A_0 - A_{eq}}{A - A_{eq}} = \frac{k_f + k_r}{2.303} t \qquad (14\text{–}38)$$

Equation (14–38) has the advantage that the approach of $A$ to equilibrium can be followed over a much wider range of concentrations than if an attempt is made to obtain the first-order rate constant, $k_f$, in the early stages of the reaction when $B \approx 0$. The equation corresponds to a straight line intersecting at zero and having a slope given by $\frac{k_f + k_r}{2.303}$. Because the equilibrium constant of the reaction is given by

$$K = \frac{k_f}{k_r} - \frac{B_{eq}}{A_{eq}} \qquad (14\text{–}39)$$

both the forward and reverse rate constants can be evaluated once the slope of the line and the equilibrium constant have been determined.

The tetracyclines and certain of their derivatives undergo a reversible isomerization at a pH in the range of 2 to 6. This isomerization has been shown to be an epimerization, resulting in *epi*tetracyclines, which show much less therapeutic activity than the natural form. Considering only that part of the tetracycline molecule undergoing change, we can represent the transformation by the scheme

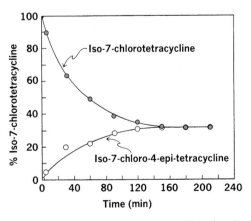

The natural configuration of tetracycline has the $N(CH_3)_2$ group above the plane and the H group below the plane of the page. Under acidic conditions, the natural compound A is converted reversibly to the *epi* isomer B.

McCormick et al.[7] followed the epimerization of iso-7-chlorotetracycline and its *epi* isomer and noted that each isomer led to the same equilibrium distribution of isomers (**Fig. 14–3**). In the solvent dimethylformamide containing 1 M aqueous $NaH_2PO_4$ at $25°C$, the equilibrium distribution consisted of 32% iso-7-chlorotetracycline and 68% iso-7-chloro-4-*epi*-tetracycline, which gives an equilibrium constant

$$K = \frac{B_{eq}}{A_{eq}} = \frac{68}{32} = 2.1$$

The data used to arrive at **Figure 14–3**, when plotted according to equation (14–38), give the line shown in **Figure 14–4**. The slope of this line is $0.010 \text{ min}^{-1}$. Because from equation (14–38) the slope $S$ is

$$S = \frac{k_f + k_r}{2.30} = 0.010 \text{ min}^{-1}$$

and from equation (14–39)

$$K = \frac{B_{eq}}{A_{eq}} = \frac{k_f}{k_r} = 2.1$$

**Fig. 14–3.** Approach to equilibrium in the reversible epimerizations of iso-7-chloro-*epi*-tetracycline (○—○—○) and iso-7-chlorotetracycline (●—●—●). (From J. D. McCormick, J. R. D. et al., J. Am. Chem. Soc. **79,** 2849, 1957.)

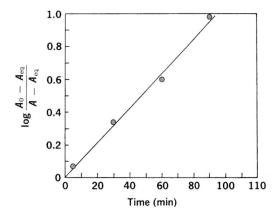

**Fig. 14–4.** Reversible epimerization of iso-7-chlorotetracycline in dimethylformamide containing 1 M $NaH_2PO_4$ at 25°C.

the elimination of $k_f$ from these equations affords a value for $k_r$. Thus, it is found that

$$\frac{2.1 k_r + k_r}{2.30} = 0.010 \text{ min}^{-1}$$

or

$$k_r = \frac{(0.010)(2.30)}{2.1 + 1} = 0.007 \text{ min}^{-1}$$

From this value, $k_f$ is found to be

$$k_f = 2.30 S - k_r = (2.30)(0.010) - 0.007$$
$$= 0.016 \text{ min}^{-1}$$

**Parallel or Side Reactions.** Parallel reactions are common in drug systems, particularly when organic compounds are involved. General acid–base catalysis, to be considered later, belongs to this class of reactions.

The base-catalyzed degradation of prednisolone will be used here to illustrate the parallel-type process. Guttman and Meister[8] investigated the degradation of the steroid prednisolone in aqueous solutions containing sodium hydroxide as a catalyst. The runs were carried out at 35°C, and the rate of disappearance of the dihydroxyacetone side chain was followed by appropriate analytic techniques. The decomposition of prednisolone was found to involve parallel pseudo–first-order reactions with the appearance of acidic and neutral steroidal products:

Prednisolone

The mechanism of the reaction can be represented as

$$P \quad \xrightarrow{k_1} \quad A \qquad \qquad (14\text{–}40)$$
$$P \quad \xrightarrow{k_2} \quad N \qquad \qquad (14\text{–}41)$$

where $P$, $A$, and $N$ are the concentrations of prednisolone, an acid product, and a neutral product, respectively.

The corresponding rate equation is

$$-\frac{dP}{dt} = k_1 P + k_2 P = kP \qquad (14\text{–}42)$$

where $k = k_1 + k_2$. This first-order equation is integrated to give

$$\ln (P_0/P) = kt \qquad \qquad (14\text{–}43)$$

or

$$P = P_0 e^{-kt} \qquad \qquad (14\text{–}44)$$

The rate of formation of the acidic product can be expressed as

$$\frac{dA}{dt} = k_1 P = k_1 P_0 e^{-kt} \qquad (14\text{–}45)$$

Integration of equation (14–45) yields

$$A = A_0 + \frac{k_1}{k} P_0 (1 - e^{-kt}) \qquad (14\text{–}46)$$

where $A$ is the concentration of the acid product at time, $t$, and $A_0$ and $P_0$ are the initial concentrations of the acid and prednisolone, respectively. Actually, $A_0$ is equal to zero because no acid is formed before the prednisolone begins to decompose. Therefore,

$$A = \frac{k_1}{k} P_0 (1 - e^{-kt}) \qquad (14\text{–}47)$$

Likewise, for the neutral product,

$$N = \frac{k_2}{k} P_0 (1 - e^{-kt}) \qquad (14\text{–}48)$$

Equations (14–47) and (14–48) suggest that for the base-catalyzed breakdown of prednisolone, a plot of the concentration $A$ or $N$ against $(1 - e^{-kt})$ should yield a straight line. At $t = 0$, the curve should pass through the origin, and at $t = \infty$, the function should have a value of unity. The value for $k$, the overall first-order rate constant, was obtained by a plot of log[Prednisolone] against the time at various concentrations of sodium hydroxide. It was possible to check the validity of expression (14–47) using the $k$ values that were now known for each level of hydroxide ion concentration. A plot of the acidic material formed against $(1 - e^{-kt})$ yielded a straight line passing through the origin as predicted by equation (14–47). The value of $k_1$, the rate constant for the formation of the acidic product, was then calculated from the slope of the line,

$$k_1 = \text{Slope} \times k / P_0 \qquad (14\text{–}49)$$

TABLE 14–3
**RATE CONSTANTS FOR THE BASE-CATALYZED DEGRADATION OF PREDNISOLONE IN AIR AT 35°C**

| NaOH (Normality) | $k$ (hr$^{-1}$) | $k_1$ (hr$^{-1}$) | $k_2$ (hr$^{-1}$) |
|---|---|---|---|
| 0.01 | 0.108 | 0.090 | 0.018 |
| 0.02 | 0.171 | 0.137 | 0.034 |
| 0.03 | 0.233 | 0.181 | 0.052 |
| 0.04 | 0.258 | 0.200 | 0.058 |
| 0.05 | 0.293 | 0.230 | 0.063 |

and the value of $k_2$, the rate constant for the formation of the neutral degradation product, was obtained by subtracting $k_1$ from $k$. The data, as tabulated by Guttman and Meister,[8] are given in **Table 14–3**.

The stability of hydrocortisone,

Hydrocortisone

was explored by Allen and Gupta[9] in aqueous and oil vehicles, water-washable ointment bases, and emulsified vehicles in the presence of other ingredients, at elevated temperatures and at various degrees of acidity and basicity. Hydrocortisone was unstable at room temperature in aqueous vehicles on the basic side of neutrality; alcohol and glycerin appeared to improve the stability. The decomposition in water and propylene glycol was a pseudo–first-order reaction. In highly acidic and basic media and at elevated temperatures, the decomposition of hydrocortisone was of a complex nature, following a parallel scheme.

**Series or Consecutive Reactions.** Consecutive reactions are common in radioactive series in which a parent isotope decays by a first-order process into a daughter isotope and so on through a chain of disintegrations. We take a simplified version of the degradation scheme of glucose as illustrative of consecutive-type reactions. The depletion of glucose in acid solution can be represented by the following scheme,[10] where 5-HMF is 5-hydroxymethylfurfural:

5-Hydroxymethylfurfural (5-HMF)

The scheme is seen to involve all of the complex-type reactions, reversible, parallel, and consecutive processes. At low concentrations of glucose and acid catalyst, the formation of polysaccharides can be neglected. Furthermore, owing to the indefinite nature of the breakdown products of 5-HMF, these can be combined together and referred to simply as constituent C. The simplified mechanism is therefore written as the series of reactions

$$A \xrightarrow{k_1} B \xrightarrow{k_2} C$$

where $A$ is glucose, $B$ is 5-HMF, and $C$ is the final breakdown products. The rate of decomposition of glucose is given by the equation

$$-dA/dt = k_1 A \qquad (14\text{--}50)$$

The rate of change in concentration of 5-HMF is

$$dB/dt = k_1 A - k_2 B \qquad (14\text{--}51)$$

and that of the breakdown products is

$$dC/dt = k_2 B \qquad (14\text{--}52)$$

When these equations are integrated and proper substitutions made, we obtain

$$A = A_0 e^{-k_1 t} \qquad (14\text{--}53)$$

$$B = \frac{A_0 k_1}{k_2 - k_1}(e^{-k_1 t} - e^{-k_2 t}) \qquad (14\text{--}54)$$

and

$$C = A_0 \left[ 1 + \frac{1}{k_1 - k_2}(k_2 e^{-k_1 t} - k_1 e^{-k_2 t}) \right] \qquad (14\text{--}55)$$

By the application of equations (**14–53**) through (**14–55**), the rate constants $k_1$ and $k_2$ and the concentration of breakdown products $C$ can be determined. Glucose is found to decompose by a first-order reaction. As glucose is depleted, the concentration of 5-HMF increases rapidly at the beginning of the reaction and then increases at a slower rate as time progresses. The decomposition products of 5-HMF increase slowly at first, indicating an induction or lag period, and then increase at a greater rate. These later products are responsible for the discoloration of glucose solutions that occurs when the solutions are sterilized at elevated temperatures.

Kinetic studies such as these have considerable practical application in pharmacy. When the mechanism of the breakdown of parenteral solutions is better understood, the manufacturing pharmacist should be able to prepare a stable product having a long shelf life. Large supplies of glucose injection and similar products can then possibly be stock-piled for use in times of emergency.

Mauger et al.[11] studied the degradation of hydrocortisone hemisuccinate at 70°C over a narrow pH range and found the reaction to be another example of the consecutive first-order type. At pH 6.9, the rate constant $k_1$ was 0.023 $hr^{-1}$ and $k_2$ was 0.50 $hr^{-1}$.

### The Steady-State Approximation

Michaelis–Menten Equation. A number of kinetic processes cannot have their rate laws integrated exactly. In situations such as these, it is useful to postulate a reasonable reaction sequence and then to derive a rate law that applies to the postulated sequence of steps. If the postulated sequence is reasonably accurate and reflects the actual steps involved in the reaction, the observed kinetics for the reaction should match the curve given by the derived rate law.

The *steady-state approximation* is commonly used to reduce the labor in deducing the form of a rate law. We illustrate this approximation by deriving the Michaelis–Menten equation.

Michaelis and Menten[12] assumed that the interaction of a substrate, $S$, with an enzyme, $E$, to yield a product, $P$, followed a reaction sequence given by

$$E + S \underset{k_2}{\overset{k_1}{\rightleftharpoons}} (E \cdot S) \overset{k_3}{\longrightarrow} P$$

According to this scheme, the rate of product formation is

$$\frac{dP}{dt} = k_3 (E \cdot S) \tag{14–56}$$

We have no easy means of obtaining the concentration of enzyme–substrate complex, so it is necessary that this concentration be expressed in terms of easily measurable quantities. In an enzyme study, we can usually measure $S$, $P$, and $E_0$, the total concentration of enzyme.

The rate of formation of $(E \cdot S)$ is

$$\frac{d(E \cdot S)}{dt} = k_1(E)(S) - k_2(E \cdot S) - k_3(E \cdot S) \tag{14–57}$$

or

$$\frac{d(E \cdot S)}{dt} = k_1(E)(S) - (k_2 + k_3)(E \cdot S) \tag{14–58}$$

If the concentration of $E \cdot S$ is constant throughout most of the reaction and is always much less than the concentrations of $S$ and $P$, we can write

$$\frac{d(E \cdot S)}{dt} = 0 \tag{14–59}$$

It follows from equations (14–58) and (14–59) that

$$(E \cdot S)_{ss} = \frac{k_1(E)(S)}{k_2 + k_3} \tag{14–60}$$

where the subscript ss is used to designate the concentration referred to as the *steady-state* value.

The total concentration of enzyme, $E_0$, is the sum of the concentrations of enzyme both free, $E$, and bound, $E \cdot S$,

$$E_0 = E + (E \cdot S)_{ss} \tag{14–61}$$

Eliminating $E$ from equations (14–60) and (14–61), we obtain

$$(E \cdot S)_{ss} = \frac{k_1 S E_0}{(k_2 + k_3) + k_1 S} \tag{14–62}$$

or

$$(E \cdots S)_{ss} = \frac{S E_0}{K_m + S} \tag{14–63}$$

where

$$K_m = \frac{k_2 + k_3}{k_1} \tag{14–64}$$

Thus, under steady-state conditions, the rate of product formation is given by

$$\frac{dP}{dt} = \frac{k_3 S E_0}{K_m + S} \tag{14–65}$$

which can be recognized as the Michaelis–Menten equation. The Michaelis–Menten constant, $K_m$, indicates the tendency of the enzyme–substrate complex to decompose to starting substrate or to proceed to product, relative to the tendency of the complex to be formed.

It is useful to introduce a maximum velocity for the Michaelis–Menten scheme, namely $(dP/dt)_{maximum}$, which is usually written as $V_m$. When $S$ is very large, all enzyme $E_0$ is present as $E \cdot S$, that is, all enzyme is combined with the substrate and the reaction proceeds at maximum velocity. From equation (14–56), $dP/dt$ becomes $V_m$ and $V_m = k_3 E_0$ because $E \cdot S$ is equivalent to $E_0$. Accordingly, from equation (14–65),

$$V = V_m \frac{S}{k_m + S} \tag{14–66}$$

Equation (14–66) can be inverted to obtain a linear expression known as the *Lineweaver–Burk equation*:

$$\frac{1}{V} = \frac{K_m + S}{V_m \cdot S} \tag{14–67}$$

$$\frac{1}{V} = \frac{1}{V_m} + \frac{K_m}{V_m} \frac{1}{S} \tag{14–68}$$

From equation (14–68) we see that a plot of $1/V$ versus $1/S$ yields a straight line with an intercept on the vertical axis of $1/V_m$ and a slope of $K_m/V_m$ (**Fig. 14–5**). Knowing $V_m$ from the intercept and obtaining $K_m/V_m$ as the slope, we can calculate $K_m$, the *Michaelis constant*.

### EXAMPLE 14–6

**Linear Transformations of the Michaelis–Menten Equation**

The velocity, $V$, of an enzymatic reaction at increasing substrate concentration [S] was experimentally determined to be as follows:

| $V[\mu g/(\text{liters min})]$ | 0.0350 | 0.0415 | 0.0450 | 0.0490 | 0.0505 |
|---|---|---|---|---|---|
| [S] (molarity, M) | 0.0025 | 0.0050 | 0.0100 | 0.0167 | 0.0333 |

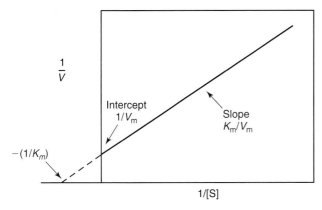

**Fig. 14–5.** A Lineweaver–Burk plot of Michaelis–Menten kinetics showing the calculation of $K_m$ by two means.

(*a*) Following the Lineweaver–Burk form of the Michaelis–Menten equation, plot $1/V$ versus $1/[S]$ using the following data and calculate $V_m$ and $K_m$ using linear regression analysis. The data for the Lineweaver–Burk plot and the regression analysis are as follows:

| $1/V$ [min/($\mu$g/liter)] | 28.57 | 24.10 | 22.22 | 20.41 | 19.80 |
|---|---|---|---|---|---|
| $1/[S]$ (liters/mole) | 400 | 200 | 100 | 59.88 | 30.0 |

(*b*) Extrapolate the line to the horizontal axis (*x* axis), where the intercept is $-1/K_m$. Read $-1/K_m$ as accurately as possible by eye and obtain $K_m$ as its reciprocal. Compare this value with that obtained by linear regression in (*a*).

*Answer:* (*a*) Linear regression analysis yields

$1/V = 19.316 + 0.0234 \, (1/[S]); \; r^2 = 0.990$

Intercept, $1/V_m = 19.316; \; V_m = 0.0518 \; \mu\text{g/(liter} \cdot \text{min)}$

Slope $= K_m/V_m = 0.0234 \; (\text{liter} \cdot \text{min}/\mu\text{g)} \; \text{M}$

$K_m = 0.0234 \; (\text{liter} \cdot \text{min}/\mu\text{g}) \; \text{M} \times 0.0518 \; \mu\text{g/liter} \cdot \text{min}$

$= 0.0012 \; \text{M}$

(*b*) By extrapolation,

$-1/K_m = -823 \; \text{M}^{-1}$

$K_m = 0.0012 \; \text{M}$

Michaelis–Menten kinetics is used not only for enzyme reactions but also for biochemical processes in the body involving carriers that transport substances across membranes such as blood capillaries and the renal tubule. It is assumed, for example, that L-tyrosine is absorbed from the nasal cavity into systemic circulation by a carrier-facilitated process, and Michaelis–Menten kinetics is applied to this case.

### Rate-Determining Step

In a reaction sequence in which one step is much slower than all the subsequent steps leading to the product, the rate at which the product is formed may depend on the rates of all the steps preceding the slow step but does not depend on any of the steps following. The slowest step in a reaction sequence is called, somewhat misleadingly, the *rate-determining step* of the reaction.

Consider the mechanistic pathway

$$A \underset{k_2}{\overset{k_1}{\rightleftharpoons}} B \quad \text{(step 1 and step 2)}$$

$$B + C \overset{k_3}{\longrightarrow} D \quad \text{(step 3)}$$

$$D \overset{k_4}{\longrightarrow} P \quad \text{(step 4)}$$

which can be postulated for the observed overall reaction

$$A + C \rightarrow P$$

If the concentrations of the intermediates $B$ and $D$ are small, we can apply the steady-state approximation to evaluate their steady-state concentrations. These are given by

$$B_{ss} = \frac{k_1 A}{k_2 + k_3 C}$$

and

$$D_{ss} = \frac{k_1 k_3 A C}{k_4(k_2 + k_3 C)}$$

For the rate of formation of the product, we can write

$$\frac{dP}{dt} = k_4 D_{ss}$$

or

$$\frac{dP}{dt} = \frac{k_1 k_3 A C}{k_2 + k_3 C} \quad (14\text{–}69)$$

If, in the mechanistic sequence, step 3 is the slow step (the rate-determining step), we can say that $k_2 \gg k_3 \, C$, and equation (14–69) is simplified to a second-order expression,

$$\frac{dP}{dt} = \frac{k_1 k_3 A C}{k_2} = k_0 A C \quad (14\text{–}70)$$

On the other hand, if step 2, the reverse reaction, is the slow step, then $k_3 C \gg k_2$, and equation (14–69) reduces to a first-order expression,

$$\frac{dP}{dt} = \frac{k_1 k_3 A C}{k_3 C} = k_1 A \quad (14\text{–}71)$$

Thus, we see that reactions may exhibit a simple first- or second-order behavior, yet the detailed mechanism for these reactions may be quite complex.

## TEMPERATURE EFFECTS

A number of factors other than concentration may affect the reaction velocity. Among these are temperature, solvents, catalysts, and light. This section discusses the effect of temperature covered.

### Collision Theory

Reaction rates are expected to be proportional to the number of collisions per unit time. Because the number of collisions increases as the temperature increases, the reaction rate is expected to increase with increasing temperature. In fact, the

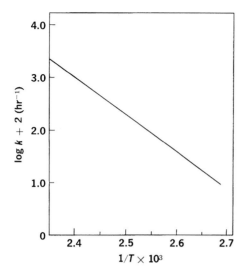

**Fig. 14–6.** A plot of log $k$ against $1/T$ for the thermal decomposition of glucose.

speed of many reactions increases about two to three times with each 10° rise in temperature. As a reaction proceeds from reactants to products, the system must pass through a state whose energy is greater than that of the initial reactants. This "barrier" is what prevents the reactants from immediately becoming products. The activation energy, $E_a$, is a measure of this barrier. The effect of temperature on reaction rate is given by the equation, first suggested by Arrhenius,

$$k = Ae^{-E_a/RT} \qquad (14\text{–}72)$$

or

$$\log k = \log A - \frac{E_a}{2.303} \frac{1}{RT} \qquad (14\text{–}73)$$

where $k$ is the specific reaction rate, $A$ is a constant known as the *Arrhenius factor* or the *frequency factor*, $E_a$ is the *energy of activation*, $R$ is the gas constant, 1.987 calories/deg mole, and $T$ is the absolute temperature. The constants $A$ and $E_a$ will be considered further in later sections of the chapter. They can be evaluated by determining $k$ at several temperatures and plotting $1/T$ against log $k$. As seen in equation (14–73), the slope of the line so obtained is $-E_a/2.303\ R$, and the intercept on the vertical axis is log $A$, from which $E_a$ and $A$ can be obtained.

Data obtained from a study of the decomposition of glucose solutions between 100°C and 140°C in the presence of 0.35 N hydrochloric acid are plotted in this manner in **Figure 14–6.** It should be observed that because the *reciprocal* of the absolute temperature is plotted along the horizontal axis, the temperature is actually *decreasing* from left to

right across the graph. It is sometimes advantageous to plot log $t_{1/2}$ instead of log $k$ on the vertical axis. The half-life for a first-order reaction is related to $k$ by equation (14–18), $t_{1/2} = 0.693/k$, and in logarithmic form

$$\log k = \log 0.693 - \log t_{1/2} \qquad (14\text{–}74)$$

Substituting equation (14–74) into equation (14–73) gives

$$\log t_{1/2} = \log 0.693 - \log A + \frac{E_a}{2.303 R} \frac{1}{T}$$

or

$$\log t_{1/2} = \frac{E_a}{2.303 R} \frac{1}{T} + \text{constant}$$

and $E_a/2.303R$ is obtained as the slope of the line resulting from plotting log $t_{1/2}$ against $1/T$. Higuchi et al.[13] plotted the results of the alkaline hydrolysis of procaine in this manner, as shown in **Figure 14–7.**

$E_a$ can also be obtained by writing equation (14–73) for a temperature $T_2$ as

$$\log k_2 = \log A - \frac{E_a}{2.303 R} \frac{1}{T_2}$$

and for another temperature $T_1$ as

$$\log k_1 = \log A - \frac{E_a}{2.303 R} \frac{1}{T_1}$$

Subtracting these two expressions yields

$$\log \frac{k_2}{k_1} = \frac{E_a}{2.303 R} \left( \frac{T_2 - T_1}{T_2 T_1} \right)$$

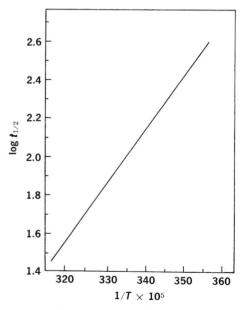

**Fig. 14–7.** A plot of log $t_{1/2}$ against $1/T$ for the alkaline hydrolysis of procaine. (From T. Higuchi, A. Havinga, and L. W. Busse, J. Am. Pharm. Assoc. Sci. Ed. **39**, 405, 1950.)

---

*Notice that log $k + 2$ is plotted on the vertical axis of Figure 14–6. This is a convenient way of eliminating negative values along the axis. For example, if $k = 1.0 \times 10^{-2}$, $2.0 \times 10^{-2}$, etc., the logarithmic expressions are log 1.0 + log $10^{-2}$, log 2.0 + log $10^{-2}$, etc., or $0.0 - 2 = -2$, $0.3 - 2 = -1.7$, etc. The negative signs can be eliminated along the vertical axis if 2 is added to each value; hence the label, log $k + 2$.

EXAMPLE 14–7

**Decomposition of 5-HMF**

The rate constant $k_1$ for the decomposition of 5-hydroxymethylfurfural at 120°C (393 K) is 1.173 hr$^{-1}$ or $3.258 \times 10^{-4}$ sec$^{-1}$ and $k_2$ at 140°C (413 K) is 4.860 hr$^{-1}$. What is the activation energy, $E_a$, in kcal/mole and the frequency factor, $A$, in sec$^{-1}$ for the breakdown of 5-HMF within this temperature range?

We have

$$\log \frac{4.860}{1.173} = \frac{E_a}{2.303 \times 1.987} \frac{413 - 393}{413 \times 393}$$

$$E_a = 23 \, \text{kcal/mole}$$

At 120°C, using equation (14–73), we obtain

$$\log(3.258 \times 10^{-4} \, \text{sec}^{-1}) = \log A - \frac{23,000 \, \text{cal}}{2.303 \times 1.987} \frac{1}{393}$$

$$A = 2 \times 10^9 \, \text{sec}^{-1}$$

## Classic Collision Theory of Reaction Rates

The Arrhenius equation is largely an empirical relation giving the effect of temperature on an observed rate constant. Relations of this type are observed for unimolecular and bimolecular reactions and often are also observed for complex reactions involving a number of bimolecular and unimolecular steps. Although it is extremely difficult, in most cases, to attach significance to the temperature dependence of complex reactions, the temperature dependence of uni- and bimolecular reactions appears to reflect a fundamental physical requirement that must be met for a reaction to occur.

The manner by which temperature affects molecular motion can be understood by considering a hypothetical situation in which all the molecules of a substance are moving in the same direction at the same velocity. If a molecule deviates from its course, it will collide with another molecule, causing both molecules to move off in different directions with different velocities. A chain of collisions between molecules can then occur, which finally results in random motion of all the molecules. In this case, only a certain fraction of the molecules have a velocity equivalent to the initial velocity of the ordered system. The net result is that for a fixed number of molecules at a given temperature, and therefore at a definite total energy, a distribution of molecular velocities varying from zero upward is attained. Because kinetic energy is proportional to the square of velocity, the distribution of molecular velocities corresponds to the distribution of molecular energies, and the fraction of the molecules having a given kinetic energy can be expressed by the *Boltzmann distribution law*,

$$f_i = \frac{N_i}{N_T} = e^{-Ei/RT} \qquad (14\text{–}75)$$

From the Boltzmann distribution law we note that of the total number of moles, $N_T$, of a reactant, $N_i$ moles have a kinetic energy given by $E_i$. The collision theory of reaction rates postulates that a collision must occur between molecules for a reaction to occur and, further, that a reaction between molecules does not take place unless the molecules are of a certain energy. By this postulate, the rate of a reaction can be considered proportional to the number of moles of reactant having sufficient energy to react, that is,

$$\text{Rate} = PZN_i \qquad (14\text{–}76)$$

The proportionality constant in this relation is divided into two terms: the collision number, $Z$, which for a reaction between two molecules is the number of collisions per second per cubic centimeter, and the steric or probability factor, $P$, which is included to take into account the fact that not every collision between molecules leads to reaction. That is, $P$ gives the probability that a collision between molecules will lead to product.

Substituting for $N_i$ in equation (**14–76**) yields

$$\text{Rate} = (PZe^{-Ei/RT})N_T \qquad (14\text{–}77)$$

which, when compared with the general rate law

$$\text{Rate} = k \times \text{Concentration of reactants} \qquad (14\text{–}78)$$

leads to the conclusion that

$$k = (PZ)e^{-Ei/RT} \qquad (14\text{–}79)$$

Thus, collision-state theory interprets the Arrhenius factor $A$ in terms of the frequency of collision between molecules,

$$A = PZ \qquad (14\text{–}80)$$

and the Arrhenius activation energy, $E_a$, as the minimum kinetic energy a molecule must possess in order to undergo reaction,

$$E_a = E_i \qquad (14\text{–}81)$$

Yang[14] showed the error possible in determining the activation energy, $E_a$, and the predicted shelf life when the kinetic order in an accelerated stability test is incorrectly assigned, for example, when an actual zero-order reaction can equally well be described by a first-order degradation.

## $Q_{10}$ Calculations

In an excellent reference text for pharmacists, Connors et al.[1] described a straightforward calculation that facilitates a practical understanding of temperature effects. Using this method, one can estimate the effect of a 10° rise in temperature on the stability of pharmaceuticals. Just as was done in *Example 14–7*, this so-called $Q_{10}$ method relies on the ratio of reaction rate constants at two different temperatures. The quantity $Q_{10}$ was originally defined by Simonelli and Dresback[15] as

$$Q_{10} = \frac{k_{(T+10)}}{k_T} \qquad (14\text{–}82)$$

$Q_{10}$ is the factor by which the rate constant increases for a 10°C temperature increase. The $Q_{10}$ factor can be calculated from the following equation:

$$Q_{10} = \exp\left[-\frac{E_a}{R}\left(\frac{1}{T+10} - \frac{1}{T}\right)\right] \qquad (14\text{–}83)$$

If the activation energy is known, the corresponding $Q_{10}$ value can be obtained from equation (**14–83**). The $Q_{10}$ approximation method is useful for making quick approximations. As noted by Connors et al.,[1] the activation energies for drug decompositions usually fall in the range of 12 to 24 kcal/mole, with typical values 19 to 20 kcal/mole. To make approximations when $E_a$ is unknown, it is reasonable to use these typical values to calculate $Q_{10}$ values. For example, using Equation (**14–83**), we have $Q_{10} = 2, 3$, and 4 when $E_a = 12.2, 19.4$, and 24.5, respectively, when the temperature rises from 20°C to 30°C. This simple calculation demonstrates that the degradation rate of most pharmaceutical agents will increase by two to four times, with an average of three times, for a 10°C rise in temperature in a range (from 20°C to 30°C) that typical consumers will experience. The more advanced student may be interested in generalizing the $Q_{10}$ approach to estimate the effect of increasing or decreasing the temperature by variable amounts. To do this, use the following equation:

$$Q_{\Delta T} = \frac{k_{(T+\Delta T)}}{k_T} = Q_{10}^{(\Delta T/10)} \qquad (14\text{–}84)$$

### EXAMPLE 14–8

**Effect of Temperature Increase/Decrease on Rate Constants**

Calculate the factors by which rate constants may change for (*a*) a 25°C to 50°C temperature change and (*b*) a 25°C to 0°C temperature change.

*Answer*:

(*a*) Using equation (14–84), with $\Delta T = +25$, we obtain

$$Q_{+25} = Q_{10}^{25/10}$$
$$= 5.7, 15.6, 32 \quad \text{for} \quad Q_{10} = 2, 3, 4, \text{ respectively.}$$

Thus, the rate increases between 6-fold and 32-fold, with a probable average increase of about 16-fold.

(*b*) When $\Delta T = -25$, we have

$$Q_{-25} = Q_{10}$$
$$= \frac{1}{5.7}, \frac{1}{15.6}, \frac{1}{32} \quad \text{for}$$
$$Q_{10} = 2, 3, 4, \text{ respectively}$$

Thus, the rate decreases to between 1/6 and 1/32 of the initial rate.

## Shelf-Life Calculations

The following examples illustrate situations that pharmaceutical scientists and practicing pharmacists are likely to encounter.

### EXAMPLE 14–9

**Increased Shelf Life of Aspirin (Connors et al.,[1] pp. 12–18)**

Aspirin is most stable at pH 2.5. At this pH the apparent first-order rate constant is $5 \times 10^{-7}$ sec$^{-1}$ at 25°C. The shelf life of aspirin in solution under these conditions can be calculated as follows:

$$t_{90} = \frac{0.10^5}{5 \times 10^{-7}} = 2.1 \times 10^5 \text{ sec} = 2 \text{ days}$$

As one can see, aspirin is very unstable in aqueous solution. Would making a suspension increase the shelf life of aspirin?

The solubility of aspirin is 0.33 g/100 mL. At pH 2.5, the apparent zero-order rate constant for an aspirin suspension is

$$k_0 = 5 \times 10^{-7} \text{ sec}^{-1} \times 0.33 \text{ g/100 mL}$$
$$= 1.65 \times 10^{-7} \text{ g/mL} \cdot \text{sec}$$

If one dose of aspirin at 650 mg per teaspoonful is administered, then one has 650 mg/5 mL = 13 g/100 mL.

For this aspirin suspension,

$$t_{90} = \frac{(0.1)(13)}{1.65 \times 10^{-7}} = 7.9 \times 10^6 \text{ sec} = 91 \text{ days}$$

The increase in the shelf life of suspensions as compared to solutions is a result of the interplay between the solubility and the stability of the drug. In the case of aspirin, the solid form of the drug is stable, whereas when aspirin is in solution it is unstable. As aspirin in solution decomposes, the solution concentration is maintained as additional aspirin dissolves up to the limit of its aqueous solubility.

### EXAMPLE 14–10

**How Long Can a Product Be Left Out at Room Temperature?**

Reconstituted ampicillin suspension is stable for 14 days when stored in the refrigerator (5°C). If the product is left at room temperature for 12 hr, what is the reduction in the expiration dating?

To solve this problem we must use the following equation:

$$t_{90}(T_2) = \frac{t_{90}(T_1)}{Q_{10}^{(\Delta T/10)}} \qquad (14\text{–}85)$$

The estimate of $t_{90}(T_2)$ is independent of order. In other words, it is not necessary to know the reaction order to make this estimate.

## OTHER FACTORS—A MOLECULAR VIEWPOINT

### Transition-State Theory

An alternative to the collision theory is the *transition-state theory* or absolute rate theory, according to which an equilibrium is considered to exist between the normal reactant molecules and an activated complex of these molecules. Decomposition of the activated complex leads to product. For an elementary bimolecular process, the reaction can be written as

$$A + B \quad \rightleftharpoons \quad [A \cdots B]^{\ddagger} \quad \rightarrow \quad P$$

| Normal reactant molecules | Activated molecules in the transition state (activated complex) | Product molecules |

$$(14\text{–}86)$$

A double dagger is used to designate the activated state, namely $[A \ldots B]^{\ddagger}$.

The rate of product formation in this theory is given by

$$\text{Rate} = v[A \cdots B]^{\ddagger} \qquad (14\text{–}87)$$

where $v$ is the frequency with which an activated complex goes to product. Because an equilibrium exists between the reactants and the activated complex,

$$K^{\ddagger} = \frac{[A \cdots B]^{\ddagger}}{[A][B]} \qquad (14\text{–}88)$$

and this expression can be rearranged to

$$[A \cdots B]^{\ddagger} = K^{\ddagger}[A][B] \qquad (14\text{–}89)$$

Hence,

$$\text{Rate} = \nu K^{\ddagger}[A][B] \qquad (14\text{–}90)$$

The general rate law for a bimolecular reaction is

$$\text{Rate} = k[A][B] \qquad (14\text{–}91)$$

so it follows that

$$k = \nu K^{\ddagger} \qquad (14\text{–}92)$$

It will be recalled from previous thermodynamic considerations that

$$\Delta G^{\circ} = -RT \quad \ln \quad K \qquad (14\text{–}93)$$

or

$$K = e^{-\Delta G^{\circ}/RT} \qquad (14\text{–}94)$$

and

$$\Delta G^{\circ} = \Delta H^{\circ} - T \, \Delta S^{\circ} \qquad (14\text{–}95)$$

Replacing the ordinary $K$ for present purposes with $K^{\ddagger}$ and making similar substitutions for the thermodynamic quantities, we obtain

$$k = \nu e^{-\Delta G^{\ddagger}/RT} \qquad (14\text{–}96)$$

and

$$k = \left(\nu e^{-\Delta S^{\ddagger}/R}\right) e^{-\Delta H^{\ddagger}/RT} \qquad (14\text{–}97)$$

where $\Delta G^{\ddagger}$, $\Delta S^{\ddagger}$, and $\Delta H^{\ddagger}$ are the respective differences between the standard free energy, entropy, and enthalpy in the transition state and in the normal reactant state.

In this theory, the Arrhenius factor $A$ is related to the entropy of activation of the transition state:

$$A = \nu e^{\Delta S^{\ddagger}/R} \qquad (14\text{–}98)$$

and the Arrhenius activation energy, $E_{a}$, is related to the entropy of activation of the transition state:

$$E_{a} = \Delta H^{\ddagger} = \Delta E^{\ddagger} + P \, \Delta V^{\ddagger} \qquad (14\text{–}99)$$

For most practical purposes, $\Delta V^{\ddagger} = 0$; hence,

$$E_{a} = \Delta E^{\ddagger} \qquad (14\text{–}100)$$

In principle, the transition-state theory gives the influence of temperature on reaction rates by the general equation

$$k = \left(\nu e^{\Delta S^{\ddagger}/R}\right) e^{-\Delta E^{\ddagger}/RT} \qquad (14\text{–}101)$$

where the frequency of decomposition of the transition-state complex, $\nu$, may vary depending on the nature of the reactants. Eyring[16] showed that the quantity $\nu$ can be considered, to

a good approximation, as a universal factor for reactions, depending only on temperature, and that it can be written as

$$\nu = \frac{RT}{Nh} \qquad (14\text{–}102)$$

where $R$ is the molar gas constant, $T$ is the absolute temperature, $N$ is Avogadro's number, and $h$ is Planck's constant. The factor $RT/Nh$ has a value of about $10^{12}$ to $10^{13}$ sec$^{-1}$ at ordinary temperatures ($\sim 2 \times 10^{10}\ T$). In many unimolecular gas reactions in which $\Delta S^{\ddagger}$ is zero so that $e^{\Delta S^{\ddagger}/R} = 1$, the rate constant ordinarily has a value of about $10^{13} e^{-E_{a}/RT}$, or

$$k \cong \frac{RT}{Nh} e^{-\Delta H^{\ddagger}/RT} \cong 10^{13} e^{-E_{a}/RT} \qquad (14\text{–}103)$$

When the rate deviates from this value, it can be considered as resulting from the $e^{\Delta S^{\ddagger}/R}$ factor. When the activated complex represents a more probable arrangement of molecules than found in the normal reactants, $\Delta S^{\ddagger}$ is positive and the reaction rate will be greater than normal. Conversely, when the activated complex results only after considerable rearrangement of the structure of the reactant molecules, making the complex a less probable structure, $\Delta S^{\ddagger}$ is negative, and the reaction will be slower than predicted from equation (14–103). The collision theory and the transition-state theory are seen to be related by comparing equations (14–80), (14–98), and (14–102). One concludes that

$$PZ = \frac{RT}{Nh} e^{\Delta S^{\ddagger}/R} \qquad (14\text{–}104)$$

The collision number, $Z$, is identified with $RT/Nh$ and the probability factor, $P$, with the entropy term $\Delta S^{\ddagger}/R$.

**EXAMPLE 14–11**

**Acid-Catalyzed Hydrolysis of Procaine**

In the study of the acid-catalyzed hydrolysis of procaine, Marcus and Baron[17] obtained the first-order reaction rate, $k$, from a plot of log $c$ versus $t$ and the activation energy, $E_{a}$, from an Arrhenius plot of log $k$ versus $1/T$. The values were $k = 38.5 \times 10^{-6}$ sec$^{-1}$ at 97.30°C and $E_{a} = 16.8$ kcal/mole.

Compute $\Delta S^{\ddagger}$ and the frequency factor, $A$, using equations (14–97) and (14–98), and the probability factor $P$. It is first necessary to obtain $RT/Nh$ at 97.30°C or about 371 K:

$$\nu = \frac{RT}{Nh} = \frac{8.31 \times 10^{7}\ \text{ergs/mole deg} \times 371\ \text{deg}}{6.62 \times 10^{-27}\text{erg sec/molecule}}$$

$$\times\ 6.02 \times 10^{23}\ \text{molecules/mole}$$

$$= 7.74 \times 10^{12}\ \text{sec}^{-1}$$

Then, from equation (14–97), in which

$$\Delta H^{\ddagger} \cong E_{a}$$

$$38.5 \times 10^{-6} = 7.74 \times 10^{12} e^{\Delta S^{\ddagger}/1.987} \times e^{-16,800/(1.987 \times 371)}$$

$$\Delta S^{\ddagger} = -24.73\ \text{cal/mole}$$

and from equation (14–98),

$$A = 7.74 \times 10^{12} e^{-33.9/1.987} = 3.05 \times 10^{7}\ \text{sec}^{-1}$$

Finally, from the discussion accompanying equation (14–104),

$$P = e^{-33.9/1.987} = 3.9 \times 10^{-6}$$

Marcus and Baron[17] compared the kinetics of the acid-catalyzed hydrolyses of procainamide, procaine, and benzocaine. They found that the frequency factors for procainamide and procaine were considerably lower than the values expected for compounds of this type. Procainamide and procaine are diprotonated species in acid solution, that is, they have taken on two protons, and hydrolysis in the presence of an acid involves the interaction of positively charged ions, namely the diprotonated procaine molecule and the hydronium ion:

Diprotonated procaine under attack by a
hydronium ion during acid hydrolysis

According to these authors, the two positively charged protonated centers on the procaine molecule exert a considerable repulsive effect on the attacking hydronium ions. This repulsion results in a low frequency factor. The $\Delta S^{\ddagger}$ is unusually negative (*Example 14–8*) perhaps for the following reason. When the third proton finally attaches itself, the activated complex that results is a highly charged ion. The activated molecule is markedly solvated, reducing the freedom of the solvent and decreasing the entropy of activation. This effect also tends to lower the frequency factor.

## Medium Effects: Solvent, Ionic Strength, Dielectric Constant

### Effect of the Solvent

The influence of the solvent on the rate of decomposition of drugs is a topic of great importance to the pharmacist. Although the effects are complicated and generalizations cannot usually be made, it appears that the reaction of nonelectrolytes is related to the relative internal pressures or solubility parameters of the solvent and the solute. The effects of the ionic strength and the dielectric constant of the medium on the rate of ionic reactions are also significant and will be discussed in subsequent sections.

Solutions are ordinarily nonideal, and equation (14–88) should be corrected by including activity coefficients. For the bimolecular reaction,

$$A + B \rightleftharpoons [A \cdots B]^{\ddagger} \rightarrow \text{Products}$$

the thermodynamic equilibrium constant should be written in terms of activities as

$$K^{\ddagger} = \frac{a^{\ddagger}}{a_A a_B} = \frac{C^{\ddagger}}{C_A C_B} \frac{\gamma^{\ddagger}}{\gamma_A \gamma_B} \quad (14\text{–}105)$$

where $a^{\ddagger}$ is the activity of the species in the transition state and $a_A$ and $a_B$ are the activities of the reactants in their normal

state. Then the following expressions, analogous to equations (14–87) and (14–90), are obtained:

$$\text{Rate} = \frac{RT}{Nh} C^{\ddagger} = \frac{RT}{NH} K^{\ddagger} C_A C_B \frac{\gamma_A \gamma_B}{\gamma^{\ddagger}} \quad (14\text{–}106)$$

or

$$k = \frac{\text{Rate}}{C_A C_B} = \frac{RT}{NH} K^{\ddagger} \frac{\gamma_A \gamma_B}{\gamma^{\ddagger}}$$

or

$$k = k_0 \frac{\gamma_A \gamma_B}{\gamma^{\ddagger}} \quad (14\text{–}107)$$

where $k_0 = RTK^{\ddagger}/Nh$ is the rate constant in an infinitely dilute solution, that is, one that behaves ideally. It will be recalled from previous chapters that the activity coefficients may relate the behavior of the solute in the solution under consideration to that of the solute in an infinitely dilute solution. When the solution is ideal, the activity coefficients become unity and $k_0 = k$ in equation (14–107). This condition was tacitly assumed in equation (14–90).

Now, the activity coefficient $\Delta_2$ of a not too highly polar nonelectrolytic solute in a dilute solution is given by the expression

$$\log \gamma_2 = \frac{V_2}{2.303 RT} (\delta_1 - \delta_2)^2 \quad (14\text{–}108)$$

where $V_2$ is the molar volume of the solute and $\Delta_1$ and $\Delta_2$ are the solubility parameters for the solvent and solute, respectively. The volume fraction term, $F^2$, is assumed here to have a value of unity.

Writing equation (14–107) in logarithmic form,

$$\log k = \log k_0 + \log \gamma_A + \log \gamma_B - \log \gamma^{\ddagger} \quad (14\text{–}109)$$

and substituting for the activity coefficients from (14–108) gives

$$\log k = \log k_0 + \frac{V_A}{2.303 RT} (\delta_1 - \delta_A)^2$$
$$+ \frac{V_B}{2.303 RT} (\delta_1 - \delta_B)^2$$
$$- \frac{V^{\ddagger}}{2.303 RT} (\delta_1 - \delta^{\ddagger})^2 \quad (14\text{–}110)$$

where $V_A$, $V_B$, $V^{\ddagger}$, and the corresponding $\delta_A$, $\delta_B$, and $\delta^{\ddagger}$ are the molar volumes and solubility parameters of reactant $A$, reactant $B$, and the activated complex $(A \cdots B)^{\ddagger}$, respectively. The quantity $\Delta_1$ is the solubility parameter of the solvent.

Thus, it is seen that the rate constant depends on the molar volumes and the solubility parameter terms.

Because these three squared terms, $(\delta_1 - \delta_A)^2$, $(\delta_1 - \delta_B)^2$, and $(\delta_1 - \delta^{\ddagger})^2$ represent the differences between solubility parameters or internal pressures of the solvent and the reactants, and the solvent and the activated complex, they can be symbolized, respectively, as $\Delta \delta_A$, $\Delta \delta_B$, and $\Delta \delta^{\ddagger}$. The molar volumes do not vary significantly, and the rate constant therefore depends primarily on the difference between

$(\Delta \delta_A + \Delta \delta_B)$ and $\Delta \delta^{\ddagger}$. This is readily seen by writing equation (14–110) as

$$\log k = \log k_0 + \frac{V}{2.303 RT}(\Delta \delta_A + \Delta \delta_B - \Delta \delta_s^{\ddagger})$$

It is assumed that the properties of the activated complex are quite similar to those of the products so that $\Delta \delta^{\ddagger}$ can be taken as a squared term expressing the internal pressure difference between the solvent and the products. This equation indicates that if the internal pressure or "polarity" of the products is similar to that of the solvent, so that $\Delta \delta^{\ddagger} \cong 0$, and the internal pressures of the reactants are unlike that of the solvent, so that $\Delta \delta_A$ and $\Delta \delta_B > 0$, then the rate will be large in this solvent relative to the rate in an ideal solution. If, conversely, the reactants are similar in "polarity" to the solvent so that $\Delta \delta_A$ and $\Delta \delta_B \cong 0$, whereas the products are not similar to the solvent, that is, $\Delta \delta^{\ddagger} > 0$, then $(\Delta \delta_A + \Delta \delta_B) - \Delta \delta^{\ddagger}$ will have a sizable negative value and the rate will be small in this solvent.

As a result of this analysis, it can be said that polar solvents, those with high internal pressures, tend to accelerate reactions that form products having higher internal pressures than the reactants. If, on the other hand, the products are less polar than the reactants, they are accelerated by solvents of low polarity or internal pressure and retarded by solvents of high internal pressure. To illustrate this principle, we can use the reaction between ethyl alcohol and acetic anhydride:

$$C_2H_5OH + (CH_3CO)_2O \rightarrow CH_3COOC_2H_5 + CH_3COOH$$

The activated complex, resembling ethyl acetate, is less polar than the reactants, and, accordingly, the reaction should be favored in a solvent having a relatively low solubility parameter. The rate constants for the reaction in various solvents are given in **Table 14–4** together with the solubility parameters of the solvents.[18] The reaction slows down in the more polar solvents as predicted.

### Influence of Ionic Strength

In a reaction between ions, the reactants $A$ and $B$ have charges $z_A$ and $z_B$, respectively, and the activated complex $(A \cdots B)^{\ddagger}$ has a charge of $z_A + z_B$. A reaction involving ions can be represented as

$$A^{z_A} + B^{z_B} \rightleftharpoons [A \cdots B]^{\ddagger(z_A + z_B)} \rightarrow \text{Products}$$

### TABLE 14–4
### INFLUENCE OF SOLVENTS ON RATE CONSTANTS

| Solvent | Solubility Parameter, $\delta$ | $k$ at 50°C |
|---|---|---|
| Hexane | 7.3 | 0.0119 |
| Carbon tetrachloride | 8.6 | 0.0113 |
| Chlorobenzene | 9.5 | 0.0053 |
| Benzene | 9.2 | 0.0046 |
| Chloroform | 9.3 | 0.0040 |
| Nitrobenzene | 10.0 | 0.0024 |

The activity coefficient, $\gamma_i$, of an ion in a dilute aqueous solution ($<0.01$ M) at 25°C is given by the Debye–Hückel equation as

$$\log \gamma_i = -0.51 z_i^2 \sqrt{\mu} \qquad (14\text{–}111)$$

where $\mu$ is the ionic strength. Therefore, we can write

$$\log \gamma_A + \log \gamma_B - \log \gamma^{\ddagger}$$
$$= -0.51 z_A^2 \sqrt{\mu} - 0.51 z_B^2 \sqrt{\mu} + 0.51(z_A + z_B)^2 \sqrt{\mu}$$
$$= -0.51 \sqrt{\mu} \, [z_A^2 + z_B^2 - (z_A^2 + 2z_A z_B + z_B^2)]$$
$$= 0.51 \times 2 z_A z_B \sqrt{\mu} = 1.02 z_A z_B \sqrt{\mu} \qquad (14\text{–}112)$$

Substituting into equation (14–109) results in the expression, at 25°C,

$$\log k = \log k_0 + 1.02 z_A z_B \sqrt{\mu} \qquad (14\text{–}113)$$

where $k_0$ is the rate constant in an infinitely dilute solution in which $\mu = 0$. It follows from equation (14–113) that a plot of $\log k$ against $\sqrt{\mu}$ should give a straight line with a slope of $1.02 z_A z_B$. If one of the reactants is a neutral molecule, $z_A z_B = 0$, and the rate constant, as seen from equation (14–113), should then be independent of the ionic strength in dilute solutions. Good agreement has been obtained between experiment and theory as expressed by equation (14–113).

If the reacting molecules are uncharged in a solution having a reasonable ionic strength, the rate expression is

$$\log k = \log k_0 + b\mu \qquad (14\text{–}114)$$

where $b$ is a constant obtained from experimental data. Carstensen[19] considered the various ionic strength effects in pharmaceutical solutions.

### Influence of Dielectric Constant

The effect of the dielectric constant on the rate constant of an ionic reaction, extrapolated to infinite dilution where the ionic strength effect is zero, is often a necessary piece of information in the development of new drug preparations. One of the equations by which this effect can be determined is

$$\ln k = \ln k_{\epsilon=\infty} - \frac{N z_A z_B e^2}{RT r^{\ddagger}} \frac{1}{\epsilon} \qquad (14\text{–}115)$$

where $k_{\epsilon=\infty}$ is the rate constant in a medium of infinite dielectric constant, $N$ is Avogadro's number, $z_A$ and $z_B$ are the charges on the two ions, $e$ is the unit of electric charge, $r^{\ddagger}$ is the distance between ions in the activated complex, and $\epsilon$ is the dielectric constant of the solution, equal approximately to the dielectric constant of the solvent in dilute solutions. The term $\ln k_{\epsilon=\infty}$ is obtained by plotting $\ln k$ against $1/\epsilon$ and extrapolating to $1/\epsilon = 0$, that is, to $\epsilon = \infty$. Such a plot, according to equation (14–115), should yield a straight line with a positive slope for reactant ions of opposite sign and a negative slope for reactants of like sign. For a reaction between ions of opposite sign, an increase in dielectric constant of the solvent results in a decrease in the rate constant. For ions of like charge, on the other hand, an increase in dielectric constant results in an increase in the rate of the reaction.

When a reaction occurs between a dipole molecule and an ion $A$, the equation is

$$\ln \ k = \ln \ k_{\in=\infty} + \frac{N z_A{}^2 e^2}{2RT} \left( \frac{1}{r_A} - \frac{1}{r^{\ddagger}} \right) \frac{1}{\in} \quad (14\text{–}116)$$

where $z_A$ is the charge on the ion $A$, $r_A$ is the radius of the ion, and $r^{\ddagger}$ is the radius of the activated complex. Equation (14–116) predicts that a straight line should be obtained when $\ln \ k$ is plotted against $1/\in$, the reciprocal of the dielectric constant. Because $r^{\ddagger}$, the radius of the combined ion and neutral molecule in the transition state, will be larger than $r_A$, the radius of the ion, the second term on the right side of the equation will always be positive, and the slope of the line will consequently be positive. Therefore, $\ln \ k$ will increase with increasing values of $1/\in$, that is, the rate of reaction between an ion and a neutral molecule will increase with *decreasing* dielectric constant of the medium. This relationship, however, does not hold if different solvents are used or if the solutions are not dilute, in which case ionic strength effects become significant.

The orientation of the solvent molecules around the solute molecules in solution will result in an effect that has not been accounted for in the equations given previously. When a solvent mixture is composed of water and a liquid of low dielectric constant, water molecules will be oriented about the ions in solution, and the dielectric constant near the ion will be considerably greater than that in the bulk of the solution. Thus, when $\ln \ k$ is plotted against the reciprocal of the dielectric constant of the solvent mixture, deviations from the straight line predicted by equations (14–115) and (14–116) will frequently result.

A number of studies relating the dielectric constant of the solvent medium to the rate of reactions have been undertaken. Several investigations involving compounds of pharmaceutical interest are briefly reviewed here.

Amis and Holmes[20] studied the effect of the dielectric constant on the acid inversion of sucrose. When the dielectric constant was reduced by adding dioxane to the aqueous solvent, the rate of the reaction was found to increase in accord with the theory of ion–dipole reactions as expressed by equation (14–116).

To determine the effect of dielectric constant on the rate of glucose decomposition in acidic solution, Heimlich and Martin[10] carried out tests in dioxane–water mixtures. The results shown in **Table 14–5** are those expected for a reaction between a positive ion and a dipole molecule. As observed in the table, the dielectric constant of the medium should be an important consideration in the stabilization of glucose solutions because replacing water with a solvent of lower dielectric constant markedly increases the rate of breakdown of glucose. Marcus and Taraszka[21] studied the kinetics of the hydrogen-ion–catalyzed degradation of the antibiotic chloramphenicol in water–propylene glycol systems. The decrease in dielectric constant resulted in an increase in the rate of the reaction, a finding that agrees with the requirements for an ion–dipole reaction.

**TABLE 14–5**

**DECOMPOSITION OF 0.278 M SOLUTIONS OF GLUCOSE AT pH 1.27 AND 100°C IN DIOXANE–WATER MIXTURES***

| Dioxane (% by Weight) | Dielectric Constant of the Solvent at 100°C | Rate Constant $k(\times 10^5 \ \text{hr}^{-1})$ |
|---|---|---|
| 0 | 55 | 4.58 |
| 9.98 | 48 | 4.95 |
| 29.74 | 35 | 6.34 |
| 49.32 | 22 | 10.30 |

*Dioxane is toxic and cannot be used in pharmaceutical preparations.

These findings have considerable pharmaceutical significance. The replacement of water with other solvents is often used in pharmacy as a means of stabilizing drugs against possible hydrolysis. The results of the investigations reviewed here suggest, however, that the use of a solvent mixture of lowered dielectric constant actually may increase rather than decrease the rate of decomposition. On the other hand, as pointed out by Marcus and Taraszka,[21] a small increase in decomposition rate due to the use of nonaqueous solvents may be outweighed by enhancement of solubility of the drug in the solvent of lower dielectric constant. Thus, there is a need for thorough kinetic studies and cautious interpretation of the results before one can predict the optimum conditions for stabilizing drug products.

## Catalysis: Specific and General Acid–Base and pH Effects

As already noted, the rate of a reaction is frequently influenced by the presence of a catalyst. Although the hydrolysis of sucrose in the presence of water at room temperature proceeds with a decrease in free energy, the reaction is so slow as to be negligible. When the hydrogen ion concentration is increased by adding a small amount of acid, however, inversion proceeds at a measurable rate.

A *catalyst* is therefore defined as a substance that influences the speed of a reaction without itself being altered chemically. When a catalyst decreases the velocity of a reaction, it is called a *negative catalyst*. Actually, negative catalysts often may be changed permanently during a reaction and should be called *inhibitors* rather than catalysts.

Because a catalyst remains unaltered at the end of a reaction, it does not change the overall $\Delta G^{\circ}$ of the reaction, and, hence, according to the relationship

$$\Delta G^{\circ} = -RT \ln K$$

it cannot change the position of the equilibrium of a reversible reaction. The catalyst increases the velocity of the reverse reaction to the same extent as the forward reaction so that although the equilibrium is reached more quickly in the presence of the catalyst, the equilibrium constant,

$$K = k_{\text{forward}}/k_{\text{reverse}}$$

remains the same and the product yield is not changed.

Catalysis is considered to operate in the following way. The catalyst combines with the reactant known as the *substrate* and forms an intermediate known as a *complex*, which then decomposes to regenerate the catalyst and yield the products. In this way, the catalyst decreases the energy of activation by changing the mechanism of the process, and the rate is accordingly increased.

Alternatively, a catalyst may act by producing free radicals such as $\dot{C}H_3$, which bring about fast *chain reactions*. Chain reactions are reactions consisting of a series of steps involving free atoms or radicals that act as intermediates. The chain reaction is begun by an initiating step and stopped by a chain-breaking or terminating step. Negative catalysts, or inhibitors, frequently serve as chain breakers in such reactions. Antiknock agents act as inhibitors in the explosive reactions attending the combustion of motor fuels.

Catalytic action may be homogeneous or heterogeneous and may occur in either the gaseous or the liquid state. *Homogeneous catalysis* occurs when the catalyst and the reactants are in the same phase. Acid–base catalysis, the most important type of homogeneous catalysis in the liquid phase, will be discussed in some detail in the next section.

*Heterogeneous catalysis* occurs when the catalyst and the reactants form separate phases in the mixture. The catalyst may be a finely divided solid such as platinum or it may be the walls of the container. The catalysis occurs at the surface of the solid and is therefore sometimes known as *contact catalysis*. The reactant molecules are adsorbed at various points or *active centers* on the rough surface of the catalyst. Presumably, the adsorption weakens the bonds of the reactant molecules and lowers the activation energy. The activated molecules then can react, and the products diffuse away from the surface.

Catalysts may be *poisoned* by extraneous substances that are strongly adsorbed at the active centers of the catalytic surface where the reactants would normally be held during reaction. Carbon monoxide is known to poison the catalytic action of copper in the hydrogenation of ethylene. Other substances, known as *promoters*, are found to increase the activity of a catalyst. For example, cupric ions promote the catalytic action of ferric ions in the decomposition of hydrogen peroxide. The exact mechanism of promoter action is not understood, although the promoter is thought to change the properties of the surface so as to enhance the adsorption of the reactants and thus increase the catalytic activity.

### Specific Acid–Base Catalysis

Solutions of a number of drugs undergo accelerated decomposition on the addition of acids or bases. If the drug solution is buffered, the decomposition may not be accompanied by an appreciable change in the concentration of acid or base so that the reaction can be considered to be catalyzed by hydrogen or hydroxyl ions. When the rate law for such an accelerated decomposition is found to contain a term involving the concentration of hydrogen ion or the concentration of hydroxyl

ion, the reaction is said to be subject to *specific acid–base catalysis*.

As an example of specific acid–base catalysis, consider the pH dependence for the hydrolysis of esters. In acidic solution, we can consider the hydrolysis to involve an initial equilibrium between the esters and a hydrogen ion followed by a rate-determining reaction with water, R:

$$S + H^+ \rightleftharpoons SH^+$$

$$SH^+ + R \rightarrow P$$

This general reaction scheme assumes that the products, $P$, of the hydrolysis reaction do not recombine to form ester.

For the generalized reaction, the rate of product formation is given by

$$\frac{dP}{dt} = k[SH^+][R] \qquad (14\text{–}117)$$

The concentration of the conjugate acid, $SH^+$, can be expressed in terms of measurable quantities because the pre-equilibrium requires that

$$K = \frac{[SH^+]}{[S][H^+]} \qquad (14\text{–}118)$$

Thus,

$$[SH^+] = K[S][H^+] \qquad (14\text{–}119)$$

and it follows that

$$\frac{dP}{dt} = kK[S][H^+][R] \qquad (14\text{–}120)$$

Because water, R, is present in great excess, equation (14–120) reduces to the apparent rate law

$$\frac{dP}{dt} = k_1[S][H^+] \qquad (14\text{–}121)$$

where

$$k_1 = kK[R] \qquad (14\text{–}122)$$

The hydrogen ion concentration term in equation (14–121) indicates that the process is a specific hydrogen-ion–catalyzed reaction.

By studying the acid-catalyzed hydrolysis of an ester at various concentrations of hydrogen ion, that is, by hydrolyzing the ester in buffer solutions of differing pH, we can obtain a rate–pH profile for the reaction. At a given pH, an apparent first-order reaction is observed:

$$\frac{dP}{dt} = k_{obs}[S] \qquad (14\text{–}123)$$

where

$$k_{obs} = k_1[H^+] \qquad (14\text{–}124)$$

Taking logarithms of equation (14–124) gives

$$\log k_{obs} = \log[H^+] + \log k_1 \qquad (14\text{–}125)$$

or, equivalently,

$$\log k_{obs} = -(-\log[H^+]) + \log k_1 \qquad (14\text{–}126)$$

We finally arrive at the expression

$$\log k_{obs} = -pH + \log k_1 \qquad (14\text{–}127)$$

Thus, a plot of log $k_{obs}$ against the pH of the solution in which the reaction is run gives a line of slope equal to –1.

Consider now the specific hydroxide-ion–catalyzed decomposition of an ester, S. We can write the general reaction as

$$S + OH^- \rightarrow P$$

and the rate of product, $P$, formation is therefore given by

$$\frac{dP}{dt} = k_2[S][OH^-] \qquad (14\text{–}128)$$

Under buffer conditions, an apparent first-order reaction is again observed:

$$\frac{dP}{dt} = k_{obs}[S] \qquad (14\text{–}129)$$

where now

$$k_{obs} = k_2[OH^-] \qquad (14\text{–}130)$$

or, because

$$K_w = [H^+][OH^-] \qquad (14\text{–}131)$$

$$k_{obs} = \frac{k_2 K_w}{[H^+]} \qquad (14\text{–}132)$$

Taking the logarithm of equation (14–132),

$$\log k_{obs} = -\log[H^+] + \log k_2 K_w \qquad (14\text{–}133)$$

we find that

$$\log k_{obs} = pH + \log k_2 K_w \qquad (14\text{–}134)$$

In this case, a plot of log $k_{obs}$ against pH should be linear with a slope equal to +1.

**Figure 14–8** shows the rate–pH profile for the specific acid–base–catalyzed hydrolysis of methyl-*dl-o*-phenyl-2-piperidylacetate.[22] Note that an increase in pH from 1 to 3 results in a linear decrease in rate, as expected from equation (14–127), for specific hydrogen ion catalysis, whereas

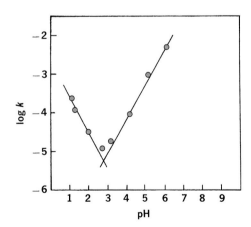

**Fig. 14–8.** Rate–pH profile for the specific acid–base–catalyzed hydrolysis of methyl-*dl-o*-phenyl-2-piperidylacetate. (From S. Siegel, L. Lachmann, and L. Malspeis, J. Pharm. Sci. **48**, 431, 1959. With permission.)

a further increase in pH from about 3 to 7 results in a linear increase in rate, as expected from equation (14–134), for specific hydroxide ion catalysis. Near pH 3, a minimum is observed that cannot be attributed to either hydrogen ion or hydroxyl ion participation in the reaction. This minimum is indicative of a solvent catalytic effect, that is, un-ionized water may be considered as the reacting species. Because of the pH independence of this reaction, the rate law is given by

$$\frac{dP}{dt} = k_0[S] \qquad (14\text{–}135)$$

so that

$$k_{obs} = k_0 \qquad (14\text{–}136)$$

Sometimes a minimum plateau extends over a limited pH region, indicating that solvent catalysis is the primary mode of reaction in this region.

Solvent catalysis may occur simultaneously with specific hydrogen ion or specific hydroxide ion catalysis, especially at pH values that are between the pH regions in which definitive specific ion and solvent catalytic effects are observed. Because each catalytic pathway leads to an increase in the same product, the rate law for this intermediate pH region can be written as:

$$\frac{dP}{dt} = (k_0 + k_1[H^+])[S] \qquad (14\text{–}137)$$

or

$$\frac{dP}{dt} = (k_0 + k_2[OH^-])[S] \qquad (14\text{–}138)$$

depending, respectively, on whether the pH is slightly lower or slightly higher than that for the solvent catalyzed case.

We can now summarize the pH dependence of specific acid–base–catalyzed reactions in terms of the general rate law,

$$\frac{dP}{dt} = (k_0 + k_1[H^+] + k_2[OH^-])[S] \qquad (14\text{–}139)$$

for which

$$k_{obs} = k_0 + k_1[H^+] + k_2[OH^-] \qquad (14\text{–}140)$$

At low pH, the term $k_1[H^+]$ is greater than $k_0$ or $k_2[OH^-]$ because of the greater concentration of hydrogen ions, and specific hydrogen ion catalysis is observed. Similarly, at high pH, at which the concentration of [OH$^-$] is greater, the term $k_2[OH^-]$ outweighs the $k_0$ and $k_1[H^+]$ terms, and specific hydroxyl ion catalysis is observed. When the concentrations of H$^+$ and OH$^-$ are low, or if the products $k_1[H^+]$ and $k_2[OH^-]$ are small in value, only $k_0$ is important, and the reaction is said to be *solvent catalyzed*. If the pH of the reaction medium is slightly acidic so that $k_0$ and $k_1[H^+]$ are important and $k_2[OH^-]$ is negligible, both solvent and specific hydrogen ion catalysis operate simultaneously. A similar result is obtained when the pH of the medium is slightly alkaline, a condition that could allow concurrent solvent and specific hydroxide ion catalysis.

### General Acid–Base Catalysis

In most systems of pharmaceutical interest, buffers are used to maintain the solution at a particular pH. Often, in addition to the effect of pH on the reaction rate, there may be catalysis by one or more species of the buffer components. The reaction is then said to be subject to *general acid* or *general base catalysis* depending, respectively, on whether the catalytic components are acidic or basic.

The rate–pH profile of a reaction that is susceptible to general acid–base catalysis exhibits deviations from the behavior expected on the basis of equations (14–127) and (14–134). For example, in the hydrolysis of the antibiotic streptozotocin, rates in phosphate buffer exceed the rate expected for specific base catalysis. This effect is due to a general base catalysis by phosphate anions. Thus, the alkaline branch of the rate–pH profile for this reaction is a line whose slope is different from 1 (Fig. 14–9).[23]

Other factors, such as ionic strength or changes in the $pK_a$ of a substrate, may also lead to apparent deviations in the rate–pH profile. Verification of a general acid or general base catalysis may be made by determining the rates of degradation of a drug in a series of buffers that are all at the same pH (i.e., the ratio of salt to acid is constant) but that are prepared with an increasing concentration of buffer species. Windheuser and Higuchi,[24] using acetate buffer, found that the degradation of thiamine is unaffected at pH 3.90, where the buffer is principally acetic acid. At higher pH values, however, the rate increases in direct proportion to the concentration of acetate. In this case, acetate ion is the general base catalyst.

Webb et al.[25] demonstrated the general catalytic action of acetic acid, sodium acetate, formic acid, and sodium formate in the decomposition of glucose. The equation for the overall rate of decomposition of glucose in water in the presence

of acetic acid, HAc, and its conjugate base, Ac⁻, can be written as

$$-\frac{dG}{dt} = k_0[G] + k_H[H^+][G] + k_A[HAc][G] + k_{OH}[OH^-][G] + k_B[Ac^-][G] \quad (14\text{–}141)$$

where [G] is the concentration of glucose, $k_0$ is the specific reaction rate in water alone, and the other $k$ values, known as *catalytic coefficients*, represent the specific rates associated with the various catalytic species. The overall first-order rate constant, $k$, which involves all effects, is written as follows:

$$k = \frac{dG/dt}{[G]} = k_0 + k_H[H^+] + k_A[HAc] + k_{OH}[OH^-] + k_B[Ac^-] \quad (14\text{–}142)$$

or, in general,

$$k = k_0 + \Sigma k_i c_i \quad (14\text{–}143)$$

where $c_i$ is the concentration of the catalytic species i, and $k_i$ is the corresponding catalytic coefficient. In reactions in which only specific acid–base effects occur, that is, in which only [H⁺] and [OH⁻] act as catalysts, the equation is

$$k = k_0 + k_H[H^+] + k_{OH}[OH^-] \quad (14\text{–}144)$$

### EXAMPLE 14–12

**Catalytic Coefficient of Glucose Decomposition**

A sample of glucose was decomposed at 140°C in a solution containing 0.030 M HCl. The velocity constant, $k$, was found to be 0.0080 hr⁻¹. If the spontaneous rate constant, $k_0$, is 0.0010 hr⁻¹, compute the catalytic coefficient, $k_H$. The catalysis due to hydroxyl ions in this acidic solution can be considered as negligible. The data are substituted in equation (14–144):

$$0.0080\ hr^{-1} = 0.0010\ hr^{-1} + k_H M^{-1}\ hr^{-1}(0.030)\ M$$

$$k_H = \frac{0.0080\ hr^{-1} - 0.0010\ hr^{-1}}{0.030\ M} = 0.233\ M^{-1}hr^{-1}$$

In 1928, Brönsted[26] showed that a relationship exists between the catalytic power as measured by the catalytic coefficients and the strength of general acids and bases as measured by their dissociation constants. The catalytic coefficient for a weak acid is related to the dissociation constant of the acid by the expression

$$k_A = aK_a^{\alpha} \quad (14\text{–}145)$$

and the corresponding equation for catalysis by a weak base is

$$k_B = bK_a^{-\beta} \quad (14\text{–}146)$$

Here $K_a$ is the dissociation constant of the weak acid, and $a$, $b$, $\alpha$, and $\beta$ are constants for a definite reaction, solvent, and temperature. From this relationship, the catalytic effect of a Brönsted–Lowry acid or base on the specific reaction rate can be predicted if the dissociation constant of the weak electrolyte is known. The relationships in equations (14–145) and (14–146) hold because both the catalytic power and the dissociation constant of a weak electrolyte depend on the ability of a weak acid to donate a proton or a weak base to accept a proton.

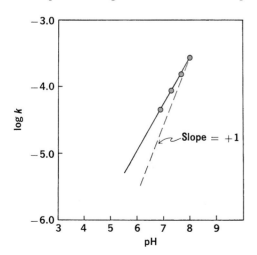

**Fig. 14–9.** Rate–pH profile of a reaction susceptible to general base catalysis. (From E. R. Garrett, J. Pharm. Sci. **49**, 767, 1960. With permission.)

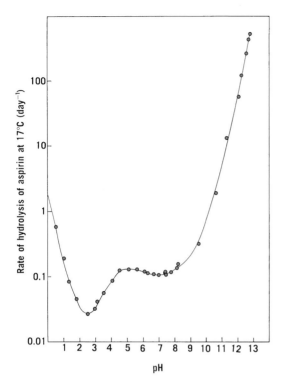

**Fig. 14–10.** Rate–pH profile for the hydrolysis of acetylsalicylic acid at 17°C. (From I. J. Edwards, Trans. Faraday Soc. **46,** 723, 1950.)

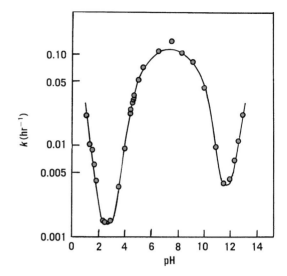

**Fig. 14–11.** The pH profile for the hydrolysis of hydrochlorothiazide. (From J. A. Mollica, C. R. Rohn, and J. B. Smith, J. Pharm. Sci. **58,** 636, 1969. With permission.)

Noncatalytic salts can affect the rate constant directly through their influence on ionic strength as expressed by equation **(14–113)**. Second, salts also affect the catalytic action of some weak electrolytes because, through their ionic strength effect, they change the classic dissociation constant, $K_a$, of equations **(14–145)** and **(14–146)**. These two influences, known respectively as the *primary* and *secondary salt effects*, are handled in a kinetic study by carrying out the reaction under conditions of constant ionic strength, or by obtaining a series of $k$ values at decreasing ionic strengths and extrapolating the results to $\mu = 0$.

An interesting rate–pH profile is obtained for the hydrolysis of acetylsalicylic acid (**Fig. 14–10**). In the range of pH from 0 to about 4, there is clearly specific acid–base catalysis and a pH-independent solvolysis, as first reported by Edwards.[27] Above pH 4, there is a second, pH-independent region, the plateau extending over at least 3 pH units. Fersht and Kirby[28] and others have provided suggestions for the presence of this plateau.

The hydrolysis of hydrochlorothiazide,

was studied by Mollica et al.[29] over a pH range from 1 to 13. The reaction was found to be reversible, the fraction that had reacted at equilibrium, $X_e$, being about 0.4. The pH profile provides a complex curve (**Fig. 14–11**), indicating multiple steps and an intermediate involved in the reaction.

## STABILITY OF PHARMACEUTICALS

### Decomposition and Stabilization of Medicinal Agents

Pharmaceutical decomposition can be classified as hydrolysis, oxidation, isomerization, epimerization, and photolysis, and these processes may affect the stability of drugs in liquid, solid, and semisolid products. Mollica et al.[30] reviewed the many effects that the ingredients of dosage forms and environmental factors may have on the chemical and physical stability of pharmaceutical preparations.

Hou and Poole[31] investigated the kinetics and mechanism of hydrolytic degradation of ampicillin in solution at 35°C and 0.5 ionic strength. The decomposition observed over a pH range of 0.8 to 10.0 followed first-order kinetics and was influenced by both specific and general acid–base catalysis. The pH–rate profile exhibited maximum stability in buffer solutions at pH 4.85 and in nonbuffered solutions at pH 5.85. The degradation rate is increased by the addition of various carbohydrates such as sucrose to the aqueous solution of ampicillin.[32] The Arrhenius plot shows the activation energy, $E_a$, to be 18 kcal/mole at pH 5 for the hydrolysis of ampicillin.

Alcohol is found to slow slow hydrolysis because of the decrease in the dielectric constant of the solvent. The half-life for the degradation of ampicillin in an acidified aqueous solution at 35°C is 8 hr; in a 50% alcohol solution the half-life is 13 hr.

Higuchi et al.[33] reported that chloramphenicol decomposed through hydrolytic cleavage of the amide linkage according to the reaction.

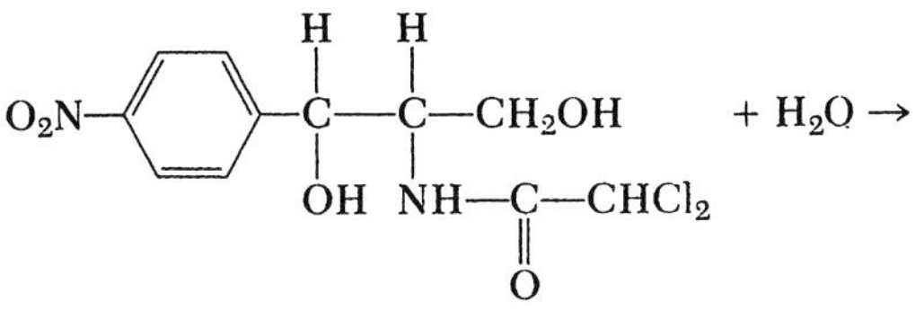

Chloramphenicol

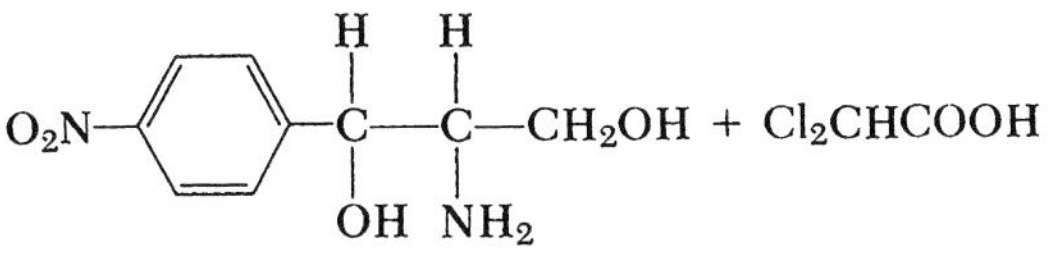

The rate of degradation was low and independent of pH between 2 and 7 but was catalyzed by general acids and bases, including $HPO_4^{2-}$ ions, undissociated acetic acid, and a citrate buffer. Its maximum stability occurs at pH 6 at room temperature, its half-life under these conditions being approximately 3 years. Below pH 2 the hydrolysis of chloramphenicol is catalyzed by hydrogen ions. In alkaline solution the breakdown is affected by both specific and general acid–base catalysis.[34]

The activation energy for the hydrolysis at pH 6 is 24 kcal/mole, and the half-life of the drug at pH 6 and 25°C is 2.9 years.

Beijnen et al.[35] investigated the stability of doxorubicin,

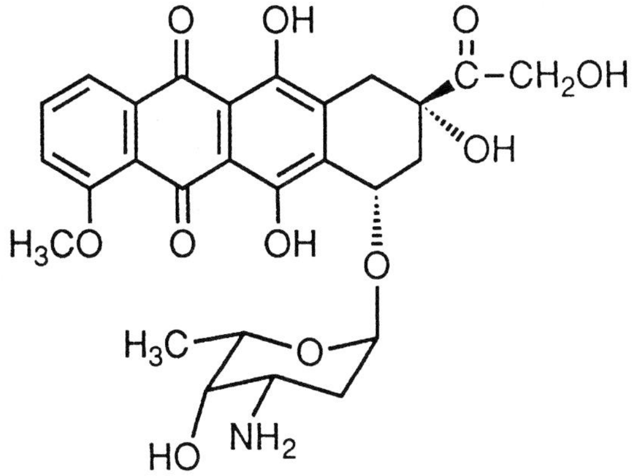

Doxorubicin

in aqueous solution using a stability-indicating high-performance liquid chromatographic assay procedure. Doxorubicin has been used with success against various human neoplasms for the last 20 years. The decomposition of the drug has not been studied in depth because it presents difficulties in analysis. It chelates with metal ions, self-associates in concentrated solutions, adsorbs to surfaces such as glass, and undergoes oxidative and photolytic decomposition.

Beijnen and associates[35] studied the degradation kinetics of doxorubicin as a function of pH, buffer effects, ionic strength, temperature, and drug concentration. The decomposition followed pseudo–first-order kinetics at constant temperature and ionic strength at various pH values. The pH–rate

profile showed maximum stability of the drug at about pH 4.5. Some study was made of the degradation in alkaline solution, other systematic work having been done only with degradation of doxorubicin in acid solution below pH 3.5. Work has also been reported on the stability of doxorubicin infusions used in clinical practice.

Steffansen and Bundgaard[36] studied the hydrolysis of erythromycin and erythromycin esters in aqueous solution:

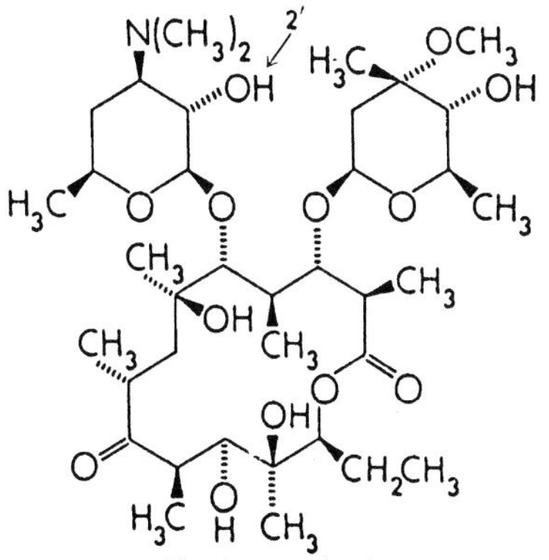

Erythromycin A

Erythromycin is an antibiotic that acts against gram-positive and some gram-negative bacteria. It has the disadvantage of degradation in an acidic environment, as found in the stomach; various methods have been suggested to protect the drug as it passes through the gastrointestinal tract. Most recent among these protective actions is the conversion of erythromycin into esters at the 2' position. These are known as *prodrugs* because they are inactive until erythromycin is released from the esters by enzymatic hydrolysis in the body.

Vinckier et al.[37] studied the decomposition kinetics of erythromycin as a function of buffer type and concentration, ionic strength, pH, and temperature. Erythromycin was found to be most stable in a phosphate buffer and least stable in a sodium acetate buffer. Changes in ionic strength showed only a negligible effect on the kinetics of erythromycin. Log $k$–pH profiles were obtained over the pH range of about 2 to 5 and showed linearity with a slope of approximately 1, indicating specific acid catalysis in the decomposition of erythromycin at 22°C. Specific base catalysis occurs at higher pH values. Erythromycin base is most stable at pH 7 to 7.5.[38]

Atkins et al.[39] also studied the kinetics of erythromycin decomposition in aqueous acidic and neutral buffers. They concluded that pH is the most important factor in controlling the stability of erythromycin A in acidic aqueous solutions.

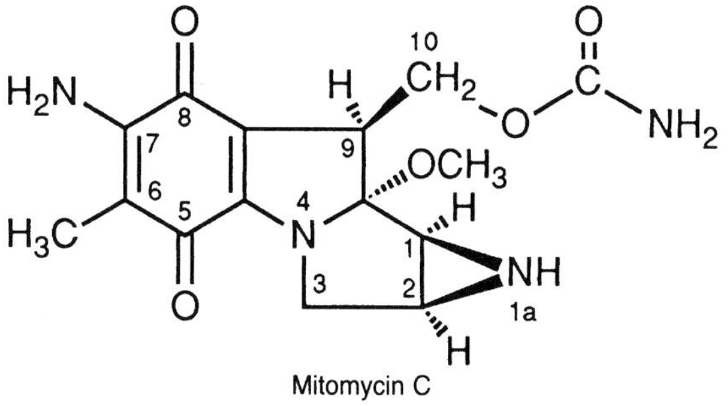

Mitomycin C

The degradation of mitomycin C in acid solution was studied by Beijnen and Underberg.[40] Mitomycin C shows both strong antibacterial and antitumor activity. Degradation in alkaline solution involves the removal of an amino group and replacement by a hydroxyl group, but the breakdown of mitomycin C is more complicated in acid solution, involving ring opening and the formation of two isomers, namely *trans* and *cis* mitosene:

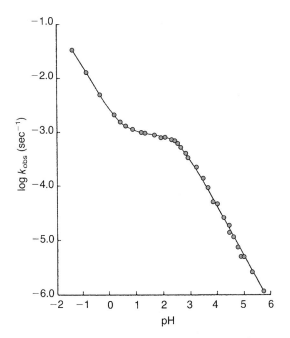

Fig. 14–12. The pH–rate constant profile for mitomycin C decomposition at 20°C. (From J. H. Beijnen and W. J. M. Underberg, Int. J. Pharm. **24,** 219, 1985. With permission.)

To study the mechanism of degradation the authors designed a high-performance liquid chromatographic assay that allows quantitative separation of the parent drug and its decomposition products. The kinetics of mitomycin C in acid solution was studied at 20°C. To obtain pH values below 3, the solutions were acidified with aqueous perchloric acid, and for the pH range of 3 to 6, they were buffered with an acetic acid–acetate buffer. The degradation of mitomycin C shows first-order kinetics over a period of more than three half-lives.

The influence of pH and buffer species on the decomposition of mitomycin C is expressed as

$$k = k_0 + k_H[H^+] + k_A[HAc] + k_B[Ac^-] \qquad (14\text{--}147)$$

where $k_0$ is the first-order constant for decomposition in water alone and $k_H$ is a second-order rate constant (catalytic coefficient) associated with catalysis due to the $[H^+]$. The second-order rate constants $k_A$ and $k_B$ are catalytic coefficients for catalysis by the buffer components, $[HAc]$ and $[Ac^-]$, respectively [equation (14–142)]. The term $k_{OH}[OH^-]$ is neglected because this study is conducted only in the acid region of the pH scale.

The log(rate constant)–pH profile for the decomposition of mitomycin C at 20°C is shown in **Figure 14–12**. In other work, Beijnen and associates[40] showed that the inflection point in the curve is associated with the $pK_a = 2.6$ for mito-

mycin C. The straight-line portions of the curve, that is, below pH = 0 and above pH = 3, both exhibit slopes of approximately –1. Slopes of –1 in this region of the profile are an indication of specific acid catalysis for decomposition of the neutral form of mitomycin C (MMC) and for the protonated form (MMCH$^+$).

Procaine decomposes mainly by hydrolysis, the degradation being due primarily to the breakdown of the uncharged and singly charged forms.[13] The reaction of procaine is catalyzed by hydrogen and hydroxyl ions. Both the free base and the protonated form are subject to specific base catalysis. Marcus and Baron[17] obtained an activation energy, $E_a$, of 16.8 kcal/mole for procaine at 97.30°C. Garrett[41] reviewed the degradation and stability of procaine.

Triamcinolone acetonide

Triamcinolone acetonide, a glucocorticoid (adrenal cortex) hormone, is a potent anti-inflammatory agent when applied topically as a cream or suspension. Gupta[42] studied the stability of water–ethanol solutions at various pH values, buffer

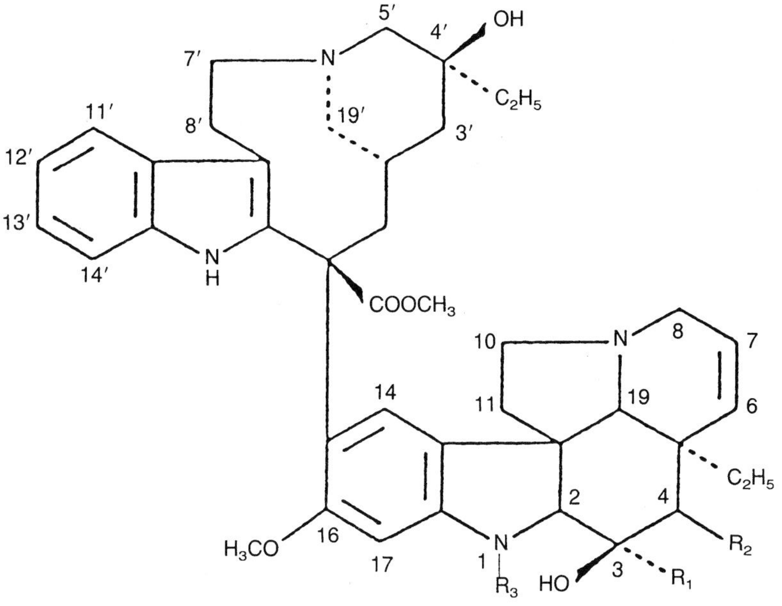

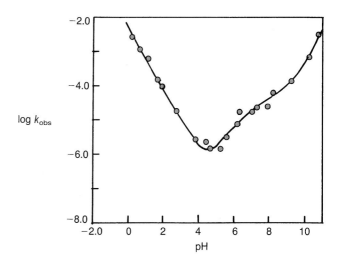

|            | $R_1$   | $R_2$   | $R_3$ |
|------------|---------|---------|-------|
| VINBLASTINE | COOCH$_3$ | OCOCH$_3$ | CH$_3$ |
| VINCRISTINE | COOCH$_3$ | OCOCH$_3$ | CHO |
| VINDESINE | CONH$_2$ | OH | CH$_3$ |

**Fig. 14–13.** Chemical structures of the closely related antineoplastic agents vinblastine and vincristine, isolated from *Vinca rosea*, and vindesine, a synthetic derivative of vinblastine. (From D. Vendrig, J. H. Beijnen, O. van der Houwen, and J. Holthuis, Int. J. Pharm, **50**, 190, 1989. With permission.)

concentrations, and ionic strengths. The decomposition of triamcinolone acetonide followed first-order kinetics, the rate constant, $k_{obs}$, varying with the pH of phosphate, sodium hydroxide, and hydrochloric acid buffer solutions. The optimum pH for stability was found from a pH–rate profile to be about 3.4 and to be related to the concentration of the phosphate buffer. In the hydrochloric acid buffer solution, triamcinolone acetonide underwent hydrolysis to form triamcinolone and acetone. A study of the reaction in solvents of varying ionic strength showed that log $k_{obs}$ decreased linearly with increasing values of $\sqrt{\mu}$, suggesting that reaction occurs between the protonated, [H$^+$], form of the drug and the phosphate buffer species, H$_2$PO$_4^-$/HPO$_4^{2-}$.

Vincristine and vinblastine are natural alkaloids used as cytotoxic agents in cancer chemotherapy (**Fig. 14–13**). Vendrig et al.[43] investigated the degradation kinetics of vincristine sulfate in aqueous solution within the pH range of −2.0 to 11 at 80°C. The drug exhibited first-order kinetics under these conditions; the rate constant, $k_{obs}$, was calculated using the first-order equation [equation (**14–14**)] at various pH values to plot the pH profile as seen in **Figure 14–14**. The degradation rates were found to be independent of buffer concentration and ionic strength within the pH range investigated. Vincristine appears to be most stable in aqueous solution between pH 3.5 and 5.5 at 80°C.

The effect of temperature on the degradation of vincristine at various pH values from 1.2 to 8.2 and within the temperature range of 60°C to 80°C was assessed using the Arrhenius equation [equation (**14–72**) or (**14–73**)]. The

**Fig. 14–14.** Log $k$–pH profile for the decomposition of vincristine. (From D. Vendrig, J. H. Beijnen, O. van der Houwen, and J. Holthuis, Int. J. Pharm. **50**, 194, 1989. With permission.)

**TABLE 14–6**

**ACTIVATION ENERGIES AND ARRHENIUS FACTORS FOR VINCRISTINE AT VARIOUS pH VALUES AT 80°C***

| pH | $E_a$ (cal/mole $\times 10^{-4}$) | $A$ (sec$^{-1}$) |
|-----|-----|-----|
| 1.2 | 1.482 | $1 \times 10^6$ |
| 3.5 | 2.008 | $9 \times 10^6$ |
| 5.2 | 1.745 | $4 \times 10^5$ |
| 7.0 | 2.534 | $9 \times 10^{10}$ |
| 8.2 | 2.773 | $9 \times 10^{12}$ |

*Based on D. E. M. M. Vendrig, J. H. Beijnen, O. A. G. J. van der Houwen, and J. J. M. Holthuis, Int. J. Pharm. **50**, 189, 1989.

activation energy, $E_a$, and the Arrhenius factor, $A$, are given in **Table 14–6**.

**EXAMPLE 14–13**

**Vincristine**

Vendrig et al.[43] listed the activation energies in kJ/mole for vincristine from pH 1.2 to 8.2. Convert the following values for $E_a$ to quantities expressed in cal/mole, as found in Table 14–6:

| pH | 1.2 | 3.5 | 5.2 | 7.0 | 8.2 |
|-----|-----|-----|-----|-----|-----|
| $E_a$(kJ/mole) | 62 | 84 | 73 | 106 | 116 |

The conversion of units is obtained by writing a sequence of ratios so as to change SI to cgs units. For the first value, that of $E_a$ at pH 1.2,

$$62 \frac{kJ}{mole} \times \frac{1000\ J}{kJ} \times \frac{10^7\ ergs}{J} \times \frac{1\ cal}{4.184 \times 10^7\ ergs}$$

or

$$62\ mole^{-1} \times 1000 \times 10^7 \times (1\ cal/4.184 \times 10^7) = 14818\ cal/mole$$

or

$$E_a = 1.4818 \times 10^4\ cal/mole = 15\ kcal/mole$$

In the *CRC Handbook of Chemistry and Physics*, we find the conversion factor 1 joule = 0.239045 cal; therefore, we can make the direct conversion

$$62000\ joules/mole \times 0.239045\ cal/joule = 14821\ cal/mole$$

or

$$E_a = 1.4821 \times 10^4\ cal/mole$$

The kinetic study of the autoxidation of ascorbic acid is an interesting research story that began about 50 years ago. Some of the reports are reviewed here as an illustration of the difficulties encountered in the study of free radical reactions. Although the decomposition kinetics of ascorbic acid probably has been studied more thoroughly than that of any other drug, we are only now beginning to understand the

mechanism of the autoxidation. The overall reaction can be represented as

Ascorbic acid

Dehydroascorbic acid

One of the first kinetic studies of the autoxidation of ascorbic acid to dehydroascorbic acid was undertaken in 1936 by Barron et al.[44] These investigators measured the oxygen consumed in the reaction, using a Warburg type of vessel and a manometer to obtain the rate of decomposition of ascorbic acid. They found that when great care was taken to free the solution of traces of copper, ascorbic acid was not oxidized by atmospheric oxygen at a measurable rate except in alkaline solutions. Cupric ion was observed to oxidize ascorbic acid rapidly to dehydroascorbic acid, and KCN and CO were found to break the reaction chain by forming stable complexes with copper.

Dekker and Dickinson[45] suggested a scheme for oxidation of ascorbic acid by the cupric ion and obtained the following equations for the decomposition:

$$-\frac{d[H_2A]}{dt} = k \frac{[Cu^{2+}][H_2A]}{[H^+]^2} \tag{14–148}$$

and in the integrated form,

$$k = \frac{2.303[H^+]^2}{[Cu^{2+}]t} \log \frac{[H_2A]_0}{[H_2A]} \tag{14–149}$$

where $[H_2 A]_0$ is the initial concentration and $[H_2A]$ is the concentration of ascorbic acid at time $t$. The experimental results compared favorably with those calculated from equation (14–149), and it was assumed that the initial reaction involved a slow oxidation of the ascorbate ion by cupric ion to a semiquinone, which was immediately oxidized by oxygen to dehydroascorbic acid. As the reaction proceeded, however, the specific reaction rate, $k$, was found to increase gradually.

Dekker and Dickinson[45] observed that the reaction was retarded by increasing the initial concentration of ascorbic acid, presumably because ascorbic acid depleted the free oxygen. When oxygen was continually bubbled through the mixture, the specific rate of decomposition did not decrease with increasing ascorbic acid concentration.

Weissberger et al.[46] showed that the autoxidation of ascorbic acid involved both a singly and a doubly charged anion of L-ascorbic acid. Oxygen was found to react with the divalent ion at atmospheric pressure about $10^5$ times as fast as with the monovalent ion of the acid at ordinary temperatures

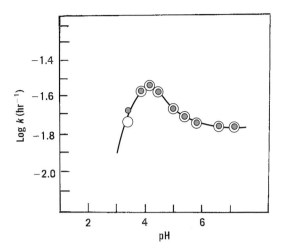

**Fig. 14–15.** The pH profile for the oxidative degradation of ascorbic acid. Key: ● = calculated rate constant; ○ = rate constant extrapolated to zero buffer concentration where only the effect of hydrogen and/or hydroxyl ions is accounted for. (From S. M. Blaug and B. Hajratwala, J. Pharm. Sci. **61**, 556, 1972; **63**, 1240, 1974. With permission.)

when metal catalysis was repressed. When copper ions were added to the reaction mixture, however, it was found that only the singly charged ion reaction was catalyzed. Copper was observed to be an extremely effective catalyst because $2 \times 10^{-4}$ mole/liter increased the rate of the monovalent ion reaction by a factor of 10,000.

Nord[47] showed that the rate of the copper-catalyzed autoxidation of ascorbic acid was a function of the concentrations of the monovalent ascorbate anion, the cuprous ion, the cupric ion, and the hydrogen ion in the solution. The kinetic scheme proposed by Nord appears to compare well with experimental findings.

Blaug and Hajratwala[48] observed that ascorbic acid degraded by aerobic oxidation according to the log(rate constant)–pH profile of **Figure 14–15**. The effects of buffer species were eliminated so that only the catalysis due to hydrogen and hydroxyl ions was considered. Dehydroascorbic acid, the recognized breakdown product of ascorbic acid, was found to decompose further into ketogulonic acid, which then formed threonic and oxalic acids.

According to Rogers and Yacomeni,[49] ascorbic acid exhibits maximum degradation at pH 4 and minimum degradation at pH 5.6 in citric acid–phosphate buffers in the presence of excess oxygen at 25°C. The pH–rate profile can be fit closely to the experimental points using first- and second-order rate constants $k_1 = 5.7 \times 10^{-6}$ $M^{-1}$ $sec^{-1}$, $k_2 = 1.7$ $sec^{-1}$, and $k_3 = 7.4 \times 10^{-5}$ $M^{-1}$ $sec^{-1}$ in the rate expression

$$k = k_1[H^+] + k_2 + k_3[OH^-] \qquad (14\text{-}150)$$

where $k_2$ is the first-order solvent catalysis term, ordinarily written as $k_0$, and $k_1$ and $k_3$ are the catalytic coefficients.

Takamura and Ito[50] studied the effect of metal ions and flavonoids on the oxidation of ascorbic acid, using polarography at pH 5.4. Transition metal ions increased the rate of

first-order oxidation; the rate was increased by 50% in the presence of $Cu^{2+}$. Flavonoids are yellow pigments found in higher plants. The flavonoid constituents rutin and hesperidan were used in the past to reduce capillary fragility and bleeding.[51] Takamura and Ito[50] found that flavonoids inhibited the $Cu^{2+}$-catalyzed oxidation in the following order of effectiveness: 3-hydroxyflavone < rutin < quercitin. This order of inhibition corresponded to the order of complexation of $Cu^{2+}$ by the flavonoids, suggesting that the flavonoids inhibit $Cu^{2+}$-catalyzed oxidation by tying up the copper ion in solution.

Oxidation rates under conditions similar to those in pharmaceutical systems were examined by Fyhr and Brodin.[52] They investigated the iron-catalyzed oxidation of ascorbic acid at 35°C at pH values of 4 to 6 and partial pressures of oxygen of 21 kilopascal (kPa) and at iron concentrations between 0.16 and 1.25 ppm. These workers found the oxidation of ascorbic acid to be first order with respect to the total ascorbic acid concentration. Trace-element analysis was used to follow changes in iron concentration.

Akers[53] studied the *standard oxidation potentials* of antioxidants in relation to stabilization of epinephrine in aqueous solution. He found that ascorbic acid or a combination of 0.5% thiourea with 0.5% acetylcysteine was the most effective in stabilizing parenteral solutions of epinephrine.

Thoma and Struve[54] attempted to protect epinephrine solutions from oxidative degradation by the addition of redox stabilizers (antioxidants) such as ascorbic acid. Sodium metabisulfite, $Na_2S_2O_5$, prevented discoloration of epinephrine solutions but improved the stability only slightly. The best stabilization of epinephrine in solution was provided by the use of nitrogen.

The decomposition of a new antiasthmatic agent (2-[(4-hydroxyphenyl)amino]-5-methoxybenzenemethanol or HPAMB), which acts therapeutically by contraction of vascular and pulmonary smooth muscles, was investigated in the presence and absence of the antioxidant ascorbic acid in phosphate buffer (pH 7.9) and in aqueous solution (pH 7.1).[55] As shown in **Figure 14–16**, the drug broke down

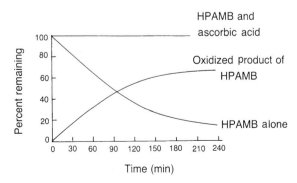

**Fig. 14–16.** Decomposition of HPAMB alone and in the presence of ascorbic acid. The curve for the oxidized product resulting from HPAMB breakdown is also shown. (From A. B. C. Yu and G. A. Portman, J. Pharm. Sci. **79**, 913, 1990. With permission.)

rapidly at 25°C in water in the absence of ascorbic acid, whereas no loss in drug concentration occurred in the presence of 0.1% ascorbic acid. In two nonaqueous solvents, ethanol and dimethyl sulfoxide, the oxidative decomposition rate of HPAMB was much slower than in aqueous solution.

## Photodegradation

Light energy, like heat, may provide the activation necessary for a reaction to occur. Radiation of the proper frequency and of sufficient energy must be absorbed to activate the molecules. The energy unit of radiation is known as the *photon* and is equivalent to one *quantum* of energy. Photochemical reactions do not depend on temperature for activation of the molecules; therefore, the rate of activation in such reactions is independent of temperature. After a molecule has absorbed a quantum of radiant energy, however, it may collide with other molecules, raising their kinetic energy, and the temperature of the system will therefore increase. The initial photochemical reaction may often be followed by thermal reactions.

The study of photochemical reactions requires strict attention to control of the wavelength and intensity of light and the number of photons actually absorbed by the material. Reactions that occur by photochemical activation are usually complex and proceed by a series of steps. The rates and mechanisms of the stages can be elucidated through a detailed investigation of all factors involved, but in this elementary discussion of the effect of light on pharmaceuticals, we will not go into such considerations.

Examples of photochemical reactions of interest in pharmacy and biology are the irradiation of ergosterol and the process of photosynthesis. When ergosterol is irradiated with light in the ultraviolet region, vitamin D is produced. In photosynthesis, carbon dioxide and water are combined in the presence of a photosensitizer, chlorophyll. Chlorophyll absorbs visible light, and the light then brings about the photochemical reaction in which carbohydrates and oxygen are formed.

Some studies involving the influence of light on medicinal agents are reviewed here.

Moore[56] described the kinetics of photooxidation of benzaldehyde as determined by measuring the oxygen consumption with a polarographic oxygen electrode. Photooxidation of drugs is initiated by ultraviolet radiation according to one of two classes of reactions. The first is a free radical chain process in which a sensitizer, for example, benzophenone, abstracts a hydrogen atom from the drug. The free radical drug adds a molecule of oxygen, and the chain is propagated by removing a hydrogen atom from another molecule of oxidant, a hydroperoxide, which may react further by a nonradical mechanism. The scheme for initiation, propagation, and termination of the chain reaction is shown in **Figure 14–17**.

The second class of photooxidation is initiated by a dye such as methylene blue.

A manometer is usually used to measure the rate of absorption of oxygen from the gas phase into a stirred solution of the

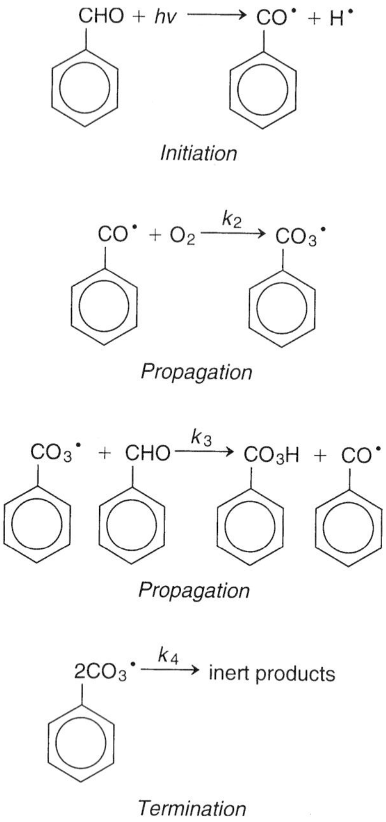

**Fig. 14–17.** Steps in the photooxidation of benzaldehyde. (From D. E. Moore, J. Pharm. Sci. **65**, 1449, 1976. With permission.)

oxidizing drug. In some cases, as in the oxidation of ascorbic acid, spectrophotometry may be used if the absorption spectra of the reactant and product are sufficiently different. An oxygen electrode or galvanic cell oxygen analyzer has also been used to measure the oxygen consumption.

Earlier studies of the photooxidation of benzaldehyde in *n*-decane solution showed that the reaction involved a free radical mechanism. Moore proposed to show whether a free radical process also occurred in a dilute aqueous solution and to study the antioxidant efficiency of some polyhydric phenols. The photooxidation of benzaldehyde was found to follow a free radical mechanism, and efficiency of the polyhydric phenolic antioxidants ranked as follows: catechol > pyrogallol > hydroquinone > resorcinol > *n*-propyl gallate. These antioxidants could be classified as retarders rather than inhibitors because they slowed the rate of oxidation but did not inhibit the reaction.

Asker et al.[57] investigated the photostabilizing effect of DL-methionine on ascorbic acid solution. A 10-mg% concentration of DL-methionine was found to enhance the stability of a 40-mg% solution of ascorbic acid buffered by phosphate but not by citrate at pH 4.5.

Uric acid was found to produce a photoprotective effect in buffered and unbuffered solutions of sulfathiazole sodium.[58] The addition of 0.1% sodium sulfite assisted in preventing the discoloration of the sulfathiazole solution prepared in either a borate or a phosphate buffer.

Furosemide (Lasix) is a potent diuretic, available as tablets and as a sterile solution for injection. It is fairly stable in alkaline solution but degrades rapidly in acid solution.

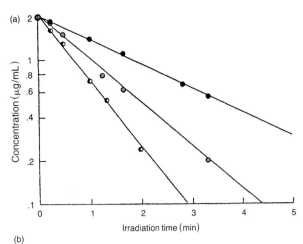

Furosemide

Irradiation of furosemide with 365 nm of ultraviolet light in alkaline solutions and in methanol results in photooxidation and reduction, respectively, to yield a number of products. The drug is relatively stable in ordinary daylight or under fluorescent (room) lighting but has a half-life of only about 4 hr in direct sunlight. Bundgaard et al.[59] discovered that it is the un-ionized acid form of furosemide that is most sensitive to photodegradation. In addition to investigating the photoliability of furosemide, these workers also studied the degradation of the ethyl, dimethylglycolamide, and diethylglycolamide esters of furosemide and found them to be very unstable in solutions of pH 2 to 9.5 in both daylight and artificial room lighting. The half-lives of photodegradation for the esters were 0.5 to 1.5 hr.

Andersin and Tammilehto[60] noted that apparent first-order photokinetics had been shown by other workers for adriamycin, furosemide, menadione, nifedipine, sulfacetamide, and theophylline. Photodegradation of the tromethamine (TRIS buffer, aminohydroxymethylpropanediol) salt of ketorolac, an analgesic and anti-inflammatory agent, appeared in ethanol to be an exception[60]; it showed apparent first-order kinetics at low concentrations, 2.0 $\mu$g/mL or less, of the drug (Fig. 14–18a). When the concentration of ketorolac tromethamine became 10 $\mu$g/mL or greater, however, the kinetics exhibited non–first-order rates. That is, the plots of drug concentration versus irradiation time were no longer linear but rather were bowed at these higher concentrations (Fig. 14–18b).[61]

Nifedipine is a calcium antagonist used in coronary artery disease and in hypertension; unfortunately, it is sensitive to light both in solution and in the solid state. Matsuda et al.[62] studied the photodegradation of nifedipine in the solid state when exposed to the radiation of mercury vapor and fluorescent light sources. The drug decomposed into four compounds, the main photoproduct being a nitrosopyridine. It readily degraded in ultraviolet and visible light, with maximum decomposition occurring at a wavelength of about 380 nm ($3.80 \times 10^{-7}$ m). The rate of degradation of nifedipine was much faster when exposed to a mercury vapor lamp than when subjected to the rays of a fluorescent lamp; however, the degradation in the presence of both light sources exhibited first-order kinetics. The drug is more sensitive to light when in solution. The photodecomposition of nifedipine

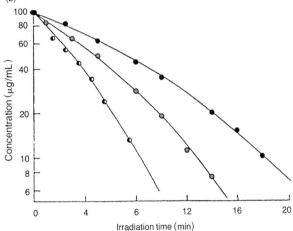

Fig. 14–18. A semilogarithmic plot of the photolysis of ketorolac tromethamine in ethyl alcohol. Key: ◐ = under argon; ○ = under air; ◑ = under oxygen. (a) At low drug concentrations; (b) at high drug concentrations. (From L. Gu, H. Chiang, and D. Johnson, Int. J. Pharm. **41**, 109, 1988. With permission.)

in the crystalline solid state was found to be directly related to the *total irradiation intensity*. The total intensity was used as a convenient parameter to measure accelerated photodecomposition of nifedipine in the solid state and thus to estimate its photostability under ordinary conditions of light irradiation.

The photosensitivity of the dye FD&C Blue No. 2 causes its solution to fade and gradually to become colorless. Asker and Collier[63] studied the influence of an ultraviolet absorber, uric acid, on the photostability of FD&C Blue No. 2 in glycerin and triethanolamine. They found that the greater the concentration of uric acid in triethanolamine, the more photoprotection was afforded the dye. Glycerin was not a suitable solvent for the photoprotector because glycerin accelerates the rate of color fading, possibly owing to its dielectric constant effect.

As would be expected for a reaction that is a function of light radiation and color change rather than concentration, these reactions follow zero-order kinetics. Photodegradation reactions of chlorpromazine, menadione, reserpine, and colchicine are also kinetically zero order.

Asker and Colbert[64] assessed the influence of various additives on the photostabilizing effect that uric acid has on solutions of FD&C Blue No. 2. The agents tested for their synergistic effects belong to the following classes: antioxidants, chelating agents, surfactants, sugars, and preservatives. It was found that the antioxidants DL-methionine and DL-leucine accelerated the photodegradation of the FD&C Blue No. 2 solutions. The addition of the surfactant Tween 80 (polysorbate 80) increased the photodegradation of the dye, as earlier reported by Kowarski[65] and other workers. Lactose has been shown by these authors and others to accelerate the color loss of FD&C Blue No. 2, and the addition of uric acid retards the photodegradation caused by the sugar. Likewise, methylparaben accelerates the fading of the blue color, and the addition of uric acid counteracts this color loss. Chelating agents, such as disodium edetate (EDTA disodium), significantly increased the rate of color loss of the dye. EDTA disodium has also been reported to increase the rate of degradation of epinephrine, physostigmine, and isoproterenol, and it accelerates the photodegradation of methylene blue and riboflavine. Acids, such as tartaric and citric, tend to increase the fading of dye solutions.

Asker and Jackson[66] found a photoprotective effect by dimethyl sulfoxide on FD&C Red No. 3 solutions exposed to long- and short-wave ultraviolet light. Fluorescent light was more detrimental to photostability of the dye solution than were the ultraviolet light sources.

## Accelerated Stability and Stress Testing

The Federal Food, Drug, and Cosmetic Act requires that manufacturers establish controls for the manufacture, processing, packing, and holding of drug products to ensure their safety, identity, strength, quality, and purity [§501(a)(2)(B)]. Requirements for these controls, also known as current good manufacturing practices, are established and monitored by the Food and Drug Administration (FDA). Stability studies should include testing of those attributes of the drug substance or drug product that are susceptible to change during storage and are likely to influence quality, safety, and/or efficacy. The testing should cover, as appropriate, the physical, chemical, biologic, and microbiologic attributes, preservative content (e.g., antioxidant, antimicrobial preservative), and function-

ality tests (e.g., for a dose delivery system). As part of the current good manufacturing practice regulations, the FDA requires that drug products bear an expiration date determined by appropriate stability testing (21 Code of Federal Regulations 211.137 and 211.166). The stability of drug products needs to be evaluated over time in the same container-closure system in which the drug product is marketed. In some cases, accelerated stability studies can be used to support tentative expiration dates in the event that full shelf-life studies are not available. When a manufacturer changes the packaging of a drug product (e.g., from a bottle to unit dose), stability testing must be performed on the product in its new packaging, and expiration dating must reflect the results of the new stability testing. Accelerated stability studies are designed to increase the rate of chemical degradation or physical change of a drug substance or drug product by using exaggerated storage conditions as part of the formal stability studies. Data from these studies, in addition to long-term stability studies, can be used to assess longer-term chemical effects at nonaccelerated conditions and to evaluate the effect of short-term excursions outside the label storage conditions such as might occur during shipping. Results from accelerated testing studies are not always predictive of physical changes. Stress testing of the drug substance or drug product can help identify the likely degradation products, which in turn can help establish the degradation pathways and the intrinsic stability of the molecule and validate the stability-indicating power of the analytical procedures used. The nature of the stress testing will depend on the individual drug substance and the type of drug product involved.

The method of accelerated testing of pharmaceutical products based on the principles of chemical kinetics was demonstrated by Garrett and Carper.[3] According to this technique, the $k$ values for the decomposition of a drug in solution at various elevated temperatures are obtained by plotting some function of concentration against time, as shown in **Figure 14–19** and already discussed in the early sections of this chapter. The logarithms of the specific rates of decomposition are then plotted against the reciprocals of the absolute temperatures as shown in **Figure 14–20**, and the resulting line is extrapolated to room temperature. The $k_{25}$ is used to obtain a measure of the stability of the drug under ordinary shelf conditions.

## ▅▅▅▅KEY CONCEPT▅ STRESS TESTING

Stress testing to elucidate the intrinsic stability of the *drug substance* is part of the development strategy and is normally carried out under more severe conditions than those used for accelerated testing. The testing typically includes the effects of temperature [in 10°C increments (e.g., 50°C–60°C) above that for accelerated testing], humidity (e.g., 75% relative humidity or greater) where appropriate, oxidation, and photolysis on the drug substance.

Stress testing of the *drug product* is undertaken to assess the effect of severe conditions on the drug product. Such studies include photostability testing and specific testing of certain products (e.g., metered-dose inhalers, creams, emulsions, refrigerated aqueous liquid products).

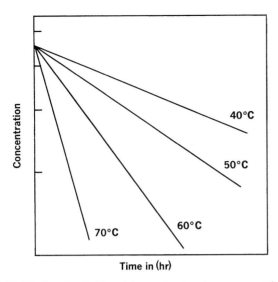

**Fig. 14–19.** Accelerated breakdown of a drug in aqueous solution at elevated temperature.

## EXAMPLE 14–14

**Expiration Dating**

The initial concentration of a drug decomposing according to first-order kinetics is 94 units/mL. The specific decomposition rate, $k$, obtained from an Arrhenius plot is $2.09 \times 10^{-5}$ hr$^{-1}$ at room temperature, 25°C. Previous experimentation has shown that when the concentration of the drug falls below 45 units/mL it is not sufficiently potent for use and should be removed from the market. What expiration date should be assigned to this product?
We have

$$t = \frac{2.303}{k} \log \frac{c_0}{c}$$

$$t = \frac{2.303}{2.09 \times 10^{-5}} \log \frac{94}{45} = 3.5 \times 10^4 \text{ hr} \cong 4 \text{ years}$$

Free and Blythe and, more recently, Amirjahed[67] and his associates suggested a similar method in which the fractional life period (*Example 14–2*) is plotted against reciprocal temperatures and the time in days required for the drug to decompose to some fraction of its original potency at room temperature is obtained. The approach is illustrated in **Figures**

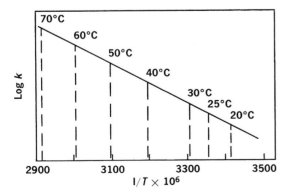

**Fig. 14–20.** Arrhenius plot for predicting drug stability at room temperatures.

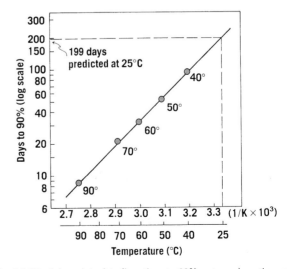

**Fig. 14–21.** Time in days required for drug potency to fall to 90% of original value. These times, designated $t_{90}$, are then plotted on a log scale in **Figure 14–22**.

14–21 and 14–22. As observed in **Figure 14–21**, the log percent of drug remaining is plotted against time in days, and the time for the potency to fall to 90% of the original value (i.e., $t_{90}$) is read from the graph. In **Figure 14–22**, the log time to 90% is then plotted against $1/T$, and the time at 25°C gives the shelf life of the product in days. The decomposition data illustrated in **Figure 14–21** result in a $t_{90}$ value of 199 days. Shelf life and expiration dates are estimated in this way; Baker and Niazi[68] pointed out limitations of the method.

By either of these methods, the *overage*, that is, the excess quantity of drug that must be added to the preparation to maintain at least 100% of the labeled amount during the expected

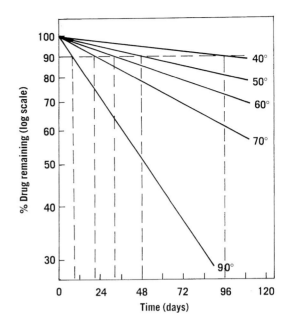

**Fig. 14–22.** A log plot of $t_{90}$ (i.e., time to 90% potency) on the vertical axis against reciprocal temperature (both Kelvin and centigrade scales are shown) on the horizontal axis.

shelf life of the drug, can be easily calculated and added to the preparation at the time of manufacture.

An improved approach to stability evaluation is that of nonisothermal kinetics, introduced by Rogers[69] in 1963. The activation energy, reaction rates, and stability predictions are obtained in a single experiment by programming the temperature to change at a predetermined rate. Temperature and time are related through an appropriate function, such as

$$1/T = 1/T_0 + at \qquad (14\text{--}151)$$

where $T_0$ is the initial temperature and $a$ is a reciprocal heating rate constant. At any time during the run, the Arrhenius equation for time zero and time $t$ can be written as

$$\ln k_t = \ln k_0 - \frac{E_a}{R}\left(\frac{1}{T_t} - \frac{1}{T_0}\right) \qquad (14\text{--}152)$$

and substituting **(14–151)** into **(14–152)** yields

$$\ln k_t = \ln k_0 - \frac{E_a}{R}at \qquad (14\text{--}153)$$

Because temperature is a function of the time, $t$, a measure of stability, $k_t$, is directly obtained over a range of temperatures. A number of variations have been made on the method,[70–73] and it is now possible to change the heating rate during a run or combine a programmed heating rate with isothermal studies and receive printouts of activation energy, order of reaction, and stability estimates for projected times and at various temperatures.

Although kinetic methods need not involve detailed studies of mechanism of degradation in the prediction of stability, they do demand the application of sound scientific principles if they are to be an improvement over extended room-temperature studies. Furthermore, before an older method, although somewhat less than wholly satisfactory, is discarded, the new technique should be put through a preliminary trial period and studied critically. Some general precautions regarding the use of accelerated testing methods are appropriate at this point.

In the first place, it should be reemphasized that the results obtained from a study of the degradation of a particular component in a vehicle cannot be applied arbitrarily to other liquid preparations in general. As Garrett[74] pointed out, however, once the energy of activation is known for a component, it probably is valid to continue to use this value although small changes of concentration (e.g., addition of overage) or slight formula changes are made. The known activation energy and a single-rate study at an elevated temperature may then be used to predict the stability of that component at ordinary temperatures.

Testing methods based on the Arrhenius law are valid only when the breakdown is a thermal phenomenon with an activation energy of about 10 to 30 kcal/mole. If the reaction rate is determined by diffusion or photochemical reactions, or if the decomposition is due to freezing, contamination by microorganisms, excessive agitation during transport, and so

on, an elevated temperature study is obviously of little use in predicting the life of the product. Nor can elevated temperatures be used for products containing suspending agents such as methylcellulose that coagulate on heating, proteins that may be denatured, and ointments and suppositories that melt under exaggerated temperature conditions. Emulsion breaking involves aggregation and coalescence of globules, and some emulsions are actually more stable at elevated temperatures at which Brownian movement is increased. Lachman et al.[75] reviewed the stability testing of emulsions and suspensions and the effects of packaging on the stability of dosage forms.

Statistical methods should be used to estimate the errors in rate constants, particularly when assays are based on biologic methods; this is accomplished by the method of least squares as discussed by Garrett[74] and Westlake.[76]

The investigator should be aware that the order of a reaction may change during the period of the study. Thus, a zero-order degradation may subsequently become first order, second order, or fractional order, and the activation energy may also change if the decomposition proceeds by several mechanisms. At certain temperatures, autocatalysis (i.e., acceleration of decomposition by products formed in the reaction) may occur so as to make room-temperature stability predictions from an elevated-temperature study impractical.

In conclusion, the investigator in the product development laboratory must recognize the limitations of accelerated studies, both the classic and the more recent kinetic type, and must distinguish between those cases in which reliable prediction can be made and those in which, at best, only a rough indication of product stability can be obtained. Where accelerated methods are not applicable, extended aging tests must be employed under various conditions to obtain the desired information.

## Containers and Closures

The information for this section is largely taken from the FDA Guidances for Containers and Closures. The interested student should refer to the specific guidances for additional information. A container closure or packaging system refers to the sum of packaging components that together contain and protect the dosage form. This includes primary packaging components and secondary packaging components, if the latter are intended to provide additional protection to the drug product. Packaging components are typically made from glass, high-density polyethylene resin, metal, or other materials. Typical components are containers (e.g., ampules, vials, bottles), container liners (e.g., tube liners), closures (e.g., screw caps, stoppers), closure liners, stopper overseals, container inner seals, administration ports (e.g., on large-volume parenterals), overwraps, administration accessories, and container labels. A package or market package refers to the container closure system and labeling, associated components (e.g., dosing cups, droppers, spoons), and external packaging (e.g., cartons or shrink wrap). A market package is the article provided to a pharmacist or retail customer upon purchase and does not include packaging used solely for the purpose of shipping such articles. There are many issues that relate to container closure systems, including protection, compatibility, safety, and performance of packaging components and/or systems. The purpose of this section is to raise the student's awareness of the stability aspects related to container closure systems.

The United States Pharmacopeial Convention has established requirements for containers that are described in many of the drug product monographs in *United States Pharmacopeia* (USP). For capsules and tablets, these requirements generally relate to the design characteristics of the container (e.g., tight, well closed, or light-resistant). For injectable products, materials of construction are also addressed (e.g., "Preserve in single-dose or in multiple-dose containers, preferably of Type I glass, protected from light"). These requirements are defined in the General Notices and Requirements (Preservation, Packaging, Storage, and Labeling) section of the USP. The requirements for materials of construction are defined in the General Chapters of the USP.

The type and extent of stability information required for a particular drug product depends on the dosage form and the route of administration. For example, the kind of information that should be provided about a packaging system for an injectable dosage form or a drug product for inhalation is often more detailed than that which should be provided about a packaging system for a solid oral dosage form. More detailed information usually should be provided for a liquid-based dosage form than for a powder or a solid because a liquid-based dosage form is more likely to interact with the packaging components. The suitability of a container closure system for a particular pharmaceutical product is ultimately proven by full shelf-life stability studies. A container closure system should provide the dosage form with adequate protection from factors (e.g., temperature, light) that can cause a reduction in the quality of that dosage form over its shelf life. As discussed earlier in this chapter, there are numerous causes of degradation such as exposure to light, loss of solvent, exposure to reactive gases (e.g., oxygen), absorption of water vapor, and microbial contamination. A drug product can also suffer an unacceptable loss in quality if it is contaminated by filth. Not every drug product is susceptible to degradation by all of these factors. Not all drug products are light sensitive. Not all tablets are subject to loss of quality due to absorption of moisture. Sensitivity to oxygen is most commonly found with liquid-based dosage forms. Light protection is typically provided by an opaque or amber-colored container or by an opaque secondary packaging component (e.g., cartons or overwrap). The USP test for light transmission is an accepted standard for evaluating the light transmission properties of a container. Situations exist in which solid- and liquid-based oral drug products have been exposed to light during storage because the opaque secondary packaging component was removed, contrary to the approved labeling and the USP monograph recommendation. Loss of solvent

can occur through a permeable barrier (e.g., a polyethylene container wall), through an inadequate seal, or through leakage. Leaks can develop through rough handling or from inadequate contact between the container and the closure (e.g., due to the buildup of pressure during storage). Leaks can also occur in tubes due to a failure of the crimp seal. Water vapor or reactive gases (e.g., oxygen) may penetrate a container closure system either by passing through a permeable container surface (e.g., the wall of a low-density polyethylene bottle) or by diffusing past a seal. Plastic containers are susceptible to both routes. Although glass containers would seem to offer better protection because glass is relatively impermeable, glass containers are more effective only if there is a good seal between the container and the closure.

## Biotechnology Products

Biotechnological/biologic products have distinguishing characteristics to which consideration should be given in any well-defined testing program designed to confirm their stability during the intended storage period. For such products in which the active components are typically proteins and/or polypeptides, maintenance of molecular conformation and, hence, of biologic activity is dependent on noncovalent as well as covalent forces. Examples of these products are cytokines (interferons, interleukins, colony-stimulating factors, tumor necrosis factors), erythropoietins, plasminogen activators, blood plasma factors, growth hormones and growth factors, insulins, monoclonal antibodies, and vaccines consisting of well-characterized proteins or polypeptides. These products are particularly sensitive to environmental factors such as temperature changes, oxidation, light, ionic content, and shear. To ensure maintenance of biologic activity and to avoid degradation, stringent conditions for their storage are usually necessary. The evaluation of stability may necessitate complex analytical methodologies. Assays for biologic activity, where applicable, should be part of the pivotal stability studies. Appropriate physicochemical, biochemical, and immunochemical methods for the analysis of the molecular entity and the quantitative detection of degradation products should also be part of the stability program whenever purity and molecular characteristics of the product permit their use. The shelf lives of biotechnological/biologic products may vary from days to several years. With only a few exceptions, the shelf lives for existing products and potential future products will be within the range of 0.5 to 5 years. This takes into account the fact that degradation of biotechnological/biologic products may not be governed by the same factors during different intervals of a long storage period. Therefore, if the expected shelf life is within this range, the FDA makes certain recommendations in their Guidance to Industry. When shelf lives of 1 year or less are proposed, the real-time stability studies should be conducted monthly for the first 3 months and at 3-month intervals thereafter. For products with proposed shelf lives of greater than 1 year, the studies should be conducted every 3 months during the first year of storage, every 6 months during the second year, and annually thereafter.

## Solid-State Stability

The breakdown of drugs in the solid state is an important topic, but it has not been studied extensively in pharmacy. The subject has been reviewed by Garrett,[77] Lachman,[78] and Carstensen,[79] and is discussed here briefly.

### Pure Solids

The decomposition of pure solids, as contrasted with the more complex mixture of ingredients in a dosage form, has been studied, and a number of theories have been proposed to explain the shapes of the curves obtained when decomposition of the compound is plotted against time. Carstensen and Musa[80] described the decomposition of solid benzoic acid derivatives, such as aminobenzoic acid, which broke down into a liquid, aniline, and a gas, carbon dioxide. The plot of concentration of decomposed drug versus time yielded a sigmoidal curve (**Fig. 14–23**). After liquid begins to form, the decomposition becomes a first-order reaction in the solution. Such single-component pharmaceutical systems can degrade by either zero-order or first-order reaction, as observed in **Figure 14–23**. It is often difficult to determine which pattern is being followed when the reaction cannot be carried through a sufficient number of half-lives to differentiate between zero and first order.

### Solid Dosage Forms

The decomposition of drugs in solid dosage forms is more complex than decay occurring in the pure state of the individual compound. The reactions may be zero or first order, but in some cases, as with pure compounds, it is difficult to distinguish between the two. Tardif[81] observed that ascorbic acid decomposed in tablets by a pseudo–first-order reaction.

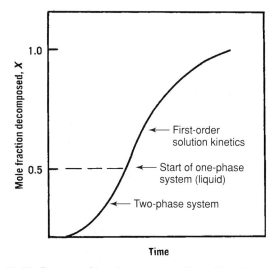

**Fig. 14–23.** Decomposition of a pure crystalline solid such as potassium permanganate, which involves gaseous reaction products. (From J. T. Carstensen, J. Pharm. Sci. **63**, 4, 1974. With permission.)

In tablets and other solid dosage forms, the possibility exists for solid–solid interaction. Carstensen et al.[82] devised a program to test for possible incompatibilities of the drug with excipients present in the solid mixture. The drug is blended with various excipients in the presence and absence of 5% moisture, sealed in vials, and stored for 2 weeks at 55°C. Visual observation is done and the samples are tested for chemical interaction by thin-layer chromatography. The method is qualitative but, in industrial preformulation, provides a useful screening technique for uncovering possible incompatibilities between active ingredient and pharmaceutical additives before deciding on a suitable dosage form.

Lach and associates[83] used diffuse reflectance spectroscopy to measure interactions of additives and drugs in solid dosage forms. Blaug and Huang[84] used this spectroscopic technique to study the interaction of spray-dried lactose with dextroamphetamine sulfate.

Goodhart and associates[85] studied the fading of colored tablets by light (photolysis reaction) and plotted the results as color difference at various light-energy values expressed in foot-candle hours.

Lachman, Cooper, and their associates[86] conducted a series of studies on the decomposition of FD&C colors in tablets and established a pattern of three separate stages of breakdown. The photolysis was found to be a surface phenomenon, causing fading of the tablet color to a depth of about 0.03 cm. Interestingly, fading did not occur further into the coating with continued light exposure, and the protected contents of the color-coated tablets were not adversely affected by exposure to light.

As noted by Monkhouse and Van Campen,[87] solid-state reactions exhibit characteristics quite different from reactions in the liquid or gaseous state because the molecules of the solid are in the crystalline state. The quantitative and theoretical approaches to the study of solid-state kinetics are at their frontier, which, when opened, will probably reveal a new and fruitful area of chemistry and drug science. The authors[87] classify solid-state reactions as *addition* when two solids, A and B, interact to form the new solid, AB. For example, picric acid reacts with naphthols to form what are referred to as *picrates*. A second kind of solid-state reaction is an *exchange* process, in which solid A reacts with solid BC to form solid AB and release solid C. Solid–gas reactions constitute another class, of which the oxidation of solid ascorbic acid and solid fumagillin are notable examples. Other types of solid-state processes include polymorphic transitions, sublimation, dehydration, and thermal decomposition.

Monkhouse and Van Campen[87] reviewed the experimental methods used in solid-state kinetics, including reflectance spectroscopy, x-ray diffraction, thermal analysis, microscopy, dilatometry, and gas pressure–volume analysis. Their review closes with sections on handling solid-state reaction data, temperature effects, application of the Arrhenius plot, equilibria expressions involved in solid-state degradation, and use of the van't Hoff equation for, say, a solid drug hydrate in equilibrium with its dehydrated form.

# CHAPTER SUMMARY

The purpose of stability testing is to provide evidence on how the quality of a drug substance or drug product varies with time under the influence of a variety of environmental factors, such as temperature, humidity, and light, and to establish a retest period for the drug substance or a shelf life for the drug product and recommended storage conditions. This fundamental topic was covered in this chapter. This chapter introduces the rates and mechanisms of reactions with particular emphasis on decomposition and stabilization of drug products. It is essential for pharmacists and pharmaceutical scientists to study, understand, and interpret conditions of instability of pharmaceutical products as well as to be able to offer solutions for the stabilization of these products. It is also essential for them to define reaction rate, reaction order, and molecularity, while understanding and applying apparent zero-order kinetics to the practice of pharmacy. By the conclusion of this chapter and some practice, the student should be able to calculate half-life and shelf life of pharmaceutical products and drugs as well as interpret pH–rate profiles and kinetic data. You should also be able to describe the influence of temperature, ionic strength, solvent, pH, and dielectric constant on reaction rates. Be familiar with $Q_{10}$ calculations as they aid in the understanding of the relationship between reaction rate constant and temperature. Finally, stabilizing pharmaceutical agents is critical for making acceptable products in the industrial and community pharmacy setting. Therefore, you should understand stabilization techniques, stability testing protocols, and regulatory requirements.

 **Practice problems for this chapter can be found at thePoint.lww.com/Sinko6e.**

# References

1. K. A. Connors, G. L. Amidon, and V. J. Stella, *Chemical Stability of Pharmaceuticals*, 2nd Ed., Wiley, New York, 1986, pp. 764–773.
2. S. Glasstone, *Textbook of Physical Chemistry*, Van Nostrand, New York, 1946, pp. 1051–1052.
3. E. R. Garrett and R. F. Carper, J. Am. Pharm. Assoc. Sci. Ed. **44**, 515, 1955.
4. K. A. Connors, G. L. Amidon, and V. J. Stella, *Chemical Stability of Pharmaceuticals*, 2nd Ed., Wiley, New York, 1986, p. 15.
5. H. S. Harned, J. Am. Chem. Soc. **40**, 1462, 1918.
6. J. Walker, Proc. Royal. Soc. London **78**, 157, 1906.
7. J. R. D. McCormick, et al., J. Am. Chem. Soc. **79**, 2849, 1957.
8. D. E. Guttman and P. D. Meister, J. Am. Pharm. Assoc. Sci. Ed. **47**, 773, 1958.
9. A. E. Allen and V. D. Gupta, J. Pharm. Sci. **63**, 107, 1974; V. D. Gupta, J. Pharm. Sci. **67**, 299, 1978.
10. K. R. Heimlich and A. Martin, J. Am. Pharm. Assoc. Sci. Ed. **49**, 592, 1960.
11. J. W. Mauger, A. N. Paruta, and R. J. Gerraughty, J. Pharm. Sci. **58**, 574, 1969.
12. L. Michaelis and M. L. Menten, Biochem. Z. **49**, 333, 1913.
13. T. Higuchi, A. Havinga, and L. W. Busse, J. Am. Pharm. Assoc. Sci. Ed. **39**, 405, 1950.
14. W. Yang, Drug Dev. Ind. Pharm. **7**, 717, 1981.
15. A. P. Simonelli and D. S. Dresback, in D. E. Francke and H. A. K. Whitney (Eds.), *Perspectives in Clinical Pharmacy*, Drug Intelligence Publications, Hamilton, IL, 1972, Chapter 19.
16. H. Eyring, Chem. Rev. **10**, 103, 1932; Chem. Rev. **17**, 65, 1935.
17. A. D. Marcus and S. Baron, J. Am. Pharm. Assoc. Sci. Ed. **48**, 85, 1959.
18. M. Richardson and F. G. Soper, J. Chem. Soc. 1873, 1929; F. G. Soper and E. Williams, J. Chem. Soc. 2297, 1931.
19. J. T. Carstensen, J. Pharm. Sci. **59**, 1141, 1970.
20. E. S. Amis and C. Holmes, J. Am. Chem. Soc. **63**, 2231, 1941.
21. A. D. Marcus and A. J. Taraszka, J. Am. Pharm. Assoc. Sci. Ed. **48**, 77, 1959.
22. S. Siegel, L. Lachman, and L. Malspeis, J. Pharm. Sci. **48**, 431, 1959.
23. E. R. Garrett, J. Pharm. Sci. **49**, 767, 1960; J. Am. Chem. Soc. **79**, 3401, 1957.
24. J. J. Windheuser and T. Higuchi, J. Pharm. Sci. **51**, 354, 1962.
25. N. E. Webb, Jr., G. J. Sperandio, and A. Martin, J. Am. Pharm. Assoc. Sci. Ed. **47**, 101, 1958.
26. J. N. Brönsted and K. J. Pedersen, Z. Physik. Chem. **A108**, 185, 1923; J. N. Brönsted, Chem. Rev. **5**, 231, 1928; R. P. Bell, *Acid–Base Catalysis*, Oxford University Press, Oxford, 1941, Chapter 5.
27. L. J. Edwards, Trans. Faraday Soc. **46**, 723, 1950; Trans. Faraday Soc. **48**, 696, 1952.
28. A. R. Fersht and A. J. Kirby, J. Am. Chem. Soc. **89**, 4857, 1967.
29. J. A. Mollica, C. R. Rehm, and J. B. Smith, J. Pharm. Sci. **58**, 636, 1969.
30. J. A. Mollica, S. Ahuja, and J. Cohen, J. Pharm. Sci. **67**, 443, 1978.
31. J. P. Hou and J. W. Poole, J. Pharm. Sci. **58**, 447, 1969; J. Pharm. Soc. **58**, 1510, 1969.
32. K. A. Connors and J. A. Mollica, J. Pharm. Sci. **55**, 772, 1966; S. L. Hem, E. J. Russo, S. M. Bahal, and R. S. Levi, J. Pharm. Sci. **62**, 267, 1973.
33. T. Higuchi and C. D. Bias, J. Am. Pharm. Assoc. Sci. Ed. **42**, 707, 1953; T. Higuchi, A. D. Marcus, and C. D. Bias, J. Am. Pharm. Assoc. Sci. Ed. **43**, 129, 530, 1954.
34. K. C. James and R. H. Leach, J. Pharm. Pharmacol. **22**, 607, 1970.
35. J. H. Beijnen, O. A. G. J. van der Houwen, and W. J. M. Underberg, Int. J. Pharm. **32**, 123, 1986.
36. B. Steffansen and H. Bundgaard, Int. J. Pharm. **56**, 159, 1989.
37. C. Vinckier, R. Hauchecorne, Th. Cachet, G. Van den Mooter, and J. Hoogmartens, Int. J. Pharm. **55**, 67, 1989; Th. Cachet, G. Van den Mooter, R. Hauchecorne, et al., Int. J. Pharm. **55**, 59, 1989.
38. K. A. Connors, G. L. Amidon, and V. J. Stella, *Chemical Stability of Pharmaceuticals*, 2nd Ed., Wiley, New York, pp. 457–462.
39. P. Atkins, T. Herbert, and N. Jones, Int. J. Pharm. **30**, 199, 1986.
40. J. H. Beijnen and W. J. M. Undenberg, Int. J. Pharm. **24**, 219, 1985.
41. E. R. Garrett, J. Pharm. Sci. **51**, 811, 1962.
42. V. Das Gupta, J. Pharm. Sci. **72**, 1453, 1983.
43. D. E. M. M. Vendrig, J. H. Beijnen, O. A. G. J. van der Houwen, and J. J. M. Holthuis, Int. J. Pharm. **50**, 189, 1989.
44. E. S. Barron, R. H. De Meio, and F. Klemperer, J. Biol. Chem. **112**, 624, 1936.
45. A. O. Dekker and R. G. Dickinson, J. Am. Chem. Soc. **62**, 2165, 1940.
46. A. Weissberger, J. E. Lu Valle, and D. S. Thomas, Jr., J. Am. Chem. Soc. **65**, 1934, 1943; A. Weissberger and J. E. Lu Valle, J. Am. Chem. Soc. **66**, 700, 1944.
47. H. Nord, Acta Chem. Scand. **9**, 442, 1955.
48. S. M. Blaug and B. Hajratwala, J. Pharm. Sci. **61**, 556, 1972; J. Pharm. Sci. **63**, 1240, 1974.
49. A. R. Rogers and J. A. Yacomeni, J. Pharm. Pharmacol. **23 S**, 218 S, 1971.
50. K. Takamura and M. Ito, Chem. Pharm. Bull. **25**, 3218, 1977.
51. V. E. Tyler, L. R. Brady, and J. E. Robbers, *Pharmacognosy*, 7th Ed., Lea & Febiger, Philadelphia, p. 97.
52. P. Fyhr and A. Brodin, Acta Pharm. Suec. **24**, 26, 1987; Chem. Abs. **107**, 46, 202y, 1987.
53. M. J. Akers, J. Parenteral Drug Assoc. **33**, 346, 1979.
54. K. Thoma and M. Struve, Pharm. Acta Helv. **61**, 34, 1986; Chem. Abs. **104**, 174, 544 m, 1986.
55. A. B. C. Yu and G. A. Portmann, J. Pharm. Sci. **79**, 913, 1990.
56. D. E. Moore, J. Pharm. Sci. **65**, 1447, 1976.

57. A. F. Asker, D. Canady, and C. Cobb, Drug Dev. Ind. Pharm. **11,** 2109, 1985.

58. A. F. Asker and M. Larose, Drug Dev. Ind. Pharm. **13,** 2239, 1987.

59. H. Bundgaard, T. Norgaard, and N. M. Neilsen, Int. J. Pharm. **42,** 217, 1988.

60. R. Andersin and S. Tammilehto, Int. J. Pharm. **56,** 175, 1989.

61. L. Gu, H.-S. Chiang, and D. Johnson, Int. J. Pharm. **41,** 105, 1988.

62. Y. Matsuda, R. Teraoka, and I. Sugimoto, Int. J. Pharm. **54,** 211, 1989.

63. A. F. Asker and A. Collier, Drug Dev. Ind. Pharm. 7, 563, 1981.

64. A. F. Asker and D. Y. Colbert, Drug Dev. Ind. Pharm. **8,** 759, 1982.

65. C. R. Kowarski, J. Pharm. Sci. **58,** 360, 1969.

66. A. F. Asker and D. Jackson, Drug Dev. Ind. Pharm. **12,** 385, 1986.

67. S. M. Free, *Considerations in sampling for stability*. Presented at American Drug Manufacturers Association, November 1955; R. H. Blythe, *Product formulation and stability prediction*. Presented at the Production Section of the Canadian Pharmaceutical Manufacturers Association, April 1957; A. K. Amirjahed, J. Pharm. Sci. **66,** 785, 1977.

68. S. Baker and S. Niazi, J. Pharm. Sci. **67,** 141, 1978.

69. A. R. Rogers, J. Pharm. Pharmacol. **15,** 101 T, 1963.

70. S. P. Eriksen and H. Stalmach, J. Pharm. Sci. **54,** 1029, 1965.

71. H. V. Maulding and M. A. Zoglio, J. Pharm. Sci. **59,** 333, 1970; M. A. Zoglio, H. V. Maulding, W. H. Streng, and W. C. Vincek, J. Pharm. Sci. **64,** 1381, 1975.

72. B. W. Madsen, R. A. Anderson, D. Herbison-Evans, and W. Sneddon, J. Pharm. Sci. **63,** 777, 1974.

73. B. Edel and M. O. Baltzer, J. Pharm. Sci. **69,** 287, 1980.

74. E. R. Garrett, J. Am. Pharm. Assoc. Sci. Ed. **45,** 171, 470, 1956.

75. L. Lachman, P. DeLuca, and M. J. Akers, in L. Lachman, H. A. Lieberman, and J. L. Kanig (Eds.), *The Theory and Practice of Industrial Pharmacy*, 3rd Ed., Lea Febiger, Philadelphia, 1986, Chapter 26.

76. W. J. Westlake, in J. Swarbrick (Ed.), *Current Concepts in the Pharmaceutical Sciences: Dosage Form Design and Bioavailability,* Lea Febiger, Philadelphia, 1973, Chapter 5.

77. E. R. Garrett, J. Pharm. Sci. **51,** 811, 1962; in H. S. Bean, A. H. Beckett, and J. E. Carless (Eds.), *Advances in Pharmaceutical Sciences*, Vol. 2, Academic Press, New York, 1967, p. 77.

78. L. Lachman, J. Pharm. Sci. **54,** 1519, 1965.

79. J. T. Carstensen, *Theory of Pharmaceutical Systems*, Vol. 2, Academic Press, New York, 1973, Chapter 5; J. Pharm. Sci. **63,** 1, 1974.

80. J. Carstensen and M. Musa, J. Pharm. Sci. **61,** 1112, 1972.

81. R. Tardif, J. Pharm. Sci. **54,** 281, 1965.

82. J. Carstensen, J. Johnson, W. Valentine, and J. Vance, J. Pharm. Sci. **53,** 1050, 1964.

83. J. L. Lach and M. Bornstein, J. Pharm. Sci. **54,** 1731, 1965; M. Bornstein and J. L. Lach, J. Pharm. Sci. **55,** 1033, 1966; J. L. Lach and M. Bornstein, J. Pharm. Sci. **55,** 1040, 1966; M. Bornstein, J. P. Walsh, B. J. Munden, and J. L. Lach, J. Pharm. Sci. **56,** 1419, 1967; M. Bornstein, J. L. Lach, and B. J. Munden, J. Pharm. Sci. **57,** 1653, 1968; W. Wu, T. Chin, and J. L. Lach, J. Pharm. Sci. **59,** 1122, 1234, 1970; J. J. Lach and L. D. Bigley, J. Pharm. Sci. **59,** 1261, 1970; J. D. McCallister, T. Chin, and J. L. Lach, J. Pharm. Sci. **59,** 1286, 1970.

84. S. M. Blaug and W.-T. Huang, J. Pharm. Sci. **61,** 1770, 1972.

85. M. Everhard and F. Goodhard, J. Pharm. Sci. **52,** 281, 1963; F. Goodhard, M. Everhard, and D. Dickcius, J. Pharm. Sci. **53,** 388, 1964; F. Goodhard, H. Lieberman, D. Mody, and F. Ninger, J. Pharm. Sci. **56,** 63, 1967.

86. R. Kuramoto, L. Lachman, and J. Cooper, J. Am. Pharm. Assoc. Sci. Ed. **47,** 175, 1958; T. Urbanyi, C. Swartz, and L. Lachman, J. Pharm. Sci. **49,** 163, 1960; L. Lachman et al., J. Pharm. Sci. **50,** 141, 1961; C. Swartz, L. Lachman, T. Urbanyi, and J. Cooper, J. Pharm. Sci. **50,** 145, 1961; C. Swartz and J. Cooper, J. Pharm. Sci. **51,** 89, 1962; J. Cooper and C. Swartz, J. Pharm. Sci. **51,** 321, 1962; C. Swartz, et al., J. Pharm. Sci. **51,** 326, 1962.

87. D. C. Monkhouse and L. Van Campen, Drug Dev. Ind. Pharm. **10,** 1175, 1984.

## Recommended Readings

K. A. Connors, G. L. Amidon, and V. J. Stella (Eds.), *Chemical Stability of Pharmaceuticals*, 2nd Ed., Wiley, New York, 1986.

K. Huynh-Ba (Ed.), *Handbook of Stability Testing in Pharmaceutical Development: Regulations, Methodologies and Best Practices*, 1st Ed., Springer, New York, 2008.

### CHAPTER LEGACY

**Fifth Edition:** published as Chapter 15 (Chemical Kinetics and Stability). Updated by Patrick Sinko.

**Sixth Edition:** published as Chapter 14 (Chemical Kinetics and Stability). Updated by Patrick Sinko.

# 15 INTERFACIAL PHENOMENA

Several types of interface can exist, depending on whether the two adjacent phases are in the solid, liquid, or gaseous state (**Table 15–1**). For convenience, these various combinations are divided into two groups, namely, *liquid interfaces* and *solid interfaces*. In the former group, the association of a liquid phase with a gaseous or another liquid phase will be discussed. The section on solid interfaces will deal with systems containing solid–gas and solid–liquid interfaces. Although solid–solid interfaces have practical significance in pharmacy (e.g., the adhesion between granules, the preparation of layered tablets, and the flow of particles), little information is available to quantify these interactions. This is due, at least in part, to the fact that the surface region of materials in the solid state is quiescent, in sharp contrast to the turbulence that exists at the surfaces of liquids and gases. Accordingly, solid–solid systems will not be discussed here. A final section will outline the electric properties of interfaces.

The term *surface* is customarily used when referring to either a gas–solid or a gas–liquid interface. Although this terminology will be used in this chapter, the reader should appreciate that every surface is an interface. Thus, a tabletop forms a gas–solid interface with the atmosphere above it, and the surface of a raindrop constitutes a gas–liquid interface.

The symbols for the various interfacial tensions are shown in the second column of **Table 15–1**, where the subscript L stands for liquid, V for vapor or gas, and S for solid. Surface and interfacial tensions are defined later.

Because every physical entity, be it a cell, a bacterium, a colloid, a granule, or a human, possesses an interface at its boundary with its surroundings, the importance of the present topic is self-evident. Interfacial phenomena in pharmacy and medicine are significant factors that affect adsorption of drugs onto solid adjuncts in dosage forms, penetration of molecules through biologic membranes, emulsion formation and stability, and the dispersion of insoluble particles in liquid media to form suspensions. The interfacial properties of a surface-active agent lining the alveoli of the lung are responsible for the efficient operation of this organ.[1–3] Several authors[4–6] have reviewed the relationship between surface properties of drugs and their biologic activity.

## LIQUID INTERFACES

### Surface and Interfacial Tensions

In the liquid state, the cohesive forces between adjacent molecules are well developed. Molecules in the bulk liquid are surrounded in all directions by other molecules for which they have an equal attraction, as shown in **Figure 15–1**. On

## KEY CONCEPT INTERFACES

When phases exist together, the boundary between two of them is known as an *interface*. The properties of the molecules forming the interface are often sufficiently different from those in the bulk of each phase that they are referred to as forming an *interfacial phase*. If a liquid and its vapors exist together in the same container, the liquid takes the bottom part of the container. The remainder of the container is filled up by the liquid vapor, which, as with any gas, has a tendency to take all available space.

Molecules in both the liquid and the gas are in constant motion and can move from the liquid into the vapor and back from the vapor to the liquid. However, the distinct boundary between the vapor and the liquid is preserved under constant temperature, and the exchange of molecules does not destroy the equilibrium between these two phases due to the labile (i.e., dynamic) character of this boundary.

### TABLE 15–1
### CLASSIFICATION OF INTERFACES

| Phase | Interfacial Tension | Types and Examples of Interfaces |
|---|---|---|
| Gas–Gas | — | No interface possible |
| Gas–liquid | $\gamma_{LV}$ | Liquid surface, body of water exposed to atmosphere |
| Gas–solid | $\gamma_{SV}$ | Solid surface, table top |
| Liquid–liquid | $\gamma_{LL}$ | Liquid–liquid interface, emulsion |
| Liquid–solid | $\gamma_{LS}$ | Liquid–solid interface, suspension |
| Solid–solid | $\gamma_{SS}$ | Solid–solid interface, powder particles in contact |

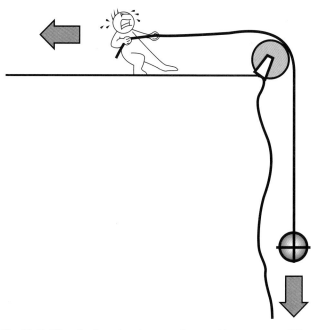

**Fig. 15–2.** Visualization of surface tension as akin to a person lifting a weight up the side of a cliff by pulling the rope in a horizontal direction.

the other hand, molecules at the surface (i.e., at the liquid–air interface) can only develop attractive cohesive forces with other liquid molecules that are situated below and adjacent to them. They can develop adhesive forces of attraction with the molecules constituting the other phase involved in the interface, although, in the case of the liquid–gas interface, this adhesive force of attraction is small. The net effect is that the molecules at the surface of the liquid experience an inward force toward the bulk, as shown in **Figure 15–1**. Such a force pulls the molecules of the interface together and, as a result, contracts the surface, resulting in a *surface tension*.

This "tension" in the surface is the force per unit length that must be applied *parallel* to the surface so as to counterbalance the net inward pull. This force, the surface tension, has the units of dynes/cm in the cgs system and of N/m in the SI system. It is similar to the situation that exists when an object dangling over the edge of a cliff on a length of rope is pulled upward by a man holding the rope and walking away from

the edge of the top of the cliff. This analogy to surface tension is sketched in **Figure 15–2**.

*Interfacial tension* is the force per unit length existing at the interface between two immiscible liquid phases and, like surface tension, has the units of dynes/cm. Although, in the general sense, all tensions may be referred to as *interfacial tensions*, this term is most often used for the attractive force between immiscible liquids. Later, we will use the term *interfacial tension* for the force between two liquids, $\gamma_{LL}$, between two solids, $\gamma_{SS}$, and at a liquid–solid interface, $\gamma_{LS}$. The term *surface tension* is reserved for liquid–vapor, $\gamma_{LV}$, and solid–vapor, $\gamma_{SV}$, tensions. These are often written simply as $\gamma_L$ and $\gamma_S$, respectively. Ordinarily, interfacial tensions are less than surface tensions because the adhesive forces between two liquid phases forming an interface are greater than when a liquid and a gas phase exist together. It follows that if two liquids are completely miscible, no interfacial tension exists between them. Some representative surface and interfacial tensions are listed in **Table 15–2**.

Surface tension as a force per unit length can also be illustrated by means of a three-sided wire frame across which a movable bar is placed (**Fig. 15–3**). A soap film is formed over the area ABCD and can be stretched by applying a force *f* (such as a hanging mass) to the movable bar, length *L*, which acts against the surface tension of the soap film. When the mass is removed, the film will contract owing to its surface tension. The surface tension, $\gamma$, of the solution forming the film is then a function of the force that must be applied to break the film over the length of the movable bar in contact with the film. Because the soap film has two liquid–gas interfaces (one above and one below the plane of the paper), the

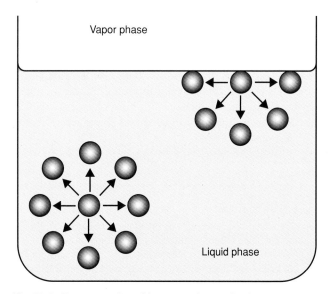

**Fig. 15–1.** Representation of the unequal attractive forces acting on molecules at the surface of a liquid as compared with molecular forces in the bulk of the liquid.

## TABLE 15–2
## SURFACE TENSION AND INTERFACIAL TENSION (AGAINST WATER) AT 20°C*

| Substance | Surface Tension (dynes/cm) | Substance | Interfacial Tension (dynes/cm) |
|---|---|---|---|
| Water | 72.8 | Mercury | 375 |
| Glycerin | 63.4 | n-Hexane | 51.1 |
| Oleic acid | 32.5 | Benzene | 35.0 |
| Benzene | 28.9 | Chloroform | 32.8 |
| Chloroform | 27.1 | Oleic acid | 15.6 |
| Carbon tetrachloride | 26.7 | n-Octyl alcohol | 8.52 |
| Caster oil | 39.0 | Caprylic acid | 8.22 |
| Olive oil | 35.8 | Olive oil | 22.9 |
| Cottonseed oil | 35.4 | Ethyl ether | 10.7 |
| Liquid petrolatum | 33.1 | | |

*From P. Becher, *Emulsions: Theory and Practice*, 2nd Ed., Reinhold, New York, 1962, and other sources.

total length of contact is in fact equal to twice the length of the bar.

Thus,

$$\gamma = f_b/2L \qquad (15\text{–}1)$$

where $f_b$ is the force required to break the film and $L$ is the length of the movable bar.

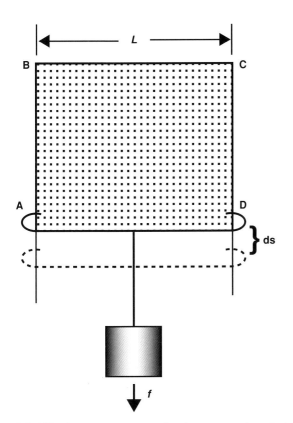

**Fig. 15–3.** Wire-frame apparatus used to demonstrate the principle of surface tension.

## EXAMPLE 15–1

**Calculating the Surface Tension of Water**

If the length of the bar, $L$, is 5 cm and the mass required to break a soap film is 0.50 g, what is the surface tension of the soap solution?

Recall that the downward force is equal to the mass multiplied by the acceleration due to gravity. Then

$$\gamma = \frac{0.50 \text{ g} \times 981 \text{ cm/sec}^2}{10 \text{ cm}} = 49 \text{ dynes/cm}$$

## Surface Free Energy

To move a molecule from the inner layers to the surface, work needs to be done against the force of surface tension. In other words, each molecule near the surface of liquid possesses a certain excess of potential energy as compared to the molecules in the bulk of the liquid. The higher the surface of the liquid, the more molecules have this excessive potential energy. Therefore, if the surface of the liquid increases (e.g., when water is broken into a fine spray), the energy of the liquid also increases. Because this energy is proportional to the size of the free surface, it is called a surface free energy. Each molecule of the liquid has a tendency to move inside the liquid from the surface; therefore, the liquid takes form with minimal free surface and with minimal surface energy. For example, liquid droplets tend to assume a spherical shape because a sphere has the smallest surface area per unit volume.

To increase the surface of the liquid without any additional changes in the liquid state, in particular without changes in liquid temperature, work must be done against the surface tension. To evaluate the amount of work in increasing the surface area, we can write equation (**15–1**) as $\gamma \times 2L = f$. When the bar is at a position AD in **Figure 15–3** and a mass is added to extend the surface by a distance ds, the work $dW$ (force multiplied by distance) is

$$dW = f \times \text{ds} = \gamma \times 2L \times \text{ds}$$

and, because $2L \times ds$ is equal to the increase in surface area, $dA$, produced by extending the soap film,

$$dW = \gamma \, dA$$

For a finite change,

$$W = \gamma \, \Delta A \qquad (15\text{-}2)$$

where $W$ is the work done, or *surface free energy* increase, expressed in ergs, $\gamma$ is the surface tension in dynes/cm, and $\Delta A$ is the increase in area in cm$^2$. Any form of energy can be divided into an intensity factor and a capacity factor (see Chapter 3). Surface tension is the intensity factor, and a change in area is the capacity factor of surface free energy. Surface tension can be defined as the *surface free energy change per unit area increase* consistent with equation (15-2).

### EXAMPLE 15-2

**Calculation of Work against Surface Tension**

What is the work required in *Example 15-1* to pull the wire down 1 cm as shown in Figure 15-3?

Because the area is increased by 10 cm$^2$, the work done is given by the equation

$$W = 49 \text{ dynes/cm} \times 10 \text{ cm}^2 = 490 \text{ ergs}$$

Repeat the calculations using SI units. We have

> 1 dyne = $10^{-5}$ N, or 49 dynes = $49 \times 10^{-5}$ N
> 49 dynes/cm = $49 \times 10^{-3}$ N/m = $49 \times 10^{-3}$ Nm/m$^2$
> = $49 \times 10^{-3}$ joule/m$^2$

Also, 1 joule = $10^7$ ergs. Therefore, $W = 49 \times 10^{-3}$ Nm/m$^2 \times 10^{-3}$ m$^2 = 490 \times 10^{-7}$ joule = 490 ergs.

Equation **15-2** defines surface tension as the work per unit area required to produce a new surface. From thermodynamics, at $T$ and $P$ constant, the surface tension can also be viewed as the increment in Gibbs free energy per unit area (see Hiemenz[7]). Thus, equation **(15-2)** can be written as

$$\gamma = \left( \frac{dG}{dA} \right)_{T,P} \qquad (15\text{-}3)$$

This definition has the advantage that the path-dependent variable $W$ is replaced by a thermodynamic function $G$, which is independent of the path. Many of the general relationships that apply to $G$ also serve for $\gamma$. This fact enables us to compute the enthalpy and entropy of a surface:

$$G^S = \gamma = H^S - TS^S \qquad (15\text{-}4)$$

and

$$\left( \frac{\partial G^S}{\partial T} \right)_P = \left( \frac{\partial \gamma}{\partial T} \right)_P = -S^S \qquad (15\text{-}5)$$

Combining equations **(15-4)** and **(15-5)**, we obtain

$$\gamma = H^S + T \left( \frac{\partial \gamma}{\partial T} \right)_P \qquad (15\text{-}6)$$

Thus, from a plot of surface tension against absolute temperature, we can obtain the slope of the line, $\partial \gamma / \partial T$, and thus find $-S^S$ from equation **(15-5)**. If $H^s$ does not change appreciably over the temperature range considered, the intercept gives the $H^s$ value. It should be noted that the units for $S^s$ and $H^s$ are given in two dimensions, ergs/cm$^2$ deg for $S^s$ and ergs/cm$^2$ for $H^s$ in the cgs system. In the SI system, $S^s$ is given in units of joule/m$^2$ deg and $H^s$ in units of joule/m$^2$.

### EXAMPLE 15-3

**Calculation of $S^s$ and $H^s$**

The surface tension of methanol in water (10% by volume) at 20°C, 30°C, and 50°C (293.15, 303.15, and 323.15 K, respectively) is 59.04, 57.27, and 55.01 dynes/cm (or ergs/cm$^2$), respectively.[8] Compute $S^s$ and $H^s$ over this temperature range.

Using linear regression of $\gamma$ versus $T$ according to equation (15-6), we find the slope to be $-0.131$ erg/cm$^2$ deg = $(\partial \gamma / \partial T)_p = -S^s$; hence, $S^s = 0.131$ and the intercept is 97.34 erg/cm$^2 = H^s$. The equation is therefore

$$G^S = \gamma = 97.34 + 0.131T$$

If we compute $H^s$ at each temperature from equation (15-6) and if $S^s$ remains constant at $-0.131$, we find

> At 20°C: $H^S = 59.04 + (0.131 \times 293.15) = 97.44$ erg/cm$^2$
> At 30°C: $H^S = 57.27 + (0.131 \times 303.15) = 96.98$ erg/cm$^2$
> At 50°C: $H^S = 55.01 + (0.131 \times 323.15) = 97.34$ erg/cm$^2$

$H^s$ appears to be practically constant, very similar to the intercept from the regression equation, $H^s = 97.34$ erg/cm$^2 = 97.34$ mJ/m$^2$. Note that the numerical value of surface tension in the cgs system, like that for $H^s$ in the cgs system, is the same as that in the SI system when the units mJ are used. Thus, one can convert surface tension readily from cgs to SI units.[7] For example, if the surface tension of methanol in water (10% by volume) at 20°C is 59.04 erg/cm$^2$ in the cgs system, we can write without carrying out the conversion calculation that $\gamma$ for the methanol-in-water mixture at 20°C is 59.04 mJ/m$^2$ in SI units.

## Pressure Differences Across Curved Interfaces

Another way of expressing surface tension is in terms of the pressure difference that exists across a curved interface.

## ◢►◤KEY CONCEPT◣ SURFACE FREE ENERGY AND SURFACE TENSION

The surface layer of a liquid possesses additional energy as compared to the bulk liquid. This energy increases when the surface of the same mass of liquid increases and is therefore called *surface free energy*. Surface free energy per unit of surface of the liquid is defined as *surface tension* and is often denoted as $\gamma$. This means that for an increase of liquid surface of $S$ units without any other changes in the liquid state and without changes in its temperature, the work equal to $\gamma S$ must be done.

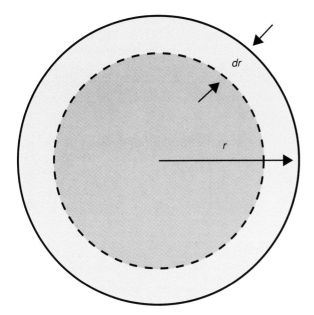

**Fig. 15–4.** Schematic representation of the pressure difference across the curved surface of a soap bubble.

Consider a soap bubble (**Fig. 15–4**) having a radius $r$. The total surface free energy, $W$, is equal to $4\pi r^2\gamma$, where $4\pi r^2$ is the area of the spherical bubble. (See the formulas, bottom, inside front cover.) Suppose that the bubble is caused to shrink so that its radius decreases by $dr$. The final surface free energy is now

$$W = 4\pi\gamma(r - dr)^2 \qquad \textbf{(15–7)}$$

$$W = 4\pi\gamma r^2 - 8\pi\gamma\, dr + 4\pi\gamma(dr)^2 \qquad \textbf{(15–8)}$$

Because $dr$ is small compared to $r$, the term containing $(dr)^2$ in equation (**15–8**) can be disregarded.

The *change* in surface free energy is therefore $-8\pi\gamma r\, dr$, negative because the surface area has shrunk. Opposing this change is an equal and opposite energy term that depends on the pressure difference, $\Delta P$, across the wall of the bubble. Because pressure is a force per unit area, or force = pressure × area, the work change brought about by a decrease in radius $dr$ is

$$W = \Delta P \times 4\pi r^2 \times -dr \qquad \textbf{(15–9)}$$

At equilibrium, this must equal the change in surface free energy, and so

$$-8\pi\gamma r\, dr = -4\,\Delta P r^2 dr \qquad \textbf{(15–10)}$$

or

$$\Delta P = \frac{2\gamma}{r} \qquad \textbf{(15–11)}$$

Therefore, as the radius of a bubble decreases, the pressure of the air inside increases relative to that outside. Equation (**15–11**) is a simplification of the Young–Laplace equation and can be used to explain capillary rise, as seen in the following section.

## Measurement of Surface and Interfacial Tensions

Of the several methods that exist for obtaining surface and interfacial tensions, only the *capillary rise* and the *DuNoüy ring* methods will be described here. For details of the other methods, such as drop weight, bubble pressure, pendent drop, sessile drop, Wilhelmy plate, and oscillating drop, refer to the treatises by Adamson,[9] Harkins and Alexander,[10] Drost-Hansen,[11] Hiemenz,[7] and Matsumoto et al.[12] It is worth noting, however, that the choice of a particular method often depends on whether surface or interfacial tension is to be determined, the accuracy and convenience desired, the size of sample available, and whether the effect of time on surface tension is to be studied. There is no one best method for all systems.

The surface tensions of most liquids decrease almost linearly with an increase in temperature, that is, with an increase in the kinetic energy of the molecules. In the region of its critical temperature, the surface tension of a liquid becomes zero. The surface tension of water at 0°C is 75.6, at 20°C it is 72.8, and at 75°C it is 63.5 dynes/cm. It is therefore necessary to control the temperature of the system when carrying out surface and interfacial tension determinations.

### Capillary Rise Method

When a capillary tube is placed in a liquid contained in a beaker, the liquid generally rises up the tube a certain distance. Because the force of adhesion between the liquid molecules and the capillary wall is greater than the cohesion between the liquid molecules, the liquid is said to *wet* the capillary wall, spreading over it and rising in the tube (spreading is discussed in some detail later). By measuring this rise in a capillary, it is possible to determine the surface tension of the liquid. It is not possible, however, to obtain interfacial tensions using the capillary rise method.

Consider a capillary tube with an inside radius $r$ immersed in a liquid that wets its surface, as seen in **Figure 15–5a**. Because of the surface tension, the liquid continues to rise in the tube, but because of the weight of the liquid, the upward movement is just balanced by the downward force of gravity.

The upward vertical component of the force resulting from the surface tension of the liquid at any point on the circumference is given by

$$a = \gamma\cos(\theta)$$

as seen in the enlarged sketch (**Fig. 15–5b**). The total upward force around the inside circumference of the tube is

$$2\pi r\gamma\cos(\theta)$$

where $\theta$ is the *contact angle* between the surface of the liquid and the capillary wall and $2\pi r$ is the inside circumference of the capillary. For water and other commonly used liquids, the angle $\theta$ is insignificant, that is, the liquid wets the capillary wall so that $\cos\theta$ is taken as unity for practical purposes (see left side of **Fig. 15–5b**).

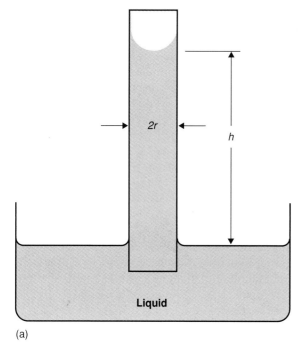

(a)

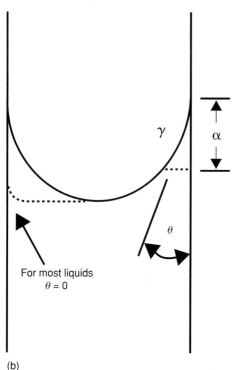

(b)

**Fig. 15–5.** (*a*) Measuring surface tension by means of the capillary-rise principle. (*b*) Enlarged view of the force components and contact angle at the meniscus of a liquid. For many liquids, the contact angle $\theta$ (exaggerated in the figure) is nearly zero, as shown on the left-hand side of the diagram.

The counteracting force of gravity (mass × acceleration) is given by the product of the cross-sectional area, $\pi r^2$, the height, $h$, of the liquid column to the lowest point of the meniscus, the difference in the density of the liquid, $\rho$, and its vapor, $\rho_0$, and the acceleration of gravity: $\pi r^2 h(\rho - \rho_0)g + w$.

The last term, $w$, is added to account for the weight of liquid above $h$ in the meniscus. When the liquid has risen to its maximum height, which can be read from the calibrations on the capillary tube, the opposing forces are in equilibrium, and accordingly the surface tension can be calculated. The density of the vapor, the contact angle, and $w$ can usually be disregarded; hence,

$$2\pi r\gamma = \pi r^2 h\rho g$$

and finally

$$\gamma = \frac{1}{2} r h\rho g \qquad (15\text{--}12)$$

---

**EXAMPLE 15–4**

**Calculation of the Surface Tension of Chloroform by the Capillary Rise Method**

A sample of chloroform rose to a height of 3.67 cm at 20°C in a capillary tube having an inside radius of 0.01 cm. What is the surface tension of chloroform at this temperature? The density of chloroform is 1.476 g/cm³. We write

$$\gamma = \frac{1}{2} \times 0.01\,\text{cm} \times 3.67\,\text{cm} \times 1.476\,\text{g/cm}^3 \times 981\,\text{cm/sec}^2$$
$$\gamma = 26.6\,\text{g/sec}^2 = 26.6\,\text{dynes/cm}$$

---

Capillary rise can also be explained as being due to the pressure difference across the curved meniscus of the liquid in the capillary. We have already seen in equation **(15–11)** that the pressure on the concave side of a curved surface is greater than that on the convex side. This means that the pressure in the liquid immediately below the meniscus will be less than that outside the tube at the same height. As a result, the liquid will move up the capillary until the hydrostatic head produced equals the pressure drop across the curved meniscus.

Using the same symbols as before and neglecting contact angles, we obtain

$$\Delta P = 2\gamma/r = \rho g h \qquad (15\text{--}13)$$

where $\rho g h$ is the hydrostatic head. Rearranging equation **(15–13)** gives

$$\gamma = r\rho g h/2$$

which is identical with equation **(15–12)** derived on the basis of adhesive forces versus cohesive forces.

### The DuNoüy Ring Method

The *DuNoüy tensiometer* is widely used for measuring surface and interfacial tensions. The principle of the instrument depends on the fact that the force necessary to detach a platinum–iridium ring immersed at the surface or interface is proportional to the surface or interfacial tension. The force required to detach the ring in this manner is provided by a torsion wire and is recorded in dynes on a calibrated dial.

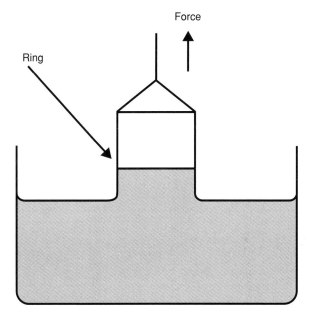

**Fig. 15–6.** Schematic of the tensiometer ring pulling a column of water above the surface before it brakes away.

The surface tension is given by the formula [compare with equation **(15–1)**]

$$\gamma = \frac{\text{Dial reading in dynes}}{2 \times \text{Ring circumference}} \times \text{Correction factor, } \beta$$
$$(15\text{–}14)$$

In effect, the instrument measures the weight of liquid pulled out of the plane of the interface immediately before the ring becomes detached (**Fig. 15–6**). A correction factor is necessary in equation **(15–15)** because the simple theory does not take into account certain variables such as the radius of the ring, the radius of the wire used to form the ring, and the volume of liquid raised out of the surface. Errors as large as 25% may occur if the correction factor is not calculated and applied. The method of calculating the correction factor has been described[13,14]; with care, a precision of about 0.25% can be obtained.

**EXAMPLE 15–5**

**DuNoüy Ring Method**

The published surface tension of water at 18°C is 73.05 dynes/cm, and the density, $\rho_1$, of water at this temperature is 0.99860 g/cm³. The density, $\rho_2$, of moist air that is, air saturated with the vapor of the liquid, water, at 18°C is 0.0012130. Therefore, $\rho_1 - \rho_2$, the density of water overlaid with air, is 0.99739 g/cm³. The dial reading in dynes or newtons on the tensiometer is equal to the mass, $M$, of the liquid lifted by the ring multiplied by the gravity constant, 980.665 cm/sec²; that is,

$$\text{Dial reading} = \text{M(g)} \times 980.665 \text{ cm/sec}^2$$

It is thus possible to obtain the mass $M$ of liquid lifted with the ring ($M = 0.7866$ g) before it breaks away from the water surface. The ring must be kept absolutely horizontal for accurate measurement. The volume, $V$, of water lifted above the free surface of water is calculated from the mass of water lifted and the density at 18°C, or

$$V = \frac{M}{\rho_1 - \rho_2} = \frac{0.7866}{0.99739} = 0.78866 \text{ cm}^3$$

**TABLE 15–3**
**SOME HARKINS AND JORDAN $\beta$ VALUES***

| $R^3 V$ | $\beta$ | | | |
| | $R/r = 30$ | $R/r = 40$ | $R/r = 50$ | $R/r = 60$ |
|---|---|---|---|---|
| 0.50 | 0.9402 | 0.9687 | 0.9876 | 0.9984 |
| 1.00 | 0.8734 | 0.9047 | 0.9290 | 0.9438 |
| 2.00 | 0.8098 | 0.8539 | 0.8798 | 0.9016 |
| 3.00 | 0.7716 | 0.8200 | 0.8520 | 0.8770 |
| 3.50 | 0.7542 | 0.8057 | 0.8404 | 0.8668 |

*From W. D. Harkins and H. F. Jordan, J. Am. Chem. Soc. **52,** 1751, 1930; H. L. Cupples, J. Phys. Chem. **51,** 1341, 1947.

The ring of the tensiometer has a radius, $R$, of 0.8078 cm, and $R^3 = 0.527122$ cm³. The radius, $r$, of the wire that forms the ring is 0.01877 cm. Two values, $R^3/V$ and $R/r$, are needed to enter the tables of Harkins and Jordan[13] to obtain the correction factor, $\beta$, by interpolation.

An abbreviated table of $R^3/V$ and $R/r$ values needed to obtain $\beta$ is given in Table 15–3. In this example $R^3/V = 0.52712/0.78866 = 0.66838$ and $R/r = 0.8078/0.01877 = 43.0368$. Introducing these values into Table VIII-C of Harkins and Jordan[13] and by interpolation, one obtains $\beta = 0.9471$ (18°C).

Finally, using equation (15–15), we obtain the surface tension for water at 18°C:

$$\gamma = \frac{M \times g}{4\pi R} \times \beta = \frac{(0.7866 \text{ g})(980.665 \text{ cm/sec}^2)}{4\pi(0.8078 \text{ cm})} \times 0.9471$$
$$= 71.97 \text{ dynes/cm} \quad \text{or} \quad 71.97 \text{ ergs/cm}^2$$

Without the correction factor, $\beta$, $\gamma$ is calculated here to be 75.99 dynes/cm. The value of $\gamma$ for water at 18°C is recorded in handbooks as approximately 73.05 dynes/cm. The error relative to the published value at 18°C is (73.05 – 71.97/73.05) × 100 = 1.48%.

The correction factor $\beta$ can be calculated from an equation rather than obtaining it from tabulated values of $R/r$ and $R^3/V$ as done in *Example 15–4*. Zuidema and Waters[15] suggested an equation for calculating $\beta$, as discussed by Shaw[16]:

$$(\beta - \alpha)^2 = \frac{4b}{\pi^2} \cdot \frac{1}{R^2} \cdot \frac{M \times g}{4\pi R(\rho_1 - \rho_2)} + c \quad (15\text{–}15)$$

**EXAMPLE 15–6**

**Surface Tension Correction Factor Calculation**

Use equation (15–15) to calculate $\beta$, the surface tension correction at 20°C, for $a = 0.7250$, $b = 0.09075$ m⁻¹ sec² for all tensiometer rings, and $c = 0.04534 - 1.6790 \, r/R$, with $R$ the radius of the ring in m, $r$ the radius in m of the wire from which the ring is constructed, $M$ the mass in kg of the liquid lifted above the liquid surface as the ring breaks away from the surface, $g$ the acceleration due to gravity in m/sec², $\rho_1$ the density of the liquid in kg/m⁻³, and $\rho_2$ the density of the air saturated with the liquid in kg/m³; that is, the upper phase of an interfacial system. With the following data, which must be expressed in SI units for use in equation (15–15), $\beta$ is calculated and used in equation (15–15) to obtain the corrected surface tension. The terms of equation (15–15) in SI units are $R = 0.012185$ m, $r = 0.0002008$ m, $M = 0.0012196$ kg, $g = 9.80665$ m/sec², $\rho_1 = 998.207$

kg/m³, and $\rho_2 = 1.2047$ kg/m³. Finally, $c = 0.04534 - 1.6790 \, r/R = 0.017671$. Substituting into equation (15–15), we have

$$(\beta - \alpha)^2 = \frac{4(0.09075 \text{ m}^{-1}\text{sec}^2)}{9.869604} \cdot \frac{1}{0.00014847 \text{ m}^2}$$
$$\cdot \frac{(0.0012196 \text{ kg})(9.80665 \text{ m/sec}^2)}{4(3.14159)(0.012185 \text{ m})(998.207 - 1.2047 \text{ kg/m}^3)}$$
$$+ \, 0.04534 - \frac{(1.6790)(0.0002008 \text{ m})}{0.012185 \text{ m}} = 0.0194077$$
$$+ \, 0.0176713 = 0.0370790$$
$$\beta - 0.7250 = \sqrt{0.0370790}$$
$$\beta = 0.7250 + 0.192559 = 0.918 \text{ (dimensionless) at } 20°\text{C}$$

The literature value of $\gamma$ for water at 20°C is 72.8 dynes/cm (or ergs/cm²) in cgs units. Using the uncorrected equation $\gamma = Mg/(4\pi R)$ and SI units, we obtain for water at 20°C

$$\gamma = (0.0012196 \text{ kg} \times 9.80665 \text{ m/sec}^2)/(4\pi \times 0.012185 \text{ m})$$
$$= 0.078109 \text{ kg/sec}^2$$

Multiplying numerator and denominator by m² yields the result 0.07811 joule/m², and expressing the value in mJ/m², we have 78.11 mJ/m². This is a useful way to express surface tension in SI units because the value 78.11 is numerically the same as that in the cgs system; namely, 78.11 ergs/cm² (see *Example 15–3*). To correct the value $\gamma = Mg/(4\pi R)$ expressed either in cgs or SI units, we multiply by the Harkins and Jordan[13] or the Zuidema and Waters[15] value for $\beta$ at a given liquid density and temperature, $M$ value, and ring dimensions.

For the particular case in this example,

$$\gamma = \frac{Mg}{4\pi R} \times \beta = 78.11 \text{ ergs/cm}^2 \text{ (or mJ/m}^2) \times 0.918$$
$$= 71.7 \text{ ergs/cm}^2$$

The error in the Zuidema and Waters[15] value of 71.7 mJ/m² relative to the literature value, 72.8 mJ/m², at 20°C is $(72.8 - 71.7)/72.8 \times 100 = 1.51\%$.

The modern variant of the surface tensiometer, the Sigma 70 Surface Tensiometer from KSV Instruments (Monroe, Conn.), offers advanced, microprocessor-based measurement using either the Wilhelmy plate or DuNoüy ring method. The Wilhelmy plate method is based on the measurement of the force necessary to detach a plate from the surface of a liquid.

## Spreading Coefficient

When a substance such as oleic acid is placed on the surface of water, it will spread as a film if the force of adhesion between the oleic acid molecules and the water molecules is greater than the cohesive forces between the oleic acid molecules themselves. The term *film* used here applies to a *duplex film* as opposed to a monomolecular film. Duplex films are sufficiently thick (100 Å or more) so that the surface (boundary between oleic acid and air) and interface (boundary between water and oleic acid) are independent of one another.

The *work of adhesion*, which is the energy required to break the attraction between the unlike molecules, is obtained by reference to **Figure 15–7**. **Figure 15–7a** shows a hypothetical cylinder (cross-sectional area 1 cm²) of the sublayer liquid, S, overlaid with a similar section of the spreading liquid, L.

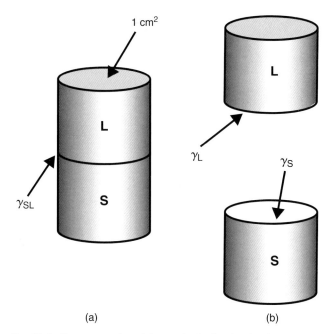

**Fig. 15–7.** Representation of the work of adhesion involved in separating a substrate and an overlaying liquid.

By equation (**15–2**), surface or interfacial work is equal to surface tension multiplied by the area increment. The work required to separate the two sections of liquid in **Figure 15–7**, each with a cross-sectional area of 1 cm², is therefore numerically related to the surface or interfacial tension involved, the area increment being unity:

$$\text{Work} = \text{Surface tension} \times \text{Unit area change}$$

Accordingly, it is seen in **Figure 15–7b** that the work done is equal to the newly created surface tensions, $\gamma_L$ and $\gamma_S$, minus the interfacial tension, $\gamma_{LS}$, that has been destroyed in the process. The work of adhesion is thus

$$W_a = \gamma_L + \gamma_S - \gamma_{LS} \qquad (15\text{--}16)$$

The *work of cohesion*, required to separate the molecules of the spreading liquid so that it can flow over the sublayer, is obtained by reference to **Figure 15–8**. Obviously, no interfacial tension exists between the like molecules of the liquid, and when the hypothetical 1-cm² cylinder in **Figure 15–8a** is divided, two new surfaces are created in **Figure 15–8b**, each with a surface tension of $\gamma_L$. Therefore, the work of cohesion is

$$W_c = 2\gamma_L \qquad (15\text{--}17)$$

With reference to the spreading of oil on a water surface, spreading occurs if the work of adhesion (a measure of the force of attraction between the oil and the water) is greater than the work of cohesion. The term $(W_a - W_c)$ is known as the *spreading coefficient*, $S$; if it is positive, the oil will spread over a water surface. Equations (**15–16**) and (**15–17**) can be written as

$$S = W_a - W_c = (\gamma_L + \gamma_S - \gamma_{LS}) - 2\gamma_L \quad (15\text{--}18)$$

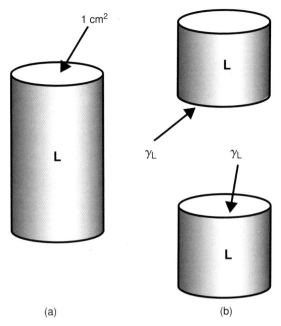

**Fig. 15–8.** Representation of the work of cohesion involved in separating like molecules in a liquid.

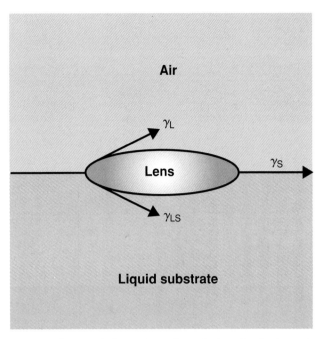

**Fig. 15–9.** Forces existing at the surface of a lens floating in a substrate liquid.

where $\gamma_S$ is the surface tension of the sublayer liquid, $\gamma_L$ is the surface tension of the spreading liquid, and $\gamma_{LS}$ is the interfacial tension between the two liquids. Rearranging equation (15–18) gives

$$S = \gamma_S - \gamma_L - \gamma_{LS} \qquad (15–19)$$

or

$$S = \gamma_S - (\gamma_S + \gamma_{LS}) \qquad (15–20)$$

Figure 15–9 shows a lens of material placed on a liquid surface (e.g., oleic acid on water). From equation (15–20), one sees that spreading occurs ($S$ is positive) when the surface tension of the sublayer liquid is greater than the sum of the surface tension of the spreading liquid and the interfacial tension between the sublayer and the spreading liquid. If $(\gamma_L + \gamma_{LS})$ is larger than $\gamma_S$, the substance forms globules or a floating lens and fails to spread over the surface. An example of such a case is mineral oil on water.

Spreading can also be thought of in terms of surface free energy. Thus, the added substance will spread if, by so doing, it reduces the surface free energy of the system. Put another way, if the surface free energy of the new surface and the new interface is less than the free energy of the old surface, spreading will take place.

Up to this point, the discussion has been restricted to *initial* spreading. Before equilibrium is reached, however, the water surface becomes saturated with the spreading material, which in turn becomes saturated with water. If we use a prime (′) to denote the values following equilibration (i.e., final rather than initial values), then the new surface tensions are $\gamma_{S'}$ and $\gamma_{L'}$. When mutual saturation has taken place, the spreading coefficient may be reduced or may even become

negative. This means that although initial spreading of the material may occur on the liquid substrate, it can be followed by coalescence of the excess material into a lens if $S'$ becomes negative in value. This reversal of spreading takes place when $\gamma_{S'}$ becomes less than $(\gamma_{LS} + \gamma_{L'})$. Note that the value of $\gamma_{LS}$ does not change because the interfacial tension is determined under conditions of mutual saturation.

### EXAMPLE 15–7

**Spreading Benzene over Water**

If the surface tension of water $\gamma_S$ is 72.8 dynes/cm at 20°C, the surface tension of benzene, $\gamma_L$, is 28.9 dynes/cm, and the interfacial tension between benzene and water, $\gamma_{LS}$, is 35.0 dynes/cm, what is the initial spreading coefficient? Following equilibration, $\gamma_{S'}$ is 62.2 dynes/cm and $\gamma_{L'}$ is 28.8 dynes/cm. What is the final spreading coefficient? We have

$S = 72.8 - (28.9 + 35.0) = 8.9$ dynes/cm (or 8.9 ergs/cm$^2$)
$S' = 62.2 - (28.8 + 35.0) = -1.6$ dynes/cm

Therefore, although benzene spreads initially on water, at equilibrium there is formed a saturated monolayer with the excess benzene (saturated with water) forming a lens.

In the case of organic liquids spread on water, it is found that although the initial spreading coefficient may be positive or negative, the final spreading coefficient always has a negative value. Duplex films of this type are unstable and form monolayers with the excess material remaining as a lens on the surface. The initial spreading coefficients of some organic liquids on water at 20°C are listed in **Table 15–4**.

It is important to consider the types of molecular structures that lead to high spreading coefficients. Oil spreads over water because it contains polar groups such as COOH or OH.

**TABLE 15–4**
**INITIAL SPREADING COEFFICIENT, S, AT 20°C***

| Substance | S (dynes/cm) |
|---|---|
| Ethyl alcohol | 50.4 |
| Propionic acid | 45.8 |
| Ethyl ether | 45.5 |
| Acetic acid | 45.2 |
| Acetone | 42.4 |
| Undecylenic acid | 32 (25°C) |
| Oleic acid | 24.6 |
| Chloroform | 13 |
| Benzene | 8.9 |
| Hexane | 3.4 |
| Octane | 0.22 |
| Ethylene dibromide | −3.19 |
| Liquid petrolatum | −13.4 |

*From W. D. Harkins, *The Physical Chemistry of Surface Films*, Reinhold, New York, 1952, pp. 44–45.

Hence, propionic acid and ethyl alcohol should have high values of $S$, as seen in **Table 15–4**. As the carbon chain of an acid, oleic acid, for example, increases, the ratio of polar–nonpolar character decreases and the spreading coefficient on water decreases. Many nonpolar substances, such as liquid petrolatum ($S = -13.4$), fail to spread on water. Benzene spreads on water not because it is polar but because the cohesive forces between its molecules are much weaker than the adhesion for water.

The applications of spreading coefficients in pharmacy should be fairly evident. The surface of the skin is bathed in an aqueous–oily layer having a polar–nonpolar character similar to that of a mixture of fatty acids. For a lotion with a mineral oil base to spread freely and evenly on the skin, its polarity and hence its spreading coefficient should be increased by the addition of a surfactant. The relation between spreading, HLB (*hydrophile–lipophile balance*), and emulsion stability has been studied.[17] Surfactant blends of varying HLBs were added to an oil, a drop of which was then placed on water. The HLB of the surfactant blend that caused the oil drop to spread was related to the required HLB of the oil when used in emulsification. (See hydrophile-lipophile classification in this chapter).

# ADSORPTION AT LIQUID INTERFACES

Surface free energy was defined previously as the work that must be done to increase the surface by unit area. As a result of such an expansion, more molecules must be brought from the bulk to the interface. The more work that has to be expended to achieve this, the greater is the surface free energy.

Certain molecules and ions, when dispersed in the liquid, move of their own accord to the interface. Their concentration at the interface then exceeds their concentration in the bulk of the liquid. Obviously, the surface free energy and the surface tension of the system are automatically reduced. Such a phenomenon, where the added molecules are partitioned in favor of the interface, is termed *adsorption*, or, more correctly, *positive adsorption*. Other materials (e.g., inorganic electrolytes) are partitioned in favor of the bulk, leading to *negative adsorption* and a corresponding increase in surface free energy and surface tension. Adsorption, as will be seen later, can also occur at solid interfaces. Adsorption should not be confused with *absorption*. The former is solely a surface effect, whereas in absorption, the liquid or gas being absorbed penetrates into the capillary spaces of the absorbing medium. The taking up of water by a sponge is absorption; the concentrating of alkaloid molecules on the surface of clay is adsorption.

## Surface-Active Agents

It is the amphiphilic nature of surface-active agents that causes them to be absorbed at interfaces, whether these are liquid–gas or liquid–liquid interfaces. Thus, in an aqueous dispersion of amyl alcohol, the polar alcoholic group is able to associate with the water molecules. The nonpolar portion is rejected, however, because the adhesive forces it can develop with water are small in comparison to the cohesive forces between adjacent water molecules. As a result, the amphiphile is adsorbed at the interface. The situation for a fatty acid at the air–water and oil–water interface is shown in **Figure 15–10**. At the air–water interface, the lipophilic chains are directed upward into the air; at the oil–water interface, they are associated with the oil phase. For the amphiphile to be concentrated at the interface, it must be balanced with the proper amount of water- and oil-soluble groups. If the molecule is too hydrophilic, it remains within the body of the aqueous phase and exerts no effect at the interface. Likewise, if it is too lipophilic, it dissolves completely in the oil phase and little appears at the interface.

## Systems of Hydrophile–Lipophile Classification

Griffin[18] devised an arbitrary scale of values to serve as a measure of the hydrophilic–lipophilic balance of surface-active agents. By means of this number system, it is possible to establish an HLB range of optimum efficiency for each class of surfactant, as seen in **Figure 15–11**. The higher the HLB of an agent, the more hydrophilic it is. The Spans, sorbitan esters manufactured by ICI Americas Inc., are lipophilic and have low HLB values (1.8–8.6); the Tweens, polyoxyethylene derivatives of the Spans, are hydrophilic and have high HLB values (9.6–16.7).

The HLB of a nonionic surfactant whose only hydrophilic portion is polyoxyethylene is calculated by using the formula

$$HLB = E/5 \qquad (15\text{–}21)$$

## KEY CONCEPT  SURFACTANTS

Molecules and ions that are adsorbed at interfaces are termed *surface-active agents* or *surfactants*. An alternative term is *amphiphile*, which suggests that the molecule or ion has a certain affinity for both polar and nonpolar solvents. Depending on the number and nature of the polar and nonpolar groups present, the amphiphile may be predominantly *hydrophilic* (water-loving), *lipophilic* (oil-loving), or reasonably well balanced between these two extremes. For example, straight-chain alcohols, amines, and acids are amphiphiles that change from being predominantly hydrophilic to lipophilic as the number of carbon atoms in the alkyl chain is increased. Thus, ethyl alcohol is miscible with water in all proportions. In comparison, the aqueous solubility of amyl alcohol, $C_5H_{11}OH$, is much reduced, whereas cetyl alcohol, $C_{16}H_{33}OH$, may be said to be strongly lipophilic and insoluble in water.

where $E$ is the percentage by weight of ethylene oxide. A number of polyhydric alcohol fatty acid esters, such as glyceryl monostearate, can be estimated by using the formula

$$HLB = 20 \left(1 - \frac{S}{A}\right) \qquad (15\text{--}22)$$

where $S$ is the saponification number of the ester and $A$ is the acid number of the fatty acid. The HLB of polyoxyethylene sorbitan monolaurate (Tween 20), for which $S = 45.5$ and $A = 276$, is

$$HLB = 20 \left(1 - \frac{45.5}{276}\right) = 16.7$$

The HLB values of some commonly used amphiphilic agents are given in **Table 15–5**.

The oil phase of an oil-in-water (O/W) emulsion requires a specific HLB, called the *required hydrophile–lipophile balance* (RHLB). A different RHLB is required to form a water-in-oil (W/O) emulsion from the same oil phase. The RHLB values for both O/W and W/O emulsions have been determined empirically for a number of oils and oil-like substances, some of which are listed in **Table 15–6**.

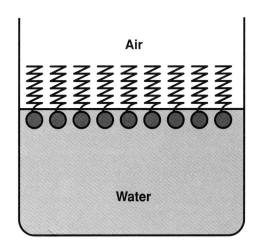

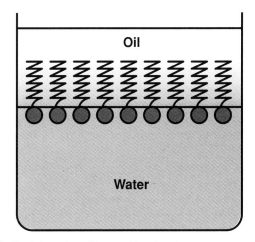

**Fig. 15–10.** Adsorption of fatty acid molecules at a water–air interface (upper panel) and a water–oil interface (lower panel).

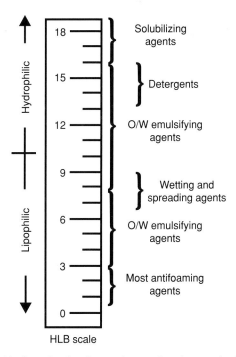

**Fig. 15–11.** A scale showing surfactant function on the basis of hydrophilic–lipophilic balance (HLB) values. Key: O/W = oil in water.

**TABLE 15–5**

## HYDROPHILIC–LIPOPHILIC BALANCE (HLB) VALUES OF SOME AMPHIPHILIC AGENTS

| Substance | HLB |
|---|---|
| Oleic acid | 1 |
| Polyoxyethylene sorbitol beeswax derivative (G-1706) | 2.0 |
| Sorbitan tristearate | 2.1 |
| Glyceryl monostearate | 3.8 |
| Sorbitan monooleate (Span 80) | 4.3 |
| Diethylene glycol monostearate | 4.7 |
| Glyceryl monostearate, self-emulsifying (Tegin) | 5.5 |
| Diethylene glycol monolaurate | 6.1 |
| Sorbitan monolaurate (Span 20) | 8.6 |
| Polyethylene lauryl ether (Brij 30) | 9.5 |
| Gelatin (Pharmagel B) | 9.8 |
| Methyl cellulose (Methocel 15 cps) | 10.5 |
| Polyoxyethylene lauryl ether (G-3705) | 10.8 |
| Polyoxyethylene monostearate (Myrj 45) | 11.1 |
| Triethanolamine oleate | 12.0 |
| Polyoxyethylene alkyl phenol (Igepal Ca-630) | 12.8 |
| Polyethylene glycol 400 monolaurate | 13.1 |
| Polyoxyethylene sorbitan monooleate (Tween 80) | 15.0 |
| Polyoxyethylene sorbitan monolaurate (Tween 20) | 16.7 |
| Polyoxyethylene lauryl ether (Brij 35) | 16.9 |
| Sodium oleate | 18.0 |
| Potassium oleate | 20 |
| Sodium lauryl sulfate | 40 |

**TABLE 15–6**

## REQUIRED HYDROPHILIC–LIPOPHILIC BALANCE (HLB) FOR SOME OIL-PHASE INGREDIENTS FOR OIL-IN-WATER (O/W) AND WATER-IN-OIL (W/O) EMULSIONS*

| | O/W | W/O |
|---|---|---|
| Cottonseed oil | 6–7 | — |
| Petrolatum | 8 | — |
| Beeswax | 9–11 | 5 |
| Paraffin wax | 10 | 4 |
| Mineral oil | 10–12 | 5–6 |
| Methyl silicone | 11 | — |
| Lanolin, anhydrous | 12–14 | 8 |
| Carnauba wax | 12–14 | — |
| Lauryl alcohol | 14 | — |
| Caster oil | 14 | — |
| Kerosene | 12–14 | — |
| Cetyl alcohol | 13–16 | — |
| Stearyl alcohol | 15–16 | — |
| Carbon tetrachloride | 16 | — |
| Lauric acid | 16 | — |
| Oleic acid | 17 | — |
| Stearic acid | 17 | — |

*From *The Atlas HLB System,* Atlas Chemical Industries, Wilmington, DE; P. Becher, *Emulsions, Theory and Practice,* 2nd Ed., Reinhold, New York, 1966, p. 249.

**EXAMPLE 15–8**

### Calculation of HLB Value for Oil-in-Water Emulsions

For the oil-in-water emulsion,

| Ingredient | Amount | RHLB (O/W) |
|---|---|---|
| 1. Beeswax | 15 g | 9 |
| 2. Lanolin | 10 g | 12 |
| 3. Paraffin wax | 20 g | 10 |
| 4. Cetyl alcohol | 5 g | 15 |
| 5. Emulsifier | 2 g | |
| 6. Preservative | 0.2 g | |
| 7. Color | As required | |
| 8. Water, purified q.s. | 100 g | |

Key: RHLB = required hydrophilic–lipophilic balance value.

One first calculates the overall RHLB of the emulsion by multiplying the RHLB of each oil-like component (items 1–4) by the weight fraction that each oil-like component contributes to the oil phase. The total weight of the oil phase is 50 g. Therefore,

| | |
|---|---|
| Beeswax | $15/50 \times 9 = 2.70$ |
| Lanolin | $10/50 \times 12 = 2.40$ |
| Paraffin | $20/50 \times 10 = 4.00$ |
| Cetyl alcohol | $5/50 \times 15 = 1.50$ |
| Total RHLB for the emulsion | $20/50 \times 10 = 10.60$ |

Next, one chooses a blend of two emulsifying agents, one with an HLB above and the other with an HLB below the required HLB of the emulsion (RHLB = 10.6 in this example). From Table 15–5, we choose Tween 80, with an HLB of 15, and Span 80, with an HLB of 4.3.

The formula for calculating the weight percentage of Tween 80 (surfactant with the higher HLB) is

$$\% \text{ Tween } 80 = \frac{\text{RHLB} - \text{HLB low}}{\text{HLB high} - \text{HLB low}} \qquad (15\text{–}23)$$

where HLB high is for the higher value, 15, and HLB low is for the lower value, 4.3. We have

$$\% \text{ Tween } 80 = \frac{10.6 - 4.3}{15.0 - 4.3} = 0.59$$

Two grams of emulsifier has been estimated as proper protection for the O/W emulsion. Therefore, 2.0 g × 0.59 = 1.18 g of Tween 80 is needed and the remainder, 0.82 g, must be supplied by Span 80 for the 100-g emulsion.

The choice of the mixture of emulsifiers and the total amount of the emulsifier phase is left to the formulator, who determines these unknowns over time by preparation and observation of the several formulas chosen.

A mathematical formula for determining the minimum amount of surfactant mixture was suggested by Bonadeo[19]:

$$Q_s = \frac{6(\rho_s/\rho)}{10 - 0.5 \cdot \text{RHLB}} + \frac{4Q}{1000} \qquad (15\text{–}24)$$

where $\rho_s$ is the density of the surfactant mixture, $\rho$ is the density of the dispersed (internal) phase, and $Q$ is the percentage of the dispersant (continuous phase) of the emulsion. The required HLB, written as RHLB, is the HLB of the oil phase needed to form an O/W or W/O emulsion.

**W/O and O/W Formulations**

We wish to formulate two products, (a) a W/O and (b) an O/W emulsion, containing 40 g of a mixed oil phase and 60 g of water.

(a) The oil phase consists of 70% paraffin and 30% beeswax. The density of the oil phase is 0.85 g/cm³ and the density of the aqueous phase is about 1 g/cm³ at room temperature. The density of the mixture of surfactants for the W/O emulsion is 0.87 g/cm³. The required HLB values of paraffin and of beeswax for a W/O emulsion are 4.0 and 5.0, respectively.

The amount $Q_s$ in grams of a mixture of sorbitan tristearate (HLB = 2.1) and diethylene glycol monostearate (HLB = 4.7) to obtain a *water-in-oil emulsion* is obtained by the use of equation (15–24), first calculating the RHLB of the oil phase:

$$RHLB = (4 \times 0.70) + (5 \times 0.30) = 4.3$$

$$Q_s = \frac{6(0.87/1)}{10 - (0.5 \times 4.3)} + \frac{4 \times 40}{1000} = 0.82 \text{ g}$$

Note that for a W/O emulsion we used the density of the internal phase, $\rho_{water} \cong 1$, and the percentage of dispersant, oil, 40%.

(b) The RHLB of the oil phase, 70% paraffin and 30% beeswax, for an O/W emulsion is

$$RHLB = (0.70 \times 10) + (0.3 \times 9) = 9.7$$

and the total amount of surfactant mixture is

$$Q_s + \frac{6(1.05/0.85)}{10 - (0.5 \times 9.7)} + \frac{4 \times 60}{1000} = 1.68 \text{ g}$$

For an O/W emulsion, we used the density $\rho$ of the oil as the internal phase and the percentage of dispersant as the aqueous phase.

For the amount of surfactant mixture in the W/O emulsion we can raise the value $Q_s$ roughly to 1.0 g and for the O/W emulsion to about 2.0 g. We can then calculate the weights of the two emulsifying agents for each emulsion, using the equation

$$\% \text{ Surfactant of higher HLB} = \frac{RHLB - HLB \text{ low}}{HLB \text{ high} - HLB \text{ low}} \quad (15\text{–}25)$$

For the W/O emulsion, the percentage by weight of diethylene glycol monostearate (HLB = 4.7) combined with sorbitan tristearate (HLB = 2.1) is

$$\% \text{ Diethylene glycol monostearate} = \frac{4.3 - 6.1}{4.7 - 2.1} = 0.85 \text{ g}$$

or   85% of 1 g

The fraction or percentage of sorbitan monostearate is therefore 0.15 g, or 15% of the 1 g of mixed emulsifier.

For the O/W emulsion, the percentage by weight of Tween 80 (HLB = 15) combined with diethylene glycol monolaurate (HLB = 6.1) is

$$\% \text{ Tween } 80 = \frac{9.7 - 6.1}{15 - 6.1} = 0.40 \quad \text{or} \quad 40\%$$

The fraction or percentage of diethylene glycol monolaurate is therefore 0.60, or 60%, and 0.40, or 40%, of a 2-g mixture of emulsifier phase = 0.8 g of Tween 80. The remainder, 1.2 g, is the amount of diethylene glycol monolaurate in the 2-g emulsifier phase.

---

Other scales of HLB have been developed, although none has gained the acceptance afforded the HLB system of Griffin. A titration method and other techniques for determining the hydrophile–lipophile character of surfactants have been proposed.[20–22]

## Types of Monolayer at Liquid Surfaces

For convenience of discussion, absorbed materials are divided into two groups: those that form "soluble" monolayers and those that form "insoluble" films. The distinction is made on the basis of the solubility of the adsorbate in the liquid subphase. Thus, amyl alcohol may be said to form a soluble monolayer on water, whereas cetyl alcohol would form an insoluble film on the same sublayer. It must be emphasized that this is really only an arbitrary distinction, for the insoluble films are, in effect, the limiting case of those compounds that form soluble monolayers at liquid interfaces. There are, however, important practical reasons why such a classification is made.

It will become apparent in the following sections that three interrelated parameters are important in studying liquid interfaces: (a) surface tension, $\gamma$; (b) surface excess, $\Gamma$, which is the amount of amphiphile per unit area of surface in excess of that in the bulk of the liquid; and (c) $c$, the concentration of amphiphile in the bulk of the liquid. As we shall see, it is relatively easy with soluble monolayers to measure surface tension and $c$ to compute the surface excess. With insoluble monolayers, $c$ is taken to be zero, whereas surface tension and surface excess can be obtained directly. Materials that lie on the borderline between soluble and insoluble systems can be studied by either approach, and, invariably, similar results are obtained.

Data obtained from such studies are of increasing biologic and pharmaceutical interest. For example, emulsions are stabilized by the presence of an interfacial film between the oil and water phases. Knowledge of the area occupied by each amphiphilic molecule at the interface is important in achieving optimum stability of the emulsion. The efficiency of wetting and detergent processes depends on the concentration of material adsorbed. Monolayers of adsorbed amphiphiles can be used as in vitro models for biologic membranes that are thought to consist of two monolayers placed back to back with the hydrocarbon chains intermeshed. Consequently, these model systems are finding increasing application for in vitro studies of drug absorption across biologic membranes. Studies of interfacial adsorption also provide valuable information on the dimensions of molecules because it is possible to calculate the areas occupied by amphiphilic molecules.

## Soluble Monolayers and the Gibbs Adsorption Equation

The addition of amphiphiles to a liquid system leads to a reduction in surface tension owing to these molecules or ions being adsorbed as a monolayer. Adsorption of amphiphiles in these binary systems was first expressed quantitatively by Gibbs[23] in 1878:

$$\Gamma = -\frac{c}{RT}\frac{d\gamma}{dc} \quad (15\text{–}26)$$

$\Gamma$ is the surface excess or surface concentration, that is, the amount of the amphiphile per unit area of surface in excess of that in the bulk of the liquid, $c$ is the concentration of amphiphile in the liquid bulk, $R$ is the gas constant, $T$ is the absolute temperature, and $d\gamma/dc$ is the change in surface tension of the solution with change of bulk concentration of the substance. The derivation of equation (15–26) is given in the following paragraphs.

Recall that the free energy change of a bulk phase containing two components is written as

$$dG = -S\,dT + V\,dp + \mu_1\,dn_1 + \mu_2\,dn_2$$

Two immiscible bulk phases can be considered to be separated by an interface or "surface phase" in which the contribution to the volume is ignored, and a new energy term, $\gamma\,dA$ [equation (15–2)], is introduced to account for the work involved in altering the surface area, $A$. The surface tension, $\gamma$, is the work done at a constant temperature and pressure per unit increase of surface area. The new work done on the surface phase is equal to the surface free energy increase, $dG^s$. Therefore, we can write

$$dG^S = -S^S\,dT + \gamma\,dA + \mu_1^S\,dn_1^S + \mu_2^S\,dn_2^S \quad (15\text{–}27)$$

At equilibrium, the free energy of the entire system is zero under the conditions of constant temperature, pressure, and surface area. Because no matter passes in or out of the system as a whole, the chemical potential of a component $i$ is the same in the two bulk phases as it is in the surface phase, $s$:

$$\mu_{i\alpha} = \mu_{i\beta} = \mu_{is} \quad\quad (15\text{–}28)$$

Such a system consisting of two immiscible liquids, water, $\alpha$, and oleic acid, $\beta$, separated by the surface phase, $s$, is shown in **Figure 15–12a**. Equation (15–27) can be integrated at constant temperature and composition to give the surface free energy,

$$G^S = \gamma A + n_1^S \mu_1^S + n_2^S \mu_2^S \quad\quad (15\text{–}29)$$

Because the surface free energy depends only on the state of the system, $dG^s$ is an exact differential and can be obtained by general differentiation of equation (15–29) under the condition of variable composition,

$$dG^S = \gamma\,dA + A\,d\gamma + n_1^S\,d\mu_1^S + n_2^S\,\mu_2^S + \mu_1^S\,dn_1^S + \mu_2^S\,dn_2^S$$
$$(15\text{–}30)$$

Comparing this result with equation (15–27) shows that

$$A\,d\gamma + S^S\,dT + n_1^S\,d\mu_1^S + n_2^S\,d\mu_2^S = 0 \quad (15\text{–}31)$$

and at constant temperature,

$$A\,d\gamma + n_1^S\,d\mu_1^S + n_2^S\,d\mu_2^S = 0 \quad\quad (15\text{–}32)$$

When equation (15–32) is divided through by the surface area $A$, and $n_1{}^s/A$ and $n_2{}^s/A$ are given the symbols $\Gamma_1$ and $\Gamma_2$, respectively, we obtain

$$d\gamma + \Gamma_1\,d\mu_1^S + \Gamma_2\,d\mu_2^S = 0 \quad\quad (15\text{–}33)$$

As expressed by equation (15–33), the chemical potentials of the components in the surface are equal to those in the bulk

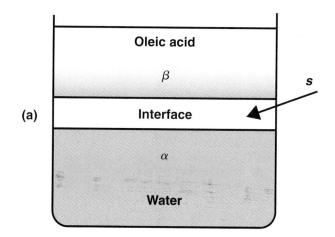

**(a)**

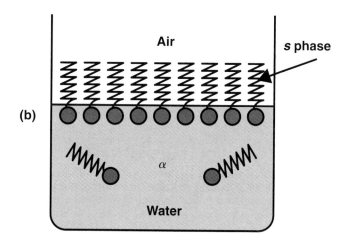

**(b)**

**Fig. 15–12.** A system consisting of oleic acid and water. (*a*) Graphic description of the two bulk phases, $\alpha$ and $\beta$, and the interface, *s*. (*b*) Condition where only the $\alpha$ phase and the surface or *s* phase need to be considered.

phases, provided that the system is in equilibrium at constant temperature, pressure, and surface area.

Now consider a single-phase solution of oleic acid (solute or component 2) in water (solvent or component 1) as shown in **Figure 15–12b**. Under these circumstances, it is possible to drop the superscripts on the chemical potentials and write

$$d\gamma + \Gamma_1\,d\mu_1 + \Gamma_2\,d\mu_2 = 0 \quad\quad (15\text{–}34)$$

where $\Gamma_1$ and $\Gamma_2$ are the number of moles of the components per unit area in the surface and $\mu_1$ and $\mu_2$ are the chemical potentials of the two components in the solution.

It is possible to make an arbitrary choice of the surface, and we do so in a manner that makes $\Gamma_1$ equal to zero, that is, we arrange the boundary so that none of the solvent is present in the surface (**Fig. 15–12b**). Then, equation (15–34) becomes

$$d\gamma + \Gamma_2\,d\mu_2 = 0 \quad\quad (15\text{–}35)$$

and

$$\Gamma_2 = -\left(\frac{\partial\gamma}{\partial\mu_2}\right)_T \quad\quad (15\text{–}36)$$

The chemical potential of the solute can be expressed in terms of the activity using the equation

$$\mu_2 = \mu^\circ + RT \ln \alpha_2$$

By differentiating at constant temperature, one obtains

$$\partial\mu_2 = RT\partial \quad \ln \quad \alpha_2 \qquad (15\text{-}37)$$

Substituting this value in equation (15–36) produces the result

$$\Gamma_2 = -\frac{1}{RT}\left(\frac{\partial\gamma}{\partial \ln \alpha_2}\right)_T \qquad (15\text{-}38)$$

From differential calculus, if $y = \ln a_2$, then $d \ln a_2 = da_2/a_2$. Substituting this result in equation (15–38) results in the *Gibbs adsorption equation*,

$$\Gamma_2 = -\frac{\alpha_2}{RT}\left(\frac{\partial\gamma}{\partial \alpha_2}\right)_T \qquad (15\text{-}39)$$

This is equation (15–26), which was given in terms of concentration, $c$, instead of activity. If the solution is dilute, $a_2$ can be replaced by $c$ without introducing a significant error.

When the surface tension, $\gamma$, of a surfactant is plotted against the logarithm of the surfactant activity or concentration, $\log c_2$, the plot takes on the shape shown in **Figure 15–13**. The initial curved segment A–B is followed by a linear segment, B–C, along which there is a sharp decrease in surface tension as $\log c_2$ increases. The point C corresponds to the critical micelle concentration (CMC), the concentration at which micelles form in the solution. Beyond the CMC, the line becomes horizontal because further additions of surfactant are no longer being accompanied by a decrease in surface tension. Along the linear segment B–C, the surface excess $\Gamma$ is constant because from equation (15–38), replacing activity with concentration, we find

$$\Gamma_2 = -\frac{1}{RT}\left(\frac{\partial\gamma}{\partial \ln c_2}\right)_T \qquad (15\text{-}40)$$

The slope $\partial\gamma/\partial \ln c_2$ reaches a limiting value and remains constant. Saturation adsorption of the surfactant has been

reached at point B; that is, $\Gamma_2$ does not increase further as the bulk concentration increases. However, the surface tension decreases greatly until point C is reached. Within the segment B–C of the curve, the surfactant molecules are closely packed at the surface and the surface area occupied per molecule is constant. Both the surface excess $\Gamma_2$ and the area per surfactant molecule can be calculated using equation (15–40).

**Calculation of Area per Molecule of a Surfactant**

The limiting slope of a plot of $\gamma$ versus $\ln c_2$ for a nonionic surfactant, $C_{12}H_{25}O(CH_2CH_2O)_{12}H_2$, is $\partial\gamma/\partial \ln c_2 = -5.2937$ dynes/cm at 23.0°C. Calculate $\Gamma_2$ and the area per molecule of this surfactant.

From the Gibbs adsorption equation (15–40),

$$\Gamma_2 = \left(\frac{1}{8.3143 \times 10^7 \text{ergs/deg mole} \times 296.15 \text{ K}}\right)(-5.2937 \text{ dyne/cm})$$

$$\Gamma_2 = 2.15 \times 10^{-10} \text{ mole/cm}^2$$

The surface excess, $2.15 \times 10^{-10}$ mole/cm$^2$, is multiplied by $6.0221 \times 10^{23}$ mole$^{-1}$, Avogadro's number, to obtain molecules/cm$^2$. The reciprocal then gives the area per molecule:

$$\text{Area/molecule} = \frac{1}{6.0221 \times 10^{23} \text{ molecule/mole} \times 2.15 \times 10^{-10} \text{ mole/cm}^2}$$

$$= 7.72 \times 10^{-15} \text{cm}^2/\text{molecule} = 77 \text{ Å}^2/\text{molecule}$$

The validity of the Gibbs equation has been verified experimentally. One of the more ingenious methods is due to McBain and Swain,[24] who literally fired a small microtome blade across a liquid surface so as to collect the surface layer. Analysis of the liquid scooped up and collected by the speeding blade agreed closely with that predicted by the Gibbs equation. Radioactive techniques using weak beta emitters also have been successfully used.[25]

## Insoluble Monolayers and the Film Balance

Insoluble monolayers have a fascinating history that goes back to before the American Revolution. During a voyage to England in 1757, Benjamin Franklin observed, as had seamen for centuries before him, that when cooking grease was thrown from the ship's galley onto the water, the waves were calmed by the film that formed on the surface of the sea. In 1765, Franklin followed up this observation with an experiment on a half-acre pond in England and found that the application of 1 teaspoonful of oil was just sufficient to cover the pond and calm the waves. In 1899, Lord Rayleigh showed that when small amounts of certain slightly soluble oils were placed on a clean surface of water contained in a trough, they spread to form a layer one molecule thick (monomolecular layer). Prior to Rayleigh's work, a woman named Agnes Pockels, from Lower Saxony, Germany, who had no formal scientific training, developed a "film balance" for studying insoluble monolayers. She carried out a series of experiments, which she summarized in a letter to Lord Rayleigh in January 1881. In fact, she invented the film balance in

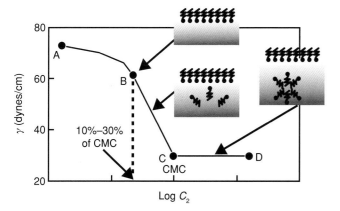

**Fig. 15–13.** Decrease in the surface tension of water when a strait-chain amphiphile is added. Key: CMC = critical micelle concentration. (Replotted from H. Schott, J. Pharm. Sci. **69**, 852, 1980.)

1883, more than 30 years before Langmuir, whose name is normally associated with this type of apparatus. These and other early contributions in the area of surface phenomena are described in a series of papers by Giles and Forrester.[26]

Knowing the area of the film and the volume of the spreading liquid, it should be possible to compute the thickness of such films. The film thickness is equal to the length of the molecules standing in a vertical position on the surface when the molecules are packed in closest arrangement. Furthermore, if the molecular weight and the density of the spreading oil are known, the cross-sectional area available to the molecules should be easily computed.

## EXAMPLE 15–11

### Calculation of the Length and the Cross-Sectional Area of a Fatty Acid Molecule

We noted that Benjamin Franklin placed 1 teaspoonful ($\sim$5 cm$^3$) of a fatty acid "oil" on a half-acre ($\sim$2 $\times$ 10$^7$ cm$^2$) pond. Assume that the acid, having a molecular weight of 300 and a density of 0.90 g/cm$^3$, was just sufficient to form a condensed monomolecular film over the entire surface. What was the length and the cross-sectional area of the fatty acid molecule?

(a) The thickness of oil on the pond is approximately equal to the length of the vertically oriented fatty acid molecule:

$$\frac{5 \text{ cm}}{2 \times 10^7 \text{ cm}^2} = 25 \times 10^{-8} \text{ cm} = 25 \text{ Å}$$

(b)

$$5 \text{ cm}^3 \times 0.9 \text{ g/cm}^3 = 4.5 \text{ g}$$

$$\frac{4.5 \text{ g}}{300 \text{ g/mole}} = 0.015 \text{ mole}$$

$$0.015 \text{ mole} \times 10^{23} \text{ molecules/mole} = 9 \times 10^{21} \text{ molecules}$$

$$\frac{2 \times 10^7 \text{ (pond area)}}{9 \times 10^{21} \text{ molecules}} = 22 \times 10^{16} \text{ cm}^2/\text{molecule}$$

$$= 22 \text{ Å}^2/\text{molecule}$$

We can readily see from this example that the area of cross section per molecule is given by

$$\text{Cross-sectional area/molecule} = \frac{MS}{V_\rho N} \quad (15\text{–}41)$$

where $M$ is molecular weight of the spreading liquid, $S$ is the surface area covered by the film, $V$ is the volume of the spreading liquid, $\rho$ is its density, and $N$ is Avogadro's number.

Langmuir, Adam, Harkins, and others made quantitative studies of the properties of films that are spread over a clear surface of the substrate liquid (usually water) contained in a trough. The film can be compressed against a horizontal float by means of a movable barrier. The force exerted on the float is measured by a torsion-wire arrangement similar to that employed in the ring tensiometer. This apparatus is called a *film balance*. The compressive force per unit area on the float is known as the *surface* or *film pressure*, $\pi$; it is the difference in surface tension between the pure substrate, $\gamma_0$, and that with a film spread on it, $\gamma$, and is written as

$$\pi = \gamma_0 - \gamma \quad (15\text{–}42)$$

Surface tension (interfacial tension) is the resistance of the surface (interface) to an expansion in area, and film pressure, $\pi$, is the lowering of this resistance to expansion, as expressed quantitatively in equation (15–42). Schott[27] stated that the film pressure, $\pi$, is an expansion pressure exerted on the monolayer that opposes the surface tension, $\gamma_0$, or contraction of the clean (water) surface. The surface-active molecules of the monolayer are thought to insert themselves into the surface of the water molecules of a film balance to reduce the resistance of the water surface to expansion. The presence of the surfactant molecules increases the ease of expansion, presumably by breaking or interfering with hydrogen bonding, van der Waals interaction, and other cohesive forces among the water molecules. These attractive forces produce the "springlike" action in the water surface, as measured by the surface tension, $\gamma_0$, and the introduction of surfactant molecules into the clean water surface reduces the springiness of the interacting water molecules and decreases the surface tension $\gamma_0$ to $\gamma_0 - \gamma$ or $\pi$ [equation (15–42)].

In carrying out an experiment with the film balance, the substance under study is dissolved in a volatile solvent (e.g., hexane) and is placed on the surface of the substrate, which has previously been swept clean by means of a paraffined or Teflon strip. The liquid spreads as a film, and the volatile solvent is permitted to evaporate. A cross-sectional view of the interface after spreading is shown in **Figure 15–14**. The movable barrier is then moved to various positions in the direction of the float. The area of the trough available to the film at each position is measured, and the corresponding film pressure is read from the torsion dial. The film pressure is then plotted against the area of the film or, more conveniently, against the cross-sectional area per molecule, $A^2$ [see *Example 15–11* and equation (15–41) for computing the molecule's cross-sectional area from the area of the film]. The results for stearic acid and lecithin are shown in **Figure 15–15**.

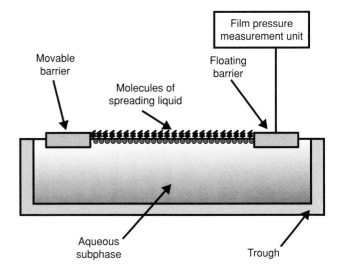

**Fig. 15–14.** Cross-sectional view of a spreading liquid on the surface of a film balance.

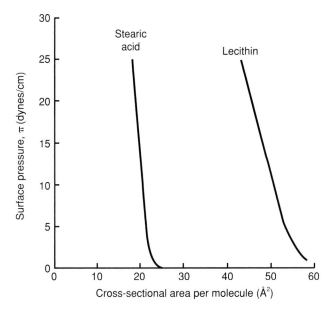

**Fig. 15–15.** Surface film pressure, $\pi$, for stearic acid and lecithin plotted as a function of cross-sectional area per molecule.

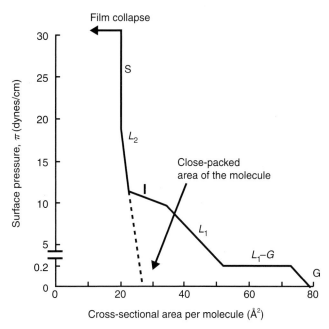

**Fig. 15–16.** Phase changes that occur when a liquid film is spread at an interface and then compressed. Key: G = two-dimensional gas; $L_1 - G$ = liquid phase in equilibrium with two-dimensional gas; $L_1$ = liquid expanded or two-dimensional bulk liquid state; I = intermediate state; $L_2$ = condensed liquid state; S = two-dimensional solid state. When compressed by a force greater than required to form a solid surface, the film collapses, as shown by the arrow at the top of the figure. (Replotted from P. C. Heimenz, *Principles of Colloid and Surface Chemistry*, 2nd Ed., Marcel Dekker, New York, 1986, p. 364.)

Frequently, a variety of phase changes are observed when an insoluble film is spread at an interface and then compressed. A representation of what can occur with a straight-chain saturated aliphatic compound at the air–water interface is shown in **Figure 15–16**. When the film is spread over an area greater than 50 to 60 $\text{Å}^2$/molecule (region G), it exerts little pressure on the floating barrier. The film acts like a gas in two dimensions. As the film begins to be compressed (region $L_1 - G$), a liquid phase, $L_1$, appears that coexists in equilibrium with the gas phase. This occurs at a low surface pressure (e.g., 0.2 dyne/cm or less). The liquid expanded state (region $L_1$) can be thought of as a bulk liquid state, but in two dimensions. Further compression of the film often leads to the appearance of an intermediate phase (region I) and then a less compressible condensed liquid state, region $L_2$. This then gives way to the least compressible state, region S, where the film can be regarded as being in a two-dimensional solid state. In these latter stages of film compression, the film or surface pressure, $\pi = \gamma_0 - \gamma_1$, rises rapidly as the curve passes through the regions $L_2$ and S in **Figure 15–16**. This increase in $\pi$ with compression of the surfactant film results from surface-active molecules being forcibly inserted and crowded into the surface. This process opposes the natural tendency of the water surface to contract, and the surface tension decreases from $\gamma_0$ to $\gamma$. Finally, the molecules slip over one another, and the film breaks when it is greatly compressed.

The regions marked along the plot in **Figure 15–16** can be represented schematically in terms of the positioning of the spreading molecules in the surface, as shown in **Figure 15–17**. In region G of **Figure 15–16**, the molecules in the monolayer lie on the surface with great distances between them, as in a three-dimensional gas. In the part of the curve marked $L_1$ and $L_2$ in **Figure 15–16**, the molecules are forced closer together, and, as shown schematically in **Figure**

15–17*b*, are beginning to stand erect and interact with one another, analogous to a three-dimensional liquid. In region S of **Figure 15–16**, the spreading molecules are held together by strong forces; this condition, analogous to the solid state in three-dimensional chemistry, shows little compressibility

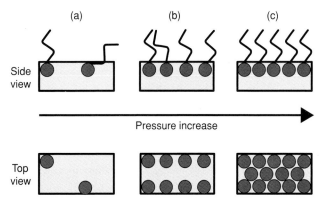

**Fig. 15–17.** Insoluble monolayers. Insoluble monolayer films exhibit characteristics that can be equated to those of the solid, liquid, and gaseous states of matter. (*a*) Gaseous film. Molecules are apart and have significant surface mobility. The molecules essentially act independently. (*b*) Liquid film. Monolayer is coherent and relatively densely packed but is still compressible. (*c*) Condensed film. Monolayer is coherent, rigid, essentially incompressible, and densely packed, with high surface viscosity. The molecules have little mobility and are oriented perpendicular to the surface.

**TABLE 15–7**
## DIMENSIONS OF ORGANIC MOLECULES DETERMINED BY MEANS OF THE FILM BALANCE

| Substance | Formula | Length of Molecule ($\mathring{A}$) | Cross-Sectional Area ($\mathring{A}^2$) |
|---|---|---|---|
| Stearic acid | $C_{17}H_{35}COOH$ | 25 | 22 |
| Tristearin | $(C_{128}H_{35}O_2)_3C_3H_5$ | 25 | 66 |
| Cetyl alcohol | $C_{16}H_{33}OH$ | 22 | 21 |
| Myricyl alcohol | $C_{30}H_{61}OH$ | 41 | 27 |

relative to that of a gas or a liquid. The S state is shown schematically in **Figure 15–17c**, where the molecules on the surface of the film balance are compressed together as far as possible. Further compression of the film by a movement from right to left on the horizontal axis of the graph in **Figure 15–14**, that is, a movement from left to right of the movable barrier, brings about a collapse of the monolayer film, one part sliding over the other, as depicted in **Figure 15–16**.

The cross-sectional area per molecule of the close-packed film at zero surface pressure is obtained by extrapolating the linear portion of the curve to the horizontal axis, as seen in **Figure 15–16**. The values for some organic molecules determined in this way by Langmuir[28] are listed in **Table 15–7**. It is seen that myricyl alcohol, with 30 carbons in the chain, has a length almost twice that of the other molecules. Its cross-sectional area at the interface is not markedly different from that of other single-chain molecules, however, confirming that it is the cross-sectional area of the alkyl chain, rather than the length, that is being measured. Tristearin, with three fatty acid chains, has a cross-sectional area about three times that of the molecules with only one aliphatic chain.

The electric potential and viscosity of monomolecular films can be studied by means of the film balance, and the molecular weight of high polymers such as proteins can be estimated by its use. The film-balance technique also has considerable significance in the study of biologic systems. Because some protein molecules unfold from a spherical configuration into a flat film when spread on the surface of the film trough, the relationship between unfolding and biologic activity can be studied. The sizes and shapes of molecules of steroids, hormones, and enzymes and their interaction with drugs at interfaces can also be investigated by means of the film balance. The interaction between insulin injected under the surface layer and several lipids spread at constant surface pressure on a film balance was studied by Schwinke et al.[29] The film balance and its applications are discussed in the books of Adam,[30] Harkins,[31] Sobotka,[32] and Gaines.[33]

Mention has been made of the fact that materials forming an insoluble monolayer can be thought of as being in the gaseous, liquid, or solid state, depending on the degree of compression to which the film is subjected. Thus, the surface pressure for molecules in the gaseous state at an interface is comparable to the pressure, $P$, that molecules in three-dimensional gaseous systems exert on the walls of their containers. Just as the equation of state for an ideal gas in three dimensions is $PV = nRT$ (see the ideal gas law in States of Matter), that for a monolayer is

$$\pi A = nRT \qquad (15\text{--}43)$$

where $\pi$ is the surface pressure in dynes/cm and $A$ is the area that each mole of amphiphile occupies at the interface.

Equation **(15–43)**, the two-dimensional ideal gas law, can be derived as follows. When the concentration of amphiphile at the interface is small, solute–solute interactions are unimportant. Under these conditions, surface tension decreases in a linear fashion with concentration. We can therefore write

$$\gamma = bc + \gamma_0 \qquad (15\text{--}44)$$

where $\gamma_0$ is the surface tension of the pure substrate, $\gamma$ is the surface tension produced by the addition of $c$ moles/liter of adsorbate, and $b$ is the slope of the line. Because the slope of such a plot is negative, and because $\pi = \gamma_0 - \gamma$, equation **(15–44)** can be rewritten as

$$\pi = -bc \qquad (15\text{--}45)$$

The Gibbs adsorption equation **(15–26)** can be expressed in the following form:

$$-(d\gamma/dc) - b = \Gamma RT/c \qquad (15\text{--}46)$$

because $d\gamma/dc$ is the slope of the line.

Substituting for equation **(15–45)** in equation **(15–46)** and canceling $c$, which is common to both sides, we obtain

$$\pi = \Gamma RT \qquad (15\text{--}47)$$

Surface excess has the dimensions of moles/cm$^2$ and can be represented by $n/A$, where $n$ is the number of moles and $A$ is the area in cm$^2$. Thus,

$$\pi = nRT/A$$

or

$$\pi A = nRT$$

which is equation **(15–43)**.

As with the three-dimensional gas law, equation **(15–43)** can be used to compute the molecular weights of materials adsorbed as gaseous films at an interface. Nonideal behavior also occurs, and plots of $\pi A$ versus $\pi$ for monolayers give results comparable to those in three-dimensional systems when $PV$ is plotted against $P$. Equations similar to van der Waal's equation (see van der Waals Equation for Real Gases in States of Matter) for nonideal behavior have been developed.

The relation between the Gibbs adsorption equation and equation **(15–43)** emphasizes the point made earlier that the distinction between soluble and insoluble films is an arbitrary one, made on the basis of the experimental techniques used rather than any fundamental differences in physical properties.

The variation of the surface pressure, $\pi$, with temperature at the several "phase changes" observed in the two-dimensional isotherm $\pi$-area (see **Fig. 15–16**) can be analyzed by a relationship analogous to the Clapeyron equation:

$$\frac{d\pi}{dT} = \frac{\Delta H}{T(A_1 - A_2)} \qquad (15\text{–}48)$$

where $A_1$ and $A_2$ are the molar areas ($cm^2$/mole) of the two phases and $T$ and $\Delta H$ are, respectively, the temperature and enthalpy for the phase change.[33] Note that $\pi$, $\Delta H$, and $(A_1 - A_2)$ are the two-dimensional equivalents of pressure, enthalpy, and change of volume, respectively, in the Clapeyron equation.

### EXAMPLE 15–12

**Calculation of Enthalpy Change**

Consideration of monolayers of insoluble amphiphilic compounds with a polymerizable group serves to investigate the polymerization behavior at the gas–water interface. The $\pi$–$A$ isotherms resulting from film balance experiments with $n$-hexadecyl acrylate monolayers in the temperature range 13°C to 28°C showed two breaks corresponding to phase transitions (changes in state).

Compute $\Delta H$, the enthalpy change of transition from the condensed liquid state, $L_2$, to the liquid expanded state, $L_1$. The areas per molecule at $L_1$ and $L_2$ are 0.357 and 0.265 $nm^2$/molecule, respectively. The change of surface pressure with temperature, $d\pi/dt$, is 0.91 mN/mK, and the temperature of transition is 24.2°C.[34]

From equation (15–48),

$$\Delta H = T(A_1 - A_2)\frac{d\pi}{dT}$$

$$\Delta H = 297.2\,\text{K}\,(0.357 - 0.265)$$
$$\times 10^{-18}\,\text{m}^2/\text{molecule} \times 0.91 \times 10^{-3}\,\frac{\text{N}}{\text{mK}}$$
$$= 2.49 \times 10^{-20}\,\text{joule/molecule}$$
$$2.49 \times 10^{-20} \times 6.022 \times 10^{23}$$
$$= 15.995\,\text{joule/mole} \approx 15\,\text{kJ/mole}$$

# ADSORPTION AT SOLID INTERFACES

Adsorption of material at solid interfaces can take place from either an adjacent liquid or gas phase. The study of adsorption of gases arises in such diverse applications as the removal of objectionable odors from rooms and food, the operation of gas masks, and the measurement of the dimensions of particles in a powder. The principles of solid–liquid adsorption are used in decolorizing solutions, adsorption chromatography, detergency, and wetting.

In many ways, the adsorption of materials from a gas or a liquid onto a solid surface is similar to that discussed for liquid surfaces. Thus, adsorption of this type can be considered as an attempt to reduce the surface free energy of the solid. The surface tensions of solids are invariably more difficult to obtain, however, than those of liquids. In addition, the solid interface is immobile in comparison to the turbulent liquid interface. The average lifetime of a molecule at the water–gas interface is about 1 $\mu$ sec, whereas an atom in the surface

of a nonvolatile metallic solid may have an average lifetime of $10^{37}$ sec.[35] Frequently, the surface of a solid may not be homogeneous, in contrast to liquid interfaces.

## The Solid–Gas Interface

The degree of adsorption of a gas by a solid depends on the chemical nature of the *adsorbent* (the material used to adsorb the gas) and the *adsorbate* (the substance being adsorbed), the surface area of the adsorbent, the temperature, and the partial pressure of the adsorbed gas. The types of adsorption are generally recognized as physical or van der Waals adsorption and chemical adsorption or chemisorption. *Physical adsorption*, associated with van der Waals forces, is reversible, the removal of the adsorbate from the adsorbent being known as *desorption*. A physically adsorbed gas can be desorbed from a solid by increasing the temperature and reducing the pressure. *Chemisorption*, in which the adsorbate is attached to the adsorbent by primary chemical bonds, is irreversible unless the bonds are broken.

The relationship between the amount of gas physically adsorbed on a solid and the equilibrium pressure or concentration at constant temperature yields an *adsorption isotherm* when plotted as shown in **Figure 15–18**. The term *isotherm* refers to a plot at constant temperature. The number of moles, grams, or milliliters, $x$, of gas adsorbed on, $m$, grams of adsorbent at standard temperature and pressure is plotted on the vertical axis against the equilibrium pressure of the gas in mm Hg on the horizontal axis, as seen in **Figure 15–18a**.

One method of obtaining adsorption data is by the use of an apparatus similar to that shown in **Figure 15–19**, which consists essentially of a balance contained within a vacuum system. The solid, previously degassed, is placed on the pan, and known amounts of gas are allowed to enter. The increase in weight at the corresponding equilibrium gas pressures is recorded. This can be achieved by noting the extension of a calibrated quartz spring used to suspend the pan containing the sample. The data are then used to construct an isotherm on the basis of one or more of the following equations.

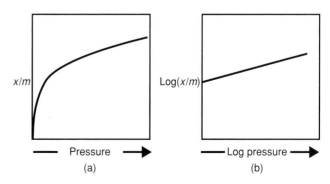

**Fig. 15–18.** Adsorption isotherms for a gas on a solid. (*a*) Amount, $x$, of gas adsorbed per unit mass, $m$, of adsorbent plotted against the equilibrium pressure. (*b*) Log of the amount of gas adsorbed per unit mass of adsorbent plotted against the log of the pressure.

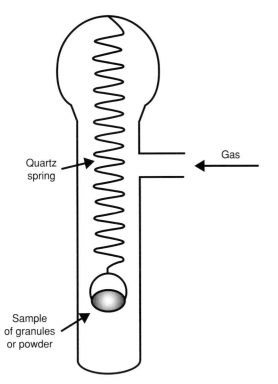

Quartz spring

Gas

Sample of granules or powder

**Fig. 15–19.** Schematic of apparatus used to measure the absorption of gases on solids.

Freundlich[36] suggested a relationship, the *Freundlich isotherm*,

$$y = \frac{x}{m} = kp^{1/n} \qquad (15\text{–}49)$$

where $y$ is the mass of gas, $x$, adsorbed per unit mass, $m$, of adsorbent, and $k$ and $n$ are constants that can be evaluated from the results of the experiment. The equation is handled more conveniently when written in the logarithmic form,

$$\log \frac{x}{m} = \log k + \frac{1}{n} \log p \qquad (15\text{–}50)$$

which yields a straight line when plotted as seen in **Figure 15–18b**. The constant, $\log k$, is the intercept on the ordinate, and $1/n$ is the slope of the line.

Langmuir[37] developed an equation based on the theory that the molecules or atoms of gas are adsorbed on active sites of the solid to form a layer one molecule thick (monolayer). The fraction of centers occupied by gas molecules at pressure $p$ is represented by $\theta$, and the fraction of sites not occupied is $1 - \theta$. The rate, $r_1$, of adsorption or condensation of gas molecules on the surface is proportional to the unoccupied spots, $1 - \theta$, and to the pressure, $p$, or

$$r_1 = k_1(1 - \theta)p \qquad (15\text{–}51)$$

The rate, $r_2$, of evaporation of molecules bound on the surface is proportional to the fraction of surface occupied, $\theta$, or

$$r_2 = k_2\theta \qquad (15\text{–}52)$$

and at equilibrium $r_1 = r_2$, or

$$k_1(1 - \theta) = k_2\theta \qquad (15\text{–}53)$$

By rearrangement, we obtain

$$\theta = \frac{k_1 p}{k_2 + k_1 p} = \frac{(k_1/k_2)p}{1 + (k_1/k_2)p} \qquad (15\text{–}54)$$

We can replace $k_1/k_2$ by $b$ and $\theta$ by $y/y_m$, where $y$ is the mass of gas adsorbed per gram of adsorbent at pressure $p$ and at constant temperature and $y_m$ is the mass of gas that 1 g of the adsorbent can adsorb when the monolayer is complete. Inserting these terms into equation **(15–54)** produces the formula

$$y = \frac{y_m b p}{1 + bp} \qquad (15\text{–}55)$$

which is known as the *Langmuir isotherm*. By inverting equation **(15–55)** and multiplying through by $p$, we can write this for plotting as

$$\frac{p}{y} = \frac{1}{by_m} + \frac{p}{y_m} \qquad (15\text{–}56)$$

A plot of $p/y$ against $p$ should yield a straight line, and $y_m$ and $b$ can be obtained from the slope and intercept.

Equations **(15–49)**, **(15–50)**, **(15–55)**, and **(15–56)** are adequate for the description of curves only of the type shown in **Figure 15–18a**. This is known as the type I isotherm. Extensive experimentation, however, has shown that there are four other types of isotherms, as seen in **Figure 15–20**, that are not described by these equations. Type II isotherms are sigmoidal in shape and occur when gases undergo physical adsorption onto nonporous solids to form a monolayer followed

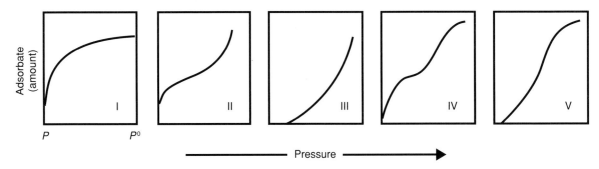

**Fig. 15–20.** Various types of adsorption isotherms.

by multilayer formation. The first inflection point represents the formation of a monolayer; the continued adsorption with increasing pressure indicates subsequent multilayer formation. Type II isotherms are best described by an expression derived by Brunauer, Emmett, and Teller[23] and termed for convenience the *BET equation*. This equation can be written as

$$\frac{p}{y(p_0 - p)} = \frac{1}{y_m b} + \frac{b-1}{y_m b} \frac{p}{p_0} \qquad (15\text{--}57)$$

where $p$ is the pressure of the adsorbate in mm Hg at which the mass, $y$, of vapor per gram of adsorbent is adsorbed, $p_0$ is the vapor pressure when the adsorbent is saturated with adsorbate vapor, $y_m$ is the quantity of vapor adsorbed per unit mass of adsorbent when the surface is covered with a monomolecular layer, and $b$ is a constant proportional to the difference between the heat of adsorption of the gas in the first layer and the latent heat of condensation of successive layers. The saturated vapor pressure, $p_0$, is obtained by bringing excess adsorbate in contact with the adsorbent. For the case of simple monolayer adsorption, the BET equation reduces to the Langmuir isotherm.

Isotherms of the shape shown as IV in **Figure 15–20** are typical of adsorption onto porous solids. The first point of inflection, when extrapolated to zero pressure, again represents the amount of gas required to form a monolayer on the surface of the solid. Multilayer formation and condensation within the pores of the solid are thought to be responsible for the further adsorption shown, which reaches a limiting value before the saturation vapor pressure, $p_0$, is attained. Type III and type V isotherms are produced in a relatively few instances in which the heat of adsorption of the gas in the first layer is *less* than the latent heat of condensation of successive layers. As with type IV isotherms, those of type V show capillary condensation, and adsorption reaches a limiting value before $p_0$ is attained. The type II isotherm results when $b$ is greater than 2.0 and type III when $b$ is less than 2.0 in the BET expression **(15–57)**. Types IV and V frequently involve hysteresis and appear as shown in **Figures 15–21** and **15–22**, respectively.

The total surface area of the solid can be determined from those isotherms in which formation of a monolayer can be detected, that is, Types I, II, and IV. This information is obtained by multiplying the total number of molecules in the volume of gas adsorbed by the cross-sectional area of each molecule. The surface area per unit weight of adsorbent, known as the *specific surface*, is important in pharmacy because the dissolution rates of drug particles depend, in part, on their surface area.

## The Solid–Liquid Interface

Drugs such as dyes, alkaloids, fatty acids, and even inorganic acids and bases can be absorbed from solution onto solids such as charcoal and alumina. The adsorption of solute molecules from solution can be treated in a manner analogous

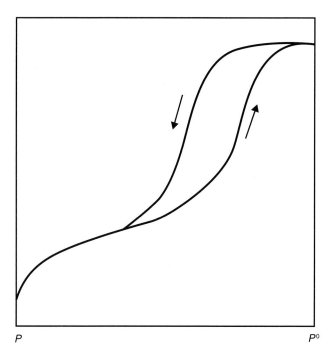

**Fig. 15–21.** Type IV isotherm showing hysteresis.

to the adsorption of molecules at the solid–gas interface. Isotherms that fit one or more of the equations mentioned previously can be obtained by substituting solute concentration for the vapor pressure term used for solid–gas systems. For example, the adsorption of strychnine, atropine, and quinine from aqueous solutions by six different clays[38] was capable of being expressed by the Langmuir equation in the form

$$\frac{c}{y} = \frac{1}{b y_m} + \frac{c}{y_m} \qquad (15\text{--}58)$$

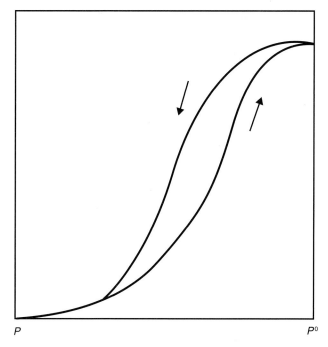

**Fig. 15–22.** Type V isotherm showing hysteresis.

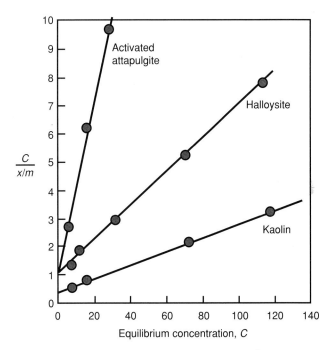

**Fig. 15–23.** Adsorption of strychnine on various clays. (Modified from M. Barr and S. Arnista, J. Am. Pharm. Assoc. Sci. Ed. **46**, 486–489, 1957.)

where $c$ is the equilibrium concentration in milligrams of alkaloidal base per 100 mL of solution, $y$ is the amount of alkaloidal base, $x$, in milligrams adsorbed per gram, $m$, of clay (i.e., $y = x/m$), and $b$ and $y_m$ are constants defined earlier. In later studies, Barr and Arnista[39] investigated the adsorption of diphtheria toxin and several bacteria by various clays. They concluded that attapulgite, a hydrous magnesium aluminum silicate, was superior to kaolin as an intestinal adsorbent. The results of the adsorption of strychnine on activated attapulgite, halloysite (similar to kaolinite), and kaolin, all washed with gastric juice, are shown in **Figure 15–23**.

The smaller the slope, the better is the adsorption. Thus, it can be calculated from **Figure 15–23** that an equilibrium concentration of, say, 400 mg of strychnine/100 mL of solution, $x/m$, gives approximately 40, 20, and 6.7 mg/g for attapulgite, halloysite, and kaolin, respectively. When an orally administered drug causes gastrointestinal disturbances, commercial adsorbent, antacid, or antidiarrheal preparations are often taken by the patient, and these preparations may interact with the drug to reduce its absorption. The absorption of quinidine salts (an antiarrhythmic agent), for example, is impaired by combining with kaolin, pectin, montmorillonite, and similar adsorbents. Moustafa et al.[40] found that the adsorption of quinidine sulfate by antacid and antidiarrheal preparations, loperamide, Kaopectate, Simeco, magnesium trisilicate, and bismuth subnitrate was well expressed by both Freundlich and Langmuir adsorption isotherms.

Nikolakakis and Newton[41] studied the solid–solid adsorption of a fine cohesive powder onto the surface of coarse free-flowing particles to form what is called an "ordered" mixture. These systems provide very homogeneous mixtures

of powders ordinarily having good physical stability. Examples of "ordered" mixtures are dry blends of sucrose and antibiotics that can be reconstituted with water to provide antibiotic syrup formulations. Sorbitol can replace sucrose to prepare sucrose-free formulations for patients with diabetes. During blending, a fine powder of an antibiotic is adsorbed onto the surface of coarse particles of sorbitol. Nikolakakis and Newton[41] obtained an apparent Langmuir or type I isotherm when the weight of drug adsorbed per unit weight of sorbitol, $x/m$, was plotted against the concentration, $c$, of nonadsorbed drug at equilibrium. Thus, using the linear form, equation (15–58), one can find the $b$ and $y_m$ values. The $y_m$ value is the amount of antibiotic per unit weight of sorbitol required to form a monolayer on the surface of sorbitol particles. This can be considered as a measure of the adsorption capacity or number of binding sites of sorbitol for the antibiotic. The quantity $b$ is an empirical affinity or binding constant that is given thermodynamic significance by some workers (see Hiemenz,[7] see Fundamentals and Concentration Effects in Chemical Kinetics and Stability).

**EXAMPLE 15–13**

**Solid–Solid Adsorption of Cephalexin**

The values of $c/y$ against $c$ for the solid–solid adsorption of cephalexin monohydrate onto sorbitol are as follows:

Calculate $b$ and $y_m$.

Using a regression analysis of $c/y$ ($y$ axis) against $c$ ($x$ axis), we find the $c/y = 25.2 + 5.93c$. Thus,

$$\text{Slope} = \frac{1}{y_m} = 5.93; \qquad y_m = 0.169 \text{ g/g (dimensionless)}$$

$$\text{Intercept} = \frac{1}{by_m} = 25.2\% \text{ (w/w)}$$

$$b = \frac{1}{25.2\% \text{ (w/w)} \times 0.169 \text{ g/g}} = 0.235\% \text{ (w/w)}^{-1}$$

| c (% w/w*) | 5 | 10 | 15 | 20 |
|---|---|---|---|---|
| c/y (% w/w) | 54.85 | 84.5 | 114.15 | 143.8 |

g adsorbate/g adsorbent

*Note that we express $c$ as percent w/w on both the $x$ and $y$ axes. We express $y = x/m$ as gram adsorbate/gram adsorbent, which is dimensionless. Therefore, the units for $c/y$ on the $x$ axis are simply % w/w. Like $y$, $y_m$ is dimensionless, and $b$ has the units 1 (% w/w).

## Activated Charcoal

An example of a substance that can adsorb enormous amounts of gases or liquids is *activated charcoal*, the residue from destructive distillation of various organic materials, treated to increase its adsorptive power. To adsorb more adsorbate, an adsorbent of a given mass should have the greatest possible surface area. This might be achieved by the use of porous or milled adsorbents. Consider the following example: A sphere with a diameter of 1.2 cm has a volume of 1 $cm^3$ and a surface area of 5 $cm^2$. If the sphere is divided into two spheres each with a diameter of 1 cm, together they will have the same volume of 1 $cm^3$ but an increased surface area of 6 $cm^2$. Particles with a diameter of about 0.01 cm and a summary volume of 1 cm$^3$ will have a total surface of about 500 cm$^3$. If we continue to divide the spheres and finally mill them to particles with diameters of about 10$^{-6}$ cm, the total surface area will increase to hundreds of square meters. Modern activated charcoal has thousands of square meters of active surface area per 1 g of mass. It is used as an antidote for poisonings due to many substances, including drugs (sulfonylureas, acetaminophen, phenobarbital, etc.).

Activated charcoal is made from material burnt in a superheated high-oxygen atmosphere, creating small holes in the range of 100 to 800 Å in diameter throughout the grain of the charcoal. This effectively increases the charcoal's surface area so that the surface area of 1 g of charcoal is approximately 1000 m$^2$. The usual dose for activated charcoal treatment is 50 to 100 g for adults and 1 to 2 g/kg for children. Activated charcoal is frequently administered to poisoned patients. The assumption is that toxin absorption is prevented and that toxicity (as defined by morbidity and mortality) of the poisoning is decreased. Yet, there is no evidence that activated charcoal improves outcome.[42]

Sorptive uptake of lignin and tannin from an aqueous phase by activated charcoal was investigated.[43] The sorption reaction was found to be of a first order. The influence on the rate of sorption of various factors, such as amount of sorbent and pH of the system, has been investigated. Sorption data fit well into the Langmuir adsorption isotherm, indicating formation of a monolayer over a homogeneous sorbent surface. Desorption studies indicate the irreversible nature of the sorption reaction, whereas interruption studies suggest film diffusion to be rate limiting.[43]

Shadnia et al.[44] described the treatment and successful outcome of a patient who had taken a dose of strychnine that would normally be fatal. A 28-year-old man was admitted 2 hr after ingestion of 1 to 1.5 g of strychnine. He was severely agitated and in mild respiratory distress; blood pressure was 90/60 mm Hg, pulse 110/min, and peripheral pulses weak. He had generalized hyperactive reflexes and had several generalized tonic–clonic convulsions in the emergency department. Treatment consisted of gastric lavage with water, oral administration of activated charcoal and sorbitol solution, continuous intravenous administration of midazolam, and then sodium thiopental, furosemide, sodium bicarbonate, and hemodialysis for acute renal failure. His clinical course included respiratory distress, agitation, generalized tonic–clonic convulsions, hyperactivity, oliguria, and acute tubular necrosis prior to recovery in 23 days.

Tanaka et al.[45] reported a case of impaired absorption of orally administered phenobarbital associated with the concomitant administration of activated charcoal, and recovery of the absorption after administration of the two drugs was separated by a 1.5-hr interval. A 78-year-old woman weighing 50 kg who had undergone brain surgery was prescribed phenobarbital 120 mg/day for postoperative convulsions. Her serum phenobarbital concentration reached 24.8 μg/mL (therapeutically effective level is 10–30 μg/mL). Thereafter, her renal function worsened, and activated charcoal 6 g/day

was started. Four months after the start of activated charcoal, blood analysis revealed that the serum phenobarbital concentration was as low as 4.3 μg/mL. The phenobarbital dose was increased to 150 mg/day. Further evaluation revealed that activated charcoal and phenobarbital had been administered concomitantly. The dosage regimen was altered to separate the administration of the agents by at least 1.5 hr. Subsequently, the patient's serum phenobarbital concentration increased to 11.9 μ/mL within 3 weeks. Her serum phenobarbital concentration was measured monthly thereafter and remained stable in the range of 15.8 to 18.6 μ/mL. The patient's low serum phenobarbital concentration was considered likely to have been due to impaired gastrointestinal absorption of phenobarbital as a result of adsorption of phenobarbital on the activated charcoal. An objective causality assessment showed that the interaction was probable. Therefore, administration of activated charcoal and phenobarbital should be separated by an interval of at least 1.5 hr.

Fourteen adsorbent materials were tested in the pH range of 3 to 8 for deoxynivalenol and nivalenol-binding ability.[46] Only activated carbon was effective, with binding capacities of 35.1 and 8.8 μ mole of deoxynivalenol and nivalenol per gram of adsorbent, respectively, calculated from the adsorption isotherms. A dynamic laboratory model simulating the gastrointestinal tract of healthy pigs was used to evaluate the small-intestinal absorption of deoxynivalenol and nivalenol and the efficacy of activated carbon in reducing the relevant absorption. The in vitro intestinal absorptions of deoxynivalenol and nivalenol were 51% and 21%, respectively, as referred to 170 μg of deoxynivalenol and 230 μg of nivalenol ingested through contaminated (spiked) wheat. Most absorption occurred in the jejunal compartment for both mycotoxins. The inclusion of activated carbon produced a significant reduction in the intestinal mycotoxin absorption. At 2% inclusion, the absorption with respect to the intake was lowered from 51% to 28% for deoxynivalenol and from 21% to 12% for nivalenol. The binding activity of activated carbon for these trichothecenes was lower than that observed for zearalenone, a mycotoxin frequently co-occurring with them in naturally contaminated cereals.

The adsorption of three barbiturates—phenobarbital, mephobarbital, and primidone from simulated intestinal fluid, without pancreatin, by activated carbon was studied using the rotating-bottle method.[46] The concentrations of each drug remaining in solution at equilibrium were determined with the aid of a high-performance liquid chromatography system employing a reversed-phase column. The competitive Langmuir-like model, the modified competitive Langmuir-like model, and the LeVan–Vermeulen model were each fit to the data. Excellent agreement was obtained between the experimental and predicted data using the modified competitive Langmuir-like model and the LeVan–Vermeulen model. The agreement obtained from the original competitive Langmuir-like model was less satisfactory. These observations are not surprising because the competitive Langmuir-like model assumes that the capacities of the adsorbates are

## ⊨▰▰▰**KEY CONCEPT** WETTING AGENT

A *wetting agent* is a surfactant that, when dissolved in water, lowers the advancing contact angle, aids in displacing an air phase at the surface, and replaces it with a liquid phase. Examples of the application of wetting to pharmacy and medicine include the displacement of air from the surface of sulfur, charcoal, and other powders for the purpose of dispersing these drugs in liquid vehicles; the displacement of air from the matrix of cotton pads and bandages so that medicinal solutions can be absorbed for application to various body areas; the displacement of dirt and debris by the use of detergents in the washing of wounds; and the application of medicinal lotions and sprays to the surface of the skin and mucous membranes.

equal, whereas the other two models take into account the differences in the capacities of the components. The results of these studies indicate that the adsorbates employed are competing for the same binding sites on the activated carbon surface. The results also demonstrate that it is possible to accurately predict multicomponent adsorption isotherms using only single-solute isotherm parameters.[47]

Hill et al.[48] studied extracorporeal liver support for episodic (acute) type C hepatic encephalopathy (AHE) failing to respond to medical therapy. A series of patients with cirrhosis and AHE failing to respond to at least 24 hr of medical therapy underwent a maximum of three 6-hr charcoal-based hemodiabsorption treatments. It was found that a charcoal-based hemodiabsorption treatment in which a standardized anticoagulation protocol is used is safe and effective treatment for AHE not responding to standard medical therapy.

Although activated charcoal is useful in the management of poisonings, it should not be considered as harmless, especially in children. Donoso et al.[49] reported the case of a patient who developed obstructive laryngitis secondary to aspiration of activated charcoal with a protected airway. This case shows that nasogastric administration of activated charcoal presents a significant degree of risk. Vomiting also frequently complicates the administration of activated charcoal. Little is known about the patient, poison, or procedure-specific factors that contribute to emesis of charcoal. Osterhoudt et al.[50] estimated the incidence of vomiting subsequent to therapeutic administration of charcoal to poisoned children 18 years or less of age and examined the relative contributions of several risk factors to the occurrence of vomiting. One of every 5 children given activated charcoal vomited. Children with previous vomiting or nasogastric tube administration were at highest risk, and these factors should be accounted for in future investigation of antiemetic strategies. Sorbitol content of charcoal was not a significant risk factor for emesis.

## Wetting

Adsorption at solid surfaces is involved in the phenomena of wetting and detergency. The tendency of molecules of liquids to move from the surface into the bulk and decrease the surface of the liquid–gas interface is explained by the fact that molecules of liquid undergo very weak attraction from the molecules of gas on the interface. There is a limited number of gas molecules in a unit of volume in the gaseous phase as compared with that in liquid phase. When a liquid comes into contact with the solid, the forces of attraction between the liquid and the solid phases begin to play a significant role. In this case, the behavior of the liquid will depend on the balance between the forces of attraction of molecules in the liquid and the forces of attraction between the liquid and the solid phases. In the case of mercury and glass, attractive forces between molecules of mercury and glass are much smaller than the forces of attraction between molecules of mercury themselves. As a result, mercury will come together as a single spherical drop. In contrast, for water and glass (or mercury and zinc), attractive forces between the solid and liquid molecules are greater than the forces between molecules of liquid themselves, and so the liquid is able to wet the surface of the glass.

The most important action of a wetting agent is to lower the *contact angle* between the surface and the wetting liquid. The contact angle is the angle between a liquid droplet and the surface over which it spreads. As shown in **Figure 15–24**, the contact angle between a liquid and a solid may be 0°,

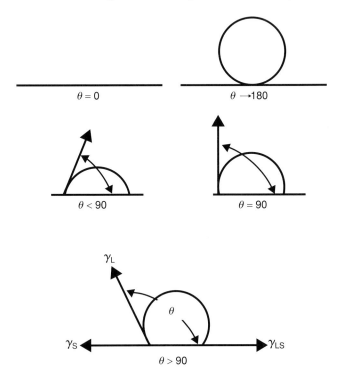

**Fig. 15–24.** Contact angles from 0° to 180°.

signifying complete wetting, or may approach 180°, at which wetting is insignificant. The contact angle may also have any value between these limits, as illustrated in the Figure. At equilibrium, the surface and interfacial tensions can be resolved into

$$\gamma_S = \gamma_{SL} + \gamma_L \cos\theta \qquad (15\text{–}59)$$

which is known as Young's equation.

When $\gamma_S$ of equation (15–59) is substituted into equation (15–19), we have

$$S = \gamma_L(\cos\theta - 1) \qquad (15\text{–}60)$$

and combining equation (15–59) with equation (15–16) results in

$$W_a = W_{SL} = \gamma_L(1 + \cos\theta) \qquad (15\text{–}61)$$

which is an alternative form of Young's equation. Equations (15–60) and (15–61) are useful expressions because they do not include $\gamma_S$ or $\gamma_{SL}$, neither of which can be easily or accurately measured. The contact angle between a water droplet and a greasy surface results when the applied liquid, water, wets the greasy surface incompletely. When a drop of water is placed on a scrupulously clean glass surface, it spreads spontaneously and no contact angle exists. This result can be described by assigning to water a high spreading coefficient on clean glass, or by stating that the contact angle between water and glass is zero. If the appropriate wetting agent is added to water, the solution will spread spontaneously on a greasy surface. For a wetting agent to function efficiently—in other words, to exhibit a low contact angle—it should have an HLB of about 6 to 9 (see **Fig. 15–11**).

## EXAMPLE 15–14

### Comparison of Different Tablet Binders

Wettability of tablet surfaces influences disintegration and dissolution and the subsequent release of the active ingredient(s) from the tablet.

A *tablet binder* is a material that contributes cohesiveness to a tablet so that the tablet remains intact after compression. The influence of tablet binders on wettability of acetaminophen tablets was studied by Esezobo et al.[51] The effect of the contact angle of water on the acetaminophen tablets, the surface tension of the liquid, and the disintegration time of the tablets is given in the following table. The water on the tablet surface is saturated with the basic formulation ingredients excluding the binder. The concentration of the tablet binders, povidone (polyvinylpyrrolidone, PVP), gelatin, and tapioca, is constant at 5% w/w.

| Binder | $\gamma$ (N/m)* | Cos $\theta$ | $t$ (min) |
|---|---|---|---|
| Povidone | 71.23 | 0.7455 | 17.0 |
| Gelatin | 71.23 | 0.7230 | 23.5 |
| Tapioca | 71.33 | 0.7570 | 2.0 |

*The surface tension, $\gamma$, is given in joules/m, or newtons, the SI force unit, divided by meters. In the cgs system, $\gamma$ is expressed in the force unit of dynes divided by centimeters, or in ergs/cm².

Using equations (15–60) and (15–61), compute $S$, the spreading coefficient, and $W_{SL}$, the work of adhesion, for water on the tablet surface, comparing the influence of the three binders in the formulation. Observe the disintegration times found in the table and use them to refute or corroborate the $S$ and $W_{SL}$ results.

Spreading coefficient, $S = \gamma(\cos\theta - 1)$

| | |
|---|---|
| PVP | $S = 71.23(0.7455 - 1) = -18.13$ |
| Gelatin | $S = 71.23(0.7230 - 1) = -19.73$ |
| Tapioca | $S = 71.33(0.7570 - 1) = -17.33$ |

Work of adhesion, $W = \gamma(1 + \cos\theta)$

| | |
|---|---|
| PVP | $W_{SL} = 71.23(1 + 0.7455) = 124.33$ N/m |
| Gelatin | $W_{SL} = 71.23(1 + 0.7230) = 122.73$ N/m |
| Tapioca | $W_{SL} = 71.33(1 + 0.7570) = 125.33$ N/m |

The spreading coefficient is negative, but the values are small. Tapioca shows the smallest negative value, $S = -17.33$, followed by PVP and finally gelatin. These results agree with the work of adhesion, tapioca > PVP > gelatin. When the work of adhesion is higher, the bond between water and tablet surface is stronger, and the better is the wetting.

From the table, we observe the tablet disintegration times to be on the order tapioca < PVP < gelatin, which agrees qualitatively with the $S$ and $W_{SL}$ values. That is, the better the wetting, reflected in a larger work of adhesion and a smaller negative spreading coefficient, the shorter is the tablet disintegration time. Other factors, such as tablet porosity, that were not considered in the study cause the relationship to be only qualitative.

## EXAMPLE 15–15

### The Influence of Additives on the Spreading Coefficient of a Film Coating Formulation to a Model Tablet Surface

One of the requirements of tablet film coating is that good adhesion of the coat to the tablet must be achieved. The properties of the coating formulation as well as those of the tablet can influence adhesion. The prerequisite for good adhesion is the spreading of the atomized droplets over the surface of the tablet and limited penetration of the coating solution into the pores of the tablet. Both of these are controlled by the surface energetics of the tablet and the coating solution. Khan et al.[52] determined the spreading coefficients of hydroxypropyl methylcellulose (HPMC) containing additives on a model tablet surface. Four formulations were studied. The formulations contained (a) 9% HPMC, 1% polyethylene glycol 400 (PEG 400); (b) 9% HPMC, 1% PEG 400, 2% microcrystalline cellulose (MCC); (c) 9% HPMC, 1% PEG 400, 2% MCC, 2% lactose; and (d) 9% HPMC, 1% PEG 400, 0.5% Tween 20. Tablets consisted of 75.2% MCC, 24.2% lactose, 0.4% magnesium atearate, and 0.2% colloidal silicon dioxide (all w/w). Two batches of tablets with average breaking loads of 127 and 191 N were produced.

Contact angles (degrees) and spreading coefficients (SC, mJ/m²) for the coating formulations of the tablets ($N = 10$; means ± SD) are as follows:

| Coating Formulations | 127-N Tablet Contact Angle | SC | 191-N Tablet Contact Angle | SC |
|---|---|---|---|---|
| (a) HPMC | 43 ± 0.48 | 41.2 | 46 ± 0.47 | 39.6 |
| (b) MCC | 54 ± 0.34 | 40.2 | 54 ± 0.32 | 38.6 |
| (c) Lactose | 50 ± 0.37 | 39.0 | 51 ± 0.52 | 37.4 |
| (d) Tween | 53 ± 0.93 | 41.8 | 52 ± 0.47 | 40.2 |

The inclusion of additives changes the contact angle of the coating formulations to a limited extent. The spreading coefficients are all high and positive, indicating effective spreading of the coating formulations on the surface of the tablets.

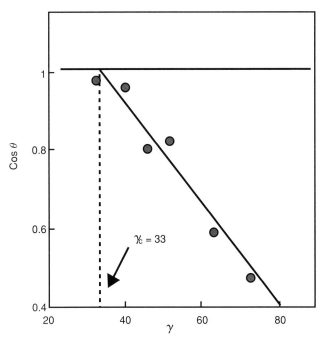

**Fig. 15–25.** Critical surface tension (Zisman) plot for a model skin. (Replotted from J. C. Charkoudian, J. Soc. Cosmet. Chem. **39**, 225, 1988.)

Zisman and his associates[53] found that when the cosine of the contact angle, $\cos \theta$, was plotted versus the surface tension for a homogeneous series of liquids spread on a surface such as Teflon (polytetrafluoroethylene), a straight line resulted. The line can be extrapolated to $\cos \theta = 1$, that is, to a contact angle of zero, signifying complete wetting. The surface tension at $\cos \theta = 1$ was given the term *critical surface tension* and the symbol $\gamma_c$. Various series of liquids on a given solid surface were all found to have about the same value of $\gamma_c$, as observed in **Figure 15–25**. Zisman concluded that $\gamma_c$ was characteristic for each solid, Teflon, for example, having a value of about 18 ergs/cm$^2$. Because the surface of Teflon consists of $-CF_2-$ groups, Zisman reasoned that all surfaces of this nature would have critical surface tensions of about 18 ergs/cm$^2$, and any liquid with a surface tension less than 18 ergs/cm$^2$ would wet a surface composed of $-CF_2-$ groups.[54,55]

**EXAMPLE 15–16**

**Wetting of Model Skin**

Charkoudian[56] designed a model skin surface with physical and chemical properties approximating those reported for human skin. The model skin consisted of a protein (cross-linked gelatin), a synthetic lipidlike substance, and water, with the protein and lipid in a ratio of 3 to 1. To further characterize the artificial skin, the surface tensions of several liquids and their contact angles on the model skin surface were determined at 20°C, as given in the following table:

| Liquid | Water | Glycerin | Diiodo methane | Ethylene glycol | Benzyl alcohol | Mineral oil |
|---|---|---|---|---|---|---|
| $\gamma$ (dynes/cm) | 72.8 | 63.4 | 50.8 | 48.3 | 39.2 | 31.9 |
| $\cos \theta$ | 0.45 | 0.56 | 0.79 | 0.77 | 0.96 | 0.97 |

Plot $\cos \theta$ versus $\gamma$ and compute the critical surface tension, $\gamma_c$, for complete wetting of the artificial skin surface. The value of $\gamma_c$ for in vivo human skin is about 26 to 28 dynes/cm.

From the results obtained, which liquid in the table would be expected to best wet the model skin surface?

The plot is shown in Figure 15–25. Although the liquids in the table do not constitute a homologous series, they appear to fit nicely the Zisman[53] principle, producing a straight line that extrapolates to $\cos \theta$ corresponding to a critical surface tension of $\gamma_c = 33$ dynes/cm.

Mineral oil, with a surface tension of 31.9 dynes/cm, most closely approximates the critical surface tension, $\gamma_c = 33$ dynes/cm, of the model skin surface. For a more exact calculation of $\gamma_c$, least squares linear regression analysis can be applied to yield

$$\cos \theta = -0.0137\gamma + 1.450, \qquad r^2 = 0.972$$

For the specific value of $\cos \theta = 1$, we obtain $\gamma_c = 33.0$ dynes/cm. It is noted that the critical surface tension, $\gamma_c$, for the artificial skin used in this study is somewhat higher ($\gamma_c = 33.0$ dynes/cm) than values reported elsewhere in the literature for human skin ($\gamma_c = 26–28$ dynes/cm). This is believed to be due in part to the absence of sweat and sebaceous secretions, which lower the $\gamma_c$ value of viable human skin.

Although one frequently desires to determine the relative efficiencies of wetting agents, it is difficult to measure the contact angle. Nor are spreading coefficients usually available, because no convenient method is known for directly measuring the surface tension of a solid surface. As a result of these difficulties, empirical tests are used in industry, one of the best-known wetting tests being that of Draves[23]. The *Draves test* involves measuring the time for a weighted skein of cotton yarn to sink through the wetting solution contained in a 500-mL graduate. No method has been suggested for estimating the ability of a wetting agent to promote spreading of a lotion on the surface of the skin, and the application properties of such products are ordinarily determined by subjective evaluation.

*Detergents* are surfactants that are used for the removal of dirt. Detergency is a complex process involving the removal of foreign matter from surfaces. The process includes many of the actions characteristic of specific surfactants: initial wetting of the dirt and of the surface to be cleaned; deflocculation and suspension; emulsification or solubilization of the dirt particles; and sometimes foaming of the agent for entrainment and washing away of the particles. Because the detergent must possess a combination of properties, its efficiency is best ascertained by actual tests with the material to be cleaned.

Other dispersion stabilizers, including deflocculating, suspending, and emulsifying agents, are considered in Chapter 17.

## APPLICATIONS OF SURFACE-ACTIVE AGENTS

In addition to the use of surfactants as emulsifying agents, detergents, wetting agents, and solubilizing agents, they find application as antibacterial and other protective agents and as aids to the absorption of drugs in the body.

A surfactant may affect the activity of a drug or may itself exert drug action. As an example of the first case, the penetration of hexylresorcinol into the pinworm, *Ascaris*, is increased by the presence of a low concentration of surfactant. This potentiation of activity is due to a reduction in interfacial tension between the liquid phase and the cell wall of the organism. As a result, the adsorption and spreading of hexylresorcinol over the surface of the organism is facilitated. When the concentration of surface-active agent present exceeds that required to form micelles, however, the rate of penetration of the anthelmintic decreases nearly to zero. This is because the drug is now partitioned between the micelles and the aqueous phase, resulting in a reduction in the effective concentration. Quaternary ammonium compounds are examples of surface-active agents that in themselves possess antibacterial activity.[55] This may depend in part on interfacial phenomena, but other factors are also important. The agents are adsorbed on the cell surface and supposedly bring about destruction by increasing the permeability or "leakiness" of the lipid cell membrane. Death then occurs through a loss of essential materials from the cell. Both gram-negative and gram-positive organisms are susceptible to the action of the cationic quaternary compounds, whereas anionic agents attack gram-positive organisms more easily than gram-negative bacteria. Nonionic surfactants are least effective as antibacterial agents. In fact, they often aid rather than inhibit the growth of bacteria, presumably by providing long-chain fatty acids in a form that is easily metabolized by the organism.

Miyamoto et al.[58] studied the effects of surfactants and bile salts on the gastrointestinal absorption of antibiotics using an in situ rat-gut perfusion technique. Polyoxyethylene lauryl ether reduced the absorption of propicillin in the stomach and increased it in the small intestine. Some surfactants increase the rate of intestinal absorption, whereas others decrease it. Some of these effects may result from alteration of the membrane by the surfactant. The effects of surfactants on the solubility of drugs and their bioabsorption have been reviewed by Mulley[59] and Gibaldi and Feldman.[60]

## Foams and Antifoaming Agents

Any solutions containing surface-active materials produce stable foams when mixed intimately with air. A foam is a relatively stable structure consisting of air pockets enclosed within thin films of liquid, the gas-in-liquid dispersion being stabilized by a *foaming agent*. The foam dissipates as the liquid drains away from the area surrounding the air globules, and the film finally collapses. *Agents* such as alcohol, ether, castor oil, and some surfactants can be used to break the foam and are known as *antifoaming agents*. Foams are sometimes useful in pharmacy (e.g., vaginal contraceptive or antimicrobial formulations) but are usually a nuisance and are prevented or destroyed when possible. The undesirable foaming of solubilized liquid preparations poses a problem in formulation.

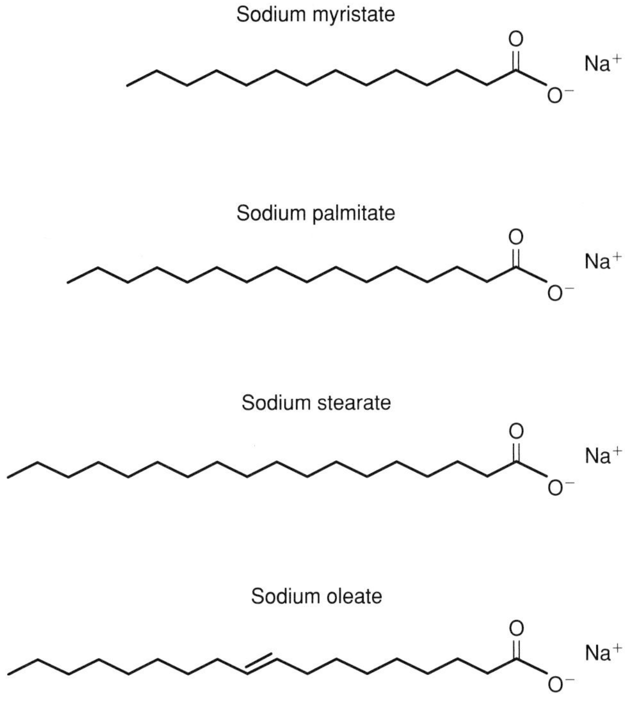

**Fig. 15–26.** Common soaps.

All of the *soaps* (sodium oleate, etc.) are fatty acid salts (anionic surfactant). They are characterized by (*a*) a long hydrocarbon chain, which may be monounsaturated (i.e., have one double bond, like sodium oleate), polyunsaturated (i.e., have more than one double bond), or saturated (i.e., no double bonds), and (*b*) a carboxylate group at the end (**Fig. 15–26**). Any surfactant that is not a soap is a *detergent*. The cleaning action of soaps and detergents is based on the property known as detergency. Possibly, the most important industrial role for surfactants is the formation of emulsions. An emulsion is a dispersion of one liquid in a second, immiscible liquid. Salad dressings, milk, and cream are emulsions, as are medicinal creams such as moisturizers. Emulsions are multiphase systems, even though they often look like they are just one phase. The phases in an emulsion are normally called the continuous phase and the dispersed phase. Detergency is a complex process involving the removal of foreign matter from surfaces. The process includes the following main steps (**Fig. 15–27**): (*a*) The hydrocarbon tails of the detergent anions dissolve in the grease; (*b*) the grease spot gradually breaks up and becomes pincushioned by the detergent anions; and (*c*) small bits of grease are held in colloidal suspension by the detergent. The anionic heads keep the grease from coalescing because the particles carry the same electric charge.

## Lung Surfactant

Lung surfactant is surface-active agent that covers the surface of alveoli contacted with air. It decreases the surface tension at the air–alveoli interface almost to zero and therefore

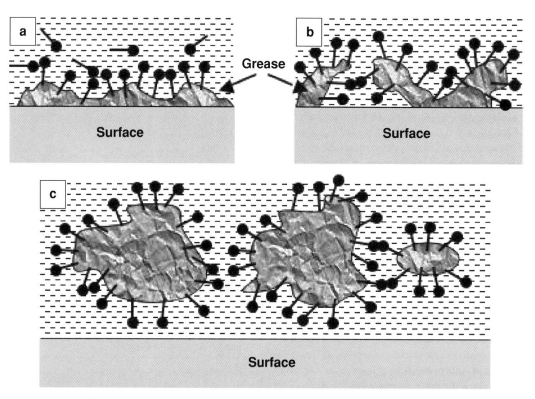

**Fig. 15–27.** Mechanism of detergent action. (*a*) The hydrocarbon tails of the detergent anions dissolve in the grease; (*b*) the grease spot gradually breaks up and becomes pincushioned by the detergent anions, and (*c*) small bits of grease are held in colloidal suspension by the detergent.

accomplishes two main tasks. First, it prevents the collapse of alveoli. Second, main surfactant function is to decrease the pressure inside the alveoli. Thus, lung surfactant allows us to breathe and prevents pulmonary edema. Lung surfactant is a complex mixture of proteins and lipids but the major component is phosphatidylcholine. Some pathologic conditions were found to decrease the activity of lung surfactant. In the United States, 40,000 premature infants per year are born without enough lung surfactant, resulting in thousands of deaths. The typical premature infant has only 1/20 of the lung surfactant needed to breathe. Fortunately, additional artificial lung surfactant can be administered. Calfactant (Infasurf; Forest Pharmaceuticals, St. Louis, MO) is one of the available artificial surfactants. Infasurf (calfactant) Intratracheal Suspension is a sterile, nonpyrogenic lung surfactant intended for intratracheal instillation only. It is an extract of natural surfactant from calf lungs and includes phospholipids, neutral lipids, and hydrophobic surfactant-associated proteins B and C (SP-B and SP-C). It contains no preservatives. Infasurf is an off-white suspension of calfactant in 0.9% aqueous sodium chloride solution. It has a pH of 5.0 to 6.0. Each milliliter of Infasurf contains 35 mg of total phospholipids (including 26 mg of phosphatidylcholine, of which 16 mg is disaturated phosphatidylcholine) and 0.65 mg of proteins, including 0.26 mg of SP-B. Treatment with calfactant often rapidly improves oxygenation and lung compliance.

# ELECTRIC PROPERTIES OF INTERFACES

This section deals with some of the principles involved with surfaces that are charged in relation to their surrounding liquid environment. Discussion of the applications arising from this phenomenon is given in the chapters dealing with colloidal systems (Chapter 16) and suspensions (Chapter 17).

Particles dispersed in liquid media may become charged mainly in one of two ways. The first involves the selective adsorption of a particular ionic species present in solution. This may be an ion added to the solution or, in the case of pure water, it may be the hydronium or hydroxyl ion. The majority of particles dispersed in water acquire a negative charge due to preferential adsorption of the hydroxyl ion. Second, charges on particles arise from ionization of groups (such as COOH) that may be situated at the surface of the particle. In these cases, the charge is a function of p$K$ and pH. A third, less common origin for the charge on a particle surface is thought to arise when there is a difference in dielectric constant between the particle and its dispersion medium.

## The Electric Double Layer

Consider a solid surface in contact with a polar solution containing ions, for example, an aqueous solution of an electrolyte. Furthermore, let us suppose that some of the cations are adsorbed onto the surface, giving it a positive charge. Remaining in solution are the rest of the cations plus the total number of anions added. These anions are attracted to the positively charged surface by electric forces that also serve to repel the approach of any further cations once the initial adsorption is complete. In addition to these electric forces, thermal motion tends to produce an equal distribution of all the ions in solution. As a result, an equilibrium situation is set up in which *some* of the excess anions approach the surface, whereas the remainder are distributed in decreasing amounts as one proceeds away from the charged surface. At a particular distance from the surface, the concentrations of anions and cations are equal, that is, conditions of electric neutrality prevail. It is important to remember that the system *as a whole* is electrically neutral, even though there are regions of unequal distribution of anions and cations.

Such a situation is shown in **Figure 15–28**, where $aa'$ is the surface of the solid. The adsorbed ions that give the surface its positive charge are referred to as the *potential-determining ions*. Immediately adjacent to this surface layer is a region of tightly bound solvent molecules, together with some negative ions, also tightly bound to the surface. The limit of this region is given by the line $bb'$ in **Figure 15–28**. These ions, having a charge opposite to that of the potential-determining ions, are known as *counterions* or *gegenions*. The degree of attraction of the solvent molecules and counterions is such that if the surface is moved relative to the liquid, the shear plane is $bb'$ rather than $aa'$, the true surface. In the region bounded by the lines $bb'$ and $cc'$, there is an excess of negative ions. The potential at $bb'$ is still positive because, as previously mentioned, there are fewer anions in the tightly bound layer than cations adsorbed onto the surface of the solid. Beyond $cc'$, the distribution of ions is uniform and electric neutrality is obtained.

Thus, the electric distribution at the interface is equivalent to a double layer of charge, the first layer (extending from $aa'$ to $bb'$) tightly bound and a second layer (from $bb'$ to $cc'$) that

is more diffuse. The so-called diffuse double layer therefore extends from $aa'$ to $cc'$.

Two situations other than that represented by **Figure 15–28** are possible: (*a*) If the counterions in the tightly bound, solvated layer equal the positive charge on the solid surface, then electric neutrality occurs at the plane $bb'$ rather than $cc'$. (*b*) Should the total charge of the counterions in the region $aa'$–$bb'$ exceed the charge due to the potential-determining ions, then the net charge at $bb'$ will be negative rather than less positive, as shown in **Figure 15–28**. This means that, in this instance, for electric neutrality to be obtained at $cc'$, an excess of positive ions must be present in the region $bb'$–$cc'$.

The student should appreciate that if the potential-determining ion is negative, the arguments just given still apply, although now positive ions will be present in the tightly bound layer.

## Nernst and Zeta Potentials

The changes in potential with distance from the surface for the various situations discussed in the previous section can be represented as shown in **Figure 15–29**. The potential at the solid surface $aa'$ due to the potential-determining ion is the *electrothermodynamic (Nernst) potential*, $E$, and is defined as the difference in potential between the actual surface and the

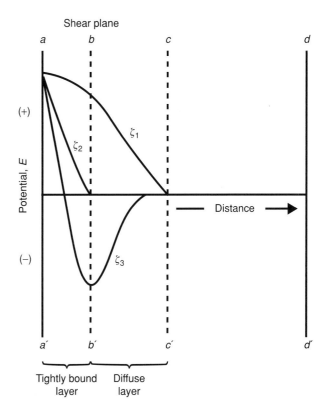

**Fig. 15–29.** Electrokinetic potential, $E$, at solid–liquid boundaries. Curves are shown for three cases characteristic of the ions or molecules in the liquid phase. Note that although $E$ is the same in all three cases, the zeta potentials are positive ($\zeta_1$), zero ($\zeta_2$), and negative ($\zeta_3$).

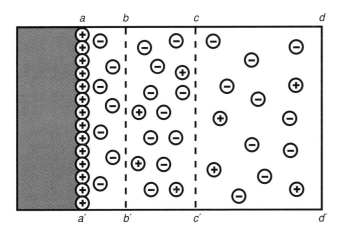

**Fig. 15–28.** The electric double layer at the surface of separation between two phases, showing distribution of ions. The system as a whole is electrically neutral.

electroneutral region of the solution. The potential located at the shear plane $bb'$ is known as the *electrokinetic*, or *zeta, potential*, $\zeta$. The zeta potential is defined as the difference in potential between the surface of the tightly bound layer (shear plane) and the electroneutral region of the solution. As shown in **Figure 15–29**, the potential initially drops off rapidly, followed by a more gradual decrease as the distance from the surface increases. This is because the counterions close to the surface act as a screen that reduces the electrostatic attraction between the charged surface and those counterions further away from the surface.

The zeta potential has practical application in the stability of systems containing dispersed particles because this potential, rather than the Nernst potential, governs the degree of repulsion between adjacent, similarly charged, dispersed particles. If the zeta potential is reduced below a certain value (which depends on the particular system being used), the attractive forces exceed the repulsive forces, and the particles come together. This phenomenon is known as *flocculation* and is discussed in the chapters dealing with colloidal and coarse dispersions.

## Effect of Electrolytes

As the concentration of electrolyte present in the system is increased, the screening effect of the counterions is also increased. As a result, the potential falls off more rapidly with distance because the thickness of the double layer shrinks. A similar situation occurs when the valency of the counterion is increased while the total concentration of electrolyte is held constant. The overall effect frequently causes a reduction in zeta potential.

# CHAPTER SUMMARY

Several types of interface can exist, depending on whether the two adjacent phases are in the solid, liquid, or gaseous state. For convenience, this chapter divided these various combinations into two groups, namely, *liquid interfaces* and *solid interfaces*. In the former group, the association of a liquid phase with a gaseous or another liquid phase was discussed. The section on solid interfaces dealt with systems containing solid–gas and solid–liquid interfaces. Although solid–solid interfaces have practical significance in pharmacy (e.g., the adhesion between granules, the preparation of layered tablets, and the flow of particles), little information is available to quantify these interactions. This is due, at least in part, to the fact that the surface region of materials in the solid state is quiescent, in sharp contrast to the turbulence that exists at the surfaces of liquids and gases. Accordingly, solid–solid systems were not discussed. Finally, the electric properties of interfaces were outlined. By the conclusion of this chapter you should understand the terms surface tension and interfacial tension and their application in pharmaceutical sciences as well as appreciate the various methods of measuring surface and interface tensions. The student should also be able to classify surface-active agents and appreciate their applications in pharmacy.

 **Practice problems for this chapter can be found at thePoint.lww.com/Sinko6e.**

## References

1. S. I. Said, Med. Clin. North Am. **51,** 391, 1967.
2. S. E. Poynter and A. M. LeVine, Crit. Care Clin. **19,** 459, 2003.
3. J. R. Wright, J. Clin Invest. **111,** 1553, 2003.
4. A. Felmeister, J. Pharm. Sci. **61,** 151, 1972.
5. P. Seeman, Pharmacol. Rev. **24,** 583, 1972.
6. G. Viscardi, P. Quagliotto, C. Barolo, P. Savarino, E. Barni, and E. Fisicaro, J. Org. Chem. **65,** 8197, 2000.
7. P. C. Hiemenz, *Principles of Colloid and Surface Chemistry,* 2nd Ed., Marcel Dekker, New York, 1986.
8. R. C. Weast (Ed.), *CRC Handbook of Chemistry and Physics*, 63rd Ed. CRC, Boca Raton, FL, p. F-35.
9. A. W. Adamson, *Physical Chemistry of Surfaces*, 4th Ed., Wiley, New York, 1982, Chapter 2.
10. W. D. Harkins and A. E. Alexander, in A. Weissberger (Ed.), *Physical Methods of Organic Chemistry*, **Vol. 1,** 3rd Ed., Interscience, New York, 1959, Chapter 15.
11. W. Drost-Hansen, in *Chemistry and Physics of Interfaces,* American Chemical Society, Washington, D.C., 1965, Chapters 2 and 3.
12. T. Matsumoto, T. Nakano, H. Fujii, M. Kamai, and K. Nogi, Phys. Rev. E Stat. Nonlinear Soft Matter Phys. **65,** 031201, 2002.
13. W. D. Harkins and H. F. Jordan, J. Am. Chem. Soc. **52,** 1751, 1930.
14. H. L. Cupples, J. Phys. Chem. **51,** 1341, 1947.
15. H. H. Zuidema and G. W. Waters, Ind. Eng. Chem. (Anal.) **13,** 312, 1941.
16. D. J. Shaw, *Introduction to Colloid and Surface Chemistry*, 2nd Ed., Butterworths, Boston, 1970, p. 62.
17. P. Becher, J. Soc. Cosmetic Chem. **11,** 325, 1960.
18. W. C. Griffin, J. Soc. Cosmetic Chem. **1,** 311, 1949.
19. I. Bonadeo, *Cosmética, Ciencía y Technologia,* Editorial Ciencia, Madrid, 1988, p. 123.
20. H. L. Greenwald, G. L. Brown, and M. N. Fineman, Anal. Chem. **28,** 1693, 1956.
21. P. Becher, *Emulsions: Theory and Practice*, 2nd Ed., Reinhold, New York, 1965.
22. K. Koga, Y. Ishitobi, M. Iwata, M. Murakami, and S. Kawashima, Biol. Pharm. Bull. **25,** 1642, 2002.
23. P. Atkins, J. D. Paula, Physical Chemistry, W. H. Freeman, NY and Company, 7 Ed., 2006.
24. J. W. McBain and R. C. Swain, Proc. Royal Soc. London A **154,** 608, 1936.
25. J. K. Dixon, A. J. Weith, A. A. Argyle, and D. J. Salby, Nature **163,** 845, 1949.
26. C. H. Giles, Chem. Ind. 1616, 1969; C. H. Giles and S. D. Forrester, Chem. Ind. 80, 1970; and 43, 1971.
27. H. Schott, J. Pharm. Sci. **69,** 852, 1980.
28. I. Langmuir, J. Am. Chem. Soc. **40,** 1361, 1918.
29. D. L. Schwinke, M. G. Ganesan, and N. D. Weiner, J. Pharm. Sci. **72,** 244, 1983.
30. N. K. Adam, *The Physics and Chemistry of Surfaces,* Oxford University Press, London, 1941, Chapter 2.
31. W. D. Harkins, *The Physical Chemistry of Surface Films*, Reinhold, New York, 1952, p. 119.
32. H. Sobotka, *Monomolecular Layers,* American Association for the Advancement of Science, Washington, D.C., 1954.
33. G. L. Gaines, Jr., *Insoluble Monolayers at Gas–Liquid Interfaces*, Interscience, New York, 1966.
34. W. Rettig and F. Kuschel, Colloid Polymer Sci. **267,** 151, 1989.

35. A. W. Adamson, *Physical Chemistry of Surfaces*, 1st Ed., Interscience, New York, 1960, p. 231.

36. H. Freundlich, *Colloid and Capillary Chemistry*, Methuen, London, 1926.

37. I. Langmuir, J. Am. Chem. Soc. **39,** 1855, 1917.

38. N. Evcim and M. Barr, J. Am. Pharm. Assoc. Sci. Ed. **44,** 570, 1955.

39. M. Barr and S. Arnista, J. Am. Pharm. Assoc. Sci. Ed. **46,** 486, 490, 493, 1957.

40. M. A. Moustafa, H. I. Al-Shora, M. Gaber, and M. W. Gouda, Int. J. Pharm. **34,** 207, 1987.

41. I. Nikolakakis and J. M. Newton, J. Pharm. Pharmacol. **41,** 155, 1989.

42. D. Seger, J. Toxicol. Clin. Toxicol. **42,** 101, 2004.

43. S. V. Mohan and J. Karthikeyan, Environ. Pollut. **97,** 183, 1997.

44. S. Shadnia, M. Moiensadat, and M. Abdollahi, Vet. Hum. Toxicol. **46,** 76, 2004.

45. C. Tanaka, H. Yagi, M. Sakamoto, Y. Koyama, T. Ohmura, H. Ohtani, and Y. Sawada, Ann. Pharmacother. **38,** 73, 2004.

46. G. Avantaggiato, R. Havenaar, and A. Visconti, Food Chem. Toxicol. **42,** 817, 2004.

47. D. E. Wurster, K. A. Alkhamis, and L. E. Matheson, AAPS Pharm-SciTech. **1,** E25, 2000.

48. K. Hill, K. Q. Hu, A. Cottrell, S. Teichman, and D. J. Hillebrand, Am. J. Gastroenterol. **98,** 2763, 2003.

49. A. Donoso, M. Linares, J. Leon, G. Rojas, C. Valverde, M. Ramirez, and B. Oberpaur, Pediatr. Emerg. Care **19,** 420, 2003.

50. K. C. Osterhoudt, D. Durbin, E. R. Alpern, and F. M. Henretig, Pediatrics **113,** 806, 2004.

51. S. Esezobo, S. Zubair, and N. Pilben, J. Pharm. Pharmacol. **41,** 7, 1989.

52. H. Khan, J. T. Fell, and C. S. Macleod, Int. J. Pharm. **227,** 113, 2001.

53. W. Z. Zisman, in F. M. Fowkes (Ed.), *Contact Angle Wettability and Adhesion*, American Chemical Society, Washington, D.C., 1964, p. 1.

54. A. W. Adamson, *Physical Chemistry of Surfaces*, 4th Ed., Wiley, New York, 1982, pp. 350–351.

55. D. Attwood and A. T. Florence, *Surfactant Systems*, Chapman and Hall, London, New York, 1983, pp. 32–33.

56. J. C. Charkoudian, J. Soc. Cosmet. Chem. **39,** 225, 1988.

57. M. Pavlikova-Moricka, I. Lacko, F. Devinsky, L. Masarova, and D. Milynarcik, Folia Microbiol. (Praha) **39,** 176, 1994.

58. E. Miyamoto, A. Tsuji, and T. Yamana, J. Pharm. Sci. **72,** 651, 1983.

59. B. A. Mulley, in H. S. Bean, A. H. Beckett, and J. E. Carless (Eds.), *Advances in Pharmaceutical Sciences*, **Vol. 1,** Academic Press, New York, 1964, p. 87.

60. M. Gibaldi and S. Feldman, J. Pharm. Sci. **59,** 579, 1970.

## Recommended Reading

C. A. Miller and P. Neogi (Eds.), *Interfacial Phenomena: Equilibrium and Dynamic Effects*, 2nd Ed., CRC Press, Boca Raton, FL, 2007.

### CHAPTER LEGACY

**Fifth Edition:** published as Chapter 16 (Interfacial Phenomena). Updated by Tamara Minko.

**Sixth Edition:** published as Chapter 15 (Interfacial Phenomena). Updated by Patrick Sinko.

**At the conclusion of this chapter the student should be able to:**

**1** Differentiate between different types of colloidal systems and their main characteristics.

**2** Understand the main optical properties of colloids and applications of these properties for the analysis of colloids.

**3** Know the main types of microscopic systems used for analysis of colloids.

**4** Appreciate the major kinetic properties of colloids.

**5** Understand the main electrical properties of colloids and their application for the stability, sensitization, and protective action of colloids.

**6** Recognize the benefits of solubilization by colloids.

**7** Understand the benefits and know the main types of modern colloidal drug delivery systems.

## INTRODUCTION

It is important that the pharmacist understand the theory and technology of dispersed systems. Knowledge of interfacial phenomena and a familiarity with the characteristics of colloids and small particles are fundamental to an understanding of the behavior of pharmaceutical dispersions. There are three types of dispersed systems encountered in the pharmaceutical sciences: molecular, colloidal, and coarse dispersions. Molecular dispersions are homogeneous in character and form true solutions. The properties of these systems were discussed in earlier chapters. Colloidal dispersions will be considered in the present chapter. Powders and granules and coarse dispersions are discussed in other chapters; all are examples of heterogeneous systems. It is important to know that the only difference between molecular, colloidal, and coarse dispersions is the size of the dispersed phase and not its composition.

Dispersions consist of at least one internal phase that is dispersed in a dispersion medium. Sometimes putting these systems into one of the three categories is a bit tricky. So, we will start by looking at an example of a complex dispersed system that we are all very familiar with—blood. Blood is a specialized fluid that delivers vital substances such as oxygen and nutrients to various cells and tissues in the body. The dispersion medium in blood is plasma, which is mostly water (~90% or so). Blood is composed of more than one dispersed phase. Nutrients such as peptides, proteins, and glucose are dissolved in plasma forming a molecular dispersion or true solution. Oxygen, however, is carried to cells and tissues by red blood cells. Given the size of red blood cells (~6 $\mu$m in diameter and 2 $\mu$m in width) they would be considered to form a coarse dispersion in blood. White blood cells such as leukocytes and platelets are the other major cells types carried in blood. The last major component of blood is serum albumin. Serum albumin forms a true solution in water. However, the size of the individual serum albumin particles in solution is $>1$ nm, which puts them into the colloidal dispersion group. As you can now see, blood is a complex bodily fluid that is an example of the three types of dispersed systems that you will encounter in the pharmaceutical sciences.

### Size and Shape of Colloidal Particles

Particles in the colloidal size range possess a surface area that is enormous compared with the surface area of an equal volume of larger particles. Thus, a cube having a 1-cm edge and a volume of 1 cm$^3$ has a total surface area of 6 cm$^2$. If the same cube is subdivided into smaller cubes each having an edge of 100 $\mu$m, the total volume remains the same, but the total surface area increases to 600,000 cm$^2$. This represents a $10^5$-fold increase in surface area. To compare the surface areas of different materials quantitatively, the term *specific surface* is used. This is defined as the surface area per unit weight or volume of material. In the example just given, the first sample had a specific surface of 6 cm$^2$/cm$^3$, whereas the second sample had a specific surface of 600,000 cm$^2$/cm$^3$. The possession of a large specific surface results in many of the unique properties of colloidal dispersions. For example, platinum is effective as a catalyst only when in the colloidal form as platinum black. This is because catalysts act by adsorbing the reactants onto their surface. Hence, their catalytic activity is related to their specific surface. The color of colloidal dispersions is related to the size of the particles present. Thus, as the particles in a red gold sol increase in size, the dispersion takes on a blue color. Antimony and arsenic trisulfides change from red to yellow as the particle size is reduced from that of a coarse powder to that within the colloidal size range.

Because of their size, colloidal particles can be separated from molecular particles with relative ease. The technique of separation, known as *dialysis*, uses a semipermeable membrane of collodion or cellophane, the pore size of which will prevent the passage of colloidal particles, yet permit small molecules and ions, such as urea, glucose, and sodium chloride, to pass through. The principle is illustrated in **Figure 16–1**, which shows that, at equilibrium, the colloidal

# KEY CONCEPT DISPERSED SYSTEMS

Dispersed systems consist of particulate matter, known as the *dispersed phase*, distributed throughout a *continuous* or *dispersion medium*. The dispersed material may range in size from particles of atomic and molecular dimensions to particles whose size is measured in millimeters. Accordingly, a convenient means of classifying dispersed systems is on the basis of the mean particle diameter of the dispersed material. Based on the size of the dispersed phase, three types of dispersed systems are generally considered: (*a*) *molecular* dispersions, (*b*) *colloidal* dispersions, and (*c*) *coarse* dispersions. The size ranges assigned to these classes, together with some of the associated characteristics, are shown in the accompanying table. The size limits are somewhat arbitrary, there being no distinct transition between either molecular and colloidal dispersions or colloidal and coarse dispersions. For example, certain *macro* (i.e., large) molecules, such as the polysaccharides, proteins, and polymers in general, are of sufficient size that they may be classified as forming both molecular and colloidal dispersions. Some suspensions and emulsions may contain a range of particle sizes such that the smaller particles lie within the colloidal range, whereas the larger ones are classified as coarse particles.

## CLASSIFICATION OF DISPERSED SYSTEMS BASED ON PARTICLE SIZE

| Class | Particle Size* | Characteristics of System | Examples |
|---|---|---|---|
| Molecular dispersion | Less than 1 nm | Invisible in electron microscope<br>Pass through ultrafilter and semipermeable membrane<br>Undergo rapid diffusion | Oxygen molecules, ordinary ions, glucose |
| Colloidal dispersion | From 1 nm to 0.5 $\mu$m | Not resolved by ordinary microscope (although may be detected under ultramicroscope)<br>Visible in electron microscope<br>Pass through filter paper<br>Do not pass semipermeable membrane<br>Diffuse very slowly | Colloidal silver sols, natural and synthetic polymers, cheese, butter, jelly, paint, milk, shaving cream, etc. |
| Coarse dispersion | Greater than 0.5 $\mu$m | Visible under microscope<br>Do not pass through normal filter paper<br>Do not dialyze through semipermeable membrane<br>Do not diffuse | Grains of sand, most pharmaceutical emulsions and suspensions, red blood cells |

* 1 nm (nanometer) = $10^{-9}$ m; 1 $\mu$m (micrometer) = $10^{-6}$ m.

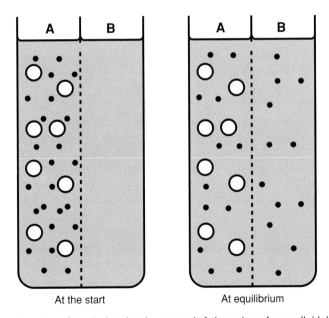

At the start          At equilibrium

**Fig. 16–1.** Sketch showing the removal of electrolytes from colloidal material by diffusion through a semipermeable membrane. Conditions on the two sides, A and B, of the membrane are shown at the start and at equilibrium. The open circles are the colloidal particles that are too large to pass through the membrane. The solid dots are the electrolyte particles that pass through the pores of the membrane.

material is retained in compartment A, whereas the subcolloidal material is distributed equally on both sides of the membrane. By continually removing the liquid in compartment B, it is possible to obtain colloidal material in A that is free from subcolloidal contaminants. Dialysis can also be used to obtain subcolloidal material that is free from colloidal contamination—in this case, one simply collects the effluent. *Ultrafiltration* has also been used to separate and purify colloidal material. According to one variation of the method, filtration is conducted under negative pressure (suction) through a dialysis membrane supported in a Büchner funnel. When dialysis and ultrafiltration are used to remove charged impurities such as ionic contaminants, the process can be hastened by the use of an electric potential across the membrane. This process is called *electrodialysis*.

Dialysis has been used increasingly in recent years to study the binding of materials of pharmaceutical significance to colloidal particles. Dialysis occurs in vivo. Thus, ions and small molecules pass readily from the blood, through a natural semipermeable membrane, to the tissue fluids; the colloidal components of the blood remain within the capillary system. The principle of dialysis is utilized in the artificial kidney, which removes low–molecular-weight impurities from the body by passage through a semipermeable membrane.

The shape adopted by colloidal particles in dispersion is important because the more extended the particle, the greater

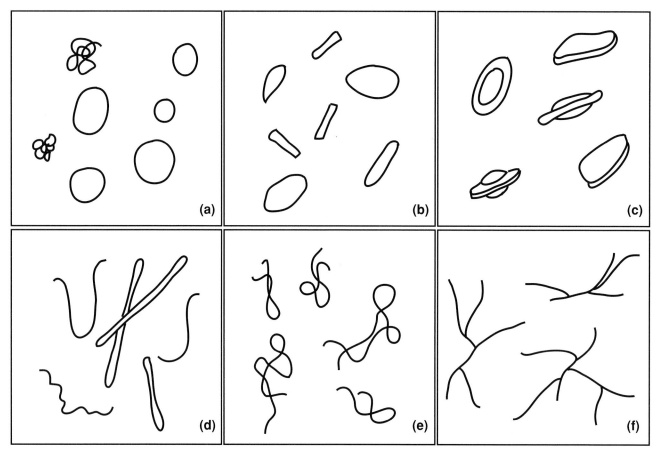

**Fig. 16–2.**  Some shapes that can be assumed by colloidal particles: (*a*) spheres and globules, (*b*) short rods and prolate ellipsoids, (*c*) oblate ellipsoids and flakes, (*d*) long rods and threads, (*e*) loosely coiled threads, and (*f*) branched threads.

is its specific surface and the greater is the opportunity for attractive forces to develop between the particles of the dispersed phase and the dispersion medium. A colloidal particle is something like a hedgehog—in a friendly environment, it unrolls and exposes maximum surface area. Under adverse conditions, it rolls up and reduces its exposed area. Some representative shapes of spherocolloids and fibrous colloids are shown in **Figure 16–2**. As will be seen in later discussions, such properties as flow, sedimentation, and osmotic pressure are affected by changes in the shape of colloidal particles. Particle shape may also influence pharmacological action.

## TYPES OF COLLOIDAL SYSTEMS

### Lyophilic Colloids

Systems containing colloidal particles that interact to an appreciable extent with the dispersion medium are referred to as *lyophilic* (solvent-loving) colloids. Owing to their affinity for the dispersion medium, such materials form colloidal dispersions, or *sols*, with relative ease. Thus, lyophilic colloidal sols are usually obtained simply by dissolving the material in the solvent being used. For example, the dissolution of acacia

or gelatin in water or celluloid in amyl acetate leads to the formation of a sol.

The various properties of this class of colloids are due to the attraction between the dispersed phase and the dispersion medium, which leads to *solvation*, the attachment of solvent molecules to the molecules of the dispersed phase. In the case of hydrophilic colloids, in which water is the dispersion medium, this is termed *hydration*. Most lyophilic colloids are organic molecules, for example, gelatin, acacia, insulin, albumin, rubber, and polystyrene. Of these, the first four produce lyophilic colloids in aqueous dispersion media (hydrophilic sols). Rubber and polystyrene form lyophilic colloids in nonaqueous, organic solvents. These materials accordingly are referred to as *lipophilic* colloids. These examples illustrate the important point that the term *lyophilic* has meaning only when applied to the material dispersed in a specific dispersion medium. A material that forms a lyophilic colloidal system in one liquid (e.g., water) may not do so in another liquid (e.g., benzene).

### Lyophobic Colloids

The second class of colloids is composed of materials that have little attraction, if any, for the dispersion medium. These

## ▶️KEY CONCEPT COLLOIDAL SYSTEMS

All kinds of dispersed phases might form colloids in all possible kinds of media, except for a gas–gas combination. Because all gases mix uniformly at the molecular level, gases only form solutions with each other. Possible types of colloidal dispersions are shown in the accompanying table. Colloidal systems are best classified into three groups—lyophilic, lyophobic, and association—on the basis of the interaction of the particles, molecules, or ions of the dispersed phase with the molecules of the dispersion medium.

### TYPES OF COLLOIDAL DISPERSIONS*

| Dispersion Medium | Dispersed Phase | Colloid Type | Examples |
|---|---|---|---|
| Solid | Solid | Solid sol | Pearls, opals |
| Solid | Liquid | Solid emulsion | Cheese, butter |
| Solid | Gas | Solid foam | Pumice, marshmallow |
| Liquid | Solid | Sol, gel | Jelly, paint |
| Liquid | Liquid | Emulsion | Milk, mayonnaise |
| Liquid | Gas | Foam | Whipped cream, shaving cream |
| Gas | Solid | Solid aerosols | Smoke, dust |
| Gas | Liquid | Liquid aerosols | Clouds, mist, fog |

* A gas in a gas always produces a solution.

are the *lyophobic* (solvent-hating) colloids and, predictably, their properties differ from those of the lyophilic colloids. This is primarily due to the absence of a solvent sheath around the particle. Lyophobic colloids are generally composed of inorganic particles dispersed in water. Examples of such materials are gold, silver, sulfur, arsenous sulfide, and silver iodide.

In contrast to lyophilic colloids, it is necessary to use special methods to prepare lyophobic colloids. These are (*a*) dispersion methods, in which coarse particles are reduced in size, and (*b*) condensation methods, in which materials of subcolloidal dimensions are caused to aggregate into particles within the colloidal size range. Dispersion can be achieved by the use of high-intensity ultrasonic generators operating at frequencies in excess of 20,000 cycles per second. A second dispersion method involves the production of an electric arc within a liquid. Owing to the intense heat generated by the arc, some of the metal of the electrodes is dispersed as vapor, which condenses to form colloidal particles. Milling and grinding processes can be used, although their efficiency is low. So-called colloid mills, in which the material is sheared between two rapidly rotating plates set close together, reduce only a small amount of the total particles to the colloidal size range.

The required conditions for the formation of lyophobic colloids by condensation or aggregation involve a high degree of initial supersaturation followed by the formation and growth of nuclei. Supersaturation can be brought about by change in solvent or reduction in temperature. For example, if sulfur is dissolved in alcohol and the concentrated solution is then poured into an excess of water, many small nuclei form in the supersaturated solution. These grow rapidly to form a colloidal sol. Other condensation methods depend on a chemical reaction, such as reduction, oxidation, hydrolysis, and double decomposition. Thus, neutral or slightly alkaline solutions of the noble metal salts, when treated with a reducing agent such as formaldehyde or pyrogallol, form atoms

that combine to form charged aggregates. The oxidation of hydrogen sulfide leads to the formation of sulfur atoms and the production of a sulfur sol. If a solution of ferric chloride is added to a large volume of water, hydrolysis occurs with the formation of a red sol of hydrated ferric oxide. Chromium and aluminum salts also hydrolyze in this manner. Finally, the double decomposition between hydrogen sulfide and arsenous acid results in an arsenous sulfide sol. If an excess of hydrogen sulfide is used, $HS^-$ ions are adsorbed onto the particles. This creates a large negative charge on the particles, leading to the formation of a stable sol.

## Association Colloids: Micelles and the Critical Micelle Concentration

*Association* or *amphiphilic* colloids form the third group in this classification. As shown in the Interfacial Phenomena chapter, certain molecules or ions, termed *amphiphiles* or *surface-active agents*, are characterized by having two distinct regions of opposing solution affinities within the same molecule or ion. When present in a liquid medium at low concentrations, the amphiphiles exist separately and are of such a size as to be subcolloidal. As the concentration is increased, aggregation occurs over a narrow concentration range. These aggregates, which may contain 50 or more monomers, are called *micelles*. Because the diameter of each micelle is of the order of 50 Å, micelles lie within the size range we have designated as colloidal. The concentration of monomer at which micelles form is termed the *critical micelle concentration* (*CMC*). The number of monomers that aggregate to form a micelle is known as the *aggregation number* of the micelle.

The phenomenon of micelle formation can be explained as follows. Below the CMC, the concentration of amphiphile undergoing adsorption at the air–water interface increases as the total concentration of amphiphile is raised. Eventually,

Critical concentration

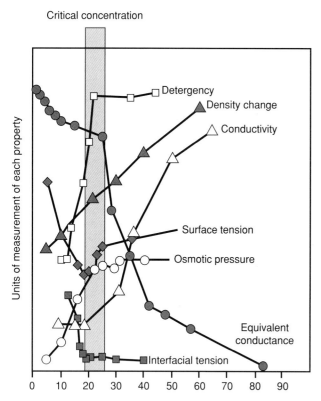

**Fig. 16–3.** Properties of surface-active agents showing changes that occur sharply at the critical micelle concentration. (Modified from W. J. Preston, Phys. Coll. Chem. **52**, 85, 1948.)

a point is reached at which both the interface and the bulk phase become saturated with monomers. This is the CMC. Any further amphiphile added in excess of this concentration aggregates to form micelles in the bulk phase, and, in this manner, the free energy of the system is reduced. The effect of micellization on some of the physical properties of solutions containing surface-active agents is shown in **Figure 16–3**. Note particularly that surface tension decreases up to the CMC. From the Gibbs' adsorption equation, this means increasing interfacial adsorption. Above the CMC, the surface tension remains essentially constant, showing that the interface is saturated and micelle formation has taken place in the bulk phase.

In the case of amphiphiles in water, the hydrocarbon chains face inward into the micelle to form, in effect, their own hydrocarbon environment. Surrounding this hydrocarbon core are the polar portions of the amphiphiles associated with the water molecules of the continuous phase. Aggregation also occurs in nonpolar liquids. The orientation of the molecules is now reversed, however, with the polar heads facing inward while the hydrocarbon chains are associated with the continuous nonpolar phase. These situations are shown in **Figure 16–4**, which also shows some of the shapes postulated for micelles. It seems likely that spherical micelles exist at concentrations relatively close to the CMC. At higher concentrations, laminar micelles have an increasing tendency to form and exist in equilibrium with spherical micelles. The

student is cautioned against regarding micelles as solid particles. The individual molecules forming the micelle are in dynamic equilibrium with those monomers in the bulk and at the interface.

As with lyophilic sols, formation of association colloids is spontaneous, provided that the concentration of the amphiphile in solution exceeds the CMC.

Amphiphiles may be anionic, cationic, nonionic, or ampholytic (zwitterionic), and this provides a convenient means of classifying association colloids. A typical example of each type is given in **Table 16–1**. Thus, **Figure 16–4a** represents the micelle of an anionic association colloid. A certain number of the sodium ions are attracted to the surface of the micelle, reducing the overall negative charge somewhat. These bound ions are termed counter ions or *gegenions*.

Mixtures of two or more amphiphiles are usual in pharmaceutical formulations. Assuming an ideal mixture, one can predict the CMC of the mixture from the CMC values of the pure amphiphiles and their mole fractions, $x$, in the mixture, according to the expression[1]

$$\frac{1}{\text{CMC}} = \frac{x_1}{\text{CMC}_1} + \frac{x_2}{\text{CMC}_2} \qquad (16\text{–}1)$$

**EXAMPLE 16–1**

**Critical Micelle Concentration**

Compute the CMC of a mixture of *n*-dodecyl octaoxyethylene glycol monoether ($C_{12}E_8$) and *n*-dodecyl $\beta$-D-maltoside (DM). The CMC of $C_{12}E_8$ is $\text{CMC}_1 = 8.1 \times 10^{-5}$ M (mole/liter) and its mole fraction is $x_1 = 0.75$; the CMC of DM is $\text{CMC}_2 = 15 \times 10^{-5}$ M.

We have

$$x_2 = (1 - x_2) = (1 - 0.75) = 0.25$$

From equation (16–1),

$$\frac{1}{\text{CMC}} = \frac{0.75}{8.1 \times 10^{-5}} + \frac{0.25}{15 \times 10^{-5}} = 10,926$$

$$\text{CMC} = \frac{1}{10,926} = 9.15 \times 10^{-5}\text{M}$$

The experimental value is $9.3 \times 10^{-5}$ M.

The properties of lyophilic, lyophobic, and association colloids are outlined in Table 16–2. These properties, together with the relevant methods, will be discussed in the following sections.

# OPTICAL PROPERTIES OF COLLOIDS

## The Faraday–Tyndall Effect

When a strong beam of light is passed through a colloidal sol, a visible cone, resulting from the scattering of light by the colloidal particles, is formed. This is the *Faraday–Tyndall effect*.

The *ultramicroscope*, developed by Zsigmondy, allows one to examine the light points responsible for the *Tyndall cone*. An intense light beam is passed through the sol against a dark background at right angles to the plane of observation, and, although the particles cannot be seen directly, the

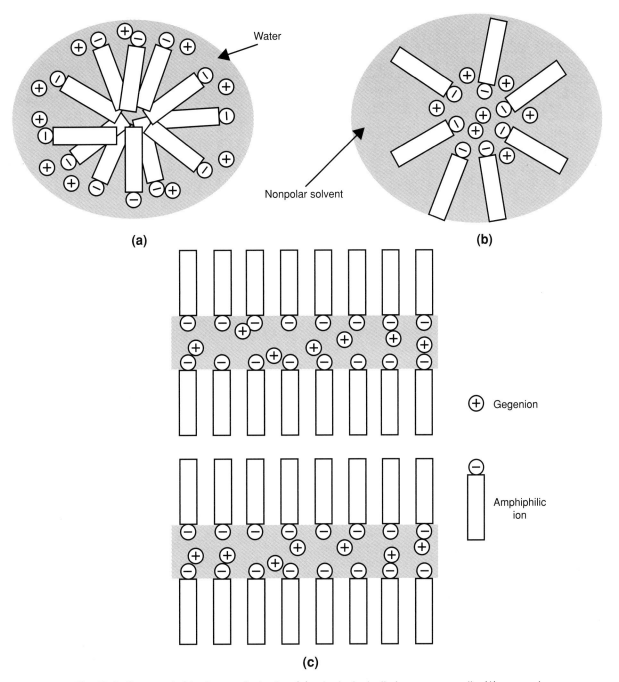

**Fig. 16–4.** Some probable shapes of micelles: (*a*) spherical micelle in aqueous media, (*b*) reversed micelle in nonaqueous media, and (*c*) laminar micelle, formed at higher amphiphile concentration, in aqueous media.

bright spots corresponding to particles can be observed and counted.

## Electron Microscope

The *electron microscope*, capable of yielding pictures of the actual particles, even those approaching molecular dimensions, is now widely used to observe the size, shape, and structure of colloidal particles.

The success of the electron microscope is due to its high resolving power, which can be defined in terms of *d*, the smallest distance by which two objects are separated and yet remain distinguishable. The smaller the wavelength of the radiation used, the smaller is *d* and the greater is the resolving power. The optical microscope uses visible light as its radiation source and is able to resolve only two particles separated by about 20 nm (200 Å). The radiation source of the electron microscope is a beam of high-energy electrons having wavelengths in the region of 0.01 nm (0.1 Å). With current instrumentation, this results in *d* being approximately 0.5 nm (5 Å), a much-increased power of resolution over the optical microscope.

**TABLE 16–1**

**CLASSIFICATION AND TYPICAL EXAMPLES OF ASSOCIATION COLLOIDS**

| Type | Compound | Example Amphiphile | Gegenions |
|------|----------|--------------------|-----------|
| Anionic | Sodium lauryl sulfate | $CH_3(CH_2)_{11}OSO_3^-$ | $Na^+$ |
| Cationic | Cetyl trimethyl-ammonium bromide | $CH_3(CH_2)_{15}N^+(CH_3)_3$ | $Br^-$ |
| Nonionic | Polyoxyethylene lauryl ether | $CH_3(CH_2)_{10}CH_2O(CH_2OCH_2)_{23}H$ | — |
| Ampholytic | Dimethyldodecylammonio-propane sulfonate | $CH_3(CH_2)_{11}N^+(CH_3)_2(CH_2)_3OSO_2^-$ | — |

## Light Scattering

This property depends on the Faraday–Tyndall effect and is widely used for determining the molecular weight of colloids. It can also be used to obtain information on the shape and size of these particles. Scattering can be described in terms of the turbidity, $\tau$, the fractional decrease in intensity due to scattering as the incident light passes through 1 cm of solution. It can be expressed as the intensity of light scattered in all directions, $I_s$, divided by the intensity of the incident light, $I$. At a given concentration of dispersed phase, the turbidity is proportional to the molecular weight of the lyophilic colloid.

Because of the low turbidities of most lyophilic colloids, it is more convenient to measure the scattered light (at a particular angle relative to the incident beam) rather than the transmitted light.

The turbidity can then be calculated from the intensity of the scattered light, provided that the dimensions of the particle are small compared with the wavelength of the light used. The molecular weight of the colloid can be obtained from the following equation:

$$\frac{Hc}{\tau} = \frac{1}{M} + 2Bc \qquad (16-2)$$

**TABLE 16–2**

**COMPARISON OF PROPERTIES OF COLLOIDAL SOLS\***

| Lyophilic | Association (Amphiphilic) | Lyophobic |
|-----------|---------------------------|-----------|
| Dispersed phase consists generally of large organic *molecules* lying within colloidal size range | Dispersed phase consists of aggregates (*micelles*) of small organic molecules or ions whose size *individually* is below the colloidal range | Dispersed phase ordinarily consists of inorganic particles, such as gold or silver |
| Molecules of dispersed phase are solvated, i.e., they are associated with the molecules comprising the dispersion medium | Hydrophilic or lipophilic portion of the molecule is solvated, depending on whether the dispersion medium is aqueous or nonaqueous | Little if any interaction (solvation) occurs between particles and dispersion medium |
| Molecules disperse spontaneously to form colloidal solution | Colloidal aggregates are formed spontaneously when the concentration of amphiphile exceeds the critical micelle concentration | Material does not disperse spontaneously, and special procedures therefore must be adopted to produce colloidal dispersion |
| Viscosity of the dispersion medium ordinarily is increased greatly by the presence of the dispersed phase; at sufficiently high concentrations, the sol may become a gel; viscosity and gel formation are related to solvation effects and to the shape of the molecules, which are usually highly asymmetric | Viscosity of the system increases as the concentration of the amphiphile increases, as micelles increase in number and become asymmetric | Viscosity of the dispersion medium is not greatly increased by the presence of lyophobic colloidal particles, which tend to be unsolvated and symmetric |
| Dispersions are stable generally in the presence of electrolytes; they may be salted out by high concentrations of very soluble electrolytes; effect is due primarily to desolvation of lyophilic molecules | In aqueous solutions, the critical micelle concentration is reduced by the addition of electrolytes; salting out may occur at higher salt concentrations | Lyophobic dispersions are unstable in the presence of even small concentrations of electrolytes; effect is due to neutralization of the charge on the particles; lyophilic colloids exert a protective effect |

\*From J. Swarbick and A. Martin, *American Pharmacy*, 6th Ed., Lippincott, Philadelphia, 1966, p. 161.

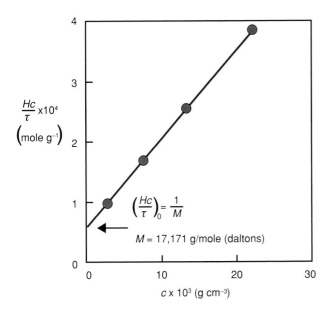

**Fig. 16–5.** A plot of $Hc/\tau$ against the concentration of a polymer (colloid).

where $\tau$ is the turbidity in cm$^{-1}$, $c$ is the concentration of solute in g/cm$^3$ of solution, $M$ is the weight-average molecular weight in g/mole or daltons, and $B$ is an interaction constant (see osmotic pressure). $H$ is constant for a particular system and is written as

$$H = \frac{32\pi^3 n^2 (dn/dc)^2}{3\lambda^4 N}$$

where $n$ (dimensionless) is the refractive index of the solution of concentration $c$(g/cm$^3$) at a wavelength $\lambda$ in cm$^{-1}$, $dn/dc$ is the change in refractive index with concentration at $c$, and $N$ is Avogadro's number. A plot of $Hc/\tau$ against concentration (**Fig. 16–5**) results in a straight line with a slope of $2B$. The intercept on the $Hc/\tau$ axis is $1/M$, the reciprocal of which yields the molecular weight of the colloid.

When the molecule is asymmetric, the intensity of the scattered light varies with the angle of observation. Data of this kind permit an estimation of the shape and size of the particles. Light scattering has been used to study proteins, synthetic polymers, association colloids, and lyophobic sols.

Chang and Cardinal[2] used light scattering to study the pattern of self-association in aqueous solution of the bile salts sodium deoxycholate and sodium taurodeoxycholate. Analysis of the data showed that the bile salts associate to form dimers, trimers, and tetramers and a larger aggregate of variable size.

Racey et al.[3] used quasielastic light scattering, a new light-scattering technique that uses laser light and can determine diffusion coefficients and particle sizes (Stokes's diameter) of macromolecules in solution. Quasielastic light scattering allowed the examination of heparin aggregates in commercial preparations stored for various times and at various temperatures. Both storage time and refrigeration caused an increase in the aggregation state of heparin solutions. It has not yet been determined whether the change in aggregation has any effect on the biologic activity of commercial preparations.

## Light Scattering and Micelle Molecular Weight

Equation (**16–2**) can be applied after suitable modification to compute the molecular weight of colloidal aggregates and micelles. When amphiphilic molecules associate to form micelles, the turbidity of the micellar dispersion differs from the turbidity of the solution of the amphiphilic molecules because micelles are now also present in equilibrium with the monomeric species. Below the CMC, the concentration of monomers increases linearly with the total concentration, $c$; above the CMC, the monomer concentration remains nearly constant; that is, $c_{\text{monomer}} \cong \text{CMC}$. The concentration of micelles can therefore be written as

$$c_{\text{micelle}} = c - c_{\text{monomer}} \cong c - c_{\text{CMC}} \qquad (16\text{–}3)$$

The corresponding turbidity of the solution due to the presence of micelles is obtained by subtracting the turbidity due to monomers, $\tau_{\text{monomer}} = \tau_{\text{CMC}}$, from the total turbidity of the solution:

$$\tau_{\text{micelle}} = \tau - \tau_{\text{CMC}} \qquad (16\text{–}4)$$

Accordingly, equation (**16–2**) is modified to

$$\frac{H(c - c_{\text{CMC}})}{(\tau - \tau_{\text{CMC}})} = \frac{1}{M} + 2B(c - c_{\text{CMC}}) \qquad (16\text{–}5)$$

where the subscript CMC indicates the turbidity or concentration at the critical micelle concentration, and $B$ and $H$ have the same meaning as in equation (**16–2**). Thus, the molecular weight, $M$, of the micelle and the second virial coefficient, $B$, are obtained from the intercept and the slope, respectively, of a plot of $H(c - c_{\text{CMC}})/(\tau - \tau_{\text{CMC}})$ versus $(c - c_{\text{CMC}})$. Equation (**16–5**) is valid for two-component systems, that is, for a micelle and a molecular surfactant in this instance.

When the micelles interact neither among themselves nor with the molecules of the medium, the slope of a plot of equation (**16–5**) is zero; that is, the second virial coefficient, $B$, is zero and the line is parallel to the horizontal axis, as seen in **Figure 16–6**. This behavior is typical of nonionic and zwitterionic micellar systems in which the size distribution is narrow. However, as the concentration of micelles increases, intermicellar interactions lead to positive values of $B$, the slope of the line having a positive value. For ionic micelles the plots are linear with positive slopes, owing to repulsive intermicellar interactions that result in positive values of the interaction coefficient, $B$. A negative second virial coefficient is usually an indication that the micellar system is polydisperse.[4,5]

**EXAMPLE 16–2**

**Computation of the Molecular Weight of Micelles**

Using the following data, compute the molecular weight of micelles of dimethylalkylammoniopropane sulfonate, a zwitterionic surfactant investigated by Herrmann[5]:

| $(c - c_{\text{CMC}}) \times 10^3$ g/mL) | 0.98 | 1.98 | 2.98 | 3.98 | 4.98 |
|---|---|---|---|---|---|
| $\left\|\dfrac{H(c - c_{\text{CMC}})}{(\tau - \tau_{\text{CMC}})}\right\| \times 10^5$ (mole/g) | 1.66 | 1.65 | 1.66 | 1.69 | 1.65 |

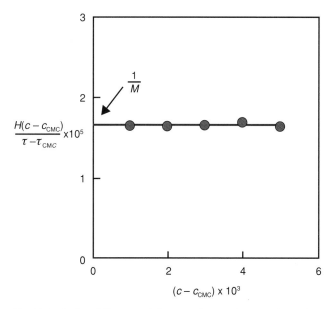

**Fig. 16–6.** A plot of $H(c - c_{CMC})/(\tau - \tau_{CMC})$ versus $(c - c_{CMC}) \times 10^3$ for a zwitterionic surfactant in which $B$ is zero. (From K. W. Hermann, J. Colloid Interface Sci. **22**, 352, 1966.)

Using equation (16–5), we obtain the micellar molecular weight from a plot of $H(c - c_{CMC})/\tau - \tau_{CMC}$ versus $(c - c_{CMC})$ (see Fig. 16–6); the intercept is $1/M = 1.66 \times 10^{-5}$ mole/g; therefore, $M = 60, 241$ g/mole. The slope is zero, that is, $2B$ in equation (16–5) is zero.

**EXAMPLE 16–3**

**Why is the Sky Blue?**

When a beam of light passes through a colloid, colloidal particles scatter the light. The intensity of scattered, $I_s$, light is inversely proportional to the fourth power of the wavelength, $\lambda$ (Rayleigh law):

$$I_s \sim \frac{1}{\lambda^4}$$

Thus, shorter-wavelength light (blue) is scattered more intensely than longer-wavelength light (yellow and red), and so the scattered light is mostly blue, whereas transmitted light has a yellow or reddish color (Fig. 16–7). Because of the constant motion of molecules, the atmosphere is inhomogeneous and constantly forms clusters with higher density of air. These inhomogeneities may be considered as colloidal particles. The scattering of short-wavelength light gives the sky its blue color. In contrast, transmitted light has a yellow color. At sunrise and sunset, sunlight has to travel a longer distance through the atmosphere than at noon. This is especially important in the lower atmosphere because it has a higher density (i.e., more gas molecules). Because of this longer distance, the yellow light also scatters. Sunsets can be more spectacular than sunrises because of an increase in the number of particles in the atmosphere due to pollution or natural causes (wind, dust), throughout the day.

## KINETIC PROPERTIES OF COLLOIDS

Grouped under this heading are several properties of colloidal systems that relate to the motion of particles with

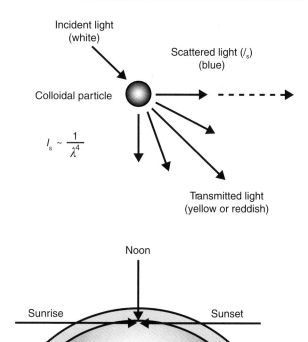

**Fig. 16–7.** Because of the constant motion of molecules, the atmosphere is inhomogeneous and constantly forms clusters with higher density of air. These inhomogeneities can be considered as colloidal particles, which scatter the light. The intensity of scattered light is inversely proportional to the fourth power of the wavelength, $(\lambda)$ (Rayleigh law). The scattering of short-wavelength light gives the sky its blue color. In contrast, transmitted light has a yellow or reddish color.

respect to the dispersion medium. The motion may be thermally induced (Brownian movement, diffusion, osmosis), gravitationally induced (sedimentation), or applied externally (viscosity). Electrically induced motion is considered in the section on electrical properties of colloids.

## Brownian Motion

Brownian motion describes the random movement of colloidal particles. The erratic motion, which may be observed with particles as large as about 5 $\mu$m, was explained as resulting from the bombardment of the particles by the molecules of the dispersion medium. The motion of the molecules cannot be observed, of course, because the molecules are too small to see. The velocity of the particles increases with decreasing particle size. Increasing the viscosity of the medium, which may be accomplished by the addition of glycerin or a similar agent, decreases and finally stops the Brownian movement.

## Diffusion

Particles diffuse spontaneously from a region of higher concentration to one of lower concentration until the concentration of the system is uniform throughout. Diffusion is a direct result of Brownian movement.

According to *Fick's first law*, the amount, $dq$, of substance diffusing in time, $dt$, across a plane of area, $S$, is directly proportional to the change of concentration, $dc$, with distance traveled, $dx$.

Fick's law is written as

$$dq = -DS \frac{dc}{dx} dt \qquad (16\text{–}6)$$

$D$ is the *diffusion coefficient*, the amount of material diffusing per unit time across a unit area when $dc/dx$, called the *concentration gradient*, is unity. $D$ thus has the dimensions of area per unit time. The coefficient can be obtained in colloidal chemistry by diffusion experiments in which the material is allowed to pass through a porous disk, and samples are removed and analyzed periodically. Another method involves measuring the change in the concentration or refractive index gradient of the free boundary that is formed when the solvent and colloidal solution are brought together and allowed to diffuse.

If the colloidal particles can be assumed to be approximately spherical, the following equation, suggested by Sutherland and Einstein[6], can be used to obtain the radius of the particle and the particle weight or molecular weight:

$$D = \frac{kT}{6\pi \eta r}$$

or

$$D = \frac{RT}{6\pi \eta r N} \qquad (16\text{–}7)$$

where $D$ is the diffusion coefficient obtained from Fick's law as already explained, $k$ is the Boltzmann constant, $R$ is the molar gas constant, $T$ is the absolute temperature, $\eta$ is the viscosity of the solvent, $r$ is the radius of the spherical particle, and $N$ is Avogadro's number. Equation (16–7) is called the *Sutherland–Einstein* or the *Stokes–Einstein* equation. The measured diffusion coefficient can be used to obtain the molecular weight of approximately spherical molecules, such as egg albumin and hemoglobin, by use of the equation

$$D = \frac{RT}{6\pi \eta N} \sqrt[3]{\frac{4\pi N}{3M\bar{v}}} \qquad (16\text{–}8)$$

where $M$ is molecular weight and $\bar{v}$ is the partial specific volume (approximately equal to the volume in $cm^3$ of 1 g of the solute, as obtained from density measurements).

Analysis of equations (16–6) and (16–7) allows us to formulate the following three main rules of diffusion: (a) the velocity of the molecules increases with decreasing particle size; (b) the velocity of the molecules increases with increasing temperature; and (c) the velocity of the molecules decreases with increasing viscosity of the medium.

**EXAMPLE 16–4**

### The Computation of Protein Properties from its Diffusion Coefficient

The diffusion coefficient for a spherical protein at 20°C is $7.0 \times 10^{-7}$ $cm^2$/sec and the partial specific volume is 0.75 $cm^3$/g. The viscosity of the solvent is 0.01 poise (0.01 g/cm sec). Compute (a) the molecular weight and (b) the radius of the protein particle.

(a) By rearranging equation (16–8), we obtain

$$M = \frac{1}{162\bar{v}} \left( \frac{1}{\pi N} \right)^2 \left( \frac{RT}{D\eta} \right)^3$$

$$M = \frac{1}{162 \times 0.75} \left( \frac{1}{3.14 \times (6.02 \times 10^{23})} \right)^2 \left( \frac{(8.31 \times 10^7) \times 293}{(7.0 \times 10^{-7}) \times 0.01} \right)^3$$

$$\cong 100{,}000 \text{ g/mole}$$

(b) From equation (16–7),

$$r = \frac{RT}{6\pi \eta N D} = \frac{(8.31 \times 10^7) \times 293}{6 \times 3.14 \times 0.01 \times (6.02 \times 10^{23}) \times (7.0 \times 10^{-7})}$$

$$= 31 \times 10^{-8} \text{ cm} = 31 \text{ Å} = 3.1 \text{ nm}$$

## Osmotic Pressure

The osmotic pressure, $\pi$, of a dilute colloidal solution is described by the van't Hoff equation:

$$\pi = cRT \qquad (16\text{–}9)$$

where $c$ is molar concentration of solute. This equation can be used to calculate the molecular weight of a colloid in a dilute solution. Replacing $c$ with $c_g/M$ in equation (16–9), in which $c_g$ is the grams of solute per liter of solution and $M$ is the molecular weight, we obtain

$$\pi = \frac{c_g}{M} RT \qquad (16\text{–}10)$$

Then,

$$\frac{\pi}{c_g} = \frac{RT}{M} \qquad (16\text{–}11)$$

which applies in a very dilute solution. The quantity $\pi/c_g$ for a polymer having a molecular weight of, say, 50,000 is often a linear function of the concentration, $c_g$, and the following equation can be written:

$$\frac{\pi}{c_g} = RT \left( \frac{1}{M} + B c_g \right) \qquad (16\text{–}12)$$

where $B$ is a constant for any particular solvent/solute system and depends on the degree of interaction between the solvent and the solute molecules. The term $B c_g$ in equation (16–12) is needed because equation (16–11) holds only for ideal solutions, namely, those containing low concentrations of spherocolloids. With linear lyophilic molecules, deviations occur because the solute molecules become solvated, leading to a reduction in the concentration of "free" solvent and an apparent increase in solute concentration. The role of $B$ in estimating the asymmetry of particles and their interactions with solute was discussed by Hiemenz.[7]

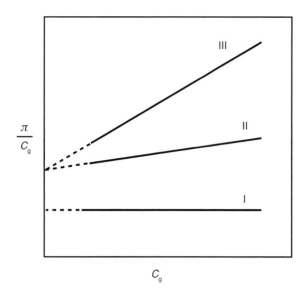

**Fig. 16–8.** Determination of molecular weight by means of the osmotic pressure method. Extrapolation of the line to the vertical axis where $c_g = 0$ gives $RT/M$, from which $M$ is obtained. Refer to text for significance of lines I, II, and III. Lines II and III are taken to represent two samples of a species of hemoglobin.

A plot of $\pi/c_g$ against $c_g$ generally results in one of three lines (**Fig. 16–8**), depending on whether the system is ideal (line I) or real (lines II and III). Equation (**16–11**) applies to line I and equation (**16–12**) describes lines II and III. The intercept is $RT/M$, and if the temperature at which the determination was carried out is known, the molecular weight of the solute can be calculated. In lines II and III, the slope of the line is $B$, the interaction constant. In line I, $B$ equals zero and is typical of a dilute spherocolloidal system. Line III is typical of a linear colloid in a solvent having a high affinity for the dispersed particles. Such a solvent is referred to as a "good" solvent for that particular colloid. There is a marked deviation from ideality as the concentration is increased and $B$ is large. At higher concentrations, or where interaction is marked, type III lines can become nonlinear, requiring that equation (**16–12**) be expanded and written as a power series:

$$\frac{\pi}{c_g} = RT \left( \frac{1}{M} + Bc_g + Cc_g^2 + \cdots \right) \quad (16\text{–}13)$$

where $C$ is another interaction constant. Line II depicts the situation in which the same colloid is present in a relatively poor solvent having a reduced affinity for the dispersed material. Note, however, that the extrapolated intercept on the $\pi/c_g$ axis is identical for both lines II and III, showing that the calculated molecular weight is independent of the solvent used.

**EXAMPLE 16–5**

**Calculation of Molecular Weight of Hemoglobin**

Let us assume that the intercept $(\pi/c_g)_0$ for line III in Figure 16–8 has the value $3.623 \times 10^{-4}$ liter atm/g, and the slope of the line is $1.80 \times 10^{-6}$ liter$^2$ atm/g$^2$. What is the molecular weight and the second virial coefficient, $B$, for a sample of hemoglobin using the data given here?

In Figure 16–8, line III crosses the vertical intercept at the same point as line II. These two samples of hemoglobin have the same *limiting reduced osmotic pressure*, as $(\pi/c_g)_0$ is called, and therefore have the same molecular weight. The $B$ values, and therefore the shape of the two samples and their interaction with the medium, differ as evidenced by the different slopes of lines II and III.

At the intercept, $(\pi/c_g)_0 = RT/M$. Therefore,

$$M = \frac{RT}{\pi/c_g} = \frac{(0.08206 \text{ liter atm/deg mole})(298 \text{ K})}{3.623 \times 10^{-4} \text{ liter atm/g}}$$

$$M = 67,498 \text{ g/mole (daltons) for both hemoglobins}$$

The slope of line III, representing one of the hemoglobin samples, is divided by $RT$ to obtain $B$, as observed in equation (16–12):

$$B = \frac{1.80 \times 10^{-6} \text{ liter}^2 \text{ atm/g}^2}{(0.08206 \text{ liter atm/mole deg})(298 \text{ K})}$$

$$= 7.36 \times 10^{-8} \text{ liter mole/g}^2$$

The other hemoglobin sample, represented by line II, has a slope of $4.75 \times 10^{-9}$ liter$^2$ atm/g$^2$, and its $B$ value is therefore calculated as follows:

$$B = \frac{4.75 \times 10^{-9} \text{ liter}^2 \text{ atm/g}^2}{(0.08206 \text{ liter atm/mole deg})(298 \text{ K})}$$

$$= 1.94 \times 10^{-10} \text{ liter mole/g}^2$$

Estimate the $B$ value for the protein represented by line I. Is its molecular weight larger or smaller than that of samples II and III? Refer to equations (16–11) and (16–12) in arriving at your answers.

## Sedimentation

The velocity, $v$, of sedimentation of spherical particles having a density $\rho$ in a medium of density $\rho_0$ and a viscosity $\eta_0$ is given by *Stokes's law*:

$$v = \frac{2r^2(\rho - \rho_0)g}{9\eta_0} \quad (16\text{–}14)$$

where $g$ is the acceleration due to gravity. If the particles are subjected only to the force of gravity, then the lower size limit of particles obeying Stokes's equation is about 0.5 $\mu$m. This is because Brownian movement becomes significant and tends to offset sedimentation due to gravity and promotes mixing instead. Consequently, a stronger force must be applied to bring about the sedimentation of colloidal particles in a quantitative and measurable manner. This is accomplished by use of the *ultracentrifuge*, developed by Svedberg in 1925,[8] which can produce a force one million times that of gravity.

In a centrifuge, the acceleration of gravity is replaced by $\omega^2 x$, where $\omega$ is the angular velocity and $x$ is the distance of the particle from the center of rotation. Equation (**16–14**) is accordingly modified to

$$v = \frac{dx}{dt} = \frac{2r^2(\rho - \rho_0)\omega^2 x}{9\eta_0}$$

The speed at which a centrifuge is operated is commonly expressed in terms of the number of revolutions per minute (rpm) of the rotor. It is frequently more desirable to express the rpm as angular acceleration ($\omega^2 x$) or the number of times that the force of gravity is exceeded.

## EXAMPLE 16–6

### Calculation of Centrifuge Force

A centrifuge is rotating at 1500 rpm. The midpoint of the cell containing the sample is located 7.5 cm from the center of the rotor (i.e., $x = 7.5$ cm). What is the average angular acceleration and the number of $g$'s on the suspended particles?

We have

$$\text{Angular acceleration} = \omega^2 x$$

$$= \left( \frac{1500 \text{ revolutions}}{\text{minute}} \times \frac{2\pi}{60} \right)^2 \times 7.5 \text{ cm}$$

$$= 1.851 \times 10^5 \text{ cm/sec}^2$$

$$\text{Number of } g\text{'s} = \frac{1.851 \times 10^5 \text{ cm/sec}^2}{981 \text{ cm/sec}^2} = 188.7 \text{ } g\text{'s}$$

that is, the force produced is 188.7 times that due to gravity.

The instantaneous velocity, $v = dx/dt$, of a particle in a unit centrifugal field is expressed in terms of the *Svedberg sedimentation coefficient s*,

$$s = \frac{dx/dt}{\omega^2 x} \qquad (16\text{–}15)$$

Owing to the centrifugal force, particles having a high molecular weight pass from position $x_1$ at time $t_1$ to position $x_2$ at time $t_2$, and the sedimentation coefficient is obtained by integrating equation (16–15) to give

$$s = \frac{\ln(x_2/x_1)}{\omega^2(t_2 - t_1)} \qquad (16\text{–}16)$$

The distances $x_1$ and $x_2$ refer to positions of the boundary between the solvent and the high–molecular-weight component in the centrifuge cell. The boundary is located by the change of refractive index, which can be attained at any time during the run and translated into a peak on a photographic plate. Photographs are taken at definite intervals, and the peaks of the *schlieren patterns*, as they are called, give the position $x$ of the boundary at each time $t$. If the sample consists of a component of a definite molecular weight, the schlieren pattern will have a single sharp peak at any moment during the run. If components with different molecular weights are present in the sample, the particles of greater weight will settle faster, and several peaks will appear on the schlieren patterns. Therefore, ultracentrifugation not only is useful for determining the molecular weight of polymers, particularly proteins, but also can be used to ascertain the degree of homogeneity of the sample. Gelatin, for example, is found to be a polydisperse protein with fractions of molecular weight 10,000 to 100,000. (This accounts in part for the fact that gelatin from various sources is observed to have variable properties when used in pharmaceutical preparations.) Insulin, on the other hand, is a monodisperse protein composed of two polypeptide chains, each made up of a number of amino acid molecules. The two chains are attached together by disulfide (S—S) bridges to form a definite unit having a molecular weight of about 6000.

The sedimentation coefficient, $s$, can be computed from equation (16–16) after the two distances $x_1$ and $x_2$ are measured on the schlieren photographs obtained at times $t_1$ and $t_2$; the angular velocity $\omega$ is equal to $2\pi$ times the speed of the rotor in revolutions per second. Knowing $s$ and obtaining $D$ from diffusion data, it is possible to determine the molecular weight of a polymer, such as a protein, by use of the expression

$$M = \frac{RT_s}{D(1 - \bar{v}\rho_0)} \qquad (16\text{–}17)$$

where $R$ is the molar gas constant, $T$ is the absolute temperature, $\bar{v}$ is the partial specific volume of the protein, and $\rho_0$ is the density of the solvent. Both $s$ and $D$ must be obtained at, or corrected to, 20°C for use in equation (16–17).

## EXAMPLE 16–7

### Molecular Weight of Methylcellulose Based on the Sedimentation Coefficient

The sedimentation coefficient, $s$, for a particular fraction of methylcellulose at 20°C (293 K) is $1.7 \times 10^{-13}$ sec, the diffusion coefficient, $D$, is $15 \times 10^{-7}$ cm$^2$/sec, the partial specific volume, $\bar{v}$, of the gum is 0.72 cm$^3$/g, and the density of water at 20°C is 0.998 g/cm$^3$. Compute the molecular weight of methylcellulose. The gas constant $R$ is $8.31 \times 10^7$ erg/(deg mole).

We have

$$M = \frac{(8.31 \times 10^7) \times 293 \times (1.7 \times 10^{-13})}{15 \times 10^{-7}[1 - (0.72 \times 0.998)]} = 9800 \text{ g/mole}$$

Kirschbaum[9] reviewed the usefulness of the analytic ultracentrifuge and used it to study the micellar properties of drugs (**Fig. 16–9**). Richard[10] determined the apparent micellar molecular weight of the antibiotic fusidate sodium by ultracentrifugation. He concluded that the primary micelles composed of five monomer units are formed, followed by aggregation of these pentamers into larger micelles at higher salt concentrations.

The sedimentation method already described is known as the *sedimentation velocity* technique. A second method, involving *sedimentation equilibrium*, can also be used. Equilibrium is established when the sedimentation force is just balanced by the counteracting diffusional force and the boundary is therefore stationary. In this method, the diffusion coefficient need not be determined; however, the centrifuge may have to be run for several weeks to attain equilibrium throughout the cell. Newer methods of calculation have been developed recently for obtaining molecular weights by the equilibrium method without requiring these long periods of centrifugation, enabling the protein chemist to obtain molecular weights rapidly and accurately.

Molecular weights determined by sedimentation velocity, sedimentation equilibrium, and osmotic pressure determinations are in good agreement, as can be seen from **Table 16–3**.

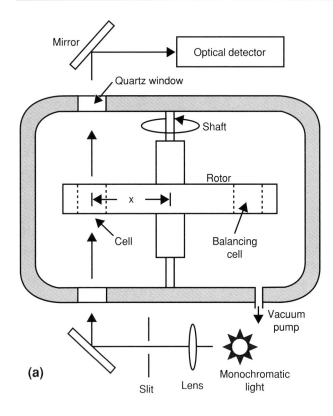

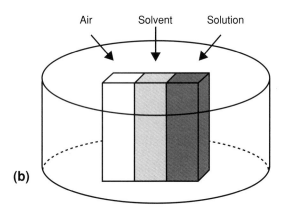

**Fig. 16–9.** (*a*) Schematic of an ultracentrifuge. (*b*) Centrifuge cell. (From H. R. Allok and F. W. Lampe, *Contemporary Polymer Chemistry*, Prentice-Hall, Englewood Cliffs, N. J., 1981, pp. 366, 367. With permission.)

## Viscosity

Viscosity is an expression of the resistance to flow of a system under an applied stress. The more viscous a liquid is, the greater is the applied force required to make it flow at a particular rate. The fundamental principles and applications of viscosity are discussed in detail in Chapter 19. This section is concerned with the flow properties of dilute colloidal systems and the manner in which viscosity data can be used to obtain the molecular weight of material comprising the

### TABLE 16–3
### MOLECULAR WEIGHTS OF PROTEINS IN AQUEOUS SOLUTION DETERMINED BY DIFFERENT METHODS*

| Material | Molecular Weight | | |
| --- | --- | --- | --- |
| | Sedimentation Velocity | Sedimentation Equilibrium | Osmotic Pressure |
| Ribonuclease | 12,700 | 13,000 | – |
| Myoglobin | 16,900 | 17,500 | 17,000 |
| Ovalbumin | 44,000 | 40,500 | 45,000 |
| Hemoglobin (horse) | 68,000 | 68,000 | 67,000 |
| Serum albumin (horse) | 70,000 | 68,000 | 73,000 |
| Serum globulin (horse) | 167,000 | 150,000 | 175,000 |
| Tobacco mosaic virus | 59,000,000 | – | – |

*From D. J. Shaw, *Introduction to Colloidal and Surface Chemistry*, Butterworths, London, 1970, p. 32. For and extensive listing of molecular weights of macromolecules, see C. Tanford, *Physical Chemistry of Macromolecules*, Wiley, New York, 1961.

disperse phase. Viscosity studies also provide information regarding the shape of the particles in solution.

Einstein developed an equation of flow applicable to dilute colloidal dispersions of spherical particles, namely,

$$\eta = \eta_0(1 + 2.5\phi) \qquad (16\text{–}18)$$

In equation (**16–18**), which is based on hydrodynamic theory, $\eta_0$ is the viscosity of the dispersion medium and $\eta$ is the viscosity of the dispersion when the volume fraction of colloidal particles present is $\phi$. The volume fraction is defined as the volume of the particles divided by the total volume of the dispersion; it is therefore equivalent to a concentration term. Both $\eta_0$ and $\eta$ can be determined using a capillary viscometer.

Several viscosity coefficients can be defined with respect to this equation. These include *relative viscosity* ($\eta_{\text{rel}}$), *specific viscosity* ($\eta_{\text{sp}}$), and *intrinsic viscosity* ($\eta$). From equation (**16–18**),

$$\eta_{\text{rel}} = \frac{\eta}{\eta} = 1 + 2.5\phi \qquad (16\text{–}19)$$

and

$$\eta_{\text{sp}} = \frac{\eta}{\eta_0} - 1 = \frac{\eta - \eta_0}{\eta_0} = 2.5\phi \qquad (16\text{–}20)$$

or

$$\frac{\eta_{\text{sp}}}{\phi} = 2.5 \qquad (16\text{–}21)$$

Because volume fraction is directly related to concentration, equation (**16–21**) can be written as

$$\frac{\eta_{\text{sp}}}{c} = k \qquad (16\text{–}22)$$

where $c$ is expressed in grams of colloidal particles per 100 mL of total dispersion. For highly polymeric materials

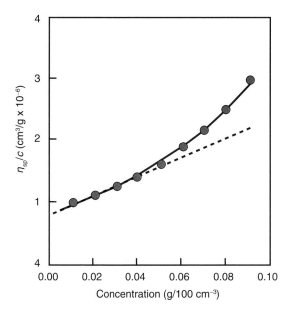

**Fig. 16–10.** Determination of molecular weight using viscosity data. (Replotted from D. R. Powell, J. Swarbrick, and G. S. Banker, J. Pharm. Sci. **55**, 601, 1966. With permission.)

dispersed in the medium at moderate concentrations, the equation is best expressed as a power series:

$$\frac{\eta_{sp}}{c} = k_1 + k_2c + k_2c + k_3c^2 \qquad (16\text{–}23)$$

By determining $\eta$ at various concentrations and knowing $\eta_0$, one can calculate $\eta_{sp}$ from equation (16–20). If $\eta_{sp}/c$ is plotted against $c$ (**Fig. 16–10**) and the line extrapolated to infinite dilution, the intercept is $k_1$ [equation (16–23)]. This constant, commonly known as the intrinsic viscosity, $[\eta]$, is used to calculate the approximate molecular weights of polymers. According to the so-called Mark–Houwink equation,

$$[\eta] = KM^a \qquad (16\text{–}24)$$

where $K$ and $a$ are constants characteristic of the particular polymer–solvent system. These constants, which are virtually independent of molecular weight, are obtained initially by determining $[\eta]$ experimentally for polymer fractions whose molecular weights have been determined by other methods such as light scattering, osmotic pressure, or sedimentation. Once $K$ and $a$ are known, measurement of $[\eta]$ provides a simple yet accurate means of obtaining molecular weights for fractions not yet subjected to other methods. Intrinsic viscosity $[\eta]$, together with an interaction constant, $k'$, provides the equation, $\eta_{sp}/c = [\eta] + k'[\eta]^2c$, which is used in choosing solvent mixtures for tablet film coating polymers such as ethyl cellulose.[11]

The shapes of particles of the disperse phase affect the viscosity of colloidal dispersions. Spherocolloids form dispersions of relatively low viscosity, whereas systems containing linear particles are more viscous. As we saw in previous sections, the relationship of shape and viscosity reflects the degree of solvation of the particles. If a linear colloid is placed

in a solvent for which it has a low affinity, it tends to "ball up," that is, to assume a spherical shape, and the viscosity falls. This provides a means of detecting changes in the shape of flexible colloidal particles and macromolecules.

The characteristics of polymers used as substitutes for blood plasma (plasma extenders) depend in part on the molecular weight of the material. These characteristics include the size and shape of the macromolecules and the ability of the polymers to impart the proper viscosity and osmotic pressure to the blood. The methods described in this chapter are used to determine the average molecular weights of hydroxyethyl starch, dextran, and gelatin preparations used as plasma extenders. Ultracentrifugation, light scattering, x-ray analysis (small-angle x-ray scattering[12]), and other analytic tools[13] were used by Paradies to determine the structural properties of tyrothricin, a mixture of the peptide antibiotics gramicidin and tyrocidine B. The antibiotic aggregate has a molecular weight of 28,600 daltons and was determined to be a rod 170 Å in length and 30 Å in diameter.

## ELECTRICAL PROPERTIES OF COLLOIDS

The properties of colloids that depend on, or are affected by, the presence of a charge on the surface of a particle are discussed under this heading. The various ways in which the surfaces of particles dispersed in a liquid medium acquire a charge were outlined in the Interfacial Phenomena chapter. Mention was also made of the *zeta* (*electrokinetic*) potential and how it is related to the *Nernst* (*electrothermodynamic*) potential. The potential versus distance diagram for a spherical colloidal particle can be represented as shown in **Figure 16–11**. Such a system can be formed, for example, by adding a dilute solution of potassium iodide to an equimolar solution of silver nitrate. A colloidal precipitate of silver iodide

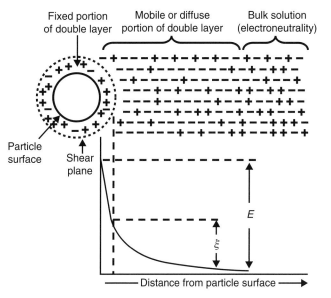

**Fig. 16–11.** Diffuse double layer and the zeta potential.

particles is produced, and, because the silver ions are in excess and are adsorbed, a positively charged particle is produced. If the reverse procedure is adopted, that is, if silver nitrate is added to the potassium iodide solution, iodide ions are adsorbed on the particles as the potential-determining ion and result in the formation of a negatively charged sol.

## Electrokinetic Phenomena

The movement of a charged surface with respect to an adjacent liquid phase is the basic principle underlying four electrokinetic phenomena: *electrophoresis, electroosmosis, sedimentation potential,* and *streaming potential.*

*Electrophoresis* involves the movement of a charged particle through a liquid under the influence of an applied potential difference. An electrophoresis cell fitted with two electrodes contains the dispersion. When a potential is applied across the electrodes, the particles migrate to the oppositely charged electrode. **Figure 16–12** illustrates the design of a commercially available instrument. The rate of particle migration is observed by means of an ultramicroscope and is a function of the charge on the particle. Because the shear plane of the particle is located at the periphery of the tightly bound layer, the rate-determining potential is the zeta potential. From knowledge of the direction and rate of migration, the sign and magnitude of the zeta potential in a colloidal system can be determined. The relevant equation,

$$\zeta = \frac{v}{E} \times \frac{4\pi\eta}{\varepsilon} \times (9 \times 10^4) \qquad \textbf{(16–25)}$$

which yields the zeta potential, $\zeta$, in volts, requires a knowledge of the velocity of migration, $v$, of the sol in cm/sec in an electrophoresis tube of a definite length in cm, the viscosity of the medium, $\eta$, in poises (dynes sec/cm$^2$), the dielectric constant of the medium, $\varepsilon$, and the potential gradient, $E$, in volts/cm. The term $v/E$ is known as the *mobility.*

It is instructive to carry out the dimensional analysis of equation (**16–25**). In one system of fundamental electric units, $E$, the electric field strength, can be expressed in electrostatic units of statvolt/cm (1 coulomb $= 3 \times 10^9$ statcoulombs, and 1 statvolt $= 300$ practical volts). The dielectric constant is not dimensionless here, but rather from Coulomb's law may be assigned the units of statcoulomb$^2$/(dyne cm$^2$). The

$$\zeta = \frac{v}{E}\frac{4\pi\eta}{\varepsilon} \qquad \textbf{(16–26)}$$

equation can then be written dimensionally, recognizing that statvolts $\times$ statcoulombs $=$ dyne cm, as

$$\zeta = \frac{\text{cm/sec}}{\text{statvolts/cm}} \times \frac{\text{dyne sec/cm}^2}{\text{statcoulomb}^2/(\text{dyne cm}^2)} = \text{statvolts}$$
$$\textbf{(16–27)}$$

It is more convenient to express the zeta potential in practical volts than in statvolts. Because 1 statvolt $= 300$ practical volts, equation (**16–27**) is multiplied by 300 to make this conversion, that is, statvolts $\times$300 practical volts/statvolt

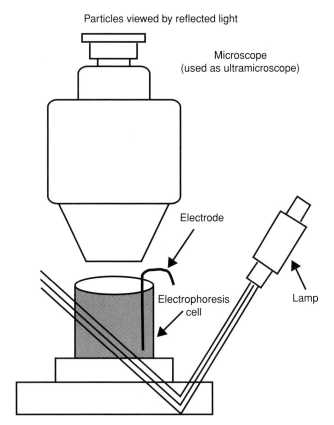

Particles viewed by reflected light

Microscope (used as ultramicroscope)

Electrode

Electrophoresis cell

Lamp

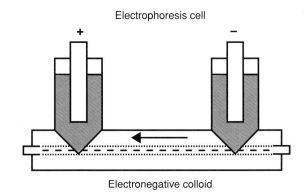

Electrophoresis cell

Electronegative colloid

**Fig. 16–12.** Principle of zeta potential measurement (based on the zeta meter) showing the ultramicroscope and the flow cell.

$= 300$ practical volts. Furthermore, $E$ is ordinarily measured in practical volts/cm and not in statvolt/cm, and this conversion is made by again multiplying the right-hand side of equation (**16–27**) by 300. The final expression is equation (**16–25**), in which the factor $300 \times 300 = 9 \times 10^4$ converts electrostatic units to volts.

For a colloidal system at 20°C in which the dispersion medium is water, equation (**16–25**) reduces approximately to

$$\zeta \cong 141\frac{v}{E} \qquad \textbf{(16–28)}$$

The coefficient of 141 at 20°C becomes 128 at 25°C.

**EXAMPLE 16–8**

**Determination of the Zeta Potential from Electrophoretic Data**

The velocity of migration of an aqueous ferric hydroxide sol was determined at 20°C using the apparatus shown in Figure 16–12 and was found to be $16.5 \times 10^{-4}$ cm/sec. The distance between the electrodes in the cell was 20 cm, and the applied emf was 110 volts. What is (a) the zeta potential of the sol and (b) the sign of the charge on the particles?

(a)

$$\frac{v}{E} = \frac{16.5 \times 10^{-4} \text{ cm/sec}}{110/20 \text{ volts/cm}} = 3 \times 10^{-4} \text{ cm}^2/\text{volt sec}$$

$$\zeta = 141 \times (3 \times 10^{-4}) = 0.042 \text{ volt}$$

(b) The particles were seen to migrate toward the negative electrode of the electrophoresis cell; therefore, the colloid is positively charged. The zeta potential is often used to estimate the stability of colloids, as discussed in a later section.

*Electroosmosis* is essentially opposite in principle to electrophoresis. In the latter, the application of a potential causes a charged particle to move relative to the liquid, which is stationary. If the solid is rendered immobile (e.g., by forming a capillary or making the particles into a porous plug), however, the liquid now moves relative to the charged surface. This is electroosmosis, so called because liquid moves through a plug or a membrane across which a potential is applied. Electroosmosis provides another method for obtaining the zeta potential by determining the rate of flow of liquid through the plug under standard conditions.

*Sedimentation potential*, the reverse of electrophoresis, is the creation of a potential when particles undergo sedimentation. The *streaming potential* differs from electroosmosis in that forcing a liquid to flow through a plug or bed of particles creates the potential.

Schott[14] studied the electrokinetic properties of magnesium hydroxide suspensions that are used as antacids and laxatives. The zero point of charge occurred at pH $\cong 10.8$, the zeta potential, $\zeta$, of magnesium hydroxide being positive below this pH value. Increasing the pH or hydroxide ion concentration produced a change in the sign of $\zeta$ from positive to negative, with the largest negative $\zeta$ value occurring at pH 11.5.

Takenaka and associates[15] studied the electrophoretic properties of *microcapsules* of sulfamethoxazole in droplets of a gelatin–acacia coacervate as part of a study to stabilize such drugs in microcapsules.

## Donnan Membrane Equilibrium

If sodium chloride is placed in solution on one side of a semipermeable membrane and a negatively charged colloid together with its counterions $R^-Na^+$ is placed on the other side, the sodium and chloride ions can pass freely across the barrier but not the colloidal anionic particles. The system at equilibrium is represented in the following diagram, in which $R^-$ is the nondiffusible colloidal anion and the vertical line separating the various species represents the semipermeable membrane. The volumes of solution on the two sides of the membrane are considered to be equal.

| Outside (o) | Inside (i) |
|:---:|:---:|
| | $R^-$ |
| $Na^+$ | $Na^+$ |
| $Cl^-$ | $Cl^-$ |

After equilibrium has been established, the concentration in dilute solutions (more correctly the activity) of sodium chloride must be the same on both sides of the membrane, according to the principle of escaping tendencies. Therefore,

$$[Na^+]_o[Cl^-]_o = [Na^+]_i[Cl^-]_i \qquad (16\text{–}29)$$

The condition of electroneutrality must also apply. That is, the concentration of positively charged ions in the solutions on either side of the membrane must balance the concentration of negatively charged ions. Therefore, on the outside,

$$[Na^+]_o = [Cl^-]_o \qquad (16\text{–}30)$$

and inside,

$$[Na^+]_i = [R^-]_i + [Cl^-]_i \qquad (16\text{–}31)$$

Equations (16–30) and (16–31) can be substituted into equation (16–29) to give

$$[Cl^-]_o^2 = ([Cl^-]_i + [R^-]_i)[Cl^-]_i = [Cl^-]_i^2 \left(1 + \frac{[R^-]}{[Cl^-]_i}\right) \qquad (16\text{–}32)$$

$$\frac{[Cl^-]_o}{[Cl^-]_i} = \sqrt{1 + \frac{[R^-]_i}{[Cl^-]_i}} \qquad (16\text{–}33)$$

Equation (16–33), the *Donnan membrane equilibrium*, gives the ratio of concentrations of the diffusible anion outside and inside the membrane at equilibrium. The equation shows that a negatively charged polyelectrolyte inside a semipermeable sac would influence the equilibrium concentration ratio of a diffusible anion. It tends to drive the ion of like charge out through the membrane. When $[R^-]_i$ is large compared with $[Cl^-]_i$, the ratio roughly equals $\sqrt{[R^-]_i}$. If, on the other hand, $[Cl^-]_i$ is quite large with respect to $[R^-]_i$, the ratio in equation (16–33) becomes equal to unity, and the concentration of the salt is thus equal on both sides of the membrane.

The unequal distribution of diffusible electrolyte ions on the two sides of the membrane will obviously result in erroneous values for osmotic pressures of polyelectrolyte solutions. If, however, the concentration of salt in the solution is made large, the Donnan equilibrium effect can be practically eliminated in the determination of molecular weights of proteins involving the osmotic pressure method.

Higuchi et al.[16] modified the Donnan membrane equilibrium, equation (16–33), to demonstrate the use of the polyelectrolyte sodium carboxymethylcellulose for enhancing the absorption of drugs such as sodium salicylate and potassium benzylpenicillin. If $[Cl^-]$ in equation (16–33) is replaced by the concentration of the diffusible drug, anion $[D^-]$ at equilibrium, and $[R^-]$ is used to represent the concentration of sodium carboxymethylcellulose at equilibrium, we have a modification of the *Donnan membrane equilibrium* for a diffusible drug anion,

$[D^-]$:

$$\frac{[D^-]_o}{[D^-]_i} = \sqrt{1 + \frac{[R^-]_i}{[D^-]_i}} \qquad (16\text{–}34)$$

It will be observed that when $[R^-]_i/[D^-]_i = 8$, the ratio $[D^-]_o/[D^-]_i = 3$, and when $[R^-]_i/[D^-]_i = 99$, the ratio $[D^-]_o/[D^-]_i = 10$. Therefore, the addition of an anionic polyelectrolyte to a diffusible drug anion should enhance the diffusion of the drug out of the chamber. By kinetic studies, Higuchi et al.[16] showed that the presence of sodium carboxymethylcellulose more than doubled the rate of transfer of the negatively charged dye scarlet red sulfonate.

Other investigators have found by in vivo experiments that ion-exchange resins and even sulfate and phosphate ions that do not diffuse readily through the intestinal wall tend to drive anions from the intestinal tract into the bloodstream. The opposite effect, that of retardation of drug absorption, may occur if the drug complexes with the macromolecule.

### EXAMPLE 16–9

**Donnan Membrane Expression**

A solution of dissociated nondiffusible carboxymethylcellulose is equilibrated across a semipermeable membrane with a solution of sodium salicylate. The membrane allows free passage of the salicylate ion. Compute the ratio of salicylate on the two sides of the membrane at equilibrium, assuming that the equilibrium concentration of carboxymethylcellulose is $1.2 \times 10^{-2}$ g equivalent/liter and the equilibrium concentration of sodium salicylate is $6.0 \times 10^{-3}$ g equivalent/liter. Use the modified Donnan membrane expression, equation (16–34):

$$\frac{[D^-]_o}{[D^-]_i} = \sqrt{1 + \frac{[R^-]_i}{[D^-]_i}}$$

$$= \sqrt{1 + \frac{12 \times 10^{-3}}{6 \times 10^{-3}}} = 1.73$$

## Stability of Colloid Systems

The presence and magnitude, or absence, of a charge on a colloidal particle is an important factor in the stability of colloidal systems. Stabilization is accomplished essentially by two means: providing the dispersed particles with an electric charge, and surrounding each particle with a protective solvent sheath that prevents mutual adherence when the particles collide as a result of Brownian movement. This second effect is significant only in the case of lyophilic sols.

A lyophobic sol is thermodynamically unstable. The particles in such sols are stabilized only by the presence of electric charges on their surfaces. The like charges produce a repulsion that prevents coagulation of the particles. If the last traces

of ions are removed from the system by dialysis, the particles can agglomerate and reduce the total surface area, and, owing to their increased size, they may settle rapidly from suspension. Hence, addition of a small amount of electrolyte to a lyophobic sol tends to stabilize the system by imparting a charge to the particles. Addition of electrolyte beyond that necessary for maximum adsorption on the particles, however, sometimes results in the accumulation of opposite ions and reduces the zeta potential below its *critical value*. The critical potential for finely dispersed oil droplets in water (oil hydrosol) is about 40 millivolts, this high value signifying relatively great instability. The critical zeta potential of a gold sol, on the other hand, is nearly zero, which suggests that the particles require only a minute charge for stabilization; hence, they exhibit marked stability against added electrolytes. The valence of the ions having a charge opposite to that of the particles appears to determine the effectiveness of the electrolyte in coagulating the colloid. The precipitating power increases rapidly with the valence or charge of the ions, and a statement of this fact is known as the *Schulze–Hardy rule*.

These observations permitted Verwey and Overbeek[17] and Derjaguin and Landau[18] to independently develop a theory that describes the stability of lyophobic colloids. According to this approach, known as the DLVO theory, the forces on colloidal particles in a dispersion are due to electrostatic repulsion and London-type van der Waals attraction. These forces result in potential energies of repulsion, $V_R$, and attraction, $V_A$, between particles. These are shown in **Figure 16–13** together with the curve for the composite potential energy, $V_T$. There is a deep potential "well" of attraction near the origin and a high potential barrier of repulsion at moderate distances. A shallow secondary trough of attraction (or minimum) is sometimes observed at longer distances of separation. The presence of a secondary minimum is significant in

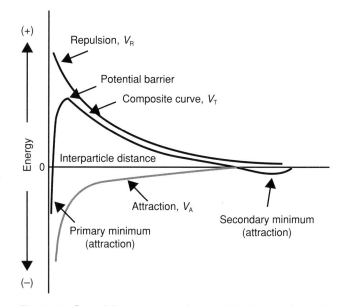

**Fig. 16–13.** Potential energy versus interparticle distance for particles in suspension.

the controlled flocculation of coarse dispersions. Following this principle, one can determine somewhat quantitatively the amount of electrolyte of a particular valence type required to precipitate a colloid.

Not only do electrolytes bring about coagulation of colloidal particles, but the mixing of oppositely charged colloids can also result in mutual agglomeration.

Lyophilic and association colloids are thermodynamically stable and exist in true solution so that the system constitutes a single phase. The addition of an electrolyte to a lyophilic colloid in moderate amounts does not result in coagulation, as was evident with lyophobic colloids. If sufficient salt is added, however, agglomeration and sedimentation of the particles may result. This phenomenon, referred to as "salting out," was discussed in the chapter on solubility.

Just as the Schulze–Hardy rule arranges ions in the order of their capacity to coagulate hydrophobic colloids, the *Hofmeister* or *lyotropic series* ranks cations and anions in order of coagulation of hydrophilic sols. Several anions of the Hofmeister series in decreasing order of precipitating power are citrate, tartrate, sulfate, acetate, chloride, nitrate, bromide, and iodide. The precipitating power is directly related to the hydration of the ion and hence to its ability to separate water molecules from the colloidal particles.

Alcohol and acetone can also decrease the solubility of hydrophilic colloids so that the addition of a small amount of electrolytes may then bring about coagulation. The addition of the less polar solvent renders the solvent mixture unfavorable for the colloid, and electrolytes can then salt out the colloid with relative ease. We can thus regard flocculation on the addition of alcohol, followed by salts, as a gradual transformation from a sol of a lyophilic nature to one of a more lyophobic character.

When negatively and positively charged hydrophilic colloids are mixed, the particles may separate from the dispersion to form a layer rich in the colloidal aggregates. The colloid-rich layer is known as a *coacervate*, and the phenomenon in which macromolecular solutions separate into two liquid layers is referred to as *coacervation*. As an example, consider the mixing of gelatin and acacia. Gelatin at a pH below 4.7 (its isoelectric point) is positively charged; acacia carries a negative charge that is relatively unaffected by pH in the acid range. When solutions of these colloids are mixed in a certain proportion, coacervation results. The viscosity of the upper layer, now poor in colloid, is markedly decreased below that of the coacervate, and in pharmacy this is considered to represent a physical incompatibility. Coacervation need not involve the interaction of charged particles; the coacervation of gelatin may also be brought about by the addition of alcohol, sodium sulfate, or a macromolecular substance such as starch.

## Sensitization and Protective Colloidal Action

The addition of a small amount of hydrophilic or hydrophobic colloid to a hydrophobic colloid of opposite charge tends

### TABLE 16–4
### THE GOLD NUMBER OF PROTECTIVE COLLOIDS

| Protective Colloid | Gold Number |
|---|---|
| Gelatin | 0.005–0.01 |
| Albumin | 0.1 |
| Acacia | 0.1–0.2 |
| Sodium oleate | 1–5 |
| Tragacanth | 2 |

to sensitize or even coagulate the particles. This is considered by some workers to be due to a reduction of the zeta potential below the critical value (usually about 20–50 millivolts). Others attribute the instability of the hydrophobic particles to a reduction in the thickness of the ionic layer surrounding the particles and a decrease in the coulombic repulsion between the particles. The addition of large amounts of the *hydrophile* (hydrophilic colloid), however, stabilizes the system, the hydrophile being adsorbed on the hydrophobic particles. This phenomenon is known as *protection*, and the added hydrophilic sol is known as a *protective colloid*. The several ways in which stabilization of hydrophobic colloids can be achieved (i.e., protective action) have been reviewed by Schott.[19]

The protective property is expressed most frequently in terms of the *gold number*. The gold number is the minimum weight in milligrams of the protective colloid (dry weight of dispersed phase) required to prevent a color change from red to violet in 10 mL of a gold sol on the addition of 1 mL of a 10% solution of sodium chloride. The gold numbers for some common protective colloids are given in **Table 16–4**.

A pharmaceutical example of sensitization and protective action is provided when bismuth subnitrate is suspended in a tragacanth dispersion; the mixture forms a gel that sets to a hard mass in the bottom of the container. Bismuth subcarbonate, a compound that does not dissociate sufficiently to liberate the bismuth ions, is compatible with tragacanth.

These phenomena probably involve a sensitization and coagulation of the gum by the $Bi^{3+}$ ions. The flocculated gum then aggregates with the bismuth subnitrate particles to form a gel or a hard cake. If phosphate, citrate, or tartrate is added, it protects the gums from the coagulating influence of the $Bi^{3+}$ ions, and, no doubt, by reducing the zeta potential on the bismuth particles, partially flocculates the insoluble material. Partially flocculated systems tend to cake considerably less than deflocculated systems, and this effect is significant in the formulation of suspensions.[20]

## SOLUBILIZATION

An important property of association colloids in solution is the ability of the micelles to increase the solubility of materials that are normally insoluble, or only slightly soluble, in the dispersion medium used. This phenomenon, known as *solubilization*, has been reviewed by many authors, including Mulley,[21] Nakagawa,[22] Elworthy et al.,[23] and Attwood and Florence.[24] Solubilization has been used with advantage in pharmacy for many years; as early as 1892, Engler and Dieckhoff[25] solubilized a number of compounds in soap solutions.

Knowing the location, distribution, and orientation of solubilized drugs in the micelle is important to understanding the kinetic aspect of the solubilization process and the interaction of drugs with the different elements that constitute the micelle. These factors may also affect the stability and bioavailability of the drug. The location of the molecule undergoing solubilization in a micelle is related to the balance between the polar and nonpolar properties of the molecule. Lawrence[26] was the first to distinguish between the various sites. He proposed that nonpolar molecules in aqueous systems of ionic surface-active agents would be located in the hydrocarbon core of the micelle, whereas polar solubilizates would tend to be adsorbed onto the micelle surface. Polar–nonpolar molecules would tend to align themselves in an intermediate position within the surfactant molecules forming the micelle. Nonionic surfactants are of most pharmaceutical interest as solubilizing agents because of their lower toxicity. Their micelles show a gradient of increased polarity from the core to the polyoxyethylene–water surface. The extended interfacial region between the core and the aqueous solution, that is, the polar mantle, is greatly hydrated. The anisotropic distribution of water molecules within the polar mantle favors the inclusion (solubilization) of a wide variety of molecules.[27] Solubilization may therefore occur in both the core and the mantle, also called the *palisade layer*. Thus, certain compounds (e.g., phenols and related compounds with a hydroxy group capable of bonding with the ether oxygen of the polyoxyethylene group) are held between the polyoxyethylene chains. Under these conditions, such compounds can be considered as undergoing inclusion within the polyoxyethylene exterior of the micelle rather than adsorption onto the micelle surface.

**Figure 16–14** depicts a spherical micelle of a nonionic, polyoxyethylene monostearate, surfactant in water. The figure is drawn in conformity with Reich's suggestion[28] that such a micelle may be regarded as a hydrocarbon core, made up of the hydrocarbon chains of the surfactant molecules, surrounded by the polyoxyethylene chains protruding into the continuous aqueous phase. Benzene and toluene, nonpolar molecules, are shown solubilized in the hydrocarbon interior of the micelle. Salicylic acid, a more polar molecule, is oriented with the nonpolar part of the molecule directed toward the central region of the micelle and the polar group toward the hydrophilic chains that spiral outward into the aqueous medium. Parahydroxybenzoic acid, a predominantly polar molecule, is found completely between the hydrophilic chains.

The pharmacist must give due attention to several factors when attempting to formulate solubilized systems successfully. It is essential that, at the concentration employed, the

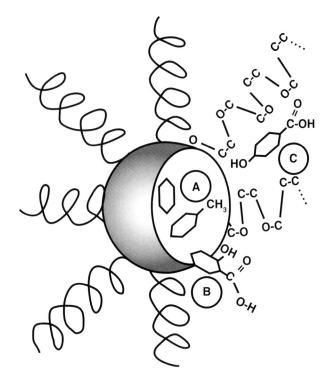

**Fig. 16–14.** A spherical micelle of nonionic surfactant molecules. (A) A nonpolar molecule solubilized in the nonpolar region of the micelle. (B) A more polar molecule found partly embedded in the central region and partially extending into the palisade region. (C) A polar molecule found lying well out in the palisade layer attracted by dipolar forces to the polyoxyethylene chains.

surface-active agent, if taken internally, be nontoxic, miscible with the solvent (usually water), compatible with the material to be solubilized, free from disagreeable odor and taste, and relatively nonvolatile. Toxicity is of paramount importance, and, for this reason, most solubilized systems are based on nonionic surfactants. The amount of surfactant used is important: A large excess is undesirable, from the point of view of both possible toxicity and reduced absorption and activity; an insufficient amount can lead to precipitation of the solubilized material. The amount of material that can be solubilized by a given amount of surfactant is a function of the polar–nonpolar characteristics of the surfactant (commonly termed the *hydrophile–lipophile balance* %HLB) and of the molecule being solubilized.

It should be appreciated that changes in absorption and biologic availability and activity may occur when the material is formulated in a solubilized system. Drastic changes in the bactericidal activity of certain compounds take place when they are solubilized, and the pharmacist must ensure that the concentration of surface-active agent present is optimum for that particular system. The stability of materials against oxidation and hydrolysis may be modified by solubilization.

Solubilization has been used in pharmacy to bring into solution a wide range of materials, including volatile oils, coal tar and resinous materials, phenobarbital, sulfonamides, vitamins, hormones, and dyes.[23,29]

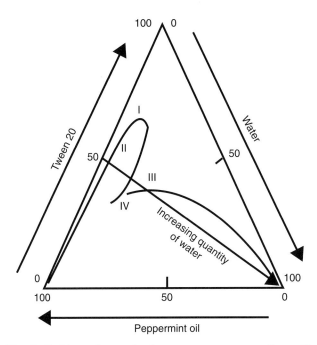

**Fig. 16–15.** Phase diagram for the ternary system water, Tween 20, and peppermint oil.

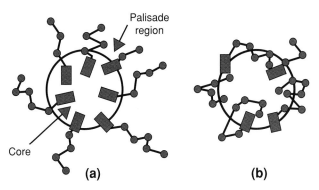

**Fig. 16–16.** Schematic of nonionic micelle of *n*-polyoxyethylene glycol monoether showing the intrusion of polyoxyethylene chains into the micelle core. (*a*) Micelle with palisade environment intact. (*b*) Palisade layer partially destroyed by loss of polyoxyethylene groups into the hydrophobic core.

O'Malley et al.[30] investigated the solubilizing action of Tween 20 on peppermint oil in water and presented their results in the form of a ternary diagram as shown in **Figure 16–15**. They found that on the gradual addition of water to a 50:50 mixture of peppermint oil and Tween 20, polysorbate 20, the system changed from a homogeneous mixture (region I) to a viscous gel (region II). On the further addition of water, a clear solution (region III) again formed, which then separated into two layers (region IV). This sequence of changes corresponds to the results one would obtain by diluting a peppermint oil concentrate in compounding and manufacturing processes. Analyses such as this therefore can provide important clues for the research pharmacist in the formulation of solubilized drug systems.

Determination of a phase diagram was also carried out by Boon et al.[31] to formulate a clear, single-phase liquid vitamin A preparation containing the minimum quantity of surfactant needed to solubilize the vitamin. Phase equilibrium diagrams are particularly useful when the formulator wishes to predict the effect on the phase equilibria of the system of dilution with one or all of the components in any desired combination or concentration.

## Factors Affecting Solubilization

The solubilization capacity of surfactants for drugs varies greatly with the chemistry of the surfactants and with the location of the drug in the micelle. If a hydrophobic drug is solubilized in the micelle core, an increase of the lipophilic alkyl chain length of the surfactant should enhance solubilization. At the same time, an increase in the micellar radius by increasing the alkyl chain length reduces the Laplace pressure, thus

favoring the entry of drug molecules into the micelle (see *Example 16–10*).

For micelles consisting of ionic surfactants, an increase in the radius of the hydrocarbon core is the principal method of enhancing solubilization,[32] whereas for micelles built up from nonionic surfactants, evidence of this effect is not well grounded. Attwood et al.[33] showed that an increase of carbon atoms above 16 in an *n*-polyoxyethylene glycol monoether— a nonionic surfactant—increases the size of the micelle, but, for a number of drugs, does not enhance solubilization. Results from NMR imaging, viscosity, and density testing[34] suggested that some of the polar groups of the micelle, that is, some polyoxyethylene groups outside the hydrocarbon core of the micelle, double back and intrude on the core, depressing its melting point and producing a fluid micellar core (**Fig. 16–16**). However, this movement of polyethylene groups into the hydrocarbon core disrupts the palisade layer and tends to destroy the region of solubilization for polar–nonpolar compounds (semipolar drugs). Patel et al.[35] suggested that the solubilizing nature of the core be increased with a more polar surfactant that would not disrupt the palisade region. Attwood et al.[33] investigated the manner in which an ether or a keto group introduced into the hydrophobic region of a surfactant, octadecylpolyoxyethylene glycol monoether, affects the solubilization and micellar character of the surfactant. It was observed that the ether group lowered the melting point of the hydrocarbon and thus was able to create a liquid core without the intrusion phenomenon, which reduced the solubilizing nature of the surfactant for semipolar drugs.

The principal effect of pH on the solubilizing power of nonionic surfactants is to alter the equilibrium between the ionized and un-ionized drug (solubilizate). This affects the solubility in water and modifies the partitioning of the drug between the micellar and the aqueous phases. As an example, the more lipophilic un-ionized form of benzoic acid is solubilized to a greater extent in polysorbate 80 than the more hydrophilic ionized form.[36] However, solubilization of drugs having hydrophobic parts in the molecule and more than one

> ## ▬▬▶ KEY CONCEPT ▮ PHARMACEUTICAL APPLICATIONS OF COLLOIDS
>
> Colloids are extensively used for modifying the properties of pharmaceutical agents. The most common property that is affected is the solubility of a drug. However, colloidal forms of many drugs exhibit substantially different properties when compared with traditional forms of these drugs. Another important pharmaceutical application of colloids is their use as drug delivery systems. The most often-used colloid-type drug delivery systems include hydrogels, microspheres, microemulsions, liposomes, micelles, nanoparticles, and nanocrystals.

dissociation constant may not correlate with the lipophilicity of the drug.[37]

## PHARMACEUTICAL APPLICATIONS OF COLLOIDS

Certain medicinals have been found to possess unusual or increased therapeutic properties when formulated in the colloidal state. Colloidal silver chloride, silver iodide, and silver protein are effective germicides and do not cause the irritation that is characteristic of ionic silver salts. Coarsely powdered sulfur is poorly absorbed when administered orally, yet the same dose of colloidal sulfur may be absorbed so completely as to cause a toxic reaction and even death. Colloidal copper has been used in the treatment of cancer, colloidal gold as a diagnostic agent for paresis, and colloidal mercury for syphilis.

Many natural and synthetic polymers are important in contemporary pharmaceutical practice. Polymers are macromolecules formed by the polymerization or condensation of smaller, noncolloidal molecules. Proteins are important natural colloids and are found in the body as components of muscle, bone, and skin. The plasma proteins are responsible for binding certain drug molecules to such an extent that the pharmacologic activity of the drug is affected. Naturally occurring plant macromolecules such as starch and cellulose that are used as pharmaceutical adjuncts are capable of existing in the colloidal state. Hydroxyethyl starch is a macromolecule used as a plasma substitute. Other synthetic polymers are applied as coatings to solid dosage forms to protect drugs that are susceptible to atmospheric moisture or degradation under the acid conditions of the stomach. Colloidal electrolytes (surface-active agents) are sometimes used to increase the solubility, stability, and taste of certain compounds in aqueous and oily pharmaceutical preparations.

In addition to mentioned pharmaceutical application, colloids are used as delivery systems for therapeutics. Seven main types of colloidal drug delivery systems in use are: hydrogels, microparticles, microemulsions, liposomes, micelles, nanoparticles, and nanocrystals (**Table 16–5**[38]). A more detailed description of different drug delivery systems is given in Chapter 23. Here, we mention the main characteristics of each colloidal delivery system.

### Hydrogels

Whereas a gel is a colloid with a liquid as dispersion medium and a solid as a dispersed phase (see Key Concept, Colloidal Systems), a hydrogel is a colloidal gel in which water is the dispersion medium. Natural and synthetic hydrogels are now used for wound healing, as scaffolds in tissue engineering, and as sustained-release delivery systems. Wound gels

## TABLE 16–5
### COLLOID-BASED DELIVERY SYSTEMS FOR THERAPEUTICS*

| Typical Mean Particle Diameter | Delivery System Type | Representative Systems of Each Type | Characteristic Applications |
|---|---|---|---|
| 0.5–20 $\mu$m | Microspheres, hydrogels | Alginate, gelatin, chitosan, polymeric microspheres, synthetic, biodegradable, polymeric hydrogels | Sustained release of therapeutics, scaffolds for cell delivery in tissue engineering |
| 0.2–5 $\mu$m | Microparticles | Polystyrene, poly(lactide) microspheres | Targeted delivery of therapeutics |
| 0.15–2 $\mu$m | Emulsions, microemulsions | Oil-in-water, water-in-oil, lipid emulsions, oil-in-water microemulsions | Controlled and targeted delivery of therapeutics |
| 30–1000 nm | Liposomes | Phospholipid and polymer-based bilayer vesicles | Targeted delivery of therapeutics |
| 3–80 nm | Micelles | Natural and synthetic surfactant micelles | Targeted delivery of therapeutics |
| 2–100 nm | Nanoparticles | Lipid, polymer, inorganic nanoparticles | Targeted delivery of therapeutics, in vivo navigational devices |
| 2–100 nm | Nanocrystals | Quantum dots | Imaging agents |

*Based on K. Kostarelos, Adv. Colloid Interface Sci. **106,** 147, 2003.

are excellent for helping create or maintain a moist environment. Some hydrogels provide absorption, desloughing, and debriding capacities to necrotic and fibrotic tissue. When used as scaffolds for tissue engineering, hydrogels may contain human cells to stimulate tissue repair.[39] Because they are loaded with pharmaceutical ingredients, hydrogels provide a sustained release of drugs. Special attention has been given to environmentally sensitive hydrogels.[40] These hydrogels have the ability to sense changes in pH, temperature, or the concentration of a specific metabolite and release their load as a result of such a change. These hydrogels can be used as site-specific controlled drug delivery systems. Hydrogels that are responsive to specific molecules, such as glucose or antigens, can be used as biosensors as well as drug delivery systems. Light-sensitive, pressure-responsive, and electrosensitive hydrogels also have the potential to be used in drug delivery. Although the concepts of these environment-sensitive hydrogels are sound, the practical applications require significant improvements in the hydrogel properties. The most important challenges that should be addressed in designing useful environmentally sensitive hydrogels include slow response time, limited biocompatibility, and biodegradability. However, if the achievements of the past can be extrapolated into the future, it is highly likely that responsive hydrogels with a wide array of desirable properties will be forthcoming.[40]

## Microparticles

Microparticles are small (0.2–5 $\mu$m), loaded microspheres of natural or synthetic polymers. Microparticles were initially developed as carriers for vaccines and anticancer drugs. More recently, novel properties of microparticles have been developed to increase the efficiency of drug delivery and improve release profiles and drug targeting. Several investigations have focused on the development of methods of reducing the uptake of the nanoparticles by the cells of the reticuloendothelial system and enhance their uptake by the targeted cells. For instance, functional surface coatings of nonbiodegradable carboxylated polystyrene or biodegradable poly(D,L-lactide-co-glycolide) microspheres with poly(L-lysine)-g-poly(ethylene glycol) (PLL-g-PEG) were investigated in attempts to shield them from nonspecific phagocytosis and to allow ligand-specific interactions via molecular recognition.[41] It was found that coatings of PLL-g-PEG-ligand conjugates provided for the specific targeting of microspheres to human blood–derived macrophages and dendritic cells while reducing nonspecific phagocytosis. Microparticles can also be used to facilitate nontraditional routes of drug administration. For example, it was found that microparticles can be used to improve immunization using the mucosal route of administration of therapeutics.[42] It was found in this study that after mucosal delivery, microparticles can translocate to tissues in the systemic compartment of the immune system and provoke immunologic reactions.

## Emulsions and Microemulsions

Microemulsions are excellent candidates as potential drug delivery systems because of their improved drug solubilization, long shelf life, and ease of preparation and administration. Three distinct microemulsions—oil external, water external, and middle phase—can be used for drug delivery, depending upon the type of drug and the site of action.[43,44] In contrast to microparticles, which demonstrate distinct differences between the outer shell and core, microemulsions are usually formed with more or less homogeneous particles. Microemulsions are used for controlled release and targeted delivery of different pharmaceutic agents. For instance, microemulsions were used to deliver oligonucleotides (small fragments of DNA) specifically to ovarian cancer cells.[45] In contrast to microemulsions, nanoemulsions consist in very fine oil-in-water dispersions, having droplet diameter smaller than 100 nm. Compared to microemulsions, they are in a metastable state, and their structure depends on the history of the system. Nanoemulsions are very fragile systems. The nanoemulsions can find an application in skin care due to their good sensorial properties (rapid penetration, merging textures) and their biophysical properties (especially their hydrating power).[46]

## Liposomes

Liposomes consist of an outer uni- or multilaminar membrane and an inner liquid core. In most cases, liposomes are formed with natural or synthetic phospholipids similar to those in cellular plasma membrane. Because of this similarity, liposomes are easily utilized by cells. Liposomes can be loaded by pharmaceutical or other ingredients by two principal ways: lipophilic compounds can be associated with liposomal membrane, and hydrophilic substances can be dissolved in the inner liquid core of liposomes. To decrease uptake by the cells of the reticuloendothelial system and/or enhance their uptake by the targeted cells, the membrane of liposomes can be modified by polymeric chains and/or targeting moieties or antibodies specific to the targeted cells. Because they are relatively easy to prepare, biodegradable, and nontoxic, liposomes have found numerous applications as drug delivery systems.[47,48]

## Micelles

Micelles are structures similar to liposomes but do not have an inner liquid compartment. Therefore, they can be used as water-soluble biocompatible microcontainers for the delivery of poorly soluble hydrophobic pharmaceuticals.[49] Similar to liposomes, their surface can be modified with antibodies (immunomicelles) or other targeting moieties providing the ability of micelles to specifically interact with their antigens.[50] One type of micelles, Pluronic block copolymers, are recognized pharmaceutical excipients listed in the US and British Pharmacopoeia. They have been used extensively in a variety of pharmaceutical formulations including delivery of low–molecular-mass drugs, polypeptides, and DNA.[51] Furthermore, Pluronic block copolymers are versatile molecules that can be used as structural elements of polycation-based gene delivery systems (polyplexes).

## Nanoparticles

Nanocapsules are submicroscopic colloidal drug carrier systems composed of an oily or an aqueous core surrounded by a thin polymer membrane. Two technologies can be used to obtain such nanocapsules: the interfacial polymerization of a monomer or the interfacial nanodeposition of a preformed polymer.[52] Solid lipid nanoparticles were developed at the beginning of the 1990s as an alternative carrier system to emulsions, liposomes, and polymeric nanoparticles. They were used, in particular, in topical cosmetic and pharmaceutical formulations.[53] A novel nanoparticle-based drug carrier for photodynamic therapy has been developed by Roy et al.[54] This carrier can provide stable aqueous dispersion of hydrophobic photosensitizers, yet preserve the key step of photogeneration of singlet oxygen, necessary for photodynamic action. Nanoparticles have also found applications as nonviral gene delivery systems.[55]

## Nanocrystals

Inorganic nanostructures that interface with biologic systems have recently attracted widespread interest in biology and medicine.[56] Larson et al.[57] set out to explore the feasibility of in vivo targeting by using semiconductor quantum dots (qdots), which are small (<10 nm) inorganic nanocrystals that possess unique luminescent properties; their fluorescence emission is stable and tuned by varying the particle size or composition. By adding a targeting moiety, one can direct these qdots specifically to the targeted organs and tissues. In particular, it was found that ZnS-capped CdSe qdots coated with a lung-targeting peptide accumulate in the lungs of mice after intravenous injection, whereas two other peptides specifically direct qdots to blood vessels or lymphatic vessels in tumors.[57] As in case of liposomes, adding polyethylene glycol to the qdot coating prevents nonselective accumulation of qdots in reticuloendothelial tissues. All these make qdots promising imaging agents. The use of semiconductor quantum dots as fluorescent labels for multiphoton microscopy enables multicolor imaging in demanding biologic environments such as living tissue.[57]

## CHAPTER SUMMARY

Although colloidal dispersion have been important in the pharmaceutical sciences for decades, with the advent of nanotechnology, they are now becoming a driving force behind drug delivery systems and technology. This chapter provided basic information on colloidal dispersions such as basic definitions, the types of colloidal systems, electric, kinetic, and optical properties, their role in solubilization, and applications of colloids in the pharmaceutical sciences. The drug delivery aspects of colloids are discussed in Chapter 23 as well.

 **Practice problems for this chapter can be found at thePoint.lww.com/Sinko6e.**

## References

1. C. J. Drummond, G. G. Warr, F. Grieser, B. W. Nihaen and D. F. Evans, J. Phys. Chem. **89,** 2103, 1985.
2. Y. Chang and J. R. Cardinal, J. Pharm Sci. **67,** 994, 1978.
3. T. J. Racey, P. Rochon, D. V. C. Awang and G. A. Neville, J. Pharm. Sci. **76,** 314, 1987; T. J. Racey, P. Rochon, F. Mori and G. A. Neville, J. Pharm. Sci. **78,** 214, 1989.
4. P. Mukerjee, J. Phys. Chem. **76,** 565, 1972.
5. K. W. Herrmann, J. Colloid Interface Sci. **22,** 352, 1966.
6. P. Atkins, J. D. Paula, Physical Chemistry, W. H. Freeman, NY and Company, 7 Ed., 2006.
7. P. C. Hiemenz, *Principles of Colloid and Surface Chemistry*, 2nd Ed., Marcel Dekker, New York, 1986, pp. 127, 133, 148.
8. T. Svedberg and J. B. Nicholas, J. Am. Chem. Soc. **49,** 2920, 1927.
9. J. Kirschbaum, J. Pharm. Sci. **63,** 981, 1974.
10. A. J. Richard, J. Pharm. Sci. **64,** 873, 1975.
11. H. Arwidsson and M. Nicklasson, Int. J. Pharm. **58,** 73, 1990.
12. H. H. Paradies, Eur. J. Biochem. **118,** 187, 1981.
13. H. H. Paradies, J. Pharm. Sci. **78,** 230, 1989.
14. H. Schott, J. Pharm. Sci. **70,** 486, 1981.
15. H. Takenaka, Y. Kawashima and S. Y. Lin, J. Pharm. Sci. **70,** 302, 1981.
16. T. Higuchi, R. Kuramoto, L. Kennon, T. L. Flanagan and A. Polk, J. Am. Pharm. Assoc. Sci. **43,** 646, 1954.
17. E. J. W. Verwey and J. Th. G. Overbeek, *Theory of the Stability of Lyophobic Colloids*, Elsevier, Amsterdam, 1948.
18. B. Derjaguin and L. Landau, Acta Phys. Chim. USSR **14,** 663, 1941; J. Exp. Theor. Physics USSR **11,** 802, 1941.
19. H. Schott, in *Remington's Pharmaceutical Sciences*, 16th Ed., Mack, Easton, PA, 1980, Chapter 20.
20. B. Haines and A. N. Martin, J. Pharm. Sci. **50,** 228, 753, 756, 1961.
21. B. A. Mulley, in *Advances in Pharmaceutical Sciences*, Academic Press, New York, 1964, Vol. 1, pp. 87–194.
22. T. Nakagawa, in *Nonionic Surfactants*, M. J. Schick, Marcel Dekker, New York, 1967.
23. P. H. Elworthy, A. T. Florence and C. B. Macfarlane, *Solubilization by Surface-Active Agents*, Chapman & Hall, London, 1968.
24. D. Attwood and A. T. Florence, *Surfactant Systems*, Chapman & Hall, London, 1983.
25. C. Engler and E. Dieckhoff, Arch. Pharm. **230,** 561, 1892.
26. A. S. C. Lawrence, Trans. Faraday Soc. **33,** 815, 1937.
27. E. Keh, S. Partyka and S. Zaini, J. Colloid Interface Sci. **129,** 363, 1989.
28. I. Reich, J. Phys. Chem. **60,** 260, 1956.
29. B. W. Barry and D. I. El Eini, J. Pharm. Pharmacol. **28,** 210, 1976.
30. W. J. O'Malley, L. Pennati and A. Martin, J. Am. Pharm. Assoc. Sci. **47,** 334, 1958.
31. P. F. G. Boon, C. L. J. Coles and M. Tait, J. Pharm. Pharmacol. **13,** 200T, 1961.
32. T. Anarson and P. H. Elworthy, J. Pharm. Pharmacol. **32,** 381, 1980.
33. D. Attwood, P. H. Elworthy and M. J. Lawrence, J. Pharm. Pharmacol. **41,** 585, 1989.
34. P. H. Elworthy and M. S. Patel, J. Pharm. Pharmacol. **36,** 565, 1984; J. Pharm. Pharmacol. **36,** 116, 1984.
35. M. S. Patel, P. H. Elworthy and A. K. Dewsnup, J. Pharm. Pharmacol. **33,** 64P, 1981.
36. J. H. Collett and L. Koo, J. Pharm. Sci. **64,** 1253, 1975.
37. K. Ikeda, H. Tomida and T. Yotsuyanagi, Chem. Pharm. Bull. **25,** 1067, 1977.

38. K. Kostarelos, Adv. Colloid Interface Sci. **106,** 147, 2003.

39. J. Kisiday, M. Jin, B. Kurz, H. Hung, C. Semino, S. Zhang and A. J. Grodzinsky, Proc. Natl. Acad. Sci. USA **99,** 9996, 2002.

40. Y. Qiu and K. Park, Adv. Drug Deliv. Rev. **53,** 321, 2001.

41. S. Faraasen, J. Voros, G. Csucs, M. Textor, H. P. Merkle and E. Walter, Pharm. Res. **20,** 237, 2003.

42. J. E. Eyles, V. W. Bramwell, E. D. Williamson and H. O. Alpar, Vaccine **19,** 4732, 2001.

43. R. P. Bagwe, J. R. Kanicky, B. J. Palla, P. K. Patanjali and D. O. Shah, Crit. Rev. Ther. Drug Carrier Syst. **18,** 77, 2001.

44. S. Benita (Ed.), *Submicron Emulsion in Drug Targeting and Delivery,* Harwood Academic Publishers, Amsterdam, 1998.

45. H. Teixeira, C. Dubernet, H. Chacun, L. Rabinovich, V. Boutet, J. R. Deverre, S. Benita and P. Couvreur, J. Controlled Release **89,** 473, 2003.

46. O. Sonneville-Aubrun, J. T. Simonnet and F. L'Alloret, Adv. Colloid Interface Sci. **108–109,** 145, 2004.

47. D. J. Crommelin and G. Storm, J. Liposome Res. **13,** 33, 2003.

48. D. D. Lasic and D. Papahadjopoulos (Eds.), *Medical Applications of Liposomes,* Elsevier Health Sciences, Amsterdam, 1998.

49. R. Savic, L. Luo, A. Eisenberg and D. Maysinger, Science **300,** 615, 2003.

50. V. P. Torchilin, A. N. Lukyanov, Z. Gao and B. Papahadjopoulos-Sternberg, Proc. Natl. Acad. Sci. USA **100,** 6039, 2003.

51. A. V. Kabanov, P. Lemieux, S. Vinogradov and V. Alakhov, Adv. Drug Deliv. Rev. **54,** 223, 2002.

52. P. Couvreur, G. Barratt, E. Fattal, P. Legrand and C. Vauthier, Crit. Rev. Ther. Drug Carrier Syst. **19,** 99, 2002.

53. R. H. Muller, M. Radtke and S. A. Wissing, Adv. Drug Deliv. Rev. **54**(Suppl 1), S131, 2002.

54. I. Roy, T. Y. Ohulchanskyy, H. E. Pudavar, E. J. Bergey, A. R. Oseroff, J. Morgan, T. J. Dougherty and P. N. Prasad, J. Am. Chem. Soc. **125,** 7860, 2003.

55. F. Scherer, M. Anton, U. Schillinger, J. Henke, C. Bergemann, A. Kruger, B. Gansbacher and C. Plank, Gene Ther. **9,** 102, 2002.

56. M. E. Akerman, W. C. Chan, P. Laakkonen, S. N. Bhatia and E. Ruoslahti, Proc. Natl. Acad. Sci. USA **99,** 12617, 2002.

57. D. R. Larson, W. R. Zipfel, R. M. Williams, S. W. Clark, M. P. Bruchez, F. W. Wise and W. W. Webb, Science **300,** 1434, 2003.

## Recommended Readings

T. Cosgrove, *Colloid Science: Principles, Methods and Applications,* Blackwell, Oxford, UK, 2005.

P. C. Hiemenz, R. Rajagopalan, *Principles of Colloid and Surface Chemistry,* 3rd Ed., Marcel Dekker, New York, 1997.

### CHAPTER LEGACY

**Fifth Edition**: published as Chapter 17 (Colloids). Updated by Tamara Minko.

**Sixth Edition**: published as Chapter 16 (Colloidal Dispersions). Updated by Patrick Sinko.

# 17 COARSE DISPERSIONS

Particulate systems have been classified on the basis of size into molecular dispersions (Chapter 5), colloidal systems (Chapter 16), and coarse dispersions (this chapter). This chapter attempts to provide the pharmacist with an insight into the role of physics and chemistry in the research and development of the several classes of coarse dispersions. The theory and technology of these important pharmaceutical classes are based on interfacial and colloidal principles, micromeritics, and rheology (Chapters 15, 16, 18, and 19, respectively).

## SUSPENSIONS

A pharmaceutical suspension is a coarse dispersion in which insoluble solid particles are dispersed in a liquid medium. The particles have diameters for the most part greater than 0.1 $\mu$m, and some of the particles are observed under the microscope to exhibit Brownian movement if the dispersion has a low viscosity.

Examples of oral suspensions are the oral antibiotic syrups, which normally contain 125 to 500 mg per 5 mL of solid material. When formulated for use as pediatric drops, the concentration of suspended material is correspondingly greater. Antacid and radiopaque suspensions generally contain high concentrations of dispersed solids. Externally applied suspensions for topical use are legion and are designed for dermatologic, cosmetic, and protective purposes. The concentration of dispersed phase may exceed 20%. Parenteral suspensions contain from 0.5% to 30% of solid particles. Viscosity and particle size are significant factors because they affect the ease of injection and the availability of the drug in depot therapy.

An acceptable suspension possesses certain desirable qualities, including the following. The suspended material should not settle rapidly; the particles that do settle to the bottom of the container must not form a hard cake but should be readily redispersed into a uniform mixture when the container is shaken; and the suspension must not be too viscous to pour freely from the orifice of the bottle or to flow through a syringe needle. In the case of an external lotion, the product must be fluid enough to spread easily over the affected area and yet must not be so mobile that it runs off the surface to which it is applied; the lotion must dry quickly and provide an elastic protective film that will not rub off easily; and it must have an acceptable color and odor.

It is important that the characteristics of the dispersed phase be chosen with care so as to produce a suspension having optimum physical, chemical, and pharmacologic properties. Particle-size distribution, specific surface area, inhibition of crystal growth, and changes in polymorphic form are of special significance, and the formulator must ensure that these and other properties[1−3] do not change sufficiently

## KEY CONCEPT SUSPENSIONS

Suspensions contribute to pharmacy and medicine by supplying insoluble and what often would otherwise be distasteful substances in a form that is pleasant to the taste, by providing a suitable form for the application of dermatologic materials to the skin and sometimes to the mucous membranes, and for the parenteral administration of insoluble drugs. Therefore, pharmaceutical suspensions can be classified into three groups: orally administered mixtures, externally applied lotions, and injectable preparations.

during storage to adversely affect the performance of the suspension. Finally, it is desirable that the product contain readily obtainable ingredients that can be incorporated into the mixture with relative ease by the use of standard methods and equipment.

The remainder of this section will be devoted to a discussion of some of the properties that provide the desirable characteristics just enumerated.

For pharmaceutical purposes, *physical stability* of suspensions may be defined as the condition in which the particles do not aggregate and in which they remain uniformly distributed throughout the dispersion. Because this ideal situation is seldom realized, it is appropriate to add that if the particles do settle, they should be easily resuspended by a moderate amount of agitation.

# INTERFACIAL PROPERTIES OF SUSPENDED PARTICLES

Little is known about energy conditions at the surfaces of solids, yet knowledge of the thermodynamic requirements is needed for the successful stabilization of suspended particles.

Work must be done to reduce a solid to small particles and disperse them in a continuous medium. The large surface area of the particles that results from the comminution is associated with a surface free energy that makes the system *thermodynamically unstable*, by which we mean that the particles are highly energetic and tend to regroup in such a way as to decrease the total area and reduce the surface free energy. The particles in a liquid suspension therefore tend to *flocculate*, that is, to form light, fluffy conglomerates that are held together by weak van der Waals forces. Under certain conditions—in a compacted cake, for example—the particles may adhere by stronger forces to form what are termed *aggregates*. Caking often occurs by the growth and fusing together of crystals in the precipitates to produce a solid aggregate.

The formation of any type of agglomerate, either floccules or aggregates, is taken as a measure of the system's tendency to reach a more thermodynamically stable state. An increase in the work, $W$, or surface free energy, $\Delta G$, brought about by dividing the solid into smaller particles and consequently increasing the total surface area, $\Delta A$, is given by

$$\Delta G = \gamma SL \cdot \Delta A \qquad (17\text{--}1)$$

where $\gamma_{SL}$ is the interfacial tension between the liquid medium and the solid particles.

### EXAMPLE 17–1

**Surface Free Energy**

Compute the change in the surface free energy of a solid in a suspension if the total surface is increased from $10^3$ to $10^7$ cm$^2$. Assume that the interfacial tension between the solid and the liquid medium, $\gamma_{SL}$, is 100 dynes/cm.

The initial free energy is

$$G_1 = 100 \times 10^3 = 10^5 \text{ ergs/cm}^2$$

When the surface area is $10^7$ cm$^2$,

$$G_2 = 100 \times 10^7 = 10^9 \text{ ergs/cm}^2$$

The change in the free energy, $\Delta G_{21}$, is $10^9 - 10^5 \cong 10^9$ erg/cm$^2$. The free energy has been increased by $10^9$, which makes the system more thermodynamically unstable.

To approach a stable state, the system tends to reduce the surface free energy; equilibrium is reached when $\Delta G = 0$. This condition can be accomplished, as seen from equation (17–1), by a reduction of interfacial tension, or it can be approached by a decrease of the interfacial area. The latter possibility, leading to flocculation or aggregation, can be desirable or undesirable in a pharmaceutical suspension, as considered in a later section.

The interfacial tension can be reduced by the addition of a surfactant but cannot ordinarily be made equal to zero. A suspension of insoluble particles, then, usually possesses a finite positive interfacial tension, and the particles tend to flocculate. An analysis paralleling this one could also be made for the breaking of an emulsion.

The forces at the surface of a particle affect the degree of flocculation and agglomeration in a suspension. Forces of attraction are of the London–van der Waals type; the repulsive forces arise from the interaction of the electric double layers surrounding each particle. The formation of the electric double layer is considered in detail in Chapter 15, which deals with interfacial phenomena. The student is advised to review, at this point, the section dealing with the electrical properties of interfaces because particle charge, electric double-layer formation, and zeta potential are all relevant to the present topic.

The potential energy of two particles is plotted in **Figure 17–1** as a function of the distance of separation. Shown are the curves depicting the energy of attraction, the energy of repulsion, and the net energy, which has a peak and two minima. When the repulsion energy is high, the potential barrier is also high, and collision of the particles is opposed. The system remains deflocculated, and, when sedimentation is complete, the particles form a close-packed arrangement with the smaller particles filling the voids between the larger ones. Those particles lowest in the sediment are gradually pressed together by the weight of the ones above; the energy barrier is thus overcome, allowing the particles to come into close contact with each other. To resuspend and redisperse these particles, it is again necessary to overcome the high-energy barrier. Because this is not easily achieved by agitation, the particles tend to remain strongly attracted to each other and form a hard cake. When the particles are flocculated, the energy barrier is still too large to be surmounted, and so the approaching particle resides in the second energy minimum, which is at a distance of separation of perhaps 1000 to 2000 Å. This distance is sufficient to form the loosely structural flocs. These concepts evolve from the Derjaguin and Landau, Verwey and Overbeek (DLVO) theory for the stability of lyophobic sols. Schneider et al.[4] prepared a computer program for

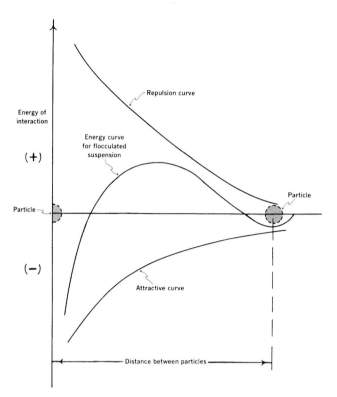

**Fig. 17–1.** Potential energy curves for particle interactions in suspension. (From A. Martin, J. Pharm. Sci. **50**, 514, 1961. With permission.)

calculating the repulsion and attraction energies in pharmaceutical suspensions. They showed the methods of handling the DLVO equations and the careful consideration that must be given to the many physical units involved. Detailed examples of calculations were given.

To summarize, flocculated particles are weakly bonded, settle rapidly, do not form a cake, and are easily resuspended; deflocculated particles settle slowly and eventually form a sediment in which aggregation occurs with the resultant formation of a hard cake that is difficult to resuspend.

## SETTLING IN SUSPENSIONS

As mentioned earlier, one aspect of physical stability in pharmaceutical suspensions is concerned with keeping the particles uniformly distributed throughout the dispersion. Although it is seldom possible to prevent settling completely over a prolonged period of time, it is necessary to consider the factors that influence the velocity of sedimentation.

### Theory of Sedimentation

The velocity of sedimentation is expressed by Stokes's law:

$$v = \frac{d^2(\rho_s - \rho_o)g}{18\eta_o} \qquad (17\text{--}2)$$

where $v$ is the terminal velocity in cm/sec, $d$ is the diameter of the particle in cm, $\rho_s$ and $\rho_o$ are the densities of the dispersed

phase and dispersion medium, respectively, $g$ is the acceleration due to gravity, and $\eta_o$ is the viscosity of the dispersion medium in poise.

Dilute pharmaceutical suspensions containing less than about 2 g of solids per 100 mL of liquid conform roughly to these conditions. (Some feel that the concentration must be less than 0.5 g/100 mL before Stokes's equation is valid.) In dilute suspensions, the particles do not interfere with one another during sedimentation, and *free settling* occurs. In most pharmaceutical suspensions that contain dispersed particles in concentrations of 5%, 10%, or higher percentages, the particles exhibit *hindered settling*. The particles interfere with one another as they fall, and Stokes's law no longer applies.

Under these circumstances, some estimation of physical stability can be obtained by diluting the suspension so that it contains about 0.5% to 2.0% w/v of dispersed phase. This is not always recommended, however, because the stability picture obtained is not necessarily that of the original suspension. The addition of a diluent may affect the degree of flocculation (or deflocculation) of the system, thereby effectively changing the particle-size distribution.

To account for the nonuniformity in particle shape and size invariably encountered in real systems. We can write Stokes's equation in other forms. One of the proposed modifications is[5]

$$v' = v\epsilon^n \qquad (17\text{--}3)$$

where $v'$ is the rate of fall at the interface in cm/sec and $v$ is the velocity of sedimentation according to Stokes's law. The term $\varepsilon$ represents the initial porosity of the system, that is, the initial volume fraction of the uniformly mixed suspension, which varies from zero to unity. The exponent $n$ is a measure of the "hindering" of the system. It is a constant for each system.

### EXAMPLE 17–2

The average particle diameter of calcium carbonate in aqueous suspension is 54 $\mu$m. The densities of $CaCO_3$ and water, respectively, are 2.7 and 0.997 g/cm³. The viscosity of water is 0.009 poise at 25°C. Compute the rate of fall $v'$ for $CaCO_3$ samples at two different porosities, $\epsilon_1 = 0.95$ and $\epsilon_2 = 0.5$. The $n$ value is 19.73.

From Stokes's law, equation (17–2),

$$v = \frac{(54 \times 10^{-4})^2(2.7 - 0.997)981}{18 \times 0.009} = 0.30 \text{ cm/sec}$$

Taking logarithms on both sides of equation (17–3), we obtain ln $v' = \ln v + n \ln \epsilon$.

For $\epsilon_1 = 0.95$,

$$\ln v' = -1.204 + [19.73(-0.051)] = -2.210$$

$$v' = 0.11 \text{ cm/sec}$$

Analogously, for $\epsilon_2 = 0.5$, $v' = 3.5 \times 10^{-7}$ cm/sec. Note that at low porosity values (i.e., 0.5, which corresponds to a high concentration of solid in suspension), the sedimentation is hindered, leading to small $v'$ values. On the other hand, when the suspension becomes infinitely diluted (i.e., $\epsilon = 1$), the rate of fall is given by $v' = v$. In the present example, if $\epsilon = 1$,

$$v' = 0.3 \times 1^{19.73} = 0.3 \text{ cm/sec}$$

which is the Stokes-law velocity.

## Effect of Brownian Movement

For particles having a diameter of about 2 to 5 $\mu$m (depending on the density of the particles and the density and viscosity of the suspending medium), Brownian movement counteracts sedimentation to a measurable extent at room temperature by keeping the dispersed material in random motion. The *critical radius, r,* below which particles will be kept in suspension by kinetic bombardment of the particles by the molecules of the suspending medium (Brownian movement) was worked out by Burton.[6]

It can be seen in the microscope that Brownian movement of the smallest particles in a field of particles of a pharmaceutical suspension is usually eliminated when the sample is dispersed in a 50% glycerin solution, having a viscosity of about 5 centipoise. Hence, it is unlikely that the particles in an ordinary pharmaceutical suspension containing suspending agents are in a state of vigorous Brownian motion.

## Sedimentation of Flocculated Particles

When sedimentation is studied in flocculated systems, it is observed that the flocs tend to fall together, producing a distinct boundary between the sediment and the supernatant liquid. The liquid above the sediment is clear because even the small particles present in the system are associated with the flocs. Such is not the case in deflocculated suspensions having a range of particle sizes, in which, in accordance with Stokes's law, the larger particles settle more rapidly than the smaller particles. No clear boundary is formed (unless only

one size of particle is present), and the supernatant remains turbid for a considerably longer period of time. Whether the supernatant liquid is clear or turbid during the initial stages of settling is a good indication of whether the system is flocculated or deflocculated, respectively.

According to Hiestand,[7] the initial rate of settling of flocculated particles is determined by the floc size and the porosity of the aggregated mass. Subsequently, the rate depends on compaction and rearrangement processes within the sediment. The term *subsidence* is sometimes used to describe settling in flocculated systems.

## Sedimentation Parameters

Two useful parameters that can be derived from sedimentation (or, more correctly, subsidence) studies are *sedimentation volume*, V, or *height*, H, and *degree of flocculation*.

The sedimentation volume, F, is defined as the ratio of the final, or ultimate, volume of the sediment, $V_u$, to the original volume of the suspension, $V_o$, before settling. Thus,

$$F = V_u / V_o \qquad (17-4)$$

The sedimentation volume can have values ranging from less than 1 to greater than 1. F is normally less than 1, and in this case, the ultimate volume of sediment is smaller than the original volume of suspension, as shown in **Figure 17–2a**, in which $F = 0.5$. If the volume of sediment in a flocculated suspension equals the original volume of suspension, then $F = 1$ (**Fig. 17–2b**). Such a product is said to be in "flocculation equilibrium" and shows no clear supernatant on standing. It is therefore pharmaceutically acceptable. It is possible for F to have values greater than 1, meaning that the final volume of sediment is greater than the original suspension volume. This comes about because the network of flocs formed in the suspension is so loose and fluffy that the volume they

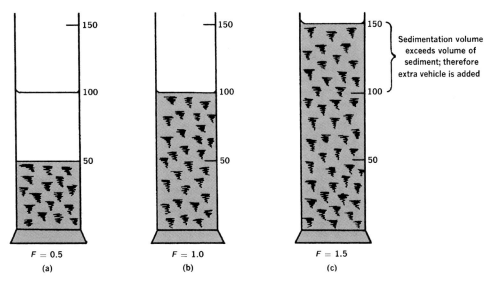

**Fig. 17–2.** Sedimentation volumes produced by adding varying amounts of flocculating agent. Examples (b) and (c) are pharmaceutically acceptable.

are able to encompass is greater than the original volume of suspension. This situation is illustrated in **Figure 17–2c**, in which sufficient extra vehicles have been added to contain the sediment. In example shown, $F = 1.5$.

The sedimentation volume gives only a qualitative account of flocculation because it lacks a meaningful reference point.[7] A more useful parameter for flocculation is $\beta$, the *degree of flocculation*.

If we consider a suspension that is completely deflocculated, the ultimate volume of the sediment will be relatively small. Writing this volume as $V_\infty$, based on equation **(17–4)**, we have

$$F_\infty = V_\infty / V_0 \qquad (17\text{–}5)$$

where $F_\infty$ is the sedimentation volume of the deflocculated, or peptized, suspension. The degree of flocculation, $\beta$, is therefore defined as the ratio of $F$ to $F_\infty$, or

$$\beta = F/F_\infty \qquad (17\text{–}6)$$

Substituting equations **(17–4)** and **(17–5)** in equation **(17–6)**, we obtain

$$\beta = \frac{V_u/V_0}{V_\infty/V_0} = V_u/V_\infty \qquad (17\text{–}7)$$

The degree of flocculation is a more fundamental parameter than $F$ because it relates the volume of flocculated sediment to that in a deflocculated system. We can therefore say that

$$\beta = \frac{\text{Ultimate sediment volume of } \textit{flocculated} \text{ suspension}}{\text{Ultimate sediment volume of } \textit{deflocculated} \text{ suspension}}$$

**EXAMPLE 17–3**

Compute the sedimentation volume of a 5% w/v suspension of magnesium carbonate in water. The initial volume is $V_0 = 100$ mL and the final volume of the sediment is $V_u = 30$ mL. If the degree of flocculation is $\beta = F/F_\infty = 1.3$, what is the deflocculated sedimentation volume, $F_\infty$?

We have

$$F = 30/100 = 0.30$$
$$F_\infty = F/\beta = 0.30/1.3 = 0.23$$

# FORMULATION OF SUSPENSIONS

The approaches commonly used in the preparation of physically stable suspensions fall into two categories—the use of a structured vehicle to maintain deflocculated particles in suspension, and the application of the principles of flocculation to produce flocs that, although they settle rapidly, are easily resuspended with a minimum of agitation.

*Structured vehicles* are pseudoplastic and plastic in nature; their rheologic properties are discussed in Chapter 19. As we shall see in a later section, it is frequently desirable that thixotropy be associated with these two types of flow. Structured vehicles act by entrapping the particles (generally deflocculated) so that, ideally, no settling occurs. In reality, some degree of sedimentation will usually take place. The "shear-thinning" property of these vehicles does, however, facilitate the re-formation of a uniform dispersion when shear is applied.

A disadvantage of deflocculated systems, mentioned earlier, is the formation of a compact cake when the particles eventually settle. It is for this reason that the formulation of flocculated suspensions has been advocated.[8] Optimum physical stability and appearance will be obtained when the suspension is formulated with flocculated particles in a structured vehicle of the hydrophilic colloid type. Consequently, most of the subsequent discussion will be concerned with this approach and the means by which controlled flocculation can be achieved. Whatever approach is used, the product must (*a*) flow readily from the container and (*b*) possess a uniform distribution of particles in each dose.

## Wetting of Particles

The initial dispersion of an insoluble powder in a vehicle is an important step in the manufacturing process and requires further consideration. Powders sometimes are added to the vehicle, particularly in large-scale operations, by dusting on the surface of the liquid. It is frequently difficult to disperse the powder owing to an adsorbed layer of air, minute quantities of grease, and other contaminants. The powder is not readily wetted, and although it may have a high density, it floats on the surface of the liquid. Finely powdered substances are particularly susceptible to this effect because of entrained air, and they fail to become wetted even when forced below the surface of the suspending medium. The *wettability* of a powder can be ascertained easily by observing the contact angle that powder makes with the surface of the liquid. The angle is approximately $90°$ when the particles are floating well out of the liquid. A powder that floats low in the liquid has a lesser angle, and one that sinks obviously shows no contact angle. Powders that are not easily wetted by water and accordingly show a large contact angle, such as sulfur, charcoal, and magnesium stearate, are said to be *hydrophobic*. Powders that are readily wetted by water when free of adsorbed contaminants are called *hydrophilic*. Zinc oxide, talc, and magnesium carbonate belong to the latter class.

Surfactants are quite useful in the preparation of a suspension in reducing the interfacial tension between solid particles and a vehicle. As a result of the lowered interfacial tension, the advancing contact angle is lowered, air is displaced from the surface of particles, and wetting and deflocculation are promoted. Schott et al.[9] studied the deflocculating effect of octoxynol, a nonionic surfactant, in enhancing the dissolution rate of prednisolone from tablets. The tablets break up into fine granules that are deflocculated in suspension. The deflocculating effect is proportional to the surfactant concentration. However, at very high surfactant concentration, say, 15 times the critical micelle concentration, the surfactant produces extensive flocculation. Glycerin and similar

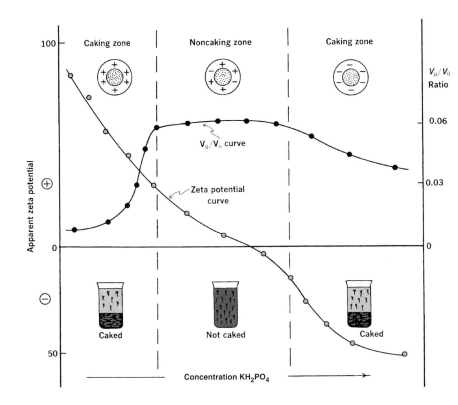

**Fig. 17–3.** Caking diagram, showing the flocculation of a bismuth subnitrate suspension by means of the flocculating agent monobasic potassium phosphate. (From A. Martin and J. Swarbrick, in *Sprowls' American Pharmacy*, 6th Ed., Lippincott, Philadelphia, 1966, p. 205. With permission.)

hygroscopic substances are also valuable in levigating the insoluble material. Apparently, glycerin flows into the voids between the particles to displace the air and, during the mixing operation, coats and separates the material so that water can penetrate and wet the individual particles. The dispersion of particles of colloidal gums by alcohol, glycerin, and propylene glycol, allowing water to subsequently penetrate the interstices, is a well-known practice in pharmacy.

To select suitable wetting agents that possess a well-developed ability to penetrate the powder mass, Hiestand[7] used a narrow trough, several inches long and made of a hydrophobic material, such as Teflon, or coated with paraffin wax. At one end of the trough is placed the powder and at the other end the solution of the wetting agent. The rate of penetration of the latter into the powder can then be observed directly.

## Controlled Flocculation

Assuming that the powder is properly wetted and dispersed, we can now consider the various means by which controlled flocculation can be produced so as to prevent formation of a compact sediment that is difficult to redisperse. The topic, described in detail by Hiestand,[7] is conveniently discussed in terms of the materials used to produce flocculation in suspensions, namely electrolytes, surfactants, and polymers.

*Electrolytes* act as flocculating agents by reducing the electric barrier between the particles, as evidenced by a decrease in the zeta potential and the formation of a bridge between adjacent particles so as to link them together in a loosely arranged structure.

If we disperse particles of bismuth subnitrate in water, we find that, based on electrophoretic mobility studies, they possess a large positive charge, or zeta potential. Because of the strong forces of repulsion between adjacent particles, the system is peptized or deflocculated. By preparing a series of bismuth subnitrate suspensions containing increasing concentrations of monobasic potassium phosphate, Haines and Martin[10] were able to show a correlation between apparent zeta potential and sedimentation volume, caking, and flocculation. The results are summarized in **Figure 17–3** and are explained in the following manner.

The addition of monobasic potassium phosphate to the suspended bismuth subnitrate particles causes the positive zeta potential to decrease owing to the adsorption of the negatively charged phosphate anion. With the continued addition of the electrolyte, the zeta potential eventually falls to zero and then increases in the negative direction, as shown in **Figure 17–3**. Microscopic examination of the various suspensions shows that at a certain positive zeta potential, maximum flocculation occurs and will persist until the zeta potential has become sufficiently negative for deflocculation to occur once again. The onset of flocculation coincides with the maximum sedimentation volume determined. *F* remains reasonably constant while flocculation persists, and only when the zeta potential becomes sufficiently negative to effect repeptization does the sedimentation volume start to fall. Finally, the absence of caking in the suspensions correlates with the maximum sedimentation volume, which, as stated previously,

## KEY CONCEPT WHAT IS A POLYMER?

*Polymers* are long-chain, high–molecular-weight compounds containing active groups spaced along their length. These agents act as flocculating agents because part of the chain is adsorbed on the particle surface, with the remaining parts projecting out into the dispersion medium. Bridging between these latter portions leads to the formation of flocs.

reflects the amount of flocculation. At less than maximum values of $F$, caking becomes apparent.

These workers[10] also demonstrated a similar correlation when aluminum chloride was added to a suspension of sulfamerazine in water. In this system, the initial zeta potential of the sulfamerazine particles is negative and is progressively reduced by adsorption of the trivalent aluminum cation. When sufficient electrolyte is added, the zeta potential reaches zero and then increases in a positive direction. Colloidal and coarse dispersed particles can possess surface charges that depend on the pH of the system. An important property of the pH-dependent dispersions is the zero point of charge, that is, the pH at which the net surface charge is zero. The desired surface charge can be achieved through adjusting the pH by the addition of HCl or NaOH to produce a positive, zero, or negative surface charge. The negative zeta potential of nitrofurantoin decreases considerably when the pH values of the suspension are charged from basic to acidic.[11]

*Surfactants*, both ionic and nonionic, have been used to bring about flocculation of suspended particles. The concentration necessary to achieve this effect would appear to be critical because these compounds can also act as wetting and deflocculating agents to achieve dispersion.

Felmeister and others[12] studied the influence of a xanthan gum (an anionic heteropolysaccharide) on the flocculation characteristics of sulfaguanidine, bismuth subcarbonate, and other drugs in suspension. Addition of xanthan gum resulted in increased sedimentation volume, presumably by a polymer-bridging phenomenon. Hiestand[13] reviewed the control of floc structure in coarse suspensions by the addition of polymeric materials.

Hydrophilic polymers also act as protective colloids, and particles coated in this manner are less prone to cake than are uncoated particles. These polymers exhibit pseudoplastic flow in solution, and this property serves to promote physical stability within the suspension. Gelatin, a polyelectrolytic polymer, exhibits flocculation that depends on the pH and ionic strength of the dispersion medium. Sodium sulfathiazole, precipitated from acid solution in the presence of gelatin, was shown by Blythe[14] to be free-flowing in the dry state and not to cake when suspended. Sulfathiazole normally carries a negative charge in aqueous vehicles. The coated material, precipitated from acid solution in the presence of gelatin, however, was found to carry a positive charge. This is due to gelatin being positively charged at the pH at which precipitation was carried out. It has been suggested[8] that the improved properties result from the positively charged gelatin-coated particles being partially flocculated in suspension, presumably because the high negative charge has been replaced by a smaller, albeit positive, charge. Positively charged liposomes have been used as flocculating agents to prevent caking of negatively charged particles. Liposomes are vesicles of phospholipids having no toxicity and that can be prepared in various particle sizes.[15] They are adsorbed on the negatively charged particles.

## Flocculation in Structured Vehicles

Although the controlled flocculation approach is capable of fulfilling the desired physical chemical requisites of a pharmaceutical suspension, the product can look unsightly if $F$, the sedimentation volume, is not close or equal to 1. Consequently, in practice, a suspending agent is frequently added to retard sedimentation of the flocs. Such agents as carboxymethylcellulose, Carbopol 934, Veegum, tragacanth, and bentonite have been employed, either alone or in combination.

This can lead to incompatibilities, depending on the initial particle charge and the charge carried by the flocculating agent and the suspending agent. For example, suppose we prepare a dispersion of positively charged particles that is then flocculated by the addition of the correct concentration of an anionic electrolyte such as monobasic potassium phosphate. We can improve the physical stability of this system by adding a minimal amount of one of the hydrocolloids just mentioned. No physical incompatibility will be observed because the majority of hydrophilic colloids are themselves negatively charged and are thus compatible with anionic flocculating agents. If, however, we flocculate a suspension of negatively charged particles with a cationic electrolyte (aluminum chloride), the subsequent addition of a hydrocolloid may result in an incompatible product, as evidenced by the formation of an unsightly stringy mass that has little or no suspending action and itself settles rapidly.

Under these circumstances, it becomes necessary to use a protective colloid to change the sign on the particle from negative to positive. This is achieved by the adsorption onto the particle surface of a fatty acid amine (which has been checked to ensure its nontoxicity) or a material such as gelatin, which is positively charged below its isoelectric point. We are then able to use an anionic electrolyte to produce flocs that are compatible with the negatively charged suspending agent.

This approach can be used regardless of the charge on the particle. The sequence of events is depicted in **Figure 17–4**, which is self-explanatory.

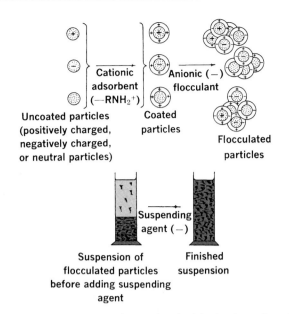

**Fig. 17–4.** The sequence of steps involved in the formation of a stable suspension. (From A. Martin and J. Swarbrick, in *Sprowls' American Pharmacy*, 6th Ed., Lippincott, Philadelphia, 1966, p. 206. With permission.)

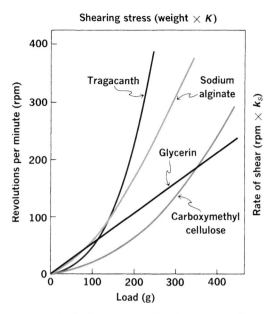

**Fig. 17–5.** Rheologic flow curves of various suspending agents analyzed in a modified Stormer viscometer.

## Rheologic Considerations

The principles of rheology can be applied to a study of the following factors: the viscosity of a suspension as it affects the settling of dispersed particles, the change in flow properties of the suspension when the container is shaken and when the product is poured from the bottle, and the spreading qualities of the lotion when it is applied to an affected area. Rheologic considerations are also important in the manufacture of suspensions.

The only shear that occurs in a suspension in storage is due to a settling of the suspended particles; this force is negligible and may be disregarded. When the container is shaken and the product is poured from the bottle, however, a high shearing rate is manifested. As suggested by Mervine and Chase,[16] the ideal suspending agent should have a *high* viscosity at negligible shear, that is, during shelf storage; and it should have a *low* viscosity at high shearing rates, that is, it should be free-flowing during agitation, pouring, and spreading. As seen in **Figure 17–5**, pseudoplastic substances such as tragacanth, sodium alginate, and sodium carboxymethylcellulose show these desirable qualities. The Newtonian liquid glycerin is included in the graph for comparison. Its viscosity is suitable for suspending particles but is too high to pour easily and to spread on the skin. Furthermore, glycerin shows the undesirable property of tackiness (stickiness) and is too hygroscopic to use in undiluted form. The curves in **Figure 17–5** were obtained by use of the modified Stormer viscometer.

A suspending agent that is thixotropic as well as pseudoplastic should prove to be useful because it forms a gel on standing and becomes fluid when disturbed. **Figure 17–6** shows the consistency curves for bentonite, Veegum

(Vanderbilt Co.), and a combination of bentonite and sodium carboxymethylcellulose. The hysteresis loop of bentonite is quite marked. Veegum also shows considerable thixotropy, both when tested by inverting a vessel containing the dispersion and when analyzed in a rotational viscometer. When bentonite and carboxymethylcellulose dispersions are mixed, the resulting curve shows both pseudoplastic and thixotropic characteristics. Such a combination should produce an excellent suspending medium.

## Preparation of Suspensions

The factors entering into the preparation and stabilization of suspensions involve certain principles of interest to physical pharmacy and are briefly discussed here. The physical

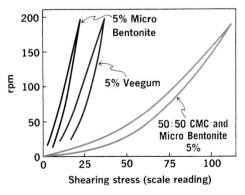

**Fig. 17–6.** Flow curves for 5% suspending agents in water, showing thixotropy. The curves were obtained with the Ferranti–Shirley cone–plate viscometer.

principles involved in the dispersion of solids by different types of equipment were discussed by Oldshue.[17]

A suspension is prepared on the small scale by grinding or levigating the insoluble material in the mortar to a smooth paste with a vehicle containing the dispersion stabilizer and gradually adding the remainder of the liquid phase in which any soluble drugs may be dissolved. The slurry is transferred to a graduate, the mortar is rinsed with successive portions of the vehicle, and the dispersion is finally brought to the final volume.

On a large scale, dispersion of solids in liquids is accomplished by the use of ball, pebble, and colloid mills. Dough mixers, pony mixers, and similar apparatus are also employed. Only the colloid mill is described here; a discussion of the other mills can be found in the book by Fischer.[18a] Dry grinding in ball mills is treated by Fischer,[18a] Berry and Kamack,[18b] and Prasher.[18c]

The colloid mill is based on the principle of a high-velocity, cone-shaped rotor that is centered with respect to a stator at a small adjustable clearance. The suspension is fed to the rotor by gravity through a hopper, sheared between the rotor and the stator, and forced out below the stator, where it may be recycled or drawn off.

The efficiency of the mill is based on the clearance between the disks, the peripheral velocity of the rotor, and the non-Newtonian viscosity of the suspension. The mill breaks down the large aggregates and flocs so that they can be dispersed throughout the liquid vehicle and then protected by the dispersion stabilizer. The shearing action that leads to disaggregation occurs at the surfaces of the rotating and stationary disks and between the particles themselves in a concentrated suspension. If the yield value is too great, the material fails to flow; if the viscosity is low, a loss in effectiveness of shearing action occurs. Therefore, the yield value should be low, and the plastic or apparent viscosity of the material should be at a maximum consistent with the optimum rate of flow through the mill. If the material is highly viscous or if the plates are adjusted to a clearance that is too narrow, the temperature rises rapidly, and cooling water must be circulated around the stator to dissipate the heat that is produced. Dilatant materials—for example, deflocculated suspensions containing 50% or more of solids—are particularly troublesome. They flow freely into the mill but set up a high shearing rate and produce overheating and stalling of the motor. Beginning any milling process with the plates set at a wide clearance minimizes this danger. If this technique fails, however, the material must be milled in another type of equipment or the paste must be diluted with a vehicle until dilatancy is eliminated.

## Physical Stability of Suspensions

Raising the temperature often leads to flocculation of *sterically stabilized* suspensions, that is, suspensions stabilized by nonionic surfactants. Repulsion due to steric interactions depends on the nature, thickness, and completeness of the surfactant-adsorbed layers on the particles. When the suspension is heated, the energy of repulsion between the particles can be reduced owing to dehydration of the polyoxyethylene groups of the surfactant. The attractive energy is increased and the particles flocculate.[19] Zapata et al.[20] studied the mechanism of freeze–thaw instability in aluminum hydrocarbonate and magnesium hydroxide gels as model suspensions because of their well-known sensitivity to temperature changes. During the freezing process, particles can overcome the repulsive barrier caused by ice formation, which forces the particles close enough to experience the strong attractive forces present in the primary minimum and form aggregates according to the DLVO theory. When the ice melts, the particles remain as aggregates unless work is applied to overcome the primary energy peak. Aggregate size was found to be inversely related to the freezing rate. The higher the freezing rate, the smaller is the size of ice crystals formed. These small crystals do not result in the aggregation of as many suspension particles as do large ice crystals.

In addition to particle aggregation, particle growth is also a destabilizing process resulting from temperature fluctuations or *Ostwald ripening* during storage. Fluctuations of temperature can change the particle size distribution and polymorphic form of a drug, altering the absorption rate and drug bioavailability.[21] Particle growth is particularly important when the solubility of the drug is strongly dependent on the temperature. Thus, when temperature is raised, crystals of drug may dissolve and form supersaturated solutions, which favor crystal growth. This can be prevented by the addition of polymers or surfactants. Simonelli et al.[3] studied the inhibition of sulfathiazole crystal growth by polyvinylpyrrolidone. These authors suggested that the polymer forms a noncondensed netlike film over the sulfathiazole crystal, allowing the crystal to grow out only through the openings of the net. The growth is thus controlled by the pore size of the polymer network at the crystal surface. The smaller the pore size, the higher is the supersaturation of the solution required for the crystals to grow. This can be shown using the Kelvin equation as applied to a particle suspended in a saturated solution[3]:

$$\ln \frac{c}{c_0} = \frac{2\gamma M}{NkT\rho R} \tag{17–8}$$

where $c$ is the solubility of a small particle of radius $R$ in an aqueous vehicle and $c_0$ is the solubility of a very large crystalline particle; $\gamma$ is the interfacial tension of the crystal, $\rho$ is the density of the crystal, and $M$ is the molecular weight of the solute. $N$ is Avogadro's number, $k$ is the Boltzmann constant, and $N \times k = 8.314 \times 10^7$ ergs$^{-1}$ mole$^{-1}$. The ratio $c/c_0$ defines the supersaturation ratio that a large crystal requires in the aqueous solution saturated with respect to the small particle. According to equation (17–8), as the radius of curvature of a protruding crystal decreases, the protrusion will require a correspondingly larger supersaturation ratio before it can grow. The radius of curvature of a protrusion must equal that of the pore of the polymer on the crystal surface.

**EXAMPLE 17–4**

**Supersaturation Ratio**

Assume that the interfacial tension of a particle of drug in an aqueous vehicle is 100 ergs/cm$^2$, its molecular weight is 200 g/mole, and the temperature of the solution is 30°C or 303 K. (*a*) Compute the supersaturation ratio, $c/c_0$, that is required for the crystal to grow. The radius, $R$, of the particle is 5 $\mu$m, or $5 \times 10^{-4}$ cm, and its density is 1.3 g/cm$^3$. (*b*) Compute the supersaturation ratio when the particle is covered by a polymer and the pore radius, $R$, of the polymer at the crystal surface is $6 \times 10^{-7}$ cm.

Using the Kelvin equation, we obtain

(*a*)

$$\ln \frac{c}{c_0} = \frac{2 \times 100 \times 200}{8.314 \times 10^7 \times 1.3 \times 303 \times 5 \times 10^{-4}} = 0.0024$$

(*b*)

$$\ln \frac{c}{c_0} = \frac{2 \times 100 \times 200}{8.314 \times 10^7 \times 1.3 \times 303 \times 6 \times 10^{-7}} = 2.036$$

$$c/c_0 = \text{antiln}(2.036) = 7.66$$

Notice that $c/c_0$ in part (*a*) represents slight oversaturation, whereas in (*b*) the supersaturation concentration must be 7.6 times larger than the solubility of the drug molecule for the crystalline particle to grow. In other words, the addition of a polymer greatly increases the point at which supersaturation occurs and makes it more difficult for the drug crystal to grow.

Ziller and Rupprecht[22] designed a control unit to monitor crystal growth and studied the inhibition of growth by poly (vinylpyrrolidone) (PVP) in acetaminophen suspensions. According to these workers, some of the segments of the polymer PVP attach to the free spaces on the drug crystal lattice and the polymer is surrounded by a hydration shell (**Fig. 17–7**). The adsorbed segments of the polymer inhibit crystal growth of acetaminophen because they form a barrier that impedes the approach of the drug molecules from the solution to the crystal surface. High–molecular-weight polymers of PVP are more effective than low–molecular-weight polymers because the adsorption of the polymer on the crystal surface becomes more irreversible as the chain length increases.

The stability of suspensions may also decrease owing to interaction with excipients dissolved in the dispersion medium. Zatz and Lue[19] studied the flocculation by sorbitol in sulfamerazine suspensions containing nonionic surfactants as wetting agents. The flocculation by sorbitol depends on the cloud point of the surfactant. Thus, the lower the cloud point, the less sorbitol was needed to induce flocculation. The fact that the cloud point can be lowered by preservatives such as methylparaben shows that the choice of additives may change the resistance to caking of a suspension containing nonionic surfactants. Zatz and Lue[19] suggested that the cloud point can be used to estimate the critical flocculation concentration of sorbitol. Lucks et al.[23] studied the adsorption of preservatives such as cetylpyridinium chloride on zinc oxide particles in suspension. Increasing amounts of this preservative led to charge reversal of the suspension. Cetylpyridinium chloride, a cationic surfactant, has a posi-

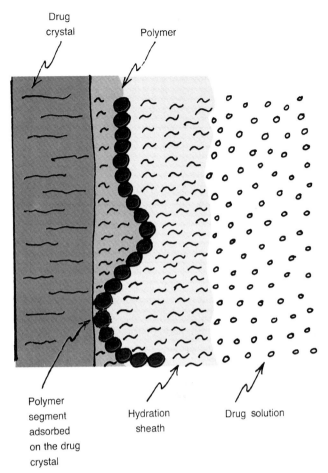

**Fig. 17–7.** Dissolution and crystallization of a drug in the presence of a polymer adsorbed on the drug crystal. (From H. K. Ziller and H. Rupprecht, Drug Dev. Ind. Pharm. **14**, 2341, 1988. With permission.)

tive charge and is strongly adsorbed at the particle surface. The positive end of the preservative molecule adsorbs on the negatively charged surface of the zinc oxide particles, forming a layer with the hydrocarbon chains oriented outward toward the dispersion medium. A second layer of preservative adsorbs at this monolayer, with the positively charged groups now directed toward the dispersion medium. Thus, the physical stability of the suspension may be enhanced owing to the repulsion of like-charged particles. However, the strong adsorption of the preservative on the zinc oxide particles reduces the biologically active free fraction of preservative in the dispersion medium, and the microbiologic activity is diminished.

# EMULSIONS

An emulsion is a thermodynamically unstable system consisting of at least two immiscible liquid phases, one of which is dispersed as globules (the dispersed phase) in the other liquid phase (the continuous phase), stabilized by the presence of an *emulsifying agent*. The various types of emulsifying agents are discussed later in this section. Either the dispersed phase or the continuous phase may range in consistency from that of a mobile liquid to a semisolid. Thus, emulsified systems range from lotions of relatively low viscosity to ointments and creams, which are semisolid in nature. The particle diameter of the dispersed phase generally extends from about 0.1 to 10 $\mu$m, although particle diameters as small as 0.01 $\mu$m and as large as 100 $\mu$m are not uncommon in some preparations.

## Emulsion Types

Invariably, one liquid phase in an emulsion is essentially polar (e.g., aqueous), whereas the other is relatively nonpolar (e.g., an oil). When the oil phase is dispersed as globules throughout an aqueous continuous phase, the system is referred to as an *oil-in-water* (o/w) emulsion. When the oil phase serves as the continuous phase, the emulsion is spoken of as a *water-in-oil* (w/o) product. Medicinal emulsions for oral administration are usually of the o/w type and require the use of an o/w emulsifying agent. These include synthetic nonionic surfactants, acacia, tragacanth, and gelatin. Not all emulsions that are consumed, however, belong to the o/w type. Certain foods such as butter and some salad dressings are w/o emulsions.

Externally applied emulsions may be o/w or w/o, the former employing the following emulsifiers in addition to the ones mentioned previously: sodium lauryl sulfate, triethanolamine stearate, monovalent soaps such as sodium oleate, and self-emulsifying glyceryl monostearate, that is, glyceryl monostearate mixed with a small amount of a monovalent soap or an alkyl sulfate. Pharmaceutical w/o emulsions are used almost exclusively for external application and may contain one or several of the following emulsifiers: polyvalent soaps such as calcium palmitate, sorbitan esters (Spans), cholesterol, and wool fat.

Several methods are commonly used to determine the type of an emulsion. A small quantity of a water-soluble dye such as methylene blue or brilliant blue FCF may be dusted on the surface of the emulsion. If water is the external phase (i.e., if the emulsion is of the o/w type), the dye will dissolve and uniformly diffuse throughout the water. If the emulsion is of the w/o type, the particles of dye will lie in clumps on the surface. A second method involves dilution of the emulsion with water. If the emulsion mixes freely with the water, it is of the o/w type. Another test uses a pair of electrodes connected to an external electric source and immersed in the emulsion. If the external phase is water, a current will pass through the emulsion and can be made to deflect a voltmeter needle or cause a light in the circuit to glow. If the oil is the continuous phase, the emulsion fails to carry the current.

## Pharmaceutical Applications

An o/w emulsion is a convenient means of orally administering water-insoluble liquids, especially when the dispersed phase has an unpleasant taste. More significant in contemporary pharmacy is the observation that some oil-soluble compounds, such as some vitamins, are absorbed more completely when emulsified than when administered orally as an oily solution. The use of intravenous emulsions has been studied as a means of maintaining debilitated patients who are unable to assimilate materials administered orally. Tarr et al.[24] prepared emulsions of taxol, a compound with antimitotic properties, for intravenous administration as an alternative method to the use of cosolvents in taxol administration. Davis and Hansrani[25] studied the influence of droplet size and emulsifying agents on the phagocytosis of lipid emulsions. When the emulsion is administered intravenously, the droplets are normally rapidly taken up by the cells of the reticuloendothelial system, in particular the fixed macrophages in the liver. The rate of clearance by the macrophages increases as the droplet size becomes larger or the surface charge, either positive or negative, increases. Therefore, emulsion droplets stabilized by a nonionic surfactant (zero surface charge) were cleared much more slowly than the droplets stabilized by negatively charged phospholipids. Radiopaque emulsions have found application as diagnostic agents in x-ray examinations.

Emulsification is widely used in pharmaceutical and cosmetic products for external use. This is particularly so with dermatologic and cosmetic lotions and creams because a product that spreads easily and completely over the affected area is desired. Such products can now be formulated to be water washable and nonstaining and, as such, are obviously more acceptable to the patient and the physician than some of the greasy products used a decade or more ago. Emulsification is used in aerosol products to produce foams. The propellant that forms the dispersed liquid phase within the container vaporizes when the emulsion is discharged from the container. This results in the rapid formation of a foam.

# THEORIES OF EMULSIFICATION

There is no universal theory of emulsification because emulsions can be prepared using several different types of emulsifying agent, each of which depends for its action on a different principle to achieve a stable product. For a theory to be meaningful, it should be capable of explaining (*a*) the stability of the product and (*b*) the type of emulsion formed. Let us consider what happens when two immiscible liquids are agitated together so that one of the liquids is dispersed as small droplets in the other. Except in the case of very dilute oil-in-water emulsions (oil hydrosols), which are somewhat stable, the liquids separate rapidly into two clearly defined layers. Failure of two immiscible liquids to remain mixed is explained by the fact that the *cohesive* force between the molecules of each separate liquid is greater than the *adhesive* force between the two liquids. The cohesive force of the individual phases is manifested as an interfacial energy or tension at the boundary between the liquids, as explained in Chapter 15.

When one liquid is broken into small particles, the interfacial area of the globules constitutes a surface that is enormous

**TABLE 17–1**
## SOME TYPICAL EMULSIFYING AGENTS*

| Name | Class | Type of Emulsion Formed |
|---|---|---|
| Triethanolamine oleate | Surface-active agent (anionic) | o/w (HLB = 12) |
| N-cetyl N-ethyl morpholinium ethosulfate (Atlas G-263) | Surface-active agent (cationic) | o/w (HLB = 25) |
| Sorbitan monooleate (Atlas Span 80) | Surface-active agent (nonionic) | w/o (HLB = 4.3) |
| Polyoxyethylene sorbitan monooleate (Atlas Tween 80) | Surface-active agent (nonionic) | o/w (HLB = 15) |
| Acacia (salts of d-glucuronic acid) | Hydrophilic colloid | o/w |
| Gelatin (polypeptides and amino acids) | Hydrophilic colloid | o/w |
| Bentonite (hydrated aluminum silicate) | Solid particle | o/w (and w/o) |
| Veegum (magnesium aluminum silicate) | Solid particle | o/w |
| Carbon black | Solid particle | w/o |

*Key: o/w = oil in water; w/o = water in oil; HLB = hydrophilic–lipophilic balance value.

compared with the surface area of the original liquid. If 1 cm$^3$ of mineral oil is dispersed into globules having a volume–surface diameter, $d_{vs}$ of 0.01 $\mu$m ($10^{-6}$ cm) in 1 cm$^3$ of water so as to form a fine emulsion, the surface area of the oil droplets becomes 600 m$^2$. The surface free energy associated with this area is about $34 \times 10^7$ ergs, or 8 calories. The total volume of the system, however, has not increased; it remains at 2 cm$^3$. The calculations are made by use of equations (18–15) and (18–17) from which

$$S_v = \frac{6}{d_{vs}}$$

$$S_v = \frac{6}{10^{-6}} = 6 \times 10^6 \text{ cm}^2 = 600 \text{ m}^2$$

The work input or surface free energy increase is given by the equation $W = \gamma_{ow} \times \Delta A$, and the interfacial tension, $\gamma_{ow}$, between mineral oil and water is 57 dynes/cm (erg/cm$^2$). Thus,

$$W = 57 \text{ ergs/cm}^2 \times (6 \times 10^6 \text{ cm}^2)$$

$$= 34 \times 10^7 \text{ ergs} = 34 \text{ joules}$$

and because 1 cal = 4.184 joules,

$$34 \text{ joules } 4.184 = 8 \text{ calories}$$

In summary, if 1 cm$^3$ of mineral oil is mixed with 1 cm$^3$ of water to produce fine particles ($d_{vs} = 0.01$ $\mu$m), the total surface is equivalent to an area slightly greater than that of a basketball court, or about 600 m$^2$. (In real emulsions, the particles are ordinarily about 10 to 100 times larger than this, and the surface area is proportionately smaller.) The increase in energy, 8 calories, associated with this enormous surface is sufficient to make the system thermodynamically unstable, hence the droplets have a tendency to coalesce.

To prevent coalescence or at least to reduce its rate to negligible proportions, it is necessary to introduce an emulsifying agent that will form a film around the dispersed globules.

Emulsifying agents can be divided into three groups, as follows:

(a) Surface-active agents, which are adsorbed at oil–water interfaces to form monomolecular films and reduce interfacial tension. These agents are discussed in detail in Chapter 15, dealing with interfacial phenomena.

(b) Hydrophilic colloids (discussed in Chapter 16), which form a *multi*molecular film around the dispersed droplets of oil in an o/w emulsion.[26,27]

(c) Finely divided solid particles, which are adsorbed at the interface between two immiscible liquid phases and form what amounts to a film of particles around the dispersed globules. The factor common to all three classes of emulsifying agent is the formation of a film, whether it be monomolecular, multimolecular, or particulate.

On this basis, we can now discuss some of the more important theories relating to the stability and type of emulsion formed.

Examples of typical emulsifying agents are given in **Table 17–1**.

## Monomolecular Adsorption

Surface-active agents, or amphiphiles, reduce interfacial tension because of their adsorption at the oil–water interface to form monomolecular films. Because the surface free energy increase, $W$, equals $\gamma_{o/w} \times \Delta A$ and we must, of necessity, retain a high surface area for the dispersed phase, any reduction in $\gamma_{o/w}$, the interfacial tension, will reduce the surface free energy and hence the tendency for coalescence. It is not unusual for a good emulsifying agent of this type to reduce the interfacial tension to 1 dyne/cm; we can therefore reduce the surface free energy of the system to approximately 1/60 of that calculated earlier.

The reduction in surface free energy is of itself probably not the main factor involved. Of more likely significance is

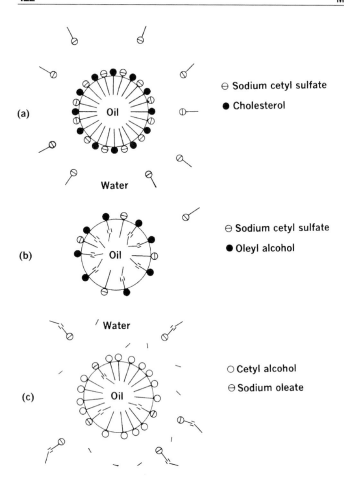

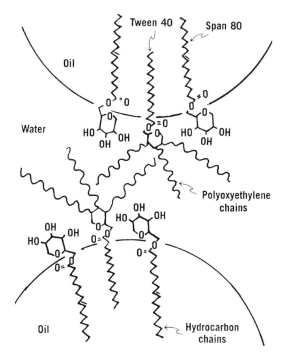

**Fig. 17–8.** Representations of combinations of emulsifying agents at the oil–water interface of an emulsion. (After J. H. Schulman and E. G. Cockbain, Trans. Faraday Soc. **36**, 651, 1940.)

**Fig. 17–9.** Schematic of oil droplets in an oil–water emulsion, showing the orientation of a Tween and a Span molecule at the interface. (From J. Boyd, C. Parkinson, and P. Sherman, J. Coll. Interface Sci. **41**, 359, 1972. With permission.)

the fact that the dispersed droplets are surrounded by a coherent monolayer that helps prevent coalescence between two droplets as they approach one another. Ideally, such a film should be flexible so that it is capable of reforming rapidly if broken or disturbed. An additional effect promoting stability is the presence of a surface charge, which will cause repulsion between adjacent particles.

In practice, combinations of emulsifiers rather than single agents are used most frequently today in the preparations of emulsions. In 1940, Schulman and Cockbain[28] first recognized the necessity of a predominantly hydrophilic emulsifier in the aqueous phase and a hydrophobic agent in the oil phase to form a complex film at the interface. Three mixtures of emulsifying agents at the oil–water interface are depicted in **Figure 17–8**. The combination of sodium cetyl sulfate and cholesterol leads to a complex film (**Fig. 17–8a**) that produces an excellent emulsion. Sodium cetyl sulfate and oleyl alcohol do not form a closely packed or condensed film (**Fig. 17–8b**), and, consequently, their combination results in a poor emulsion. In **Figure 17–8c**, cetyl alcohol and sodium oleate produce a close-packed film, but complexation is negligible, and again a poor emulsion results.

A hydrophilic Tween can be combined with a lipophilic Span, varying the proportions so as to produce the desired o/w or w/o emulsion.[29] Boyd et al.[30] discussed the molecular association of Tween 40 and Span 80 in stabilizing emulsions. In **Figure 17–9**, the hydrocarbon portion of the Span 80 (sorbitan monooleate) molecule lies in the oil globule and the sorbitan radical lies in the aqueous phase. The bulky sorbitan heads of the Span molecules prevent the hydrocarbon tails from associating closely in the oil phase. When Tween 40 (polyoxyethylene sorbitan monopalmitate) is added, it orients at the interface such that part of its hydrocarbon tail is in the oil phase and the remainder of the chain, together with the sorbitan ring and the polyoxyethylene chains, is located in the water phase. It is observed that the hydrocarbon chain of the Tween 40 molecule is situated in the oil globule between the Span 80 chains, and this orientation results in effective van der Waals attraction. In this manner, the interfacial film is strengthened and the stability of the o/w emulsion is increased against particle coalescence. The same principle of mixed emulsifying agents can be applied in the use of combinations such as sodium stearate and cholesterol, sodium lauryl sulfate and glyceryl monostearate, and tragacanth and Span. Chun et al.[31] determined the hydrophile–lipophile balance (HLB) of some natural agents and further discussed the principle of mixed emulsifiers.

The type of emulsion that is produced, o/w or w/o, depends primarily on the property of the emulsifying agent. This characteristic is referred to as the *hydrophile–lipophile* balance, that is, the polar–nonpolar nature of the emulsifier. In fact, whether a surfactant is an emulsifier, wetting agent, detergent, or solubilizing agent can be predicted from a knowledge of the HLB, as discussed in a previous chapter. In an emulsifying agent such as sodium stearate, $C_{17}H_{35}COONa$, the nonpolar hydrocarbon chain, $C_{17}H_{35}-$, is the lipophilic or "oil-loving" group; the carboxyl group, $-COONa$, is the *hydrophilic* or "water-loving" portion. The balance of the hydrophilic and lipophilic properties of an emulsifier (or combination of emulsifiers) determines whether an o/w or w/o emulsion will result. In general, o/w emulsions are formed when the HLB of the emulsifier is within the range of about 9 to 12, and w/o emulsions are formed when the range is about 3 to 6. An emulsifier with a high HLB, such as a blend of Tween 20 and Span 20, will form an o/w emulsion. On the other hand, Span 60 alone, having an HLB of 4.7, tends to form a w/o emulsion.

It would appear, therefore, that the type of emulsion is a function of the relative solubility of the surfactant, the phase in which it is more soluble being the continuous phase. This is sometimes referred to as the *rule of Bancroft*, who observed this phenomenon in 1913. Thus, an emulsifying agent with a high HLB is preferentially soluble in water and results in the formation of an o/w emulsion. The reverse situation is true with surfactants of low HLB, which tend to form w/o emulsions. Beerbower, Nixon, and Hill[32] suggested an explanation for emulsion type and stability and devised a general scheme for emulsion formulation based on the Hildebrand and Hansen solubility parameters.

## Multimolecular Adsorption and Film Formation

Hydrated lyophilic colloids have been used for many years as emulsifying agents, although their use is declining because of the large number of synthetic surfactants now available. In a sense, they can be regarded as surface active because they appear at the oil–water interface. They differ, however, from the synthetic surface-active agents in that (*a*) they do not cause an appreciable lowering of interfacial tension and (*b*) they form a multi- rather than a monomolecular film at the interface. Their action as emulsifying agents is due mainly to the latter effect because the films thus formed are strong and resist coalescence. An auxiliary effect promoting stability is the significant increase in the viscosity of the dispersion medium. Because the emulsifying agents that form multilayer films around the droplets are invariably hydrophilic, they tend to promote the formation of o/w emulsions.

## Solid-Particle Adsorption

Finely divided solid particles that are wetted to some degree by both oil and water can act as emulsifying agents. This results from their being concentrated at the interface, where they produce a particulate film around the dispersed droplets so as to prevent coalescence. Powders that are wetted preferentially by water form o/w emulsions, whereas those more easily wetted by oil form w/o emulsions.

# PHYSICAL STABILITY OF EMULSIONS

Probably the most important consideration with respect to pharmaceutical and cosmetic emulsions is the stability of the finished product. The stability of a pharmaceutical emulsion is characterized by the absence of coalescence of the internal phase, absence of creaming, and maintenance of elegance with respect to appearance, odor, color, and other physical properties. Some workers define instability of an emulsion only in terms of agglomeration of the internal phase and its separation from the product. Creaming, resulting from flocculation and concentration of the globules of the internal phase, sometimes is not considered as a mark of instability. An emulsion is a dynamic system, however, and flocculation and resultant creaming represent potential steps toward complete coalescence of the internal phase. Furthermore, in the case of pharmaceutical emulsions, creaming results in a lack of uniformity of drug distribution and, unless the preparation is thoroughly shaken before administration, leads to variable dosage. Certainly, the visual appeal of an emulsion is affected by creaming, and this is just as real a problem to the pharmaceutical compounder as is separation of the internal phase.

Another phenomenon important in the preparation and stabilization of emulsions is *phase inversion*, which can be an aid or a detriment in emulsion technology. Phase inversion involves the change of emulsion type from o/w to w/o or vice versa. Should phase inversion occur following preparation, it may logically be considered as an instance of instability.

In the light of these considerations, the instability of pharmaceutical emulsions may be classified as follows:

(*a*) Flocculation and creaming
(*b*) Coalescence and breaking
(*c*) Miscellaneous physical and chemical changes
(*d*) Phase inversion

## Creaming and Stokes's Law

Those factors that find importance in the creaming of an emulsion are related by Stokes's law, equation (**17–2**). The limitations of this equation to actual systems have been discussed previously for suspensions, and these apply equally to emulsified systems.

Analysis of the equation shows that if the dispersed phase is less dense than the continuous phase, which is generally the case in o/w emulsions, the velocity of sedimentation becomes negative, that is, an upward *creaming* results. If the internal phase is heavier than the external phase, the globules settle, a phenomenon customarily noted in w/o emulsions in which the internal aqueous phase is denser than the continuous oil phase. This effect can be referred to as *creaming in a downward direction*. The greater the difference between the density of the two phases, the larger the oil globules, and the less viscous the external phase, the greater is the rate of creaming.

By increasing the force of gravity through centrifugation, the rate of creaming can also be increased. The diameter of the globules is seen to be a major factor in determining the rate of creaming. Doubling the diameter of the oil globules increases the creaming rate by a factor of 4.

### EXAMPLE 17–5

**Velocity of Creaming**

Consider an o/w emulsion containing mineral oil with a specific gravity of 0.90 dispersed in an aqueous phase having a specific gravity of 1.05. If the oil particles have an average diameter of 5 $\mu$m, or $5 \times 10^{-4}$ cm, the external phase has a viscosity of 0.5 poise (0.5 dyne sec/$cm^2$ or 0.5 g/cm sec), and the gravity constant is 981 $cm/sec^2$, what is the velocity of creaming in cm/day?

We have

$$v = \frac{(5 \times 10^{-4})^2 \times (0.90 - 1.05) \times 981}{18 \times 0.5}$$

$$= -4.1 \times 10^{-6} \text{ cm/sec}$$

and because a 24-hr day contains 86,400 sec, the rate of upward creaming, $-v$, is

$$-v = 4.1 \times 10^{-6} \text{ cm/sec} \times 86,400 \text{ sec/day} = 0.35 \text{ cm/day}$$

The factors in Stokes's equation can be altered to reduce the rate of creaming in an emulsion. The viscosity of the external phase can be increased without exceeding the limits of acceptable consistency by adding a *viscosity improver* or *thickening agent* such as methylcellulose, tragacanth, or sodium alginate. The particle size of the globules can be reduced by homogenization; this, in fact, is the basis for the stability against creaming of homogenized milk. If the average particle size of the emulsion in the example just given is reduced to 1 $\mu$m, or one fifth of the original value, the rate of creaming is reduced to 0.014 cm/day or about 5 cm/year. Actually, when the particles are reduced to a diameter below 2 to 5 $\mu$m, Brownian motion at room temperature exerts sufficient influence so that the particles settle or cream more slowly than predicted by Stokes's law.

Little consideration has been given to the adjustment of densities of the two phases in an effort to reduce the rate of creaming. Theoretically, adjusting the external and internal phase densities to the same value should eliminate the tendency to cream. This condition is seldom realized, however, because temperature changes alter the densities. Some research workers have increased the density of the oil phase by the addition of oil-soluble substances such as $\alpha$-bromonaphthalene, bromoform, and carbon tetrachloride, which, however, cannot be used in medicinal products. Mullins and Becker[33] added a food grade of a brominated oil to adjust the densities in pharmaceutical emulsions.

Equation (17–2) gives the rate of creaming of a single droplet of the emulsion, whereas one is frequently interested in the rate of creaming at the center of gravity of the mass of the disperse phase. Greenwald[34] developed an equation for the mass creaming rate, to which the interested reader is referred for details.

## Coalescence and Breaking

Creaming should be considered as separate from breaking because creaming is a reversible process, whereas breaking is irreversible. The cream floccules can be redispersed easily, and a uniform mixture is reconstituted from a creamed emulsion by agitation because the oil globules are still surrounded by a protective sheath of emulsifying agent. When breaking occurs, simple mixing fails to resuspend the globules in a stable emulsified form because the film surrounding the particles has been destroyed and the oil tends to coalesce. Considerable work has been devoted to the study of breaking instability. The effects of certain factors on breaking are summarized in the following paragraphs.

King[35] showed that reduction of particle size does not necessarily lead to increased stability. Rather, he concluded that an optimum degree of dispersion for each particular system exists for maximum stability. As in the case of solid particles, if the dispersion is nonuniform, the small particles wedge between larger ones, permitting stronger cohesion so that the internal phase can coalesce easily. Accordingly, a moderately coarse dispersion of uniform-sized particles should have the best stability. Viscosity alone does not produce stable emulsions; however, viscous emulsions may be more stable than mobile ones by virtue of the retardation of flocculation and coalescence. Viscous or "tacky" emulsifiers seem to facilitate shearing of the globules as the emulsion is being prepared in the mortar, but this bears little or no relationship to stability. Knoechel and Wurster[36] showed that viscosity plays only a minor role in the gross stability of o/w emulsions. Probably an *optimum* rather than a *high* viscosity is needed to promote stability.

The *phase–volume* ratio of an emulsion has a secondary influence on the stability of the product. This term refers to the relative volumes of water and oil in the emulsion. As shown in the section on powders, uniform spherical particles in loose packing have a porosity of 48% of the total bulk volume. The volume occupied by the spheres must then be 52%.

If the spheres are arranged in closest packing, theoretically they cannot exceed 74% of the total volume regardless of their size. Although these values do not consider the distortions of size and shape and the possibility of small particles lying between larger spheres, they do have some significance with respect to real emulsions. Ostwald and Kolloid[37] showed that if one attempts to incorporate more than about 74% of oil in an o/w emulsion, the oil globules often coalesce and the emulsion breaks. This value, known as the *critical point*, is defined as the concentration of the internal phase above which the emulsifying agent cannot produce a stable emulsion of the desired type. In some stable emulsions, the value may be higher than 74% owing to the irregular shape and size of the globules. Generally speaking, however, a phase–volume ratio of 50:50 (which approximates loose packing) results in about the most stable emulsion. This fact was discovered empirically by pharmacists many years ago, and most

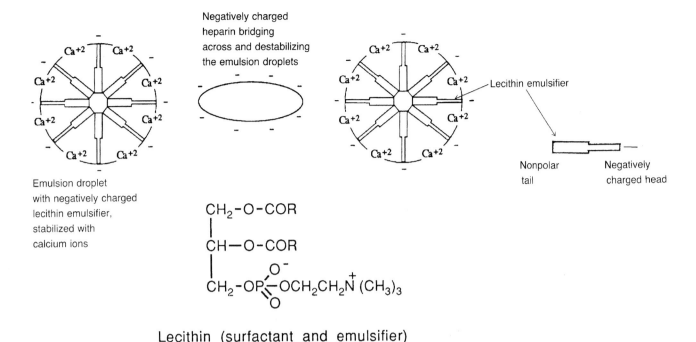

Lecithin (surfactant and emulsifier)

**Fig. 17–10.** Parental emulsion droplets in the presence of the negatively charged emulsifier lecithin and stabilized by electrostatic repulsion by calcium ions. The emulsion may be flocculated and destabilized by the bridging effect of heparin, a negatively charged polyelectrolyte, which overcomes the stabilizing electrostatic repulsion of the $Ca^{2+}$ ions. (From O. L. Johnson, C. Washington, S. S. Davis, and K. Schaupp, *Int. J. Pharm.* **53,** 237, 1989. With permission.)

medicinal emulsions are prepared with a volume ratio of 50 parts of oil to 50 parts of water.

Emulsions can be stabilized by electrostatic repulsion between the droplets, that is, by increasing their zeta potential. Magdassi and Siman-Tov[38] used lecithin to stabilize perfluorocarbon emulsions, which appear to be a good blood substitute. Lecithin is a mixture of phospholipids having a negative charge at physiologic pH. The stabilizing effect is due to the adsorption of lecithin at the droplet surface, which creates a negative charge and consequently electrostatic repulsion. Lecithin produces very stable emulsions of triglyceride acids in water for intravenous administration. However, the stability of these emulsions may be poor because in clinical practice they are mixed with electrolytes, amino acids, and other compounds for total parenteral nutrition. The addition of positively charged species such as sodium and calcium ions or cationic amino acids—the charge on the latter depending on the pH—reduces the zeta potential and may cause flocculation. Johnson et al.[39] studied the effect of heparin and various electrolytes, frequently used clinically, on the stability of parenteral emulsions. Heparin, an anticoagulant, is a negatively charged polyelectrolyte that causes rapid flocculation in emulsions containing calcium and lecithin. The critical flocculation concentration occurs at a specific zeta potential. The value of this zeta potential can be determined by plotting the flocculation rate against the surface potential and extrapolating to zero flocculation rate.[40] Johnson et al.[39] explained the destabilizing effect of heparin as follows. Divalent electrolytes such as calcium bind strongly to

the surface of droplets stabilized with lecithin to form 1:2 ion–lipid complexes. This causes a charge reversal on the droplets, leading to positively charged particles. The droplets are then flocculated by a bridging of the negatively charged heparin molecules across the positively charged particles, as depicted in **Figure 17–10.**

When the oil particles, which usually carry a negative charge, are surrounded in an o/w emulsion by a film of emulsifier, particularly a nonionic agent, the electrokinetic effects are probably less significant than they are in suspensions in maintaining the stability of the system. The effect of electrolytes in these systems has been studied by Schott and Royce.[41] Probably the most important factors in the stabilization of an emulsion are the physical properties of the emulsifier film at the interface. To be effective, an emulsifier film must be both tough and elastic and should form rapidly during emulsification. Serrallach et al.[42] measured the strength of the film at the interface. They found that a good emulsifying agent or emulsifier combination brings about a preliminary lowering of the interfacial tension to produce small uniform globules and forms rapidly to protect the globules from reaggregation during manufacture. The film then slowly increases in strength over a period of days or weeks.

## Evaluation of Stability

According to King and Mukherjee,[43] the only precise method for determining stability involves a size–frequency analysis of the emulsion from time to time as the product ages. For

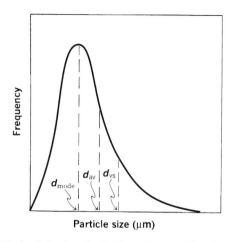

**Fig. 17–11.** Particle-size distribution of an emulsion. Such curves ordinarily are skewed to the right as shown in the figure, and the mode diameter, that is, the highest point on the curve or the most frequent value, is seen to occur at the lower end of the scale of diameters. The arithmetic mean diameter, $d_{av}$, will be found somewhat to the right of the mode in a right-skewed distribution, and the mean volume–surface diameter, $d_{vs}$, is to the right of the arithmetic mean.

rapidly breaking emulsions, macroscopic observation of separated internal phase is adequate, although the separation is difficult to read with any degree of accuracy. In the microscopic method, the particle diameters are measured, and a size–frequency distribution of particles ranging from 0.0 to 0.9 $\mu$m, 1.0 to 1.9 $\mu$m, 2.0 to 2.9 $\mu$m, and so on, is made as shown in **Figure 17–11**. The particle size or diameter of the globules in micrometers is plotted on the horizontal axis against the frequency or number of globules in each size range on the vertical axis. Finkle et al.[44] were probably the first workers to use this method to determine the stability of emulsions. Since that time, many similar studies have been made. Schott and Royce[45] showed that the experimental problems involved in microscopic size determinations are Brownian motion, creaming, and field flow. Brownian motion affects the smallest droplets, causing them to move in and out of focus so that they are not consistently counted. Velocity of creaming is proportional to the square of the droplet diameter, and creaming focuses attention on the largest droplets because they move faster toward the cover glass than do smaller ones. *Field flow* is the motion of the entire volume of emulsion in the field due to the pressure exerted by the immersion objective on the cover glass, evaporation of the continuous phase, or convection currents resulting from heating by the light source. These workers[45] described an improved microscopic technique that overcomes these experimental problems and gives a more accurate measure of the droplet size.

An initial frequency distribution analysis on an emulsion is not an adequate test of stability because stability is not related to initial particle size. Instead, one should perhaps consider the coalescence of the dispersed globules of an aging emulsion or the separation of the internal phase from the emulsion over a period of time. Boyd et al.,[30] however, deemed

this method unsatisfactory because the globules can undergo considerable coalescence before the separation becomes visible. These workers conducted particle-size analyses with a Coulter centrifugal photosedimentometer. Mean volume diameters were obtained, and these were converted to number of globules per milliliter. King and Mukherjee[43] determined the specific interfacial area, that is, the area of interface per gram of emulsified oil, of each emulsion at successive times. They chose the reciprocal of the decrease of specific interfacial area with time as a measure of the stability of an emulsion.

Other methods used to determine the stability of emulsions are based on accelerating the separation process, which normally takes place under storage conditions. These methods employ freezing, thaw–freeze cycles, and centrifugation.

Merrill[46] introduced the centrifuge method for evaluating the stability of emulsions. Garrett, Vold, and others[47] used the ultracentrifuge as an analytic technique in emulsion technology. Coulter counting, turbidimetric analysis, and temperature tests have also been used in an effort to evaluate new emulsifying agents and determine the stability of pharmaceutical emulsions. Garti et al.[48] developed a method for evaluating the stability of oil–water viscous emulsions (ointments and cosmetic creams) containing nonionic surfactants. The method is based on electric conductivity changes during nondestructive short heating–cooling–heating cycles. Conductivity curves are plotted during the temperature cycling. A stability index is defined as $\Delta/h$, where $h$ is the change in the conductivity between 35°C and 45°C and $\Delta$ is the conductivity interval within the two heating curves at 35°C, as shown in **Figure 17–12**. The *stability index* indicates the relative change in conductivity between two cycles. The smaller the conductivity, the greater is the stability of the emulsion. The method was applied in a series of emulsions at different HLBs, emulsifier concentrations, and oil-phase

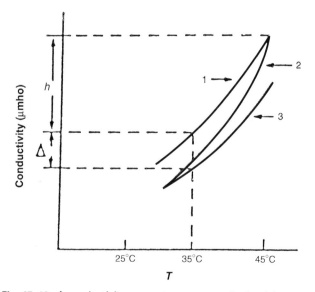

**Fig. 17–12.** A conductivity versus temperature plot involving successively (*a*) heating, (*b*) cooling, and (*c*) heating. (From N. Garti and S. Magdassi, Drug Dev. Ind. Pharm. **8**, 475, 1982. With permission.)

concentrations. The authors reviewed earlier work on electric conductivity of emulsions as related to stability.

## Phase Inversion

When controlled properly during the preparation of an emulsion, phase inversion often results in a finer product, but when it gets out of hand during manufacturing or is brought about by other factors after the emulsion is formed, it can cause considerable trouble.

An o/w emulsion stabilized with sodium stearate can be inverted to the w/o type by adding calcium chloride to form calcium stearate. Inversion can also be produced by alterations in phase-volume ratio. In the manufacture of an emulsion, one can mix an o/w emulsifier with oil and then add a small amount of water. Because the volume of the water is small compared with that of the oil, the water is dispersed by agitation in the oil even though the emulsifier preferentially forms an oil-in-water system. As more water is slowly added, the inversion point is gradually reached and the water and emulsifier envelope the oil as small globules to form the desired o/w emulsion. This procedure is sometimes used in the preparation of commercial emulsions, and it is the principle of the *continental method* used in compounding practice. The preparation of emulsions is discussed in books on general pharmacy and on compounding and dispensing.

## PRESERVATION OF EMULSIONS

Although it is not always necessary to achieve sterile conditions in an emulsion, even if the product is for topical or oral use, certain undesirable changes in the properties of the emulsion can be brought about by the growth of microorganisms. These include physical separation of the phases, discoloration, gas and odor formation, and changes in rheologic properties.[49] Emulsions for parenteral use obviously must be sterile.

The propagation of microorganisms in emulsified products is supported by one or more of the components present in the formulation. Thus, bacteria have been shown to degrade nonionic and anionic emulsifying agents, glycerin, and vegetable gums present as thickeners, with a consequent deterioration of the emulsion. As a result, it is essential that emulsions be formulated to resist microbial attack by including an adequate concentration of preservative in the formulation. Given that the preservative has inherent activity against the type of contamination encountered, the main problem is obtaining an *adequate* concentration of preservative in the product. Some of the factors that must be considered to achieve this end are presented here.

Emulsions are heterogeneous systems in which partitioning of the preservative will occur between the oil and water phases. In the main, bacteria grow in the aqueous phase of emulsified systems, with the result that a preservative that is partitioned strongly in favor of the oil phase may be virtually useless at normal concentration levels because of the low concentration remaining in the aqueous phase. The phase–volume ratio of the emulsion is significant in this regard. In addition, the preservative must be in an un-ionized state to penetrate the bacterial membrane. Therefore, the activity of weak acid preservatives decreases as the pH of the aqueous phase rises. Finally, the preservative molecules must not be "bound" to other components of the emulsion, because the complexes are ineffective as preservatives. Only the concentration of free, or unbound, preservative is effective. These points have been discussed in some detail in earlier sections. In addition to partitioning, ionization, and binding, the efficacy of a particular preservative is also influenced by emulsion type, nutritive value of the product, degree of aeration, and type of container used. These factors are discussed by Wedderburn.[49]

## RHEOLOGIC PROPERTIES OF EMULSIONS

Emulsified products may undergo a wide variety of shear stresses during either preparation or use. In many of these processes, the flow properties of the product will be vital for the proper performance of the emulsion under the conditions of use or preparation. Thus, spreadability of dermatologic and cosmetic products must be controlled to achieve a satisfactory preparation. The flow of a parenteral emulsion through a hypodermic needle, the removal of an emulsion from a bottle or a tube, and the behavior of an emulsion in the various milling operations employed in the large-scale manufacture of these products all indicate the need for correct flow characteristics. Accordingly, it is important for the pharmacist to appreciate how formulation can influence the rheologic properties of emulsions.

The fundamentals of rheology are discussed in Chapter 19. Most emulsions, except dilute ones, exhibit non-Newtonian flow, which complicates interpretation of data and quantitative comparisons among different systems and formulations. In a comprehensive review, Sherman[50] discussed the principal factors that influence the flow properties of emulsions. The material of this section outlines some of the viscosity-related properties of the dispersed phase, the continuous phase, and the emulsifying agent. For a more complete discussion of these and other factors that can modify the flow properties of emulsions, the reader is referred to the original article by Sherman[50] and Sherman's book.[51]

The factors related to the dispersed phase include the phase–volume ratio, the particle-size distribution, and the viscosity of the internal phase itself. Thus, when volume concentration of the dispersed phase is low (less than 0.05), the system is Newtonian. As the volume concentration is increased, the system becomes more resistant to flow and exhibits pseudoplastic flow characteristics. At sufficiently high concentrations, plastic flow occurs. When the volume concentration approaches 0.74, inversion may occur, with a marked change in viscosity; reduction in mean particle size increases the viscosity; and the wider the particle size distribution, the lower is the viscosity when compared with a system having a similar mean particle size but a narrower particle-size distribution.

The major property of the continuous phase that affects the flow properties of an emulsion is not, surprisingly, its own viscosity. The effect of the viscosity of the continuous phase may be greater, however, than that predicted by determining the bulk viscosity of the continuous phase alone. There are indications that the viscosity of a thin liquid film, of say 100 to 200 Å, is several times the viscosity of the bulk liquid. Higher viscosities may therefore exist in concentrated emulsions when the thickness of the continuous phase between adjacent droplets approaches these dimensions. Sherman pointed out that the reduction in viscosity with increasing shear may be due in part to a decrease in the viscosity of the continuous phase as the distance of separation between globules is increased.

Another component that may influence the viscosity of an emulsion is the emulsifying agent. The type of agent will affect particle flocculation and interparticle attractions, and these in turn will modify flow. In addition, for any one system, the greater the concentration of emulsifying agent, the higher will be the viscosity of the product. The physical properties of the film and its electric properties are also significant factors.

## MICROEMULSIONS

The term *microemulsion* may be a misnomer because microemulsions consist of large or "swollen" micelles containing the internal phase, much like that found in a solubilized solution. Unlike the common macroemulsions, they appear as clear, transparent solutions, but unlike micellar solubilized systems, microemulsions may not be thermodynamically stable. They appear to represent a state intermediate between thermodynamically stable solubilized solutions and ordinary emulsions, which are relatively unstable. Microemulsions contain droplets of oil in a water phase (o/w) or droplets of water in oil (w/o) with diameters of about 10 to 200 nm, and the volume fraction of the dispersed phase varies from 0.2 to 0.8.

As often recommended in the formation of ordinary emulsions or macroemulsions, an emulsifying adjunct or cosurfactant is used in the preparation of microemulsions. An anionic surfactant, sodium lauryl sulfate or potassium oleate, can be dispersed in an organic liquid such as benzene, a small measured amount of water is added, and the microemulsion is formed by the gradual addition of pentanol, a lipophilic cosurfactant, to form a clear solution at 30°C. The addition of pentanol temporarily reduces the surface tension to approximately zero, allowing spontaneous emulsification. The surfactant and cosurfactant molecules form an adsorbed film on the microemulsion particles to prevent coalescence.

Shinoda and Kunieda[52] showed that by choosing a surfactant and cosurfactant that have similar HLB values, one can increase the solubilization of an organic liquid in water

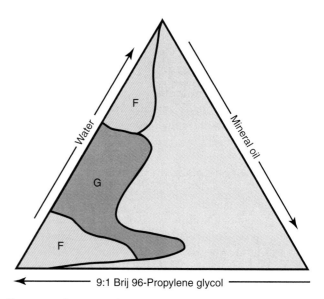

**Fig. 17–13.** A ternary-phase diagram of water, mineral oil, and a mixture of surfactants showing the boundary of the microemulsion region. The zones within the microemulsion region are labeled F for fluid and G for gel. (From N. J. Kate and L. V. Allen, Jr., Int. J. Pharm. **57,** 87, 1989. With permission.)

and enlarge the microemulsion droplet size without affecting stability. With ionic surfactants at normal temperatures, one expects o/w microemulsions to be formed when the phase volume ratio favors water, analogous to the rule for macroemulsions.

The microemulsion region is usually characterized by constructing ternary-phase diagrams, as shown in **Figure 17–13**, the axes representing water, mineral oil, and a mixture of surfactant and cosurfactant at different ratios.[53] The phase diagrams allow one to determine the ratios oil:water: surfactant–cosurfactant at the boundary of the microemulsion region. The microemulsion appears by visual observation as an isotropic, optically clear liquid system. Kale and Allen[53] studied water-in-oil microemulsions consisting of the system Brij 96-cosurfactant–mineral oil–water. Brij 96 [polyoxyethylene(10) oleyl ether] is a nonionic surfactant commonly used in the preparation of macro- and microemulsions. The cosurfactants studied were ethylene glycol, propylene glycol, and glycerin. **Figure 17–13** shows the phase diagram for the system upon varying the ratio Brij 96:propylene glycol. Within the microemulsion region, zones of different viscosity, labeled as fluid (F) or gel (G), can be observed. The microemulsion region becomes smaller as the cosurfactant concentration increases. According to the researchers, the transition from fluid microemulsion to gel-like microemulsion may be due to the change in the nature and shape of the internal oil phase. Thus, at low water content the internal phase consists of spherical structures, whereas at higher water concentration the interfacial film expands to form gel-like cylindrical and laminar structures. As the water content is further increased, aqueous continuous systems of low viscosity with internal phases of spherical structures (droplets) are again formed.

The droplet average molecular weight of a microemulsion can be measured by light-scattering techniques. Because the internal phase is not usually very dilute, the droplets interact with one another, resulting in a decrease in the turbidity. Thus, the effective diameter obtained is smaller than the actual droplet diameter. The latter can be obtained from a plot of the effective diameter (obtained at various dilutions of the microemulsion) against the concentration of the internal phase. Extrapolation to zero concentration gives the actual diameter.[53] Attwood and Ktistis[54] showed that the extrapolation procedure often cannot be applied because many microemulsions exhibit phase separation on dilution. They described a procedure for overcoming these difficulties and obtaining true particle diameter using light scattering.

Microemulsions have been studied as drug delivery systems. They can be used to increase the bioavailability of drugs poorly soluble in water by incorporation of the drug into the internal phase. Halbert et al.[55] studied the incorporation of both etoposide and a methotrexate diester derivative in water-in-oil microemulsions as potential carriers for cancer chemotherapy. Etoposide was rapidly lost from the microemulsion particles, whereas 60% of the methotrexate diester remained incorporated in the internal phase of the microemulsion. The methotrexate diester microemulsions showed an in vitro cytotoxic effect against mouse leukemia cells. Microemulsions have also been considered as topical drug delivery systems. Osborne et al.[56] studied the transdermal permeation of water from water-in-oil microemulsions formed from water, octanol, and dioctyl sodium sulfosuccinate, the latter functioning as the surfactant. These kinds of microemulsions can be used to incorporate polar drugs in the aqueous internal phase. The skin used in the experiments was fully hydrated so as to maximize the water permeability. The delivery of the internal phase was found to be highly dependent on the microemulsion water content: The diffusion of water from the internal phase increased tenfold as the water amount in the microemulsion increased from 15% to 58% by weight. Linn et al.[57] compared delivery through hairless mouse skin of cetyl alcohol and octyl dimethyl para-aminobenzoic acid (PABA) from water-in-oil microemulsions and macroemulsions. The delivery of these compounds from microemulsions was faster and showed deeper penetration into the skin than delivery from the macroemulsions. The authors reviewed a number of studies on the delivery of drugs from the microemulsions. These reports, including several patents, dealt with the incorporation of fluorocarbons as blood substitutes and for the topical delivery of antihypertensive and anti-inflammatory drugs. Microemulsions are used in cosmetic science,[58] foods, and dry cleaning and wax-polishing products.[59]

# SEMISOLIDS

## Gels

A gel is a solid or a semisolid system of at least two constituents, consisting of a condensed mass enclosing and interpenetrated by a liquid. When the coherent matrix is rich in liquid, the product is often called a *jelly*. Examples are ephedrine sulfate jelly and the common table jellies. When the liquid is removed and only the framework remains, the gel is known as a *xerogel*. Examples are gelatin sheets, tragacanth ribbons, and acacia tears.

Hydrogels retain significant amounts of water but remain water-insoluble and, because of these properties, are often used in topical drug design. The diffusion rate of a drug depends on the physical structure of the polymer network and its chemical nature. If the gel is highly hydrated, diffusion occurs through the pores. In gels of lower hydration, the drug dissolves in the polymer and is transported between the chains.[60] Cross-linking increases the hydrophobicity of a gel and diminishes the diffusion rate of the drug. The fractional release, F, of a drug from a gel at time t can be expressed in general as

$$F = \frac{M_t}{M_0} = kt^n \qquad (17\text{--}9)$$

where $M_t$ is the amount released at time $t$, $M_0$ is the initial amount of drug, $k$ is the rate constant, and $n$ is a constant

---

## ◄══ KEY CONCEPT ■ CLASSIFICATION OF GELS

Gels can be classified as two-phase or single-phase systems. The gel mass may consist of floccules of small particles rather than large molecules, as found in aluminum hydroxide gel, bentonite magma, and magnesia magma, and the gel structure in these two-phase systems is not always stable (Fig. 17–14a and b). Such gels may be thixotropic, forming semisolids on standing and becoming liquids on agitation.

On the other hand, a gel may consist of macromolecules existing as twisted, matted strands (Fig. 17–14c). The units are often bound together by stronger types of van der Waals forces so as to form crystalline and amorphous regions throughout the entire system, as shown in Figure 17–14d. Examples of such gels are tragacanth and carboxymethylcellulose. These gels are considered to be one-phase systems because no definite boundaries exist between the dispersed macromolecules and the liquid.

Gels can be classified as *inorganic* and *organic*. Most inorganic gels can be characterized as two-phase systems, whereas organic gels belong to the single-phase class because the condensed matrix is dissolved in the liquid medium to form a homogeneous gelatinous mixture. Gels may contain water, and these are called *hydrogels*, or they may contain an organic liquid, in which case they are called *organogels*. Gelatin gel belongs to the former class, whereas petrolatum falls in the latter group.

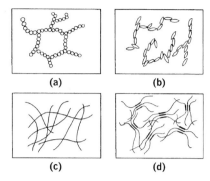

**Fig. 17–14.** Representations of gel structures. (*a*) Flocculated particles in a two-phase gel structure. (*b*) Network of elongated particles or rods forming a gel structure. (*c*) Matted fibers as found in soap gels. (*d*) Crystalline and amorphous regions in a gel of carboxymethylcellulose. (From H. R. Kruyt, *Colloid Science*, Vol. II, Elsevier, New York, 1949.)

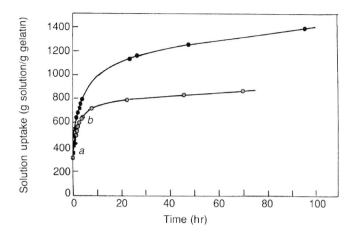

**Fig. 17–15.** Swelling isotherms of gelatin at (●) 25°C and (○) at 20°C. Swelling is measured as the increase in weight of gelatin strips in buffer solution at various times. The points *a* and *b* are discussed in the text. (From C. M. Ofner III and H. Schott, J. Pharm. Sci. **75**, 790, 1986. With permission.)

called the *diffusional exponent*. When $n = 0$, $t^n = 1$ and the release $F$ is of zero order; if $n = 0.5$, Fick's law holds and the release is represented by a square root equation. Values of $n$ greater than 0.5 indicate anomalous diffusion due generally to the swelling of the system in the solvent before the release takes place.[61] Morimoto et al.[62] prepared a polyvinyl alcohol hydrogel for rectal administration that has a porous, tridimensional network structure with high water content. The release of indomethacin from the gel followed Fickian diffusion over a period of 10 hr.

**EXAMPLE 17–6**

**Diffusional Component**

The release fraction, $F$, of indomethacin is 0.49 at $t = 240$ min. Compute the diffusional exponent, $n$, knowing that $k = 3.155\%$ $min^{-n}$.

Because the rate constant $k$ is expressed as percentage, the fractional release, $F$, is also expressed in percentage units in equation (17–9), that is, 49%. Taking the ln on both sides of equation (17–9), we obtain

$$\ln F = \ln k + n \ln t$$

$$n = \frac{\ln F - \ln k}{\ln t} = \frac{\ln 49 - \ln 3.155}{\ln 240}$$

$$n = \frac{3.892 - 1.149}{5.481} = 0.5$$

Therefore, with the exponent of $t$ equal to 0.5, equation (17–9) becomes $F = kt^{1/2}$, which is a Fickian diffusion.

## Syneresis and Swelling

When a gel stands for some time, it often shrinks naturally, and some of its liquid is pressed out. This phenomenon, known as *syneresis*, is thought to be due to the continued coarsening of the matrix or fibrous structure of the gel with a consequent squeezing-out effect. Syneresis is observed in table jellies and gelatin desserts. The "bleeding" in connec-

tion with the liberation of oil or water from ointment bases usually results from a deficient gel structure rather than from the contraction involved in syneresis.

The opposite of syneresis is the taking up of liquid by a gel with an increase in volume. This phenomenon is known as *swelling*. Gels may also take up a certain amount of liquid without a measurable increase in volume, and this is called *imbibition*. Only those liquids that solvate a gel can bring about swelling. The swelling of protein gels is influenced by pH and the presence of electrolytes.

Ofner and Schott[63] studied the kinetics of swelling of gelatin by measuring the increase in weight of short rectangular strips of gelatin films after immersion in buffer solutions as a function of time, $t$. A plot of the weight, $W$, in grams of aqueous buffer absorbed per gram of dry gelatin against $t$ in hours gives the swelling isotherms (**Fig. 17–15**). The horizontal portions of the two isotherms correspond to equilibrium swelling. To obtain a linear expression, $t/W$ is plotted against $t$ (the plot is not shown here) according to the equation

$$\frac{t}{W} = A + Bt \qquad (17\text{–}10)$$

Rearranging and differentiating equation (17–10), we obtain

$$\frac{dW}{dt} = \frac{A}{(A + Bt)^2} \qquad (17\text{–}11)$$

As $t \to 0$, equation (17–11) gives the *initial swelling rate*, $dW/dt = 1/A$, which is the reciprocal of the intercept of equation (17–10). The reciprocal of the slope, $1/B = W_\infty$, is the *equilibrium swelling*, that is, the theoretical maximum uptake of buffer solution at $t_\infty$.

## EXAMPLE 17–7

### Initial Swelling Rate

The increase in weight of 330 mg for a 15% gelatin sample 0.27 mm thick was measured in 0.15 M ammonium acetate buffer at 25°C. The $t/W$ values at several time periods are as follows:[*]

| $t$ (hr) | 0.5 | 1 | 1.5 | 2 | 3 | 4 |
|---|---|---|---|---|---|---|
| $t/W \dfrac{\text{hr}}{\text{g(buffer)/g(gelatin)}}$ | 0.147 | 0.200 | 0.252 | 0.305 | 0.410 | 0.515 |

Compute the initial swelling rate and the equilibrium swelling.

A regression of $t/W$ against $t$ gives

$$\frac{t}{W} = 0.0946 + 0.1051\, t$$

The initial swelling rate, $1/A$, is the reciprocal of the intercept,

$$\frac{1}{A} = \frac{1}{0.0946} = 10.57 \text{ g(buffer solution)/hr g(gelatin)}$$

The equilibrium swelling is

$$W_\infty = \frac{1}{B} = \frac{1}{0.1051} = 9.513 \frac{\text{g(buffer solution)}}{\text{g(gelatin)}}$$

Equation (17–10) represents a second-order process. When the constants $A$ and $B$ are used to backcalculate the swelling, $W$, at several times and are compared with the experimental data, the higher deviations are found in the region of maximum curvature of the isotherms (Fig. 17–15). Ofner and Schott[63] attributed the deviations to the partially crystalline structure of gelatin. Thus, the first part of curve $a$ in Figure 17–15 corresponds to the swelling of the amorphous region, which is probably complete at times corresponding to maximum curvature, namely 6 to 10 hr at 20°C. The penetration of the solvent into the crystalline region is slower and less extensive because this region is more tightly ordered and has a higher density (part $b$ of the curve in Fig. 17–15).

Gelatin is probably the most widely employed natural polymer in pharmaceutical products; it is used in the preparation of soft and hard gelatin capsules, tablet granulations and coatings, emulsions, and suppositories. Gelatin may interact with gelatin-encapsulated drugs or excipients by absorbing significant amounts of them, and some compounds may change the dissolution rate of soft gelatin capsules. Ofner and Schott[64] studied the effect of six cationic, anionic, and nonionic drugs or excipients on the initial swelling rate and equilibrium swelling in gelatin. The cationic compounds reduced the equilibrium swelling, $W_\infty$, substantially, whereas the nonionic and anionic compounds increased it. The researchers suggested that the cationic additives such as quaternary ammonium compounds may cause disintegration and dissolution problems with both hard and soft gelatin capsules.

Cross-linked hydrogels with ionizable side chains swell extensively in aqueous media. The swelling depends on the nature of the side groups and the pH of the medium. This property is important because diffusion of drugs in hydrogels depends on the water content in the hydrogel. Kou et al.[65] used phenylpropanolamine as a model compound to study its diffusion in copolymers of 2-hydroxyethyl methacrylate and methacrylic acid cross-linked with tetraethylene glycol dimethacrylate. The drug diffusivity, $D$, in the gel matrix is related to the matrix hydration by the relation

$$\ln D = \ln D_0 - K_f \left(\frac{1}{\mathbf{H}} - 1\right) \qquad (17\text{–}12)$$

where $D_0$ is the diffusivity of the solute in water and $K_f$ is a constant characteristic of the system. The term $\mathbf{H}$ represents the matrix hydration and is defined as

$$\mathbf{H} = \frac{\text{Equilibrium swollen gel weight } - \text{ Dry gel weight}}{\text{Equilibrium swollen gel weight}}$$

According to equation (17–12), a plot of $\ln D$ against $1/(\mathbf{H} - 1)$ should be linear with slope $K_f$ and intercept $\ln D_0$.

## EXAMPLE 17–8

### Diffusion Coefficients

Compute the diffusion coefficients of phenylpropanolamine in a gel for two gel hydrations, H = 0.4 and H = 0.9. The diffusion coefficient of the solute in water is $D_0 = 1.82 \times 10^{-6}$ cm²/sec, and $K_f$, the constant of equation (17–12), is 2.354.

For H = 0.4,

$$\ln D = \ln(1.82 \times 10^{-6}) - 2.354\left(\frac{1}{0.4} - 1\right) = -16.748$$

$$D = 5.33 \times 10^{-8} \text{ cm}^2/\text{sec}$$

For H = 0.9,

$$\ln D = \ln(1.82 \times 10^{-6}) - 2.354\left(\frac{1}{0.9} - 1\right) = -13.479$$

$$D = 1.4 \times 10^{-6} \text{ cm}^2/\text{sec}$$

The swelling (hydration) of the gel favors drug release because it enhances the diffusivity of the drug, as shown in the example.

## Classification of Pharmaceutical Semisolids

Semisolid preparations, with special reference to those used as bases for jellies, ointments, and suppositories, can be classified as shown in Table 17–2. The arrangement is arbitrary and suffers from certain difficulties, as do all classifications.

Some confusion of terminology has resulted in recent years, partly as a result of the rapid development of the newer types of bases. Terms such as "emulsion-type," "water-washable," "water-soluble," "water-absorbing," "absorption base," "hydrophilic," "greaseless," and others have appeared in the literature as well as on the labels of commercial bases where the meaning is obscure and sometimes misleading. The title "greaseless" has been applied both to water-dispersible bases that contain no grease and to o/w bases because they feel greaseless to the touch and are easily removed from the skin and clothing. The terms "cream" and "paste" are also often used ambiguously. Pectin paste is a jelly, whereas zinc oxide paste is a semisolid suspension. And what does the term

---

[*]The data are calculated from the slope and intercept given in Table III in Ofner and Schott.[64]

## TABLE 17–2
## A CLASSIFICATION OF SEMISOLID BASES

| | Examples |
|---|---|
| I. Organogels | |
| A. Hydrocarbon type | Petrolatum, mineral oil–polyethylene gel* |
| B. Animal and vegetable fats | Lard, hydrogenated vegetable oils, Theobroma oil |
| C. Soap base greases | Aluminum stearate, mineral oil gel |
| D. Hydrophilic organogels | Carbowax bases, polyethylene glycol ointment |
| II. Hydrogels | |
| A. Organic hydrogels | Pectin paste, tragacanth jelly |
| B. Inorganic hydrogels | Bentonite gel, colloidal magnesium aluminum silicate gels |
| III. Emulsion-type semisolids | |
| A. Emulsifiable bases | |
| 1. Water-in-oil (absorption) | Hydrophilic petrolatum, wool fat |
| 2. Oil-in-water | Anhydrous Tween base† |
| B. Emulsified bases | |
| 1. Water-in-oil | Hydrous wool fat, rose water ointment |
| 2. Oil-in-water | Hydrophilic ointment, vanishing cream |

*Plastibase (E. R. Squibb). J. Am. Pharm. Assoc. Sci. Ed. **45,** 104, 1956.
† White petrolatum, stearyl alcohol, glycerin, Tween 60 (Atlas-ICI).

"absorption base" mean? Does it imply that the base is readily absorbed into the skin, that drugs incorporated in such a base are easily released and absorbed percutaneously, or that the base is capable of absorbing large quantities of water? These few examples point out the difficulties that arise when different titles are used for the same product or when different definitions are given to the same term.

### Organogels

Petrolatum is a semisolid gel consisting of a liquid component together with a "protosubstance" and a crystalline waxy fraction. The crystalline fraction provides rigidity to the gel structure, whereas the protosubstance or gel former stabilizes the system and thickens the gel. Polar organogels include the polyethylene glycols of high molecular weight known as Carbowaxes (Union Carbide Corp., New York). The Carbowaxes are soluble to about 75% in water and therefore are completely washable, although their gels look and feel like petrolatum.

### Hydrogels

Bases of this class include organic and inorganic ingredients that are colloidally dispersible or soluble in water. Organic hydrogels include the natural and synthetic gums such as tragacanth, pectin, sodium alginate, methylcellulose, and sodium carboxymethylcellulose. Bentonite mucilage is an inorganic hydrogel that has been used as an ointment base in about 10% to 25% concentration.

### Emulsion-Type Bases

Emulsion bases, as might be expected, have much greater affinity for water than do the oleaginous products.

The o/w bases have an advantage over the w/o bases in that the o/w products are easily removed from the skin and do not stain clothing. These bases are sometimes called *water wash-*

*able*. They have the disadvantage of water loss by evaporation and of possible mold and bacterial growth, thus requiring preservation. Two classes of emulsion bases are discussed: emulsifiable and emulsified.

(*a*) *Emulsifiable bases.* We choose to call these bases *emulsifiable* because they initially contain no water but are capable of taking it up to yield w/o and o/w emulsions. The w/o types are commonly known as *absorption bases* because of their capacity to absorb appreciable quantities of water or aqueous solutions without marked changes in consistency.

(*b*) *Emulsified bases.* Water-in-oil bases in which water is incorporated during manufacture are referred to in this book as *emulsified* w/o *bases* to differentiate them from the emulsifiable w/o bases (absorption bases), which contain no water. The emulsified *oil-in-water bases* are formulated as is any emulsion with an aqueous phase, an oil phase, and an emulsifying agent. The components of emulsified ointments, however, differ in some ways from the ingredients of liquid emulsions.

The oil phase of the ointment may contain petrolatum, natural waxes, fatty acids or alcohols, solid esters, and similar substances that increase the consistency of the base and provide certain desirable application properties.

### Comparison of Emulsion Bases

The absorption bases have the advantage over oleaginous products in absorption of large amounts of aqueous solution. Furthermore, they are compatible with most drugs and are stable over long periods. When compared with o/w bases, the w/o preparations are superior in that they do not lose water readily by evaporation because water is the internal phase. Although emulsified o/w or washable bases do have the undesirable property of drying out when not stored properly and of losing some water during compounding operations, they are more acceptable than the nonwashable absorption bases because they are easily removed with water from the skin and clothing.

## Hydrophilic Properties of Semisolids

Petrolatum is hydrophilic to a limited degree, taking up about 10% to 15% by weight of water through simple incorporation.

The water-absorbing capacity of oleaginous and water-in-oil bases can be expressed in terms of the *water number*, first defined in 1935 by Casparis and Meyer[66] as the maximum quantity of water that is held (partly emulsified) by 100 g of a base at 20°C. The test consists in adding increments of water to the melted base and triturating until the mixture has cooled. When no more water is absorbed, the product is placed in a refrigerator for several hours, removed, and allowed to come to room temperature. The material is then rubbed on a slab until water no longer exudes, and, finally, the amount of water remaining in the base is determined. Casparis and Meyer found the water number of petrolatum to be about 9 to 15; the value for wool fat is about 185.

## Rheologic Properties of Semisolids

Manufacturers of pharmaceutical ointments and cosmetic creams have recognized the desirability of controlling the consistency of non-Newtonian materials.

Probably the best instrument for determining the rheologic properties of pharmaceutical semisolids is some form of a rotational viscometer. The cone–plate viscometer is particularly well adapted for the analysis of semisolid emulsions and suspensions. The Stormer viscometer, consisting of a stationary cup and rotating bob, is also satisfactory for semisolids when modified, as suggested by Kostenbauder and Martin.[67]

Consistency curves for the emulsifiable bases hydrophilic petrolatum and hydrophilic petrolatum in which water has been incorporated are shown in **Figure 17–16**. It will be observed that the addition of water to hydrophilic petrolatum has lowered the yield point (the intersection of the extrapolated downcurve and the load axis) from 520 to 340 g. The plastic viscosity (reciprocal of the slope of the downcurve) and the thixotropy (area of the hysteresis loop) are increased by the addition of water to hydrophilic petrolatum.

The effect of temperature on the consistency of an ointment base can be analyzed by use of a properly designed rotational viscometer. **Figures 17–17** and **17–18** show the changes of plastic viscosity and thixotropy, respectively, of petrolatum and Plastibase as a function of temperature.[68] The modified Stormer viscometer was used to obtain these curves. As observed in **Figure 17–17**, both bases show about the same temperature coefficient of plastic viscosity. These results account for the fact that the bases have about the same degree of "softness" when rubbed between the fingers. Curves of yield value versus temperature follow approxi-

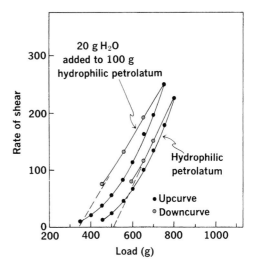

**Fig. 17–16.** Flow curves for hydrophilic petrolatum and hydrophilic petrolatum containing water. (After H. B. Kostenbauder and A. Martin, J. Am. Pharm. Assoc. Sci. Ed. **43**, 401, 1954.)

mately the same relationship. The curves of **Figure 17–18** suggest strongly that it is the alternation of thixotropy with temperature that differentiates the two bases. Because thixotropy is a consequence of gel structure, **Figure 17–18** shows that the waxy matrix of petrolatum is probably broken down considerably as the temperature is raised, whereas the resinous structure of Plastibase withstands temperature changes over the ranges ordinarily encountered in its use.

Based on data and curves such as these, the pharmacist in the development laboratory can formulate ointments with more desirable consistency characteristics, the worker in the

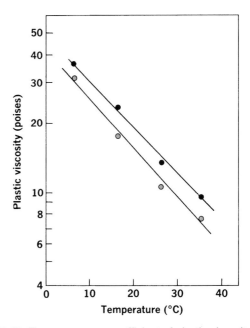

**Fig. 17–17.** The temperature coefficient of plastic viscosity of (●) Plastibase (E. R. Squibb and Sons, New Brunswick, NJ) and (○) petrolatum. (From A. H. C. Chun, M. S. Thesis, Purdue University, Purdue, Ind., June 1956.)

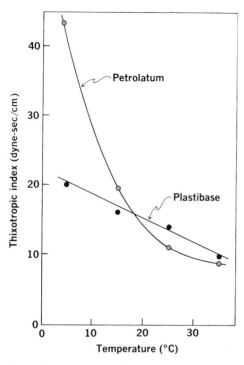

**Fig. 17–18.** The temperature coefficient of thixotropy of Plastibase (E. R. Squibb and Sons) and petrolatum. (After A. H. C. Chun, M. S. Thesis, Purdue University, Purdue, Ind., June 1956.)

production plant can better control the uniformity of the finished product, and the dermatologist and the patient can be assured of a base that spreads evenly and smoothly in various climates, yet adheres well to the affected area and is not tacky or difficult to remove.

Rigidity and viscosity are two separate parameters used to characterize the mechanical properties of gels. Ling[69] studied the effect of temperature on rigidity and viscosity of gelatin. He used a *rigidity index, f,* which is defined as the force required to depress the gelatin surface a fixed distance. To measure rigidity, a sample of gelatin solution or gel mass is subjected to penetrative compression by a flat-ended cylindrical plunger that operates at a constant speed. In this method, the strain rate (rate of deformation of the gel) is constant and independent of stress (force applied). Ling found that thermal degradation with respect to rigidity followed second-order kinetics,

$$-df/dt = k_f f^2 \qquad (17\text{--}13)$$

The integrated form of equation **(17–13)** is

$$\frac{1}{f} - \frac{1}{f_0} = k_f t \qquad (17\text{--}14)$$

where *f* is the *rigidity index* of the gelatin solution or gelatin gel at time $t$, $f_0$ is the rigidity index at time zero, $k_f$ is the rate constant ($g^{-1}$ $hr^{-1}$), and $t$ is the heating time in hours. The quantities $f_0$ and $k_f$ can be computed from the intercept and the slope of equation **(17–14)** at a given temperature.

---

**EXAMPLE 17–9**

**Rigidity Index**

The rigidity degradation of a 6% pharmaceutical-grade gelatin USP was studied[69] at 65°C. The rigidity index values at several times are as follows:

| $t$ (hr) | 10 | 20 | 30 | 40 | 50 |
|---|---|---|---|---|---|
| $\dfrac{1}{f}(g^{-1})$ | 0.0182 | 0.0197 | 0.0212 | 0.0227 | 0.0242 |

Compute the rigidity index, $f_0$, at time zero and the rate constant, $k_f$, at 65°C.

The regression of $1/f$ versus $t$ gives the equation

$$\frac{1}{f} = 1.5 \times 10^{-4}\, t + 0.0167$$

At $t = 0$ we have intercept $1/f_0 = 0.0167$ $g^{-1}$; $f_0 = 59.9$ g. The slope is $k_f = 1.5 \times 10^{-4}$ $g^{-1}$ $hr^{-1}$. Using the regression equation, we can compute the rigidity index, $f$, at time $t$, say 60 hr:

$$\frac{1}{f} = (1.5 \times 10^{-4} \times 60) + 0.0167 = 0.0257\ g^{-1}$$

$$f = \frac{1}{0.0257} = 38.9\ \text{g}$$

The force needed to depress the gelatin surface has decreased from its original value, $f_0 = 59.9$ g. Therefore, gelatin lost rigidity after heating for 60 hr.

---

The effect of temperature on the rate constant, $k_f$, can be expressed using the Arrhenius equation,

$$k_f = Ae^{-E_a/RT} \qquad (17\text{--}15)$$

Thus, a plot of ln $k_f$ against $1/T$ gives the Arrhenius constant, $A$, and the energy of activation, $E_a$.

Fassihi and Parker[70] measured the change in the rigidity index, $f$, of 15% to 40% gelatin gel, USP type B, before and after gamma irradiation (which is used to sterilize the gelatin). They found that the rigidity index diminished with irradiation and that the kinetics of rigidity degradation is complex. For gels containing more than 20% gelatin, the rigidity index follows a sigmoidal curve at increasing radiation doses, as shown in **Figure 17–19.** Gelatin is widely used in tablet manufacturing as a binder to convert fine powders into granules. The loss of rigidity index reduced the binding properties of gelatin and decreased the hardness of lactose granules prepared with irradiated gelatin. These workers suggested that doses of gamma radiation should be held to less than 2 megarad (Mrad) to obtain gelatins of acceptable quality for pharmaceutical applications.

## Universe of Topical Medications

Katz[71] devised a "universe of topical medications" (**Fig. 17–20**) by which one can consider the various topical medications such as pastes, absorption bases, emulsified products, lotions, and suspensions. The basic components of most dermatologic preparations are powder, water, oil, and emulsifier. Beginning at *A* on the "universal wheel" of **Figure 17–20,** one is confronted with the simple powder medication, used

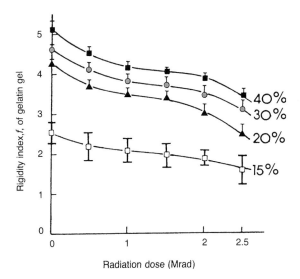

**Fig. 17–19.** Rigidity index of gelatin gel as a function of gamma irradiation at various concentrations (15%–40%) of the gel. (From A. R. Fassihi and M. S. Parker, J. Pharm. Sci. **77**, 876, 1988. With permission.)

as a protective, drying agent and lubricant and as a carrier for locally applied drugs. Passing counterclockwise around the wheel, we arrive at the paste, B, which is a combination of powder from segment A and an oleaginous material such as mineral oil or petrolatum. An oleaginous ointment for lubrication and emolliency and devoid of powder is shown in segment C.

The next section, D, is a waterless absorption base, consisting of oil phase and w/o emulsifier and capable of absorbing aqueous solutions of drugs. At the next region of the wheel, E, water begins to appear along with oil and emulsifier, and a w/o emulsion results. The proportion of water is increased at

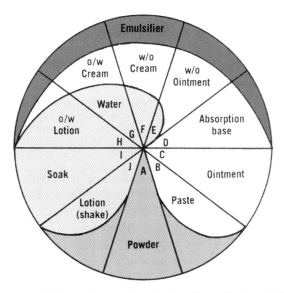

**Fig. 17–20.** Universe of topical medication. (From M. Katz, in E. J. Ariens (Ed.), *Drug Design*, Academic Press, New York, 1973. With permission.)

F to change the ointment into a w/o cream. At G, the base is predominantly water, and an o/w emulsifier is used to form the opposite type of emulsion, that is, an o/w cream. Still more water and less oil converts the product into an o/w lotion at H. At point I on the universal wheel, only water remains, both oil and surfactant being eliminated, and this segment of the wheel represents an aqueous liquid preparation, a soak, or a compress.

Finally, at section J, the powder from A is incorporated, and the aqueous product becomes a shake preparation, as represented by calamine lotion. Accordingly, this ingenious wheel classifies nearly all types of topical preparations from solid pastes and ointments, through w/o and o/w emulsions, to liquid applications and shake lotions. It serves as a convenient way to discuss the various classes of dermatologic and toiletry products that are prepared by the manufacturer or practicing pharmacist and applied topically by the patient.

# DRUG KINETICS IN COARSE DISPERSE SYSTEMS

The kinetics of degradation of drugs in suspension[72] can be described as a pseudo–zero-order process (see Chapter 14),

$$M = M_0 - k_1 V C_s t \qquad (17\text{–}16)$$

where $k_1$ is the first-order constant of the dissolved drug, $V$ is the volume of the suspension, and $C_s$ is the solubility of the drug. If the solubility is very low, the kinetics can be described as found in the section on solid-state kinetics (see Chapter 14). For very viscous dispersed systems, the kinetics of degradation can be partially controlled by the dissolution rate as given by the Noyes–Whitney equation,

$$dc/dt = KS(C_s - C) \qquad (17\text{–}17)$$

where $C_s$ is the solubility of the drug, $C$ is the concentration of solute at time $t$, $S$ is the surface area of the expanded solid, and $K$ is the dissolution rate constant. It is assumed that as a molecule degrades in the liquid phase it is replaced by another molecule dissolving. The overall decrease in concentration in the liquid phase can be written as

$$dc/dt = -kC + KS(C_s - C) \qquad (17\text{–}18)$$

where $-kC$ expresses the rate of disappearance at time $t$ due to degradation, and $KS(C_s - C)$ is the rate of appearance of the drug in the liquid phase due to dissolution of the particles. The solution of this differential equation is

$$C = [C_s KS/(k_1 + KS)]e^{-(k_1 + SK)t} \qquad (17\text{–}19)$$

At large $t$ values, $C$ becomes

$$C = C_s KS/(k_1 + KS) \qquad (17\text{–}20)$$

and the amount of drug remaining in suspension at large values of $t$ is

$$M = M_0 - [k_1 SKC_s V/(k_1 + KS)]t \qquad (17\text{–}21)$$

where $M_0$ is the initial amount of drug in suspension. Equation (17–21) is an expression for a zero-order process, as is equation (17–16), but the slopes of the two equations are different. Because the dissolution rate constant, $K$, in equation (17–21) is proportional to the diffusion coefficient, $D$, $K$ is inversely proportional to the viscosity of the medium; therefore, the more viscous the preparation, the greater is the stability.

### EXAMPLE 17–10

**Particles and Decomposition**

The first-order decomposition rate of a drug in aqueous solution is $5.78 \times 10^{-4}$ sec$^{-1}$ and the dissolution rate constant, $K$, is $3.35 \times 10^{-6}$ cm$^{-2}$ sec$^{-1}$. What is the amount of drug remaining in 25 cm$^3$ of a 5% w/v suspension after 3 days? Assume spherical particles of mean volume diameter, $d_{vn}$, $2 \times 10^{-4}$ cm. The density of the powder is 3 g/cm$^3$ and the solubility of the drug is $2.8 \times 10^{-4}$ g/cm$^3$.

The initial amount of drug is

$$\frac{5}{100} = \frac{M_0}{25}, M_0 = 1.25 \text{ g/25 cm}^3$$

The number of particles, $N$, in 25 cm$^3$ can be computed from equation (18–4):

$$N = \frac{6}{\pi(d_{vn})^3 \rho} = \frac{6}{3.1416 \times (2 \times 10^{-4})^3 \times 3}$$

$$= 7.96 \times 10^{10} \frac{\text{particles}}{\text{gram}}$$

The number of particles in 1.25 g is $N = 7.96 \times 10^{10} \times 1.25 = 9.95 \times 10^{10}$ particles.

The total surface area is

$$S = N\pi d^2 = 9.95 \times 10^{10} \times 3.1416 \times (2 \times 10^{-4})^2$$
$$= 1.25 \times 10^4 \text{ cm}^2$$

From equation (17–21),

$$M = 1.25 - \frac{5.78 \times 10^{-4} \times 1.25 \times 10^4 \times 3.35}{\times 10^{-6} \times 2.8 \times 10^{-4} \times 25}{5.78 \times 10^{-4} + (3.35 \times 10^{-6} \times 1.25 \times 10^4)}$$

$$\times (2.6 \times 10^5 \text{ sec})$$

$$= 1.25 - [(3.99 \times 10^{-6})(2.6 \times 10^5)] = 1.25 - 1.0374 = 0.213 \text{ g}$$

Kenley et al.[73] studied the kinetics of degradation of fluocinolone acetonide incorporated into an oil-in-water cream base. The degradation followed a pseudo–first-order constant at pH values from 2 to 6 and at several temperatures. The observed rate constants increased with increasing temperature, and acid catalysis at low pH values and basic catalysis at pH above 4 were observed. The observed rate constant for the degradation process can be written as

$$k = k_o + k_H[H^+] + k_{OH}[OH^-] \quad (17\text{–}22)$$

Figure 17–21 compares the degradation of fluocinolone acetonide from oil-in-water creams with that of triamcinolone acetonide, a related steroid, in aqueous solution. From the figure, both creams and solution share a similar log(Rate)−pH

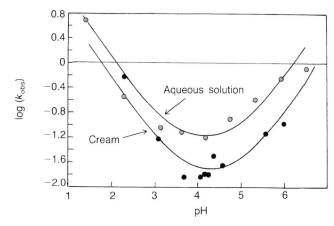

**Fig. 17–21.** The pH–log($k_{obs}$) profile for degradation of fluocinolone acetonide and triamcinolone acetonide at 50°C. Key: ● = experimentally determined $k_{obs}$ (month$^{-1}$) for fluocinolone acetonide cream; ○ = triamcinolone acetonide solution. The solid lines were obtained from the calculated values of $k_{obs}$ using equation (17–22). (From A. Kenley, M. O. Lee, L. Sukumar, and M. Powell, Pharm. Res. **4**, 342, 1987. With permission.)

profile over the pH range of 2 to 6, with a minimum rate near pH 4. This may indicate that the degradation in oil-in-water creams is confined to an aqueous environment, the nonaqueous components of the cream having little influence.[73]

Because $\ln k = \ln A - E_a/RT$, where $A$ is the Arrhenius factor and $E_a$ is the energy of activation, equation (17–22) can be rewritten in terms of activation parameters, $A$ and $E_a$, for each of the catalytic coefficients, $k_o$, $k_H$, and $k_{OH}$:

$$k = \exp[\ln A_o - (E_{ao}/RT)]$$
$$+ \exp[\ln A_H - (E_{aH}/RT)][H^+] \quad (17\text{–}23)$$
$$+ \exp[\ln A_{OH} - (E_{aOH}/RT)][OH^-]$$

Equation (17–23) allows one to compute the degradation rate constant $k$ at several temperatures and pH values.

### EXAMPLE 17–11

**Degradation Rate Constant**

The natural logarithm of the Arrhenius parameters for neutral-, acid-, and base-catalyzed hydrolysis of fluocinolone acetonide in oil-in-water creams are $\ln A_o = 22.5$, $\ln A_H = 38.7$, and $\ln A_{OH} = 49.5$. The corresponding energies of activation are $E_{aO} = 17,200$, $E_{aH} = 22,200$, and $E_{aOH} = 21,100$ cal/mole. The H$^+$ and OH$^-$ concentrations in equation (17–23) are expressed, as usual, in moles per liter, and the first-order rate constant, $k$, is expressed in this example in month$^{-1}$. Compute the degradation rate constant, $k$, at 40°C and pH 4.

From equation (17–23),

$$k = \exp[22.5 - (17,200/1.9872 \times 313)]$$
$$+ \exp[38.7 - (22,200/1.9872 \times 313)] \times (1 \times 10^{-4})$$
$$+ \exp[49.5 - (21,100/1.9872 \times 313)] \times (1 \times 10^{-10})$$
$$= (5.782 \times 10^{-3}) + (2.025 \times 10^{-3}) + (5.820 \times 10^{-4})$$
$$k = 8.39 \times 10^{-3} \text{ month}^{-1}$$

Teagarden et al.[74] determined the rate constant, $k$, for the degradation of prostaglandin $E_1$ (PGE$_1$) in an oil-in-water emulsion. At acidic pH values, the degradation of PGE$_1$ showed large rate constants. This fact was attributed to the greater effective concentration of hydrogen ions at the oil–water interface, where PGE$_1$ is mainly located at low pH values.

# DRUG DIFFUSION IN COARSE DISPERSE SYSTEMS

The release of drugs suspended in ointment bases can be calculated from the Higuchi equation:

$$Q = [D(2A - C_s)C_s t]^{1/2} \qquad (17\text{–}24)$$

where $Q$ is the amount of drug released at time $t$ per unit area of exposure, $C_s$ is the solubility of the drug in mass units per cm$^3$ in the ointment, and $A$ is the total concentration, both dissolved and undissolved, of the drug. $D$ is the diffusion coefficient of the drug in the ointment (cm$^2$/sec).

Iga et al.[75] studied the effect of ethyl myristate on the release rate of 4-hexylresorcinol from a petrolatum base at pH 7.4 and temperature 37°C. They found that the release rate was proportional to the square root of time, according to the Higuchi equation. Increasing concentrations of ethyl myristate enhanced the release rate of the drug owing to the increase of drug solubility, $C_s$, in the ointment [see equation (17–24)]. This behavior was attributed to formation of 1:1 and 1:2 complexes between hexylresorcinol and ethyl myristate.

## EXAMPLE 17–12

### Calculate $Q$

The solubility of hexylresorcinol in petrolatum base is 0.680 mg/cm$^3$. After addition of 10% ethyl myristate, the solubility, $C_s$, of the drug is 3.753 mg/cm$^3$. Compute the amount, $Q$, of drug released after 10 hr. The diffusion coefficient, $D$, is $1.31 \times 10^{-8}$ cm$^2$/sec and the initial concentration, $A$, is 15.748 mg/cm$^3$.

We have

$$Q = \{(1.31 \times 10^{-8}\,\text{cm}^2/\text{sec})[(2 \times 15.748\,\text{mg/cm}^3)$$
$$- 0.68\,\text{mg/cm}^3)]\}^{\frac{1}{2}}$$
$$\times [0.68\,\text{mg/cm}^3 \times (10 \times 3600)\,\text{sec}]^{\frac{1}{2}} = 0.099\,\text{mg/cm}^3$$

After addition of 10% ethyl myristate, we find

$$Q = \{(1.31 \times 10^{-8}\,\text{cm}^2/\text{sec})\,[(2 \times 15.748\,\text{mg/cm}^3)$$
$$- 3.753\,\text{mg/cm}^3)]\}^{\frac{1}{2}}$$
$$\times [3.753\,\text{mg/cm}^3 \times (10 \times 3600)\,\text{sec}]^{\frac{1}{2}} = 0.222\,\text{mg/cm}^2$$

The release of a solubilized drug from emulsion-type creams and ointments depends on the drug's initial concentration. It is also a function of the diffusion coefficient of the drug in the external phase, the partition coefficient between the internal and external phases, and the volume fraction of the internal phase. If the drug is completely solubilized in a minimum amount of solvent, the release from the vehicle is faster than it is from a suspension-type vehicle.

Ong and Manoukian[76] studied the delivery of lonapalene, a nonsteroidal antipsoriatic drug, from an ointment, varying the initial concentration of drug and the volume fraction of the internal phase. In the study, lonapalene was completely solubilized in the ointment systems. Most of the drug was dissolved in the internal phase, consisting of propylene carbonate–propylene glycol, but a fraction was also solubilized in the external phase of a petrolatum base consisting of glyceryl monostearate, white wax, and white petrolatum. The data were treated by the approximation of Higuchi,[77]

$$Q = 2C_0 \sqrt{\frac{D_e t}{\pi}} \qquad (17\text{–}25)$$

where $Q$ is the amount of drug released per unit area of application, $C_0$ is the initial concentration in the ointment, $D_e$ is the effective diffusion coefficient of the drug in the ointment, and $t$ is the time after application. For a small volume of the internal phase,

$$D_e = \frac{D_1}{\phi_1 + K\phi_2} \left[ 1 + 3\phi_2 \left( \frac{KD_2 - D_1}{KD_2 + 2D_1} \right) \right] \qquad (17\text{–}26)$$

where the subscripts 1 and 2 refer to the external and internal phases, respectively, and $K$ is the partition coefficient between the two phases. When $D_2$ is much greater than $D_1$,

$$D_e = \frac{D_1(1 + 3\phi_2)}{\phi_1 + K\phi_2} \qquad (17\text{–}27)$$

$D_e$, the effective diffusion coefficient, is obtained from the release studies [equation (17–25)], and $D_1$ can be computed from equation (17–27) if one knows the volume fraction of the external and internal phases, $\phi_1$ and $\phi_2$, respectively. The drug is released according to two separate rates: an initial nonlinear and a linear, diffusion-controlled rate (Fig. 17–22). The initial rates extending over a period of 30 min are higher than the diffusion-controlled rates owing to the larger transference of drug directly to the skin from the surface globules. The high initial rates provide immediate availability of the drug for

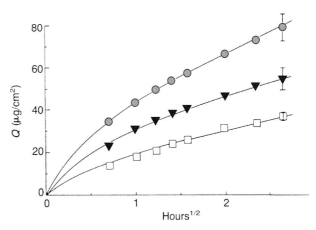

**Fig. 17–22.** Amount per unit area, $Q$, of lonapalene at time $t$ from an emulsion-type ointment. Key: $\square$ = 0.5%; $\blacktriangledown$ = 1.0%; and $\bullet$ = 2.0% drug. (From J. T. H. Ong and E. Manoukian, Pharm. Res. **5**, 16, 1988. With permission.)

absorption. In addition, the release of drug from the external phase contributes to the initial rates. Equation (17–25) is applicable only to the linear portion of the graph, where the process becomes diffusion controlled (**Fig. 17–22**).

### EXAMPLE 17–13

#### Amount Released

Compute the amount of lonapalene released per $cm^2$ after $t = 24$ hr from a 0.5% w/v emulsified ointment. The internal phase of the ointment consists of the drug solubilized in a propylene carbonate–propylene glycol mixture and the external phase is a white petrolatum–glyceryl monostearate–white wax mixture. The volume fraction of the internal phase, $\phi_2$, is 0.028, the diffusion coefficient of the drug in the external phase, $D_1$, is $2.60 \times 10^{-9}$ $cm^2$/sec, and the partition coefficient, $K$, between the internal and external phases is 69.

From equation (17–27), the effective diffusion coefficient is

$$D_e = \frac{(2.60 \times 10^{-9} \text{ cm}^2/\text{sec})[1 + (3 \times 0.028)]}{(1 - 0.028) + (69 \times 0.028)}$$

$$= 0.97 \times 10^{-9} \text{ cm}^2/\text{sec}$$

Note that the sum of the volume fractions of internal and of external phases is equal to 1; therefore, knowing the external volume fraction to be $\phi_2 = 0.028$, one simply has the internal volume fraction, $\phi_1 = 1 - 0.028$. The initial concentration of drug is 0.5 g per 100 $cm^3$, that is, 5 mg/mL. From equation (17–25), the amount of lonapalene released after 24 hr is

$$Q = 2 \times (5 \text{ mg/cm}^3) \sqrt{\frac{(0.97 \times 10^{-9} \text{cm}^2/\text{sec}) \times (24 \times 3600) \text{ sec}}{3.1416}}$$

$$= 0.05 \text{ mg/cm}^2$$

The rate of release also depends on the solubility of the drug as influenced by the type of emulsion. Rahman et al.[78] studied the in vitro release and in vivo percutaneous absorption of naproxen from anhydrous ointments and oil-in-water and water-in-oil creams. The results fitted equation (17–25), the largest release rates being obtained when the drug was incorporated into the water phase of the creams by using the soluble sodium derivative of naproxen. After application of the formulations to rabbit skin, the absorption of the drug followed first-order kinetics, showing a good correlation with the in vitro release.

Chiang et al.[79] studied the permeation of minoxidil, an antialopecia (antibaldness) agent, through the skin from anhydrous, oil-in-water, and water-in-oil ointments. The rate of permeation was higher from water-in-oil creams.

Drug release from fatty suppositories can be characterized by the presence of an interface between the molten base and the surrounding liquid. The first step is drug diffusion into the lipid–water interface, which is influenced by the rheologic properties of the suppository. In a second step, the drug dissolves at the interface and is then transported away from the interface.[80] Because the dissolution of poorly water-soluble drugs on the aqueous side of the lipid–water interface is the rate-limiting step, the release is increased by the formation of a water-soluble complex. Arima et al.[80] found that the release of ethyl 4-biphenyl acetate, an anti-inflammatory drug, from a lipid suppository base was

enhanced by complexation of the drug with a hydrosoluble derivative of $\beta$-cyclodextrin. The increase in solubility and wettability as well as the decrease in crystallinity due to an inclusion-type complexation may be the cause of the enhanced release. On the other hand, complexation of flurbiprofen with methylated cyclodextrins, which are oil soluble and surface active, enhances the release from hydrophilic suppository bases. This is due to the decreased interaction between the drug complex and the hydrophilic base.[81] Coprecipitation of indomethacin with PVP also enhances the release from lipid suppository bases because it improves wetting, which avoids the formation of a cake at the oil–aqueous suppository interface.[82]

Nyqvist-Mayer et al.[83] studied the delivery of a eutectic mixture of lidocaine and prilocaine (two local anesthetics) from emulsions and gels. Lidocaine and prilocaine form eutectic mixtures at approximately a 1:1 ratio. The eutectic mixture has a eutectic temperature of 18°C, meaning that it is a liquid above 18°C and can therefore be emulsified at room temperature. The mechanism of release from this emulsion and transport through the skin is complex owing to the presence of freely dissolved species, surfactant-solubilized species, and emulsified species of the local anesthetic mixture. The passage of these materials across the skin membrane is depicted in **Figure 17–23**. The solute lost due to transport across the membrane is replenished by dissolution of droplets as long as a substantial number of droplets are present. Micelles of surfactant with a fraction of the solubilized drug may act as carriers across the aqueous diffusion layer, diminishing the diffusion layer resistance. Droplets from the bulk are also transported to the boundary layer and supply solute, which diffuses through the membrane, thus decreasing the limiting effect of the aqueous layer to diffusion of solute. Because the oil phase of this emulsion is formed by the eutectic mixture itself, there is no transport of drug between the inert oil and water, as occurs in a conventional emulsion and which would result in a decreased thermodynamic activity, $a$, or "escaping tendency." The system actually resembles a suspension that theoretically has high thermodynamic activity owing to the saturation of the drug in the external phase. In a suspension, the dissolution rate of the particles could be a limiting factor. In contrast, the fluid state of the eutectic mixture lidocaine–prilocaine

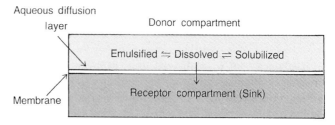

**Fig. 17–23.** Delivery of a eutectic mixture of lidocaine–prilocaine from an emulsion into a receptor compartment. (From A. A. Nyqvist-Mayer, A. F. Borodin, and S. G. Frank, J. Pharm. Sci. **75,** 365, 1986. With permission.)

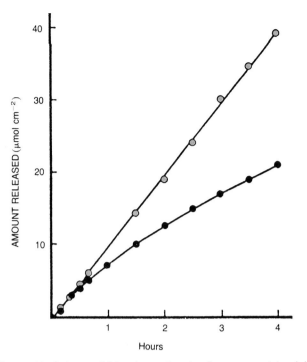

**Fig. 17–24.** Release of lidocaine–prilocaine from an emulsion (○) and from a gel (●). (From A. A. Nyqvist-Mayer, A. F. Borodin, and S. G. Frank, J. Pharm. Sci. **75,** 365, 1986. With permission.)

may promote a higher dissolution rate. The total resistance, $R_T$, to the skin permeation of the free dissolved fraction of prilocaine is given by the sum of the resistances of the aqueous layer, $R_a$, and the resistance of the membrane, $R_m$:

$$R_T = R_a + R_m \qquad (17\text{–}28)$$

or

$$R_T = \frac{1}{P} = \frac{h_m}{D_m K} + \frac{h_a}{D_a} \qquad (17\text{–}29)$$

where $D$ is the diffusion coefficient of the drug, $h_a$ is the thickness of the aqueous layer, $h_m$ is the thickness of the membrane, and $P$ is the permeability coefficient associated with the membrane and the aqueous layer; $K$ is the partition coefficient between the membrane and the aqueous layer. The subscripts a and m stand for aqueous layer and membrane, respectively. Equation **(17–29)** is analogous to equation **(11–30)**, except that the constant 2 in the denominator has been eliminated in this case because we consider only one aqueous layer **(Fig. 17–24)**.

**EXAMPLE 17–14**

Compute the total permeability, $P$, of a 1:1.3 ratio of lidocaine–prilocaine in the form of a eutectic mixture. The thicknesses of the aqueous and membrane layers are 200 and 127 $\mu$m, respectively. The diffusion coefficient and the partition coefficient of the drugs at the membrane–aqueous layers are as follows: lidocaine, $D_a = 8.96 \times 10^{-6}$ cm²/sec, $D_m = 2.6 = 10^{-7}$ cm²/sec, and $K = 9.1$; prilocaine, $D_a = 9.14 \times 10^{-6}$ cm²/sec, $D_m = 3 \times 10^{-7}$ cm²/sec, and $K = 4.4$.

For lidocaine, according to equation (17–29),

$$\frac{1}{P} = \frac{127 \times 10^{-4}\text{ cm}}{(2.6 \times 10^{-7}\text{ cm}^2/\text{sec}) \times 9.1} + \frac{200 \times 10^{-4}\text{ cm}}{8.96 \times 10^{-6}\text{ cm}^2/\text{sec}}$$
$$= 7599.8\text{ sec/cm}$$
$$P = 1/7599.8 = 1.32 \times 10^{-4}\text{ cm/sec}$$

For prilocaine,

$$\frac{1}{P} = \frac{127 \times 10^{-4}\text{ cm}}{(3 \times 10^{-7}\text{ cm}^2/\text{sec}) \times 4.4} + \frac{200 \times 10^{-4}\text{ cm}}{9.14 \times 10^{-6}\text{ cm}^2/\text{sec}}$$
$$= 11809.2\text{ sec/cm}$$
$$P = 1/11809.2 = 8.47 \times 10^{-5}\text{ cm/sec}$$

The permeability of the mixture $P_T$ can be calculated from the proportion of each component.[83] Because the proportion of lidocaine is 1 and that of prilocaine 1.3, the total amount is $1 + 1.3 = 2.3$. Therefore, the permeability of the mixture is

$$P_T = \frac{(1 \times 1.32 \times 10^{-4}) + (1.3 \times 8.47 \times 10^{-5})}{2.3}$$
$$= 1.05 \times 10^{-4}\text{ cm/sec}$$

The total amount released from the emulsion consists of an initial steady-state portion, from which the release rate can be computed. When the formulation is thickened with carbomer 934P (carbopol), a gel results. The release rates from the gel and the emulsion are compared in **Figure 17–24**. In the gel, the release rate continuously decreases owing to the formation of a depletion zone in the gel. The thickness of the stagnant diffusion layer next to the membrane increases to such a degree that the release process becomes *vehicle controlled*. After 1 hr, the amount delivered is a function of the square root of time, and the apparent diffusion coefficient in the gel can be computed from the Higuchi equation **(17–24)**. The release process is both membrane layer and aqueous layer controlled for nongelled systems (emulsions). For gelled systems the initial release is also membrane layer and aqueous layer controlled, but later, at $t > 1$ hr, the release becomes formulation or vehicle controlled, that is, the slowest or rate-determining step in the diffusion of the drug is passage through the vehicle.

## CHAPTER SUMMARY

Particulate systems have been classified on the basis of size into molecular dispersions, colloidal systems, and coarse dispersions. This chapter attempts to provide the pharmacist with an insight into the role of physics and chemistry in the research and development of the several classes of coarse dispersions. The theory and technology of these important pharmaceutical classes are based on interfacial and colloidal principles, micromeritics, and rheology (Chapters 15, 16, 18 and 19, respectively). Pharmaceutical suspensions were introduced and the roles they play in the pharmaceutical sciences were described. In addition, the desirable qualities of pharmaceutical suspensions and the factors that affect the stability of suspensions were also discussed. The concepts of flocculation, settling and sedimentation theory were introduced and the student was shown how to calculate sedimentation rates. Two useful sedimentation parameters, sedimentation volume and degree of flocculation were discussed. The student should be aware of the approaches commonly used in the preparation of physically stable suspensions. Pharmaceutical emulsions and emulsifying agents were introduced and the main types of emulsions discussed. The student should be able to classify pharmaceutical semisolids as well as understand thixotropic properties, syneresis, and swelling. Finally, examples of coarse dispersions were given.

 **Practice problems for this chapter can be found at thePoint.lww.com/Sinko6e.**

## References

1. K. J. Frederick, J. Pharm. Sci. **50,** 531, 1961.
2. J. C. Samyn, J. Pharm. Sci. **50,** 517, 1961.
3. A. P. Simonelli, S. C. Mehta, and W. I. Higuchi, J. Pharm. Sci. **59,** 633, 1970.
4. W. Schneider, S. Stavchansky, and A. Martin, Am. J. Pharm. Educ. **42,** 280, 1978.
5. K. S. Alexander, J. Azizi, D. Dollimore, and V. Uppala, J. Pharm. Sci. **79,** 401, 1990.
6. E. F. Burton, in A. E. Alexander (Ed.), *Colloid Chemistry,* Vol. I, Reinhold, New York, 1926, p. 165.
7. E. N. Hiestand, J. Pharm. Sci. **53,** 1, 1964.
8. A. Martin, J. Pharm. Sci. **50,** 513, 1961; R. A. Nash, Drug Cosmet. Ind. **97,** 843, 1965.
9. H. Schott, L. C. Kwan, and S. Feldman, J. Pharm. Sci. **71,** 1038, 1982.
10. B. A. Haines and A. Martin, J. Pharm. Sci. **50,** 228, 753, 756, 1961.
11. A. Delgado, V. Gallardo, J. Salcedo, and F. Gonzalez-Caballero, J. Pharm. Sci. **79,** 82, 1990.
12. A. Felmeister, G. M. Kuchtyak, S. Kozioi, and C. J. Felmeister, J. Pharm. Sci. **62,** 2027, 1973; J. S. Tempio and J. L. Zaps, J. Pharm. Sci. **69,** 1209, 1980; J. Pharm. Sci. **70,** 554, 1981; J. L. Zaps et al., Int. J. Pharm. **9,** 315, 1981.
13. E. N. Hiestand, J. Pharm. Sci. **61,** 269, 1972.
14. R. H. Blythe, U. S. Patent 2,369,711, 1945.
15. S.-L. Law, W.-Y. Lo, and G.-W. Teh, J. Pharm. Sci. **76,** 545, 1987.
16. C. K. Mervine and G. D. Chase, Presented at the American Pharm. Association Meeting, 1952.
17. J. Y. Oldshue, J. Pharm. Sci. **50,** 523, 1961.
18. (a) E. K. Fischer, *Colloidal Dispersions,* Wiley, New York, 1950; (b) C. E. Berry and H. J. Kamack, *Proceedings 2nd International Congress of Surface Activity,* Vol. IV, Butterworths, London, 1957, p. 196; (c) C. L. Prasher, *Crushing and Grinding Process Handbook,* Wiley, New York, 1987, Chapter 6.
19. J. L. Zatz and R.-Y. Lue, J. Pharm. Sci. **76,** 157, 1987.
20. M. I. Zapata, J. R. Feldkamp, G. E. Peck, J. L. White, and S. L. Hem, J. Pharm. Sci. **73,** 3, 1984.
21. S. C. Mehta, P. D. Bernardo, W. I. Higuchi, and A. P. Simonelli, J. Pharm. Sci. **59,** 638, 1970.
22. K. H. Ziller and H. Rupprecht, Drug Dev. Ind. Pharm. **14,** 2341, 1988.
23. J. S. Lucks, B. W. Müller, and R. H. Müller, Int. J. Pharm. **58,** 229, 1990.
24. B. D. Tarr, T. G. Sambandan, and S. H. Yalkowsky, Pharm. Res. **4,** 162, 1987.
25. S. S. Davis and P. Hansrani, Int. J. Pharm. **23,** 69, 1985.
26. E. Shotton and R. F. White, in P. Sherman (Ed.), *Rheology of Emulsions,* Pergamon Press, Oxford, 1963, p. 59.
27. J. A. Serrallach and G. Jones, Ind. Eng. Chem. **23,** 1016, 1931.
28. J. H. Schulman and E. G. Cockbain, Trans. Faraday Soc. **36,** 651, 661, 1940.
29. Atlas Powder Co., *A Guide to Formulation of Industrial Emulsions with Atlas Surfactants,* Atlas Powder Co., Wilmington, Del., 1953.
30. J. Boyd, C. Parkinson, and P. Sherman, J. Coll. Interface Sci. **41,** 359, 1972.
31. A. H. C. Chun, R. S. Joslin, and A. Martin, Drug Cosmet. Ind. **82,** 164, 1958.
32. A. Beerbower and J. Nixon, Am. Chem. Soc. Div. Petroleum Chem. Preprints **14**(1), 62, 1969; A. Beerbower and M. W. Hill, in *McCutcheon's Detergents and Emulsifiers,* MC Publishing Company, Princeton, WI. 1971, p. 223.
33. J. Mullins and C. H. Becker, J. Am. Pharm. Assoc. Sci. Ed. **45,** 110, 1956.
34. H. L. Greenwald, J. Soc. Cosmet. Chem. **6,** 164, 1955.
35. A. King, Trans. Faraday Soc. **37,** 168, 1941.
36. E. L. Knoechel and D. E. Wurster, J. Am. Pharm. Assoc. Sci. Ed. **48,** 1, 1959.
37. W. Ostwald, Kolloid Z. **6,** 103, 1910; **7,** 64, 1910.
38. S. Magdassi and A. Siman-Tov, Int. J. Pharm. **59,** 69, 1990.
39. O. L. Johnson, C. Washington, S. S. Davis, and K. Schaupp, Int. J. Pharm. **53,** 237, 1989.
40. C. Washington, A. Chawla, N. Christy, and S. S. Davis, Int. J. Pharm. **54,** 191, 1989.
41. H. Schott and A. E. Royce, J. Pharm. Sci. **72,** 1427, 1983.
42. J. A. Serrallach, G. Jones, and R. J. Owen, Ind. Eng. Chem. **25,** 816, 1933.
43. A. King and L. N. Mukherjee, J. Soc. Chem. Ind. **58,** 243T, 1939.
44. P. Finkle, H. D. Draper, and J. H. Hildebrand, J. Am. Chem. Soc. **45,** 2780, 1923.
45. H. Schott and A. E. Royce, J. Pharm. Sci. **72,** 313, 1983.
46. R. C. Merrill, Jr., Ind. Eng. Chem. Anal. Ed. **15,** 743, 1943.
47. E. R. Garrett, J. Pharm. Sci. **51,** 35, 1962; R. D. Vold and R. C. Groot, J. Phys. Chem. **66,** 1969, 1962; R. D. Vold and K. L. Mittal, J. Pharm. Sci. **61,** 869, 1972; S. J. Rehfeld, J. Coll. Interface Sci. **46,** 448, 1974.
48. N. Garti, S. Magdassi, and A. Rubenstein, Drug Dev. Ind. Pharm. **8,** 475, 1982.
49. D. L. Wedderburn, in *Advances in Pharmaceutical Sciences,* Vol. 1, Academic Press, London, 1964, p. 195.
50. P. Sherman, J. Pharm. Pharmacol. **16,** 1, 1964.
51. P. Sherman (Ed.), *Rheology of Emulsions,* Pergamon Press, Oxford, 1963.
52. K. Shinoda and H. Kunieda, J. Coll. Interface Sci. **42,** 381, 1973; K. Shinoda and S. Friberg, Adv. Coll. Interface Sci. **44,** 281, 1975.
53. N. J. Kale and J. V. Allen, Jr., Int. J. Pharm. **57,** 87, 1989.
54. D. Attwood and G. Ktistis, Int. J. Pharm. **52,** 165, 1989.
55. G. W. Halbert, J. F. B. Stuart, and A. T. Florence, Int. J. Pharm. **21,** 219, 1984.
56. D. W. Osborne, A. J. I. Ward, and K. J. O'Neill, Drug Dev. Ind. Pharm. **14,** 1203, 1988.
57. E. E. Linn, R. C. Pohland, and T. K. Byrd, Drug Dev. Ind. Pharm. **16,** 899, 1990.
58. H. L. Rosano, J. Soc. Cosmet. Chem. **25,** 601, 1974.
59. L. M. Prince (Ed.), *Microemulsions, Theory and Practice,* Academic Press, New York, 1977.
60. J. M. Wood and J. H. Collett, Drug Dev. Ind. Pharm. **9,** 93, 1983.
61. C. Washington, Int. J. Pharm. **58,** 1, 1990.
62. K. Morimoto, A. Magayasu, S. Fukanoki, K. Morisaka, S.-H. Hyon, and Y. Ikada, Pharm. Res. **6,** 338, 1989.
63. C. M. Ofner III and H. Schott, J. Pharm. Sci. **75,** 790, 1986.
64. C. M. Ofner III and H. Schott, J. Pharm. Sci. **76,** 715, 1987.
65. J. H. Kou, G. L. Amidon, and P. L. Lee, Pharm. Res. **5,** 592, 1988.
66. P. Casparis and E. W. Meyer, Pharm. Acta Helv. **10,** 163, 1935.
67. H. B. Kostenbauder and A. Martin, J. Am. Pharm. Assoc. Sci. Ed. **43,** 401, 1954.
68. A. H. C. Chun, M. S. Thesis, Purdue University, Purdue, Ind., June 1956.
69. W. C. Ling, J. Pharm. Sci. **67,** 218, 1978.
70. A. R. Fassihi and M. S. Parker, J. Pharm. Sci. **77,** 876, 1988.
71. M. Katz, in E. J. Ariens (Ed.), *Drug Design,* Academic Press, New York, 1973.
72. J. T. Carstensen, Drug Dev. Ind. Pharm. **10,** 1277, 1984.
73. R. A. Kenley, M. O. Lee, L. Sukumar, and M. F. Powell, Pharm. Res. **4,** 342, 1987.

74. D. L. Teagarden, B. D. Anderson, and W. J. Petre, Pharm. Res. **6,** 210, 1989.

75. K. Iga, A. Hussain, and T. Kashihara, J. Pharm. Sci. **70,** 939, 1981.

76. J. T. H. Ong and E. Manoukian, Pharm. Res. **5,** 16, 1988.

77. W. I. Higuchi, J. Pharm. Sci. **51,** 802, 1962; J. Pharm. Sci. **56,** 315, 1967.

78. M. M. Rahman, A. Babar, N. K. Patel, and F. M. Plakogiannis, Drug Dev. Ind. Pharm. **16,** 651, 1990.

79. C.-M. Chiang, G. L. Flynn, N. D. Weiner, W. J. Addicks, and G. J. Szpunar, Int. J. Pharm. **49,** 109, 1989.

80. H. Arima, T. Irie, and K. Uekama, Int. J. Pharm. **57,** 107, 1989.

81. K. Uekama, T. Imai, T. Maeda, T. Irie, F. Hirayama, and M. Otagiri, J. Pharm. Sci. **74,** 841, 1985.

82. M. P. Oth and A. J. Moës, Int. J. Pharm. **24,** 275, 1985.

83. A. A. Nyqvist-Mayer, A. F. Borodin, and S. G. Frank, J. Pharm. Sci. **74,** 1192, 1985; J. Pharm. Sci. **75,** 365, 1986.

## Recommended Reading

E. K. Fischer, *Colloidal Dispersions*, National Bureau of Standards, Washington, DC John Wiley & Sons, Inc., New York, Chapman & Hall, Limited, London, 1950.

### CHAPTER LEGACY

**Fifth Edition:** published as Chapter 18 (Coarse Dispersions). Updated by Patrick Sinko.

**Sixth Edition:** published as Chapter 17 (Coarse Dispersions). Updated by Patrick Sinko.

**At the conclusion of this chapter the student should be able to:**

**1** Understand the concept of particle size as it applies to the pharmaceutical sciences.

**2** Discuss the common particle sizes of pharmaceutical preparations and their impact on pharmaceutical processing/preparation.

**3** Be familiar with the units for particle size, area, and volume and typical calculations.

**4** Describe how particles can be characterized and why these methods are important.

**5** Discuss the methods for determining particle size.

**6** Discuss the role and importance of particle shape and surface area.

**7** Understand the methods for determining particle surface area.

**8** State the two fundamental properties for any collection of particles.

**9** Describe what a derived property of a powder is and identify the important derived properties.

Knowledge and control of the size and the size range of particles are of profound importance in pharmacy. Thus, size, and hence surface area, of a particle can be related in a significant way to the physical, chemical, and pharmacologic properties of a drug. Clinically, the particle size of a drug can affect its release from dosage forms that are administered orally, parenterally, rectally, and topically. The successful formulation of suspensions, emulsions, and tablets, from the viewpoints of both physical stability and pharmacologic response, also depends on the particle size achieved in the product. In the area of tablet and capsule manufacture, control of the particle size is essential in achieving the necessary flow properties and proper mixing of granules and powders. These and other factors reviewed by Lees[1] make it apparent that a pharmacist today must possess a sound knowledge of micromeritics.

## PARTICLE SIZE AND SIZE DISTRIBUTION

In a collection of particles of more than one size (in other words, in a polydisperse sample), two properties are important, namely, (a) the shape and surface area of the individual particles and (b) the size range and number or weight of particles present and, hence, the total surface area. Particle size and size distributions will be considered in this section; shape and surface area will be discussed subsequently.

The size of a sphere is readily expressed in terms of its diameter. As the degree of asymmetry of particles increases, however, so does the difficulty of expressing size in terms of a meaningful diameter. Under these conditions, there is no one unique diameter for a particle. Recourse must be made to the use of an *equivalent spherical diameter*, which relates the size of the particle to the diameter of a sphere having the same surface area, volume, or diameter. Thus, the surface diameter, $d_s$, is the diameter of a sphere having the same surface area as the particle in question. The diameter of a sphere having the same volume as the particle is the volume diameter, $d_v$, whereas the projected diameter, $d_p$, is the diameter of a sphere having the same observed area as the particle when viewed normal to its most stable plane. The size can also be expressed as the Stokes diameter, $d_{st}$, which describes an equivalent sphere undergoing sedimentation at the same rate as the asymmetric particle. Invariably, the type of diameter used reflects the method employed to obtain the diameter. As will be seen later, the projected diameter is obtained by microscopic techniques, whereas the Stokes diameter is determined from sedimentation studies on the suspended particles.

## KEY CONCEPT MICROMERITICS

The science and technology of small particles was given the name *micromeritics* by Dalla Valle.[2] Colloidal dispersions are characterized by particles that are too small to be seen in the ordinary microscope, whereas the particles of pharmaceutical emulsions and suspensions and the "fines" of powders fall in the range of the optical microscope. Particles having the size of coarser powders, tablet granulations, and granular salts fall within the sieve range. The approximate size ranges of parti-

cles in pharmaceutical dispersions are listed in Table 18–1. The sizes of other materials, including microorganisms, are given in Tables 18–2 and 18–3. The unit of particle size used most frequently in micromeritics is the micrometer, $\mu$m, also called the micron, $\mu$, and equal to $10^{-6}$ m, $10^{-4}$ cm, and $10^{-3}$ mm. One must not confuse $\mu$m with m$\mu$, the latter being the symbol for a millimicron or $10^{-9}$ m. The millimicron now is most commonly referred to as the nanometer (nm).

## TABLE 18–1
## PARTICLE DIMENSIONS IN PHARMACEUTICAL DISPERSE SYSTEMS

**Particle Size, Diameter**

| Micrometers ($\mu$m) | Millimeters | Approximate Sieve Size | Examples |
|---|---|---|---|
| 0.5–10 | 0.0005–0.010 | – | Suspensions, fine emulsions |
| 10–50 | 0.010–0.050 | – | Upper limit of subsieve range, coarse emulsion particles; flocculated suspension particles |
| 50–100 | 0.050–0.100 | 325–140 | Lower limit of sieve range, fine powder range |
| 150–1000 | 0.150–1.000 | 100–18 | Coarse powder range |
| 1000–3360 | 1.000–3.360 | 18–6 | Average granule size |

Any collection of particles is usually polydisperse. It is therefore necessary to know not only the size of a certain particle but also how many particles of the same size exist in the sample. Thus, we need an estimate of the size range present and the number or weight fraction of each particle size. This is the particle-size distribution, and from it we can calculate an average particle size for the sample.

If a drug product formulator desires to work with particles of approximately uniform size (i.e., *monodisperse* rather than *polydisperse*), he or she may obtain batches of latex particles as small as 0.060 $\mu$m (60 nm) in diameter with a standard deviation, $\sigma$, of $\pm 0.012$ $\mu$m and particles as large as 920 $\mu$m (0.920 nm) with $\sigma = \pm 32.50$. Such particles of uniform size[3] are used in science, medicine, and technology for various diagnostic tests; as particle-size standards for particle analyzers; for the accurate determination of pore sizes in filters; and as uniformly sized surfaces upon which antigens can be coated for effective immunization. Nanosphere Size Standards[4] are available in 22 sizes, from 21 nm (0.021 $\mu$m) to 900 nm (0.9 $\mu$m or 0.0009 mm) in diameter for instrument calibration and quality control in the manufacture of submicron-sized products such as liposomes, nanoparticles, and microemulsions.

## Average Particle Size

Suppose we have conducted a microscopic examination of a sample of a powder and recorded the number of particles lying within various size ranges. Data from such a determination are shown in **Table 18–4**. To compare these values with those from, say, a second batch of the same material, we

## TABLE 18–2
## A SCALE OF THE RANGES OF VARIOUS SMALL PARTICLES, TOGETHER WITH THE WAVELENGTH OF LIGHT AND OTHER ELECTROMAGNETIC WAVES THAT ILLUMINATE MATERIALS FOUND IN THESE SIZE RANGES

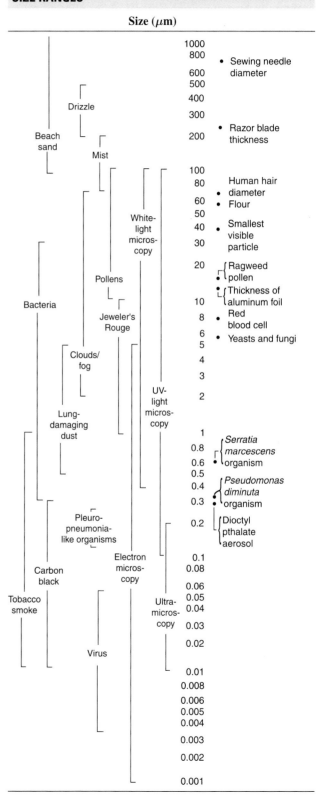

**TABLE 18–3**
### ROD LENGTH AND DIAMETER OF VARIOUS MICROORGANISMS

| Organism | Rod Length ($\mu$m) | Rod or Coccus Diameter ($\mu$m) | Significance |
|---|---|---|---|
| *Acetobacter melanogenus* | 1.0–2.0 | 0.4–0.8 | Strong beer/vinegar bacterium |
| *Alcaligenes viscolactis* | 0.8–2.6 | 0.6–1.0 | Causes ropiness in milk |
| *Bacillus anthracis* | 3.0–10.0 | 1.0–1.3 | Causes anthrax in mammals |
| *B. stearothermophilus* | 2.0–5.0 | 0.6–1.0 | Biologic indicator for steam sterilization |
| *B. subtilis* | 2.0–3.0 | 0.7–0.8 | Biologic indicator for ethylene oxide sterilization |
| *Clostridium botulinum* (B) | 3.0–8.0 | 0.5–0.8 | Produces exotoxin causing botulism |
| *C. perfringens* | 4.0–8.0 | 1.0–1.5 | Produces toxin causing food poisoning |
| *C. tetani* | 4.0–8.0 | 0.4–0.6 | Produces exotoxin causing tetanus |
| *Diplococcus pneumoniae* | | 0.5–1.25 | Causes lobar pneumonia |
| *Erwinia aroideae* | 2.0–3.0 | 0.5 | Causes soft rot in vegetables |
| *Escherichia coli* | 1.0–3.0 | 0.5 | Indicator of fecal contamination in water |
| *Haemophilus influenzae* | 0.5–2.0 | 0.2–0.3 | Causes influenza and acute respiratory infections |
| *Klebsiella pneumoniae* | 5.0 | 0.3–0.5 | Causes pneumonia and other respiratory inflammations |
| *Lactobacillus delbrueckii* | 2.0–9.0 | 0.5–0.8 | Causes souring of grain mashes |
| *Leuconostoc mesenteroides* | | 0.9–1.2 | Causes slime in sugar solutions |
| *Mycoplasma pneumoniae* (PPLO) | | 0.3–0.5 | Smallest known free-living organism |
| *Pediococcus acidilactici* | | 0.6–1.0 | Causes mash spoilage in brewing |
| *P. cerevisiae* | | 1.0–1.3 | Causes deterioration in beer |
| *Pseudomonas diminuta* | 1.0 | 0.3 | Test organism for retention of 0.2-$\mu$m membranes |
| *Salmonella enteritidis* | 2.0–3.0 | 0.6–0.7 | Causes food poisoning |
| *S. hirschfeldii* | 1.0–2.5 | 0.3–0.5 | Causes enteric fever |
| *S. typhimurium* | 1.0–1.5 | 0.5 | Causes food poisoning in humans |
| *S. typhosa* | 2.0–3.0 | 0.6–0.7 | Causes typhoid fever |
| *Sarcina maxima* | | 4.0–4.5 | Isolated from fermenting malt mash |
| *Serratia marcescens* | 0.5–1.0 | 0.5 | Test organism for retention of 0.45-$\mu$m membranes |
| *Shigella dysenteriae* | 1.0–3.0 | 0.4–0.6 | Causes dysentery in humans |
| *Staphylococcus aureus* | | 0.8–1.0 | Causes pus-forming infections |
| *Streptococcus lactis* | | 0.5–1.0 | Contaminant in milk |
| *S. pyogenes* | | 0.6–1.0 | Causes pus-forming infections |
| *Vibrio percolans* | 1.5–1.8 | 0.3–0.4 | Test organism for retention of 0.2-$\mu$m membranes |

**TABLE 18–4**
### CALCULATION OF STATISTICAL DIAMETERS FROM DATA OBTAINED BY USE OF THE MICROSCOPIC METHOD (NORMAL DISTRIBUTION)

| Size Range ($\mu$m) | Mean of Size Range ($\mu$m) | Number of Particles in Each Size Range, $n$ | $nd$ | $nd^2$ | $nd^3$ | $nd^4$ |
|---|---|---|---|---|---|---|
| 0.50–1.00 | 0.75 | 2 | 1.50 | 1.13 | 0.85 | 0.64 |
| 1.00–1.50 | 1.25 | 10 | 12.50 | 15.63 | 19.54 | 24.43 |
| 1.50–2.00 | 1.75 | 22 | 38.50 | 67.38 | 117.92 | 206.36 |
| 2.00–2.50 | 2.25 | 54 | 121.50 | 273.38 | 615.11 | 1384.00 |
| 2.50–3.00 | 2.75 | 17 | 46.75 | 128.56 | 353.54 | 972.24 |
| 3.00–3.50 | 3.25 | 8 | 26.00 | 84.50 | 274.63 | 892.55 |
| 3.50–4.00 | 3.75 | 5 | 18.75 | 70.31 | 263.66 | 988.73 |
| | | $\sum n = 118$ | $\sum nd = 265.50$ | $\sum nd^2 = 640.89$ | $\sum nd^3 = 1645.25$ | $\sum nd^4 = 4468.95$ |

## TABLE 18–5
## STATISTICAL DIAMETERS*

| $\left(\dfrac{\sum nd^{p+f}}{\sum nd^{f}}\right)^{1/p}$ | $p$ | $f$ | Type of Mean | Size Parameter | Frequency | Mean Diameter | Value for Data in Table 18–4 ($\mu$m) | Comments |
|---|---|---|---|---|---|---|---|---|
| $\dfrac{\sum nd}{\sum n}$ | 1 | 0 | Arithmetic | Length | Number | Length-number mean, $d_{\mathrm{ln}}$ | 2.25 | Satisfactory if size range is narrow and distribution is normal; these conditions are rarely found in pharmaceutical powders. |
| $\sqrt{\dfrac{\sum nd^{2}}{\sum n}}$ | 2 | 0 | Arithmetic | Surface | Number | Surface-number mean, $d_{\mathrm{sn}}$ | 2.33 | Refers to particle having average surface area |
| $\sqrt{\dfrac{\sum nd^{3}}{\sum n}}$ | 3 | 0 | Arithmetic | Volume | Number | Volume-number mean, $d_{\mathrm{vn}}$ | 2.41 | Refers to particle having average weight and is related inversely to $N$, the number of particles per gram of material |
| $\dfrac{\sum nd^{2}}{\sum nd}$ | 1 | 1 | Arithmetic | Length | Length | Surface-length or length-weighted mean, $d_{\mathrm{sl}}$ | 2.41 | No practical significance |
| $\dfrac{\sum nd^{3}}{\sum nd^{2}}$ | 1 | 2 | Arithmetic | Length | Surface | Volume-surface or surface-weighted mean, $d_{\mathrm{vs}}$ | 2.57 | Important pharmaceutically because inversely related to $S_{\mathrm{w}}$, the specific surface |
| $\dfrac{\sum nd^{4}}{\sum nd^{3}}$ | 1 | 3 | Arithmetic | Length | Weight | Weight-moment or volume-weighted mean, $d_{\mathrm{wm}}$ | 2.72 | Limited pharmaceutical significance |

*Modified from I. C. Edmundson, in H. S. Bean, J. E. Carless, and A. H. Beckett (Eds.), *Advances in Pharmaceutical Sciences,* Vol. 2, Academic Press, London, 1967, p. 950. With permission.

usually compute an average or mean diameter as our basis for comparison.

Edmundson[5] derived a general equation for the average particle size, whether it be an arithmetic, a geometric, or a harmonic mean diameter:

$$d_{\mathrm{mean}} = \left(\frac{\sum nd^{p+f}}{\sum nd^{f}}\right)^{1/p} \qquad (18\text{–}1)$$

In equation (18–1), $n$ is the number of particles in a size range whose midpoint, $d$, is one of the equivalent diameters mentioned previously. The term $p$ is an index related to the size of an individual particle, because $d$ raised to the power $p = 1$, $p = 2$, or $p = 3$ is an expression of the particle length, surface, or volume, respectively. The value of the index $p$ also decides whether the mean is arithmetic ($p$ is positive), geometric ($p$ is zero), or harmonic ($p$ is negative). For a collection of particles, the frequency with which a particle in a certain size range occurs is expressed by $nd^{f}$. When the frequency index, $f$, has values of 0, 1, 2, or 3, then the size frequency distribution is expressed in terms of the total number, length, surface, or volume of the particles, respectively.

Some of the more significant arithmetic ($p$ is positive) mean diameters are shown in **Table 18–5**. These are based on the values of $p$ and $f$ used in equation (18–1). The diameters calculated from the data in **Table 18–4** are also included. For a more complete description of these diameters, refer to the work of Edmundson.[5]

## Particle-Size Distribution

When the number, or weight, of particles lying within a certain size range is plotted against the size range or mean particle size, a so-called *frequency distribution curve* is obtained. Typical examples are shown in **Figures 18–1** (based on **Table 18–4**) and **18–2** (based on **Table 18–6**). Such plots give a visual representation of the distribution that an average diameter cannot achieve. This is important because it is possible to have two samples with the same average diameter but different distributions. Moreover, it is immediately apparent from a frequency distribution curve what particle size occurs most frequently within the sample. This is termed the *mode*.

An alternative method of representing the data is to plot the cumulative percentage over or under a particular size versus

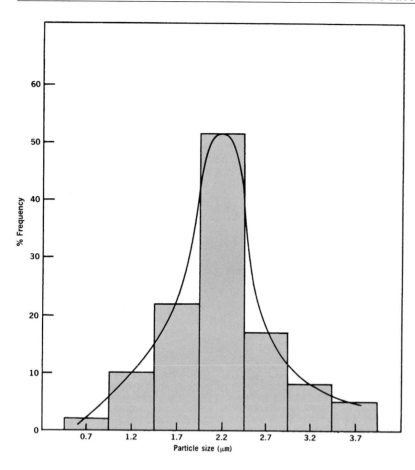

**Fig. 18–1.** A plot of the data of Table 18–4 so as to yield a size–frequency distribution. The data are plotted as a bar graph or *histogram*, and a super-imposed smooth line or frequency curve is shown drawn through the histogram.

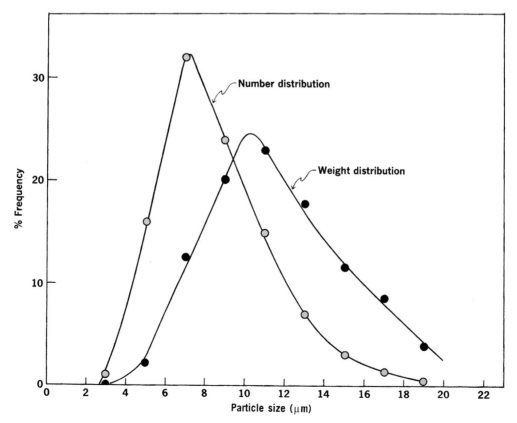

**Fig. 18–2.** Frequency distribution plot of the data in Table 18–6.

**TABLE 18–6**
## CONVERSION OF NUMBER DISTRIBUTION TO WEIGHT DISTRIBUTION (LOG-NORMAL DISTRIBUTION)

| (1)<br>Size<br>Range ($\mu$m) | (2)<br>Mean of Size<br>Range, $d$ ($\mu$m) | (3)<br>Number of<br>Particles<br>in Each Size<br>Range, $n$ | (4)<br>Percent, $n$ | (5)<br>Cumulative<br>Percent<br>Frequency<br>Undersize<br>(Number) | (6)<br>$nd$ | (7)<br>$nd^2$ | (8)<br>$nd^3$ | (9)<br>Percent<br>$nd^3$<br>(Weight) | (10)<br>Cumulative<br>Percent<br>Frequency<br>Undersize<br>(Weight) |
|---|---|---|---|---|---|---|---|---|---|
| 2.0–4.0 | 3.0 | 2 | 1.0 | 1.0 | 6 | 18 | 54 | 0.03 | 0.03 |
| 4.0–6.0 | 5.0 | 32 | 16.0 | 17.0 | 160 | 800 | 4000 | 2.31 | 2.34 |
| 6.0–8.0 | 7.0 | 64 | 32.0 | 49.0 | 448 | 3136 | 21952 | 12.65 | 14.99 |
| 8.0–10.0 | 9.0 | 48 | 24.0 | 73.0 | 432 | 3888 | 34992 | 20.16 | 35.15 |
| 10.0–12.0 | 11.0 | 30 | 15.0 | 88.0 | 330 | 3630 | 39930 | 23.01 | 58.16 |
| 12.0–14.0 | 13.0 | 14 | 7.0 | 95.0 | 182 | 2366 | 30758 | 17.72 | 75.88 |
| 14.0–16.0 | 15.0 | 6 | 3.0 | 98.0 | 90 | 1350 | 20250 | 11.67 | 87.55 |
| 16.0–18.0 | 17.0 | 3 | 1.5 | 99.5 | 51 | 867 | 14739 | 8.49 | 96.04 |
| 18.0–20.0 | 19.0 | 1 | 0.5 | 100.0 | 19 | 361 | 6859 | 3.95 | 99.99 |
| | | $\sum n = 200$ | | | | | | | |

particle size. This is done in **Figure 18–3**, using the cumulative percent undersize (column 5, **Table 18–6**). A sigmoidal curve results, with the mode being that particle size at the greatest slope.

The reader should be familiar with the concept of a *normal* distribution. As the name implies, the distribution is symmetric around the mean, which is also the mode.

The standard deviation, $\sigma$, is an indication of the distribution about the mean.* In a normal distribution, 68% of the population lies $\pm 1 \sigma$ from the mean, 95.5% lies within the mean $\pm 2 \sigma$, and 99.7% lies within the mean $\pm 3 \sigma$. The normal distribution, shown in **Figure 18–1**, is not commonly found in pharmaceutical powders, which are frequently processed by milling or precipitation.[6] Rather, these systems tend to have an nonsymmetric, or skewed, distribution of the type depicted in **Figure 18–2**. When the data in **Figure 18–2** (taken from **Table 18–6**) are plotted as frequency versus the *logarithm* of the particle diameter, a typical bell-shaped curve is frequently obtained. This is depicted in **Figure 18–4**. A size distribution fitting this pattern is spoken of as a *log-normal distribution*, in contrast to the normal distribution shown in **Figure 18–1**.

A log-normal distribution has several properties of interest. When the logarithm of the particle size is plotted against the cumulative percent frequency on a probability scale, a linear relationship is observed (**Fig. 18–5**). Such a linear plot has the distinct advantage that we can now characterize a log-normal distribution *curve* by means of two parameters—the slope of the line and a reference point. Knowing these two

parameters, we can reproduce **Figure 18–5** and, by working back, can come up with a good approximation of **Figure 18–2**, **Figure 18–3**, or **Figure 18–4**. The reference point used is the logarithm of the particle size equivalent to 50% on the probability scale, that is, the 50% size. This is known as the *geometric mean diameter* and is given the symbol $d_g$. The slope is given by the geometric standard deviation, $\sigma_g$, which is the quotient of the ratio (84% undersize or 16% oversize)/(50% size) or (50% size)/(16% undersize or 84% oversize). This is simply the slope of the straight line. In **Figure 18–5**, for the number distribution data, $d_g = 7.1$ $\mu$m and $\sigma_g = 1.43$. Sano et al.[7] used a spherical agglomeration technique with soluble polymers and surfactants to increase the dissolution rate of the poorly soluble crystals of tolbutamide. The spherical particles were free flowing and yielded log probability plots as shown in **Figure 18–5**. The dissolution of the tolbutamide agglomerates followed the Hixon–Crowell cube root equation, as did the dissolution rate of tolbutamide crystals alone.

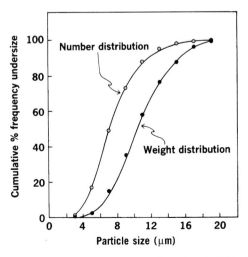

**Fig. 18–3.** Cumulative frequency plot of the data in Table 18–6.

---

*The statistic $\sigma$ is the standard deviation of a very large number of measurements approximating the total population or universe of particles. Because the particle sample measured in pharmaceutical systems ordinarily is small relative to the universe, the statistic used to express the variability of a sample is usually written as $s$ rather than $\sigma$. Authors of works on particle-size analysis frequently do not make a distinction between $\sigma$ and $s$, a practice that is followed in this chapter.

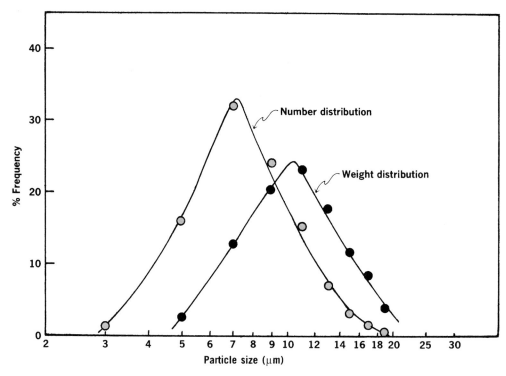

**Fig. 18–4.** Frequency distribution plot of the data in Table 18–6 showing log-normal relation.

## Number and Weight Distributions

The data in **Table 18–6** are shown as a number distribution, implying that they were collected by a counting technique such as microscopy. Frequently, we are interested in obtaining

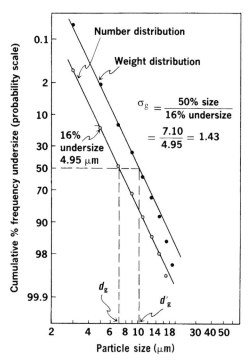

**Fig. 18–5.** Log probability plots of the data in Table 18–6.

data based on a weight, rather than a number, distribution. Although this can be achieved by using a technique such as sedimentation or sieving, it will be more convenient, if the number data are already at hand, to convert the number distribution to a weight distribution and vice versa.

Two approaches are available. Provided the general shape and density of the particles are independent of the size range present in the sample, an estimate of the weight distribution of the data in **Table 18–6** can be obtained by calculating the values shown in columns 9 and 10. These are based on $nd^3$ in column 8. These data have been plotted alongside the number distribution data in **Figures 18–2** and **18–3**, respectively.

The significant differences in the two distributions are apparent, even though they relate to the same sample. For example, in **Figure 18–3**, only 12% of the sample by number is greater than 11 $\mu$m, yet these same particles account for 42% of the total weight of the particles. For this reason, it is important to distinguish carefully between size distributions on a weight and a number basis. Weight distributions can also be plotted in the same manner as the number distribution data, as seen in **Figures 18–4** and **18–5**. Note that in **Figure 18–5** the slope of the line for the weight distribution is identical with that for the number distribution. Thus, the geometric standard deviation on a weight basis, $\sigma'_g$, also equals 1.43. Customarily, the prime is dropped because the value is independent of the type of distribution. The geometric mean diameter (the particle size at the 50% probability level) on a weight basis, $d'_g$, is 10.4 $\mu$m, whereas $d_g = 7.1$ $\mu$m.

Provided the distribution is log-normal, the second approach is to use one of the equations developed by Hatch

## HATCH–CHOATE EQUATIONS FOR COMPUTING STATISTICAL DIAMETERS FROM NUMBER AND WEIGHT DISTRIBUTIONS

| Diameter | Number Distribution | Weight Distribution |
|---|---|---|
| Length-number mean | $\log d_{\mathrm{ln}} = \log d_{\mathrm{g}} + 1.151 \log^2 \sigma_{\mathrm{g}}$ | $\log d_{\mathrm{ln}} = \log d_{\mathrm{g}}' - 5.757 \log^2 \sigma_{\mathrm{g}}$ |
| Surface-number mean | $\log d_{\mathrm{sn}} = \log d_{\mathrm{g}} + 2.303 \log^2 \sigma_{\mathrm{g}}$ | $\log d_{\mathrm{sn}} = \log d_{\mathrm{g}}' - 4.606 \log^2 \sigma_{\mathrm{g}}$ |
| Volume-number mean | $\log d_{\mathrm{vn}} = \log d_{\mathrm{g}} + 3.454 \log^2 \sigma_{\mathrm{g}}$ | $\log d_{\mathrm{vn}} = \log d_{\mathrm{g}}' - 3.454 \log^2 \sigma_{\mathrm{g}}$ |
| Volume-surface mean | $\log d_{\mathrm{vs}} = \log d_{\mathrm{g}} + 5.757 \log^2 \sigma_{\mathrm{g}}$ | $\log d_{\mathrm{vs}} = \log d_{\mathrm{g}}' - 1.151 \log^2 \sigma_{\mathrm{g}}$ |
| Weight-moment mean | $\log d_{\mathrm{wm}} = \log d_{\mathrm{g}} + 8.059 \log^2 \sigma_{\mathrm{g}}$ | $\log d_{\mathrm{wm}} = \log d_{\mathrm{g}}' + 1.151 \log^2 \sigma_{\mathrm{g}}$ |

and Choate.[8] By this means, it is possible to convert number distributions to weight distributions with a minimum of calculation. In addition, a particular average can be readily computed by use of the relevant equation. The Hatch–Choate equations are listed in **Table 18–7**.

**EXAMPLE 18–1**

**Using Distribution Data**

From the number distribution data in Table 18–6 and Figure 18–5, it is found that $d_{\mathrm{g}} = 7.1$ $\mu$m and $\sigma_{\mathrm{g}} = 1.43$, or $\log \sigma_{\mathrm{g}} = 0.1553$. Using the relevant Hatch–Choate equation, calculate $d_{\mathrm{ln}}$ and $d_{\mathrm{g}}'$. The equation for the length-number mean, $d_{\mathrm{ln}}$, is

$$\log d_{\mathrm{ln}} = \log d_{\mathrm{g}} + 1.151 \log^2 \sigma_{\mathrm{g}}$$
$$= 0.8513 + 1.151(0.1553)^2$$
$$= 0.8513 + 0.0278$$
$$= 0.8791$$
$$d_{\mathrm{ln}} = 7.57 \, \mu\mathrm{m}$$

To calculate $d_{\mathrm{g}}'$, we must substitute into the following Hatch–Choate equation:

$$\log d_{\mathrm{ln}} = \log d_{\mathrm{g}}' - 5.757 \log^2 \sigma_{\mathrm{g}}$$
$$8.791 = \log d_{\mathrm{g}}' - 5.757(0.1553)^2$$

or

$$\log d_{\mathrm{g}}' = 0.8791 + 0.1388$$
$$= 1.0179$$
$$d_{\mathrm{g}}' = 10.4 \, \mu\mathrm{m}$$

One can also use an equation suggested by Rao,[9]

$$d_{\mathrm{g}}' = d_{\mathrm{g}} \sigma_{\mathrm{g}}^{(3 \ln \sigma_{\mathrm{g}})} \tag{18–2}$$

to readily obtain $d_{\mathrm{g}}'$ knowing $d_{\mathrm{g}}$ and $\sigma_{\mathrm{g}}$. In this example,

$$d_{\mathrm{g}}' = 7.1(1.43)^{(3 \ln 1.43)}$$
$$= 10.42$$

The student should confirm that substitution of the relevant data into the remaining Hatch–Choate equations in Table 18–7 yields the following statistical diameters:

$$d_{\mathrm{sn}} = 8.07 \, \mu\mathrm{m}; \qquad d_{\mathrm{vn}} = 8.60 \, \mu\mathrm{m};$$
$$d_{\mathrm{vs}} = 9.78 \, \mu\mathrm{m}; \qquad d_{\mathrm{wm}} = 11.11 \, \mu\mathrm{m}$$

## Particle Number

A significant expression in particle technology is the *number of particles per unit weight, N*, which is expressed in terms of $d_{\mathrm{vn}}$.

The number of particles per unit weight is obtained as follows. Assume that the particles are spheres, the volume of a single particle is $\pi d_{\mathrm{vn}}^3 / 6$, and the mass (volume × density) is $(\pi d_{\mathrm{vn}}^3 \rho)/6$ g per particle. The number of particles per gram is then obtained from the proportion

$$\frac{(\pi d_{\mathrm{vn}}^3 \rho)/6 \text{ g}}{1 \text{ particle}} = \frac{1 \text{ g}}{N} \tag{18–3}$$

and

$$N = \frac{6}{\pi d_{\mathrm{vn}}^3 \rho} \tag{18–4}$$

**EXAMPLE 18–2**

**Number of Particles**

The mean volume number diameter of the powder, the data for which are given in Table 18–4, is 2.41 $\mu$m, or $2.41 \times 10^{-4}$ cm. If the density of the powder is 3.0 g/cm$^3$, what is the number of particles per gram?

We have

$$N = \frac{6}{3.14 \times (2.41 \times 10^{-4})^3 \times 3.0} = 4.55 \times 10^{10}$$

## METHODS FOR DETERMINING PARTICLE SIZE

Many methods are available for determining particle size. Only those that are widely used in pharmaceutical practice and are typical of a particular principle are presented. For a detailed discussion of the numerous methods of particle size analysis, consult the work by Edmundson[5] and Allen[10] and the references given there to other sources. The methods available for determining the size characteristics of submicrometer particles were reviewed by Groves.[11]

Microscopy, sieving, sedimentation, and the determination of particle volume are discussed in the following section. None of the measurements are truly direct methods. Although the microscope allows the observer to view the actual particles, the results obtained are probably no more "direct" than those resulting from other methods because only two of the three particle dimensions are ordinarily seen. The sedimentation methods yield a particle size relative to the rate at which particles settle through a suspending medium, a measurement important in the development of emulsions and suspensions. The measurement of particle volume, using an

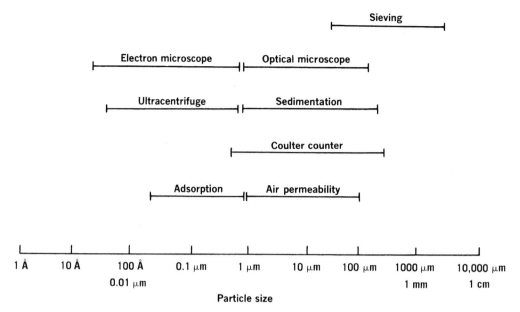

**Fig. 18–6.** Approximate size ranges of methods used for particle-size and specific-surface analysis.

apparatus called the Coulter counter, allows one to calculate an equivalent volume diameter.

However, the technique gives no information as to the shape of the particles. Thus, in all these cases, the size may or may not compare with that obtained by the microscope or by other methods; the size is most directly applicable to the analysis for which it is intended. A guide to the range of particle sizes applicable to each method is given in **Figure 18–6.**

## Optical Microscopy

It should be possible to use the ordinary microscope for particle-size measurement in the range of 0.2 to about 100 $\mu$m. According to the microscopic method, an emulsion or suspension, diluted or undiluted, is mounted on a slide or ruled cell and placed on a mechanical stage. The microscope eyepiece is fitted with a micrometer by which the size of the particles can be estimated. The field can be projected onto a screen where the particles are measured more easily, or a photograph can be taken from which a slide is prepared and projected on a screen for measurement.

The particles are measured along an arbitrarily chosen fixed line, generally made horizontally across the center of the particle. Popular measurements are the *Feret diameter*, the *Martin diameter*,[12] and the *projected area diameter*, all of which can be defined by reference to **Figure 18–7**, as suggested by Allen.[13] Martin's diameter is the length of a line that bisects the particle image. The line can be drawn in any direction but must be in the same direction for all particles measured. The Martin diameter is identified by the number 1 in **Figure 18–7.** Feret's diameter, corresponding to the number 2 in the figure, is the distance between two tangents on opposite sides of the particle parallel to some fixed direction, the *y* direction in the figure. The third measurement, number 3 in **Figure 18–7**, is the projected area diameter. It is the diameter

of a circle with the same area as that of the particle observed perpendicular to the surface on which the particle rests.

A size–frequency distribution curve can be plotted as in **Figure 18–1** for the determination of the statistical diameters of the distribution. Electronic scanners have been developed to remove the necessity of measuring the particles by visual observation.

Prasad and Wan[14] used video recording equipment to observe, record, store, and retrieve particle-size data from a microscopic examination of tablet excipients, including microcrystalline cellulose, sodium carboxymethylcellulose, sodium starch glycolate, and methylcellulose. The projected area of the particle profile, Feret's diameter, and various shape factors (elongation, bulkiness, and surface factor) were

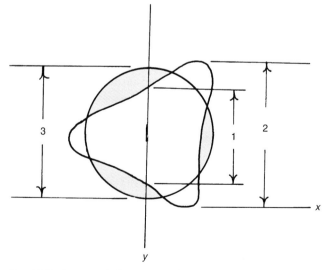

**Fig. 18–7.** A general diagram providing definitions of the Feret, Martin, and projected diameters. (From T. Allen, *Particle Size Measurements*, 2nd Ed., Chapman Hall, London, 1974, p. 131. With permission.)

determined. The video recording technique was found to be simple and convenient for microscopic examination of excipients.

A disadvantage of the microscopic method is that the diameter is obtained from only two dimensions of the particle: length and breadth. No estimation of the depth (thickness) of the particle is ordinarily available. In addition, the number of particles that must be counted (300–500) to obtain a good estimation of the distribution makes the method somewhat slow and tedious. Nonetheless, microscopic examination (photomicrographs) of a sample should be undertaken even when other methods of particle-size analysis are being used, because the presence of agglomerates and particles of more than one component may often be detected.

## Sieving

This method uses a series of standard sieves calibrated by the National Bureau of Standards. Sieves are generally used for grading coarser particles; if extreme care is used, however, they can be employed for screening material as fine as 44 $\mu$m (No. 325 sieve). Sieves produced by photoetching and electroforming techniques are available with apertures from 90 $\mu$m to as low as 5 $\mu$m. According to the method of the *U. S. Pharmacopeia* for testing powder fineness, a mass of sample is placed on the proper sieve in a mechanical shaker. The powder is shaken for a definite period of time, and the material that passes through one sieve and is retained on the next finer sieve is collected and weighed.

Another approach is to assign the particles on the lower sieve with the arithmetic or geometric mean size of the two screens. Arambulo and Deardorff[15] used this method of size classification in their analysis of the average weight of compressed tablets. Frequently the powder is assigned the mesh number of the screen through which it passes or on which it is retained. King and Becker[16] expressed the size ranges of calamine samples in this way in their study of calamine lotion.

When a detailed analysis is desired, the sieves can be arranged in a nest of about five with the coarsest at the top. A carefully weighed sample of the powder is placed on the top sieve, and after the sieves are shaken for a predetermined period of time, the powder retained on each sieve is weighed. Assuming a log-normal distribution, one plots the cumulative percent by weight of powder retained on the sieves on the probability scale against the logarithm of the arithmetic mean size of the openings of each of two successive screens. As illustrated in **Figure 18–5**, the geometric mean weight diameter, $d'_g$, and the geometric standard deviation, $\sigma_g$, can be obtained directly from the straight line.

According to Herdan,[17] sieving errors can arise from a number of variables including sieve loading and duration and intensity of agitation. Fonner et al.[18] demonstrated that sieving can cause attrition of granular pharmaceutical materials. Care must be taken, therefore, to ensure that reproducible techniques are employed so that different particle-size distributions between batches of material are not due simply to different sieving conditions.

## Sedimentation

The application of ultracentrifugation to the determination of the molecular weight of high polymers has already been discussed. The particle size in the subsieve range can be obtained by gravity sedimentation as expressed in Stokes's law,

$$v = \frac{h}{t} = \frac{d_{st}^2(\rho_s - \rho_0)g}{18\eta_0} \qquad (18\text{--}5)$$

or

$$d_{st} = \sqrt{\frac{18\eta_0 h}{(\rho_s - \rho_0)gt}} \qquad (18\text{--}6)$$

where $v$ is the rate of settling, $h$ is the distance of fall in time $t$, $d_{st}$ is the mean diameter of the particles based on the velocity of sedimentation, $\rho_s$ is the density of the particles and $\rho_0$ that of the dispersion medium, $g$ is the acceleration due to gravity, and $\eta_0$ is the viscosity of the medium. The equation holds exactly only for spheres falling freely without hindrance and at a constant rate. The law is applicable to irregularly shaped particles of various sizes as long as one realizes that the diameter obtained is a relative particle size equivalent to that of sphere falling at the same velocity as that of the particles under consideration. The particles must not be aggregated or clumped together in the suspension because such clumps would fall more rapidly than the individual particles and erroneous results would be obtained. The proper deflocculating agent must be found for each sample that will keep the particles free and separate as they fall through the medium.

### EXAMPLE 18–3

**Stokes Diameter**

A sample of powdered zinc oxide, density 5.60 g/cm$^3$, is allowed to settle under the acceleration of gravity, 981 cm/sec$^2$, at 25°C. The rate of settling, $v$, is 7.30 $\times$ 10$^{-3}$ cm/sec; the density of the medium is 1.01 g/cm$^3$, and its viscosity is 1 centipoise = 0.01 poise or 0.01 g/cm sec. Calculate the Stokes diameter of the zinc oxide powder. We have

$$d_{st} = \sqrt{\frac{(18 \times 0.01 \text{ g/cm sec}) \times (7.30 \times 10^{-3} \text{ cm/sec})}{(5.60 - 1.01 \text{ g/cm}^3) \times (981 \text{ cm/sec}^2)}}$$

$$= 5.40 \times 10^{-4} \text{ cm or } 5.40 \,\mu\text{m}$$

For Stokes's law to apply, a further requirement is that the flow of dispersion medium around the particle as it sediments is *laminar* or *streamline*. In other words, the rate of sedimentation of a particle must not be so rapid that turbulence is set up, because this in turn will affect the sedimentation of the particle. Whether the flow is turbulent or laminar is indicated by the dimensionless *Reynolds number*, $R_e$, which is defined as

$$R_e = \frac{v \, d\rho_0}{\eta_0} \qquad (18\text{--}7)$$

where the symbols have the same meaning as in equation (18–5). According to Heywood,[19] Stokes's law cannot be used if

$R_e$ is greater than 0.2 because turbulence appears at this value. On this basis, the limiting particle size under a given set of conditions can be calculated as follows.

Rearranging equation (18–7) and combining it with equation (18–5) gives

$$v = \frac{R_e \eta}{d \rho_0} = \frac{d^2 (\rho_s - \rho_0) g}{18 \eta} \qquad (18\text{–}8)$$

and thus

$$d^3 = \frac{18 R_e \eta^2}{(\rho_s - \rho_0) \rho_0 g} \qquad (18\text{–}9)$$

Under a given set of density and viscosity conditions, equation (18–9) allows calculation of the maximum particle diameter whose sedimentation will be governed by Stokes's law, that is, when $R_e$ does not exceed 0.2.

### EXAMPLE 18–4

**Largest Particle Size**

A powdered material, density 2.7 g/cm$^3$, is suspended in water at 20°C. What is the size of the largest particle that will settle without causing turbulence? The viscosity of water at 20°C is 0.01 poise, or g/cm sec, and the density is 1.0 g/cm$^3$.

From equation (18–9),

$$d^3 = \frac{(18)(0.2)(0.01)^2}{(2.7 - 1.0)1.0 \times 981}$$

$$d = 6 \times 10^{-3} \text{ cm} = 60 \, \mu m$$

### EXAMPLE 18–5

**Particle Size, Setting, and Viscosity**

If the material used in *Example 18–4* is now suspended in a syrup containing 60% by weight of sucrose, what will be the critical diameter, that is, the maximum diameter for which $R_e$ does not exceed 0.2? The viscosity of the syrup is 0.567 poise, and the density is 1.3 g/cm$^3$.

We have

$$d^3 = \frac{(18)(0.2)(0.567)^2}{(2.7 - 1.3)1.3 \times 981}$$

$$d = 8.65 \times 10^{-2} \text{ cm} = 865 \, \mu m$$

Several methods based on sedimentation are used. Principal among these are the pipette method, the balance method, and the hydrometer method. Only the first technique is discussed here because it combines ease of analysis, accuracy, and economy of equipment.

The Andreasen apparatus is shown in **Figure 18–8**. It usually consists of a 550 mL vessel containing a 10 mL pipette sealed into a ground-glass stopper. When the pipette is in place in the cylinder, its lower tip is 20 cm below the surface of the suspension.

The analysis is carried out in the following manner. A 1% or 2% suspension of the particles in a medium containing a suitable deflocculating agent is introduced into the vessel and brought to the 550 mL mark. The stoppered vessel is shaken to distribute the particles uniformly throughout the suspension, and the apparatus, with pipette in place, is clamped securely in a constant-temperature bath. At various time intervals,

**Fig. 18–8.** Andreasen apparatus for determining particle size by the gravity sedimentation method.

10 mL samples are withdrawn and discharged by means of the two-way stopcock. The samples are evaporated and weighed or analyzed by other appropriate means, correcting for the deflocculating agent that has been added.

The particle diameter corresponding to the various time periods is calculated from Stokes's law, with $h$ in equation (18–6) being the height of the liquid above the lower end of the pipette at the time each sample is removed. The residue or dried sample obtained at a particular time is the weight fraction having particles of sizes less than the size obtained by the Stokes-law calculation for that time period of settling. The weight of each sample residue is therefore called the *weight undersize*, and the sum of the successive weights is known as the *cumulative weight undersize*. It can be expressed directly in weight units or as percentage of the total weight of the final sediment. Such data are plotted in **Figures 18–2** through **18–4**. The cumulative percentage by weight undersize can then be plotted on a probability scale against the particle diameter on a log scale, as in **Figure 18–5**, and the statistical diameters obtained as explained previously.

## Particle Volume Measurement

A popular instrument for measuring the volume of particles is the Coulter counter (**Fig. 18–9**). This instrument operates on the principle that when a particle suspended in a conducting liquid passes through a small orifice on either side of which are electrodes, a change in electric resistance occurs. In practice, a known volume of a dilute suspension is pumped through the orifice. Provided the suspension is sufficiently dilute, the particles pass through essentially one at a time. A constant voltage is applied across the electrodes to produce a current. As the particle travels through the orifice, it

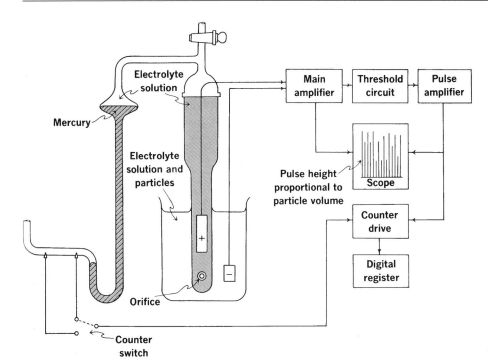

**Fig. 18–9.** Schematic diagram of a Coulter counter, used to determine particle volume.

displaces its own volume of electrolyte, and this results in an increased resistance between the two electrodes. The change in resistance, which is related to the particle volume, causes a voltage pulse that is amplified and fed to a pulse-height analyzer calibrated in terms of particle size. The instrument records electronically all those particles producing pulses that are within two threshold values of the analyzer. By systematically varying the threshold settings and counting the number of particles in a constant sample size, it is possible to obtain a particle-size distribution. The instrument is capable of counting particles at the rate of approximately 4000 per second, and so both gross counts and particle-size distributions are obtained in a relatively short period of time. The data may be readily converted from a volume distribution to a weight distribution.

The Coulter counter has been used to advantage in the pharmaceutical sciences to study particle growth and dissolution[20,21] and the effect of antibacterial agents on the growth of microorganisms.[22]

The use of the Coulter particle-size analyzer together with a digital computer was reported by Beaubien and Vanderwielen[23] for the automated particle counting of milled and micronized drugs. Samples of spectinomycin hydrochloride and a micronized steroid were subjected to particle-size analysis together with polystyrene spheres of 2.0 to 80.0 $\mu$m diameter, which were used to calibrate the apparatus. The powders showed log-normal distributions and were well characterized by geometric volume mean diameters and geometric standard deviations. Accurate particle sizes were obtained between 2 and 80 $\mu$m diameter with a precision of about 0.5 $\mu$m. The authors concluded that the automated Coulter counter was quite satisfactory for quality control of pharmaceutical powders. The Coulter particle counter was used by Ismail and Tawashi[24] to obtain size distributions of the min-

eral part of human kidney (urinary) stones and to determine whether there is a critical size range for stone formation. The study provided a better understanding of the clustering process and the packing of the mineral components of renal stones.

Beckman Coulter also manufactures a *submicron*-particle sizing instrument, the Beckman Coulter Model N5, for analyzing particles in the size range of 0.0033 to 0.3 $\mu$m. By the use of photon correlation spectroscopy, the instrument senses the Brownian motion of the particles in suspension. The smaller a particle, the faster it moves by Brownian motion. A laser beam passes through the sample and a sensor detects the light scattered by the particles undergoing Brownian motion. The Beckman Coulter Model N5 instrument provides not only particle-size and size distribution data but also molecular weights and diffusion coefficients. Submicron size determination is important in pharmacy in the analysis of microemulsions, pigments and dyes, colloids, micelles and solubilized systems, liposomes, and microparticles.

An investigation of contaminant particulate matter in parenteral solutions for adherence to the standards set by the 1986 Italian *Pharmacopóeia* was conducted by Signoretti et al.[25] They studied the number and nature of the particulates in 36 large-volume injectable solutions using scanning electron microscopy and x-ray analysis. About one fifth of the samples showed a considerable number of particles of sizes greater than 20 $\mu$m in diameter. The particles were identified as textile fibers, cellulose, plastic material, and contaminants from the manufacturing and packaging processes, such as pieces of rubber and bits of metal. Because of their number, size, shape, surface properties, and chemical nature, these contaminants can cause vascular occlusions and inflammatory, neoplastic, and allergic reactions. Embolisms may occur with particles larger than 5 $\mu$m.

According to the standards of the Italian *Pharmacopóeia* for parenteral solutions of greater than 100 mL, no more than 100 particles 5 $\mu$m and larger and no more than four particles 20 $\mu$m in diameter and larger may be present in each milliliter of solution. These workers found that a considerable number of the manufacturers failed to produce parenteral preparations within the limits of the Pharmacopóeia, the contaminants probably occurring in most cases from filters, clothing, and container seals.

In the preparation of indomethacin sustained-release pellets, Li et al.[26] used a Microtrac particle-size analyzer (Leeds and Northrup Instruments) to determine the particle size of indomethacin as obtained from the manufacturer and as two types of micronized powder. The powders were also examined under a microscope with a magnification of 400×, and photomicrographs were taken with a Polaroid SX-70 camera. Pellets (referred to as IS pellets) containing indomethacin and Eudragit S-100 were prepared using a fluid bed granulator or a Wurster column apparatus. Eudragit (Röhm Pharma) is an acrylic polymer for the enteric coating of tablets, capsules, and pellets. Its surface properties and chemical structure as a film coating polymer were reviewed by Davies et al.[27] Sieve analysis with U.S. standard sieves Nos. 12, 14, 16, 18, 20, 25, and 35 was used to determine the particle-size distribution of the IS pellets. The yield of IS pellets depended greatly on the particle size of the indomethacin powder. Batches using two micronized powders (average diameter of 3.3 and 6.4 $\mu$m, respectively) produced a higher yield of the IS pellets than did the original indomethacin powder (40.6 $\mu$m) obtained directly from the drug manufacturer. Davies et al.[27] concluded that both the average particle diameter and the particle-size distribution of the indomethacin powder must be considered for maximum yield of the sustained-release pellets.

Carli and Motta[28] investigated the use of microcomputerized mercury porosimetry to obtain particle-size and surface area distributions of pharmaceutical powders. Mercury porosimetry gives the volume of the pores of a powder, which is penetrated by mercury at each successive pressure; the pore volume is converted into a pore-size distribution. The total surface area and particle size of the powder can also be obtained from the mercury porosimetry data.

# PARTICLE SHAPE AND SURFACE AREA

Knowledge of the shape and the surface area of a particle is desirable. The shape affects the flow and packing properties of a powder as well as having some influence on the surface area. The surface area per unit weight or volume is an important characteristic of a powder when one is undertaking surface adsorption and dissolution rate studies.

## Particle Shape

A sphere has minimum surface area per unit volume. The more asymmetric a particle, the greater is the surface area

per unit volume. As discussed previously, a spherical particle is characterized completely by its diameter. As the particle becomes more asymmetric, it becomes increasingly difficult to assign a meaningful diameter to the particle—hence, as we have seen, the need for equivalent spherical diameters. It is a simple matter to obtain the surface area or volume of a sphere because for such a particle

$$\text{Surface area} = \pi d^2 \qquad (18\text{–}10)$$

and

$$\text{Volume} = \frac{\pi d^3}{6} \qquad (18\text{–}11)$$

where $d$ is the diameter of the particle. The surface area and volume of a spherical particle are therefore proportional to the square and cube, respectively, of the diameter. To obtain an estimate of the surface or volume of a particle (or collection of particles) whose shape is not spherical, however, one must choose a diameter that is characteristic of the particle and relate this to the surface area or volume through a correction factor. Suppose the particles are viewed microscopically, and it is desired to compute the surface area and volume from the projected diameter, $d_p$, of the particles. The square and cube of the chosen dimension (in this case, $d_p$) are proportional to the surface area and volume, respectively. By means of proportionality constants, we can then write

$$\text{Surface area} = \alpha_s d_p^2 = \pi d_s^2 \qquad (18\text{–}12)$$

where $\alpha_s$ is the surface area factor and $d_s$ is the equivalent surface diameter. For volume we write

$$\text{Volume} = \alpha_v d_p^3 = \frac{\pi d_v^2}{6} \qquad (18\text{–}13)$$

where $\alpha_v$ is the volume factor and $d_v$ is the equivalent volume diameter. The surface area and volume "shape factors" are, in reality, the ratio of one diameter to another. Thus, for a sphere, $\alpha_s = \pi d_s^2/d_p^2 = 3.142$ and $\alpha_v = \pi d_v^3/6d_p^3 = 0.524$. There are as many of these volume and shape factors as there are pairs of equivalent diameters. The ratio $\alpha_s/\alpha_v$ is also used to characterize particle shape. When the particle is spherical, $\alpha_s/\alpha_v = 6.0$. The more asymmetric the particle, the more this ratio exceeds the minimum value of 6.

## Specific Surface

The specific surface is the surface area per unit volume, $S_v$, or per unit weight, $S_w$, and can be derived from equations (18–12) and (18–13). Taking the general case, for asymmetric particles where the characteristic dimension is not yet defined,

$$S_v = \frac{\text{Surface area of particles}}{\text{Volume of particles}}$$

$$= \frac{n\alpha_s d^2}{n\alpha_v d^3} = \frac{\alpha_s}{\alpha_v d} \qquad (18\text{–}14)$$

where $n$ is the number of particles. The surface area per unit weight is therefore

$$S_w = \frac{S_v}{\rho} \tag{18–15}$$

where $\rho$ is the true density of the particles. Substituting for equation (18–14) in (18–15) leads to the general equation

$$S_w = \frac{\alpha_s}{\rho d_{vs} \alpha_v} \tag{18–16}$$

where the dimension is now defined as $d_{vs}$, the volume–surface diameter characteristic of specific surface. When the particles are spherical (or nearly so), equation (18–16) simplifies to

$$S_w = \frac{6}{\rho d_{vs}} \tag{18–17}$$

because $\alpha_s/\alpha_v = 6.0$ for a sphere.

**EXAMPLE 18–6**

**Surface Area**

What are the specific surfaces, $S_w$ and $S_v$, of particles assumed to be spherical in which $\rho = 3.0$ g/cm$^3$ and $d_{vs}$ from Table 18–5 is 2.57 $\mu$m?
   We have

$$S_w = \frac{6}{3.0 \times 2.57 \times 10^{-4}} = 7.78 \times 10^3 \text{ cm}^2/\text{g}$$

$$S_v = \frac{6}{2.57 \times 10^{-4}} = 2.33 \times 10^4 \text{ cm}^2/\text{cm}^3$$

# METHODS FOR DETERMINING SURFACE AREA

The surface area of a powder sample can be computed from knowledge of the particle-size distribution obtained using one of the methods outlined previously. Two methods are commonly available that permit direct calculation of surface area. In the first, the amount of a gas or liquid solute that is *adsorbed* onto the sample of powder to form a monolayer is a direct function of the surface area of the sample. The second method depends on the fact that the rate at which a gas or liquid *permeates* a bed of powder is related, among other factors, to the surface area exposed to the permeant.

## Adsorption Method

Particles with a large specific surface are good adsorbents for the adsorption of gases and of solutes from solution. In determining the surface of the adsorbent, the volume in cubic centimeters of gas adsorbed per gram of adsorbent can be plotted against the pressure of the gas at constant temperature to give a type II *isotherm* as shown in **Figure 18–10**.

The adsorbed layer is monomolecular at low pressures and becomes multimolecular at higher pressures. The completion of the monolayer of nitrogen on a powder is shown as point $B$ in **Figure 18–10**. The volume of nitrogen gas, $V_m$, in cm$^3$ that

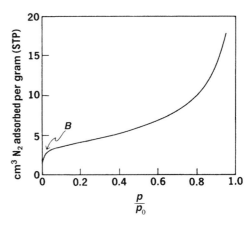

**Fig. 18–10.** Isotherm showing the volume of nitrogen adsorbed on a powder at increasing pressure ratio. Point $B$ represents the volume of adsorbed gas corresponding to the completion of a monomolecular film. Key: STP = standard temperature and pressure.

1 g of the powder can adsorb when the monolayer is complete is more accurately given by using the Brunaver, Emmett, and Teller (BET) equation, which can be written as

$$\frac{p}{V(p_0 - p)} = \frac{1}{V_m b} + \frac{(b-1)p}{V_m b p_0} \tag{18–18}$$

where $V$ is the volume of gas in cm$^3$ adsorbed per gram of powder at pressure $p$, $p_0$ is the saturation vapor pressure of liquefied nitrogen at the temperature of the experiment, and $b$ is a constant that expresses the difference between the heat of adsorption and heat of liquefaction of the adsorbate (nitrogen). Note that at $p/p_0 = 1$, the vapor pressure, $p$, is equal to the saturation vapor pressure.

   An instrument used to obtain the data needed to calculate surface area and pore structure of pharmaceutical powders is the Quantasorb QS-16, manufactured by the Quantachrome Corporation (Boynton, FL). Absorption and desorption of nitrogen gas on the powder sample is measured with a thermal conductivity detector when a mixture of helium and nitrogen is passed through a cell containing the powder. Nitrogen is the absorbate gas; helium is inert and is not adsorbed on the powder surface. A Gaussian or bell-shaped curve is plotted on a strip-chart recorder, the signal height being proportional to the rate of absorption or desorption of nitrogen and the area under the curve being proportional to the gas adsorbed on the particles. Quantasorb and similar instruments have replaced the older vacuum systems constructed of networks of glass tubing. These required long periods of time to equilibrate and were subject to leakage at valves and breaks in the glass lines. The sensitivity of the new instrument is such that small powder samples can be analyzed. Quantasorb's versatility allows the use of a number of individual gases or mixtures of gases as adsorbates over a range of temperatures. The instrument can be used to measure the true density of powdered material and to obtain pore-size and pore-volume distributions. The characteristics of porous materials and the method of analysis are discussed in the following sections.

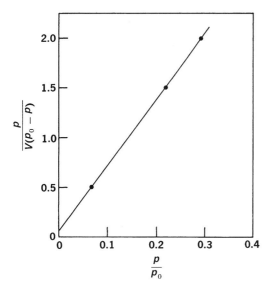

**Fig. 18–11.** A linear plot of the Brunaver, Emmett, and Teller (BET) equation for the adsorption of nitrogen on a powder.

Instead of the graph shown in **Figure 18–10**, a plot of $p/V(p_0 - p)$ against $p/p_0$, as shown in **Figure 18–11**, is ordinarily used to obtain a straight line, the slope and intercept of which yield the values $b$ and $V_m$. The specific surface of the particles is then obtained from

$$S_w = \frac{A_m N}{M/\rho} \times V_m \ cm^3/g$$

$$S_w = \frac{(16.2 \times 10^{-16})(6.02 \times 10^{23})}{22.414 \times 10^4} \times V_m$$

$$S_w = 4.35 \ m^2/cm^3 \times V_m \ cm^3/g \qquad \textbf{(18–19)}$$

where $M/\sigma$ is the molar volume of the gas, 22,414 $cm^3$/mole at standard temperature and pressure (STP), and the factor $10^4$ is included in the denominator to convert square centimeters to square meters. $N$ is Avogadro's number, $6.02 \times 10^{23}$ molecules/mole, and $A_m$ is the area of a single close-packed nitrogen molecule adsorbed as a monolayer on the surface of the particles. Emmett and Brunauer[29] suggested that the value of $A_m$ for nitrogen be calculated from the formula

$$A_m = 1.091 \left( \frac{M}{\rho N} \right)^{2/3} \qquad \textbf{(18–20)}$$

where $M$ is the molecular weight, 28.01 g/mole, of $N_2$; $\sigma$ is the density, 0.81 $g/cm^3$, of $N_2$ at its boiling point, 77 K ($-196°C$); and $N$ is Avogadro's number. The quantity 1.091 is a packing factor for the nitrogen molecules on the surface of the adsorbent. We have

$$A_m = 1.091 \left( \frac{28.01 \ g/mole}{(0.81 \ g/cm^3)(6.02 \times 10^{23} \ molecules/mole)} \right)^{2/3}$$

$$= 16.2 \times 10^{-16} \ cm^2 = 16.2 \ \mathring{A}^2$$

$A_m$ for liquid nitrogen has been obtained by several methods and is generally accepted as 16.2 $\mathring{A}^2$, or $16.2 \times 10^{-16} \ cm^2$.

The specific surface is calculated from equation (**18–19**) and is expressed in square meters per gram.

Experimentally, the volume of nitrogen that is adsorbed by the powder contained in the evacuated glass bulb of the Quantasorb or similar surface-area apparatus is determined at various pressures and the results are plotted as shown in **Figure 18–11**. The procedure was developed by Brunauer, Emmett, and Teller[30] and is commonly known as the *BET method*. It is discussed in some detail by Hiemenz and by Allen.[31] Swintosky et al.[32] used the procedure to determine the surface area of pharmaceutical powders. They found the specific surface of zinc oxide to be about 3.5 $m^2$/g; the value for barium sulfate is about 2.4 $m^2$/g.

**EXAMPLE 18–7**

**Specific Surface**

Using the Quantasorb apparatus, a plot of $p/V(p_0 - p)$ versus $p/p_0$ was obtained as shown in Figure 18–11 for a new antibiotic powder. Calculate $S_w$, the specific surface of the powder, in $m^2$/g. The data can be read from the graph to obtain the following values:

| $p/V(p_0 - p)$ | 0.05 | 0.150 | 0.20 |
|---|---|---|---|
| $p/p_0$ | 0.07 | 0.220 | 0.290 |

Following the BET equation (18–18) and using linear regression, the intercept, $1/(V_m b)$, is $I = 0.00198$ and the slope $(b - 1)/(V_m b)$ is $S = 0.67942$. By rearranging equation (18–18), we find $V_m$:

$$V_m = \frac{1}{1 + S} = \frac{1}{0.00198 + 0.67942}$$

$$= 1.46757 \ cm^3/g$$

The specific surface, $S_w$, is obtained using equation (18–19):

$$S_w = 4.35 \ m^2/cm^3 \times V_m \ cm^3/g = 6.38 \ m^2 g^{-1}$$

Assuming that the particles are spherical, we can calculate the mean volume–surface diameter by use of equation (**18–17**):

$$d_{vs} = \frac{6}{\rho S_w}$$

where $\rho$ is the density of the adsorbent and $S_w$ is the specific surface in square centimeters per gram of adsorbent. Employing this method, Swintosky et al.[32] found the mean volume–surface diameter of zinc oxide particles to be 0.3 $\mu$m.

## Air Permeability Method

The principal resistance to the flow of a fluid such as air through a plug of compacted powder is the surface area of the powder. The greater is the surface area per gram of powder, $S_w$, the greater is the resistance to flow. Hence, for a given pressure drop across the plug, permeability is inversely proportional to specific surface; measurement of the former provides a means of estimating this parameter. From equation (**18–16**) or (**18–17**), it is then possible to compute $d_{vs}$.

A plug of powder can be regarded as a series of capillaries whose diameter is related to the average particle size. The

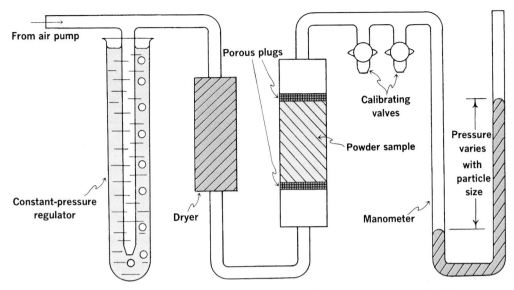

**Fig. 18–12.** The Fisher subsieve sizer. An air pump generates air pressure to a constant head by means of the pressure regulator. Under this head, the air is dried and conducted to the powder sample packed in the tube. The flow of air through the powder bed is measured by means of a calibrated manometer and is proportional to the surface area or the average particle diameter.

internal surface of the capillaries is a function of the surface area of the particles. According to Poiseuille equation,

$$V = \frac{\pi d^4 \Delta P t}{128 l \eta} \qquad (18\text{--}21)$$

where $V$ is the volume of air flowing through a capillary of internal diameter $d$ and length $l$ in $t$ seconds under a pressure difference of $\Delta P$. The viscosity of the fluid (air) is $\eta$ poise.

In practice, the flow rate through the plug, or bed, is also affected by (*a*) the degree of compression of the particles and (*b*) the irregularity of the capillaries. The more compact the plug, the lower is the *porosity*, which is the ratio of the total space between the particles to the total volume of the plug. The irregularity of the capillaries means that they are longer than the length of the plug and are not circular.

The Kozeny–Carman equation, derived from the Poiseuille equation, is the basis of most air permeability methods. Stated in one form, it is

$$V = \frac{A}{\eta S_w^2} \cdot \frac{\Delta P t}{K l} \cdot \frac{\varepsilon^3}{(1-\varepsilon)^2} \qquad (18\text{--}22)$$

where $A$ is the cross-sectional area of the plug, $K$ is a constant (usually $5.0 \pm 0.5$) that takes account of the irregular capillaries, and $\varepsilon$ is the porosity. The other terms are as defined previously.

A commercially available instrument is the Fisher subsieve sizer. The principle of its operation is illustrated in **Figure 18–12**. This instrument was modified by Edmundson[33] to improve its accuracy and precision.

Equation **(18–22)** apparently takes account of the effect of porosity on $S_w$ or $d_{vs}$. It is frequently observed, however, that $d_{vs}$ decreases with decreasing porosity. This is especially true of pharmaceutical powders that have diameters of a few

micrometers. It is customary, therefore, in these cases to quote the minimum value obtained over a range of porosities as the diameter of the sample. This noncompliance with equation **(18–22)** probably arises from initial bridging of the particles in the plug to produce a nonhomogeneous powder bed.[5] It is only when the particles are compacted firmly that the bed becomes uniform and $d_{vs}$ reaches a minimum value.

Because of the simple instrumentation and the speed with which determinations can be made, permeability methods are widely used pharmaceutically for specific-surface determinations, especially when the aim is to control batch-to-batch variations. When using this technique for more fundamental studies, it would seem prudent to calibrate the instrument.

Bephenium hydroxynaphthoate, official in the 1973 *British Pharmaceutical Codex*, is standardized by means of an air permeability method. The drug, used as an anthelmintic and administered as a suspension, must possess a surface area of not less than $7000 \text{ cm}^2/\text{g}$. As the specific surface of the material is reduced, the activity of the drug also falls.

Seth et al.[34a] studied the air permeability method of the *U. S. Pharmacopeia*, 20th edition, which used a Fisher subsieve sizer for determining the specific surface area of griseofulvin (also see *U. S. Pharmacopeia*[34b]). The authors suggested improvements in the method, principal among which was the use of a defined porosity, such as 0.50. This specified value is used in the ASTM Standard C-204–79 (1979) for measuring the fineness of Portland cement.

The volume surface diameter, $d_{vs}$, and therefore the specific surface, $S_w$, or surface area per unit weight in grams [equation **(18–19)**] of a powder can be obtained by use of this instrument (see **Fig. 18–12**). It is based on measuring the flow rate of air through the powder sample. If the sample weight is made exactly equal to the density of the powder

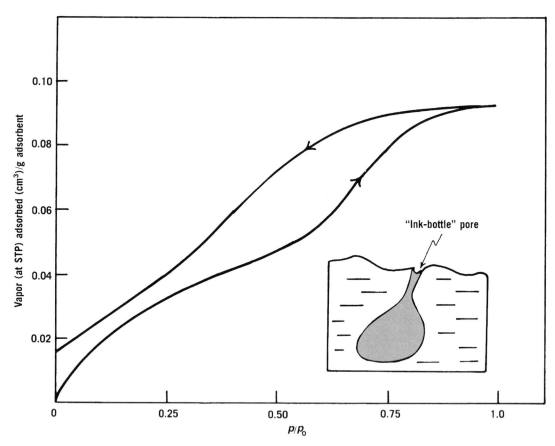

**Fig. 18–13.** Open hysteresis loop of an adsorption isotherm, presumably due to materials having "ink-bottle" pores, as shown in the inset. Key: STP = standard temperature and pressure.

sample, a more elaborate equation[35,36] for the average particle diameter, $d_{vs}$, is reduced to the simple expression

$$d_{vs} = \frac{cL}{[(AL) - 1]^{3/2}} \cdot \sqrt{\frac{F}{P - F}} \qquad (18\text{–}23)$$

where $c$ is an instrument constant, $L$ is the sample height in cm, $A$ is the cross-sectional area of the sample holder in cm$^2$, $F$ is the pressure drop across a flowmeter resistance built into the instrument, and $P$ is the air pressure as it enters the sample. The pressure (in cm of water) is measured with a water manometer rather than the better-known mercury manometer.

## PORE SIZE

Materials of high specific area may have cracks and pores that adsorb gases and vapors, such as water, into their interstices. Relatively insoluble powdered drugs may dissolve more or less rapidly in aqueous medium depending on their adsorption of moisture or air. Other properties of pharmaceutical importance, such as the dissolution rate of drug from tablets, may also depend on the adsorption characteristics of drug powders. The adsorption isotherms for porous solids display hysteresis, as seen in **Figures 18–13** and **18–14**, in which the desorption or downcurve branch lies above and to the left of the adsorption or upcurve. In **Figure 18–13**, the open hystere-

sis loop is due to a narrow-neck or "ink-bottle" type of pore (see the inset in **Fig. 18–13**) that traps adsorbate, or to irreversible changes in the pore when adsorption of the gas has occurred so that desorption follows a different pattern than adsorption. The curve of **Figure 18–14** with its closed hysteresis loop is more difficult to account for. Notice in **Figures 18–13** and **18–14** that at each relative pressure $p/p_0$, there are two volumes (at points $a$ and $b$ in **Fig. 18–14**) corresponding to a relative pressure $c$.

The upcurves of **Figures 18–13** and **18–14** correspond to gas adsorption into the capillaries and the downcurve to desorption of the gas. A smaller volume of gas is adsorbed during adsorption (point $a$ of **Fig. 18–14**) than is lost during desorption (point $b$). Vapor condenses to a liquid in small capillaries at a value less than $p_0$, the saturation vapor pressure, which can be taken as the vapor pressure at a flat surface. If the radius of the pore is $r$ and the radius of the meniscus is $R$ (**Fig. 18–15**, point $a$), $p/p_0$ can be calculated using expression known as the *Kelvin equation*,[*]

$$NkT \ln(p/p_0) = -\frac{2M\gamma}{\rho R} \qquad (18\text{–}24)$$

---

[*]Note that the solubility-of-solids equation has essentially the same form as equation (**18–24**), modified as necessary to deal with liquid or solid particles. Note further that $R$ in equation (**18–24**) is not the gas constant but rather the radius of a meniscus.

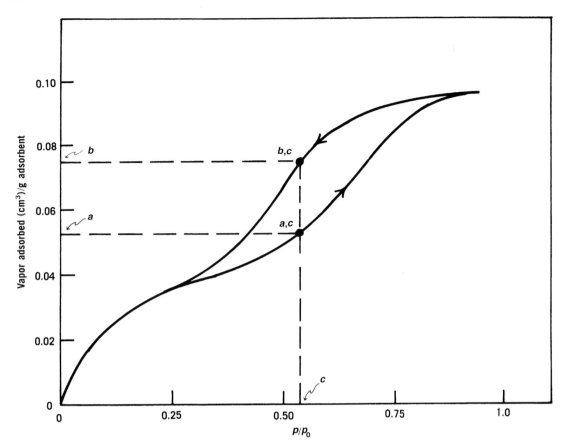

**Fig. 18–14.** Closed hysteresis loop of the adsorption isotherm of a porous material. At $p/p_0 = c$ on the curve of the loop, the volume of the pore is given by point *a*. At relative pressure *c* on the downcurve of the pore, volume is given by point *b*.

where $M$ is molecular weight of the condensing gas and $\rho$ is its density at a particular temperature, $M/\sigma$ is the molar volume of the fluid and $\gamma$ is its surface tension, $N$ is Avogadro's number, and $k$ is the Boltzmann constant, $1.381 \times 10^{-16}$ erg/deg

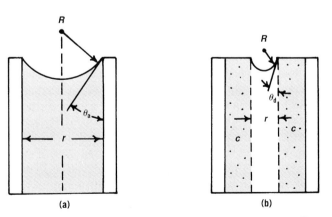

**Fig. 18–15.** (*a*) Pore into which vapor is condensing, corresponding to point *a, c* on the upcurve of Figure 18–14. Key: $\theta_a$ = advancing contact angle; $r$ = pore radius; $R$ = radius of meniscus. (*b*) Pore from which the liquid is vaporizing, corresponding to point *b, c* on the downcurve of Figure 18–14. Key: $\theta_d$ = receding or desorption contact angle; $R$ and $R$ are defined as in (*a*); $c$ = condensed vapor on walls of the capillary.

molecule. If the condensing vapor is water with a density of 0.998 at 20°C and a surface tension of 72.8 ergs/cm$^2$ and if the radius of the meniscus in the capillary $R$ is $1.67 \times 10^{-7}$ cm, we can calculate $p/p_0$ to be

$$\ln \frac{p}{p_0} = -\frac{2(18.015 \text{ g/mole})(72.8 \text{ ergs/cm}^2)}{(6.022 \times 10^{23} \text{ molecules/mole})}$$

$$\times 1/[(1.381 \times 10^{-16} \text{ erg/deg molecules})]$$

$$\times 1/[(0.998 \text{ g/cm}^3)(1.67 \times 10^{-7} \text{ cm})(293.18 \text{ K})]$$

$$\ln \frac{p}{p_0} = -0.6455$$

$$\frac{p}{p_0} = 0.5244$$

During adsorption, the capillary is filling (point *a, c* in **Fig. 18–14**) and the contact angle, $\theta_a$ (advancing contact angle), is greater than that during desorption, $\theta_d$, at which time the capillary is emptying. The radius of the meniscus will be smaller in the receding stage than in the advancing stage because the capillary is partly filled with fluid from multilayer adsorption. This smaller receding contact angle means a smaller radius of the meniscus, as seen in **Figure 18–15***b*, and $p/p_0$ will decrease because $R$ is in the denominator of the Kelvin equation, the right-hand side of which is negative.

**WATER ADSORPTION AND DESORPTION ON A CLAY AS A FUNCTION OF RELATIVE PRESSURE, $p/p_0$**

| (1) $p/p_0$ | (2) $V_1$ (Absorption) (mL/g) | (3) $V_2$ (Desorption) (mL/g) | (4) Radius (Å) | (5) Cumulative Pore Volume (%) |
|---|---|---|---|---|
| 1   0.20 | 0.079 | 0.123 | <6.7 | 54.9 |
| 2   0.31 | 0.109 | 0.147 | <9.2 | 65.6 |
| 3   0.40 | 0.135 | 0.165 | <11.7 | 73.7 |
| 4   0.49 | 0.141 | 0.182 | <15.1 | 81.3 |
| 5   0.66 | 0.152 | 0.191 | <30 | 85.3 |
| 6   0.80 | 0.170 | 0.200 | <48.4 | 89.3 |
| 7   0.96 | 0.224 | 0.224 | <265 | 100 |

The Kelvin equation gives a reasonable explanation for the differences of $p/p_0$ on adsorption and desorption and consequently provides for the existence of the hysteresis loop. The Kelvin equation, together with the hysteresis loops in adsorption–desorption isotherms (**Fig. 18–14**), can be used to compute the pore-size distribution.[37] During desorption, at a given $p/p_0$ value, water will *condense* only in pores of radius equal to or below the value given by the Kelvin equation. Water will *evaporate* from pores of larger radius. Thus, from the desorption isotherm, the volume of water retained at a given pressure $p/p_0$ corresponds to the volume of pores having radius equal to or below the radius calculated from the Kelvin equation at this $p/p_0$ value.

### EXAMPLE 18–8

**Pore Radius**

Yamanaka et al.[38] obtained experimental values for a water adsorption–desorption isotherm at 20°C on a clay. These values, which are given in Table 18–8, are selected from Figure 7 of their work.[38]

(*a*) Compute the radius of pores corresponding to the relative pressures $p/p_0$ given in Table 18–8.

(*b*) Assuming that all pores are of radius less than 265 Å, compute the cumulative percentage of pore volume with radii less than those found in part (*a*).

(*c*) Compute the percentage of pore volume at 20°C with radii between 40 and 60 Å.

(*a*) Using the Kelvin equation for $p/p_0 = 0.2$, we obtain

$$R = -2\gamma V([RT \ln(p/p_0)]$$

$$= -\frac{2 \times 72.8 \times 18}{(8.3143 \times 10^7 \times 293)(\ln 0.2)}$$

$$= (+)6.7 \times 10^{-8} \text{ cm} = 6.7 \text{ Å}$$

The results for the several $p/p_0$ values are shown in the fourth column of Table 18–8.

(*b, c*) The total cumulative pore volume is 0.224 mL/g, corresponding to the intersection of the adsorption and desorption curves (row 7 in Table 18–8). It corresponds to 100% cumulative pore

volume. Therefore, for, say, pores of radius less than 48.4 Å (Table 18–8, column 4, row 6), the cumulative percentage of pore volume is

$$\% = \frac{0.200}{0.224} \times 100 = 89.3\%$$

where the value 0.200 mL/g is taken from the desorption isotherm (Table 18–8, column 3, row 6). The results are given in column 5 of Table 18–8.

Christian and Tucker[39] made a careful and extensive study of pore models and concluded that a model that included a combination of cylindrical and slit-shaped pores provided the best quantitative fit of the data obtained on both the adsorption and desorption branches of the pore distribution plots. A modification of the BET equation assuming multilayer adsorption at the capillary walls has also been found to provide a satisfactory model for the hysteresis that occurs with porous solids.[40]

The adsorption of water vapor, flavoring agents, perfumes, and other volatile substances into films, containers, and other polymeric materials used in pharmacy is important in product formulation and in the storage and use of drug products. Sadek and Olsen[41] showed that the adsorption isotherms for water vapor on methylcellulose, povidone, gelatin, and polymethylmethacrylate all exhibited hysteresis loops. Hydration of gelatin films was observed to be lowered by treatment with formaldehyde, which causes increased cross-linking in gelatin and a decrease in pore size. Povidone showed increased water adsorption on treatment with acetone, which enlarged pore size and increased the number of sites for water sorption. In a study of the action of tablet disintegrants, Lowenthal and Burress[42] measured the mean pore diameter of tablets in air permeability apparatus. A linear correlation was observed between log mean pore diameter and tablet porosity, allowing a calculation of mean pore diameter from the more easily obtained tablet porosity. Gregg and Sing[43] discussed pore size and pore-size distribution in some detail.

# DERIVED PROPERTIES OF POWDERS

The preceding sections of this chapter have been concerned mainly with size distribution and surface areas of powders. These are the two *fundamental* properties of any collection of particles. There are, in addition, numerous *derived* properties that are based on these fundamental properties. Those of particular relevance to pharmacy are discussed in the remainder of this chapter. Very important properties, those of particle dissolution and dissolution rate, are subjects of separate chapters.

## Porosity

Suppose a powder, such as zinc oxide, is placed in a graduated cylinder and the total volume is noted. The volume occupied is known as the *bulk volume*, $V_b$. If the powder is nonporous, that is, has no internal pores or capillary spaces, the bulk volume of the powder consists of the true volume of the solid particles plus the volume of the spaces between the particles. The volume of the spaces, known as the *void volume*, $v$, is given by the equation

$$v = V_b - V_p \qquad (18\text{-}25)$$

where $V_p$ is the *true volume* of the particles. The method for determining the volume of the particles will be given later.

The *porosity* or *voids* $\varepsilon$ of the powder is defined as the ratio of the void volume to the bulk volume of the packing:

$$\varepsilon = \frac{V_b - V_p}{V_b} = 1 - \frac{V_p}{V_b} \qquad (18\text{-}26)$$

Porosity is frequently expressed in percent, $\varepsilon \times 100$.

### EXAMPLE 18–9

**Calculate Porosity**

A sample of calcium oxide powder with a true density of 3.203 and weighing 131.3 g was found to have a bulk volume of 82.0 cm³ when placed in a 100 mL graduated cylinder. Calculate the porosity. The volume of the particles is

$$131.3 \text{ g}/(3.203 \text{ g/cm}^3) = 41.0 \text{ cm}^3$$

From equation (18–25), the volume of void space is

$$v = 82.0 \text{ cm}^3 - 41.0 \text{ cm}^3 = 41.0 \text{ cm}^3$$

and from equation (18–26) the porosity is

$$\varepsilon = \frac{82 - 41}{82} = 0.5 \text{ or } 50\%$$

## Packing Arrangements

Powder beds of uniform-sized spheres can assume either of two ideal packing arrangements: (*a*) *closest* or *rhombohedral* and (*b*) *most open, loosest,* or *cubic packing*. The theoretical porosity of a powder consisting of uniform spheres in closest packing is 26% and for loosest packing is 48%. The arrangements of spherical particles in closest and loosest packing are shown in **Figure 18–16**.

**Fig. 18–16.** Schematic representation of particles arranged in (*a*) closest packing and (*b*) loosest packet. The dashed circle in (*a*) shows the position taken by a particle in a plan above that of the other three particles.

The particles in real powders are neither spherical in shape nor uniform in size. It is to be expected that the particles of ordinary powders may have any arrangement intermediate between the two ideal packings of **Figure 18–16**, and most powders in practice have porosities between 30% and 50%. If the particles are of greatly different sizes, however, the smaller ones may shift between the larger ones to give porosities below the theoretical minimum of 26%. In powders containing flocculates or aggregates, which lead to the formation of bridges and arches in the packing, the porosity may be above the theoretical maximum of 48%. In real powder systems, then, almost any degree of porosity is possible. Crystalline materials compressed under a force of 100,000 lb/in.² can have porosities of less than 1%.

## Densities of Particles

Because particles may be hard and smooth in one case and rough and spongy in another, one must express densities with great care. Density is universally defined as weight per unit volume; the difficulty arises when one attempts to determine the volume of particles containing microscopic cracks, internal pores, and capillary spaces.

For convenience, three types of densities can be defined[43,44]: (*a*) *the true density* of the material itself, exclusive of the voids and intraparticle pores larger than molecular or atomic dimensions in the crystal lattices, (*b*) the *granule density* as determined by the displacement of mercury, which does not penetrate at ordinary pressures into pores smaller than about 10 $\mu$m, and (*c*) the *bulk density* as determined from the bulk volume and the weight of a dry powder in a graduated cylinder.*

When a solid is nonporous, true and granule densities are identical, and both can be obtained by the displacement of helium or a liquid such as mercury, benzene, or water. When the material is porous, having an internal surface, the true density is best approximated by the displacement of helium, which penetrates into the smallest pores and is not adsorbed by the material. The density obtained by liquid displacement is considered as approximately equal to true density but may

---

*The term *apparent density* has been used by various authors to mean granular or bulk density, and some have used it to mean true density obtained by liquid displacement. Because of this confusion, the use of the term *apparent density* is discouraged.

differ from it somewhat when the liquid does not penetrate well into the pores.

The methods for determining the various densities are now discussed. *True density*, $\rho$, is the density of the actual solid material. Methods for determining the density of nonporous solids by displacement in liquids in which they are insoluble are found in general pharmacy books. If the material is porous, as is the case with most powders, the true density can be determined by use of a helium densitometer, as suggested by Franklin.[45] The volume of the empty apparatus (dead space) is first determined by introducing a known quantity of helium. A weighed amount of powder is then introduced into the sample tube, adsorbed gases are removed from the powder by an outgassing procedure, and helium, which is not adsorbed by the material, is again introduced. The pressure is read on a mercury manometer, and by application of the gas laws, the volume of helium surrounding the particles and penetrating into the small cracks and pores is calculated. The difference between the volume of helium filling the empty apparatus and the volume of helium in the presence of the powder sample yields the volume occupied by the powder. Knowing the weight of the powder, one is then able to calculate the true density. The procedure is equivalent to the first step in the BET method for determining the specific surface area of particles.

The density of solids usually listed in handbooks is often determined by liquid displacement. It is the weight of the body divided by the weight of the liquid it displaces, in other words, the loss of weight of the body when suspended in a suitable liquid. For solids that are insoluble in the liquid and heavier than it, an ordinary pycnometer can be used for the measurement. For example, if the weight of a sample of glass beads is 5.0 g and the weight of water required to fill a pycnometer is 50.0 g, then the total weight would be 55.0 g. When the beads are immersed in the water and the weight is determined at 25°C, the value is 53.0 g, or a displacement of 2.0 cm$^3$ of water, and the density is 5.0 g/2.00 cm$^3$ = 2.5 g/cm$^3$. The true density determined in this manner may differ slightly depending on the ability of the liquid to enter the pores of the particles, the possible change in the density of the liquid at the interface, and other complex factors.

Because helium penetrates into the smallest pores and crevices (**Fig. 18–17**), it is generally conceded that the helium method gives the closest approximation to true density. Liquids such as water and alcohol are denied entrance into the smallest spaces, and liquid displacement accordingly gives a density somewhat smaller than the true value. True densities are given in **Table 18–9** for some powders of pharmaceutical interest.

*Granule density*, $\rho_g$, can be determined by a method similar to the liquid displacement method. Mercury is used because it fills the void spaces but fails to penetrate into the internal pores of the particles. The volume of the particles together with their *intraparticle spaces* then gives the granule volume, and from a knowledge of the powder weight, one finds the granule density. Strickland et al.[46] determined

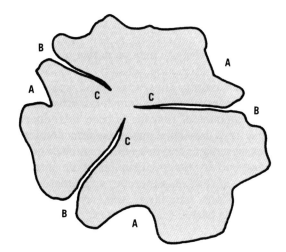

**Fig. 18–17.** Pores and crevices of a pharmaceutical granule. Water or mercury surrounds such a particle and rests only in the surface irregularities such as regions A and B. Helium molecules may enter deep into the cracks at points C, allowing calculation of true rather than granule density.

the granule density of tablet granulations by the mercury displacement method, using a specially designed pycnometer. A measure of true density was obtained by highly compressing the powders. The samples were compressed to 100,000 lb/in$^2$, and the resulting tablets were weighed. The volumes of the tablets were computed after measuring the tablet dimensions with calipers. The weight of the tablet divided by the volume then gave the "true" or high-compression density.

The *intraparticle porosity* of the granules can be computed from a knowledge of the true and granule density. The porosity is given by the equation

$$\varepsilon_{\text{intraparticle}} = \frac{V_g - V_p}{V_g} = 1 - \frac{V_p}{V_g}$$

$$= 1 - \frac{\text{Weight/True density}}{\text{Weight/Granule density}} \quad (18\text{–}27)$$

or

$$\varepsilon_{\text{intraparticle}} = 1 - \frac{\text{Granule density}}{\text{True density}} = 1 - \frac{\rho_g}{\rho} \quad (18\text{–}28)$$

where $V_p$ is the true volume of the solid particles and $V_g$ is the volume of the particles together with the intraparticle pores.

### EXAMPLE 18–10

**Intraparticle Porosity**

The granule density, $\rho_g$, of sodium bicarbonate is 1.450 and the true density, $\rho$, is 2.033. Compute the intraparticle porosity.

We have

$$\varepsilon_{\text{intraparticle}} = 1 - \frac{1.450}{2.033} = 0.286 \text{ or } 28.6\%$$

The granule densities and internal porosity or percent pore spaces in the granules, as obtained by Strickland et al.,[46] are shown in **Table 18–10**. The difference in porosity depends on the method of granulation, as brought out in the table.

**TABLE 18–9**
**TRUE DENSITY IN g/cm³ OF SOLIDS COMMONLY USED IN PHARMACY**

| | | | |
|---|---|---|---|
| Aluminum oxide | 4.0 | Mercuric chloride | 5.44 |
| Benzoic acid | 1.3 | Mercuric iodide | 6.3 |
| Bismuth subcarbonate | 6.86 | Mercuric oxide | 11.1 |
| Bismuth subnitrate | 4.9 | Mercurous chloride | 7.15 |
| Bromoform | 2.9 | Paraffin | 0.90 |
| Calcium carbonate (calcite) | 2.72 | Potassium bromide | 2.75 |
| Calcium oxide | 3.3 | Potassium carbonate | 2.29 |
| Chalk | 1.8–2.6 | Potassium chloride | 1.98 |
| Charcoal (air free) | 2.1–2.3 | Potassium iodide | 3.13 |
| Clay | 1.8–2.6 | Sand, fine dry | 1.5 |
| Cork | 0.24 | Silver iodide | 5.67 |
| Cotton | 1.47 | Silver nitrate | 4.35 |
| Gamboge | 1.19 | Sodium borate, borax | 1.73 |
| Gelatin | 1.27 | Sodium bromide | 3.2 |
| Glass beads | 2.5 | Sodium chloride | 2.16 |
| Graphite | 2.3–2.7 | Sucrose | 1.6 |
| Kaolin | 2.2–2.5 | Sulfadiazine | 1.50 |
| Magnesium carbonate | 3.04 | Sulfur, precipitated | 2.0 |
| Magnesium oxide | 3.65 | Talc | 2.6–2.8 |
| Magnesium sulfate | 1.68 | Zinc oxide (hexagonal) | 5.59 |

*Bulk density*, $\rho_b$, is defined as the mass of a powder divided by the bulk volume. A standard procedure for obtaining bulk density or its reciprocal, *bulk specific volume*, has been established.[47] A sample of about 50 cm³ of powder that has previously been passed through a U. S. Standard No. 20 sieve is carefully introduced into a 100 mL graduated cylinder. The cylinder is dropped at 2 sec intervals onto a hard wood surface three times from a height of 1 in. The bulk density is then obtained by dividing the weight of the sample in grams by the final volume in cm³ of the sample contained in the cylinder. The bulk density does not actually reach a maximum until the container has been dropped or tapped some 500 times; however, the three-tap method has been found to give the most consistent results among various laboratories. The bulk density of some pharmaceutical powders is compared with true and apparent densities in **Table 18–11**. The term "light" as

applied to pharmaceutical powders means low bulk density or large bulk volume, whereas "heavy" signifies a powder of high bulk density or small volume. It should be noted that these terms have no relationship to granular or true densities.

The bulk density of a powder depends primarily on particle-size distribution, particle shape, and the tendency of the particles to adhere to one another. The particles may pack in such a way as to leave large gaps between their surfaces, resulting in a light powder or powder of low bulk density. On the other hand, the smaller particles may shift between the larger ones to form a heavy powder or one of high bulk density.

The *interspace* or *void porosity* of a powder of porous granules is the relative volume of interspace voids to the bulk volume of the powder, exclusive of the intraparticle pores. The interspace porosity is computed from a knowledge of

**TABLE 18–10**
**DENSITIES AND POROSITIES OF TABLET GRANULATIONS***

| Granulation | "True" or High-Compression Density (g/cm³) | Granule Density by Mercury Displacement (g/cm³) | Pore Space (Porosity) |
|---|---|---|---|
| Sulfathiazole[†] | 1.530 | 1.090 | 29 |
| Sodium bicarbonate[†] | 2.033 | 1.450 | 29 |
| Phenobarbital[†] | 1.297 | 0.920 | 29 |
| Aspirin[‡] | 1.370 | 1.330 | 2.9 |

*From W. A. Strickland, Jr., L. W. Busse, and T. Higuchi, J. Am. Pharm. Assoc. Sci. Ed. **45,** 482, 1956. With permission.
[†] Granulation prepared by wet method using starch paste.
[‡] Granulation prepared by dry method (slugging process).

## COMPARISON OF BULK DENSITIES WITH TRUE DENSITIES

| | Bulk Density (g/cm³) | True Density (g/cm³) |
|---|---|---|
| Bismuth subcarbonate heavy | 1.01 | 6.9* |
| Bismuth subcarbonate light | 0.22 | 6.9* |
| Magnesium carbonate heavy | 0.39 | 3.0* |
| Magnesium carbonate light | 0.07 | 3.0* |
| Phenobarbital | 0.34 | 1.3† |
| Sulfathiazole | 0.33 | 1.5† |
| Talc | 0.48 | 2.7* |

*Density obtained by liquid displacement.
†True density obtained by helium displacement.

the bulk density and the granule density and is expressed by the equation

$$\varepsilon_{\text{interspace}} = \frac{V_b - V_g}{V_b} = 1 - \frac{V_g}{V_b}$$
$$= 1 - \frac{\text{Weight/Granule density}}{\text{Weight/Bulk density}} \quad (18-29)$$

$$\varepsilon_{\text{interspace}} = 1 - \frac{\text{Bulk density}}{\text{Granule density}}$$
$$= 1 - \frac{\rho_b}{\rho_g} \quad (18-30)$$

where $V_b = w/\rho_b$ is the bulk volume and $V_g = w/\rho_g$ is the granule volume, that is, the volume of the particles plus pores.

The *total porosity* of a porous powder is made up of voids between the particles as well as pores within the particles. The total porosity is defined as

$$\varepsilon_{\text{total}} = \frac{V_b - V_p}{V_b} = 1 - \frac{V_p}{V_b} \quad (18-31)$$

where $V_b$ is the bulk volume and $V_p$ is the volume of the solid material. This equation is identical with that for nonporous powders [equation **(18–26)**]. As in the previous cases, $V_p$ and $V_b$ can be expressed in terms of powder weights and densities:

$$V_p = \frac{w}{\rho}$$

and

$$V_b = \frac{w}{\rho_b}$$

where $w$ is the mass ("weight") of the powder, $\rho$ is the true density, and $\rho_b$ is the bulk density. Substituting these relationships into equation **(18–31)** gives for the total porosity

$$\varepsilon_{\text{total}} = 1 - \frac{w/\rho}{w/\rho_b} \quad (18-32)$$

or

$$\varepsilon_{\text{total}} = 1 - \frac{\rho_b}{\rho} \quad (18-33)$$

**Bulk Density and Total Porosity**

The weight of a sodium iodide tablet was 0.3439 g and the bulk volume was measured by use of calipers and found to be 0.0963 cm³. The true density of sodium iodide is 3.667 g/cm³. What is the bulk density and the total porosity of the tablet?

We have

$$\rho_b = \frac{0.3439}{0.0963} = 3.571 \text{ g/cm}^3$$

$$\varepsilon_{\text{total}} = 1 - \frac{3.571}{3.667}$$

$$= 0.026 \text{ or } 2.6\%$$

In addition to supplying valuable information about tablet porosity and its evident relationship to tablet hardness and disintegration time, bulk density can be used to check the uniformity of bulk chemicals and to determine the proper size of containers, mixing apparatus, and capsules for a given mass of the powder. These topics are considered in subsequent sections of this chapter.

In summary, the differences among the three densities (true, granule, and bulk) can be understood better by reference to their reciprocals: specific true volume, specific granule volume, and specific bulk volume.

The specific true volume of a powder is the volume of the solid material itself per unit mass of powder. When the liquid used to measure it does not penetrate completely into the pores, the specific volume is made up of the volume per unit weight of the solid material itself and the small part of the pore volume within the granules that is not penetrated by the liquid. When the proper liquid is chosen, however, the discrepancy should not be serious. Specific granule volume is the volume of the solid and essentially all of the pore volume within the particles. Finally, specific bulk volume constitutes the volume per unit weight of the solid, the volume of the *intra*particle pores, and the void volume or volume of *inter*particle spaces.

**Total Porosity**

The following data apply to a 1 g sample of a granular powder:

Volume of the solid alone = 0.3 cm³/g

Volume of intraparticle pores = 0.1 cm³/g

Volume of spaces between particles = 1.6 cm³/g

(*a*) What are the specific true volume, $V$, the specific granule volume, $V_g$, and the specific bulk volume, $V_b$?

$$V = 0.3 \text{ cm}^3$$

$$V_g = V + \text{intraparticle pores}$$

$$= 0.3 + 0.1 = 0.4 \text{ cm}^3/\text{g}$$

$$V_b = V + \text{intraparticle pores}$$
$$+ \text{spaces between particles}$$

$$= 0.3 + 0.1 + 1.6$$

$$= 2.0 \text{ cm}^3/\text{g}$$

(*b*) Compute the total porosity, $\varepsilon_{\text{total}}$, the interspace porosity, $\varepsilon_{\text{interspace}}$, or void spaces between the particles, and the intraparticle porosity, $\varepsilon_{\text{intraparticle}}$, or pore spaces within the particles.

We have

$$\varepsilon_{\text{total}} = \frac{V_b - V_p}{V_b} = \frac{2.0 - 0.3}{2.0}$$

$$= 0.85 \text{ or } 85\%$$

$$\varepsilon_{\text{interspace}} = \frac{V_b - V_g}{V_b} = \frac{2.0 - 0.4}{2.0}$$

$$= 0.80 \text{ or } 80\%$$

$$\varepsilon_{\text{intraparticle}} = \frac{V_g - V_p}{V_g} = \frac{0.4 - 0.3}{0.4}$$

$$= 0.25 \text{ or } 25\%$$

Thus, the solid constitutes 15% of the total bulk and 85% is made up of void space; 80% of the bulk is contributed by the voids between the particles and 5% of the total bulk by the pores and crevices within the particles. These pores, however, contribute 25% to the volume of granules, that is, particles plus pores.

## Bulkiness

Specific bulk volume, the reciprocal of bulk density, is often called *bulkiness* or *bulk*. It is an important consideration in the packaging of powders. The bulk density of calcium carbonate can vary from 0.1 to 1.3, and the lightest or bulkiest type would require a container about 13 times larger than that needed for the heaviest variety. Bulkiness increases with a decrease in particle size. In a mixture of materials of different sizes, however, the smaller particles shift between the larger ones and tend to reduce the bulkiness.

## Flow Properties

A bulk powder is somewhat analogous to a non-Newtonian liquid, which exhibits plastic flow and sometimes dilatancy, the particles being influenced by attractive forces to varying degrees. Accordingly, powders may be *free-flowing* or *cohesive* ("sticky"). Neumann[48] discussed the factors that affect the flow properties of powders. Of special significance are particle size, shape, porosity and density, and surface texture. Those properties of solids that determine the magnitude of particle–particle interactions were reviewed by Hiestand.[49]

With relatively small particles (less than 10 $\mu$m), particle flow through an orifice is restricted because the cohesive forces between particles are of the same magnitude as gravitational forces. Because these latter forces are a function of the diameter raised to the third power, they become more significant as the particle size increases and flow is facilitated. A maximum flow rate is reached, after which the flow decreases as the size of the particles approaches that of the orifice.[50] If a powder contains a reasonable number of small particles, the powder's flow properties may be improved be removing the "fines" or adsorbing them onto the larger particles. Occasionally, poor flow may result from the presence of moisture, in which case drying the particles will reduce the cohesiveness. A review by Pilpel[51] deals with the various apparatus for the measurement of the properties of cohesive powders and the effects on cohesive powders of particle size, moisture, glidants, caking, and temperature.

Dahlinder et al.[52] reviewed the methods for evaluating flow properties of powders and granules, including the *Hausner ratio* or packed bulk density versus loose bulk density, the rate of tamping, the flow rate and free flow through an orifice, and a "drained" angle of repose. The Hausner ratio, the free flow, and the angle of repose correlated well with one another and were applicable even for fairly cohesive tablet granulations.

Elongated or flat particles tend to pack, albeit loosely, to give powders with a high porosity. Particles with a high density and a low internal porosity tend to possess free-flowing properties. This can be offset by surface roughness, which leads to poor flow characteristics due to friction and cohesiveness.

Free-flowing powders are characterized by "dustibility," a term meant to signify the opposite of stickiness. Lycopodium shows the greatest degree of dustibility; if it is arbitrarily assigned a dustibility of 100%, talcum powder has value of 57%, potato starch 27%, and fine charcoal 23%. Finely powdered calomel has a relative dustibility of 0.7%.[48] These values should have some relation to the uniform spreading of dusting powders when applied to the skin, and stickiness, a measure of the cohesiveness of the particles of a compacted powder, should be of some importance in the flow of powders through filling machines and in the operation of automatic capsule machines.

Poorly flowing powders or granulations present many difficulties to the pharmaceutical industry. The production of uniform tablet dosage units has been shown to depend on several granular properties. Arambulo and coworkers[53] observed that as the granule size was reduced, the variation in tablet weight fell. The minimum weight variation was attained with granules having a diameter of 400 to 800 $\mu$m. As the granule size was reduced further, the granules flowed less freely and the tablet weight variation increased. The particle-size distribution affects the internal flow and segregation of a granulation.

Raff et al.[54] studied the flow of tablet granulations. They found that internal flow and granule demixing (i.e., the tendency of the powder to separate into layers of different sizes) during flow through the hopper contribute to a decrease in tablet weight during the latter portion of the compression period. Hammerness and Thompson[55] observed that the flow rate of a tablet granulation increased with an increase in the quantity of fines added. An increase in the amount of lubricant also raised the flow rate, and the combination of lubricant and fines appeared to have a synergistic action.

The frictional forces in a loose powder can be measured by the *angle of repose*, $\phi$. This is the maximum angle possible between the surface of a pile of powder and the horizontal plane. If more material is added to the pile, it slides down the sides until the mutual friction of the particles, producing a surface at an angle $\phi$, is in equilibrium with the gravitational force. The tangent of the angle of repose is equal to the coefficient of friction, $\mu$, between the particles:

$$\tan \phi = \mu \qquad (18\text{--}34)$$

Hence, the rougher and more irregular the surface of the particles, the higher will be the angle of repose. This situation was observed by Fonner et al.,[56] who, in studying granules prepared by five different methods, found the repose angle to be primarily a function of surface roughness. Ridgeway and Rupp[57] studied the effect of particle shape on powder properties. Using closely sized batches of sand separated into different shapes, they showed that, with increasing departure from the spherical, the angle of repose increased while bulk density and flowability decreased.

To improve flow characteristics, materials termed *glidants* are frequently added to granular powders. Examples of commonly used glidants are magnesium stearate, starch, and talc. Using a recording powder flowmeter that measured the weight of powder flowing per unit time through a hopper orifice, Gold et al.[58] found the optimum glidant concentration to be 1% or less. Above this level, a decrease in flow rate was usually observed. No correlation was found between flow rate and repose angle. By means of a shear cell and a tensile tester, York[59] was able to determine an optimum glidant concentration for lactose and calcium hydrogen phosphate powders. In agreement with the result of Gold et al.,[58] the angle of repose was found to be unsuitable for assessing the flowability of the powders used.

Nelson[60] studied the repose angle of a sulfathiazole granulation as a function of average particle size, presence of lubricants, and admixture of fines. He found that, in general, the angle increased with decreasing particle size. The addition of talc in low concentration decreased the repose angle, but in high concentration it increased the angle. The addition of fines—particles smaller than 100 mesh—to coarse granules resulted in a marked increase of the repose angle.

The ability of a powder to flow is one of the factors involving in mixing different materials to form a powder blend. Mixing, and the prevention of unmixing, is an important pharmaceutical operation involved in the preparation of many dosage forms, including tablets and capsules.[61] Other factors affecting the mixing process are particle aggregation, size, shape, density differences, and the presence of static charge. Train[62] and Fischer[63] described the theory of mixing.

## Compaction: Compressed Tablets

Neumann[64] found that when powders were compacted under a pressure of about 5 $kg/cm^2$, the porosities of the powders composed of rigid particles (e.g., sodium carbonate) were higher than the porosities of powders in closest packing, as determined by tapping experiments. Hence, these powders were *dilatant*, that is, they showed an unexpected expansion, rather than contraction, under the influence of stress. In the case of soft and spongy particles (e.g., kaolin), however, the

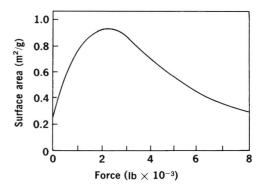

**Fig. 18–18.** The influence of compressional force on the specific surface of a sulfathiazole granulation. (From T. Higuchi et al., J. Am. Pharm. Assoc. Sci. Ed. **41**, 93, 1952; J. Am. Pharm. Assoc. Sci. Ed. **42**, 1944, 1953; J. Am. Pharm. Assoc. Sci. Ed. **43**, 344, 596, 685, 718, 1954; J. Am. Pharm. Assoc. Sci. Ed. **44**, 223, 1955; J. Am. Pharm. Assoc. Sci. Ed. **49**, 35, 1960; J. Pharm. Sci. **52**, 767, 1963.)

particles deformed on compression, and the porosities were lower than after tapping the powder down to its condition of closest packing. Similar experiments might be conducted to determine the optimum condition for packing powders into capsules on the manufacturing scale.

The behavior of powders under compression is significant in pharmaceutical tableting. Although basic information can be obtained from the literature on powder metallurgy and the compression of metallic powders, Train,[65] who performed some of the fundamental work in this area, pointed out that not all the theories developed for the behavior of metals will necessarily hold when applied to nonmetals.

Much of the early work was carried out by Higuchi and associates,[66] who studied the influence of compression force on the specific surface area, granule density, porosity, tablet hardness, and disintegration time of pharmaceutical tablets. As illustrated in **Figure 18–18**, the specific surface of a sulfathiazole tablet granulation as determined by the BET method increased to a maximum and then decreased. The initial increase in surface area can be attributed to the formation of new surfaces as the primary crystalline material is fragmented, whereas the decrease in specific surface beyond a compression force of 2500 lb is presumably due to cold bonding between the unit particles. It was also observed that porosity decreased and density increased as a linear function of the logarithm of the compression force, except at the higher force levels. As the compression increased, the tablet hardness and fracture resistance also rose. Typical results obtained using an instrumented rotary tablet machine[67] are shown in **Figure 18–19**.

The strength of a compressed tablet depends on a number of factors, the most important of which are compression force and particle size. The literature dealing with the effect of particle size has been outlined by Hersey et al.,[68] who, as a result of their studies, concluded that, over the range 4 to 925 $\mu$m, there is no simple relationship between strength and particle size. These workers did find that for simple crystals,

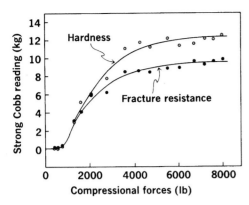

**Fig. 18–19.** Effect of compressional force on tablet hardness and fracture strength. (From E. L. Knoechel, C. C. Sperry, and C. J. Lintner, J. Pharm. Sci. **56**, 116, 1967.)

the strength of the tablet increased with decreasing particle size in the range of 600 to 100 $\mu$m.

The work initiated by Higuchi and coworkers[66] involved the investigation of other tablet ingredients, the development of an instrumented tablet machine, and the evaluation of tablet lubricants. The reader who desires to follow this interesting work should consult the original reports, as well as other studies[69–73] in this area. Tableting research and technology were comprehensively reviewed in 1972 by Cooper and Rees.[74] Deformation processes during decompression may be the principal factors responsible for the success or failure of compact formation.[75]

# CHAPTER SUMMARY

Knowledge and control of the size and the size range of particles are of profound importance in pharmacy. At this point you should understand that particle size is related in a significant way to the physical, chemical, and pharmacologic properties of a drug. Clinically, the particle size of a drug can affect its release from dosage forms that are administered orally, parenterally, rectally, and topically. The successful formulation of suspensions, emulsions, and tablets, from the viewpoints of both physical stability and pharmacologic response, also depends on the particle size achieved in the product. The student should have an understanding of the common particle sizes of pharmaceutical preparations and their impact on pharmaceutical processing/preparation; be familiar with the units for particle size, area, and volume and typical calculations; and be able to describe how particles can be characterized and why these methods are important. In the area of tablet and capsule manufacture, control of the particle size is essential in achieving the necessary flow properties and proper mixing of granules and powders.

**Practice problems for this chapter can be found at thePoint.lww.com/Sinko6e.**

# References

1. K. A. Lees, J. Pharm. Pharmacol. **15**, 43T, 1963.
2. J. M. Dalla Valle, *Micromeritics*, 2nd Ed., Pitman, New York, 1948, p. xiv.
3. Seradyn, Particle Technology Division, 1200 Madison Ave., Box 1210, Indianapolis, In., 46206.
4. Duke Scientific, 1135D San Antonio Road, Palo Alto, Calif., 94303.
5. I. C. Edmundson, in H. S. Bean, J. E. Carless, and A. H. Beckett (Eds.), *Advances in Pharmaceutical Sciences*, **Vol. 2**, Academic Press, New York, 1967, p. 95.
6. E. L. Parrott, J. Pharm. Sci. **63**, 813, 1974.
7. A. Sano, T. Kuriki, T. Handa, H. Takeuchi, and Y. Kawashima, J. Pharm. Sci. **76**, 471, 1987.
8. T. Hatch, J. Franklin. Inst. **215**, 27, 1933; T. Hatch and S. P. Choate, J. Franklin. Inst. **210**, 793, 1930.
9. H. L. Rao, *Private Communication*, Manipal, India, 1986,
10. T. Allen, *Particle Size Measurement*, 2nd Ed., Chapman Hall, London, 1974.
11. M. J. Groves, Pharm. Tech. **4**, 781, 1980.
12. G. Martin, Br. Ceram. Soc. Trans. **23**, 61, 1926; Br. Ceram. Soc. Trans. **25**, 51, 1928; Br. Ceram. Soc. Trans. **27**, 285, 1930.
13. T. Allen, *Particle Size Measurement*, 2nd Ed., Chapman Hall, London, 1974, p. 131.
14. K. P. P. Prasad and L. S. C. Wan, Pharm. Res. **4**, 504, 1987.
15. A. S. Arambulo and D. L. Deardorff, J. Am. Pharm. Assoc. Sci. Ed. **42**, 690, 1953.
16. L. D. King and C. H. Becker, Drug Standards **21**, 1, 1953.
17. G. Herdan, *Small Particle Statistics*, Elsevier, New York, 1953, p. 72.
18. D. E. Fonner, Jr., G. S. Banker, and J. Swarbrick, J. Pharm. Sci. **55**, 576, 1966.
19. H. J. Heywood, J. Pharm. Pharmacol. **15**, 56T, 1963.
20. W. I. Higuchi et al., J. Pharm. Sci. **51**, 1081, 1962; J. Pharm. Sci. **52**, 162, 1963; J. Pharm. Sci. **53**, 405, 1964; J. Pharm. Sci. **54**, 74, 1205, 1303, 1965.
21. S. Bisaillon and R. Tawashi, J. Pharm. Sci. **65**, 222, 1976.
22. E. R. Garrett, G. H. Miller, and M. R. W. Brown, J. Pharm. Sci. **55**, 593, 1966; G. H. Miller, S. Khalil, and A. Martin, J. Pharm. Sci. **60**, 33, 1971.
23. L. J. Beaubien and A. J. Vanderwielen, J. Pharm. Sci. **69**, 651, 1980.
24. S. I. Ismail and R. Tawashi, J. Pharm. Sci. **69**, 829, 1980.
25. E. C. Signoretti, A. Dell'Utri, L. Paoletti, D. Bastisti, and L. Montanari, Drug Dev. Ind. Pharm. **14**, 1, 1988.
26. S. P. Li, K. M. Feld, and C. R. Kowarski, Drug Dev. Ind. Pharm. **15**, 1137, 1989.
27. M. C. Davies, I. B. Wilding, R. D. Short, M. A. Khan, J. F. Watts, and C. D. Melia, Int. J. Pharm. **57**, 183, 1989.
28. F. Carli and A. Motta, J. Pharm. Sci. **73**, 197, 1984.
29. P. H. Emmett and S. Brunauer, J. Am. Chem. Soc. **59**, 1553, 1937; S. J. Gregg and K. S. W. Sing, *Adsorption, Surface Area and Porosity*, 2nd Ed., Academic Press, New York, 1982, p. 62.
30. S. Brunauer, P. H. Emmett, and F. Teller, J. Am. Chem. Soc. **60**, 309, 1938.
31. P. C. Hiemenz, *Principles of Colloid and Surface Chemistry*, 2nd Ed., Marcel Dekker, New York, 1986, pp. 513–529; T. Allen, *Particle Size Measurement*, Chapman Hall, London, 1975, pp. 358–366.
32. J. V. Swintosky, S. Riegelman, T. Higuchi, and L. W. Busse, J. Am. Pharm. Assoc. Sci. Ed. **38**, 210, 308, 378, 1949.
33. I. C. Edmundson, Analyst (Lond.), **91**, 1082, 1966.
34. a: P. Seth, N. Moller, J. C. Tritsch, and A. Stamm, J. Pharm. Sci. **72**, 971, 1983; and b: *United States Pharmacopeia*, 22nd Ed., US Pharmacopeial Convention, Rockville, MD, 1990, pp. 616.
35. P. C. Carman, J. Soc. Chem. Ind. **57**, 225, 1938; F. M. Lea and R. W. Nurse, J. Soc. Chem. Ind. **58**, 277, 1939.
36. E. L. Gooden and C. M. Smith, Ind. Eng. Chem. Anal. Ed. **12**, 479, 1940.
37. J. T. Carstensen, *Solid Pharmaceutics: Mechanical Properties and Rate Phenomena*, Academic Press, New York, 1980, pp. 130–131.
38. S. Yamanaka, P. B. Malla, and S. Komarheni, J. Colloid Interface Sci. **134**, 51, 1990.

39. S. D. Christian and E. E. Tucker, Am. Lab. **13,** 42, 1981; Am. Lab. **13,** 47, 1981.

40. P. C. Hiemenz, *Principles of Colloid and Surface Chemistry*, 2nd Ed., Marcel Dekker, New York, 1986, pp. 534–539; A. W. Adamson, *Physical Chemistry of Surfaces*, 4th Ed., Wiley, New York, 1982, pp. 584, 594.

41. H. M. Sadek and J. Olsen, Pharm. Technol. **5,** 40, 1981.

42. W. Lowenthal and R. Burress, J. Pharm. Sci. **60,** 1325, 1971; W. Lowenthal, J. Pharm. Sci. **61,** 303, 1972.

43. S. J. Gregg and K. S. W. Sing, *Adsorption, Surface Area and Porosity,* Academic Press, New York, 1982; S. J. Gregg, *The Surface Chemistry of Solids*, 2nd Ed., Reinhold, New York, 1982, pp. 132–152.

44. J. J. Kipling, Q. Rev. **10,** 1, 1956.

45. R. E. Franklin, Trans. Faraday Soc. **45,** 274, 1949.

46. W. A. Strickland, Jr., L. W. Busse, and T. Higuchi, J. Am. Pharm. Assoc. Sci. Ed. **45,** 482, 1956.

47. A. Q. Butler and J. C. Ransey, Jr., Drug Standards **20,** 217, 1952.

48. B. S. Neumann, in H. S. Bean, J. E. Carless, and A. H. Beckett (Eds.), *Advances in Pharmaceutical Sciences*, **Vol. 2**, Academic Press, New York, 1967, p. 181.

49. E. N. Hiestand, J. Pharm. Sci. **55,** 1325, 1966.

50. C. F. Harwood and N. Pilpel, Chem. Process Eng. **49,** 92, 1968.

51. N. Pilpel, in H. S. Bean, A. H. Beckett, and J. E. Carless (Eds.), *Advances in Pharmaceutical Sciences*, **Vol. 3**, Academic Press, New York, 1971, pp. 173–219.

52. L.-E. Dahlinder, M. Johansson, and J. Sjögren, Drug Dev. Ind. Pharm. **8,** 455, 1982.

53. A. S. Arambulo, H. Suen Fu, and D. L. Deardorff, J. Am. Pharm. Assoc. Sci. Ed. **42,** 692, 1953.

54. A. M. Raff, A. S. Arambulo, A. J. Perkins, and D. L. Deardorff, J. Am. Pharm. Assoc. Sci. Ed. **44,** 290, 1955.

55. F. C. Hammerness and H. O. Thompson, J. Am. Pharm. Assoc. Sci. Ed. **47,** 58, 1958.

56. D. E. Fonner, Jr., G. S. Banker, and J. Swarbrick, J. Pharm. Sci. **55,** 181, 1966.

57. K. Ridgway and R. Rupp, J. Pharm. Pharmacol. **21 Suppl.**, 30S, 1969.

58. G. Gold, R. N. Duvall, B. T. Palermo, and J. G. Slater, J. Pharm. Sci. **55,** 1291, 1966.

59. P. York, J. Pharm. Sci. **64,** 1216, 1975.

60. E. Nelson, J. Am. Pharm. Assoc. Sci. Ed. **44,** 435, 1955.

61. E. L. Parrott, Drug Cosmetic Ind. **115,** 42, 1974.

62. D. Train, J. Am. Pharm. Assoc. Sci. Ed. **49,** 265, 1960.

63. J. J. Fischer, Chem. Eng. **67,** 107, 1960.

64. B. S. Neumann, in J. J. Hermans (Ed.), *Flow Properties of Disperse Systems,* Interscience, New York, 1953, Chapter 10.

65. D. Train, J. Pharm. Pharmacol. **8,** 745, 1956; Trans. Inst. Chem. Eng. **35,** 258, 1957.

66. T. Higuchi et al., J. Am. Pharm. Assoc. Sci. Ed. **41,** 93, 1952; J. Am. Pharm. Assoc. Sci. Ed. **42,** 1944, 1953; J. Am. Pharm. Assoc. Sci. Ed. **43,** 344, 596, 685, 718, 1954; J. Am. Pharm. Assoc. Sci. Ed. **44,** 223, 1955; J. Am. Pharm. Assoc. Sci. Ed. **49,** 35, 1960; J. Pharm. Sci. **52,** 767, 1963.

67. E. L. Knoechel, C. C. Sperry, and C. J. Lintner, J. Pharm. Sci. **56,** 116, 1967.

68. J. A. Hersey, G. Bayraktar, and E. Shotton, J. Pharm. Pharmacol. **19,** 24S, 1967.

69. E. Shotton and D. Ganderton, J. Pharm. Pharmacol. **12,** 87T, 93T, 1960; J. Pharm. Pharmacol. **13,** 144T, 1961.

70. J. Varsano and L. Lachman, J. Pharm. Sci. **55,** 1128, 1966.

71. S. Leigh, J. E. Carless, and B. W. Burt, J. Pharm. Sci. **56,** 888, 1967.

72. J. E. Carless and S. Leigh, J. Pharm. Pharmacol. **26,** 289, 1974.

73. P. York and N. Pilpel, J. Pharm. Pharmacol. **24** (Suppl.), 47P, 1972.

74. J. Cooper and J. E. Rees, J. Pharm. Sci. **61,** 1511, 1972.

75. E. N. Hiestand, J. E. Wells, C. B. Peot, and J. F. Ochs, J. Pharm. Sci. **66,** 510, 1977.

CHAPTER LEGACY

**Fifth Edition:** published as Chapter 19 (Micromeritics). Updated by Patrick Sinko.

**Sixth Edition:** published as Chapter 18 (Micromeritics). Updated by Patrick Sinko.

## CHAPTER OBJECTIVES

**At the conclusion of this chapter the student should be able to:**

**1** Define rheology, provide examples of fluid pharmaceutical products exhibiting various rheologic behaviors, and describe the application of rheology in the pharmaceutical sciences and practice of pharmacy.

**2** Understand and define the following concepts: shear rate, shear stress, viscosity, kinematic viscosity, fluidity, plasticity, yield point, pseudoplasticity, shear thinning, dilatancy, shear thickening, thixotropy, hysteresis, antithixotropy, rheopexy, plug flow, and viscoelasticity.

**3** Define and understand Newton's law of flow and its application.

**4** Differentiate flow properties and corresponding rheograms between Newtonian and non-Newtonian materials.

**5** Understand and calculate the effects of temperature on viscosity and recognize similarities between viscous flow and diffusion relative to temperature.

**6** Recognize and identify specific rheologic behaviors with their corresponding rheograms.

**7** Appreciate the fundamentals of the practical determination of rheologic properties and describe four types of viscometers and their utility and limitations in determining rheologic properties of various systems.

**8** Appreciate the differences between continuous or steady shear rheometry and oscillatory and creep measurements in determining the consistency of viscoelastic materials.

---

The term "rheology," from the Greek *rheo* ("to flow") and *logos* ("science"), was suggested by Bingham and Crawford (as reported by Fischer[1]) to describe the flow of liquids and the deformation of solids. *Viscosity* is an expression of the resistance of a fluid to flow; the higher the viscosity, the greater is the resistance. As will be seen later, simple liquids can be described in terms of absolute viscosity. Rheologic properties of heterogeneous dispersions are more complex, however, and cannot be expressed by a single value.

Fundamental principles of rheology are used to study paints, inks, doughs, road-building materials, cosmetics, dairy products, and other materials. An understanding of the viscosity of liquids, solutions, and dilute and concentrated colloidal systems has both practical and theoretical value. Scott-Blair[2] recognized the importance of rheology in pharmacy and suggested its application in the formulation and analysis of such pharmaceutical products as emulsions, pastes, suppositories, and tablet coatings. Manufacturers of medicinal and cosmetic creams, pastes, and lotions must be capable of producing products with acceptable consistency and smoothness and reproducing these qualities each time a new batch is prepared. In many industries, a trained person with extensive experience handles in-process material periodically during manufacture to determine its "feel" and "body" and judge proper consistency. The variability of such subjective tests at different times under varying environmental conditions is, however, well recognized. A more serious objection, from a scientific standpoint, is the failure of subjective tests to distinguish various properties that make up the total consistency of the product. If these individual physical characteristics are delineated and studied objectively according to the analytic methods of rheology, valuable information can be obtained for use in formulating better pharmaceutical products.

Rheology is involved in the mixing and flow of materials, their packaging into containers, and their removal prior to use, whether this is achieved by pouring from a bottle, extrusion from a tube, or passage through a syringe needle. The rheology of a particular product, which can range in consistency from fluid to semisolid to solid, can affect its patient acceptability, physical stability, and even biologic availability. For example, viscosity has been shown to affect absorption rates of drugs from the gastrointestinal tract.

Rheologic properties of a pharmaceutical system can influence the selection of processing equipment used in its manufacture. Inappropriate equipment from this perspective may result in an undesirable product, at least in terms of its flow characteristics. These and other aspects of rheology that apply to pharmacy are discussed by Martin et al.[3]

When classifying materials according to types of flow and deformation, it is customary to place them in one of two categories: Newtonian or non-Newtonian systems. The choice depends on whether or not their flow properties are in accord with Newton's law of flow.

## NEWTONIAN SYSTEMS

### Newton's Law of Flow

Consider a "block" of liquid consisting of parallel plates of molecules, similar to a deck of cards, as shown in **Figure 19–1**. If the bottom layer is fixed in place and the top plane of liquid is moved at a constant velocity, each lower layer will move with a velocity directly proportional to its distance

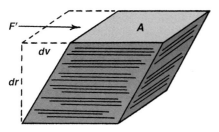

**Fig. 19–1.** Representation of the shearing force required to produce a definite velocity gradient between the parallel planes of a block of material.

from the stationary bottom layer. The difference of velocity, $dv$, between two planes of liquid separated by an infinitesimal distance, $dr$, is the *velocity gradient* or *rate of shear*, $dv/dr$. The force per unit area, $F'/A$, required to bring about flow is called the *shearing stress* and is given the symbol $F$. Newton was the first to study flow properties of liquids in a quantitative way. He recognized that the higher the viscosity of a liquid, the greater is the force per unit area (*shearing stress*) required to produce a certain rate of shear. Rate of shear is given the symbol $G$. Hence, rate of shear should be directly proportional to shearing stress, or

$$\frac{F'}{A} = \eta \frac{dv}{dr} \qquad (19\text{–}1)$$

where $\eta$ is the *coefficient of viscosity*, usually referred to simply as *viscosity*.

Equation **(19–1)** is frequently written as

$$\eta = \frac{F}{G} \qquad (19\text{–}2)$$

where $F = F'/A$ and $G = dv/dr$. A representative flow curve, or *rheogram*, obtained by plotting $F$ versus $G$ for a Newtonian system is shown in **Figure 19–2a**. As implied by equation **(19–2)**, a straight line passing through the origin is obtained.

The unit of viscosity is the *poise*, defined with reference to **Figure 19–1** as the shearing force required to produce a velocity of 1 cm/sec between two parallel planes of liquid each 1 cm$^2$ in area and separated by a distance of 1 cm. The cgs units for poise are dyne sec cm$^{-2}$ (i.e., dyne sec/cm$^2$) or g cm$^{-1}$ sec$^{-1}$ (i.e., g/cm sec). These units are readily obtained by a dimensional analysis of the viscosity coefficient. Rearranging equation **(19–1)** to

$$\eta = \frac{F' dr}{A\, dv} = \frac{\text{dynes} \times \text{cm}}{\text{cm}^2 \times \text{cm/sec}} = \frac{\text{dyne sec}}{\text{cm}^2}$$

gives the result

$$\frac{\text{dyne sec}}{\text{cm}^2} = \frac{\text{g} \times \text{cm/sec}^2 \times \text{sec}}{\text{cm}^2} = \frac{\text{g}}{\text{cm sec}}$$

A more convenient unit for most work is the *centipoise* (cp, plural cps), 1 cp being equal to 0.01 poise. *Fluidity*, $\phi$, a term sometimes used, is defined as the reciprocal of viscosity:

$$\phi = \frac{1}{\eta} \qquad (19\text{–}3)$$

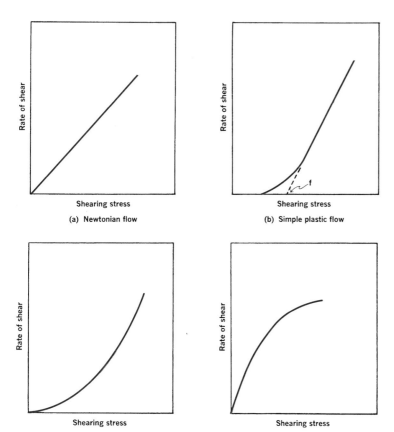

(a) Newtonian flow

(b) Simple plastic flow

(c) Simple pseudoplastic flow

(d) Dilatant flow

**Fig. 19–2.** Representative flow curves for various materials.

## ABSOLUTE VISCOSITY OF SOME NEWTONIAN LIQUIDS AT 20°C

| Liquid | Viscosity (cp) |
|---|---|
| Castor oil | 1000 |
| Chloroform | 0.563 |
| Ethyl alcohol | 1.19 |
| Glycerin, 93% | 400 |
| Olive oil | 100 |
| Water | 1.0019 |

## Kinematic Viscosity

*Kinematic viscosity* is the absolute viscosity [as defined in equation (19–1)] divided by the density of the liquid at a specific temperature:

$$\text{Kinematic viscosity} = \frac{\eta}{\rho} \qquad (19–4)$$

The units of kinematic viscosity are the *stoke* (s) and the *centistoke* (cs). Arbitrary scales (e.g., Saybolt, Redwood, Engler, and others) for measurement of viscosity are used in various industries; these are sometimes converted by use of tables or formulas to absolute viscosities and vice versa.

Viscosities of some liquids commonly used in pharmacy are given in **Table 19–1** at 20°C. A number of viscosity-increasing agents are described in the *United States Pharmacopeia*.

**EXAMPLE 19–1**

**Measuring Viscosity**

(a) An Ostwald viscometer (see Fig. 19–11) was used to measure acetone, which was found to have a viscosity of 0.313 cp at 25°C. Its density at 25°C is 0.788 g/cm$^3$. What is the kinematic viscosity of acetone at 25°C?

(b) Water is ordinarily used as a standard for viscosity of liquids. Its viscosity at 25°C is 0.8904 cp. What is the viscosity of acetone relative to that of water (relative viscosity, $\eta_{rel}$) at 25°C?

Solutions:

(a) Kinematic viscosity = 0.313 cp ÷ 0.788 g/cm$^3$ = 0.397 poise/ (g/cm$^3$), or 0.397 cs.

(b) $\eta_{rel(acetone)}$ = 0.313 cp/0.8904 cp = 0.352 (dimensionless).

## Temperature Dependence and the Theory of Viscosity

Whereas the viscosity of a gas increases with temperature, that of a liquid decreases as temperature is raised, and the fluidity of a liquid (the reciprocal of viscosity) increases with temperature. The dependence of the viscosity of a liquid on temperature is expressed approximately for many substances by an equation analogous to the Arrhenius equation of chemical kinetics:

$$\eta = Ae^{E_v RT} \qquad (19–5)$$

where $A$ is a constant depending on the molecular weight and molar volume of the liquid and $E_v$ is an "activation energy" required to initiate flow between molecules.

The energy of vaporization of a liquid is the energy required to remove a molecule from the liquid, leaving a "hole" behind equal in size to that of the molecule that has departed. A hole must also be made available in a liquid if one molecule is to flow past another. The activation energy for flow has been found to be about one-third that of the energy of vaporization, and it can be concluded that the free space needed for flow is about one-third the volume of a molecule. This is presumably because a molecule in flow can back, turn, and maneuver in a space smaller than its actual size, like a car in a crowded parking lot. More energy is required to break bonds and permit flow in liquids composed of molecules that are associated through hydrogen bonds. These bonds are broken at higher temperatures by thermal movement, however, and $E_v$ decreases markedly. Diffusional phenomena (Chapter 12) exhibit a similar dependence on temperature; like fluidity (the reciprocal of viscosity), rates of diffusion increase exponentially with temperature.

**EXAMPLE 19–2**

**Temperature Dependence of Viscosity**

The modified Arrhenius equation (19–5) is used to obtain the dependence of viscosity of liquids on temperature. Use equation (19–5) and the viscosity versus temperature data for glycerin (Table 19–2) to obtain the constant $A$ and $E_v$ (activation energy to initiate flow). What is the value of $r^2$, the square of the correlation coefficient?

Equation (19–5) is written in logarithmic form

$$\ln \eta = \ln A + \frac{E_v}{R} \frac{1}{T} \qquad (19–6)$$

## VISCOSITY OF GLYCERIN AT SEVERAL TEMPERATURES*

| Temperature (°C) | −42 | −20 | 0 | 6 | 15 | 20 | 25 | 30 |
|---|---|---|---|---|---|---|---|---|
| Temperature (K) | 231 | 253 | 273 | 279 | 288 | 293 | 298 | 303 |
| $1/T$ (K$^{-1}$) | 0.00432 | 0.00395 | 0.00366 | 0.00358 | 0.00347 | 0.00341 | 0.00336 | 0.00330 |
| $\eta$ (cp) | $6.71 \times 10^6$ | $1.34 \times 10^5$ | 12110 | 6260 | 2330 | 1490 | 954 | 629 |
| $\ln \eta$ | 15.719 | 11.806 | 9.402 | 8.742 | 7.754 | 7.307 | 6.861 | 6.444 |

*Data from *CRC Handbook of Chemistry and Physics*, 63rd Ed., CRC Press, Boca Raton, Fla., 1982, p. F–44.

According to equation (19–6), a regression of ln $\eta$ against $1/T$ gives $E_v$ from the slope and ln $A$ from the intercept. Using the values given in Table 19–2, we obtain

$$\ln \eta = -23.4706 + 9012\frac{1}{T}$$

Slope $= 9012 = E_v/R; E_v = 9012 \times 1.9872 = 17,909$ cal/mole
Intercept $= -23.4706 = \ln, A; A = 6.40985 \times 10^{-11}; r^2 = 0.997$

# NON-NEWTONIAN SYSTEMS

The majority of fluid pharmaceutical products are not simple liquids and do not follow Newton's law of flow. These systems are referred to as *non-Newtonian*. Non-Newtonian behavior is generally exhibited by liquid and solid heterogeneous dispersions such as colloidal solutions, emulsions, liquid suspensions, and ointments. When non-Newtonian materials are analyzed in a rotational viscometer and results are plotted, various consistency curves, representing three classes of flow, are recognized: *plastic*, *pseudoplastic*, and *dilatant*.

## Plastic Flow

In **Figure 19–2b**, the curve represents a body that exhibits plastic flow; such materials are known as *Bingham bodies* in honor of the pioneer of modern rheology and the first investigator to study plastic substances in a systematic manner.

Plastic flow curves do not pass through the origin but rather intersect the shearing stress axis (or will if the straight part of the curve is extrapolated to the axis) at a particular point referred to as the *yield value*. A Bingham body does not begin to flow until a shearing stress corresponding to the yield value is exceeded. At stresses below the yield value, the substance acts as an elastic material. The rheologist classifies Bingham bodies, that is, those substances that exhibit a yield value, as solids, whereas substances that begin to flow at the smallest shearing stress and show no yield value are defined as liquids. Yield value is an important property of certain dispersions.

The slope of the rheogram in **Figure 19–2b** is termed the *mobility*, analogous to fluidity in Newtonian systems, and its reciprocal is known as the *plastic viscosity*, *U*. The equation describing plastic flow is

$$U = \frac{F - f}{G} \tag{19–7}$$

where *f* is the yield value, or intercept, on the shear stress axis in dynes/cm$^2$, and *F* and *G* are as previously defined.

Plastic flow is associated with the presence of flocculated particles in concentrated suspensions. As a result, a continuous structure is set up throughout the system. A yield value exists because of the contacts between adjacent particles (brought about by van der Waals forces), which must be broken down before flow can occur. Consequently, the yield value is an indication of force of flocculation: The more flocculated the suspension, the higher will be the yield value. Frictional forces between moving particles can also contribute to

yield value. As shown in the following example, once the yield value has been exceeded, any further increase in shearing stress (i.e., $F - f$) brings about a directly proportional increase in $G$, rate of shear. In effect, a plastic system resembles a Newtonian system at shear stresses above the yield value.

**EXAMPLE 19–3**

**Calculating Plastic Viscosity**

A plastic material was found to have a yield value of 5200 dynes/cm$^2$. At shearing stresses above the yield value, $F$ was found to increase linearly with $G$. If the rate of shear was 150 sec$^{-1}$ when $F$ was 8000 dynes/cm$^2$, calculate $U$, the plastic viscosity of the sample.

Substituting into equation (19–7), we find

$$U = (8000 - 5200)/150$$
$$= 2800/150$$
$$= 18.67 \text{ poise}$$

## Pseudoplastic Flow

Many pharmaceutical products, including liquid dispersions of natural and synthetic gums (e.g., tragacanth, sodium alginate, methylcellulose, and sodium carboxymethyl cellulose) exhibit *pseudoplastic flow*. Pseudoplastic flow is typically exhibited by polymers in solution, in contrast to plastic systems, which are composed of flocculated particles in suspension. As seen in **Figure 19–2c**, the consistency curve for a pseudoplastic material begins at the origin (or at least approaches it at low rates of shear). Therefore, there is no yield value as there is in a plastic system. Furthermore, because no part of the curve is linear, the viscosity of a pseudoplastic material cannot be expressed by any single value.

The viscosity of a pseudoplastic substance decreases with increasing rate of shear. An apparent viscosity can be obtained at any rate of shear from the slope of the tangent to the curve at the specified point. The most satisfactory representation for a pseudoplastic material, however, is probably a graphic plot of the entire consistency curve.

The curved rheogram for pseudoplastic materials results from a shearing action on long-chain molecules of materials such as linear polymers. As shearing stress is increased, normally disarranged molecules begin to align their long axes in the direction of flow. This orientation reduces internal resistance of the material and allows a greater rate of shear at each successive shearing stress. In addition, some of the solvent associated with the molecules may be released, resulting in an effective lowering of both the concentration and the size of the dispersed molecules. This, too, will decrease apparent viscosity.

Objective comparisons between different pseudoplastic systems are more difficult than with either Newtonian or plastic systems. For example, Newtonian systems are completely described by viscosity, $\eta$, and plastic systems are adequately described by yield value, *f*, and plastic viscosity, *U*. However, several approaches have been used to obtain meaningful parameters that will allow different pseudoplastic materials

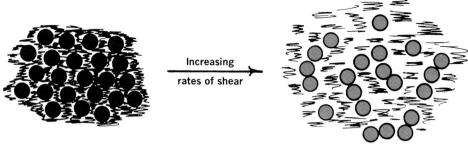

Close-packed particles;
minimum void volume;
sufficient vehicle;
relatively low consistency

Increasing
rates of shear

Open-packed (dilated) particles;
increased void volume;
insufficient vehicle;
relatively high consistency

**Fig. 19–3.** Explanation of dilatant flow behavior.

to be compared. Of those discussed by Martin et al.,[3] the exponential formula

$$F^N = \eta' G \qquad (19\text{--}8)$$

has been used most frequently for pseudoplastics. The exponent $N$ rises as flow becomes increasingly non-Newtonian. When $N = 1$, equation **(19–8)** reduces to equation **(19–2)** and flow is Newtonian. The term $\eta'$ is a viscosity coefficient. Following rearrangement, we can write equation **(19–8)** in the logarithmic form

$$\log G = N \log F - \log \eta' \qquad (19\text{--}9)$$

This is an equation for a straight line. Many pseudoplastic systems fit this equation when $\log G$ is plotted as a function of $\log F$.[4] Several of the more important pseudoplastic suspending agents used in pharmacy, however, do not conform to equation **(19–9)**.[5] Modified equations were suggested by Shangraw et al. and Casson and Patton.[6] Pseudoplastic systems have been characterized on the basis of the assumption that the typical rheogram of a pseudoplastic substance is composed of a first-order segment and a zero-order segment.[7]

### Dilatant Flow

Certain suspensions with a high percentage of dispersed solids exhibit an increase in resistance to flow with increasing rates of shear. Such systems actually increase in volume when sheared and are hence termed *dilatant*; **Figure 19–2d** illustrates their flow properties. This type of flow is the inverse of that possessed by pseudoplastic systems. Whereas pseudoplastic materials are frequently referred to as "shear-thinning systems," dilatant materials are often termed "shear-thickening systems." When stress is removed, a dilatant system returns to its original state of fluidity.

Equation **(19–8)** can be used to describe dilatancy in quantitative terms. In this case, $N$ is always less than 1 and decreases as degree of dilatancy increases. As $N$ approaches 1, the system becomes increasingly Newtonian in behavior.

Substances possessing dilatant flow properties are invariably suspensions containing a high concentration (about 50% or greater) of small, deflocculated particles. As discussed previously, particulate systems of this type that are flocculated would be expected to possess plastic, rather than dilatant, flow characteristics. Dilatant behavior can be explained as follows. At rest, particles are closely packed with minimal interparticle volume (voids). The amount of vehicle in the suspension is sufficient, however, to fill voids and permits particles to move relative to one another at low rates of shear. Thus, a dilatant suspension can be poured from a bottle because under these conditions it is reasonably fluid. As shear stress is increased, the bulk of the system expands or dilates; hence the term *dilatant*. The particles, in an attempt to move quickly past each other, take on an open form of packing, as depicted in **Figure 19–3**. Such an arrangement leads to a significant increase in interparticle void volume. The amount of vehicle remains constant and, at some point, becomes insufficient to fill the increased voids between particles. Accordingly, resistance to flow increases because particles are no longer completely wetted, or lubricated, by the vehicle. Eventually, the suspension will set up as a firm paste.

Behavior of this type suggests that appropriate precaution be used during processing of dilatant materials. Conventionally, processing of dispersions containing solid particles is facilitated by the use of high-speed mixers, blenders, or mills. Although this is advantageous with all other rheologic systems, dilatant materials may solidify under these conditions of high shear, thereby overloading and damaging processing equipment.

## THIXOTROPY

As described in the previous sections, several types of behavior are observed when rate of shear is progressively increased and plotted against resulting shear stress. It may have been assumed that if the rate of shear were reduced once the desired maximum had been reached, the downcurve would be identical with, and superimposable on, the upcurve. Although this is true for Newtonian systems, the downcurve for non-Newtonian systems can be displaced relative to the upcurve. With shear-thinning systems (i.e., pseudoplastic), the downcurve is frequently displaced to the left of the upcurve (as in

## ◤KEY CONCEPT◢ RHEOGRAMS

A *rheogram* is a plot of *shear rate*, *G*, as a function of *shear stress*, *F*. Rheograms are also known as consistency curves or flow curves. The rheologic properties of a given material are most completely described by its unique rheogram.

The simplest form of a rheogram is produced by Newtonian systems, which follow the equation for a straight line passing through the origin:

$$G = fF$$

The slope, *f*, is known as *fluidity* and is the reciprocal of *viscosity*, $\eta$:

$$f = 1/\eta$$

Therefore, the greater the slope of the line, the greater is the fluidity or, conversely, the lower is the viscosity. The rheogram of Newtonian systems can easily be obtained with a single-point determination.

*Plasticity* is the simplest type of non-Newtonian behavior in which the curve is linear only at values of *F*, beyond its yield value. If the curve is nonlinear throughout the entire range of *F* values tested, then the system is non-Newtonian and either *pseudoplastic* (shear thinning; slope increases with *F*) or *dilatant* (shear thickening; slope decreases with *F*) or a combination of the two. If the curve shows a hysteresis loop, that is, the curve obtained on increasing shear stress is not superimposable with that obtained on decreasing shear stress, then the system is *thixotropic*.

A non-Newtonian system can only be fully characterized by generating its complete rheogram, which requires use of a multipoint rheometer.

Fig. 19–4), showing that the material has a lower consistency at any one rate of shear on the downcurve than it had on the upcurve. This indicates a breakdown of structure (and hence shear thinning) that does not reform immediately when stress is removed or reduced. This phenomenon, known as *thixotropy*, can be defined[8] as "an isothermal and comparatively slow recovery, on standing of a material, of a consistency lost through shearing." As so defined, thixotropy can be applied only to shear-thinning systems. Typical rheograms for plastic and pseudoplastic systems exhibiting this behavior are shown in **Figure 19–4**.

Thixotropic systems usually contain asymmetric particles that, through numerous points of contact, set up a loose three-dimensional network throughout the sample. At rest, this structure confers some degree of rigidity on the system, and it resembles a gel. As shear is applied and flow starts, this structure begins to break down as points of contact are disrupted and particles become aligned. The material undergoes a gel-to-sol transformation and exhibits shear thinning. On removal of stress, the structure starts to reform. This process is not instantaneous; rather, it is a progressive restoration of consistency as asymmetric particles come into contact with

one another by undergoing random Brownian movement. Rheograms obtained with thixotropic materials are therefore highly dependent on the rate at which shear is increased or decreased and the length of time a sample is subjected to any one rate of shear. In other words, the previous history of the sample has a significant effect on the rheologic properties of a thixotropic system. For example, suppose that in **Figure 19–5** the shear rate of a thixotropic material is increased in a constant manner from point *a* to point *b* and is then decreased at the same rate back to *e*. Typically, this would result in the so-called *hysteresis loop abe*. If, however, the sample was taken to point *b* and the shear rate held constant for a certain period of time (say, $t_1$ seconds), shearing stress, and hence consistency, would decrease to an extent depending on time of shear, rate of shear, and degree of structure in the sample. Decreasing the shear rate would then result in the *hysteresis loop abce*. If the sample had been held at the same rate of shear for $t_2$ seconds, the loop *abcde* would

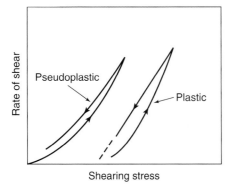

**Fig. 19–4.** Thixotropy in plastic and pseudoplastic flow systems.

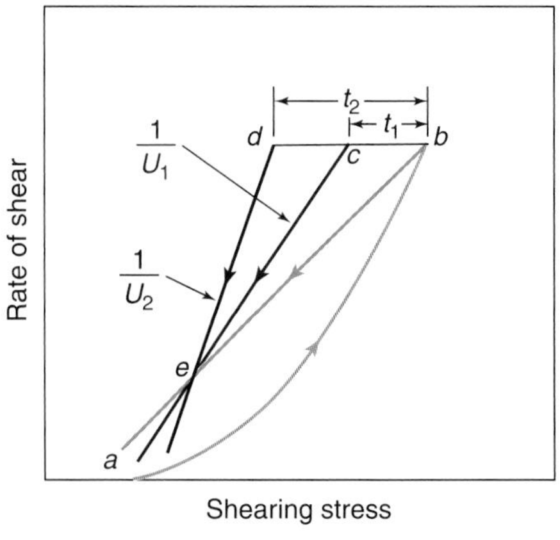

**Fig. 19–5.** Structural breakdown with time of a plastic system possessing thixotropy when subjected to a constant rate of shear for $t_1$ and $t_2$ seconds. See text for discussion.

have been observed. Therefore, the rheogram of a thixotropic material is not unique but will depend on rheologic history of the sample and approach used in obtaining the rheogram. This is an important point to bear in mind when attempting to obtain a quantitative measure of thixotropy. This will become apparent in the next section.

## Measurement of Thixotropy

A quantitative measurement of thixotropy can be attempted in several ways. The most apparent characteristic of a thixotropic system is the hysteresis loop formed by the upcurves and downcurves of the rheogram. This *area of hysteresis* has been proposed as a measure of thixotropic breakdown; it can be obtained readily by means of a planimeter or other suitable technique.

With plastic (Bingham) bodies, two approaches are frequently used to estimate degree of thixotropy. The first is to determine structural breakdown with time at a *constant* rate of shear. The type of rheogram needed for this estimation is shown in **Figure 19–5**; the steps necessary to obtain it have already been described. Based on such a rheogram, a thixotropic coefficient, $B$, the rate of breakdown with time at constant shear rate, is calculated as follows:

$$B = \frac{U_1 - U_2}{\ln\frac{t_2}{t_1}} \qquad (19\text{--}10)$$

where $U_1$ and $U_2$ are the plastic viscosities of the two downcurves, calculated from equation **(19–7)**, after shearing at a constant rate for $t_1$ and $t_2$ seconds, respectively. The choice of shear rate is arbitrary. A more meaningful, though time-consuming, method for characterizing thixotropic behavior is to measure fall in stress with time at several rates of shear.

The second approach is to determine the structural breakdown due to *increasing* shear rate. The principle involved in this approach is shown in **Figure 19–6**, in which two hysteresis loops are obtained having different maximum rates of shear, $v_1$ and $v_2$. In this case, a thixotropic coefficient, $M$,

the loss in shearing stress per unit increase in shear rate, is obtained from

$$M = \frac{U_1 - U_2}{\ln(v_2/v_1)} \qquad (19\text{--}11)$$

where $M$ is in dynes sec/cm$^2$ and $U_1$ and $U_2$ are the plastic viscosities for two separate downcurves having maximum shearing rates of $v_1$ and $v_2$, respectively. A criticism of this technique is that the two rates of shear, $v_1$ and $v_2$, are chosen arbitrarily; the value of $M$ will depend on the rate of shear chosen because these shear rates will affect the downcurves and hence the values of $U$ that are calculated.

## Bulges and Spurs

Dispersions employed in pharmacy may yield complex hysteresis loops when sheared in a viscometer in which shear rate (rather than shear stress) is increased to a point, then decreased, and the shear stress is read at each shear rate value to yield appropriate rheograms. Two such complex structures are shown in **Figures 19–7** and **19–8**. A concentrated aqueous bentonite gel, 10% to 15% by weight, produces a hysteresis loop with a characteristic *bulge* in the upcurve. It is presumed that the crystalline plates of bentonite form a "house-of-cards structure" that causes the swelling of bentonite magmas. This three-dimensional structure results in a bulged hysteresis loop as observed in **Figure 19–7**. In still more highly structured systems, such as a procaine penicillin gel formulated by Ober et al.[9] for intramuscular injection, the bulged curve may actually develop into a spurlike protrusion (**Fig. 19–8**). The structure demonstrates a high yield or *spur value*, $\Upsilon$, that traces out a bowed upcurve when the three-dimensional structure breaks in the viscometer, as observed in **Figure 19–8**. The spur value represents a sharp point of structural breakdown at low shear rate. It is difficult to produce the spur, and it may not be observed unless a sample of the gel is allowed to age undisturbed in the cup-and-bob assembly for some time before the rheologic run is made. The

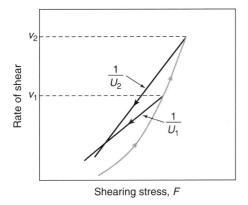

**Fig. 19–6.** Structural breakdown of a plastic system possessing thixotropy when subjected to increasing shear rates. See text for discussion.

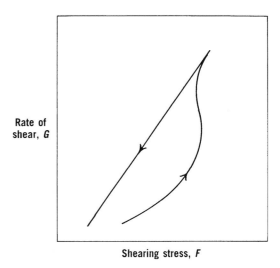

**Fig. 19–7.** Rheogram of a thixotropic material showing a bulge in the hysteresis loop.

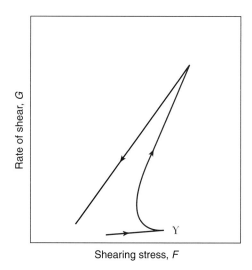

**Fig. 19–8.** Rheogram of a thixotropic material showing a spur value $\gamma$ in the hysteresis loop.

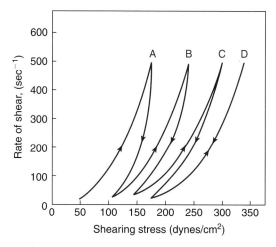

**Fig. 19–9.** Rheogram of magnesia magma showing antithixotropic behavior. The material is sheared at repeated increasing and then decreasing rates of shear. At stage *D*, further cycling no longer increased the consistency, and the upcurves and downcurves coincided. (From C. W. Chong, S. P. Eriksen, and J. W. Swintosky, J. Am. Pharm. Assoc. Sci. Ed. **49,** 547, 1960. With permission.)

spur value is obtained by using an instrument in which the rate of shear can be slowly and uniformly increased, preferably automatically, and the shear stress read out or plotted on an *X–Y* recorder as a function of shear rate. Ober et al.[9] found that penicillin gels having definite $\Upsilon$ values were very thixotropic, forming intramuscular depots upon injection that afforded prolonged blood levels of the drug.

## Negative Thixotropy

From time to time in the measurement of supposedly thixotropic materials, one observes a phenomenon called *negative thixotropy* or *antithixotropy*, which represents an increase rather than a decrease in consistency on the downcurve. This increase in thickness or resistance to flow with increased time of shear was observed by Chong et al.[10] in the rheologic analysis of magnesia magma. It was detected at shear rates of greater than 30 sec$^{-1}$; below 30 sec$^{-1}$ the magma showed normal thixotropy, the downcurve appearing to the left of the upcurve. As pointed out by Chong et al., antithixotropy had been reported by other investigators but not in pharmaceutical systems.

It was observed that when magnesia magma was alternately sheared at increasing and then at decreasing rates of shear, the magma continuously thickened (an increase in shearing stress per unit shear rate) but at a decreasing rate, and it finally reached an equilibrium state in which further cycles of increasing–decreasing shear rates no longer increased the consistency of the material. The antithixotropic character of magnesia magma is demonstrated in **Figure 19–9**. The equilibrium system was found to be gel-like and to provide great suspendability, yet it was readily pourable. When allowed to stand, however, the material returned to its sol-like properties.

Antithixotropy or negative thixotropy should not be confused with dilatancy or rheopexy. Dilatant systems are deflocculated and ordinarily contain greater than 50% by volume of solid dispersed phase, whereas antithixotropic systems have

low solids content (1%–10%) and are flocculated, according to Samyn and Jung.[11] *Rheopexy* is a phenomenon in which a solid forms a gel more readily when gently shaken or otherwise sheared than when allowed to form the gel while the material is kept at rest.[12] In a rheopectic system, the gel is the equilibrium form, whereas in antithixotropy, the equilibrium state is the sol. Samyn and Jung noted that magnesia magma and clay suspensions may show a negative rheopexy, analogous to negative thixotropy. It is believed that antithixotropy results from an increased collision frequency of dispersed particles or polymer molecules in suspension, resulting in increased interparticle bonding with time. This changes an original state consisting of a large number of individual particles and small floccules to an eventual equilibrium state consisting of a small number of relatively large floccules. At rest, the large floccules break up and gradually return to the original state of small floccules and individual particles.

As more rheologic studies are done with pharmaceuticals, negative thixotropy no doubt will be observed in other materials.

## Thixotropy in Formulation

Thixotropy is a desirable property in liquid pharmaceutical systems that ideally should have a high consistency in the container, yet pour or spread easily. For example, a well-formulated thixotropic suspension will not settle out readily in the container, will become fluid on shaking, and will remain long enough for a dose to be dispensed. Finally, it will regain consistency rapidly enough so as to maintain the particles in a suspended state. A similar pattern of behavior is desirable with emulsions, lotions, creams, ointments, and parenteral suspensions to be used for intramuscular depot therapy.

With regard to suspension stability, there is a relationship between degree of thixotropy and rate of sedimentation;

the greater the thixotropy, the lower the rate of settling. Concentrated parenteral suspensions containing from 40% to 70% w/v of procaine penicillin G in water were found to have a high inherent thixotropy and were shear thinning.[9] Consequently, breakdown of the structure occurred when the suspension was caused to pass through the hypodermic needle. Consistency was then recovered as rheologic structure reformed. This led to formation of a depot of drug at the site of intramuscular injection where drug was slowly removed and made available to the body. The degree of thixotropy was related to the specific surface of the penicillin used.

Degree of thixotropy may change over time and result in an inadequate formulation. Thixotropic systems are complex, and it is unrealistic to expect that rheologic changes can be meaningfully followed by the use of one parameter. Thus, in a study concerned with the aging effects of thixotropic clay, Levy[13] found it necessary to follow changes in plastic viscosity, area of hysteresis, yield value, and spur value.

# DETERMINATION OF RHEOLOGIC PROPERTIES

## Choice of Viscometer

Successful determination and evaluation of rheologic properties of any particular system depend, in large part, on choosing the correct instrumental method. Because shear rate in a Newtonian system is directly proportional to shearing stress, instruments that operate at a single shear rate can be used. These "single-point" instruments provide a single point on the rheogram; extrapolation of a line through this point to the origin will result in a complete rheogram. Implicit in the use of a single-point instrument is prior knowledge that the flow characteristics of the material are Newtonian. Unfortunately, this is not always the case, and, if the system is non-Newtonian, a single-point determination is virtually useless in characterizing its flow properties. It is therefore essential that, with non-Newtonian systems, the instrument can operate at a variety of shear rates. Such multipoint instruments are capable of producing a complete rheogram for non-Newtonian systems. For example, multipoint evaluation of pseudoplastic materials would allow assessment of viscosity of a suspending agent at rest (negligible shear rate), while being agitated, poured from a bottle, or applied to the skin (moderately high shear rate). Single-point instruments are unable to describe these changes. As illustrated in **Figure 19–10**, single-point instruments can lead to erroneous results if used to evaluate non-Newtonian systems because flow properties could vary significantly despite identical measured viscosities. Even multipoint instruments, unless properly designed, will not give satisfactory results.

The important conclusion, therefore, is that although all viscometers can be used to determine viscosity of Newtonian

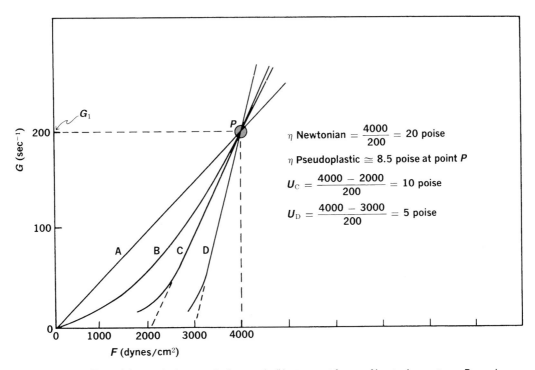

**Fig. 19–10.** Errors inherent in the use of a "one-point" instrument for non-Newtonian systems. Regardless of the fact that A is Newtonian, B is pseudoplastic, and C and D are two different plastic systems, a "one-point" instrument could indicate a common viscosity of 20 poise ($F = 4000$ dynes/cm$^2$ and $G = 200$ sec$^{-1}$). Use of a "one-point" instrument is proper only in the case of the Newtonian systems. (From A. Martin, G. S. Banker, and A. H. C. Chun, in H. S. Bean, A. H. Beckett, and J. E. Carless (Eds.), *Advances in Pharmaceutical Sciences*, Academic Press, London, 1964, Chapter 1. With permission.)

systems, only those with variable–shear-rate controls can be used for non-Newtonian materials. Many types of viscometers have been discussed in detail.[3,14–16] This discussion will be limited to four instruments: capillary, falling-sphere, cup-and-bob, and cone-and-plate viscometers. The first two are single–shear-rate instruments suitable for use only with Newtonian materials, whereas the latter two (multipoint, rotational instruments) can be used with both Newtonian and non-Newtonian systems.

Other rheologic properties such as tackiness or stickiness, "body," "slip," and "spreadability" are difficult to measure by means of conventional apparatus and, in fact, do not have precise meanings. However, the individual factors—viscosity, yield value, thixotropy, and the other properties that contribute to the total consistency of non-Newtonian pharmaceuticals—can be analyzed to some degree of satisfaction in reliable apparatus. An attempt must be made to express these properties in meaningful terms if rheology is to aid in the development, production, and control of pharmaceutical preparations.

## Capillary Viscometer

The viscosity of a Newtonian liquid can be determined by measuring the time required for the liquid to pass between two marks as it flows by gravity through a vertical capillary tube known as an *Ostwald viscometer*. A modern adaptation of the original Ostwald viscometer is shown in **Figure 19–11**. The time of flow of the liquid under test is compared with the time required for a liquid of known viscosity (usually water) to pass between the two marks. If $\eta_1$ and $\eta_2$ are the viscosities

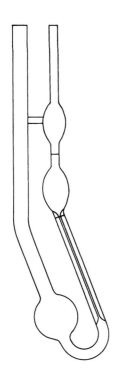

**Fig. 19–11.** Ostwald–Cannon–Fenske viscometer.

of the unknown and the standard liquids, respectively, $\eta_1$ and $\eta_2$ are the respective densities of the liquids, and $t_1$ and $t_2$ are the respective flow times in seconds, the absolute viscosity of the unknown liquid, $\eta_1$, is determined by substituting the experimental values in the equation

$$\frac{\eta_1}{\eta_2} = \frac{\rho_1 t_1}{\rho_2 t_2} \tag{19–12}$$

The value $\eta_1/\eta_2 = \eta_{\text{rel}}$ is known as the *relative viscosity* of the liquid under test.

### EXAMPLE 19–4

**Viscosity of Acetone**

Consider the viscosity measurement of acetone discussed in *Example 19–1*. Assume that the time required for acetone to flow between the two marks on the capillary viscometer was 45 sec and for water the time was 100 sec, at 25°C. The density of acetone is 0.786 g/cm$^3$ and that of water is 0.997 g/cm$^3$ at 25°C. The viscosity of water is 0.8904 cp at this temperature. The viscosity of acetone at 25°C can be calculated using equation (19–12):

$$\frac{\eta_1}{0.8904} = \frac{0.786 \times 45.0}{0.997 \times 100}$$

$$\eta_1 = 0.316 \text{ cp}$$

Equation **(19–12)** is based on *Poiseuille's law* for a liquid flowing through a capillary tube,

$$\eta = \frac{\pi r^4 t \Delta P}{8 l V} \tag{19–13}$$

where $r$ is the radius of the inside of the capillary, $t$ is the time of flow, $\Delta P$ is the pressure head in dyne/cm$^2$ under which the liquid flows, $l$ is the length of the capillary, and $V$ is the volume of liquid flowing. Equation **(19–12)** is obtained from Poiseuille's law, equation **(19–13)**, as follows. The radius, length, and volume of a given capillary viscometer are invariants and can be combined into a constant, $K$. Equation **(19–13)** can then be written as

$$\eta = K t \Delta P \tag{19–14}$$

The pressure head $\Delta P$ depends on density the $\rho$ of the liquid being measured, the acceleration of gravity, and the difference in heights of liquid levels in the two arms of the viscometer. Acceleration of gravity is a constant, however, and if the levels in the capillary are kept constant for all liquids, these terms can be incorporated in the constant and the viscosities of the unknown and the standard liquids can be written as

$$\eta_1 = K' t_1 \rho_1 \tag{19–15}$$

$$\eta_2 = K' t_2 \rho_2 \tag{19–16}$$

Therefore, when flow periods for two liquids are compared in the same capillary viscometer, the division of **(19–15)** by **(19–16)** gives equation **(19–12)**. The *United States Pharmacopeia* suggests a capillary apparatus for determining the viscosity of high-viscosity types of methylcellulose solutions.

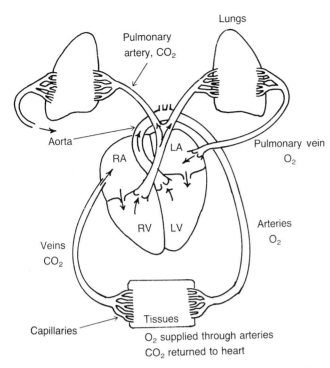

**Fig. 19–12.** Blood flow through the heart, lungs, arteries, veins, and capillaries. Blood with oxygen bound to hemoglobin is pumped through the left ventricle (LV) of the heart to the arteries and is released in the tissues. Carbon dioxide is taken up by the venous blood and is pumped to the right ventricle (RV) of the heart by way of the right atrium (RA). The blood then passes to the lungs, where carbon dioxide is released and oxygen is taken up. The blood, now rich in oxygen, passes from the lungs to the left atrium (LA) and through the left ventricle (LV) to complete the cycle.

**Clinical Correlate: Poiseuille's Law**

Poiseuille's law can be used to calculate the pressure difference in the arteries and capillaries: **Figure 19–12** depicts blood circulation in the body.[17] The systolic pressure is normally about 120 mm Hg and the diastolic pressure about 80 mm Hg. Therefore, at rest the average blood pressure is about 100 mm Hg.

The Poiseuille equation **(19–13)** can be written as

$$r = \left( \frac{8\eta l(V/t)}{\pi \Delta P} \right)^{1/4}$$

where the viscosity, $\eta$, of the blood at normal body temperature is 4 cp or 0.04 poise = 0.04 dyne sec/cm$^2$, and $l$ is the distance, say 1 cm, along an artery. The average rate of blood flow, $V/t$, at rest is 80 cm$^3$/sec, and the pressure drop, $\Delta P$, over a distance of 1 cm along the artery is 3.8 mm Hg (1 dyne/cm$^2$ = 7.5 × 10$^{-4}$ mm Hg). The radius, $r$ (cm), of the artery can be calculated as follows:

$$r(\text{cm}) = \left( \frac{8(0.04 \text{ dyne sec/cm}^2)(1 \text{ cm})(80 \text{ cm}^3/\text{sec})}{\pi(3.8 \text{ mm Hg}) \times (1 \text{ dyne cm}^{-2}/7.50 \times 10^{-4} \text{ mm Hg})} \right)^{1/4}$$

The radius is $(0.001608)^{1/4} = 0.200$ cm.

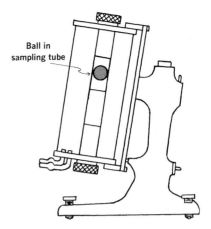

**Fig. 19–13.** Hoeppler falling-ball viscometer.

## Falling-Sphere Viscometer

In this type of viscometer, a glass or steel ball rolls down an almost vertical glass tube containing the test liquid at a known constant temperature. The rate at which a ball of a particular density and diameter falls is an inverse function of the viscosity of the sample. The Hoeppler viscometer, shown in **Figure 19–13**, is a commercial instrument based on this principle. The sample and ball are placed in the inner glass tube and allowed to reach temperature equilibrium with the water in the surrounding constant-temperature jacket. The tube and jacket are then inverted, which effectively places the ball at the top of the inner glass tube. The time for the ball to fall between two marks is accurately measured and repeated several times. The viscosity of a Newtonian liquid is then calculated from

$$\eta = t(S_b - S_f)B \qquad (19\text{–}17)$$

where $t$ is the time interval in seconds for the ball to fall between the two points and $S_b$ and $S_f$ are the specific gravities of the ball and fluid, respectively, at the temperature being used. $B$ is a constant for a particular ball and is supplied by the manufacturer. Because a variety of glass and steel balls of different diameters are available, this instrument can be used over the range 0.5 to 200,000 poise. For best results, a ball should be used such that $t$ is not less than 30 sec.

## Cup-and-Bob Viscometer

In cup-and-bob viscometers, the sample is sheared in the space between the outer wall of a bob and the inner wall of a cup into which the bob fits. The principle is illustrated in **Figure 19–14**. The various instruments available differ mainly in whether the torque results from rotation of the cup or of the bob. In the *Couette* type of viscometer, the cup is rotated. The viscous drag on the bob due to the sample causes it to turn. The resultant torque is proportional to the viscosity of the sample. The MacMichael viscometer is an example of such an instrument. The *Searle* type of viscometer uses a stationary cup and a rotating bob. The torque resulting from

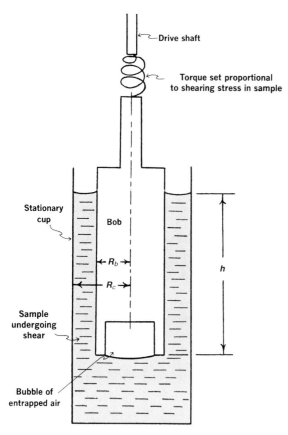

**Fig. 19–14.** Principle of rotational cup-and-bob viscometer (Searle type). See text for explanation.

the viscous drag of the system under examination is generally measured by a spring or sensor in the drive to the bob. The Rotovisco viscometer, shown in **Figure 19–15**, is an example of this type; it can also be modified to operate as a cone and plate instrument.

### EXAMPLE 19–5

#### Shear Stress and Rate of Shear

The Haake Rotovisco apparatus uses interchangeable measuring heads, MK-50 and MK-500. The shear stress, $F$, in dyne/cm$^2$ is obtained from a dial reading $S$ and is calculated using the formula

$$F \text{ (dynes/cm}^2) = K_F \cdot S \qquad (19\text{–}18)$$

where $K_F$ is a shear stress factor.

The shear rate, $G$, in sec$^{-1}$, is proportional to the adjustable speed, $n$, in revolutions per minute of the rotating cylinder in the cup containing the sample. The formula for shear rate is

$$G \text{ (sec}^{-1}) = K_G \cdot n \qquad (19\text{–}19)$$

where $K_G$ is a shear rate factor that varies with the particular rotating cylinder used. Three cups and cylinders (sensor systems) are supplied with the instrument, MVI, MVII, and MVIII. For the measuring head MK-50 and the sensor system MVI, the values for the constants $K_F$ and $K_G$ are 2.95 dyne/cm$^2$ and = 2.35 min/sec, respectively.

In the analysis of a solution of a new glucose derivative that is found to be Newtonian, the following data were obtained in a typical experimental run at 25°C using the Haake viscometer with the MK-50 head and the MVI sensor system. With the cylinder rotating at 180 rpm, the dial reading $S$ was obtained as 65.5 scale divisions.[18] Calculate the Newtonian viscosity of the new glucose derivative. What are the values of shear stress, $F$, and the rate of shear, $G$?

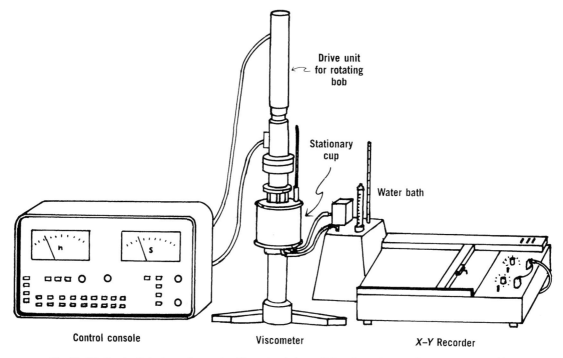

**Fig. 19–15.** Haake Rotovisco viscometer. The rate of shear, $G$, is selected manually or programmed for automatic plotting of upcurves and downcurves. Its value in sec$^{-1}$ is proportional to the speed of the bob shaft, dialed in and read as $n$ on the console. The shear stress is read on the scale $S$ or obtained from the rheogram, plotted on the X–Y recorder.

Using equations (19–18) and (19–19), we obtain

$$F = 2.95 \times 65.5 = 193.2 \text{ dynes/cm}^2$$
$$G = 2.35 \times 180 = 423.0 \text{ sec}^{-1}$$

Now, the Newtonian viscosity is readily obtained as

$$\eta = \frac{F}{G} = \frac{193.2}{423.0} = 0.457 \text{ poise, or } 45.7 \text{ cp}$$

A popular viscometer based on the Searle principle is the Stormer instrument. This viscometer, a modification of that described by Fischer,[19] is shown in **Figure 19–16**. In operation, the test system is placed in the space between the cup and the bob and allowed to reach temperature equilibrium. A weight is placed on the hanger, and the time required for the bob to make 100 revolutions is recorded. These data are then converted to revolutions per minute (rpm). The weight is increased and the whole procedure repeated. In this way, a rheogram can be constructed by plotting rpm versus weight added. By the use of appropriate constants, the rpm values can be converted to actual shear rates in sec$^{-1}$. Similarly, the weights added can be transposed into the units of shear stress, namely, dyne/cm$^2$. According to Araujo,[20] the Stormer instrument should not be used with systems having a viscosity below 20 cp.

It can be shown that, for a rotational viscometer, equation (19–1) becomes

$$\Omega = \frac{1}{\eta} \frac{T}{4\pi h} \left( \frac{1}{R_b^2} - \frac{1}{R_c^2} \right) \qquad (19\text{–}20)$$

where $\Omega$ is the angular velocity in radians/sec produced by $T$, the torque in dynes cm. The depth to which the bob is immersed in the liquids is $h$, and $R_b$ and $R_c$ are the radii of the bob and cup, respectively (see **Fig. 19–14**). The viscous drag of the sample on the base of the bob is not taken into account by equation (19–20). Either an "end correction" must be applied or, more usually, the base of the bob is recessed,

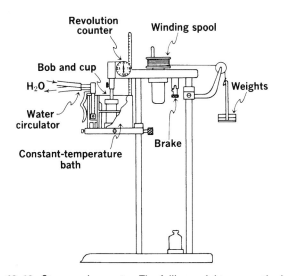

**Fig. 19–16.** Stormer viscometer. The falling weights cause the bob to rotate in the stationary cup. The velocity of the bob is obtained by means of a stopwatch and the revolution counter.

as shown in **Figure 19–14**. In this case, a pocket of air is entrapped between the sample and the base of the bob, rendering the contribution from the base of the bob negligible. It is frequently more convenient to combine all the constants in equation (**19–20**), with the result that

$$\eta = K_v \frac{T}{\Omega} \qquad (19\text{–}21)$$

where $K_v$ is a constant for the instrument. With the modified Stormer viscometer, $\Omega$ is a function of $v$, the rpm generated by the weight, $w$, in grams, which is proportional to $T$. Equation (**19–21**) can then be written as

$$\eta = K_v \frac{w}{v} \qquad (19\text{–}22)$$

The constant $K_v$ can be determined by analyzing an oil of known viscosity in the instrument; reference oils for this purpose are obtained from the National Bureau of Standards.

The equation for plastic viscosity when employing the Stormer viscometer is

$$U = K_v \frac{w - w_f}{v} \qquad (19\text{–}23)$$

where $U$ is the plastic viscosity in poises, $w_f$ is the yield value intercept in grams, and the other symbols have the meaning given in equation (**19–22**).

The yield value of a plastic system is obtained by use of the expression

$$f = K_f \times w_f \qquad (19\text{–}24)$$

where $K_f$ is equal to

$$K_v \times \frac{2\pi}{60} \times \frac{1}{2.303 \log(R_c/R_b)}$$

where $R_c$ is the radius of the cup and $R_b$ is the radius of the bob.

### EXAMPLE 19–6

**Plastic Viscosity of a Gel**

A sample of a gel was analyzed in a modified Stormer viscometer (see Fig. 19–16). A driving weight, $w$, of 450 g produced a bob velocity, $v$, of 350 rpm. A series of velocities was obtained using other driving weights, and the data were plotted as shown in Figure 19–2$b$. The yield value intercept, $w_f$, was obtained by extrapolating the curve to the shearing stress axis where $v = 0$, and the value of $w_f$ was found to be 225 g. The instrumental constant, $K_v$, is 52.0, and $K_f$ is 20.0. What is the plastic viscosity and the yield value of the sample?

We have

$$U = 52.0 \times \frac{450 - 225}{350} = 33.4 \text{ poise}$$
$$f = 20 \times 225 = 4500 \text{ dynes/cm}^2$$

The Brookfield viscometer is a rotational viscometer of the Searle type that is popular in the quality-control laboratories of pharmaceutical manufacturers. A number of spindles (bobs) of various geometries, including cylinders, t-bars, and a cone–plate configuration, are available to provide scientific rheologic data for Newtonian and non-Newtonian liquids and for empirical viscosity measurements on pastes and other

## ▶️ KEY CONCEPT  PLUG FLOW

One potential disadvantage of cup-and-bob viscometers is variable shear stress across the sample between the bob and the cup. In contrast to Newtonian systems, the apparent viscosity of non-Newtonian systems varies with shear stress. With plastic materials, the apparent viscosity below the yield value can be regarded as infinite. Above the yield value, the system possesses a finite viscosity $U$, the plastic viscosity. In a viscometer of the Searle type, the shear stress close to the rotating bob at relatively low rates of shear may be sufficiently high so as to exceed the yield value. The shear stress at the inner wall of the cup could (and frequently does), however, lie below the yield value. Material in this zone would therefore remain as a solid plug and the measured viscosity would be in error. A major factor determining whether or not *plug flow* occurs is the gap between the cup and the bob. The operator should always use the largest bob possible with a cup of a definite circumference so as to reduce the gap

and minimize the chances of plug flow. In a system exhibiting plug flow in the viscometer, more and more of the sample is sheared at a stress above the yield value as the speed of rotation of the bob is increased. It is only when the shear stress at the wall of the cup exceeds the yield value, however, that the system as a whole undergoes laminar, rather than plug, flow and the correct plastic viscosity is obtained.

The phenomenon of plug flow is important in the flow of pastes and concentrated suspensions through an orifice (e.g., the extrusion of toothpaste from a tube). High-shear conditions along the inner circumference of the tube aperture cause a drop in consistency. This facilitates extrusion of the material in the core as a plug. This phenomenon is, however, undesirable when attempting to obtain the rheogram of a plastic system with a cup-and-bob viscometer. Cone-and-plate viscometers do not suffer from this drawback.

semisolid materials. Various models of the Brookfield viscometer are available for high-, medium-, and low-viscosity applications. **Figure 19–17** depicts a cone-and-plate type of Brookfield viscometer.

### Cone-and-Plate Viscometer

The Ferranti–Shirley viscometer is an example of a rotational cone-and-plate viscometer. The measuring unit of the apparatus is shown in **Figure 19–18**; the indicator unit and speed control amplifier are not shown. In operation, the sample is placed at the center of the plate, which is then raised into position under the cone, as shown in **Figure 19–19**. A variable-speed motor drives the cone, and the sample is sheared in

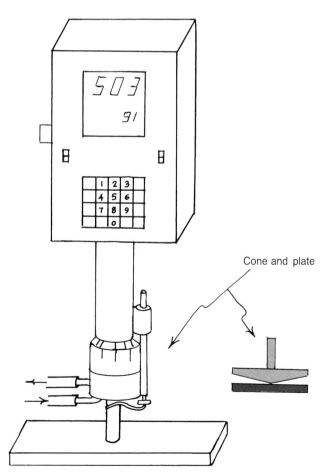

**Fig. 19–17.** A digital-type cone-and-plate Brookfield viscometer.

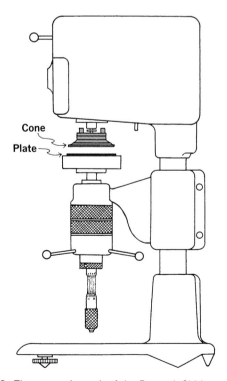

**Fig. 19–18.** The measuring unit of the Ferranti–Shirley cone–plate viscometer.

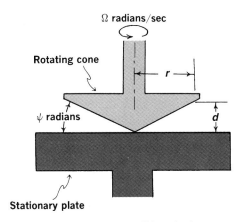

**Fig. 19–19.** Constant–shear-rate conditions in the cone and plate viscometer. The cone-to-plate angle, $\psi$, is greatly exaggerated here; it is ordinarily less than 1° (<0.02 rad).

the narrow gap between the stationary plate and the rotating cone. The rate of shear in revolutions per minute is increased and decreased by a selector dial and the viscous traction or torque (shearing stress) produced on the cone is read on the indicator scale. A plot of rpm or rate of shear versus scale reading or shearing stress can thus be constructed in the ordinary manner.

The viscosity in poise of a Newtonian liquid measured in the cone–plate viscometer is calculated by use of the equation

$$\eta = C \frac{T}{v} \qquad (19\text{–}25)$$

where $C$ is an instrumental constant, $T$ is the torque reading, and $v$ is the speed of the cone in revolutions per minute. For a material showing plastic flow, the plastic viscosity is given by the equation

$$U = C \frac{T - T_f}{v} \qquad (19\text{–}26)$$

and the yield value is given by

$$f = C_f \times T_f \qquad (19\text{–}27)$$

where $T_f$ is the torque at the shearing stress axis (extrapolated from the linear portion of the curve) and $C_f$ is an instrumental constant.

### EXAMPLE 19–7

**Plastic Viscosity of an Ointment Base**

A new ointment base was designed and subjected to rheologic analysis at 20°C in a cone–plate viscometer with an instrumental constant, $C$, of 6.277 cm$^{-3}$. At a cone velocity of $v = 125$ rpm the torque reading, $T$, was 1287.0 dyne cm. The torque, $T_f$, at the shearing stress axis was found to be 63.5 dyne cm.

The plastic viscosity of the ointment base at 20°C was thus calculated using equation (19–26) to be

$$U = \frac{1287 - 63.5}{125} \times 6.277 = 61.44 \text{ poise}$$

The yield value, $f$, is obtained using equation (19–27), where $C_f = 113.6$ cm$^{-3}$ for a medium-size cone (radius of 2.007 cm):

$$f = 113.6 \times 63.5 = 7214 \text{ dynes/cm}^2$$

A cone-and-plate viscometer possesses several significant advantages over the cup-and-bob type of instrument. Most important is the fact that the rate of shear is constant throughout the entire sample being sheared. As a result, any chance of plug flow is avoided. The principle is illustrated in **Figure 19–19**, from which it can be seen that $G$, the rate of shear at any diameter, is the ratio of the linear velocity, $\Omega r$, to the gap width, $d$. Thus,

$$G = \frac{\Omega r}{d} \frac{\text{cm/sec}}{\text{cm}} \qquad (19\text{–}28)$$

The ratio $r/d$ is a constant and is proportional to $\psi$, the angle between the cone and the plate in radians. Thus,

$$G = \frac{\Omega}{\psi} \text{sec}^{-1} \qquad (19\text{–}29)$$

and is independent of the radius of the cone. The cone angle generally ranges from 0.3° to 4°, with the smaller angles being preferred. Other advantages of a cone-and-plate viscometer are the time saved in cleaning and filling and the temperature stabilization of the sample during a run. Whereas a cup-and-bob viscometer may require 20 to 50 mL of a sample for a determination, the cone-and-plate viscometer requires a sample volume of only 0.1 to 0.2 mL. By means of a suitable attachment, it is also possible to increase and then decrease the rate of shear in a predetermined, reproducible manner. At the same time, the shear stress is plotted as a function of the rate of shear on an X–Y recorder. This is a valuable aid when determining the area of hysteresis or thixotropic coefficients because it allows comparative studies to be run in a consistent manner. The use of this instrument in the rheologic evaluation of some pharmaceutical semisolids has been described by Hamlow,[21] Gerding,[22] and Boylan.[23]

## VISCOELASTICITY

A number of methods have been used to measure the consistency of pharmaceutical and cosmetic semisolid products. The discussion in this chapter has centered on the fundamentals of continuous or steady shear rheometry of non-Newtonian materials. Oscillatory and creep measurements are also of considerable importance for investigating the properties of semisolid drug products, foods, and cosmetics that are classified as viscoelastic materials.

Continuous shear mainly employs the rotational viscometer and is plotted as flow curves (see **Fig. 19–2**), which provide useful information by which to characterize and control products in industry. Continuous shear does not keep the material being tested in its rheologic "ground state" but resorts to gross deformation and alteration of the material during measurement. Analysis of viscoelastic materials is designed instead not to destroy the structure, so that measurements can provide information on the intermolecular and interparticle forces in the material.

Viscoelastic measurements are based on the mechanical properties of materials that exhibit both viscous properties of liquids and elastic properties of solids. Many of the systems studied in pharmacy belong to this class, examples being creams, lotions, ointments, suppositories, suspensions, and the colloidal dispersing, emulsifying, and suspending agents. Biologic materials such as blood, sputum, and cervical fluid also show viscoelastic properties. Whereas steady shear in rotational viscometers and similar flow instruments yields large deformations and may produce false results, oscillatory and creep methods allow the examination of rheologic materials under nearly quiescent equilibrium conditions. Davis[24] described creep and oscillatory methods for evaluating the viscoelastic properties of pharmaceutical materials, and Barry[25] reviewed these methods for pharmaceutical and cosmetic semisolids.

A semisolid is considered to demonstrate both solid and liquid characteristics. The flow of a Newtonian fluid is expressed by using equation (19–2),

$$\eta = F/G$$

relating shear stress, $F$, and shear rate, $G$. A solid material, on the other hand, is not characterized by flow but rather by elasticity, and its behavior is expressed by the equation for a spring (derived from Hooke's law):

$$E = F/\gamma \qquad (19\text{–}30)$$

where $E$ is the elastic modulus (dyne/cm$^2$), $F$ is the stress (dyne/cm$^2$), and $\gamma$ is the strain (dimensionless). Using a mechanical model, we can represent a viscous fluid as movement of a piston in a cylinder (or *dashpot*, as it is called) filled with a liquid, as seen in **Figure 19–20a**. An example of a dashpot is an automobile shock absorber. An elastic solid is modeled by the movement of a Hooke spring (**Fig. 19–20b**). The behavior of a semisolid as a viscoelastic body can therefore be described by the combination of the dashpot and spring, as observed in **Figure 19–20c**. The combination of spring and shock absorber in a car, which provides a rela-

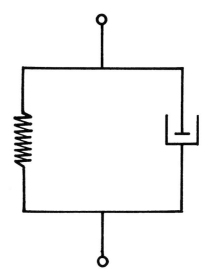

**Fig. 19–21.** Spring and dashpot combined in parallel as a mechanical model of a viscoelastic material, known as a *Voigt element.*

tively smooth ride over rough roads, is analogous to the spring and dashpot of **Figure 19–20c**.

This mechanical model of a viscoelastic material, a non-Newtonian material showing both viscosity of the liquid state and elasticity of the solid state, and combined in *series* is called a *Maxwell element*. The spring and dashpot can also be combined in a *parallel* arrangement as seen in **Figure 19–21**. This second model for viscoelasticity is known as a *Voigt element*.

As a constant stress is applied to the Maxwell unit, there is a strain on the material that can be thought of as a displacement of the spring. The applied stress can be thought of as also producing a movement of the piston in the dashpot due to viscous flow. Removal of the stress leads to complete recovery of the spring, but the viscous flow shows no recovery, that is, no tendency to return to its original state. In the Voigt model, in which the spring and the dashpot are attached in parallel rather than in series, the drag of the viscous fluid in the dashpot simultaneously influences the extension and compression of the spring that characterizes the solid nature of the material, and the strain will vary in an exponential manner with time. Strain is expressed as a deformation or *compliance*, $J$, of the test material, in which $J$ is strain per unit stress. The compliance of a viscoelastic material following the Voigt model is given as a function of time, $t$, by the expression

$$J = J_{\infty}(1 - e^{-t/\tau}) \qquad (19\text{–}31)$$

where $J_{\infty}$ is the compliance or strain per unit stress at infinite time and $\tau$ is viscosity per unit modulus, $\eta/E$ (dyne sec cm$^{-2}$/dyne cm$^{-2}$), which is called *retardation time* and has the unit of seconds.

The Maxwell and Voigt mechanical models representing viscoelastic behavior in two different ways can be combined into a generalized model to incorporate all possibilities of flow and deformation of non-Newtonian materials. One of

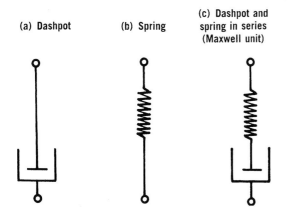

**Fig. 19–20.** Mechanical representation of a viscoelastic material using a dashpot and spring. The dashpot and spring in series is called a *Maxwell element* or *unit.*

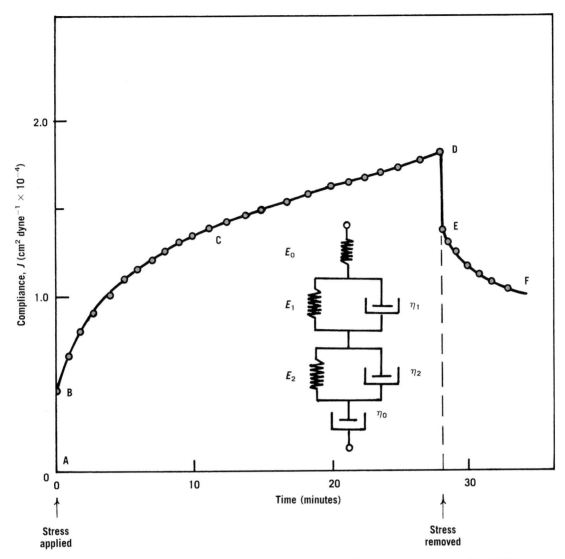

**Fig. 19–22.** A creep curve obtained by analyzing a sample of wool fat in a creep viscometer (Fig. 19–23) at 30°C. The creep curve results from a plot of compliance, $J$, equation (19–31), against the time in minutes during which a stress is applied to the sample. The inset shows the combination of Maxwell and Voigt elements required to represent the viscoelasticity of the wool fat sample. $E_0$, $E_1$, and $E_2$, the spring moduli, can be calculated from the plot and by use of equation (19–32) and the three viscosities $\eta_1$, $\eta_2$, and $\eta_0$. (From S. S. Davis, Pharm. Acta Helv. **49**, 161, 1974. With permission.)

several Voigt units can be combined with Maxwell elements to represent the changes that a pharmaceutical solid, such as an ointment or a cream, undergoes as it is stressed. As observed in **Figure 19–22**, two Voigt elements are combined with a Maxwell element to reproduce the behavior of a sample of wool fat[24] at 30°C. The compliance, $J$, as a function of time is measured with an instrument known as a *creep viscometer* (**Fig. 19–23**) and is plotted in **Figure 19–22** to obtain a *creep curve*. The creep curve is observed to be constructed of three parts, first a sharply rising portion $AB$ corresponding to the elastic movement of the uppermost spring; second, a curved portion $BC$, a viscoelastic region representing the action of the two Voigt units; and third, a linear portion $CD$ corresponding to movement of the piston in the dashpot at the bottom of the Maxwell–Voigt model representing viscous flow.

The compliance equation corresponding to the observed behavior of wool fat (**Fig. 19–22**), as simulated by the Maxwell–Voigt model (inset in **Fig. 19–22**), is

$$J = \frac{\gamma_0}{F} + J_m(1 - e^{-t/\tau_m}) + J_n(1 - e^{-t/\tau_n}) + \frac{\gamma}{F} \quad \textbf{(19–32)}$$

where $\gamma_0$ is the instantaneous strain and $F$ is the constant applied shear stress.[24,25] The quantity $\gamma_0/F$ is readily obtained from the experimental curve (region $AB$) in **Figure 19–22**. The viscoelastic region of the curve ($BC$) is represented by the intermediate term of equation **(19–32)**, where $J_m$ and $J_n$ are the mean compliance of bonds in the material and $\tau_m$ and $\tau_n$ are the mean retardation times for the two Voigt units of **Figure 19–22**. It is sometimes found that three or more Voigt units are needed in the model to reflect the

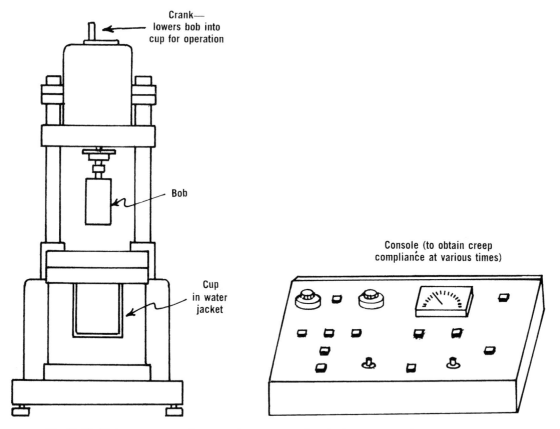

**Fig. 19–23.** Main components of a creep viscometer used to obtain creep compliance curves such as those found in Figures 19–22 and 19–24.

observed behavior of the material. The final term of equation **(19–32)** corresponds to the linear portion, *CD*, of the creep curve. This section represents a condition of Newtonian compliance in which the rupture of bonds leads to the flow of the material, where *F* is the constant applied stress and $\gamma$ is the shear strain in this region of the curve.

When stress is removed by the operator of the creep rheometer (**Fig. 19–23**), a recovery (*DEF*) of the sample is obtained. It is composed of an instantaneous elastic recovery, *DE*, equivalent to *AB*, followed by an elastic recovery region, *EF*, equivalent to *BC*. In the creep compliance curve of **Figure 19–22**, flow occurs in region *CD*, irreversibly destroying the structure, and in the recovery curve this portion is not reproduced. By such an analysis, Davis[24] obtained the elastic moduli (inset of **Fig. 19–22**) $E_0 = 2.7 \times 10^4$ dynes/cm, $E_1 = 5.4 \times 10^4$ dynes/cm, and $E_2 = 1.4 \times 10^4$ dynes/cm, and the three viscosities $\eta_1 = 7.2 \times 10^5$ poise, $\eta_2 = 4.5 \times 10^6$ poise, and $\eta_0 = 3.1 \times 10^7$ poise for wool fat.

The creep curve used to measure the viscoelasticity of non-Newtonian pharmaceutical, dermatologic, and cosmetic materials can shed some light on the molecular structure of the materials and therefore provide information for modification and improvement of these vehicles. Creep compliance curves were used by Barry[25] to study the changes with temperature in samples of white petrolatum (White Soft Paraffin, British *Pharmacopoeia*) as observed in **Figure 19–24**.

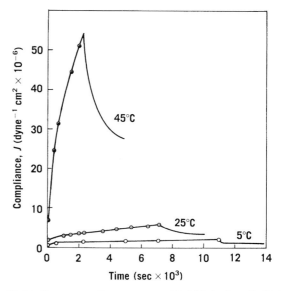

**Fig. 19–24.** Creep compliance curves of Soft White Paraffin (British Pharmacopoeia) at three temperatures. (From B. W. Barry, in H. S. Bean, A. H. Beckett, and J. E. Carless (Eds.), *Advances in Pharmaceutical Sciences*, Vol. 4, Academic Press, New York, 1974, p. 36. With permission.)

The behavior was complex, requiring five Voigt units and one Maxwell element to describe the observed creep compliance curves at 5°C and 25°C and three Voigt units at 45°C, where some of the structure had been destroyed by melting. Three curves are characteristic of the crystalline bonding and the interaction of crystalline and amorphous material that constitute petrolatum. The curves were automatically plotted on an *X–Y* recorder as the material was stressed in the creep viscometer. The circles plotted along the lines of **Figure 19–24** were obtained by use of an equation similar to equation **(19–32)**, showing the accuracy with which the creep curves can be reproduced by a theoretical model of Voigt and Maxwell units.

Another dynamic rheologic method that does not disturb the structure of a material is that of oscillatory testing.[24–27] A thin layer of material is subjected to an oscillatory driving force in an apparatus known as a rheogoniometer, such as that shown in **Figure 19–25**. The shearing stress produced by the oscillating force in the membrane of the apparatus results in a shear rate proportional to the surface velocity of the material. The viscoelastic behavior of materials obtained by oscillatory shear measurements can be analyzed by an extension of the Maxwell spring-and-dashpot model.

Steady shear methods involving rotational viscometers tend to break down materials under analysis, and although they yield useful data on thixotropy and yield stress, for exam-

ple, they do not provide information about the original structure and bonding in pharmaceutical and cosmetic semisolids. Viscoelastic analysis performed by creep or oscillatory methods is particularly useful for studying the structure of liquid and semisolid emulsions and gels.[26] Viscoelastic measurements can also be used to measure the rheologic changes occurring in a cream after it is broken down in various stages by milling, incorporation of drugs, or spreading on the skin.

Radebaugh and Simonelli[28] studied the viscoelastic properties of anhydrous lanolin, which were found to be a function of strain, shear frequency, shear history, and temperature. The energy of activation, $E_v$, was calculated for the structural changes of the lanolin sample, which was found to undergo a major mechanical transition between 10°C and 15°C. The $E_v$ for the transition was about 90 kcal that expected for glass transition. Rather than a sharp change from a rubbery to a glasslike state, however, anhydrous lanolin appeared to change to a state less ordered than glass. The glass–rubber transition and the glass transition temperature are discussed in Chapter 20. The viscoelastic properties were determined using a Rheometrics mechanical spectrometer (RMS 7200; Rheometrics, Inc., Union, N.J.). The rheometer introduces a definite deformation into the sample at a specified rate and at a chosen temperature.

For the design of mucolytic agents in the treatment of bronchitis, asthma, and cystic fibrosis, viscoelastic methods are also of value in the analysis of sputum. Other biologic fluids such as blood, vaginal material, and synovial fluids may be analyzed by viscoelastic test methods. The unsteady shear to which synovial fluids are subjected in the body during the movement of leg and arm joints requires the elastic properties of these fluids, in addition to viscous properties that are observed only in steady shear. Thurston and Greiling[29] used oscillatory shear to analyze cases of noninflammatory and inflammatory joint disease associated with arthritis. The macromolecule hyaluronic acid is primarily responsible for the high viscosity and non-Newtonian character of synovial fluid and gives it simple Newtonian rather than the desired non-Newtonian properties. Changes in viscoelasticity of synovial fluids, measured in the oscillatory instrument shown in **Figure 19–25**, can therefore serve as sensitive indicators of joint disease.

## PSYCHORHEOLOGY

In addition to desirable pharmaceutical and pharmacologic properties, topical preparations must meet criteria of feel, spreadability, color, odor, and other psychologic and sensory characteristics. Workers in the food industry have long tested products such as butter, chocolate, mayonnaise, and bread dough for proper consistency during manufacture, packaging, and end use. Sensations in the mouth, between the fingers, and on the skin are important considerations for manufacturers of foods, cosmetics, and dermatologic products. Scott-Blair[30] discussed *psychorheology* (as this subject is called)

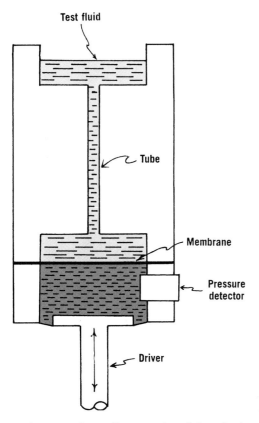

**Fig. 19–25.** Apparatus for oscillatory testing of viscoelastic materials. (From G. B. Thurston and A. Martin, J. Pharm. Sci. **67**, 1499, 1978. With permission.)

in the food industry. Kostenbauder and Martin[31] assessed the spreadability of ointments in relation to their rheologic properties. In consultation with dermatologists, they divided the products into three classes. Class I products were soft, mainly for ophthalmic use; class II products included common medicated ointments of intermediate consistency; and class III products involved stiff protective products for use in moist ulcerative conditions. The yield values and plastic viscosity for each class of product were reported.

Boylan[23] showed that the thixotropy, consistency, and yield value of bacitracin ointment, USP, decreased markedly as the temperature was raised from 20°C to 35°C. Thus, although a product may be sufficiently thixotropic in its container, this property can be lost following application to the skin.

Barry et al.[32] carried out sensory testing on topical preparations. They used a panel to differentiate textural parameters and established rheologic methods for use in industry as control procedures for maintaining uniform skin feel and spreadability of dermatologic products. Cussler et al.[33] studied the texture of non-Newtonian liquids of widely different rheologic properties applied to the skin. They found that a panel of untrained subjects could accurately assess the consistency of a material by the use of only three attributes: smoothness, thinness, and warmth. Smoothness was related to a coefficient of friction and thinness to non-Newtonian viscous parameters that could be measured with appropriate instruments. The characteristic of warmth was found to be sufficiently complex to require further study.

## APPLICATIONS TO PHARMACY

The rheologic behavior of poloxamer vehicles was studied as a function of concentration over a temperature range of 5°C to 35°C using a cone–plate viscometer.[34] Poloxamers are block polymers from BASF Wyandotte Corp. that have the chemical structure

$$HO(CH_2CH_2O)_a(CH[CH_3]CH_2O)_b(CH_2CH_2O)_cH$$

Poloxamers with a wide range of molecular weights are available as Pluronics. Some of the poloxamers are used in dermatologic bases or topical ophthalmic preparations because of their low toxicity and their ability to form clear water-based gels.

The aqueous solubility of the poloxamers decreases with an increase in temperature, the hydration of the polymer being reduced by the breaking of hydrogen bonds at higher temperatures. The desolvation that results, together with the entanglement of the polymer chains, probably accounts for the gel formation of the poloxamers.

A linear relationship was found between shear rate and shear stress (Newtonian behavior) for the poloxamer vehicles in the sol state, which exists at low concentrations and low temperatures. As the concentration and temperature were increased, some of the poloxamers exhibited a sol–

gel transformation and became non-Newtonian in their rheologic character. The addition of sodium chloride, glycerin, or propylene glycol resulted in increased apparent viscosities of the vehicles.

Polymer solutions can be used in ophthalmic preparations, as wetting solutions for contact lenses, and as tear replacement solutions for the condition known as *dry eye syndrome*. Both natural (e.g., dextran) and synthetic (e.g., polyvinyl alcohol) polymers are used with the addition of various preservatives. A high-molecular-weight preparation of sodium hyaluronate at concentrations of 0.1% to 0.2% has been introduced to overcome the dry eye condition.

For high-polymer solutions, the viscosity levels off to a *zero-shear viscosity* (a high viscosity) at low shear rates. The viscosity decreases as the shear rate is increased because the normally twisted and matted polymer molecules align in the streamlined flow pattern and exhibit pseudoplasticity or shear thinning.

Bothner et al.[35] suggested that a suitable tear substitute should have shear-thinning properties as do natural tears to conform to the low shear rate during nonblinking and the very high shear rate during blinking. The low viscosity at high shear rates produces lubrication during blinking, and the high viscosity at zero shear rate prevents the fluid from flowing away from the cornea when the lids are not blinking. Using a computer-controlled Couette viscometer, they studied the rheologic properties of eight commercial tear substitutes, together with 0.1% and 0.2% solutions of sodium hyaluronate. For five of the commercial products, the viscosity was independent of shear rate; thus, these products behaved as Newtonian liquids. Two products showed slight shear thinning at high shear rates. Only the commercial product Neo-Tears and the two noncommercial sodium hyaluronate solutions showed the desired pseudoplastic behavior. For Neo-Tears the viscosity at high shear rate, $1000 \ sec^{-1}$, was 3-fold that at zero shear. For 0.1% sodium hyaluronate the value was 5-fold and for 0.2% sodium hyaluronate it was 30-fold. Therefore, sodium hyaluronate appears to be an excellent candidate as a tear replacement solution.

The rheologic properties of suppositories at rectal temperatures can influence the release and bioabsorption of drugs from suppositories, particularly those having a fatty base. Grant and Liversidge[36] studied the characteristics of triglyceride suppository bases at various temperatures, using a rotational rheometer. Depending on the molten (melted) character of the base, it behaved either as Newtonian material or as a plastic with thixotropy.

Fong-Spaven and Hollenbeck[37] studied the rheologic properties as a function of the temperature of mineral oil–water emulsions stabilized with triethanolamine stearate (TEAS). The stress required to maintain a constant rate of shear was monitored as temperature increased from 25°C to 75°C. Unexpected, but reproducible discontinuities in the plots of temperature versus apparent viscosity were obtained using a Brookfield digital viscometer and were attributed

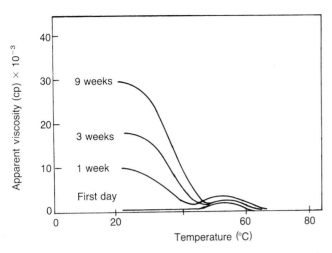

**Fig. 19–26.** Viscosity versus temperature plots of an oil–water emulsion over a period of 9 weeks. (From F. Fong-Spaven and R. G. Hollenbeck, Drug Dev. Ind. Pharm. **12,** 289, 1986. With permission.)

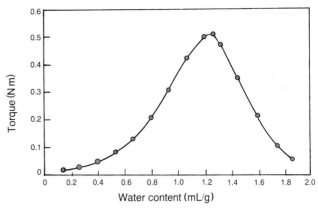

**Fig. 19–27.** Changes of torque in a mixer torque rheometer as water is added to a mixture of powders. (From R. C. Rowe and G. R. Sadeghnejad, Int. J. Pharm. **38,** 227, 1987. With permission.)

possibly to shifts in the liquid crystalline structures. As seen in **Figure 19–26,** where apparent viscosity is plotted versus temperature for a 5% TEAS mineral oil–water emulsion, viscosity decreases as temperature is raised to about 48°C. The viscosity reverses and increases to a small peak at 54°C and then decreases again with increasing temperature. This unusual behavior is considered to result from gel formation, which stabilizes the internal phase. Liquid-crystalline structures of TEAS exist, and at higher temperatures the structures disintegrate or "melt" to form a large number of TEAS molecules in a gel-like arrangement that exhibits increased resistance to flow. As the temperature rises above 54°C the gel structure is gradually destroyed and viscosity again decreases, as shown in **Figure 19–26.**

Patterned after the manufacture and use of cosmetic sticks, solidified sodium stearate–based sticks were prepared and tested for topical application using a Ferranti–Shirley cone–plate viscometer. The sticks contained propylene glycol, polyethylene glycol 400, and polyethylene glycol 600 as humectants and the topically active drugs panthenol, chlorphenesin, and lignocaine. Thixotropic breakdown was much lower in these medicated sticks than in comparable bases. The addition of the three topical drugs to the stearate-based sticks caused changes in yield values, thixotropy, and plastic viscosity; possible reasons for the changes were advanced.[38]

Rowe and Sadeghnejad[39] studied the rheologic properties of microcrystalline cellulose, an ingredient incorporated into wet powder masses to facilitate granulation in the manufacture of tablets and granules. The authors designed a *mixer torque rheometer* to measure the torque changes as water was added to the powder mixture (*torque* is the force acting to produce rotation of a body). As the mixture became wetter, torque increased until the mass was saturated, then decreased with further addition of water as a slurry (suspension) was formed. A plot of torque in Newton meters (1 N m = 1 joule) against increasing water content produced a

bell-shaped curve, as shown in **Figure 19–27.** This behavior was explained, according to the authors, by the three states of liquid saturation of a powder mass, as described by Newitt and Conway-Jones.[40]

With the early addition of liquid, a *pendular state* exists (see **Fig. 19–28**) with lenses of liquid at the contact points of the particles. The liquid forces out some of the air originally filling the spaces between particles. As more liquid is added, a mixture of liquid and air exists between the particles to produce the *funicular state.* The torque on the mixer increases for these two conditions until the end of the funicular state. The pores are then filled with liquid to yield the *capillary state,* and with the addition of more liquid the torque decreases as a slurry (suspension) is produced (*liquid-droplet state*). These stages of saturation are depicted schematically in **Figure 19–28.**

The three microcrystalline celluloses from different sources[39] exhibited essentially the same plot of torque versus water added (see **Fig. 19–27**). Yet the curves, only one of three shown here, rose to slightly different heights and the maxima occurred at different amounts of water added.

An account of the rheology of suspensions, emulsions, and semisolids is presented in Chapter 19, and the flow properties of powders are dealt with in Chapter 20. Consideration is given in Chapter 18 to the rheology of colloid materials,

(a) Pendular    (b) Funicular    (c) Capillary    (d) Liquid droplet

**Fig. 19–28.** The states of liquid saturation of a powder. (*a*) *Pendular* state with lenses of liquid at the contact points of the particles. (*b*) A mixture of liquid and air between the particles, producing the *funicular* state. (*c*) Pores filled with liquid to yield the capillary state. (*d*) Liquid droplets completely enveloping particles (the *liquid-droplet* state). (Modified from D. M. Newitt and J. M. Conway-Jones, Trans. Inst. Chem. Eng. **36,** 422, 1958.)

**TABLE 19–3**

## PHARMACEUTICAL AREAS IN WHICH RHEOLOGY IS SIGNIFICANT*

1. Fluids
   a. Mixing
   b. Particle-size reduction of disperse systems with shear
   c. Passage through orifices, including pouring, packaging in bottles, and passage through hypodermic needles
   d. Fluid transfer, including pumping and flow through pipes
   e. Physical stability of disperse systems
2. Quasisolids
   a. Spreading and adherence on the skin
   b. Removal from jars or extrusion from tubes
   c. Capacity of solids to mix with miscible liquids
   d. Release of the drug from the base
3. Solids
   a. Flow of powders from hoppers and into die cavities in tabletting or into capsules during encapsulation
   b. Packagability of powdered or granular solids
4. Processing
   a. Production capacity of the equipment
   b. Processing efficiency

*From A. Martin, G. S. Banker, and A. H. C. Chun, in H. S. Bean, A. H. Beckett, and J. E. Carless (Eds.), *Advances in Pharmaceutical Sciences*, Academic Press, London, 1964, Chapter 1. With permission.

which find wide application in pharmacy as suspending agents. Boylan[23] considered some of the rheologic aspects of parenteral suspensions and emulsions.

A summary of the major areas of product design and processing in which rheology is significant is given in **Table 19–3**. Although the effects of processing can affect the flow properties of pharmaceutical systems, a detailed discussion of this area is outside the scope of this text. For an account of this topic as well as a comprehensive presentation of the theoretical and instrumental aspects of rheology, refer to the review by Martin et al.[3] The theory and application of viscoelasticity were briefly reviewed in the previous section. Detailed discussions of this approach are given in the references cited.

# CHAPTER SUMMARY

In this chapter the basics of rheology were presented. An understanding of the viscosity of liquids, solutions, and dilute and concentrated colloidal systems has both practical and theoretical values in the pharmaceutical sciences. Rheology is involved in the mixing and flow of materials, their packaging into containers, and their removal prior to use, whether this is achieved by pouring from a bottle, extrusion from a tube, or passage through a syringe needle. The rheology of a particular product, which can range in consistency from fluid to semisolid to solid, can affect its patient acceptability, physical stability, and even biologic availability. Materials are classi-

fied according to types of flow and deformation. Finally, the theory and methods for determining the rheologic properties of pharmaceutical materials as well as the application to the pharmaceutical sciences were discussed.

**Practice problems for this chapter can be found at thePoint.lww.com/Sinko6e.**

## References

1. E. K. Fischer, J. Colloid. Sci. **3**, 73, 1948.
2. G. W. Scott-Blair, Pharm. J. **154**, 3, 1945.
3. A. Martin, G. S. Banker, and A. H. C. Chun, in H. S. Bean, A. H. Beckett, and J. E. Carless (Eds.), *Advances in Pharmaceutical Sciences*, Academic Press, London, 1964, Chapter 1.
4. S. P. Kabre, H. G. DeKay, and G. S. Banker, J. Pharm. Sci. **53**, 492, 1964.
5. E. E. Hamlow, *Correlation of Rheological Methods for Measuring Newtonian and Non-Newtonian Materials*, Ph.D. Thesis, Purdue University, Purdue, Ind., 1958.
6. R. Shangraw, W. Grim, and A. M. Mattocks, Trans. Soc. Rheol. **5**, 247, 1961; N. Casson, in C. C. Mill (Ed.), *Rheology of Disperse Systems*, Pergamon Press, New York, 1959, pp. 84–104; T. C. Patton, *Paint Flow and Pigment Dispersion*, 2nd Ed., Wiley-Interscience, New York, 1979, Chapter 16.
7. G. J. Yakatan and O. E. Araujo, J. Pharm. Sci. **57**, 155, 1968.
8. M. Reiner and G. W. Scott-Blair, in F. E. Eirich (Ed.), *Rheology*, Vol. 4, Academic Press, New York, 1967, Chapter 9.
9. S. S. Ober, H. C. Vincent, D. E. Simon, and K. J. Frederick, J. Am. Pharm. Assoc. Sci. Ed. **47**, 667, 1958.
10. C. W. Chong, S. P. Eriksen, and J. W. Swintosky, J. Am. Pharm. Assoc. Sci. Ed. **49**, 547, 1960.
11. J. C. Samyn and W. Y. Jung, J. Pharm. Sci. **56**, 188, 1967.
12. H. Freundlich and F. Juliusburger, Trans. Faraday Soc. **31**, 920, 1935.
13. G. Levy, J. Pharm. Sci. **51**, 947, 1962.
14. E. Hatschek, *Viscosity of Liquids*, Bell and Sons, London, 1928.
15. P. Sherman, in *Emulsion Science*, Academic Press, London, 1968, p. 221.
16. J. R. Van Wazer, J. W. Lyons, K. Y. Kim, and R. E. Colwell, *Viscosity and Flow Measurement—A Laboratory Handbook of Rheology*, Interscience, New York, 1963.
17. R. Chang, *Physical Chemistry with Applications to Biological Systems*, 2nd Ed., Macmillan, New York, 1981, pp. 76, 77, 93.
18. B. Millan-Hernandez, *Properties and Design of Pharmaceutical Suspensions*, M. S. Thesis, University of Texas, 1981.
19. E. K. Fischer, *Colloidal Dispersions*, Wiley, New York, 1950, Chapter 5.
20. O. E. Araujo, J. Pharm. Sci. **56**, 1023, 1967.
21. E. E. Hamlow, Ph.D. Thesis, Purdue University, Purdue, Ind., 1958.
22. T. G. Gerding, Ph.D. Thesis, Purdue University, Purdue, Ind., 1961.
23. J. C. Boylan, Bull. Parenter. Drug Assoc. **19**, 98, 1965; J. Pharm. Sci. **55**, 710, 1966.
24. S. S. Davis, Pharm. Acta Helv. **9**, 161, 1974.
25. B. W. Barry, in H. S. Bean, A. H. Beckett, and J. E. Carless (Eds.), *Advances in Pharmaceutical Sciences*, Vol. 4, Academic Press, New York, 1974, pp. 1–72.
26. G. B. Thurston and S. S. Davis, J. Colloid Interface Sci. **69**, 199, 1979.
27. G. B. Thurston and A. Martin, J. Pharm. Sci. **67**, 1949, 1978.
28. G. W. Radebaugh and A. P. Simonelli, J. Pharm. Sci. **72**, 415, 422, 1983; J. Pharm. Sci. **73**, 590, 1984.
29. G. B. Thurston and H. Greiling, Rheol. Acta **17**, 433, 1978.
30. G. W. Scott-Blair, *Elementary Rheology*, Academic Press, New York, 1969.
31. H. B. Kostenbauder and A. Martin, J. Am. Pharm. Assoc. Sci. Ed. **43**, 401, 1954.
32. B. W. Barry and A. J. Grace, J. Pharm. Sci. **60**, 1198, 1971; J. Pharm. Sci. **61**, 335, 1972; B. W. Barry and M. C. Meyer, J. Pharm. Sci. **62**, 1349, 1973.

33. E. L. Cussler, S. J. Zlolnick, and M. C. Shaw, Percept. Psychophys. **21,** 504, 1977.

34. S. C. Miller and B. R. Drabik, Int. J. Pharm. **18,** 269, 1984.

35. H. Bothner, T. Waaler, and O. Wik, Drug Dev. Ind. Pharm. **16,** 755, 1990.

36. D. J. W. Grant and G. G. Liversidge, Drug Dev. Ind. Pharm. **9,** 247, 1983.

37. F. Fong-Spaven and R. G. Hollenbeck, Drug Dev. Ind. Pharm. **12,** 289, 1986.

38. A. G. Mattha, A. A. Kassem, and G. K. El-Khatib, Drug Dev. Ind. Pharm. **10,** 111, 1984.

39. R. C. Rowe and G. R. Sadeghnejad, Int. J. Pharm. **38,** 227, 1987.

40. D. M. Newitt and J. M. Conway-Jones, Trans. Inst. Chem. Eng. **36,** 422, 1958.

## Recommended Readings

J. W. Goodwin and R. W. Hughes, *Rheology for Chemists: An Introduction, Royal Society of Chemistry*, 2nd Ed., London, UK, 2008.

C. W. Macosko, *Rheology: Principles, Measurements, and Applications*, 1st Ed., Wiley-VCH, Inc., New York, 1994.

### CHAPTER LEGACY

**Fifth Edition:** published as Chapter 20 (Rheology). Updated by Patrick Sinko.

**Sixth Edition:** published as Chapter 19 (Rheology). Updated by Patrick Sinko.

# 20 ■ PHARMACEUTICAL POLYMERS

## INTRODUCTION

Synthetic and natural-based polymers have found their way into the pharmaceutical and biomedical industries and their applications are growing at a fast pace. Understanding the role of polymers as ingredients in drug products is important for a pharmacist or pharmaceutical scientist who deals with drug products on a routine basis. Having a basic understanding of polymers will give you the opportunity to not only familiarize yourself with the function of drug products but also possibly develop new formulations or better delivery systems. This chapter will provide the basis for understanding pharmaceutical polymers. The basic concepts of polymer chemistry, polymer properties, types of polymers, polymers in pharmaceutical and biomedical industries, and reviews of some polymeric products in novel drug delivery systems and technologies will be covered.

## HISTORY OF POLYMERS

Polymers have a wide-ranging impact on modern society. Polymers are more commonly referred to as "plastics" since people are more familiar with plastic products that they encounter around the house than any other type of polymeric product. Plastics have the ability to be molded, cast, extruded, drawn, thermoformed, or laminated into a final product such as plastic parts, films, and filaments. The first semisynthetic polymer ever made was guncotton (cellulose nitrate) by Christian F. Schönbein in 1845. The manufacturing process for this polymer was changed over the years due to its poor solubility, processability, and explosivity resulting in a variety of polymers such as Parkesine, celluloid (plasticized cellulose nitrate), cellulose acetate (cellulose treated with acetic acid), and hydrolyzed cellulose acetate soluble in acetone. In 1872, Bakelite, a strong and durable synthetic polymer based on phenol and formaldehyde, was invented. Polycondensation-based polymeric products such as Bakelite

and those based on phenoxy, epoxy, acrylic, and ketone resins were used as cheap substitutes for many parts in the auto and electronic industries. Other synthetic polymers were invented later including polyethylene (1933), poly (vinyl chloride) (1933), polystyrene (1933), polyamide (1935), Teflon (1938), and synthetic rubbers (1942). Polyethylene was used to make radar equipment for airplanes. The British air force used polyethylene to insulate electrical parts of the radars in their airplanes. Synthetic rubber, which could be made in approximately 1 hr as compared to 7 years for natural rubbers, was used to make tires and other military supplies. Teflon was used in atomic bombs to separate the hot isotopes of uranium. Nylon was used to make parachutes, replacing silk, which had to be imported from Japan.

The plastics revolution advanced technologies in the 20th century and opened new fields of application in the pharmaceutical and biomedical sectors. In recent years, polymers have been used to develop devices for controlling drug delivery or for replacing failing natural organs. In oral delivery, polymers are used as coatings, binders, taste maskers, protective agents, drug carriers, and release controlling agents. Targeted delivery to the lower part of the gastrointestinal tract (e.g., in the colon) was made possible by using polymers that protect drugs during their passage through the harsh environment of the stomach. Transdermal patches use polymers as backings, adhesives, or drug carriers in matrix or membrane products (these are described later in the book). Controlled delivery of proteins and peptides has been made possible using biodegradable polymers. In many drug products you may find at least one polymer that enhances product performance. The key difference between early polymers and pharmaceutical polymers is biocompatibility.

## POLYMERS IN GENERAL[1-5]

The word "polymer" means "many parts." A polymer is a large molecule made up of many small repeating units. In the

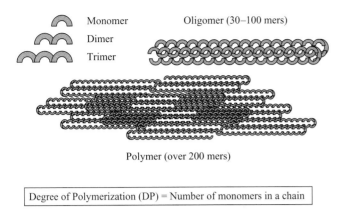

Degree of Polymerization (DP) = Number of monomers in a chain

**Fig. 20–1.** Polymer anatomy.

early days of polymer synthesis, little was known about the chemical structures of polymers. Herman Staudinger, who received the Nobel Prize in Chemistry in 1953, coined the term "macromolecule" in 1922 and used it in reference to polymers. The difference between the two is that polymers are made of repeating units, whereas the term macromolecule refers to any large molecule, not necessarily just those made of repeating units. So, polymers are considered to be a subset of macromolecules.

A *monomer* is a small molecule that combines with other molecules of the same or different types to form a polymer. Since drawing a complete structure of a polymer is almost impossible, the structure of a polymer is displayed by showing the repeating unit (the monomer residue) and an "*n*" number that shows how many monomers are participating in the reaction. From the structural prospective, monomers are generally classified as olefinic (containing double bond) and functional (containing reactive functional groups) for which different polymerization methods are utilized. If two, three, four, or five monomers are attached to each other, the product is known as a dimer, trimer, tetramer, or pentamer, respectively. An *oligomer* contains from 30 to 100 monomeric units. Products containing more than 200 monomers are simply called a polymer (**Fig. 20–1**). From a thermodynamic perspective, polymers cannot exist in the gaseous state because of their high molecular weight. They exist only as liquids or high solid materials.

**EXAMPLE 20–1**

**Molecular Weight**

A polyethylene with molecular weight of 100,000 g/mol is made of almost 3570 monomer units ($-CH_2CH_2-$) with the molecular weight of 28 g/mol.

Since polymers originate from oil, they are generally cheap materials. Unlike other materials such as metals or ceramics, polymers are large molecular weight materials and their molecular weight can be adjusted for a given application. For example, silicone polymers are supplied as vacuum grease (low molecular weight) and as durable implants (very high molecular weight). By changing the molecular weight,

the physical and mechanical properties of the polymer can be tailor-made. This can be achieved by changing the structure of the monomer building blocks or by blending them with other polymers. Blending is a process intended to achieve superior properties that are unattainable from a single polymer. For example, polystyrene is not resistant against impact, so a polystyrene cup can be easily smashed into pieces if compressed between your fingers. However, polystyrene blended with polybutadiene is an impact resistant product. Alternatively, monomers of styrene and butadiene can be copolymerized to make a new copolymer of styrene–butadiene.

# POLYMER SYNTHESIS

To make polymers, monomers have to interact with each other. Let us consider a simple scenario in which just one monomer type is going to be polymerized. The structure of the monomer molecule will tell us how we should polymerize it. A monomer may be unsaturated; in other words it may contain a double bond of $\sigma$ (sigma) and $\pi$ (pi) between a pair of electrons. The $\pi$ bond generally requires low energy to break; therefore, polymerization starts at this site by the addition of a free radical on the monomer. On the other hand, if a monomer does not contain a double bond but possesses functional groups such as hydroxyl, carboxyl, or amines, they can interact via condensation. These two types of polymerization processes are described in the next two sections.

## Addition Polymerization

Free-radical polymerization is also known as chain or addition polymerization. As the name implies, a radical-generating ingredient induces an initiator triggering polymerization. The initiator is an unstable molecule that is cleaved into two radical-carrying species under the action of heat, light, chemical, or high-energy irradiation. Each initiating radical has the ability to attack the double bond of a monomer. In this way, the radical is transferred to the monomer and a monomer radical is produced. This step in polymerization is called *initiation*. The monomer radical is also able to attack another monomer and then another monomer, and so on and so forth. This step is called *propagation* by which a *macroradical* is formed. Macroradicals prepared in this way can undergo another reaction with another macroradical or with another inert compound (e.g., an impurity in the reaction) which terminates the macroradical. **Figure 20–2** shows the free-radical polymerization of styrene, a monomer, to polystyrene. Monomers such as acrylic acid, acrylamide, acrylic salts (such as sodium acrylate), and acrylic esters (methyl acrylate) contain double bonds and they can be polymerized via addition reactions.

## Condensation Polymerization

In condensation polymerization, also called step polymerization, two or more monomers carrying different reactive

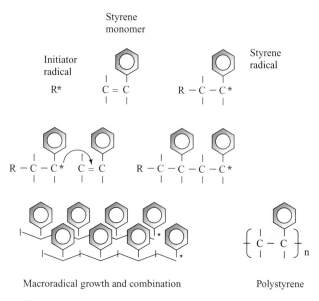

**Fig. 20–2.** Addition or free-radical polymerization of styrene.

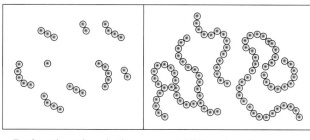

**Fig. 20–4.** Condensation versus addition polymerizations.

functional groups interact with each other as shown in **Figure 20–3**. For example, a monomer containing a reactive hydrogen from the amine residue can react with another monomer containing a reactive hydroxyl group (a residue of carboxyl group) to generate a new functional group (amide) and water as a side product. If a monomer containing the reactive hydrogen reacts with a monomer containing reactive chlorine, the side product will be hydrochloric acid. Since each monomer is bifunctional (in other words, it contains two reactive hydrogens or two reactive chlorines), the reaction product can grow by reacting with another monomer generating a macromonomer. Nylon is prepared via condensation polymerization of a diamine and diacid chloride. The diamine and diacid chloride are dissolved in water and tetrachloroethylene, respectively. Since the two solutions do not mix with each other, they form two immiscible separate layers, with tetrachloroethylene at the bottom. At the interface of the two solutions, the two monomers interact and form the polymer. The polymers can then be gently removed from the interface as fiber. There are no radicals involved in this polymerization reaction.

Free-radical polymerization is an addition reaction that is characterized by fast growth of macroradicals. There is a high chance that high–molecular-weight chains are formed

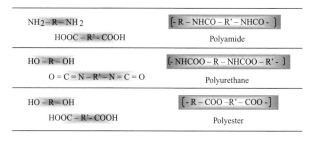

**Fig. 20–3.** Examples of condensation polymerization.

at the beginning of the reaction. On the other hand, condensation polymerization is a stepwise reaction in which smaller species are initially formed first and then combined to make higher-molecular-weight species. This reaction tends to be slow generally lasting for several hours. **Figure 20–4** shows the concept.

## POLYMERIZATION METHODS

Now, the question is how can polymers be made from monomers? Reactions may be carried out in homogeneous or heterogeneous systems. The former includes bulk and solution polymerizations, whereas the latter includes any dispersed system such as suspensions, emulsions, and their reverse phase counterparts; in other words, inverse suspensions and inverse emulsions.

### Homogeneous Polymerization

Bulk polymerization occurs when no other materials except the monomer and initiator are used. If the monomer is water-soluble, a linear water-soluble polymer is theoretically prepared. With oil-soluble monomers, the polymer will be linear and soluble in oil. Surprisingly, sometimes when an olefinic water-soluble monomer is polymerized in bulk, a water-swellable polymer is prepared. This is due to excessive exothermic heat resulting in hydrogen abstraction from the polymer backbone, which promotes cross-linking reactions at the defective site. The cross-linked polymer obtained without using any chemical cross-linker is called a popcorn polymer and the reaction is called "popcorn polymerization." Crospovidone, a superdisintegrant in solid dose formulations, is a cross-linked polymer of vinyl pyrrolidone which is produced by popcorn polymerization.

In certain circumstances when the monomer is very temperature sensitive, a popcorn polymer can be obtained even without using an initiator. The monomer acrylic acid is glacial with a melting point around 13°C. If the monomer was stored at freezing temperature, the polymerization stabilizer will be unevenly distributed between the liquid and thawing phases. This results in poor protection of the monomer and sudden polymerization that generates tremendous amounts of heat.

To solve the problems associated with exothermic heat in bulk polymerization, polymerization can alternatively be conducted in solution. Depending on the monomer solubility, water or organic solvents can be used as diluents or solvents. Again, a water-soluble or an oil-soluble polymer is obtained if monomers are water-soluble or oil-soluble, respectively. The solvent or diluent molecules reside in between the monomer molecules and they reduce the amount of interaction between the two neighboring monomers. In this way, less amounts of heat are generated in a given period of time and a less exothermic but controllable reaction is conducted. Polymers prepared accordingly are generally soluble in their corresponding solvents, but they are swellable if a cross-linker is used during their polymerization. The cross-linker can be water-soluble or oil-soluble. Swellability of a polymer can be modified by the simultaneous use of water-soluble and oil-soluble cross-linkers.

## Dispersion Polymerization

Dispersion polymerization occurs in suspensions, emulsions, inverse suspensions, and inverse emulsions. In dispersion polymerization, two incompatible phases of water and oil are dispersed into each other. One phase is known as the minor (dispersed) phase and the other as the major (continuous) phase. The active material (monomer) can be water-soluble or oil-soluble. To conduct polymerization in a dispersed system, the monomer (in the dispersed phase) is dispersed into the continuous phase using a surface-active agent. The surfactant is chosen on the basis of the nature of the continuous phase. Generally, a successful dispersion polymerization requires that the surfactant be soluble in the continuous phase. Therefore, if the continuous phase is water, the surfactant should have more hydrophilic groups. On the other hand, if the continuous phase is oil, a more hydrophobic (lipophilic) surfactant would be selected. Generally, two basic factors control the nature of the dispersion system. These are surfactant concentration and the surface tension of the system (nature of the dispersed phase) as shown in **Figure 20–5**. Dispersed systems were discussed in earlier chapters. The surfactant concentration determines the size of the polymer particles. The system will be a suspension or inverse suspension with

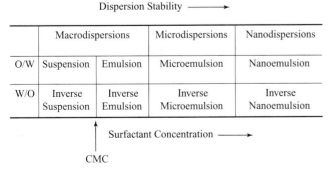

**Fig. 20–5.** Dispersion systems.

particle sizes around 0.2 to 0.8 mm below the critical micelle concentration. Above the critical micelle concentration, 10 to 100 $\mu$m particles are formed. Nanosize particles can be made if a sufficient amount of surfactant is used. Nanoemulsion or inverse nanoemulsion systems are rarely used in the pharmaceutical industry because of the amount of surfactant required to stabilize the system. Surfactants represent undesirable impurities that affect drug stability and formulation acceptability. Water-insoluble polymers based on acrylic or methacrylic esters are prepared via suspension or emulsion polymerization. Eudragit L30D is a copolymer of methacrylic acid and ethyl acrylate which is manufactured using an emulsion technique. Eudragit NE30D is also a copolymer of ethyl acrylate and methyl methacrylate which can be manufactured in an emulsion system. On the other hand, water-soluble polymers based on acrylic or methacrylic salts as well as acrylamide can be prepared using inverse suspension or inverse emulsion systems. Emulsion systems that use water as a continuous phase are known as latex. **Table 20–1** summarizes important polymerization methods that are potentially used to prepare pharmaceutical polymers.

## COPOLYMERS AND POLYMER BLENDS

If one polymer system cannot address the needs of a particular application, its properties need to be modified. For this reason, polymer systems can be physically blended or chemically reacted. With the former, a two-phase system generally exists, whereas with the latter a monophase system exists. This can clearly be seen in a differential scanning calorimeter by monitoring the glass transition temperature ($T_g$) of the individual polymers. With polymer blends, two $T_g$ values are observed while one single $T_g$ is detected for copolymers. Thermal analysis is discussed in detail in Chapter 2.

*Copolymerization* refers to a polymerization reaction in which more than one type of monomer is involved. Generally, copolymerization includes two types of monomers. If one monomer is involved, the process is called polymerization and the product is a *homopolymer*. For example, polyethylene is a homopolymer since it is made of just one type of monomer. Depending on their structure, monomers display different reactivities during the polymerization reaction. If the reactivities of two monomers are similar, there will be no preference for which monomer is added next, so the polymer that is formed is called a *random copolymer*. When one monomer is preferentially added to another monomer, the monomers are added to each other alternatively and the polymer product is called an *alternate copolymer*. Sometimes, monomers preferentially add onto themselves and a *block copolymer* is formed. This happens when one monomer has a very high reactivity toward itself. Once more reactive monomers have participated in the reaction, the macroradical of the first monomer will attack the second monomer with the lower activity, and the second monomer will then grow as a block. Pluronic surfactants (EO-PO-EO terpolymers) are

**POLYMERIZATION METHODS**

|  | Bulk | Solution | Suspension | Inverse Suspension | Emulsion | Inverse Emulsion | Pseudo latex |
|---|---|---|---|---|---|---|---|
| Monomer | WS or OS | WS or OS | OS | WS | OS | WS | |
| Initiator | WS or OS | WS or OS | Generally OS | Generally WS | Generally OS | Generally WS | |
| Cross-linker* | WS or OS | WS or OS | Generally OS | Generally WS | Generally OS | Generally WS | |
| Water | | If WS monomer is used | CP | DP | CP | DP | CP |
| Organic solvent | | If OS monomer is used | DP | CP | DP | CP | DP |
| Surfactant | | | WS; $C_{surf}$ <CMC; high HLB | OS; $C_{surf}$ <CMC; low HLB | WS; $C_{surf}$ >CMC; high HLB | OS; $C_{surf}$ >CMC; low HLB | WS |
| Polymer | | | | | | | OS[†] |

*Cross-linker is added if a swellable polymer is desired.
[†]Polymer is soluble in organic solvents, but the latex itself is water dispersible.
*Key*: WS = water-soluble; OS = organic-soluble; DP = dispersed phase; CP = continuous phase; CMC = critical micelle concentration; HLB = hydrophilic lipophilic balance; $C_{surf}$ = surfactant concentration.

composed of block units of ethylene oxide and propylene oxides attached to each other. The major difference between graft copolymer and the other copolymer types is the nature of their building blocks. Other copolymer types are made of two or more monomer types, while a monomer and a polymer are generally used to make graft copolymers. For example, the physical chemical properties of carboxymethyl cellulose (CMC) can be changed by grafting various monomers such as acrylic acid, acrylamide, and acrylonitrile onto the cellulose backbone. Although not very common, a terpolymer will be obtained when three monomers participate in the polymerization reaction. Different types of copolymer products are shown in **Figure 20–6**.

**Pharmaceutical Polymers**

In pharmaceutical solid oral dosage forms, the Eudragit polymers are used for sustained release, drug protection, and taste-masking

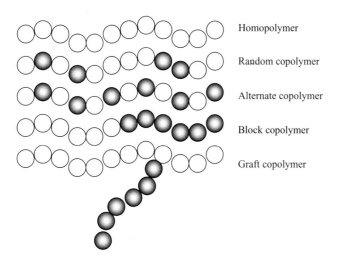

**Fig. 20–6.** Polymers made of two or more monomer units.

applications. These polymers are made of acrylic esters (methyl methacrylate, ethyl acrylate). Their solubility, swellability, and pH dependent properties have been modified by incorporating anionic and cationic monomers such as methacrylic acid and dimethylaminoethyl acrylate.

From a commercial standpoint, polymer properties can be simply changed by mixing or blending one or two polymer systems. Polymer blends are simply made by physical blending of two different polymers in molten or in solution state. The blend is either solidified at lower temperature if prepared by melting or recovered at higher temperature if prepared in solution. Some thermoplastic polymers are not resistant to sudden stresses. Once impacted, the craze (microcrack) and macrocracks will grow very quickly within their structure and the polymer will simply and suddenly break apart. These polymers have rigid structures with high $T_g$ values. Adding a low $T_g$ polymer (in other words, a flexible polymer) such as rubber particles improves the impact resistance of these polymers by preventing the cracks from growing.

# INTERPENETRATING POLYMER NETWORKS

Interpenetrating polymer networks (IPNs) are also composed of two or more polymer systems but they are not a simple physical blend. Semi-IPNs or semi-interpenetrating polymer networks are prepared by dissolving a polymer into a solution of another monomer. An initiator as well as a cross-linker is added into the solution and the monomer is polymerized and cross-linked in the presence of the dissolved polymer. The result will be a structure in which one cross-linked polymer interpenetrates into a non–cross-linked polymer system. With fully interpenetrated structures, two different monomers and their corresponding cross-linkers are polymerized and cross-linked simultaneously. This results in a doubly cross-linked polymer system that interpenetrates into one another. Alternatively, conducting the cross-linking reaction on a semi-interpenetrated product can form a full-IPN structure. The non–cross-linked phase of the semi-IPN product will be further cross-linked with a chemical cross-linker or via physical complexation.[6,7]

### EXAMPLE 20–3

**IPN Polymer Structure**

Elastic superporous hydrogels have been developed for oral gastric retention of the drugs with a narrow absorption window. These hydrogels are prepared using a two-step process. First, a semi-IPN structure is prepared by polymerizing and cross-linking a synthetic monomer (such as acrylamide) in the presence of a water-soluble polymer (e.g., alginate). Although the cross-linked acrylamide polymer is not soluble in water, the alginate component is. In the second step, the prepared semi-IPN is further treated with cations (such as calcium) to provide insolubility to the alginate component via ion-complexation. This results in a full IPN structure with a balanced swelling and mechanical properties.

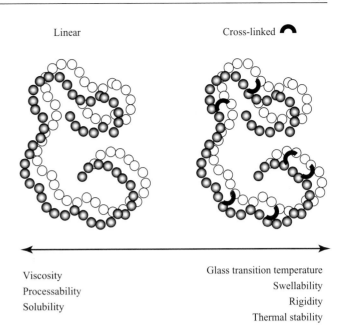

**Fig. 20–7.** Polymer topology and properties.

## TOPOLOGY AND ISOMERISM

The topology of a polymer describes whether the polymer structure is linear, branched, or cross-linked. Topology can affect polymer properties in its solid or solution states. With a linear polymer, the polymer chains are not chemically attached to each other, instead weaker intermolecular forces hold the polymer chains together. A linear polymer can show dual behavior. Chains in a linear polymer can freely move, which offers the polymer a low melting temperature. On the other hand, linear chains have a higher chance of approaching each other in their solid state, which increases their crystallinity and melting temperature. The same holds true for branched polymers in which short or long side groups are attached to the backbone of the polymer. Branched polymer chains move with difficulty because of the steric hindrance induced by the side groups but they presumably possess weaker intermolecular forces, which apparently help them move freely. With cross-linked polymers, the chains are chemically linked and will be restricted from moving to a sensible extent depending on the level of cross-linking. Very highly cross-linked polymers are very rigid structures that degrade at high temperatures before their chains start to move.

In solution, a branched polymer might display a better solvent permeability compared to its linear counterpart due to its side groups. Gum Arabic is a highly branched polymer with very high solubility in water. If a linear polymer is cross-linked, its solubility will be sacrificed at the expense of swellability. Therefore, a cross-linked polymer can swell in a solvent to an extent that is inversely related to the amount of cross-linker. **Figure 20–7** summarizes and correlates polymer topology to its solution and melt properties.

Isomerism can be classified as structural isomerism (**Fig. 20–8a**), sequence isomerism (**Fig. 20–8b**), and

stereoisomerism (**Fig. 20–8c**). Gutta Percha natural rubber (*trans*-polyisoprene) and its synthetic counterpart (*cis*-polyisoprene) are similar in structure but their *trans* and *cis* nature results in a medium-crystal and amorphous behavior, respectively. This important feature can be accounted for in terms of the position of a methyl group. The *cis* and *trans* isomers of a same polymer display different $T_g$ and $T_m$ values, for example, polyisoprene ($T_g$ of $-70°C$ versus $-50°C$; $T_m$ of $39°C$ versus $80°C$), polybutadiene ($T_g$ of $-102°C$ versus $-50°C$; $T_m$ of $12°C$ versus $142°C$).[1] With sequence isomerism, monomers with pendant groups can attach to each other in head-to-tail, head-to-head, or tail-to-tail conformation. Stereoisomerism applies to polymers with chiral centers, which results in three different configurations—isotactic (pendant groups located on one side), syndiotactic (pendant groups located alternatively on both sides), and atactic (pendant groups located randomly on both sides) configurations.

### EXAMPLE 20–4

**Stereoisomerism**

The isotactic and atactic polypropylenes display glass transition temperatures of 100°C and −20°C, respectively. While the isotactic one is used for special packaging purposes, the atactic one is commonly used as a cheap excipient in general adhesive formulations.

## THERMOPLASTIC AND THERMOSET POLYMERS

Polymers with a linear or branched structure generally behave as thermoplastics. Thermoplastic polymers can undergo melting, which is potentially useful in processes such as compression molding, injection molding, and thermoforming. In other words, a polymer that is originally a solid can flow upon application of heat. The process of thermomelting and

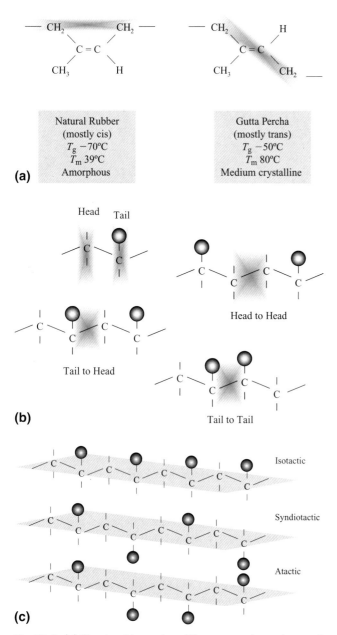

**Fig. 20–8.** (*a*) Structural isomerism; (*b*) sequence isomerism; and (*c*) stereoisomerism.

solidification can be repeated indefinitely with thermoplastic polymers. Examples include polystyrene, polyethylene, and poly (vinyl chloride). On the other hand, thermosetting polymers are cross-linked polymers, which are formed upon combined application of a cross-linker and heat or combined application of heat and reaction of internal functional groups. In some cross-linking reactions such as in curing rubbers, the reaction is assisted by simultaneous application of heat and pressure. Therefore, these polymers assume a different status than thermoplastic polymers as their flow behavior is temperature independent. Once a thermoset polymer is formed, it does not soften upon heating and decomposes with further application of heat. Since there is no reversible melting and solidifying in thermoset polymers, this feature is very useful when a thermoresistant polymer is desirable. Processing of polymers is generally favored by increasing temperature. A more processable polymer is one that requires a lower temperature to move its chains. A cross-linked polymer loses its processability as chain movement is hindered with the addition of cross-linker. On the other hand, linear and branched polymers gain more freedom to move as temperature increases.

For a polymer in its solution state, solubility in a solvent is also an entropy-favored process. In other words, a linear or branched polymer generally dissolves in an appropriate solvent. Addition of cross-links to their structure will hinder chain movement and reduce their solubility in that solvent. This is why cross-linked polymers swell when they are placed in a compatible solvent.

## POLYMER PROPERTIES

### Crystalline and Amorphous Polymers

Polymers display different thermal, physical, and mechanical properties depending on their structure, molecular weight, linearity, intra- and intermolecular interactions. If the structure is linear, polymer chains can pack together in regular arrays. For example, polypropylene chains fit together in a way that intermolecular attractions stabilize the chains into a regular lattice or crystalline state. With increased temperature, the crystal cells (crystallites) start to melt and the whole

---

**◆━━ KEY CONCEPT ━ CROSS-LINKING**

A ladder is composed of two long legs and multiple short pieces that are used to connect them. When a ladder is used, you do not want its legs to move or even worse, to separate from each other. A ladder with more connection points on the two legs is more secure and more stable than a ladder with less. One cannot climb on a ladder without a connector. In polymer terms, cross-linked polymers are long linear chains (ladder legs) that are cross-linked using a functional or an olefin cross-linker

(ladder legs connector). Cross-linked polymers are also intended for applications where a certain amount of load is applied. Examples of this are tires (made of cross-linked rubbers) and hydrogels (made of cross-linked hydrophilic polymers) that are expected to function and to survive under the service load of mechanical and swelling pressure, respectively. When you drive your car, the last thing you want is to have your tire melt away.

polymer mass suddenly melts at a certain temperature. Above the melting temperature, polymer molecules are in continuous motion and the molecules can slip past one another.

In many cases, the structure of a polymer is so irregular that crystal formation is thermodynamically infeasible. Such polymers form glass instead of crystal domains. A glass is a solid material existing in a noncrystalline (i.e., amorphous) state. Amorphous structure is formed due to either rapid cooling of a polymer melt in which crystallization is prevented by quenching or due to the lack of structural regularity in the polymer structure. Rotation around single bonds of the polymer chains becomes very difficult at low temperatures during rapid cooling; therefore, the polymer molecules forcedly adopt a disordered state and form an amorphous structure. Amorphous or glassy polymers do not generally display a sharp melting point; instead, they soften over a wide temperature range.[8]

### EXAMPLE 20–5

**Crystalline and Amorphous**

Polystyrene and poly (vinyl acetate) are amorphous with melting range of 35°C to 85°C and 70°C to 115°C, respectively. On the other hand, poly (butylene terephthalate) and poly (ethylene terephthalate) are very crystalline with sharp melting range of 220 and 250°C to 260°C, respectively.

Polymer strength and stiffness increases with crystallinity as a result of increased intermolecular interactions. With an increase in crystallinity, the optical properties of a polymer are changed from transparent (amorphous) to opaque (semicrystalline). This is due to differences in the refractive indices of the amorphous and crystalline domains, which lead to different levels of light scattering. From a pharmaceutical prospective, good barrier properties are needed when polymers are used as a packaging material or as a coating. Crystallinity increases the barrier properties of the polymer. Small molecules like drugs or solvents usually cannot penetrate or diffuse through crystalline domains. Therefore, crystalline polymers display better barrier properties and durability in the presence of attacking molecules. Diffusion and solubility are two important terms that are related to the level of crystallinity in a polymer. On the other hand, a less crystalline

or an amorphous polymer is preferred when the release of a drug or an active material is intended. Crystallinity in a given polymer depends on its topology and isomerism (linear versus branched; isotactic versus atactic), polymer molecular weight, intermolecular forces, pendant groups (bulky versus small groups), rate of cooling, and stretching mode (uniaxial versus biaxial). Another unique property of a crystalline polymer or a polymer-containing crystalline domains is anisotropy. A crystal cell displays different properties along longitudinal and transverse directions. This causes the polymer to behave like an anisotropic material.

## Thermal Transitions

Thermal transitions in polymers can occur in different orders. In other words, the volume of a polymer can change with temperature as a first- or second-order transition. When a crystal melts, the polymer volume increases significantly as the solid turns to a liquid. The melting temperature ($T_m$) represents a first-order thermal transition in polymers. On the other hand, the volume of an amorphous polymer gradually changes over a wide temperature range or so-called glass transition temperature. This behavior represents a second-order thermal transition in polymers. As shown in **Figure 20–9**, $T_m$ and $T_g$ of a given polymer can be detected by differential scanning calorimetry (DSC) as an endothermic peak and a baseline shift, respectively. These two thermal transitions reflect the structural movement of the crystalline and amorphous regions of a polymer chain.

## Glass Transition Temperature

$T_g$ is an expression of molecular motion, which is dependent on many factors. Therefore, the $T_g$ is not an absolute property of a material and is influenced by the factors affecting the movement of polymer chains. At temperatures well below the $T_g$, amorphous polymers are hard, stiff, and glassy although they may not necessarily be brittle. On the other hand, at temperatures well above the $T_g$, polymers are rubbery and might flow. The $T_g$ values for linear organic polymers range from about $-100°C$ to above $300°C$. Even though some organic polymers are expected to have $T_g$ values above

## KEY CONCEPT  ANISOTROPY

Take a roll of toilet paper from your bathroom and try to tear it apart from two directions perpendicular to each other. What will you observe?

If you tear it along the roll direction (its length), it will easily tear apart and the tear line will be smooth and even. On the other hand, tearing in the other direction would be very difficult and the tear line will appear as a random irregular corrugated line. Why is this?

Toilet paper is manufactured using a process that applies a force along the roll direction. Because of the applied force, the

chains are aligned in the direction of the force. When you try to tear the tissue in this direction, there is no barrier to the force and the material does not resist. On the other hand, tearing the tissue requires cutting the chains in the perpendicular direction that implies resistance from the material. This is **anisotropy,** which means material properties are different in different directions. Pharmaceutical tablets are generally compressed in one direction, which might affect drug release or tablet properties throughout.

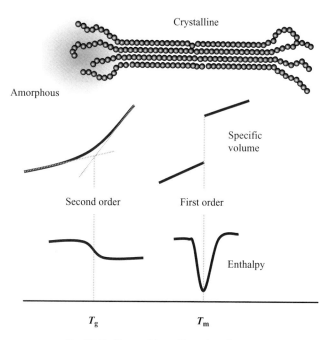

**Fig. 20–9.** Thermal transitions in polymers.

300°C, they decompose at temperatures below their transition temperatures. From a pharmaceutical standpoint, $T_g$ is an important factor for solid dosage forms. For example, a chewable dosage form needs to be soft and flexible at mouth temperature of about 37°C. This means the polymer used as a chewable matrix should be softened at this temperature. Pharmaceutically acceptable polymers with their $T_g$ values close to the service temperature of 37°C would be the best candidates. Nicotine gum (Nicorette) gum is used as an aid in smoking cessation. It works by providing low levels of nicotine, which lessens the physical signs of withdrawal symptoms. The nicotine is released into the mouth as the patient chews the gum. After placing a piece of the Nicorette gum into the mouth the patient should chew it slowly several times. The patient should stop chewing it once he or she notices a tingling sensation or a peppery taste in the mouth. At this point the nicotine is being released and the patient should "park" the gum in the buccal area (in other words, between the cheek and the gum) and leave it there until the taste or tingling sensation is almost gone. The patient can resume chewing a few more times and then stop once the taste comes back. The patient should repeat this for about 30 min or until the taste or tingling sensation does not return. The reason that the patient "parks" the gum is because the release of nicotine should be slow and constant chewing will release the nicotine too quickly resulting in nausea, hiccups, or stomach problems. The patient should also avoid drinking or eating at this time. You have most likely experienced how gum behaves differently when you drink cold water or hot tea. This is all reflected in glass–rubber transition of the chewable matrix.

As mentioned before, the $T_g$ of a polymer is dependent on many factors and the most important ones are discussed here. Segmental motion in polymers is facilitated by the empty space in between the polymer chain ends, also called the *free volume*. As the free volume increases, polymer segments gain more freedom to move and this affects the temperature at which the movement occurs. For example, low- and high-density polyethylenes are different in terms of the size of the free volume inside their structures. At a given weight, a low-density polymer occupies more volume as compared with its high-density counterpart. This means the polymer chain in general and the chain segments in particular can move with more ease resulting in a lower $T_g$ value.

$T_g$ **and the length of the polymer chain:** Long polymer chains provide smaller free volume than their shorter counterparts. Since more free volume corresponds to lower $T_g$ values, polymers containing short chains or having lower molecular weight possess lower $T_g$ values.

$T_g$ **and polymer chain side group:** A side group may be bulky or polar. Because of its steric hindrance, higher temperature is needed to induce segmental motion in polymers containing bulky groups. For example, polystyrene and polypropylene are only different in terms of their side groups, phenyl versus methyl, respectively. The larger size of the phenyl group results in the much higher $T_g$ value of polystyrene, 100°C as opposed to −20°C for polypropylene. On the other hand, polar side groups provide stronger intermolecular interactions that significantly affect the segmental motion of the polymer chains. Poly (vinyl chloride) is similar to polyethylene except hydrogen is replaced by one chlorine atom. Since chlorine is more polar than hydrogen, the PVC polymer displays a much higher $T_g$ of 100°C compared to −120°C for polyethylene.

$T_g$ **and polymer chain flexibility:** Flexible polymer chains display higher entropy (desire to move) than rigid chains. Flexible and rigid chains behave similar to liquid and solid, respectively. Groups such as phenyl, amide, sulfone, and carbonyl either inside the backbone or as a side group hanging on the backbone affect the overall polymer flexibility. For example, poly (ethylene adipate) and poly (ethylene terephthalate) are structurally very similar except for the phenylene residue in phthalate versus the butylene residue in adipate. This results in almost a 100°C difference in $T_g$ values of the two polymers (−70°C versus 69°C, respectively).

$T_g$ **and polymer chain branching:** Linear polymer chains possess smaller free volume as opposed to their branched counterparts. Therefore, higher $T_g$ values are expected for linear polymers. On the other hand, branches in branched polymers impose hindrance or restriction to segmental motion, for which higher $T_g$ values are expected. Therefore, branching has no obvious effect on the $T_g$ unless the whole structure of the polymer is known.

$T_g$ **and polymer chain cross-linking:** Compared to cross-linked chains, linear chains have a higher entropy and the desire to move; hence, they display low $T_g$ values. Adding cross-links to linear polymer chains limits chain movement resulting in less entropy at a given temperature and hence a higher $T_g$ value. For very highly cross-linked polymers, $T_g$ values are expected to be very high to the extent that the polymer starts to decompose before it shows any segmental motion.

$T_g$ **and processing rate:** In order to prepare polymer products, the polymer needs to be processed at different temperatures or pressures that can significantly affect the molecular motion in polymers. Therefore, the rate of processes such as heating, cooling, loading, and so on and so forth might be considered when evaluating the $T_g$ value of a given polymer. Kinetically speaking, if the rate of the process is high (fast cooling, fast loading), the polymer chains cannot move to the extent that they are expected to. They virtually behave like rigid chains with lower tendency to move, which results in reading high $T_g$ values. For instance, when a differential scanning calorimeter is used to measure the $T_g$ of a polymer, different $T_g$ values may be observed if the same polymer is heated up at different heating rates. This implies that the heating rate has to be very realistic and should be consistent with the conditions in which the polymer is expected to serve.

$T_g$ **and plasticizers:** Plasticizer molecules can increase the entropy and mobility of the polymer chains. This is translated to lower $T_g$ values for plasticized polymers compared with their nonplasticized counterparts.

## Plasticized Polymers

A plasticizer is added to a polymer formulation to enhance its flexibility and to help its processing. It facilitates relative movement of polymer chains against each other. The addition of a plasticizer to a polymer results in a reduction in the glass transition temperature of the mixture. Since plasticizers increase molecular motion, drug molecules can diffuse through the plasticized polymer matrix at a higher rate depending on the plasticizer concentration.

### EXAMPLE 20–6

**Plasticized Polymers**

Fluoxetine (Prozac Weekly) (fluoxetine hydrochloride) capsules contain hydroxypropyl methylcellulose and hydroxypropyl cellulose acetate succinate plasticized with sodium lauryl sulfate and triethyl citrate. Omeprazole magnesium (Prilosec), a delayed release oral suspension, contains hydroxypropyl cellulose, hydroxypropyl methylcellulose, and methacrylic acid copolymer plasticized with glyceryl monostearate, triethyl citrate, and polysorbate. Triacetin can be found in ranitidine HCl (Zantac) 150-tablet formulations, which contains hydroxypropyl methylcellulose as its polymer matrix. Dibutyl sebacate is found in methylphenidate HCl (Metadate) CD which contains polymers such as povidone, hydroxypropyl methylcellulose, and ethyl cellulose.

## Molecular Weight

Addition of a monomer to a growing macroradical during polymer synthesis occurs by a diffusion or a random walk process. Monomers may or may not be added equally to the growing macroradicals. As a result, a polymer batch may contain polymer chains with different lengths (molecular weights) and hence different molecular weight distributions. A very narrow molecular weight distribution is very much desired for a polymer that is intended to be mechanically strong. On the other hand, a polymeric adhesive may have wide distribution of molecular sizes. In general, a given polymer cannot be identified as a molecule with a specific molecular weight. Since chains are different, the molecular weight of all chains should be considered and must be averaged to have a more realistic figure for molecular weight of a given polymer. There are different ways that molecular weights of a polymer can be expressed; by the number of the chains, by the weight of the chains (the chain size), or by viscosity. However, the two most common ways are number ($M_n$) and weight ($M_w$) average calculations. If all polymer chains are similar in size, then the number and weight average values will be equivalent. If chains are of different sizes, then weight average is distancing itself from the number average value. The term polydispersity (PD) indicates how far the weight average can distance itself from the number average. A PD value closer to 1 means the polymer system is close to monodispersed and all of the polymer chains are almost similar in size. The farther the value from 1 indicates that the polymer system is polydispersed and chains are different in size. **Figure 20–10** shows the concept.

Consider that you have received two different batches of a same polymer as shown in **Table 20–2**. The first batch contains 2 chains of 50,000 g/mol and 10 chains of 20,000 g/mol in size. The second batch contains 2 chains of 100,000 g/mol and 10 chains of 10,000 g/mol in size. Calculations show both batches have the same number averages of 25,000 g/mol. Should you, as a pharmaceutical scientist, claim that the two batches are similar and you can use them interchangeably within your formulation? You continue with the calculation to find out the weight average values for the two batches. Surprisingly, two very different numbers, 30,000 g/mol and 70,000 g/mol, are found for the batch 1 and the batch 2,

Monodispersed
Polydispersity = 1
$M_w = M_n$

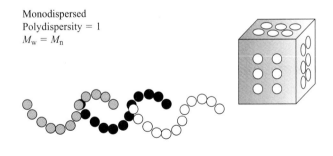

Polydispersed
Polydispersity $\gg 1$
$M_w \gg M_n$

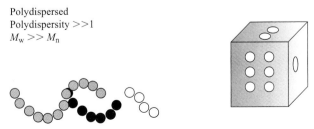

**Fig. 20–10.** Molecular weight distribution in polymer systems.

## KEY CONCEPT NUMBER AND WEIGHT AVERAGES

A research institute is planning to hire a good scientist who can publish a tangible number of high-quality manuscripts per year. The institute receives two resumes in which both scientists have claimed 20 publications a year. The applicants were then asked to submit more details about the journals in which they have published. Now, the institute knows not only the total number of publications but also the type and the number of journals they have published in. Journals were then categorized on the basis of their impact factor (IF) as very high, high, medium, and low. If the total number of publications is divided by the number of journals, both scientists score the same with an average of five publications per journal per year. The number will change if the impact factors of the journals are also considered in the calculation. So, the new calculation shows average numbers of 1.6 as opposed to 3.4 for the scientists 1 and 2, respectively. It looks like the insti-

tute has found a tool to discriminate between the achievements of the two scientists. These new numbers show that the scientist 2 is more capable in publishing high-quality manuscripts. A similar discussion is valid for different polymer chains (journals) with different molecular weights (impact factors). With number average, all chains are considered similar and the effect of their size is simply overlooked.

| Journal | IF | Scientist 1 | Scientist 2 | $N_1$ | $W_1$ | $N_2$ | $W_2$ |
|---|---|---|---|---|---|---|---|
| Very high IF | 20 | 2 | 8 | | | | |
| High IF | 10 | 4 | 6 | 5 | 1.6 | 5 | 3.4 |
| Medium IF | 4 | 6 | 4 | | | | |
| Low IF | 1 | 8 | 2 | | | | |
| Total | 35 | 20 | 20 | | | | |

respectively. This shows that the two batches are beyond a doubt different. Another important piece of data that can help you with your decision is PD which is the ratio of weight to number averages. *Polydispersity* of 2.8 versus 1.2 indicates that the batch 2 contains very different chains. If both polymer batches are soluble in water, they will definitely show different solubility behavior in the presence of water. The shorter chains are dissolved faster in water than longer chains. Drug release from these batches will certainly be different as they assume different PD values.

Different techniques are used to calculate different averages. Since the number average relies on the number of polymer chains, technique to measure this should also rely on the number of species such as number of particles, and so on and so forth. It is well-known that colligative properties such as osmotic pressure and freezing point depression are dependent on the number of particles in the solution. Colligative properties were introduced earlier in the book. These techniques are very appropriate for calculating the average $M_n$ of

a given polymer. On the other hand, the weight average relies on the size of the molecules. Techniques such as light scattering are also reliant on the size of the molecules. Large- and small-sized molecules scatter light in a very different way. Therefore, it is reasonable to use a light-scattering technique to calculate the average $M_w$ of a polymer.

## Mechanical Properties

Depending on their structure, molecular weight, and intermolecular forces, polymers resist differently when they are stressed. They can resist against stretching (tensile strength), compression (compressive strength), bending (flexural strength), sudden stress (impact strength), and dynamic loading (fatigue). With increasing molecular weight and hence the level of intermolecular forces, polymers display superior properties under an applied stress. As far as structure is concerned, a flexible polymer can perform better under stretching whereas a rigid polymer is better under compression.

## TABLE 20–2
### AVERAGE MOLECULAR WEIGHTS AND POLYDISPERSITY

| Number of Chains | Batch 1 | Batch 2 |
|---|---|---|
| 2 | 50,000 g/mol | 100,000 g/mol |
| 10 | 20,000 g/mol | 10,000 g/mol |

| Batch 1 | | | Batch 2 | | |
|---|---|---|---|---|---|
| $M_n$ | $M_w$ | PD | $M_n$ | $M_w$ | PD |
| 25,000 g/mol | 30,000 g/mol | 1.2 | 25,000 g/mol | 70,000 g/mol | 2.8 |

$$\overline{M_n} = \frac{\sum M_i N_i}{N_i} = \frac{(M_1 N_1) + (M_2 N_2) + (M_3 N_3) + (M_4 N_4) + \cdots}{N_1 + N_2 + N_3 + N_4 + \cdots}$$

$$\overline{M_w} = \frac{\sum M_i^2 N_i}{\sum M_i N_i} = \frac{(M_1^2 N_1) + (M_2^2 N_2) + (M_3^2 N_3) + (M_4^2 N_4) + \cdots}{(M_1 N_1) + (M_2 N_2) + (M_3 N_3) + (M_4 N_4) + \cdots}$$

## ▶ KEY CONCEPT MOLECULAR WEIGHT DISTRIBUTION

If you are planning to hire individuals for a cheerleading team, you may impose very strict requirements. For example they should all be 6-feet tall with a body mass index of 20. In the same sense, a soccer team might need a goalkeeper as well as a defense and forward, for each of which you may have different requirements. For example, height is very important for the goalkeeper position, whereas speed and accuracy is the most important requirement

for the forward. In polymer terms, chain size distribution per se is not a bad or a good thing. Depending on the application, the polymer needs to have a sharp or wide size distribution. A polymer for an engineering application like the ones they use in aircrafts or spacecrafts may need to have a narrow size distribution, whereas for a general-purpose application you may use a polymer with a wider size distribution.

A polymer is loaded and its deformation is monitored to measure its strength. **Figure 20–11** shows the stress–strain behavior of different materials. For elastic materials such as metals and ceramics, the stress and strain (deformation) correlation is linear up to the failure point. Generally, these materials show high stress and very low elongation (deformation, strain) at their breaking point. Polymeric materials such as fibers and highly cross-linked polymers display elastic behavior, in other words, a linear stress/strain correlation up to their breaking point. With an increase in intermolecular forces within a fibrous product or cross-link density of a cross-linked polymer, the slope of the stress/strain line will become steeper. The sharper the slope, the higher the modulus. Modulus and stiffness are two terms that can be used interchangeably to demonstrate the strength of a polymer. Some polymeric materials do not display a sharp or abrupt breaking point. Instead, they yield at certain stresses and continue to deform under lower stresses before they finally break apart. Tough plastics show this typical behavior. Rubbers or elastomers on the other hand display completely different behavior, which depends on the level of cross-linking or curing. Generally, under very small stresses, they deform to a large extent to more than 10 to 15 times their original lengths. You may recall a rubber band when you stretch it from both ends. Highly cross-linked rubbers show very low deformation at their breaking point. In fact, cross-linking is the process by which properties of a rubber can be enhanced to a very tough plastic or even a fiber. Regardless of the polymer type

(fiber, tough plastic, or rubber), certain amounts of energy are needed to break the polymer apart (in other words, toughness) and the area under the stress/strain curve measures it. The larger the area is, the tougher the polymer.

### Viscoelastic Properties

Mechanical properties of a given polymer are generally measured at a fixed rate of loading, certain temperature or relative humidity, and so on and so forth. Polymers are neither a pure elastic nor a pure fluid material. They have the ability to store energy (display elastic behavior) and to dissipate it (display viscous behavior). For this reason, most polymers are viscoelastic materials. For example, poly (vinyl chloride) has a glass transition temperature of about $100°C$. This means, it behaves like a solid at temperatures below its $T_g$ and like a fluid at temperatures above its $T_g$. Since a typical PVC product is generally used at room temperature, its $T_g$ is supposed to be well above the temperature of the environment in which it is expected to serve. In other words, a PVC product behaves like a solid or glass at any temperature (including its service temperature) below its $T_g$. Now, assume that your PVC product is expected to serve under a certain load (thermal, mechanical, etc.) and at certain temperature below its $T_g$, but for various periods of time. Such a loaded polymer, which originally behaves as a solid, or elastic may change its behavior upon a long-term loading. Over time, the polymer intermolecular forces will essentially become weaker and hence, the polymer becomes softer. This can be seen in the glass windows used in the old churches as they show different thicknesses from top to the bottom.

There are generally two methods to evaluate the viscoelasticity in polymers; the creep test and the stress relaxation test. With the former, the polymer is first loaded with a certain weight and its deformation is then monitored over the time. With the latter, the polymer is first deformed to a certain extent, and then its stress relaxation (internal stress) is monitored with the time.

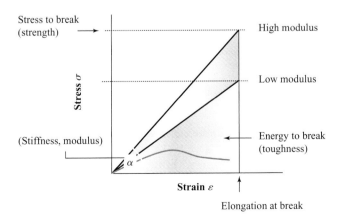

**Fig. 20–11.** Mechanical properties of polymers.

### Molecular Weight and Polymer Properties

Mechanical properties of a given polymer generally increase with an increase in molecular weight. Polymer melts and

## KEY CONCEPT ENTANGLEMENT

Let us say that there are two laundry machines with the total capacity of each 20 lb and you separate your clothes into two small (shorts) and large (pants) groups, each weighing 20 lb. Once the laundry step is completed, the clothes are to be transferred into a dryer. An important observation to make is that more time will be spent to separate the large clothes from each other, which is not the case with the small clothes. This happens because large clothes have a tendency to tie into each other. Because

of this, the washer should be loaded with a smaller number of large articles of clothing as it makes it easier to wash and dry them. In polymer terms, large molecular-weight polymers (large clothes) have a better affinity to tie into each other as opposed to their smaller molecular-weight counterparts (small clothes). This is called *entanglement*. This occurs after a certain molecular weight and affects the polymer properties in both the solution and solid states.

polymer solutions are handled with more difficulty as their molecular weight increases. This is due to a phenomenon called entanglement, which affects the flow of the polymer chains. As molecular weight increases, polymer chains are more likely entangled into each other at certain molecular weights. This results in poor polymer flow either in solid state (as a melt) or in solution state (as a solution). For many applications, there is a working range of molecular weights that a given polymer in solid or solution state can successfully be processed.

## VARIETY OF POLYMERS

Depending on their applications, polymers may be classified as rubbers, plastics, fibers, adhesives, and coatings. Each application requires a polymer to possess certain properties.

### Rubbers

Rubbers are mostly used in tire manufacturing. A tire is a dynamic service environment that experiences friction with the ground surface; has to carry a heavy load of car weight and its passengers; and is exposed to ultraviolet radiation, ozone, oxygen (inside and outside of the tire), weathering conditions (wind, rain), and fatigue (dynamic loading and unloading). From a processing prospective, a tire is a composite of a few rubbers, metal, fiber, particulate fillers, and more. This requires rubber components of a tire to have excellent cohesive (strength) and adhesive (adhesion) properties. Rubbers have unique elongation properties, they can be stretched without failure, and they can be loaded with static and dynamic loads under very severe conditions. Just imagine for a moment, landing of a fully loaded cargo plane or a commercial aircraft. Different rubbers offer different properties. Those with double bonds (e.g., isoprene, butadiene) offer resiliency but are very susceptible to oxidation and ozonation. Those without double bonds (e.g., ethylene–propylene rubber) are very durable against weathering conditions. Some are very resistant to oil (e.g., chloroprene and nitrile) and some have excellent impermeability (e.g., isobutylene–isoprene rubber). Tube-in tires are still used in which the tube part

is basically made of an air-impermeable rubber called butyl. Silicone is a very inert rubber with almost no affinity to any material. Therefore, silicone rubber is an excellent candidate for very durable parts such as implants in biomedical applications. Rubbers in general are not very strong in their raw form but they have a potential to be cross-linked and cured. None of the rubbers used in tires can serve this application without undergoing a curing process. Rubber is loaded with certain chemicals (curing agents) and is cured or cross-linked at high pressure and temperature. Generally speaking, the glass transition temperatures of the rubbery polymers (elastomers) are below the room temperature.

### Plastics

Plastics on the other hand possess completely different properties. Their glass transition temperature is generally above the room temperature as opposed to elastomers as shown in **Figure 20–12**. Plastic parts are manufactured by techniques such as injection molding, extrusion, and thermoforming that require the plastic to be in its molten state. Plastics that are used in general applications such as packaging are generally cheap and are structurally weak. Polymers such as polyethylene, polypropylene, and polystyrene have only carbon in their backbone. The other groups of plastics which are used in engineering applications are required to be impact resistant, weather resistant, solvent resistant, and so on and so forth. These are generally heterogeneous plastics, which have elements other than carbon such as N, Si, and O in their backbone. Polyesters, polyamides, and polyacetals are engineering plastics with very high intermolecular forces and hence high melting point.

### Fibers

Polymers for fibrous products are required to have a crystalline structure with a very sharp melting point. For this application, polymers need to be meltable and spinnable. Polypropylene fibers are used for plastic baskets, they are weak, and do not possess any specific properties. On the other hand, Kevlar fibers are used for bulletproof jackets. This application requires the fiber to have very strong

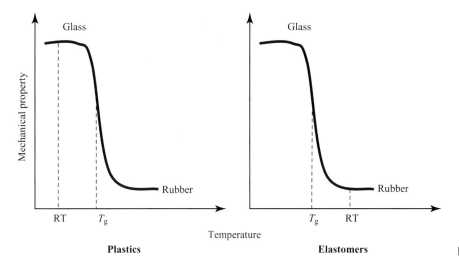

Fig. 20–12. Plastics and elastomers.

intermolecular forces. In manufacturing fibers, both general and engineering plastics are used. Examples of fiber-forming materials are cellulose acetate, rayon, polypropylene, nylon, polyester, polyamide, and polyacrylonitrile.

## Adhesives and Coatings

The required properties of polymers for adhesive and coating applications are tackiness and adhesiveness. This means that adhesive forces (interaction with a second material) should be in balance with cohesive forces (interaction with itself). Both forces increase with the molecular weight of the polymer as molecular interactions increase between the same or different molecules due to increased surface area. Structurally speaking, the cohesive forces within a polymer can be modulated by changing its molecular weight, crystallinity, or addition of a second material such as plasticizers or oils. The adhesive intended for a nonpolar adherent should be nonpolar as well. On the other hand, very polar adhesive materials such as epoxy and cyanoacrylate are suggested for very polar adherents including metals. Generally speaking, the rule of thumb "like dissolves like" is simply applied to polymers for adhesive and coating applications. Like plastics, adhesives can be categorized as general and engineering (structural). The difference is the level of intermolecular forces within the adhesive structure. Structural adhesives are generally used for engineering application such as in air and aerospace industries where high quality, durability, and strength are the basic requirements. To ensure that these requirements are met, the adhesive undergoes special treatment such as curing. Cyanoacrylate-based adhesives or silicone adhesives are generally cured by absorbing moisture from the air. Epoxy adhesives are generally supplied as two components and cured in the presence of a third component (primary, secondary, and tertiary amines). Polyester adhesives are cured using peroxides and catalyzed by amines. The curing process increases the cohesive forces at the expense of adhesive forces. Since an adhesive should possess a balance of cohesive and adhesive properties, the curing process should also be optimized.

Coating and adhesive applications rely on similar concepts. A successful adhesive or coating process requires that the matrix onto which the adhesive is applied to be fully covered by the polymer material, which is generally applied in an emulsion form. Coatings are used for protection purposes. A successful adhesive application requires careful understanding of the properties of the adhesive and adherents since an adhesive is generally trapped in between two or more materials. For coating applications, the coating polymer is generally exposed to a second environment such as air, oxygen, water, stomach fluid, intestinal fluid, solvents, and so on and so forth. This requires a thorough understanding of the coated matrix, coating material, as well as the service environment in which the material is expected to serve. Examples of coating materials are poly (vinyl acetate), acrylate esters, ethyl cellulose, and so on and so forth.

## POLYMERS AS RHEOLOGY MODIFIERS

Polymer chains are in a coiled conformation at rest, and they assume extended conformation once they are loaded. In applications where increased viscosity of the solution is desirable, the goal is to increase the chain end-to-end distance under a given load. In dissolution of a polymer and polymer swelling, the load originates from the interaction of a polymer and a solvent as well as concentration gradient of ions inside the polymer structure and the solution. Apparently longer end-to-end distances are potentially obtained if the polymer chains are longer and have more interaction with the solvent. In case of water as a solvent, the more hydrophilic polymer will be better. On the other hand, a more lipophilic polymer would be more desirable when the dissolution or swelling medium is organic. **Figure 20–13** shows how different polymer chains and solutions display different rheological behavior, which is characterized by the volume occupied by the polymer chains. Because of their hydrophilicity and high molecular weight, gums are the candidate of choice for increasing the viscosity of the aqueous solutions or dispersions.

## KEY CONCEPT  POLYMER STRUCTURE AND SOLUTION VISCOSITY

Your cotton-based clothes get wetter on a rainy day as compared to plastics. So, if you go for a daylong trip on a rainy day, you might want to wear a poncho which is 100% plastic in order to repel the water. In polymer terms, polymers with water-loving functional groups make more and closer contacts with water, which causes molecules of water to move slower, which means they generate more viscosity.

## HYDROGELS

The concept of an end-to-end distance is also applied in swellable polymers. As mentioned earlier, the driving force for the dissolution and swelling processes are similar. Certain materials, when placed in excess water, are able to swell rapidly and retain large volumes of water in their structures. Such aqueous gel networks are called hydrogels. These are usually made of a hydrophilic polymer that is cross-linked either by chemical bonds or by other cohesion forces such as ionic interaction, hydrogen bonding, or hydrophobic interactions. Hydrogels behave like an elastic solid in a sense that they can return to their original conformation even after a long-term loading.

A hydrogel swells for the same reason as its linear polymer dissolves in water to form a polymer solution or hydrosol. From a general physicochemical standpoint, a hydrosol is simply an aqueous solution of a polymer. Many polymers can undergo reversible transformation between hydrogel and hydrosol. When a hydrogel is made by introducing gas (air, nitrogen, or carbon dioxide) during its formation, it is called a porous hydrogel.

A hydrogel swells in water or in any aqueous medium because of positive forces (polymer–solvent interaction, osmotic, electrostatic) and negative forces (elastic) acting upon the polymer chains as shown in **Figure 20–14**. Dissolution of a polymer in a solvent is an entropy-driven process that happens spontaneously. A dry hydrogel is in its solid state and has the tendency to obtain more freedom as it goes into solution. If a polymer structure is nonionic, the major driving force of swelling will be polymer–solvent interactions. As the ion content of a hydrogel increases, two very strong osmotic and electrostatic forces are generated within the hydrogel structure. The presence of ions inside the gel and the absence of the same ions in the solvent trigger a diffusion process (osmosis) by which water enters the polymer structure until the concentration of the ion inside the gel and the solvent becomes equivalent. In fact, the polymer diffuses into water to balance its ion content with the surrounding solution. Polymer chains carrying ions are charged either negatively (anionic) or positively (cationic). In either case, similar charges on the polymer backbone will repel each other upon ionization in an aqueous medium. This creates more spaces inside the hydrogel and more water can be absorbed into its structure. Since swelling in many applications is a desirable property (e.g., in superabsorbent baby diapers or superdisintegrants in pharmaceutical solid dosage forms), the infinite dilution of the polymer needs to be restricted. Linking polymer chains to each other can do this, generating elastic forces and causing less entropy.

### EXAMPLE 20–7

**Superdisintegrants**

In pharmaceutical solid dosage forms, a superdisintegrant is generally used to help the dosage form with a proper disintegration. The concept behind this is the osmotic pressure that is generated by either hydrophilicity (as in vinyl pyrrolidone) or ionic (as in

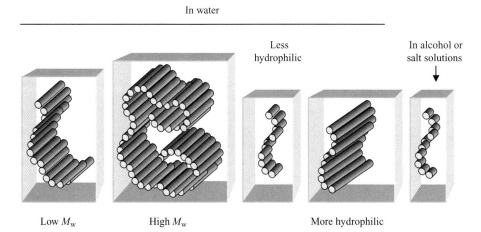

In water

Less hydrophilic

In alcohol or salt solutions

Low $M_w$          High $M_w$                    More hydrophilic

**Fig. 20–13.** Polymers as rheology modifier.

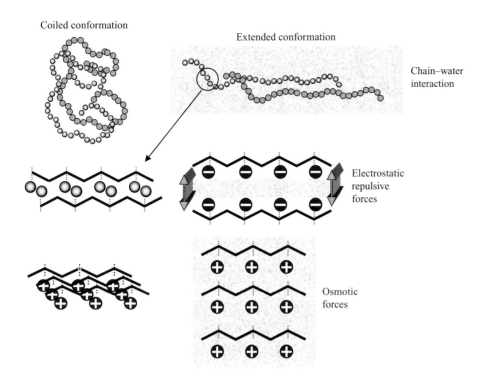

Coiled conformation

Extended conformation

Chain–water interaction

Electrostatic repulsive forces

Osmotic forces

**Fig. 20–14.** Swelling forces in hydrogels.

carboxymethyl cellulose) nature of the structure. Sodium starch glycolate (Explotab, Primojel, Vivastar P), cross-linked poly (vinyl pyrrolidone) (Crospovidone), and cross-linked sodium salt of carboxymethyl cellulose (Ac-Di-Sol, Croscarmelose) are widely used as a tablet and capsule disintegrant in oral dosage forms.

---

**EXAMPLE 20–8**

**Osmotic Tablet and Pump**

Alza's Oros and Duros technologies are based on an osmosis concept. Oros provides 24 hr controlled drug release that is independent of many factors such as diet status. Tablets using Oros technology as shown in Figure 20–15 are made of two sections coated with a semipermeable material. The upper section contains drug and the lower section contains the osmotic agent either a salt or a water-soluble/swellable polymer. The membrane allows water or the aqueous medium to enter into the osmotic agent compartment. In the presence of water, osmotic pressure pushes the bottom compartment upward which in turn forces the drug through a laser-

drilled orifice on top of the tablet. Since 1983, this technology has been used in a number of prescription and over-the-counter products marketed in the United States, including nifedipine (Procardia XL), glipizide (Glucotrol XL), methylphenidate, oxybutynin, and pseudoephedrine (Sudafed 24 Hour). Duros technology is utilized in implants that deliver drugs over a very long period. Leuprolide implant (Viadur) osmotic implant is based on Alza's Duros pump technology which delivers leuprolide acetate over a year long period.

---

Depending on the nature of cross-linking, a hydrogel is classified as chemical or physical.

## Chemical Gels

Chemical gels are those that are covalently cross-linked. Therefore, chemical gels will not dissolve in water or other organic solvent unless the covalent cross-links are broken apart. At least two different approaches can be used to form chemical gels, either by adding an unsaturated olefinic monomer carrying more than one double bond (e.g., $N,N'$-methylene bisacrylamide, ethylene glycol dimethacrylate) or by reacting the functional groups on the polymer backbone. The first approach is used to make water swellable gels or hydrogels. In general, cross-linking through double bond is energetically favored as less energy is required to break a double bond than to react the functional groups. Cross-linked polymers of acrylic acid, sodium acrylate, and acrylamide have found extensive application in hygiene and agricultural industries as water absorbent polymers. These can absorb urine in diapers or can retain the water in the soil.

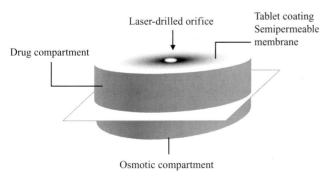

Laser-drilled orifice

Tablet coating
Semipermeable membrane

Drug compartment

Osmotic compartment

**Fig. 20–15.** An osmotic tablet based on Oros technology.

As far as the swelling is concerned, temperature has often a positive effect on the swelling process. Most chemical gels especially those made of hydrophilic chains can swell more in warmer solutions. These gels are so-called thermoswelling chemical gels. On the other hand, some hydrogels made of relatively hydrophobic monomers shrink upon increase in temperature, and they are known as thermoshrinking chemical gels. The thermoshrinking hydrogels undergo thermally reversible swelling and deswelling. The temperature at which this sharp transition occurs is corresponded to a lower critical solution temperature of the non–cross-linked polymer.

## Physical Gels

Hydrogen bonding, hydrophobic interaction, and complexation are three major tools in preparing a physical gel. A hydrogen bond is formed when two electronegative atoms, such as nitrogen and oxygen interact with the same hydrogen, $N–H \cdots O$. The hydrogen is covalently attached to one atom, the donor, but interacts electrostatically with the other, the acceptor. This type of interaction occurs extensively in poly (vinyl alcohol), for example. Although its structure suggests an easy dissolution in water, a poly (vinyl alcohol) at molecular weight more than 100,000 g/mol is water insoluble. In order to dissolve the polymer, the hydrogen bonds need to be broken and that requires the solution to be heated up to 80°C to 90°C.

Hydrophobic interactions are considered to be the major driving force for the folding of thermoresponsive hydrogels and globular proteins. The existence of hydrophobic groups will change the hydrophilic lipophilic balance (HLB) of the polymer that in turn affects its solubility in water. The more hydrophobic groups within the hydrogel structure, the more temperature dependent the swelling will be. As the number of hydrophobic group increases, the solubility–insolubility transition or swelling–deswelling transition shifts to a lower temperature. Polymers, such as methylcellulose, hydroxypropyl methylcellulose, or certain PEO/PPO/PEO triblock copolymers, dissolve only in cold water and form a viscous solution. Once the solution temperature increases up to a certain point, these solutions become thicker by forming a gel.

Complexation may happen between two oppositely charged groups of different polymer structures or via metal ions. In water, alginic acid with negatively charged groups and chitosan with positively charged groups can form a complex. The solubility of the complex is generally dependent on the pH of the dissolution medium and the $pK_a$ of the polymers. On the other hand, alginic acid carrying negatively charged carboxyl groups can form insoluble complexes with divalent and trivalent ions such as calcium, aluminum, and iron. These complexes are also reversible and pH dependent. Hydrogels either chemical or physical are also known as smart, intelligent, or responsive as they react to the environmental changes such as pH, temperature, salt concentration, salt type, solvent composition, or pressure. The unique properties of responsive hydrogels are ideal for making sensors and modulated drug delivery systems.

# POLYMERS FOR PHARMACEUTICAL APPLICATIONS

In a traditional pharmaceutics area, such as tablet manufacturing, polymers are used as tablet binders to bind the excipients of the tablet. Modern or advanced pharmaceutical dosage forms utilize polymers for drug protection, taste masking, controlled release of a given drug, targeted delivery, increase drug bioavailability, and so on and so forth.

Apart from solid dosage forms, polymers have found application in liquid dosage forms as rheology modifiers. They are used to control the viscosity of an aqueous solution or to stabilize suspensions or even for the granulation step in preparation of solid dosage forms. Major application of polymers in current pharmaceutical field is for controlled drug release, which will be discussed in detail in the following sections. In the biomedical area, polymers are generally used as implants and are expected to perform long-term service. This requires that the polymers have unique properties that are not offered by polymers intended for general applications. **Table 20–3** provides a list of polymers with their applications in pharmaceutical and biomedical industries.

In general, the desirable polymer properties in pharmaceutical applications are film forming (coating), thickening (rheology modifier), gelling (controlled release), adhesion (binding), pH-dependent solubility (controlled release), solubility in organic solvents (taste masking), and barrier properties (protection and packaging).

From the solubility standpoint, pharmaceutical polymers can be classified as water-soluble and water-insoluble (oil-soluble or organic soluble). The cellulose ethers with methyl and hydroxypropyl substitutions are water-soluble, whereas ethyl cellulose and a group of cellulose esters such as cellulose acetate butyrate or phthalate are organic soluble. Hydrocolloid gums are also used when solubility in water is desirable. The synthetic water-soluble polymers have also found extensive applications in pharmaceutical industries, among them polyethylene glycol, polyethylene glycol vinyl alcohol polymers, polyethylene oxide, polyvinyl pyrrolidone, and polyacrylate or polymethacrylate esters containing anionic and cationic functionalities are well-established.

## Cellulose-Based Polymers

Although cellulose itself is insoluble in water, its water-soluble derivatives have found extensive applications in pharmaceutical dosage forms. The structure of cellulose is shown in **Figure 20–16**. Methyl cellulose, CMC, and hydroxypropyl methylcellulose are the most common cellulose-based polymers with methyl, carboxymethyl, and

**TABLE 20-3**

## POLYMERS IN PHARMACEUTICAL AND BIOMEDICAL APPLICATIONS

**Water-Soluble Synthetic Polymers**

| | |
|---|---|
| Poly (acrylic acid) | Cosmetic, pharmaceuticals, immobilization of cationic drugs, base for Carbopol polymers |
| Poly (ethylene oxide) | Coagulant, flocculent, very high molecular-weight up to a few millions, swelling agent |
| Poly (ethylene glycol) | $M_w < 10,000$; liquid ($M_w < 1000$) and wax ($M_w > 1000$), plasticizer, base for suppositories |
| Poly (vinyl pyrrolidone) | Used to make betadine (iodine complex of PVP) with less toxicity than iodine, plasma replacement, tablet granulation |
| Poly (vinyl alcohol) | Water-soluble packaging, tablet binder, tablet coating |
| Polyacrylamide | Gel electrophoresis to separate proteins based on their molecular weights, coagulant, absorbent |
| Poly (isopropyl acrylamide) and poly (cyclopropyl methacrylamide) | Thermogelling acrylamide derivatives, its balance of hydrogen bonding, and hydrophobic association changes with temperature |

**Cellulose-Based Polymers**

| | |
|---|---|
| Ethyl cellulose | Insoluble but dispersible in water, aqueous coating system for sustained release applications |
| Carboxymethyl cellulose | Superdisintegrant, emulsion stabilizer |
| Hydroxyethyl and hydroxypropyl celluloses | Soluble in water and in alcohol, tablet coating |
| Hydroxypropyl methyl cellulose | Binder for tablet matrix and tablet coating, gelatin alternative as capsule material |
| Cellulose acetate phthalate | Enteric coating |

**Hydrocolloids**

| | |
|---|---|
| Alginic acid | Oral and topical pharmaceutical products; thickening and suspending agent in a variety of pastes, creams, and gels, as well as a stabilizing agent for oil-in-water emulsions; binder and disintegrant |
| Carrageenan | Modified release, viscosifier |
| Chitosan | Cosmetics and controlled drug delivery applications, mucoadhesive dosage forms, rapid release dosage forms |
| Hyaluronic acid | Reduction of scar tissue, cosmetics |
| Pectinic acid | Drug delivery |

**Water-Insoluble Biodegradable Polymers**

| | |
|---|---|
| (Lactide-co-glycolide) polymers | Microparticle–nanoparticle for protein delivery |

**Starch-Based Polymers**

| | |
|---|---|
| Starch | Glidant, a diluent in tablets and capsules, a disintegrant in tablets and capsules, a tablet binder |
| Sodium starch glycolate | Superdisintegrant for tablets and capsules in oral delivery |

**Plastics and Rubbers**

| | |
|---|---|
| Polyurethane | Transdermal patch backing (soft, comfortable, moderate moisture transmission), blood pump, artificial heart, and vascular grafts, foam in biomedical and industrial products |
| Silicones | Pacifier, therapeutic devices, implants, medical grade adhesive for transdermal delivery |
| Polycarbonate | Case for biomedical and pharmaceutical products |
| Polychloroprene | Septum for injection, plungers for syringes, and valve components |
| Polyisobutylene | Pressure sensitive adhesives for transdermal delivery |
| Polycyanoacrylate | Biodegradable tissue adhesives in surgery, a drug carrier in nano- and microparticles |
| Poly (vinyl acetate) | Binder for chewing gum |
| Polystyrene | Petri dishes and containers for cell culture |
| Polypropylene | Tight packaging, heat shrinkable films, containers |
| Poly (vinyl chloride) | Blood bag, hoses, and tubing |
| Polyethylene | Transdermal patch backing for drug in adhesive design, wrap, packaging, containers |
| Poly (methyl methacrylate) | Hard contact lenses |
| Poly (hydroxyethyl methacrylate) | Soft contact lenses |
| Acrylic acid and butyl acrylate copolymer | High $T_g$ pressure–sensitive adhesive for transdermal patches |
| 2-Ethylhexyl acrylate and butyl acrylate copolymer | Low $T_g$ pressure–sensitive adhesive for transdermal patches |
| Vinyl acetate and methyl acrylate copolymer | High cohesive strength pressure–sensitive adhesive for transdermal patches |
| Ethylene vinyl acetate and polyethylene terephthalate | Transdermal patch backing (occlusive, heat sealable, translucent) |
| Ethylene vinyl acetate and polyethylene | Transdermal patch backing (heat sealable, occlusive, translucent) |
| Polyethylene and polyethylene terephthalate | Transdermal patch backing (when ethylene vinyl acetate copolymer is incompatible with the drug) |

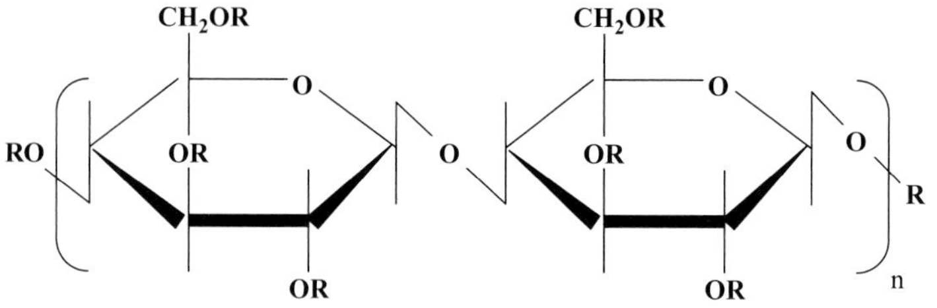

**Fig. 20–16.** Cellulose structure.

hydroxypropyl/methyl substitution, respectively. **Table 20–4** shows how functional group substitution results in different cellulose-based polymers with different properties.

Methocel polymers including pure methylcellulose and hydroxypropyl-substituted methylcellulose display thermo-gelling property in water. As the temperature of the solution increases, the hydrophobic groups of these polymers start to aggregate, as a result the polymer solution will assume a cloudy appearance. The cloud point temperature for the pure methyl cellulose (with no hydroxypropyl substitution) is about 50°C. As more methyl groups are substituted with hydroxypropyl groups, which have better solubility in water, the cloud point temperature shifts to higher temperature (60°C–85°C for the Methocel E, F, and K). Generally speaking, cloud point temperature is critically dependent on the methyl substitution. On the other hand, aqueous viscosity of the Methocel polymers is more dependent on the polymer molecular weight than its methyl/hydroxypropyl content.

## Hydrocolloids

Various hydrocolloids or polysaccharide gums are originated from a variety of sources as summarized in **Table 20–5**.

Most gums are hydrophilic and contain very long poly-meric chains as well as different functional groups. These features are very attractive in many pharmaceutical processes such as coating, stabilization, thickening, binding, solubiliza-tion, and disintegration. Gums behave differently in water and aqueous solutions. Almost all display thickening property, whereas some show gelling property. Although thickening is a desirable property for solution, suspension, and emul-sion dosage forms, gelling property is utilized in drug encap-sulation for controlled delivery applications. Gums such as guar gum can provide excellent thickening property, whereas gums including alginate and chitosan can offer a gelling prop-erty in the presence of ions. Similar to synthetic polymers, gums can be blended to provide superior properties through

## TABLE 20–4
## CELLULOSE-BASED POLYMERS

| R | Polymer | Characteristics |
|---|---|---|
| H | Cellulose | Water-insoluble due to excessive hydrogen bonding |
| H and $CH_3$ | Methyl cellulose (MC) | Soluble in cold water only; commercially available as Methocel A (Dow Chemical); swells and disperses slowly in cold water to form a colloidal dispersion; practically insoluble in ethanol, saturated salt solutions, and hot water; soluble in glacial acetic acid, displays thermogelling property |
| H and $CH_2CH_3$ | Ethyl cellulose (EC) | Water-insoluble; aqueous coating system for sustained release applications; impermeable barrier; plasticized EC composed of dibutyl sebacate and oleic acid; Ethocel is commercially available from Dow; Ethyl cellulose latex, Aquacoat, is also available from FMC Corp |
| H and $CH_2COOH$ | Carboxymethyl cellulose (CMC) | Water-soluble; variable degree of substitution; cross-linked CMC is water-swellable and known as croscarmellose sodium in National Formulary (NF); FMC Corp. supplies cross-linked CMC (Ac-Di-Sol; Accelerated Dissolution) as tablet superdisintegrant |
| H and $CH_2CH_2OH$ | Hydroxyethyl cellulose (HEC) | Soluble in water and in alcohol |
| H and $CH_2CHOHCH_3$ | Hydroxypropyl cellulose (HPC) | Water-soluble at low temperature; film-coating application |
| H and $CH_3$ and $CH_2CHOHCH_3$ | Hydroxypropyl methylcellulose (HPMC) | Soluble in water below 60°C and in organic solvents; Dow Chemical supplies HPMC as Methocel (such as E, F, K) for tablet coating; HPMC coating replaced sugar coating with the advantage of much shorter coating time; possess thermogelling property; is also used as capsule material to substitute the animal-based gelatin |

**TABLE 20–5**

**HYDROCOLLOIDS FROM DIFFERENT SOURCES**

| Plant Exudates | Seed Gum | Seaweed Extract | Microbial | Plant Extract | Animal-Based |
|---|---|---|---|---|---|
| Gum arabic | Guar | Agar | Xanthan | Pectin | Chitosan |
| Karaya | Locust bean | Carrageenan | Gellan | Konjac | Gelatin |
| Tragacanth | Psyllium | Alginate | Curdlan | | |
| Ghatti | | | | | |

synergy, which cannot be achieved by individual gum alone. On the negative side, gums are obtained from natural sources with different assay and qualities. Therefore, as opposed to synthetic polymers, the batch-to-batch consistency and quality would be a major challenge for pharmaceutical suppliers. Besides, gums are a good platform for bacteria growth, which limits their shelf life and requires sterilization.

Polysaccharides and their derivatives can be used as a rate controller in sustained release formulations due to their gelling property.[9] Gums can easily be derivatized to change their solution properties. For instance, chitosan is only soluble in acidic water, but its carboxymethyl derivative is soluble at a wider pH range. Gums offer a wide range of molecular weights that also affect their dissolution properties. They are biodegradable and their chemical composition varies greatly.[10] The physicochemical properties of polysaccharide solutions and gels have recently been reviewed for pharmaceutical and food applications.[11] Polysaccharides are claimed to effectively treat local colon disorders if they are used in colon-targeting delivery systems, which utilize the colonic microflora.[12] Inulin, amylase, guar gum, and pectin are specifically degraded by the colonic microflora and used as polymer drug conjugates and coating. It has been shown that drug release in the colon can be maximized if the hydrophobicity of the gums is modified chemically or physically using other conventional hydrophobic polymers.[13] In cancer therapy, polysaccharides are used as immune-modulators. A few fungal polysaccharides, either alone or in combination with chemotherapy and/or radiotherapy, have been used clinically in the treatment of various cancers.[9] It was suggested that iron stabilized into a polysaccharide structure can be used to treat anemia. The product can also be used in resonance imaging as well as in separation of cells and proteins utilizing magnetic fields due to its magnetic properties.[14]

Alginic acid is a linear polysaccharide that mainly consists of two building blocks of mannuronic (M) and guluronic (G) acid residues. Alginic acid and its salts are anionic polymers that can offer gelling properties. Since they contain carboxyl group, they can easily undergo a complexation reaction in the presence of ions. Depending on its source, the mannuronic and guluronic contents of the alginate product can be different. Between the two building blocks, the G blocks are responsible for the gelling property; as such, a product rich in G block offers stronger gel in the presence of ions, in particular, calcium. Excipient suppliers provide different grades of alginate with broad range of G/M ratios. Apparently, the mechanical property of the alginate gel is determined by the G/M contents of the product, the type of ion (monovalent, divalent, trivalent), the concentration of ion in the solution, as well as the duration of the gelling process. For the encapsulation purposes, all these factors have to be considered in designing a tailor-made delivery system. Alginic acid and its derivatives have found applications as a stabilizing agent, binding agent, drug carrier, and so on and so forth. The antibiotic griseofulvin, which is supplied as oral suspension, contains sodium alginate stabilized with methylparaben. Alginic acid and ammonium calcium alginate can be found in metaxalone tablets. Alginate microbeads can be used to entrap drugs, macromolecules, and biological cells. For this application, parameters such as calcium salt, other hardening agents, and different drying methods have been studied.[15] Islets of Langerhans embedded into alginate encapsulates can be transplanted without the risk of protein contamination and immune system suppression.[16]

Carrageenan is a sulfated linear polysaccharide of galactose and anhydrogalactose. It carries a half-ester sulfate group with the ability to react with proteins. If carrageenan is used in a solution containing proteins, the solution becomes gel or viscous due to a complex formation between carrageenan and charged amino acids of the protein. Therefore, a formulation scientist should be aware of any incompatibility issues which might jeopardize the stability of the drug solution, suspension, or emulsion. Depending on the concentration and location of the sulfated ester groups, carrageenan can be found in three different grades of kappa, iota, and lambda. Kappa carrageenan can form a strong and brittle gel, especially in the presence of potassium ions or if blended with locust bean gum. If a drug formulation requires a softer and more elastic gel, iota carrageenan can be used. Both kappa and iota carrageenan can be used for controlled delivery application as they display gelling properties under certain circumstances. If a drug formulation needs to be thickened and does not contain proteins of any source, a lambda carrageenan can be utilized. Donepezil hydrochloride orally disintegrating tablets and cefpodoxime proxetil oral suspension contain carrageenan. Carrageenan is shown to increase the permeation of sodium fluorescein through porcine skin as it changes the rheological properties of the drug solution.[17] In capillary electrophoresis, a sulfated polysaccharide such as carrageenan can be used to separate the enantiomers of weakly basic pharmaceutical compounds. Different enantiomers of racemic compounds such as propranolol and pindolol have been separated using carrageenan.[18,19]

Chitosan is obtained from chitin, the second most abundant natural polymer after cellulose, which can be found in shrimp, crab, and lobster shells. Chitosan is a cationic polymer and has been investigated as an excipient in controlled delivery formulations and mucoadhesive dosage forms because of its gelling and adhesive properties. The bitter taste of natural extracts such as caffeine has been masked using chitosan. Chitosan can potentially be used as a drug carrier, a tablet excipient, delivery platform for parenteral formulations, disintegrant, and tablet coating.[20,21] From toxicity and safety standpoint, lower–molecular-weight chitosan (as an oligosaccharide) has been shown to be safer with negligible cytotoxicity on Caco-2 cells.[22] During the encapsulation process using synthetic polymers, the protein is generally exposed to the conditions which might cause their denaturation or deactivation. Therefore, a biocompatible alternative such as chitosan is desirable for such applications.[23] Because of its cationic nature, chitosan can make complexes with negatively charged polymers such as hyaluronic acid (HA) to make a highly viscoelastic polyelectrolyte complex. The complex has the potential to be used as cell scaffold and as a drug carrier for gene delivery.[24] Gels based on chitosan and ovalbumin protein have been suggested for pharmaceutical and cosmetic use.[25] In veterinary area, chitosan can be used in the delivery of chemotherapeutics such as antibiotics, antiparasitics, anesthetics, painkillers, and growth promotants.[26] As an absorption enhancer, a protonated chitosan is able to increase paracellular permeability of peptide drugs across mucosal epithelia. While trimethyl derivative of chitosan is believed to enhance the permeation of neutral and cationic peptide analogs, the carboxymethyl derivative of chitosan offers gelling properties in an aqueous environment or with anionic macromolecules at neutral pHs.[27] Chitosan can also be mixed with nonionic surfactants such as sorbitan esters to make emulsion like solutions or creams.[28] To prepare chitosan beads or microspheres, the chitosan matrix needs to be treated with an anionic compound like pentasodium tripolyphosphate. A sustained release dosage form of salbutamol sulfate bead can be prepared using chitosan in phosphate buffer.[29]

Pectin is a ripening product of green fruits such as lemon and orange skin. Protopectin is an insoluble pectin precursor present in such fruits, which is converted to pectinic acid and subsequently to pectin. The main component of pectin is D-galacturonic acid, which is in part esterified via methylation. Depending on its methoxyl content, pectin is classified as high methoxyl (HM, 50% or more esterification) and low methoxyl (LM, less than 50% esterification). Pectins can form a gel in an aqueous solution if certain conditions are existed. For instance, high methoxyl pectins require a minimum of 65% soluble solids and low pH (<3.5) to form a gel, whereas low methoxyl pectins require calcium and may form a gel at a much lower solid content, that is, 20%. Gelation and association of pectin chains in the presence of pH-reducing additives has also been reported.[30] While pectin is generally known as a suspending and thickening agent, it is also claimed to reduce blood cholesterol levels and to treat gastrointestinal disorders.[31] Pectin can be found in amlexanox oral paste.

Xanthan gum is found in a number of drug formulations including cefdinir oral suspension and nitazoxanide tablets. It is a highly branched glucomannan polysaccharide with excellent stability under acidic conditions. Xanthan is generally used in solution and suspension products for its thickening property. Because of its very rigid structure, its aqueous solution is significantly stable over a wide pH range. Similar to locust bean gum, xanthan gum is also nongelling but forms a gel once it is mixed with the locust bean gum. Concentrated xanthan gum solutions resist flow due to excessive hydrogen bonding in the helix structure, but they display shear-thinning rheology under the influence of shear flow. This feature of xanthan gum solutions is critical in food, pharmaceutical, and cosmetic manufacturing processes.[32] Oxymorphone hydrochloride extended-release tablets contain TIMERx, which consists of xanthan gum, and locust bean gum for controlled delivery.[33] Rectal delivery of morphine can be controlled using cyclodextrins as an absorption enhancer and xanthan as a swelling hydrogel. This combination produces a sustained plasma level of morphine and increases its rectal bioavailability.[34]

HA consists of $N$-acetyl-D-glucosamine and beta-glucoronic acid and has been used as fluid supplement in arthritis, in eye surgery, and to facilitate healing of surgical wounds. Solaraze gel for the topical treatment of actinic keratoses is composed of 3% sodium diclofenac in 2.5% HA gel.[35] Sodium hyaluronate and its derivatives have been evaluated in vitro and in vivo for optimized delivery of a variety of active molecules such as antibiotic gentamicin and cytokine interferon.[36] Hyaluronan is biocompatible and nonimmunogenic and has been suggested as a drug carrier for ophthalmic, nasal, pulmonary, parenteral, and dermal routes.[37] Sodium hyaluronate topical (Seprafilm), which is used to reduce scar tissue as a result of abdominal or pelvic surgery, is a bioresorbable membrane containing sodium hyaluronate.

Gum arabic or gum acacia is best known for its emulsifying property and its solution viscosity at very high solid concentration. Locust bean gum consists of mannose and galactose sugar units at a ratio of 4:1. Like almost all gum solutions, an aqueous solution of this gum displays shear-thinning rheology. It shows synergistic effect with xanthan and kappa carrageenan. Gellan gum has been used in pharmaceutical dosage forms as a swelling agent,[38] as a tablet binder,[39] and as a rheology modifier.[40] Drug delivery behavior of scleroglucan hydrogels has been examined using theophylline as a model drug.[41] As an alternative tablet binder to starch and polyvinyl pyrrolidone, seed galactomannan of Leucaena, a natural polysaccharide comparable to guar gum, has been evaluated for poorly compressible drugs such as paracetamol.[42] Schizophyllan, which is secreted by fungus, has been evaluated as an immunostimulant (to suppress tumor growth), antitumor, antihepatitis, anti-HIV, and antiviral agent.[43,44] Sustained release formulations based on guaran,[45] in situ gel-forming ability of xyloglucan,[46] and borax–guar gum complexes for colon specific drug delivery have also been studied.[47]

One of the most recent applications of gums is in film forming. Recent concepts and products such as breath films, cough strips, flu, and sore throat strips have all been realized on the basis of film-forming ability of gums. Generally speaking, as opposed to branched gums, linear gums have more sites available for intermolecular hydrogen bonding; as a result, they offer better film-forming properties. Individual and blended gum products based on agar, alginate, $\kappa$-carrageenan, methyl cellulose, pectin, CMC, and guar can potentially be used in film dosage forms.

# POLYMERS IN DRUG DELIVERY

## Introduction

Pharmaceutical polymers are widely used to achieve taste masking; controlled release (e.g., extended, pulsatile, and targeted), enhanced stability, and improved bioavailability. Monolithic delivery devices are systems in which a drug is dispersed within a polymer matrix and released by diffusion. The rate of the drug release from a matrix product depends on the initial drug concentration and relaxation of the polymer chains, which overall displays a sustained release characteristic. Extended release alprazolam tablet is an example of monolithic products, in which extended or sustained delivery is provided by swelling and erosion of the polymer matrix. Alternatively, a drug can be released from a drug core through a porous or nonporous membrane. While drug release through a nonporous membrane is essentially driven by diffusion, porous membrane generates an extra path for the drug release, that is, through pores. The status of drug release from membrane systems can generally be modified via membrane thickness, use of plasticizers, pore structure (size, size distribution, and morphology), and filler tortuosity (filler orientation). Membrane systems have found applications in drug stability, enteric release, taste masking, and sustained release. Enteric-coated products are the ones that pass the stomach environment safely and release the drug at a higher pH environment of the intestine. These have to be coated with a pH-operative coating such as an anionic polymer. Examples of enteric-coated products are duloxetine, mesalazine, naproxen, omeprazole, and amino salicylic acid. Drugs such as lutein and lycopene are more stable in membrane dosage forms. Reservoir systems have been utilized to taste mask acetaminophen and caffeine. Potassium chloride and diltiazem are also offered sustained release property if formulated in a membrane system.

## Synthetic Polymers

Synthetic polymers based on acrylic or methacrylic acids have found extensive applications in the drug delivery area to protect the drug or to release the drug in a controlled manner. These are classified as cationic, anionic, and neutral (nonionic) polymers. Despite the different solubility and swellability across the GI tract, the drug release from these matrices occurs through a diffusion process.

Cationic polymers: One of the most widely used cationic polymers for protective coating applications has dimethyl aminoethyl methacrylate for a functional group. As far as its purity is concerned, the polymer contains less than 3000 ppm of residual monomers including butyl methacrylate (<1000 ppm), methyl methacrylate (<1000 ppm), and dimethyl aminoethyl methacrylate (<1000 ppm). These are used as pH-dependent drug delivery platforms to protect sensitive drugs, to mask unpleasant tastes and odor, to protect the active ingredient from moisture, and also to improve drug storage stability. Eudragit E 100 is, for instance, supplied as a granule and is used in taste-masking applications where a low pH solubility (<5) is desirable.

Anionic polymers: Anionic polymers have methacrylic acid as a functional group and are generally used for drug delivery past the stomach into the duodenum, jejunum, ileum, or colon. As discussed in Chapter 10, since the pH of the fasted stomach is below pH 3 in nearly every healthy person and below pH 2 in most people, the stomach represents a harsh environment for many drugs. Since the methacrylic group dissociates at the higher pH of the small intestine and colon, anionic polymers such as Eudragit L 100-55 (with dissolution at pH 5.5) and Eudragit S 100 (with dissolution at pH >7.0) are highly desirable. Aqueous dispersions of these polymers (Eudragit L 30 D-55) are generally available for direct use in enteric coating applications. Kollicoat MAE 30 DP (**Fig. 20–17**), a combination of methacrylic acid and ethyl acrylate (1:1) monomers, is supplied as a 30% aqueous dispersion. The polymer is used as a film-former in enteric coating of solid dosage forms.

Neutral polymers: Acrylate or methacrylate polymers with or without aminoethyl functionality are generally insoluble or have pH-independent swelling property. These are neutral acrylic polymers which are used for sustained release applications where insolubility of the polymeric drug carrier is very much desirable. Neutral polymers with added functionality are supplied as powder (e.g., Eudragit RS PO), granule (e.g., Eudragit RS 100), and aqueous dispersions (e.g., Eudragit

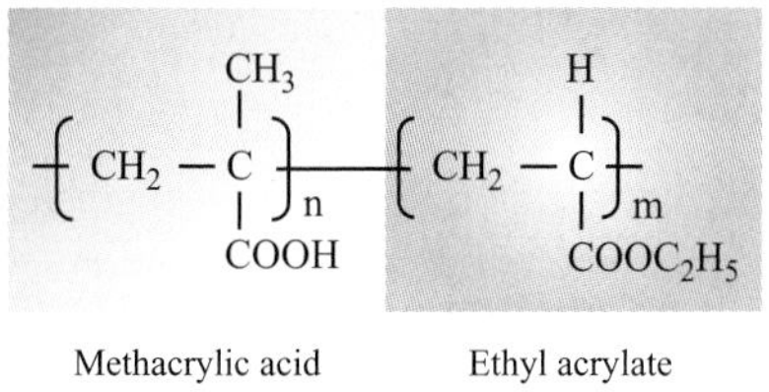

Fig. 20–17. A copolymer for an enteric tablet-coating application.

RS 30 D). Neutral polymers with no added functionality are supplied as aqueous dispersions (Eudragit NE30D, NM30D, and NE40D).

## Biodegradable Polymers

Alternatively the drug can be released from a dosage form as a result of polymer erosion. Erosion occurs because of biodegradation or swelling and might happen within the bulk of the polymer or may be limited to the polymer surface at a time. Porous and nonporous platforms can provide bulk and surface erosion, respectively. Polymers with ester and amide functional groups are susceptible to a hydrolytic degradation in strong acidic and basic environment. When a polymer starts to erode from its surface, the drug imbedded within the polymer matrix will be released at a rate depending on the erosion rate of the polymer. If erosion happens in bulk, a much faster release is expected as an enormous number of hydrolysable sites are simultaneously cleaved up in water.

Biodegradable polymers are classified as natural-based and synthetic. Polysaccharides and protein-based polymers are obtained from the natural sources. Polyesters or copolyesters of lactic acid and glycolic acid, polycaprolactone, polyanhydrides, and polyethylene glycol are the most common synthetic biodegradable polymers, which are used for variety of pharmaceutical applications.

### EXAMPLE 20–9

**Injectable Implant**

Injectable implants have been developed on the basis of biodegradable polymers. Leuprolide (Eligard), a delivery system for prostate cancer, is supplied as an injectable suspension that utilizes the Atrigel technology for delivering the hormone leuprolide acetate. The delivery system consists of a biodegradable (lactide-co-glycolide) copolymer dissolved in a biocompatible solvent. The polymer gradually loses its organic solubility once it is injected subcutaneously. Doxycycline (Atridox), a bioabsorbable delivery system for the treatment of periodontal disease, also uses Atrigel technology to deliver an antibiotic, doxycycline hyclate.

Alternatively, a drug may be released as a result of matrix swelling. The matrix is made of nonbiodegradable but erodible polymers, which control the drug delivery due to its swelling. A polymer in its swollen form is mechanically weak and can be eroded at different rates depending on the swelling feature of the matrix.[48] A fast swelling hydrogel may undergo faster erosion and provide faster drug release compared with a slow swelling hydrogel. The release kinetics from a swellable matrix is generally zero-order.

## Ion-Exchange Resins

These are polymeric materials with two characteristics; they swell in an aqueous medium and they contain complexable and ionizable groups. Ion-exchange resins are generally made of methacrylic acid, sulfonated styrene, and divinyl benzene (DVB). The acidic resins can be weak or strong. Weak acid resins are produced on the basis of methacrylic acid (containing COOH) monomer cross-linked with DVB. The counter-ion of the acidic carboxyl group is hydrogen (as in Polacrilex resin, Amberlite IRP64) or potassium (as in Polacrilin potassium, Amberlite IRP88). To make an ion-exchange resin with stronger acidity, the water-insoluble styrene is used as a monomer, which is sulfonated to become water compatible. Similarly, DVB is used to cross-link the polymer and the counter-ion of the sulfate group ($SO_3$) is generally sodium. The commercial product of sodium polystyrene sulfonate (Amberlite IRP69) is used to treat hyperkalemia. Cation exchangers are anionic polymers which contain carboxyl or sulfate groups with hydrogen, potassium, and sodium as counter-ions. On the other hand, cationic ion-exchange resins with the ability to exchange anions carry quaternary ammonium groups, $-N^+(R)^3$ with chlorine as a counter-ion. Cholestyramine resin (Duolite AP143) is cationic styrene DVB polymer which is an anion exchanger and used to reduce cholesterol or to sequestrate the bile acid. Because of their unique properties, the ion-exchange resins are generally used for taste masking, drug stabilization, tablet disintegration, and sustained release applications.

Cationic or anionic drugs can be complexed into the structure of an ion-exchange resin due to its ionic structure. The stability of the complex inside the mouth masks the taste of the drug since the drug will not be free to interact with taste buds. The drug will be released in the gastric medium once the complex becomes unstable. Certain drugs have a poor shelf life due to their instability against moisture, light, heat, and so on and so forth. Shelf stability and bioavailability of these drugs increase when formulated with ion-exchange resins. The DVB–cross-linked potassium methacrylate copolymer possesses a very high swelling capacity in water. Although this polymer generally swells fast and to a high degree, its swelling properties are significantly affected by pH, the presence of salts, and the ionic strength of the aqueous solution. Nevertheless, the swelling pressure generated by this polymer is sufficient to disintegrate tablet dosage forms in an aqueous medium. These polymers can also provide a sustained or a zero-order release due to their high swelling capacity.

## CHAPTER SUMMARY

Polymers have been used as a main tool to control the drug release rate from the formulations. They are also increasingly used as taste-masking agent, stabilizer, and protective agent in oral drug delivery. Polymers can bind the particles of a solid dosage form and also change the flow properties of a liquid dosage form. Extensive applications of polymers in drug delivery have been realized because polymers offer unique properties which so far have not been attained by any other materials. Polymers are macromolecules having very large chains, contain a variety of functional groups, can be blended with other low- and high–molecular-weight materials, and can be tailored for any applications. Understanding the basic concepts of polymers provides a foundation for further understanding of drug products and designing of better delivery systems.

This chapter provides basic concepts behind the behavior of polymers in the solid and solution states. Chapter begins with a general introduction on polymers and continues with different types of polymer structure and polymerization methods to make them. The major polymer concepts and properties, such as synthesis, topology, crystallinity, thermal transitions, molecular weight (averages and distribution), swelling, entanglement, rheology, and mechanical properties, are discussed in detail. The chapter continues with a variety of polymer products including rubbers, plastics, fibers, adhesives, and coatings, and also highlights important properties of each group. The chapter concludes with major applications of polymers in pharmaceutical industry, such as ion-exchange resins. Many examples, pictures, and concept boxes have been added to better appreciate these topics. This chapter can serve as a valuable source of information for those with little or no background in polymers, researchers in the polymer, pharmaceutics and biomedical areas, as well as pharmacy students.

 **Practice problems for this chapter can be found at thePoint.lww.com/Sinko6e.**

## References

1. F. Rodriguez, *Principles of Polymer Systems*, Taylor & Francis, Philadelphia, PA, 1996.
2. G. Odian, *Principles of Polymerization*, 3rd Ed., Wiley, New York, 1991.
3. P. J. Flory, *Principles of Polymer Chemistry*, Cornell University, Ithaca, New York, 1953.
4. F. W. Billmeyer, *Textbook of Polymer Science*, 3rd Ed., Wiley, New York, 1984.
5. L. H. Sperling, *Introduction to Physical Polymer Science*, Wiley, New York, 1992.
6. H. Omidian, et al., Hydrogels having enhanced elasticity and mechanical strength properties. US patent 6,960,617. 2005.
7. H. Omidian, J. G. Rocca, and K. Park, Macromol. Biosci. **6**(9), 703–710, 2006.
8. D. Braun, *Simple Methods for Identification of Plastics*, 2nd Ed., Hanser Publishers, Munchen, Wien, 1986.
9. S. C. Jong, J. M. Birmingham, and S. H. Pai, Eos-Rivista Di Immunologia Ed Immunofarmacologia, **11**(3), 115–122, 1991.
10. L. Hovgaard and H. Brondsted, Crit. Rev. Ther. Drug Carrier Syst. **13**(3–4), 185–223, 1996.
11. K. Nishinari and R. Takahashi, Curr. Opin. Colloid Interface Sci. **8**(4–5), 396–400, 2003.
12. A. W. Basit, Drugs, **65**(14), 1991–2007, 2005.
13. T. F. Vandamme, et al., Carbohydr. Polym. **48**(3), 219–231, 2002.
14. K. I. Shingel and R. H. Marchessault, in Robert H. Marchessault, Francois Ravenelle, Xiao Xia Zhu (Eds.), *Polysaccharides for Drug Delivery and Pharmaceutical Applications*, ACS Symposium Series 934, American Chemical Society, 2006, pp. 271–287.
15. L. W. Chan, H. Y. Lee, and P. W. S. Heng, Int. J. Pharm. **242**(1–2), 259–262, 2002.
16. J. Dusseault, et al., J. Biomed. Mater. Res. A **76**(2), 243–251, 2006.
17. C. Valenta and K. Schultz, J. Control. Release, **95**(2), 257–265, 2004.
18. G. M. Beck and S. H. Neau, Chirality, **8**(7), 503–510, 1996.
19. G. M. Beck and S. H. Neau, Chirality, **12**(8), 614–620, 2000.
20. O. Felt, P. Buri, and R. Gurny, Drug Dev. Ind. Pharm. **24**(11), 979–993, 1998.
21. A. K. Singla and M. Chawla, J. Pharm. Pharmacol. **53**(8), 1047–1067, 2001.
22. S. Y. Chae, M. K. Jang, and J. W. Nah, J. Control. Release, **102**(2), 383–394, 2005.
23. M. George and T. E. Abraham, J. Control. Release, **114**(1), 1–14, 2006.
24. J. Wu, et al., J. Biomed. Mater. Res. A **80**(4), 800–812, 2007.
25. S. Y. Yu, et al., Langmuir, **22**(6), 2754–2759, 2006.
26. S. Senel and S. J. McClure, Adv. Drug Deliv. Rev. **56**(10), 1467–1480, 2004.
27. M. Thanou, J. C. Verhoef, and H. E. Junginger, Adv. Drug Deliv. Rev. **52**(2), 117–126, 2001.
28. J. Grant, J. Cho, and C. Allen, Langmuir, **22**(9), 4327–4335, 2006.
29. E. A. El Fattah, et al., Drug Dev. Ind. Pharm. **24**(6), 541–547, 1998.
30. P. S. Holst, et al., Polym Bull. **56**(2–3), 239–246, 2006.
31. B. R. Thakur, R. K. Singh, and A. K. Handa, Crit. Rev. Food Sci. Nutr. **37**(1), 47–73, 1997.
32. K. W. Song, Y. S. Kim, and G. S. Chang, Fibers Polym. **7**(2), 129–138, 2006.
33. M. J. Tobyn, et al., Int. J. Pharm. **128**(1–2), 113–122, 1996.
34. K. Uekama, et al., J. Pharm. Sci. **84**(1), 15–20, 1995.
35. M. B. Brown and S. A. Jones, J. Eur. Acad. Dermatol. Venereol. **19**(3), 308–318, 2005.
36. N. E. Larsen and E. A. Balazs, Adv. Drug Deliv. Rev. **7**(2), 279–293, 1991.
37. Y. H. Liao, et al., Drug Deliv. **12**(6), 327–342, 2005.
38. P. J. Antony and N. M. Sanghavi, Drug Dev. Ind. Pharm. **23**(4), 413–415, 1997.
39. P. J. Antony and N. M. Sanghavi, Drug Dev. Ind. Pharm. **23**(4), 417–418, 1997.
40. P. B. Deasy and K. J. Quigley, Int. J. Pharm. **73**(2), 117–123, 1991.
41. M. E. Daraio, N. Francois, and D. L. Bernik, Drug Deliv. **10**(2), 79–85, 2003.
42. U. P. Deodhar, A. R. Paradkar, and A. P. Purohit, Drug Dev. Ind. Pharm. **24**(6), 577–582, 1998.
43. T. Fuchs, W. Richtering, and W. Burchard, Macromol. Symp. **99**, 227–238, 1995.
44. J. Munzberg, U. Rau, and F. Wagner, Carbohydr. Polym. **27**(4), 271–276, 1995.
45. B. P. Nagori and N. K. Mathur, Ind. J. Chem. Technol. **3**(5), 279–281, 1996.
46. N. Kawasaki, et al., Int. J. Pharm. **181**(2), 227–234, 1999.
47. A. Rubinstein and I. Glikokabir, Stp. Pharma. Sci. **5**(1), 41–46, 1995.
48. H. Omidian and K. Park, J. Drug Deliv. Sci. Technol. **18**(2), 83–93, 2008.

## Recommended Readings

A. T. Florence and D. Attwood, *Physicochemical Principles of Pharmacy*, 4th Ed., Pharmaceutical Press, London, 2006, Chapter 8, pp. 273–322.

D. Jones, Rapra Rev. Rep. **174**, 124, 2004.

M. Saltzman, *Tissue Engineering: Engineering Principles for the Design of Replacement Organs and Tissues*, Appendix A: Introduction to Polymers, Oxford University Press, Inc., New York, 2004, pp. 453–480.

A. M. Stephen, *Food Polysaccharides and Their Applications*, Marcel Dekker, Inc., New York, 1995.

## CHAPTER LEGACY

**Fifth Edition:** published as Chapter 21 (Biomaterials). Updated by Bozena Michniak-Kohn.

**Sixth Edition:** published as Chapter 20 (Pharmaceutical Polymers). Rewritten de novo by Hossein Omidian, Kinam Park, and Patrick Sinko.

# 21 PHARMACEUTICAL BIOTECHNOLOGY

## INTRODUCTION

Up to this point, the text has primarily been concerned with drugs of molecular weight under a few thousand ("small molecules"). Nevertheless, much larger molecules, such as proteins, DNA, and carbohydrates as well as macromolecular assemblies including viruses and bacteria, have been used as drugs and vaccines for quite some time and many are currently in development. For example, animal-derived hormones such as insulin and somatotropin (growth hormone) as well as human blood–derived proteins such as coagulation factors and immunoglobulin (antibody) preparations have saved millions of lives during the last century. The availability of synthetic versions of such materials fall into a subclass of pharmaceutical products derived by a general series of procedures known as biotechnology.

Here, in keeping with the general theme of physical pharmacy, the focus will be on what is often called "pharmaceutical biotechnology." There are many comprehensive texts[3–8] that discuss the more general aspects of biotechnology with extended discussions of fermentation (the major emphasis of biotechnology until fairly recently), industrial enzymes, and related topics. In this chapter, the analysis, preformulation and formulation of large molecules intended for pharmaceu-

tical use, generally focusing on proteins and nucleic acids, as well as vaccines will be introduced. Methods of their production and delivery will also be briefly discussed but the interested student should refer to more detailed discussions in these areas.[3–8]

The basic idea in pharmaceutical biotechnology is to employ biological processes and biological molecules to create drugs and vaccines. Our ability to do this has arisen from a dramatic increase in our understanding of the molecular and cellular basis of life. This has included the ability to create and manipulate both nucleic acids and proteins through an extensive series of procedures that is usually referred to as molecular biology. These methods will be very briefly discussed below. One consequence of this dramatic technology expansion has been the rise of the biotechnology industry. This is directly manifested in the creation of a series of companies such as Genentech, Amgen, Genzyme, Biogen-Idec, Med-Immune, and many more. In addition, there also exist many hundreds of smaller biotech businesses spread throughout the world focusing on an extensive variety of human diseases employing a diverse array of biotechnology-related technologies. Furthermore, previously small molecule focused large pharmaceutical companies such as Pfizer, Merck, GlaxoSmithKline, Wyeth, and Novartis all contain within

## KEY CONCEPT | DEFINING PHARMACEUTICAL BIOTECHNOLOGY

The Oxford American Dictionary defines this as: "the exploitation of biological processes for industrial and other purposes, esp. the genetic manipulation of microorganisms for the production of antibiotics, hormones, etc."[1] The Oxford Dictionary of Biochemistry and Molecular Biology somewhat more elaborately defines it as: "the integration of natural sciences and engineering sciences in order to achieve the application of organisms, cells, parts thereof and molecular analogues for products and services (European Federation of Biotechnology General Assembly, 1989);

a field of technological activity in which biochemical, genetic, microbiological, and engineering techniques are combined for the pursuit of technical and applied aspects of research into biological materials and, in particular, into biological processing. It includes traditional technologies such as fermentation processes, antibiotic production, and sewage treatment, as well as newer ones such as biomolecular engineering and single-cell protein production."[2]

themselves large biotechnology divisions. Currently, there are more than a hundred approved peptide and protein pharmaceuticals (see http://www.drugbank.ca) with hundreds more in clinical trials. It seems fair to say that the clinical importance of large molecules is approaching that of their smaller cousins. The recent mapping of the human genome has further opened up new opportunities with the identification of several tens of thousands of genes providing both new targets and potential new macromolecular drugs. Opportunities for both improved and novel diagnostic procedures have also appeared as a result of the advances mentioned above.

# TYPES OF BIOTECHNOLOGY-DERIVED PRODUCTS

## Peptides and Proteins[9–11]

Peptides and proteins are formed by the creation of a peptide bond between combinations of the 20 naturally occurring amino acids. The primary distinction between the two is one of size with polymer lengths less than 30 to 40 residues defining peptides and longer sequences, proteins. A second distinction is one of levels of structure. While both peptides and proteins contain defined orders of their amino acids (their primary structure or sequence), proteins also usually contain additional higher-order structures (**Fig. 21–1**). For example, most proteins contain regions of regular, local chain interactions known as secondary structure. The two most common types of secondary structure in proteins are the $\alpha$ helix and $\beta$ sheet. Various types of turns (reversals of chain direction) and more disordered regions are also commonly present. Recently, it has been recognized that some proteins also exist in fairly disordered states, at least in their purified forms. Large peptides may also contain significant regions of secondary structure.

Proteins can also bring their various elements of secondary structure together to form what we refer to as tertiary structure. This creates a distinct three-dimensional structure for most proteins. Modern methods of x-ray crystallography and nuclear magnetic resonance (NMR) spectroscopy often allow the location in 3-D space of the individual atoms in a protein to be determined with a resolution of 1 to 3 Å. This has resulted in very detailed pictures of thousands of proteins with more appearing each day (**Fig. 21–1**). One conclusion that has been reached from such work is that distinct, common "domains" (small compact regions of various secondary structure combinations) exist in most proteins, reflecting both evolutionary and functional relationships among many proteins. It is important to realize, however, that the static picture of proteins, seen by crystallography, fails to provide a complete representation of protein structure. Proteins exist in a highly dynamic equilibrium of various conformational states. This will be considered in more detail below.

Individual proteins can also associate into defined multisubunit assemblies forming what is known as quaternary structure. This can involve either multiple copies of the same proteins or heterogeneous mixtures of different types of subunits. There also exists another way in which proteins can associate with themselves, which is referred to as aggregation (**Fig. 21–2**). This type of structure is especially important to the pharmaceutical scientist since it constitutes a major pathway of physical degradation for many protein pharmaceuticals. This process can be highly ordered (as seen in crystallization or the assembly of fibers) or highly disordered forming amorphous protein particles. We will further consider this latter process below.

Peptides and proteins of pharmaceutical utility can conveniently be placed into a number of different classes. A wide variety of peptide-based drugs are now available. These include antitumor agents such as leuprolide, diabetes drugs

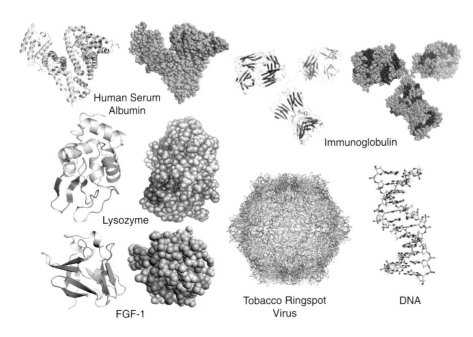

Fig. 21–1. Some typical structures of biopharmaceuticals. Several of these (human serum albumin (HSA), IgG, lysozyme, a fibroblast growth factor (FGF), and DNA) are used to produce data in later figures. Note the highly helical nature of HSA and lysozyme and the extensive $\beta$-structure present in IgG and FGF. In the first four cases, the structures are shown in both ribbon and space-filling forms.

Human Serum Albumin

Immunoglobulin

Lysozyme

FGF-1

Tobacco Ringspot Virus

DNA

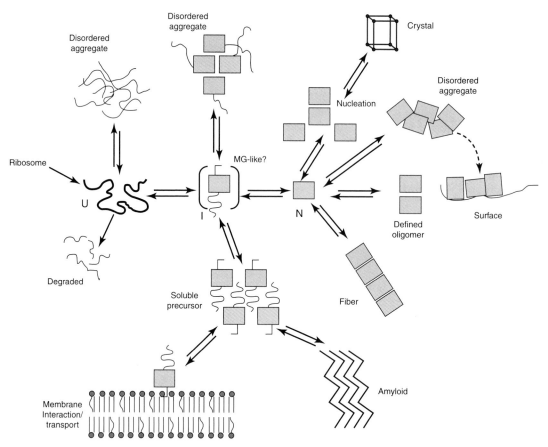

**Fig. 21–2.** Aggregation of proteins is a major pathway of their physical degradation. The native state of a protein can associate into ordered species that are crystalline or fibrous in nature and assemble into defined oligomeric species such as dimers and tetramers or into amorphous aggregates. Under various forms of stress, some structure can be lost, often into forms known as molten globule states, which can also form disordered aggregates or associate into soluble aggregates which form amyloid materials or become surface active and bind to membrane or container surfaces. Native assemblies can also frequently be surface active. These intermediate states are the most common origin of aggregation problems in protein pharmaceuticals. More complete unfolding is rarely encountered because these intermediate states dominate but unfolded (denatured) protein can aggregate as well. (Modified from C. M. Dobson and M. Karplus. The fundamentals of protein folding: bringing together theory and experiment. Curr. Opin. Struct. Biol. **9**, 92–101, 1999.)

such as insulin and exenatide (Byetta), immunosuppressants like cyclosporine, and labor-inducing agents like oxytocin. Peptides can be either isolated from natural sources such as animals, bacteria, or fungi, or chemically synthesized. They are usually sufficiently small that they can in many ways be treated like lower–molecular-weight pharmaceuticals.

Proteins that are used as drugs currently cover such a wide range of diseases that it is difficult to summarize their activities and applications. Only a few examples are provided here. Many of the early pharmaceutical proteins were derived from human or animal blood or tissue. Insulin and somatotropin (growth hormone) were originally derived from animal pancreases and human brains, respectively. Serum albumin, coagulation factor XIII, and the hepatitis B surface antigen were all obtained from human blood. Various antivenoms against snake and spider venoms were (and still are in many cases) obtained from the blood of large animals like horses

and goats. Most of these are now, however, obtained recombinantly, reducing or eliminating problems of immunological reactions, contamination, expense, and supply.

Many other proteins of potential therapeutic use were initially impossible to obtain from natural sources in sufficient amounts or quality (purity) to make them realistic candidates for use as therapeutic agents. Many such proteins are now produced recombinantly. While in some cases still expensive, the recombinant forms are both safer and available in abundant supply.

## Currently Available Recombinant Protein-Based Pharmaceuticals

Currently available protein-based drugs can be subdivided into several categories. These are (based on 2008 sales) hematopoietics (23%), monoclonal antibodies (20%), cytokines (19%), vaccines (13%), antithrombins (11%), plasma proteins (6%), insulin (5%), and growth hormones (2%). (Information cited in genengnews.com, January 1, 2009.) A brief discussion of a few representative examples follows.

Tissue plasminogen activator (t-PA) is a protease, which can dissolve thrombi which form at sites of coronary vessel occlusion and can induce myocardial infarction and stroke. Removal of such blockages within the first few hours of a coronary event can be lifesaving. Previous therapy using proteins such as urokinase and streptokinase was less specific, often leading to internal bleeding problems. Because t-PA binds to fibrin and is a natural human protein, it is more specific with generally less side effects. The protein must be produced in mammalian cell culture because it is glycosylated in its natural form (see below). Erythropoietin (EPO) is another example of a protein drug with a dramatic pharmacological effect. This protein stimulates the production of red blood cells (erythrocytes) and has a variety of uses including treatment of kidney dialysis patients to prevent anemia. Like t-PA, the protein is produced in mammalian cells because it must be correctly glycosylated for full biological activity. Similarly, granulocyte macrophage colony stimulating factor (GM-CSF) and granulocyte colony stimulating factor (G-CSF) are used to enhance the proliferation of hematopoietic progenitor cells important for immune responses and are used during bone marrow reconstitution as well as some cancers.

A variety of proteins known as interferons and cytokines are now available in recombinant form and are used for a wide variety of disorders. Interferon-$\beta$ is used to effectively treat multiple sclerosis (MS) and $\alpha$-interferon as an antiviral agent. The interleukins also belong to the cytokine family. The best known of these is interleukin-2 and is most commonly used for the treatment of renal cell cancers. A wide variety of interleukins (more than 20 are currently known) are being explored for immune-related applications. Some other notable therapeutic proteins include the Factor VIII and IX coagulation factors, DNAse for the treatment of cystic fibrosis, and glucocerebrosidase for enzyme replacement therapy in Gaucher disease. These examples are fairly typical in that they either target receptors or their ligand or are themselves ligands or receptors or fragments or mutants, thereof. Alternatively, they are proteins with enzymatic activities of therapeutic consequences. Recently, however, the biotechnology industry has begun to be dominated by the class of proteins known as immunoglobulins (Igs). When these proteins have defined affinity for specific ligands they are known as antibodies. All Igs consist of two heavy chains and two light chains bound covalently by disulfide bonds (**Fig. 21–1**). The chains themselves are composed of two globular domains in the case of light chains and four or five in the case of the heavy chains. Each domain comprises a beta-sandwich-like structure with the two flat sides of the sandwich also held together by disulfide bonds. The five different classes of immunoglobulins (IgG, IgA, IgD, IgE, and IgM) are defined

by differences in their heavy chains. There also exist two different types of light chains known as lambda and kappa. The N-terminal part of the heavy and light chains varies significantly in three regions of their sequence known as hypervariable regions, a phenomenon which arises because of controlled genetic recombination events at the DNA level combined with somatic mutation. These highly variable regions are brought together in space to form millions of different binding clefts sufficient to recognize with both low and high affinity most substances, which are referred to as antigens. This generates a huge library of receptorlike molecules that can be used to create pharmaceutical proteins that can interact with virtually any chosen target. Not surprisingly, this has resulted in the use of Ig as therapeutic agents for amazingly diverse applications. Of the 5 Ig classes, it is the IgG type that is generally used. IgG itself consists of several subclasses with differing biological properties. By using cellular cloning methods, it is possible to create unique (monoclonal) recombinant antibodies as well as their fragments for virtually any ligand. Although the original technology was developed for mouse antibodies, it is now routinely possible to produce entirely human antibodies or animal antibodies in which the nonvariable parts of the antibodies are converted to human form (humanized monoclonal antibodies, hmAbs). Although IgG molecules are large ($\sim$150 kDa) and glycosylated (usually necessitating their production in mammalian cell culture), their unique specificity and long serum half-life contribute to their expanding use as therapeutic agents. A list of some currently marketed antibodies including their target and use is shown in **Table 21–1**. This diversity of applications ensures their continued development as therapeutic agents well into the immediate future.

## Vaccines[12,13]

Outside of public health measures, there is little doubt that vaccines have had the greatest positive effect on human health. Although vaccines have now been used for several hundred years, there has been a resurgence of interest in their use in the last decade due to the new technologies available for their creation and an improvement in their financial viability.[14] Vaccines function by exposing our immune systems to attenuated pathogens, pieces of pathogens, or other agents (all referred to as "antigens"). Under the right conditions (still incompletely understood), this can produce a "memory" response which results in a very robust immune response that can protect the immunized individual against later exposure to actual disease causing pathogens. There are three divisions of the human immune response generally recognized[15,16] (**Fig. 21–3**). The first is the innate response which primarily involves the recognition of repetitive structures on the surface of pathogens by receptors on immune cells known as "toll" receptors. This leads to a complex series of cellular responses including activation of the adaptive immune response involving the production of antibodies (the humoral response) and the cellular response

### CURRENTLY MARKETED ANTIBODIES TARGET AND USE

| Name | Target | Typical Uses |
| --- | --- | --- |
| Avastin | VEGF | Cancer (multiple) |
| Bexxar | CD20 | Non–Hodgkin lymphoma |
| Campath | CD52 | Leukemia |
| Erbitux | EGFR | Colorectal cancer |
| Hercetin | HER-2 | Breast cancer |
| Humira | TNF-alpha | Arthritis |
| Mylotarg | CD33 | Leukemia |
| Orthoclone OKT3 | CD3 | Prevent transplant rejection |
| Raptiva | CD11alpha | Psoriasis |
| Remicade | TNF-alpha | Immune inflammatory disorders |
| Reopro | GPIIb IIIa | Inhibits platelet aggregation (thrombux formation) |
| Rituxan | CD20 | Non–Hodgkin lymphoma/arthritis |
| Simulect | CD25 | Inhibits allograft rejection |
| Synagis | RSV | Treats RSV infections |
| Tysabri | VLA4 | MS/Crohn disease |
| Xolair | Ig E | Allergic asthma |
| Zenapax | IL-2 | Inhibits allograft rejection |
| Zevalin | CD20 | Non–Hodgkin lymphoma |

which among other activities produces cells which can kill infected host cells. When testing vaccines, the production of antibodies is the most common event measured although it is becoming common to quantitate the production of cytokines as an indicator of cellular responses.

A wide variety of different antigens have been employed as vaccines.[12] Among the most effective are those that employ attenuated viruses. Vaccines such as measles, mumps, rubella, varicella, rotavirus, and one form of the polio vaccine are all examples of such vaccines and have had a dramatic impact on human health. Effective vaccines can also be created by inactivating viruses and bacteria through chemical or radiation methods, with the hepatitis A, rabies, and some forms of the polio and influenza vaccines representative examples. From early in the history of vaccinology there has been the hope that individual components of organisms could also be used as the active components of vaccines. Today, proteins, carbohydrates, and nucleic acids are all important antigens with the former two currently employed in marketed vaccines. Originally, proteins for vaccine use were purified from actual pathogens. These include inactivated proteins such as typhoid and cholera toxins, partially purified influenza proteins (primarily the hemagglutinin and neuraminadase), as well as viruslike particles from the serum of hepatitis B infected individuals. More recently, recombinant methods (see below) have been employed, although there has been only limited success in these efforts. Nevertheless, the hepatitis B surface antigen (HBV), cholera toxin B, and surface proteins from the human papillomavirus (HPV) are all successfully used in highly effective vaccines. The OspA protein from the organism causing Lyme disease was also employed but the vaccine was eventually removed from the market. It should be noted that two of these recombinant vaccines (HBV and HPV) are in the form of viruslike particles which significantly enhance their immunogenicity. Several vaccines that employ carbohydrates as antigens are also available. In the case of adults, a vaccine against pneumonia has been developed which contains 23 different purified polysaccharide chains. Polysaccharides are only weakly immunogenic in children, however, and must be conjugated to protein-based carriers for them to be effective in infants. These carriers include diphtheria and tetanus toxoids as well

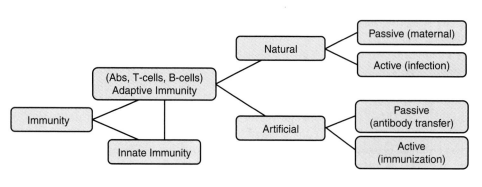

**Fig. 21–3.** A basic diagram of the mammalian immune system (see text).

as an outer membrane protein complex from *Neisseria meningitides*. Important childhood vaccines based on polysaccharide/protein complexes include those for *Haemophilus influenzae*, meningoccal disease, and pneumococcus. Finally, it has recently become apparent that it is possible to make vaccines using DNA plasmids in which antigenic proteins are encoded within appropriate expression sequences. Through mechanisms that are still not entirely understood, expression of these antigenic proteins by cells of the immune system or their secretion by other cells can lead to significant cellular and humoral immune responses. Although successful human vaccines have yet to be developed on the basis of this technology, two veterinary vaccines are on the market. In another emerging technology, the integration of specific protein sequences into virus delivery vehicles such as adenoviruses is also being explored. Perhaps the most important points to note here are that (a) there are a wide variety of different approaches to vaccines and (b) vaccines themselves are usually quite complex presenting significant challenges for their formulation and delivery.

## Nucleic Acids[17–20]

As indicated in the previous section by the example of DNA vaccines, nucleic acids in both their DNA and RNA forms can also be used as pharmaceuticals. Historically, an approach known as antisense RNA was used to either alter gene expression or interfere with the translation of RNA into protein. The idea is simple in principle but difficult in execution. A piece of single stranded RNA that has a sequence complementary to a gene or RNA of interest (the "sense" target) is introduced into appropriate cells. The highly specific binding of the antisense drug to a specific mRNA or gene may result in its destruction through an enzymatic process or blocking of transcription, respectively. Various chemical analogues of RNA are often used for this purpose to improve their stability. The delivery of antisense RNA pharmaceuticals into cells, however, has presented a formidable challenge which remains unsolved despite a wide range of attempts to overcome the barriers that exist to this problem. So far, only a single antisense drug has reached market (for treatment of cytomegalovirus retinitis). Currently, the enthusiasm for this approach is rather low with approaches based on RNA interference (RNAi) (see below) more actively pursued.

A second approach relies on the introduction of a gene (DNA) coding for a protein of potential therapeutic activity into cells (i.e., gene therapy). This gene (or genes) is usually introduced either incorporated into the genome of a virus[21] or as a part of a bacterial plasmid, the latter often complexed to a polymeric, (usually) cationic carrier to facilitate entry into cells.[22,23] Initially, RNA retroviruses with the ability to integrate their genes into host chromosomes were the most actively employed viral vectors. Problems of toxicity, however, have reduced their use. More recently, the DNA viruses, adenovirus and adeno-associated virus have been used. For safety reasons, there is also significant interest in using bacterial plasmids to delivery genes for therapeutic applications. Plasmids are circular, double-stranded, facilitated self-replicating pieces of DNA that can contain multiple genes as well as auxiliary sequences that can aid in replication and gene expression. They are easily produced in high numbers in host bacteria or other cell types making their manipulation and manufacture at high concentration easier than viruses. In most cases, plasmids are complexed to cationic polymers such as positively changed lipids or polyethyleneimines to facilitate cellular entry and to enhance their stabilization. The use of plasmids provides an especially flexible approach to gene therapy given their large genomic capacity and ease of synthesis, but so far no human therapeutics have been directly derived from this technology. Again, problems with their delivery, potency, and safety have inhibited their development.

The most recent approach to the use of nucleic acids as therapeutics has come from the discovery that many genes are naturally regulated by the presence of small RNA molecules.[24,25] This is a rapidly evolving field with new discoveries being made almost weekly. From a pharmaceutical perspective, this new technology clearly holds significant promise. Initially, efforts have been focused on the use of 21 to 23 nucleotide double-stranded RNA molecules which are complementary to target mRNA species. In a manner similar to but distinctly different from antisense activity, the complexes formed are subject to destruction by a naturally occurring catalytic activity. The biochemistry and cell biology of RNA interference is both complex and fascinating.[19] We leave it to the interested student to pursue this topic further on their own.[20] Promising effects have been seen in animal disease models and human clinical trials are underway.[24,25]

The potential use of "aptamers" as therapeutic agents will be briefly mentioned. These are small DNA or RNA molecules (typically 15–60 nucleotides in length) that have been specifically selected for their specificity and affinity for proteins or other biological targets.[26–28] In fact, a drug for age-related macular degeneration is already available based on the use of aptamers. A number of clinical trails are also underway employing these unique molecules. Peptide-based aptamers are also under investigation. These generally consist of a variable peptide loop that is incorporated into a protein matrix.

## Discovery of Biotech Drugs

Small molecule drugs are typically discovered by screening libraries of natural or synthetic compounds (combined with rational structure optimization) against protein-based targets. Targets are selected on the basis of our current fairly extensive understanding of metabolic and hormonal pathways and extrapolated therapeutic effects. It is this same modern understanding of cell and molecular biology[29,30] accompanied by corresponding pathologies that is used to design biotechnology-based drugs. One difference, however, is that in many cases components of these pathways or their analogues are used as actual drug substances. Thus, protein drugs such as insulin, human growth hormone, and EPO are simply used to supplement their naturally occurring counterparts. With the availability of the sequence of the human genome of approximately 20,000 to 30,000 genes (the exact number remains quite controversial), in principle at least, all such genes are now available for therapeutic use. In addition, the many variants of each gene (sometimes in the hundreds and known as single-nucleotide polymorphisms or SNPs) are being increasingly defined and offer the potential for more specific therapeutic use. In addition, the use of antisense RNA and more recently RNA interference offers the possibility of gaining fairly detailed functional information. This can often be used to create cellular models that provide useful analogues of specific biochemical pathways or even disease states that can be employed to further develop protein and nucleic acid therapeutics. In ideal cases, animal models either natural or transgenic in nature in which specific genetic changes have been made to simulate human diseases can be used. This subject is a vast one and rapidly changing, but it seems certain that the information necessary to develop recombinant pharmaceutics will become increasingly available.

## Cloning[31,32]

Approximately 30 years ago, Paul Berg, Herbert Boyer, and their colleagues realized that it should be possible to manipulate DNA in such a way that specific genes could be inserted in cellular systems and their expression induced. This has led to a now routine series of procedures to accomplish this task. Although the details of these methods can be quite complex and an art form in themselves, the basic idea is straightforward.[31,32] First, a specific gene (e.g., one that is selected as a potential protein pharmaceutical) must be isolated. This is typically done by screening a large library of genes that have been inserted into circular pieces of bacterial DNA (plasmids). This gene is then inserted into a plasmid which has been specifically designed for expression of the gene into protein (an "expression vector"). To accomplish this, the gene is removed from the library plasmid by cutting with a highly specific protein known as a "restriction enzyme." The expression plasmid is then opened with the same enzyme and combined with the desired gene. The gene hybridizes to the sticky ends of the plasmid in a highly specific manner. The host plasmid is then covalently closed with another enzyme known as a ligase. The "recombinant" plasmid can then be inserted into a host cell for reproduction (we say the cell is transformed, a process not to be confused with the transformation of cancer cells).

A wide variety of different cells types are available for expression purposes. These include bacteria such as *Escherichia coli*, yeast cells, baculovirus, animal cells in culture, plants such as corn and tobacco, and transgenic animals like goats, sheep, and cattle. It is also possible to express proteins in cell-free systems using extracts of mammalian cell cytoplasm. There are advantages and disadvantages to

each expression system. For example, low cost and high levels of expression often dictate the use of bacterial and yeast cells. If posttranslation modifications such as glycosylation are necessary for the proper functioning of the protein, then eukaryotic, yeast, or baculovirus systems are typically used although each may produce a uniquely modified protein. If larger proteins are being expressed, then eukaryotic cells are usually employed.

The vectors used to transform the expression system of choice must meet a number of requirements. They must contain one or more appropriate promoter sequences (binding sites for RNA polymerase), an origin of replication for DNA polymerase, a ribosomal binding site, appropriate restriction sites, and termination sequences. In addition, selectable genes that when expressed allow cellular survival when stressed and affinity tags to aid in the isolation of the expressed protein are often included (see **Fig. 21–4**).

Common promoters include those from phages, viruses, and inducible systems such as the arabinose system. A wide variety of tags are available for purification purposes. The most common is the His6 sequence, which can be used for Nickel affinity chromatography, and the FLAG epitope sequence of DYKDDDDK. Others include the cellulose-binding domain and glutathione-S-transferase tags as well as the c-myc epitope (EQKLISEEDL), poly E 35 and poly K35, and CBP-calmodulin-binding domain tags.

The next step is to get the DNA into the host cell (transformation). In the case of bacteria, this is usually accomplished by making the plasmid particulate through the addition of a positively charged ion such as calcium, followed by rapidly lowering the temperature to damage the bacterial membrane, facilitating plasmid entry. A more modern and efficient method is through the use of a process known as electroporation. This uses a pulse of electromagnetic energy to open up pores in the bacterial membrane. Although this procedure is

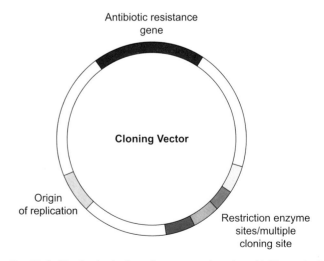

**Fig. 21–4.** The basic design of an expression plasmid. The actual plasmids used in the productions of biopharmaceuticals are usually much more complex (see references 31 and 32).

still poorly understood, it works for many types of cells and is even being considered for human use and gene therapy. Genes can also be introduced into cells by using viruses as cloning vectors (which have natural receptors) or by complexing the recombinant plasmid to a positively charged agent such as a cationic lipid or other positively charged polymer. To some extent, the process is an empirical one with the extent and stability of expression the criteria for success.

Transformation of yeast and mammalian cell lines ("transfection") have some similarities to bacterial systems but some key differences as well. The student is referred to the more technical literature to explore these differences, which include selection procedures, plasmid design, and cellular growth requirements among others.[3-8,31,32]

Once cellular transformation has been performed, it is necessary to grow these cells to a high mass to produce sufficient material for isolation of the target product (e.g., protein, virus, DNA plasmid). This is accomplished by a process known as fermentation. This procedure varies depending on the cell type, but in general, variables such as $CO_2$ requirement, $O_2$ levels, and necessary nutrients must be individually optimized. Typically, microbial cells grow much more rapidly than animal cells and grow to higher cell densities. Growth is usually performed in large fermentors or bioreactors, which permit the growth process to be continuously monitored and the amounts of $O_2$, $CO_2$, and nutrients to be maintained at appropriate levels. The use of incubators and fermentors is a highly specialized activity and requires extensive knowledge and training for maximum effectiveness.

## Purification of Macromolecular Therapeutics[33,34]

After sufficient growth of cells and subsequent expression of the target protein or other macromolecular drug agent, the product can be found either inside the cells, secreted into the surrounding medium, or perhaps associated with the cells' surface. If the protein (or DNA, or virus) is secreted into the growth medium, its purification can usually begin immediately. If it is found inside the cell, however, its isolation is usually much more difficult. The cells must initially be broken open. This is usually done by using a French pressure cell (shear), sonication (sound), or disruption with glass beads (mechanical stress). All three methods are relatively rough, so gentler methods are also often used. These include freeze-thaw stress, lysis with detergents (dissolution of cell membranes), and enzymatic or osmotic lysis. Once the macromolecular drug substance is released from its association with the transformed cells, it must then be isolated from the contaminating proteins, lipids, carbohydrates, and nucleic acids. This is usually a relatively complex multistep process that is primarily empirically based although the physical properties of the protein/nucleic acid/virus can often provide important clues to the most effective steps.

Initial steps in the purification of macromolecular substances typically involve crude separations based on the gross solution behavior of the material. For example, differential precipitation by salts, organic solvents or solutes, pH, or temperature is frequently used for this purpose. In some cases, this may be followed by filtration steps using filters containing micron or submicron-sized pores that selectively pass or retain the macromolecular drug.

The primary class of higher resolution approaches used today involves a variety of types of chromatography.[35,36] Most generally, these procedures rely on passing a mixture of molecules in a mobile (liquid) phase, through a stationary (solid) phase, resulting in a partial (usually) or complete separation of the components. This occurs because of an interaction or lack thereof between the molecules to be separated and the solid phase. There are currently many different types of chromatography, some used analytically (to be discussed below) and some to separate molecules at preparative levels. The four most widely used approaches for the latter purpose are described below (**Fig. 21–5**).

If it can be successfully performed, some version of "affinity" chromatography is usually an optimal choice. This is

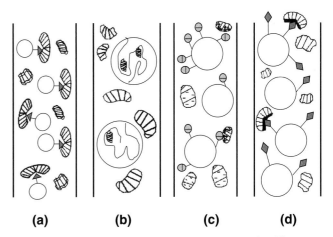

**(a)**     **(b)**     **(c)**     **(d)**

Fig. 21–5. The basic forms of liquid chromatography. The most common forms of chromatography used to purify and characterize biomolecules and their complexes are based on interactions between the biopharmaceuticals and a solid support (indicated by large open circles) which are usually derivatized with a molecular moiety to induce interactions with target molecules. (a) In affinity chromatography the immobilized ligand (▲) binds to a specific site on the macromolecule (all shown with lined interiors) and is usually eluted with a related soluble ligand or a change in pH. (b) In size exclusion chromatography, no ligand is used but smaller molecules are able to diffuse into particle support interiors slowing their progress through the column resulting in separations based on size. (c) In ion-exchange chromatography, macromolecules and their complexes are separated on the basis of charge/charge interactions with the support with elution induced by either changing the pH or increasing salt concentrations. (d) Hydrophobic interaction chromatography separates target macromolecules based on the interactions between apolar (hydrophobic) ligands and apolar sites on proteins. Elution is usually performed by lowering salt concentrations (which weakens apolar interactions). The same principle is used in reversed-phase HPLC but elution is typically induced by increasing concentration of an apolar solvent.

because of the degree of purification (often greater than a thousandfold) frequently obtained by this method. The basic principle behind affinity chromatography is that highly specific interactions often exist between proteins and protein complexes like viruses and other molecules (referred to as "ligands"). If a ligand can be fixed (usually covalently) to an insoluble matrix (often in the form of small beads), then this can be used as a basis for separation since most or all of the other components in the mixture will not undergo this specific interaction. It is, of course, necessary to remove the molecule of interest from the support, but this can often be achieved by some type of competitive interaction with another substance or alteration of the solution conditions by a variable such as (low) pH or (high) ionic strength. The trick is not to alter the desired macromolecule by the elution process. Ligands are typically attached covalently to a resin such as sepharose while retaining an affinity in the range of $K_D$ $10^{-4}$ – $10^{-8}$ M. A wide variety of ligands have been used for protein affinity chromatography. For example, proteins with polyhistidine tags will bind to columns containing immobilized metal ions such as $Zn^{+2}$ and $Cu^{+2}$. They can generally be eluted with either imidazole which competes for the binding or low pH (e.g., ~4.5). The problem with this approach is that the protein drug now contains the His-tag, often an undesirable modification of its native structure (His-tags can be "sticky" because of their positive charge at pH <5). This problem can be overcome by using a cleavable linker between the proteins and the tag, but in general, His-tag affinity chromatography is primarily used in the discovery and basic research stages of pharmaceutical development. In contrast, a method that is commonly used to purify large amounts of immunoglobins uses protein A and G. These proteins bind to Igs with high affinity. They have been attached to various resins and are routinely used in immunoglobulin purification with elution performed by lowering the pH. Another commonly used ligand is heparin or other highly polyanionic polymer. Many proteins contain high affinity (although not necessarily highly specific) polycationic sites that can interact electrostatically with polyanions.[37,38] For example, many growth factors, cytokines and blood proteins, contain such sites. High ionic strength is often sufficient to elute such proteins from heparin columns. Other ligands that have been used to isolate proteins of pharmaceutical interest include calmodulin (kinases, phosphatases, cyclases), adenosine monophosphate (AMP) (adenosine 5'-triphosphate (ATP)-dependent proteins, nicotinamide-adenine dinucleotide (NAD) cofactor utilizing proteins), and cibacron Blue dye (proteins with large hydrophobic binding sites such as the serum albumins).

A chromatographic method which separates proteins on an entirely different physical basis is gel filtration (also commonly known as size exclusion chromatography or SEC). In this case, proteins, nucleic acids, and even viruses can be separated on the basis of size. The method employs small beads of various porosities. Smaller proteins can enter into the highly channeled interior of these matrices and consequently their progression through a column of such material is slowed.

In contrast, larger proteins can less efficiently enter into the beads or pass between them entirely. The overall result is a time-dependent separation of mixtures of proteins, nucleic acids, or even viruses from one another based on their hydrodynamic size. As discussed below, this method can also be used as an analytical tool to estimate the size and molecular weight of macromolecules and to detect aggregated material.

A very commonly used method to separate macromolecules and especially proteins is ion-exchange chromatography. This technique separates molecules on the basis of their charge rather than their size or specific affinity for ligands. To perform this method, either positively or negatively charged groups are covalently attached to a polymer, which permits the free flow of macromolecules. The most commonly employed groups are the negatively charged carboxymethyl or sulphopropyl moieties and the positively charged quaternary aminoethyl and diethylaminoethyl side chains. The negatively and positively charged groups are referred to as cation and anion exchangers, respectively. Once adsorbed to the charged surface, the macromolecule is eluted by either increasing the ionic strength with salt (in either a step or gradient fashion) or altering the pH to minimize the electrostatic attraction. Ion-exchange resins are classified as either weak or strong. Weak exchangers are typically effective over a more limited pH range than the strong exchangers. The selection of an appropriate ion exchange resin is determined by the macromolecules isoelectric point (the pH at which it has zero net charge), size, pH stability, and the scale at which the separation is conducted.

The fourth method is known as hydrophobic interaction chromatography. In this case, the group which is derivatized to the support resin is hydrophobic (more correctly "apolar") rather than charged or a specific ligand. Apolar sites on molecular surfaces can bind to these apolar groups. To increase the strength of this interaction, the molecules are usually loaded at high salt concentrations. In contrast to electrostatic interactions, which are weakened at higher salt concentration, hydrophobic interactions are increased under these conditions. Thus, macromolecules are eluted by lowering the salt concentration. This technique bears a resemblance to high-performance liquid chromatography (HPLC), which will be discussed below as an analytical technique. Both employ resin-attached apolar side chains of length C1-C18. HPLC, however, usually elutes protein with an organic solvent like acetylnitrile or propanol. Unfortunately, this usually alters macromolecular 3-D structure (although often, reversibly), often negating its use as an isolation technique. The exception is peptides, which may not require a defined 3-D structure for their biological activity.

## CHARACTERIZATION[39,40]

A key to the successful development of a macromolecular pharmaceutical is to obtain a detailed picture of its structure and the response of this structure to environmental stresses such as temperature, pH, ionic strength, agitation, and freeze/thaw exposure. Although a detailed structural picture of a molecule like that obtained by x-ray crystallography or NMR spectroscopy can be very helpful, this is not absolutely necessary for the successful formulation and development of a biopharmaceutical. The usual approach is to use a variety of different lower resolution experimental approaches to obtain pictures of the molecule from a wide range of perspectives. Some of the most common methods are discussed below.

## Biology-Based Assays (Bioassays)[41]

Although often lacking a high degree of precision and accuracy, assays based on the response of animals or cells to biotherapeutics are generally considered to be the ultimate arbiter of the retention of pharmaceutical activity. Well-known examples of the use of animals for this purpose include the lowering of blood glucose in test animals by insulin, an increase in weight upon the injection of growth hormone, and the production of specific antibodies when animals are exposed to vaccines. In some cases, transgenic animals have been developed as disease models.

Thus, diabetic mice, mice with Alzheimer-like disease, and many animals with genetic defects similar to those found in humans have all been created and provide a basis with which to check the effectiveness of a particular biotherapeutic. The advantages of such approaches are obvious. They permit a direct evaluation of the critical properties of the pharmaceutical and may be sensitive to changes in its structure. It is possible, however, for structural changes to occur that are not detected by such methods because either the structural alteration does not affect activity or the usually fairly wide experimental variability in such measurements does not permit detection of relevant structural changes, even if the biological activity is perturbed. For these reasons (among others), additional assays based upon a variety of different physical and chemical properties of the target biopharmaceutical are usually also employed.

Assays based on the response of cells to the presence of drugs are being increasingly used to check for structural integrity and biological activity. Many if not most biopharmaceuticals act on one or more cell types by either binding to receptors on their surface or entering cells and producing a consequent molecular response. These cellular responses may involve changes in the level of important cellular messenger molecules such as cyclic AMP, ATP, and inositol phosphates or alterations in cellular properties like membrane potential, rate of cell division, or even cell death (toxicity). This is a large subject that we cannot explore further here, but an increasing number of experimental approaches for such measurements are becoming available facilitating this approach. A question that has yet to be unambiguously answered is to what extent cellular-based assays can replace animal studies. The speed, simplicity, and precision of the former, however, point to their increased usage.

## Immunoassays[42,43]

Immunoassays are methods that employ antibodies to detect the amount of an antigen (e.g., protein, DNA, polysaccharides). These are solution or solid state assays that can employ either monoclonal antibodies obtained by a variety of methods or antisera obtained from the blood of animals injected with the macromolecule or macromolecular complex of interest. Monoclonals have the advantage that they are highly specific for a single site but antisera can be more easily obtained and their multispecificity can be advantageous under certain circumstances. In the case of vaccines, immunoassays can be used to measure antibodies produced in response to the vaccine. Alternatively, the presence and amount of a biopharmaceutical can be determined. In both cases, either the antibody or the antigen must be labeled in a way that the amount of that component can be easily quantitated. Common forms of labeling include attaching a fluorescent or colored group, a radioisotope (a radioimmunoassay or RIA), an enzyme (an enzyme-liked immunoassay or ELISA), or a magnetic particle (a magnetic immunoassay or MIA). A number of other approaches are also frequently used. For example, addition of antibody to antigen often produces insoluble complexes due to the multivalent nature of the antibody. This can be measured by light scattering or turbidity (see below). The latter is often referred to as nephelometry. The antibody or antigen can also be attached to cells (usually red blood cells) and the aggregation of the cells measured by a method known as agglutination.

Many immunoassays are conducted in a competitive manner. Either a labeled antibody or antigen is used and the unlabeled molecule or virus of interest is used to compete or displace the labeled component in the antigen–antibody interaction (**Fig. 21–6**).

## Electrophoresis[44,45]

The fundamental theory of electrophoresis is described earlier in the book and should be reviewed prior to reading this section. Here the focus is on the versions of electrophoresis that are primarily used for biomolecules, gel, and capillary electrophoresis. Gel electrophoresis is today usually performed in thin gels of cross-linked polymers. When the method is used for proteins, the gel usually consists of cross-linked polyacrylamide, often in gradient form. When nucleic acids such as DNA are being electrophoresed, this is usually performed in agarose. In both cases, a semisolid medium of porous properties is used and an electromagnetic field is employed to move the macromolecule through the matrix. The relative "mobility" of the macromolecule is determined by a number of factors including its size and charge as well as in some cases, its 3-D structure. The most common form of protein electrophoresis is known as sodium dodecyl sulfate (SDS) polyacrylamide gel electrophoresis (SDS-PAGE). In this case, the presence of SDS, a negatively charged detergent, causes proteins to unfold into rod-shaped structures.

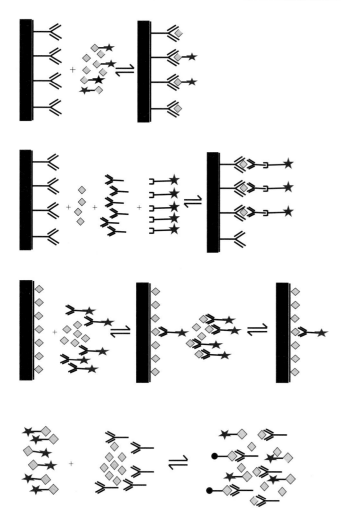

**Fig. 21–6.** Types of immunoassays, the symbols represent antibodies (➤–), antigens (◆) labels (★, ●), and a solid support such as the well of a microtiter plate (—). (*a*) A heterogeneous, simple competitive assay in which labeled and unlabeled antigen compete with one another. (*b*) A sandwich assay in which an excess of a labeled reagent antibody competes for a site on the antigen which is captured on the support by another antibody which binds to a different site. (*c*) In the third method, detection is indirect through the use of the labeled antibody. (*d*) This illustrates a homogeneous competitive method in which the properties of the label itself (i.e., the signal) changes. (Adapted from G. Kersten and J. Westdijk, Immunoassays, in W. Jiskoot and D. J. A. Crommelin (Eds.), *Methods for Structural Analysis of Protein Pharmaceuticals*, AAPS Press, Arlington, 2005.)

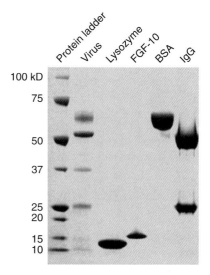

**Fig. 21–7.** SDS-PAGE of representative biopharmaceuticals under reducing conditions employing a gradient gel. As expected, the virus produces multiple protein components and the IgG, two bands corresponding to the heavy and light chains. Molecular weights can be estimated by comparison to the known standards present in the left-most lane.

Because the SDS binds relatively uniformly to the protein, the rodlike complexes migrate in proportion to their molecular weight (**Fig. 21–7**). If protein molecular weight standards are included in lanes of the gel not containing the protein of interest, a fairly accurate estimate of the latter's molecular weight can be made from a plot of the logarithm of the molecular weight of the standards versus their position on the gel and comparison of the unknown's migration behavior. If a reducing agent such as $\beta$-mercaptoethanol or dithiothreitol is present, any disulfide bonds present will be broken and individual bands for any subunit present will be seen. For example, when immunoglobulins are subjected to SDS-PAGE in the presence of a reducing agent, bands for both the heavy and light chains can be clearly distinguished and their molecular weights estimated (**Fig. 21–7**). If no SDS is used, then it is much more difficult to estimate molecular weights but relative size information about proteins can still be obtained, including separation of proteins in mixtures. Because proteins do not usually absorb visible light (with the exception of those containing chromophores such as the heme groups of hemoglobin and the cytochromes), the gels are usually stained with a visible dye. Alternatively, gels can be stained with silver nitrate, a method approximately $50\times$ more sensitive than the most commonly used dye, coomassie blue. The amount of protein present in each band is roughly proportional to the amount of associated dye. This can be quantitated by scanning the gel with a laser or by using some other type of imaging technique. There are a large variety of other types of electrophoresis that can also be used with proteins. For example, the separated protein can be blotted onto another piece of material and this stained with labeled antibodies to identify specific proteins of interest (a "western" blot). This permits a highly reliable identification of proteins to be made. Gradients of a protein unfolding agent like urea or a gradient of temperature can also be imposed. Such techniques permit protein unfolding and subunit dissociation to be studied.

Because nucleic acids carry a strong negative charge due to their phosphate groups, they can also be electrophoresed. In this case, however, a gel based on agarose is usually employed. The migration of DNA and RNA is dependent on their structure. Small pieces of double-stranded DNA or RNA generally migrate as a function of their radius of gyration, but larger species such as supercoiled DNA or single-stranded

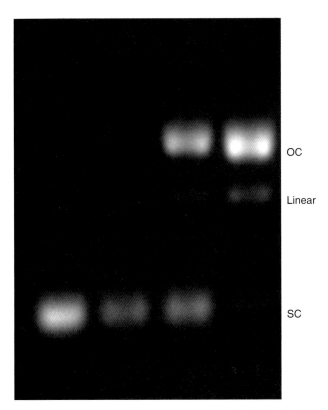

OC

Linear

SC

**Fig. 21–8.** An agarose gel of a supercoiled plasmid (sc) (left) which as been subjected to long-term storage with consequent development of open circle (oc) and linear forms (right).

nucleic acids which can internally base pair and therefore form tertiary structure can behave in quite complex ways (**Fig. 21–8**). Nucleic acid agarose gels are usually stained with a dye such as ethidium bromide which intercalates between the bases, causing it to become highly fluorescent. The high separatory capacity of agarose gel electrophoresis (polyacrylamide can also be used) makes this an especially important analytical tool.

A third variant of electrophoresis of great importance is isoelectric focusing. Here, compounds known an ampholytes (mixtures of charged peptides) are included during electrophoresis. The externally applied voltage causes these charged agents to form a pH gradient within the gel. This causes the target protein to migrate to the point in the ampholyte gradient when its charge is neutralized, that is, its isoelectric point. Thus, this method allows one to measure the pH at which a protein is charge neutral. Proteins differing by a single charge (such as seen during a deamidation event, see below) can be distinguished. High concentrations of urea are often included during gel isoelectric focusing to reduce extraneous protein/protein interactions.

Electrophoresis can also be conducted in small capillaries containing an electrolyte (capillary zone electrophoresis or CZE). Like gel electrophoresis, species are generally separated on the basis of their size to charge ratio. In this type of analysis, the motion of the buffer solution itself under the influence of the electromagnetic field (electroosmotic flow)

often exceeds that of the electrophoretic migration of the sample and this contributes significantly to the separation. A variety of detection methods are available and are usually instituted through a clear area of the capillary. The most commonly used methods are UV/VIS absorption and fluorescence (either through the intrinsic fluorescence of the macromolecular such as protein tryptophan fluorescence (see below) or an extrinsic fluorescent group introduced through chemical modification). The output of the capillary can also be fed to a mass spectrometer for molecular weight analysis. Isoelectric focusing can also be performed in a capillary format. Although more commonly used for proteins, capillary electrophoresis can also be used for nucleic acids. In a combination of the gel and capillary approaches, a separatory gel matrix can be included within the capillary. Electrophoresis has also been used to measure the zeta potential of macromolecules. This is discussed in Chapters 15 and 16.

## High-Performance Liquid Chromatography[46]

In addition to electrophoresis, HPLC in its myriad forms is the most common analytical technique used to characterize macromolecules. The various types of separations available for HPLC are similar to those described above for the different types of preparative chromatography. The major differences are the use of small diameter columns packed with smaller particles that permit the use of high pressure (more than 1000 atmospheres) and consequent rapid flow rates and separation times (typically minutes). A wide array of detection methods is available including UV/VIS absorption, fluorescence, vibrational spectroscopies, electrochemical methods, and mass spectrometry.

Probably the most frequently used form of HPLC is the reversed-phase mode (RP-HPLC). In this method, the packing material is derivatized with alkyl or other apolar substituents. The molecules to be separated interact with the apolar moieties through hydrophobic sites on their surfaces. They are then eluted with an apolar solvent like acetylnitrile or propanol, often in a step or gradient mode. Under optimized conditions, the time of elution (the "retention time") may be taken as highly characteristic of the analyte and can be used to establish its identity and structural integrity. A common use of RP-HPLC with proteins is the construction of "peptide" maps (**Fig. 21–9**). In this important procedure, a protein is digested into peptide fragments of defined size by one or more proteases with their molecular weight determined by online mass spectrometry. This permits the chemical (primary) structure of a protein to be rigorously established and is routinely used to confirm a protein's identity and the presence of chemically altered residues. In general, peptides tend to elute on the basis of their relative polarity. The elution behavior of proteins themselves is much more complex due to their secondary, tertiary, and quaternary structure, but it is assumed that the presence of apolar binding sites plays a key role in their elution behavior as well as conformational changes produced by the eluting solvent.[47]

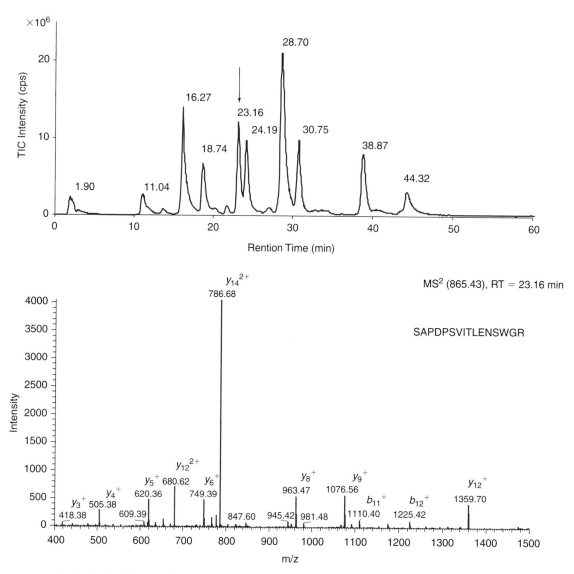

**Fig. 21–9.** Peptide map of a mutated Ricin-A Chain after digestion by trypsin and analyzed with LC-MS/MS. The peptides were separated on a reverse-phase column $C_{18}$ with gradient elution from 0% to 80% acetonitrile in water (containing 0.06% formic acid) at 10 $\mu$L/min, and the separated peptides were detected by mass spectrometry (upper figure). The lower figure shows a product ion spectrum which can be used to identify the sequence of the peptide shown in the upper right corner. The fragments with chargers on the N-Terminal side are called "b" ions and the fragments with chargers on the C-terminal end are called "y" ions.

There are many additional forms of HPLC that are used as analytical tools in macromolecular analysis. Especially important are size exclusion (SEC), ion exchange and bioaffinity chromatography. SEC is of exceptional importance since it is used to detect changes in size. Thus, degradation that results in significant decreases in size or increases in mass such as that produced by limited association or more extensive aggregation is often first detected by this sensitive technique. One problem with its use is that dilution occurs during separation and thus the concentration at which separation occurs is not that of the initial sample. Furthermore, interactions of proteins with a column's matrix can signifi-

cantly alter elution behavior. This makes estimates of size sometimes difficult. Inclusion of agents such as urea and high salt concentration can be used to minimize this problem. Nevertheless, this method provides the most frequently used criteria for detection of the crucial phenomenon of protein aggregation. The method can also be used for nucleic acids and even viral particles if highly porous resins are used. With the advent of high pressures and appropriate packing materials, it is now possible to obtain analytical separations in a few minutes (and even seconds in certain circumstances) making HPLC-based methods of both widespread use and significant analytical importance.

## Ultraviolet/Visible Spectroscopy

A variety of spectroscopic methods are widely used to characterize biomolecules. An introduction to the general topic of spectroscopy is presented in Chapter 4 and it is highly recommended that this section be reviewed before perusing this section. This chapter is restricted to aspects of these techniques that are particularly relevant to biomolecules in a pharmaceutical context.

Absorption spectroscopy in the ultraviolet and visible regions is a very versatile technique widely used with both proteins and nucleic acids.[48,49] The technique has shown a recent resurgence due to the availability of diode array detection. Conventionally, absorption spectra were produced by scanning a moveable monochromator (a light dispensing element) through a sample contained in a cuvette (a visible and/or UV transparent rectangular sample holder of path length 0.01–10 cm, most commonly 0.1–1 cm) with detection by a photomultiplier tube. In contrast, in a diode array instrument, all wavelengths of light are put through the sample simultaneously and a spectrum created after absorption of the light. The resultant spectrum is then projected onto a diode array for detection purposes. Most importantly, by using mathematical fitting techniques to interpolate between the individual wavelengths detected by the diodes (typical spaced at 0.5–2 nm), highly resolved spectra can be obtained after derivative analysis to produce a resolution on the order of ± 0.01 nm.

In proteins, there are two major intrinsic chromophoric groups to consider. In the far UV region (175–220 nm), there are three electronic transitions observed because of peptide bonds with a broad peak seen at 185 to 195 nm. Although this region contains information about a protein's secondary structure, it has rarely been used for this purpose due to optical interference by most substances. The high wavelength tail is often used, however, to detect proteins during various forms of chromatography.

In contrast, the near UV region is used for a wide variety of purposes. This portion of a protein absorption spectrum (240–310 nm) is dominated by the $\pi \rightarrow \pi^*$ transitions of the three aromatic amino acid side chains. Phenylalanine (Phe) manifests a weak peak with marked vibrational structure between 250 and 270 nm, tyrosine (Tyr) a stronger, pH-dependent multicomponent peak from 250 to 290 nm, and tryptophan, the strongest absorbing side chain (another multicomponent peak) from 250 to 300 nm. In the case of proteins, the broad overlapping nature of these three contributions results in a broad peak centered between 277 and 287 nm (primarily from Trp and Tyr) with weak undulating bumps in the spectrum below 270 nm (due to the vibrational fine structure of the Phe contribution) and a marked shoulder for Trp at approximately 290 nm. Weak contributions from His and disulfide bonds can also occasionally be seen. The second or fourth derivative of a protein's spectrum dramatically brings out the underlying contributions usually in the form of 6 peaks (3 Phe, 1 Tyr, 1 Tyr/Trp, and 1 Trp) (**Fig. 21–10**). As indi-

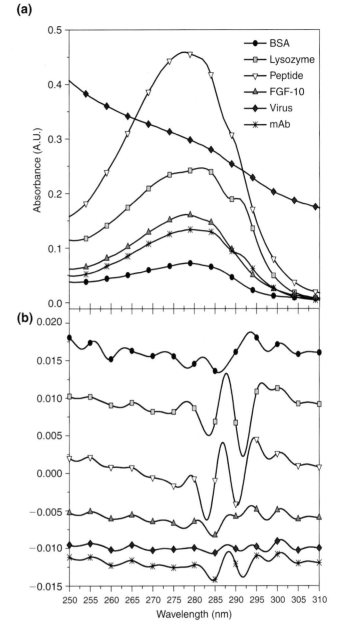

**Fig. 21–10.** Zero-order absorption spectra of four proteins, a peptide and a viral particle (upper panel). Note that all display rather broad featureless spectra. When their second derivative is calculated, however, usually six peaks are seen corresponding to the three classes of aromatic side chains (bottom panel). The second derivative spectra in the lower panel are displaced for clarity. See text for further discussion of the interpretation of such spectra.

cated above, these peaks can often be localized to within 0.01 nm, providing a highly distinctive spectrum for each protein. Some additional information is possible since the Phe residues are usually buried in a protein's interior, most Tyr are interfacial, and the Trp residues dispersed throughout the protein's matrix. The important point here is that the position (wavelength) of each of these peaks is sensitive to the polarity of their environment. The general rule is that as the environment of an aromatic amino acid's side chain becomes less polar (more hydrophobic), its wavelength is shifted to a higher wavelength (note that this effect is opposite to that seen in a protein intrinsic fluorescence experiment as will be discussed below). Thus, as the structure of a protein is altered, to the extent that one or more aromatic side chains experiences a subsequent change in its immediate environment, changes in the position of the derivative absorption peaks provide a measure of conformational change (and potentially physical degradation). The folding and unfolding of a protein as induced by temperature, pH, or a potential unfolding agent like urea, guanidinium hydrochloride, alcohols, chaotropic salts, and detergents can be simply followed by this method if their optical properties do not interfere with the measurements. Data obtained by absorption measurements as a function of multiple variables such as temperature and pH can be summarized and visualized by a method known as the "empirical phase diagram (EPD)." In this approach, the six peak positions are used as components of a vector at any particular set of variables (e.g., T, pH). If colors are then assigned to the major components, a map of temperature versus pH displays different colored regions corresponding to different physical states of the protein (**Fig. 21–11**). For more information about this method, see reference (50). By far the most common use of protein UV spectroscopy, however, is to determine their concentrations using Beer's law (Chapter 4), which states that the concentration is linearly proportional to the absorbance with a constant of proportionality known as the extinction coefficient ($\varepsilon$, $A = \varepsilon cl$). Extinction coefficients, which are characteristic of each molecule, can be determined by dry weight or amino acid analysis or calcu-

lated from the aromatic amino acid content from a number of empirical algorithms.[51,52]

A number of extrinsic chromophores may also be present in proteins. These include metals such as copper or iron as well as lanthanides (with Tb and Co the most commonly employed), which can be used as calcium analogues. Prosthetic groups such as flavin-adenine dinucleotide (FAD), flavin mononucleotide (FMN), NAD, rhodopsin, pyridoxal phosphate, and heme groups among others, all provide strong spectra in the UV/VIS region and are sensitive to their local environment. They can therefore be used as biological sensors of a variety of phenomena such as redox state, $O_2/CO_2$ binding, and light effects.

## Fluorescence[53,54]

Probably the most versatile spectral technique for macromolecular structural analysis is fluorescence spectroscopy. When a chromophore such as the indole side chain of tryptophan in proteins is raised to an excited (singlet) state, rather than return to the ground state through internal conversion processes as in an absorbance measurement, it can do this by the emission of a photon. This constitutes the phenomenon of fluorescence. If this emission is from a triplet state, the process is known as phosphorescence (the combination of the two is called luminescence). For brevity, we will not be concerned with the latter here although the technique of phosphorescence can be quite useful in the analysis of proteins. In the case of fluorescence, the (relatively) long periods of time spent in the excited states ($10^{-3} - 10^{-9}$s versus $<10^{-15}$s for absorbance) allow various types of interactions with this state. This makes fluorescence usually quite sensitive to the fluorophores' immediate environment. The spectrum of this emission is always at longer wavelengths than the absorption band(s) because prior to emission of fluorescence photons, energy is lost as the excited state returns to its lowest vibrational energy level. As a first approximation, the absorption and emission spectra are mirror images of one another. In addition to emission spectra, the lifetime of the excited state ($\tau$; the time it takes fluorescence to fall to $1/e$ of its initial value) can also be measured. The amount of emission can be measured either in terms of its quantum yield (the number of photons emitted divided by the number absorbed) or by simple intensity changes at a fixed wavelength. The latter is approximately proportional to the quantum yield.

Fluorescence is usually measured at right angles to the exciting light, with the emission monochromoter scanned. Alternatively, the excitation monochromoter can be varied and emission monitored at a fixed wavelength to produce a version of the absorption spectrum known as an excitation spectrum. Lifetimes are measured by either exciting with single pulses of light and measuring their emission decay or determining shifts in the phase of the emitted photons after modulation of the exciting light. Both techniques can measure more than one lifetime component (up to three or four) by deconvolution methods or display distributions of

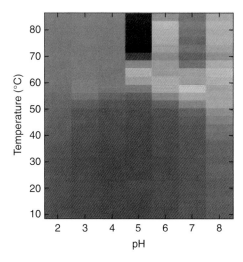

**Fig. 21–11.** An empirical phase diagram (EPD) based on the peak positions of second derivative UV absorption spectra. At each T/pH condition, the protein is represented as a six-dimensional vector which is truncated to the three largest contributions. The regions with different shaded characteristics represent different structural states of the protein. Such diagrams are usually shown in color as described in the text.

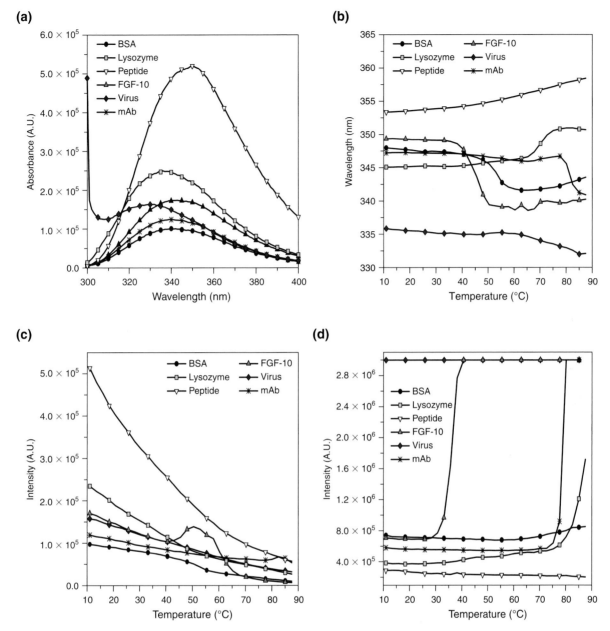

**Fig. 21–12.** Intrinsic fluorescence spectra of biopharmaceuticals. (*a*) The position of the emission maximum (upon excitation at 295 nm) varies from approximately 330 to 350 nm in these examples. Emission at 350 nm indicates that the tryptophan residues are on average highly exposed to the solvent. As the emission maximum decreases, increased burial of the indole side chains in indicated. (*b*) The effect of temperature on the position of the emission maximum and (*c*) its intensity. Note the appearances of transitions in all the macromolecules with the exception of the peptide. (*d*) Light scattering at 295 nm as a function of temperature acquired simultaneously with the intrinsic fluorescence data. The high value seen with the viral particle directly reflects its large size compared to the individual proteins and the peptide.

lifetimes. Most lifetimes of natural fluorphores are on the order of a few nanoseconds (Trp [∼2.6 nsec], Tyr [∼3.6], Phe [∼6.4], NADH [∼0.4], etc.). In general, low quantum yields correlate with short lifetimes.

The intrinsic fluorescence spectra of proteins tend to be dominated by tryptophan emission unless this residue is absent. Under such conditions, the emission of tyrosine and or phenylalanine can often be seen. Of special importance is the observation that this fluorescence is very sensitive to the environment of the endogenous indoles (**Fig. 21–12**). Thus, various kinds of conformational changes that range from the very subtle to complete unfolding can usually be followed by this method. The position of protein Trp containing emission spectra varies from 310 to 320 nm (completely buried indole side chains) to more than 350 nm (totally exposed to the aqueous solvent). Other phenomena such as ligand binding and subunit association and dissociation can also often be detected by changes in intrinsic fluorescence. In a particularly elegant use of this approach, single Trp residues can be placed at many positions throughout the structure of a protein by site-directed mutagenesis (it may be necessary to remove some tryptophans if the target protein contains more than one) and by measuring peak positions to estimate Trp exposure. This information can then be used to generate actual 3-D structures using additional modeling considerations.

Studies of the fluorescence of proteins are not limited to the aromatic amino acids. It is also possible to add an extrinsic probe, which either covalently or noncovalently binds to a particular site on a protein. Common covalent fluorescent labels include molecules like dansyl chloride or fluorescein isothiocyanate. Covalent labels are available which bind to a variety of protein reactive sites such as amino, carboxyl, and sulfhydryl groups. Such probes are typically highly fluorescent and their spectral properties highly sensitive to their environment. Examples of fluorescent probes which are used to bind noncovalently to proteins include 8-anilino naphthalene sulfonic acid (ANS) and its dimeric analogue bis-ANS. Probes such as this are usually assumed to bind to apolar sites on proteins although the negative charge on ANS may also result in electrostatic interactions. One use of ANS is to detect molten-globule (MG) states in proteins. In such states, as the tertiary structure begins to be disrupted, compounds like ANS can interact with the protein and their normally solvent quenched emission can be relieved with enhanced fluorescence expressed as well as blue shifts in wavelength emission maxima. Another use of noncovalent probes involves the detection of protein aggregation. When proteins self-associate and then aggregate, they often form intermolecular $\beta$-structure. Certain dyes (e.g., Congo Red, Thioflavin T) can bind to such structures with a change in their fluorescence or absorption spectra.

Nucleic acids lack significance fluorescence (with the exception of a modified base in tRNA). Noncovalent probes, however, have been extensively used in their characterization and analysis. Some planar dyes can intercalate between nucleic acid bases, whereas others can bind within the grooves of the helix. Upon such interactions, their fluorescence can dramatically increase. This phenomenon has been used to measure DNA and RNA concentration, analyze the binding of other substances to nucleic acids through dye displacement (**Fig. 21–13**), and follow the behavior of DNA microscopically among other applications. Fluorescent dyes are also commonly used to study lipid bilayers and cell membranes. Using probes that enter into the bilayer, the fluidity of membrane interiors and associated phase changes can be monitored by changes in the spectral properties of certain dyes, often using temperature to induce phase transitions.

There are many additional applications of fluorescence to macromolecular systems. Only three are briefly considered here. Quenching of protein Trp fluorescence by extrinsically added solutes is a commonly used approach to analyze the

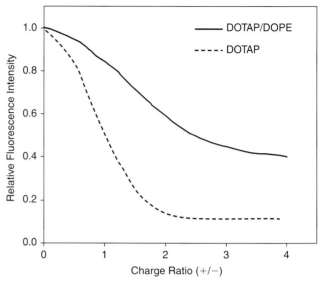

**Fig. 21–13.** An example of the use of a fluorescent dye to measure complex formation between DNA and cationic lipids. The dye ethidium bromide is intercalated between the bases of a DNA plasmid where it becomes highly fluorescent. When cationic lipids are added, the dye is displaced and its fluorescence reduced. If the charge density of the cationic lipid is reduced by the addition of a neutral helper lipid (DOPE, dioeyl phosphatidylcholine), its efficiency in displacing the dye is reduced. Formulations containing such helper lipids are actually more efficient at transfecting target cells than those with cation lipids alone.

accessibility of these residues to the aqueous solvent. Certain low molecular solutes can quench indole fluorescence either by a dynamic collisional process (dynamic quenching) or by forming a complex with the side chain, preventing it from reaching an excited state (static quenching). Some commonly used quenchers are $O_2$, acrylamide, $I^-$, and $Cs^+$. The different charge characteristics of the quenchers permit them to probe different electrostatic environments. The dynamic quenching process can be described by a relationship known as the Stern-Volmer equation:

$$\frac{F_o}{F} = 1 + K[Q]$$

or

$$\frac{F_o}{F} - 1 = k_q \tau [Q]$$

where $F_o$ is the fluorescence (or quantum yield) in the absence of quencher, $F$ the fluorescence in the presence of quencher, $Q$ is the concentration of quencher, and $K$ is a constant known as the Stern-Volmer constant. If the quenching process is entirely dynamic, a plot of $F_o/F$ versus $[Q]$ will yield a straight line of slope $K$ or $k_q\tau$. Curvature in such plots can be introduced by heterogeneity in the environment of multiple Trp residues or the presence of static quenching. If the lifetime $(\tau)$ of the indole is known, the rate constant $(k_q)$ for the collisional process can be evaluated. This can be used to characterize changes in structure or interestingly, the dynamic behavior of proteins in terms of the ability of the protein matrix to permit the transport of the quencher to the indole side chains. We will consider below other ways, including some that involve fluorescence, of studying the intramolecular dynamics of proteins. If no Trp is present in a protein, this method may also be used to study the quenching of Tyr residues. An ultraviolet absorbance technique can be used to acquire related information. In this case, simple cations of various sizes are added in increasing amounts to proteins and the positions of the derivative peaks from Trp, Tyr, and Phe are monitored. In this case, the shifts are due to the formation of cation-$\pi$ interactions of cations such as $K^+$, $Cs^+$, and $Li^+$ with the negatively changed $\pi$ electrons of the aromatic rings. Since smaller cations diffuse more easily into the protein interiors, this method can also be used to analyze protein dynamics.[55]

Another method in wide use is known as singlet–singlet energy transfer (or fluorescence resonance energy transfer or FRET). If the emission spectrum of a fluorophore (the "donor") overlaps the absorption spectrum of a second fluorophore (the "acceptor"), then when the donor is excited and the donor and acceptor are close to one another, under certain circumstances the acceptor will emit radiation as the donor is quenched. When this is *not* due to the trivial reemission of an absorbed photon, the process occurs as a result of a resonant interaction between the emission process of the donor and the absorption process of the acceptor. The efficiency of this transfer process is a function of the spectral overlap, the relative orientation of the two fluorophores and the distance between them. This distance ($R$) is given by:

$$R = R_o \left( \frac{1 - E}{E} \right)^{1/6}$$

where $E$ is the efficiency of the transfer process and $R_o$ is the characteristic transfer distance that corresponds to $R$ where $E = 50\%$. $R_o$ is a function of the degree of spectral overlap, the refractive index between the donor and acceptor, an orientation factor ($\kappa^2$), which depends on the relative orientation of D and A, and the quantum yield of the donor. Methods exist to estimate the orientation factor but a value of 2/3 for a random orientation is usually used.

This method is most commonly used by placing either covalently or noncovalently specific fluorophores with the proper spectral properties (especially spectral overlap) at single locations either within a macromolecule or at sites in different molecules that are sufficiently close (<80 Å) that efficient resonance energy transfer can occur. Thus, relatively accurate distance estimates can be determined by this method. Numerous systems have been examined by variations of this technique and its utility is well-established for mapping a wide variety of structural features of single molecules, molecular complexes such as viruses and ribosomes, and the surface of cell membranes. FRET has also frequently been used in both static and kinetic modes to study nucleic acids and their complexes.

A third important application of fluorescence involves the use of polarized radiation. This is based on the principle that there is preferential absorption of light when a chromophore has its transition dipole(s) parallel to that of the exciting light. Thus, if polarized light is used to excite a randomly oriented collection of fluorophores, those transiently oriented parallel to the exciting light will preferentially absorb light. If the "photoselected" molecules rotate within their excited state lifetime, the emitted light will be depolarized to some extent. Thus, the motion of fluorophores and the molecules to which they are attached can be analyzed by this method. Depolorization is typically measured in terms of a quantity known as the fluorescence anisotropy ($r$), which is defined as:

$$r = \frac{I_\parallel - I_\perp}{I_\parallel + 2I_\perp}$$

where $I_\parallel$ and $I_\perp$ are the intensities of the emitted light oriented parallel and perpendicular to the exciting light. Experiments can usually be performed in one of two ways. In steady state studies, the anisotropy or the depolarization ratio ($I_\perp - I_\parallel / I_\perp + I_\parallel$) is measured. By varying the temperature or viscosity, it is possible to calculate the rotational correlation time ($\rho$) of the macromolecule to which the fluorophore is attached. Alternatively, the target fluorophores can be excited with single photons of polarized light and the anisotropy of the emitted light detected. Because the anisotropy ($A(t)$) decays exponentially with time, molecules that emit later have more time to rotate:

$$A(t) = A_o e^{-3t/\rho}$$

where the rotation correlation time $\rho$ is just the time it takes a molecule to rotate 1/e of a complete rotation. Multiple rotation modes can often be resolved by deconvoluting the experimental data into individual exponential components. Thus, it is frequently possible to resolve internal dynamic motions of macromolecules from their overall motion. Polarization measurements have also been widely used to study ligand binding and protein/protein interactions. If a fluorescent label is attached to the smaller component, its polarization increases as it becomes part of the larger complex. There are many variations on such experiments, but the results can be quite sensitive to the presence of the interactions.

We have only touched on the use of fluorescence techniques for biological molecules here. A wide variety of experimental methods are based on this principle. These include fluorescence microscopy, single molecule fluorescence, fluorescence photobleaching and recovery, and fluorescence correlation spectroscopy. The interested student is referred to the comprehensive text by Lakowicz for further information.[54]

## Circular Dichroism

The physical basis of circular dichroism (CD; see references 56 and 57) is different from that of simple absorption and fluorescence and as such can provide somewhat different information: light can be thought of as composed of two opposite circularly polarized components. If one of these components is greater or less than the other due to differential absorption and they are combined, light that is elliptically polarized is produced. It is the angle of rotation of the long axis of such an ellipse that is measured in a CD experiment (i.e., the "ellipticity"). CD is seen only in absorption bands, thus requiring appropriate chromophores. CD signals are produced by the interaction (technically, through the "dot product") of electronic and magnetic absorption processes. This is, in fact, the general origin of optical activity. Electronic absorption can be thought of as a unidirectional displacement of charge, whereas magnetic absorption can be represented as a light-induced current loop. When a vertical motion of charge acts on such a circular displacement, a helical charge distribution is produced. As we will see in a moment, this is especially important for CD analysis of proteins and nucleic acids.

According to the above, CD can only be produced when the local environment of a chromophore is asymmetric. In biomolecular systems, there are at least three situations in which this is seen. In the first, transitions could involve electrons near the $\alpha$-carbon atoms of amino acids in proteins. Because there are no major absorptive chromophores here, however, any such signals are quite weak. Second, the tertiary structure of a macromolecule could place relatively symmetric absorptive molecular groups into asymmetric environments. Third, helices can facilitate a helical flow of charge thus producing relatively large optical activity if there is an appropriate absorptive chromophore. It is this last situation that has received the most attention and we will begin our discussion of the CD of proteins and DNA here.

Proteins have several chromophores of potential interest from a CD perspective. These are the peptide bond, certain side chains (especially the aromatic side chains and disulfide bonds), endogenous chromophores such as heme groups, and extrinsically added chromophores whose optical activity changes or is induced when they are added to proteins. The CD of the peptide bond consists of a band near 222 nm (n $\rightarrow$ $\pi*$) and a signal, which is split into two parts through interactions between transitions at 200 to 210 nm ($\pi \rightarrow \pi*$, $\parallel$) and 191 to 193 nm ($\pi \rightarrow \pi*$, $\perp$) (**Fig. 21–14**). Right-handed alpha helices produce a distinct CD spectrum with negative peaks at approximately 208 and 222 nm and a positive peak near 192 nm. $\beta$-sheets manifest a weaker signal ($\beta$ structure can be thought of as distorted helices) at 215 to 218 nm and a positive peak at about 195 nm. Beta turns give a number of weak signals in the same region depending on their type. A left-handed $\alpha$-helix results in a spectrum that is approximately a mirror image of the right-handed version. Cross-$\beta$-structure and beta-trefoils produce negative signals at 210 to 215 and 203 to 208 nm, respectively. Lastly, disordered structure typically gives a peak with a 195 to 200 nm minimum and often a weak positive signal near 230 nm. Because of the distinct nature of the CD spectra of these different types of secondary structure, a protein's CD spectrum can be used to estimate fairly accurately its secondary structure content. The basic idea is a simple one. One can extract inherent values for the ellipticity of the various types of secondary structure from a library of proteins of known secondary structure content (determined by x-ray crystallography or NMR). These values will reflect typical effects of

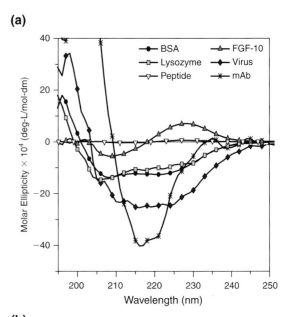

**(a)**

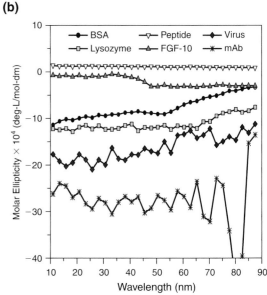

**(b)**

**Fig. 21–14.** Circular dichroism (CD) spectra of representative biopharmaceuticals (*a*) with (*b*) accompanying thermal melting curves. A strong double minimum is seen at 208 nm and 222 nm for the $\alpha$-helix rich BSA and lysozyme. The weak $\beta$-structure minimum at 217 to 218 nm for IgG is more difficult to see on the same scale. Note the positive peak for the peptide between 225 and 230 nm which is often assigned to loosely coiled peptides.

3-D structure, the contribution of side chains, and the length and distortion of regions of secondary structure as well as other factors. This data serves as a basis from which to fit unknown spectra and provide secondary structure content estimates. Such an analysis does not provide absolute values, but the fractional content determined and is often good to 2% to 3%. The method is especially powerful when used to monitor changes in secondary structure with very subtle changes in helix and $\beta$-sheet content detectable. Note the specificity of analysis of secondary structure when CD spectra are obtained in the 180 and 250 nm range. In contrast, UV absorption and fluorescence in the near UV region are primarily sensitive to changes in tertiary structure although they will reflect indirectly secondary structure alterations. CD does, however, offer tertiary structure information when used between 250 and 300 nm where the aromatic side chains and disulfide bonds become the absorbing chromophores. The protein spectrum in this region is quite complex, consisting of a series of positive and negative peaks. These are primarily produced by the induction of optical activity in these side chains although they do possess some weak, intrinsic optical asymmetry. Attempts to derive specific structure information from the details of such spectra have largely been unsuccessful. Changes in the CD spectra of proteins in the 250 to 300 nm region, however, are quite useful in the same way that UV absorbance and intrinsic fluorescence spectra are employed to detect changes in tertiary structure. These signals are, however, significantly weaker than the far UV peaks and either higher concentrations or longer cuvette path lengths must be used. In a number of cases, other intrinsic chromophores such as heme or rhodopsin groups or extrinsically added dyes can produce strong CD signals associated with their absorption bands. If detailed structural information is available, it may be possible to further use such spectra to provide additional structural information. There also exist a number of hybrid versions of CD such as vibrational, magnetic, and fluorescence-detected circular dichroism, which are useful in specific indications. In general, however, the most common use of protein CD involves the monitoring of protein conformational changes (in terms of changes in either secondary or tertiary structure, or both). Alternatively, the effect of ligand binding either through the induction of conformational changes in proteins or induced optical activity in the ligand is another common use. Using the latter method, binding constants and stoichiometries can often be determined. If temperature is varied, the enthalpy, entropy, and heat capacity of the binding process may also be determined using a Van't Hoff analysis.

Because of their helical nature, nucleic acids also produce strong CD signals. In this case, the CD arises from the nucleotide bases and their absorption between 200 and 300 nm. The CD spectra of the different forms of the nucleic acids are quite distinct from one another due to differences in interactions between the bases (**Fig. 21–15**). For example, the A form of DNA (11 base pairs bp/turn) produces a spectrum with a maximum near 270 nm and a minimum near 210 nm. The spectrum of A-RNA has a spectrum similar to A-DNA but is shifted to 10 nm lower wavelengths. B-DNA (10 bp/turn) has a less intense spectrum that is similar in shape but the peaks are slightly shifted. Z-DNA (12 bp/turn, but left-handed in contrast to the A and B forms) is roughly a mirror image of the A and B forms. Depending upon the actual sequence of the nucleic acids, the spectra are subtly different but most importantly highly sensitive to changes in structure. Thus, the melting of ds DNA or RNA can easily be followed by this method. If small pieces of DNA (oligonucleotides) which contain specific binding sites

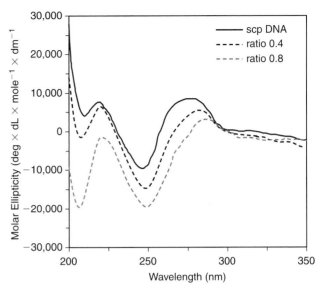

**Fig. 21–15.** CD spectra of pDNA and the DNA in the presence of a cationic polymer complex at different charge ratios (+/−). The CD spectrum of the DNA alone shows that it is in the B-form. The presence of the cationic polymers causes the structure of the DNA to change. A change in the CD spectrum of the DNA is seen because the interaction between the nucleic acids bases is altered, thus perturbing the helical nature of the DNA.

are examined, the binding of proteins to DNA can be analyzed by CD changes. When dyes, drugs, and delivery agents bind to DNA, they often display quite marked induced CD or changes in the spectrum of the nucleic acid (**Fig. 21–15**). This can be used to study their interaction. If DNA becomes highly compacted which can occur when its charges are neutralized, quite unusual spectra can be produced which are diagnostic of unique forms of condensed DNA.[58]

## Vibrational Spectroscopy

The secondary and tertiary structure of both proteins and nucleic acids can also be analyzed by vibrational spectroscopy. Both infrared[59–63] and Raman[64,65] spectroscopies have been employed for this purpose. The former is an absorptive method depending on a change in permanent dipole moment during excitation, whereas the latter is based on small shifts in the frequency of scattered light due to interactions with vibrational states and requires a change in bond polarizability. Today, infrared spectroscopy is almost always performed in a Fourier transform mode and is therefore (somewhat inappropriately) referred to as "FTIR."

FTIR spectroscopy is the more commonly used of the two techniques at least partially due to the wider availability and lower expense of the instrumentation. The theory of IR absorption is briefly discussed in Chapter 4. The method has a number of advantages over CD including the ability to more easily monitor various states of matter (solid, liquid, gas, suspensions), an increased number of secondary structure sensitive signals, and the ability to perform experiments such as isotope exchange (see below), dichroism measurements,

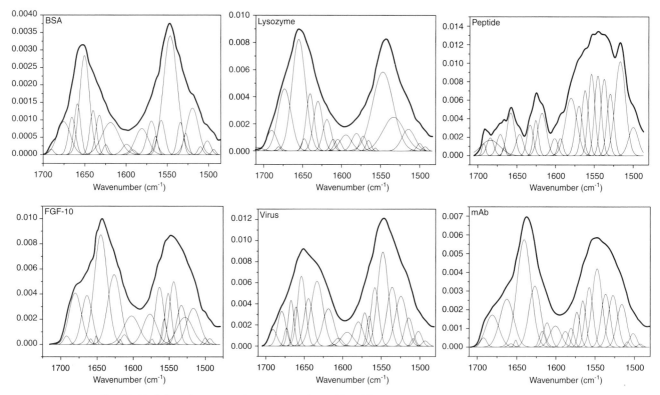

**Fig. 21–16.** Infrared spectra of proteins and peptides. The amide I region extends from 1700 to 1600 cm$^{-1}$. The spectra have been deconvoluted to show the relative contributions of the individual secondary structure types (and in the case of the peptide certain amino acid side chains) to the zero-order spectrum. Most commonly, the amide I region is used to estimate secondary structure content by assigning the origin of the individual bonds to different structural types (helix, $\beta$-sheet, turns, and disordered) which are ratioed to the total band area (see the text).

and two-dimensional correlation analyses. Furthermore, in recent years it has begun to approach the sensitivity of CD.

Infrared spectroscopy of proteins has been most frequently applied to the spectral signals known as the amide bands. There are many such vibrational absorption peaks with the most commonly examined the amide I, II, and III bands. The most frequently used band for secondary structure analysis in proteins is the amide I, which arises primarily from carbonyl stretching and to a lesser extent from NH wag. When proteins are examined, however, a very broad band is seen between 1600 and 1700$^{-1}$ cm in this region (**Fig. 21–16**). A careful examination of this peak usually shows a number of small bumps and inflections suggesting the existence of underlying component signals. Several "band narrowing" procedures are available to deconvolute this broad peak into its constituents. These include derivative analysis as well as a procedure known as "Fourier self-deconvolution." These component peaks are known to arise from different types of protein secondary structure. Assignments are based on studies of model amino acid polymers as well as analysis of proteins of known (secondary) structure. Examples of assignments include $\alpha$-helix (1650–1658 cm$^{-1}$), intramolecular $\beta$-sheet (1670–1680 and 1620–1640, multiple components), turns (1680–1700 cm$^{-1}$), and loops and disordered regions

(1645–1655 cm$^{-1}$). In an observation of special importance to pharmaceutical scientists, aggregated proteins (manifesting intermolecular $\beta$-sheet) often produce distinct bands at either 1610 to 1620 or 1690 to 1695 cm$^{-1}$. Some amino acid side chains (Asn, Gln, Arg, Lys, Tyr, His) also produce signals in the amide I region but unless they are present at very high levels, their absorption bands are usually ignored. The major problem with the use of FTIR to analyze proteins in the past has been the strong absorption bands of water and water vapor in the region of the amide I band. For this reason, $D_2O$ was often used as a solvent since its spectrum does not significantly overlap the amide I signals. Recent improved subtraction techniques, however, now permit the use of normal water for such studies. Thus, FTIR is now routinely used to estimate the relative secondary structure content of proteins with a precision and accuracy similar to that of CD.

As indicated previously, one of the advantages of FTIR is its ability to obtain spectra from samples in a variety of physical states. This is often accomplished through the use of different sampling techniques. For example, in addition to conventional transmittance geometry, spectra of both solids and liquids can also be obtained by attenuated total reflectance (ATR). In this technique, the sample is placed on a transparent plate of appropriate material which permits infrared radiation to penetrate to a small extent into the sample allowing an absorption spectrum to be obtained. Solids can also be monitored by pressing them into a disc with a medium like potassium bromide (KBr) or illuminated from above and a reflective absorption spectrum obtained by a technique known as diffuse reflectance (DRIFT) spectroscopy. Infrared spectra can also be obtained through a microscope or a diamond-based cell, both of which permit very small areas to have their infrared spectra measured.

Nucleic acids can also be usefully examined by FTIR spectroscopy. In this case, the major signals of interest originate from the phosphate, base, and sugar moieties. All three groups are sensitive to nucleic acid conformation and to the binding of various ligands. Thus, B-form DNA can be distinguished from the other forms by the position of base carbonyl bands above $1700 \text{ cm}^{-1}$ and the various phosphate-stretching vibrations in the 1000 to $1300 \text{ cm}^{-1}$ region. Furthermore, the IR spectra of nucleic acids are quite sensitive to hydrogen bonding. Thus, the unwinding of DNA or RNA as measured in a thermal melting experiment can be monitored by FTIR spectroscopy.

Plasmid DNA is frequently used to deliver genes for gene therapy or as a vaccine. Often, the DNA is complexed to cationic lipids or positively charged polymers of various types. The cationic partners in such delivery complexes also produce distinct IR spectra. When the complexes are formed, IR signals from both components typically are altered as the various groups interact.[66,67] This permits complex formation and stability to be directly analyzed by this technique. For example, the $CH_2$ asymmetric stretching vibration of lipids can be used to measure the fluidization of their acyl chains in lipids bound to nucleic acids, providing a quantitative measure of their thermal stability and the effect of DNA upon lipid structure.[68]

Raman spectroscopy is also routinely used for all types of macromolecular-based systems.[64,65] This method has the advantages that the water bands are quite weak. This substantially decreases interference and many side chain vibrations are much better seen. Like FTIR, it can also be used to examine samples in multiple physical states. Its major disadvantage is that it is generally less sensitive than infrared spectroscopy. Two exceptions to this rule exist, however. If an absorption band can be directly excited, a spectral coupling process can produce dramatically enhanced vibrational signals from the specific chromophore (resonance Raman spectroscopy). In addition, if the molecule of interest is absorbed to certain types of materials such as silver, a much enhanced vibrational spectrum is again seen (surface-enhanced Raman spectroscopy or SERS). The mechanism of this enhancement is still debated. A major increase in use of Raman spectroscopy and its enhanced varieties is due to the availability of tunable lasers of wide wavelength range and the potential to reduce the amount of material necessary for such measurements.

In the case of proteins, the most common application is the analysis of amide bands to obtain secondary structure information, analogues to applications of FTIR. It is more common

with Raman spectroscopy, however, to use the amide III band because it is better resolved. General assignments are 1260 to $1305 \text{ cm}^{-1}$ ($\alpha$-helix), 1230 to $1245 \text{ cm}^{-1}$ ($\beta$-sheet), 1258 to $1300 \text{ cm}^{-1}$ ($\beta$-turn), and 1242 to $1255 \text{ cm}^{-1}$ (disordered). Tyrosine side chains give well-resolved signals near 850 and $830 \text{ cm}^{-1}$ and the ratio of these intensities has been used to analyze the relative exposure of phenolic side chains. Tryptophan (multiple peaks) and disulfide bands ($500–550 \text{ cm}^{-1}$) also produce conformationally sensitive signals. Thus, secondary and tertiary structure changes can be simultaneously examined by Raman spectroscopy, a particularly attractive aspect of the technique.

Raman spectroscopy of nucleic acids has also frequently been used to explore nucleic acid structure, perhaps more than infrared absorption spectroscopy. The two major forms of DNA manifest differences in the $800 \text{ cm}^{-1}$ region (phosphodiester antisymmetric stretching) with this signal present in the A form but absent in the B form. The left-handed Z-form produces a unique peak near $625 \text{ cm}^{-1}$ (shifted from $675 \text{ cm}^{-1}$ in the A and B forms). A variety of other conformationally sensitive signals are available in other regions of the spectrum. Raman spectroscopy has also been widely used to characterize viruses in which distinct signals from both protein and nucleic acid components can be easily resolved.[65] Furthermore, the interaction between these two components can be analyzed as well as various aspects of viral structure and stability. Few studies of this type have yet been performed with IR spectroscopy.

## Scattering,[69–73] Hydrodynamic,[74,75] and Calorimetric[76–81] Techniques[39]

The theory behind many of the methods described in this section is discussed in Chapter 16 and should be reviewed accordingly. Specific applications to biopharmaceuticals and topics not reviewed previously will be focused on here.

Light scattering is an extremely useful technique for analyzing the size and shape of biomolecules. For macromolecules which are much smaller ($<\lambda/50$) than the wavelength ($\lambda$) of light used in a scattering experiment, it can be shown that:

$$\frac{I_\theta}{I_o} = \frac{2\pi n_o^2 \left(\frac{dn}{dc}\right)^2}{N\lambda^4 r^2}(1 + \cos^2\theta)M_w c = KM_w c$$

where $I_\theta$ is the intensity of the scattered light at some angle $\theta$, $I_o$ is the incident intensity, $n_o$ is the refractive index of the solvent, dn/dc is the variation in refractive index of the solution with variation in concentration of the scattered (the refractive index increment), $N$ is Avogadro's number, $M_w$ is the weight average molecular weight, and $c$ the concentration. Thus, in this case the intensity of scattered light is proportional to the molecular weight. The concentration dependence of this expression can also be used to calculate virial coefficients, which can be used to characterize the interaction between molecules. In the case of large molecules, which possess multiple scattering centers within themselves, it is found that:

$$\frac{K_c}{R_0} = \left[1 + \frac{16\pi^2 R_g^2}{3\lambda^2}\sin^2\frac{\theta}{2}\right]\left[\frac{1}{M_w} + 2Bc\right]$$

where $R_\theta = \frac{r^2}{1+\cos^2\theta}\frac{I_\theta}{I_o}$ (the "Raleigh ratio"), and

$$K = \frac{2\pi n_o^2\left(\frac{dn}{dc}\right)^2}{N\lambda^4}, \qquad M_w = \frac{\sum n_i m_i^2}{\sum n_i m_i}$$

where $B$ is the second virial coefficient and $R_G$ is the radius of gyration which is defined as $(\sum m_i r_i^2)/(m_i)$, where $m_i$ is the mass of the ith element at distance $r_i$ from the center of mass of the scattering particle. Inspection of this somewhat complex expression, however, shows that a plot of $\frac{Kc}{R_\theta}$ vs $\sin^2\frac{\theta}{2} + Kc$ yields an intercept equal to the molecular weight ($M_w$) and a slope proportion to $M$ and $R_G$. The physical meaning of $R_G$ is not obvious but it has a fairly simple relationship to various shapes. For example, for a sphere, $R_G = R\sqrt{3/5}$ where $R$ is the radius of a sphere, a rod, $R_G = L\sqrt{1/12}$ where $L$ is the length of a rod and $R_G = \frac{\sqrt{\overline{h}^2}}{\sqrt{6}}$ for a random coil, where $\overline{h}^2 =$ mean square end-to-end distance. More generally, shapes can be modeled as prolate or oblate ellipsoids. For example, for a prolate ellipsoid, $R_G = a\sqrt{2 - \gamma^{-2}}$ where $2a$, $2a$, and $\Upsilon 2a$ are the axes of the ellipsoid. Given these types of relationships, the shape of a particle can be estimated from the ratio of the value of the observed $R_G$ to the calculated $R_G$ for a sphere. Globular proteins typically give a value near 1, while more elongated molecules like myosin and DNA produce values greater than 10. If greatly elongated molecules like DNA give observed values of the $R_G$ much less than those calculated for an equivalent rigid rod, this is direct evidence for their flexibility. Another way to think about scattering by large particles is in terms of their turbidity ($\tau = \frac{1}{\ell}\ln\frac{I_o}{I}$ and $\tau = H_c M_w$). This is discussed in more detail in Chapter 16. Although this approach is less sensitive than studying scattering at other angles, it can be simply obtained in an absorption spectrometer since it is simply related to decreases in transmittance. Turbidity measurements are widely used in biopharmaceutics for this reason and have been especially widely applied in kinetic studies of macromolecular aggregation (see below).

A second type of light scattering experiment that has been widely used in biopharmaceutics is "dynamic" or "quasielastic" light scattering (DLS or QELS). This involves an analysis of the fluctuations in intensity of scattered light due to Brownian motion of the scatterers. This analysis is in terms of what is known as an "autocorrelation function," $G(\tau)$ which is defined as $\langle I(t) \times I(t+\tau)\rangle$ where $I(t)$ is the intensity of scattered light at time $t$ and $I(t+\tau)$ is the intensity at some short time ($\tau$) later. Inspection of this function finds that if the intensity remains high as $\tau$ is increased, its value will be high. If $\tau$ increases, however, and the value of $I(t+\tau)$ changes rapidly, it will quickly time average to zero. Thus, the value of the autocorrelation function falls toward zero more

**(a)**

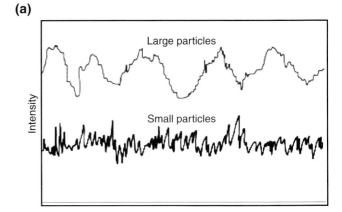

**(b)**

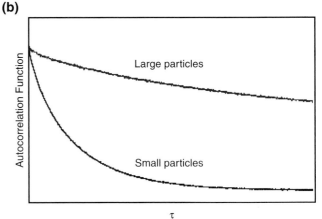

**Fig. 21–17.** Dynamic light scattering. The size of particles can be determined by measuring the fluctuations in the intensity of their scattered light (*a*). Because of their Brownian motion, small particles produce more rapid changes in scattered intensity than larger ones (*b*). If one measures the change in intensity at very short time intervals, the intensity changes less quickly for light scattered by the larger particles and we say that these intensity changes are more "auto-correlated." Analysis of autocorrelation functions (see text) allows the size (more specifically the hydrodynamic radius or diameter) of the scattering particle to be determined.

rapidly for smaller, faster moving molecules (**Fig. 21–17**). This decrease is exponential in form and is given by:

$$G(\tau) = A\exp(-2DQ^2\tau) + B$$

where $D$ is the diffusion coefficient of the scattering molecule, $B$ is a constant, and $Q$ is given by:

$$Q = \frac{4\pi n}{\lambda}\sin\frac{\theta}{2}$$

where $n$ is the refractive index and $\theta$ the angle at which the scattered light is observed. Thus, the angular dependence is small for smaller molecules ($d < \lambda/10$) but larger at lower angles for larger molecules. $D$ can be related to $R$ (the Stoke's radius of the scatterer) through the Stokes–Einstein equation in the following convenient form (see equation **17–7**):

$$D = \frac{kT}{6\pi\eta R}$$

where $\eta$ is the solution viscosity. Thus, the size of a large molecule like a protein or nucleic acid can be directly measured by this method. If the solution is homogenous, an accurate size is easily obtained. If it is not, two approaches can be taken. In the first, the data are fit to a function of the form $\ln[g(\tau) - k]$ to yield two parameters, a weight average mean diameter and a polydispersity parameter (a measure of the width of the size distribution). This is known as the method of cumulants. In the second, the data are fit to a sum of exponential functions to yield a multimodal distribution of sizes. If the different species differ by more than a factor of two in diameter, they can usually be resolved if distinct populations are present. Analysis can be in terms of weight, number and intensity distributions, with the number distribution usually the most intuitively useful. If there are large, internal fluctuations within a macromolecule as might be seen by a large, flexible molecule like DNA, DLS may also be able to detect and resolve these motions as contributions to autocorrelation functions. Dynamic light scattering has become an increasingly important tool to the biopharmaeutical scientist as highly convenient commercial instrumentation has become increasingly available and changes in size and aggregation state are recognized as important degradation pathways for biotechnology products.

A number of other methods often prove useful in the analysis of macromolecular size and shape. Osmotic pressure and viscosity measurements have previously been discussed in earlier chapters and will not be further considered here. More frequently used in the last few years, however, is analytical ultracentrifugation. This is primarily due to the introduction of modern instrumentation to perform such studies. Recall that in a velocity sedimentation experiment, the experimentally measured sedimentation coefficient ($s$, the velocity of the sedimenting particles divided by the unit centrifugal acceleration) is directly proportional to the molecular weight, allowing this quantity to be determined if the diffusion coefficient ($D$) and partial specific volume ($\bar{v}$) are known ($M = \frac{RTs}{D(1-\bar{v}\rho)}$). If a mixture of macromolecules has multiple components which differ significantly in $S$, they can sometimes be resolved by this technique (**Fig. 21–18**). It is possible to perform sedimentation analysis in gradients of substances such as sucrose and cesium chloride. This has the advantage that differences in densities of particles are exploited. Therefore, mixtures containing proteins, nucleic acids, and viruses which differ in density can be separated and analyzed, and to some extent at a preparative scale.

If instead of measuring the velocity of sedimenting particles or using solute gradients, one spins a solution of macromolecules into an equilibrium gradient, one can also calculate molecular weight from the resultant distribution of mass. This is described by:

$$M_w = \frac{2RT}{\omega^2(1-\bar{v}\rho)}\frac{d\log c}{d(x^2)}$$

where $\omega^2$ is the angular velocity of the rotor at equilibrium and $d\log c/d(x^2)$ describes the gradient of concentration as a function of the distance from the center of the rotor. Thus, the slope of a plot of $d\log c$ versus $dx^2$ allows the molecular weight to be determined. Note that the need to know the diffusion coefficient has disappeared and therefore one can obtain an absolute estimate of the molecular weight from an equilibrium sedimentation study. A quite large molecular weight range (less than 100 to more than 10 million Daltons) can be characterized by this method. A very powerful application of equilibrium sedimentation involves the analysis of associating or dissociating systems. This is accomplished by fitting the data to various models of such behavior. The sensitivity and accuracy of this method permits both stoichiometries and equilibrium constants of associating and dissociating macromolecules to be obtained.

Two versions of microcalorimetry are also widely used in the analysis of biopharmaceuticals. As described in Chapter 2, differential scanning calorimetry (DSC) measures the excess heat capacity of a molecule as a function of temperature. If there is an absorption or release of heat due to a structural change in a macromolecular system (an endothermic or exothermic transition), a peak is usually seen in a DSC experiment. A plot of Cp versus $T$ is known as a thermogram. If the process is reversible, the area under such a curve corresponds to the enthalpy ($\Delta H$) of the change in state. Structural changes in biopharmaceuticals are often detected by this technique (**Fig. 21–19**). Protein unfolding, the melting of nucleic acids, and phase changes in lipid bilayers are all routinely studied by this method. If transitions are not reversible, the temperature at which the peak of the transition occurs (the "$T_m$" or melting temperature) is used as a measure of thermal stability. In many cases, DSC thermograms can be quite complex. This can occur for a variety of reasons. For example, samples such as membranes or viral particles which contain multiple components can correspondingly manifest several transitions. Furthermore, individual structural domains within individual proteins can also often be resolved as isolated or overlapping thermal events. For example, the multiple structural domains present in immunoglobulins usually produce multiple peaks in the thermograms of these molecules (**Fig. 21–19**). When ligands bind to macromolecules, they often perturb their stability and thus can be detected as a change in $T_m$. Furthermore, protein aggregation can sometimes be seen as exothermic transitions, in contrast to the endothermic events seen as bonds are broken in other processes. Because it is not dependent on the presence of specific chromophores, is reasonably sensitive (sample concentrations as low as 10 $\mu$g/mL have been employed) and is now available in a high throughput (HTP) (autosampling) mode, DSC is widely used as a routine tool in the characterization and formulation of biopharmaceuticals (see below).

A second calorimetry method of importance to the pharmaceutical analysis of biopolymers is isothermal titration calorimetry (ITC). In such experiments, small amounts of one component are incrementally introduced to a second. A common application is to study the interaction of a small molecule (ligand) such as an enzyme effector or excipient

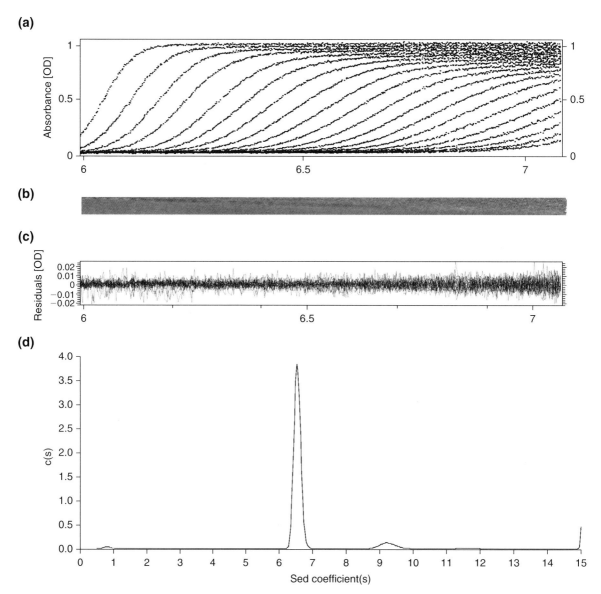

**Fig. 21–18.** (*a*) Sedimentation velocity profiles of a monoclonal antibody. The data was acquired from the absorbance at 280 nm. (*b*) and (*c*) are, respectively, the residue bitmap and residues of the best data fit. The fitting was done with the program Sedfit at a confidence level of 0.68 with sedimentation coefficients seen from 0.5 ~ 15, the best fit friction ratio at 1.491 and partial specific volume at 0.728. (*d*) Continuous sedimentation coefficient distribution of the monoclonal antibody finds the sedimentation coefficient of the antibody monomer of the antibody at 6.53 and the dimer at 9.20.

stabilizer to a protein or nucleic acid. The small heats produced during the binding interaction can be plotted as a function of the molar ratio of ligand to receptor and the data fit to various binding models (see Chapter 11). If a good fit can be obtained, quantitative analysis of such data can yield the binding constant (free energy), enthalpy, and entropy of the interaction as well as the stoichiometry of the binding event. Thus, this approach has been widely used to study the interaction of plasmids with their delivery partners.[82,83] The heats of dilution produced as a consequence of the titration process must be subtracted from the experimental binding heats. The sample requirements for ITC analysis are modest

(0.1–1 mg) and the method has also been adopted for HTP applications.

### Analysis of Macromolecular Dynamics

The previous techniques discussed for use in macromolecular analysis are in general time averaged methods in that they see an averaged property that is smeared out over the time of the measurements which usually take many seconds to hours. Molecular systems like proteins, nucleic acids, and viral particles, however, display a wide variety of much faster motions that play an important role in their structure, function, and stability.[84,85] In the case of proteins, these internal

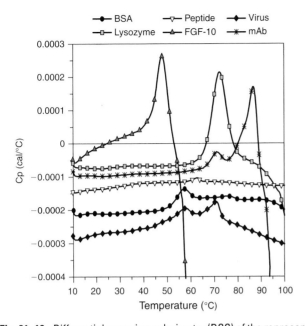

**Fig. 21–19.** Differential scanning calorimetry (DSC) of the representative biopharmaceuticals. The sudden decrease in heat capacity seen with FGF-10 and the IgG is due to their aggregation and precipitation as the temperature is raised. Note the multiple transitions seen with the IgG and BSA, both of which are multidomain proteins (see **Fig. 21–1**). The multicomponent nature of the viral particle also produces at least two transitions.

motions range from the rotations and flexing of individual side chains through movements and telescoping of regions of secondary structure to large-scale motions of entire domains. The presence of such motions means that probably the best picture of a protein molecule in solution is that of a large, Boltzman-like distribution of rapidly interconverting conformational states, with the true native state of any protein best described by such a distribution (see **Fig. 21–21**, later in this chapter). The importance of this view of protein structure is just beginning to be recognized in the world of biopharmaceuticals because of its relationship to physical and chemical stability. Nucleic acids also display marked internal motions, the most common characterized as "breathing modes." This involves the rapid breaking and remaking of hydrogen bonds between the bases as well as changes in the stacking interactions between the bases. Although these fluctuations are quite small in large DNA molecules, they can be quite significant in regions of stress or at the ends of duplexes or at ss/ds junctions. Similarly, lipid bilayers are also subject to significant thermal motions that play a key role in their structure and functional properties.

A number of experimental methods that can detect these types of rapid motions have already been described. These additionally include isotope exchange measurements as detected by NMR, FTIR, and mass spectrometry as well as molecular dynamic simulations. The availability of an array of novel multidimensional NMR methods also provides the ability to directly monitor the dynamic behavior of individual residues. The results of isotope-exchange studies (employ-

ing deuterons) usually reveal at least three different classes of exchangeable protons from peptide bonds: (*a*) a rapidly exchanging group that exchanges too fast to be detected (i.e., in less than a few seconds), (*b*) a class that exchange over seconds to many hours, and (*c*) a small number of buried protons that do not exchange over the lifetime of the experiments. The binding of ligands or alterations in protein/protein interactions typically produces changes in the relative number of each class of exchangeable protons. This then provides one picture of the dynamic aspects of protein structure in terms of the accessibility of the peptide backbone to solvent water.

A wide variety of other methods are also available to probe similar and different aspects of protein dynamic behavior. As discussed above, several fluorescence-based methods provide alternative pictures of internal protein motions. Instead of the use of proton exchange, the quenching of tryptophan (and to a lesser extent tyrosine) residues can be used to study protein motions that permit the diffusion of various solutes into different protein regions. Similarly, as mentioned previously, cations of various sizes can be used in the same manner by measuring shifts in the derivative absorption peaks of Trp, Tyr, and Phe due to cation/pi interactions. The increased number and type of residues in the latter approach offers several advantages. In a new method, the slopes in the shifts of these same derivative absorption peaks with temperature can be used as a measure of protein motions.[86] This is based on the well-understood temperature dependence of the dielectric constant of water and solvent penetration into protein interiors with highly buried aromatic side chains producing little or no temperature-dependent slopes in contrast to more exposed ones.

Time-resolved fluorescence anisotropy methods can be used to sample very rapid motions in the picosecond to nanosecond range of times. In this technique, polarized photons are used to excite fluorophores and their depolarization upon emission is used to characterize the motion of individual molecular groups in terms of their rotational correlation times. Because the fluorescence lifetimes of indole are so short, only rapid motions can be seen in this case. But if extrinsic fluorophores with longer lifetimes are either covalently attached or noncovalently bound to specific sites on a protein or nucleic acid, larger scale motions can be sampled. For example, if a long lifetime fluorophore is placed in the antigen-binding site of an antibody or is attached to a cysteine residue at a defined location, motions such as the flexing of the arms of a Y-shaped antibody can be measured. Another dynamics-sensitive fluorescence-based technique is red-edge excitation in which slow dipole relaxation and photoselection are on the same (or longer) timescale than fluorophore lifetimes. Because these processes are solvent dependent, they can be related to the rigidity of the local environment. If this local matrix is not altered, the emission wavelength of a subensemble present may be uniquely excited and be of lower energy than the mean distribution. Thus, the fluorescence emission spectrum will be excitation dependent and shifted to longer wavelength.

Two relatively simple techniques can also be used to measure the expansivity and contractibility of proteins, both parameters related to their dynamic behavior. In pressure-perturbation isothermal titration calorimetry, the heat emitted or absorbed when pulses of pressure are applied differentially to a sample and reference is measured. This heat difference can be used to determine the coefficient of thermal expansion of the partial volume of the target macromolecule. Such studies also permit the accessible surface area and solvation to be obtained. In complementary measurements, ultrasonic spectroscopy can be used to obtain the adiabatic and isothermal compressibility of a sample of any type. High-frequency sound waves are sensitive to intramolecular interactions because they produce compressions (and subsequent relaxations) of highly structured polymeric systems. By measuring the speed of sound through such materials, the attenuation produced by the pressure-induced compressions and decompressions can be related to the presence of cavities in macromolecular interiors. This is, in turn, related to fluctuation in volumes and their coupling to the local solvent and thus protein dynamic behavior (**Fig. 21–20**).

A large number of other techniques are available to probe the dynamics of higher molecular weight systems. These include neutron diffraction, single molecule fluorescence spectroscopy, three pulse photon echo peak shift spectroscopy, ultrafast two-dimensional vibration echo, and correlation spectroscopy among others. We will not discuss them here, but the interested student needs to be aware that this is a rapidly expanding field with new approaches routinely becoming available.

Breathing modes of nucleic acids can also be measured by many of the above methods. A number of unique methods are available as well. For example, chemical probes such as formaldehyde or dimethylsulfate that specifically react with single-stranded sequences can be used to measure the fluctuations in duplexes that are responsible for their reactivity. It is also possible to replace adenine bases in DNA and RNA with 2-aminopurine, which possesses unique CD, and fluorescence spectral properties that can be used to sense local dynamic behavior. A wide variety of fluorescence and electron spin resonance probes can be used to study membrane dynamics along with NMR and variations on some of the methods described above.

What exactly is the utility of the many dynamics-sensitive methods briefly indicated here? Although such studies are still in their infancy, it is clear that an intimate relationship exists between the dynamics of biopharmaceutical systems and their stability. It was initially thought that this correlation might be a simple one in which increased rigidity (reduced dynamic motions) was related in increased stability. Although this relationship has sometimes been observed, it has also been found that local decreases in stability can be observed upon ligand binding and macromolecule/macromolecule interactions. This is probably due to increases in rigidity and stability in one region of a molecule being relieved by decreases in stability (and increases in dynamic behavior) in other parts. It is clear, however, that the role of internal dynamics is becoming better understood in macromolecular system and that these phenomena will play an increasingly important role in the stabilization and formulation of biopharmaceuticals.

# PREFORMULATION[87–90]

The insertion of a molecule into a chemical system in which it possesses sufficient solubility, stability, and deliverability such that it can be used as a drug or vaccine is commonly referred to as "formulation." The final form of this system containing the molecule itself as well as its accompanying excipients used to achieve these acceptable properties is also described as its formulation (noun). The formulation of biomolecules follows a process generally similar to that used for smaller molecules except that the physical nature of these much larger molecules necessitates the use of many different experiment methods (see above) as well a variety of other considerations based on their unique properties. It is conventional to consider the degradation of biomolecules as either physical or chemical in nature. In general, by degradation we mean change in structure. This may or may not be accompanied by a loss in biological activity as described below. This initial analysis of a macromolecular system prior to the preparation of the formulation is referred to as "preformulation."

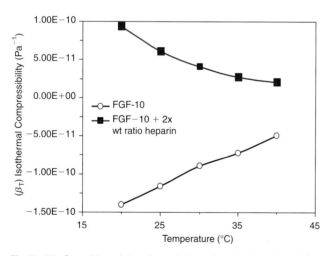

**Fig. 21–20.** Some idea of the effect of a ligand on the dynamics of the protein behavior can be obtained by measuring its compressibility. This is done by measuring the attenuation of sound when it is passed through a protein-containing solution (ultrasonic spectroscopy). As the sound-induced compression of the solvent squeezes a protein, a certain amount of energy is lost as the protein is compressed. This loss can be used to estimate the compressibility. In this example, when heparin is added to fibroblast growth factor-10 (FGF-10) (square), the protein appears to become more compressible, that is, its range of dynamic motion increases. The difference in compressibility is decreased, however, as the temperature is raised. See reference 84 for further discussion.

## Physical Degradation

Physical degradation is characterized by changes in the non-covalent interactions within and between biomolecules. The relationship between physical and chemical degradation will be considered in a later section. Physical degradation is usually discussed in terms of a catalogue of the various types of noncovalent interactions. Here, however, these phenomena will be considered in a somewhat different manner. Imagine a protein in solution of average stability at moderate concentration (0.01–10 mg/mL) at a fixed, near neutral pH (5–8). What happens to the macromolecule when we raise and lower the temperature starting under ambient conditions (15°C–30°C)? The effect of temperature will be considered from two different perspectives: changes in the structure of an individual protein and alterations in the distribution of the microstates of a population of such molecules (**Fig. 21–21**). Similar comments are applicable to nucleic acids and other macromolecular systems. Temperature is chosen as the "stressing" variable here because of its general nature and critical role in the storage stability of biopharmaceuticals. The reader is reminded that the primary effect of temperature is on the thermal motion of the solvent (the water molecules) and the internal motions of the various molecular entities within the protein. Increasing the temperature, of course, increases the rate and magnitude of such motions while lowering it does the opposite.

As the temperature gradually begins to rise the interior motions of the protein will begin to gradually increase. Thus, the local motion of side chains, larger scale motions of elements of secondary structure, and the translational movement of the protein all begin to gradually increase in magnitude. At least at first, these effects on the structure of a protein are usually difficult to detect although they can often be seen in

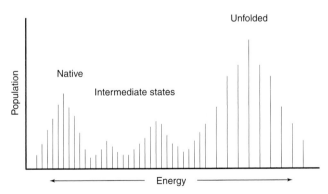

**Fig. 21–21.** The behavior of proteins and other macromolecular-based systems can be described in terms of a distribution of different structural states, each possessing certain energies. As the environment of the macromolecules changes as a result of alterations in solution variables such as pH and temperature, the distribution itself will alter. For example, at lower temperatures, lower energy state may be more populated (i.e., "native states"). At higher temperatures, as bonds are broken and the structure is altered, the population of these states (i.e., "unfolded states") will increase. The existence of intermediate states is equally important due to their unique, individual behavior.

continuous changes in parameters like UV-absorption derivation peak positions. Most importantly, they generally have little immediate, obvious effect on the structure and stability of most proteins. This also results, however, in an alteration in the distribution of microstates with a shift to higher energy. As the temperature is further increased, however, these increases in internal motions can lead to a significant weakening of many forces that stabilize protein structure such as hydrogen bonding, electrostatic, and Van der Waals interactions.[91] Note that the strength of apolar (hydrophobic) interactions tends to increase as the temperature is raised because of favorable entropic ($-T\Delta S$) and heat capacity ($\Delta Cp$) effects. In some proteins, this can lead to actual conformational alterations with the distribution of states splitting into two or more peaks. This does not necessarily imply that the protein must unfold (denature) under these conditions, but rather that the weakening of key intramolecular interactions can produce local alterations in structure. This could be due to unfolding-like changes in a particular region of the protein, a change in structure of one or more domains in a multidomain protein, or a consequent dissociation of a subunit-containing protein, among other possibilities. Such alterations could result in a change in a protein's biological activity or its immunogenicity. It could also lead to aggregation of the protein, especially if an apolar region becomes exposed. The type of conformational changes indicated may or may not be detected by methods like CD, intrinsic or ANS fluorescence, or DSC, depending on their magnitude and the exact nature of the structural change. As mentioned previously, one special form of these types of structural changes produces an important class of altered protein configurations known as MG states.[92]

MG states have several distinguishing characteristics. They display a dramatic decrease in tertiary structure as seen by exposure of their aromatic side chains as monitored by intrinsic fluorescence, ultraviolet absorption, near UV CD or related techniques. In contrast, their secondary structure remains substantially intact as seen by far UV CD, infrared, or Raman spectroscopies. Thus, when proteins that display this state are heated, the contacts between secondary structure elements and other distant contacts within the polypeptide chain are broken prior to major alterations of secondary structure. One consequence of this is that dyes such as ANS usually bind to MG-states with a dramatic increase in fluorescence, aiding in their identification. The reason MG states are so important to the pharmaceutical scientist is their tendency to aggregate. It is now generally thought that many cases of aggregation are due to the population of such states (**Fig. 21–2**). They are also commonly seen in proteins at low pH and high salt concentration, but their transient presence is probably responsible for many if not most cases of protein aggregation. If a protein continues to be heated, a much more comprehensive disruption of structure may take place. Although these so called "unfolded" or "denatured" states usually still contain some structure (especially in thermally induced unfolding), a loss in biological activity is typically produced. When proteins begin to unfold, they often interact with themselves to form intermolecular $\beta$-structure. These aggregative states can usually be identified by unique FTIR signals and dye binding as mentioned above and appear to be involved in a variety of disease states such as Alzheimer and Parkinson disease in which precipitated protein is present in vivo. Complete unfolding does not necessarily (and probably rarely) produce aggregation. Although this was once thought to be the case, passage through MG states is more likely responsible for most examples of commonly observed thermal aggregation. Although thermally induced aggregation is usually (but not always) irreversible, use of denaturing agents such as urea, guanidinium hydrochloride, or a chaotropic salt (e.g., $LiClO_4$, $NaSCN_4$) can often be used to produce reversibly and more extensively unfolded protein. Frequently, data obtained from such experiments can be modeled as a simple, reversible, two state unfolding transition in which intermediates do not play a significant role[93,94] (**Fig. 21–22**):

$$N \geq U$$

where N and U refer to native and unfolded states, respectively. The fraction unfolded (fu) as induced by temperature or unfolding agent is then given by

$$f_u = \frac{X_N - X}{X_N - X_U}$$

where $X$ is the experimental value determined from a method such as CD, fluorescence, and so forth for the native ($X_N$), unfolded ($X_U$), and fractionally unfolded ($X$) state. The equilibrium constant for unfolding ($K_u$) is then given by

$$K_u = \frac{[U]}{[N]} = \frac{f_u}{1 - f_u}$$

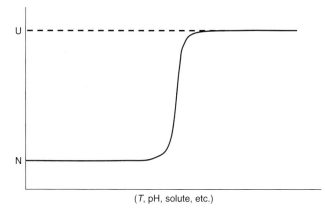

(T, pH, solute, etc.)

**Fig. 21–22.** A generic "unfolding curve" for a simple two state system. If only two states, such as a native and unfolded form, are detectably present, a very sharp, highly cooperative transition is usually seen. Such transitions can be induced in macromolecular systems by a wide variety of different variables including temperature, pH, and the presence of solutes such as urea and guanidine hydrochloride. The midpoint of such curves when temperature is used as the perturbing variable is known as the melting temperature or the $T_m$.

and the free energy ($\Delta G_u$) by

$$\Delta G_u = RT \ln K_u$$

The relationship $\Delta G_u = \Delta H_u - T\Delta S_u$ and the Van't Hoff equation permit estimates of the associated thermodynamic parameters for the unfolding process.

If a denaturant is used[93–95]:

$$\Delta G_u = \Delta G_{H_2O} - m[\text{denaturant}]$$

where $\Delta G_{H_2O}$ is the extrapolated valve of the free energy at zero denaturant concentration and $m$ (the dependence of $\Delta G_u$ on denaturant concentration) is a parameter related to the amount of protein surface area that becomes solvent exposed upon the induced unfolding. Thus, at least in these special circumstances, a fairly comprehensive quantitative picture of the unfolding process can be obtained.

Let us return to more moderate temperature conditions. It is also possible for proteins to aggregate without any conformational change in their structure. The highly amphipathic nature of protein surfaces (i.e., they possess both polar and apolar regions) means that they have a significant potential to interact with themselves. Such interactions are usually temperature dependent with increased aggregation seen as the temperature is lowered. As discussed above, however, increases in temperature can also lead to aggregation through conformational changes or be due to the increasing strength of apolar interactions at higher temperatures. A variety of other forms of environmental stress important to the pharmaceutical scientist such as shaking and freeze/thaw events can also lead to protein conformational changes and/or aggregation although these phenomena are less well understood despite their pharmaceutical relevance.

Low temperature can also promote the unfolding of proteins due to the reversed temperature dependence of the hydrophobic effect and the large heat capacity (and surface area) changes, which are usually associated with protein conformational alterations. Maximum destabilization is often seen at temperatures below freezing but destabilization per se can frequently be a factor in the behavior of proteins at quite moderate temperatures. Another form of physical degradation also arises from the amphipathic nature of protein surfaces. Proteins in solution must be resident in some type of container. This presents a variety of different types of surfaces with which a protein can interact. In fact, the plastic and glass containers commonly used to store proteins and other biopharmaceuticals may themselves possess some charged or apolar characteristics, enhancing the potential for protein/surface interactions. The air/water interface produced during shaking can also be considered an example of such a surface. Rubber stoppers and various types of syringes are frequently siliconized to reduce such interactions but often with only limited success. In fact, at very low protein concentration ($<10\ \mu g/mL$), a substantial portion of most proteins may be adsorbed to container surfaces. As proteins remain adsorbed to surfaces, they may undergo conformational changes that optimize their interaction with the surfaces, leading to additional problems. Potential solutions to such problems are described below.

The effect of environmental factors other than temperature can also play an important role in the physical degradation of proteins. The highly charged nature of protein surfaces (and to a much lesser extent their interiors) makes them very pH sensitive. Important approximate $pK_a$'s are Asp (3.0), Glu (4.2), Lys (10.0), Arg (12.5), His (6.0), Tyr (10.0), and Lys (9.1). These valves can vary quite significantly (by several pH units) in proteins because of local environmental effects.[96,97] Rough estimates of the total change on a protein can be made from these values or from average valves of the individual residues based on actual measurement of a large number of proteins. The charge density of a protein ($P_c$) can be crudely estimated from[39]:

$$P_c = \frac{(pI - pH)}{M_w}$$

where the pI is the isoelectric point of the protein (the pH at which the charges sum to neutrality) and $M_w$ is the molecular weight of the protein. Proteins tend to display their minimum solubility near their pI although many exceptions to this rule exist because of the potential for specific interactions among protein molecules.

Charged residues often provide key interactions in the stabilization of protein structure. In addition to direct interactions between oppositely charged side chains (sometimes called salt bridges or ionic interactions), ion–dipole and cation–pi interactions are commonly seen in proteins. Thus, as pH is varied, any of these (and many more) types of interactions can be altered leading to structural changes and changes in the stability of proteins.

The presence of salt has a major effect on the electrostatic behavior of proteins. At low salt concentrations ($<0.2$ M for a simple salt like NaCl), this is successfully modeled by a simple charge-shielding model as originally described by Debye–Hückel theory. This also works well for the interactions of ions like $Mg^{2+}$ with the phosphate backbone of DNA. As salt concentrations are raised, however, a variety of other phenomena are observed. In the case of proteins, solubility can be dramatically decreased to the extent that certain ions can produce precipitation from solution (so called "salting out" salts) due to preferential hydration of the salt ions. In contrast, certain anions such as thiocyanate and perchloride and cations such as $Li^+$ can increase the solubility as well as reduce the stability of a wide variety of macromolecules.

All of the above phenomena can also be considered in the light of the "distribution of microstates" picture. Thus, particular microstates or populations of microstates can be considered to produce distinct surface properties, solubilities or aggregative tendencies that are responsible for their physical degradation. Whichever description is used, however, the challenge for formulation science is to reduce the rate (and extent) at which they appear.

Nucleic acids also undergo a variety of physical changes that can be damaging to their use as vaccines, gene therapy agents, and RNA-based therapeutics.[58,98] Physical pathways of degradation of DNA and RNA are at least partially a function of their size. DNA is often used in the form of supercoiled (sc) plasmids. If the ends of a large piece of double-stranded DNA are covalently joined, it is possible for one strand to pass through the other. Assuming no breakage of either strand, the number of times this occurs is a constant known as the linking number (Lk). One strand can, however, be twisted about the other (Tw) or a writhing about the duplex axis (Wr) can occur. The relationship between these three quantities is simple:

$$Lk = Tw + Wr$$

Changes in these parameters can occur as a result of strand breakage and reclosure (often due to enzymatic processes by topoisomerases or by physical processes as well), producing a variety of topological forms that can be easily seen as individual bands on agarose gels (**Fig. 21–8**) or peaks in HPLC analysis. If cleavage of a single strand occurs, a closed circular form of the DNA (oc) will be produced. If cleavage of both strands occurs near one another, linear strands of ds DNA (l) result. These forms are also easily detected by various electrophoretic methods. Because sc, oc, and l forms of DNA may well have distinct (and possibly altered) stabilities and biological activities, they represent important degradation products.

If heat is applied to either single or double stranded oligonucleotides, changes in state will also occur. In the case of large DNA molecules like plasmids, thermal melting (i.e., complete disruption of base hydrogen bonding and stacking interactions) appears only at higher temperatures ($>90°C$) and is probably of limited pharmaceutical relevance. An exception is longer single-stranded nucleic acids which may contain extensive regions of internal base pairing which lead to stem-loop type–structures. These regions can often melt in a highly cooperative manner with simple UV absorbance as well as other types of spectroscopic and calorimetric measurements able to detect their presence as lower temperature melting events. Shorter DNA and RNA oligonucleotides, of course, melt at much lower temperatures and this can also constitute important physical degradation events. Furthermore, "breathing phenomena" (see above) before melting and destacking of bases in single-stranded (unbase-paired) nucleic acids also comprise physical events of potential relevance to stability. It is now recognized that many RNA molecules also contain biologically significant tertiary structure due to distant contacts within polynucleotide chains (often mediated by metal ions). Although this has yet to constitute a major pharmaceutical issue, it can be expected to become so in the near future.

Nucleic acids are also capable of aggregation when the charges on the phosphate backbone of polynucleotides are partially or fully neutralized by a variety of cationic molecules such as polyamines and metals. Under such conditions, the nucleic acids may begin to associate and ultimately extensively aggregate. This may also be a result of structural changes in the nucleic acid. Related phenomena are often seen when DNA is complexed to polycations to create gene and vaccine delivery complexes when the zeta potential of the complex approaches neutrality. The conformational properties and stability of the DNA within gene delivery vehicles are also subject to conformational alterations, which can be accessed by methods such as CD, FTIR, and DSC. One form of such change is known as condensation. When there is a reduction in the repulsive electrostatic forces between phosphate groups, large DNA molecules can often collapse into highly condensed structures, which are of both dramatically minimized volume and high density. Such structures can possess unique physical and spectral properties. The former makes these candidates for various types of delivery applications and the latter fairly easily recognized by a variety of experimental methods (especially CD).

## Chemical Degradation[99,100]

Protein drugs are often also chemically unstable with chemical degradation events occurring at specific amino acids within proteins. These degradation reactions are influenced by intrinsic factors such as primary, secondary, tertiary, and quaternary structures as well as extrinsic factors such as pH, temperature, buffer, and excipients. Common chemical degradation reactions include deamidation, hydrolysis, oxidation, N,O-acyl migration, and beta-elimination. This section describes the characteristics of these degradation reactions.

## Deamidation Reactions

The deamidation reaction is one of the most studied and best understood reactions in peptides and proteins (**Fig. 21–23**). It occurs primarily in asparagine (Asn) residues. In this reaction, asparagine residues can be converted to succinimide (Asu), aspartic acid (Asp), and iso-aspartic (Iso-Asp) acid[101–104] moieties. The rate and mechanism of deami-

dation reactions are strongly influenced by the pH of the solution.[102–104] At pH < 4.0, the amide group on the side chain of the Asn residue undergoes direct hydrolysis to release ammonia to generate Asp. At pH > 6.0, the deamidation reaction proceeds via a cyclic imide intermediate (Asu) due to the attack on the carbonyl carbon of the Asn side chain by the backbone amide nitrogen of the C-terminal of the Asn residue to expel an ammonia molecule. The cyclic imide can be hydrolyzed at two different sites to produce Asp or Iso-Asp residues in proteins (**Fig. 21–26**). The Asp-containing protein can be further hydrolyzed at the backbone to give two protein fragments (Asp-mediated degradation). The hydrolysis of cyclic imide intermediates generates the Iso-Asp- and the Asp-containing proteins with a ratio around 4:1, suggesting that the hydrolysis reaction favors the backbone carbonyl over the side chain carbonyl group. The bulkiness of the side chain of the amino acid C-terminal to the Asn residue (i.e., the $n + 1$ residue) affects the rate of deamidation; the presence of a bulky hydrophobic side chain at the $n + 1$ residue impedes the deamidation reaction and changes the Iso-Asp:Asp ratio from 4:1 to 2.5:1. The conversion of Asn to the Iso-Asp or Asp residue adds an additional negative charge to the protein, which can potentially influence the structural and physical stability of the protein and its biological activity. This change can be detected by both peptide mapping and isoelectric focusing.

The impact of secondary structure on the deamidation reaction has been elucidated in both peptides and proteins; for example, bovine growth releasing factor (bGRF) peptide (Leu[27]bGRF) contains 32 amino acids with two segments of $\alpha$-helix at Phe6-Gly15 and Arg20-Leu27 (**Fig. 21–24**).[105] The Asn8 residue on Leu[27]bGRF undergoes a deamidation reaction to produce isoAsp and Asp peptides. To test the effect of secondary structure on deamidation, the Gly15 residue was mutated to Ala15 and Pro15 to give the Ala[15]Leu[27]bGRF and Pro[15]Leu[27]bGRF peptides, respectively. Because alanine is a strong $\alpha$-helix inducer, the Ala[15]Leu[27]bGRF peptide

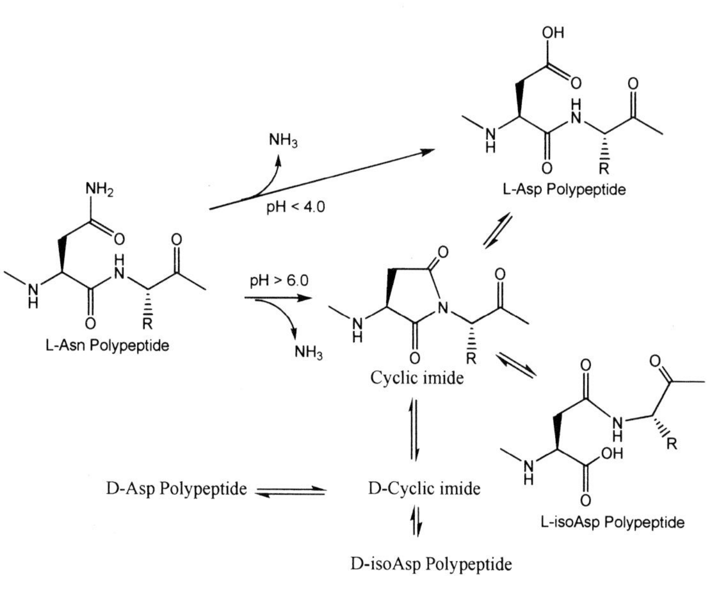

**Fig. 21–23.** Diagram of a deamidation reaction at Asn residues. The formation of a cyclic imide intermediate is shown followed by formation of Asp and Iso-Asp.

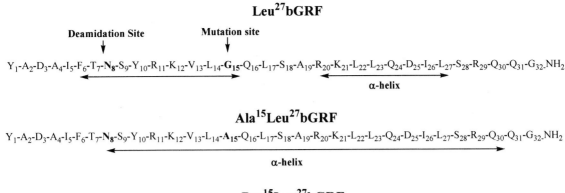

## Leu²⁷bGRF

**Deamidation Site**     **Mutation site**

Y₁-A₂-D₃-A₄-I₅-F₆-T₇-**N₈**-S₉-Y₁₀-R₁₁-K₁₂-V₁₃-L₁₄-**G₁₅**-Q₁₆-L₁₇-S₁₈-A₁₉-R₂₀-K₂₁-L₂₂-L₂₃-Q₂₄-D₂₅-I₂₆-L₂₇-S₂₈-R₂₉-Q₃₀-Q₃₁-G₃₂-NH₂

α-helix

## Ala¹⁵Leu²⁷bGRF

Y₁-A₂-D₃-A₄-I₅-F₆-T₇-**N₈**-S₉-Y₁₀-R₁₁-K₁₂-V₁₃-L₁₄-**A₁₅**-Q₁₆-L₁₇-S₁₈-A₁₉-R₂₀-K₂₁-L₂₂-L₂₃-Q₂₄-D₂₅-I₂₆-L₂₇-S₂₈-R₂₉-Q₃₀-Q₃₁-G₃₂-NH₂

α-helix

## Pro¹⁵Leu²⁷bGRF

Y₁-A₂-D₃-A₄-I₅-F₆-T₇-**N₈**-S₉-Y₁₀-R₁₁-K₁₂-V₁₃-L₁₄-**P₁₅**-Q₁₆-L₁₇-S₁₈-A₁₉-R₂₀-K₂₁-L₂₂-L₂₃-Q₂₄-D₂₅-I₂₆-L₂₇-S₂₈-R₂₉-Q₃₀-Q₃₁-G₃₂-NH₂

**Random Structure**

**Fig. 21–24.** The sequence of Leu²⁷bGRF and its mutants Ala¹⁵Lue²⁷bGRF and Pro¹⁵Leu²⁷bGRF. Asn8 is the deamidation site on these peptides. The mutations discussed in the text affect the secondary structure of the peptides.

possesses helical structure from Asn8 to Gln30, which is a greater helical content than in the parent Leu²⁷bGRF.[105,106] In contrast, Pro¹⁵Leu²⁷bGRF does not have any α-helical structure (i.e., it is random) because proline is a strong α-helix breaker.[105] Evaluation of the deamidation reaction at the Asn8 residue showed that the rate of deamidation of Ala¹⁵Leu²⁷bGRF was slower ($t_{1/2} = 21.53 \pm 2.83$ hr) than the parent peptide ($t_{1/2} = 15.74 \pm 2.45$ hr); this is presumably due to the greater α-helical structure in Ala¹⁵Leu²⁷bGRF than in Leu²⁷bGRF. Conversely, the rate of deamidation of Pro¹⁵Leu²⁷bGRF ($t_{1/2} = 10.78 \pm 2.95$ hr) is faster than the parent peptide,[105,106] indicating that the secondary structure of this peptide has an accelerating effect on the deamidation rate.

The rate of deamidation reactions is also affected by the position of Asn residues in β-turns. Linear (AcNG and AcGN) and cyclic (cNG and cGN) peptides have a β-turn structure at the Asn-Gly or Gly-Asn sequence in which the Asn residue is at the $n + 1$ or $n + 2$ position of the β-turn (**Fig. 21–25**).[107] At pH 8.8, the rates of degradation of the cyclic peptides cGN ($<2.2 \times 10^4 s^{-1}$) and cNG ($9.36 \times 10^7 s^{-1}$) are slower than the respective linear peptide counterparts, AcGN ($20.1 \times 10^7 s^{-1}$) and Ac-NG ($42.2 \times 10^7 s^{-1}$) [107]. This result suggests that the molecular rigidity of the cyclic peptides hampers the deamidation reaction. In addition, the cGN peptide with Asn at $n + 2$ is more stable than the cNG peptide with Asn at $n + 1$. The Asn residue at $n + 1$ of a β-turn undergoes deamidation reaction more readily than the Asn residue at the $n + 2$ position; this is due to a more favorable formation of the cyclic imide when the Asn residue is at the $n + 1$ position than when it is at the $n + 2$ residue of a β-turn. For the Asn residue at the $n + 1$ position, the shortest distance between the carbonyl carbon of the Asn side chain and the backbone nitrogen atom of the $n + 2$ residue for forming the cyclic imide intermediate is 1.89 Å (**Fig. 21–23**). In contrast, the shortest distance between reactive atoms when the Asn residue is at position $n + 2$ is 4.8 Å

(**Fig. 21–23**). In conclusion, the secondary structure and the position of the Asn residue in a peptide or protein affect the rate of their deamidation reactions. Furthermore, it has been demonstrated that increased mobility of deamidation sites in proteins (e.g., when they are located on a protein's surface) also facilitates the deamidation reaction.[108]

## Asp-Mediated Backbone Hydrolysis

Aspartic acid residues can catalyze backbone hydrolysis in peptides and proteins.[109–113] The deamidation products of a protein containing susceptible Asn residues are Asp- and

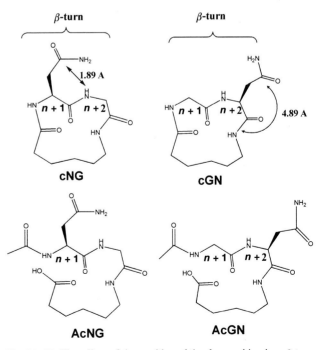

**Fig. 21–25.** The effect of the position of the Asn residue in a β-turn on a deamidation reaction.

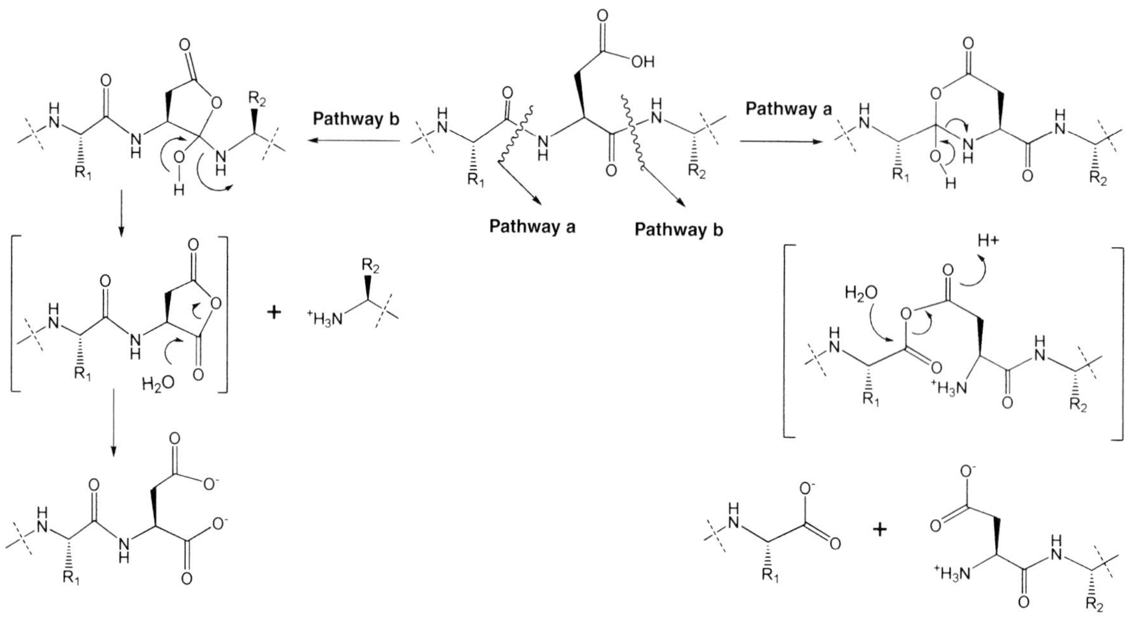

**Fig. 21–26.** The degradation pathways of Asp residues to induce peptide bond hydrolysis.

iso-Asp–containing proteins that can undergo peptide bond hydrolysis at both the N- and C-terminal of the Asp residue via pathways a and b, respectively (**Fig. 21–26**). In pathway a, the carbonyl carbon of the $n-1$ residue is attacked by the carboxyl group of Asp to form a six-membered ring intermediate.[112] Upon rearrangement, the six-membered ring is open to the anhydrate intermediate that immediately hydrolyzes to form two fragments of the peptide or protein. Pathway b proceeds via the attack of the C-terminal of the carbonyl of the Asp residue by the side chain carboxylic acid oxygen to form a five-membered ring. The rearrangement of the five-membered ring hydrolyzes the peptide bond to separate the two portions of the protein. Like the Asn residue, Asp can also form a cyclic imide intermediate that can further rearrange to form iso-Asp. Comparison of the stability of

Asp-containing linear and cyclic peptides has shown that the rate of peptide bond hydrolysis mediated by the Asp residue in a cyclic peptide is slower than the rate of degradation in a linear one.[112] Molecular dynamics simulations show that it is more difficult to form cyclic imide intermediates in a cyclic peptide than in a linear peptide; this is due to the rigidity of the cyclic peptide backbone and a favorable distance between the reactive atoms which can be easily accommodated in the linear peptide compared to that of cyclic peptides.[113]

## N,O-Acyl Migration in Ser or Thr Residues

Ser and Thr residues are prone to undergo N,O-acyl migration reactions which rearrange the protein/peptide backbone in acidic conditions (**Fig. 21–27**).[114] This reaction can occur

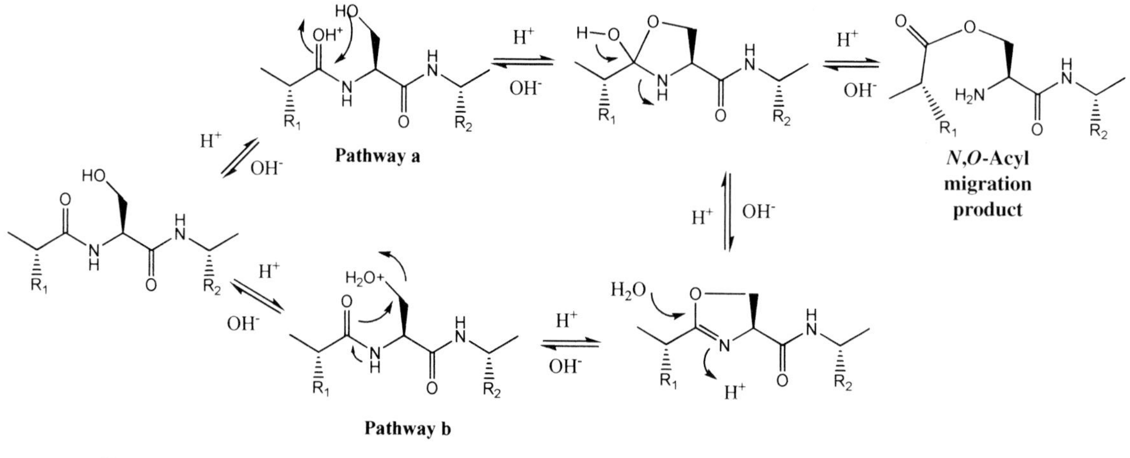

**Fig. 21–27.** The N,O-acyl migration reaction occurs at Ser and Thr residues. This reaction produces a rearrangement of the peptide backbone at the N-terminus of the side chain of the Ser or Thr residue to make an ester bond.

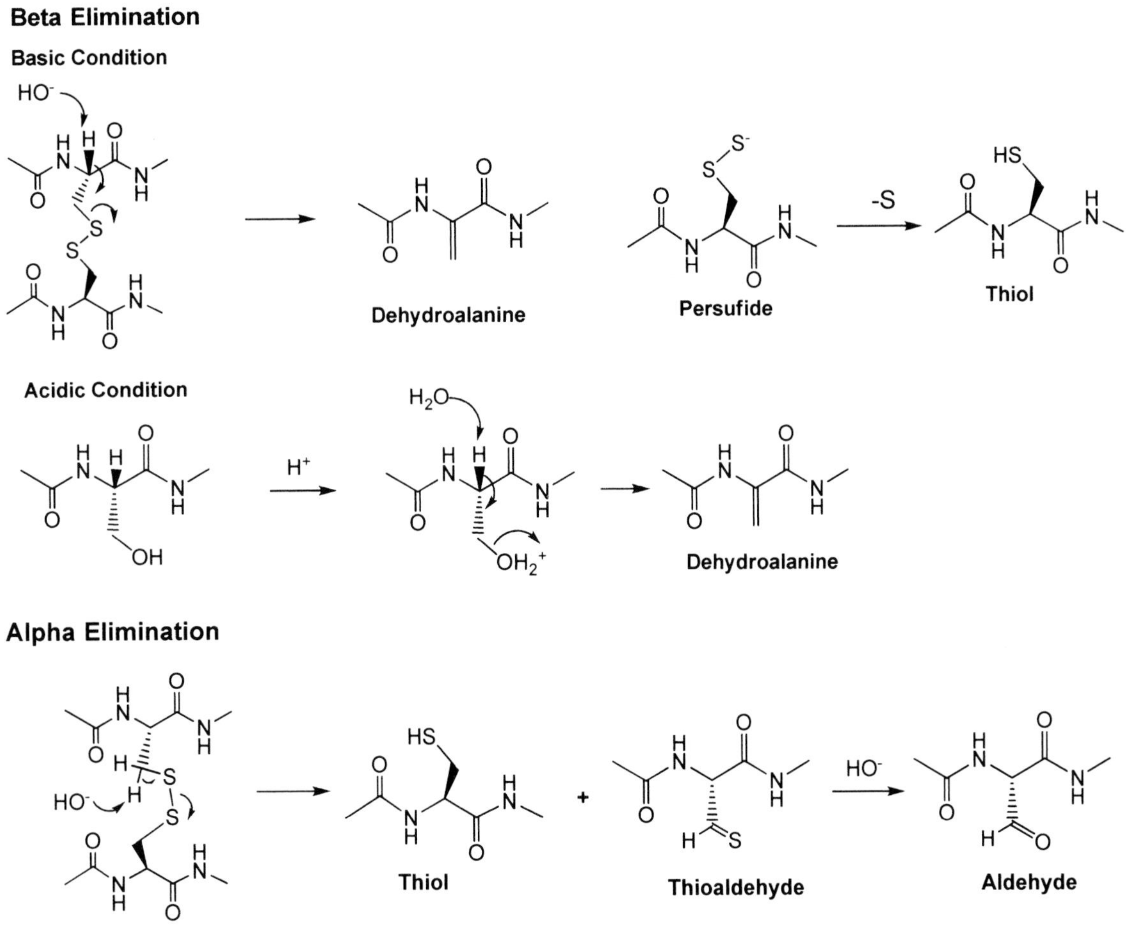

## Beta Elimination

### Basic Condition

Dehydroalanine          Persufide          Thiol

### Acidic Condition

Dehydroalanine

## Alpha Elimination

Thiol          +          Thioaldehyde          Aldehyde

**Fig. 21–28.** The beta- and alpha-elimination reactions in Ser and Cys residues under acidic and basic conditions.

via two possible mechanisms (pathways a and b). Pathway **a** is initiated by protonation of the carbonyl oxygen of the residue $n - 1$ to the reactive Ser residue followed by an attack of its carbon by the Ser residue's oxygen to form a five-membered ring intermediate. The opening of the five-membered ring upon cleavage of the C–N bond generates an ester bond from the carbonyl of the $(n - 1)$ residue to the side chain oxygen of Ser. The second possible mechanism is via protonation of the hydroxyl group of the Ser residue followed by an attack of the beta-carbon of the Ser residue by the carbonyl oxygen of the $n - 1$ residue to produce a five-membered oxazoline, which upon the nucleophilic attack of water on the double bond produces the five-membered ring oxazolidine intermediate. As in pathway a, opening of the five-membered ring intermediate generates the *N,O*-Acyl migration product.

## Beta- and Alpha-Elimination Reactions

Disulfide bonds have important roles in stabilizing the folded structure of proteins. They are normally formed by two Cys residues that are in close proximity due to tertiary structure constraints. The destruction or reduction of disulfide bonds may frequently have an impact on the structure and biological

activity of a protein. The degradation of disulfide bonds can occur in mild to strong alkaline conditions when hydroxide ions abstract the alpha-proton of the Cys residue to generate dehydroalanine and persulphide ion (**Fig. 21–28**).[112,115] Extrusion of the sulfur atom from the persulfide ion produces the thiol group of a Cys residue. In basic conditions, beta-elimination is often observed in Cys residues that are involved in disulfide bonds. This reaction is frequently observed in proteins that contain disulfide bonds. In acidic conditions, beta-elimination can take place in Ser residues; upon protonation of the side chain OH group, the alpha proton of Ser is abstracted by a water molecule to produce the dehydroalanine residue (**Fig. 21–28**).

Alpha elimination can also occur in a disulfide bond under basic conditions to form thioaldehyde and aldehyde products. The alpha-elimination reaction proceeds via proton abstraction of the beta-carbon of the Cys residue to form thioaldehyde and releases the thiolate anion from the other Cys residue. The thio-aldehyde can further react with the hydroxide anion to produce aldehyde. The presence of a reactive aldehyde may further react with amino groups (i.e., Lys side chains) within a protein or with another protein to form an imine bond.

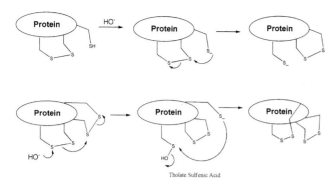

**Fig. 21–29.** Disulfide bond exchange in proteins under basic conditions.

The presence of multiple disulfide bonds in a protein can also lead to a disulfide bond exchange reaction under basic conditions.[116] At high protein concentrations, disulfide bond exchange can form dimers and higher oligomers that can precipitate the protein from solution. The exchange reaction can be initiated by the attack of thiolate anion on the sulfur atom of a disulfide bond (**Fig. 21–29**). Presumably, this exchange reaction occurs when the thiolate anion is in close proximity to the disulfide bond. The exchange reaction could also occur via the attack of the hydroxyl anion of the sulfur atom of the disulfide bond to produce thiolate anion and sulfenic acid. Further reaction of the thiolate anion with the sulfur of sulfenic acid to release hydroxyl anion can produce the disulfide bond exchange reaction.

## Oxidative Reactions (Met, His, Trp)

Oxidation reactions of methionine (Met), histidine (His), and tryptophan (Trp) residues are often observed during protein

production and formulation. Such oxidation reactions are due to reactive oxygen species (e.g., $^{\bullet}OH$, $O_2^{\bullet-}$, $H_2O_2$, $O_3$, $^1O_2$). The formation of reactive oxygen species can be catalyzed by metals (i.e., ferryl, perferryl) and can be produced by ionizing radiation and photochemical reactions. To prevent protein drug oxidation, reducing agents (i.e., glutathione, dithiothreitol, thioacetic acid, and cysteine) have been added to protein formulations. Methionine oxidation produces the sulfoxide amino acid and further oxidation generates a sulfone group on the side chain of Met (**Fig. 21–30a**).[117–120] The sensitivity of the Met residue to oxidation depends on its location within the tertiary structure of a protein. In protein formulations, the oxidation reaction can occur because of the presence of residual hydrogen peroxide that is used to sterilize containers and vials for storage. Oxidation of methionine can significantly reduce the half-life of protein drugs and generate major problems in protein purification and formulation. Oxidation may also alter the physical stability of proteins through the production of oxygen radicals.[120,121]

The aromatic rings of His, Trp, Tyr, and Phe residues are also prone to oxidation. The imidazole ring of His in serum albumin is oxidized by ascorbic acid/$Cu^{2+}$ or $H_2O_2$/$Cu^{2+}$, producing an oxo-dihydro-imidazol ring (**Fig. 21–30b**).[122] This oxidation is via a hydroxyl radical attack at the C2 position of the imidazole ring followed by the removal of the hydrogen radical to produce the 2-oxo-imidazol moiety. Further oxidation of His can produce an Asp residue.[122]

The oxidation of Trp residues in peptides and protein with hydrogen peroxide generates $N$-formylkynurenine (NFK), kynurenine (Kyn), oxindolylalanine (Oia), dioxindolylalanine (DiOia), and 5-hydroxytryptophan (5-OH-Trp).[123–125] Similarly, oxidation of the Trp residue in di- and tripeptides (i.e., Ile-Trp, Trp-Leu, Gly-Trp-Leu, and Ala-Trp-Ile)

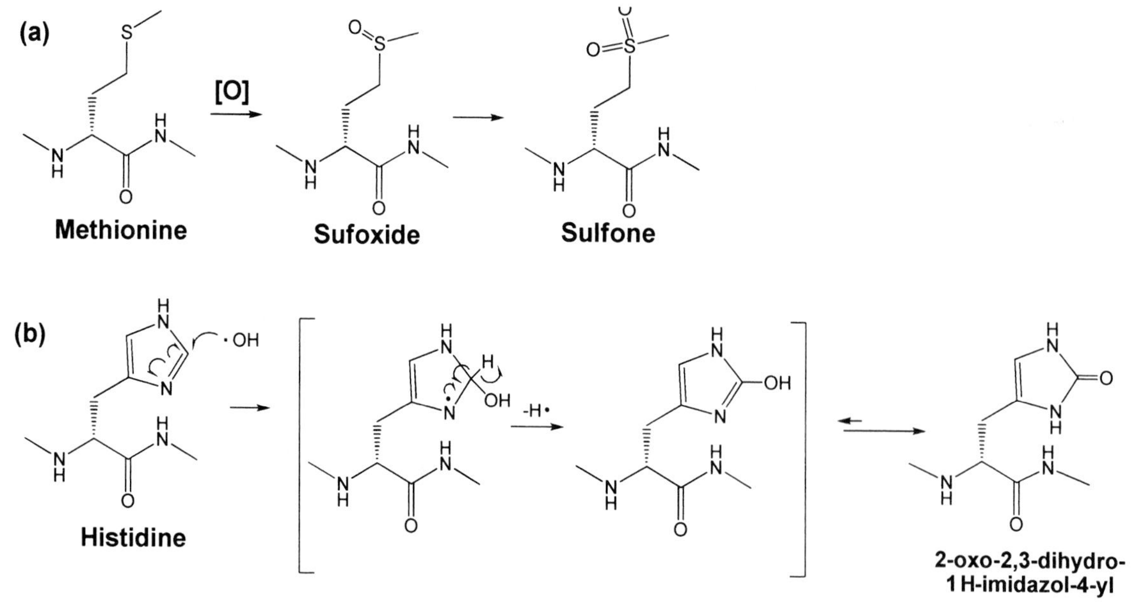

**(a)**

**Methionine**        [O]        **Sufoxide**        **Sulfone**

**(b)**

**Histidine**        · OH        –H·        **2-oxo-2,3-dihydro-1H-imidazol-4-yl**

**Fig. 21–30.** (*a*) The oxidation of methionine to form sulfoxide and sulfone. (*b*) Oxidation of a histidine residue by hydroxyl radicals to form oxo-imidazole.

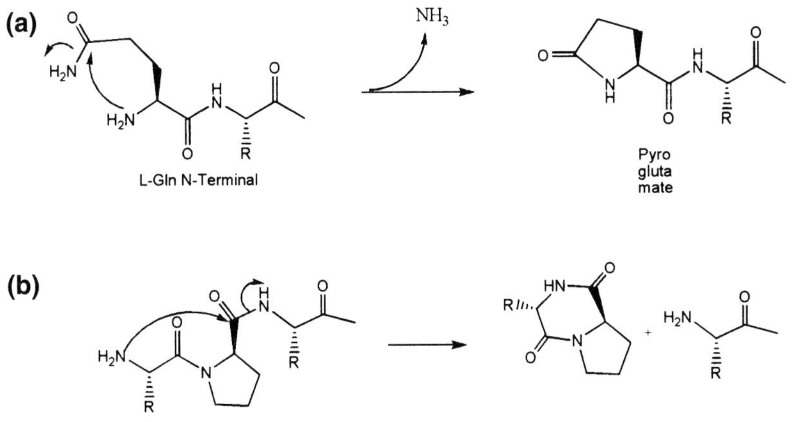

**Fig. 21–31.** (*a*) The conversion reaction of N-terminal glutamine to pyroglutamine. (*b*) The reaction to form diketopeperazine at the N-terminal of a protein or peptide.

by a superoxide-generating system such as hypoxanthine/xanthine oxidase in the presence of Iron(III) and ethylenediaminetetraacetic acid (EDTA) generates NFK and Oia as the major products. The hypoxanthine/xanthine oxidase/Fe(III)-EDTA system generates the reactive oxygen species hydroxyl radical (•OH) produced via a Fenton reaction that oxidizes the Trp residue.[124]

Reactive oxygen species can oxidize the aromatic ring of Tyr to produce 3,4-OH-Phe as the major product; this reaction can also occur upon exposure of proteins to ionizing radiation. Tyrosine-tyrosine cross-linking has been observed in proteins upon UV and gamma irradiation as well as in low-density lipoproteins found in vivo.[126] Similarly, the aromatic ring of Phe can be oxidized to form Tyr, 2-OH-Phe, 3-OH-Phe, and 2,3-OH-Phe with *ortho* tyrosine (2-OH-Phe) as the major product.

## Other Reactions

Other side reactions in peptides and proteins include the formation of pyroglutamate and diketopiperazine. Peptides and proteins that have glutamine and glutamic acid residues at their N-terminus can form pyroglutamate degradation products (**Fig. 21–31**). The deamidation of glutamine at the N-terminus is more rapid than the deamidation of this residue when it is located in the middle of protein sequences. The driving force for this reaction is the formation of a stable five-membered ring when the Gln or Glu residue is present at the N-terminus. This reaction is not observed in N-terminal Asn residues because the ring product is an unfavorable four-membered ring. Pyroglutamate formation was observed during a stability study of a decapeptide vaccine (ELAGIGILTV) containing an N-terminal glutamic acid.[127]

Peptides and proteins that have an Xaa-Pro residue at the N-terminus may be prone to diketopiperazine formation with release of the rest of the protein product with deficient Xaa-Pro residues. This reaction was first observed in a Gly-Pro peptide (**Fig. 21–31**). This reaction has been observed in recombinant human vascular endothelial growth factor.

As mentioned previously, oligonucleotides (i.e., DNA and RNA) have also been investigated as potential therapeutic

agents. These molecules are also subject to a variety of chemical degradation reactions.[100,98] DNA and RNA molecules can both undergo various chemical changes via hydrolysis or oxidation reactions. RNA is generally less stable than DNA. The hydrolysis reaction can cause the breakup of the oligonucleotide chain and isomerization of the phophoester group on the ribose ring. As expected, these degradation reactions are strongly influenced by the intrinsic properties of the solution such as pH, buffer, and ionic strength. Similar to proteins, external conditions such as temperature and light can also have dramatic effects on the stability of oligonucleotide-based drugs. Physical instability (i.e., conformational changes, aggregation, and precipitation) of oligonucleotides and plasmids can often be induced by their chemical degradation.

The hydrolysis reactions of oligonucleotides are catalyzed by acid or base as illustrated for RNA degradation in **Figure 21–32**. Acid-catalyzed degradation produces two degradation pathways. The first results in a phosphoester bond shift from C3′ to C2′ to make oligonucleotide 5; this bond shift will affect the higher-order structure of the RNA. In this case, the acid-catalyzed degradation is initiated by protonation of the oxygen of the phophoester group (compound 2) followed by a nucleophilic attack of the 2′-OH group on the phosphorous atom to produce intermediate 3. The proton transfer from the 2′-oxygen to the 3′-oxygen generates intermediate 4, which upon a five-membered ring opening reaction produces degradation product 5 with an oligonucleotide chain shift. The second pathway generates two fragments of RNA (i.e., 7 and 8 in **Fig. 21–32**). In this route, the proton on the 2′-oxygen in compound 3 can shift to the oxygen attached to the methylene group of the next nucleic acid to generate intermediate 6. The unstable intermediate 6 undergoes a fragmentation reaction and produces the two smaller pieces of RNAs, 7 and 8.

Fragmentation of RNA can also be catalyzed by base. In this case, the reaction is initiated by deprotonation of the 2′-hydroxyl group as shown in intermediate 9 followed by the nucleophilic attack of the phosphorous atom by the 2′-oxy-anion to yield intermediate 10. The base abstraction of the hydroxyl proton of the phosphate group in intermediate

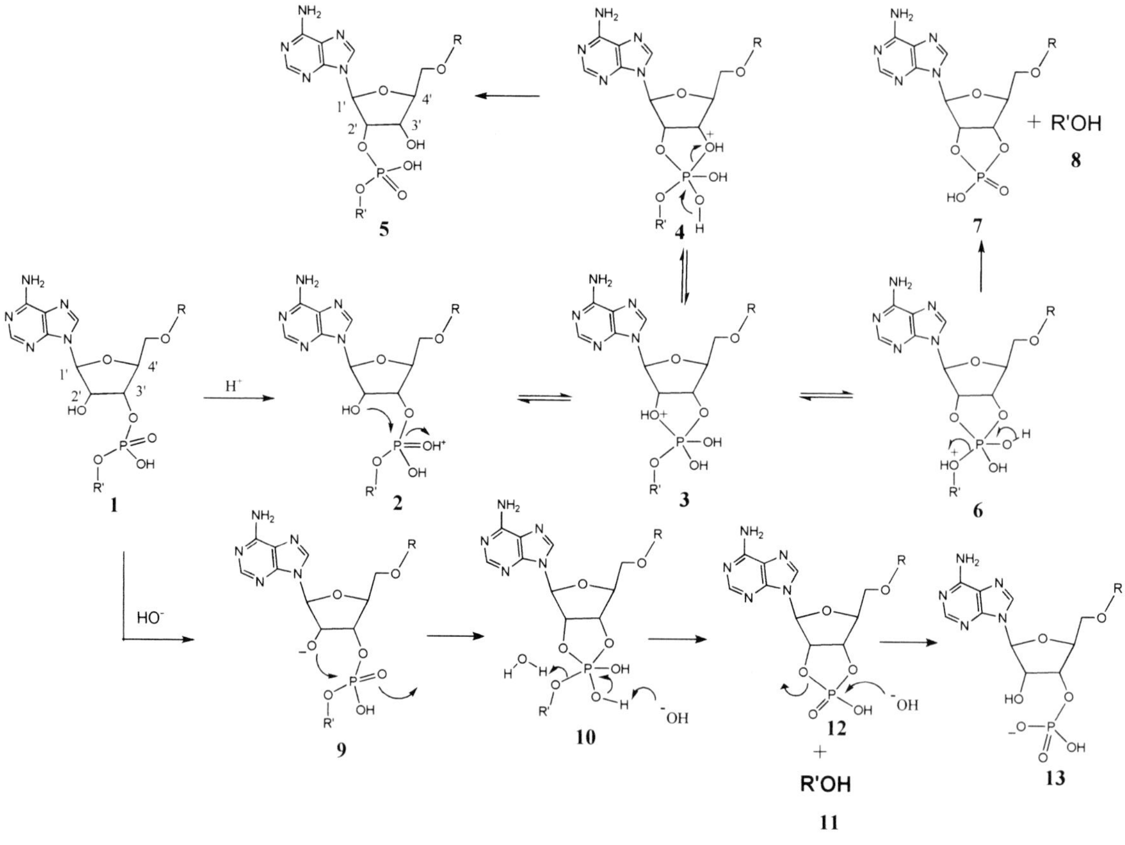

**Fig. 21–32.** Schematic diagram of some degradation reactions of RNA catalyzed by acid or base. These degradation reactions cause RNA fragmentations and phosphoester bond shift.

10 leads to the fragmentation of the RNA to give a smaller RNA 11 and intermediate 12. Opening of the five-membered phophoester in 12 produces another RNA fragment 13.

Oligonucleotides undergo other hydrolysis reactions at different locations within the molecule to produce other products, including the release of a base (e.g., purines) as well as the ring opening of the ribose group. Furthermore, modification of the base groups can be catalyzed by base. Oxidation reactions of RNA produce RNA with an open ribose ring as well as RNA with modified base groups. For a further description of these other reactions and related changes in DNA, readers are encouraged to read the appropriate reviews.[100,98]

## FORMULATION

In the previous sections the production, characterization, and the most common physical and chemical pathways of degradation of biotechnology-based products were described. How are these macromolecules and their complexes formulated? The goal is to take the various types of information that was gathered to create actual drugs and vaccines.

The initial major concerns with solution state biopharmaceuticals usually involve conformational stabilization, prevention of aggregation, and inhibition of chemical degradation reactions. Thus, the first step in most formulation

procedures is to identify such events. This is most commonly done using "accelerated-stability" protocols. Thus, various types of stress are applied to the system of interest and the methods described previously are used to detect changes in their physical and chemical structure including the association state of the molecule/macromolecular complex. The most common forms of stress applied in rough order of their utility are temperature, pH, redox potential, solute, shear (shaking), and freeze/thaw cycles. Ideally, one could employ actual, selected storage conditions (e.g., 2°C–8°C, 24–48 months) but this is generally not possible due to temporal constraints in product development timelines. Ultimately, however, real-time stability studies must be the ultimate arbitrator of successful stabilization. Given the molecular complexity of biopharmaceuticals and the wide variety of methods available for their analysis, the decision on how to proceed is often a difficult one.

One frequent approach is to pick one or several techniques that are expected to be sensitive to major degradation pathways. For example, one might use DSC to evaluate thermal stability at several different pH values and monitor stress-induced aggregation with SEC, oxidation induced by $H_2O_2$, and deamidation by high pH with HPLC-MS. Then, as described below, potential excipients can be tested for their ability to inhibit any degradation processes observed. The problem with such an approach is that important degradation events might be missed because of the limited ability of a small number of conditions, events, and methods to detect all potential problems. There are a large number of variations on this approach based on both the techniques and types of stress employed. Alternatively, attempts can be made to cover a much wider formulation space by using many techniques and a wide range of solution conditions. One widely documented approach to the analysis of physical degradation makes measurements at 0.5 or 1.0 pH intervals from 3 to 8 and from low to high temperatures (e.g., 0°C–100°C). A series of methods such as CD (to detect secondary structure changes), intrinsic fluorescence (to monitor tertiary structure), ANS fluorescence (to detect alterations in apolar surface exposure), and static and/or dynamic light scattering to measure association (aggregation)/dissociation phenomena are used to characterize the response of the physical state of the system to stress. Changes in dynamic behavior can be analyzed in a similar manner using methods such as isotope exchange, US spectroscopy, PP-DSC, fluorescence anisotropy, or solute quenching. Similarly, the presence of chemical alterations of individual residues can be described in terms of their rate constants as detected by LC-MS. All of the above approaches have recently been facilitated by the availability of HTP technology for their execution. The major weakness in all of the above is that it is still possible to miss key degradation events. Furthermore, the results obtained may not always be extrapolated to actual pharmaceutical storage conditions although this does not often appear to be the case. Variations of this HTP approach have been successfully applied to therapeutic peptides, proteins, VLPs, viruses, and bacterial cells as well as various vaccines types and accompanying adjuvants. It can also be used with solid-state formulations (see below). How does one analyze the vast array of information that is obtained by such an extensive collection of data and experimental conditions? One method is to use the EPD method described earlier for the analysis of high-resolution UV absorption data. In the multiple techniques version, however, normalized data from all of the techniques employed are used to construct the EPD. The resultant colored summary of the effect of the chosen variables (T, pH, drug concentration, ionic strength, agitation, freeze/thaw, etc.) on the physical and chemical structure and behavior of the biopharmaceutical can then be used to guide formulation development (**Fig. 21–33**). We should also mention that it is possible to redesign the macromolecular system if the exact mechanism of degradation is known. For example, residues in a protein that undergo deamidation or oxidation could be replaced with nondegrading analogues or the interior packing of the protein's amino acid residues could be improved through modern protein design methods to improve stability. Although in many ways an ideal solution (although immunogenicity can become a problem), this has yet to become a routine approach to stability problems. It may become so in the future.

Often the first thing one does with a biopharmaceutical solution formulation is to select a buffer. A number of considerations are necessary for an optimal choice. First and

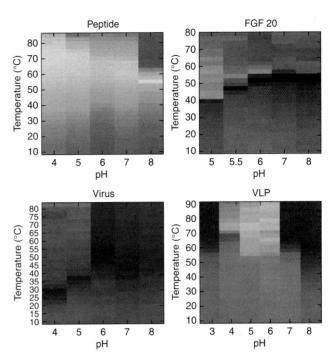

**Fig. 21–33.** Representative EPDs using multiple methods. For example, CD spectra (secondary structure), intrinsic fluorescence (tertiary structure), ANS fluorescence (appearance of apolar binding sites), and light scattering (association and aggregation) are obtained after normalization as vectors and represented as colors (or in this case regions of different shadings). Apparent phases of different colors then represent different structure states. See **Figure 21–11** and text.

foremost is that the buffer supports optimum structural stability and solubility. This can often be achieved through the methods outlined above. For example, a pH can be selected to place the formulation as far as possible from structural changes produced by changes in pH and temperature using a single method (e.g., CD, fluorescence, DSC) or as distant as possible from the apparent phase boundaries of an EPD. One caveat, however, is that a number of proteins undergo pH-dependent conformational changes as a part of their normal biological functions. For example, the lowering of the pH inside endosomes (sometimes the immediate destination of a protein taken up into a cell) might induce a structure change that exposes apolar regions, which permits interaction with the endosomal membrane and subsequent release into the cytoplasm. Furthermore, because so many degradation processes are pH dependent, it may be necessary to compromise in the selection of pH. Tables of physiologically acceptable buffers and their $pK_a$'s are readily available. Among the many popular buffers used for biopharmaceuticals are phosphate ($pK_a$'s 2.12, 7.21, 12.32), citrate (3.06, 4.74, 5.40), and imidazole (7.00). By combining buffers, it is also possible to obtain a very broad buffering range.

A variety of other factors may be involved in buffer choice. Buffer ions may specifically interact with proteins due to their charged nature. They may also chelate or be contaminated with metals, a potentially important phenomenon. One may also need to consider the effect of temperature on a buffer's $pK_a$ values. Some buffers, such as the commonly used Tris (tris-(hydroxymethyl) aminomethane) species shift their $pK_a$ by as much as $-0.03/^\circ C$.

The next step in the formulation of macromolecules (if necessary, as it frequently is) is the selection of excipients to control critical degradation processes. This is usually done by screening a group of compounds and polymers usually referred to as GRAS (Generally Regarded as Safe) materials. This can be accomplished by using any one or a combination of the methods discussed above. Potential excipients are usually initially tested at high concentrations with concentration dependence studies employed later to define the minimum concentration that can be used to obtain the desired effect.

The GRAS excipient list is based on compounds currently used in marketed formulations of drugs. It consists of a collection of carbohydrates, polysaccharides, amino acids, small molecules, detergents, and polymers among other agents. A number of different mechanisms mediated by these compounds can stabilize macromolecular systems. High concentrations (i.e., $>0.3$ M) of sugars, amino acids, and some salts appear to stabilize through a mechanism known as preferential exclusion.[128] This is based on the greater surface area of a macromolecule in its structurally disrupted (unfolded) state. The presence of the stabilizing agent causes the chemical potential (free energy) of the macromolecule to be increased in a manner proportional to its surface area. Because this is an unfavorable process, the effect is to differentially stabilize the native state (**Fig. 21–34**). In contrast, some stabilizers bind better directly to the native state. This shifts the $N \leftrightarrow U$ equilibrium to the native form resulting in stabilization. For example, the presence of extended polyanion binding sites on many proteins such as growth factors and coagulation factors means that polymers like heparin and dextran sulfate can often have dramatic stabilizing abilities.[129–131] Some compounds act by either directly or indirectly inhibiting aggregation. Those that act indirectly generally do so by stabilizing the native state which delays formation of aggregation competent species such as MG forms. Direct effects occur through blocking of the protein/protein interactions that are responsible for association processes. Inhibition of protein aggregation by detergents is thought to occur through one or both mechanisms. Because of the presence of disulfide bonds and free thiol groups in proteins, it is sometimes possible to stabilize proteins by the inclusion of a reducing agent to either maintain free thiols in their reduced (and active) form or prevent the formulation of nonnative inter- or intramolecular disulfides which leads to inactive forms. Since metals can inactivate macromolecules through a variety of mechanisms such as oxidation, the presence of a chelating agent can be used to minimize such problems.

As mentioned previously, the amphipathic nature of proteins means they are usually quite surface active. By this we mean that they have a strong tendency to bind to surfaces such as air/water interfaces as produced by agitation or to the inner surfaces of storage devices such as vials and syringes. In fact, at low concentration ($<10$ $\mu g/mL$), a substantial portion of the macromolecules may be resident on a surface.[132] At least three common approaches have been used to minimize these problems. In the first, the design or nature of the surface itself can be altered. This is accomplished most frequently by the use of different materials or the addition of a coating that lowers interactions with proteins. The second method often employs the presence of a proteinaceous material such as serum albumin, casein, or gelatin to competitively prevent or displace the macromolecular drug substance or vaccine from the surface. Initially, animal-derived versions of these proteins were used but they are being replaced by recombinant forms of the same or similar proteins. Third, detergents are often used for a related purpose due to their affinity for both proteins and/or container surfaces. It has also recently been recognized that particles originating from various sources such as the plastic or metal materials produced by the degradation of vial filling pumps and the tungsten used in syringes can result in particulate matter that must be removed for clean formulations to result. Many of the above phenomena can also result from conditions produced during the shipping of biopharmaceuticals. Solutions to such problems are similar to those described above, but careful shipping studies are essential to identify and minimize their occurrence.

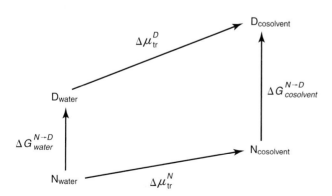

**Fig. 21–34.** The thermodynamic mechanism of stabilization of macromolecules by solutes which cause their preferential hydration. The solute (cosolvent) causes a greater difference in free energy of the unfolding reaction than in its absence leading to a destabilization of the unfolded (D) state (and therefore a relative stabilization of the native (N) form).

## Drying of Biopharmaceuticals[133]

Despite recent advances in the preparation of stable, solution-based formulations of peptides, proteins, nucleic acids, and viruses, it is still frequently necessary to employ dried formulations of biopharmaceuticals. By removing most (but not all) of the water and reducing inter- and intramolecular mobility, it is usually possible to dramatically stabilize such systems. Dried formulations are often considered less desirable because of significantly increased expense and the need for reconstitution, but the dramatic improvement in stability obtained by such technologies is thought to be more than adequate reason for their use. In fact, many currently marketed biopharmaceuticals and vaccines employ such methods. By far, the most common method employed is that of lyophilization, which is also known as freeze-drying. The techniques of spray drying and spray–freeze drying have, however, also been used to dry biopharmaceuticals. There exist a variety of other potentially useful technologies such as foam drying, which will not be described here.

Freeze-drying is far and away the most commonly used procedure.[134] It is essentially a batch process in which water is removed directly from the solid state. It is generally performed in three distinct phases. In the first, the water in a solution of the biopharmaceutical agent is converted to ice. This results in the production of a concentrated frozen macromolecular solid. This freezing step is typically performed in the temperature range of $-45°C$ to $-10°C$ for 2 to 5 hr. The second stage is referred to as primary drying. This involves the removal of some unfrozen water (ca. 15%) and sublimation of ice at $-10°C$ to $-40°C$ for 5 hr to 5 days (this stage is highly variable in time). The final procedure (secondary drying) involves removal of most of the remainder of the unfrozen water down to 1% to 4% as the temperature is increased from the previous process to $4°C$ to $50°C$ for 5 to 15 hr.

In contrast, spray drying is a continuous process and involves drying from the liquid state.[135] Its initial step consists of atomization of a macromolecule containing aqueous solution into small droplets. This is generally considered to be the most problematic aspect of the procedure given the presence of the air/water interface, a potential site of protein degradation and aggregation. The droplets are mixed with hot air (ca. $120°C$) which rapidly (in seconds) removes most of the water resulting in concentrated solute. After further cooling at $40°C$ to $50°C$, only a low water content (3%–5%) remains. The latter water content is typically somewhat greater than that produced by lyophilization and may result in greater degradation of the macromolecular drug or vaccine upon storage. In spray-freeze drying, the atomization and freezing is carried out in a solvent such as liquid nitrogen followed by macromolecule/vaccine drying.

Because it is so much more commonly employed, we will consider lyophilization in more detail. In the initial freezing step, most biomolecules form amorphous solids (in contrast to many small molecules solutes which may crystallize). In general, primary drying is performed $2°C$ to $3°C$ below $T_g'$ (Tee Gee prime), the glass transition temperature of the freeze concentrate. The glass transition temperature ($T_g$ for a pure solid) refers to the softening of a glasslike solid to form a viscous liquid state, which permits increased molecular mobility and subsequently enhanced degradation. One potential problem occurs because of the concentration that occurs during freezing. A buffer-like sodium phosphate may crystallize causing a shift of several units to lower pH, a potentially degradative condition. Thus, phosphates are usually not

employed (although small amounts may be acceptable). In general, a minimal weight ratio of buffer to other solutes is used to minimize crystallization-induced pH shifts as well as prevent large reductions in $T_g'$ and therefore increased solute mobility.

What is the significance of glass transitions[136]? During primary drying (ice sublimation), $T_g'$ reflects the temperature at which the conversion from a glassy to rubbery solid state occurs. As the material is dried it can also undergo a loss of structure that is referred to as cake "collapse" with the temperature at which this occurs designated the collapse temperature ($T_c$). This can be detected by a special form of microscopy known as "freeze-drying microscopy"[137] or by DSC[138] as a district thermal event or by a change in electrical resistance. The collapse temperature is usually a few degrees higher than $T_g'$.

During secondary drying as the unfrozen water is removed, this phase change is seen hear the $T_g$. These temperatures are very formulation dependent. As one moves above a glass transition temperature, the mobility and reactivity of the macromolecule or its complexes increase. During storage, the $T_g$ can have a controlling impact on the drug's stability. In general, as the amount of residual water increases, the $T_g$ is lowered. Thus, knowledge of this property is one key to preparing a stable, lyophilized formulation.

To create a stable dry formulation of a biotechnology-based drug, all of the above must be considered in its creation. Besides optimization of the lyophilization cycle, such formulations almost always contain excipients.[139,140] These are used to facilitate stabilization during freezing stress ("cryoprotection"), stabilization during freezing and drying ("lyoprotection"), and stabilization in the dry state to enhance integrity during storage. In addition, excipients are used for a number of other reasons. Bulking agents such as mannitol or glycine are often employed for "elegance" and to prevent "blow-out" in which the dry cake can be expelled into the freeze dryer. Bulking agents are often chosen for their crystallinity and their high eutectic temperature to facilitate rapid, easy drying. Crystallinity is typically evaluated by a combination of polarized light microscopy (to detect birefringence), x-ray powder diffraction, and calorimetry. Buffers are often included for pH control although care must be taken that their crystallization does not produce large and potentially destructive pH shifts. As a general rule, however, they are used in minimal amounts. Isotonicity modifiers such as glycerol and NaCI are also often used although they may be present in the diluent rather than the formulation itself. In addition, compounds such as hydroxyethyl starch can be used to raise the $T_c$ of the product (i.e., to increase $T_g'$). Especially critical for biopharmaceuticals, stabilizers are often necessary to provide a sufficiently robust formulation. The complex relationship between water content, molecular mobility, and the physical and chemical degradation of dried macromolecular systems can make the selection and optimization of stabilizers especially challenging. There are, however, several generally accepted principles for successful stabilization of biomolecules in the solid state.[140] First, it is well established that the stabilizer must remain amorphous and in the same phase as the drug. Conversely, physical mixtures do not effectively stabilize.[141] Second, the stabilizer should be chemically and physically inert. A well-known example of this problem involves the use of sucrose, often a highly effective stabilizer. At low pH, this disaccharide can be hydrolyzed to reducing sugars, which can covalently interact with proteins. Third, as mentioned previously, the formulation should not permit selective buffer crystallization and consequent pH shifts. Many macromolecules and viruses and other biological entities are often pH sensitive with losses in biological properties upon exposure to extremes of pH. Finally, it is very clear that low water content is often essential for optimal stabilization in dried formulations. It should also be mentioned that ice is a major stress during freezing. The formulation of an ice/water interface may result in adsorption of proteins and other amphipathic macromolecules which can be significantly destabilizing. This is at least partially due to the forces exerted by the surface on macromolecules due to the multipoint nature of the contacts between the surface and the drug or vaccine. The presence of stabilizers may reduce such destabilizing effects but the mechanisms are incompletely understood. The preferential hydration (solute exclusion) mechanism discussed previously may be operative at this level. In the solid state, several factors are considered to play critical roles in stabilization. Chief among these is mobility. It has long been thought that dried formulations are most stable in the glassy, solid state. The existence of such a state, however, does not guarantee long-term stability, even at moderate temperatures. In addition, the presence of the "native" state of a protein is generally considered necessary. This is most frequently analyzed by FTIR spectroscopy which can conveniently analyze protein, nucleic acid, and viral structure in the solid state. A common observation is that without the presence of stabilizers, lyophilized proteins are not in their native state in their solid forms, producing accelerated chemical as well as physical degradation.

A number of mechanisms have been proposed to explain how stabilizers are able to maintain macromolecular structure under conditions of low moisture.[140] The water substitution hypothesis argues that many stabilizers interact with proteins and other biological entities in a manner similar to water. This is proposed to support the native state of such molecules and provide stabilization during freezing and drying by providing an appropriate physical environment. Two main lines of evidence in support of this hypothesis are that (a) many stabilizers are sugars and due to their multiple hydroxyl groups are able to hydrogen bond to macromolecular systems in a manner similar to water and (b) spectroscopic studies demonstrate water-like interactions between stabilizers and biomolecules in the solid state. The water substitute hypothesis has primarily been used to explain stabilization during drying rather than during storage. The second major hypothesis postulates that by creation of a glassy state, there results in a reduction in macromolecular mobility which leads to

a decrease in the rate of degradative events. Two types of motions are recognized in glasses. The more global dynamic behavior is known as $\alpha$-relaxation. It is directly related to viscosity and involves long time and length scales. Conversion to this behavior occurs when solids are converted to liquid-like states and are measured by $T_g$. Fast dynamic behavior is designated $\beta$-relaxation. This involves local motions on a much shorter length and time scale and can be measured by a variety of methods including dielectric, neutron scattering and NMR relaxation techniques. The relationship between the effects of stabilizers on these different types of motions is not simple, however. While it might be expected that stabilizers would simply decrease the amplitude of such processes, both increases and decreases have been seen similar to observations in solution. Furthermore, lowering the $T_g$ does not always destabilize. This remains a very active area of current research with a consensus that dynamics are important, but their precise role yet to be definitively defined.

Some tentative conclusions about the mechanisms of excipient stabilization can, however, be advanced. Cryoprotection may involve solvent exclusion if instability occurs early in freezing. If it occurs later, immobilization by vitrification is more likely. If surfaces are involved, the coating of such surfaces by a surface-active agent such as a surfactant or protein (serum albumin, gelatin, etc.) may be helpful. Lyoprotectants are usually amorphous, chemically inert glass formers. They form single phases with macromolecules and "moderately" interact with their surfaces. They should couple all relevant modes of motion, both local and global, to the matrix and preserve native structure during freezing and drying. The requirements for storage stabilization are similar but specific considerations are usually necessary for each individual biomaterial based on its specific sensitivity to their unique degradation pathways.

Here are a few general rules to guide formulation of macromolecules for freeze-drying. (a) The amount of buffer should be minimized (avoid phosphates). (b) Employ other salts only if needed (i.e., for solubility) and minimize their amounts. (c) Maximize $T_g'$ ($< -35°C$ is usually a problem; $< -40°C$ is typically unacceptable). Lyophilization should be performed below $T_g'$ (or at least below the collapse temperature) in primary drying and below $T_g$ in secondary drying.

If stability problems arise, the following approaches have often proven successful to minimize such difficulties. If the problem occurs during freeze-drying, it should be isolated to either freeze/thaw or freeze dry stability. If the problem is during freezing, the addition of surfactants (Tweens and Pluronics) or high levels of "excluded solutes" (e.g., amino acids, nonreducing carbohydrates, polyethylene glycols) may be useful. If the problem is seen during drying only, the use of nonreducing carbohydrates such as sucrose or trehalose is often successful. If one has a storage stability problem, the moisture level needs to be carefully controlled. It is not unusual, however, for this to be inadequate to completely solve the problem. In general, the best additional step has been to use a nonreducing carbohydrate like sucrose at neutral or basic pH (if moisture control is not a problem). At low pH and higher moisture content, trehalose often proves to be a better choice as a storage stabilizer.

## Formulation of Vaccines[87,90,142,143]

The formulation of vaccines is complicated by their varied nature (peptide, protein, VLP, virus, DNA, bacteria, polysaccharide) as well as the need for adjuvants[142,143] in their less immunogenic forms. Nevertheless, the procedures employed are very similar to those described above for individual macromolecules. One apparently important difference is the ultimate goal. In the case of biotherapeutics, one wishes to keep immunogenicity to a minimum. In contrast, we desire to maintain an optimal immunogenicity for vaccines. Fortunately, however, both states can be achieved in the same manner, namely, by maintaining a particular structure (typically the native one) both in vivo and during long-term storage. Thus, the immediate relevance of the discussion in the preceding sections should be evident. What about complex vaccines such as attenuated viruses[87]? In such cases, it is still often possible to treat them as physiochemical systems. One must first isolate the virus in purified (i.e., >90%) form. They can then be subjected to the various stresses as described above and analyzed by the same physical and chemical methods. In this case, however, the resultant signals are the sum of the signals from all of the component viral proteins and nucleic acids weighted by the relative amount of each macromolecule and their individual signal intensities. In the case of many viruses, the measured experimental results will primarily reflect viral coat proteins as well as the integrity of the viral particle. If the rate-limiting degradation events are reflected by changes in the properties of components with large contributions to observed signals, the formulation methods described in the previous sections may well be effective. In many cases, however, live attenuated viruses are needed for efficacious vaccines. In fact, these live entities may constitute a very small minority of the total viral particles in a preparation. It is frequently the case, however, that all of the viral particles in such a mixture undergo the destabilizing changes of interest so that physical and chemical methods can still be used in accelerated stability studies. This is because the events that originally inactivated the majority of the viruses are distinct from those that are relevant to long-term stability studies (the initial inactivating events often occur during the cell culture process as is evident from kinetic studies). A second problem is that many vaccines are now used in combination form (measles, mumps, rubella (MMR); diphtheria, tetanus, pertussis (DPT); etc.).[144] In such cases, the individual components must be studied separately and stabilizers identified for each individual component. Mixtures of stabilizers can then be used to address individual problems. Such approaches can also be applied to VLPs, DNA vaccines containing cationic delivery vehicles, and even entire bacterial cells. The latter is again possible if a critical event in bacterial integrity can be detected during accelerated stability testing. Never-

theless, critical physical and/or chemical degradation events may not be detectable in complex vaccines. In such cases, a trial and error (empirical) method must be used. One then usually employs an animal model (most commonly mice) with maintenance of immunogenicity as judged by stimulation of specific antibody levels (less commonly cytokine production). This is usually done by various types of ELISA assays. Selection of potential excipients is based on the principles outlined above with GRAS agents screened as potential stabilizers. In the case of live antigens such as attenuated viruses, cellular responses may also be measured because of the essential nature of the replicative states. Thus, vaccines such as those for measles can have their ability to kill sensitive cells used as a criterion for efficacy using plague assays.

Another complication is the use of adjuvants with weakly immunogenic antigens such as monomeric recombinant proteins. The most commonly used adjuvants are the aluminum salts. Although only effective at enhancing humoral (antibody) responses, their well-established safety and efficacy profiles have resulted in their widespread use. Aluminum salts are usually used in the form of aluminum hydroxide and aluminum phosphate, the former positively charged at neutral pH, the latter negatively charged under these conditions. A general principle in the use of these adjuvants is that antigens must be adsorbed to the surface of these particulate salts for them to be effective at enhancing immune responses. Thus, the first step in their formulation is to perform binding studies of antigens to their surface (**Fig. 21–35**). This is simply done by adding incremental amounts of antigens to aluminum salts, incubating for a short period, followed by separation of unbound antigen from antigen/aluminum salt complexes by centrifugation. The amount of unbound antigen is then measured (typically by optical absorbance or a dye-binding method) and the amount of antigen bound determined by subtraction from the amount added. The general rule is that negatively charged antigens bind to the positively charged aluminum hydroxide and positively charged antigens

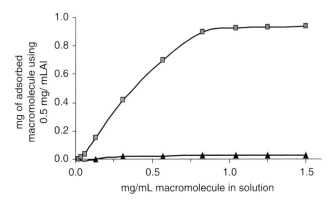

**Fig. 21–35.** A representative binding isotherm of a vaccine protein to an aluminum salt adjuvant. In this case, the protein is negatively charged (it has a low pI) and binds well to positively charged aluminum hydroxide (squares). In contrast, it binds poorly to negatively charged aluminum phosphate (triangles).

to the negatively charged aluminum phosphate. Typical formations contain a milligram or less of aluminum with the amount of antigen more variable but typically in the range of a few micrograms to hundreds of micrograms. The charge on the antigen is usually estimated from its isoelectric point.

Suspensions of aluminum salts are optically opaque. Thus, methods that can be used to examine adsorbed proteins are more limited than those available for solution studies. Several methods are available, however, to examine proteins and other macromolecular systems adsorbed to aluminum salts adjuvants. For example, fluorescence methods can still be used. Sometimes enough light can still penetrate and exit such suspensions, sufficient for emission spectra to be obtained. If this is not the case, emission can be measured off the surface of the sample by examining emission at a lower angle (e.g., $45°$–$60°$). This is known as front face fluorescence. As discussed earlier, both FTIR (in ATR and DRIFT modes) and Raman spectroscopy can be used in highly scattering samples. Thus, secondary structure information can be obtained by analysis of amide bands by both techniques. The problem of light scattering can be avoided entirely by the use of DSC since this is a thermal technique. The latter methods may not be sensitive enough to detect small amounts of absorbed antigens but fluorescence is usually sensitive enough for such applications. Effects of temperature and pH can be analyzed but it must be shown that the antigen remains adsorbed to the adjuvant surface under these conditions.

Alternatively, the antigen can be removed from the surface and then examined. This can often be done by low pH, high salt concentration, or treatment with an agent that dissolves the adjuvant (e.g., citrate) or a reagent that weakens antigen/adjuvant interactions such as low concentrations of urea or guanidine hydrochloride (high concentrations cannot be used because they usually disrupt antigen structure). Although the interactions between antigen and adjuvant often contain a major electrostatic component, they may also involve apolar and van der Waal forces among other types of weak noncovalent interactions. A mechanism known as ligand exchange may also occur.[145] Unfortunately, it is not uncommon for antigens to be difficult or impossible to remove. This is especially the case after long storage times where the antigen may undergo structural changes as it optimizes its interaction with the aluminum salt surface. In fact, it is often found that antigens are destabilized when they interact with aluminum salts.[146] This is typically manifested by a lowering of the $T_m$ measured by DSC, temperature-dependent fluorescence, or vibrational spectroscopic methods. Fortunately, it turns out that compounds that stabilize macromolecules and viruses in solution often also stabilize them on aluminum salt surfaces, although usually to a lesser extent.[147] One problem with aluminum salt formulations deserves special mention. Because aluminum hydroxide is positively charged, it attracts hydroxide anions to its surface. This increases surface pH leading to enhanced deamidation of protein antigens. This can often be prevented, however, by the inclusion of small (millimolar) amounts of phosphate in the formulation, which

lowers the surface pH by converting small amounts of the aluminum hydroxide to aluminum phosphate. Until recently, it was thought that aluminum salt adjuvants could not be lyophilized. Recent work, however, suggests that the presence of carbohydrate stabilizers will permit this to be done[148] and opens up the possibility of using drying technologies to improve the stability of aluminum salt vaccine formulations.

Although the use of other adjuvants is still in its infancy, it is clear that this is unlikely to be the case in the near future. Recent discoveries have found that all mammals possess a series of proteins on the surface of immune cells that recognize highly repetitive structures on the surface of pathogens. These are now known as "toll" receptors.[149] Concomitant with these findings, it was determined that many of the adjuvant materials that had been empirically discovered actually acted through binding to these receptors. This is in contrast to aluminum salt adjuvants, which appear to work through a variety of other mechanisms including depot effects, facilitation of antigen entry into cells, and other specific immune effects. A number of toll receptor–based adjuvants have now been tested in human clinical studies. While safety still remains an issue, it seems highly probable that many of these as well as non–toll receptor adjuvants will become available for human use. For example, a number of synthetic lipid A analogues (e.g., monophosphoryl lipid A), saponins, oil-in-water, and water-in-oil emulsions used alone and in combination appear quite promising.[142,143] In fact, an oil-in-water emulsion containing squalene and two surfactants (MF59) is already available in a commercial flu vaccine in Europe. Novel adjuvant containing vaccines will present unique formulation problems due to their diversity. It seems probable, however, that the methods currently developed in conjunction with new technologies should be able to meet these challenges.

## CHAPTER SUMMARY

Despite the sophistication of modern biotechnology, significant problems still exist from a pharmaceutical perspective. Like all drugs, biotechnology-based pharmaceuticals and vaccines produce side effects in their recipients. Mechanism-based toxicity as well as adverse effects due to the general physical properties of macromolecules and their complexes remain poorly understood. It is clear that animal models are currently inadequate to address such problems. Improved animal models (especially disease based) are presently an area of great interest as are cell culture systems that might be used to elucidate mechanism-based toxicity. One phenomenon of great concern is the immunogenicity of biopharmaceuticals. This can lead to both loss of activity through neutralization by antibodies as well as pathological immune responses such as allergic reactions. Although immune responses to therapeutic proteins often have little if any negative effects, there is a growing concern that this issue must be more aggressively addressed. Conversely, both recombinant protein and

DNA-based vaccines are usually insufficiently immunogenic. Thus, the development of novel adjuvants as well as improvements in the delivery of DNA vaccines is clearly required. With regard to the latter, nucleic acid–based therapeutics remain inadequately developed and understood, necessitating a greater emphasis on pharmaceutical aspects of their behavior and delivery.

In general, the precise relationships between the structure and behavior of biomolecules in both the solution and solid state are still poorly understood. In many ways, this remains a key to the successful development of biopharmaceuticals from a process, analytical, and formulation perspective. As discussed above, the role of protein dynamics in each of these areas has yet to be definitively explored. One aspect of this poor understanding is a lack of availability of potential stabilizers for use as excipients in biopharmaceutical and vaccine formulations. The GRAS list is, in fact, rather limited and offers a quite restricted number of options to the formulation scientist. As our understanding of biomolecular structure increases, however, we can expect that a combination of rational design and HTP screening methods should allow us to greatly expand stabilizer options after appropriate safety considerations.

A practical problem with biopharmaceuticals is their manufacture at a scale sufficient for use in large populations. For example, in the case of monoclonal antibodies it appears that there is insufficient manufacturing capability if a significant number of such proteins currently in clinical trials come to fruition as marketed pharmaceuticals. The manufacture of live agents such as viruses has always proven to be challenging at an industrial scale. Thus, the development of new technologies to aid in the high-level manufacture of biopharmaceuticals is an important goal of modern biotechnology.

As biotechnology-based products begin to go off patent, the possibility and then the reality of less expensive versions of these drugs and vaccines has become apparent.[150,151] The word "generic" is not generally applied to these agents because they are unlikely to be physically, biologically, and functionally equivalent to the original product. Both the terms "follow-on biologics" and "biosimilars" have been applied to such drugs. The major area of controversy with follow-on recombinant proteins has been the extent to which extensive clinical trials are necessary to ensure their safety and efficacy. In particular, are physical and chemical comparisons of biosimilars to the original innovator drugs sufficient to ensure these critical properties? A recent detailed study comparing the properties of EPO products from a wide variety of different sources[152] emphasizes striking differences based on manufacturing process and company of origin raising significant scientific, legal, and regulatory concerns.

Although at its beginnings, the use of biotechnology (e.g., fermentation, industrial enzymes, etc.) produced little public concern, this changed dramatically when it began to be used to genetically manipulate plants and animals. Scientists themselves expressed concerns that it might be difficult to predict the result of the alteration and insertion of new genes into novel cellular environments. At the public level, this went so far as to imagine the creation of genetically altered organisms with unique pathological characteristics and plants, which might spread deleterious genes into non-target plants. All of the proceeding can and have occurred, but so far without any significant disasters. Initially, we perhaps forgot that plant and animal breeders have been doing the same thing for hundreds if not thousands of years with essentially positive results. While the potential for problems and negative public perceptions remain real, the success of modern biopharmaceuticals and vaccines has maintained forward momentum in the use of these technologies. Several recent problems, however, illustrate continued negative perceptions in a minority of the population. For example, the use of stem cells derived from embryos to treat various diseases has raised much controversy due to the source of the cells.[153] This is, however, more of an ethical issue rather than a scientific one. Claims that the measles vaccine or the compound thimerosal (a mercury-containing preservative) in certain vaccines causes autism in children are completely unsubstantiated by scientific evidence but have resulted in significant public concern, nevertheless.[154,155] Such issues will no doubt continue to be raised but do not seem a major impediment to progress in biotechnology.

 **Practice problems for this chapter can be found at thePoint.lww.com/Sinko6e.**

## References

1. E. J. Jewell and F. Abate (Eds.), *The New Oxford American Dictionary*, Oxford University Press, New York, 2001.
2. A. D. Smith et al., (Eds.), *Oxford Dictionary of Biochemistry and Molecular Biology*, Rev ed., Oxford University Press, Oxford, 2000.
3. C. Ratledge and B. Kristiansen (Eds.), *Basic Biotechnology*, 3rd Ed., Cambridge University Press, Cambridge, 2006.
4. D. L. Oxender and L. E. Post (Eds.), *Novel Therapeutics from Modern Biotechnology: From Laboratory To Human Testing*, Springer-Verlag, Heidelberg, 1999.
5. W. J. Thieman and M. A. Palladino, *Introduction to Biotechnology*, Benjamin Cummings, San Francisco, 2004.
6. J. E. Smith, *Biotechnology*, 4th Ed., Cambridge University Press, Cambridge, 2004.
7. V. Moses, R. E. Cape, and D. G. Springham (Eds.), *Biotechnology—The Science and The Business*, 2nd Ed., Harwood Academic Publishers, Singapore, 1999.
8. A. L. Demain (Ed.), *Biotechnology for Beginners*, Academic Press, Heidelberg, 2008.
9. G. A. Petsko and D. Ringe, *Protein Structure and Function*, New Science Press Ltd, London, 2004.
10. J. Kyte, *Structure in Protein Science*, Garland Publishing, Inc., New York, 1995.
11. T. E. Creighton, *Proteins: Structures and Molecular Properties*, 2nd Ed., W. H. Freeman and Company, New York, 1984.
12. S. A. Plotkin, W. A. Orenstein, and P. A. Offit, *Vaccines*, 5th Ed., Saunders, Philadelphia, 2008.
13. B. R. Bloom and P.-H. Lambert (Eds.), *The Vaccine Book*, Academic Press, San Diego, 2003.
14. A. Allen, *Vaccine: The Controversial Story of Medicine's Greatest Lifesaver*, W. W. Norton & Company, New York, 2007.
15. C.-A. Siegrist, in S. A. Plotkin, W. A. Orenstein, and P. A. Offit (Eds.), *Vaccines*, 5th Ed., Saunders, China, 2008, p. 17.

16. R. A. Seder and J. R. Mascola, in B. R. Bloom and P.-H. Lambert (Eds.), *The Vaccine Book*, Academic Press, San Diego, 2003, p. 51.

17. D. V. Schaffer and W. Zhou (Eds.), *Gene Therapy and Gene Delivery Systems (Advances in Biochemical Engineering/Biotechnology)*, Springer-Verlag, Heidelberg, 2006.

18. N. S. Templeton (Ed.), *Gene and Cell Therapy: Therapeutic Mechanisms and Strategies*, 3rd Ed., CRC Press, Boca Raton, 2008.

19. T. A. Cooper, L. Wan, and G. Dreyfuss, Cell, **136**, 777, 2009.

20. D. Castanotto and J. J. Rossi, Nature, **457**, 426, 2009.

21. D. V. Schaffer, J. T. Koerber, and K. I. Lim, Annu. Rev. Biomed Eng. **10**, 169, 2008.

22. D. Schaffert and E. Wagner, Gene Ther. **15**, 1131, 2008.

23. M. A. Mintzer and E. E. Simanek, Chem. Rev. **109**, 259, 2009.

24. J. Kurreck, Angew. Chem. Int. Ed. Engl. **48**, 1378, 2009.

25. K. A. Whitehead, R. Langer, and D. G. Anderson, Nat. Rev. Drug Discov. **8**, 129, 2009.

26. L. Cerchia, P. H. Giangrande, J. O. McNamara, and V. de Franciscis, Methods Mol. Biol. **535**, 59, 2009.

27. G. Mayer, Angew. Chem. Int. Ed. Engl. **48**, 2672, 2009.

28. E. Levy-Nissenbaum, A. F. Radovic-Moreno, A. Z. Wang, R. Langer, and O. C. Farokhzad, Trends Biotechnol. **26**, 442, 2008.

29. H. Lodish, A. Berk, C. A. Kaiser, M. Krieger, M. P. Scott, A. Bretscher, H. Ploegh, and P. Matsudaira, *Molecular Cell Biology*, 6th Ed., W. H. Freeman, New York, 2007.

30. B. Alberts, A. Johnson, J. Lewis, M. Raff, K. Roberts, and P. Walter, *Molecular Biology of the Cell*, 5th Ed., Garland Science, New York, 2007.

31. J. Sambrook and D. W. Russell, *Condensed Protocols from Molecular Cloning: A Laboratory Manual*, Cold Spring Harbor Laboratory Press, Cold Spring Harbor, 2006.

32. B. Lewin, *Genes IX*, Jones & Bartlett Publishers, Sudbury, 2007.

33. R. J. Simpson, P. D. Adams, and E. A. Golemis, *Basic Methods in Protein Purification and Analysis: A Laboratory Manual*, Cold Spring Harbor Laboratory Press, Cold Spring Harbor, 2008.

34. S. A. Doyle (Ed.), *High Throughput Protein Expression and Purification: Methods and Protocols (Methods in Molecular Biology)*, Humana Press, New York, 2008.

35. J. M. Miller, *Chromatography: Concepts and Contrasts*, 2nd Ed., Wiley-Interscience, New York, 2004.

36. L. Hagel, G. Jagschies, and G. K. Sofer, *Handbook of Process Chromatography: Development, Manufacturing, Validation and Economics*, 2nd Ed., Academic Press, Amsterdam, 2007.

37. N. Salamat-Miller, J. Fang, C. W. Seidel, A. M. Smalter, Y. Assenov, M. Albrecht, and C. R. Middaugh, Mol. Cell Proteomics, **5**, 2263, 2006.

38. N. Salamat-Miller, J. Fang, C. W. Seidel, Y. Assenov, M. Albrecht, and C. R. Middaugh, J. Biol. Chem. **282**, 10153, 2007.

39. K. E. van Holde, W. C. Johnson, and P. S. Ho, *Principles of Physical Biochemistry*, 2nd Ed., Pearson Education Inc., Upper Saddle River, 2006.

40. W. Jiskoot and D. J. A. Crommelin, *Methods for Structural Analysis of Protein Pharmaceuticals*, AAPS Press, Arlington, 2005.

41. S. Peng, C. Cui, M. Zhao, and G. Cui, *Pharmaceutical Bioassays: Methods and Applications*, Wiley, New York, 2009.

42. D. Wild (Ed.), *The Immunoassay Handbook*, 3rd Ed., Elsevier Science, Amsterdam, 2005.

43. J. P. Gosling, *Immunoassays: A Practical Approach*, Oxford University Press, New York, 2000.

44. R. Westermeier, *Electrophoresis in Practice: A Guide to Methods and Applications of DNA and Protein Separations*, 4th Ed., Wiley-VCH, Weinheim, 2005.

45. S. Ahuja and M. Jimidar (Eds.), *Capillary Electrophoresis Methods for Pharmaceutical Analysis: Separation Science and Technology*, **Vol. 9**, Academic Press, San Diego, 2008.

46. W. J. Lough and I. W. Wainer (Eds.), *High Performance Liquid Chromatography: Fundamental Principles and Practice*, Springer, New York, 2008.

47. G. E. Katzenstein, S. A. Vrona, R. J. Wechsler, B. L. Steadman, R. V. Lewis, and C. R. Middaugh, Proc. Natl. Acad. Sci. U.S.A. **83**, 4268, 1986.

48. L. A. Kueltzo and C. R. Middaugh, in W. Jiskoot and D. J. A. Crommelin (Eds.), *Methods for Structural Analysis of Protein Pharmaceuticals*, AAPS Press, Arlington, 2005, p. 1.

49. C. S. Braun, L. A. Kueltzo, and C. R. Middaugh, in M. A. Findeis (Ed.), *Nonviral Vectors for Gene Therapy: Methods and Protocols*, Humana Press, Totowa, 2001, p. 253.

50. L. A. Kueltzo, B. Ersoy, J. P. Ralston, and C. R. Middaugh, J. Pharm. Sci. **92**, 1805, 2003.

51. H. Mach, C. R. Middaugh, and R. V. Lewis, Anal. Biochem. **200**, 74, 1992.

52. C. N. Pace, F. Vajdos, L. Fee, G. Grimsley, and T. Gray, Protein Sci. **4**, 2411, 1995.

53. W. Jiskoot, A. J. W. G. Visser, J. N. Herron, and M. Sutter, in W. Jiskoot and D. J. A. Crommelin (Eds.), *Methods for Structural Analysis of Protein Pharmaceuticals*, AAPS Press, Arlington, 2005, p. 27.

54. J. R. Lakowicz, *Principles of Fluorescence Spectroscopy*, 3rd Ed., Springer, Singapore, 2006.

55. L. H. Lucas, B. A. Ersoy, L. A. Kueltzo, S. B. Joshi, D. T. Brandau, N. Thyagarajapuram, L. J. Peek, and C. R. Middaugh, Protein Sci. **15**, 2228, 2006.

56. N. Berova, K. Nakanishi, and R. W. Woody (Eds.), *Circular Dichroism*, 2nd Ed., Wiley-VCH, New York, 2000.

57. G. D. Fasman (Ed.), *Circular Dichroism and the Conformational Analysis of Biomolecules*, Plenum Press, New York, 1996.

58. V. A. Bloomfield, D. M. Crothers, and I. Tinoco, Jr., *Nucleic Acids*, University Science Books, Sausalito, 2000.

59. Z. Ganim, H. S. Chung, A. W. Smith, L. P. Deflores, K. C. Jones, and A. Tokmakoff, Acc. Chem. Res. **41**, 432, 2008.

60. A. Barth, Biochim Biophys. Acta. **1767**, 1073, 2007.

61. M. C. Manning, Expert. Rev. Proteomics **2**, 731, 2005.

62. C. R. Middaugh, H. Mach, J. A. Ryan, G. Sanyal, and D. B. Volkin, in B. A. Shirley (Ed.), *Protein Stability and Folding: Theory and Practice*, Humana Press, Totowa, 1995, p. 137.

63. M. van de Weert, J. A. Hering, and P. I. Haris, in W. Jiskoot and D. J. A. Crommelin (Eds.), *Methods for Structural Analysis of Protein Pharmaceuticals*, AAPS Press, Arlington, 2005, p. 63.

64. Z. Q. Wen, J. Pharm. Sci. **96**, 2861, 2007.

65. G. J. Thomas, Jr., Annu. Rev. Biophys. Biomol. Struct. **28**, 1, 1999.

66. S. Choosakoonkriang, C. M. Wiethoff, G. S. Koe, J. G. Koe, T. J. Anchordoquy, and C. R. Middaugh, J. Pharm. Sci. **92**, 115, 2003.

67. S. Choosakoonkriang, C. M. Wiethoff, T. J. Anchordoquy, G. S. Koe, J. G. Smith, and C. R. Middaugh, J. Biol. Chem. **276**, 8037, 2001.

68. B. A. Lobo, S. A. Rogers, S. Choosakoonkriang, J. G. Smith, G. Koe, and C. R. Middaugh, J. Pharm. Sci. **91**, 454, 2002.

69. S. E. Harding, D. B. Sattelle, and V. A. Bloomfield, *Laser Light Scattering in Biochemistry*, Science and Behavior Books, Palo Alto, 1992.

70. B. Chu, *Laser Light Scattering*, Academic Press, New York, 1974.

71. B. J. Berne and R. Pecora, *Dynamic Light Scattering: With Applications to Chemistry, Biology, and Physics*, John Wiley & Sons, Inc., New York, 1976.

72. J. Demeester, S. S. de Smedt, N. N. Sanders, and J. Haustraete, in W. Jiskoot and D. J. A. Crommelin, *Methods for Structural Analysis of Protein Pharmaceuticals*, AAPS Press, Arlington, 2005, p. 245.

73. C. M. Wiethoff and C. R. Middaugh, in M. A. Findeis (Ed.), *Nonviral Vectors for Gene Therapy: Methods and Protocols*, Humana Press, Totowa, 2001, p. 349.

74. J. S. Philo, in W. Jiskoot and D. J. A. Crommelin (Eds.), *Methods for Structural Analysis of Protein Pharmaceuticals*, AAPS Press, Arlington, 2005, p. 379.

75. J. Liu and S. J. Shire, J. Pharm. Sci. **88**, 1237, 1999.

76. A. Schön and A. Velázquez-Campoy, in W. Jiskoot and D. J. A. Crommelin (Eds.), *Methods for Structural Analysis of Protein Pharmaceuticals*, AAPS Press, Arlington, 2005, p. 573.

77. G. W. H. Höhne, W. F. Hemminger, and H.-J. Flammersheim, *Differential Scanning Calorimetry*, 2nd rev ed., Springer, Berlin, 2003.

78. S. Gaisford and M. A. A. O'Neill, *Pharmaceutical Isothermal Calorimetry*, Informa Health Care, New York, 2006.

79. J. E. Ladbury and M. L. Doyle (Eds.), *Biocalorimetry 2*, John Wiley & Sons, Ltd., Chichester, 2004.

80. C. H. Spink, Methods Cell Biol. **84**, 115, 2008.

81. P. L. Privalov and A. I. Dragan, Biophys. Chem. **126,** 16, 2007.
82. B. A. Lobo, S. A. Rogers, C. M. Wiethoff, S. Choosakoonkriang, S. Bogdanowich-Knipp, and C. R. Middaugh, in M. A. Findeis (Ed.), *Nonviral Vectors for Gene Therapy: Methods and Protocols*, Humana Press, Totowa, 2001, p. 319.
83. B. A. Lobo, G. S. Koe, J. G. Koe, and C. R. Middaugh, Biophys. Chem. **104,** 67, 2003.
84. T. J. Kamerzell and C. R. Middaugh, J. Pharm. Sci. **97,** 3494, 2008.
85. V. J. Hilser, E. B. Garcia-Moreno, T. G. Oas, G. Kapp, and S. T. Whitten, Chem. Rev. **106,** 1545, 2006.
86. R. Esfandiary, J. S. Hungan, G. H. Lushington, S. B. Joshi, and C. R. Middaugh, Protein Sci., In Press, 2009.
87. C. J. Burke, T. A. Hsu, and D. B. Volkin, Crit. Rev. Ther. Drug Carrier Syst. **16,** 1, 1999.
88. D. B. Volkin, G. Sanyal, C. J. Burke, and C. R. Middaugh, Pharm. Biotechnol. **14,** 1, 2002.
89. J. F. Carpenter and M. C. Manning (Eds.), *Rational Design of Stable Protein Formulations: Theory and Practice (Pharmaceutical Biotechnology)*, Springer, New York, 2002.
90. D. T. Brandau, L. S. Jones, C. M. Wiethoff, J. Rexroad, and C. R. Middaugh, J. Pharm. Sci. **92,** 218, 2003.
91. K. A. Dill and S. Bromberg, *Molecular Driving Forces: Statistical Thermodynamics in Chemistry and Biology*, Garland Science, New York, 2003.
92. A. L. Fink, in B. A. Shirley (Ed.), *Protein Stability and Folding: Theory and Practice*, Humana Press, Totowa, 1995, p. 343.
93. B. A. Shirley, in B. A. Shirley (Ed.), *Protein Stability and Folding: Theory and Practice*, Humana Press, Totowa, 1995, p. 177.
94. C. N. Pace and K. L. Shaw, Proteins, Suppl 4, 1, 2000.
95. J. K. Myers, C. N. Pace, and J. M. Scholtz, Protein Sci. **4,** 2138, 1995.
96. G. R. Grimsley, J. M. Scholtz, and C. N. Pace, Protein Sci. **18,** 247, 2009.
97. C. N. Pace, G. R. Grimsley, and J. M. Scholtz, J. Biol. Chem. **284,** 13285, 2009.
98. C. R. Middaugh, R. K. Evans, D. L. Montgomery, and D. R. Casimiro, J. Pharm. Sci. **87,** 130, 1998.
99. T. J. Ahern and M. C. Manning (Eds.), *Stability of Protein Pharmaceuticals*, Plenum Press, New York, 1992.
100. D. Pogocki and C. Schöneich, J. Pharm. Sci. **89,** 443, 2000.
101. S. Clarke, Int. J. Pept. Protein Res. **30,** 808, 1987.
102. K. Patel and R. T. Borchardt, J. Parenter Sci. Technol. **44,** 300, 1990.
103. K. Patel and R. T. Borchardt, Pharm. Res. **7,** 787, 1990.
104. K. Patel and R. T. Borchardt, Pharm. Res. **7,** 703, 1990.
105. C. L. Stevenson, M. E. Donlan, A. R. Friedman, and R. T. Borchardt, Int. J. Pept. Protein Res. **42,** 24, 1993.
106. C. L. Stevenson, A. R. Friedman, T. M. Kubiak, M. E. Donlan, and R. T. Borchardt, Int. J. Pept. Protein Res. **42,** 497, 1993.
107. M. Xie, J. Aube, R. T. Borchardt, M. Morton, E. M. Topp, D. Vander Velde, and R. L. Schowen, J. Pept. Res. **56,** 165, 2000.
108. N. E. Robinson and A. B. Robinson, Proc. Natl. Acad. Sci. U.S.A. **98,** 4367, 2001.
109. C. Oliyai and R. T. Borchardt, Pharm. Res. **10,** 95, 1993.
110. C. Oliyai and R. T. Borchardt, Pharm. Res. **11,** 751, 1994.
111. C. Oliyai, J. P. Patel, L. Carr, and R. T. Borchardt, Pharm. Res. **11,** 901, 1994.
112. S. J. Bogdanowich-Knipp, S. Chakrabarti, T. D. Williams, R. K. Dillman, and T. J. Siahaan, J. Pept. Res. **53,** 530, 1999.
113. S. J. Bogdanowich-Knipp, D. S. Jois, and T. J. Siahaan, J. Pept. Res. **53,** 523, 1999.
114. Y. Sohma, Y. Hayashi, M. Skwarczynski, Y. Hamada, M. Sasaki, T. Kimura, and Y. Kiso, Biopolymers, **76,** 344, 2004.
115. H. T. He, R. N. Gursoy, L. Kupczyk-Subotkowska, J. Tian, T. Williams, and T. J. Siahaan, J. Pharm. Sci. **95,** 2222, 2006.
116. D. B. Volkin and A. M. Klibanov, J. Biol. Chem. **262,** 2945, 1987.
117. S. Li, C. Schöneich, and R. T. Borchardt, Biotechnol. Bioeng. **48,** 490, 1995.
118. S. Li, T. H. Nguyen, S. Schöneich, and R. T. Borchardt, Biochemistry, **34,** 5762, 1995.
119. S. Li, C. Schöneich, and R. T. Borchardt, Pharm. Res. **12,** 348, 1995.
120. S. Li, C. Schöneich, G. S. Wilson, and R. T. Borchardt, Pharm. Res. **10,** 1572, 1993.
121. C. Schöneich, F. Zhao, G. S. Wilson, and R. T. Borchardt, Biochim. Biophys. Acta. **1158,** 307, 1993.
122. K. Uchida and S. Kawasaki, Biochem Biophys. Res. Commun. **138,** 659, 1986.
123. T. J. Simat and H. Steinhart, J. Agric. Food Chem. **46,** 490, 1998.
124. K. Itakura, K. Uchida, and S. Kawakishi, Chem. Res. Toxicol. **7,** 185, 1994.
125. E. L. Finley, J. Dillon, R. K. Crouch, and K. L. Schey, Protein Sci. **7,** 2391, 1998.
126. C. Giulivi, N. J. Traaseth, and K. J. Davies, Amino Acids, **25,** 227, 2003.
127. A. Beck, M. C. Bussat, C. Klinguer-Hamour, L. Goetsch, J. P. Aubry, T. Champion, E. Julien, J. F. Haeuw, J. Y. Bonnefoy, and N. Corvaia, J. Pept. Res. **57,** 528, 2001.
128. S. N. Timasheff, Biochemistry, **41,** 13473, 2002.
129. H. Fan, S. N. Vitharana, T. Chen, D. O'Keefe, and C. R. Middaugh, Mol. Pharm. **4,** 232, 2007.
130. T. J. Kamerzell, S. B. Joshi, D. McClean, L. Peplinskie, K. Toney, D. Papac, M. Li, and C. R. Middaugh, Protein Sci. **16,** 1193, 2007.
131. S. F. Ausar, M. Espina, J. Brock, N. Thyagarajapuram, R. Repetto, L. Khandke, and C. R. Middaugh, Hum. Vaccin. **3,** 94, 2007.
132. C. J. Burke, B. L. Steadman, D. B. Volkin, p. K. Tsai, M. W. Bruner, and C. R. Middaugh, Int. J. Pharm. **86,** 89, 1992.
133. H. R. Costantino and M. J. Pikal, *Lyohilization of Biopharmaceuticals*, AAPS Press, Arlington, 2004.
134. B. S. Chang and S. Y. Patro, in H. R. Costantino and M. J. Pikal (Eds.), *Lyohilization of Biopharmaceuticals*, AAPS Press, Arlington, 2004, p. 113.
135. Y.-F. Maa and H. R. Costantino, in H. R. Costantino and M. J. Pikal (Eds.), *Lyohilization of Biopharmaceuticals*, AAPS Press, Arlington, 2004, p. 519.
136. C. A. Angell and J. L. Green, in H. R. Costantino and M. J. Pikal (Eds.), *Lyohilization of Biopharmaceuticals*, AAPS Press, Arlington, 2004, p. 367.
137. D. E. Overcashier, in H. R. Costantino and M. J. Pikal (Eds.), *Lyohilization of Biopharmaceuticals*, AAPS Press, Arlington, 2004, p. 337.
138. D. Lechuga-Ballesteros, D. P. Miller, and S. P. Duddu, in H. R. Costantino and M. J. Pikal (Eds.), *Lyohilization of Biopharmaceuticals*, AAPS Press, Arlington, 2004, p. 271.
139. H. R. Costantino, in H. R. Costantino and M. J. Pikal (Eds.), *Lyohilization of Biopharmaceuticals*, AAPS Press, Arlington, 2004, p. 139.
140. J. F. Carpenter, B. S. Chang, and T. W. Randolph, in H. R. Costantino and M. J. Pikal (Eds.), *Lyohilization of Biopharmaceuticals*, AAPS Press, Arlington, 2004, p. 423.
141. P. O. Souillac, H. R. Costantino, C. R. Middaugh, and J. H. Rytting, J. Pharm. Sci. **91,** 206, 2002.
142. C. J. Hackett and D. A. Harn, Jr. (Eds.), *Vaccine Adjuvants: Immunological and Clinical Principles*, Humana Press, Totowa, 2006.
143. M. Singh (Ed.), *Vaccine Adjuvants and Delivery Systems*, John Wiley & Sons, Inc., Hoboken, 2007.
144. M. D. Decker, K. M. Edwards, and H. H. Bogaerts, in S. A. Plotkin, W. A. Orenstein, and P. A. Offit (Eds.), *Vaccines*, 5th Ed., Saunders, Philadelphia, 2008, p. 1069.
145. F. R. Vogel and S. L. Hem, in S. A. Plotkin, W. A. Orenstein, and P. A. Offit (Eds.), *Vaccines*, 5th Ed., Saunders, Philadelphia, 2008, p. 59.
146. L. S. Jones, L. J. Peek, J. Power, A. Markham, B. Yazzie, and C. R. Middaugh, J. Biol. Chem. **280,** 13406, 2005.
147. L. J. Peek, T. T. Martin, C. Elk Nation, S. A. Pegram, and C. R. Middaugh, J. Pharm. Sci. **96,** 547, 2007.
148. A. L. Clausi, S. A. Merkley, J. F. Carpenter, and T. W. Randolph, J. Pharm. Sci. **97,** 2049, 2008.
149. B. Beutler, in C. J. Hackett and D. A. Harn, Jr. (Eds.), *Vaccine Adjuvants: Immunological and Clinical Principles*, Humana Press, Totowa, 2006, p. 1.
150. D. M. Dudzinski and A. S. Kesselheim, N. Engl. J. Med. **358,** 843, 2008.
151. J. Woodcock, J. Griffin, R. Behrman, B. Cherney, T. Crescenzi, B. Fraser, D. Hixon, C. Joneckis, S. Kozlowski, A. Rosenberg,

L. Schrager, E. Shacter, R. Temple, K. Webber, and H. Winkle, Nat. Rev. Drug Discov. **6**, 437, 2007.

152. S. S. Parks, J. Park, J. Ko, L. Chen, D. Meriage, J. Crouse-Zeineddini, W. Wong, and B. A. Kerwin, J. Pharm. Sci. **98**, 1688, 2009.
153. J. M. Wilson, Science, **324**, 727, 2009.
154. P. A. Offit, *Autism's False Prophets: Bad Science, Risky Medicine, and the Search for a Cure*, Columbia Press, New York, 2008.
155. G. A. Poland, R. M. Jacobson, and I. G. Ovsyannikova, Vaccine, **27**, 3240, 2009.

## Recommended Reading

It is recommended that the interested student view three excellent films, which provide a nice overview of the history of biotechnology. These are "Glory Enough for All" (the discovery of insulin), "Double Helix" (the discovery of the structure of DNA), and "And the Band Played On" (the early days of the AIDS epidemic). In dramatic form, these three films well illustrate the promise and problems of biotechnology.

CHAPTER LEGACY

**Sixth Edition:** published as Chapter 21 (Pharmaceutical Biotechnology). This is a new chapter written by Charles Russell Middaugh and Teruna J. Siahaan.

# CHAPTER OBJECTIVES

**At the conclusion of this chapter the student should be able to:**

**1** Understand the basic concepts and challenges associated with the development of an oral solid dosage form.

**2** Describe the Biopharmaceutics Classification System (BCS) for drugs and understand how it may be applied to oral dosage form development.

**3** Understand the importance of solubility and permeability in oral drug delivery.

**4** Describe preformulation development activities and their importance in developing a drug product.

**5** Apply basic physicochemical principles to active pharmaceutical ingredients.

**6** Identify the roles that pharmaceutical excipients play in product development.

**7** Understand the important physical, chemical, and mechanical properties of pharmaceutical materials and their relevance in formulation development.

**8** Describe the common unit processes used to manufacture oral solid dosage forms.

**9** Understand the importance and role of oral dosage form performance tests in ensuring product quality and performance.

## INTRODUCTION

This chapter, in many ways, is the culmination of those that preceded it. Physical pharmacy and pharmaceutical science is the science of the delivery of active pharmaceutical ingredients (APIs) to the target site to achieve the desired pharmacological effect. For the drug to exert its biological effect, it must be released from the dosage form, permeate through biological membranes, and reach the site of action. Successful design and delivery of APIs requires a sound fundamental understanding of the diverse array of scientific topics presented in this text. The goal of this chapter is to provide an introduction to how these topics are integrated into dosage form design, product development, and manufacturing activities. The focus of this chapter is on oral drug delivery, and in particular solid dosage forms. **Table 22–1** shows that a majority of pharmaceutical products, 60% or more, are offered as solid dosage forms. However, many of the basic principles apply to the design and manufacture of all types of pharmaceutical dosage forms. The pharmaceutical industry is, after all, a drug product industry, not a drug industry.

## GASTROINTESTINAL ABSORPTION

As the focus of this chapter is on oral dosage forms, a brief review and understanding of the gastrointestinal tract and drug absorption is beneficial. Additional discussion of details of the physiology and absorption of drugs from the gastrointestinal (GI) tract has been presented in chapters on Biopharmaceutics (Chapter 12) and Drug Delivery Systems (Chapter 23).

The gastrointestinal tract is depicted in **Figure 22–1** and some details of the dimensions and volumes and residence time are shown in **Table 22–2**. The oral cavity provides the first contact with biological fluids where mastication and mixing with saliva takes place and digestion begins. As ingested components are swallowed, they move through the esophagus into the stomach. The stomach provides several major functions. It processes food into chyme with vigorous contractions that mix the ingested contents with gastric secretions that continue digestion. It also regulates the input of these liquefied components into the intestinal tract and serves as a major site of chemical and enzymatic breakdown. As stomach contents empty, the chyme enters the small intestine where the absorption of a majority of drugs and nutrients takes place.

Absorption of drugs and nutrients can occur from each section of the small intestine and colon. The small intestine is partitioned into three sections: the duodenum, the jejunum, and the ileum. For most drugs, the duodenum and the proximal jejunum are the best sites of absorption as they have the highest absorptive surface area and often the highest concentration of dissolved drug is achieved in the lumen of this region. Small intestinal absorption is now understood to be dramatically affected by regional differences in the distribution of transporters, enzymes, and greater detail is provided on these aspects in Chapter 12 on Biopharmaceutics. Significant drug absorption from the colon may also occur although the absorptive surface area is substantially less than that of the small intestine.[1,2] However, drug may remain in the colon for 12 to 72 hr and this longer residence time makes the colon an effective site of drug absorption in some cases. Drug absorption may also occur from the oral cavity[3–5] or, rarely, the stomach depending upon the drug and dosage form properties, which must be conducive to absorption from these sites.[6] The low absorptive surface area and typically short residence time of the stomach limits absorption from this site.

## BIOPHARMACEUTICS CLASSIFICATION SYSTEM

An important goal of pharmaceutical formulation development is to "facilitate" drug absorption and ensure that an

## ◀━━━KEY CONCEPT BIOPHARMACEUTICS CLASSIFICATION SYSTEM

The Biopharmaceutics Classification System (BCS) is a scientific framework for classifying drug substances based on their intestinal permeability and aqueous solubility. When combined with drug product dissolution, the BCS takes into account three of the most important factors that influence the rate and extent of drug absorption for immediate release dosage forms: intestinal permeability, solubility, and dissolution. The framework of the BCS may be used as a drug development tool to improve product development efficiency, identify necessary clinical testing, and establish useful in vitro evaluation strategies.

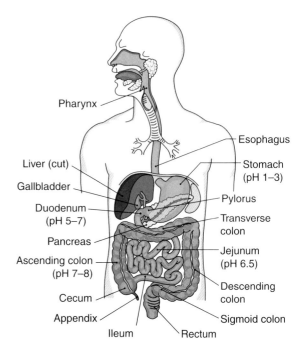

**Fig. 22–1.** Human digestive system.

adequate amount of drug reaches the systemic circulation. Many orally administered drugs enter systemic circulation via a passive diffusion process through the small intestine, although paracellular and transport-mediated absorption also occurs and our understanding of these absorption mechanisms continuous to grow. The Biopharmaceutics Classification System (BCS) is a tool to categorize compounds according to two key parameters: solubility and permeability.[7] Although the BCS does not address other important factors such as the drug absorption mechanism and presystemic degradation, it nonetheless provides a useful framework for identifying potential drug delivery challenges. It also facilitates the identification of appropriate oral dosage forms and strategies to consider that provide opportunities to overcome physicochemical limitations. According to the BCS, compounds are grouped into four classes according to their solubility and permeability as shown in **Table 22–3**. A detailed analysis of the transport and absorption of drugs is described earlier in this book and the student is directed there for a detailed discussion of the various aspects to relate to intestinal absorption.

The basis of the BCS is rooted in the understanding that two very critical parameters affecting drug absorption are solubility and permeability. The importance of these two properties in determining oral absorption can be seen from the following equations describing the flux of drug across the intestinal membrane.

From Fick's first law (also see equation **11–2**), the flux of drug through a unit cross section (in other words, a $cm^2$ surface area) of intestinal membrane can be described by the following equation:

$$\text{Flux} = -D\frac{dC}{dx} = -D_m \left( \frac{C_2 - C_1}{x_2 - x_1} \right) \quad (22\text{--}1)$$

where $D_m$ is the diffusion coefficient of the drug, $C$ is the concentration on the luminal side (1) and serosal side (2) of the membrane, and $x$ is the distance of movement perpendicular to the membrane surface. This equation can be simplified further as (see also equation **11–11**):

$$\text{Flux} = D_m \frac{C_1 - C_2}{h_m} \quad (22\text{--}2)$$

### TABLE 22–1
**MOST COMMONLY AVAILABLE PHARMACEUTICAL DOSAGE FORMS[1,2]**

| Dosage Form | WHO List of Essential Medicines (2007) | Top 100 Best Selling Drugs in 2007 |
|---|---|---|
| Tablet | 48% | 63% |
| Capsule | 11% | 3% |
| Injection | 38% | 27% |
| Oral liquid | 13% | 2% |
| Topical | 4% | 3% |

### TABLE 22–2
**APPROXIMATE VOLUME, RESIDENCE TIME, AND DIMENSIONS OF THE HUMAN GASTROINTESTINAL TRACT[8–12]**

| | Fluid Volume (mL) | Residence Time (hr) | Diameter (cm) | Length (cm) |
|---|---|---|---|---|
| Oral cavity | 1 | | | |
| Stomach | 15–250 | 0.25–3 | | 15 × 30 |
| Duodenum | | | 3–4 | 25–30 |
| Jejunum | 22–300 | 2–4 | 3–4 | 200–250 |
| Ilium | | | 2–3.5 | 300–350 |
| Cecum | | | 7–9 | 9–12 |
| Colon | 2–100 | 12–72 | 5–6 | 85 |

## TABLE 22–3
## THE BIOPHARMACEUTICS CLASSIFICATION SYSTEM[7]

| Class I | Class II | Class III | Class IV |
|---------|----------|-----------|----------|
| High solubility | Low solubility | High solubility | Low solubility |
| High permeability | High permeability | Low permeability | Low permeability |

where $h_m$ is the membrane thickness (also see equation 12–11).

Note that $C_1$ and $C_2$ are the concentrations of drug inside the membrane, but since these are very rarely known or measured, a distribution coefficient, $K$, is typically introduced into this equation to transform the concentrations to the respective aqueous concentrations on the bulk aqueous donor, $C_d$, and receiver, $C_r$, sides. The distribution coefficient reflects the tendency of the drug to partition into the membrane and is the ratio of the drug concentration in the membrane ($C_1$, $C_2$) to that in the aqueous phase immediately adjacent to the membrane ($C_d$, $C_r$). A lipophilic drug would have a distribution coefficient greater than 1 since biological membranes tend to be lipophilic.

$$K = \frac{C_1}{C_d} = \frac{C_2}{C_r} \qquad (22\text{--}3)$$

Equation (22–2) can then be rewritten as:

$$\text{Flux} = \frac{KD_m}{h_m} \cdot (C_d - C_r) = P_m(C_d - C_r) \qquad (22\text{--}4)$$

where $P_m$ is the permeability of the biological membrane, and $C_d$ and $C_r$ are the aqueous concentrations of drug in the intestinal lumen (donor side) and serosal side (e.g., blood), respectively.

Finally, if the drug concentration is much lower on the serosal side of the intestinal membrane as is usually the case (often referred to as sink condition), equation (22–4) can be approximated by the following:

$$\text{Flux} \approx P_m(C_d) \qquad (22\text{--}5)$$

Equation (22–5) represents the essential point of the BCS that drug absorption (i.e., the flux) is determined by two factors, the membrane permeability, $P_m$, and the concentration of drug in the lumen of the intestine, $C_d$.

With the presence of solid drug in the intestine, the concentration of drug dissolved in the intestinal tract, $C_d$, may approach or equal its aqueous solubility if dissolution of drug from the dosage form is sufficiently rapid that it is not rate limiting. From equation (22–5), it is apparent that the flux of drug across the intestine is proportional to the aqueous solubility in the lumen, $C_d$. For drugs that have high intestinal membrane permeability, $P_m$, the aqueous solubility may be the limiting factor for adequate drug flux (BCS Class II). Where the membrane permeability is low, it may be the factor limiting drug absorption (BCS Class III). BCS Class I compounds are the least problematic; both dissolution and oral absorption are generally not major challenges. Finally, Class IV compounds with poor solubility and poor permeability are very difficult compounds to develop using conventional oral dosage form strategies.

Utilization of BCS has led to extensive evaluation of drugs and drug products that now impact regulatory decisions on the type and level of testing necessary as, for example, to ensure equivalence of dosage forms. The BCS has evolved over the past decade to provide additional guidance on classification of products with respect to solubility, permeability, and even dissolution. In recent guidelines issued by the Food and Drug Administration (FDA), a drug substance is considered highly soluble when the highest dose strength is soluble in <250 mL water (e.g., a glass of water) over a pH range of 1–7.5. Since dissolution rate is closely tied to solubility, FDA also provides additional guidance on dissolution criteria: a drug product is considered to be rapidly dissolving when >85% of the labeled amount of drug substance dissolves within 30 min using United States Pharmacopeia (USP) apparatus I or II in a volume of ≤900 mL. Finally, a drug is considered highly permeable when the extent of absorption in humans is determined to be >90% of an administered dose, based on mass-balance or in comparison to an intravenous reference dose.[13] Early in product development the extent of human absorption may not be known and alternative methods of characterizing intestinal permeability may be considered. These include in vivo intestinal perfusion studies in humans, in vivo or in situ intestinal perfusion studies in animals, in vitro permeation experiments with excised human or animal intestinal tissue, or in vitro permeation experiments across epithelial cell monolayers.[7,13–17]

Even though the BCS was designed to guide decisions with respect to in vivo and in vitro correlations and the need for bioequivalence studies[13] (see Chapter 12), it can also be used to categorize the types of formulation strategies that might be pursued.[18] Table 22–4 summarizes some dosage form options that may be considered for each biopharmaceutics class. Each class of compound, and especially Classes II, III, and IV, requires different dosage forms to deal with the challenges associated with solubility or permeability limitations. Characterizing the properties of the drug, also known as preformulation characterization, provides the information necessary to classify drugs and identify suitable dosage forms to address drug delivery issues. Back in an era when local pharmacies offered a delivery service, drug delivery was described as "a boy on a bicycle." (J. Robinson, Oral Communication, 1995) In a way, drug delivery has not changed much. The goal of drug delivery today is still to efficiently and

**TABLE 22–4**

## ORAL DOSAGE FORM OPTIONS BASED ON BIOPHARMACEUTICS CLASSIFICATION SYSTEM[18]

Class I: High Solubility, High Permeability
- No major challenges for immediate-release dosage forms
- Controlled release dosage forms may be needed to slow drug release from the dosage form and reduce absorption rate.

Class II: Low Solubility, High Permeability
Formulations designed to overcome solubility or dissolution rate problems
- Particle size reduction
- Salt formation
- Precipitation inhibitors
- Metastable forms
- Solid dispersion
- Complexation
- Lipid technologies
- Cocrystals

Class III: High Solubility, Low Permeability
Approaches to improve permeability
- Prodrugs
- Permeation enhancers
- Ion pairing
- Bioadhesives
- Lipid technologies

Class IV: Low Solubility, Low Permeability
- Formulations often use a combination of approaches identified in Class II and Class III to overcome dissolution and permeability problems.
- Strategies for oral administration are not often feasible.
- Often use alternative delivery methods, such as intravenous administration.

effectively provide the medicine where it is needed and when it is needed.

**EXAMPLE 22–1**

Chloroquine phosphate has the following physicochemical and biological properties.[19] Although the FDA has required in vivo documentation of bioavailability (BA) and bioequivalence (BE) for many drug products, in some cases FDA has allowed the use of in vitro methods for documenting BA and BE. Obtaining a biowaiver for a drug product based on in vitro BA and BE very often simplifies the application process and shortens the time to market. An FDA guidance describes recommendations for requesting waivers of in vivo BA/BE studies on the basis of the solubility and intestinal permeability of the drug substance and dissolution characteristics of the drug product, based on the biopharmaceutics classification system.[13]

Using chloroquine phosphate as an example, is it a BCS Class I compound and therefore a suitable candidate for a BCS Biowaiver?

- Aqueous solubility: Greater than 100 mg/mL in water
- Dose: 150 mg
- Human oral absorption: Rapid and almost complete. Bioavailability = 89% with high variability (67%–114%)

The aqueous solubility is high, although data over the entire pH range of interest (pH = 1–7.5) are lacking. One dose of 150 mg will dissolve in less than 2 mL of water. This suggests that chloroquine phosphate can be classified as a BCS high-solubility compound. The human absorption data from commercially available products indicate that the drug is well absorbed since the bioavailability is 89%. The FDA guidance defines "high permeability" as not less than 90% absorbed. While this falls slightly below the FDA guidance criteria, recent discussions have indicated that a minimum value of bioavailability can be lowered to 85%.[17,20] This information supports the classification of chloroquine phosphate as a BCS Class I compound with high solubility and permeability and it would be a suitable candidate for a Biowaiver.

An oral solid dosage form of chloroquine phosphate should conform to the following:[19]

- Utilize standard excipients.
- Comply with the requirements for "rapidly dissolving" at pH 1.0, pH 4.5, and pH 6.8.[13]
- Comply with the similarity requirements for comparative dissolution testing versus the reference product at pH 1.0, pH 4.5, and pH 6.8.[13]

## PREFORMULATION CHARACTERIZATION

While hundreds of thousands of compounds are synthesized and evaluated every year in the pharmaceutical industry, very few make it to clinical testing and fewer still make it to the market. There are many reasons for failure. Because of the challenges associated with drug discovery and development, the opportunity to identify and develop a safe and effective product benefits greatly from the integration of pharmacology, chemistry, toxicology, metabolism, clinical research, thorough physicochemical characterization and, very importantly, dosage form development. The ability to identify a suitable dosage form is critical to success. The dosage form must deliver the drug to the desired site at the desired concentration (often considered the blood) for the desired duration. Finally, the dosage form must be robust and manufacturable!

To initiate formulation development activities, that is, the identification of an effective drug delivery system— important physical, chemical, and even mechanical properties (physicochemical or physicomechanical properties) as

##  KEY CONCEPT  THE PHARMACEUTICAL INDUSTRY

The pharmaceutical industry is a drug product industry, not a drug industry. The pharmaceutical industry is highly regulated and much of that regulation is focused on ensuring a safe, effective, high quality, and consistently performing product. The active ingredient (the drug) is obviously critically important, but it is only in the context of a drug product (a dosage form) that the drug can be safely administered to the patient with the confidence that it will have the desired performance. It is the drug product that is of true value to the patient.

well as drug absorption (permeability) properties need to be determined.

Evaluation of these properties during the drug discovery and development process, known as preformulation, help identify the most promising molecules for development and also provide key information for scientific dosage form design and development. Dosage forms that make sense to consider are dictated to a large extent by the molecular, particle, and bulk powder properties.

Typically, for oral dosage forms, crystalline drug forms are preferred. Common solid forms include crystalline polymorphs, hydrates, and crystalline salts of the active ingredient. This is especially true for solid dosage forms such as tablets and capsules since the solid form of the active ingredient may be a significant component in the dosage form and will impact manufacturing, dosage form performance and stability. Ideally, the most thermodynamically stable form is chosen as it will generally provide the greatest physical and chemical stability. Therefore, early identification and selection of the solid form to be used in development becomes paramount as it has a direct impact on physicochemical and drug delivery attributes.

Many of the physicochemical properties that have been discussed in this text are, in fact, dependent on the solid form. Aqueous solubility, hygroscopicity, and chemical stability are three obvious examples where very large differences may exist between solid forms of the same drug molecule. It is therefore important to rigorously characterize solid forms early in the discovery/development process. Selection of the right solid form can allow the pharmaceutical scientist to design the dosage form with optimal physicochemical, manufacturing, and dosage form performance properties. A thorough understanding of solid forms maximizes the opportunity to understand, control, and predict the behavior of a compound in the solid state, identify the appropriate dosage forms to consider, and develop a marketable product. The reader is directed to *Chapter 2: States of Matter* and the literature for additional discussion of crystalline solids and polymorphism.[21–24]

### EXAMPLE 22–2

An antiviral compound, ritonavir, marketed in a semisolid capsule as Norvir (Abbott) began to demonstrate physical instability and dissolution failures in 1998.[21,25] Upon investigation, the failures were shown to be caused by the crystallization of a new and previously unknown polymorph (Form II) in the semisolid capsule matrix that was approximately half the solubility of the original polymorphic form (Form I). The new form II, with lower solubility, was supersaturated in the formulation and upon storage, precipitated out in the capsule. The formation of this new polymorph was surprising since the semisolid dosage form using Form I showed no evidence of Form II formation on stability even after 24 months.[25] The lower solubility polymorph exhibited slower dissolution which compromised dosage form performance. The semisolid capsule was withdrawn from the market and an alternative dosage form, a soft-gel capsule formulation with adequate stability, was developed and marketed.

The physicochemical characterization described in this chapter and indeed throughout this text can be applied to each of the forms that have been identified and isolated as each solid form will have a unique set of physicochemical and mechanical properties. Careful consideration of these properties will inevitably lead to the identification of better lead compounds and forms with which to enter development. Some of the key physical, chemical, mechanical, and biological properties that should be of interest to the development team and the pharmaceutical scientist are listed in **Table 22–5**. Each of these physical, chemical, mechanical, and biological properties can have a significant effect on the final dosage form design, performance, manufacturing, or stability and these are discussed in greater detail in the following sections. It should be kept in mind that many of these properties are dependent on the solid form and complete characterization of each of the most relevant solid forms is needed to

##  KEY CONCEPT  PREFORMULATION CHARACTERIZATION

Preformulation characterization is the evaluation of those properties of the drug substance, and the solid forms in which it exists, that can impact drug delivery and drug product performance. Every form of a drug substance has unique physical and chemical properties that must be evaluated and understood to ensure the successful development of a safe and effective drug product with consistent drug delivery performance.

**TABLE 22–5**
**TABLE 22–5**

**IMPORTANT PHYSICAL, CHEMICAL, MECHANICAL, AND BIOLOGICAL PROPERTIES FOR ORAL DRUG DELIVERY**

| *Physical properties* | *Chemical properties* |
|---|---|
| Polymorphic form(s) | Ionization constant ($pK_a$) |
| Crystallinity | Solubility product ($K_{sp}$) of salt forms |
| Melting point | Chemical stability in solution |
| Particle size, shape, surface area | Chemical stability in solid state |
| Density | Photolytic stability |
| Hygroscopicity | Oxidative stability |
| Aqueous solubility as a function of pH | Incompatibility with formulation additives |
| Solubility in organic solvents | Complexation with formulation additives |
| Solubility in presence of surfactants (e.g., bile acids) | |
| Dissolution rate | |
| Wettability | |
| Partition coefficient (octanol–water) | |
| | |
| *Mechanical properties* | *Biological properties* |
| Elasticity | Membrane permeability |
| Plasticity (hardness) | Absorption, distribution, metabolism, excretion (ADME) |
| Bonding | |
| Brittleness | Metabolism: Gut, first pass, systemic |
| Viscoelasticity | |

provide a complete physicochemical picture. Because of its importance in product development, extensive discussion of physicochemical characterization to support product development is available in many reference books.[26]

## Physical Properties

### Melting Point

Melting point is defined as the temperature at which the solid phase exists in equilibrium with its liquid phase. As such, the melting point is a measure of the "energy" required to overcome the attractive forces that hold the crystal together. Melting point determination is of great value and can successfully be accomplished by any of several commonly used methods including visual observation of the melting of material packed in a capillary tube (Thiele arrangement), by hot-stage microscopy, or other thermal analysis methods such as differential scanning calorimetry. Careful characterization of thermal properties such as that possible with differential scanning calorimetry provides an opportunity to assess and quantify the presence of impurities as well as the presence or interconversion of polymorphs and pseudopolymorphs. Melting points and the energetics of desolvation can also be evaluated, as can the enthalpies of fusion for different solid forms. Chapter 2, States of Matter, provides additional discussion of melting point and thermal analysis.

As a practical matter, low melting materials tend to be more difficult to handle in conventional solid dosage forms. Melting points below about 60°C are generally considered to be problematic and melting points above 100°C are

considered desirable. Temperatures in conventional manufacturing equipment such as high shear granulation and conventional tablet machine equipment may exceed 40°C, whereas fluid bed granulation and drying can involve temperatures approaching or exceeding 80°C. While amorphous solids do not have a distinct melting point, they undergo softening as temperatures approach the glass transition temperature. Furthermore, common handling procedures (e.g., weighing, processing) can be difficult for low melting materials. Alternative dosage forms (liquid type) may be required for liquid or low melting materials. A comparison of melting points of polymorphs also provides a perspective on the relative stability of polymorphic forms.[24]

### Aqueous Solubility

The importance of aqueous solubility in determining oral absorption can be seen from equation (22–5). From this equation it is apparent that the flux of drug across the intestinal membrane is proportional to the concentration of drug in solution, and more specifically, the nonionized drug concentration in solution. The aqueous solubility reflects this and an understanding of aqueous solubility, pH dependence, and the impact of biological fluid components is important in the physicochemical characterization of APIs.

Drug solubility may be determined experimentally by adding excess solid drug to well-defined aqueous media and agitating until equilibrium is achieved. Appropriate temperature control, solute purity, agitation rate, and time as well as monitoring of solid phase at equilibration are needed to ensure high-quality solubility data is obtained.[27] A wide variety of techniques have been proposed for estimating aqueous solubility. They can broadly be classified as (*a*) methods based on group contributions, (*b*) techniques based on experimental or predicted physicochemical properties (e.g., partition coefficient, melting point), (*c*) methods based on molecular structure (e.g., molar volume, molecular surface area, topological indices), and (*d*) methods which use a combination of approaches.[27–29] While all of the methods have some theoretical basis, their use in predicting aqueous solubility is largely empirical. Detailed discussions on solubility predictions may be found in the literature and in Chapter 9 of this book. Each predictive approach has advantages and has been successfully applied to a variety of classes of compounds to develop and test the accuracy of solubility predictions. Usually, approaches that are developed from structurally related analogues yield more accurate predictions.[29]

Aqueous solubility is, in a simple sense, determined by the interaction of solute molecules in the crystal lattice, interactions in solution, and the entropy changes that occur as solute passes from the solid phase to the solution phase. For example, the pioneering work of Yalkowsky and Valvani[30] illustrates the importance of two physical properties (melting point and lipophilicity) on solubility. They successfully estimated the solubility of rigid short-chain nonelectrolytes with the following equation:

$$\text{Log}(S) = -\log(P) - 0.01 \times (MP) + 1.05 \quad \textbf{(22–6)}$$

where S is the molar solubility, P is the octanol–water partition coefficient, and MP is the melting point of the compound. Equation **(22–6)** provides insight into the relative importance of crystal energy (melting point) and lipophilicity (partition coefficient). Increasing either lipophilicity, P, or melting point, MP, results in decreased predicted aqueous solubility, S. This semiempirical approach has been applied and refined for a variety of solutes and classes of compounds.[31–35] From equation **(22–6)**, one can see that the octanol–water partition coefficient is a significant predictor of aqueous solubility. A 1-log unit change in aqueous solubility can be expected for each log unit change in partition coefficient. By comparison, a melting point change of $100°C$ is required to have the same 1-log unit change on solubility. The Yalkowsky–Valvani and similar equations can be used to predict aqueous solubility often within a factor of 2, using physical, chemical, and molecular properties.

Caffeine (log $P = -0.2$, MP $= 238°C$) and cortisone (log $P = 1.47$, MP $= 222°C$) have similar melting points but substantially different log $P$ values. Use equation (22–6) to estimate the molar aqueous solubility of each.

For caffeine: Log(S) $= -(-0.2) - 0.01 \times (238) + 1.05 = -1.13$
  and S $= 0.074$ mol/L

For cortisone: Log(S) $= -(1.47) - 0.01 \times (222) + 1.05 = -2.64$
  and S $= 0.0023$ mol/L

These two compounds illustrate the impact of partition coefficient on aqueous solubility.

Triazolam (log $P = 2.42$, MP $= 224°C$) and ethyl-p-hydroxybenzoate (log $P = 2.47$, MP $= 116°C$) have similar log $P$ values but substantially different melting points. Use equation (22–6) to estimate the molar aqueous solubility of each.

For triazolam: Log(S) $= -(2.42) - 0.01 \times (224) + 1.05 = -3.61$
  and S $= 0.00025$ mol/L

For cortisone: Log(S) $= -(2.47) - 0.01 \times (116) + 1.05 = -2.58$
  and S $= 0.0026$ mol/L

These two compounds illustrate the impact of melting point on aqueous solubility.

Aqueous solubility prediction continues to be an active area of research with a wide variety of approaches being applied to this important and challenging area and additional discussion of solubility can be found in Chapter 9 of this book.

### Dissolution Rate

Aqueous solubility can also play a critical role in the rate of dissolution of drug and release from dosage forms. The rate at which a solute dissolves was described in quantitative terms by Noyes and Whitney in 1897[36] and the equation can be written in the following way (see also equation **13–2**):

$$\text{Dissolution Rate} = \frac{dM}{dt} = \frac{D \cdot S}{h} \cdot (C_s - C) \quad \textbf{(22–7)}$$

where $M$ is the mass of solute dissolved in time $t$, $dM/dt$ is the rate of dissolution, $D$ is the aqueous diffusion coefficient, $S$ is the surface area of the exposed solid, $h$ is the aqueous diffusion layer thickness which is dependent on viscosity and agitation rate, $C_s$ is the aqueous drug solubility at the surface of the dissolving solid, and $C$ is the concentration of drug in the bulk aqueous phase. When $C \sim 0$, this is commonly referred to as sink conditions and equation **(22–7)** can be simplified to the following (see also equation **13–7**).

$$\text{Dissolution Rate} = \frac{D \cdot S}{h} \cdot C_s \quad \textbf{(22–8)}$$

From the Noyes Whitney equation, dissolution rate is seen to be directly proportional to the aqueous solubility, $C_s$, as well as the surface area, $S$, of drug exposed to the dissolution medium. It is common practice, especially for low-solubility drugs, to increase dissolution rate by increasing the surface area of a drug. This can be done through particle size reduction. If drug surface area is too low, the dissolution rate may be too slow and absorption can become dissolution rate limited. For high-solubility drugs, the dissolution rate is generally fast enough that a high drug concentration is achieved in the lumen and extensive particle size reduction is not needed. Use of high-solubility salts is commonly undertaken to facilitate rapid dissolution in the GI tract.

Although the mathematics becomes somewhat more complicated, dissolution of particles may also be modeled and this provides greater insight into the interplay of solubility and drug particle size on dissolution rate. For a drug powder consisting of uniformly sized, spherical particles, it is possible

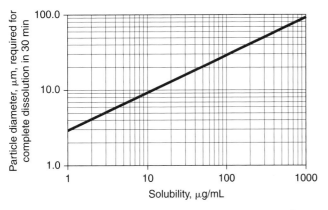

**Fig. 22–2.** Relationship between aqueous solubility and maximum spherical particle diameter that will dissolve in 30 min.

to derive an equation that expresses the rate of dissolution. A detailed discussion and derivation of the following equation is provided in Chapter 13 (equation **13–20**) and will not be repeated here. The resulting equation that predicts the change in particle radius with time is:

$$r^2 = r_0^2 - \frac{2DC_s t}{\rho} \qquad (22\text{–}9)$$

where $r$ is the radius of the dissolving particle at time $t$, $r_0$ is the initial radius of the particle, $D$ is the aqueous diffusion coefficient, $C_s$ is the aqueous solubility, and $\rho$ is the particle density.

The time for complete dissolution, $\tau$, is the time it takes for the initial particle radius to be reduced to zero (i.e., set $r = 0$ in equation **22–9**) and is given by:

$$\tau = \frac{\rho \cdot r_0^2}{2DC_s} \qquad (22\text{–}10)$$

These equations may be used to make useful predictions about dissolution and the relationship between particle size (diameter) and solubility is shown in **Figure 22–2**. Based on these considerations, the need for particle size reduction to achieve adequate dissolution can be made.

Because pharmaceutical powders are not monodispersed, that is, not all the same size, it is important to consider the particle size distribution as well. A few large particles may seriously affect the dosage form dissolution rate of a material under some circumstances and more sophisticated mathematical models have been developed to address these issues.[37,38] As a rough "rule of thumb," if the particle diameter in $\mu$m is greater than the solubility in $\mu$g/mL, further particle size reduction may be needed to achieve adequate dissolution for an immediate release dosage form.

**EXAMPLE 22–5**

A new drug is under development and the pharmaceutical scientist responsible for designing the first clinical formulation must identify the particle size necessary to achieve an acceptable rate of dissolution (e.g., complete dissolution in 30 min or less). Based on the physicochemical data available, the drug has a constant aqueous solubility of 10 $\mu$g/mL in the physiological pH range of 1 to 7. Additional

information available includes the density of the crystalline drug ($\rho = 1.52$ g/cm$^3$) and the aqueous diffusion coefficient (estimated $D = 9 \times 10^{-6}$ cm$^2$/sec).

Estimate the time it would take for particles of 1 $\mu$m, 10 $\mu$m, and 100 $\mu$m in diameter to dissolve.

Use equation (22–10). Note: use consistent units for mass, time, and volume.

For 1 $\mu$m diameter particle $\tau = (1.52 \times (0.5 \times 10^{-4})^2)/(2 \times (9 \times 10^{-6}) \times (10 \times 10^{-6})) = 21$ sec
For 10 $\mu$m diameter particle $\tau = 2211$ sec $= 35.2$ min
For 100 $\mu$m diameter particle $\tau = 2.1 \times 10^5$ sec $= 3518$ min

Based on these calculations, the particle size of the drug should be 10 $\mu$m or less to achieve rapid dissolution. Certainly, a particle size of 100 $\mu$m would be too large to achieve rapid dissolution.

## Ionization Constant

Knowledge of acid–base ionization properties is essential to an understanding of solubility properties, partitioning, complexation, chemical stability, and drug absorption, and an extensive discussion of ionic equilibria is given in Chapter 7. The ionized molecule exhibits markedly different properties from the nonionized form. For weak acids, the equilibrium between the free acid, HA, and its conjugate base, A$^-$, is described by the following equilibrium equation (see also equation **7–10**):

$$HAc = H_3O^+ + Ac^- \qquad (22\text{–}11)$$

and the corresponding acid dissociation constant is given by:

$$K_a = \frac{[H_3O^+][Ac^-]}{HAc} \qquad (22\text{–}12)$$

The equation for a weak base, B, and its conjugate acid, BH$^+$, is described by (also see equation **7–21**):

$$K_a = \frac{[H_3O^+][B]}{[BH^+]} \qquad (22\text{–}13)$$

Of particular interest to the pharmaceutical scientist is the impact of p$K_a$ on aqueous solubility and partitioning (see Chapter 9). Taking a weak acid as an example, the total aqueous solubility, $S_T$, is equal to the sum of the ionized and nonionized species concentrations in solution. Assuming that the solution is saturated with respect to free acid, the total solubility, $S_T$, can be written (see also equation **10–61**):

$$S_T = S_a(1 + (K_a/[H^+])) \qquad (22\text{–}14)$$

where the intrinsic solubility of the free acid is $S_a$. The solubility equation for a weak base is given by:

$$S_T = S_b(1 + ([H^+]/K_a)) \qquad (22\text{–}15)$$

These equations can be written in log form respectively as:

$$(S_T) = (S_a)(1 + 10^{(pH - pK_a)}) \qquad (22\text{–}16)$$

$$(S_T) = (S_b)(1 + 10^{(pK_a - pH)}) \qquad (22\text{–}17)$$

Based on these equations, typical solubility curves are shown in **Figure 22–3** for a weak acid and a weak base and

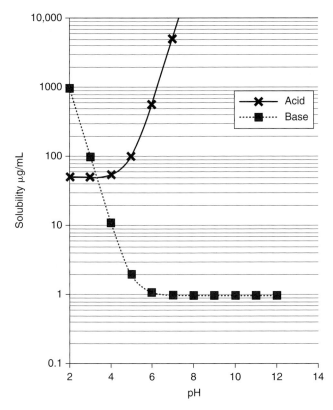

**Fig. 22–3.** Solubility ($\mu$g/mL) versus pH for an acid (intrinsic solubility = 50 ($\mu$g/mL) and base (intrinsic solubility = 1 ($\mu$g/mL) with $pK_a = 5.0$.

several significant conclusions and implications are worth pointing out. Taking the free base as an example, at pH greater than the $pK_a$, the predominant form present in solution is the nonionized form (free base) and the total solubility is essentially equal to the intrinsic solubility, that is, the free-base solubility. As the pH decreases below the $pK_a$, a rapid increase in total solubility is observed since the ionized form, $BH^+$, is dramatically increasing. In fact, for each unit decrease in pH, the total aqueous solubility will increase 10-fold in this region as shown in **Figure 22–3**. The total solubility will continue to increase as long as the ionized form continues to be soluble. Such dramatic increases in solubility as a function of pH demonstrate the importance of understanding and controlling solution pH and also offer the pharmaceutical scientist a number of possible opportunities to modify the dosage form and factors leading to oral absorption properties. It is important to recall, however, that often only the nonionized drug is well absorbed.

For weak acids, one will observe a rapid increase in total solubility as the pH exceeds the $pK_a$ (**Fig. 22–3**) since the ionized species concentration, $Ac^-$, will increase with increasing pH. The pharmaceutical scientist must understand the solubility properties of both the nonionized species and its corresponding conjugate, ionized, form(s) since each may limit solubility.

From the solubility curves, one can draw conclusions regarding which solid form will exist at equilibrium as a

function of pH. This basic principle is of significance in vivo since one might imagine dosing patients with a soluble salt of a base, which could rapidly dissolve in the low pH of the stomach, but as drug in the gastric contents enters the intestine where solution pH approaches neutral, precipitation of the insoluble free base could occur. Such changes have been proposed as an explanation for the poor bioavailability of highly soluble salts of weak bases.

### EXAMPLE 22–6

Kramer and Flynn[39] investigated the aqueous solubility of 2-ethyl-2-phenyl-4-(2′-piperidyl)-1,3-dioxolane and its hydrochloride salt as a function of pH at 30°C in 0.05 M succinate buffer. The $pK_a$ of the amine functional group was determined to be 8.5, the intrinsic solubility of the free base was 2.87 mg/mL, and the solubility of the hydrochloride salt corresponded to 60 mg/mL. Calculate the solubility of the free base and the hydrochloride salt at pH = 4, 6, 8, and 10.

Using equation (22–17), the solubility of the free base is:

$$(S_T) = (2.87)(1 + 10^{(8.5-4)}) = 90760 \text{ mg/mL at pH} = 4$$
$$(S_T) = (2.87)(1 + 10^{(8.5-6)}) = 910 \text{ mg/mL at pH} = 6$$
$$(S_T) = (2.87)(1 + 10^{(8.5-8)}) = 11.9 \text{ mg/mL at pH} = 8$$
$$(S_T) = (2.87)(1 + 10^{(8.5-10)}) = 2.96 \text{ mg/mL at pH} = 10$$

Using equation (22–16), the solubility of the hydrochloride salt is:

$$(S_T) = (60)(1 + 10^{(4-8.5)}) = 60 \text{ mg/mL at pH} = 4$$
$$(S_T) = (60)(1 + 10^{(6-8.5)}) = 60.1 \text{ mg/mL at pH} = 6$$
$$(S_T) = (60)(1 + 10^{(8-8.5)}) = 79 \text{ mg/mL at pH} = 8$$
$$(S_T) = (60)(1 + 10^{(10-8.5)}) = 1957 \text{ mg/mL at pH} = 10$$

### Hygroscopicity

Moisture uptake or sorption is a significant concern for pharmaceutical powders. The extent of sorption of water depends on the chemical nature of the drug. Two types of moisture sorption are generally recognized: physical sorption and chemical sorption. Physical asorption is that which is associated with van der Waals forces and is reversible. A graph of the amount of water that is physically sorbed to the surface of a solid material as a function of equilibrium water vapor pressure yields an sorption isotherm. Greater detail on physical and chemical sorption is provided in Chapter 15. In addition to surface sorption, water may condense in pores and the reader is referred to Chapter 18 for additional discussion of this topic.

Moisture has been shown to have a significant impact on the physical, chemical, and manufacturing properties of drugs, excipients, and formulations. It is also a key factor in decisions related to packaging, storage, handling, and shelf life, and successful development requires a sound understanding of hygroscopic properties. Moisture sorption isotherms can yield an abundance of information regarding the physical state of the solid and the conditions under which significant changes may occur. Conversion from an anhydrous form to a hydrated form may be observed when the relative humidity exceeds a critical level and moisture content rapidly increases in the solid. Quantitative measurement of

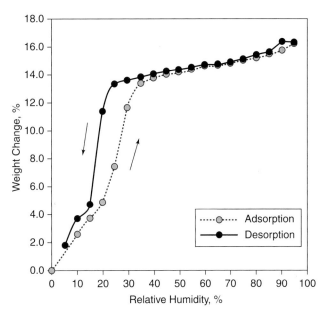

**Fig. 22–4.** Moisture sorption (% weight change) as a function of % Relative Humidity for an active pharmaceutical ingredient (API).

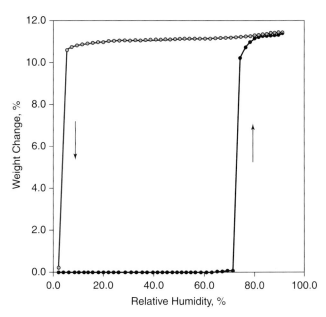

**Fig. 22–5.** Moisture sorption (% weight change) as a function of % Relative Humidity for an API demonstrating hydrate formation above 70% Relative humidity.

moisture content also provides valuable information on the type of hydrate that has formed.

Measurement of moisture uptake is typically done by either of two general methods. The classical approach involves equilibration of solid at several different humidities and the subsequent determination of water content either by gravimetric or by analytical methods such as Karl Fischer titration or loss on drying. Moisture adsorption or desorption may be measured using this method and the process is effective but tedious and time-consuming. An automated controlled atmosphere system in conjunction with an electronic microbalance is now commonly used.[40–42] Such systems can generate an atmosphere with well-controlled humidity passing over a sample (often only a few milligrams are needed) and weight change is monitored and can be programmed to carry out a series of humidity increments to generate the adsorption and desorption curves. In this way, hysteresis may be observed as well as phase or form changes that are associated with moisture sorption. Examples of a moisture sorption curve are shown in **Figure 22–4** and **Figure 22–5**.

In general, water adsorption to only the surface of crystalline materials will result in very limited moisture uptake. Only 0.1% water uptake is predicted to be needed to achieve monolayer coverage of a crystalline material with an average particle size of 1 $\mu$m.[31] Amorphous phases tend to be much more prone to moisture sorption and high moisture uptake by a solid is likely to reflect either the presence of significant amorphous regions or a change in solid form such as the formation of a hydrate. Moisture sorption has, in fact, been used to quantitate the amorphous content of predominantly crystalline materials.

Dynamic moisture sorption, in particular, provides an excellent opportunity to study solid form conversion and Figure 22–5 depicts a typical sorption curve of an antiarrhythmic compound that shows the conversion of an anhydrous form to a hydrate. Moisture uptake by the anhydrous form is very small on the moisture uptake curve until a critical humidity of about 70% is achieved. At this point, rapid moisture uptake occurs and a hydrate form containing about 10% moisture is formed.

Subsequent reduction in the humidity (desorption) shows the hydrate to remain until approximately 5% RH when it spontaneously converts to the anhydrous form. It is important to recognize, however, that conversion between solid forms is very time dependent. The relative humidities at which conversion was seen in Figure 22–5 are very dependent upon the length of time the solid material was equilibrated. For the material shown, conversion from the anhydrous to the hydrate "at equilibrium" will occur somewhere between 10% and 70% RH.

## Particle Size

Understanding a pharmaceutical powder's particle size, shape, and distribution is an important component of formulation development. When working with the API, a few large or small particles in a batch can alter the final tablet's content uniformity (potency, segregation), dissolution profile, and/or processing (e.g., flow, compression pressure profile, granulating properties). Yalkowsky, Bolton, and others have developed a model to estimate the API particle size needed to pass United States Pharmacopeia (USP) content uniformity criteria.[32,33] A plot of the particle size needed to pass content uniformity as a function of particle size and size distribution is shown in **Figure 22–6**. It is useful for estimating particle size requirements and determining whether additional drug

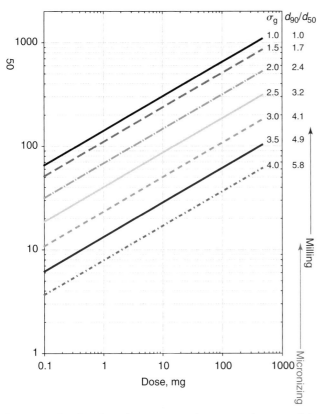

**Fig. 22–6.** Predicted maximum geometric mean volume particle diameter needed to pass USP content uniformity requirements as a function of dose (mg) and geometric standard deviation ($\sigma_g$) assuming log normal particle size distribution of active pharmaceutical ingredient.[33]

particle size processing, such as milling, is necessary. Drug particle size and size distribution information can also help decide whether a direct compression formulation or dry granulation approach is more appropriate. Examination of drug particle size can also reveal inter and intrabatch differences or trends. If the particle size distribution has changed from one batch of API to the next, this could significantly impact the processing properties of the final formulation, as well as dissolution, leading to inconsistent dosage form performance.

Particle size and size distribution are also important as they are critical parameters in assuring that the desired dissolution rate is achieved. Several theoretical models for dissolution of powders have been developed and discussed in previous chapters of this book. Flow characteristics of formulations are also influenced by particle shape, size, and size distribution.

### EXAMPLE 22–8

The drug molecule described in *Example* 22–5 will be manufactured as a tablet dosage form for Phase I clinical testing. Using Figure 22–6, identify the particle size necessary to achieve acceptable content uniformity for tablet strengths of 1 mg, 3 mg, and 10 mg assuming a monodispersed particle size distribution (i.e., all particles are the same size, $\sigma_g = 1.0$).

From Figure 22–6, for a 1-mg dose approximately a 130-$\mu$m particle size is needed to achieve content uniformity. For a 3-mg dose, approximately a 200-$\mu$m particle size is needed, whereas the high dose of 10 mg would require approximately 300 $\mu$m particle size. For this particular drug molecule, achieving the particle size needed to obtain adequate dissolution (*Example* 22–5, approximately 10 $\mu$m) will ensure that content uniformity can be achieved.

Following milling of the drug to achieve a mean particle size diameter of 10 $\mu$m for clinical supply manufacture, the drug is found to have an extremely wide particle size distribution which is log normal with a geometric standard deviation, $\sigma_g$, of 4.0. Will content uniformity likely be achieved with the 1-mg dose tablet?

From Figure 22–6, the particle size necessary to achieve content uniformity for this extremely wide particle size distribution is estimated to be approximately 8 $\mu$m. These theoretical estimates indicate that content uniformity could be a problem during clinical supply manufacture of the low-dose tablet. Further processing the drug may be appropriate to better ensure success in clinical manufacturing.

## Chemical Properties

### Stability

Both solution and solid-state stability are key considerations for oral delivery. Chemical stability is addressed in detail in Chapter 15. The drug molecule must be adequately stable in the dosage form to ensure a satisfactory shelf life. For oral dosage forms, it is generally considered that 2 years is an acceptable shelf life. This allows sufficient time for the manufacture and storage of the active ingredient, the manufacture of the dosage form, shipping, storage, and finally sale

## KEY CONCEPT  SHELF LIFE

The shelf life of a pharmaceutical product is the time period during which the product is expected to remain within the acceptance criteria established by the manufacturer for the critical physical, chemical, and aesthetic properties when stored according to the manufacturer's recommendations. The shelf life depends on the drug molecule, the dosage form, packaging, and the environmental conditions to which the dosage form is exposed.

According to the FDA,[43] there shall be a written testing program designed to assess the stability characteristics of drug products. The results of such stability testing must be used in

determining appropriate storage conditions and expiration dates and include the following: sample size and test intervals based on statistical criteria for each attribute; storage conditions of samples; reliable, meaningful, and specific test methods; testing of the drug product in the same container-closure system as that in which the drug product is marketed; and testing of drug products for reconstitution at the time of dispensing (as directed in the labeling) as well as after they are reconstituted. An adequate number of batches of each drug product must be tested to determine an appropriate expiration date.

to and use by the consumer. Loss of potency is an obvious consideration and, generally, stability guidelines require that at least 90% of the drug remains at the end of the shelf life. More often though, shelf life is determined by the appearance of relatively low levels of degradation products. While perhaps a 5% loss of drug may be considered acceptable, the appearance of a degradation product or impurity of unknown toxicity at a level of 0.1% to 1% will likely require identification or qualification. Detailed guidance regarding stability has been provided by regulatory agencies such as those in the FDA Guidance for Industry and the International Conference on Harmonization.[34,35,43]

### Solution Stability

Solution stability is important for oral products because the drug generally has to dissolve in the gastric or intestinal fluids prior to absorption. Residence time in the stomach varies between 15 min and a couple of hours depending on fasting/fed state. In addition, the stomach is generally acidic in a majority of subjects but may depend on disease state. In this context, stability under acidic conditions over a period of a couple of hours at 37°C may be satisfactory. Residence time in the small intestine is approximately 3 hr where the pH may range from approximately 5 to 7, whereas residence in the large intestine ranges up to 24 hr or more. Stability studies for up to 24 hr in the pH range of 5 to 7 at 37°C with no significant appearance of degradation products of unknown toxicity generally indicates that significant decomposition in the intestine will not occur. Other intestinal components such as microorganisms, enzymes, and surfactants can dramatically alter in vivo solution stability however.

Buffered aqueous solution stability studies are typically done at pH of 1.2 to 2 and in the range of 5 to 7. A complete degradation rate profile can provide valuable information regarding the degradation mechanism and degradation products. A complete study and understanding of solution stability is particularly critical for aqueous and cosolvent solution formulations which may be developed for pediatric or geriatric populations.

### Solid-State Stability

Adequate solid-state stability is often critical for many drugs since solid dosage forms (tablets, capsules) are generally the preferred delivery system. Stability of the drug in the dosage form for several years at room temperature is desirable. Unstable drugs may be developed, but the time and resources needed are generally greater and the chances of failure also greater. Accelerated stability studies are often carried out early in development on pure drug to assess stability and identify degradation products and mechanism. Testing at 50°C, 60°C, or even 70°C under dry and humid conditions (75% RH) for 1 month is often sufficient to provide an initial assessment. More quantitative assessments of drug and formulation stability are carried out to support regulatory filings and generally follow regulatory guidances.[34,35,44]

The field of solution and solid-state stability is large, varied, and beyond the scope of this chapter. Stability stud-

ies described above at a variety of conditions provide the perspective and understanding needed to make meaningful predictions of long-term stability and shelf life. Typically, solid-state decomposition occurs either by zero-order or first-order processes and additional discussion is provided in Chapter 14. Arrhenius analysis and extrapolation to room temperature provide additional confidence that the dosage form will have acceptable stability. Generally though, regulatory guidance allows for New Drug Applications to project shelf life based on accelerated conditions but data at the recommended storage temperature is required to support the actual shelf life of marketed products.

## Mechanical Properties

Many investigations have demonstrated the importance and impact of the physical and chemical properties of materials on powder processing, oral dosage form design, and manufacturing. Physical properties such as particle size and shape clearly influence powder flow, for example. The previous sections of this chapter provide some perspective on characterization. However, mechanical properties (i.e., properties of a material under the influence of an applied stress) are also of great importance for oral dosage form development and manufacturing—particularly for solid dosage forms such as tablets. This section describes the importance of the mechanical properties of materials. For the purposes of this discussion, physical properties are considered to be those properties that are "perceptible especially through the senses"[45] (i.e., properties such as particle size and shape). In contrast, mechanical properties are those properties of a material under an applied load: for example, elasticity, plasticity, viscoelasticity, bonding, and brittleness.

**Table 22–5** lists some of the physical and mechanical properties that influence powder properties and compaction. For example, surface energy and elastic deformation properties influence individual particle true areas of contact as particles are compressed together. Plastic deformation likely occurs to some extent in powders and depends on the applied load and almost certainly it occurs during the compaction of powders into tablets. At asperities, local regions of high pressure can lead to localized plastic yielding. Electrostatic forces can also play a role in powder flow depending on the insulating characteristics of the material and environmental conditions. Particle size, shape, and size distribution have all been shown to influence flow and compaction as well. A number of environmental factors such as humidity, adsorbed impurities (air, water, etc.), consolidation load and time, direction and rate of shear, and storage container properties are also important. With so many variables, it is not surprising that a wide variety of methods have been developed to characterize materials.

What holds particles together in a tablet? A detailed discussion is beyond the scope of this chapter and excellent references are available in the literature.[46,47] However, it is important to realize that the forces that hold particles together

**TABLE 22-6**

**MECHANICAL PROPERTIES OF COMPACTS OF SELECTED EXCIPIENTS DETERMINED AT A COMPACT SOLID FRACTION OF 0.9**

| Excipient | Compression Pressure (MPa) | Tensile Strength (MPa) | Permanent Deformation Pressure (MPa) | Brittle Fracture Index | Bonding Index |
|---|---|---|---|---|---|
| Calcium phosphate, dibasic dihydrate[48] | 395 | 5.6 | 667 | 0.10 | |
| Microcrystalline cellulose[48] | 98 | 8.7 | 153 | 0.08 | 0.06 |
| Croscarmellose, sodium[48] | 200 | 13.6 | 300 | 0.10 | 0.05 |
| Lactose, anhydrous[48] | 178 | 2.6 | 520 | 0.04 | 0.005 |
| Lactose, monohydrate[48] | 191 | 2.5 | 485 | 0.09 | 0.005 |
| Lactose, monohydrate, spray process[48] | 155 | 2.4 | 543 | 0.17 | 0.004 |
| Sucrose[46] | 180 | 2.0 | 473 | 0.68 | 0.004 |
| Corn starch[46] | – | 0.8 | 105 | 0.8 | 0.008 |
| Sorbitol[46] | – | 1.9 | 410 | 0.03 | 0.005 |
| Calcium sulfate, dihydrate[46] | – | 1.9 | 235 | 0.08 | 0.008 |

in a tablet are the very same forces discussed in detail in introductory physical chemistry texts and in this book. There is nothing magical about particle–particle interactions; the forces involved are London dispersion forces, dipole interactions, surface energy considerations, and hydrogen bonding. The compression of powders into tablets brings particles into close proximity and these fundamental forces can then act effectively to produce strong particle–particle interactions and bonding. Particle rearrangement, elastic, and plastic deformation of material can establish large areas of true contact between particles; if the resulting particle–particle bonds are strong, a strong and intact tablet is produced.

Materials used in the pharmaceutical industry can be elastic, plastic, viscoelastic, hard, or brittle in the same sense that metals, plastics, or wood are. The same concepts that mechanical engineers use to explain or characterize tensile, compressive, or shear strength are relevant to pharmaceutical materials. These mechanical properties of materials can have a profound effect on solids processing.

The mechanical properties of a material play an important role in powder flow and compaction by influencing particle–particle interaction, cohesion, and adhesion. They are critical properties that influence the true areas of contact between particles. Therefore, it is essential to be able to quantitatively characterize them. **Table 22–6** provides some mechanical property information for a number of pharmaceutical excipients. One can see that there are a wide range of values and it is important to take these material properties into consideration when developing tablet dosage forms since these mechanical properties determine how the formulation will behave during tablet compaction. There are a wide range of methodologies available for mechanical property characterization and it is important to realize that experimental results are very dependent on the methods used. The reader is directed to comprehensive reviews of this branch of science for additional information.[18,49–56]

Reliable mechanical property information can be useful in helping choose a processing method such as granulation or direct compression, selecting excipients with properties that will mask the poor mechanical properties of the drug, or helping document what went wrong, for example, when a tableting process is being scaled up or when a new bulk drug process is being tested. Since all of these can influence the quality of the final product, it is to the pharmaceutical scientist's advantage to understand the importance of mechanical properties of the active and inactive ingredients and quantitate them.

**Elastic Deformation**

In general, during the initial stages of compression, a material will be deformed elastically and a change in shape caused by an applied pressure is completely reversible and the specimen will return to its original shape on release of the pressure. During elastic deformation, the stress–strain relationship for a specimen is described by Hooke's law:

$$\sigma = E \cdot \varepsilon \qquad (22-18)$$

where

$$\varepsilon = \frac{l - l_{o}}{l_{o}} \qquad (22-19)$$

$E$ is referred to as Young's modulus of elasticity, $\sigma$ is the applied pressure, and $\varepsilon$ is strain where $l_{o}$ is the initial length of the specimen and $l$ is the final length. The region of elastic deformation of a specimen is shown graphically in **Figure 22–7**. As long as the elastic limit is not exceeded, only elastic deformation occurs. The elastic properties of materials can be understood by considering the attractive and repulsive forces between atoms and molecules. Elastic strain results from a change in the intermolecular spacing and, at least for small deformations, is reversible.

**Plastic Deformation**

Plastic deformation is the permanent change in shape of a specimen due to applied stress. The onset of plastic deformation is seen as the change of curvature in the stress–strain curve shown in **Figure 22–7**. Plastic deformation is important

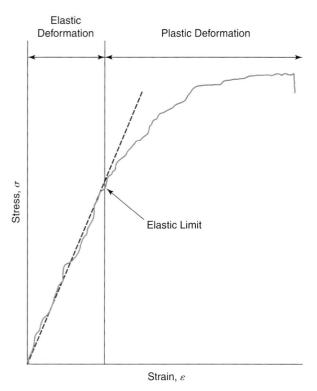

**Fig. 22–7.** Stress–strain curve depicting elastic and plastic deformation regions.

because it permits pharmaceutical excipients and drugs to establish large true areas of contact during compaction. In this way, strong tablets can be prepared.

Plastic deformation, unlike elastic deformation, is generally not accurately predicted from atomic or molecular properties. Rather, plastic deformation is often determined by the presence of crystal defects such as dislocations, grain boundaries, and slip planes within crystals. The formation of dislocations and grain boundaries, and hence the mechanical properties of materials, is influenced by factors such as the rate of crystallization, particle size, the presence of impurities, and the type of crystallization solvent used. Slip planes may exist within crystals due to molecular packing arrangements that result in weak interplanar forces. Processes that influence these (e.g., crystallization rate, solvent, temperature) can be expected to influence the plastic deformation properties of materials and processing properties.

### Brittle and Ductile Fracture

In addition to plastic deformation, materials may fail by either brittle fracture or ductile fracture; fracture being the separation of a body into two or more parts. Brittle fracture occurs by the rapid propagation of a crack throughout the specimen. Conversely, ductile fracture is characterized by extensive plastic deformation followed by fracture. Ductile failure is not typically seen with compacts of pharmaceutical materials. The characteristic snap of a tablet when pressed between the fingers as it breaks in half is indicative of brittle fracture. Brittle fracture of tablets experienced during normal

processing and handling is not acceptable and selection of formulation components can allow pharmaceutical scientists to obtain tablets with acceptable properties.

## Viscoelastic Properties

Viscoelastic properties can be important; viscoelasticity reflects the time-dependent nature of stress–strain. A basic understanding of viscoelasticity can be gained by considering processes that occur at a molecular level when a material is under stress. An applied stress, even when in the elastic region, effectively moves atoms or molecules from their equilibrium energy state. With time, the permanent rearrangement of atoms or molecules can occur.

The stress–strain relationship can therefore depend on the time frame over which the test is conducted. In compacting tablets, for example, it is frequently noted that higher compaction forces are required to make a tablet when the compaction speed is fast. All pharmaceutical materials are viscoelastic; the degree to which their mechanical properties are influenced by the rate of application of stress depends on the material. Appropriate selection of additional pharmaceutical ingredients is needed to address these problems.

## Biological Properties

### Partition Coefficient

The basic principle of the distribution of solute between immiscible solvents has been described in some detail in Chapter 9. The partition coefficient is defined for dilute solutions as the molar concentration ratio of a single, neutral species between two phases at equilibrium:

$$P = \frac{[A]_o}{[A]_w} \qquad (22\text{–}20)$$

Usually the logarithm (base 10) of the partition coefficient (logP) is used because partition coefficient values may range over 8 to 10 orders or magnitude. Indeed, the partition coefficient, typically the octanol–water partition coefficient, has become a widely used and studied physicochemical parameter in a variety of fields including medicinal chemistry, physical chemistry, pharmaceutics, environmental science, and toxicology. While $P$ is the partition coefficient notation generally used in the pharmaceutical and medicinal chemistry literature, environmental and toxicological sciences have more traditionally used the term $K$ or $K_{ow}$. One of the earliest applications of oil/water partitioning to explain pharmacological activity was the work of Overton[57] and Meyer[58] over a century ago, which demonstrated that narcotic potency tended to increase with oil/water partition coefficient. The estimation and application of partition coefficient data to drug delivery began to grow rapidly in the 1960s to become one of the most widely used and studied physicochemical parameters in medicinal chemistry and pharmaceutics.[59]

Selection of the octanol–water system is often justified in part because, like biological membrane components, octanol is flexible and contains a polar head and a nonpolar tail. Hence, the tendency of a drug molecule to leave the aqueous phase and partition into octanol is viewed as a measure of how efficiently drug will partition into and diffuse across biological barriers such as the intestinal membrane. While the octanol–water partition coefficient is, by far, most commonly used, other solvent systems such as cyclohexane–water and chloroform–water systems offer additional insight into partitioning phenomena. Partition coefficients are relatively simple, at least in principle, to measure. However, the devil is in the details and certain aspects demand sufficient attention that rapid throughput methodologies have not yet been successfully developed.[60,61]

As mentioned above, partition coefficient refers to the distribution of the neutral species. For ionizable drugs where the ionized species does not partition into the organic phase, the apparent partition coefficient, $D$, can be calculated from the following:

$$\text{Acids: } \log D = \log P - \log(1 + 10^{(\text{pH}-\text{p}K_a)}) \quad \textbf{(22–21)}$$

$$\text{Bases: } \log D = \log P - \log(1 + 10^{(\text{p}K_a-\text{pH})}) \quad \textbf{(22–22)}$$

## Permeability

New chemical compounds generated in today's pharmaceutical research environment often have unfavorable biopharmaceutical properties. These compounds are generally more lipophilic, less soluble, and are of higher molecular weight.[62] Indeed, permeability, solubility, and dose have been referred to as the "triad" that determines whether a drug molecule can be developed into a commercially viable product with the desired properties. As seen in equation **(22–5)** above, intestinal permeability can be critically important in controlling the rate and extent of absorption and to achieving desired plasma levels.

With the difficulties associated with accurate estimation of permeability based only on physicochemical properties, a variety of methods of measuring permeability have been developed and used. Among them are (*a*) cultured monolayer cell systems such as Caco-2 or MDCK, (*b*) diffusion cell systems which utilize small sections of intestinal mucosa between two chambers, (*c*) in situ intestinal perfusion experiments performed in anesthetized animals such as rats, and (*d*) intestinal perfusion studies performed in humans. All of these methods offer opportunities to study transport of drug across biological membranes under well-controlled conditions.

## ORAL SOLID DOSAGE FORMS

Oral administration is the most frequently utilized route of drug delivery and solid dosage forms are the most commonly available (see **Table 22–1**). To successfully develop oral solid formulations, the important physical, chemical, biological, and mechanical properties of the API need to be assessed and integrated into a suitable strategy that will lead to a dosage form that meets the necessary drug delivery requirements. The focus of this section is to provide an overview of the most commonly available oral solid dosage forms and manufacturing technologies used today. The principles of formulation development and manufacturing apply to any pharmaceutical dosage form though. Often, the decision to manufacture a product is influenced by the cost of manufacturing, packaging, storage, and shipping as well as the drug delivery requirements of the active ingredient. The properties of the drug may require alternative dosage form technologies such as liquid preparations (oral solutions, suspensions), liquid-filled soft gelatin capsules, and so on. Preformulation characterization influences the proper selection of the dosage form technology needed and a wide range of options exist (**Table 22–4**). Every dosage form requires a thorough characterization and understanding of formulation components, manufacturing processes, and product performance requirements. What follows is a discussion of these considerations focusing on the two most common oral dosage forms: tablets and capsules. The reader is also referred to the extensive literature available on this topic.[63–66]

## DRUG RELEASE FROM ORAL DOSAGE FORMS

Drug release is the process by which a drug leaves the drug product and is available for absorption, distribution, metabolism, and excretion (ADME), eventually becoming available for pharmacologic action (**Fig. 22–8**). The selection of the appropriate drug release profile is dependent upon the drug ADME properties and the desired pharmacological effect. Proper selection of excipients and manufacturing methods for the dosage form permits a wide range of drug release profiles to be achieved when properly matched with drug properties. Some of the more common drug release profiles for solid oral dosage forms are immediate release, modified release, delayed release, extended release, and pulsatile

## KEY CONCEPT FORMULATION DEVELOPMENT

Formulation development is the process of identifying the materials and methods necessary to manufacture a stable dosage form that consistently meets specified performance requirements throughout the product's shelf life. Dosage form efficacy, safety, quality, and manufacturability must be ensured. Chemical stability considerations, drug release characteristics, phys-

ical stability, absence of undesirable impurities or degradation products, aesthetic considerations, and the ability to consistently manufacture the dosage form in an environment that meets product supply demand are important factors that must be addressed in formulation development.

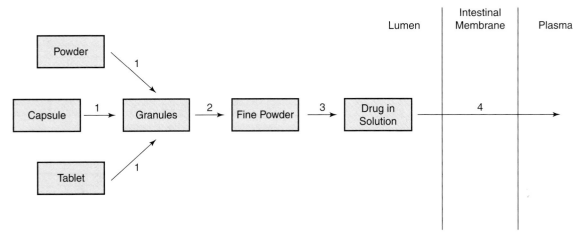

**Fig. 22–8.** Summary of processes associated with the administration of an immediate-release oral solid dosage form: (1) wetting and disintegration, (2) deagglomeration, (3) dissolution, and (4) absorption across the intestinal membrane.

release. Immediate-release drug products allow drugs to dissolve with no intention of delaying or prolonging dissolution or absorption of the drug. Delayed-release is defined as the release of a drug at a time other than immediately following administration. An excellent example of a delayed release dosage form is an enteric-coated tablet. Enteric-coated tablets protect the dosage form from the very acidic environment of the stomach and prevent tablet disintegration until it enters the upper GI tract where the less acidic intestinal fluid can dissolve the enteric polymer and facilitate the disintegration and dissolution of the tablet. Extended-release products are formulated to make the drug available over an extended period after administration. Typically, extended-release products reduce the dosing frequency required. Modified-release dosage forms is a term that applies to both delayed- and extended-release drug products and describes dosage forms whose drug-release characteristics are chosen to accomplish therapeutic or convenience objectives that are not offered by immediate release dosage forms. Pulsatile release involves the release of finite amounts (or pulses) of drug at distinct time intervals that are programmed into the drug product. Finally, controlled-release dosage forms is an inclusive term that includes extended-release and pulsatile-release products. Additional details on these topics and the relevant scientific principles are presented in chapters on "Drug Release and Dissolution" and "Drug Delivery Systems."

## Tablets

A wide variety of tablet dosage forms are available. Compressed tablets as a dosage form originated in the mid-19th century and are still the most commonly available dosage form. The technology and the science of tablet compression has advanced substantially making it a convenient and effective manufacturing approach for a wide variety of drugs. Compressed tablets are manufactured by mechanically compressing the pharmaceutical formulation using a tablet punch

and die system as shown in **Figure 22–9**. In a production environment, high-speed tablet presses can produce tablets at a very high rate, often at a rate that exceeds several thousand tablets per minute. Excipients are incorporated into a formulation with the API using a variety of manufacturing processes to ensure satisfactory manufacturing, stability, and dosage form performance. Tablets are compacted sufficiently hard to ensure that they will withstand normal handling during manufacturing, transport, and patient use but will perform as required to deliver the active ingredient when administered. For immediate release tablets, this involves the rapid disintegration of the compressed tablet into particulate material with subsequent dissolution of the drug substance in the gastrointestinal tract. Tablets are the most frequently prescribed dosage form and can provide the patient with a stable, elegant, effective, and convenient dosage form. However,

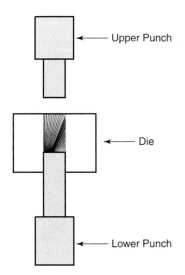

**Fig. 22–9.** Schematic drawing of typical punch and die system for tablet compaction.

tablets are available only in fixed dosage strengths and offer a limited range of doses for the patients. Tablets may be scored to facilitate breaking them to provide fractional doses. Although uncommon, tablets may be formed using molding methods and it is now possible for compound pharmacies to prepare small quantities of compressed or molded tablets for patients. Tablets must be uniform in weight and appearance, contain the proper amount of active ingredient, and consistently achieve the overall drug release properties required to ensure effective administration of the drug for the entire shelf life of the product. For tablet dosage forms, the shelf life is often 2 years or more.

## Capsules

Capsule products have also been used for well over a century and they continue to be used in today's high-speed product manufacturing environment. They have an important role in drug delivery as they are quite flexible, relatively easy to manufacture, and they are amenable to small-scale manufacturing by the compounding pharmacist. Capsule shells consist of two parts: the base, or body, which is longer and has a smaller diameter, and the cap which is shorter and has a slightly larger diameter allowing it to slide over the base portion and form a snug seal. Capsule products are generally prepared by filling formulated material into the base and slipping the cap over the base to seal it. Capsule manufacturing can be done by manual, semimanual, or fully automatic methods. In today's large manufacturing environment, capsule products can be manufactured at high speed, though not as fast as compressed tablets. For this reason, tablets are generally a more cost-effective dosage form. Capsule products are generally dosed in their entirety.

Like tablets, capsules must be uniform in weight and appearance, contain the proper amount of active ingredient, and consistently achieve the overall drug release properties required to ensure effective administration of the drug. Gelatin is still the most common material used to manufacture capsule shells though newer polymeric materials such as hypromellose (HPMC) (hydroxypropyl methylcellulose) are becoming more commonly available and used.

## Powders

Historically, powders have been used for both oral and external applications. Unlike standard tablets and capsules, powders enable physicians and pharmacists to more easily alter the quantity of active ingredient that is administered in a dose. Powders can also be useful in clinical studies because of the flexibility in dosing. Powders are not often made in mass quantities, however, and the application of powdered dosage forms is now largely limited to small clinical studies and compounding pharmacy practices. Powder formulations must contain the proper amount of active ingredient in each dose (e.g., that portion of the powder that is to be dosed) and consistently achieve the overall drug release property requirements.

## Formulation Development

In addition to the APIs, dosage forms contain a number of other pharmaceutical additives called excipients. A formulation is a combination of excipients and active ingredient processed using one or more manufacturing processes to yield a pharmaceutical dosage form. According to USP, "Excipients are substances, other than the active drug substance or finished dosage form, that have been appropriately evaluated for safety and are included in drug delivery systems: 1) to aid in the processing of the drug delivery system during its manufacture; 2) to protect, support, or enhance stability, bioavailability, or patient acceptability; 3) to assist in product identification; or 4) to enhance any other attribute of the overall safety, effectiveness, or delivery of the drug during storage or use."[67] Excipients are used for a reason and they play critical roles in a dosage form. A number of common excipient functions are listed in **Table 22–7**. Some of the more important excipients are described below. All excipients used in approved products are well studied and shown to be safe for human and veterinary use.

There are a number of pharmacopeias worldwide such as United States Pharmacopeia, the European Pharmacopeia, and the Japanese Pharmacopeia, that provide public standards for excipients.

As our understanding of drug absorption and intestinal physiology has increased, it has become clear that some excipients may serve a more active role of enhancing drug absorption by influencing intestinal transporters or other membrane properties. Such "active" excipients are the topic of a number of research investigations as a way to improve the oral delivery of what has traditionally been considered "difficult to deliver" drugs. These active excipients offer new opportunities for pharmaceutical scientists but caution is also warranted as indiscriminate permeability enhancement can lead to unwanted consequences.[68–70]

**TABLE 22–7**

**COMMON PHARMACEUTICAL TABLET AND CAPSULE EXCIPIENT FUNCTIONAL CATEGORIES**

Anticaking agent
Antioxidant
Binder (for wet granulation)
Coating agent
Coloring agent
Diluent
Disintegrant
Dissolution retardant (polymers)
Flavoring agent
Glidant
Lubricant
Preservative
Solubilizing agent
Sweetening agent
Wetting agent

## TABLE 22–8
### COMMON PHARMACEUTICAL TABLET AND CAPSULE DILUENTS

Calcium carbonate
Dicalcium phosphate
Lactose anhydrous
Lactose monohydrate
Lactose spray process
Mannitol
Microcrystalline cellulose
Sorbitol
Starch
Sucrose

### Diluents

Diluents are ingredients incorporated into formulations to increase dosage form volume or weight. They are sometimes referred to as fillers and they often comprise a significant proportion of the dosage form. The quantity and type of diluent selected depends upon its physical and chemical properties and it must be matched to the active ingredient to ensure satisfactory stability and performance. Because the diluent may comprise a large portion of the dosage form, successful and robust manufacturing and dosage form performance is very dependent upon its properties. Among the most important functional roles diluents play is to impart desirable manufacturing properties such as good powder flow, tablet compaction strength, and desired performance including content uniformity, disintegration, dissolution, tablet integrity, friability, and physical and chemical stability. A number of commonly used solid dosage form diluents are listed in **Table 22–8**. Among the most commonly used diluents are lactose, dicalcium phosphate, and microcrystalline cellulose.

### Binder

Tablet binders (**Table 22–9**) are incorporated into formulations to facilitate the agglomeration of powder into granules during mixing with a granulating fluid such as water, hydroalcoholic mixture, or other solvent. In a wet granulation process, the binder may be either dissolved or dispersed in the granulation liquid or blended in a dry state with other components and the granulation liquid added separately during agitation. Following evaporation of the granulation liquid, binders typically produce dry granules that achieve desirable

## TABLE 22–9
### COMMON PHARMACEUTICAL TABLET AND CAPSULE BINDERS

Hypromellose (HPMC)
Povidone
Pregelatinized starch
Sodium carboxymethylcellulose
Starch

## TABLE 22–10
### COMMON PHARMACEUTICAL TABLET AND CAPSULE DISINTEGRANTS

Alginic acid
Crospovidone
Microcrystalline cellulose
Pregelatinized starch
Sodium croscarmellose
Sodium starch glycolate
Starch

manufacturing properties such as granule size and size distribution, shape, content, mass, active ingredient content, and compaction properties. Wet granulation facilitates the further processing of the granules by improving one or more granule properties such as flow, handling, strength, resistance to segregation, dustiness, appearance, solubility, compaction, or drug release. Tablet binders are soluble or partially soluble in the granulating solvent. Upon addition of liquid, binders typically facilitate the production of moist granules (agglomerates) by altering interparticle adhesion. During drying, solid bridges are produced that result in significant granule strength.

### Disintegrant

For most tablets and capsules, it is necessary to incorporate a disintegrant to overcome the cohesive strength of the tablet that was generated during compression. Disintegrants (**Table 22–10**) facilitate the uptake of water into the tablet or swell in contact with water producing an expansion of the tablet and the breakup of the bonds that hold the tablet together. So-called superdisintegrates perform both of these functions and cause tablets to disintegrate very rapidly upon exposure to water.

### Lubricant

Lubricants typically are used to reduce frictional forces between formulation components and metal contact surfaces of manufacturing equipment such as tablet punches and dies (**Table 22–11**). The most commonly used lubricant in solid dosage forms is magnesium stearate. It is a solid powder that can be blended with other formulation components. Lubricants adhere to solid surfaces (formulation components and

## TABLE 22–11
### COMMON PHARMACEUTICAL TABLET AND CAPSULE LUBRICANTS

Magnesium stearate
Calcium stearate
Stearic acid
Sodium steryl fumerate
Polyethylene glycol
Sodium lauryl sulfate
Starch

**TABLE 22–12**

## COMMON PHARMACEUTICAL TABLET AND CAPSULE GLIDANTS AND ANTICAKING AGENTS

Colloidal silicon dioxide
Calcium silicate
Magnesium silicate
Talc

**TABLE 22–14**

## COMMON PHARMACEUTICAL TABLET AND CAPSULE COATING AGENTS

Hypromellose
Ethylcellulose
Methylcellulose
Ammonio methacrylate copolymer
Cellulose acetate
Cellulose acetate phthalate
Methacrylic acid copolymer
Sucrose

equipment parts) and reduce the particle–particle friction or the particle–equipment surface friction.

Lubricants are typically incorporated in very low levels—often 1% w/w or less. Caution is required, however, because excessive lubricant levels may retard tablet disintegration or dissolution by creating large hydrophobic surfaces that will not wet or dissolve.

### Glidants and Anticaking Agents

Glidants and anticaking agents (**Table 22–12**) are used to promote powder flow and to reduce the caking or clumping that can occur when powders are stored in bulk. Glidants and anticaking agents can also reduce the incidence of bridging during the emptying of powder hoppers and during powder processing. Glidants likely work through a combination of adsorption onto the surfaces of larger particles to help reduce particle–particle adhesive and cohesive forces and also by being dispersed between the larger particles and acting to reduce the friction between those particles. Anticaking agents generally work by absorbing free moisture that may otherwise permit the formation of particle–particle bridges that can cause caking.

### Wetting and Solubilizing Agents

Surfactants, or surface-active agents, are amphiphilic molecules that contain both a polar and nonpolar region that can function as emulsifying, wetting, and solubilizing agents (see **Table 22–13**). The amphiphilic nature of surfactants is responsible for two important properties of these compounds that account for a variety of interfacial phenomena. One is the ability of surfactant molecules to adsorb at gas–liquid, liquid–liquid, and solid–liquid interfaces to reduce interfacial tension. They also have a tendency to self-associate and form aggregates or micelles once the critical micelle concentration is exceeded. The ability of surfactants to reduce interfacial

tension is critical to emulsification and wetting while micelle formation enables the solubilization of water-insoluble compounds. These excipients are added to formulations to facilitate the wetting or solubilization of the drug substance.

### Coating Agents

Pharmaceuticals may be coated for several reasons including taste masking, improving ingestion, improving appearance, ease of identification, protecting active ingredients from the environment, and controlling drug release in the GI tract. The materials used in coating systems (**Table 22–14**) include natural and synthetic or semisynthetic materials. Although more popular decades ago, sugarcoating tablets is still performed. Some coating materials are used as colloidal dispersions. Titanium dioxide, an inorganic compound, is used in coatings as an opacifier. The coating system forms a layer on the tablet and changes appearance (nonfunctional coat) or performance (functional coat). Coating materials that are used must have the ability to form a film or coating system around the tablet that is complete and stable. The coating material must be applied uniformly to ensure proper performance by spreading over the surface of the dosage form and coalescing to form a smooth film. One important functional tablet coating is enteric coating. Enteric coating polymers are insoluble in the acidic environment of the stomach and protect the drug. Once the enteric-coated dosage form enters the intestine where the pH is higher, the polymer dissolves and allows the dosage form to disintegrate and the drug to dissolve.

### Drug Release Modifying Agents

A variety of excipients, typically polymeric, may be used to delay the release of drug from a dosage form (**Table 22–15**). Common technologies used for this purpose include: matrix tablets, multiparticulate–coated particles, and osmotically controlled dosage forms. Selection of the release-modifying agent is dependent upon the drug properties and the drug release profile that is needed to optimize dosage form performance. In comparing the tables of excipients provided here, it is clear that excipients may serve different functions depending on how they are used in a formulation. For example,

**TABLE 22–13**

## COMMON PHARMACEUTICAL TABLET AND CAPSULE WETTING AND SOLUBILIZING AGENTS

Sodium lauryl sulfate
Docusate sodium
Lecithin
Poloxamer
Polysorbate 80

## TABLE 22–15
## COMMON PHARMACEUTICAL TABLET AND CAPSULE DRUG RELEASE MODIFYING AGENTS

Hypromellose
Hydroxypropyl methylcellulose—acetate succinate
Ethylcellulose
Ammonio methacrylate copolymer
Cellulose acetate
Cellulose acetate phthalate
Methacrylic acid copolymer
Polymethacrylate
Carboxymethylcellulse
Polyvinylchloride
Polyvinylacetate

hypromellose (HPMC) may be used as a tablet binder, a delayed release agent, or a tablet-coating agent depending on the quantity and processing methods used.

### Other Excipients

There are a variety of other excipients that are utilized in solid dosage forms that are not enumerated here. All excipients in a dosage form are there for a reason and current regulatory filings require a dosage form manufacturer to indicate the function of each ingredient and ensure that they meet standards for safety, efficacy, and quality. The United States Pharmacopeia (USP/NF), the European Pharmacopeia, and the Japanese Pharmacopeia provide publically available standards for these purposes. The USP,[71] the Handbook of Pharmaceutical Excipients,[72] and other standard textbooks identify the functional categories of excipients and their typical uses.

## EXAMPLE 22–9

A commercially available tablet dosage form for the treatment of Parkinson disease lists the following excipients in its list of inactive ingredients: mannitol, starch, colloidal silicon dioxide, povidone, and magnesium stearate. Identify the functional purpose of each excipient:

　Mannitol: diluent
　Starch: diluent, binder, and/or disintegrant
　Colloidal silicon dioxide: glidant and/or anticaking agent
　Povidone: binder
　Magnesium Stearate: lubricant

Note that starch and colloidal silicon dioxide may serve more than one purpose in this formulation and it is sometimes difficult to know exactly what an excipient function is without knowing more about the formulation. The function of an excipient is dependent upon the formulation, the manufacturing process, and the dosage form performance requirements.

# MANUFACTURING

## Regulatory Environment

The FDA regulates the new drug approval process in the United States and other countries have similar regulatory bodies to ensure that pharmaceutical products are manufactured and distributed in a way that ensures safety, efficacy, and quality. In 1906, President Theodore Roosevelt signed into law the Food and Drug Act that, in effect, established what is now known as the FDA. The responsibilities of the FDA were substantially expanded when President Franklin Roosevelt signed the Food, Drug, and Cosmetic (FD&C) Act into law in 1938. These changes came about as a result of the 1937 sulfanilamide elixir tragedy in which more than 100 people died after using the drug formulated in the toxic solvent ethylene glycol. The 1938 act required predistribution clearance for the safety of new drugs, authorized factory inspections, and expanded the legal authority of the FDA. Further revisions and expansion of the FDA responsibilities occurred in 1962 when the Kefauver–Harris Amendment to the FD&C Act established the requirement that all new drug applications demonstrate, for the first time, substantial evidence of efficacy for marketed claims in addition to the previous requirements of demonstrated safety.

Today's pharmaceutical industry is highly regulated and global in nature and the impact of regulatory requirements is far reaching. Regulatory agencies, including the FDA, have established good laboratory practices, good manufacturing practices, good clinical practices, good distribution practices, good regulatory practices, guidelines for new drug applications, limits on advertising, postmarketing surveillance and clinical monitoring, and a host of other guidelines and requirements to ensure product quality, safety, and efficacy. The FDA's stated mission is to protect "the public health by assuring the safety, efficacy, and security of human and veterinary drugs, biological products, medical devices, our nation's food supply, cosmetics, and products that emit radiation. The FDA is also responsible for advancing the public health by helping to speed innovations that make medicines and foods more effective, safer, and affordable; and helping

## KEY CONCEPT　CURRENT GOOD MANUFACTURING PRACTICES

Current Good Manufacturing Practices (cGMPs) are a set of regulations established by the US Food and Drug Administration that contain the minimum current good manufacturing practice for methods to be used in, and the facilities or controls to be used for, manufacturing, processing, packing, or holding of a drug to ensure that the drug meets the requirements for safety, identity, strength, and the necessary quality and purity requirements. Failure to comply with cGMPs in the manufacturing, processing, packing, or holding of a drug renders it to be adulterated and subject to regulatory action.

the public get the accurate, science-based information they need to use medicines and foods to improve their health."[73]

With respect to pharmaceutical manufacturing, current Good Manufacturing Practices (cGMPs) play a pivotal role. Originally established in 1963 and expanded upon in 1979, cGMPs present the minimum requirements for manufacturing, packaging, and storage of human and veterinary products. These cGMPs provide guidance on organization and personnel, buildings and facilities, equipment, production and process controls, packaging and labeling, holding and distribution, and laboratory controls as well as records and reports. In effect, virtually every aspect of the manufacturing, packaging, and storage of a pharmaceutical product is carefully assessed, analyzed, and controlled to ensure product quality.[74,75]

## Manufacturing

Pharmaceutical manufacturing on a large scale is carried out in facilities that conform to good manufacturing practices. In comparison, pharmaceutical compounding (medications made by a pharmacist or other healthcare provider in response to a valid prescription) comprises approximately 1% of prescriptions filled, totaling approximately 30 million prescriptions and $1 billion annually.[76] Following physical, chemical, and mechanical property characterization of the API, initial formulation development activities are undertaken to design a formulation with the desired stability, drug release, and manufacturing properties. A general outline of the overall formulation development process is provided in **Table 22–16**. Different approaches may be taken. The "plan for success" approach often front-loads formulation development activities where extensive work is done to identify robust, manufacturable formulations very early in development. If the drug being studied moves successfully through early clinical studies, product manufacturing will not be on the critical path and the development process can move as quickly as possible. An alternate approach being taken these days is a material and resource sparing one, in which only enough time and effort is expended to identify and manufacture a formulation that meets the clinical and regulatory requirements of the project. In the former approach, the formulations utilized in early clinical testing are often very representative of what the final dosage form will look and behave like. With the latter approach, extensive formulation and process development is

### TABLE 22–16
### FORMULATION DESIGN AND DEVELOPMENT

**Preliminary Formulation Development**
  Physical, chemical and mechanical property characterization of the API
  Preliminary formulation design (preliminary selection of excipients, processing)
  Preliminary formulation process selection
**Initiate Marketed Product Formulation Development**
  Excipient range-finding studies
  Identify and assess manufacturing process variables
**Final Formulation Development**
  Final process characterization
**Product Appearance**
  Tablet coating process characterization
  Tablet tooling evaluation
**Scale Up Activities**
  Prepare large-scale lots
**Stability**
  Establish final packaging and stability
**Regulatory Filings**
  File NDA
  File regulatory documents worldwide

postponed until the drug successfully passes the early clinical testing milestones. With either approach, as the drug moves through development, a wide range of studies are conducted to identify the components and quantities of the formulation that are required to achieve the desired dosage form performance. Following formulation design activities, additional effort goes into identifying the manufacturing processes and specific processing conditions that will be necessary to combine the drug and excipients into a manufacturable product.

### Unit Processes

Most pharmaceutical manufacturing today consists of a series of separate and distinct manufacturing steps called unit processes. Typically, each of these discrete steps can be viewed as an individual activity and each can be evaluated and optimized to produce a consistent material. Several examples of a series of manufacturing steps (sometimes referred to as a manufacturing or process train) are shown in **Figure 22–10**. Among the most common unit processes for oral solid dosage forms are milling, blending, granulation, tablet

## KEY CONCEPT PHARMACEUTICAL QUALITY BY DESIGN

Pharmaceutical Quality by Design (QbD) is a systematic, scientific, risk-based, and proactive approach to pharmaceutical development that begins with predefined objectives and emphasizes product and process understanding and process control. This includes designing and developing formulations and manufacturing processes to ensure that predefined product quality objectives are consistently met. QbD identifies characteristics that are critical to quality and translates them into the attributes

that the drug product should possess and establishes how the critical process parameters can be varied to consistently produce a drug product with the desired characteristics. The specifications of a drug product under QbD should be clinically relevant and generally determined by product performance. Under QbD, consistency comes from the design and control of the manufacturing process.

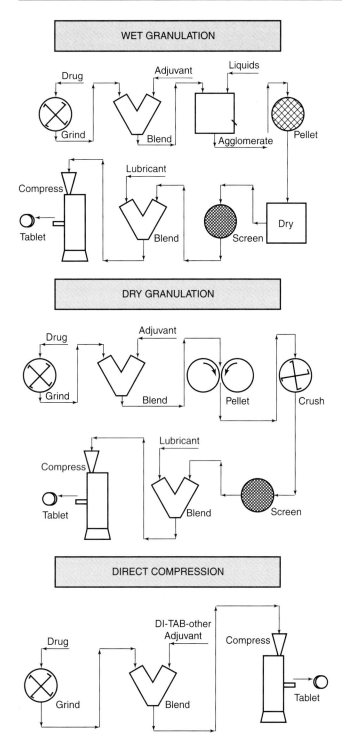

**Fig. 22–10.** The three main methods for the preparation of tablets. (Courtesy of Stauffer Chemical Co. Tarpon Springs, FL.)

compression or capsule filling, and tablet coating. Each of these unit processes offers challenges and opportunities to the pharmaceutical scientist. The physical, chemical, and mechanical properties of the materials (active ingredient, excipients, intermediate materials) that are introduced into equipment may influence the properties of the material that is obtained after processing. Those properties that have a

significant and important impact on the manufacturing or performance of the product can be considered critical material attributes. Those processing parameters that have a significant and important impact on manufacturing or performance are referred to as critical process parameters. A recent paradigm shift within the regulatory agencies has pharmaceutical manufacturers moving toward a Quality by Design or QbD approach in which well designed, controlled, and studied materials and manufacturing processes identify the critical material attributes and critical process parameters needed to achieve a final product that consistently meets the performance requirements.[75,77] Each unit process is often studied in some detail using appropriate experimental methods (e.g., design of experiments).

**Milling.** The particle size of pharmaceutical materials is often a critical material attribute that can impact processing and performance. Particle size has been shown to influence processes like blending, granulation, and compaction. Particle size also influences dosage form performance characteristics such as dissolution and content uniformity. For this reason particle size is often carefully studied and controlled. Where particle size, shape, and size distribution can be controlled by crystallization, crystal-engineering strategies are desirable. Where this is not possible, materials may be milled to achieve the desired particle size, shape, and size distribution (micromeritic properties, Chapter 18). A variety of mill types are available to the pharmaceutical scientist and some of these are shown in **Table 22–17**. Proper selection of mill type and process conditions can be used to tailor micromeritic properties. Milling is most often applied to the API since pharmaceutical grade excipients may be purchased in a range of particle sizes that usually meet development scientist's needs.

**Blending or Mixing.** The blending of solid particles in the dry state is one of the oldest industrial processes known to man. Blending or mixing is a unit process that is used at some point in virtually every oral solid dosage form manufacturing process train. Science and technology have advanced our understanding of blending and a variety of methods are available to carry out this process. Blending is a process that results in the randomization of particles within a powder system and achieves an assembly of particles that are more or less thoroughly dispersed. Blending can be described as proceeding in the following steps. A static powder must first expand before particle–particle movement is possible. Once expansion of a powder occurs, particles are able to move; shear forces are necessary to produce movement between particles. Movement of particles relative to one another requires adequate three-dimensional stresses that result in essentially random particle movement and mixing.

Diffusion mixers operate by facilitating the reorientation of particles relative to one another due to powder bed expansion and random motion of particles. Diffusion type mixers are commonly used in the pharmaceutical industry and include V-blenders, double-cone blenders, bin blenders, and drum blenders. Convection mixers facilitate mixing by

**TABLE 22–17**
**MILL TYPES AND APPROXIMATE PARTICLE SIZE ACHIEVED**

| Type | Coarse | Medium Fine | Fine | Very Fine | Super Fine | Ultra Fine |
|---|---|---|---|---|---|---|
| Range of particle size, $\mu$m | 1000–5000 | 500–1000 | 150–500 | 50–150 | 25–50 | 2–10 |
| Screening mills | X | X | X | | | |
| Impact mills | | X | X | X | | |
| Air swept impact mill | | X | X | X | | |
| Fluid energy mill | | | | X | X | X |

reorienting particles relative to one another due to mechanical movement using paddles. Examples of convective mixers are ribbon blenders, screw blenders, planetary blenders, and high-intensity mixers.

Granulation. If a simple blend of excipients and active ingredients does not have the physical, chemical, or mechanical properties needed to achieve the manufacturing and performance requirements, the blended powders may be further processed using granulation methods. A direct blend, often referred to as a direct compression formulation, is advantageous and often preferred over granulated powders because it requires fewer steps for manufacturing and is therefore more cost-effective. Granulation is the process of particle agglomeration and size enlargement of powdered ingredients to achieve desirable processing properties such as improved powder flow or compression. Within the pharmaceutical industry dry granulation and wet granulation methodologies are most often used if a direct blend is not suitable.

Wet granulation is achieved by mixing a granulating fluid, often water, together with other blended components to achieve a wet mass that forms larger agglomerates called granules. Once the desired granule growth has been achieved, commonly referred to as a granulation endpoint, the wet massing process is stopped and the granules are then dried. As the drying occurs, ingredients which were dissolved in the granulation fluid form solid bridges that hold the particles together. Generally, a pharmaceutical binder (see **Table 22–9** for examples) is added to the blend or granulating liquid which acts as the glue to permanently hold the particles together. The dried granules may then be milled to achieve the final desired particle size. The wet granulation process has a number of advantages related to improved processability but its disadvantages include exposure of the formulation components to granulating liquid and exposure to the elevated temperatures necessary to dry the wet granules. Wet granulation may be carried out in high shear equipment or alternatively utilize fluid bed technology. The properties of the granules formed depend on the properties of the individual materials used and the process and the process parameters that are used in granulation.

Dry granulation is achieved by compressing powdered materials into dense, cohesive compacts which are then milled and screened to produce a granular form of material with desirable particle size characteristics. The compaction process in dry granulation may be achieved in a continuous fashion using what is known as roller compaction. Roller compaction is the process of compressing powder blend to produce a solid ribbon between two rollers. An alternative and less commonly used method is to compress powders into large tablets, called slugs, which are then milled and screened. Among the advantages of dry granulation is that the materials are not exposed to granulation fluid or the high temperatures required to dry the granulated material.

Drying. In the manufacture of solid dosage forms, it is sometimes necessary to include a wet granulation step in the manufacturing process as described above. Drying is undertaken to remove excess water (or other granulation liquid) from the solid granules by evaporation. The drying process is designed to reduce the moisture content to an acceptable value. The final value depends upon the material being dried. There are a wide variety of drying methods. Among the most commonly used in the pharmaceutical industry are direct heating methods where heat transfer is accomplished by direct contact between the wet solid mass and heated air. An example of a static method of drying is tray-drying where the granulation is placed on a tray that is then placed in an oven and drying takes place. An alternative and more common method that is conducive to large-scale manufacturing is to physically move the moist granulation with heating to cause evaporation. The most commonly used method of drying is fluidized bed drying where the granulation is fluidized in heated air.

Lubrication. A separate blending step, called the lubrication step, is described here because it is a very frequently used unit process. The lubrication step involves a separate mixing step where a lubricant (**Table 22–11**) is incorporated into the formulation. It is very often the last step before tablet compression or capsule filling. As with the other unit processes, the properties of the lubricant and the process parameters must be carefully assessed and characterized because an inappropriately performed lubrication step can have a significant negative impact on dosage form performance. The commonly used lubricants magnesium stearate or stearic acid are very water insoluble. Incorporation of an excessive amount of one of these ingredients or excessive mixing has been shown to decrease the dissolution rate of the final dosage form. Appropriate characterization and control of the lubricant as well as

**TABLE 22–18**

## COMMON PHARMACEUTICAL COMPRESSED TABLET DOSAGE FORMS

Immediate-release tablet controlled release tablet
Bilayer tablet
Multilayer tablet
Osmotic tablet
Sugarcoated tablet
Film-coated tablet
Enteric-coated tablet
Gelatin-coated tablet
Buccal tablet
Sublingual tablet
Chewable tablet
Effervescent tablet
Molded tablet
Rapidly disintegrating tablet
Mucoadhesive tablet
Gastroretentive tablet

an understanding and control of this unit process is very often important in ensuring consistent dosage form performance.

**Compression.** Following the blending, granulation, and lubrication steps, the formulation is ready for compression into a tablet. Tablet dosage forms are manufactured using a compression process. A wide variety of tablet dosage forms with a remarkable range of performance characteristics can be prepared with the proper selection of formulation ingredients and processing. The seeming simplicity of the tablet dosage form belies the remarkable flexibility and creativity this technology offers as a sophisticated drug delivery device. Many of the available tablet dosage forms are listed in **Table 22–18**.

Powder compression into tablets is the application of pressure to the formulated powder to achieve a reduction in volume and the generation of strength within the compacted material to form an intact tablet. Tablet tooling consists of a lower punch which snugly fits into the tablet die from below and an upper punch which can enter the die from above (**Fig. 22–9**). The die serves to hold the formulated powder in place when the lower punch is in place, and the upper and lower punches are forced together to compress the powder. Powder compaction can be done using a small, hand-operated press but, of course, in a large-scale manufacturing environment, high-speed tableting machines are used to produce thousands of tablets per minute. An example of a large-scale tableting machine is shown in **Fig. 22–11**.

The process of powder compaction into tablets can be described as a six-step process as shown schematically in **Figure 22–12**. The first step (Stage 1) is the die filling step in which powder is moved into the die. The powder in this state is loosely aggregated in the die. The lower punch position holds the powder in the die and determines the amount of powder that the die will hold. The compression process begins in the second step (Stage 2) as the upper punch is

**Fig. 22–11.** Example of a production tablet press. (Courtesy of Korsch Tableting, Korsch AG, Berlin.)

pressed into the die; the applied force results in rearrangement and consolidation of the powder. In the third stage of compression (Stage 3), significant particle deformation and possibly particle fracture occur as the powder further consolidates into a cohesive mass. In this stage of compression, significant areas of contact are established between particles as they are pressed closer together and this can result in significant particle–particle bonding. The decompression stage (Stage 4) begins as the upper punch force is reduced and the upper punch is removed. During the decompression stage some of the elastic deformation that occurred during compression results in some tablet expansion. (Stage 5) involves the lower punch being pushed upward as the compacted tablet is pushed upward. If the formulation is properly designed, the final stage (Stage 6) results in the ejection of an intact tablet that has the desired strength and performance characteristics. On rotary tablet machines, multiple punches and dies are located around the outside of a circular die table and the compression process described above occurs as they are rotated under circular compression rolls that force the upper and lower punches together and punch guides pull them apart with precise timing. The entire process described above can occur on a production tablet press in less than 100 milliseconds.

The compression process has been studied in detail by a number of investigators and a variety of equations to describe the relationship between compression pressure and tablet density have been developed.[78–83] One of the most commonly utilized equations was developed by Heckel.[81,82] He proposed that there was a relationship between the yield

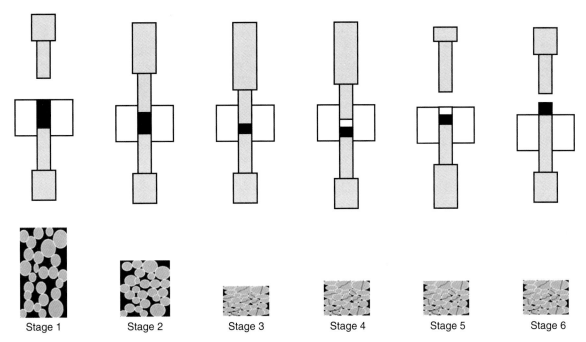

**Fig. 22–12.** Stages of tablet compaction.

strength of the material and the pressure necessary to cause compaction. The yield strength of a material is a measure of its ability to permanently or plastically deform as discussed in the previous section on mechanical properties. A high-yield pressure indicates that a material is hard; a low-yield pressure indicates a soft material. From this, he derived an equation referred to as the Heckel equation (equation **22–25**), expressing the relationship between the relative density of the compact and the compression pressure applied.

In tablet compaction, an important concept is that of relative density. The relative density, $D$, of a material is given by the following equation:

$$D = \frac{\rho_S}{\rho_A} \qquad (22\text{--}23)$$

where $\rho_S$ is the density of the powder or compact in g/cm$^3$ and $\rho_A$ is the absolute or true density of the material in g/cm$^3$. The true density of a material is its density in the absence of pores, meaning that the material contains absolutely no void space between particles. The reader is directed to Chapter 18 on Micromeritics for further discussion of density and methods of measurement.

From equation (**22–23**), the relative density, $D$, has a maximum value of 1.0 and this occurs when all of the void space is compressed out of the compressed powder and only solid material with no pores remains. Ranges of $D$ are between 0.4 and 0.95 for loose powders and highly compacted tablets, respectively. Virtually all pharmaceutical tablets have some porous structure though, and typical values for relative density are in the range of 0.7 to 0.9, meaning that between 30% and 10% of the volume of the tablet consists of pores. The

relationship between relative density and porosity, $\varepsilon$, is given by:

$$\varepsilon = 1 - D \qquad (22\text{--}24)$$

The Heckel relationship is based on the assumption that the decreasing void space within the tablet follows a first-order rate process.[82]

$$\frac{dD}{dP} = K \cdot (1 - D) \qquad (22\text{--}25)$$

where $D$ is the tablet relative density, $P$ is the applied pressure, and $K$ is a constant that reflects the ability of the powder to consolidate into a coherent mass.

Integrating equation (**22–26**), the Heckel equation is:

$$\ln\left(\frac{1}{1-D}\right) = K \cdot P + A \qquad (22\text{--}26)$$

$K$ is the slope of the Heckel equation and is a measure of the plasticity of the material. *A greater slope indicates that the material has greater plasticity and is more easily permanently deformed. A* Heckel plot obtained by plotting $\ln(1/1 - D)$ versus $P$ is shown in **Figure 22–13** for three different pharmaceutical excipients. As seen in this figure, only the terminal portion of the plot is linear and conforms to equation (**22–26**). The different terminal slopes indicate that these three materials have significantly different deformation properties. The initial nonlinear region of the plot is the region in which the Heckel equation does not apply and reflects the initial stage of consolidation where significant particle rearrangement is occurring.

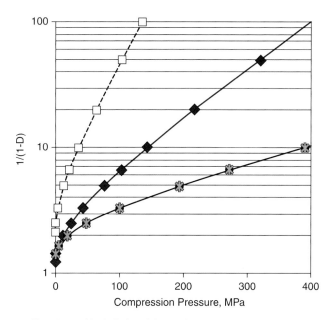

**Fig. 22–13.** Heckel plot of three pharmaceutical excipients.

Heckel and other equations have been used to interpret and predict the compaction properties of pharmaceutical materials and formulations. Because of the critical importance of the compaction process in forming tablet dosage forms, a great deal of research has revolved around understanding and modeling this process. The reader is directed to other comprehensive discussions of powder compaction for further information.[84,85]

There are several significant challenges to developing and successfully manufacturing compressed tablets in a production environment. The formulation must have the necessary properties to ensure consistent powder flow properties that allow it to move through the manufacturing equipment, including movement from any intermediate storage containers to the tablet press hopper and feeder system that directs the powder down into the tablet die. Following compression, the tablet must have acceptable aesthetic properties such as a smooth, elegant appearance without cracks or chips. It must also be robust enough to handle any remaining processing that is required such as coating and packaging, yet consistently meet all the performance characteristics that ensure satisfactory performance such as disintegration and dissolution.

**Tablet Coating.** Some tablet dosage forms may be coated. Coatings may be described as either functional or nonfunctional. Although more popular decades ago, sugarcoating tablets are still performed. The sugarcoating process seals and protects the tablet dosage form and, with the incorporation of color, adds a distinctive look and taste to the tablet. Many tablets are film-coated today as it is a more cost-efficient process than sugarcoating. A thin polymeric layer with a color added is sprayed onto the surface of the tablets to provide a distinctive appearance. Both sugarcoating and film coating may also serve to prevent the patient from experiencing the undesirable taste that some active ingredients have. An example of a functional coat is enteric coating. The enteric-coating material is insoluble in the acidic fluid of the stomach but dissolves in the relatively neutral pH of the intestine. Enteric coating is therefore a way of protecting acid labile drugs from being exposed to the harsh acidic stomach media that can degrade some active ingredients. Controlled release polymers may also be applied to dosage forms to control the rate at which drug enters the intestinal tract and is absorbed.

**Capsule Filling.** Capsules are solid dosage forms in which the medicinal agent and excipients are enclosed in a small shell of gelatin. The capsule shells may be hard two-piece capsules or a soft gelatin film. Two-piece capsules consist of a capsule body into which the formulated material can be filled and the slightly larger diameter cap that slips over the body to seal the capsule. Soft gelatin capsules are sometimes referred to as soft elastic capsules. Soft gelatin capsules may be filled with liquids or semisolid ingredients, whereas the two-piece capsules are very often filled with dried powders. Recent advancements in two-piece capsule technology now allow for liquid and semisolid fills. The vast majority of capsules are intended to be swallowed whole by the patient. While a majority of capsules are manufactured from gelatin, new polymeric, two-piece capsules prepared from HPMC and pullulan (a water-soluble polysaccharide) are now available and additional materials such as starches are being investigated. Capsule machine equipment is designed to move the formulation into the capsule body followed by positioning and closing the cap to produce the final product. One of the main advantages of capsule formulations is that they do not have to undergo the compaction process as tablets do. This can simplify the formulation process. Industrial capsule machines are capable of fast manufacturing speeds though capsule machines do not currently reach the dosage form output of tablet machines.

**Continuous Processing.** While the previous sections have covered some of the details of current pharmaceutical manufacturing processes which are done in batch mode, the future of pharmaceutical manufacturing is moving toward continuous processing. Continuous processing often combines one or more pharmaceutical processes utilizing equipment designed to allow for continuous input of starting materials, material processing, and continuous exit of final processed material. There are a variety of benefits to continuous processing. A continuous process inherently provides an opportunity for improved quality and consistency as it involves processing a much smaller quantity of material at any one time. For example, in a typical batch wet granulation unit operation involving a batch size of 300 Kg, the entire quantity is processed simultaneously. For a continuous operation with reasonable material throughput, the quantity of material being processed at any given time may be only about 400 g.

Continuous processing involves operating equipment at steady state and this makes process control strategies including online measurement of critical process parameters as well as product quality attributes more feasible. Process and statistical modeling can be used to develop control strategies consistent with the Quality by design (Qbd) initiative. These advantages provide the opportunity for simple scale-up. In many instances, it is possible to make large-scale lots with the same equipment used during small-scale development activities by running the process longer. Finally, as the unit operations are integrated into a continuous process, there may be considerable material and resource savings.

Among the commonly used unit operations in solid oral dosage form development, milling, roller compaction (dry granulation), compression (tableting), encapsulation, and packaging are inherently continuous processes. Wet granulation, drying, blending and coating are inherently batch processes. Over the last decade, a concerted effort from industry and equipment manufacturers has resulted in significant progress being done to make these unit operations continuous. As an example, for wet granulation, a modified twin screw extruder and other similar equipment have been designed for use. For drying, fluid bed and dielectric drying have been used. Several designs of low and high shear continuous dry blending equipment are also being marketed. For coating, there are several large-scale continuous coaters in use. With active research and development activities underway, significant advancement is expected to occur in the equipment engineering as well as overall process integration to make this approach for solid oral dosage from development a success in the next 5 to 10 years.

Final Dosage Form Finishing and Packaging. Following the manufacture of tablets or capsules, final finishing of the dosage forms takes place. Two-piece capsules, for example, may be polished to remove small amounts of powder that may adhere to the outside of the capsules during filling. On a large scale, many capsule and tablet machines are affixed to a cleaning vacuum that removes extraneous material from the tablets or capsules as they leave the machine. Following manufacture, the tablets and capsules may be stored in bulk containers until they are packaged. The final dosage form packaging plays a critical part in ensuring and maintaining product quality. Drugs that are adversely affected by light will be packaged in light-resistant containers, whereas moisture-sensitive drugs may be packaged with a desiccant to ensure that the dosage forms are not exposed to high moisture levels that could cause physical or chemical degradation. Properly stored dosage forms will remain stable and effective throughout the entire labeled shelf life of the product.

# DOSAGE FORM TESTING

## Tablet Hardness

The mechanical strength of tablet dosage forms is an important property and it plays a significant role in product development and manufacturing control. The mechanical strength of tablet dosage forms is sometimes referred to as tablet hardness or tablet crushing strength. Pharmaceutical ingredients which bond well together are capable of forming tablets with high strength. An old rule of thumb for tablet hardness was that the tablet should break with a sharp snap when squeezed between the fingers and thumb. Commercially available tablet hardness testers are available to provide quantitative data on tablet hardness. Tablet hardness is the force necessary to cause a tablet to fracture when compressed between two rigid platens. Tablet strength is influenced by the formulation components, the processing used to make the formulation, and the process of forming the compressed tablet. The resistance of tablets to chipping, abrasion, and breakage depends on tablet hardness. Tablet hardness is used as a manufacturing control tool and hardness values are often determined throughout a tablet manufacturing lot. If tablet hardness values vary, adjustments to the tablet machine can be made to ensure that the tablet hardness remains within the accepted range. Tablet hardness values should be high enough to ensure satisfactory appearance and tablet strength to withstand further tablet processing and handling but not so high that the dosage form will fail performance criteria such as disintegration or dissolution.

## Friability

Tablets must be hard enough to withstand the agitation and stresses that occur during manufacturing, coating, packaging, shipping, and patient use. However, tablets must also be friable enough to break up when swallowed. The pharmaceutical scientist's responsibility in developing a robust tablet formulation is to produce a dosage form that has adequate hardness and tablet strength to withstand the stresses but, when necessary, will break up and release the drug in a consistent fashion when administered to the patient. Tablet friability is a measure of the ability of the tablets to withstand stresses. The USP describes friabilator apparatus and test methodologies to evaluate tablet resistance to abrasion. Tablets are placed in a 12-inch diameter drum which rotates for a set period of revolutions, typically 100. A shaped arm lifts the tablets and drops them half the height of the drum with each revolution. At the end of this operation, tablets are removed, dedusted, and reweighed. The percent weight change is calculated and is used as a measure of friability. Tablets that remain intact without cracking or chipping (e.g., <1% weight change) typically have sufficient strength to withstand further processing and packaging.

## Disintegration

One simple measure of the ability of a compressed tablet or capsule to release drug is the disintegration test. The disintegration time is the time it takes for a dosage form to break apart upon exposure to water with mild agitation. Pharmacopeia's worldwide, including USP, provide details for carrying out standardized disintegration testing that specify the disintegration liquid, the apparatus, the number of dosage units to test, and disintegration endpoint determination. Disintegration tests, official in the USP since 1950, are only indirectly related to drug bioavailability and product performance.

For conventional immediate release tablets, disintegration times may range from less than 1 min to as much as 5 to 15 min. The disintegration time is markedly affected by formulation ingredients and processing. However, disintegration time does not necessarily bear a direct relationship to in vivo release of drug from a dosage form. To be absorbed, the drug substance must be in solution and the disintegration test only measures the time required for the tablet to break up into particles or for a capsule to disperse its contents. The test is useful as a quality assurance tool and is still used today for this purpose.

## Dissolution

Dissolution refers to the process by which a solid phase (e.g., a tablet or powder) goes into a solution phase such as water or gastrointestinal fluid. If the dosage form is intended to disintegrate, the tablet or capsule disintegrates into granules and these granules deaggregate, in turn, into fine particles that disperse in the dissolution medium. The individual particles then separate and dissolve (e.g., mix molecule by molecule) with the liquid. Disintegration, deaggregation, and dissolution may occur simultaneously with the release of a drug from its delivery form. Some kinds of controlled release dosage forms are not intended to fully disintegrate on exposure to fluid but rather to slowly release drug from the dosage form over a period of time. Drug dissolution is therefore the process by which drug molecules are liberated from a solid phase and enter into solution. If particles remain in the solid phase once they are introduced into a solution, a pharmaceutical suspension results. Suspensions are covered in Chapters 16 and 17. In the vast majority of circumstances, only drugs in solution can be absorbed, distributed, metabolized, excreted, or even produce a pharmacologic action. Thus, dissolution is an important process.

The effectiveness of a tablet in releasing its drug for systemic absorption is influenced by the rate of disintegration and the deaggregation of the granules. Ordinarily of more importance, however, is the dissolution rate of the solid drug. Dissolution is the limiting or rate-controlling step in the absorption of drugs with low solubility (see BCS discussion) when it is the slowest of the steps involved in the release of the drug from a dosage form and passage into systemic circulation.

Although there are many customized and unique dissolution testing devices reported in the literature, the United States Pharmacopeia (USP) and other pharmacopeias worldwide have established standard methodologies and equipment to perform testing of immediate- and modified-release oral dosage forms. The most commonly used methods for evaluating dissolution first appeared in the USP in the early 1970s. The two most common methods are known as the USP basket (method I) and paddle (method II) methods. The reader is referred to Chapter 13 for additional discussion of dissolution testing methods. In practice, a rotating basket or paddle provides a steady stirring motion in a large vessel with 500 to 1000 mL of fluid controlled to 37°C. The devices are rel-

atively simple and standardized. The USP basket and paddle methods are the methods of choice for dissolution testing of immediate-release oral solid dosage forms. Although water is one of the most commonly listed dissolution media found in USP monographs, it may not be physiologically relevant due to the lack of buffering capacity or other biological components. Biorelevant dissolution media are sometimes used instead of buffered aqueous solutions to more precisely simulate in vivo conditions and these are discussed in greater detail in Chapter 13.

Modified-release delivery systems are similar in size and shape to conventional immediate-release dosage forms but the mechanism of drug release is very different and depends upon the design. The mechanisms for controlling the release of the drugs is becoming very sophisticated and special consideration must be given to how drug release is evaluated. For this reason there are several alternative dissolution apparatuses that may be used for modified-release dosage forms.

## Stability

One of the most important activities of formulation development is to evaluate both the physical and chemical stability of the drug substance in the dosage form. It is essential that the drug substance have known purity (typically 97% or greater) and sufficiently low levels of impurities to ensure safety and efficacy. The presence of impurities, or the generation of degradation products as a result of decomposition on storage, must be carefully characterized and controlled and where possible, eliminated with appropriate product design, packaging, and storage. Chemical decomposition of medicinal agents may take on many forms; among the most common decomposition processes are those of hydrolysis and oxidation. Additional details on the various aspects of chemical stability are described in previous sections of this book.

Stability is defined as the extent to which a product retains the same properties and characteristics that it possessed at the time of manufacture. A stable dosage form is one that retains all of its critical physical, chemical, and dosage form performance characteristics such as chemical stability, potency, disintegration, dissolution, and drug release. Pharmaceutical scientists are interested not only in chemical stability, that is, the extent to which the active ingredient retains its chemical integrity and potency but also in physical stability. Physical stability considerations include appearance, tablet hardness or capsule integrity, disintegration, and dissolution profiles. Appropriate characterization and control of physical and chemical stability of dosage forms generally will ensure therapeutic performance.

Both physical and chemical stability considerations are important in selecting storage conditions and containers. Temperature, exposure to light, and humidity often are the critical parameters that influence dosage form physicochemical stability. Stability and expiration dating are based on reaction kinetics, that is, the study of the rate of chemical and physical change and the way the rate is influenced by storage conditions. The FDA and other regulatory bodies

## THRESHOLDS FOR DEGRADATION PRODUCTS IN NEW DRUG PRODUCTS[86],*

| Maximum Daily Dose | Threshold |
|---|---|
| **Reporting Thresholds** | |
| $\leq 1$ g | 0.1% |
| $> 1$ g | 0.05% |
| **Identification Thresholds** | |
| $< 1$ mg | 1.0% or 5 $\mu$g TDI, whichever is lower |
| 1 mg–10 mg | 0.5% or 20 $\mu$g TDI, whichever is lower |
| $> 10$ mg–2 g | 0.2% or 2 mg TDI, whichever is lower |
| $> 2$ g | 0.10% |
| **Qualification Thresholds** | |
| $< 10$ mg | 1.0% or 50 $\mu$g TDI, whichever is lower |
| 10 mg–100 mg | 0.5% or 200 $\mu$g TDI, whichever is lower |
| $> 100$ mg–2 g | 0.2% or 3 mg TDI, whichever is lower |
| $> 2$ g | 0.15% |

*The maximum daily dose is the amount of drug substance administered per day. Thresholds for degradation products are expressed either as a percentage of the drug substance or as total daily intake (TDI) of the degradation product. Lower thresholds can be appropriate if the degradation product is unusually toxic. Higher thresholds should be scientifically justified.

worldwide have provided guidance and regulations regarding stability and stability testing of pharmaceutical ingredients and products.[44,74,77,87] Stability testing during each stage of development provides the information needed to optimize product stability and performance. **Table 22–19** provides International Conference on Harmonization guidelines and working recommendations to support regulatory filings regarding the presence of impurities and degradation products. Each commercially available pharmaceutical product has a well-defined shelf life and use of the product within its shelf life assures the patient that the product will be safe and effective when stored as directed. Typically oral solid dosage forms such as tablets and capsules have shelf lives of 2 years or more from the date of manufacture when stored at room temperature in appropriate containers, which may be necessary to protect them from light and humidity. In general, kinetic studies are performed to characterize stability of the active ingredient alone (bulk drug stability study) as well as the product. Accelerated stability is done to stress the drug in the dosage form to help define the limits and critical parameters that impact stability. Accelerated stability studies may be used to extrapolate or estimate shelf life at room temperature. Accelerated stability studies are very often done in the early stages of product development and may be used to support establishing the product shelf life. In addition to the accelerated stabilities, long-term stability studies carried out under the usual conditions of transport and storage are done. Consideration of the different climate zones to which the

product may be shipped must be considered as the different climate zones experience different temperature and humidity conditions throughout the year. While the details of all that is necessary to characterize the stability of a pharmaceutical product are beyond the scope of this section, regulatory guidance is available and Chapter 14 provides additional details.

**EXAMPLE 22–10**

An antihypertensive drug under development was placed on stability and the potency was measured over a 36-month period. Graph the following data and determine the first-order decomposition rate, the half-life, and the shelf life (time to 90% of label):

| % Potency | Time (months) |
|---|---|
| 100 | 0 |
| 98.5 | 3 |
| 97.0 | 6 |
| 94.6 | 12 |
| 92.0 | 18 |
| 90.4 | 24 |
| 85.0 | 36 |

Calculate logarithm of $A/A_0$ at each timepoint and plot as a function of time. Calculate the slope of the line using linear regression.

The linear regression line for the plot of $\log(A/A_0) = -0.0019t - 0.0008$ with $R^2 = 0.997$

The rate constant from equation (14–13) related to the slope of the line: $k = -$slope $\times 2.303 = 0.0044$ mo$^{-1}$

Using equation (14–18), $t_{1/2} = 0.693/k = 158$ months

The shelf life is defined as the time required for 10% of the drug to disappear.

$t_{90\%} = 0.105/k = 23.9$ months.

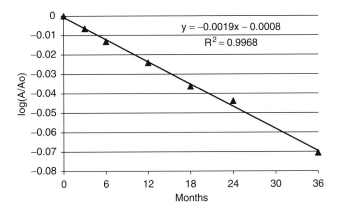

$\ln(A/A_0) = -0.058t - 0.0001$

The first-order rate constant is the slope of the linear regression line: $k = 0.058$ months$^{-1}$

The half-life can be calculated using equation (15–18) as:

$t_{1/2} = 0.693/k = 11.9$ months

The shelf life (time to reach 90% of initial potency) can be calculated using equation (15–14) as:

$t_{90\%} = (2.303/k) \times \log(100/90) = 0.105/k = 8.3$ months

## CHAPTER SUMMARY

Physical pharmacy and pharmaceutical science is the science of the delivery of APIs to the target site to achieve the desired pharmacological effect. For a drug to exert its biological effect, it must be released from the dosage form into the body, permeate through biological membranes, and reach the site of action. Successful delivery of APIs requires a sound understanding of a diverse array of scientific topics including physical and chemical properties, particle and powder properties, excipient properties and selection, dosage form manufacturing, drug absorption and transport, dosage form performance, and stability.

 **Practice problems for this chapter can be found at thePoint.lww.com/Sinko6e.**

# References

1. A. Rubinstein, Drug Discov. Today Technol. **2,** 33–37, 2005.
2. V. R. Sinha, A. Singh, R. V. Kumar, S. Singh, J. R. Bhinge, et al., Crit. Rev. Ther. Drug Carrier Syst. **24,** 63–92, 2007.
3. R. Birudaraj, R. Mahalingam, X. L. Li, and B. R. Jasti, Crit. Rev. Ther. Drug Carrier Syst. **22,** 295–330, 2005.
4. A. H. Shojaei, J. Pharm. Pharmaceut. Sci. **1,** 15–30, 1998.
5. S. Rossi, G. Sandri, and C. Caramella, Drug Discov. Today Technol. **2,** 59–64, 2005.
6. Y. Masaoka, Y. Tanaka, M. Kataoka, S. Sakuma, and S. Yamashita, Eur. J. Pharm. Sci. **29,** 240–250, 2006.
7. G. L. Amidon, H. Lennernas, V. P. Shah, and J. R. Crison, Pharm. Res. **12,** 413–420, 1995.
8. C. Schiller, C. Frohlich, T. Giessmann, W. Siegmund, H. Monnikes, N. Hosten, and W. Weitschies, Aliment. Pharmacol. Ther. **22,** 971–979, 2005.
9. A. Vander, J. Sherman, and D. Luciano, *Human Physiology*, McGraw-Hill, New York, 2001.
10. R. Carola, J. P. Harley, and C. Noback, *Human Anatomy and Physiology*, McGraw-Hill, Inc, New York, 1990.
11. R. M. Berneand and M. N. Levy, *Physiology*, Mosby, New York, 1988.
12. I. Coskun, H. Yildiz, K. Arslan, and B. Yildiz, Ital. J. Anat. Embryol. **112,** 27–36, 2007.
13. *Guidance for industry. Waiver of the In Vivo Bioavailability and Bioequivalence Studies for Immediate-Release Solid Oral Dosage Forms Based on a Biopharmaceutics Classification System*, U.S. Department of Health & Human Services, Food and Drug Administration (FDA), Center for Drug Evaluation and Research, Washington, DC, 2000.
14. U. Fagerholm, J. Pharm. Pharmacol. **59,** 751–757, 2007.
15. Annex 7: Multisource (generic) pharmaceutical products: guidelines on registration requirements to establish interchangeability, World Health Organization, Geneva, Switzerland, 2006, pp. 347–390.
16. T. Takagi, C. Ramachandran, M. Mermejo, S. Yamashita, L. Yu, and G. L. Amidon, Mol. Pharm. **3,** 631–643, 2006.
17. Annex 8: Proposal to waive in vivo bioequivalence requirements for WHO Model of List Essential Medicines immediate release, solid oral dosage forms, World Health Organization, Geneva, Switzerland, 2006.
18. G. E. Amidon, X. He, and M. J. Hageman, Physicochemical Characterization and Principles of Oral Dosage Form Selection. In D. J. Abraham (Ed.), *Burger's Medicinal Chemistry and Drug Discovery*, **Vol. 2;** *Drug Discovery and Drug Development*, **Vol. 2,** J Wiley, New York, 2003.
19. R. Verbeeck, H. Junginger, K. Midha, V. Shah, and D. Barends, J. Pharm. Sci. **94,** 1389–1395, 2005.
20. J. E. Polli, L. X. Yu, J. A. Cook, G. L. Amidon, R. T. Borchardt, B. A. Burnside, P. S. Burton, M. L. Chen, D. P. Conner, P. J. Faustino, A. A. Hawi, A. S. Hussain, H. N. Oshi, G. Kwei, V. H. Lee, L. J. Lesko, R. A. Lipper, A. E. Loper, S. G. Nerurkar, J. W. Polli, D. R. Sanvordeker, R. Taneja, R. S. Uppoor, C. S. Vattikonda, I. Wilding, and G. Zhang, J. Pharm. Sci. **93,** 1375–1381, 2004.
21. J.-O. Henckand and S. R. Byrn, Drug Discov. Today, **12,** 189–199, 2007.
22. S. R. Byrn, Solid State Chemistry of Drugs, SSCI, West Lafayette, IN, 2000.
23. J. Bernstein, *Polymorphism in Molecular Crystals*, Oxford Science Publications, Oxford University Press, Oxford, UK, 2002.
24. J. Haleblianand and W. McCrone, J. Pharm. Sci. **58,** IS 8:911, 1969.
25. J. Bauer, S. Spanton, R. Henry, J. Quick, W. Dziki, W. Porter, and J. Morris, Pharm. Res. **18,** 859–866, 2001.
26. J. Carstensen, *Pharmaceutical Preformulation*, Informa Health Care, Boca Raton, FL, 1998.
27. S. Yalkowsky and S. Banerjee (Eds.), *Aqueous Solubility: Methods of Estimation for Organic Compounds*, Marcel Dekker Inc, New York, 1992.
28. S. H. Yalkowsky, G. L. Flynn, and G. L. Amidon, J. Pharm. Sci. **61,** 983–984, 1972.
29. J. Huuskonen, Comb. Chem. High Throughput Screen. **4,** 311–316, 2001.
30. S. H. Yalkowsky and S. C. Valvani, J. Pharm. Sci. 69, 912, 1980.
31. C. Ahlneckand and G. Zografi, Int. J. Pharm. **62,** 87, 1990.
32. S. H. Yalkowsky and S. Bolton, Pharm. Res. **7,** 962, 1990.
33. B. R. Rohrs, G. E. Amidon, R. H. Meury, P. J. Secreast, H. M. King, and C. J. Skoug, J. Pharm. Sci. **95,** 1049, 2006.
34. ICH Harmonized Tripartite Guideline: Specifications: Test Procedures and Acceptance Criteria for New Drug Substances and New Drug Products: Chemical Substances Q6A, International Conference on Harmonization, 1999, pp. 1–22.
35. ICH Harmonized Tripartite Guideline: Pharmaceutical Development Q8, 2005, pp. 1–11.
36. A. A. Noyesand and W. R. Whitney, J. Am. Chem. Soc. **19,** 930, 1897.
37. R. J. Hintzand and K. C. Johnson, Int. J. Pharm. **51,** 9–17, 1989.
38. G. E. Amidonand and B. R. Rohrs, Particle Engineering: A Formulator's Perspective, *AAPS Arden House*, West Point, NY, 2005.
39. S. F. Kramer and G. L. Flynn, J. Pharm. Sci. **61,** 1986, 1972.
40. M. S. Bergren, Int. J. Pharm. **103,** 103–114, 1994.
41. G. Engel, N. Rarid, M. Faul, L. Richardson, and L. Winneroski, Int. J. Pharm. **198,** 239–247, 2000.
42. K. Morris, A. Newman, D. Bugay, S. Ranadive, and A. Serajuddin, Int. J. Pharm. **108,** 195–206, 1994.
43. Food and Drugs Chapter I: Food and Drug Administration, Department of Health and Human Services, Subchapter C, Drugs: General, **Vol. 4,** Title 22, Code of Federal Regulations, 2008.
44. *Guidance for Industry: Q1A(R2) Stability Testing of New Drug Substances and Products*, U.S. Department of Health & Human Services, Food and Drug Administration, Center for Drug Evaluation and Research (CDER), Center for Biologics Evaluation and Research (CBER), Washington, DC, 2003.
45. Merriam-Webster OnLine Dictionary. www.Merriam-Webster.com.
46. E. N. Hiestand, in G. Alderbornand and C. Nystrom (Eds.), *Pharmaceutical Powder Compaction Technology*, **Vol. 71,** Marcel Dekker Inc, New York, 1996.
47. E. N. Hiestand, *Mechanics and Physical Principles for Powders and Compacts*, SSCI Inc, West Lafayette, Ind, 2000.
48. A. H. Kibbe (Ed.), *Handbook of Pharmaceutical Exicpients*, American Pharmaceutical Association and Pharmaceutical Press, Washington, DC, 2000.
49. E. N. Hiestand, Pharm. Technol. **10,** 52, 54, 56, 58, 1986.
50. E. N. Hiestand, in G. Alderbornand and C. Nystrom (Eds.), *Pharmaceutical Powder Compaction Technology*, **Vol. 71,** Marcel Dekker, New York, 1996, pp. 229–244.
51. E. N. Hiestand, J. E. Wells, C. B. Peot, and J. F. Ochs, J. Pharm. Sci. **66,** 510–519, 1977.
52. G. E. Amidon, in H. G. Brittain (Ed.), *Physical Characterization of Pharmaceutical Solids*, **Vol. 70,** Marcel Dekker Inc, New York, 1995, p. 281.
53. R. C. Roweand and R. J. Roberts, *The Mechanical Properties of Powders*, in D. Ganderton, T. Jones, J. McGinity (Eds.), Academic Press, *Advances in Pharmaceutical Sciences*, Elsevier, Maryland Heights, MO, 1995.
54. R. C. Roweand and R. J. Roberts, in G. Alderbornand and C. Nystrom (Eds.), *Pharmaceutical Powder Compaction Technology*, **Vol. 71,** Marcel Dekker Inc, New York, 1996, p. 165.
55. G. T. Carlsonand and B. C. Hancock, in A. Katdareand and M. Chaubal (Eds.), *Excipient Development for Pharmaceutical, Biotechnology, and Drug Delivery Systems*, Healthcare, USA, Inc, New York, 2006.

56. G. E. Amidon, P. J. Secreast, and D. M. Mudie, in Y. Qiu (Ed.), *Pharmaceutical Theory and Practice in Developing Solid Oral Dosage Forms*, Academic Press, New York, 2009.

57. E. Overton, Phys. Chem. **8,** 189–209, 1891.

58. H. Meyer, Arch. Exp. Pathol. Pharmakol. **42,** 109–118, 1899.

59. P. Buchwaldand and N. Bodor, Curr. Med. Chem. **5,** 353–380, 1998.

60. P. Taylor, *Comprehensive Medicinal Chemistry*, Pergamon Press, New York, 1990, pp. 241–294.

61. A. J. Leo, J. Pharm. Sci. **76,** 166–168, 1987.

62. C. A. Lipinski, F. Lombardo, B. W. Dominy, and P. J. Feeney, Adv. Drug Deliv. Rev. **23,** 3–25, 1997.

63. Y. Qiu, Y. Chen, G. Zhang, L. Liu, and W. Porter (Eds.), *Developing Solid Oral Dosage Forms: Pharmaceutical Theory & Practice*, Academic Press, New York, 2009.

64. G. S. Banker and C. Rhodes, *Modern Pharmaceutics*, Marcel Dekker, New York, 2002.

65. H. A. Lieberman, L. Lachman, and J. Schwartz, *Pharmaceutical Dosage Forms: Tablets*, Marcel Dekker, New York, 1989.

66. L. Augsburger and S. W. Hoag, *Pharmaceutical Dosage Forms; Tablets*, Informa Healthcare, 2008.

67. USP31/NF26. <1078> Good Manufacturing Practices for Bulk Pharmaceutical Excipients. United States Pharmacopeial Convention, Washington, DC, 2008.

68. *Guidance for Industry, Nonclinical Studies for the Safety Evaluation of Pharmaceutical Excipients*, In FDA U.S. Department of Health & Human Services, Center for Drug Evaluation and Research (CDER), Center for Biologics Evaluation and Research (CBER) (Eds.), 2005.

69. *Guidance for Industry: Drug Product Chemistry, Manufacturing, and Controls Information*, In FDA U.S. Department of Health & Human Services, Center for Drug Evaluation and Research (CDER), Center for Biologics Evaluation and Research (CBER) (Eds.), 2003.

70. *Guidance for Industry: Variations in Drug Products That May Be Included in a Single ANDA*, U.S. Department of Health & Human Services, Food and Drug Administration (FDA), Center for Drug Evaluation and Research, Washington, DC, 1998.

71. USP31/NF26, United States Pharmacopeia, United States Pharmacopeial Convention, Washington, DC, 2008.

72. R. C. Rowe, P. J. Sheskey, and S. C. Owen (Eds.), *Handbook of Pharmaceutical Excipients*, Pharmaceutical Press and the American Pharmacists Association, Washington, DC, 2006.

73. U.S. Food and Drug Administration Webpage. http://www.fda.gov/opacom/morechoices/mission.html.

74. Code of Federal Regulations: Title 22, Volume 4, Part 221: Current Good Manufacturing Practices for Finished Pharmaceuticals, Food and Drug Administration, Revised April 1, 2008.

75. *Pharmaceutical CGMPs for the 22nd Century—A Risk-based Approach: Final Report*, U.S. Food and Drug Administration, 2004.

76. L. Allen, *The Art, Science, and Technology of Pharmaceutical Compounding*, American Pharmaceutical Association, Washington, DC, 2002.

77. *Guidance for Industry Q8 Pharmaceutical Development*, U.S. Department of Health & Human Services Food and Drug Administration, 2006.

78. K. Kawakita, Science, **26,** 149, 1953.

79. K. Kawakita and Y. Tsutsumi, Bull. Chem. Soc. Japan. **39,** 1364, 1966.

80. K. Kawakita and K. H. Ludde, Powder Tech. **4,** 61, 1970.

81. R. W. Heckel, Trans. Metallurgical Soc. AIME. **222,** 1001, 1961.

82. R. W. Heckel, Trans. Metallurgical Soc. AIME. **222,** 671, 1961.

83. A. R. Cooperand and L. E. Eaton, J. Am. Ceram. Soc. **45,** 97–101, 1962.

84. G. S. Banker and C. T. Rhodes, *Modern Pharmaceutics*, Marcel Dekker, Inc, 1996.

85. G. Alderborn and C. Nystrom, *Pharmaceutical Powder Compaction Technology*, Marcel Dekker, New York, 1996.

86. *Guidance for Industry, Q3 A Impurities in New Drug Substances*, In FDA U.S. Department of Health and Human Services, Center for Drug Evaluation and Research (CDER), Center for Biologics Evaluation and Research (CBER) (Eds.), Washington, DC, 2008.

87. *Guidance for Industry Q7A Good Manufacturing, Practice Guidance for Active Pharmaceutical Ingredients*, In FDA U.S. Department of Health & Human Services, Center for Drug Evaluation and Research (CDER), Center for Biologics Evaluation and Research (CBER) (Eds.), 2001.

## Recommended Readings

Gordon L. Amidon, Hans Lennernas, Vinod P. Shah, and John R. Crison, Pharm. Res. **12,** 413–420, 1995.

*Remington: The Science and Practice of Pharmacy*, Lippincott Williams & Wilkins, Philadelphia, PA, 2006.

Y. Qiu, Y. Chen, G. Zhang, L. Liu, and W. Porter (Eds.), *Developing Solid Oral Dosage Forms: Pharmaceutical Theory & Practice*, Academic Press, New York, 2009.

CHAPTER LEGACY

**Sixth Edition:** published as Chapter 22 (Oral Solid Dosage Forms). Written by Gregory E. Amidon. Contribution to the continuous processing section by Mayur P. Lodaya is gratefully acknowledged.

## INTRODUCTION

"Drugs" that are taken by a patient exert a biological effect usually by interacting with specific receptors at the site of action.[1] Unless the drug is delivered to the target site (in other words the site of action) at a rate and concentration, which minimizes the side effects and maximizes the therapeutic effect, the efficiency of a therapy is compromised.[1] Often, the delivery and targeting barriers are so great that they render an otherwise potent drug ineffective. Dosage forms serve many purposes including facilitating drug administration and improving drug delivery. Traditional dosage forms include injections, oral formulations (solutions, suspension, tablets, and capsules), and topical creams and ointments. Unfortunately, most traditional dosage forms are unable to do all of the following: facilitate adequate drug absorption and access to the target site; prevent nonspecific drug distribution (side effects) and premature metabolism and excretion; and match drug input with the dose requirement.[1] Alternative routes of drug administration and advanced drug delivery systems are therefore needed to meet these drug delivery challenges and improve drug therapy. In this chapter, the student will learn about advanced drug delivery systems.[2,3]

Advanced drug delivery systems aim to overcome limitations of conventional drug delivery using traditional dosage forms by achieving enhanced bioavailability and therapeutic index, reduced side effects, and improved patient acceptance or compliance.[3] While the first three factors are well appre-

ciated, the improved patient compliance is equally important because it has been estimated that patients take almost one billion prescriptions per year incorrectly resulting in a significant number of hospitalizations and nursing home admissions. Improved patient compliance is achieved by developing "user-friendly" delivery systems that are convenient to take and require lower dosing frequency. During the 1950s and 1960s some of the first attempts were made to transform common dosage forms into advanced delivery systems by sustaining drug release via the oral route.[4] The Spansule capsule, consisting of hundreds of tiny-coated pellets of drug substance developed by Smith Kline and French Laboratories, is considered the first such example.[1] As a pellet travels through the gastrointestinal (GI) tract, the coating dissolves to release the drug. The pellet thickness is changed to control the drug release pattern. By the 1960s, polymers began to be used to deliver drugs and scientists started using a systems approach to product development that combined an understanding of pharmacokinetics, the biological interface, and the biological compatibility.[4] Nanoparticles were introduced in drug delivery in 1970s; transdermal drug delivery system started appearing in 1980s and transepithelial delivery models were developed in 1990s. The phenomenal advances in the field of biotechnology and molecular biology during 1980s and 1990s made possible large quantity synthesis of biologics/biopharmaceuticals such as peptides, proteins, antisense oligonucleotides, and siRNA. These compounds, although highly potent, are difficult to deliver because of their large molecular size, water solubility, and instability.

## ■ KEY CONCEPT ■ ADVANCED DRUG DELIVERY SYSTEMS

*Advanced drug delivery systems* are defined as a formulation or device that delivers drug to specific site in the body at a certain rate. Advanced drug delivery systems usually represent a more sophisticated system that incorporates advanced technologies such as controlled, pulsatile, or bioresponsive drug delivery.[2] Usually some form of targeting technology may also be present.

There is an economic rationale as well for developing advanced drug delivery technologies.[5] It has been estimated that the sales of advanced drug delivery systems in the United States were $64.1 billion by the end of 2006.[6] The sales are projected to reach $153.5 billion in 2011. Similarly, the European market for advanced drug delivery systems totaled at $25 billion in 2007 and was expected to reach $47 billion by 2013.[7] In 2009, the largest market share is for targeted drug delivery systems (~$50 billion) followed by sustained-release formulations (~$45 billion). While oral drug delivery systems currently represent about half of the drug delivery market, pulmonary, transdermal, and nanodrug delivery systems are expected to show most promising growth in the future.

This chapter aims to provide an overview of advanced drug delivery and targeting technologies. Major drug delivery routes are described and both advantages and disadvantages associated with each delivery route are discussed. An introduction to the concepts of controlled drug delivery and targeting is presented and representative examples of different drug delivery systems are presented.

## TERMINOLOGY[2,3]

**Active targeting:** Targeting is achieved by binding to specific antigens or cell surface receptors.

**ADME:** Abbreviation for absorption, distribution, metabolism, and excretion.

**Bioavailability:** The rate and extent to which a drug is absorbed and becomes available at the site of action.

**Biocompatible:** The system is able to perform the desired function without eliciting toxic and immunogenic responses, either systemically or locally.

**Biodegradable:** The system degrades (chemical breakdown) either chemically or enzymatically by physiological environment.

**Bioerosion:** The gradual dissolution of the system (mostly polymer matrix).

**Bioequivalence:** Absence of significant difference in the rate and extent to which the active ingredient or moiety in two formulations becomes available at the site of action. Formulations showing superimposable drug plasma concentration (Cp) versus time (*T*) curve are said to be bioequivalent.

**Bioresponsive release:** Drug delivery is controlled by a biological stimulus.

**Blood–brain barrier (BBB):** The permeability barrier present between the brain (brain capillary endothelium) and blood, which prevents substances in blood from entering the brain tissue.

**Carrier:** Monoclonal antibodies, carbohydrates, proteins, peptides, hormones, vitamins, growth factors, immunotoxins conjugated to the drug for achieving site-specific delivery.

$C_{max}$: The maximum plasma concentration reached after the drug administration.

**Drug Delivery System (DDS):** Formulation or device that delivers drug to a specific site in the body at a certain rate.

**Drug disposition:** All processes involved in the DME of drugs in living organism.

**Half-life ($t_{1/2}$):** The time required for half of the drug to be removed from the body.

**Passive targeting:** Exploits the in vivo passive distribution pattern of a carrier for drug targeting.

**Prodrug:** Pharmacologically inert derivatives that can be converted to active drug molecule in vivo, enzymatically or nonenzymatically, to exert a therapeutic effect.

**Rate controlled delivery:** Drug delivered at predetermined rate either systemically or locally for a specific period of time.

**Spatial drug delivery:** Delivery to a specific region of the body.

**Sustained drug delivery:** Drug delivery, which prolongs or sustains the therapeutic blood or tissue levels of drug for an extended period of time.

**Targeted drug delivery:** Drug is delivered to specific sites in the body.

**Temporal drug delivery:** Control of drug delivery to produce an effect in time-dependant manner.

**Therapeutic index:** Ratio of toxic to therapeutic drug dose.

$t_{max}$: The time at which $C_{max}$ occurs.

**Variable release:** The drug is delivered at variable rate.

**Zero-order release:** Drug release does not vary with time and relatively constant drug level is maintained in the body for longer periods.

## ROUTES OF DRUG DELIVERY

The route of administration or delivery is a very important factor in designing a drug delivery system. For example, a conventional oral tablet could not be used to deliver medication in the ear since tablets require a certain amount and type of fluid to disintegrate and dissolve. The ear canal does not have the fluid or volume to be able to accommodate tablets. In addition, many tablets are simply too big to be inserted into the ear canal. Another factor is the therapeutic agent that has to be delivered. Some drugs are so poorly absorbed across the intestine that they need to be injected directly into the bloodstream through a vein (intravenous) or an artery (intra-arterial). It may be desirable to deliver a drug locally to the target organ or tissue without first entering the systemic circulation. This type of drug delivery is usually referred to as local or topical (Greek topikos, "place"). A good example of this is hydrocortisone cream. Hydrocortisone cream is applied topically to the skin where it is expected to exert its action as an anti-inflammatory and antipruritic (i.e., anti-itch) agent. Because of the potency and side effects associated with steroid drugs, avoiding systemic absorption is often desirable. In contrast, drugs can be delivered to the whole body via the general blood circulation. This is usually referred to as systemic drug delivery. In order to enter the systemic circulation, a drug has to pass through a rate-limiting membrane such as the intestine or vaginal mucosa or it may be directly injected into the body avoiding a transmembrane absorption step altogether (e.g., it may be directly injected into a vein). Often, several dosage forms of the same drug are produced that may be suitable for different routes of administration. In such cases, the selection of the route of administration for the particular drug is generally dictated by the desired onset and duration of drug effect, reliability, patient's discomfort, and compliance.

The common routes of administration are summarized in **Table 23–1**. It should be stressed that local drug delivery can result in the release of the drug into the systemic circulation and therefore provide systemic drug delivery to many other organs (**Fig. 23–1**). For example, if a drug is being delivered to the lung, it might penetrate the respiratory barrier and enter the circulation. Similarly, penetration of a drug through the oral mucosa (i.e., sublingual, buccal, or gingival) results in the systemic delivery of the drug. Fundamental differences in the various biologic routes of drug administration critically affect the onset and duration of drug action. **Table 23–2** compares different routes of administration using nitroglycerin as an example. The selection of route of administration and dosage form depends on the desired drug concentration profiles that need to be achieved, patient issues (e.g., the ability to

tolerate treatment regimens such as the frequency of administration or the ability to swallow a tablet), and the disease state that is being treated. For instance, intravenous drug administration is often used in emergency situations when

### TABLE 23–1
### COMMON ROUTES OF DRUG ADMINISTRATION*

| Route | Site of Absorption |
| --- | --- |
| Parenteral | Injected directly into body, so the absorption step is usually minimal or nonexistent |
|   Intravenous | Into a vein |
|   Intramuscular | Into a muscle |
|   Subcutaneous | Under the skin |
| Buccal | In the mouth through the oral mucosa (cheek near the gumline) |
| Inhalation | By mouth or nose and absorbed by the pulmonary (lung) mucosa |
| Nasal | In the nose through the nasal mucosa |
| Ocular | In the eye |
| Oral | Given by mouth and absorbed through the gastrointestinal mucosa |
| Rectal | In the rectum |
| Sublingual | Under the tongue |
| Topical | On the skin, local action |
| Transdermal | On the skin, systemic delivery |
| Vaginal | Into the vagina and absorbed through the vaginal mucosa |

*Most of these routes of drug administration can be used for both topical (local) and systemic drug delivery.

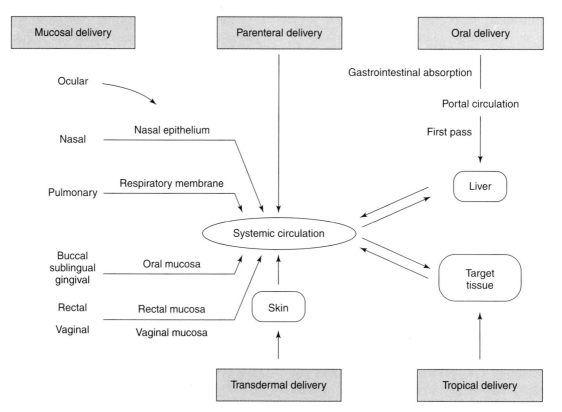

**Fig. 23–1.** Commonly used routes of drug delivery.

**TABLE 23–2**

## COMPARISON OF ROUTES OF NITROGLYCERIN ADMINISTRATION*

| Route of Administration | Drug Delivery System | Onset of Action | Duration of Action |
|---|---|---|---|
| Intravenous | Parenteral solution | Immediate | Several minutes (dose dependent) |
| Translingual | Rapidly dissolving tablets | 2–4 min | 30–60 min |
| | Extended-release capsules and tablets | 20–45 min | 8–12 hr |
| Sublingual | Tablets and drops | 1–3 min | 30 min |
| Transmucosal (buccal) | Extended-release tablets | 2–3 min | 5 hr |
| | Ointment | 20–60 min | 4–8 hr |
| Transdermal | Transdermal patches | 40–60 min | 18–24 hr |

*Nitroglycerin belongs to the class of drugs called nitrates. It dilates (widens) blood vessels (arteries and veins). Nitroglycerin is used to prevent angina attacks (oral tablets, buccal tablets) and to treat attacks once they have started (sublingual tablets, chewable tablets, spray).

fast action is critical. In contrast, extended-release oral tablets or transdermal drug delivery systems are often used to prolong the drug action. Specific drug delivery systems that are used for the various routes of delivery are discussed in detail later in this and other chapters.

## Gastrointestinal (Oral, Per Os)

Oral administration of drugs is the simplest, easiest, and most common route of drug administration. After absorption from the GI tract, the drug enters the liver through the portal circulation (**Fig. 23–1**). During this "first pass" through the intestine and liver, drugs can be metabolically deactivated unless special measures are taken to protect them. Beyond the liver, the drug enters the systemic circulation and is delivered to the target tissues as well as all other tissues. The existence of first-pass intestinal–hepatic metabolism is the most significant challenge of oral drug delivery. In addition, the oral route is not suitable for the systemic delivery of drugs that are poorly absorbed or significantly destroyed in the GI tract.

In many countries, oral administration is the most common and preferred route of administration. The most commonly used dosage forms for the oral route are liquids, dispersed systems, and solids. Liquid dosage forms include oral solutions of drugs with added substances to make the preparation pharmaceutically stable and aesthetically acceptable. Drugs administered in this form are more rapidly absorbed than other forms when administered on an empty stomach because gastric emptying is rapid and the drug is immediately available for absorption. Dispersed systems include emulsions and suspensions (see Chapters 16 and 17). The drugs in the emulsion or suspension forms are absorbed much more rapidly than in the solid forms since solid forms have to disintegrate, deaggregate, and dissolve. The solid dosage forms include the vast majority of the preparations used for oral administration. The widely used solid dosage forms are powders, tablets, caplets, and capsules. An entire chapter is devoted to this important dosage form and route of administration (see Chapter 22). Powders are administered occasionally for rapid systemic action. A tablet is a compressed form of the powdered drug along with therapeutically inactive ingredients

that enable the proper disintegration, dissolution, lubrication, and so on, of the dosage form. A well-formulated immediate-release tablet should disintegrate rapidly and make the drug available for dissolution and absorption. Exceptions are slow-release and delayed-release tablets, which are designed for the continuous delivery of drug over a defined period of time or for delayed dissolution and drug release to target or avoid a specific GI location (e.g., enteric-coated tablets to avoid gastric release and inactivation of drugs). A capsule is a solid dosage form in which the drug with or without other ingredients is enclosed in either a hard or soft soluble shell generally prepared from a suitable form of gelatin.

Many new drugs cannot be delivered in oral form because they are too large, highly charged, or are degradable by stomach acid or the various enzymes in the GI tract. Insufficient amounts of these types of drugs traverse the intestinal barrier and reach the bloodstream. Consequently, these molecules can be delivered only by injection or other nonoral means. Nevertheless, oral ingestion is regarded as the safest, most convenient, and most economical method of drug administration. When compared to other alternatives, patient acceptance and adherence to a dosing regimen is typically higher among orally delivered medicines. Prodrug approaches have been widely used to enhance the oral delivery of small-molecule therapeutics such as acyclovir or ganciclovir (GCV). As described later in this chapter, by orally administering valacyclovir, the valyl-ester prodrug of acyclovir, one obtains blood concentrations of acyclovir similar to those achieved after the intravenous administration. Although the pharmaceutical industry has been successful in delivering small-molecule therapeutic agents orally, successes have been much more limited when it comes to larger drugs with more complex secondary and tertiary structures. Controlled-release drug delivery systems suitable for oral drug delivery are discussed later in this chapter (see Controlled Drug Delivery section).

## Parenteral

The word "parenteral" (Greek para, "outside"; enteron, "intestine"), meaning outside of the intestine, denotes the

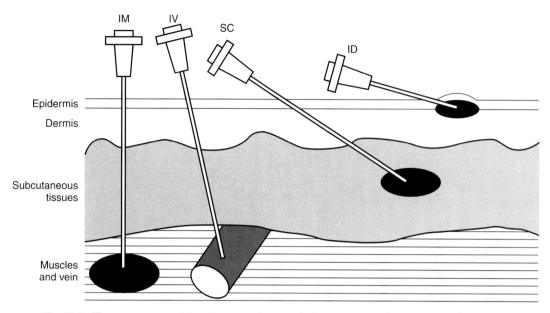

**Fig. 23–2.** The most common injection routes: into a vein (intravenous, IV); into a muscle (intramuscular, IM); under the skin (subcutaneous, SC, sub-Q, hypodermic, hypo); into the skin (intradermal, ID, intracutaneous).

routes of administration other than the oral route (mainly by injection). Parenteral routes of administration are often used when the administration of drugs through the oral route is ineffective or impractical. It is also suitable for administration of drugs that are poorly absorbed or inactivated in the GI tract. The parenteral route is also used for drugs that are too irritating to be given by mouth. Almost any organ or area of the body can be used to inject drugs. However, the most commonly used routes of injections include the intramuscular (IM), intravenous (IV), subcutaneous (SC), and intradermal (ID) routes (**Fig. 23–2**). Intravenous injection provides very rapid onset of drug action, precision of dose, and accommodation of a large volume of drug solutions. It is suitable for administration of high–molecular-weight compounds. The other common forms of parenteral administration require that the drug pass through a significant amount of tissues and blood vessels (i.e., the endothelium of capillaries) to enter the circulation. The longer the path to the systemic circulation, the more delayed is the onset of drug reaction. Bioavailability of a parenteral drug also depends significantly on physicochemical characteristics of the drug (e.g., solubility, polarity, degree of ionization, molecular size).

Protein formulations, when given as injections, show poor pharmacokinetic profiles.[8] The proteins are metabolized and rapidly cleared, which limits their therapeutic utility. PEGylation (covalent attachment of PEG to active moiety) reduces the plasma clearance of proteins by reducing their metabolic degradation and receptor-mediated uptake from systemic circulation. It also improves their safety profiles by shielding antigenic and immunogenic epitopes. An important example to consider is that of recombinant interferon-$\alpha$ (IFN), which is approved for the treatment of chronic hepatitis C, cell leukemia, malignant melanoma, non-Hodgkin lym-

phoma, and chronic myelogenous leukemia. When given subcutaneously, IFN is rapidly absorbed ($t_{1/2}$: 2.3 hr) and reaches a peak plasma level in 1 to 8 hr, which then falls rapidly (elimination $t_{1/2}$: 3–8 hr) and becomes undetectable in 24 hr.[9] PEGylation with a 12-kDa linear PEG (peginterferon alpha-2b, PEG-Intron, Schering-Plough) significantly increases the absorption ($t_{1/2}$: 7 hr) and elimination half-lives ($t_{1/2}$: 4-days) of IFN.[10] PEGylation with a 40-kDa branched PEG (peginterferon alpha-2a, Pegasys, Roche) enhances the absorption ($t_{1/2}$: 50 hr) and elimination ($t_{1/2}$: 11 days) half-lives to much higher levels.[9] The improved pharmacokinetic profile reduces the dosing from three times weekly to once a weekly subcutaneous injection (with ribavirin) for patients with hepatitis C. Both products have safety profiles similar to unmodified IFN.

The advantage of using these routes lies in reliability, precision of dosage, and timed control of the onset of action. Disadvantages of all parenteral routes of drug administration include discomfort, possibility of infection, tissue damage, administration by trained personnel, and so on. Drug delivery systems and devices suitable for parenteral use are discussed later in this chapter.

## Mucosal

Delivery of drugs via the absorptive mucosa in various easily accessible body cavities like the buccal, nasal, ocular, sublingual, rectal, and vaginal mucosae offers distinct advantages over peroral administration for systemic drug delivery. The primary advantage of using these routes is that they avoid the first-pass effect of drug clearance. Some of the numerous approaches that have been taken to facilitate mucosal drug delivery are described in the following sections.

Mucosal delivery faces several challenges such as retention on the mucosal surface so that bioavailability can be maximized. Bioadhesive polyacrylic acid nanoparticles are an example of a novel DDS designed for mucosal drug delivery.[11] They had a narrow size range, averaging approximately 50 nm, and are stable in buffer. The drug timolol maleate is loaded into the nanoparticles from aqueous drug solutions, which is then released over several hours on dispersal of drug-loaded particles in phosphate buffer solution. Another variant of a mucoadhesive drug delivery formulation is based on H-bonded complexes of poly(acrylic acid) (PAA) or poly(methacrylic acid) with the poly(ethylene glycol) (PEG) of a PEG–drug conjugate.[12,13] The PEGylated prodrugs are synthesized with degradable PEG-anhydride-drug bonds for eventual delivery of free drug from the formulation. The complexes are designed to dissociate as the formulation swells in contact with mucosal surfaces at pH 7.4, releasing PEG-bound drug, which then hydrolyzes to release free drug and PEG. It has been found that as the molecular weight of PAA increases, the dissociation rate of the complex decreases, which results in a decreased rate of drug release. On the other hand, drug release from PEG–drug conjugates alone and from a solid mixture of PEG–indomethacin +PAA was much faster than that from the H-bonded complexes. Because of the differences in thermal stability, the poly(methacrylic acid) complex exhibited slightly faster drug release than the PAA complex of comparable molecular weight. These H-bonded complexes of degradable PEGylated drugs with bioadhesive polymers may be useful for mucosal drug delivery.

## Buccal and Sublingual

The buccal and sublingual mucosae in the oral cavity provide an excellent alternative for the delivery of certain drugs. The buccal mucosa is located on the cheeks in the mouth, and the sublingual mucosa is located under the tongue and on the floor of the mouth. Both of these mucosae offer an easily accessible area for the placement of dosage forms such as adhesive tablets. The buccal and sublingual routes provide improved delivery for certain drugs that are inactivated by first-pass intestinal/hepatic metabolism or are inactivated by proteolytic enzymes in the GI tract. Although this route shows some promise, it can only be used for potent drugs, as only a small surface area of about 100 cm$^2$ is available for absorption. Delivery of drugs into the mouth is also potentially limited by the taste of the drug or components of the delivery system.

The delivery of drugs to the oral mucosal cavity can be classified into three categories: (a) *sublingual delivery*, which is systemic delivery of drugs through the mucosal membranes lining the floor of the mouth; (b) *buccal delivery*, which is drug administration through the mucosal membranes lining the cheeks (buccal mucosa); and (c) *local delivery*, which is drug delivery into the oral cavity for nonsystemic delivery.

The sublingual mucosa is relatively permeable, giving rapid absorption and onset of drug action with acceptable bioavailabilities for many drugs. This route of drug delivery is also convenient, accessible, and generally well accepted. Sublingual drug delivery systems are generally of two different designs: (a) rapidly disintegrating tablets and (b) soft gelatin capsules filled with the drug in solution. Such systems create a very high drug concentration in the sublingual region before they are absorbed across the mucosa. Because of the high permeability and the rich blood supply, the sublingual route is capable of producing a rapid onset of action, making it appropriate for drugs with short delivery period requirements and an infrequent dosing regimen.

The most commonly used dosage form for the administration of drug through this route is a small tablet. These tablets are placed under the tongue. The tablets are designed to dissolve rapidly, and the drug substances are readily absorbed to the systemic circulation. Nitroglycerin sublingual tablets are frequently used for the prompt relief from an acute angina attack. The other drug products designed for this route are hormones such as dehydroepiandrosterone, melatonin, and vitamin C, and several metal salts.

The buccal mucosa is considerably less permeable than the sublingual area and is generally unable to provide the rapid absorption properties and higher bioavailabilities as seen with sublingual administration. Two main differences between sublingual and buccal routes should be considered when designing drug delivery systems suitable for oral mucosa delivery. First, these two oral mucosa routes differ significantly in their permeability characteristics. The onset of action from the buccal mucosa is not as rapid as the sublingual mucosa because it is much less permeable and absorption is not as rapid. Therefore, it is more suitable for a sustained-release approach. Second, the buccal mucosa has an expanse of smooth muscle and is relatively immobile, whereas the sublingual region lacks both of these features and is constantly washed by a considerable amount of saliva. This makes the buccal mucosa a more desirable region for retentive delivery systems used for oral transmucosal drug delivery. Thus, the buccal mucosa is more useful for sustained-delivery applications, delivery of less permeable molecules, and perhaps peptide drugs. Similar to any other mucosal barrier, the buccal mucosa has limitations as well. One of the major disadvantages associated with buccal drug delivery is the low rate of absorption, which results in low drug bioavailability.

Because of the relative low permeability of buccal mucosa, DDSs for this route of administration usually include permeability or penetration enhancers—compounds that promote or enhance the absorption of drugs through the skin or mucosae, usually by reversibly altering the permeability of the barrier. Various compounds have been investigated for their use as buccal penetration enhancers to increase the flux of drugs across the mucosa. These compounds include but are not limited to ethers, cholates, aprotinin, azone, benzalkonium chloride, cetylpyridinium salts, cyclodextrins, dextrans, lauric

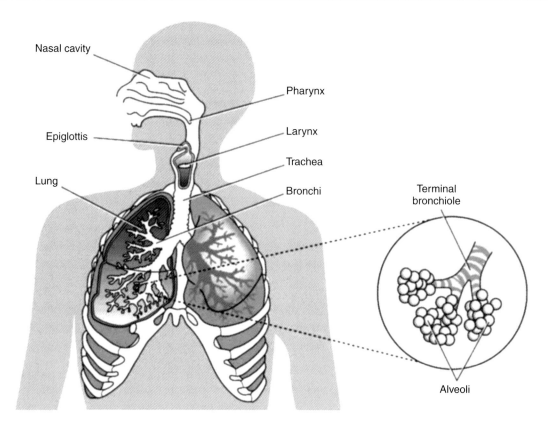

**Fig. 23–3.** Human respiratory system.

acid and its salts, propylene glycol, phospholipids, menthol, salicylates, ethylene diamine tetraacetic acid, several salts, sulfoxides, and various alkyl glycosides.[14]

## Pulmonary (Inhalation)

The respiratory tract (**Fig. 23–3**), which includes the nasal mucosa, pharynx, and large and small airway structures (trachea, bronchi, bronchioles, alveoli), provides a large mucosal surface for drug absorption. Utilization of this huge surface for drug delivery might provide a more convenient way compared with parenteral administration. The advantage of this drug delivery route was succinctly stated by J. S. Patton, "Taking advantage of the body's ability to transfer large molecules through the lungs is a better way to deliver drugs than sticking people with needles."[15] This route of administration is useful for the treatment of pulmonary conditions and for the delivery of drugs to distant target organs by means of the circulatory system.

The respiratory tract has a large surface area and therefore can be used for local and systemic drug delivery. The surface increases from the exterior region (nasopharyngeal) to the tracheobronchial and pulmonary regions, the latter consisting of bronchioles and alveoli (**Fig. 23–3**). One of the oldest examples of pulmonary administration for systemic drug delivery is inhalation anesthesia. An increasing variety of drugs such as beta-agonists, corticosteroids, mast cell stabilizers, antibiotics, and antifungal and antiviral agents are

being administered by this route to obtain a direct effect on the target tissues of the respiratory system.

The pulmonary route has been used for decades to administer drug to the lung for the treatment of asthma and other local ailments. Recently, this route has received more attention for the systemic delivery of drugs. The onset of action following the pulmonary administration of drugs is very fast and comparable to the intravenous route. The lungs offer a larger surface area (70 m$^2$) for systemic absorption of drugs than other nontraditional routes of systemic drug delivery such as the buccal, sublingual, nasal, rectal, and vaginal cavities. The major challenge is the lack of reproducibility in the deposition site of the administered dose. The rate of drug absorption is expected to vary in many regions in the lung owing to the variable thickness of the epithelial lining in the bronchial tree.

### Nasal

The uppermost portion of the human respiratory system, the nose, is a hollow air passage, which functions in breathing and in the sense of smell. The nasal cavity moistens and warms incoming air, and small hairs and mucus filter out harmful particles and microorganisms. The nose (**Fig. 23–4**) consists of two openings (nostrils) separated by a median septum. The vestibule at the entrance of each nostril is covered with hairs, which prevent the entrance of air-suspended particles. The nose cavity is divided by the septum into two chambers called fossae. They form passages for air movement from the

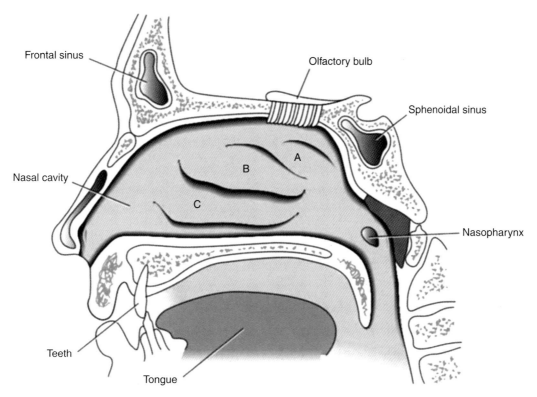

**Fig. 23–4.** The nasal anatomy. The nasal cavity is lined with mucous membrane and consists of three passageways or meatuses: A: upper; B: middle; and C: lower.

nostrils to the nasopharyngeal space at the back of the nose. Each fossa consists of two parts, an olfactory region at the front of the nose and a respiratory region that accounts for the remainder of the fossae. The nasal cavity is lined with a mucous membrane, called the membrana mucosa nasi, which is continuous with the skin of the nostrils. The respiratory portion of the nasal cavity contains ciliated (hairlike) projections consisting of columnar epithelial cells.

The nasal cavity consists of three passageways or meatuses: upper, middle, and lower meatus. The nasolacrimal duct drains into the lower meatus. The nose is connected to the middle ear through the nasopharynx or postnasal space and through the auditory canal. The compartments of the nose are connected to the conjunctiva of the eye by way of the nasolacrimal and lacrimal ducts, and through several sinuses that drain into the nose. A portion of a drug administered into the conjunctiva of the eye may enter the nose through these ducts and sinuses and may also pass into the esophagus.[16] The nasal mucosa is the only location in the body that provides a direct connection between the central nervous system (CNS) and the atmosphere. Drugs administered to the nasal mucosa rapidly traverse through the cribriform plate into the CNS by three routes: (a) directly by the olfactory neurons located in olfactory bulb (**Fig. 23–4**); (b) through supporting cells and the surrounding capillary bed; and (c) directly into the cerebrospinal fluid.[17] Therefore, in addition to local and systemic drug delivery, the nasal mucosa can be used to deliver drugs to CNS.

Traditionally, the nasal route is used for locally acting drugs. This route is getting more and more attention for the systemic delivery of protein and peptide drugs. The highly vascular nature of the nasal cavity makes it a suitable alternate route for systemic drug delivery. This route is also useful for potent drugs because of its smaller surface area of about 200 cm$^2$ available for absorption. The most commonly used dosage form for the administration of drug through this route is liquid solutions of drug. For large polar molecules such as peptides or polysaccharides in the form of drugs or vaccines, the nasal route provides a viable, noninvasive alternative to injections. For conventional molecules, the nasal route provides other clinical benefits relevant to certain drugs and patient groups: pulsatile or sustained plasma profiles, fast absorption and rapid onset of action, avoidance of first-pass metabolism, and avoidance of the effects of gastric stasis and vomiting often seen in migraine patients.

One of the major challenges is developing nasal formulations that improve the absorption of macromolecules and water-soluble drugs.[18] Another challenge is the problem of short retention time in the nasal cavity due to the efficient physiological clearance mechanisms. Good systemic bioavailability after nasal drug delivery can be achieved for molecules with a molecular weight of up to 1000 daltons when no enhancer is used. With the inclusion of enhancers, good bioavailability can be extended to a molecular weight of at least 6000 daltons. Several methods have been used to facilitate the nasal absorption of drugs:

1. **Structural modification.** The drug is chemically modified to alter its physicochemical properties to enhance its nasal absorption.
2. **Salt or ester formation.** The drug is converted to a salt with increased solubility or an ester derivative with better nasal membrane permeability for achieving better transnasal absorption.
3. **Formulation design.** Appropriate formulation excipients are selected, which could improve the stability and/or enhance the nasal absorption of drugs.
4. **Surfactants.** Surfactants are incorporated into the nasal formulations to modify the permeability of nasal mucosa, which may facilitate the nasal absorption of drugs.

Chitosan is used as an absorption enhancer in nasal delivery (as described earlier for oral mucosa drug delivery). The chitosan nasal technology can be exploited as a solution, dry powder, or microsphere formulation to further optimize the delivery system for individual drugs. For compounds requiring rapid onset of action, the nasal chitosan technology can provide a fast peak concentration compared with oral or subcutaneous administration.[18–20]

Two kinds of organic-based pharmaceuticals are used for nasal drug delivery: (*a*) Drugs with extensive presystemic metabolism (e.g., progesterone, estradiol, testosterone, hydralazine) can be rapidly absorbed through the nasal mucosa with a systemic bioavailability of approximately 100%; and (*b*) water-soluble organic-based compounds, which are well absorbed (e.g., sodium cromoglycate). Recently, nasal drug delivery has been used for systemic delivery of peptide-based pharmaceuticals.[21–23] Because of their physicochemical instability and susceptibility to hepatogastrointestinal first-pass elimination, peptide and protein pharmaceuticals generally have a low oral bioavailability and are normally administered by parenteral routes. Most nasal formulations of peptide and protein pharmaceuticals have been prepared in simple aqueous (or saline) solutions with preservatives. Another recent example is a commercially available nasal salmon calcitonin formulation. The calcitonin (Miacalcin) nasal spray is licensed for the treatment of established osteoporosis for postmenopausal women. Unlike injectable calcitonin, it is recommended for long-term rather than short-term use and has been shown to reduce the risk of new vertebral fractures. The extent of systemic delivery of peptides or proteins by transnasal permeation may depend on the structure and size of the molecules, partition coefficient, susceptibility to proteolysis by nasal enzymes, nasal residence time, and formulation variables (pH, viscosity, and osmolarity).

## Ocular

The eye is uniquely shielded from foreign substance penetration by its natural anatomic barriers, which makes effective drug delivery to the inside of the eye difficult. Two main barriers that protect the eye are (*a*) the cornea, which protects the front of the eye, and (*b*) the blood–retina barrier, which protects the back of the eye.

Topical medications are frequently impeded in reaching the targeted site due to the eye's natural protective surface. In many situations, less than 1% of the medication applied to the surface of the eye will actually reach the disease site. The solution instilled as eye drops into the ocular cavity may disappear from the precorneal area of the eye by any of the following composite routes: nasolacrimal drainage, tear turnover, productive corneal absorption, and nonproductive conjunctival uptake (**Fig. 23–5**). Traditional dosage forms for delivery of drugs into the eye are mostly solutions and ointments; however, as a consequence of its function as the visual apparatus, mechanisms exist for the clearance of applied materials from the cornea to preserve visual acuity. This presents problems in the development of formulations for ophthalmic therapy. A large proportion of the topically applied drug is immediately diluted in the tear film, excess fluid spills over the lid margin, and the remainder is rapidly drained into the nasolacrimal duct. In addition, part of the drug is not available for therapeutic action because it binds to the surrounding extraorbital tissues. These processes lead to a typical corneal contact time of about 1 to 2 min in humans for solutions and an ocular bioavailability that is commonly less than 10%.

To achieve a sufficient concentration of drug delivered to the back of the eye, medications are frequently administered systemically at very high doses to overcome the blood–retina barrier. Drug injections into the back of the eye are occasionally used, but are quickly removed by the eye's natural circulatory process, often necessitating frequent injections that can carry toxicity risks.

To optimize ocular drug delivery, the following characteristics are required: good corneal penetration, prolonged contact time with the corneal epithelium, simplicity of instillation for the patient, nonirritative and comfortable form (i.e., the system should not provoke lacrymation and reflex blinking), and appropriate rheologic properties. Several novel drug delivery systems have been developed to enhance drug delivery to the eye[24] (described later in this chapter).

## Transdermal

The skin has been used for centuries as the site for the topical administration of drugs, but only recently has it been used as a pathway for systemic drug delivery (i.e., transdermal).[25] The barrier function of the skin prevents both water loss and the entrance of external agents; however, some drugs are able to penetrate the skin in sufficient amounts to produce a systemic action. The transdermal route is of particular interest for drugs that have a systemic short elimination half-life or undergo extensive first-pass metabolism, therefore, requiring frequent dosing.

The concept of delivering drugs through the skin was first introduced in the early 1950s. However, the first commercial product was made available in the United States only in the early 1980s. These first-generation, passive, transdermal patches set the foundation for this route of delivery. These

**(a)**            **(b)**

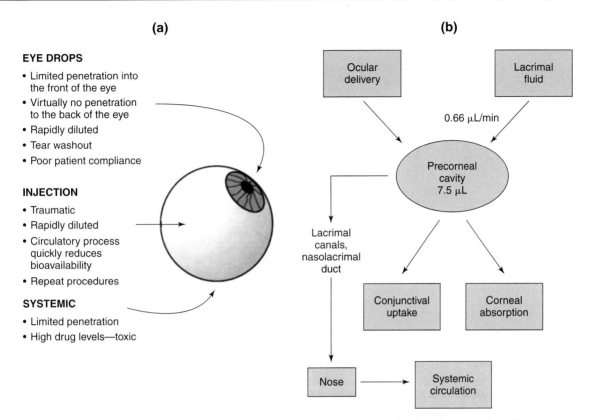

**EYE DROPS**

- Limited penetration into the front of the eye
- Virtually no penetration to the back of the eye
- Rapidly diluted
- Tear washout
- Poor patient compliance

**INJECTION**

- Traumatic
- Rapidly diluted
- Circulatory process quickly reduces bioavailability
- Repeat procedures

**SYSTEMIC**

- Limited penetration
- High drug levels—toxic

**Fig. 23–5.** (*a*) The mechanisms of drug elimination from eye after the administration of eyedrops into the ocular cavity. (*b*) The disadvantages of conventional ocular drug delivery.

transdermal patches, which were designed to control nausea, vomiting, and angina, however, failed to succeed in the market. The introduction of nicotine patches for smoking cessation gave the necessary impetus to this technology during the initial years. Today, transdermal patches are widely used to deliver hormones and pain management medications. The basic process involved in the development of transdermal drug delivery systems (i.e., patches) is percutaneous or transdermal absorption. Novel transdermal systems, including iontophoresis, thermophoresis, and phonophoresis, were also developed to enhance transdermal delivery of drugs.

The skin epidermis consists of three main layers, the stratum corneum, the granular layer, and the basal layer. The stratum corneum is considered the most important barrier to drug transfer. It is a heterogeneous nonliving structure, formed by keratinized cells, protein-rich cells, and intercellular lipid layers. The lipid composition among the epidermal layers is very different. Polar phospholipids, which are components of living cell membrane, are absent in the dead stratum corneum. These lipids form bilayers and their acyl chains can exist in "gel" and "liquid crystalline" states. The transition between these two states occurs at certain temperatures without loss of the bilayer structure.[26] The principal lipids of the stratum corneum are ceramide (50%) and fatty acids (25%). Although the stratum corneum does not contain phospholipids, the mixture of ceramides, cholesterol, and fatty acids is capable of forming bilayers. These lipid bilayers provide the barrier function of the stratum corneum.[26]

To study the percutaneous transfer of drugs, the skin can be considered as a bilaminate membrane consisting of the dead stratum corneum (lipophilic layer) and the living tissue (hydrophilic layer) that comprises the granular and basal layers of the epidermis and the dermis (**Fig. 23–6**). Diffusion of polar drugs is much faster through the viable tissue than across the stratum corneum.[27] The permeability coefficient through the skin, $P$, can be expressed as:[28]

$$P = \frac{D_v D_s}{K l_v D_s + l_s D_v} \qquad (23\text{–}1)$$

where $K$ is the partition coefficient of the drug between the stratum corneum (s) and the viable tissue (v) and $l_v$, $l_s$, $D_v$, and $D_s$ are the diffusion path lengths and diffusion coefficients, respectively. The subscripts s and v refer to the stratum corneum and the viable tissue, respectively. If the drug diffuses slowly through the stratum corneum, $K l_v D_s$ is much less than $l_s D_v$, and equation (**23–1**) becomes:

$$P = \frac{D_s}{l_s} \qquad (23\text{–}2)$$

In this case, the skin permeability is controlled by the stratum corneum alone. If the diffusion through the stratum corneum is fast, $K l_v D_s$ is much greater than $l_s D_v$, and equation (**23–1**) becomes:

$$P = \frac{D_v}{l_v K} \qquad (23\text{–}3)$$

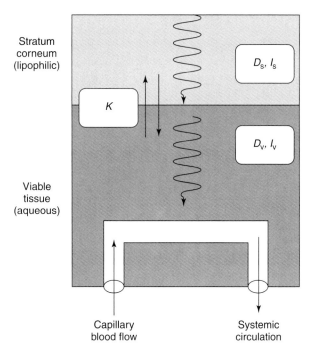

**Fig. 23–6.** The skin represented as two layers: the dead stratum corneum ($s$) and the viable or living tissue ($v$) (dermis and part of epidermis). The diffusion coefficients, $D_s$ and $D_v$, and the diffusion path lengths, $l_s$ and $l_v$, together with the partition coefficient, $K$, control the transport of the penetrant through the two layers. (Modified from R. H. Guy and J. Hadgraft, Pharm. Res. **5,** 753, 1988. With permission.)

In this case, the partition coefficient may be influential in the permeability. As $K$ increases, the transfer from the stratum corneum to viable epidermis becomes less favorable and slower. At large $K$ values, partitioning of the drug is the rate-limiting step.[28]

### EXAMPLE 23–1

**Transdermal Permeability**

Compute the permeability of a drug across the skin assuming that $D_s$ is $10^{-10}$ cm$^2$/sec and $D_v$ is $10^{-7}$ cm$^2$/sec. The path lengths are $l_s = 350$ $\mu$m and $l_v = 150$ $\mu$m. The large value for $l_s$ is due to the fact that the molecules follow a tortuous pathway through the intercellular spaces. The value for $l_v$ is the distance from the underside of the stratum corneum to the upper capillary region of the dermis. The partition coefficient is taken to be $K = 1$.

Because diffusion through the stratum corneum is very slow ($D_s = 10^{-10}$ cm$^2$/sec) and $Kl_vD_s = 1.0 \times 150 \times 10^{-4} \times 10^{-10} = 1.5 \times 10^{-12}$ is much less than $l_sD_v = 350 \times 10^{-4} \times 10^{-7} = 3.5 \times 10^{-9}$, then from equation (23–3),

$$P = \frac{10^{-10}\ \text{cm}^2/\text{sec}}{350 \times 10^{-4}\ \text{cm}} = 2.86 \times 10^{-9}\ \text{cm/sec}$$

Using equation (23–2), we arrive at a similar order of magnitude:

$$P = \frac{(10^{-7}\ \text{cm}^2/\text{sec})(10^2\ \text{cm}^2/\text{sec})}{(1.0 \times 150 \times 10^{-4} \times 10^{-10})\ \text{cm} \cdot \text{cm}^2/\text{sec}}$$
$$+ (10^{-7} \times 350 \times 10^{-4})\ \text{cm} \cdot \text{cm}^2/\text{sec}$$
$$= 2.86 \times 10^{-9}\ \text{cm/sec}$$

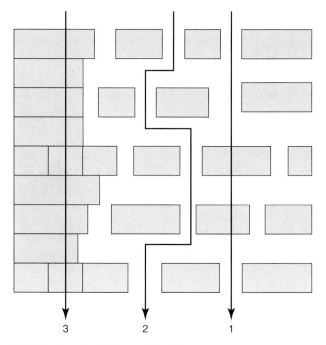

**Fig. 23–7.** Random-brick model for the stratum corneum. Arrows 1, 2, and 3 show three possible routes of drug diffusion. (Modified from K. Tojo, J. Pharm. Sci. **76,** 889, 1987. With permission.)

Tojo[29] proposed a random-brick model for the transfer of drugs across the stratum corneum. As shown in **Figure 23–7,** the cells rich in proteins separated from one another by thin-layer intercellular lipids represent the stratum corneum. The side length of the cells varies, but the total average surface area is constant. The thickness of the cells and the lipid layer are also assumed to be constant. According to this model, the transfer of a drug is divided into three parallel pathways: (*a*) across the cellular–intercellular regions in series; (*b*) across the lipid intercellular spaces; and (*c*) across thin lipid layers sandwiched between flattened protein cells of the stratum corneum.

According to the brick model, the effective diffusion coefficient, $D_{\text{eff}}$, across the skin is given by:

$$D_{\text{eff}} = 2\varepsilon(1 - \varepsilon)D_1 + \varepsilon^2 D_2 + (1 - \varepsilon)^2 D_3 \quad \textbf{(23–4)}$$

The first, second, and third terms on the right-hand side represent the three possible routes 1, 2, and 3 in **Figure 23–7,** respectively; $D_1, D_2$, and $D_3$ are the diffusivities across routes 1, 2, and 3, respectively; $\varepsilon$ and $(1 - \varepsilon)$ are the average fraction of diffusion area of the lipid and protein on the skin surface, respectively. Substituting $D_1, D_2$, and $D_3$ by their corresponding expressions, we find that equation **(23–4)** becomes:

$$D_{\text{eff}} = \frac{2\varepsilon(1 - \varepsilon)(2n + 4)}{\dfrac{n}{D_{\text{p}}} + \dfrac{n + 4}{KD_1}} + \varepsilon^2 KD_1 + \frac{(1 - \varepsilon)^2(2n + 4)}{\dfrac{2n}{D_{\text{p}}} + \dfrac{4}{KD_1}}$$
$$\textbf{(23–5)}$$

The term $n$ is related to the volume fraction of lipids in the skin and the average fraction of diffusion area of the lipids,

$\varepsilon$; $K$ is the lipid–protein partition coefficient, and $D_p$ and $D_l$ are the diffusion coefficients across the protein layer and the lipid layer, respectively.

The flux across the skin is given by:

$$J = \frac{dQ}{dt} = \frac{D_{eff}\, C_p}{h} \qquad (23\text{–}6)$$

where $C_p$ is the concentration of drug in the protein cell layer, $h$ is the thickness of the skin, and $D_{eff}$ is defined by equation (23–5).

### EXAMPLE 23–2

**Brick Model of Transdermal Penetration***

Compute the flux at the steady state, $dQ/dt$, of a new drug from the following data: $D_l = 1 \times 10^{-10}$ cm$^2$/sec, $D_p = 1 \times 10^{-7}$ cm$^2$/sec, $C_p = 10$ mg/cm$^3$, $\varepsilon = 0.02$, $h = 0.0020$ cm, $K = 0.1$, and $n = 14.3$. Use the random-brick model.

We have

$$D_{eff} = \frac{(2)(0.02)(1-0.020)(2 \times 14.30 + 4)}{\dfrac{14.3}{1 \times 10^{-7}} + \dfrac{14.3 + 4}{(0.1)(10^{-10})}} + (0.02)^2(0.1)(10^{-10})$$

$$+ \frac{(1-0.02)^2(2 \times 14.3 + 4)}{\dfrac{2 \times 14.3}{1 \times 10^{-7}} + \dfrac{4}{(0.1)(1 \times 10^{-10})}} = 7.89 \times 10^{11}\ \text{cm}^2/\text{sec}$$

The flux is

$$J = \frac{(7.89 \times 10^{-11}\ \text{cm}^3/\text{sec})(10\ \text{mg/cm}^3)}{0.0020\ \text{cm}}$$

$$= 3.95 \times 10^{-7}\ \text{mg/(cm}^2\ \text{sec)} = 3.95 \times 10^{-4}\ \mu\text{g/(cm}^2\ \text{sec)}$$

## Factors Affecting Permeability

### Hydration

The skin permeability of a drug depends on the hydration of the stratum corneum; the higher the hydration, the greater is the permeability. The dermal tissue is fully hydrated, whereas the concentration of water in the stratum corneum is much lower, depending on ambient conditions. Hydration may promote the passage of drugs in the following way. Water associates through hydrogen bonding with the polar head groups of the lipid bilayers present in the intercellular spaces. The formation of a hydration shell loosens the lipid packing so that the bilayer region becomes more fluid.[30] This facilitates the migration of drugs across the stratum corneum. From the rate of transpiration (i.e., passage of water from inner layers to the stratum corneum) and diffusivity of water in the stratum corneum, the amount of water in the tissue can be obtained.[31]

### Solubility of the Drug in Stratum Corneum

Using the experimental results obtained from intact skin and stripped skin (layers of stratum corneum removed using Scotch tape), the solubility, $C$, of a drug in the stratum

---

*This example was kindly provided by K. Tojo, Kyushu Institute of Technology, Japan.

corneum was calculated from the following expression[29]:

$$C = \frac{1 - 3\tau + 2\eta\tau}{(1 + 2\eta)(1 - \eta)} \cdot \frac{6t_2}{h_2} \left(\frac{dQ}{dt}\right)_2 \qquad (23\text{–}7)$$

where $(dQ/dt)_2$ is the steady-state permeation across the intact skin, $\tau$ is the ratio of the two time lags, $(t_1/t_2)$, and $\eta$ is the ratio $(dQ/dt)_2/(dQ/dt)_1$. The subscripts 1 and 2 refer to the stripped and intact skin, respectively; $h_2$ is the thickness of the stratum corneum.

### EXAMPLE 23–3

**Solubility of Progesterone in the Stratum Corneum**

Compute the solubility of progesterone in the stratum corneum. The lag times across the intact and stripped skin are $t_2 = 5.49$ and $t_1 = 1.55$ hr, respectively, and the permeation rates across the intact and stripped skin are $(dQ/dt)_2 = 2.37\ \mu\text{g/(cm}^2\ \text{hr)}$ and $(dQ/dt)_1 = 3.62\ \mu\text{g/(cm}^2\ \text{hr)}$, respectively. The thickness of the stratum corneum is 10 $\mu$m ($10 \times 10^{-4}$ cm).

We have

$$\tau = \frac{t_1}{t_2} = \frac{1.55}{5.49} = 0.28$$

$$\eta = \frac{(dQ/dt)_2}{(dQ/dt)_1} = \frac{2.37}{3.62} = 0.65$$

From equation (23–7),

$$C = \frac{1 - (3 \times 0.28) + (2 \times 0.65 \times 0.28)}{[1 + (2 \times 0.65)](1 - 0.65)} \cdot \frac{6 \times 5.49}{10 \times 10^{-4}} \cdot 2.37$$

$$C = 50816.8\ \mu\text{g/mL} = 50.8\ \text{mg/mL}$$

## Excipients

Common solvents and surfactants can affect penetration of drugs though the skin. Sarpotdar and Zatz[32] studied the penetration of lidocaine through hairless mouse skin in vitro from vehicles containing various proportions of propylene glycol and polysorbate 20. Propylene glycol is a good solvent for lidocaine and reduces its partitioning into the stratum corneum, lowering the penetration rate. In this study, the effect of the surfactants depended on the concentration of propylene glycol in the vehicle. The decrease of flux for 40% (w/w) propylene glycol concentration can be explained by micellar solubilization of lidocaine. It is generally assumed that only the free form of the drug is able to penetrate the skin. Thus, the micellar solubilization of lidocaine reduces its thermodynamic activity in the vehicle and retards its penetration. At higher propylene glycol concentrations, 60% and 80%, an increase in flux was observed, possibly owing to an interaction of the surfactant with propylene glycol.

### Influence of pH

According to the pH-partition hypothesis only the un-ionized form of the drug is able to cross the lipoidal membranes in significant amounts. However, in studies of isolated intestinal membranes, both the ionized and un-ionized forms of sulfonamides permeated the membrane. The diffusion of ionized drug through the skin may be nonnegligible, particularly at pH values at which a large number of ionized molecules are present.[33] Fleeker et al.[34] studied the influence of pH on the transport of clonidine, a basic drug, through hydrated shed snakeskin. The contribution to the total flux, $J$, in $\mu$g cm$^{-2}$ hr$^{-1}$ of the nonionized and ionized species for a basic drug can be written as:

$$J = J_B + J_{BH^+} \qquad (23\text{--}8)$$

where B and BH$^+$ represent the basic (nonionized) and protonated forms, respectively. The dissociation of the protonated form is represented as:

$$[BH^+] \xoverleftrightarrow{K_a} [B] + [H^+] \qquad (23\text{--}9)$$

and

$$K_a = \frac{[B][H^+]}{[BH^+]} \qquad (23\text{--}10)$$

Taking the logarithm of both sides of equation (23–10) and rearranging it, we obtain the concentration of the protonated form as:

$$[BH^+] = \frac{10^{(pK_a - pH)}}{1 + 10^{(pK_a - pH)}}[T] \qquad (23\text{--}11)$$

where [T] represents the total concentration of both the charged and uncharged species,

$$[T] = [B] + [BH^+] \qquad (23\text{--}12)$$

The concentration of nonionized form [B] can be computed from equation (23–12). The total flux, equation (23–8), can also be written in terms of the permeability coefficients times the concentrations of each species:

$$J = P_B[B] + P_{BH^+}[BH^+] \qquad (23\text{--}13)$$

From equations (23–11) and (23–12), provided PB $\cong$ PBH$^+$s, we have the following results:

**(a)** When $pK_a$ = pH, [B] = [BH$^+$], and both species B and BH$^+$ contribute to the total flux.
**(b)** When pH is much greater than $pK_a$, [B] is much greater than [BH$^+$], and the total flux, $J$, is approximately PB[B].
**(c)** When pH is much less than $pK_a$, [B] is much less than [BH$^+$], and the total flux, $J$, is approximately PBH$^+$[BH$^+$].

Equation (23–13) allows one to compute the permeabilities of both species from the total experimental flux and the values of [B] and [BH$^+$] from equations (23–11) and (23–12).

---

### EXAMPLE 23–4

**Transdermal Permeability of Clonidine**

Compute the permeability coefficients PB and PBH$^+$ corresponding to the nonionized and protonated forms of clonidine. The total fluxes at pH 4.6 and pH 7 are 0.208 and 0.563 $\mu$g/(cm$^2$ hr), respectively. The $pK_a$ of the protonated form is 7.69. The total concentration*

---
*"Concentration," expressed in $\mu$g/uv, was obtained by using 0.4 mL of an appropriate concentration of clonidine solution.

of the two species, [T], is 4 $\times$ 103 $\mu$g/uv, where $\mu$g/uv stands for microgram/unit volume.

At pH 4.6,

$$[BH^+] = \frac{10^{(7.69-4.6)}}{1 + 10^{(7.69-4.6)}}(4 \times 10^3) = 3.996 \times 10^3 \ \mu g/uv$$

$$[B] = (4 \times 10^3) - (3.996 \times 10^3) = 4.0 \ \mu g/uv$$

At pH 7,

$$[BH^+] = \frac{10^{(7.69-7)}}{1 + 10^{(7.69-7)}}(4 \times 10^3) = 3.322 \times 10^3 \ \mu g/uv$$

$$[B] = (4 \times 10^3) - (3.322 \times 10^3) = 678 \ \mu g/uv$$

From equation (23–13) with the flux, $J$, expressed in $\mu$g/cm$^2$ hr and the permeability coefficient, $P$, in units of cm/hr, at pH 4.6,

$$0.208 \ \mu g/cm^2 \ hr = P_B(4 \ \mu g/cm^3) + P_{BH^+}(3996 \ \mu g/cm^3)$$

At pH 7,

$$0.563 \ \mu g/cm^2 \ hr = P_B(678 \ \mu g/cm^3) + P_{BH^+}(3322 \ \mu g/cm^3)$$

PB and PBH$^+$ are calculated by solving the two equations simultaneously:

$$P_B = (0.208 - 3996 \ P_{BH^+})/4$$

$$0.563 = 678 \frac{0.208 - 3996 \ P_{BH^+}}{4} + 3322 \ P_{BH^+}$$

$$P_{BH^+} = \frac{-34.693}{-674000} = 5.15 \times 10^{-5} \ cm/hr$$

$$P_B = [0.208 - (5.15 \times 10^{-5} \times 3996)]/4 = 5.5 \times 10^{-4} \ cm/hr$$

It is noted that the values found for PBH$^+$ and PB do not change with pH, whereas the fluxes, $J$, are markedly different at pH 4.6 and pH 7.0.

---

Equations (23–10) and (23–13) can be combined to give:

$$\frac{J}{[B]} = \frac{P_{BH^+}}{K_a}[H^+] + P_B \qquad (23\text{--}14)$$

From equation (23–14), the permeability coefficients of the ionized [BH$^+$] and nonionized [B] forms can be computed from the slope and intercept of a plot of $J$/[B] against [H$^+$]. The corresponding equation for acids is:

$$\frac{J}{[A^-]} = \frac{P_{HA}}{K_a}[H^+] + P_{A^-} \qquad (23\text{--}15)$$

From equation (23–15), the permeability coefficients of the nonionized [HA] and ionized [A$^-$] forms are computed from the slope and the intercept of a plot of $J$/[A$^-$] against [H$^+$]. Swarbrick et al.[33] found that both the ionized and non-ionized forms of four chromone-2 carboxylic acids permeated skin, although the permeability of the nonionized form was about 104 times greater.

### Binding of Drug to the Skin

The skin can act as a reservoir for some drugs that are able to bind to macromolecules. The drug fraction bound is not able to diffuse, and binding hinders the initial permeation rate of molecules, resulting in larger lag times. Banerjee and Ritschel[35] studied the binding of vasopressin and corticotropin to rat skin. Penetration of large molecules such as collagen, used in cosmetic formulations, is questionable, but partial hydrolysates of collagen are able to reach the deeper skin layers. The sorption process can be represented by the Langmuir equation:

$$\frac{c}{x/m} = \frac{1}{bY_m} + \frac{c}{Y_m} \qquad (23\text{-}16)$$

where $c$ represents the equilibrium concentration of the drug, $x$ is the amount of drug adsorbed per amount, $m$, of adsorbent (the skin proteins in this case), $b$ is the affinity constant, and $Y_m$ is the maximum adsorption capacity, $(x/m)_{max}$. The sorption isotherm was obtained by Banerjee and Ritschel[35] by equilibration of a measured weight of rat epidermis with a known concentration of radiolabeled vasopressin solution and was analyzed by scintillation counting (measured radioactivity). The small value for the adsorption constant in equation (23-16), $b = 6.44 \times 10^{-4}$ mL/$\mu$g, suggests low affinity of vasopressin for the binding sites in the skin.

### Drug Metabolism in the Skin

The metabolism of drugs during transport through the skin affects bioavailability and can produce significant differences between in vivo and in vitro results. Oxidation, reduction, hydrolysis, and conjugation are kinetic processes that compete with the transport of drugs across the skin. Guzek et al.[36] and Potts et al.[37] found differences in the in vitro and in vivo extents of enzymatic cleavage in the skin and in the distribution of the metabolites of a diester derivative of salicylic acid. The authors suggested that the in vitro measurements overestimated the metabolism because of the increased enzymatic activity and/or decreased removal of the drug in the absence of capillaries. The fact that the skin contains esterases and other enzymes is useful for the administration of prodrugs. The solubility and absorption can be improved, and the enzymes could be used to cleave the prodrug to give the active drug in the skin.[38]

## Vaginal[39,40]

The vagina has been used for a long time for topical drug administration. The most frequently used vaginal preparations include:[39] (*a*) antimicrobials (antibacterial, antifungal, antiprotozoal, antichlamydial, and antiviral) pessaries, or creams such as metronidazole, 5-nitroimidazoles (tinadazole and ornidazole), and imidazole (clotrimazole, econazole, isoconazole, and miconazole); (*b*) estrogen creams; and (*c*) spermicidal foams, gels, and creams such as nonoxynol-9, octoxinol, and p-di-isobutylphenoxy-poly(ethoxyethanol).

Earlier, the vagina was considered as an organ incapable of absorbing drugs systemically and, therefore, systemic absorption of a drug through vagina was considered only from the standpoint of toxicity.[40] However, it was shown that a number of topical drugs are able to achieve sufficient blood levels and can achieve systemic effects. Later, it was also shown that vaginal permeability of substances such as water, 17-$\beta$-estradiol, arecoline, and arecaidine is in fact higher than the intestinal mucosa.[41] As a result, there is an interest in the design of vaginal delivery systems for systemic delivery of drugs such as estrogens, progesterones, prostaglandins, peptides, and proteins. The interest in systemic vaginal drug delivery systems is due to the ease of administration, rich blood supply facilitating rapid absorption, high permeability to certain drugs in different phases of menstrual cycle, avoidance of hepatic first-pass metabolism, reduction in GI side effects, and decrease in hepatic side effects (e.g., for steroids). The disadvantages associated with vaginal drug delivery on the other hand are that it is limited to only potent drugs; there is a possibility of adverse effects due to the low amount of fluids present, hormone-dependant changes, and possibility of leakage.

The vagina is a tubular, fibromuscular organ extending from the cervix of the uterus to the vaginal vestibule.[39,40] In an adult female, the vaginal tract is about 2 cm in width and comprises an anterior wall of ~8 cm and a posterior wall of ~11 cm in length. Histologically, it consists of four distinct layers: epithelial with underlying basement membrane, lamina propria, muscular layer, and adventicia. Although sometimes considered a mucosal tissue, the normal vagina does not have glands and the vaginal secretions present on the surface is a mixture of fluids from different sources. It must be noted that vaginal characteristics, particularly the pH, changes with the phase of the menstrual cycle. The normal vaginal pH (4.5–5.5) is maintained by *Lactobacilli* present in the vagina. There may be an atrophy of vaginal epithelium, elevation of pH (6.0–7.5), and decrease in secretions, postmenopause. The vagina is normally collapsed on itself and capable of holding about 2–3 g of fluid/gel without leakage. Drug permeation across the vaginal membrane (epithelial) occurs mainly through diffusion, where hydrophilic molecules are absorbed via the paracellular route (diffusion between adjacent cells) and hydrophobic substances are absorbed via the transcellular route (across epithelial cells by passive diffusion, carrier-mediated transport, or endocytic process).[40]

The vaginal route is used for estrogens and progesterone delivery. Controlled-release delivery devices such as suppositories, inserts, and rings are also available. Vaginal route is also being investigated for the delivery of GnRH analogues and insulin. Antiviral vaginal gels and liposomal preparations, vaginal mucosal vaccines, microspheres (starch and hyaluronan), bioadhesive polymers, and gels are under various stages of development. Penetration enhancers such as organic acids and $\alpha$-cyclodextrin are also being used to enhance the drug absorption across vaginal epithelium but they are associated with side effects.

## Central Nervous System[42,43]

The brain is not a route of drug delivery but still an important pharmaceutical target. Drugs that act on the CNS are those

## KEY CONCEPT BLOOD–BRAIN BARRIER

In the early 1900s, researchers found the first evidence that the brain had a selective barrier that protects its cells. It is now known as the *blood–brain barrier* (*BBB*) and it is responsible for regulating the entry of molecules into the brain. The BBB separates the blood compartment from the extracellular fluid compartment of the brain parenchyma and consists of monolayer of polarized endothelial cells. The brain endothelial cells are connected by tight junctions, unlike the nonbrain capillary endothelial cells comprising large fenestrations (opening). BBB acts as a selective barrier and performs following functions:

(*a*) isolate the brain from systemic influence; (*b*) provide pathway for the transport of nourishment to neurons; and (*c*) clear potentially toxic substance from brain into the blood. Besides the permeability barrier, highly selective enzymes are present in these endothelial cells, which further restrict the entry of substrates to the brain. The problem is further compounded by the presence of efflux transporters such as p-glycoprotein that are active in astrocyte membranes. All components work in tandem to form a multicomponent BBB.[42,43] Most drugs are unable to reach the brain because of their inability to penetrate the BBB.

used for psychosis, depression and mania, anxiety, epilepsies, Parkinson disease, Alzheimer disease, pain, and brain tumors. Furthermore, the AIDS virus is also known to attack neuron and glial cells causing memory loss, palsy, dementia, and paralysis.[42] Drug delivery to brain is highly difficult due to the presence of a blood–brain barrier (BBB) regulating the entry of molecules to the brain. The BBB makes the brain inaccessible to CNS-targeted drugs in the systemic circulation, more so for biotherapeutics such as peptides, proteins, and nucleic acids.

Various transport mechanisms exist in the brain endothelium for the uptake of nutrients into the CNS, which may also be utilized for the drug delivery. The transport mechanisms available for passage through the BBB are the following (**Fig. 23–8**):[42] passive diffusion, active transport, and receptor-mediated transport.

Physicochemical factors also influence the drug delivery to brain. Increasing the lipid solubility of a drug increases its permeability across the BBB. Highly lipid soluble molecules (barbiturate drugs and alcohol) rapidly cross the BBB into the brain. However, this is true only for low–molecular-weight drugs in the range of 400 to 600 Da. Presence of

p-glycoprotein efflux restricts the passive diffusion of drugs that are substrate for p-glycoprotein. Examples include vinblastine, vincristine, and cyclosporin. In addition, the transport of drugs that are highly charged or bind strongly to plasma protein across the BBB is slow.

## DELIVERY OF NUCLEIC ACID THERAPEUTICS

### Nucleic Acid Therapeutics

Exogenous nucleic acids can be used to modify gene expression.[44] Zamecnik and Stephenson[45] demonstrated that a short oligodeoxynucleotide (13-mer) that was antisense to the Rous sarcoma virus could inhibit viral replication in cell culture. The existence of natural antisense nucleic acids and their role in regulating gene expression was shown in the mid 1980s.[46] This led to the development of technologies employing synthetic oligonucleotides as therapeutics for manipulating gene expression in living cells. Synthetic oligonucleotides (ODNs) are short (<30 nucleotides) nucleic acid strands (DNA or RNA), which are chemically synthesized.

1. Between the cells
   (paracellular route)

   – Cells
     • Infection
     • Immune
   – Chemicals
     • Ions
     • Waste
   – Hydrophilic

2. Through the cells
   (transcellular route)

   – Gases
   – Water
   – Lipophilic

3. Active transport

   – D-glucose
   – Large neutral
     amino acids

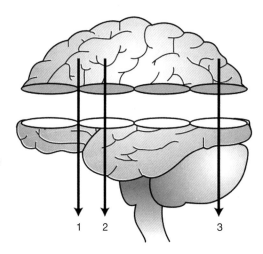

**Fig. 23–8.** Transport mechanisms available for passage across the blood–brain barrier (BBB).

## ⬛⬛⬛KEY CONCEPT⬛ NUCLEIC ACID THERAPEUTICS

The genetic information in biological systems is transferred from DNA→RNA→proteins (**Fig. 23–9**). The double helical DNA unwinds and creates a copy of itself by the process called *replication*. As a result, genetic information stored in DNA is faithfully transferred to the next generation of cells or organisms. The information stored in a part of DNA is transferred to mRNA by a process called *transcription*. In eukaryotic cells, the primary transcript (pre-mRNA) is processed further by alternative splicing (mRNA blocks are cut out and rearranged) and it migrates from nucleus to cytoplasm. The mRNA binds to ribosomes in cytoplasm, where information stored in mRNA is read as triplet codon (three base pairs for one amino acid) to assemble proteins (biological activity) by a process called *translation*. Information from proteins cannot be transferred back to either nucleic acids or proteins. Francis Crick[47] called this unidirectional flow of information the Central Dogma of Molecular Biology. The importance of *nucleic acid therapeutics* and *gene therapy* lies in the fact that they provide the capability to interfere at different stages of this process with high specificity.

Nucleic acids are polynucleotides comprising sugar (ribose and 2-deoxy-ribose), purine (adenine/A and guanine/G) and pyrimidine (cytosine/C, thymidine/T, and uracil/U) bases, and phosphate. DNA contains $2'$-deoxy-ribose sugar and A, G, C, and T bases, whereas RNA contains ribose sugar and A, G, C, and U bases. DNA is a double helical structure, whereas RNA is single stranded but may fold back to form duplex structures. Base and sugar react to form nucleoside and addition of a phosphate group to the nucleoside gives nucleotide. The nucleic acids are linear polynucleotide chains connected through phosphodiester backbone. All nucleic acid hybridizations (DNA-DNA, DNA-RNA, and RNA-RNA) are stabilized by hydrogen bonds (*Watson–Crick base pairing*); G always binds to C with three hydrogen bonds and A always binds to T or U with two hydrogen bonds. This phenomenon is called base complementarity and accounts for target specificity of nucleic acids. AT and GC base pairs in the major groove can establish additional hydrogen bonds with T and protonated C*, respectively. These are called *Hoogsteen base pairing* and used for triplex formation.

Antisense is not the only mechanism available; other mechanisms are now known to cause specific inhibition of gene expression (**Fig. 23–10**).

1. *Antigene mechanism*: Triplex-helix forming oligonucleotides[48] are synthetic single-stranded DNA, which hybridize to purine or pyrimidine-rich region in the major groove of double-helical DNA through Hoogsteen base pairing. If a stable triple helix is formed, it prevents unwinding of double helical DNA necessary for transcription of the targeted region or blocks the binding of transcription factor complexes. This mechanism is not considered efficient for clinical applications even though it provides an opportunity for therapeutic interventions at a very early stage.

2. *Antisense mechanism*: Reverse-complementary (antisense) oligonucleotides, which hybridize to the mRNA strand of the targeted gene.[49–51] After hybridization, antisense oligonucleotides block expression either sterically by obstructing the ribosomes or by forming an RNA-DNA hybrid, which is a substrate for RNase H enzyme, thereby causing the cleavage of target mRNA. RNase H is a ubiquitous enzyme that hydrolyzes the RNA strand of an RNA-DNA duplex.

3. *RNA interference*: RNA interference (RNAi) or posttranscriptional gene silencing is a natural process in eukaryotic cells by which double-stranded RNA targets mRNA for cleavage in sequence-specific manner.[52] The mechanism of RNAi involves processing of a very long (500–1000 nucleotides) double-stranded RNA, which is cleaved

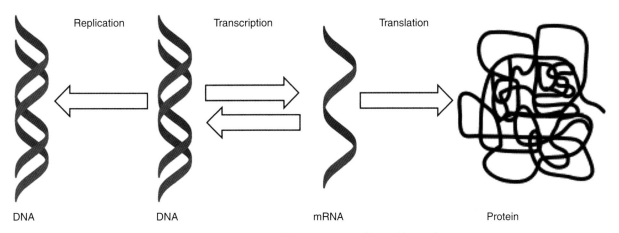

**Fig. 23–9.** Transfer of genetic information in living organisms (Central Dogma).

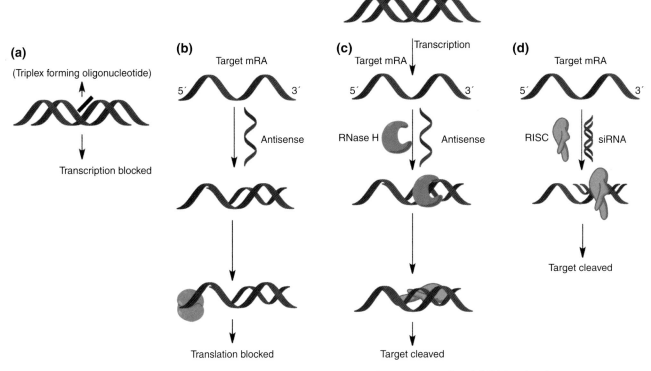

**Fig. 23–10.** Different mechanisms for selective inhibition of gene expression: (*a*) Triplex-forming oligonucleotide binds to the purine or pyrimidine rich region in the major groove of DNA (antigene); (*b*) antisense oligonucleotide sterically blocks ribosome activity (antisense); (*c*) antisense oligonucleotide forms hybrid with target mRNA, which is a substrate for RNase H (antisense); and (*d*) siRNA forms complex with RISC, which facilitate recognition of target mRNA followed by its cleavage (RNA interference).

into short double-stranded RNA (small interfering RNA or siRNA, 21–25 nucleotides) by the DICER enzyme.[53] Synthetic siRNA can be produced and directly introduced into the cells. Once inside the cytoplasma of the cell, siRNA is incorporated into a large multicomponent complex called RNA-induced silencing complex (RISC). A multifunctional protein (Argonaute 2) in the RISC unwinds the double-stranded RNA and cleaves the sense strand of siRNA. The activated RISC containing the antisense strand of siRNA selectively seeks out and degrades mRNA complementary to the antisense strand. The activated RISC then moves on to destroy additional mRNA targets.[50,54]

4. *Aptamer selection*: Aptamers are nucleic acid ligands (15–40 nucleotides) isolated from combinatorial oligonucleotide libraries by in vitro selection.[55] In solution, the oligonucleotide chain forms intramolecular interactions that fold the molecule into a complex three-dimensional shape. Aptamers have capability to tightly and specifically bind to the target molecules ranging from small molecules to complex multimeric structures. The therapeutic potential of aptamers arises from the fact that many aptamers targeted against proteins are able to interfere with their biological activity.

Other nucleic acid agents such as microRNA are also known and so are nucleic acids with catalytic activity: *ribozymes* and *DNAzymes*. *Ribozymes* are RNA molecules containing different catalytic motifs.[53] They hybridize (bind) to the substrate RNA through Watson–Crick base pairing and cause sequence-specific cleavage. *DNAzymes* on the other hand are DNA molecules containing a catalytic motif and similar to ribozyme, cleave the substrate RNA after binding to it.

The unmodified nucleic acids have phosphodiester backbone, which are rapidly degraded in biologic fluids and in cells by the nucleases. Moreover, they show extremely poor penetration (diffusion) across the cell membrane.[49–51] Most of the antisense oligonucleotides or siRNA therefore contain structural modifications. Oligonucleotides for clinical applications mostly have phosphorothioate or morpholino backbone instead of phospodiester. Several advanced antisense oligonucleotides are "gapmers" consisting of a central DNA portion that can recruit RNase H and flanking 2'-modified regions. Other important structural modifications are 2'-OH modifications (e.g., 2'-O-methyl), locked nucleic acids, peptide nucleic acids, and hexitol nucleic acids.

The only nucleic acid therapeutic currently approved for human use is fomivirsen (Vitravene, 1998, Isis Pharmaceuticals) for the treatment of inflammatory viral infection of the eye (retinitis) caused by cytomegalovirus.[53] Vitravene is a 21-mer antisense phosphorothioate oligonucleotide complementary to mRNA transcribed from the main immediate-early transcriptional unit of cytomegalovirus. The drug was later withdrawn from the market. Antisense oligonucleotides in clinical trials are alicaforsen for ulcerative colitis (Isis); ISIS113715 for diabetes (Isis); ATL1102 for multiple sclerosis (ATL/Isis); OGX-011 for prostate cancer (Onco-Gene/Isis); and Genesense for varied cancer (Genta).[50] The siRNAs in clinical trials are AEG35156 for X-linked inhibitor of apoptosis protein (i.v., Aegera Therapeutics); AGN211745 for vascular endothelial growth factor receptor (i.v., Allergan); and ALN-RSV01 for respiratory syncytial virus nucleocapsid (nasal, Alnylam Pharmaceuticals).[50,56]

*Gene therapy* is the treatment of human disease by transferring genetic material into the specific cells of the patient.[57] It must not be confused with the nucleic acid therapeutics described above. Advances in molecular biology, biotechnology, and the Human Genome Project have led to the identification of several disease-causing genes. Gene-therapy approaches are being suggested for the replacement of genes responsible for genetic diseases like hemophilia, muscular dystrophy, and cystic fibrosis. Gene replacement can also be used for altering the expression of an existing gene, inhibiting or augmenting the synthesis of a particular protein, and producing cytotoxic proteins or prodrug activating enzymes (see GDEPT). Gene therapy is also being investigated for the treatment of cardiovascular, neurological, and infectious diseases and cancer. Clinical success with gene therapy was first reported in 2000 for the treatment of severe combined immunodeficiency.

The limited success of antisense oligonucleotides or siRNA and gene therapy is attributed to the lack of efficient delivery systems.

## Systemic Delivery of Nucleic Acids

Local delivery of nucleic acid therapeutics could be achieved in eye, skin, mucus membranes, and local tumors. It is particularly well suited for the treatment of lung diseases and infections. The advantages are higher bioavailability and reduced adverse side effects, but not every tissue is amenable to local or topical delivery.

Systemic delivery of antisense or siRNA oligonucleotides requires that the antisense or siRNA oligonucleotides are able to travel throughout the body to reach the target tissue/organ while avoiding the nontarget tissues. For effective systemic delivery, ODNs must navigate the circulatory system of the body while avoiding rapid excretion by the kidney, degradation by the serum and tissue nucleases, and uptake by the phagocytes of the reticuloendothelial system. Moreover, they should be able to overcome failure to cross the capillary endothelium, slow diffusion/binding in extracellular matrix, and inefficient endocytosis by tissue cells and release from endosomes. The major systemic delivery approaches are described below.[57,58]

## Viral Delivery Systems (Viral Vectors)

A virus carries its genome from one host cell to another. It enters the new target cells, navigates to the cell's nucleus, and initiates expression of its genome, as a part of its self-replication cycle. It is possible to convert a virus into a gene-delivery vehicle by replacing a part of the virus genome with a therapeutic gene. Viruses that are used for delivering genetic material to host cells are called vectors, the process is known as transduction, and infected cells are described as being transduced.

The viral vectors could be nonreplicating or replicating and common examples are:[59] (*a*) *retroviruses*, which contain single-stranded RNA molecule as genetic material. Herpes simplex virus (HSV) is a retrovirus, whereas lentiviruses are a subclass of retroviruses; (*b*) *adenoviruses*, which contain the genetic material in the form of double-stranded DNA and cause common cold as well as respiratory, intestinal, and eye infections in humans; and (*c*) *adeno-associated viruses*, which have single-stranded DNA as genetic material. The lesser-used viruses are *baculovirus* and *vaccinia*. Although to a lesser extent, bacteria have also been used as a delivery vehicle. The most prevalent organisms used are *Salmonella* and *Clostridium*; other lesser-used examples are *Bifidobacterium* and *Escherichia coli*.

The evolution of viruses has taken place essentially as gene-delivery vehicles and therefore they are typically very efficient. Viral vectors are used in a large number of preclinical and clinical studies involving gene-delivery. Safety of viral vectors is the major concern limiting their use in clinical applications. The viral vectors are replication deficient and therefore nonpathogenic, but there is a slight risk of reversion to the wild type virus. There is also the possibility of inducing severe immune responses. The introduction of retrovirus vectors may cause mutagenesis of the host genome. The selective delivery and expression of genes (target-cell specificity) may be difficult to achieve and the manufacturing costs are high.

## Nonviral Delivery Systems (Synthetic Vectors)

Safety concerns associated with viral vectors have led to the use of synthetic materials (synthetic vectors) for oligonucleotide delivery.[49,57,58] Moreover, they are less expensive, easier and safer to make, and suitable for long-term storage. Synthetic vectors are molecules, which electrostatically bind to DNA or RNA, condensing them into nanosized particles. When nanoparticles are formed by complexation between cationic lipids (or cationic liposomes) and DNA, they are called lipoplexes, whereas when formed by complexation between cationic polymers (or polypeptides) and DNA, they are called polyplexes. This process of delivering nucleic acids into the cell by nonviral methods is called transfection (also lipofection when lipids or liposomes are used).

Cationic liposomes are among the most extensively used synthetic materials for systemic gene delivery due to their relatively higher transfection efficiencies. They electrostatically interact with the negatively charged phosphate backbone of nucleic acids, which neutralizes the charge and promotes the condensation of nucleic acids into a more compact structure. A common example of cationic liposome is transfection reagent Lipofectin, which comprises 1:1 mixture of N-[1-(2,3-dioleyloxy)propyl]-N,N,N-trimethyl ammonium chloride (DOTMA) and the colipid dioleylphosphatidyl ethanol amine (DOPE). Other examples of cationic lipids are 3β[(N,N'-dimethylaminoethane)-carbamoyl] cholesterol (DC-CHOL); 1,2-bis(oleoyloxy)-3-(trimethylamino) propane (DOTAP); and (1,2-dimyristyloxypropyl-3-dimethyl-hydroxy) ethyl ammonium bromide (DMRIE). Recently, stable nucleic acid–lipid particle (SNALP) formulations have demonstrated high efficacy in several models in vivo.

Polypeptides such as polylysine have also been investigated for gene delivery. Polyplexes of DNA and polylysine are poor gene-delivery vectors and require the addition of chloroquine.

Cationic polymers with linear, branched, or dendritic structures also serve as efficient transfection agent due to their ability to bind and condense large nucleic acids into stabilized nanoparticles. Polymers with primary, secondary, tertiary, and quarternary amines as well as other positively charged group like amidines are particularly useful for this purpose. Examples of cationic polymers investigated for gene delivery are polyethylenimine (PEI), polyamidoamine or starburst dendrimer, imidazole-containing polymers, cyclodextrin-containing polymers, and membrane-disruptive polymers (polyethylacrylic acid/PEAA, methylacrylic acid/MAA, polyacrylic acid/PAA, and polypropylacrylic acid/PPAA).

Unlike the ionic complexes, nucleic acid conjugates with cell-penetrating peptides, carbohydrates, and lipid molecules have been used for improved delivery with moderate success.

Liposomes and lipid-based formulation have been mostly used for systemic delivery of oligonucleotides in clinical applications. The disadvantages of using synthetic vectors are low transfection efficiency (polyplexes), reproducibility, toxicity to some cell type's in vitro and in vivo, and colloidal stability upon systemic administration.

# TARGETED DRUG DELIVERY

## Magic Bullet

Paul Ehrlich coined the term "Magic Bullet."[60,61] He envisaged a treatment of pathogens and toxins in the human body by means of a chemical substance, which is equipped with high affinity for the causative agent. Moreover, this substance should be efficacious in a concentration that is harmless for patients. While screening trivalent arsenic compounds for their potency on *Treponema pallidum*, the causative agent for syphilis, he discovered Salvarsan (magic bullet), which killed syphilis organisms in most cases without killing the host. Although the concept of the magic bullet was introduced 100 years ago, the challenge of making drugs with

**KEY CONCEPT** TARGETED DRUG DELIVERY

The main goal of *targeted drug delivery* is to optimize drug's therapeutic index (the ratio of the therapeutic dose to the toxic dose) by strictly localizing its activity at the target (diseased) site.[62] Drugs can be targeted to specific organs (organ targeting), systems (systemic targeting), cells (cellular targeting), or specific intracellular organelles, or molecules (molecular targeting). Drug targeting could be achieved by physical, biologic, or molecular systems that result in high concentrations of pharmaceutically active agents at the targeted site, thus lowering its concentrations in the rest of the body. Successful drug targeting results into significantly lower drug toxicities, reduced doses, and increased efficacy.

selective toxicity (in other words, targeting) has not been broadly achieved.

Drugs administered by routine parenteral administration are distributed throughout the body and reach nontarget (normal/healthy) organs/tissues leading to possible toxic side effects and low efficacy of treatment.[62] Besides, there is a possibility of drug metabolism in the liver or other organs and excretion by kidney. As a result, only small fraction of the administered drug dosage will reach the target (diseased) organ or tissues. Targeted drug delivery[62] aims to overcome limitations associated with routine drug administration by delivering drugs specifically to diseased cells and tissues while not exposing healthy tissues. Ensuring minimal drug loss during the transit to the target site, protecting the drug from metabolism and premature clearance, retaining the drug at the target site for desired period of time, facilitating the drug transport into the cell, and delivering the drug to the appropriate intracellular target site are other requirements of targeted drug delivery. Last but not least, these targeted drug delivery systems should be biocompatible, biodegradable, and nonantigenic.

## Types of Drug Targeting

Drug targeting approaches are grouped into two major categories: (*a*) active targeting and (*b*) passive targeting. A brief overview of major active and passive approaches is provided below.[63]

## Active Targeting

Active targeting is achieved by binding to cell surface antigens and receptors or membrane transporters. A brief overview of different active targeting strategies is provided below.

### Carrier-Linked Prodrug Strategy for Targeting Cell Surface Antigens and Receptors

This strategy aims to develop prodrugs by conjugating drug molecules to monoclonal antibodies (mAbs) or ligands for

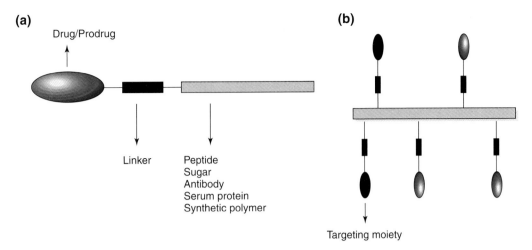

**(a)**

Drug/Prodrug

Linker

Peptide
Sugar
Antibody
Serum protein
Synthetic polymer

**(b)**

Targeting moiety

**Fig. 23–11.** General design of carrier-linked prodrug/conjugate for active targeting.

specific interaction with antigens or receptors expressed on target cell surface. These cell surface targets are distinguished into two categories: noninternalizing and internalizing. In noninternalizing systems, the drug conjugate is cleaved extracellularly, whereas in internalizing systems drug is cleaved intracellularly after endocytosis.

Active targeting to antigens/receptors is achieved by employing carrier-linked prodrug strategy[64] as shown in **Figure 23–11.** The delivery system has three components: (a) drug, (b) carrier, and (c) the homing device or the targeting moiety. Carrier-linked prodrugs are obtained by conjugating the drug molecules to low- to high–molecular-weight molecules (carriers) like sugars, growth factors, vitamins, antibodies, peptides, and synthetic polymers that can transport the drugs to the target site and subsequently release them there. The drug release in most of the prodrugs is accomplished by conjugating the drug to the carrier through a spacer that incorporates a predetermined breaking point, which allows the drug to be released at the cellular target site. It is achieved by incorporating linkages sensitive to enzymatic cleavage, acidic pH, hypoxia, or thiol-exchange reactions. Drug release is also accomplished by employing self-immolative linkers.

Several cell-specific receptors are expressed under physiological conditions, which are specific to ligands and therefore could be exploited for targeting.[62] Examples are (a) parenchymal liver cells, which are specific to galactose, polymeric IgA, cholesterol ester-VLDL, and cholesterol ester-LDL; (b) kupffer cells, which are specific to mannose-fucose, galactose, and LDL; (c) liver endothelial cells, which are specific to mannose and acetylated LDL; and (d) leucocytes, which are specific to chemotactic peptide and complement C3b. Receptors may also become available under pathological conditions. Examples include (a) antigenic sites on pathogens (bacteria, viruses, and parasites); (b) infected cells expressing specific antigens; and (c) tumor-associated antigens/receptors.[64]

Monoclonal antibodies are used for active targeting because of their high binding affinity for respective anti-

gens. Most mAbs belong to the immunoglobulins of the IgG class, which is smallest in size but most abundant antibody found in all biological fluids. Several standard chemotherapeutic agents including antifolates, vinca alkaloids, or anthracyclines have been conjugated to mAbs mostly through cathepsin-B sensitive peptide linker or disulfide bond. It was found that these antibody conjugates indeed have selectivity toward the cells that expresses the respective antigens. These conjugates, however, failed in clinical trials because the mAbs were of murine origin and invoked immune responses. The problem of immunogenicity was resolved by the development of chimeric and humanized antibodies that do not carry murine sequences. FDA has now approved five chimeric or humanized antibodies such as rituximab (Rituxan), tratuzumab (Herceptin), alemtuzumab (Campath), bevacizumab (Avastin) and cetuximab (Erbitux) for the treatment of hematological and solid cancers. Of several immunoconjugates evaluated in clinical trials, gemtuzumab ozogamicin (Mylotarg, Wyeth, NJ)[65] has been approved for the treatment of cancer (**Fig. 23–12**). This immunoconjugate consists of humanized anti-CD33 mAb linked to the cytotoxic antibiotic ozogamicin ($N$-acetyl-$\gamma$-calicheamicin). The linker consists of two cleavable bonds. Mylotarg[65] is used for the treatment of CD33+ acute myeloid leukemia in elderly patients who are not eligible for other chemotherapies and who are suffering from their first relapse. Mylotarg demonstrated clinical efficiency in pediatric patients with advanced CD33+ acute myeloid leukemia.

Immunotoxins are antibody conjugates of highly potent drugs (DOX is frequently used) or toxins. Immunotoxins contain a toxin made by plants; insects; or microorganisms and examples include Pseudomonas exotoxin A (PE), diphtheria toxin (DT), and ricin. The primary targets of immunotoxins are tumor cells. BR96-DOX conjugate in an extensively investigated example, where an average of eight molecules of DOX are linked to chimeric mAB BR 96 through an acid-sensitive hydrazone linkage.[66] Promising immunotoxins currently in clinical trials include TransMID 107 (transferrin-CRM107) and PRECISE (IL13-PEI-301-R03).[64]

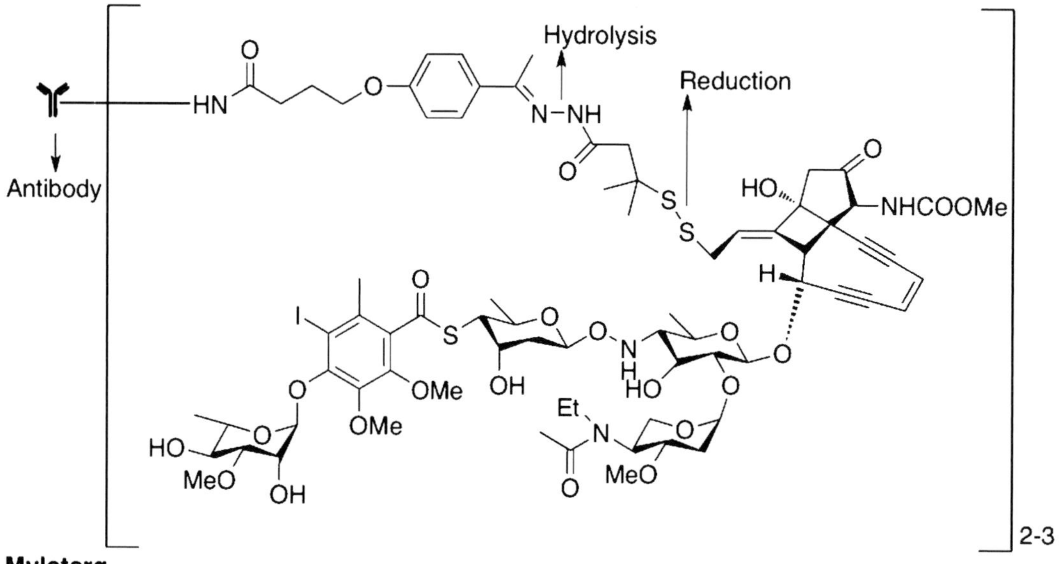

**Mylotarg**

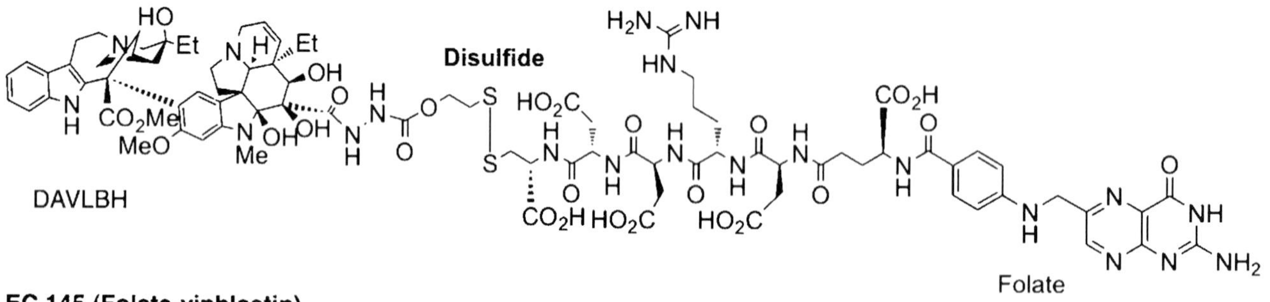

**DAVLBH**

**EC 145 (Folate-vinblastin)**

**Fig. 23–12.** Examples of actively targeted prodrugs: Mylotarg and EC 145.

Besides antigens, cellular receptors also provide targets for prodrug design. Active targeting is achieved by binding drugs to ligands that display high affinity for a particular receptor. The ligands can be low- or high–molecular-weight compounds such as vitamins, peptides, sugars, native or modified proteins, and antibodies. Prodrug is taken up by receptor-mediated endocytosis after binding. The drug is then released in endosomes or lysosomes depending on the route of cellular trafficking of the particular receptor.

Folic acid is one of the most highly used ligands because it retains high affinity for its receptor even after modification with drug/carrier molecules. It is overexpressed in many human types of cancers and a broad spectrum of low- and high–molecular-weight drug folate conjugates with alkylating agents, platinum complexes, paclitaxel, 5-fluorouracil, camptothecin, doxorubicin, and mitomycin has been investigated. Prodrug EC 145,[67] currently in clinical trials, is probably the most promising folate-targeted prodrug (**Fig. 23–12**). It is composed of vinca alkaloid desacetylvinblastine mono-

hydrazide linked to folic acid through reducible disulfide bridge. EC 145 was found to be more active and better tolerated than the free drug in in vivo preclinical studies and showed superior antitumor activity.

Cyclic peptides that bind to integrins can be used to target vascular receptors. Vascular receptor proteins are crucial for the interaction between a cell and the extracellular matrix and they are involved in tumor angiogenesis. Certain integrins ($\alpha_v\beta_3$, $\alpha_v\beta_5$) are overexpressed on proliferating endothelial cells and some tumor cells. Peptides containing the RGD sequence (Arg-Gly-Asp) that are present in extracellular matrix are used to target integrins and subsequently inhibit angiogenesis. A number of RGD-drug conjugates with cytostatic and diagnostic agents have been prepared to obtain the proof of concept.

The asialoglycoprotein receptor (ASGPR) is a membrane-bound lectin expressed on hepatocytes and liver cancer. It has been used for prodrug targeting for the treatment of hepatocellular carcinoma. ASGPR has high affinity for terminal

$\beta$-galactoside or $\beta$-$N$-acetylgalactosamine on glycoproteins and is responsible for the endocytosis of several glycoproteins. The strong interaction of glycoproteins with ASGPR receptor is attributed to the cluster effect (multivalency) in which adjacent saccharide group binds to the receptor with high binding constants. The cluster effect is mainly due to the thermodynamic property of multivalent ligands rather than the presence of multiple receptor binding sites. N-(2-hydroxylpropyl) methyl acrylamide (HPMA)-Gly-Phe-Leu-Gly containing galactosamine (PK2, FCE28069) is the only polymer–drug conjugate bearing a targeting ligand to be tested clinically. PK2 has Mw $\sim$25,000, DOX content ($\sim$7.5 wt%), and galactosamine content of 1.5 mol% to 2.5 mol%. The prodrug showed 30% delivery to the hepatic region in preclinical studies. The prodrug was found to accumulate in tumors also due to the enhanced permeability and retention (EPR) effect and the ratio of tumor tissue to normal liver uptake was 1:3 in 24 hr. The galactose-mediated liver targeting was about 15% to 20% of dose at 24 hr.

The antigens and receptors described here are only representative and there are many more that have been investigated for active drug targeting. The disadvantage associated with antibody–drug conjugates and drug modified with ligands having affinity for particular receptors is that they are not exclusively target-specific and cross-reactivity of drug conjugate with normal tissue is observed.[64]

### Antibody-Directed Enzyme Prodrug Therapy

Antibody-directed enzyme prodrug therapy (ADEPT) is a two-step mechanism for prodrug targeting, where a tumor-associated mAb linked to drug-activating enzyme (usually antibody–enzyme fusion protein) is administered intravenously in the first step, which binds to specific antigen expressed on the tumor cell surface.[68] A nontoxic prodrug is administered systemically in the second step and converted to the cytotoxic drug by the pretargeted enzyme. The enzymes used in ADEPT are divided into following three categories: Class I: enzymes of nonmammalian origin with no mammalian homologues such as alkaline phosphatase and $\alpha$-galactosidase; Class II: enzymes of nonmammalian origin with mammalian homologues such as carboxypeptidase A, $\beta$-glucuronidase, and nitroreductase; and Class III: enzymes of mammalian origin such as $\beta$-lactamase, carboxypeptidase, cytosine deaminase, benzylpenicillin amidase, and phenoxymethyl.

Currently, there are two ADEPT systems in Phase I/II clinical trials with prodrug ZD2767P or $N$-{4-[$N$,$N$-bisamino]phenyloxycarbonyl}-L-glutamic acid. The prodrug is activated by enzyme carboxypeptidase 2 (CPG2) to active drug 4-[$N$,$N$-bis(2-iodoethyl)amino]phenol or phenol bisiodide mustard and is active against colorectal tumors. The first ADEPT system in a clinical trial uses a recombinant fusion of murine anticarcinoembryonic antigen (CEA) F(ab)2 fused to CPG2 in combination with prodrug ZD2767 for the treatment of advanced colorectal cancer and the second ADEPT

study in clinical trials utilizes a recombinant fusion of CEA sFv fused to CPG2, in combination with ZD2767P for the treatment of CEA-expressing tumors.

The ideal drugs for ADEPT are small molecules with the ability to diffuse into the tumor tissues to cause a bystander effect. The bystander effect is defined as the capability to kill the surrounding nondividing/nonexpressing tumor cells and is an important requirement for this type of therapy as it amplifies the drug effect. To avoid systemic toxicity in clinical application, the time interval between enzyme and prodrug administration should be optimized so that the conjugate accumulates only in tumors and not in blood and normal tissues. The target antigen should be either expressed on tumor cell membrane or secreted into the extracellular matrix of the tumor and use of high-affinity mAb is essential. The drug should be dose dependant and cell cycle independent. For effective therapy, the antibody–enzyme conjugate should remain on the cell surface after binding to the respective antigens and it must also be cleared rapidly from the circulation to prevent toxicity.

The major advantage of ADEPT over antibody conjugates is the amplification of the cytotoxic effects due to catalytic activation of prodrug. Another benefit is the ability to kill surrounding tumor cells thereby reducing the risk of tumor evading therapy by antigen loss. The cytotoxic effects of drugs are largely confined to the tumor target and hence the side effects are reduced as compared to systemic administration of chemotherapy. A significant obstacle for ADEPT is the immunogenicity of enzymes used for prodrug activation and the targeting mAb as both were derived from nonhuman sources. This problem has been resolved by the use of human enzymes in conjunction with humanized or human mAbs.

Another analogous approach is "**Lectin-directed Enzyme-Activated Prodrug Therapy**," where glycosylated-enzyme conjugates are administered first, which binds to cells surfaces expressing specific lectin receptors.[69] The enzyme then activates the systemically administered prodrug at the site of action.

### Gene-Directed Enzyme Prodrug Therapy/Virus-Directed Enzyme Prodrug Therapy

Gene-directed enzyme prodrug therapy (GDEPT) is also known as suicide gene therapy and involves physical delivery of a gene for a foreign enzyme (not naturally expressed in the host) to tumor cells by a targeting mechanism that leaves the surrounding noncancerous cells untransformed.[59] The transformed tumor cells express the enzyme, which in turn activates the systemically delivered nontoxic prodrug. Viral vectors are mostly used for gene delivery but they suffer from limited amount and size of plasmid-DNA. There is also the possibility of inducing severe immune responses. This approach is recognized as virus-directed enzyme prodrug therapy (VDEPT) when viral vectors are used for gene delivery. Nonviral vectors such as cationic liposomes are less expensive, easier and safer to make, and suitable for long-time storage, but their gene delivery properties are far from optimal.

An earlier example of GDEPT is herpes simplex virus thymidine kinase and GCV. The drug is phosphorylated by the herpes simplex virus thymidine kinase and then by cellular kinases to produce GCV-triphosphate, which incorporates into the elongating DNA during the cell division (S-phase) and causes inhibition of DNA polymerase and single-strand breaks. These characteristics make HSV TK-GCV useful for eradicating tumor cells invading non-proliferating tissues. Other examples in clinical trials are (a) the purinenucleoside phosphorylase enzyme in combination with 6-methylpurine (prodrug); (b) carboxylesterases enzyme in combination with prodrug irinotecan (CPT11); and (c) cytochrome P450 (CYP450) enzyme in combination with prodrug cyclophosphamide.

For GDEPT to be effective, the expressed enzyme or a related protein should not be present in normal human tissues or expressed only at very low concentrations and must achieve sufficient expression in the tumors to give high catalytic activity. The prodrug should be lipophilic so that it can diffuse into the tumor cells before it can be converted into cytotoxic drug by the suicide enzyme. Alternatively, if the prodrug cleavage takes place extracellularly, the active drug should be capable of diffusing through cell membranes. The drug should be able to kill surrounding nondividing or nonexpressing tumor cells by bystander effect. The advantage of the GDEPT approach is the possibility of delivering target-specific cancer therapy with reduced systemic toxicity resulting in a better prognosis for patients. There are certain theoretical risks associated with GDPET such as insertional mutagenesis, anti-DNA antibody, local infection, and tumor nodule ulceration, which may restrict its use.

### Antibody-Targeted, Triggered, Electrically Modified Prodrug Type Strategy

Antibody-targeted, triggered, electrically modified prodrug type strategy (ATTEMPTS)[70] delivery system comprises large complex made of two components: (a) targeting component consisting of an antibody chemically linked with an anionic heparin molecule and (b) a drug component consisting of the enzyme drug modified with a cationic moiety. The two components are linked through a tight but reversible electrostatic attraction. The cationic species conjugated to the enzyme is relatively small (positively charged peptide) and hence the enzyme conjugates retain its catalytic activity. However, this enzyme conjugate is unable to exert its catalytic activity because it is bound to antibody–heparin conjugate via electrostatic bonds. The ATTEMPTS complex is delivered to the targeted site by the attached antibody and the enzyme drug is released at the site by using a triggering agent such as protamine. Protamine is a heparin antidote, which binds to heparin more strongly than most of the cationic species. The released enzyme is then concentrated at the site of action thereby maximizing its catalytic activity toward drug conjugate at the targeted site while minimizing its toxic effects toward the normal cells. The selection of an appropriate cationic moiety is key to success of ATTEMPTS strategy because retention of prodrug after administration and its conversion to active drug relies on the binding strength of modified enzyme toward heparin. An important aspect of this approach is that both chemical and biological methods can be used to insert the cationic moiety. Yang and colleagues[70] modified the tissue plasminogen activator (t-PA) with a cationic species ($Arg_7$-Cys-) and rendered it inactive by electrostatic binding with negatively charged heparin–antifibrin antibody conjugate. After targeting the complex to the target site, t-PA activity was restored by administration of protamine (heparin antidote).

The approach could be important because prodrug activation in ADEPT depends on chemical conjugation and enzyme cleavage, respectively. ADEPT is therefore restricted to small-molecule drugs only and macromolecular drugs such as proteins are not suitable candidate for ADEPT.

### Membrane Transporters

Membrane transporters are integral plasma membrane proteins that mediate the uptake of different substrates including, polar nutrients, amino acids and peptides, nucleosides, and sugars.[63,71] They fall into two major families: (a) the ATP-binding cassette family, which includes transporters responsible for drug resistance through efflux transporters like Pgp and (b) the solute carrier (SLC) transporters, which are capable of influx into the cell. An example of the solute carrier transporters is nucleoside transports, which are responsible for the uptake of nucleosides, the precursors of nucleotides. Transporters can also be selective for different classes of substrates as is the case for the transporters of purine and pyrimidines. Often, transporters require the flux of a secondary substrate, like the peptide transporters PEPT1 and PEPT2, which require the influx of $H^+$ to facilitate their uptake functions. Because many transporters have nutrients as substrates, prodrug design is manipulated to mask the drug with nutrient moiety so as to initiate prodrug uptake through these transporters (**Fig. 23–13**).

The use of membrane transporters as prodrug targets has largely stemmed from the discovery of the absorption mechanism of valacyclovir through the PEPT1 transporter.[72] The addition of the L-valyl ester to the parent drug acyclovir vastly improves the bioavailability of acyclovir. Upon absorption, valacyclovir is rapidly converted to its parent drug acyclovir via esterase cleavage. Acyclovir is then free to enter the bloodstream, where it is taken up by cells via nucleoside transporters. Valacyclovir is the first example of a novel peptide ester prodrug that actively targets the human transporters, PEPT1, for increased oral absorption. Besides the obvious benefit of increased oral absorption, this method of drug delivery has been tailored to target cancer cells over expressing certain membrane transporters. The malignant ductal pancreatic cancer cell lines AsPc-1 and Capan-2 and the human fibrosarcoma cell line HT-1080 overexpress the PEPT1 membrane transporter. It has been suggested that if

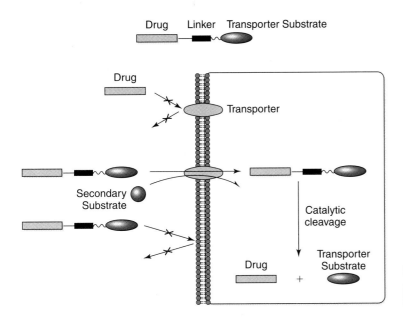

**Fig. 23–13.** Schematic representation of membrane transporter targeted prodrug uptake. (Source: H.-K. Han and G. L. Amidon, AAPS Pharm. Sci. **2,** 1, 2000.)

tumor cell uptake of hydrophilic polymer drug conjugates via a specific mechanism of internalization can be achieved; these drugs may avoid the development of or elimination due to multidrug resistance.

The advantage of using transporters like PEPT1 is the potential for increased oral absorption. Thus, prodrugs utilizing the membrane transporters may be afforded the luxury of an oral formulation, which is the gold standard of administration. Amino acid side chains could be varied to alter the physicochemical properties of prodrug. The broad specificity of these transporters for multiple substrates increases the potential flexibility in prodrug design. Also because the transporter-targeting substrates are generally sugars, vitamins, or peptides, the byproducts of the prodrug's conversion yield nontoxic nutrients. The disadvantage associated with this approach is that these transporters are also expressed in cells of the small intestine, kidney, bile duct, and pancreas. Therefore, if PEPT1 substrates are used as targeting moieties, toxicity in these cell types may occur. Another problem could arise if the prodrug upon absorption by the transporter is quickly converted to parent drug. Once this occurs, the parent drug will be passively distributed throughout the systemic circulation, which could in turn cause systemic toxicity.

## Passive Targeting

Passive targeting utilizes the natural or passive distribution characteristics of a carrier for drug targeting and no homing device is attached. Some of the major passive-targeting approaches are briefly described below.

### Mononuclear Phagocyte System

Particulate carriers are phagocytosed by the cells of mononuclear phagocyte system (MPS), leading to major accumulation in the liver and the spleen[62] (**Fig. 23–14**). After phagocytosis, the particulate drug/carrier complex is transported to lysosomes, where the complex is disintegrated to release the drug. If the complex is not broken down in the lysosomes, it may be released from the lysosomal compartment into the cytoplasm and may even escape from phagocyte causing a prolonged release systemic effect. The ability of macrophages to rapidly phagocytose particulate drug/carrier complexes could be diminished by grafting PEG chains on the surfaces of particulate material.

Passive targeting to MPS could be used for the treatment of macrophage-associated microbial, viral, or bacterial diseases and lysosomal enzyme deficiencies.

### Enhanced Permeability and Retention (EPR) Effect

Prodrugs could be passively targeted to tumors by exploiting the EPR phenomenon[73] (**Fig. 23–14**). Angiogenesis is induced in tumors to accommodate their ever-increasing demand for nutrition and oxygen as the tumor cells multiply and cluster together to reach the size of 2 to 3 nm.[74] Unlike the normal tissue, the blood vessels in tumors become irregular in shape, dilated, leaky, or defective, and the endothelial cells become poorly aligned and disorganized with large fenestrations. The perivascular cells and the smooth muscle layers are frequently absent or abnormal in the vascular wall. Tumor vessels develop wide lumens and the lymphatic drainage in tumor tissues becomes impaired. This results into extensive leakage of blood plasma components such as macromolecules, nanoparticles, and lipidic particles in tumor tissues. These macromolecules and nanoparticles are retained in tumors due to the slow venous return in tumor tissues and poor lymphatic drainage. Longer plasma residence time and neovasculature are known to influence the EPR. This phenomenon has been shown to achieve 10- to 50-fold (1%–5% and in some cases 20% of injected dose per gram of tumor) high local concentration of drugs in the tumor tissues than in normal tissues within 1 to 2 days of injection.

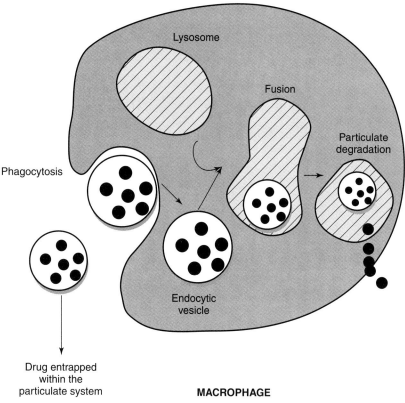

**MACROPHAGE**

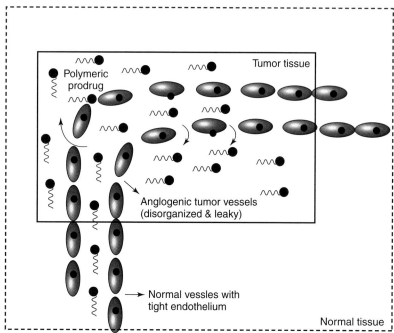

**EPR EFFECT**

**Fig. 23–14.** Passive targeting approaches: mononuclear phagocyte system (MPS) and enhanced permeability and retention (EPR) effect. (Modified from D. J. A. Crommelin et al., Drug Delivery and Targeting for Pharmacist and Pharmaceutical Scientists, CRC Press, Boca Raton, FL, 2001, pp. 117 and R. Duncan, Nat. Rev. Drug Discov. **2,** 347, 2003.)

Passive targeting through the EPR effect is achieved by attaching the drug to macromolecules (e.g., synthetic or biopolymers) or nanoparticles (e.g., liposomes, nanospheres) that act as inert carriers. These macromolecules or nanoparticles do not interact with tumor cells but strongly influence the drug biodistribution. Molecular weight is absolutely impor-

tant but not the sole criterion for predicting the molecule's biodistribution. The chemical nature of polymer, as well as shape and conformation in water, also influences its molecular size. Attachment to polymers results into improved water solubility, prolonged stay in blood circulation, and reduced toxicities. Polymer used for such applications should

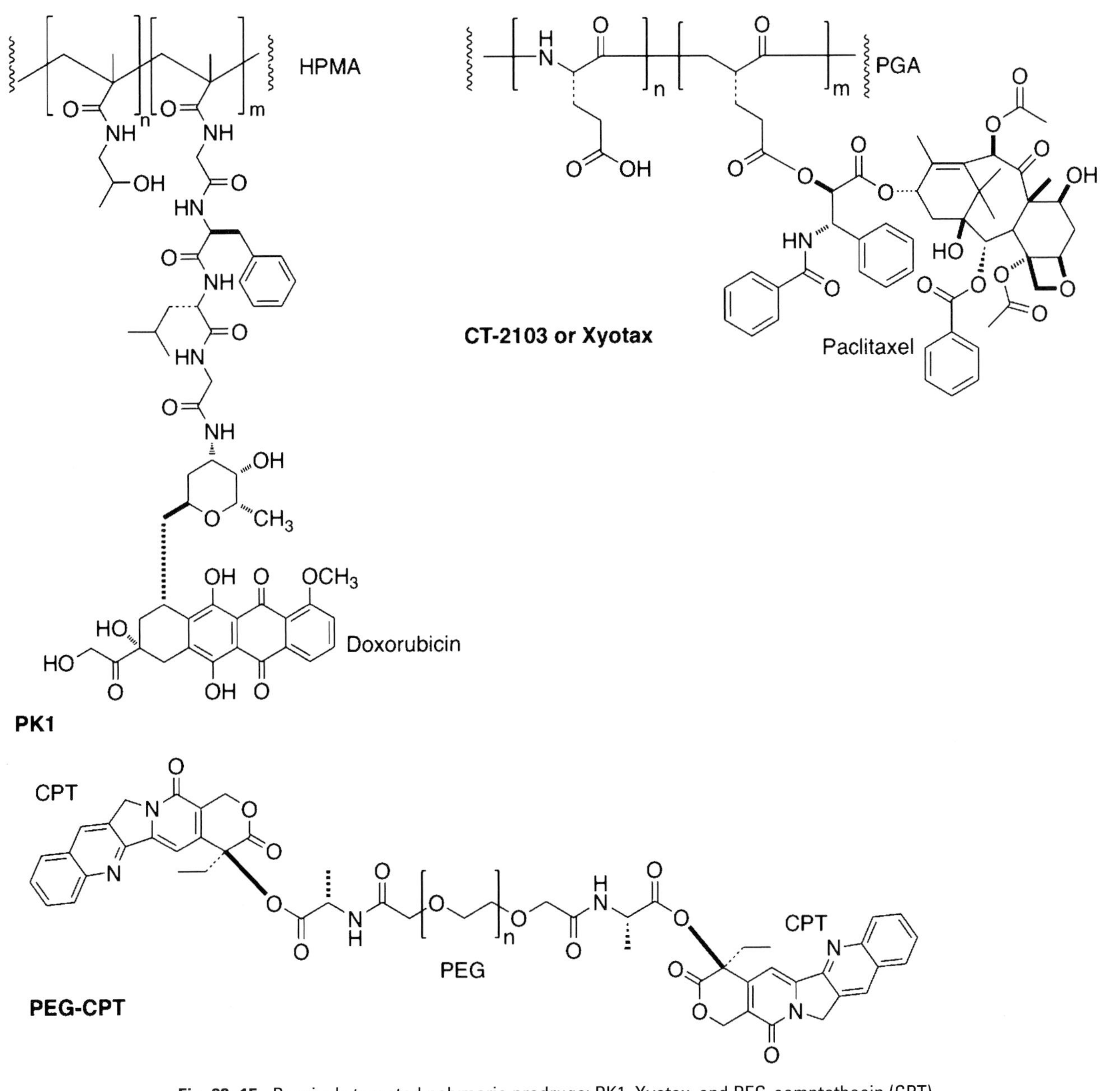

**Fig. 23–15.** Passively targeted polymeric prodrugs: PK1, Xyotax, and PEG-camptothecin (CPT).

be biocompatible (nontoxic, nonimmunogenic, preferably biodegradable), able to carry the required drug payload, able to protect the drug against premature metabolism, display active/passive targeting, able to liberate the drug at a rate appropriate to its mechanism of action, and able to enter tumor cells by endocytosis (if designed for lysosomotropic release).

Several passively targeted polymeric prodrugs are being evaluated in clinical trials[75,76] (**Fig. 23–15**). Examples are (a) PK1 or FCE28068 (HPMA linked to doxorubicin through tetrapeptide linker); (b) ProLindac (HPMA linked to diaminocyclohexane platinum [II]); (c) Xyotax or CT-2103 (polyglutamate or PGA attached to paclitaxel through ester bond); (d) CT-21006 (PGA attached to camptothecin);

and (e) EZN-2208 or PEG-SN38 (PEG attached to 7-ethyl-10-hydroxycamptothecin). Genoxol-PM is paclitaxel-loaded biodegradabe polymeric micelle of PEG-PLA.

Another example is styrene-malic-anhydride-neocarcinostatin systems (SMANCS) obtained by conjugating neocarcinostatin (NCS, ~12 kDa protein) to two poly(styrene-co-maleic anhydride copolymer. NCS is a small protein, which is rapidly cleared by the kidney and shows nonspecific cytotoxicity. Styrene-malic-anhydride-neocarcinostatin, which are targeted possibly by EPR, show improved pharmacokinetic properties. Clinical successes have been reported in patients with hepatocellular carcinomas. Dendritic structures and particulate materials are also being explored for passive targeting of prodrugs.

### Polymer-Directed Enzyme Prodrug Therapy and Polymer-Directed Enzyme Liposome Therapy

The polymer-directed enzyme prodrug therapy (PDEPT) and polymer-directed enzyme liposome therapy (PELT) also exploit EPR for drug targeting.[75,76] PDEPT is a two-step approach and involves the initial administration of the polymeric prodrug to promote tumor targeting followed by administration of the activating polymer-enzyme conjugate. The process utilizes the EPR effect to target the polymeric prodrug as well as polymer enzyme conjugate. The feasibility of PDEPT for targeted delivery is being evaluated using following: (i) PK1 prodrug and HPMA copolymer-cathepsin B enzyme conjugate; and (ii) HPMA-methacryloyl-Gly-Gly-cephalosporin-doxorubicin prodrug and HPMA-methacryloyl-Gly-Gly-$\beta$-lactamase enzyme conjugate. In another strategy, known as polymer-directed enzyme liposome therapy, liposomes (e.g., HPMA-phospholipase) are used to deliver the prodrug (improved EPR) followed by the polymer enzyme.

### Particulate Carriers in Drug Targeting

Besides the soluble prodrugs/conjugates, particulate carriers have also been used in the targeted drug delivery.[62,77] Advantages associated with particulate carriers are (a) high drug payload, (b) possibility of both covalent and ionic association between the drug and the carrier, and (c) high degree of protection available to drug after encapsulation. Several solid-particulate nanosuspensions are in the market; examples include sirolimus (Rapamune) (immunosuppressant, Wyeth) and aprepitant (Emend) (antiemetic, Merck). Several examples of particulate carriers are briefly described below.

#### Liposomes

Liposomes are vesicular structures based on one or more lipid bilayers encapsulating an aqueous core (**Fig. 23–16**).[62,77] The lipid molecules are usually naturally occurring or synthetic phospholipids, amphipathic moieties with a hydrophilic polar head group along with two nonpolar hydrophobic chains

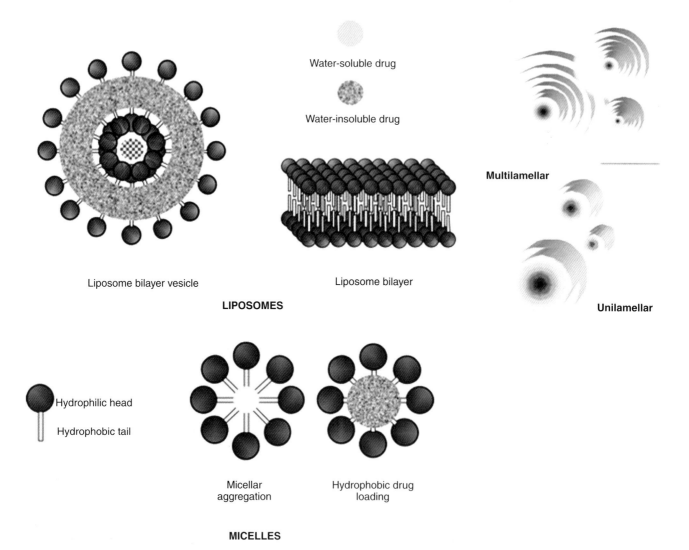

**Fig. 23–16.** Particulate carriers in drug targeting: liposomes and micelles.

(tails). These molecules spontaneously arrange themselves in water to give the thermodynamically most stable bilayer structures, where the hydrophilic head groups protrude outside into the aqueous environment, and the hydrophobic chains orient themselves inward, away from water. The flat bilayers self-close into concentric compartments around a central aqueous phase to give spherical liposomes with diameters in the range of 0.02 to 20 $\mu$m.

The liposomes are either unilamellar (one concentrically oriented bilayers around an aqueous core) or multilamellar (multitude of concentrically oriented bilayers around aqueous core) (**Fig. 23–16**). When the liposomes are multilamellar, water may be present in the aqueous core and also between the bilayers. Multilamellar vesicles are formed under low shear agitation with wide size distribution, display relatively low level of aqueous encapsulation, and have relatively short circulation half-lives. The smaller unilamellar vesicles on the other hand have narrow size distribution and therefore preferred for intravenous applications where a long circulation time is demanded. They are also useful for encapsulation of water-soluble drugs but suffer from tendency to aggregate into larger liposomes. Multilamellar vesicles are readily converted to unilamellar vesicles by employing high shear processes such as sonication, homogenization, and extrusion.

Different types of liposomes have been used in drug delivery[62]: (a) *liposomes*, which are neutral or negatively charged and used for passive targeting to the cells of MPS; (b) *stealth liposomes*, which are sterically stabilized liposomes carrying hydrophilic coating (PEG) for longer circulation times; (c) *immunoliposomes*, which contain specific antibody or antibody fragment for active targeting; and (d) *cationic liposomes*, which are positively charged and used for gene delivery.

Liposomes are capable of incorporating both water-soluble (inside the aqueous core, e.g., daunorubicin) and water-insoluble drugs (within the bilayer, e.g., amphotericin B). They are biodegradable and nontoxic and proven to improve pharmacokinetic properties of the drugs. Injectable anticancer drugs, which exploit liposome technology, are doxorubicin (Doxil, Alza, CA, 1995), daunorubicin (Daunoxome, Gilead, CA, 1996), and cytarabine (Depocyt, SKYE Pharma, UK, 1999). The technology has also been used for topical antifungal products such as Amphotericin B (Ambisome, Gilead, CA, 1997). Liposomes have been used in pulmonary drug delivery. Lung fluids have a pH of ~6.8 and low protease or lipase activities and, therefore, liposomes are expected to be stable during administration. The drug release from the liposomes depends on the nature of phospholipids, composition of the liposomes, and hydrophilic or lipophilic properties of active ingredients. Multilamellar vesicles are used for nebulization. During the nebulization, the shear force generated by extrusion through the jet orifice reduces the liposome size to 0.2 to 0.3 $\mu$m.

Liposomes suffer from drawbacks such as (a) physical and chemical instability in liquid state; (b) low encapsulation efficiency for several drugs; and (c) challenging scale-up and sterilization of the final formulation. Stability of liposomes is dependent on the lipid composition, storage conditions (light, oxygen, temperature, moisture), and stabilizers (cholesterol, $\alpha$-tocopherols and inert atmosphere) used, and in some liposomal formulations, physical and chemical stability can be improved by lyophilization. Another major concern is the short half-life of liposomes in blood circulation, which can be improved by "stealth" technology (PEGylation).

## Micelles

Surfactants molecules aggregate in aqueous solution to form micelles at certain concentrations and temperature[62] (**Fig. 23–16**). Surfactants have a hydrophilic polar head group attached to a long-chain lipophilic (nonpolar) tail. Surfactant molecule used for micelles formation could be anionic (sodium dodecyl sulfate or SDS and deoxycholic acid); cationic (hexadecyltrimethyl ammonium bromide); zwitterionic (lecithin or phosphatidylcholine); or nonionic (methyl cellulose and other lipophilic cellulose derivatives). Block copolymers comprising hydrophilic and hydrophobic segments are used to form polymeric micelles. Micelles are formed only when surfactants are present above a certain concentration, known as critical micelle concentration (CMC), which is characteristic for each surfactant. There is also a critical temperature requirement for micelle formation. A high CMC value suggests a rapid exchange of constitutive components and a fast disintegration of the micelles upon dilution, whereas a low CMC value suggests the contrary.

Micelles are used for reducing the surface tension of water, increasing the miscibility of different solvent phases, and stabilizing the emulsions. Micelles used in targeted drug delivery should be of low CMC so that it is stable in blood circulation and does not disintegrate upon contact with blood components. The diameter of the micelles could be chosen in the range where EPR effect is expected to occur (0.2 $\mu$m), to allow for accumulation of drug-loaded micelles in tumors or inflammation sites. Micelles obtained from amphipathic block copolymers consisting of hydrophilic PEG block and hydrophobic doxorubicin-conjugated poly(aspartic acid) or poly($\beta$-benzyl-L-aspartate) have been extensively investigated. These drug-loaded block copolymers form micelles in water with spherical core/shell structure with drugs present in hydrophobic core.

## Polymersomes and Dendrimers

Polymersomes are polymer vesicles with a core-shell structure similar to liposomes. They are made of diblock copolymers, which contain hydrophilic and hydrophobic portions similar to phospholipids. Since polymersomes are stronger and more stable than liposomes, they display less deformation under load and slower rate of drug leakage. The degree of polymerization and melting temperature ($T_g$) of the polymer are varied to control the vesicle-like properties such as rigidity, thickness, and permeability. However, polymersomes are not biocompatible, their degradation products are usually toxic, and the drug release from these platforms is generally too slow.

Polymeric dendrimers on the other hand are treelike or star-shaped polymers that adopt a quasispherical shape. Drugs are incorporated into the internal cavities or attached through the surface functional groups. The main use of polymeric dendrimers is to enhance aqueous solubility of the poorly soluble drugs. Toxicity of the dendrimic polymers remains a major concern. Similar to polymersomes, polymeric dendrimers too are not commercially available.

Lipoproteins, which are endogenous lipid carrier systems comprising a lipid core and a coat where apolipoprotein is found, have also been used for targeted drug delivery. Examples are chylomicrons (10–90 nm); very low-density lipoprotein (VLDL, 30–90 nm); low-density lipoprotein (LDL, ~25 nm); and high-density lipoproteins (HDL, ~10 nm). Other common examples of particulate carriers are albumin microspheres, poly(lactide-co-glycolide) or PLA microspheres, and niosomes.

## PRODRUG APPROACHES[78,79]

Conventional prodrug design aims to improve pharmacokinetic and pharmacodynamic properties of a drug by chemically altering its structure, which is usually achieved by attaching a promoiety to the drug through reversible (enzymatic/nonenzymatic) bonds (**Fig. 23–17**). The most frequently used chemical linkages[78] (bonds) in prodrug design are ester, carbonates, carbamates, amides, phosphates, and oximes. Linkages such as thioethers, thioesters, imines, and Mannich bases have also been used but to a lesser extent. The chemical groups that can produce the aforementioned linkages are carboxylic acid, hydroxyl, amine, phosphate/phosphonate, and carbonyl.

Esters are the most common prodrug linkages constituting about half of marketed prodrugs.[78] They are obtained by attaching the promoiety to the water-soluble charged group (e.g., carboxyl) on the drug through ester bonds and aim to improve the lipophilicity or membrane permeability of the parent drug. Once in the body, the ester bond is cleaved

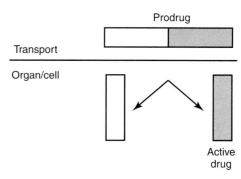

**Fig. 23–17.** The conventional prodrug design, where inactive prodrugs are transformed into active drugs inside the cell.

by ubiquitous esterases (blood, liver, and other organs) to release the active drug. Contrary to esters, phosphate ester prodrugs are prepared to enhance the aqueous solubility of parent drugs for achieving more favorable oral or parenteral administration.[78] These display adequate chemical stability and are readily converted to active drugs by phosphatases present at intestinal brush border and in the liver. The amide prodrugs are relatively more stable in vivo and hydrolyzed to active drugs by enzymes such as carboxylesterases, peptidases, and proteases. The carbonates and carbamate prodrugs are converted to active drugs by esterases, whereas oxime prodrugs are converted to active form by microsomal cytochrome P450 (CY450). Enzymatic cleavage is not the only mechanism available for prodrug activation; physicochemical environment at the target site (e.g., acidic pH, hypoxia, glutathione-based thiol-exchange reactions) is also exploited for prodrug activation. Activation is also achieved by incorporating self-immolative linker in the prodrug design.

Attachment to promoiety through reversible bonds has been used to obtain prodrugs with improved lipophilicity or membrane permeability[78,79] (e.g., enalapril, pivampicillin, adefovir dipivoxil, tenofovir disoproxil, famciclovir), aqueous solubility (e.g., sulindac, fosamprenavir, estramustine phosphate, prenisolone phosphate), and parenteral administration (fosphenytoin, foflucanazole, propofol phosphate

## ▶ KEY CONCEPT ◀ PRODRUGS

The term *prodrug* is used to characterize pharmacologically inert drug derivatives that can be converted to active drug molecules in vivo, enzymatically or nonenzymatically, to exert a therapeutic effect (**Fig. 23–17**).[78,79] It therefore implies a covalent linkage between the drug and the chemical moiety (also called *promoiety*) causing the inertness. The conventional prodrug design aims to overcome (*a*) pharmaceutical problems such as poor solubility, insufficient chemical stability, unacceptable taste or odor, and irritation or pain; (*b*) pharmacokinetic problems such as insufficient oral absorption, inadequate BBB permeability, marked presystemic metabolism, and toxicity; and (*c*) pharmacodynamic problems such as low therapeutic index and lack of selectivity at the site of action. Prodrugs have been earlier referred to as reversible or bioreversible derivatives, latentiated drugs, and

biolabile drug carrier conjugates but the term prodrug is now standard. About 10% of the drugs approved worldwide can be classified as prodrugs.

In some cases, a prodrug may consist of two pharmacologically active drugs coupled together into a single molecule so that each drug becomes a promoiety for the other; such derivatives are called *codrugs*. A *bioprecursor prodrug* is a prodrug that does not contain a carrier or promoiety but results from molecular modification of active agent itself (active metabolite). *Soft drugs*, which must not be confused with prodrug, are active drugs designed to undergo a predictable and controllable deactivation or metabolism in vivo after achieving the therapeutic effect. Prodrugs also differ from conjugates in that they are reversible.

etc.). Prodrugs for improved ophthalmic (dipivefrin [propine] and latanoprost) and dermal (tazarotene) delivery and diseases such as Parkinson disease (levodopa), viral (pradefovir), asthma (bambuterol), hypercholesterolemia (simvastatin), and cancer (e.g., capecitabine [Xeloda] and ftorafur [Tegafur]) have also been developed (**Table 23–3**).

A few specific prodrug examples are discussed below to illustrate the salient features of conventional prodrug design. Enalapril is a prodrug used for lowering the blood pressure, congestive heart failure, and kidney problems (**Fig. 23–18**). It is an ethyl ester prodrug of enalaprilat, which suffers from low oral bioavailability (36%–44%). Conversion to ester enhances the absorption from 53% to 74% and the prodrug is readily converted to active form by hydrolysis of ester in vivo, which inhibits angiostensin-converting enzyme. Many prodrugs undergo site-selective activation too, common examples being the lipid-lowering bioprecursor prodrugs simvastatin (Zocor, Merck, NJ) (**Fig. 23–18**) and lovastatin (Mevacor, Merck, NJ) used for the treatment of hypercholesterolemia. Both prodrugs are administered in their inactive hydrophobic lactone form, which are then converted to active $\beta$-hydroxyacid form by CYP450 enzymes, mainly in liver. The active form inhibits the 3-hydroxy-3-methylglutaryl coenzyme A reductase (HMG-CoA), which is involved in cholesterol biosynthesis.

Another example is capecitabine (Xeloda, Roche, Switzerland), which is an orally administered tumor-selective carbamate prodrug of 5′-deoxy-5-fluorouridine (5′-FU) and used for the treatment of breast, colorectal, and gastric cancer (**Fig. 23–18**). The conversion of capecitabine to active drug is achieved in three-steps: (a) a hepatic carboxylesterase converts it into 5′-deoxy-5-fluorocytidine, (b) 5′-deoxy-5-fluorocytidine is converted to 5′-deoxy-5-fluorouridine by cytidine deaminase enzyme in the liver/tumor, and (c) tumor-associated enzyme thymidine phosphorylase converts 5′-deoxy-5-fluorouridine to 5-fluorouracil, which in turn is converted to 5′-fluoruridine or 5′-fluoro-2-deoxyuridine. The 5′-fluoro-2-deoxyuridine is incorporated into the RNA and DNA, respectively. This prodrug demonstrates satisfactory GI absorption (~100% bioavailability) and low GI toxicity when compared to the parent drug.

Conventional prodrugs are associated with several limitations, most important being their nonspecific activation inside the body. Current prodrug designs are therefore highly focused on the development of targeted prodrugs, where targeting is achieved by employing either active or passive targeting strategies described earlier (see section Targeted Drug Delivery).

Carrier-linked design has been employed to obtain prodrug targeted to cell/tissue-specific antigens and receptors (**Fig. 23–11**). A recently approved example in this category is gemtuzumab ozogamicin (Mylotarg, Wyeth)[65] (**Fig. 23–12**). Prodrug systems designed to exploit more advanced active targeting strategies such as ADEPT, GDEPT, and ATTEMPTS are in various stages of development and have been described earlier (see section Targeted Drug Delivery). Examples of prodrugs exploiting carrier-mediated transport (**Fig. 23–13**) are valacyclovir (Valtrex, GlaxoSmithKline, UK) (**Fig. 23–18**) and valganciclovir (Valcyte, Roche). These are L-valyl esters (promoiety: amino acid valine) of acyclovir and GCV, both drugs having limited and variable bioavailability. The prodrugs on the other hand show 3 to 10 times high intestinal permeation, which is mediated by di and tripeptide (hPEPT1) membrane transporters. Following membrane transport, the active drug is released by intracellular hydrolysis. Levodopa is a substrate for neutral amino acid transporter (LAT1) at the BBB. After penetrating the BBB, levodopa is decarboxylated to dopamine, which can act locally, as it is no longer a substrate for amino acid transporter. Other prodrug examples in this category are midodrine (hPEPT1) and XP13512 (MCT1 and SMVT).

Passively targeted prodrugs are also being investigated in clinical trails.[75,76] For example, anticancer prodrugs have been obtained by conjugating (covalent attachment) larger molecules (synthetic or biopolymers) or micro/nanoparticles (liposomes, nanospheres) to active drugs. These macromolecules or particles do not interact with the target tumor cells but strongly influence the drug biodistribution due to the EPR effect. Examples of passively targeted prodrugs and prodrug systems used in advanced passive targeting strategies such as PDEPT and PLEPT have been described earlier (see section Targeted Drug Delivery).

## CONTROLLED DRUG DELIVERY

The drug concentration in the plasma does not remain constant and follows a "sawtooth" kinetic profile, where the drug concentration fluctuates between maximum and minimum (**Fig. 23–19**). As a result, the drug level may rise too high

---

**KEY CONCEPT** **CONTROLLED DRUG DELIVERY**

*Controlled drug delivery*, also known as *rate controlled drug delivery*, is defined as the delivery of drug or active agent in the body at a predetermined rate.[80] A controlled drug delivery system is therefore one that provides some control over the drug delivery in the body: temporal or spatial or both. Controlled drug delivery should not be confused with prolonged or sustained drug delivery because the controlled drug delivery attempts to control drug level in the target tissues or cells, whereas the sustained drug delivery is restricted to maintaining therapeutic blood or tissue levels of drug for extended period of time.

**TABLE 23–3**

## FOOD AND DRUG ADMINISTRATION–APPROVED PRODRUGS

| | Prodrug/Product Name | Chemical Modification | Drug Release Mechanism/Location | Drug Name | References[*] |
|---|---|---|---|---|---|
| 1 | Valacyclovir (Valtrex) | L-Valyl ester | First-pass metabolism, esterases/intestine, liver | Acyclovir | a |
| 2 | Valganciclovir (Valcyte) | L-Valyl ester | First-pass metabolism, esterases/intestine, liver | Ganciclovir | b |
| 3 | Azidothymidine (Zidovudine) | Thymidine analogue | Phosphorylation by kinases/infected and uninfected cells | Zidovudine triphosphate | c |
| 4 | Capecitabine (Xeloda) | N4-pentyloxycarbonyl-5′-deoxy-5-fluoro-cytidine | Thymidine phosphorylase/ tumor tissues | 5-Fluorouracil | d |
| 5 | Famciclovir (Famvir) | Diacetyl 6-deoxy derivative | Deacetylation followed by oxidation at the 6 position/liver | Penciclovir | e |
| 6 | Nabumetone (Relafen) | 4-(6-Methoxy-2-naphthyl)-butan-2-one | First-pass metabolism/liver | 6-Methoxy-2-naphthylacetic acid (6-MNA) | f |
| 7 | Pivampicillin (Pondocillin) | Ester | Nonspecific esterases/ gastrointestinal tract | Ampicillin | g |
| 8 | Irinotecan (Camptosar) | 7-Ethyl-10[4-(1-piperidino)-1-piperidino] carbonyloxycamp-tothecin | Human carboxylesterase/liver | SN-38 | h |
| 9 | Terfenadine (Seldane) | Alkyl derivative | Cytochrome P-450 3A4/hepatic first-pass metabolism | Fexofenadine (Allegra) | i |
| 10 | Enalapril (Vasotec) | Ester | Hydrolysis by esterase/liver | Enalaprilat | j |
| 11 | Ramipril (Altace) | Ester | Esterase/liver | Ramiprilat | k |
| 12 | Dipivefrin (Propine) | Pivalic acid ester | Esterase/human eye | Epinephrine | l |
| 13 | Omeprazole (Prilosec) | Sulfonamide | Cytochrome P-450/parietal cells | Omeprazole/hfill Sulfonamide | m |
| 14 | Sulfasalazine (Azulfidine) | Azo modification | Bacterial azo reduction/colon | 5-Amino salicylic acid | n |
| 15 | Olsalazine (Dipentum) | Azo modification | Diazo-bond cleavage by colonic microflora/colon | 5-Amino salicylic acid | o |
| 16 | Methenamine (Urex) | Hexamethylenetetramine | Chemical hydrolysis/urine | Formaldehyde | p |
| 17 | Bambuterol (Bambec) | bis-Dimethyl carbamate | Oxidative metabolism/liver | Terbutaline | q |
| 18 | Allopurinol (Zyloprim) | | Oxidative metabolism/liver | Oxypurinol | r |
| 19 | Gemcitabine (Gemzar) | Dephosphorylated form | Cellular kinases/deoxycytidine kinase | Triphosphate | s |
| 20 | Fludarabine (Fludara) | Dephosphorylated form | Cellular kinases/deoxycytidine kinase | Triphosphate | t |
| 21 | Cladribine (Leustatin) | Dephosphorylated form | Cellular kinases/deoxycytidine kinase | Triphosphate | u |
| 22 | Simvastatin (Zocor) | Lactone | Biotransformation/liver | β-hydroxy acid | v |
| 23 | Tegafur (Ftorafur) | Dehydroxylated form | Cytochrome P-450, thymidine phosphorylase/liver, cytosol | 5-Fluorouracil | w |

[*]a. P. de Miranda and T. C. Burnette, Drug. Metab. Dispos. **22,** 55, 1994; b. F. Brown, L. Banken, K. Saywell, and I. Arum, Clin. Pharmacokinet. **37,** 167, 1999; c. H. H. Chow, P. Li, G. Brookshier, and Y. Tang, Drug. Metab. Dispos. **25,** 412, 1997; d. G. Pentheroudakis and C. Twelves, Clin. Colorectal Cancer **2,** 16, 2002; e. S. E. Clarke, A. W. Harrell, and R. J. Chenery, Drug Metab. Dispos. **23,** 251, 1995; f. R. E. Haddock, D. J. Jeffery, J. A. Lloyd, and A. R. Thawley, Xenobiotica **14,** 327, 1984; g. J. C. Loo, E. L. Foltz, H. Wallick, and K. C. Kwan, Clin. Pharmacol. Ther. **16,** 35, 1974; h. L. P. Rivory, M. R. Bowles, J. Robert, and S. M. Pond, Biochem. Pharmacol. **52,** 1103, 1996; i. B. C. Jones, R. Hyland, M. Ackland, C. A. Tyman, and D. A. Smith, Drug Metab. Dispos. **26,** 875, 1998; j. T. N. Abu-Zahra and K. S. Pang, Drug Metab. Dispos. **28,** 807, 2000; k. S. Tabata, H. Yamazaki, Y. Ohtake, and S. Hayashi, Arzneimittelforschung **40,** 865, 1990; l. J. A. Anderson, W. L. Davis, and C. P. Wei, Invest. Ophthalmol. Vis. Sci. **19,** 817, 1980; m. A. Abelo, T. B. Andersson, M. Antonsson, A. K. Naudot, I. Skanberg, and L. Weidolf, Drug Metab. Dispos. **28,** 966, 2000; n. C. P. Rains, S. Noble, and D. Faulds, Drugs **50,** 137, 1995; o. A. N. Wadworth and A. Fitton, Drugs **41,** 647, 1991; p. R. Gollamudi, M. C. Meyer, and A. B. Straughn, Biopharm. Drug Dispos. **1,** 27, 1979; q. C. Lindberg, C. Roos, A. Tunek, and L. A. Svensson, Drug Metab. Dispos. **17,** 311, 1989; r. Y. Moriwaki, T. Yamamoto, Y. Nasako, S. Takahashi, M. Suda, K. Hiroishi, T. Hada, and K. Higashino, Biochem. Pharmacol. **46,** 975, 1993; s. W. Plunkett, P. Huang, Y. Z. Xu, V. Heinemann, R. Grunewald, and V. Gandhi, Semin. Oncol. **22,** 3, 1995; t. V. Gandhi and W. Plunkett, Clin. Pharmacokinet. **41,** 93, 2002; u. J. Liliemark, Clin. Pharmacokinet. **32,** 120, 1997; v. S. Vickers, C. A. Duncan, I. W. Chen, A. Rosegay, and D. E. Duggan, Drug Metab. Dispos. **18,** 138, 1990; w. T. Komatsu, H. Yamazaki, N. Shimada, S. Nagayama, Y. Kawaguchi, M. Nakajima, and T. Yokoi, Clin. Cancer Res. **7,** 675, 2001.

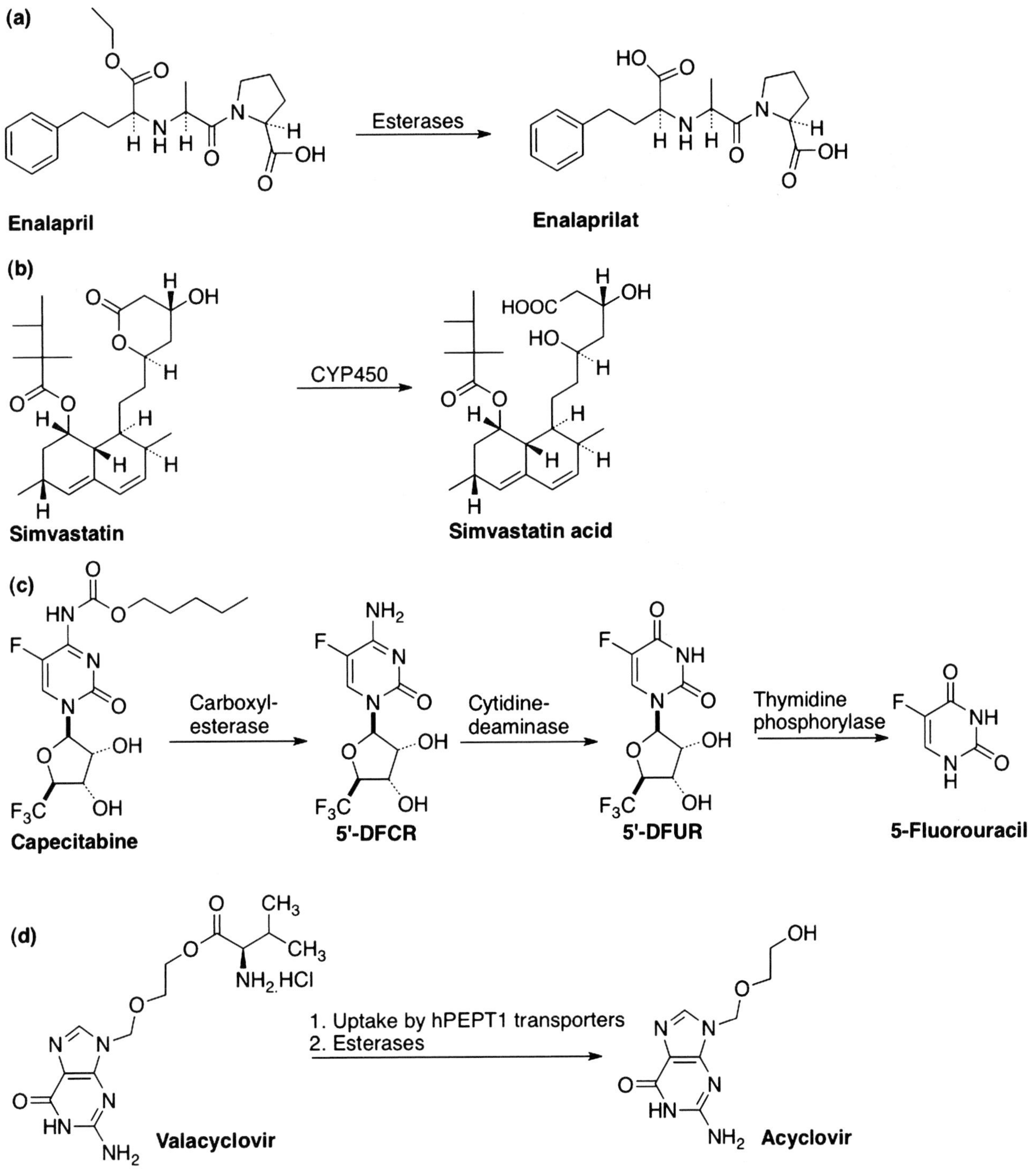

**(a)**

**Enalapril**                              **Enalaprilat**

Esterases

**(b)**

**Simvastatin**                          **Simvastatin acid**

CYP450

**(c)**

**Capecitabine**          **5'-DFCR**          **5'-DFUR**          **5-Fluorouracil**

Carboxyl-esterase          Cytidine-deaminase          Thymidine phosphorylase

**(d)**

**Valacyclovir**                              **Acyclovir**

1. Uptake by hPEPT1 transporters
2. Esterases

**Fig. 23–18.** Prodrug activation of (*a*) enalapril, (*b*) simvastatin, (*c*) capecitabine, and (*d*) valacyclovir.

leading to toxic side effects or fall too low resulting into the lack of efficacy. Frequent dosing is therefore needed to maintain therapeutically effective plasma drug level, more so for drugs with short half-lives, which is likely to result into toxic side effects and poor patient compliance. Using controlled drug delivery, which involves delivering drug either locally or systemically at a predetermined rate, undesirable fluctuation of drug levels in plasma can be avoided. Design-

ing a controlled drug delivery system requires simultaneous consideration of several factors[2b,3,80] such as the nature of disease and therapy (acute/chronic), drug property, route of drug administration, nature of delivery vehicle, mechanism of drug release, targeting ability, and biocompatibility. It is not easy to achieve all these in one system due to their extensive interdependency. Besides, reliability and reproducibility are also crucial to successful designing of delivery systems.

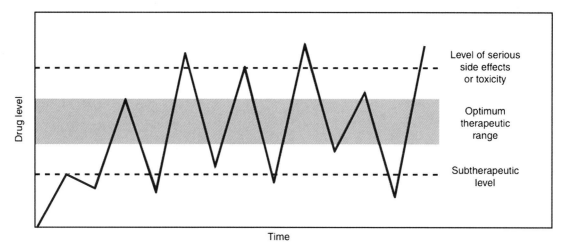

**Fig. 23–19.** The sawtooth kinetic profile obtained after normal dosing and optimum therapeutic profile obtainable with controlled-delivery devices.

Ideally the controlled drug delivery system should be inert, biocompatible, mechanically strong, convenient for the patient, capable of achieving high drug loading, safe from accidental drug release, simple to administer and remove, and easy to fabricate and sterilize.

Advantages of controlled drug delivery are that fluctuations in drug plasma level associated with conventional dosage forms are avoided and therapeutic drug concentration is maintained, which leads to more effective therapies with lesser side effects. Fewer doses are required resulting into improved patient compliance. While the therapeutic considerations are prime driving force for the development of drug delivery systems, there are economic considerations too.[2b,3,80] Once the patent on the new drug has expired (20 years), the pharmaceutical company responsible for its discovery starts loosing its market share to generic competitors, which supply the same drug at lower prices. Repackaging the drug in a new delivery system allows the company to extend the patent life of its product. The disadvantages on the other hand are higher costs compared to conventional formulations, possible toxicity or nonbiocompatibility of the material used, and undesirable by-products. More importantly, many controlled drug delivery systems are invasive and require surgical intervention for their insertion and removal from the body.

## Types of Controlled Drug Delivery

Drug release from a controlled drug delivery system is of three types: zero-order, variable, and bioresponsive[2b,80] (**Fig. 23–20**).

1. **Zero-order release.** The drug release does not vary with time and relatively constant drug level is maintained in plasma over an extended period of time. Since the typical "sawtooth" kinetic profile is not obtained, the risk of drug achieving toxic peak plasma level is abated and so is the

possibility of symptom breakthrough resulting from drop in drug plasma level.

2. **Variable release.** The drug is released at variable rates to match circadian rhythms or mimic natural biorhythms. It is characterized by an episodic increase in drug concentration followed by a "rest" period, where drug level falls below the therapeutic level. It may also be fluctuating or pulsatile (release pulses at predetermined lag times). Variable release is used in situations where changing level of response is needed. For example, in hypertension, blood pressure is lower in the night but increases in the early morning, and consequently maximum drug levels are needed in the early morning. Similarly in nocturnal asthma, bronchoconstriction is worse at night.

3. **Bioresponsive release.** The drug release is triggered by biological stimulus like changes in pH, temperature, or concentration of certain biologically active substances.

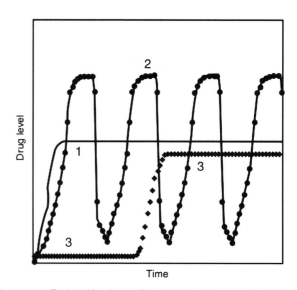

**Fig. 23–20.** Typical kinetic profiles obtained from controlled drug delivery systems: (1) zero-order, (2) variable, and (3) bioresponsive.

Blood glucose level triggering the release of insulin from a drug delivery system is an example of bioresponsive release.

## Mechanisms of Controlled Drug Delivery

Only a small number of mechanisms are involved in the drug release from a controlled drug delivery system[2b,80]: (1) diffusion controlled release mechanism, (2) dissolution controlled release mechanism, (3) osmosis controlled release mechanism, (4) mechanical controlled release mechanism, and (5) bioresponsive controlled release mechanism. Any or all of the above mechanisms may be involved in the drug release from the system.

1. **Diffusion controlled release mechanism.** The drug is released by diffusion through either a polymeric membrane or a polymeric/lipid matrix (**Fig. 23–21**). Diffusion controlled devices could be grouped into two categories: reservoir devices and matrix devices. In reservoir devices, the drug is surrounded by a rate controlling polymer membrane (nonporous, microporous). The rate of diffusion follows Fick's law and depends on partition and diffusion coefficients of the drug in the membrane, the available surface area, the membrane thickness, and the drug concentration gradient. If the drug concentration gradient remains constant, zero-order drug release is attained. Examples include Norplant subdermal implant (parenteral), Vitrasert intravitreal and Ocusert implant (ocular), Transderm-Scop transdermal patch system and Catapres-TTS transdermal system (transdermal), and Cervidil vaginal insert and Estring vaginal ring (vaginal).

In matrix (monolith) devices, the drug is distributed throughout a continuous phase composed of polymer or lipid. As release continues, the rate of drug release decreases with square root of time. This decrease in drug release is due to the fact that as the drug present at the surface is being released, the drug present in the center of the matrix has to migrate longer distances for release, which takes more time. Such devices usually do not provide zero-order release. Polymeric controlled release microspheres represent an example of a matrix-controlled release system.[81] Commercial examples are Compudose cattle growth implant (parenteral) and Deponit transdermal patch (transdermal).

2. **Dissolution controlled release mechanism.** The drug release is controlled by dissolution rate of employed polymer. Similar to diffusion controlled devices, dissolution controlled devices are also either the reservoir type or the matrix type. Since the drug release is dissolution controlled, the polymer must be water soluble and/or degradable. In reservoir devices, the release is controlled by the thickness and/or the dissolution rate of polymer membrane surrounding the drug core. Once the coating is dissolved, the drug is available for dissolution and absorption. Polymer coatings of different thickness can be employed to delay the drug release for certain period of time. Such systems are used for zero-order oral drug delivery and examples include Spansule, Sequel, and SODAS capsules.

In matrix-type devices, on the other hand, the drug release is controlled by the dissolution of matrix and decreases with time due to the decrease in the size of the matrix. Examples of matrix dissolution devices are goserelin (Zoladex) subcutaneous implant comprising

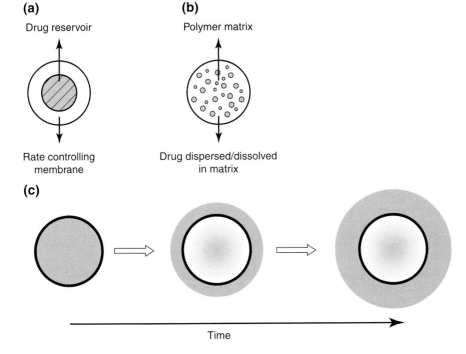

**(a)**
Drug reservoir

Rate controlling membrane

**(b)**
Polymer matrix

Drug dispersed/dissolved in matrix

**(c)**

Time

**Fig. 23–21.** Diffusion controlled controlled-delivery devices: (*a*) reservoir and (*b*) matrix. (*c*) Drug diffusion from a homogeneous controlled drug delivery system. (Modified from H. Sah et al., Drug Delivery and Targeting for Pharmacist and Pharmaceutical Scientists, CRC Press, Boca Raton, FL, 2001, pp. 83.)

poly(lactide-co-glycolide) or PLGA matrix system for goserelin delivery and leuprolide (Lupron) depot comprising PLGA microsphere for parenteral goserelin delivery.

3. **Osmosis controlled or active efflux controlled release mechanism.** Osmosis is defined as diffusion of water through a semipermeable membrane from a solution of low concentration (hypotonic) to a solution of high concentration (hypertonic) resulting into an increase in the pressure of the solution. The pressure difference is termed as osmotic pressure and defined as pressure required for maintaining equilibrium, with no net movement of water. Osmotic pressure can be used to deliver drug at a constant rate, and device and formulation parameters can be controlled to obtain zero-order release. Osmotic pressure is a colligative property and therefore depends only on molar concentration of solute and not its identity. An important consequence is that osmosis controlled devices operate independently of environmental factors. Examples include osmotic mini pumps from Alza Corporation used in experimental animal studies, DUROS implant pump for controlled delivery of peptides and proteins, and Oros osmotic pump for oral delivery. Commercial examples are nifedipine (Procardia XL) and chlorpheniramine (Efidac 24).

4. **Mechanical controlled release mechanism.** Mechanically driven pumps used for drug administration in hospital settings. These pumps can be programmed to achieve zero-order or intermittent release.

5. **Bioresponsive controlled release mechanism.** Drug is released in response to changes in the external environment. The external stimulus could be a change in pH or ionic strength, which might cause drug release by influencing the swellability of polymeric delivery systems. Similarly, there are systems incorporating enzymes, which may cause localized change in pH or substrate (e.g., glucose) concentration to trigger drug release by causing change in swelling or permeability of polymeric systems.

## Examples of Controlled Drug Delivery Systems

The characteristic features of implants and Oros osmotic pump are provided below. More examples are discussed in later sections of this chapter.

### Implant

An implant is a single unit drug delivery system designed for delivering drug at predetermined rate over an extended period of time.[2b,3,80] It is available in many forms and the two most frequently used implants are (a) polymeric implant made of either nondegradable or biodegradable polymer (available in shapes like rod, cylinder, ring, film etc.) and (b) minipumps powered by osmotic pressure or mechanical force. Usually, an implant is placed (implanted) subcutaneously into the loose interstitial tissues of the outer surface of the upper arm or the anterior surface of the thigh or the lower portion of the abdomen. Implants may also be surgically implanted in places like the vitreous cavity of the eye or intraperi-

toneally. Implants have been mostly used for sustained parenteral administration, including ocular and subcutaneous drug delivery.

Polymeric implants are made from either nondegradable or degradable polymers. Examples of nondegradable polymers used in implants include silicone rubber, silicone–carbonate copolymers, poly(ethylene-vinyl acetate), polyethylene, polyurethane, polyisoprene, polyisobutylene, polybutadiene, polyamide, polyvinyl chloride, plasticized soft nylon, hydrogels of polyhydroxyethyl methyl acrylate, polyethylene oxide, polyvinyl alcohol, polyvinyl pyrrolidone, cellulose esters, cellulose triacetate, cellulose nitrate, modified insoluble collagen, polyacrbonates, polysulfonates, polychloroethers, acetal polymers, and halogenated polyvinylidene fluoride. Implants such as Norplant subdermal and Vitrasert are made from nondegradable polymers such as dimethylsiloxane/methylvinylsiloxane copolymer (containing levonorgestrel) and poly(vinyl alcohol)/poly(ethylene-co-vinyl acetate), respectively.

Implants are also made of degradable polymers. Degradation is achieved by either biodegradation or bioerosion. Biodegradation is the degradation of polymer structure by chemical or enzymatic processes, whereas bioerosion is gradual dissolution of polymer matrix. Bioerosion can be of two types: (a) bulk erosion and (b) surface erosion. In bulk erosion the entire polymer matrix is subject to chemical or enzymatic processes whereas in surface erosion, polymer degradation is limited to the surface of implant exposed to the medium, and therefore takes place layer by layer. Polymers in biodegradable implants are either water soluble and/or degradable in water. The water-soluble polymers are PAA, PEG, and poly(vinyl pyrrolidone), whereas degradable polymers include poly(hydroxy butyrate), poly(lactide-co-glycolide), poly(orthoesters), poly(caprolactone), and polyanhydrides. Naturally occurring biodegradable polymers are proteins (albumin, casein, collagen, and gelatin) and polysaccharides (cellulose, chitin, dextran, hyaluronic acids, insulin, and starch). Zoladex implant and Lupron depot are made of PLA/PLGA, whereas Gliadel is poly[bis(p-carboxyphenoxy propane: sebacic acid in 20:80 ratio.

There are mechanical implants too and a very recent example is Medtronic IsoMed Constant-Flow Infusion system. The system was approved for (a) delivering chemotherapy (floxuridine) in hepatic arterial infusion therapy for patient with colorectal liver cancer and (b) delivering morphine to the spinal fluid for patient with chronic intractable pain. The implant comprises two basic parts: (a) pump made of an outer round titanium shell (biocompatible) with a silicone rubber septum containing a drug reservoir (20, 35, and 60 mL) and (b) catheter made of silicone rubber tube, which is tunneled under the skin to the site of action for delivering drug from the pump. The whole pump is about 3-inches wide and weighs 6 oz with standard flow rates of 0.5, 1.0, and 1.5 mL/day. The reservoir is surrounded by a propellant, which forces the drug content through the catheter to the site of the delivery and can be refilled. The pump is implanted surgically in

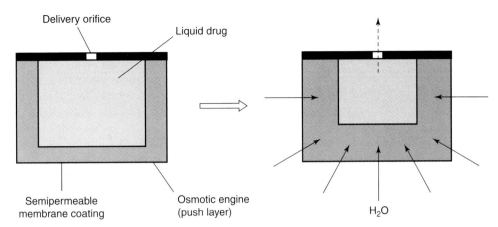

**Fig. 23–22.** Schematic representation of L-OROS osmotic drug delivery system. The drug is liberated through the small orifice due to the osmosis of fluids through the semipermeable membrane into the osmotic push layer.

the abdomen and the catheter is placed in blood in case of chemotherapy and within the sheath around the spinal cord for pain relief. The system is ideal for patients with stable dosing requirements. Other examples of mechanical implants are SynchroMed pumps for chemotherapy (floxuridine, doxorubicin, cisplatin, methotrexate), intractable cancer pain (morphine sulfate), osteomyelitis (clindamycin), and spasticity therapy (muscle relaxant baclofen); MiniMed insulin pumps; and Arrow pumps (floxuridine, morphine sulfate, baclofen, heparinized saline).

Implants are more convenient because they eliminate the need for continuous intravenous infusion or injections for maintaining the drug concentrations and they improve patient compliance by reducing or eliminating the patient-involved dosing. Implants are capable of providing controlled drug release and deliver drug either locally or directly to systemic circulation bypassing the GI tract and liver. Implants are considered new drug product and therefore provide patent exclusivity of 5 years to new drugs and 3 years to existing drugs. However, implants are invasive and require surgical interventions. If not degradable, they have to be retrieved from the body, and if degradable, it is difficult to terminate the drug delivery. Invasiveness is probably the most serious limitation associated with implants. Danger of device failure, possibility of adverse reactions, and biocompatibility are other concerns. Developing implants is highly cost-intensive and therefore limited to only potent drugs.

## Osmotic Pump

The elementary osmotic pump, also known as Oros or the GI therapeutic system, was first described by Theeuwes and Yum[82,83] and introduced by Alza Corporation. The system is composed of core tablet surrounded by a coating of semipermeable membrane containing a laser-created hole (orifice). The core tablet contains two layers, one with drug and the other with the electrolyte. When the tablet is swallowed, a semipermeable membrane permits the entry of fluid from the stomach and intestines to the tablet, which dissolves/suspends

the drug. As pressure increases due to the movement of water, drug is pumped out of the orifice. Only the drug solution is able to pass through the orifice and the system is so designed that only few drops of water are drawn in the tablet every few hours. Drug delivery is controlled by osmotic gradient between the contents of the core tablet and fluids in the GI tract. Surface area, thickness or composition of membrane, and diameter of orifices are altered to control the drug delivery rate. Other systems based on this technology are L-Oros softcap and hardcap systems (**Fig. 23–22**).

In L-Oros softcap, the drug formulation is encased in a soft gelatin capsule surrounded by a barrier layer, an osmotic engine, and a semipermeable membrane. The barrier layer separates the soft gelatin capsule from the osmotic engine, thus minimizing its hydration and mixing with the drug layer. In the L-Oros hardcap system, the drug layer and osmotic engine are encased in a hard capsule surrounded by the semipermeable membrane. Drug release of about 2 to 24 hr is obtained. Other examples in this category are nifedipine (Procardia XL) and Efidac 24.

### EXAMPLE 23–5

**Elementary Osmotic Pump**

Theeuwes[84] first tested the elementary osmotic pump for drug delivery using potassium chloride to serve as both the osmotic agent and the drug model. In a later report, Theeuwes et al.[85] designed a therapeutic system based on the principle of the osmotic pump to deliver indomethacin at a constant zero-order rate. For zero-order rate, these workers used following equation:

$$\left(\frac{dM}{dt}\right)_z = \frac{S}{h}k'\pi_s C_s \qquad (23\text{–}17)$$

where $(dM/dt)_z$ is the rate of delivery of the solute under zero-order conditions, $S$ is the semipermeable membrane area (2.2 cm$^2$), $h$ is the membrane thickness (0.025 cm), $k'$ is a permeability coefficient, $2.8 \times 10^{-6}$ cm$^2$/atm hr, and $\pi_s$ is the osmotic pressure, 245 atm, of the formulation under zero-order conditions (saturated solution) ($k'\pi_s = 0.686 \times 10^{-3}$ cm$^2$/hr). The concentration of the saturated solution, $C_s$, at 37°C is 330 mg/cm$^3$. The zero-order delivery rate for this system is calculated as follows:

$$\left(\frac{dM}{dt}\right)_z = \left(\frac{2.2\,\text{cm}^2}{0.025\,\text{cm}}\right)(0.686 \times 10^{-3}\,\text{cm}^2/\text{hr})(330\,\text{mg/cm}^3)$$

$$\left(\frac{dM}{dt}\right)_z = 19.9\,\text{mg/hr}$$

Some of the drug is released from the device by simple diffusion through the membrane. Equation (23–17) should therefore be modified as follows:

$$\left(\frac{dM}{dt}\right)_z = \frac{S}{h}k'\pi_s C_s + \frac{S}{h}PC_s$$

or

$$\left(\frac{dM}{dt}\right)_z = \frac{S}{h}(k'\pi_s + P)C_s \qquad (23\text{–}18)$$

where $P$ is the permeability coefficient for passage of KCl across the semipermeable membrane ($0.122 \times 10^{-3}$ cm$^2$/hr).

We have following:

$$\left(\frac{dM}{dt}\right)_z = \left(\frac{2.2\,\text{cm}^2}{0.025\,\text{cm}}\right)(330\,\text{mg/cm}^3)(0.686 \times 10^{-3}\,\text{cm}^2/\text{hr}$$
$$+\ 0.122 \times 10^{-3}\,\text{cm}^2/\text{hr})$$
$$\left(\frac{dM}{dt}\right)_z = 23.5\,\text{mg/hr}$$

The time, $t_z$, in which the mass of the drug, $M_z$, is delivered (disregarding the start-up time required to reach equilibrium) is

$$t_z = M_t\left(1 - \frac{C_s}{\rho}\right)\frac{1}{dM/dt} \qquad (23\text{–}19)$$

where $M_t$ is the total mass of drug in the core (500 mg KCl) and $\rho$ is the density of the drug (2 g/cm$^3$ or 2000 mg/cm$^3$):

$$t_z = (500\,\text{mg})\left(1 - \frac{330\,\text{mg/cm}^2}{2000\,\text{mg/cm}^3}\right)\frac{1}{23.5\,\text{mg/hr}}$$
$$t_z = 17.8\,\text{hr}$$

Beyond $t_z$, the drug is delivered under non–zero-order conditions.

# DRUG DELIVERY SYSTEMS

Representative examples of drug delivery systems designed for different routes of drug delivery are described below.

## Buccal Drug Delivery Systems

Most of the buccal drug delivery systems are designed to overcome two major limitations: (*a*) low flux and (*b*) the lack of drug retention at the site of absorption, which is due to the saliva produced by the salivary glands. The major determinant of the salivary composition is its flow rate, which is influenced by the time of day, the type of stimulus, and the degree of stimulation. The salivary pH ranges from 5.5 to 7, but at high flow rates, the sodium and bicarbonate concentrations increase, which further increases the salivary pH. The daily salivary volume is between 0.5 and 2 liters, which is enough to clear the released drug. Hydrophilic polymeric matrices are used for oral transmucosal drug delivery systems. Some drug delivery systems are discussed below.

## Polymers

Bioadhesive (mucoadhesive when the substrate is mucosal tissue) polymers are capable of adhering onto a biologic substrate. Diverse classes of polymers have been investigated for their potential use as mucoadhesives. Examples include synthetic polymers like polyacrylic acid, hydroxypropyl methylcellulose, polymethacrylate derivatives, polyurethanes, epoxy resins, polystyrene, and naturally occurring polymers such as cement, hyaluronic acid, and chitosan.[14] *Chitosan* is derived from a material called chitin, which is an amino polysaccharide extracted from the powdered shells of crustaceans like shrimps and crabs. Chitosan is similar to cellulose in chemical structure, a plant fiber, and has many of the same properties, except that chitosan is positively charged and actively attracts fat. Chitosan works like a "pollution magnet" to soak up pollutants and make them easier to remove. It is bioadhesive and binds to the mucosal membrane, prolonging retention time of the formulation on the mucosa.

## Gels

The dosage forms designed for buccal administration should not cause irritation and they should be small and flexible enough to be accepted by the patient. Gels meet these requirements. Gels are hydrophilic matrices that are capable of swelling in water, without loosing their shape.[86] When drug-loaded gels are placed in water, chain relaxation occurs due to the swelling, and the drug is released through the spaces or channels within the gel network. Examples include natural gums and cellulose derivatives. These "pseudohydrogels" swell, and the component molecules dissolve from the surface of the matrix. Drug release occurs through the spaces or channels within the network as well as through the dissolution and/or the disintegration of the matrix. A buccal mucoadhesive device (copolymer hydrogel disk) was developed for the controlled release of buprenorphine.[87] The device was applied for a 3-hr application time, and steady-state levels were maintained during the time of application. In general, mucoadhesive oral drug delivery systems are used for both sublingual and buccal drug delivery. They provide an onset of drug action in 1 to 3 min and duration of about 30 min to 5 hr for sublingual and buccal drug delivery systems, respectively.

### Adhesive Patches: Systemic Mucosal Delivery
Adhesive patches for mucosal sustained release consist of an impermeable backing layer and a mucoadhesive polymer layer containing the drug (**Fig. 23–23a**). The shape and size varies depending on the site of administration: the buccal, sublingual, or gingival mucosa. The duration of mucosal adhesion depends on the polymer type and the viscosity of the polymer used. The release of the drug is controlled by the dissolution kinetics of the polymer carrier rather than the drug diffusion out of the polymer.

### Adhesive Patches: Local Oral Delivery
Adhesive patches for local oral sustained release generally consist of three layers (**Fig. 23–23b**): the upper layer of a nonadhesive and flavored waxy material containing the drug, the middle layer prepared from antiadhesive material

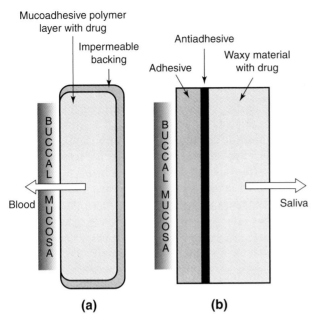

Fig. 23–23. Adhesive patches for drug delivery: (*a*) systemic mucosal and (*b*) local oral.

(magnesium stearate), and the lower layer designed for adhering to the oral mucosa. This three-level device sustains a constant saliva level of drugs, usually anti-infective drugs, over a 3-hr period.[14]

The buccal mucosa offers several advantages for controlled drug delivery: (*a*) the mucosa is well supplied with both vascular and lymphatic drainage; (*b*) first-pass intestinal/hepatic metabolism and presystemic degradation in the GI tract are avoided; (*c*) the area is well suited for a retentive device and is usually acceptable to the patient; and (*d*) with the right DDS design, the permeability and the local environment of the mucosa can be controlled and manipulated to accommodate drug permeation.

## Pulmonary Drug Delivery Systems

Aerosols are widely used to deliver drugs in the respiratory tract. The deposition mechanism of the particles depends on the inhalation regime, the particle size, shape, density, charge, and hygroscopicity. The size of solid particles or liquid droplets in aerosols normally ranges from 1 to 10 $\mu$m and expressed as the *aerodynamic diameter, $d_{ae} = \rho^{1/2}d$*, where $\rho$ is the density of the particle and $d$ is the observed diameter. The particles are delivered via mouth inhalation to bypass the nasopharyngeal cavity and the total retention of particles is only between 50% and 60% of the administered dose.

Liquid jets and ultrasonic nebulizers, metered-dose inhalers (MDIs) (**Fig. 23–24**), and breath-activated dry powder inhalers (DPIs) have proved useful in the management of asthma.

### Nebulizers

These are the device for converting drug solution or suspension into an aerosol suitable for inhalation. There are different types of nebulizers, but the most common are *air jet*

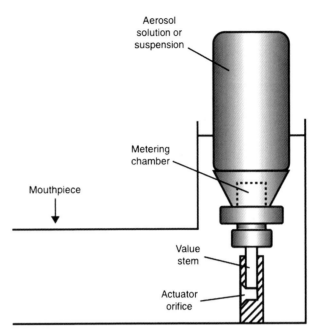

Fig. 23–24. Metered-dose inhaler (MDI) for pulmonary and nasal drug delivery. (Modified from P. R. Byron (Ed.), *Respiratory Drug Delivery*, CRC Press, Boca Raton, Fl, 1990, p. 171.)

*nebulizers*, which are connected to compressed air source that causes air and oxygen to blast through the solution at high velocity, converting them into an aerosol. Another category is *ultrasonic nebulizers* where high-frequency waves are used to create vertical capillaries of drug solution, which breaks up at high energy to provide an aerosol. Nebulizers generate small particles with high delivery capacities but suffer from inconvenience, long inhalation time, and poor dose control. They have been used for the delivery of insulin (diabetes), penicillin (lung infection), isoprenaline and hydrocortisone (asthma), and DNAse and tobramycin (cystic fibrosis).

### Pressurized Metered-Dose Inhalers (MDIs)

MDIs have drug solution or suspension in a volatile propellant. They are equipped with metering valves in conjunction with a propellant and therefore provide multiple dosing capabilities. A typical unit contains a container (10 mL), metering valve (for releasing drug volumes in the range of 25–100 $\mu$L), elastomer seal (critical to valve performance), and actuator (propellant). MDIs have been used for delivering albuterol (asthma).

### Dry Powder Inhalers (DPIs)

DPIs deliver the drugs to the airways as a dry powder aerosol. They are mostly breath-actuated; a cloud of drug powder (aerosol) is produced in response to patient's breathing. Use of DPIs avoids coordination problems and there is a lower drug loss, but they are associated with problems such as requirement of high inspiratory effort from patient and more inconvenient coughing reflexes as compared to other devices. DPIs have been used for the delivery of noradrenaline and terbutaline (asthma) and insulin (diabetes).

The use of lungs for the delivery of peptides and proteins, which are otherwise injected, is possibly the most exciting development in pulmonary delivery.[88] Lungs provide higher bioavalabilities for macromolecules than any other noninvasive route of delivery. However, unlike the smaller molecules, which undergo minimal metabolism in lungs, macromolecules (unless modified) are subject to enzymatic hydrolysis. As the molecular mass increases, possibility of enzymatic hydrolysis is decreased or even eliminated, which significantly increases the bioavailability.[88] After 15 years of development effort, inhaled human insulin was approved in Europe and United States for the treatment of diabetes in adult (insulin inhalation [Exubera, Pfizer]).[88] It has been reported that 4% of the insulin dose reaches the deep lung after the inhalation of single dose and it maintains effective glycemic control comparable to subcutaneously administered fast-acting insulin for type 1 and 2 diabetes. Exubera consists of blisters containing human insulin powder, which are administered using the Exubera inhaler (DPI) before each meal. Each unit dose blister contains 1 to 3 mg of insulin along with the excipients. After the blister is inserted into the inhaler, the patient pumps the handle of the inhaler and then presses a button, causing the blister to be pierced. The insulin inhalation powder (aerosol) is then dispersed into the chamber, which is inhaled by the patient. A major problem associated with this device is inability to delivery precise insulin doses. The product was withdrawn, reasons cited being commercial rather than safety.

The key challenge for pulmonary drug delivery is to provide drug penetration deep into the lung to the smaller airways and the alveoli. The aerosol formulations are unable to move the medication into the deep lungs. The lungs are endowed with a sophisticated defense system that protects the body from the penetration of exogenous particles (**Table 23–4**). Upper airways provide filtering mechanisms that trap and eliminate particles with size $>10$ $\mu$m. Two reflexes, sneezing and coughing, also eliminate large foreign particles. Mucovisciliar transport in conducting airways removes smaller particles out of the respiratory tract into the mouth. In addition, immunoglobulins produced by plasma cells in the submucosa help fight against infections. Alveolar macrophages and neutrophils provide a defense against the smallest foreign matter

that penetrates into the gas exchange area of the respiratory system.

Larger particles ($>10$ $\mu$m) are either filtered in the nose or impacted in the nasal and oral pharynx and cleared by coughing or sneezing. Consequently, drug delivery systems of such size do not penetrate further. Moderate-size particles ($5$–$10$ $\mu$m) are trapped in a mucous blanket in the conducting airways and move cephalad (i.e., toward the head) by ciliary action (cilia move only in the cephalad direction). At the level of the larynx they are either swallowed or expectorated. Small particles of aerodynamic diameter ($<2$ $\mu$m) can penetrate into the lower airways (e.g., bronchial and alveolar regions) and can be phagocytosed by alveolar macrophages. Therefore, submicron-size drug delivery systems can be used for drug delivery to lower airways, alveoli, and systemic circulation through the gas–blood barrier (i.e., the alveolocapillary barrier). The duration of local therapeutic activity is a complex function of particle deposition, mucociliary clearance, drug dissolution or release (for solid aerosols), absorption, tissue sequestration, and metabolism kinetics.

Byron[89] proposed a mathematical model for calculating drug residence times and dose fractions in the three functional regions of the respiratory tract: the nasopharyngeal, the tracheobronchial, and the alveolar. The deposition in the ciliated airways was largely unaffected by breath holding, and the particles showed a maximum $d_{ae}$ between 5 to 9 $\mu$m (slow inhalation) and 3 to 6 $\mu$m (fast inhalation). Alveolar deposition was dependent on the mode of inhalation and breath holding. The latter is a common practice and allows deposition of small particles that otherwise would be exhaled. To determine the effect of breath holding on the deposition of particles, the *sedimentation efficiency*, $S$, was defined as

$$S = \frac{\text{Distance the particle falls during breath holding}}{\text{Mean regional airway diameter}}$$

$$(23\text{–}20)$$

The mean regional diameters for the three areas considered are 5 cm for the mouth, 0.2 cm for the tracheobronchial region, and 0.073 cm for the alveoli. In cases where the sedimentation efficiency is greater than 1, $S$ is assigned a value of unity. A total lung volume after inhalation, $V_t$, of 3000 cm$^3$ is divided into 30, 170, and 2800 cm$^3$ for the mouth (M), tracheobronchial (TB) region, and alveolary (P) region. For each particle size, the *exhaled dose fraction, E*, is given by

$$E = 1 - (f_M + f_{TB} + f_P) \qquad (23\text{–}21)$$

where the terms $f_M$, $f_{TB}$, and $f_P$ stand for the respective fractional depositions in the three regions. During breath holding, additional fractions will sediment, depending on the sedimentation efficiency and the ratio of the regional volume to the total volume of the lungs. The *sedimentation dose fraction*, SDF, after breath holding is calculated from the expression

$$SDF = E \left( \frac{S V_r}{V_t} \right) \qquad (23\text{–}22)$$

---

**TABLE 23–4**
**DEFENSE MECHANISMS OF THE LUNG**

In the upper airways
  Filtering mechanisms in the nasal cavity trap and eliminate
    larger particles ($>10$ $\mu$m)
  Two reflexes: sneezing and coughing
In conducting airways
  Mucociliary escalator, immunoglobulin A, produced by the
    plasma cells in the submucosa
In alveoli
  Alveolar macrophages with some interplay with and by
    neutrophils
  Immunologic mechanisms: interplay between the alveolar
    macrophages and T and B lymphocytes; immunoglobulin G

where $V_r$ is the regional volume of the M, TB, or P regions and $V_t$ is the total volume after inhalation (3000 cm$^3$).

## EXAMPLE 23–6

### Aerosol Deposition in the Lung

For a 3-$\mu$m monodisperse aerosol inhaled at 22.5 L/min, the fractional depositions were found to be $f_M = 0.04$, $f_{TB} = 0.14$, and $f_P = 0.55$. Compute the undeposited or exhaled fraction, $E$, and the additional sedimentation of the undeposited fraction in the mouth and tracheobronchial and alveolary regions after 10 sec of breath holding. The velocity of sedimentation of the particles is 0.027 cm/sec.

From equation (23–21) the undeposited (exhaled) fraction is

$$E = 1 - (0.04 + 0.14 + 0.55) = 0.27$$

After 10 sec of breath holding, the distance of particle fall is 10 sec $\times$ 0.027 cm/sec $= 0.27$ cm. Substitution of this value and the mean diameter of the alveolary (P) region in equation (23–20) gives the sedimentation efficiency in the pulmonary region, $S_P$:

$$S_P = \frac{0.27}{0.073} = 3.70$$

Because $S_P > 1$, a value of $S_P = 1$ is taken. From equation (23–22) the sedimentation dose fraction becomes

$$SDF_P = 0.27 \times 1 \times \frac{2800}{3000} = 0.252$$

Analogously, for the tracheobronchial zone (mean diameter 0.2 cm), a value of $S_{TB} > 1$ is obtained, so $S_{TB}$ is taken as equal to unity, and for the sedimentation dose fraction, one obtains

$$SDF_{TB} = 0.27 \times 1 \times \frac{170}{3000} = 0.0153$$

For the mouth region, M, the mean diameter, is 5 cm, and so

$$S_M = \frac{0.27}{5} = 0.054$$

and

$$SDF_M = 0.054 \times \frac{30}{3000} = 5.4 \times 10^{-4}$$

Thus, the total additional dose deposited after 10 sec of breath holding is

$$SDF = SDF_P + SDF_{TB} + SDF_M = 0.252 + 0.0153$$
$$+ 5.4 \times 10^{-4} = 0.268$$

These results show that for 3-$\mu$m particles, breath holding is adequate for depositing particles in the alveolar region ($SDF_P = 0.252$), whereas breath holding is inadequate for depositing these particles in the mouth ($SDF_M = 0.00054$) and in the tracheobronchial region ($SDF_{TB} = 0.0153$).

Byron et al.[90] studied the deposition and absorption of disodium fluorescein from solid aerosols having an aerodynamic diameter of 3 to 4 $\mu$m. The aerosols were administered for 20 min under different inhalation regimes (respiration frequency, RF, in cycles/min). The total dose administered, $D$, was divided into a transferable amount, $A$, which diffuses into the perfusate according to a first-order rate constant $k$, and an untransferable amount, $U$, according to the scheme.

Assuming instantaneous dissolution of $A$ and first-order kinetics, the amount in the perfusate, $B$, at any time $t$ is given by the product of perfusate concentration and volume. The amount transferred to $B$ can be computed from[*]

$$B = \frac{A}{20\,RF}\left[ n - \frac{(1 - e^{-\frac{nk}{RF}})e^{-\frac{k}{RF}}}{1 - e^{-\frac{k}{RF}}} \right] \quad \text{for} \quad t \leq 20$$

(23–23)

and

$$B = B_{20} + A_{20}\left[1 - e^{-k(t-20)}\right] \quad \text{for} \quad t > 20 \quad (23–24)$$

where $B_{20}$ is computed from equation (23–23) at time $t = 20$ min, and $A_{20}$ is calculated using the following expression:

$$A_{20} = \left[ \left(\frac{A}{20\,RF}\right)\left(1 - e^{-20k}\right)e^{-\frac{k}{RF}} \right]\left(1 - e^{-\frac{k}{RF}}\right)$$

(23–25)

The term $A/20\,RF$ in equations (23–23) and (23–25) is the transferable amount deposited after each inhalation. The inhalation or dose number, $n$, is equal to $t \times RF$ for $t \leq 20$, where RF is the respiratory frequency.

The ratio of transferable amount to amount deposited increases at high respiratory frequency, RF, large tidal volume, and decreasing aerosol particle size. *Tidal volume* is the amount of air that enters the lungs with each inspiration or leaves the lungs with each expiration.

## EXAMPLE 23–7

### Small-Particle Transfer

Compute the transfer, $B$, of 3-$\mu$m particles at $t = 20$ min, knowing that the transferable amount, $A$, is 37.7 $\mu$g, the respiratory frequency RF is 28, and $k = 0.049$ min$^{-1}$.

The inhalation or dose number is $n = 20$ min $\times$ 28 cycles/min = 560 cycles; from equation (23–23)

$$B = \frac{37.7}{560}\left[ 560 - \frac{\left(1 - e^{\frac{-560 \times 0.049}{28}}\right) \times e^{\frac{-0.049}{28}}}{1 - e^{\frac{-0.049}{28}}} \right] = 13.69\,\mu g$$

The pulmonary route provides effective administration of beta-adrenergic agonists in asthma treatment. Corticosteroids have been added to the therapeutic regime, in particular triamcinolone acetonide and beclomethasone, which are safe and effective in aerosol formulations.[91] However, this route has shown to be of limited usefulness for antimicrobial drugs. The pulmonary route can be useful for controlled delivery of drugs to the respiratory tract, depending on the characteristics of the drug and the aerosol device. It is unlikely that this route will be a substitute for the administration of more conventional oral or parenteral drugs in foreseeable future.

## Nasal Drug Delivery Systems

The most suitable dosage forms for the nasal drug delivery are aerosols, gels, liquids, ointments, suspensions, and

---

[*]In personal correspondence, Byron noted that $n$ is incorrectly placed in equation (2) of Byron et al.[90] Equation (23–23) here is the correct expression.

sustained-release formulations. Dosage forms for nasal absorption must be deposited and remain in the nasal cavity long enough to allow effective absorption. The standard methods of administration are sprays and drops. The particle size in aerosols is important in determining the site of deposition. Particles $<0.5$ $\mu$m in diameter pass through the nose and reach the terminal bronchi and alveoli of the lungs. A nasal spray requires that the particles have a diameter $>4$ $\mu$m to be retained in the nose and to minimize passage into the lungs. The nasal spray deposits drug in the proximal part of the nasal atrium, whereas nasal drops are dispersed throughout the nasal cavity. A spray clears more slowly than drops because the spray is deposited in nonciliated regions. An MDI is most often used for nasal and pulmonary delivery. This device (**Fig. 23–24**), when manually compressed, delivers an accurate and reproducible dose of the nasal (or bronchial) formulation.

One of the limitations of nasal drug delivery is rapid removal of the therapeutic agent from the site of absorption. To overcome this, the addition of bioadhesive materials and mixtures with polymers has been investigated. By adding these materials to the drug in solution or powder preparations, increased drug absorption was observed because of increased residence time. Quadir et al.[92] examined the effect of microcrystalline cellulose on the bioavailability of ketorolac. They found that the bioavailability of spray formulations of ketorolac alone in rabbits was approximately 50% after the intravenous administration (**Fig. 23–25**). Nasal administration of ketorolac with microcrystalline cellulose significantly improved the absolute bioavailability (i.e., compared to intravenous injection) of the drug to 90%.

A new nasal gel drug, zinc gluconate (Zicam) (Gel-Tech LLC, Woodland Hill, CA), significantly reduces the length of the common cold.[93,94] The active ingredient in Zicam is zinc ion, which has long been used in cold lozenges. The gel formulation allows the ions to stay within the nasal cavity long

### TABLE 23–5

### ADVANTAGES AND LIMITATIONS OF NASAL MUCOSAL DRUG DELIVERY

| Advantages | Limitations |
|---|---|
| Avoidance of hepatic first-pass elimination and destruction in the gastrointestinal tract | Possible local tissue irritation |
| Rapid absorption of drug molecules across the nasal membrane | Rapid removal of the therapeutic agent from the site of absorption |
| Can be used for both local and systemic drug delivery | Pathologic conditions such as cold or allergies that may alter significantly the nasal bioavailability |
| Relative ease and convenience | |

enough to interact with the virus. Patients who took Zicam within 24 hr of the onset of three or more cold symptoms recovered in an average of 1.5 to 3.3 days, whereas patients who received a placebo recovered in an average of 9.8 days.

Major advantages and limitations of nasal drug delivery are summarized in **Table 23–5**.

## Controlled Ocular Drug Delivery Systems

The action of a drug for ocular delivery is prolonged by (*a*) reducing drainage by using viscosity-enhancing agents, suspensions, emulsions, and erodible and nonerodible matrices and (*ii*) enhancing the corneal penetration by using the prodrugs and liposomes. The optimal viscosity range for reducing drainage loss is between 12 and 15 cp when polyvinyl alcohol[95] or methyl-cellulose[96] is used as viscosity enhancer. To minimize potential irritation, ophthalmic suspensions are prepared by micronization techniques. The dissolution rate of large particles is smaller than that of small particles. To obtain the desired bioavailability, the dissolution rate of the drug must be greater than the clearance of the dose from the conjunctival sac and approximately equal to the absorption rate. Many drugs do not satisfy these requirements.

Using water-soluble matrices, where the drug is either dispersed or dissolved, increases the precorneal retention and duration of action. The delivery of drugs from hydrophilic matrices is fast because the tear fluid rapidly penetrates into the matrix. The prolonged action is not controlled by the vehicle but by the precorneal retention of the drug. The penetration of water into the matrix can be reduced by hydrophobic polymers such as alkyl half-esters of poly(methyl vinyl ether–maleic anhydride) (PVM–MA). The matrix surface is water soluble above certain pH, owing to the ionizable carboxylic groups. However, the hydrophobic alkyl ester groups avoid the penetration of water into the matrix. The diffusion of drug from the matrix is impeded and it is released at the rate at which the polymer surface is dissolved. In one study, pilocarpine was released from PVM–MA polymers according to zero-order kinetics and controlled by the erosion of the

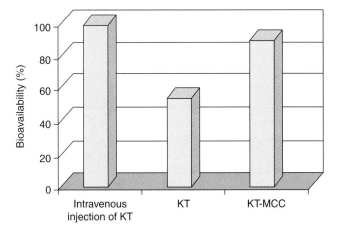

**Fig. 23–25.** Comparison of bioavailability of injection and spray formulations of ketorolac (KT) alone and with a microcrystalline cellulose (MCC) in rabbits. (Replotted from M. Quadir, H. Zia, and T. E. Needham, Drug. Deliv. **7**, 223, 2000.)

polymer surface.[97] Grass et al.[98] prepared erodible and nonerodible dry films for sustained delivery of pilocarpine. The polymers used in both cases were polyvinyl alcohol and carboxyl copolymer (carbomer 934). The release of drug from the films fitted either the Hixon–Crowell dissolution cube root equation or the diffusion-controlled dissolution equation proposed by Cobby et al.,[99,100] the latter providing the best fit.

The ocular delivery of drugs from matrices can be improved by the use of bioadhesive polymers. Johnson and Zografi[101] measured the adhesion (i.e., adhesive strength) of hydroxypropyl cellulose to solid substrates as a function of dry film thickness. A "butt adhesion test" used by them provided a constant slow rate of film detachment to maintain the viscoelastic contribution of the film relative to the adhesion measurements as a constant. For thickness less than 20 $\mu$m, there is a linear relationship between the adhesive strength, $Y$ (in g/cm$^2$) and the film thickness, $h$ (in $\mu$m). The adhesive failure, $Y_0$, can be obtained by extrapolating the adhesive strength to zero film thickness.

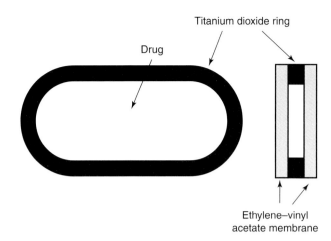

Titanium dioxide ring

Drug

Ethylene–vinyl acetate membrane

**Fig. 23–26.** The Ocusert system consisting of a pilocarpine core (drug) sandwiched between two rate-controlling ethylene–vinyl acetate copolymer membranes. When the device is placed under the upper or lower eyelid, the pilocarpine molecules dissolved in the lacrimal fluid are released at preprogrammed rates through the rate-controlling membranes.

---

**EXAMPLE 23–8**

**Adhesion Properties of Hydroxypropyl Cellulose**

The adhesion, $Y$, of hydroxypropyl cellulose to polyethylene surfaces as a function of the film thickness, $h$, of the adhesive is given as follows:

| $Y$ (g/cm$^2$) | 3850 | 2800 | 1750 | 700 |
|---|---|---|---|---|
| $h$ ($\mu$m) | 5 | 10 | 15 | 20 |

Compute the adhesive failure, $Y_0$. A regression of $Y$ (dependent variable) against $h$ (independent variable) gives

$$Y = -210\,h + 4900$$

The adhesive failure is given by the intercept, $Y_0 = 4900$ g/cm$^2$.

The *work of adhesion*, $W_a$, of the dry film on the solid surface can be computed from the surface tension of the polymer and the solid surface[101]:

$$W_a = \frac{4\gamma_s^d\gamma_p^d}{\gamma_s^d + \gamma_p^d} + \frac{4\gamma_s^p\gamma_p^p}{\gamma_s^p + \gamma_p^p} \qquad (23-26)$$

where $\gamma_s$ and $\gamma_p$ are the surface tensions of the solid and the polymer, respectively. The superscripts d and p represent the contribution to the total surface tension from nonpolar and polar portions of the molecule.

---

**EXAMPLE 23–9**

**Work of Adhesion**

Compute the work of adhesion of hydroxypropyl cellulose films to a solid surface of polyethylene from the following data: $\gamma_s^d = 34.2$ ergs/cm$^2$; $\gamma_s^p = 3.4$ ergs/cm$^2$; $\gamma_p^d = 24.7$ ergs/cm$^2$; and $\gamma_p^p = 16.3$ ergs/cm$^2$.

We have

$$W_a = \frac{4 \times 34.2 \times 24.7}{34.2 + 24.7} + \frac{4 \times 3.4 \times 16.3}{3.4 + 16.3}$$

$$W_a = 68.6\,\text{ergs/cm}^2$$

**Ocusert System**

Alza Corporation introduced the pilocarpine-containing device Ocusert. Made of an ethylene–vinyl acetate copolymer, the device (**Fig. 23–26**) has a central core or reservoir of pilocarpine between two membrane surfaces that control the rate of release of the drug. The oval device, slightly larger than a contact lens, is placed under the upper or lower lid, where pilocarpine is released at a zero-order rate and absorbed into the cornea of the eye. Two products are available, Ocusert P-20, which delivers a dose of 20 $\mu$g/hr, and Ocusert P-40, which delivers a dose of 40 $\mu$g/hr.

Because of the close contact with the eye and continuous release of drug from the Ocusert over a period of a week, only about one fourth of the pilocarpine dose is administered, when compared to drops. With drugs that are only sparingly soluble in water, such as chloramphenicol, release in the eye is calculated from a form of Fick's law:

$$M = \frac{SDKC_s}{h}t \qquad (23-27)$$

where $M$ is the accumulated amount released and $t$ is the time. $S$ is the surface area of the device in contact with the eye, $D$ is the diffusion coefficient of the Ocusert membrane, $K$ is a liquid–liquid partition coefficient between the Ocusert and the eye fluids, $C_s$ is the solubility of the drug in water, and $h$ is the Ocusert membrane thickness. As observed in **Figure 23–27**, a plot of accumulated drug release against time is linear, showing a break at point $A$ (125 hr), then becoming horizontal, indicating that chloramphenicol is depleted, and no more is released after 125 hr. A plot of *release rate*, rather than amount, versus time results in a straight horizontal line to point $A$, then tends toward zero (**Fig. 23–27, inset**). The curve does not fall vertically following point $A$ but is attenuated parabolically as observed in the inset of **Figure 23–27**. These plots indicate that the release rate of a sparingly soluble drug

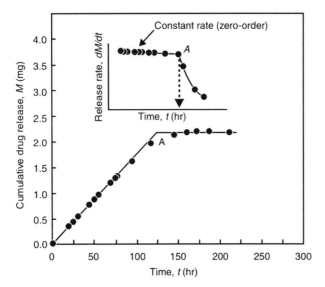

**Fig. 23–27.** Controlled drug delivery of chloramphenicol through the Ocusert membrane. The graph shows the cumulative release versus time (inset: release rate versus time).

is almost constant, that is, a zero-order release rate over most of the lifetime of the device. The flux or rate per unit time for a water-soluble compound is not horizontal (**Fig. 23–27, inset**) but becomes attenuated as the release proceeds. By the proper choice of membranes for the Ocusert, the release rate of pilocarpine can be held essentially constant (zero-order drug release) for up to 7 days of delivery.

These preparations present some disadvantages, such as noncompliance, especially in elderly people, and many patients lose the device sometimes without becoming aware of it. From the point of view of patient acceptability, a liquid dosage form is preferable.

---

**EXAMPLE 23–10**

**Chloramphenicol Analogue Release from Ocusert**

The diffusion coefficient of a new chloramphenicol derivative in the Ocusert device is $3.77 \times 10^{-5}$ cm²/hr. The surface area, $S$, of the Ocusert is 0.80 cm², the partition coefficient, $K$, between the Ocusert and ocular fluids is 1.03, the thickness of the membrane, $h$, is 0.007 cm, and the solubility, $C_s$, in water (25°C) of the new compound is 3.93 mg/cm³. By use of equation (23–27), calculate the cumulative amount of drug released in 125 hr.

We have

$$M = \frac{(0.80\, \text{cm}^2)(3.77 \times 10^{-5}\, \text{cm}^2/\text{hr})(1.03)(3.93\, \text{mg/cm}^2)(125\, \text{hr})}{0.007\, \text{cm}}$$

$$M = 2.18\, \text{mg released in 125 hr}$$

---

## Biodegradable Drug Delivery System

Oculex Pharmaceuticals (Sunnyvale, CA) developed the biodegradable drug delivery (BDD) system, which is based on a microsized polymer system that enables microencapsulated drug therapies to be implanted within the eye. Unlike any other intraocular drug delivery system, this technology is completely biodegradable. This allows biodegradable microsystems to release therapeutic agents directly into the

area requiring medication for a predetermined period of time, enabling treatment of a broad spectrum of conditions and diseases that occur within the back of the eye. The key features of this technology include programmable site-specific drug delivery, a biodegradable therapeutic solution, minimally invasive delivery, versatile drug delivery platform, and better patient compliance. Two types of BDD systems have been developed. The first system (Surodex BDD) is inserted in the front of the eye, whereas the second (Posurdex BDD) might be inserted in the back of the eye by elective surgery. BDD delivery systems are designed to provide continuous, controlled release drug therapy directly to the targeted site for periods ranging from several days to several years. Several drug delivery systems based on this technology are undergoing clinical trials.

## Mucoadhesive Drug Delivery Formulation

It is based on an ionic complex of partially neutralized PAA and a highly potent beta-blocker drug, levobetaxolol hydrochloride (LB × HCl), and used for the treatment of glaucoma.[12,13] PAA is neutralized with sodium hydroxide to varying degrees of neutralization. Aqueous solutions containing varied concentrations of LB × HCl equivalent to the degree of PAA neutralization are added to the PAA solutions to form insoluble complexes. Complexes are prepared with different drug loading, such that the same PAA chain would have free—COOH groups for mucoadhesion along with ionic complexes of LB × H⁺ with COO—groups. From thin films of the complexes, drug is released by ion exchange with synthetic tear fluid. The film thickness attenuated continuously during the release of the drug and dissolved completely in 1 hr. Solid inserts of these films could be useful as a mucoadhesive ophthalmic drug delivery system.

## Transdermal Drug Delivery Systems

### Enhancers for Percutaneous Absorption

The transport of molecules through the skin is increased by the use of adjuvants known as *enhancers*. Ionic surfactants enhance transdermal absorption by disordering the lipid layer of the stratum corneum and by denaturation of the keratin. Enhancers increase the drug penetration by causing the stratum corneum to swell and/or leach out some of the structural components, thus reducing the diffusional resistance and increasing the permeability of the skin.[102]

Nishihata et al.[103] proposed a mechanism for the enhancing effect of reducing agents such as ascorbate and dithiothreitol. The poor permeability of the skin is due to the ordered layer of intercellular lipids and the low water content. Proteins in keratinized tissue are rich in cysteine residues, and the strong disulfide bonds are possibly responsible for the insoluble nature of the protein. The reducing agents cause a decrease in the number of disulfide bridges, leading to an increase in the hydration of the proteins, which results into increased membrane permeability. Azone or laurocapram (1-dodecyl-azacycloheptan-2-one) is the most efficient enhancer for percutaneous absorption. It greatly improves the penetration of hydrophilic and hydrophobic compounds, though the latter to a smaller degree. Azone is an oily liquid, insoluble in water, but freely soluble in organic solvents. Most of the azone applied remains on the skin surface; the small fraction absorbed is located mainly in the stratum corneum.

The compound has been found to be the most effective enhancer in the percutaneous absorption of dihydroergotamine, a drug widely used in the prevention and/or treatment of migraine. The effect of azone increases in the presence of propylene glycol. A possible mechanism could be the fluidization of the intercellular lipid lamellar region of the stratum corneum by azone. Azone is a very nonpolar molecule, enters the lipid bilayers, and disrupts their structure (**Fig. 23–28**).[104] In contrast, a strongly dipolar solvent, dimethyl sulfoxide (DMSO), enters the aqueous region and interacts with the lipid polar heads to form a large solvation shell and to

expand the hydrophilic region between the polar heads. As a result, both azone and DMSO increase the lipid fluidity, thus reducing the resistance of the lipid barrier to the diffusion of drugs. Alcohol derivatives of *N,N*-disubstituted amino acids and hexamethylene lauramine also enhance the permeability of drugs.

### Membrane-Controlled Systems for the Percutaneous Absorption Route

A transdermal device is a laminated structure consisting of four layers, as shown in **Figure 23–29**. It consists of (*a*) an impermeable backing membrane, which is the mechanical support of the system; (*b*) an adjacent polymer layer, which serves as the drug reservoir; (*c*) a microporous membrane filled with a nonpolar material (e.g., paraffin); and (*d*) an adhesive film to make close contact with the skin and maintain the device in the desired position.

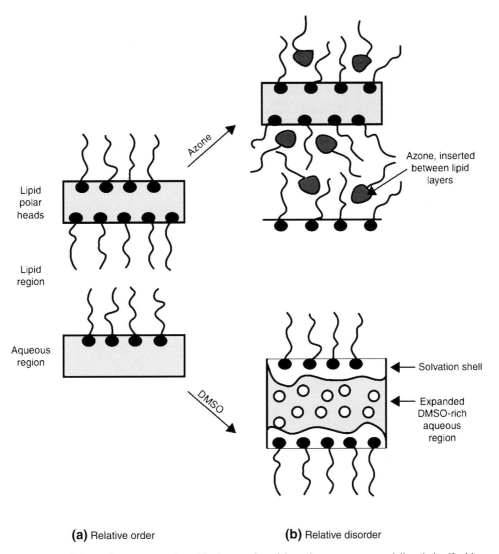

**(a)** Relative order                                    **(b)** Relative disorder

**Fig. 23–28.** Schematic representation of the interaction of the enhancers azone and dimethyl sulfoxide (DMSO) with the intercellular lipids of the stratum corneum: (*a*) Relatively ordered structure of the lipid bilayers. (*b*) Disordered lipid array due to the azone and DMSO activity. (Modified from B. W. Barry, Int. J. Cosmet. Sci. **10,** 281, 1988. With permission.)

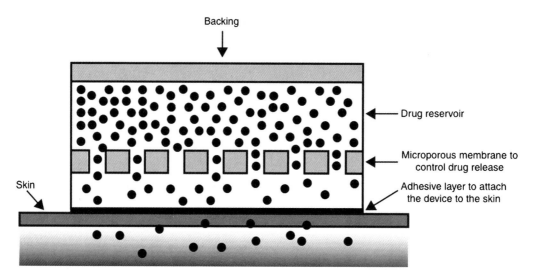

**Fig. 23–29.** A schematic representation of the transdermal therapeutic system. (Modified from K. Heilmann, *Therapeutic Systems*, Georg Thieme, Stuttgart, 1978, p. 53.)

Guy and Hadgraft[105,106] proposed a model for the transport of clonidine across the skin (**Fig. 23–30**) from a membrane-controlled adhesive system. The constant $k_0$ represents the zero-order rate constant for the membrane-controlled leaching of the drug, and $k_R$ represents the partition between the patch and the skin surface. The system should be designed so that the partitioning favors the skin and $k_R$ remains negligibly small. The first-order constants $k_1$ and $k_2$ in **Figure 23–30** are for drug transport across the stratum corneum and the living part of the epidermis. These constants are directly proportional to the diffusion coefficients for passage through the layers of the skin and therefore inversely proportional to the penetrant molecular weight as observed from the Stokes–

Einstein equation:

$$D = \frac{RT}{6\pi \eta N} \sqrt[3]{\frac{4\pi N}{3M\bar{v}}} \qquad (23\text{–}28)$$

The rate constant $k_3$ of **Figure 23–30** is included to express any tendency for reverse drug transport from epidermis to the stratum corneum and, in conjunction with $k_2$, can be considered as a partition coefficient. The authors computed the values of $k_1$ and $k_2$ for benzoic acid and used them to compute these rate constants for other drug molecules. The expressions are

$$k_1 = k_1^{BA} \left( \frac{M^{BA}}{M} \right)^{1/3} \qquad (23\text{–}29)$$

$$k_2 = k_2^{BA} \left( \frac{M^{BA}}{M} \right)^{1/3} \qquad (23\text{–}30)$$

where $k_1^{BA}$, $k_2^{BA}$, and $M^{BA}$ are the rate constants and molecular weight of benzoic acid and $M$ is the molecular weight of the drug for which the constants $k_1$ and $k_2$ are calculated. The ratio $k_3/k_2$ was found to be a function of the octanol–water partition coefficient, $K$:

$$K \cong 5(k_2/k_3) \qquad (23\text{–}31)$$

Using this model, one can predict the constants $k_1$, $k_2$, and $k_3$ from the physicochemical properties of the drug. The rate constant $k_4$ in **Figure 23–30** represents the first-order elimination of the drug from the blood and cannot be predicted. It must be measured experimentally.

Examples of membrane-controlled systems are the Transderm-Nitro (Ciba, Basel, Switzerland, and Alza, Mountain View, CA) for the delivery of nitroglycerin, Transderm-Scop (Alza and Ciba-Geigy) for scopolamine, and Catapress TTS (Alza and Boehringer Ingelheim, Germany) for clonidine.[107]

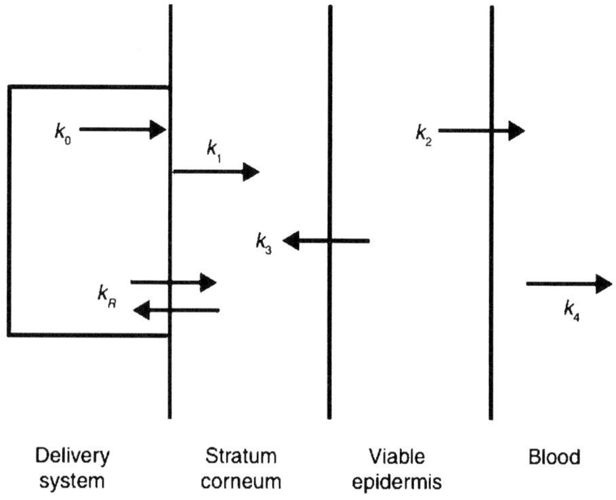

**Fig. 23–30.** Transdermal delivery of clonidine from a membrane-controlled patch. (Modified from R. H. Guy and J. Hadgraft, J Pharm. Sci. **74**, 1016, 1985. With permission.)

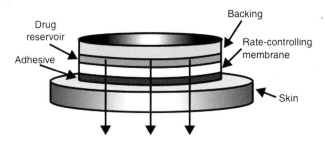

**Fig. 23–31.** A schematic representation of an adhesive diffusion-controlled transdermal drug delivery system.

---

### EXAMPLE 23–11

**Clonidine Release from a Transdermal Patch**

(*a*) Compute the rate constants $k_1$, $k_2$, and $k_3$ for the transport of clonidine from a membrane-controlled patch. The first-order rate constants $k_1^{BA}$ and $k_2^{BA}$ are $5.11 \times 10^{-5}$ and $80 \times 10^{-5}$ sec$^{-1}$, respectively. The octanol–water partition coefficient of clonidine is $K = 6.7$. The molecular weight of benzoic acid is 122.12 g/mole, and the molecular weight of clonidine is 230.10 g/mole.

We have:

$$k_1 = 5.11 \times 10^{-5}(122.12/230.10)^{1/3} = 4.14 \times 10^{-5}\,\text{sec}^{-1}$$
$$k_2 = 80 \times 10^{-5}(122.12/230.10)^{1/3} = 64.77 \times 10^{-5}\,\text{sec}^{-1}$$

From equation (23–31),

$$k_3 = \frac{Kk_2}{5} = \frac{6.7 \times 64.77 \times 10^{-5}\,\text{sec}^{-1}}{5} = 8.68 \times 10^{-4}\,\text{sec}^{-1}$$

(*b*) The steady-state plasma concentration of clonidine, $C^{SS}$, can be computed from

$$C^{SS} = Ak_0 / V_d k_4 \qquad (23\text{–}32)$$

where $A$ is the area of the patch, $k_0$ is the zero-order rate constant for the delivery of clonidine from the patch, and $V_d$ is the *volume of distribution*,* the amount of drug in the body divided by the plasma concentration. For the most efficient membrane-controlled patch, which contains 2.5 mg of clonidine, $A$, the area of the patch is 5 cm$^2$ and $k_0$ is 1.6 $\mu$g/cm$^2$ hr. The volume of distribution $V_d$ for clonidine is 147 liters and the first-order constant, $k_4$, is 0.08 hr$^{-1}$. From equation (23–31),

$$C^{SS} = \frac{5\,\text{cm}^2 \times 1.6\,\mu\text{g/cm}^2\,\text{hr}}{147{,}000\,\text{cm}^3 \times 0.08\,\text{hr}^{-1}} = 6.8 \times 10^4\,\mu\text{g/mL}$$

---

### Adhesive Diffusion-Controlled Systems

The basic difference between this system and the one previously described is the absence of microporous membrane (**Fig. 23–31**). The device consists of an impermeable plastic barrier on the top, a drug reservoir in the middle, and several rate-controlling adhesive layers next to the skin. The rate of drug release, $dQ/dt$, depends on the partition coefficient, $K$, of the drug between the reservoir (r) and the adhesive layers (a), the diffusion coefficient, $D_a$, the sum of the thicknesses

---

*To understand $V_d$, the volume of distribution, consider the following: If an amount of drug, say 1.5 $\mu$g, is distributed in the body plasma to give a drug concentration of $9.38 \times 10^{-6}$ $\mu$g/mL, the volume of plasma containing this amount (1.5 $\mu$g) of drug is 1.5 $\mu$g/($9.38 \times 10^{-6}$ $\mu$g/mL) $=$ 160,000 mL, or $V_d = 160$ liters.

---

of the adhesive layers, $h_a$, and the concentration $C_r$ of the drug in the reservoir layer[108]:

$$\frac{dQ}{dt} = \frac{KD_a C_r}{h_a} \qquad (23\text{–}33)$$

Examples of these devices are the nitroglycerin (Nitrodisc) (Searle, Chicago, IL) and glyceryl trinitrate (Deponit) (Swarz Pharma, Monheim, Germany) for the delivery of nitroglycerin.

### Matrix-Controlled Devices

In a matrix-controlled device, the drug reservoir consists of a hydrophilic or hydrophobic polymer containing the dispersed drug, attached to a plastic backing that is impermeable to the drug. The drug reservoir is in direct contact with the skin, and the release of drug is matrix controlled, that is, it is a function of the square root of time. To obtain zero-order release of the drug across the skin from such systems, the stratum corneum must control the rate of drug delivery. This can be achieved if the release rate of the drug from the device is much greater than the rate of skin uptake. Example is Nitro-Dur system (Schering-Plough Corp., Kenilworth, NJ).

### Iontophoresis

It is an electrochemical method that enhances the transport of some soluble molecules by creating a potential gradient through the skin tissue with an applied electrical current or voltage (**Fig. 23–32**). This technique is used to enhance the transdermal transport of drugs by applying a small current through a reservoir that contains ionized species of the drug. Positive or negative electrodes are placed between the drug reservoir and the skin. Positive ions are introduced in the skin from a positive electrode and negative ions from a negative electrode. **Figure 23–32** shows an iontophoresis circuit with the active electrode being negative. A second electrode, positive in this case, is placed a short distance away on the body to complete the circuit, and the electrodes are connected to a power supply. When the current flows, the negatively charged ions are transported across the skin, mainly through the pores. The isoelectric point of the skin is between 3 and 4 pH

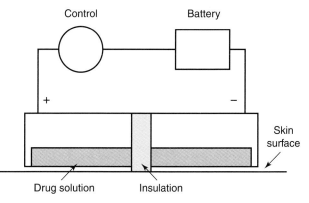

**Fig. 23–32.** A schematic representation of iontophoresis apparatus on skin.

units; below pH 3 the pores are positively charged and above pH 4 they are negatively charged. Owing to the negative charge in the upper skin layers, basic drugs are relatively easy to introduce.

In vitro systems designed to study iontophoretic transport involve the use of diffusion cells in which a skin membrane is placed vertically between the two halves of the cell. The "active" electrode, say the positive electrode for the transport of positive ions, is placed on the epidermal side. The other side of the cell contains a passive (oppositely charged) electrode in a conductive fluid. Iontophoresis enhances the transdermal absorption of insulin. At a pH below the isoelectric point of insulin (pH 5.3) the drug acts as a positive electrode, whereas at a pH above the isoelectric point, the drug reservoir acts as a negative electrode. The greatest transport of insulin was found at pH 3.68 rather than at 7 or 8 owing possibly to low aggregation and a high charge density of insulin at pH 3.68.[109]

The rate of skin permeation depends on the drug concentration, the ionic strength of the buffer solution, the magnitude of current applied, and the duration of iontophoresis.[110] Iontophoresis is a promising method for delivering peptides through the skin. Burnette and Marrero[111] showed that the flux of both ionized and nonionized species of thyrotropin-releasing hormone was greater than the flux obtained by passive diffusion alone. The increased flux was proportional to the applied current density. Transport through the pores is favored for positive ions, whereas transport of negative ions is probably smaller.

Faraday's law states that equal quantities of electricity will deposit equivalent quantities of ions at either electrode. However, the correlation between the prediction by Faraday's law and the experimental values is not good owing to several factors involved in iontophoretic transport. Kasting et al.[112] used an electrodiffusion model to study the transport of etidronate disodium, a negatively charged bone resorption agent, through excised human skin. At steady state, the flux, $J_i$, of drug through the membrane is given by the Nernst–Planck flux equation:

$$J_i = -D_i \frac{dc_i}{dx} + \frac{D_i z_i e E c_i}{kT} \tag{23–34}$$

where $D_i$ is the diffusion coefficient for the ion i (in the $x$ direction), $z_i$ is its charge, and $c_i$ is its concentration. The term $kT$ is the thermal energy of the system, where $k$ is the Boltzmann constant and $T$ is the absolute temperature. The Goldman approximation[112] provides a solution of equation (23–34):

$$J_i = \frac{-D_i K v}{h} \frac{c_i - c_0 e^{-v}}{1 - e^{-v}} \tag{23–35}$$

where $K$ is the partition coefficient, $h$ is the thickness of the membrane, and $c_i$ and $c_0$ are the concentrations at either side of the membrane. Assuming that the concentration $c_0 = 0$, in the limit as $v \rightarrow 0$, equation (23–35) becomes the flux passive diffusion, $J$:

$$J_i = \frac{-D_i K c_i}{h} \tag{23–36}$$

The term $v$ is a dimensionless driving force, defined as:

$$v = \frac{z_i e V}{kT} \tag{23–37}$$

where $e$ is the electronic charge, $z$ is the charge on the drug, $k$ is the Boltzmann constant, $T$ is the absolute temperature, and V is the applied voltage across the membrane. The *iontophoretic enhancement factor*, $J_i/J_0$, is given by:

$$\frac{J_i}{J_0} = \frac{v}{1 - e^{-v}} \tag{23–38}$$

Equation (23–38) measures the increase in transport of a drug relative to the passive diffusion due to the electrical current applied. For positive values [$z_i$ and V of the same sign in equation (23–37)], equation (23–35) predicts that the enhancement in flux is proportional to $v$. For negative $v$ values, the flux will fall exponentially with increasing magnitude of $v$.

**EXAMPLE 23–12**

**Iontophoretic Enhancement Factor**

(a) Compute the iontophoretic enhancement factor, $J_i/J_0$, across human excised skin for a 10% etidronate solution. The voltage applied is 0.25 V, the average number of charges, $z$, per ion is 2.7. The charge on the electron is 1 eV. The value of $kT$ at 25°C is 0.025 eV.

From equation (23–37),

$$v = \frac{2.7 \times 1 \times 0.25}{0.025} = 27$$

Using equation (23–38), we find the iontophoretic enhancement factor

$$\frac{J_i}{J_0} = \frac{27}{1 - e^{-27}} = 27$$

Thus, we see that the flux for the drug promoted by iontophoresis is 27 times that expected for passive diffusion.

(b) Compute the flux under the driving force of iontophoresis, knowing that the passive permeability coefficient, $P$, of the drug is $4.9 \times 10^{-6}$ cm/hr and the concentration, $c$, is $1.02 \times 10^5$ $\mu$g/cm$^3$.

Because $P = DK/h$ cm/hr from equation (23–36),

$$J_0 = 4.6 \times 10^{-6} \times 1.02 \times 10^{-5} = 0.5 \, \mu\text{g} \, (\text{cm}^2 \cdot \text{hr})$$

From equation (23–35), using the value 27 obtained in part (a) for $J_i/J_0$, we obtain

$$J_i = 0.5 \times 27 = 13.5 \, \mu\text{g} \, (\text{cm}^2 \cdot \text{hr})$$

Kasting et al.[113] found that equation (23–35) applies up to 0.25 V. At higher voltages, the flux of etidronate disodium rises much faster than the predicted values because of alteration of the membrane and because the diffusion coefficient is no longer a constant value, as assumed in equation (23–36). Burnette and Bagniefski[114] determined the skin electrochemical resistance, $R$, after iontophoresis. The decrease in resistance suggested that the current alters the ion-conducting pathways of the skin even at the clinically acceptable current densities, leading to tissue damage.[115]

The main advantages of iontophoresis include (a) controlling the delivery rates through variations of current density, pulsed voltage, drug concentration, and/or ionic strength; (b) eliminating GI incompatibility, erratic absorption, and first-pass metabolism; (c) reducing side effects and interpatient variability; (d) avoiding the risk of infection, inflammation, and fibrosis associated with continuous injection or infusion; and (e) enhancing patient compliance with a convenient and noninvasive therapeutic regimen. The main disadvantage of iontophoresis is skin irritation at high current densities, which can be eliminated by lowering the current of administration.

### Phonophoresis

It is defined as transport of drugs through the skin using ultrasound (synonyms: *ultrasound, ultrasonophoresis, ultraphonophoresis*) (**Fig. 23–33**). It is a combination of ultrasound therapy with topical drug therapy to achieve therapeutic drug concentrations at selected sites in the skin. The ultrasonic unit has a sound transducer head emitting energy at 1 MHz at 0.5 to 1 W/cm$^2$. In this technique, the drug is mixed with a coupling agent, usually a gel, but sometimes a cream or an ointment, which transfers ultrasonic energy from the phonophoresis device to the skin through this coupling agent. The exact mechanism of phonophoresis action is not known.

## Vaginal Administration, Intrauterine, and Rectal Drug Delivery Systems

Vaginal delivery systems are in use for estrogen replacement therapy, which when used alone carries the risk of endometrial cancer. Traditionally, this risk is overcome by treatment with progesterone for about 14 to 30 days but it is associated with low oral bioavailability, lack of efficacy, and high level of metabolites. Consequently progesterone tablets, suppositories, and gels have been developed for vaginal administration.[39,40] Vaginal administration pro-

vides higher and sustained plasma levels and low amount of metabolites. Various vaginal preparations of estrogens and progesterones are now available for use as contraceptives, in hormone replacement therapy, and in vitro fertilization programs.

### Suppositories

As mentioned earlier, suppositories are solid dosage forms intended for insertion into body orifices where they melt, soften, or dissolve, and exert localized or systemic drug delivery (suppository, from the Latin sup, "under," and ponere, "to place"). Once inserted, the suppository base melts, softens, or dissolves, distributing the medications it carries to the tissues of the region. Suppositories are preferred for their safety, suitability for sustained systemic and/or local drug delivery, and nonmessy, nonstaining, and convenient administration. The progesterones and estrogens vaginal suppositories are available commercially.

### Vaginal Rings

Vaginal rings containing various progesterones and estrogens are available as steroidal contraceptives (**Fig. 23–34***a*). These rings consist of a drug reservoir surrounded by a polymeric membrane. These are pliable drug delivery system that can be inserted into the vagina, where they slowly release the drug, which is absorbed into the bloodstream. The most common one being Silastic toroidal-shaped ring, which is about 2$\frac{1}{4}$″ in diameter and the size of the outer rim of a diaphragm, designed for insertion into vagina and positioned around the cervix for about 21 days. Levonorgestrel (progesterone analog) is released from the device at a concentration of 20 μg/day with nearly a zero-order release. These rings are easy to use with the advantage of reversibility, self-insertion and removal, continuous drug administration at an effective dose level, and better patient compliance. This above device was

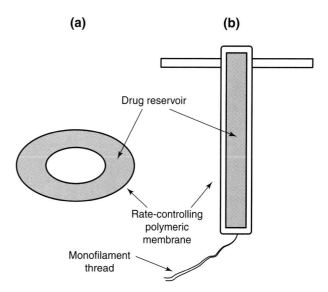

**(a)**                     **(b)**

Drug reservoir

Rate-controlling polymeric membrane

Monofilament thread

**Fig. 23–34.** A schematic representation of (*a*) vaginal ring and (*b*) intrauterine device.

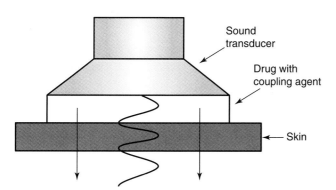

Sound transducer

Drug with coupling agent

Skin

**Fig. 23–33.** A schematic representation of phonophoresis apparatus on skin.

however associated with irregular bleeding. Another vaginal Silastic silicone ring (Estring) was launched in the United States in 1997 for treating postmenopausal women with symptoms of urogenital aging. The ring provides a constant release (6.5–9.5 $\mu$g/day) of estradiol over 3-month period and gives better results when compared to estradiol containing pessaries and creams.

### Vaginal Inserts

Vaginal inserts are in use for prostaglandin delivery. Prostaglandin E2 (PGE2) is used to ripen the cervix for induction of labor and for second trimester abortion. Prostaglandins provide benefits such as reduced time for onset of labor, reduced need for oxytocin, and shortened time for vaginal and caesarean-operated delivery. These inserts are polymeric hydrogel material, which has the ability to absorb fluid and swell without losing its physical form. As it swells, the incorporated drug is released in a controlled manner. An example is Cervidil Vaginal, which contains 10 mg of dinoprostone and provides release at a rate of 0.3 mg/hr in vivo. The retrieval system comprises Dacron polyester net, which surrounds the insert and has a long ribbon end (net plus ribbon = 31 cm). Another example is the Hycore (CeNeS Pharmaceuticals, Cambridge, UK), which exists in two main forms, Hycore-V, a hydrogel pessary used to deliver drugs locally via the vagina, and Hycore-R, a rectal delivery system used to deliver drugs systemically. Misoprostol (prostaglandin E1 analog used for terminating second trimester pregnancy) is also given through the vaginal route at a dose 100 to 200 $\mu$g every 12 hr and provides three times higher bioavailability when compared to oral administration.

### Intrauterine Device

An intrauterine device (IUD) is a small plastic device that is placed into the uterine cavity for sustained intrauterine drug release and is usually used for contraception. A typical IUD is shaped like a T and contains a drug, usually progesterone, in its vertical arm (**Fig. 23–34b**). The progesterone release causes the cervical mucus to become thicker, so sperm cannot reach the egg. It also changes the lining of the uterus so that implantation of a fertilized egg cannot occur. The IUD is inserted through the cervix and placed in the uterus. A small string hangs down from the IUD into the upper part of the vagina and is used to periodically check the device. A shorter-than-normal string can be a warning sign of an imbedded IUD.

## Central Nervous System Drug Delivery Systems

Over the years, various strategies have been developed to overcome the BBB and deliver the drugs to the CNS.[42,43]

### Invasive Strategies

The invasive strategies are:[42,43] (a) *intracerebroventricular (ICV) drug infusion*, where the drug is directly injected into the ventricles (large cavities in the middle of the brain). Following the infusion, drug diffusion to brain is

still limited by physical barriers, catabolic enzymes, high and low affinity uptake sites, and low diffusion coefficients of high–molecular-weight drugs; (b) *implants*, where either genetically engineered cells secreting a drug or polymer matrix/reservoir containing the drug is implanted within the brain. Polymeric implants such as Gliadel are commercially available (Guilford Pharmaceuticals, Baltimore, MD). Gliadel is a small, white, dime-sized wafer made of a biodegradable polymer containing chemotherapy (carmustine or BCNU). Up to eight wafers are implanted in the cavity created, when a brain tumor is surgically removed. Once implanted, they slowly dissolve over a period of 2 to 3 weeks, delivering the drug directly to the tumor site in high concentrations. First approved in 1996 for use as an adjunct to surgery for prolonging patient survival, these wafers are now approved for patients undergoing initial surgery for malignant glioma; and (c) *reversible BBB disruption*, where transient disruption or opening of BBB is achieved by the intracarotid infusion of hyperosmolar (2M) solution of mannitol, leukotrienes, or bradykinin. This approach has significant side effects. All the above-mentioned strategies are invasive and require intervention by trained professionals.

### Noninvasive Strategies

Besides the invasive strategies for local drug delivery to CNS, there are pharmacology and physiology-based strategies for systemic delivery to CNS. The most common strategy is to increase the lipophilicity (lipidization) of the drug for improved drug penetration into the brain (**Fig. 23–35**). This is achieved by either blocking the hydrogen bond forming functional groups in the drug or covalent attachment of lipophilic moieties such as long chain fatty acids. For example,[42] O-methyation of morphine to form codeine or di-O-methylation to form heroin enhances the BBB permeability. Multivesicular liposomes (diameter $<2$ $\mu$m) are retained in the brain following systemic administration and therefore used for systemic drug delivery to CNS. A further

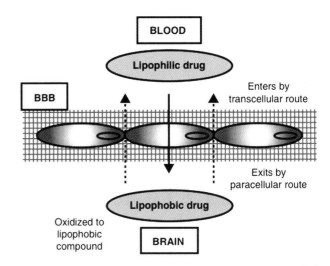

**Fig. 23–35.** Improving the drug penetration across the blood–brain barrier (BBB) by increasing its lipophilicity.

improvement in liposome-based technologies is achieved by employing immunoliposomes, where antibodies are attached to the liposomes through bifunctional PEG linker to exploit receptor-mediated transcytosis.

Carrier-mediated transport mechanism is also used for drug delivery to CNS.[42,43] As explained earlier, nutrients are transferred to the CNS by various carrier-mediated systems. Drugs containing molecular structure similar to the nutrient are designed for transport to the CNS by appropriate carrier-mediated system. Drugs such as L-dopa and $\alpha$-methyldopa are transported across the BBB by neutral amino acid carrier system. Similarly, receptor-mediated transcytosis is also exploited to achieve systemic delivery to CNS. This strategy (prodrug) involves coupling of the drug to a peptide or protein vector (insulin, insulinlike growth factor, and transferrin) through reversible linker, which undergoes receptor-mediated transcytosis. Once it has been exocytosed into the brain interstitial fluid, the linker is cleaved releasing the drug. Pep:trans peptide-derived vectors (Synt:em, Nimes, France) are examples of such vectors. This vector family is derived from the optimization of natural mammalian antimicrobial peptides involved in an ancestral immune response system. Drugs linked to Pep:trans typically show up to 100-fold enhancement in the brain uptake. This approach suffers from rapid clearance of peptide from the bloodstream. Monoclonal antibodies such as anti-insulin (mAb83–7 or 83–14) and anti-transferrin (OX26) receptor antibodies have been alternatively used as vectors. The limitation with receptor-mediated transcytosis is that it is a saturable process.

# CHAPTER SUMMARY

The objective of this chapter was to provide a perspective on advanced drug delivery. The two key characteristic features of advanced drug delivery systems are controlled delivery and targeting. Oral route remains the most preferred route of drug delivery but not always feasible, and therefore alternative routes of drug delivery are being explored. Each drug delivery route has its own merits and limitations, and the delivery systems depending on the route of drug delivery, must meet characteristic design requirements. The delivery of biotherapeutics (peptides, proteins, and nucleic acids) is more challenging because of their inherent disadvantageous delivery features.

 **Practice problems for this chapter can be found at thePoint.lww.com/Sinko6e.**

# References

1. A. M. Hillery, in A. M. Hillery, A. W. Lloyd, and J. Swarbrick (Eds.), *Drug Delivery: The Basic Concepts,* CRC Press, Boca Raton, FL, 2001, pp. 1.

2a. A. M. Hillery, in A. M. Hillery, A. W. Lloyd, and J. Swarbrick (Eds.), *Advanced Drug Delivery and Targeting: An introduction,* CRC Press, Boca Raton, FL, 2001, pp. 63.

2b. V. H. K. Li, J. R. Robinson, and V. H. L. Lee, in V. H. L. Lee and J. R. Robinson (Eds.), Controlled Drug Delivery: Fundamentals and Applications, 2 Ed, Marcel Dekker, NY, 1987, pp.

3. C. R. Gardner, in P. Johnson and J. G. Lloyd-Jones (Eds.), *Drug Delivery Systems: Fundamental and Techniques,* Ellis Horwood Ltd., UK, and VCH Verlag, 1987, pp. 11.

4. H. Rosen and T. Abribat, Nat. Rev. Drug Discov. **4,** 381, 2005.

5. P. Evers, in A. M. Hillery, A. W. Lloyd, and J. Swarbrick (Eds.), *Drug Delivery: Market Perspectives,* CRC Press, Boca Raton, FL, 2001, pp. 49.

6. Advanced drug delivery systems: new development, new technologies. Business Communication Company, Wellesley, MA, 2006.

7. European market for drug delivery products and technologies. Medtech Insight, Irvine, CA, 2009.

8. S. Frokjaer and D. E. Otzen, Nat. Rev. Drug Discov. **4,** 298, 2005.

9. K. R. Reddy, M. W. Modi, and S. Pedder, Adv. Drug Deliv. Rev. **54,** 571, 2002.

10. Y.-S. Wang, S. Youngster, M. Grace, J. Bausch, R. Bordens, and D. F. Wyss, Adv. Drug Deliv. Rev. **54,** 547, 2002.

11. T. K. De and A. S. Hoffman, Artif. Cells Blood Substit. Immobil. Biotechnol. **29,** 31, 2001.

12. B. S. Lele and A. S. Hoffman, J. Biomater. Sci. Polym. Ed. **11,** 1319, 2000.

13. B. S. Lele and A. S. Hoffman, J. C. Release, J. Control. Release, **69,** 237, 2000.

14. A. H. Shojaei, J. Pharm. Pharm. Sci. **1,** 15, 1998.

15. J. S. Patton, Chemtech. **27,** 34, 1997.

16. Y. W. Chien, K. S. E. Su, and S.-F. Chang, *Nasal Systemic Drug Delivery,* Marcel Dekker, NY, 1989, chapter 1.

17. P. A. Hilger, *Fundamentals of Otolaryngology,* W. B. Saunders, Philadelphia, PA, 1989, pp. 184.

18. L. Illum, J. Control. Release, **87,** 187, 2003.

19. P. Sinswat and P. Tengamnuay, Int. J. Pharm. **257,** 15, 2003.

20. H. Pavis, A. Wilcock, J. Edgecombe, D. Carr, C. Manderson, A. Church, and A. Fisher, J. Pain Symptom Manage. **24,** 598, 2002.

21. Y. H. Liu, M. C. Kao, Y. L. Lai, and J. J. Tsai, J. Allergy Clin. Immunol. **112,** 301, 2003.

22. M. A. Pogrel, Oral Maxillofac. Surg. **61,** 649, 2003.

23. S. Borsutzky, V. Fiorelli, T. Ebensen, A. Tripiciano, F. Rharbaoui, A. Scoglio, C. Link, F. Nappi, M. Morr, S. Butto, A. Cafaro, P. F. Muhlradt, B. Ensoli, and C. A. Guzman, Eur. J. Immunol. **33,** 1548, 2003.

24. Y. Ali and K. Lehmussaari, Adv. Drug Deliv. Rev. **58,** 1258, 2006.

25. M. R. Prausnitz, S. Mitragotri, and R. Langer, Nat. Rev. Drug Discov. **3,** 115, 2004.

26. W. Curatolo, Pharm. Res. **4,** 271, 1987.

27. R. L. Bronaugh and R. F. Stewart, J. Pharm. Sci. **75,** 487, 1986.

28. R. H. Guy and J. Hadgraft, Pharm. Res. **5,** 753, 1988.

29. K. Tojo, C. C. Chiang, and Y. W. Chien, J. Pharm. Sci. **76,** 123, 1987.

30. W. Barry, Int. J. Cosmet. Sci. **10,** 281, 1988.

31. M.-S. Wu, J. Pharm. Sci. **72,** 1421, 1983.

32. P. P. Sarpotdar and J. L. Zatz, J. Pharm. Sci. **75,** 176, 1986.

33. J. Swarbrick, G. Lee, J. Brom, and N. P. Gensmantel, J. Pharm. Sci. **73,** 1352, 1984.

34. C. Fleeker, O. Wong, and J. H. Rytting, Pharm. Res. **6,** 443, 1989.

35. P. S. Banerjee and W. A. Ritschel, Int. J. Pharm. **49,** 189, 1989.

36. D. B. Guzek, A. H. Kennedy, S. C. McNeill, E. Wakshull, and R. O. Potts, Pharm. Res. **6,** 33, 1989.

37. R. O. Potts, S. C. McNeill, C. R. Desbonnet, and E. Wakshull, Pharm. Res. **6,** 119, 1989.

38. S. Y. Chan and A. L. W. Po, Int. J. Pharm. **55,** 1, 1989.

39. J. das Neves and M. F. Bahia, Int. J. Pharm. **318,** 1, 2006.

40. H. Okada and A. M. Hillery, in A. M. Hillery, A. W. Lloyd, and J. Swarbrick (Eds.), *Vaginal Drug Delivery,* CRC Press, Boca Raton, FL, 2001, pp. 301.

41. P. van der Bijl and A. D. V. Eyk, Int. J. Pharm. **261,** 147, 2003.

42. W. M. Pardridge and P. L. Golden, in A. M. Hillery, A. W. Lloyd, and J. W. Swarbrick, (Eds.), *Drug Delivery to Central Nervous System,* CRC Press, Boca Raton, FL, 2001, pp. 355.

43. J. Temsamani, J.-M. Scherrmann, A. R. Rees, and M. Kaczorek, Pharm. Sci. Tech. Today, **3,** 155, 2000.

44. B. M. Paterson, B. E. Roberts, and E. L. Kuff, Proc. Natl. Acad. Sci. USA, **74,** 4370, 1977.

45. M. L. Stephenson and P. C. Zamecnik, Proc. Natl. Acad. Sci. USA, **75,** 285, 1978.

46. T. Mizuno, M. Y. Chou, and M. Inouye, Proc. Natl. Acad. Sci. USA, **81,** 1966, 1984.

47. F. Crick, Nature, **227,** 561, 1970.

48. M. Praseuth, A. L. Guieysse, C. Helene, Biochim. Biophys. Acta **1489,** 181, 1999.

49. R. Juliano, M. R. Alam, V. Dixit, and H. Kang, Nucleic Acids Res. **36,** 4158, 2008.

50. D. R. Corey, Nat. Chem. Biol. **3,** 8, 2007.

51. N. Dias and C. A. Stein, Mol. Cancer Ther. **1,** 347, 2002.

52. S. M. Elbashir, J. Harborth, W. Lendeckel, A. Yalchin, K. Weber, and T. Tuschl, Nature, **411,** 494, 2001.

53. J. B. Opalinska and A. M. Gewirtz, Nat. Rev. Drug Discov. **1,** 503, 2002.

54. Z. Paroo and D. Corey, Trends. Biotechnol. **22,** 390, 2004.

55. M. Famulok, G. Mayer, and M. Blind, Acc. Chem. Res. **33,** 591, 2000.

56. D. Jones, Nat. Rev. Drug Discov. **8,** 525, 2009.

57. D. W. Pack, A. S. Hoffman, S. Pun, and P. S. Stayton, Nat. Rev. Drug Discov. **4,** 581, 2005.

58. K. A. Whitehead, R. Langer, and D. G. Anderson, Nat. Rev. Drug Discov. **8,** 129, 2009.

59. D. Hedley, L. Oglivie, and C. Springer, Nat. Rev. Cancer, **7,** 870, 2007.

60. P. Ehrlich, Studies in Immunity' Wiley, NY, 1906.

61. F. Winau, O. Westphal, and R. Winau, Microbes Infect. **6,** 786, 2004.

62. D. J. A. Crommelin, W. E. Hennink, and G. Storm, in A. M. Hillery, A. W. Lloyd, and J. Swarbrick (Eds.), Drug Targeting Systems: Fundamentals and Applications to Parenteral Drug Delivery, CRC Press, Boca Raton, FL, 2001, pp. 117.

63. Y. Singh, M. Palombo, and P. J. Sinko, Curr. Med. Chem. **15,** 1802, 2008.

64. F. Kratz, I. A. Muller, C. Ryppa, and A. W. Warnecke, Chem. Med. Chem. **3,** 20, 2008.

65. P. F. Bross, J. Beitz, G. Chen, X. H. Chen, E. Duffy, L. Kieffer, S. Roy, R. Sridhara, A. Rahman, G. Williams, and R. Padzur, Clin. Cancer Res. **7,** 1490, 2001.

66. N. P. Barbour, M. Paborji, T. C. Alexander, W. P. Coppola, and J. B. Bogardus, Pharm. Res. **12,** 215, 1995.

67. J. A. Reddy, R. Dorton, E. Westrick, A. Dawson, T. Smith, L.-C. Xu, M. Vetzel, P. Kleindl, I. R. Vlahov, and C. P. Leamon, Cancer Res. **67,** 4434, 2007.

68. P. Carter, Nat. Rev. Cancer, **1,** 118, 2001.

69. M. A. Robinson, S. T. Chariton, P. Garnier, X.-T. Wang, S. S. Davies, A. C. Perkins, M. Frier, R. Duncan, T. J. Savage, D. A. Wyatt, S. A. Watson, and B. G. Davies, Proc. Natl. Acad. Sci. USA, **101,** 14527, 2004.

70. Y.-J. Park, J.-F. Liang, H. Song, Y. T. Li, S. Naik, and V. C. Yang, Adv. Drug Del. Rev. **55,** 251, 2003.

71. H.-K. Han and G. L. Amidon, AAPS Pharm. Sci. **2,** E6, 2000.

72. M. A. Jacobson, J. Med. Virol. **Suppl. 1,** 150, 1993.

73. H. Maeda, Adv. Enzyme Regul. **41,** 189, 2001.

74. A. K. Iyer, G. Khaled, J. Fang, and H. Maeda, Drug Discov. Today, **11,** 812, 2006.

75. R. Duncan, Nat. Rev. Drug Discov. **2,** 347, 2003.

76. R. Duncan, Nat. Rev. Cancer, **6,** 688, 2006.

77. B. E. Rabinow, Nat. Rev. Drug Discov. **3,** 785, 2004.

78. P. Ettmayer, G. L. Amidon, B. Clement, and B. Testa, J. Med. Chem. **47,** 2393, 2004.

79. J. Rautio, H. Kumpulainen, T. Heimbach, R. Oliyai, D. Oh, T. Jarvinen, and J. Savolainen, Nat. Rev. Drug Discov. **7,** 255, 2008.

80. H. Sah and Y. W. Chien, in A. M. Hillery, A. W. Lloyd, and J. Swarbrick (Eds.), Rate Control in Drug Delivery and Targeting: Fundamentals and Applications to Implantable Systems, CRC Press, Boca Raton, FL, 2001, pp. 83.

81. H. Kim and D. J. Burgess, J. Microencapsul. **19,** 631, 2002.

82. F. Theeuwes and S. I. Yum, Ann. Biomed. Eng. **4,** 343, 1976.

83. F. Theeuwes, in R. T. Borchardt, A. J. Repta, and V. J. Stella (Eds.), Directed Drug Delivery, Humana Press, NJ, 1985.

84. F. Theeuwes, J. Pharm. Sci. **64,** 1987, 1975.

85. F. Theeuwes, D. Swanson, P. Wong, P. Bonsen, V. Place, K. Heimlich, and K. C. Kwan, J. Pharm. Sci. **72,** 253, 1983.

86. A. S. Hoffman, Ann. N. Y. Acad. Sci. **944,** 62, 2001.

87. J. P. Cassidy, N. M. Landzert, and E. Quadros, J. Control. Release, **25,** 21, 1993.

88. J. S. Patton and P. R. Byron, Nat. Rev. Drug Discov. **6,** 67, 2007.

89. P. R. Byron, J. Pharm. Sci. **75,** 433, 1986.

90. P. R. Byron, N. S. R. Roberts, and A. R. Clark, J. Pharm. Sci. **75,** 168, 1986.

91. P. R. Byron (Ed.), Respiratory Drug Delivery, CRC Press, Boca Raton, Fl, 1990.

92. M. Quadir, H. Zia, and T. E. Needham, Drug Deliv. **7,** 223, 2000.

93. G. Eby, Am. J. Ther. **10,** 233, 2003.

94. M. Hirt, S. Nobel, and E. Barron, Ear Nose Throat J. **79,** 778, 2000.

95. T. F. Patton and J. R. Robinson, J. Pharm. Sci. **64,** 1312, 1975.

96. S. S. Chrai and J. R. Robinson, J. Pharm. Sci. **63,** 1218, 1974.

97. A. Urtti, L. Salminen, and O. Miinalainem, Int. J. Pharm. **23,** 147, 1985.

98. G. M. Grass, J. Cobby, and M. C. Makoid, J. Pharm. Sci. **73,** 618, 1984.

99. J. Cobby, M. Mayersohn, and G. C. Walker, J. Pharm. Sci. **63,** 725, 1974.

100. S. S. Jambhekar and J. Cobby, J. Pharm. Sci. **74,** 991, 1985.

101. B. A. Johnson and G. Zografi, J. Pharm. Sci. **75,** 529, 1986.

102. E. M. Niazy, A. M. Molokhia, and A. S. El-Gorashi, Int. J. Pharm. **56,** 181, 1989.

103. T. Nishihata, J. H. Rytting, K. Takahashi, and K. Sakai, Pharm. Res. **5,** 738, 1988.

104. J. W. Wiechers, B. F. H. Drenth, J. H. G. Joknman, and R. A. D. Zeeuw, Pharm. Res. **4,** 519, 1987.

105. R. H. Guy and J. Hadgraft, J. Pharm. Sci. **73,** 883, 1984.

106. R. H. Guy and J. Hadgraft, J. Pharm. Sci. **74,** 1016, 1985.

107. D. Arndts and K. Arndts, Eur. Clin. Pharmacol. **26,** 79, 1984.

108. Y. W. Chien, in J. R. Robinson and V. H. L. Lee (Eds.), Controlled Drug Delivery, Marcel Dekker, NY, 1987.

109. O. Siddiqui, Y. Sun, J.-C. Liu, and Y. W. Chien, J. Pharm. Sci. **76,** 341, 1987.

110. S. Del Terzo, C. R. Behl, and R. A. Nash, Pharm. Res. **6,** 85, 1989.

111. R. R. Burnette and D. Marrero, J. Pharm. Sci. **75,** 738, 1986.

112. B. Kasting and J. C. Keister, J. Control. Release, **8,** 195, 1989.

113. B. Kasting, E. W. Merrit, and J. C. Keister, J. Membrane Sci. **35,** 137, 1988.

114. R. R. Burnette and T. M. Bagniefski, J. Pharm. Sci. **77,** 492, 1988.

115. R. R. Burnette and B. Ongpipattanakul, J. Pharm. Sci. **77,** 132, 1988.

# Recommended Readings

## Drug Delivery and Targeting

Y. W. Chien, Novel Drug Delivery Systems, Marcel Dekker, NY, 1992.

A. M. Hillery, A. W. Lloyd, and J. Swarbrick (Eds.), Drug Delivery and Targeting for Pharmacists and Pharmaceutical Scientists, CRC Press, Boca Raton, FL, 2001.

D. A. LaVan, D. M. Lynn, and R. Langer, Nat. Rev. Drug Discov. **1,** 77, 2002.

B. E. Rabinow, Nat. Rev. Drug Discov. **3,** 785, 2004.

## Pulmonary Drug Delivery

J. S. Patton, P. R. Byron, Nat. Rev. Drug Discov. **6,** 67, 2007.

## Nasal Drug Delivery

L. Illum, J. Control. Release **87,** 187, 2003.

## Ocular Drug Delivery

Y. Ali and K. Lehmussaari, Adv. Drug. Deliv. Rev. **58,** 1258, 2006.

## Transdermal Drug Delivery

M. R. Prausnitz, S. Mitragotri, and R. Langer, Nat. Rev. Drug Discov. **3,** 115, 2004.

## Central Nervous System Drug Delivery

J. Temsamani, J.-M. Scherrmann, A. R. Rees, and M. Kaczorek, Pharm. Sci. Tech. Today **3,** 155, 2000.

## Gene, Antisense, and siRNA Delivery

R. Juliano, M. R. Alam, V. Dixit, and H. Kang, Nucleic Acids Res. **36,** 4158, 2008.

D. W. Pack, A. S. Hoffman, S. Pun, and P. S. Stayton, Nat. Rev. Drug Discov. **4,** 581, 2005.

## Prodrugs

P. Ettmayer, G. L. Amidon, B. Clement, and B. Testa, J. Med. Chem. **47,** 2393, 2004.

J. Rautio, H. Kumpulainen, T. Heimbach, R. Oliyai, D. Oh, T. Jarvinen, and J. Savolainen, Nat. Rev. Drug Discov. **7,** 255, 2008.

CHAPTER LEGACY

**Fifth Edition:** published as Chapter 22 (Drug Delivery Systems). New chapter by Tamara Minko.

**Sixth Edition:** published as Chapter 23 (Drug Delivery and Targeting). Rewritten de novo by Yashveer Singh, Hamid Omidian, and Patrick Sinko.

Page number followed by *f* indicate figures and *t* indicate tables.